CELL CULTURE TECHNOLOGY FOR PHARMACEUTICAL AND CELL-BASED THERAPIES

Biotechnology and Bioprocessing Series

1. Membrane Separations in Biotechnology, *edited by W. Courtney McGregor*
2. Commercial Production of Monoclonal Antibodies: A Guide for Scale-Up, *edited by Sally S. Seaver*
3. Handbook on Anaerobic Fermentations, *edited by Larry E. Erickson and Daniel Yee-Chak Fung*
4. Fermentation Process Development of Industrial Organisms, *edited by Justin O. Neway*
5. Yeast: Biotechnology and Biocatalysis, *edited by Hubert Verachtert and René De Mot*
6. Sensors in Bioprocess Control, *edited by John V. Twork and Alexander M. Yacynych*
7. Fundamentals of Protein Biotechnology, *edited by Stanley Stein*
8. Yeast Strain Selection, *edited by Chandra J. Panchal*
9. Separation Processes in Biotechnology, *edited by Juan A. Asenjo*
10. Large-Scale Mammalian Cell Culture Technology, *edited by Anthony S. Lubiniecki*
11. Extractive Bioconversions, *edited by Bo Mattiasson and Olle Holst*
12. Purification and Analysis of Recombinant Proteins, *edited by Ramnath Seetharam and Satish K. Sharma*
13. Drug Biotechnology Regulation: Scientific Basis and Practices, *edited by Yuan-yuan H. Chiu and John L. Gueriguian*
14. Protein Immobilization: Fundamentals and Applications, *edited by Richard F. Taylor*
15. Biosensor Principles and Applications, *edited by Loï'efc J. Blum and Pierre R. Coulet*
16. Industrial Application of Immobilized Biocatalysts, *edited by Atsuo Tanaka, Tetsuya Tosa, and Takeshi Kobayashi*
17. Insect Cell Culture Engineering, *edited by Mattheus F. A. Goosen, Andrew J. Daugulis, and Peter Faulkner*
18. Protein Purification Process Engineering, *edited by Roger G. Harrison*
19. Recombinant Microbes for Industrial and Agricultural Applications, *edited by Yoshikatsu Murooka and Tadayuki Imanaka*
20. Cell Adhesion: Fundamentals and Biotechnological Applications, *edited by Martin A. Hjortso and Joseph W. Roos*
21. Bioreactor System Design, *edited by Juan A. Asenjo and José C. Merchuk*
22. Gene Expression in Recombinant Microorganisms, *edited by Alan Smith*
23. Interfacial Phenomena and Bioproducts, *edited by John L. Brash and Peter W. Wojciechowski*
24. Metabolic Engineering, *edited by Sang Yup Lee and Eleftherios T. Papoutsakis*
25. Biopharmaceutical Process Validation, *edited by Gail Sofer and Dane W. Zabriskie*
26. Membrane Separations in Biotechnology: Second Edition, Revised and Expanded, *edited by William K. Wang*
27. Isolation and Purification of Proteins, *edited by Rajni Hatti-Kaul and Bo Mattiasson*

28. Biotransformation and Bioprocesses, *Mukesh Doble, Anil Kumar Kruthiventi, and Vilas Gajanan Gaikar*
29. Process Validation in Manufacturing of Biopharmaceuticals: Guidelines, Current Practices, and Industrial Case Studies, *edited by Anurag Singh Rathore and Gail Sofer*
30. Cell Culture Technology for Pharmaceutical and Cell-Based Therapies, *edited by Sadettin S. Ozturk and Wei-Shou Hu*

CELL CULTURE TECHNOLOGY FOR PHARMACEUTICAL AND CELL-BASED THERAPIES

edited by

Sadettin S. Ozturk

Centocor, Inc.
Malvern, Pennsylvania, U.S.A.

Wei-Shou Hu

University of Minnesota
Minneapolis, Minnesota, U.S.A.

Taylor & Francis
Taylor & Francis Group
New York London

Published in 2006 by
CRC Press
Taylor & Francis Group
6000 Broken Sound Parkway NW, Suite 300
Boca Raton, FL 33487-2742

Printed in the United States of America on acid-free paper
10 9 8 7 6 5 4 3

International Standard Book Number-10: 0-8247-5334-8 (Hardcover)
International Standard Book Number-13: 978-0-8247-5334-4 (Hardcover)

Library of Congress Cataloging-in-Publication Data

Catalog record is available from the Library of Congress

T&F informa

Taylor & Francis Group
is the Academic Division of T&F Informa plc.

Visit the Taylor & Francis Web site at
http://www.taylorandfrancis.com

and the CRC Press Web site at
http://www.crcpress.com

Preface

In the past quarter century, we have seen protein therapeutics greatly expand the horizon of health care. Many life-threatening diseases or chronic dysfunctions—once beyond the reach of medicine—are now readily treatable. Much of the impact of protein therapeutics has come about through the successful use of a new production vehicle—mammalian cells. Although the emergence of cell culture cultivation technology is relatively recent, the demand for technological advances to meet medical needs has induced a rapid industrial transformation. Cell culture processes have become an established method of creating new medicines in a short decade.

Numerous books, encyclopedias, and conference proceedings have captured important aspects of cell culture processing over the years, helping to disseminate knowledge of this technology. As we have watched the reach of protein therapeutics continue to grow and many new professionals continue to enter careers related to this technology, we have come to the realization that a comprehensive, single-volume treatise on cell culture technology would be a worthwhile addition. We envisioned the book to cover cell culture bioprocessing from A to Z, starting from cell line development to the finished product. It encompasses both fundamental principles and practical aspects of the technology to serve as a resource for information for seasoned professionals, and as comprehensive reading for those new to any specific area of cell culture technology. This book is designed to last for many years to come.

Covering a wide range of topics essential to cell culture technology in a single volume is no easy task. We are fortunate to have the participation of many esteemed experts as contributors. Their dedication, commitment, and hard work transformed our ideal into reality. Chapter 1, an overview of cell culture technology, reviews the state-of-the-art technology and provides an historical perspective. The following chapters on cell line development, medium, and process take the reader through a journey of product development. Chapter 2 focuses on recombinant DNA technology and cell line development. A detailed discussion on medium development is presented in Chapter 3. Chapter 4 outlines the current understanding of cell metabolism for optimizing the performance of cell culture systems. An extensive review on protein glycosylation is presented in Chapter 5. Chapter 6 gives an overview of cell culture bioreactors. In this chapter and Chapter 7, a number of reactor operational topics, including aeration, mixing, and hydrodynamics in bioreactors is discussed. Process instrumentation and process control are important for production of cell culture-derived products, and these topics are covered in Chapter 8. This discussion is followed by Chapter 9, which covers the utilization of cell culture kinetics and modeling. Cell culture bioreactors operate in different modes, and this book covers both fed-batch (Chapter 10) and high-density perfusion (Chapter 11) operation. Both chapters are comprehensive in their coverage and they provide extensive guidelines for the cultivation of mammalian cells for the production of recombinant proteins.

Cell separation and product capture in cell culture technology are the topics of Chapter 12. This is followed by detailed coverage of downstream processing (Chapter 13) and formulation (Chapter 14). The next two chapters focus on commercialization of cell culture processes. Chapter 15 deals with process and equipment validation, and Chapter 16 provides extensive guidelines for a commercial facility design. Two somewhat unconventional methods for protein production using cell culture technology are presented in Chapters 17 and 18, which cover transient expression and large-scale propagation of insect cells, respectively. The final two chapters of this book are dedicated to stem cells and their application for cell-based therapies.

We thank all the contributing authors for their dedication and diligence in bringing this project to fruition. Our families and many friends have been extremely supportive during the editing of this book. In addition, we would like to recognize Amy Fayette, Denny Kraichely, Subinay Ganguly, David Epstein, and Kristen Cosgrove for their help in proofreading and Anita Lekhwani for the initiation and support of this project.

Sadettin S. Ozturk
Wei-Shou Hu

Contents

Contributors

Ron Bates Bristol-Myers Squibb, Syracuse, New York, U.S.A.

Alain R. Bernard Serono Pharmaceutical Research Institute, Geneva, Switzerland

Jing-xiu Bi Institute of Process Engineering, National Laboratory of Biochemical Engineering, Chinese Academy of Sciences, Beijing, P.R. China

A. Burgener Department of Microbiology, University of Manitoba, Manitoba, Canada

M. Butler Department of Microbiology, University of Manitoba, Manitoba, Canada

Jordi Joan Cairó Department d'Enginyeria Química, ETSE, Universitat Autònoma de Barcelona, Bellaterra, Barcelona, Spain

Jeffrey J. Chalmers Department of Chemical and Biomolecular Engineering, The Ohio State University, Columbus, Ohio, U.S.A.

M. E. M. Cromwell Early Stage Formulation, Genentech, Inc., South San Francisco, California, U.S.A.

Chandra M. Dwivedi Bayer Corporation, Berkeley, California, U.S.A.

Sandino Estrada-Mondaca Departamento de Medicina Molecular y Bioprocesos, Instituto de Biotecnología, Universidad Nacional Autónoma de México, Morelos, Mexico

Christel Fenge AstraZeneca Biotech Laboratory, Södertälje, Sweden

Francesc Gódia Department d'Enginyeria Química, ETSE, Universitat Autònoma de Barcelona, Bellaterra, Barcelona, Spain

Sarah W. Harcum Department of Bioengineering, Clemson University, Clemson, South Carolina, U.S.A.

Dhinakar S. Kompala University of Colorado, Boulder, Colorado, U.S.A.

Lynne A. Krummen Genentech, Inc., Department of Cell Culture & Fermentation R&D, South San Francisco, California, U.S.A.

Elke Lüllau AstraZeneca Biotech Laboratory, Södertälje, Sweden

Ningning Ma Department of Chemical and Biomolecular Engineering The Ohio State University, Columbus, Ohio, U.S.A.

A. Mantalaris Department of Chemical Engineering and Chemical Technology, Imperial College of Science, Technology, and Medicine, London, U.K.

Mike Mollet Department of Chemical and Biomolecular Engineering The Ohio State University, Columbus, Ohio, U.S.A.

Kim Nelson CRB Consulting Engineers, Plymouth Meeting, Pennsylvania, U.S.A.

Sadettin S. Ozturk Centocor Inc., Malvern, Pennsylvania, U.S.A.

Laura A. Palomares Departamento de Medicina Molecular y Bioprocesos, Instituto de Biotecnología, Universidad Nacional Autónoma de México, Morelos, Mexico

Octavio T. Ramírez Departamento de Medicina Molecular y Bioprocesos, Instituto de Biotecnología, Universidad Nacional Autónoma de México, Morelos, Mexico

Arthur E. Reyes Genentech, Inc., Department of Cell Culture & Fermentation R&D, South San Francisco, California, U.S.A.

Mark Riley Agricultural and Biosystems Engineering, The University of Arizona, Tucson, Arizona, U.S.A.

Thomas Seewoester Amgen, Thousand Oaks, California, U.S.A.

Amy Y. Shen Genentech, Inc., Department of Cell Culture & Fermentation R&D, South San Francisco, California, U.S.A.

E. S. Tzanakakis School of Medicine, Diabetes Center University of California-San Francisco, San Francisco, U.S.A.

Jana Van de Goor Genentech, Inc., Department of Cell Culture & Fermentation R&D, South San Francisco, California, U.S.A.

C. M. Verfaillie Stem Cell Institute, University of Minnesota, Minneapolis, Minnesota, U.S.A.

J. H. D. Wu Department of Chemical Engineering, Biomedical Engineering, and Microbiology and Immunology, University of Rochester, Rochester New York, U.S.A.

Liangzhi Xie Merck Research Laboratories, Merck & Co., Inc., Sumneytown Pike, West Point, Pennsylvania, U.S.A.

An-Ping Zeng GBF National Research Institute for Biotechnology Division of Biochemical Engineering, Braunschweig, Germany

Lisa Zheng Genentech, Inc., Department of Cell Culture & Fermentation R&D, South San Francisco, California, U.S.A.

Weichang Zhou Merck Research Laboratories, Merck & Co., Inc., Sumneytown Pike, West Point, Pennsylvania, U.S.A.

1

Cell Culture Technology—An Overview

Sadettin S. Ozturk
Centocor Inc., Malvern, Pennsylvania, U.S.A.

INTRODUCTION

The last three decades have witnessed a major development in cell culture technology. The field developed substantially and became an essential part of biotechnology (1). During its evolution, cell culture technology successfully integrated various disciplines, including cell biology, genetic engineering, protein chemistry, genomics, and chemical engineering. Cell culture technology is now the established method of producing a number of important proteins, especially those that are large, complex, and glycosylated (2).

Cell culture technology–derived products are currently used as medicines to prevent and treat serious diseases such as cancer, viral infections, heredity deficiencies, and a variety of chronic diseases. The products are proven to be safe, effective, and economical. The capacity requirements for these products and the market they generated exceed the initial estimates. Some of the cell culture–derived products have a demand of 500 kg/year and generate $1–2 billion in revenue.

Cell culture technology went through a significant evolution from its origin to its commercialization. Cell culture techniques were originated as study tools to investigate cell and tissue behavior and function in vitro (3,4). Utilization of cell cultures for therapeutic purposes started with the use of cells for vaccine production. Cultured cells were successfully used as hosts to grow viruses, opening the field of large-scale vaccine manufacturing.

The next big step for cell culture technology was the acceptance of continuous cell lines by regulatory agencies. Continuous cell lines can grow indefinitely, have less stringent growth requirements, and, most importantly, they can be cultured in suspension. Elimination of solid substrate in suspension culture allows scaling up by volume and allows the cells to grow in bioreactors using well-established methods similar to those used for microbial systems.

Genetic engineering and the use of recombinant DNA technology made it possible to produce a vast number of products in cell culture. A number of vectors are now used in a variety of cell lines to produce native and modified human proteins. Engineering the machinery of the cells also allowed the modification of protein products for stability, efficacy, and biological activity. A parallel development in cell culture technology resulted in tissue engineering, where cells are produced as products for tissue replacement and gene therapy.

Advancement and development of cell culture technology required an interdisciplinary approach. As a result of close collaboration and tight integration between cell biology and biochemical engineering, cell lines with excellent productivities can now be cultivated in large-scale bioreactors and at very high cell densities. Issues related to shear sensitivity, aeration, and mixing in bioreactors are largely resolved. The processes can be scaled up successfully and cells can be cultivated consistently in 20,000 L bioreactors. Advances in medium development and cell retention technologies resulted in very high cell densities in the bioreactors. In addition to enhancing productivity in the bioreactors, the cost of the cell culture process was diminished significantly by the elimination of serum and other high-cost proteins from the medium. Now, chemically defined medium is a reality. This progress contributes significantly to the safety and reliability of raw materials in the production of biologicals. Finally, the technology now utilizes efficient separation, purification, and virus inactivation methods that result in highly purified, efficacious, and safe products.

This chapter overviews the development and evolution of cell culture technology from a historical perspective, and reviews the state of the art of cell culture technology. There are several products from cell culture technology approved for treatment of diseases; a brief description of these products is also presented. Finally, a look into the future of the biotechnology field and an assessment of current products in development are provided.

A BRIEF HISTORY OF CELL CULTURE TECHNOLOGY

Methods for growth and maintenance of primary cells in vitro opened the doors for cell culture technology. Early studies with embryos and fibroblasts provided tools for studying cell behavior and function in vitro. These studies also provided a tool for developing cell culture media. The cells used in those days had a limited life span unless they were transformed or were originated from tumors. The production of biologicals from cell culture evolved gradually over the last three decades.

Early Days of Cell Culture Technology—First Products

Utilization of cultured cells for the production of viral vaccines was the first application of cell culture technology. Viruses for vaccine production need living cells to propagate. Embryonic chicken in eggs traditionally was used for vaccine production. Due to increased demand, alternative methods were sought and cell culture technology was the answer. The production of polio vaccine using cells grown in culture started in 1954 (4). The cells were primary monkey kidney cells grown on surfaces (attachment dependent cells). A similar development took place for the vaccine against foot-and-mouth disease for veterinary use. Large-scale vaccine production for FMD virus was a major activity 36 years ago. Increased demand and process economics required more effective and scalable processes. Baby hamster kidney (BHK) cells were adopted for this process and FMD vaccine could then be produced at a 5000 L scale in suspension bioreactors. The medium used for this production was Eagle's medium with 5% adult bovine serum. Virus production continued to utilize cell culture technology, and vaccines such as measles, rabies, mumps, rubella, and varicella are produced in large quantities. In addition to these human vaccines several veterinary vaccines are manufactured by cell culture (5).

Production of biologicals for medical applications also underwent a significant evolution. Interferon (IFN) alpha was the first cell culture derived product used as a drug. IFNs were traditionally made from human white blood cells from blood donors. The production of IFN was very limited, as it was dependent on donor blood cells and was present at very low concentrations. Work at Wellcome Research Laboratories was instrumental in this area. Initially, researchers at Wellcome wanted to use BHK cells as they could be expanded in suspension and these cells have already been approved for vaccine production. However, IFN-alpha was to be used as a medicine and issues were raised for the BHK cell line. BHK cells contain C-type virus particles and they have oncogenic potential. Nawalwa lymphoblastoid cells, on the other hand, had a better biosafety profile and could be cultivated as easily as BHK cells. The production of IFN moved cell culture technology to the 10,000 L scale and the process is being used successfully for IFN production.

Progress in Cell Line Acceptability

As mentioned before, cells used for early vaccine production were primary cells such as monkey kidney cells and chicken embryo. These cells are attachment-dependent cells requiring a surface on which to grow. In addition the primary cells have stringent growth requirements (growth factors, serum) and their expansion requires a lot of manipulation (trypsinization). These requirements inhibit the use of attachment-dependent cells for large-scale production. The shift from attachment cells to suspension cells was a major advancement for cell culture technology. The use of BHK cells for veterinary vaccines came from the need to meet the demand for FMD vaccine. BHK cells are continuous cell lines and can grow in suspension. Although the BHK cells were accepted for veterinary vaccines, these cells could not be approved for human vaccines. BHK cells were considered unsafe because they contain tumorigenic agents and viruslike particles. As a result, regulatory agencies (WHO, FDA) evaluated other options and, in the end, they approved the use of human diploid cells such as WI-38 and MRC-5 (6). These cells are "normal" cells, virus-free, and contain no carcinogenic substrate. Although the biosafety profile of these cells was acceptable, these cells were still a challenge in large-scale production. They have a limited life span and are attachments-dependent cells.

The concerns of regulatory agencies about the use of transformed cells or cells derived from cancer tissue were addressed by tight monitoring and control of cell lines. In the 1980s, agencies accepted Namalwa cells for the production of IFNs based on characterization data of the master cell bank. These cells did not have any contaminants and the product did not have detectable infective agents. Even though these cells had Epstein-Barr (EB) virus, it was not a serious drawback. The product from these cells could be purified in a process that effectively removes the majority of cell-derived proteins and DNA (6).

Continuous cell lines such as BHK, Chinese hamster ovary (CHO), myeloma cells (SP2/0, NS0), and human embryonic kidney (HEK) cells were accepted gradually for use in cell culture technology. These cells can be grown in suspension and are easy to scale up. Since the cells are continuous cell lines, their growth requirements are manageable. In fact, these cells can grow in chemically defined media without serum or protein supplements. In addition, these cells can be modified genetically to produce recombinant proteins. In 1987, tissue plasminogen activator (tPA) was approved from recombinant CHO cells. In the last decade, tissue engineering was used to generate tissues for replacement therapy and the cells themselves were approved as products for tissue replacements. The latest development in tissue engineering is the use of embryonic stem cells. These cells can differentiate to different tissues that can be implanted or used as tissue replacement.

Progress in Large-Scale Cell Culture Technology

While small-scale cell culture provided sufficient products for investigational and diagnostic purposes, the demand for vaccines and therapeutic products required large-scale production. Some of the cell culture technology-based products such as monoclonal antibodies need to be manufactured at 1000 kg/year. This can only be done in large bioreactors (up to 20,000 L) by efficient processes that produce several kilograms a day (1,7).

Early cell lines used for vaccines were attachment-dependent cells. These cells required medium supplemented with serum and surfaces on which to grow. The expansion and propagation of cells required trypsinization. Scale-up of these systems was achieved by adding more surfaces into the bioreactor system. Roller bottles, T-flasks, and disk propagators are still used to scale up these cultures. A modernized version of these systems is the cell cube, which bears a similarity to a multichamber T-flask. Compared to suspension systems, attachment-dependent cell systems are more difficult to scale up because the surface-to-volume ratio gets smaller at larger scales.

The development of microcarrier cultures was a major breakthrough for attachment-dependent cell systems (8). The microcarriers can be suspended in stirred tanks similar to those used for microbial systems. Modifications to the impellers and aeration systems ensured the success of the bioreactor process. The cells, when inoculated into microcarrier-containing bioreactors, attach and grow on microspheres. The cells can be fed by fresh medium and the product can be removed from the bioreactor continuously. Microcarriers are easier to scale up; one can simply put more microcarriers in larger bioreactors. These systems are now used for vaccine production as well as for protein production at scales of 10,000 L or higher. An improvement in the microcarrier culture was the use of macroporous carriers in the bioreactors. These macroporous carriers allow cells to grow inside as well as outside, thus achieving higher densities by the extra room provided.

The acceptance of continuous cell lines made it possible to grow the cells in suspension in large scale bioreactors. Continuous cell lines are mostly suspension cells that can be agitated and aerated in bioreactors. However, culturing cells in bioreactors was not straightforward. The absence of cell membranes and the stringent growth requirement of cells required both the cultivation procedures and bioreactor operating conditions to be customized. Optimal growth conditions in the bioreactor were achieved by implementation of the results from intensive studies on shear, agitation, aeration, and medium optimization. Today, cells can be cultivated at large-scale bioreactors (up to 20,000 L), at high densities (up to 50 million cells/mL), for many months of operation (9).

High-Efficiency Cell Culture Processes

The efficiency of cell culture processes is measured by the productivity of the cultures, i.e., the amount made per volume per day. This quantity, the volumetric productivity, is the product of the cell density achieved and the productivity of each cell (specific productivity). Volumetric productivity determines the amount made in a certain production period that can be measured as accumulated product concentration in the bioreactor.

Novel transfection, amplification, and selection methods resulted in high-producing cell clones with high specific productivities. Specific production rate is a measure of how much product is secreted from the cell per unit of time. For monoclonal antibodies, the specific production rate is commonly expressed in picograms per cell per day. For commercial production of antibodies, a minimal specific productivity of 10–20 pg/cell/day is expected. Advances in cell line development and optimization of culture conditions can result in specific productivities as high as 40–100 pg/cell/day.

In addition to specific productivity, the volumetric productivity is directly proportional to the cell density in the bioreactor. Cell density attainable in the bioreactor depends on several factors such as cell growth and death rates, composition of the medium, and aeration capacity (9). In early large-scale bioreactor development studies, the aeration was a major limitation for achieving high cell density. Over the years effective aeration strategies were developed and most of the issues with sparging have been resolved. An optimized design of the sparger (gas flow rate, diameter, etc.) and some media additives (F68, antifoam) can effectively minimize the impact of sparging. Today, the cell density in the bioreactor is mostly dictated by the media composition and media exchange rate.

Mammalian cells require a highly complicated medium and a delicate formulation is required to maximize cell growth. If the aeration is not limiting, achievable cell density in the bioreactor is controlled by: (a) the nutrient levels in the medium, and (b) the medium exchange rate (9). Nutrient levels in the medium can be increased by extensive medium optimization. Media development and optimization is an art and a science. Many variables are involved in cell metabolism and the components of the medium interact with each other. Early media formulations provided cell densities of 1–2 million cells/mL. Today, a cell density of 10 million cells/mL is achievable in a batch culture. This is an order of magnitude improvement, accomplished in the last two decades. In perfusion cultures, the cell density can be even higher due to high medium exchange rates resulting in cell densities of 20–50 million cells/mL (9).

Media Development

Process economics and implementation of cost-effective processes benefited not only by increasing the productivity, but also by reducing the manufacturing cost. Medium development contributed greatly to the process economics by allowing cells to grow to high densities and by decreasing the cost of the medium components.

In the beginning, cell culture medium contained animal products such as serum, albumin, and growth factors. For attachment-dependent cells, other components were also added to stimulate attachment and to form extracellular matrices. These components added to the cost of production and complicated the material sourcing and material release processes. Over the years, serum and other animal products were eliminated from the medium and most continuous cell lines can now be grown in serum- and protein-free media formulations. The use of simple media made it possible to produce kilogram amounts of protein per year economically from volumes close to 100,000 L.

Serum-free media development was followed by the complete elimination of animal proteins from the medium. Initially, albumin and other proteins have been used in serum-free media to substitute for some of the serum components. However, concerns about viruses and animal disease such as bovine spongiform encephalopathy (BSE) required the development of animal product-free (APF) media formulations. Early APF media relied heavily on the use of plant hydrolysates such as soy peptone and yeast extracts for optimal results. Thanks to extensive media development efforts, these components could also be eliminated from the medium with minimal loss in productivity. These efforts resulted in chemically defined media (CDM) formulations. Chemically defined medium is now utilized in many biotechnology processes as the medium of choice because every component could be traced and consistently manufactured.

PRODUCTS FROM CELL CULTURE TECHNOLOGY

Viral Vaccines

Vaccination for the prevention of infectious diseases has been an effective strategy for many years. Polio vaccine was the first cell culture technology–based vaccine and was produced in cultured monkey kidney cells (7). The cell-based vaccine technology evolved in the last four decades; the production of vaccines now utilizes primary cells, human diploid cells, and continuous or even recombinant cell lines. Vaccines against hepatitis B, measles and mumps, rubella, rabies, and FMD had been very effective in preventing life-threatening diseases. New vaccines target human immunodeficiency virus (HIV), herpes simplex virus, respiratory syncytial virus, cytomegal virus (CMV), and influenza, and continue to utilize cell culture technology for production. The latest developments in vaccine development include genetically engineered vaccines and DNA vaccines that will open new frontiers in this field. Development of new vaccines for HIV and cancer is very exciting and these efforts should come to successful conclusion in the next decade.

Cytokines (Interferons and Interleukins)

Cytokines are soluble mediators or glycoproteins helping cells communicate and function as part of the immunological, hematological, and neurological systems (10). Interferons and interleukins are cytokines with enormous therapeutic potential. Alpha interferon (Wellferon®) was developed as the first cell culture-derived biological for treatment of cancer (4). This was achieved in 8000 L bioreactors using Nawalwa cells at the Wellcome facility. It was followed by the production of several other interferons and interleukins. Due to the small size and relatively simple molecular structure of cytokines, production by simpler cell systems (microbial or yeast) seemed to be adequate. However, mammalian cells express more native cytokines, and cell culture technology is the method of choice for alpha, beta, and gamma interferons. In addition, interleukins 2 to 4, 6, 11, and 12 are also produced by cell culture technology (7).

Hematopoietic Growth Factors

The formation and differentiation of hematopoietic cells to give rise to mature blood cells require a series of growth factors. These growth factors are mainly single-chain polypeptides and can be produced by microbial systems. However, cell culture technology can offer advantages when the molecule is complicated by glycosylation and when the native form of the molecule is required for therapy.

Erythropoietin (EPO) is a hormone that controls the maturation of red blood cells and it is used for clinical applications in anemia. Recombinant EPO was genetically engineered in CHO cells and the product was launched in 1989 under the names of Epogen® (Amgen), Procrit® (Johnson and Johnson), and Eprex® (Johnson and Johnson). EPO was launched in 1990 in Europe and Japan under the names of Epogin® and Recormon®.

Growth Hormones

Although conventionally expressed in *Escherichia coli,* human growth hormone can be produced efficiently using cell culture technology. Seostim® and Saizen® are produced by C127 cells and marketed by Serono S.A.

Monoclonal Antibodies

Antibodies were hailed as "magic bullets" for targeting and neutralizing their antigens as therapeutic agents. Initial application of antibodies involved in vivo and in vitro small-scale production for diagnostic kits. It took almost two decades since then for the field to mature and produce antibodies that are safe and effective therapeutics. Today, antibodies constitute more than 25% of total biotechnology production (11,12).

The first antibody was produced by murine ascites. OKT3 antibody was approved in 1987 for the treatment of transplant rejection. Since then both the molecular structure of the antibodies and their production methods evolved significantly. The mouse antibodies were replaced first by chimeric, then by humanized, and recently by fully human antibodies. This allowed the gradual elimination of immunigenicity of antibodies. Antibodies are now produced in hundreds of kilograms using stirred tank bioreactors.

Table 1 presents the antibodies approved to date. Antibodies are used for HIV, cancer, allergic diseases, arthritis, renal prophylaxis, septic shock, transplantation, asthma, CMV, and anti-idiotype vaccines. In addition to those presented in Table 1, there are a large number of antibodies in clinical trials.

Recombinant Thrombolytic Agents

Recombinant tissue plasminogen activator (tPA) was the first product from cell culture (7). Genentech obtained the approval for tPA production from CHO cells in 1987 (Table 2). tPA is produced in a large-scale (>10,000 L) cell culture process. Although the current tPA is highly efficacious, there are several attempts to improve the pharmacokinetics of the molecule. In addition, there is a new generation of recombinant thrombolytic agents under development.

Recombinant tPA from Genentech is marketed as Activase® and it is used for acute myocardial infarction (approved in 1987), acute massive pulmonary embolism (approved in 1990), acute myocardial infarction accelerated infusion (approved in 1995), and ischemic stroke within 3–5 hr of symptom onset (approved in 1996).

Table 1 Approved Monoclonal Antibodies

Product	Company	Indication	Approved
ORTHOCLONE OKT® 3	Ortho Biotech	Kidney, heart, and liver transplantation	1991
OncoScint® CR/OV	CYTOGEN	Diagnosis of colorectal and ovarian cancers	1992
ReoPro®	Centocor/Eli Lilly	Transluminal coronary angioplasty	1994
CEA-Scan®	Immunomedics	Diagnosis of colorectal cancer	1996
ProstaScint®	CYTOGEN	Diagnosis of prostate adenocarcinoma	1996
Verluma®	DuPont Merck	Diagnosis of small-cell lung cancer	1996
Neumega®	Wyeth	Thrombocytopenia	1997
Rituxan®	Genentech/IDEC	Non-Hodgkin's lymphoma	1997
Zenapax®	Roche/Protein Design	Kidney transplant rejection	1997
Thmoglobulin®	SangStat	Kidney transplant rejection	1998
Herceptin®	Genentech	Breast cancer	1998
Remicade®	Centocor	Crohn's disease, rheumatoid arthritis	1998
Simulect®	Novartis	Renal transplant rejection	1998
Synagis®	MedImmune	Respiratory synctial virus	1998
Mylotarg®	Wyeth	Acute myeloid leukemia (AML)	2000
Campath®	Berlex/ILEX Oncology	B-cell chronic lymphocytic leukemia (B-CLL)	2001
Zevalin®	IDEC Pharm	Non-Hodgkin's lymphoma	2002
Xolair®	Genentech	Astma	2003
Bexxar®	Corixa	Non-Hodgkin's lymphoma	2003
Raptiva®	Genentech	Psiorasis	2003
Tysabri®	Biogen/DEC/Elan	Multiple sclerosis	2004
Erbutux®	Imclone	Colorectal cancer	2004
Humira®	Abbott	Rheumatoid arthritis	2004
Avastin®	Genentech	Colorectal cancer	2004

Table 2 Recombinant Thrombolytic Agents Produced by Cell Culture

Product	Company	Indication	Approved
Activase®	Genentech	Myocardial infarction, ischemic stroke	1987
TNKase®	Genentech	Myocardial infarction	2000

Table 3 Recombinant Coagulation Factors Approved

Protein	Medicine	Company	Indication	Approved
rFVIII	Recombinate®	Baxter/Wyeth	Hemophilia A	1992
rFVIII	KoGENate	Bayer	Hemophilia A	1993
rFIX	BeneFIX®	Wyeth	Hemophilia B	1997
rFVII	NovoSeven®	Novo Nordisk	Hemophilia A/B	1999
rFVIII	Helixate® FS	Aventis	Hemophilia A	2000
rFVIII	KoGENate-FS	Bayer	Hemophilia A	2000
rFVIII	ReFacto®	Wyeth	Hemophilia A	2000

Medicines for myocardial infarction now include TNKase™ from Genentech, which was approved in 2000 and is produced from CHO cells.

Recombinant Coagulation Factors

Blood coagulation factors traditionally obtained from plasma fractionation were used for the treatment of genetic diseases such as hemophilia. The issues related to blood safety, HIV, and Jacob-Creutzfeld Disease required a switch to recombinant protein production using cell culture technology.

The treatment of hemophilia benefited greatly with the introduction of recombinant factor VIII in 1992 (Table 3). Baxter/Wyeth obtained the license for rFVIII product Recombinate™ for the treatment of hemophilia A. This was followed by Bayer (KoGENate and KoGENate-FS, approved in 1993 and 2000, respectively) and Wyeth (ReFacto®, approved in 2000). Other companies such as Aventis market rFVIII under the name of Helixate® and Helixate® FS.

Recombinant factors VII and IX are now also produced by cell culture technology. Wyeth developed rFIX under the name of BeneFIX™ using CHO cells. This product was approved for the treatment of hemophilia B in 1997. Recombinant factor VII NovoSeven® is a product from Novo Nordisk for treatment of bleeding episodes in hemophilia A or B patients with inhibitors to factor VIII or factor IX.

Recombinant Enzymes

CHO cells are used in the manufacture of DNAse. This drug is sold under the name of Pulmozyme® and used for the treatment of cystic fibrosis. Genentech obtained the license for manufacturing of Pulmozyme in 1993. Genzyme now manufactures another enzyme, glucocerebrosidase, using CHO cells under the name Cerezyme®. This product obtained the license in 1994 for the treatment of Gaucher disease and replaced placenta-derived Ceredase®.

Cell Therapy

The latest development in cell and tissue engineering resulted in therapies for replacement, repair, or enhancement of damaged tissue. The first tissue therapy product was

Table 4 Cell Therapy Products Approved

Product	Company	Indication	Approved
Dermagraft-TC	Advanced Tissue Sciences[a]	Skin replacement for burn patients	1997
Carticel®	Genzyme Tissue Repair	Cartilaginous defects	1997
Apligraf®	Novartis	Venous leg ulcers	1998
DACS® SC	Denderon	Chemotherapy	1999

[a] Dermagraft-TC is discontinued and Advanced Tissue Sciences is out of business.

for burn patients. Keratinocytes could be grown in vitro into large sheets of skin tissue that could be grafted. Dermagraft-TC has been approved as a temporary wound covering for partial-thickness burns. Dermagraft-TC was first approved for marketing in 1997 as a temporary wound covering for severe burns (Table 4). Dermagraft-TC was the first human, fibroblast-derived temporary skin substitute for the treatment of partial-thickness burns that has been approved for marketing by the FDA.*

Other tissue therapy products were also developed. Carticel(TM) is used to repair clinically significant, symptomatic cartilaginous defects of the femoral condyle (medial, lateral, or trochlear) caused by acute or repetitive trauma. Genzyme Tissue Repair obtained approval for this product in 1997. Another cell/tissue therapy product is Apligraf(TM) manufactured by Novartis Pharmaceuticals. This product was approved in 1998 for the treatment of venous leg ulcers. Finally, DACS(TM) SC from Dendreon was approved in 1999 for rescue therapy following high-dose chemotherapy.

Growing cells and tissues for cellular therapies have many potential applications including diabetes (islet cells), Parkinson's disease (fetal dopamine cells), cancer (heamopoitic cells, bone marrow), liver disease (hapatocytes), and cartilage damage (chrondocytes). The latest developments in stem cells will eventually allow growing, grafting, and replacing any kind of tissue of the human body.

Gene Therapy

Gene therapy offers a permanent fix for some of the genetic diseases by introducing the gene that is missing by a delivery system (13). In addition, gene therapy can be used for the treatment of cancer and viral infections. Injection of a gene and its possible integration requires a carefully designed delivery system that is safe and effective. Gene delivery can be performed in vivo, in situ, or in vitro using several vectors. The production of vectors, mostly engineered viruses, rely on systems similar to vaccine production.

The first genetic therapy targeted ADA-SCID disease in 1990 with some success. Currently, there are several gene therapy protocols executed as clinical trials for cancer, genetic diseases, and AIDS.

---

*Due to some regulatory and business concerns, the production of Dermagraft was discontinued. The company that developed the Dermagraft, Advanced Tissue Sciences closed down.

Other Products

There are several other products from cell culture technology that must be mentioned here. Recombinant activated protein C is a major product marketed by Eli Lily, approved in 2001 for the treatment of severe sepsis. HEK 293 cells are used for the production of activated protein C.

Follicle stimulating hormone is used for the treatment of infertility. Follistim® (Organon) and Gonal® (Serano) obtained approval in 1997 and 1998, respectively, for stimulation of ovulation during assisted reproduction. CHO cells are used for the production of both products.

Soluble TNF receptor, Embrel®, is used as a fusion protein TNFR (conjugated to antibody-Fc) for the treatment of moderate to severe active rheumatoid arthritis. Following initial approval in 1998, the product was later approved for juvenile rheumatoid arthritis (1999), for disease modification of active rheumatoid arthritis, and for psoriatic arthritis (January 2002). Embrel is produced from CHO cells.

FUTURE PROSPECTS

The future of biotechnology-derived medicines is bright and full of promise (2). An industrial survey of biotechnology companies revealed approximately 371 medicines under development (14). These biopharmaceutical therapeutics target about 425 diseases, as presented in Figure 1. Most of the therapeutics are for cancer, with a

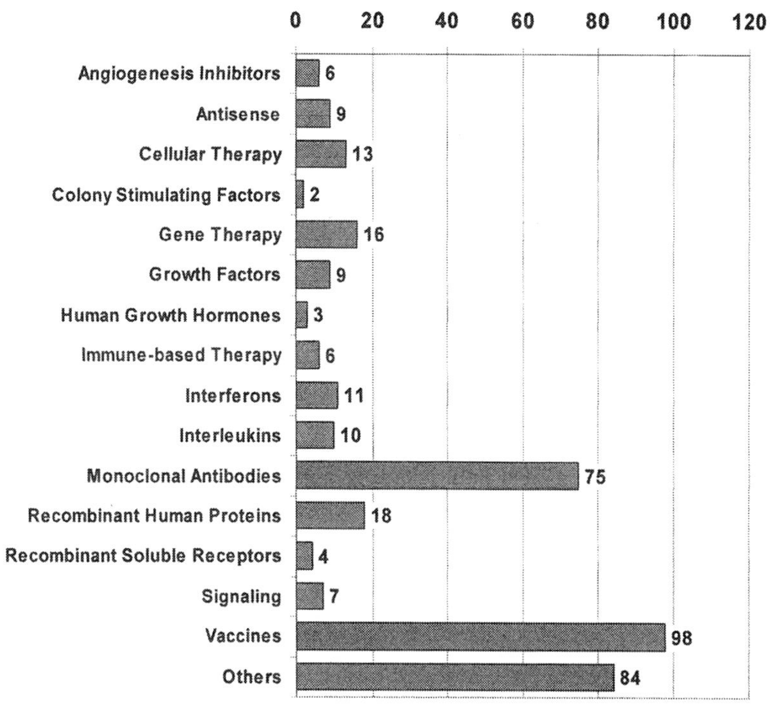

Figure 1 Biotechnology medicines under development in the United States by product category. [Data from Ref. (14).]

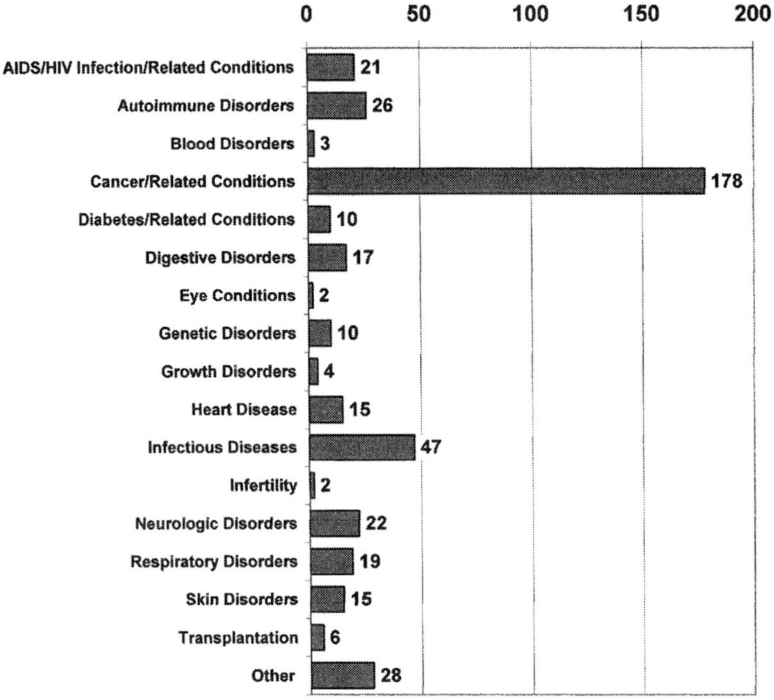

Figure 2 Biotechnology medicines under development in the United States by therapeutic category. [Data from Ref. (14).]

total of 178 cancer drugs in development by pharmaceutical companies and the National Cancer Institute. There are 47 drugs under development for infectious diseases, 26 for autoimmune diseases, and 21 for AIDS/HIV conditions (14). These drugs are at different stages of clinical trials and it is likely that some of them will get approved by the FDA. These drugs will be added to the 95 already approved biotechnology-derived products in the growing assenal against diseases.

The drugs under development are presented in Figure 2, categorized by product type (14). The majority of the drugs are vaccines (a total of 98) and monoclonal antibodies (a total of 75). There will be more medicines using recombinant human proteins, interferons and interleukins, growth factors, and recombinant soluble receptors. New categories of products such as angiogenesis inhibitors and immune-based therapy are on the horizon. In addition, there are 16 gene therapy protocols under development. Tissue engineering and cell therapy account for 13 trials.

The human genome project and sequencing the human gene will open new frontiers for biotechnology (2,15). Some researchers believe the human genome will result in instructions for recreating as many as 300,000 different proteins. This will help researchers increase the number of disease targets from 500 to 10,000. It is the hope of pharmaceutical companies that increased understanding of the human genome will cause a quantum leap in drug development, increasing not only the number of drug candidates but also the effectiveness of the drug, and the success rate in clinical trials (2). New developments in genomics and proteomics will help in translating the genetic information to protein products. Understanding of the molecular mechanism of the diseases and their relationships to the genes will be used to predict the likelihood of diseases and will offer early treatment options.

The future also looks bright for cellular therapies and gene therapy. As mentioned before, there are extensive efforts in these areas and treatments are on the horizon. The latest developments in the embryonic stem cell area could possibly generate tissues for repair and replacement therapies in the upcoming years. New gene delivery systems are being developed for gene therapy applications.

It is clear that cell culture technology has great potential for growth. The number of medicines produced from cell culture will increases, as well as the amounts to be produced. The potential requirements for monoclonal antibodies particularly will be huge. Optimizing the productivity and minimizing the cost of goods for these products will make it possible for the medicines to penetrate a larger market. The promise of biotechnology is fulfilled and; cell culture technology is here to stay.

REFERENCES

1. Das RC. Proteins and antibodies make advances as therapeutic products. Am Clin Lab 2001; 20(8):10–14.
2. PhRMA. Pharmaceutical Industry Profile. Pharmaceutical Research and Manufacturers of America, 2002.
3. Griffiths B. Animal cell products, overview. In: Spier RE, ed. Encyclopedia of Cell Technology. Vol. 1. New York: John Wiley & Sons, 2000:70–76.
4. Finter NB, Garland AJM, Telling RC. Large-scale mammalian cell culture: a perspective. Bioprocess Technology 1990; 10:1–14.
5. Lubiniecki AS, ed. Large Scale Animal Cell Culture Technology. New York: Marcel Dekker, 1990.
6. Petricciani JC. The acceptability of continuous cell lines: a personal and historical perspective. In: Beuvery EC, Griffith JB, Ziejlemaker WP, eds. Animal Cell Technology: Developments towards the 21st Century. Dordrecht: Kluwer Academic Publishers, 1995:605–609.
7. Lubiniecki AS. Historical reflections on cell culture engineering. Cytotechnology 1998; 28:139–145.
8. van der Velden-de Groot CAM. Microcarrier technology, present status and perspective. Cytotechnology 1995; 18:51–56.
9. Ozturk SS. Engineering challenges in high density cell culture systems. Cytotechnology 1996; 22:3–16.
10. Crommelin DJA, Sindelar RD. Pharmaceutical Biotechnology. 2nd ed. Philadephia, PA: Taylor and Francis, 2002.
11. Das RC. Antibodies in biotech: year of the bear. Am Biotechnol Lab 2003; 20:4–6.
12. Holliger PHH. Antibodies came back from the brink. Nat Biotechnol 1998; 16:1015–1016.
13. Smith AE. Gene therapy: where are we? Lancet 1999; 354:1–4.
14. PhRMA. Biotechnology New Medicines in Development. Pharmaceutical Research and Manufacturers of America, 2002.
15. Venter JC, et al. The sequence of human genome. Science 2001; 291:1304–1351.

2

Recombinant DNA Technology and Cell Line Development

Amy Y. Shen, Jana Van de Goor, Lisa Zheng, Arthur E. Reyes, and Lynne A. Krummen
Genentech, Inc., Department of Cell Culture & Fermentation R&D, South San Francisco, California, U.S.A.

CELL LINE DEVELOPMENT OVERVIEW

Production of recombinant proteins in animal cell culture for research, clinical development, and commercial purposes necessarily begins with the construction of a suitable cell line. The choice of expression system and mode of expression is generally based on the nature of the protein to be expressed, the quantity of material needed, and the timeframe over which maintained production is required. Mammalian expression systems are generally favored for the production of large, highly disulfide bound or glycosylated proteins. For production of small quantities of such molecules over a short timeframe, transient expression systems can be the method of choice whereas for clinical or commercial production, stable expression systems capable of continuous product expression over prolonged periods (months) generally are required. This chapter will discuss in detail the steps involved in the process of construction, selection, and screening of stable cell lines for industrial production of biotechnology products (Fig. 1).

Transient Expression

In transient expression systems or "transient transfections," DNA is introduced into the cell and maintained/replicated as an extrachromasomal unit. Besides the strength of the promoter driving expression of the product gene, transfection efficiency (percentage of cells taking-up and expressing DNA) is the most important factor in determining the overall expression level, and several methods have been recently developed which yield reproducibly high transfection efficiencies (~50%, see following text). The average production life of the transient expression system is usually limited by toxicity of the rapidly multiplying DNA, or the loss of DNA from the cell population during division. While not suitable for long-term production or commercial manufacture, expression levels in transient systems may range from >1 to 100 mg/L, making these systems extremely useful for rapid production

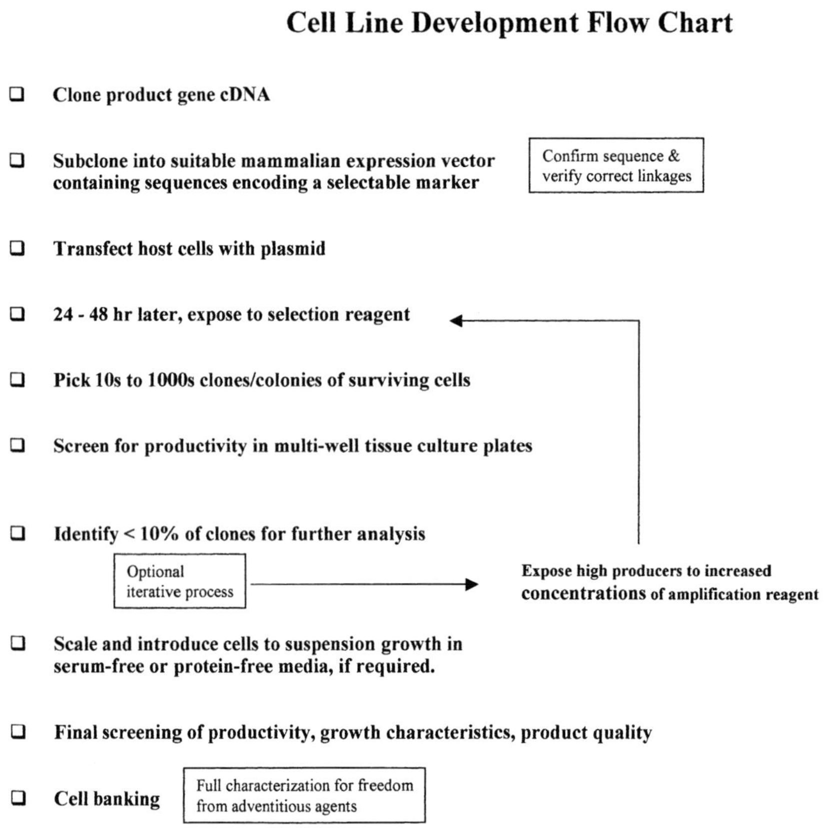

Figure 1 Cell line development flow chart.

of small quantities of research materials for drug candidate identification and in vivo evaluation. Moreover, methods for transient transfection of larger batch cultures, that will potentially enable scalability to at least the 100 L scale, are presently in development (1).

Stable Expression Systems

DNA containing the product gene must be integrated into the host cell genome for stable product production. Besides promoter strength, the frequency of DNA integration into the chromosome (copy number) as well as the position of integration, rather than the transfection efficiency are the more important factors in determining overall expression levels. There are no differences in transfection methods per se between transient and stable transfection. What does differ is the need to identify and select cells that have integrated the plasmid into the genome. This is most often accomplished by transfecting a sequence encoding a marker gene into the host cell along with the expression vector containing the product gene(s). Transfected cells are then cultured under defined conditions that allow for direct selection of cells that have incorporated the marker sequence into the genome and are expressing its gene product (2,3). Over the past decade, many vector design strategies have been developed which help ensure cointegration of the selection marker and the product gene. Thus, it is expected that most of the selected cells will also express the product gene(s).

Depending on the cell type, the selection system used, and the desired levels of product expression, the specific productivity of the initially selected stable cells may not be optimal. In this case there may be the possibility to further enhance protein expression by subjecting the cells to repetitive rounds of exposure to a selective agent that often results in gene amplification. Individual clones are generally screened for productivity at the 96-well scale, and lead candidates further evaluated for performance in conditions which model high cell density cell culture process conditions. In most instances it is also desirable to make some early evaluation of the biochemical quality of the product being produced. Based on data collected, a single cell line is chosen for continued process development. The cell line is banked and extensively characterized in accordance with specific Regulatory guidelines for freedom from adventitious agents to ensure suitability for cGMP manufacturing.

CELL SOURCE AND HOST CELL LINE SELECTION

There are several mammalian systems capable of yielding high-level expression of recombinant proteins that are suitable for use in the manufacture of protein pharmaceuticals. Cell lines derived from rodents such as Chinese hamster ovary (CHO) cells, baby hamster kidney (BHK) cells, and lymphoma cell lines such as NSO, SP2, and YB2/0 are the most commonly used. However, there are no restrictions on the type of host cell that can be used for commercial manufacturing provided the cells: (1) support high-level product expression over many months in culture, (2) can be scaled to large volumes (100–10,000 L), (3) can reach and maintain high viable cell density in an acceptable manufacturing process ($>5 \times 10^6$ cells/mL in batch culture, $>10^8$ cells/mL in perfusion), (4) provide appropriate posttranslational processing capabilities, and (5) can be appropriately characterized to assure freedom of the cell substrate from adventitious agents, especially viruses.

Reports in the scientific literature suggest that most of the cell lines commonly used for biopharmaceutical production are capable of high-level gene expression on the order of 10–100 pg/cell/day (4,5), although in practice yields >10 pg/cell/day have been more commonly achieved with monoclonal antibody products than with other types of recombinant proteins. Events at the transcriptional, translational, or posttranslational level have the potential to limit the cell specific productivity that can be achieved with any given protein. For many monoclonal antibodies, protein secretion appears to mirror mRNA expression levels up to specific productivities of 50 pg/cell/day or more (4). However, for other classes of proteins such as multimeric, cross-linked, or highly glycosylated growth factors and fusion proteins, posttranslational processing and protein folding may be the primary rate-limiting step.

The ability to achieve high-level mRNA expression appears to depend primarily on selecting and optimizing the right *combination* of host cell, expression vector, and selection strategy rather than on the attributes of any one particular host cell. For example, a panel of CHO cells deficient in dihydrofolate reductase (DHFR) are available that can be combined with a spectrum of vectors that combine promoters highly active in CHO cells with strategies to accomplish efficient transfectant selection using DHFR gene replacement (3,6). Similarly, for murine myeloma NSO cells deficient in expression of glutamine synthase (GS), use of vectors combining strong promoters/enhancers with efficient GS selection systems has allowed optimization of this system (7).

There are accumulating examples of engineering or selecting host cells for attributes that further increase the efficiency or robustness of protein production. These include coexpression of genes that enhance expression from a given promoter (*trans*- and *cis*-activators), promote posttranslational protein processing and folding, enhance ability to grow in nutritionally defined media (SUPER-CHO (8), VEGGIE-CHO (9)), improve cell cycle characteristics or provide improvements in product quality (10). Moreover, there is ongoing research directed towards marking transcriptionally active sites in the genome with a reporter gene flanked by recognition sequences that allow for the repetitive replacement of product genes into this same site. Some of these technology refinements will be touched upon briefly later in the chapter. These improvements seem equally suited to most host cells and can be applied to further enhance and optimize any well-developed expression system.

Glycosylation of recombinant glycoproteins is an attribute that has received much attention regarding the ability of various host cell lines to provide appropriate product quality (11–13). Data collected on glycoforms produced by CHO, BHK, and NSO cells indicate that the complex N-linked carbohydrate structures they synthesize contain the same basic oligosaccharide structure naturally occurring on human proteins (Fig. 2). However, each cell type has its own unique qualities. Murine cell lines, unlike hamster lines, may introduce significant quantities

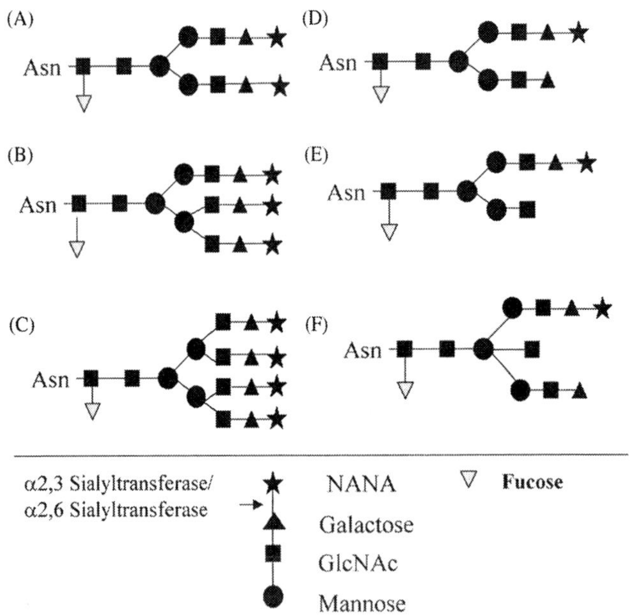

Figure 2 Examples of typical N-linked glycans found on recombinant proteinsproduced in mammalian cells. Structures generally contain a variable number of lactosamine branches containing *N*-acetylglucosamine (GlcNAc, squares) followed by a galactose (triangles) residue. These branches may be capped with a terminal *N*-acetyl neuraminic acid residue (sialic acid, NANA, stars, see bottom of figure). Biantennary (A), triantennary (B), and tetraantennary structures are common (C), although these structures may be undersialylated (D) or undergalactosylated (E). Cell line specific attributes, such as substitution of *N*-glycolyl neuraminic acid (NGNA) for NANA or addition of a bisecting GlcNAc structure (F), or variable fucosylation may occur.

of terminal *N*-glycolylneuramic acid (NGNA) in place of terminal NANA, or display an α-1,3-galactosyltransferase activity, that may lead to synthesis of glycan structures potentially immunogenic in humans. However, CHO and BHK cells lack the ability to provide certain structures normally found on some human glycoproteins, including terminal sialic acid residues attached in an α-2,6 linkage to galactose or a bisecting GlcNAc residue. Since the nature of the protein itself as well as the cell culture conditions also greatly influence the quality of the glycans attached to any given recombinant protein (14), the choice of cell line that provides acceptable product quality needs to be made in practice on a protein-by-protein basis.

VECTORS FOR EXPRESSION IN MAMMALIAN CELLS

Specialized vectors are required for expression of heterologous proteins in mammalian cells; [see Refs. 2,3 for general review]. Mammalian shuttle vectors by definition are plasmids that contain a cassette for expression of genes in mammalian cells as well as elements that allow plasmid replication in bacteria and selection of plasmid-containing bacteria. The β-lactamase gene that encodes an ampicillin-resistant protein and allows for selection of bacteria containing plasmid DNA in the presence of ampicillin is one example. Many vectors also include an M13 origin of replication that confers the ability to make single-stranded DNA useful for site-directed mutagenesis, although more efficient polymerase chain reaction (PCR)-based methodologies are now generally used to accomplish this task.

The cassette for gene expression in mammalian cells generally consists of two types of elements: promoter/enhancer elements that drive mRNA transcription, and sequences that help to stabilize or enhance translation of the primary transcript. When the vector is to be used for generation of a stable cell lines, sequences encoding a selectable marker gene are also either integrated into the product gene expression cassette directly or in a separate expression cassette on the same vector.

Promoters and Enhancers

Promoters are generally defined as upstream (5′) elements that promote the formation of a transcription complex with RNA polymerase II and accurately position the start of transcription. Elements like TATA sequences and the SP1 binding region (CCGCCC) occur frequently in eukaryotic promoters and are analogous to the −10 and −35 bp consensus sequences recognized by bacterial RNA polymerase. These regions are thought to act as binding domains for association of RNA polymerase II with the transcriptional factors required to initiate an active polymerase complex. Enhancers augment transcriptional complex assembly. They are generally, but not strictly, positioned upstream of the promoter, but their spacing from the promoter, and their sequence orientation is variable. Endogenous mammalian genes have a defined promoter coupled to one or more enhancers that operate in vivo.

For expression of recombinant proteins in mammalian cells, the product gene's endogenous promoter/enhancer elements are usually replaced by a strong viral promoter/ enhancer or by a promoter/enhancer combination known to be particularly active within the context of the chosen host cell. Many such sequences are available which work well in rodent-derived cell lines (15–18) including myeloproliferative sarcoma virus (MPSV) (16), Rous sarcoma virus (RSV) (17), VISNA (18), and the newly described CHO elongation factor-alpha (EF1-α) (19). The two most widely used promoter/enhancers are derived from the simian virus 40 (SV40) (20,21) and cytomegalovirus (CMV) (22). The SV40 promoter/enhancer sequence is about 400 bases long and is from the early transcriptional regulatory region. The CMV promoter/enhancer is about 650 bp long and derived from cytomegalovirus immediate early gene.

Cis- and *Trans*-Activators of Promoter Activity

Further augmentation of expression can be achieved by use of a host cell line that is selected or engineered to express a transcription factor capable of enhancing the activity of the chosen promoter. The viral *trans*-activating proteins Ela/Elb, for example, have been used to enhance expression driven by the CMV promoter (23,24). Alternatively, incorporation of sequences directly onto the plasmid that act in *cis* to increase promoter activity have been reported to be useful in some cases (19,25), although definitive studies confirming this effect are not yet available. These *cis*-acting sequences can themselves serve as response elements which enhance the activity of a nearby promoter through binding of an endogenous transcription factor or by imparting a structure to the chromatin in the vicinity of the promoter that allows enhanced access to nuclear regulatory factors (26). These later elements, known as matrix attachment regions (MARS) or scaffold-attachment regions (SARS) have also been hypothesized to delineate the boundary of transcriptionally active units within the genome and have been shown to neutralize the effects of integration location or "position-dependent" effects on expression of stably integrated transgenes. The elements often have little or negative effect on transient gene expression.

Elements that Stabilize and Enhance Translation of the Primary Transcript

Polyadenylation Signals

A unique property of the majority of eukaryotic mRNAs is their 3′ poly (A) tail. This poly (A) tract is added posttranscriptionally by poly A polymerase after the nascent mRNA has been released from the transcription complex. The site of addition of the poly (A) tail is signaled by the presence of the sequence AAUAAA in the 3′ untranslated region of the mRNA. The role of the poly A tail is not well understood, but is thought to be involved with export of the mRNA from the nucleus, prolonging the half-life of mRNA in the cytoplasm and enabling efficient translation (27). For construction of recombinant expression vectors, however, endogenous polyadenylation signal sequences are discarded and replaced by more generic elements. Two polyadenylation signals commonly used in construction of mammalian expression cassettes are derived from SV40 (28) and bovine growth hormone gene (29). These sequences usually contain an AATAAA consensus sequence, which is 20–30 nucleotides 5′ of a diffuse GT or T-rich region.

Kozak Sequence

A survey of sequences surrounding the initiator codon (the ATG codon that specifies methionine) of mammalian genes was carried out by Kozak to identify the optimum consensus sequence for this purpose. This sequence, referred to as the "Kozak" sequence (CCA/GCC) (30,31) is usually placed right in front of the first ATG codon of gene of interest to improve the translational initiation of mRNA.

Intervening Sequences

Mammalian genes contain an intervening sequence (IVS) or intron in their DNA. These introns are removed by a specific splicing reaction before mRNA transcripts are translocated to the cytoplasm (32,33). Although not required, many mammalian expression vectors also include an intron in the 5′ untranslated leaders of genes to be expressed. Nuclear processing of the message results in removal of the intron from the primary transcript. Studies have demonstrated that inclusion of an intron between the promoter and translational start of the expressed sequence can result in increased steady state cytoplasmic poly (A) mRNA levels (27). Available data suggest that this effect is the result of enhanced efficiency of polyadenylation of nascent transcripts in the nucleus.

Vectors for Regulated Expression of Genes in Mammalian Cells

The promoters described in the preceding text are employed to drive constitutive transcription of cDNA, leading to continuous mRNA and protein synthesis. However, there are situations where regulated or induced gene expression is desired; for instance, when an expressed protein is cytotoxic or cytostatic to host cells. To address these situations there have been several inducible mammalian expression systems developed. Early systems relied on inducers that had pleiotropic effects on host cells, for example, heat shock induction of the heat shock promoter, heavy metal induction of the metallothionine promoter, or glucocorticoid induction of steroid responsive promoters (34–36). These systems often displayed high basal or "leaky" expression and suffered from the broad effects of inducers on a spectrum of cellular genes. More recently developed inducible systems including the Lac/IPTG system (37,38), the tetracycline (tet) system (39–43), the streptagramin system (44), and the ecdysone system (45,46) appear to have solved many of these problems. Each system employs two plasmids, a regulatory plasmid for repressor or trans-activator expression and an expression plasmid containing the regulatable promoter linked to the product gene. Repressor or activator activity is regulated in each system by addition of a compound to the cell culture medium that modulates binding to the promoter's regulatory element, thus regulating the expression of the gene of interest. For example, one of the most widely used regulated systems is based on *Escherichia coli* tet repressor (Fig. 3) (39,40). In this system, a tet-responsive transactivator (tTA) fusion gene containing the tet repressor and VP16 activation domain from herpes simplex virus is expressed constitutively from a regulatory vector, which is integrated into the host cell genome. The modified host cell is then retransfected with an expression vector containing the gene of interest driven by a weak or minimal CMV promoter with an adjacent tet operator. The binding of the tTA drives the expression of the gene of interest to the tet operator. Addition of a tetracycline derivative to the media inhibits tTA binding to the operator and thus, represses transcription. When set up in this configuration, the system functions as a "tet-off" system. Modifications of this system as well as other similarly convertible on/off systems based on gene regulation by other antibiotics have also recently been described. The attractiveness of these expression systems is their ability to be used modularly to regulate both the product gene and one or more functional genes involved in complex regulatory pathways in a concerted fashion (42,43).

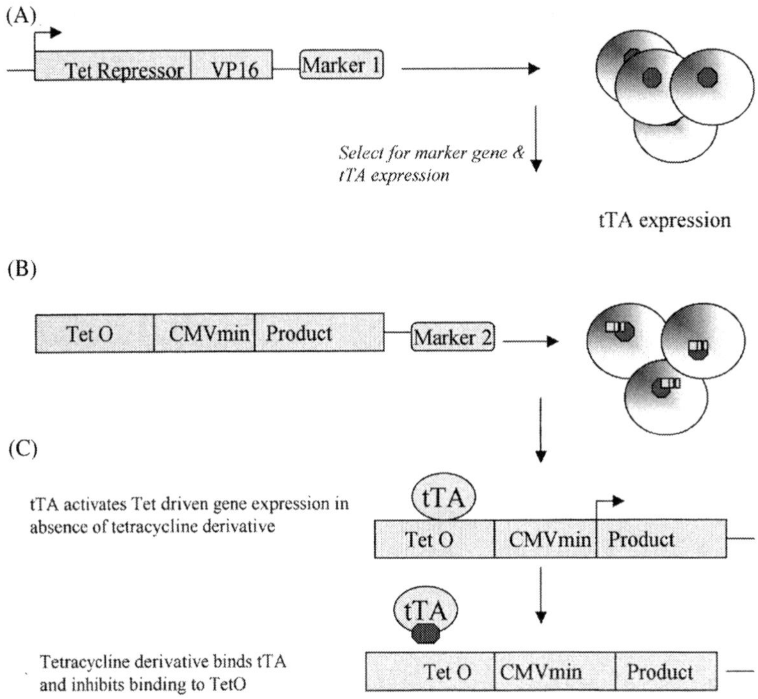

Figure 3 Diagram of expression system used to provide tetracycline-regulatable gene expression in mammalian cells. Cells are first transfected with a vector containing cDNA encoding the tet transactivator protein (tTA), which is a fusion of the tet repressor and VP16 activation domain. Cells selected for tTA expression are then used as hosts for a second transfection event with a plasmid containing the product gene driven by a minimal promoter (CMVmin) adjacent to the tet operator (Tet O). In the absence of tetracycline, tTA binds to Tet O and drives transcription of the product. Addition of tet derivatives interferes with this binding by competing for binding to tTA and thus, inhibits product transcription.

HIGH EFFICIENCY SELECTION OF STABLE CELL LINES

Selectable Markers

In order to differentiate cells that have integrated the plasmid into the host genome, a selectable marker that allows identification and selection of those cells is required. Moreover, by physically linking the expression of the marker gene and the product gene, and using selection strategies that impair the ability to express the marker gene product relative to product gene, it is possible to stringently select for cells producing high levels of product gene. There are several types of markers available for mammalian expression systems (Table 1) [see also Refs. (47–51) for additional examples]. These can be divided into dominant and recessive classes. A subset of each of these can be amplified, or the copy number of the marker gene can be further increased, by application of the proper selective agent (50,51).

Many of the dominant selectable markers are bacterial antibiotic resistant genes or genes that detoxify the effects of various protein synthesis inhibitors on the cell. For example puromycin *N*-acetyltransferase renders the cell resistant to the protein synthesis inhibitor puromycin. The "dominant" feature of these markers refers to the fact that there is no background activity of this sort expressed by the

Table 1 Selective Agents Commonly Used in Constructing Mammalian Expression Vectors

Selectable marker	Selective agent	Enzyme inhibitor used for gene amplification
Dominant		
Aminoglycoside phosphotransferase (aph)	Geneticin (G418)/Neomycin	
Hygromycin-B phosphotransferase (hph)	Hygromycin B	
Puromycin *N*-acetyltransferase (pac)	Puromycin	
Zeocin resistance gene (sh ble)	Zeocin	
Recessive		
Glutamine synthase (GS)	Glutamine-free media	Methionine sulfoximine (MSX)
Dihydrofolate reductase (DHFR)	Media lacking glycine, hypoxanthine and thymidine	Methotrexate (MTX)

cell, and as a result, selection using these markers does not rely on availability of a host cell that contains a mutant or impaired endogenous gene.

Neomycin-, hygromycin-, and puromycin-resistant genes are very commonly used dominant selectable markers. The expression level of the resistance gene per cell, and of any other cointroduced genes, is dependent on the position and number of integrations achieved and the strength of the promoters driving their expression. The expression level of associated product genes can be improved by optimizing the concentration of selective agent used to select for cells which have acquired the highest expression of the marker gene. However, further enhancement of gene expression by repeated exposure to increasing concentrations of the selective agent using these selective markers has not been demonstrated to be generally useful.

In contrast to dominant markers, recessive markers generally restore a critical enzyme or protein to the cell that is required for survival, and therefore require host cells that are mutant or deficient in that activity. Often the activity is required for cellular DNA synthesis. In the nontransfected host cell, the requirement for the activity is replaced by supplementing the media with the metabolic byproduct of the enzyme's action. Upon withdrawal of these media components cell survival is dependent on acquiring the needed activity through transfection.

For expression of product genes in CHO cells the recessive marker DHFR is often used for selection of transfected cells. DHFR is responsible for the production of tetrahydrofolate (FH4) from folic acid (Fig. 4) (3). In the absence of FH4, the primary pathway for synthesis of purines and pyrimidines is inhibited. These metabolites can be made by the cell via salvage pathways that involve the conversion of hypoxanthine to GMP and AMP and conversion of thymidine to TMP via thymidine kinase activity. As long as cells deficient in DHFR are supplied with glycine, hypoxanthine, and thymidine from an exogenous source, they can make DNA and grow. If the media is not supplemented with these precursors the cells will die unless they have acquired the ability to express DHFR through transfection. The availability of several strains of DHFR (minus) CHO cells (52), many of which provide a perfect host for this marker, are available from public culture collections. In

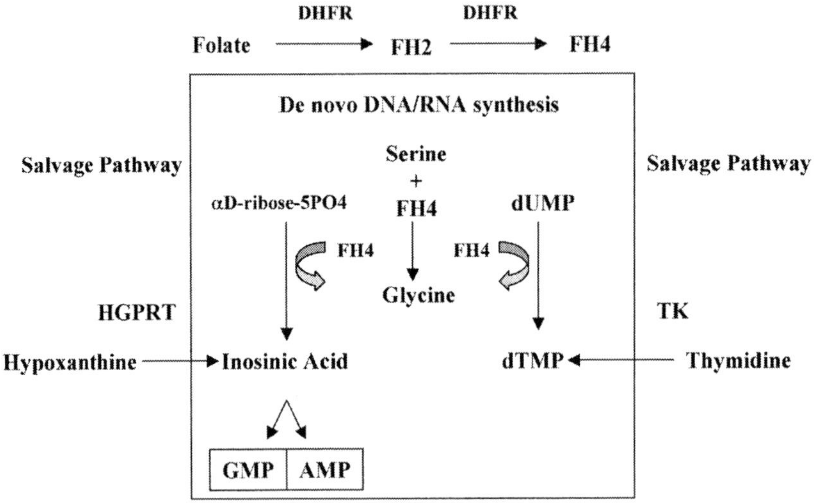

Figure 4 Diagram of folate metabolism in mammalian cells.

addition, this marker allows the option of being able to identify cells which may have increased or amplified integrated gene copy numbers and expression level after exposure to methotrexate (MTX), an agent that binds to and irreversibly inhibits DHFR (see section titled "Amplification of Transfected Sequences to Enhance Production").

Similarly, myeloma and hybridoma cell lines have an absolute requirement for glutamine due to insufficient activity of glutamine synthase (GS) (7), a cellular enzyme responsible for formation of glutamine from glutamate and ammonia. Supplementation of cellular GS level through transfection restores the cell's ability to grow in glutamine-free media, and so serves as an appropriate selectable marker in these cells. Moreover, in the presence of methionine sulfoximine (MSX) a specific inhibitor of the GS enzyme, efficient amplification of gene expression can be accomplished.

Although both MTX and MSX can be used as selective agents with wild-type cells integrating DHFR- and GS-containing plasmids, respectively, the high background of surviving nontransfected cells and the excessively high concentrations of MTX and MSX required usually undermine the success of these selections.

Selection Vector Design

When selectable markers are used to identify cells that have stably integrated the vector sequences into their genome, they are essentially being used as surrogates to identify cells that express the product. However, the linked integration of the marker gene and the product gene is not always guaranteed, and it is possible that cells which exhibit drug-resistant phenotypes, or survive in nutrient deficient media may not express the product. Initially, genes for the product and the selectable marker were introduced as independent expression cassettes on one or more plasmids.

Although this approach can be used successfully, there is an increased risk of producing drug-resistant colonies that do not coexpress the product, which increases the demand on secondary product expression screening strategies.

In order to increase the frequency of cells overexpressing both a high level of product and the marker gene, recent vector designs have emphasized optimization of product-to-selectable marker gene expression ratios and improving the efficiency of selecting cells with product gene coexpression.

At a simple level, putting the marker gene under control of a weaker promoter than that driving the product gene or reducing the efficiency of the Kozak sequence (4) has been associated with increased efficiency of selection of higher expressing cell lines. Alternatively, some investigators have created a dicistronic expression vector where the product gene, driven by a strong promoter, is followed by an internal ribosome-binding sequence derived from picornavirus (IRES) inserted in front of the marker gene (53–56). These elements allow ribosome binding and cap-independent initiation of marker gene translation. However, since the translation of genes downstream of IRES often is not very efficient, these dicistronic vectors minimize marker gene synthesis while insuring the linkage of integrated gene of interest and selectable marker.

Other approaches include impairment of selectable marker gene expression so that extremely high levels of marker gene transcription are required in order for cells to make sufficient marker gene product to survive. This is the goal of vector designs that place the selectable marker within an intron inserted between the promoter and the 5′ initiation site of the product gene (57). Imperfect splice sites that direct inefficient processing of the intron during nuclear splicing flank the marker sequence. Initially all the RNA transcripts produced will contain sequences encoding both marker and product. However, during nuclear processing, the marker sequence will be spliced out of the majority of the transcripts (95%). Since only the few transcripts containing both genes will express the selection marker, the system is biased towards tight selection of cells producing high levels of mRNA transcripts and therefore increased levels of product gene (Fig. 5A). Alternatively, the product gene expression

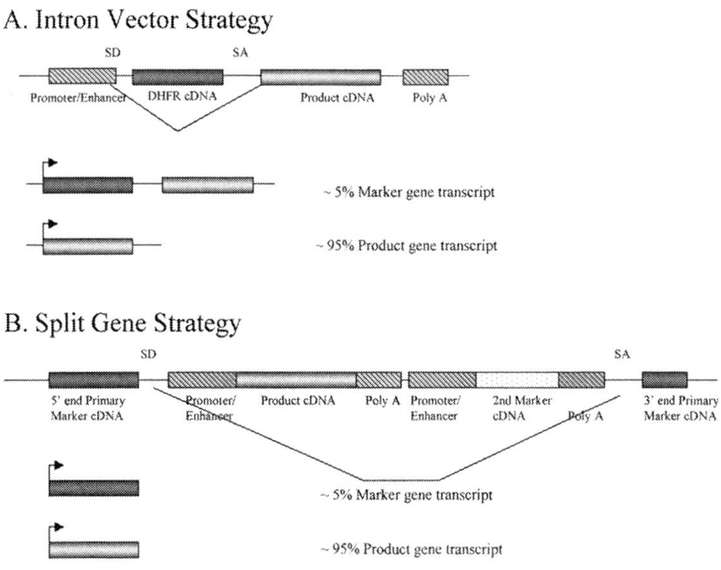

Figure 5 Examples of two vector design strategies that allow efficient selection of highly productive recombinant cell lines by physically linking expression of product to marker gene on the plasmid and increasing the product gene to marker gene mRNA expression ratio.

cassette has been placed within an artificial intron inside a selectable bacterial marker gene (a "split gene," Fig. 5B) (4). In this system only a small percentage of transcripts are correctly processed to allow expression of marker gene. Thus, the cell is required to produce large amounts transcript to provide adequate selection protection. Data presented on the mechanism of action of this later system suggests that high-level expression is achieved from only a few integrated gene copies, suggesting that the stringent selection system directs integration of the vector into chromosomal locations that provide exceptional levels of transcriptional activity.

Targeted Integration

Cell lines stably expressing a protein product are typically generated by transfection followed by integration of the expression plasmid into the host genome via nonhomologous recombination. Such integration is a random event and generates clones with a wide range of expression levels (Fig. 6), reflecting the gene copy number integrated and the transcriptional activity of the locus in which the copies were integrated (positional effect). The concept that high expression loci exist in the genome has lead to the development of several strategies for targeting gene integration specifically to these transcriptionally active loci, thus, introducing a controlled integration event.

The targeted integration approach is comprised of two steps. First, a transcriptionally active locus in the host genome is identified by transfecting a plasmid encoding a reporter gene and a selectable marker (employing vector design strategies described in Chapter 4) flanked by specific recombination target sites. Clones are selected under an appropriate selection pressure and a large screening is performed to identify a high expressing clone. In the second step, a plasmid carrying a product gene and a different selectable marker, flanked again by the recombination target sites, is transfected into the clone together with a second plasmid that transiently expresses a site-specific recombinase. This recombinase recognizes the specific sequence of the target sites and promotes a recombination event between them, termed homologous recombination. This event is highly specific and ensures integration of the product gene at the preexisting target site. All clones generated in this manner display an identical site of integration (isogenic clones) and should, therefore, express similar levels of the transgene. This approach eliminates the need for

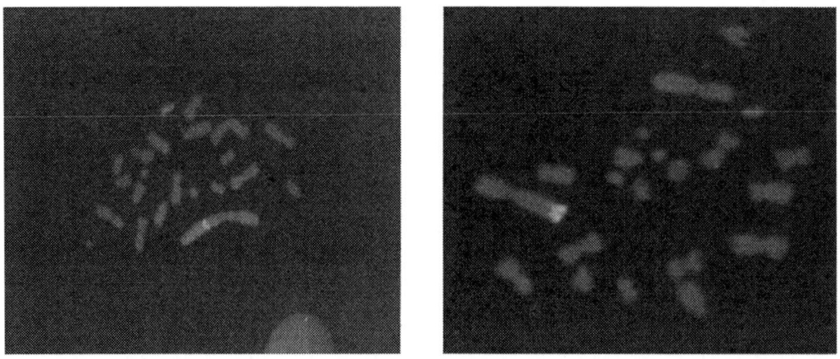

Figure 6 Fluorescent in situ hybridization of recombinant sequences integrated into the chromosomes of CHO cells. Photos are of two different cell lines resulting from random integration of the plasmid DNA during a single transfection event (*See color insert p. 1.*)

screening a large number of clones, and potentially minimizes/eliminates the need for amplification, thus shortening the time required to develop a highly productive cell line. In addition, once identified, the "marked" host cell line can be used for recombination and expression of any desired gene.

Several approaches have been investigated, using different recombinases, such as Cre recombinase from the bacteriophage PI (58), which recognizes Lox P sites, or Flp recombinase from *Saccharomyces cerevisiae*, which recognizes FRT sites (59). Both Lox P and FRT sites are 34 bp long and asymmetrical. Depending on the orientation of the recognition sites, the recombinase may catalyze excision/integration (recognition sequences are in the same orientation) or inversion (opposite orientation). For the application described here, the excision/integration reaction is utilized. Both recombinases have been shown to work efficiently in mammalian cells (58–61).

TRANSFECTION: INTRODUCTION OF PLASMID DNA INTO MAMMALIAN CELLS

A variety of methods to introduce DNA into mammalian cells have been developed. They can be divided into four classes: chemical, physical, lipid-based lipofection, and polymer or dendrimer based. Different cell types may be more amenable to one approach than to another. Detailed discussion of the methods for DNA transfection is included elsewhere in this volume. Therefore, only a brief discussion is given here.

Calcium Phosphate Method

Graham et al. (62) first described the calcium phosphate transfection method in 1973. It is inexpensive and works in a wide range of cell types. In this method, calcium chloride, phosphate buffer and DNA coprecipitate and form a fine precipitate or complex which enters mammalian cells via an endocytic vesicle (62–64). This method is still widely used and variations of this transfection protocol including a DMSO, glycerol, or chloroquine treatment posttransfection have been incorporated to improve transfection efficiency (63). Transfection efficiencies are generally <25% using this method. The primary disadvantages of this technique are that the efficiency of transfection is related to the size of the calcium phosphate precipitate and the concentration of DNA. Calcium phosphate crystals nucleate immediately upon reagent combination and continue to grow rather rapidly depending upon the relative concentrations of calcium chloride and phosphate, the pH of the reagents and temperature. Therefore, the method needs to be optimized carefully since minor alterations in procedure or environment can dramatically alter the outcome of the transfection.

Electroporation

Electroporation utilizes a pulsed electric field to disrupt the voltage gradient that exists across the plasma membrane and create reversible pores that allow DNA to enter the cell (65,66). It is a simple and rapid method. It involves a direct physical interaction between electrical field and cell membrane so that it is less cell type specific compared with other transfection methods. Electroporation parameters like peak voltage and fall time of the discharge waveform need to be empirically determined in each case. In addition, electroporation generally results in lower posttransfection cell viability.

Lipofection and Polyfection

Polyfection and lipofection are the most recently developed and probably simplest transfection methodologies for gene delivery in a variety of cell types. Lipofection refers to cationic lipid-mediated gene transfer into the cells. The cationic lipids containing a quaternary amine form positively charged liposomes that interact with negatively charged DNA to form a complex. Various models have been proposed to describe how DNA and lipids interact. The most popular model, supported by in situ optical microscopy and X-ray diffraction, proposed the complex as lipoplexes consisting of a higher-ordered multilamellar structure with DNA sandwiched between cationic layers (67–69). The lipoplexes presumably enter the cell by endocytosis or direct cell membrane fusion and DNA probably is released from lipoplexes before entry into the nucleus (70). Lipofection can be done in the presence or absence of serum with little or no toxicity. The most critical parameters for optimal lipofection are the amount of DNA and lipid used as well as the ratio between DNA and lipid (71–73). Too little DNA results in low transgene expression whereas large amounts are cytotoxic to host cell. The optimization of the ratio of DNA and lipid is probably important because net positive charge of the complex is required to facilitate interaction between lipoplex and the negatively charged cell surface.

Polyfection refers to gene transfer mediated by cationic polymers and dendrimers. The cationic polymers, especially highly branched and spherical dendritic macromolecules have a high number of surface primary amino groups that interact with DNA forming a polyplex that protects DNA from degradation before reaching nucleus (74,75). There are many lipofection and polyfection reagents commercially available and the most efficient method for a given cell culture system should be empirically determined.

AMPLIFICATION OF TRANSFECTED SEQUENCES TO ENHANCE PRODUCTION

When exposed to cytotoxic drugs, some cells can develop a drug-resistant phenotype. Although there are mechanisms to generate drug tolerance that do not employ gene amplification, amplification of DNA in response to growth-limiting conditions is a well-known phenomenon (51), and one that can be exploited in order to maximize expression of recombinant products. In this case, the gene that is targeted for amplification is the transfected marker gene. During the process of amplification, both the target gene and surrounding DNA sequences are coamplified via a process that may involve over-replication of DNA and/or recombination of sequences. As a result, amplification of the marker gene provides an indirect means of amplifying the cointegrated product gene(s) (50,51,76–78). MTX irreversibly binds to and inhibits the activity of DHFR, therefore dose-dependently limiting the cells ability to produce tetrahydrofolate. Under conditions in which the MTX concentration is limiting for growth, amplification of the DHFR gene frequently occurs. This effect has been exploited in the development of highly productive commercial cell lines. A similar effect can be obtained using the GS selection marker and MSX. The DNA replicated during the isolation of a drug-resistant phenotype is structurally referred to as an amplicon, and can contain a few to thousands of copies of the amplified gene (79). Increases in the gene copy number of a given cell line are often associated with increased product expression or secretion. However, since not all gene copies may be transcriptionally active or other bottlenecks to protein expression may occur, cell lines with high gene copy numbers may not always be highly productive (79).

The initial selection of the transfected cells in selective media or in low concentrations of the selective agent is generally thought to identify cells which have either integrated multiple copies of the plasmid or which have single or low copy number integrations at exceptionally active transcription loci. The more stringent the selection, the higher the initial expression level of the cell pool and the more frequently high expressors will occur within this pool. In cases where amplification is desired, these initial selectants are then exposed to stepwise increases in the toxic agent. The steepness of the increase in selective agent during gene amplification protocols seems to have an impact on whether drug-resistant cells have increased copy number of the integrated sequences and co-express high levels of drug product or have acquired drug resistance through mechanisms not involving gene amplification (80). Detailed analyses of integration site, gene copy number, and specific productivity of clones resulting from various types of selection and amplification strategies are in progress to better define the types of events which lead to highly productive, stable cell lines (81,82). However, at present, the optimal selection and amplification procedures appear to be highly dependent on the design of the expression construct and needs to be determined independently for each system.

Stability of Amplified Gene Expression

The resistance phenotype and amplified gene(s) can be stable or unstable. When a population of cells that have been recently subjected to an amplification event are maintained in the absence of selective pressure, the amplified gene can decline in the population over a period of weeks or months depending on the cell type. This can be due to continued rearrangements of structures within the amplicon and loss of genetic material during subsequent cell divisions. The more heterogeneous the population becomes, the more likely instability will result in significant changes in the productivity and growth rate of the population. However, though exceptions occur, cells maintained in the presence of selective agent generally exhibit stable production throughout many months in culture. For cells cultivated in batch mode, where only production culture and one or two passages preceding are non-selective, instability of expression due to loss of the gene is generally not a problem. However, for cells in continuous production cultures that run over long periods in the absence of selective agent, gene stability is more critical and care should be taken to screen the cell line for stability or to devise expression strategies that minimize the need for amplifications.

SCREENING AND DEVELOPMENT OF PRODUCTION CELL LINES

During the initial selection process and/or following the final round of amplification, individual clones are isolated from the cell pool and screened for desirable attributes such as high specific productivity (pg/cell/day), good growth rate, and suitable biochemical product quality. Generally the objective of the first stage of screening is to rapidly evaluate a large number of cells, using the smallest possible number of cells per evaluation, to find the highest producers.

Fluorescence-Activated Cell Sorting-Based Methods

The number of clones that need to be screened in order to identify members of the population with acceptably high productivity will be determined by the ability of the chemical selection method to find high producers within the transfected population and/or the availability of a rapid high throughput screen for cell line evaluation. Fluorescence-activated cell sorting (FACS) methods represent the potentially most elegant means of high-throughput screening. Most methods described below can be used successfully to separate non- or low producers from highly productive cells and as such represent an attractive way to enrich pools of transfected cells prior to more traditional clonal screening. However, to date, a practical FACS method that can be used to reliably identify and sort the highest individual viable cells from a pool of already high producing cells remains a challenge. The following methods are among the most commonly explored while trying to develop a robust high-throughput screen.

For cell culture systems where gene of interest expression is linked to DHFR expression, fluorescent MTX binding has been used as a measure of the level of DHFR expressed by transfected cells (81). This method requires the incubation of the test cells with high concentrations of FITC-MTX, but is generally easily performed on viable cells. The drawback of this technique is that DHFR expression, particularly with traditional expression constructs, does not always correlate directly to product gene expression. Although as vector strategy has improved to further link DHFR to product gene, the success of this method has improved.

Green fluorescent protein (GFP) and related proteins that exhibit autofluorescence also make attractive surrogate markers of gene expression (82,83). As with fluorescent MTX binding, this requires tight physical linkage of the gene(s) of interest with the marker to effectively identify high producers. The advantage of GFP is that there is no need to add exogenous reagents to cells. However, reported toxicity of high levels of GFP may self-limit the usefulness of this marker for development of the most highly productive cell lines (84), although a nontoxic red-shifted mutant of this marker is now commercially available which may circumvent this problem.

Other FACS-based methods developed recently employ a solid or semisolid matrix for cell suspension that contains a labeled antibody that traps secreted product in the vicinity of the cell's surface (85–88). The accumulation of cell-associated product is taken as a measure of cell specific productivity and is used to separate high secretors. Although there are technical challenges to capturing and recovering single viable cells in the matrix, these methods have the advantage that they do not use a surrogate to monitor gene of interest expression and, unlike fluorescent-MTX or GFP, they measure secreted protein rather than intracellular accumulation.

Cell Culture Performance Assays

More commonly, clones resulting from transfection and/or amplification steps are isolated by limiting dilution, by picking of well-isolated colonies on low-density plates, or by similar means and are evaluated by more traditional cell culture assays quantitating clone productivity. The end point of such assays can be specific productivity, which is the amount of product produced per cell per unit time (often pg/cell/day), or volumetric productivity, which is the product of specific productivity and the cumulative viable cell mass present in the culture over time. Ultimately, it is the volumetric productivity, or the titer (expressed in mg/L) at the end of a given manufacturing cell culture run which is important. However, at early stages of cell screening, the growth properties of cells growing in dishes can be very different from what they will be once the cells have been adapted to suspension and/or serum-free (or protein-free) growth. Therefore, at the initial stages of screening the potential of the cell lines is usually ranked based on specific productivity followed by volumetric productivity evaluations as the growth characteristics of the cell lines are developed.

In order to evaluate specific productivity among hundreds to thousands of clones efficiently, a moderately high throughput measure of product secretion or expression and a measure of cell number that can be adapted to a 96- or 384-well format is needed. Titer estimates can be derived from measurement of the accumulated product in cell culture fluid using an immunoassay (enzyme linked immunosorbent assay or ELISA) or bioassay (functional assay) for the protein. A robust ELISA is desirable because of its high throughput and high sensitivity aspects. Optical measurements of cell density, either of light transmission through the cells themselves or of a dye that identifies viable cells and can be read in a microtiter plate reader is used as a rough measure of total cell number in each well. Calcein-AM is an example of a viable cell dye that is easily used for this purpose (Fig. 7) (89). The native form of this compound is freely diffusible into cells and is converted to an impermeable, fluorescent species by non-specific cellular esterase activity. Unless there are differences in the volume of cells, the amount of fluorescence per cell is constant among cells of the same type and the intensity of the fluorescent signal per well corresponds well to the total number of viable cells.

If no immunoassay or bioassay is available, clones can be screened for mRNA expression. Development of higher throughput quantitative mRNA analyses using real-time quantitative PCR protocols coupled to higher throughput mRNA extraction technology has certainly made these types of screens much more feasible and

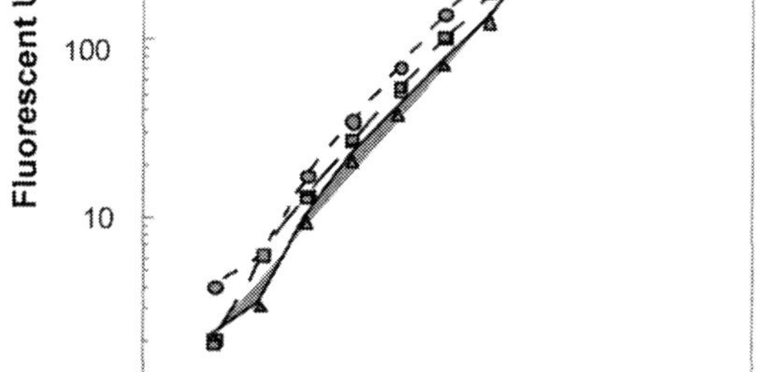

Figure 7 Correlation of fluorescence and number of cells per well for CHO cells grown in 96-well tissue culture plates. Fluorescence was measured in a 96-well plate reader after various lengths of exposure to calcein-AM.

affordable. mRNA screens may not be useful in extreme cases where protein secretion is generally limited by factors other than transcription, but for most applications, cells that produce high quantities of product mRNA are good targets to move into more complex screening paradigms.

Generally two to three rounds of microtiter scale screening are employed to segregate the top 1–5% of clones that will be expanded for further evaluation.

Scale-Up and Adaptation to Suspension, Serum-Free Growth

Clones identified based on high specific productivity or mRNA expression are generally expanded on tissue culture plates until enough cells are available to inoculate them into a suspension culture. Depending on the cell type, two T-175 tissues culture flasks usually contain sufficient cells to inoculate a spinner flask with a 200 mL working volume with $\sim$3–4 $\times$ 10^5 cells/mL. For cell lines that are maintained as adherent cultures in serum-containing media, suspension adaptation can often be accomplished quite readily in the same serum-supplemented media in spinner vessels maintained at 50–80 rpm. Altering the media or adding components that discourage adhesion of cells to solid substrates, or use of specially treated glass and plastic-ware can facilitate this process. The primary consideration in methods choice should be robustness and compatibility of the resulting condition of the cells to growth under manufacturing conditions. During this procedure it is critical to initiate or "seed" the cultures at a suitably high cell density and to monitor growth and viability closely over the first several passages. For cell lines that are resistant to culture in suspension, media optimization altering the concentrations of cations may be necessary (90). Several commercial vendors have specialized media available for this purpose.

Cell lines that are intended for commercial cell culture processes should also be adapted to serum-free or protein-free growth conditions early in development. In fact, the desire to eliminate as many animal-derived raw materials as possible from commercial cell culture operations has lead to the development in some instances of transfection and selection procedures which completely remove the need for serum in the process. Where this is not possible, adaptation of cells to serum-free growth is routinely performed. This process is not well understood physiologically and is technically more challenging than adaptation of cells to suspension growth in serum-containing media. Appropriate supplementation of the basal media powder with growth factors, trace elements, lipids and vitamins is critical for success (90,91). Some protocols call for slowly reducing the serum concentration of media, whereas others favor a complete removal of serum in a single step. In either case, cells often will show a slowed growth profile and decreased viability for several passages after the complete withdrawal of serum and need to be carefully monitored to ensure the success of this procedure. In many instances, several media exchanges using centrifugation of cells away from spent media and resuspension in fresh media need to be performed in order to remove waste products and maintain an adequate seeding density (3–10 $\times$ 10^5 cells/mL). It is important that selective pressure be maintained on the gene of interest throughout this process to ensure retention of high-level gene expression. However, the concentration of selective agent and of supplemental growth factors may need to be temporarily modified during this procedure to ensure success.

After cells have adapted to suspension and/or serum-free growth, their growth properties are much closer to what they will be in a large-scale cell culture process.

At this stage the cell lines can be compared for volumetric productivity and product quality using assays designed to model conditions encountered in the extended, high-density culture conditions typical of large-scale industrial cell culture processes. Such assays usually involve both plate-based and small-scale bioreactor formats. Plate based designs provide the ability to look at a broader number of cell lines to develop a sense of which lines perform best under defined process conditions. However, performance of cells in these assays does not always mirror cell performance under the pH and dissolved oxygen-controlled environment of the bioreactor. Bioreactors allow information to be developed on a limited number of cell lines and for key process parameters that might alter a given cell lines performance to begin to be identified. Several new high throughput systems that more closely approximate the gas exchange and control typical of fermentation systems, without the complexity of fermentor set-up and monitoring offer an attractive compromise for cell line selection studies.

Many aspects of product quality, ranging from gross estimates of molecular weight and presence or absence of glycans to detailed analysis of glycan structure or charge variants using mass spectrophotometric and chromatographic techniques can be monitored during this final selection period to ensure that the final clone selected for further process development has the most desirable cell culture and product properties.

GENETIC ENGINEERING OF HOST CELL LINES TO IMPROVE PRODUCTION CHARACTERISTICS

One of the most notable advances in recent years has been the application of genetic engineering approaches to rationally modify specific features of mammalian host cells to improve their utility in recombinant protein expression applications. One such example is in the area of glycosylation control. A breadth of work in the last decade has demonstrated how various reactions in the glycosylation pathway can be influenced by cell culture factors, host cell selection, and protein specific features (92–94) leading to production of molecules with variable or suboptimal clearance or bioactivity properties. Other chapters in this volume contain a detailed discussion of the importance of glycosylation control in maintaining a glycoprotein's efficacy and/or clearance properties. In the context of expression of recombinant glycoproteins for pharmaceutical manufacture, recent studies have demonstrated that specific manipulation of oligosaccharide structures on a recombinant protein can be achieved by overexpression of appropriate glycosyltransferases. This strategy can be used to either enhance glycan quality by increasing the homogeneity of native structures or by introducing non host cell residues to specialize glycan quality and function. For example, the overexpression of a galactosyltransferase and a sialyltransferase in CHO cells has been shown to lead to corresponding increases in the galactose and sialic acid levels on recombinant therapeutic proteins expressed by these cells (Fig. 8) (95). Other groups have successfully overexpressed *N*-acetylglucosaminyltransferase III in order to increase the fraction of bisecting *N*-acetylglucosamine residues on antibodies produced in CHO cells (96,97) or introduced sialic acid in an α-2,6 linkage to glycoproteins synthesized by CHO and BHK cells that lack the specific sialyltransferase responsible for this transfer (98–100).

Another focus of genetic manipulation has been to improve the efficiency of central carbon metabolism and the reduction of lactate accumulation. Earlier work

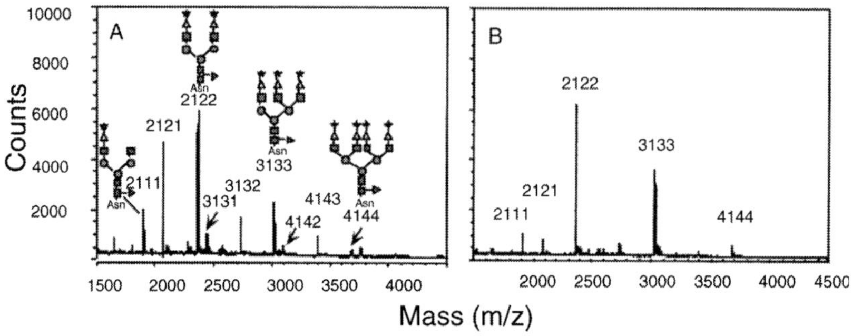

Figure 8 Remodeling of N-linked glycans on a recombinant protein through overexpression of glycosyltransferases in CHO cells. Insets show mass spectrophotometric analysis of the individual glycan structures associated with the protein. Glycan structures are identified by a nomenclature consisting of four digits that denote number of branches, presence of a core fucose sugar, number of galactose residues, and number of sialic acids. A 2122 structure is biantennary, contains core fucose and is fully galactosylated and sialylated. (A) shows the heterogeneity of the glycans produced by the unmodified cell line and (B) shows the effect of over expression of α-2,3 sialyltransferase in the same cell line. Note the increased presence of fully sialylated glycans compared to the control condition.

showed that overexpression of a pyruvate carboxylase gene in BHK cells reduced lactate accumulation and improved cell yields although it is not clear if this approach will be generically helpful (101). Partial disruption of the LDH-A gene has also been used to reduce lactate accumulation and improve cell culture performance in hybridoma cultures (102).

Perhaps the most interesting of the targets for metabolic control are cell growth and apoptosis (programmed cell death). A significant body of work performed during the last few years lends collective support to the hypothesis that control of progression through the cell cycle may significantly increase productivity in mammalian cell cultures. For example, recent reports suggest that productivity may vary dramatically in different growth phases (103,104).

Work demonstrating the effectiveness chemical inhibitors of apoptosis in extending culture viability has laid the foundation for a growing area of work directed towards genetic engineering of cell death pathway, including overexpression of bcl-2 and inhibition of caspase expression (105–108).

CONCLUSIONS

The success of cell line development efforts depends in part on understanding the optimal use of the components of a given expression system and in part on the availability of adequate high-throughput methods to identify cell lines that will perform well in the context of the process. Over the past decade, developments in host cell line selection, vector technology, selection and screening strategies have combined to enable stable, high-level expression of many recombinant genes in mammalian cells for clinical and commercial supply. Since timely development of robust, highly productive cell lines is key to maximizing the efficiency of the drug development process and ultimately minimizing the cost of manufacturing, current efforts in the field are directed towards rapid, consistent generation of highly productive cell lines and genetic engineering of host cell lines to display functional attributes that improve product consistency and quality and robustness of performance in large-scale cell culture processes.

REFERENCES

1. Girad P, Derouzari M, Baumgartner G, Bourgeois M, Jordan M, Wurm FM. 100 liter transfection. 17th ESACT Meeting, Conference Abstract, Tyleosand, Sweden, June 10–14, 2001.
2. Chisholm V. High efficiency gene transfer into mammalian cells. In: Glover DM, Hames BD, eds. DNA Cloning: A Practical Approach. Vol. IV. Mammalian Systems. Oxford: Oxford University Press, 1995:1.
3. Levinson et al. Section V. Expression in mammalian cells. In: Goedell DV, ed. Methods in Enzymology. New York: Academic Press, 1990:485.
4. Barnett RS, Limoli KL, Huynh TB, Ople EA, Reff ME. Antibody production in Chinese hamster ovary cells using an impaired selectable marker. Proceedings of the American Chemical Society, 94, San Diego, CA, 1995:27–40.
5. Turner BG. Manufacturing proves development for D2E7, a recombinant human antibody to tumor necrosis factor. Cell Culture Engineering VIII, Conference Abstract, Snowmass, CO, April 1–5, 2002.
6. Wood CR, Dorner AJ, Morris GE, Alderman EM, Wilson D, O'Hara R, Kaufman RJ. High level synthesis of immunoglobulins in Chinese hamster ovary cells. J Immunol 1990; 145:3011.
7. Bebbington CR, Renner G, Thomson S, King D, Abrams D, Yarranton GT. High-level expression of a recombinant antibody from myeloma cells using a glutamine synthetase gene as an amplifiable selectable marker. Biotechnology 1992; 10:169.
8. Pak SCO, Hunt SMN, Bridges MW, Sleigh MJ, Gray PP. Super-CHO a cell line capable of autocrine growth under fully-defined protein-free conditions. Cytotechnology 1996; 22:139–146.
9. Rasmussen B, Davis R, Thomas J, Reddy, P. Isolation, characterization and recombinant protein expression in veggie-CHO: a serum-free CHO host cell line. Cytotechnology 1998; 28:31.
10. Andersen DC, Krummen LA. Recombinant protein expression for therapeutic applications. Curr Opin Biotechnol 2002; 13:117.
11. Bhatia PK, Mukhopadhay A. Protein glycosylation: implications for in vivo functions and therapeutic applications. Adv Biochem Eng Biotech 1998; 64:157.
12. Jenkins N, Curling EMA. Glycosylation of recombinant proteins: problems and prospects. Enzyme Microb Technol 1994; 16:354.
13. Grabenhorst E, Schlenke P, Pohl S, Nimtz M, Conradt HS. Genetic engineering of recombinant glycoproteins and the glycosylation pathway in mammalian host cells. Glycoconj J 1999; 16:81.
14. Andersen D, Goochee CF. The effect of cell culture conditions on the oligosaccharide structures of recombinant glycoproteins. Curr Opin Biotechnol 1994; 5:546.
15. Colosimo A, Goncz KK, Holmes AR, Kunzelmann K, Novelli G, Malone RW, Bennett MJ, Gruenert DC. Transfer and expression of foreign genes in mammalian cells. BioTechniques 2000; 29:314.
16. Hilberg F, Stocking C, Ostertag W, Grez M. Functional analysis of a retroviral host-range mutant: altered long terminal repeat sequences allow expression in embryonal carcinoma cell. Proc Natl Acad Sci USA 1987; 84:5232.
17. Gorman CM, Merlina GD, Willingham MC, Pastan I, Howard BH. Proc Natl Acad Sci USA 1982; 79:6777.

18. Carruth LM, Morse BA, Clements JE. The leucine domain of the Visna Virus Tat protein mediates targeting of an AP-1 site in the viral long terminal repeat. J Virol 1996; 70:4338.
19. Allison D. Development and use of CHEF1 vectors for rapid isolation of mammalian cell lines expressing high levels of recombinant proteins. Cell Culture Engineering VIII, Conference Abstract, Snowmass, CO, April 1–5, 2002.
20. Tooze J. The SV40 nucleotide sequence. Tooze J, ed. DNA Tumor Virus. Part 2. Cold Spring Harbor: Cold Spring Harbor Laboratory Press, 1982:799.
21. Carswell S, Alwine J. Efficiency of utilization of the simian virus 40 late polyadenylation site: effects of upstream sequences. Mol Cell Biol 1989; 9:4248.
22. Boshart M, Weber F, Jahn G, Dorsch-Hasler K, Fleckenstein B, Schaffher W. A very strong enhancer is located upstream of an immediate early gene of human cytomegalovirus. Cell 1985; 41:521.
23. Gorman CM, Gies D, McCray G, Huang, M. The human cytomegalovirus immediate early promoter can be trans-activated by adenovirus early proteins. Virology 1989; 171:377.
24. Cockett MI, Bebbington CR, Yarranton GT. The use of engineered Ela genes to transactivate the hCMV-MIE promoter in permanent CHO cell lines. Nucleic Acids Res 1990; 19:319.
25. Morris AE, Lee C-C, Hodges K, Aldrich TL, Krantz C, Smidt PS, Thomas JN. Expression augmenting sequence element (EASE) isolated from Chinese hamster ovary cells. In: Carrondo MJT, et al, eds. Animal Cell Technology. Dordrecht, The Netherlands: Kluwer Academic Publishers, 1997:529.
26. Zahn-Zabal M, Kobr M, Girod P-A, Imhof M, Chatellard P, deJesus M, Wurm F, Mermod N. Development of stable cell lines for production or regulated expression using matrix attachment regions. J Biotechnol 2001; 87:29.
27. Huang MTF, Gorman CM. Intervening sequences increase efficiency of RNA 3′ processing and accumulation of cytoplasmic RNA. Nucleic Acids Res 1990; 18:937.
28. Carswell S, Alwine J. Efficiency of utilization of the simian virus 40 late polyadenylation siter:effects of upstream sequences. Mol Cell Biol 1989; 9:4248.
29. Miller WL, Martial JA. Molecular cloning of DNA complementary to bovine growth hormone mRNA. J Biol Chem 1980; 25:7521.
30. Kozak M. Point mutations define a sequence flanking the AUG initiator codon that modulates translation by eucaryotic ribosomes. Cell 1986; 44:283.
31. Kozak M. Adherance to the first-AUG rule when a second AUG codon follows closely upon the first. Proc Natl Acad Sci USA 1995; 92:2662.
32. Petitclerc D, Attal J, Theron MC, Bearzotti M, Bolifraud P, Kann G, Stinnakre MG, Pointu H, Puissant C, Houdebine LM. The effect of various introns and transcription terminators on the efficiency of expression vectors in various cultured cell lines and in the mammary gland of transgenic mice. J Biotechnol 1995; 40:169.
33. Abrams JM, Thotpe SM, Schimke RT. Intronic positioning maximizes co-expression and co-amplification of nonselectable heterologous genes. J Biol Chem 1989; 264:14016.
34. Schweinfest CW, Jorcyk CL, Fujiwara S, Papas TS. A heat shock-inducible eukaryotic expression vector. Gene 1998; 71:207.
35. Israel DI, Kaufman RJ. Highly inducible expression from vectors containing multiple GRES in CHO cells overexpressing the glucocorticoid receptor. Nucleic Acids Res 1989; 12:4589.
36. Yarranton GT. Inducible vectors for expression in mammalian cells. BioTechnology 1992; 3:506.
37. Hu MC, Davidson N. The inducible lac operator-repressor system is functional in mammalian cells. Cell 1987; 48:555.
38. Kamashev DE, Esipova NG, Ebralidse KK, Mirzabekov AD. Mechanism of Lac repressor switch-off: orientation of the Lac repressor DNA-binding domain is reversed upon inducer binding. FEBS Lett 1995; 375:27.

39. Gossen M, Bujard H. Tight control of gene expression in mammalian cells by tetracycline-responsive promoters. Proc Natl Acad Sci USA 1992; 89:5547.
40. Gossen M, Freundlieb S, Bender G, Muller G, Hillen W, Bujard H. Transcriptional activation by tetracyclines in mammalian cells. Science 1995; 268:1766.
41. Hoffmann A, Villalba M, Journot L, Spengler D. A novel tetracycline-dependent expression vector with low basal expression and potent regulatory properties in various mammalian cell lines. Nucleic Acids Res 1997; 25:1078.
42. Fussenegger M, Schlatter S, Datwyler D, Mazur X, Bailey JE. Controlled proliferation by multigene metabolic engineering enhances the productivity of Chinese hamster ovary cells. Nat Biotechnol 1998; 16:468.
43. Kaufmann H, Mazur X, Marone R, Bailey JE, Fussenegger M. Comparative analysis of two controlled proliferative strategies regarding product quality, influence on tetracycline-regulated gene expression and productivity. Biotechnol Bioeng 2001; 72:592.
44. Fussenegger M, Morris RP, Fux C, Rimann M, von Stockar B, Thompson CJ, Bailey JE. Streptogramin-based gene regulation systems for mammalian cells. Nat Biotechnol 2000; 18:1203.
45. Christopherson KS, Mark M, Bajaj V, Godowski P. Ecdysteroid-dependent regulation of genes in mammalian cells by a Drosophila ecdysone receptor and chimeric transactivators. Proc Natl Acad Sci USA 1992; 89:6314.
46. No D, Yao TP, Evans RM. Ecdysone-inducible gene expression in mammalian cells and transgenic mice. Proc Natl Acad Sci USA 1996; 93:3346.
47. Etcheverry T. Expression of engineered proteins in mammalian cell culture. In: Cleland JL, Craik CS, eds. Protein Engineering: Principles and Practice. New York: Wiley-Liss, 1996:163.
48. Blochlinger K, Dinggelmann H. Hygromycin B phosphotransferase as a selectable marker for DNA transfer experiments with higher eukaryotic cells. Mol Cell Biol 1984; 4:2929.
49. Bebbington CR. The use of vectors based on gene amplification for the expression of cloned genes in mammalian cells. In: Glover DM, Hames BD, eds. DNA Cloning. Vol. IV. Mammalian Systems. Oxford: Oxford University Press, 1995:163.
50. Simonsen CC, Levinson AD. Isolation and expression of an altered mouse dihydrofolate reductase cDNA. Proc Natl Acad Sci USA 1983; 80:2495.
51. Schimke RT. Gene amplification in cultured cells. J Biol Chem 1988; 263:5989.
52. Urlaub G, Chasin LA. Isolation of Chinese hamster cell mutants deficient in dihydrofolate reductase. Proc Natl Acad Sci USA 1980; 77:4216.
53. Jang SK, Davies MV, Kaufman RJ, Wimmer E. Initiation of protein synthesis by internal entry of ribosomes into the 5′ nontranslated region of encephalomyocarditis virus RNA in vitro. J Virol 1989; 63:1651.
54. Gurtu V, Yan G, Zhang G. IRES bicistronic expression vectors for efficient creation of stable mammalian cell lines. Biochem Biophys Res Commun 1996; 229:295.
55. Hennecke M, Kwissa M, Metzger K, Oumard A, Kröger A, Schirmbeck R, Reimann J, Hauser H. Composition and arrangement of genes define the strength of IRES-driven translation in bicistronic mRNSs. Nucleic Acids Res 2001; 29:3327.
56. Mielke C, Tümmler M, Schübeler D, Von Hoegen I, Hauser H. Stabilized, long-term expression of heterodimeric proteins from tricistronic mRNA. Gene 2000; 254:1.
57. Lucas BK, Giere LM, DeMarco RA, Shen A, Chisholm V, Crowley CW. High-level production of recombinant proteins in CHO cells using a dicistronic DHFR intron expression vector. Nucleic Acids Res 1996; 24:1774.
58. Sauer B, Henderson N. Site-specific DNA recombination in mammalian cells by the Cre recombinase of bacteriophage PI. Proc Natl Acad Sci USA 1988; 85:5166.
59. Dymecki SM. Flp recombinase promotes site-specific DNA recombination in embryonic stem cells and transgenic mice. Proc Natl Acad Sci USA 1996; 93:6191.
60. Fox DT, Wahl GM. Recombinase-mediated gene activation and site-specific integration in mammalian cells. Science 1991; 251:1351.

61. Karreman S, Hauser H, Karreman C. On the use of double FLP recognition targets (FRTs) in the LTR of retroviruses for the construction of high producer cell lines. Nucleic Acids Res 1996; 24:1616.
62. Graham FL, Van der Eb AJ. A new technique for the assay of infectivity of human adenovirus 5 DNA. Virology 1973; 52:456.
63. Wilson SP, Smith LA. Addition of glycerol during DNA exposure enhances calcium phosphate transfection. Anal Biochem 1997; 246:148.
64. Brandt CR, Buonaguro FM, McDougall JK, Galloway DA. Plasmid mediated mutagenesis of a cellular gene in transfected eukaryotic cells. Nucleic Acids Res 1987; 15:561.
65. Xie TD, Tsong TY. Study of mechanisms of electric field-induced DNA transfection. V. effects of DNA topology on surface binding, cell uptake, expression, and integration into host chromosomes and DNA in the mammalian cell. Biophys J 1993; 65:1684.
66. Canatella PJ, Karr JF, Petros JA, Prausnitz MR. Quantitative study of electroporation-mediated molecular uptake and cell viability. Biophys J 2001; 80:775.
67. Eastman SJ, Siegel C, Tousignant J, Smith AE, Cheng SH, Scheule RK. Biophysical characterization of cationic lipid:DNA complexes. Biochem Biophys Acta 1997:41.
68. Felgner PL, Rinegold GM. Cationic liposome mediated transfection. Nature 1989; 337:387.
69. Boussif O, Lezoualch F, Zanta MA, Mergny MD, Scherman D, Demeneix B, Behr JP. A versatile vector for gene and oligonucleotide transfer into cells in culture and in vivo: polyethylenimine. Proc Natl Acad Sci USA 1995; 92:7297.
70. Radler J, Koltover L, Salditt T, Safinya CR. Structure of DNA-cationic liposome complexes: DNA intercalation in multilamellar membranes in distinct interhelical packing regimes. Science 1997; 275:810.
71. Gao X, Huang L. Cationic liposome-mediated gene transfer. Gene Ther 1995; 2:710.
72. Gao X, Huang L. Novel cationic liposome reagent for efficient transfection of mammalian cells. Biochem Biophys Res Commun 1997; 179:280.
73. Gershon H, Ghirlando R, Guttman SB, Minsky A. Mode of formation and structural features of DNA-cationic liposome complexes used for transfection. Biochemistry 1993; 32:7143.
74. Haensler J, Szoka FC. Polyamidoamine cascade polymers mediate efficient transfection of cells in culture. Bioconjug Chem 1993; 4:372.
75. Kukowska-Latallo JF, Bielinska AU, Johnson J, Spindler R, Tomalia DA, Baker JR. Efficient transfer of genetic material into mammalian cell using Starbust polyamidoamine dendrimers. Proc Natl Acad Sci USA 1996; 93:4897.
76. Grouse GF, McEwan RN, Pearson ML. Expression and amplification of engineered dihydrofolate reductase minigenes. Mol Cell Biol 1983; 1:854.
77. Murray M, Kaufman RJ, Latt SA, Weinberg RA. Construction and use of a dominant, selectable marker: a Harvey-sarcoma virus-dihydrofolate reductase chimera. Mol Cell Biol 1983; 3:32.
78. Cockett MI, Bebbington CR, Yarranton GT. High level expression of tissue inhibitor of metalloproteinases in Chinese hamster ovary cells using glutamine synthetase gene amplification. BioTechnology 1990; 8:662.
79. Wurm, FM, Johnson A, Ryll T, Kohne C, Scherthan H, Glaab F, Lie YS, Petropoulous CJ, Arathoon R. Efficient methods for maximizing specific productivity and assessment of genetic consequences. Ann N Y Acad Sci 1996; 782:70.
80. Wurm FM. Integration, amplification and stability of plasmid sequences in CHO cell cultures. Biologicals 1990; 18:159.
81. Sherwood SW, Schimke RT. Flow cytometric analysis of gene amplification in culture mammalian cells. In: Yen A, ed. Flow Cytometry: Advanced Research and Clinical Applications. Boca Raton, FL: CRC Press, 1989:85.
82. Sabath DE, Shin M-H. Use of green fluorescent protein/Flp recombinase fusion protein and flow cytometric sorting to enrich for cells undergoing Flp-mediated recombination. Biotechniques 2000; 28:966.

83. Meng YG, Liang J, Wong WL, Chisholm V. Green florescent protein as a second selectable marker for selection of high producing clones from transfected CHO cells. Gene 2000; 242:201.
84. Liu H-S, Jan M-S, Chou C-K, Chen P-H, Ke N-J. Is green fluorescent protein toxic to the living cells? Biochem Biophys Res Commun 1999; 260:712.
85. Weaver JC, Bliss JG, Powell KT, Harrison GI, Williams GB. Rapid and clonal growth measurements at the single cell level: gel microdrops and flow cytometry. Biotechnology 1991; 9:873.
86. Gray F, Kennedy JS, Dunne JF. Secretion capture and report web: use of affinity derivatized agarose microdroplets for the selection of hybridoma cells. J Immunol Methods 1995; 182:155.
87. Holmes P, Al-Rubeai M. Improved cell line developement by a high throughput affinity capture surface display technique to select for high secretors. J Immunol Methods 1999; 230:141.
88. Borth N, Zeyda M, Katinger H. Efficient selection of high-producing subclones during gene amplification of recombinant Chinese hamster ovary cells by flow cytometry and cell sorting. Biotechnol Bioeng 2001; 71:266.
89. Ihrig TJ, Tsao M, Hilton M, Jacobson F, Sliwkowski MB. Cell culture raw materials screening by calcein-AM fluorescence using a 96 well plate format. Animal Cell Technology: Products for Today, Prospects for Tomorrow. Oxford, UK: Butterworth Heinemann, 1994:110.
90. Mather JP, Moore A, Shawley R. Optimization of growth, viability and specific productivity for expression of recombinant proteins in mammalian cells. In: Tuan R, ed. Methods in Molecular Biology. Vol. 62. Recombinant Gene Expression Protocols. Totowa, NJ: Humana Press, 1995:369.
91. Zang M, Trautmann H, Gandor C, Messi F, Asselbergs F, Leist C, Fiechter A, Reiser J. Production of recombinant proteins in Chinese hamster ovary cells using a protein-free cell culture medium. Biotechnology 1995; 13:389.
92. Goochee CF, Gramer MU, Andersen DC, Bahr JB, Rasmussen JR. The oligosaccharides of glycoproteins: bioprocess factors affecting oligosaccharide structure and their effect on glycoprotein properties. Biotechnology 1991; 9:1347.
93. Jenkins N, Parekh RB, James DC. Getting the glycosylation right: implications for the biotechnology industry. Nat Biotechnol 1996; 14:975.
94. Grabenhorst E, Schlenke P, Pohl S, Nimtz M, Conradt HS. Genetic engineering of recombinant glycoproteins and the glycosylation pathway in mammalian host cells. Glycoconj J 1999; 16:81.
95. Weikert S, Papac D, Briggs J, Cowfer D, Tom S, Gawlitzek M, Lofgren J, Mehta S, Chisholm V, Modi N, et al. Engineering Chinese hamster ovary cells to maximize sialic acid content of recombinant glycoproteins. Nat Biotechnol 1999; 17:1116.
96. Umaña P, Jean-Mairet J, Moudry R, Amstutz H, Bailey JE. Engineering glycoforms of an antineuroblastoma IgGl with optimized antibody-dependent cellular cytotoxic activity. Nat Biotechnol 1999; 17:176.
97. Davies J, Jiang LY, Pan L-Z, LaBarre MU, Andersen D, Reff M. Expression of GnTIII in recombinant anti-CD20 CHO production cell line: expression of antibodies with altered glycoforms leads to an increase in ADCC through higher affinity for FcγRIII. Biotechnol Bioeng 2001; 74:288.
98. Monaco L, Marc A, Eon-Duval A, Acerbis G, Distefano G, Lamotte D, Engasser J-E, Soria M, Jenkins N. Genetic engineering of α2,6-sialytransferase in recombinant CHO cells and its effects on the sialylation of recombinant interferon-γ. Cytotechnology 1996; 22:197.
99. Grabenhorst E, Horrman A, Nimtz M, Zettlmeissl G, Conradt HS. Construction of stable BHK-21 cells coexpressing human secretory glycoproteins and human Gal (β1-4) GlcNAc-R α2,6-sialytransferase. Eur J Biochem 1995; 232:718.

100. Lee UE, Roth J, Paulsen JC. Alteration of terminal glycosylation sequences on N-linked oligosaccharides of Chinese hamster ovary cells by expression of β-galactoside α2,6-sialytransferase. J Biol Chem 1989; 164:13848.
101. Irani N, Wirth M, van den Heuvel J, Wagner R. Improvement of primary metabolisms of cell cultures by introducing a new cytoplasmic pyruvate carboxylase reaction. Biotechnol Bioeng 1999; 66:238.
102. Chen K, Liu Q, Xie L, Sharp PA, Wang DIC. Engineering of a mammalian cell line for reduction of lactate formation and high monoclonal antibody production. Biotechnol Bioeng 2001; 72:55.
103. Miller WM, Blanch HW, Wilke CR. A kinetic analysis of hybridoma growth and metabolism in batch and continuous suspension culture: effect of nutrient concentration, dilution rate and pH. Biotechnol Bioeng 1988; 32:947.
104. Fussenegger M, Schlatter S, Dätwyler D, Mazur X, Bailey JE. Controlled proliferation by multigene metabolic engineering enhances the productivity of Chinese hamster ovary cells. Nat Biotechnol 1998; 16:468.
105. Mastrangelo AJ, Hardwick JM, Zou S, Betenbaugh MU. Part II. Overexpression of Bci-2 family members enhances survival of mammalian cells in response to various culture insults. Biotechnol Bioeng 2000; 67:555.
106. Laken HA, Leonard MW. Understanding and modulating apoptosis in industrial cell culture. Curr Opin Biotechnol 2001; 12:175.
107. Kim NS, Lee GM. Overexpression of Bci-2 inhibits sodium butyrate-induced apoptosis in Chinese hamster ovary cells resulting in enhanced humanized antibody production. Biotechnol Bioeng 2001; 71:184.
108. van de Goor J. Apoptosis pathway in CHO cells. Proceedings of the Second European Meeting on Cell Engineering, Costa Brava, Spain 2001.

3
Medium Development

A. Burgener and M. Butler
Department of Microbiology, University of Manitoba, Manitoba, Canada

INTRODUCTION

The development of culture media for mammalian cells has been studied for more than 50 years. The first attempts at culturing animal cells in vitro made use of biological fluids, such as serum and other blood or tissue extracts. This was followed by the attempt to culture animal cells in defined media through the analysis of the contents of biological fluids (1). Another approach, developed by Eagle, consisted of finding the minimum ingredients that were essential for growth, which led to the development of Eagle's minimal essential medium (EMEM) (2). This medium consisted of 13 amino acids, 8 vitamins, 6 ionic species, and dialyzed serum to provide the necessary undefined components required for growth.

As new cell lines became available in the scientific community, new formulations were developed. Many of these cell lines could be cultured in Eagle's MEM, while others required more complex formulations. These included Dulbecco's modification of Eagle's medium (DMEM) (3), F12 (4) medium, and Roswell Park Memorial Institute (RPMI) medium (5). As progress was made in the understanding of cell metabolism and growth factor requirements, various serum-free formulations were also developed.

Presently, there are many formulations available for the culture of animal cells. The decision of which formulation to use is dependent on the purpose of the culture. For the production of viruses or other nonspecific molecular studies, basic formulations such as serum-supplemented MEM are often used. However, for other studies where undefined components can affect the results, or in large-scale production systems where productivity is an issue, serum-free formulations are relied upon.

The supplementation of culture media with undefined components, such as serum, has many inherent disadvantages. This has fueled the demand for better serum-free media in both the research and the industrial communities. New formulations are being developed all the time, and there are presently many available from various commercial suppliers. However, the performance of many of these formulations remains poor as many of them contain undefined components that can affect their quality and consistency.

This chapter focuses on the composition of classic animal culture media, with emphasis on the development of serum-free formulations.

CULTURE MEDIA

Basal Media

The components of these media formulations include a complex mixture of carbohydrate, amino acids, salts, vitamins, hormones, and growth factors.

The media originally used for the growth of animal cells were based entirely on biological fluids such as plasma and embryonic extracts. However, the use of chemically undefined media suffers from the disadvantages of batch variation and vulnerability to contamination. This led to a need for chemically defined culture media, which were originally based on the analysis of plasma, and this led to complex formulations such as Medium 199, which contains over 60 synthetic ingredients.

An alternative approach in media design is to reduce the number of components to the minimum shown to be essential for cell growth. This strategy led to Eagle's basal medium (BME), which was designed for the optimal growth of mouse-L cells and HeLa cells. It should be noted that there are several versions of BME. This was later modified and improved as EMEM, which has also found wide application for the growth of a variety of cell lines.

Although there are numerous media formulations now available for cell growth, the following list indicates those that have found greatest application for the cell technologist. The choice of a particular medium for a cell type has to be empirical but those listed are the ones that would normally be a starting point for attempting the growth of any new cell line.

Simple Basal Media

Table 1 shows the composition of several commonly used basal culture media developed for mammalian cell cultures.

- *BME* was originally designed for the growth of mouse-L and HeLa cells and has been used subsequently for the growth of a wide range of cell lines including human diploid cells. The medium needs to be changed at least every other day to support continued cell growth. It should be noted that there are several versions of BME (2).
- *EMEM* was developed as an improvement of BME for the optimal growth of a wide variety of cell lines including clonal cultures. EMEM has a higher concentration of amino acids than BME. It contains a balanced salt solution (Earle's); alternatively Hanks' salt solution can be used. The medium can support cell growth for several days (6).
- *Glasgow's modification of Eagle's medium (GMEM)* is a modification of BME and contains 2× the concentrations of the amino acids and vitamins with extra glucose and bicarbonate. This medium was originally developed for the growth of baby hamster kidney 21 (BHK21)/C13 cells and is usually supplemented with 10% tryptose phosphate broth as well as serum (7–10).
- *Joklik's modified Eagle's medium* is a general-purpose growth medium for suspension cultures (10).
- *Alpha modified Eagle's medium* was originally used to study ribosomes in mouse–hamster hybrid cells. The medium, which contains both essential and nonessential amino acids, can be used as a general-purpose growth medium (11).
- *DMEM* has a high nutrient concentration and includes 4× the BME concentration of amino acids and vitamins as well as additional nonessential

Table 1 Simple Basal Media Formulations Used in Mammalian Cell Culture

	Molarity (mM)			
Components	BME	GMEM	DMEM	RPMI (1640)
Inorganic salts				
Calcium chloride ($CaCl_2$)	1.8	4.5	1.8	
Calcium nitrate ($Ca(NO_3)_2\ 4H_2O$)				0.424
Cupric sulfate ($CuSO_4 \cdot 5H_2O$)				
Ferric nitrate ($Fe(NO_3) \cdot 3–9H_2O$)		0.25	0.25	
Ferrous sulfate ($FeSO_4 \cdot 7H_2O$)				
Potassium chloride (KCl)	5.3	37.33	5.3	5.3
Potassium nitrate (KNO_3)				
Magnesium chloride ($MgCl_2$ anhydrous)	0.5	11.26		
Magnesium sulfate ($MgSO_4$)	0.81		0.813	0.407
Magnesium sulfate ($MgSO_4$ anhydrous)		11.31		
Sodium chloride (NaCl)	117	110	110.34	103.44
Sodium bicarbonate ($NaHCO_3$)	26.2	4.17	44.1	23.8
Sodium phosphate ($Na_2HPO_4 \cdot H_2O$)				5.63
Sodium phosphate ($Na_2HPO_4 \cdot H_2O$)	1.01	7.43	0.906	
Energy metabolism				
D-Glucose	5.55	3.88	25	11.1
Sucrose		78		
D-Fructose		2.22		
Fumaric acid		0.474		
α-Ketoglutaric acid		2.53		
Malic acid		5		
Succinic acid		0.508		
Amino acids				
β-Alanine		2.2		
L-Alanine		2.5		
L-Arginine-HCl	0.1	3.3	0.398	1.1
L-Asparagine		2.65		0.379
L-Aspartic acid		2.63		0.15
L-Cystine-2HCl	5.0×10^{-2}	9.2×10^{-2}	0.2	0.206
L-Glutamic acid		4.08		0.136
L-Glutamine		4.11	4	2.05
Glycine		8.67	0.399	0.133
L-Histidine	5.2×10^{-2}	16.2	0.2	9.7×10^{-2}
L-Hydroxyproline				0.153
L-Isoleucine	0.198	0.382	0.802	0.382
L-Leucine	0.198	0.573	0.802	0.382

(*Continued*)

Table 1 Simple Basal Media Formulations Used in Mammalian Cell Culture (*Continued*)

	Molarity (mM)			
Components	BME	GMEM	DMEM	RPMI (1640)
L-Lysine-HCl	0.2	3.42	0.798	0.219
L-Methionine	0.05	0.336	0.201	0.101
L-Phenylalanine	0.1	0.909	0.4	9.1×10^{-2}
L-Proline		3.04		0.174
L-Serine		10.5	0.4	0.286
L-Threonine	0.202	1.47	0.078	0.168
L-Tryptophan	2.0×10^{-2}	0.49	0.078	2.5×10^{-2}
L-Tyrosine-2Na•$2H_2O$	10.0×10^{-2}	0.276	0.398	0.11
L-Valine	0.2	0.855	0.803	0.171
Vitamins				
Biotin	4.0×10^{-3}	4.1×10^{-5}		8.0×10^{-3}
D-Calcium pantothenate	2.0×10^{-3}	4.2×10^{-5}	8.3×10^{-3}	5.0×10^{-4}
Choline chloride	7.1×10^{-3}	1.4×10^{-3}	2.9×10^{-2}	2.1×10^{-2}
Folic acid	2.2×10^{-3}	4.5×10^{-5}	9.1×10^{-3}	2.2×10^{-3}
i-Inositol	1.1×10^{-2}	1.0×10^{-4}	4.0×10^{-2}	1.9×10^{-1}
Nicotinic acid (Niacin)		1.0×10^{-4}		
Nicotinamide	8.1×10^{-3}		3.3×10^{-2}	8.1×10^{-3}
p-Aminobenzoic acid		1.0×10^{-4}		7.2×10^{-3}
Pyridoxal-HCl	4.9×10^{-3}	9.7×10^{-5}	2.0×10^{-2}	4.8×10^{-3}
Pyridoxine hydrochloride				
Riboflavin	2.7×10^{-4}	5.3×10^{-5}	1.1×10^{-3}	5.0×10^{-4}
Thiamine-HCl	2.9×10^{-3}	5.9×10^{-5}	1.2×10^{-2}	2.9×10^{-3}
Vitamin B_{12}				3.7×10^{-6}
Lipids and derivatives				
i-Inositol	1.1×10^{-2}	1.0×10^{-4}	4.0×10^{-2}	1.9×10^{-1}
Glutathione (reduced)				3.2×10^{-3}
Indicators				
Phenol red	2.5×10^{-2}		3.5×10^{-2}	1.3×10^{-2}

amino acids and trace elements. The original formulation contained 5.6 mM glucose but 25 mM has been proved to be optimal for the growth of several cell types. It was first reported for the culture of embryonic mouse cells but has since found a wide application in the culture of various cells, including primary mouse and chicken cells, and in virus production.

- *RPMI 1629 or McCoy's 5A medium* was based on BME with the amino acid and vitamin mixture from Medium 199. The medium was originally formulated by McCoy et al. (12) and later modified by Hsu and Kellogg (13) and Iwakata and Grace (14). The latter modification differs in the use of D-alanine rather than L-alanine. In another modification, Park and Terasaki added a further 25 mg/L phenol red in order to observe pH changes more readily. McCoy's medium has been used as a standard medium for cloning cells (12–16).
- *RPMI 1630* was developed in 1968 by Moore and Kitamura (17).
- *RPMI 1640* was developed for the long-term culture of peripheral blood lymphocytes. It was developed as a modification of McCoy's 5A medium. It has now been recognized as a general-purpose medium particularly for

lymphocyte and hybridoma cultures. There are two commonly used modifications of RPMI 1640 that can enhance growth promotion or lymphocyte stimulation (5).

- *Searle's modification of RPMI 1640* includes Na_2HPO_4 (399 mg/L) and $NaH_2PO_4{\cdot}2H_2O$ (440 mg/L).
- *Dutch modification of RPMI 1640* includes NaCl (6400 mg/L), $NaHCO_3$ (1000 mg/L), and Hepes (4770 mg/L).

Other Simple Media

- *Fischer's medium* was originally used to support the growth of mouse leukemia cells in suspension culture (18,19).
- *Leibovitz L-15 medium* was designed for use without the need for an enriched CO_2 atmosphere. It has been used to support the growth of human diploid fibroblasts (15,20).
- *Trowell 's T-8 medium* (15,21).
- *Williams' medium E* was originally formulated for the long-term culture of adult rat liver epithelial cells (22).

Complex Media Intended for Use as Serum-Free Formulations

Examples of these media compositions are listed in Table 2:

- *Biggers' Medium* was originally designed to support isolated hepatocytes in culture (23). Additional amino acids and vitamins led to the Fitton–Jackson modification, which has been used to support cultures of embryonic bone.
- *Connaught Medical Research Laboratories (CMRL) 1066 medium* is a complex medium designed to support cell growth without the addition of serum. The formulation represents an extensive modification of Medium 199 (24,25).
- *eRDF* is based on a mixture of RPMI 1640:DMEM:F12 (2:1:1). This has been found to be effective with a serum-free formulation for the support of hybridomas (26). The enriched form of RDF contains 3 × the original concentration of amino acids and 2 × the concentration of glucose. Such a formulation with the addition of ITES (insulin, transferrin, ethanolamine, and selenite) can be more effective for myeloma or hybridoma cell growth than a serum-based medium.
- *Ham's F10 medium* was originally developed for the clonal growth of differentiated cells from chicken embryos. Modifications from the original formulation have included a substitution of $MgCl_2$ for $MgSO_4{\cdot}7H_2O$ and a 10-fold increase in the concentration of phenol red.
- *Ham's F12 medium* has a complex composition including various trace elements and was originally designed for cloning diploid hamster ovary cells (4). It was originally designed as a serum-free formulation but now is commonly used with a serum supplement to support the growth of a variety of normal and transformed cells. F12 has been combined with DMEM (1:1) as the basis for development of serum-free formulations (27). The idea was to combine the richness of F12 with the high nutrient concentrations of DMEM.
- *Ham's F14 medium* was originally developed to allow the study of sensory neurons.
- *Iscove's modified Dulbecco's medium (IMDM)* is a modification of DMEM containing additional amino acids and vitamins, selenium, sodium

Table 2 Complex Basal Formulations Used for Serum-Free Media

	Molarity (mM)							
Components	CMRL (1066)	DMEM/ F12	F12 (Ham's)	IMDM	MCDB 153	MCDB 131	Media 199	Waymouth's MB 752/1
Inorganic salts								
Calcium chloride ($CaCl_2$)	1.8	1.05	0.299	1.49	0.03	1.6	1.8	0.816
Calcium nitrate ($Ca(NO_3)_2 \cdot 4H_2O$)		1.20×10^{-7}						
Cupric sulfate ($CuSO_4 \cdot 5H_2O$)		7.80×10^{-9}	1.0×10^{-5}		1.1×10^{-5}	5.0×10^{6}		
Ferric nitrate ($Fe(NO_3)_3 \cdot 9H_2O$)		1.2×10^{-4}					1.7×10^{-3}	
Ferrous sulfate ($FeSO_4 \cdot 7H_2O$)		1.5×10^{-3}	0.003		5.0×10^{-3}	1.0×10^{-3}		
Potassium chloride (KCl)	5.3	4.16	2.98	4.44	1.5	3.97	5.33	2
Potassium nitrate (KNO_3)				7.5×10^{-4}				
Magnesium chloride ($MgCl_2$)					0.6			1.18
Magnesium chloride ($MgCl_2$, anhydrous)			0.6					
Magnesium sulfate ($MgSO_4$)	0.814	0.407				10.02	0.397	0.813
Magnesium sulfate ($MgSO_4$, anhydrous)				0.814				
Mangenous sulfate ($MnSO_4 \cdot H_2O$)					1.0×10^{-6}	1.0×10^{-6}		
Ammonium molybdate ($(NH_4)6Mn_7O_{24} \cdot 4H_2O$)					1.0×10^{-6}	3.0×10^{-6}		
Nickelous chloride ($NiCl_2 \cdot 6H_2O$)					5.0×10^{-7}	3.0×10^{-7}		
Sodium chloride (NaCl)	116	120.61	131	77.59	120	110.86	81.89	1.0×10^{-4}
Sodium bicarbonate ($NaHCO_3$)	26.2	29	14	36	14	14	14.9	26.7
Sodium meta silicate ($Na_2SiO_3 \cdot 9H_2O$)					5.0×10^{-4}	1.0×10^{-2}		
Sodium phosphate ($Na_2HPO_4 \cdot H_2O$)		0.5	1		2			2.11
Sodium phosphate ($NaH_2PO_4 \cdot H_2O$)	1.01	0.453		0.906		0.5	1.01	0.46
Stannous chloride ($SnCl_2$)					5.0×10^{-7}			
Sodium selenite ($Na_2SeO_3 \cdot 5H_2O$)				6.5×10^{-5}				
Selenious acid (H_2SeO_3)					3.0×10^{-5}	3.0×10^{-5}		
Ammonium metavanadate ($NaVO_3$)					5.0×10^{-6}	5.0×10^{-6}		

Zinc sulfate ($ZnSO_4 \cdot 7H_2O$)		1.5×10^{-3}	3.0×10^{-3}		5.0×10^{-4}	1.0×10^{-6}		
Energy metabolism								
D-Glucose	3.33	17.51	10	25	6	5.56	5.55	27.2
Sodium acetate-$3H_2O$	0.61				3.7		0.61	
Sodium Pyruvate		0.5	1	1	0.5	1		
Nucleic acid derivatives								
Adenine sulfate					0.18		5.4×10^{-2}	
ATP							1.7×10^{-3}	
AMP							5.0×10^{-4}	
Adenine						1.0×10^{-3}		
2′-Deoxyadenosine	4.0×10^{-2}							
5-Methyl-deoxycytidine	4.0×10^{-4}							
2′-Deoxycytidine	4.4×10^{-2}							
2′-Deoxyguanosine	3.8×10^{-2}							
Diphosphopyridine nucleotide (NAD)	1.1×10^{-2}							
Flavin adenine dinucleotide (FAD)	1.3×10^{-3}							
Hypoxanthine (Na)		1.5×10^{-2}	3.0×10^{-2}				2.9×10^{-3}	0.184
Guanine hydrochloride							1.6×10^{-3}	
Ribose							3.3×10^{-3}	
2-deoxy-D-ribose							3.7×10^{-3}	
Thymidine	4.1×10^{-2}	1.5×10^{-3}	3.0×10^{-3}		3.0×10^{-3}	1.0×10^{-4}		
Thymine							2.4×10^{-3}	
Triphosphopyridine nucleotide (NADP)	1.3×10^{-3}							
Uracil							2.7×10^{-3}	
UTP	2.0×10^{-3}							
Xanthine (Na)							2.2×10^{-3}	
Amino acids								
L-Alanine	0.281	5.0×10^{-2}	0.1	0.281	0.28	0.03	0.28	
L-Arginine-HCl	0.33	0.7	1	0.398	1	0.3	0.33	0.36
L-Asparagine		5.0×10^{-2}	0.1	0.189	0.1	0.1		
L-Aspartic acid	0.23	5.0×10^{-2}	0.1	0.23	0.03	0.1	0.23	0.451
L-Cystine-2HCl	0.1083			0.381			0.108	

(*Continued*)

Table 2 Complex Basal Formulations Used for Serum-Free Media (*Continued*)

Components	Molarity (mM) CMRL (1066)	DMEM/ F12	F12 (Ham's)	IMDM	MCDB 153	MCDB 131	Media 199	Waymouth's MB 752/1
L-Cysteine		0.1						0.504
L-Cysteine-HCl • H_2O	1.48	0.1	0.2		0.24	0.2	6.0×10^{-4}	6.3×10^{-2}
L-Glutamic acid	0.51	0.05	0.1	0.51	0.1	0.03	0.454	1.02
L-Glutamine		2.5	1	4	6		0.685	2.4
Glycine	0.667	0.25	0.1	0.399	0.1	0.03	0.667	0.667
L-Histidine hydrochloride H_2O	9.5×10^{-2}	0.15	10.0×10^{-2}	0.2		0.2	0.104	0.826
L-Histidine					0.08			
L-Hydroxyproline	7.6×10^{-2}						0.763	
L-Isoleucine	0.153	0.416	3.0×10^{-2}	0.802	1.5×10^{-2}	0.504	0.305	0.191
L-Leucine	0.458	0.451	0.1	0.802	0.5	1	0.458	0.382
L-Lysine-HCl	0.383	0.499	0.199	0.798	0.1	0.995	0.383	1.31
L-Methionine	0.101	0.116	3.0×10^{-2}	0.201	3.0×10^{-2}	0.1	0.1	0.336
L-Phenylalanine	0.152	0.215	3.0×10^{-2}	0.4	3.0×10^{-2}	0.2	0.152	0.303
L-Proline	0.348	0.15	0.3	0.348	0.3	0.1	0.348	0.435
L-Serine	0.238	0.25	0.1	0.4	0.6	0.305	0.238	
L-Threonine	0.252	0.449	0.1	0.078	0.1	0.101	0.252	0.63
L-Tryptophan	0.049	0.0442	1.0×10^{-2}	0.078	1.5×10^{-2}	2.0×10^{-2}	0.049	0.196
L-Tyrosine-2Na • $2H_2O$	0.26	0.214	3.0×10^{-2}	0.462	1.5×10^{-2}	0.1	0.221	0.221
L-Valine	0.214	0.452	0.1	0.803	0.3	0.103	0.214	0.566
Vitamins								
Ascorbic acid	0.284						2.0×10^{-4}	9.9×10^{-2}
α-tocopherol phosphate							1.4×10^{-5}	
Biotin	4.1×10^{-5}	1.4×10^{-5}	3.0×10^{-5}	5.3×10^{-5}	6.0×10^{-5}	3.0×10^{-5}	4.1×10^{-5}	8.2×10^{-5}
Calciferol (vitamin D_2)							2.5×10^{-4}	
D-Calcium pantothenate	2.1×10^{-5}	4.6×10^{-3}	1.0×10^{-3}	8.3×10^{-3}		2.5×10^{-2}	2.1×10^{-5}	2.0×10^{-3}
Choline chloride	3.5×10^{-3}	6.4×10^{-2}	10.0×10^{-2}	2.9×10^{-2}	0.1	0.1	3.5×10^{-3}	1.79
Folic acid	2.3×10^{-5}	6.0×10^{-3}	2.9×10^{-3}	9.1×10^{-3}	1.8×10^{-3}	1.0×10^{-3}	2.0×10^{-4}	1.1×10^{-3}
i-Inositol	2.0×10^{-4}	7.0×10^{-2}	0.1	4.0×10^{-2}	0.1	4.0×10^{-2}	2.8×10^{-4}	5.5×10^{-3}

Menadione (vitamin K_3)							5.8×10^{-4}	
Nicotinic acid (Niacin)	2.0×10^{-4}						2.0×10^{-4}	
Nicotinamide	2.0×10^{-3}	1.7×10^{-2}	3.0×10^{-4}	3.3×10^{-2}		5.0×10^{-2}	2.0×10^{-4}	8.1×10^{-3}
p-Aminobenzoic acid	3.0×10^{-4}						3.0×10^{-4}	
Pantothenate					1.0×10^{-3}			
Pyridoxal-HCl	1.0×10^{-4}			2.0×10^{-2}			1.0×10^{-4}	
Pyridoxine hydrochloride	1.2×10^{-4}	9.7×10^{-3}	3.0×10^{-4}		3.0×10^{-4}	1.0×10^{-2}	1.0×10^{-4}	4.8×10^{-3}
Riboflavin	2.7×10^{-5}	6.0×10^{-4}	1.0×10^{-4}	1.1×10^{-3}	1.0×10^{-4}	1.0×10^{-5}	2.7×10^{-5}	2.6×10^{-3}
Thiamine-HCl	3.0×10^{-5}	6.4×10^{-3}	1.0×10^{-3}	1.2×10^{-2}	1.0×10^{-3}	1.0×10^{-2}	3.0×10^{-5}	3.0×10^{-2}
Vitamin B_{12}		5.0×10^{-4}	1.0×10^{-3}	9.6×10^{-6}	3.0×10^{-4}	1.0×10^{-5}	3.1×10^{-4}	1.0×10^{-4}
Vitamin A (acetate)							3.1×10^{-4}	
Lipids and derivatives								
Cholesterol	5.2×10^{-4}						5.2×10^{-4}	
i-Inositol	2.0×10^{-4}							
Linoleic acid		1.5×10^{-4}	3.0×10^{-4}					
Lipoic acid		5.1×10^{-4}	9.7×10^{-4}		1.0×10^{-3}	1.0×10^{-5}		
Ethanolamine					0.1			
Prostaglandin E_1					2.5×10^{-5}			
Other compounds								
EGF(ng/ml)					25			
Putrescine-2HCl		5.0×10^{-4}	1.0×10^{-3}		1.0×10^{-3}	1.0×10^{-6}		
Glutathione(reduced)	3.3×10^{-2}						1.0×10^{-4}	4.9×10^{-2}
Hydrocortisone					1.4×10^{-4}			
Sodium glucuronate-H_2O	1.8×10^{-2}							
Co-carboxylase	2.1×10^{-3}							
Coenzyme A	3.3×10^{-3}							
Tween 80® (mg/L)	5.0						20	
Insulin (μg/ml)					5.0			
Dialyzed FBS (μg P/ml)					1.0			
Buffers/indicators								
Hepes				25	28			
Phenol red	5.0×10^{-2}	2.0×10^{-2}	3.0×10^{-3}	3.5×10^{-2}	3.3×10^{-2}	3.1×10^{-2}	5.0×10^{-2}	2.5×10^{-2}

pyruvate, Hepes, and potassium nitrate instead of ferric nitrate. This medium has proved useful as a basis for serum-free formulations (28) and has been widely used for the growth of lymphocytes and hybridomas.

- *MCDB 104* was developed from Ham's F12 as a medium for the clonal growth of human diploid fibroblasts.
- *MCDB 110* was developed from MCDB 104 for the growth of human diploid fibroblasts (29). The modifications included the addition of fatty acids, phospholipids, growth factors, hormones, and other supplements to simulate the action of serum (30).
- *MCDB 153* was developed as a basal medium for the serum-free growth of human epidermal keratinocytes and other epithelial cell types (31–33).
- *Medium 199* was formulated in an attempt to produce growth medium free from added protein, although for the growth of some cell lines serum is still added. This in an extremely complex medium based on Earle's salt solution and containing an extensive range of amino acids, vitamins, nucleic acid derivatives, growth factors and lipids (15,25,34).
- *NCTC 135 medium* was originally formulated for the serum-free growth of mammalian cells over long periods. The medium is a modification of the original NCTC 109 medium (35,36).
- *Waymouth's medium MB 752/1* was originally designed for the growth of L929 cells under serum-free conditions. However, it has been used for the growth of many cell types, usually with serum supplementation (15,37).

Standard media formulations such as those outlined previously can be purchased from commercial suppliers (e.g., Sigma, Life Technologies, etc.), either as a liquid or as a powder. Liquid forms can come as 1 × or 10 × concentrates, the latter requiring dilution with sterile deionized distilled water. Alternatively, powdered media may be preferred, especially if large quantities are needed. In this case, the powder should be dissolved in distilled water and sterilized by filtration through a 0.22 μm filter under positive pressure. Sterilized supplements of sodium bicarbonate, antibiotics, and glutamine are often added after sterilization of the bulk of the medium. This is essential if the medium is autoclaved, as these supplements are unstable.

Water for Media Preparation

The quality of water for mammalian cell culture is critical in any process. Due to the delicacy of mammalian cells the presence of trace minerals and/or contaminants will drastically affect the performance of the culture medium, especially in serum-free cultures. Due to this, specialized water purification systems are used for water in media preparation.

The overall process of water purification involves three or four-stages. These are reverse osmosis (or distillation), charcoal filtration, deionization, and micropore filtration. The first stage is typically reverse osmosis but distillation can also be used. This provides a relatively pure water supply to begin the process. Charcoal filtration will remove the majority of organic and inorganic impurities from the water, followed by deionization to remove any remainder of trace metals or ions. The final product is then passed through a micropore filter to remove any possible microbial contamination. In most systems the purified water will be continuously recycled throughout the filtering system, resulting in increasingly pure water.

The purity of the water is measured by its ability to conduct an electrical charge, in this case its resistance in MΩ cm. Most purification systems will come with a resistance meter display. The ISO 3696 standard for type-I water resistance is designated at $\geq$20 MΩ cm at 25°C (38).

To ensure consistency of the water, the cartridges should be replaced on a regular basis (specified by the manufacturer). Furthermore, any tubing or storage containers used with the system should be cleaned regularly to prevent the introduction of contaminants.

For the purposes of making culture media the water should be taken directly from the purification system during media preparation. Alternatively, the water can be stored if it is sterilized before hand. This can be accomplished by autoclaving or by micropore filtration (20 μm pore size). For autoclaving the water should be placed in a suitable container, such as a pyrex (borosilicate) bottle.

Antibiotics

Antibiotics are often included in media for short-term cultures. The use of antibiotics for routine subculture or in stock cultures should be avoided as this may mask a low-level contamination that may cause problems at a later date. Furthermore, extensive use of antibiotics may cause the selective retention of antibiotic-resistant contaminants in the laboratory. When antibiotics are used, the following cocktail may be recommended:

- Penicillin (100 IU/mL) to inhibit the growth of gram-positive bacteria
- Streptomycin (50 μg/mL) to inhibit the growth of gram-negative bacteria
- Amphotericin-B (25 μg/mL) as an antifungal agent.

Nutrients

Glucose

The primary carbohydrate normally included in mammalian culture media is glucose, routinely supplemented from 5 to 25 mM. The majority of glucose is metabolized via glycolysis to form pyruvate, which is then reduced to lactate. This eventually leads to a buildup of lactic acid in the medium. Studies of metabolic pathways have revealed that only a small portion of glucose (~20–30%) enters the other pathways including the tricarboxylic acid cycle and the pentose phosphate shunt (39).

In addition to glucose evidence has shown that mammalian cells also utilize glutamine as their major energy source. Glutamine is often included at high concentrations relative to the other amino acids (2–4 mM), and has been shown to be a high requirement of many cell lines. In most cultures, glutamine and glucose are utilized particularly rapidly and can cause cell growth limitations even before their complete exhaustion. This trend is more pronounced in serum-free or low-protein media, and higher utilization rates have been observed (39).

Amino Acids

The nutrient requirement of cultured cells includes the essential amino acids (i.e., those not normally synthesized in mammals in vivo) along with others depending on the specific requirements of the individual cell line. The nonessential amino acids are usually included because some cell lines may not be able to produce their own, which would lead to a limitation in their growth potential. The effect of an amino acid limitation in the medium will be a reduction in the growth rate and/or the maximum cell density. Other nonessential amino acids can be produced by the cell but not in sufficient quantities to maintain maximal growth. Some amino acids are not stable in the medium (e.g., glutamine), thus requiring supplementation to maintain a suitable level.

Amino acids are often added as defined components into the medium. There are, however, undefined sources of amino acids, such as serum, tryptose phosphate broth or lactalbumin, or various plant and animal cell hydrolysates.

Glutamine is routinely used in culture media as it has been shown to be a major energy source for mammalian cells. However, glutamine can lead to a buildup of ammonia in the medium, which can result in growth inhibition of some ammonia-sensitive cell lines (40). In this case glutamate, or glutamine dipeptides (ala-gln and gly-gln), have been shown to be successful substitutes and will lower the accumulation of ammonia in the medium.

Branched amino acids are consumed particularly rapidly by a number of cell lines including Madin–Darby canine kidney (MDCK) cells (41), human fibroblasts (42), mouse myeloma cells (43), and BHK cells (44). It has been found that branched amino acids are more extensively oxidized by hybridoma cells when the specific glutamine utilization rate is low (45). This group also observed that the serine consumption was higher when the glutamine consumption was reduced.

Some cell lines can be more sensitive to amino acid limitations than others and can be important for feeding strategies in larger-scale systems. For example, under low glutamine concentrations, BHK cells will increase their consumption of other amino acids, especially the essential amino acids leucine, isoleucine, and valine and the nonessential amino acids serine and glutamate (46).

Salts

The media salt concentration is isotonic to prevent osmotic imbalances. The osmolality of standard growth media is approximately 300 mOsm/kg and is optimal for most cell lines. Care should be taken when supplementing a medium with extra salts as this will change the osmolality. However, many cell lines have been shown to tolerate variation of approximately 10% of this optimal value.

The salts that are primarily used are those of Na^+, K^+, Mg^{2+}, Ca^{2+}, Cl^-, SO_4^{2-}, PO_4^{3-}, and HCO_3^-. These are present in most basal media formulations.

Trace Elements

Other inorganic elements that are present in serum in trace amounts may also be included in media formulations. These include Mn, Cu, Zn, Mo, Va, Se, Fe, Ca, Mg, Si, Ni. The less common elements that are not seen as essential have also been supplemented into media, including Al, Ag, Ba, Br, Cd, Co, Cr, F, Ge, J, Rb, and Zr. Many of these elements play an important role in enzyme activity and are essential to the survival of most cells.

The most commonly added element to serum-free formulations is selenium. The major functions of selenium can be attributed to its antioxidative properties and its role in cell growth. Selenium functions with mammalian cells in the form of selenoproteins, which have a number of physiological roles, of which 11 have been identified. Most of these proteins have been proven or implicated in antioxidant activities (47), such as glutathione peroxidase and thioredoxin reductase (48).

In some cases the supplementation of certain trace elements can decrease the requirement of certain growth factors in the medium. A ferrous salt, for example, can be used to replace transferrin as an iron carrier in many cell lines, such as B-lymphocytes, MDCK, and human diploid fibroblasts (49–51).

Calcium has been shown to act as an important element involved in cell proliferation. It is important to many processes, including signal transduction, cell division, and cell adhesion. Calmodulin, a calcium-activated protein, regulates many serine/threonine kinases involved in cell division, in a calcium concentration dependent manner (52). Elevated calcium levels has been shown to be beneficial to some cell lines. These include keratinocytes (53,54), and human diploid fibroblasts (55). The reduction of calcium in culture media can be effective in reducing the degree of cell–cell adhesion. Cell clumping is a major problem for some production systems as this may reduce productivity and cell viability (56). Reducing the calcium in the medium can avoid this problem and is routinely done in many agitated-culture systems.

Vitamins

The vitamin and hormone components are present at relatively low concentrations and are utilized as cofactors, the requirements for which show considerable variations between cell lines. Thus, the content of these cofactors varies considerably among different media formulations. The requirement for extra vitamins is increased whenever the serum concentration is reduced. Cells that are limited in vitamins will vary in cell growth and survival but not in maximum cell density.

The vitamins included in each basal media formulation will vary depending on the cell line for which it was designed. The more basic media formulations, such as BME, designed for HeLa cells and mouse fibroblasts, contain biotin, folic acid, nicotinamide, pantothenate, pyridoxal, riboflavin, and thiamine. These are typically supplemented with serum, which provides the other cofactor and vitamin requirements. The more complex formulations, such as F12, DMEM, or M199, are designed for serum-free formulations and contain a greater source of vitamins. M199, for example, contains all of the vitamins in BME but also includes vitamins A, B_6, C, D_3, E, K, and *p*-aminobenzoic acid (PABA). Each of these vitamin formulations was designed empirically for a specific cell line. Therefore, the requirements of another cell line may be different and will have to be determined independently.

Buffers

Bicarbonate is often included to act as a buffer system in conjunction with the carbon dioxide environment (5–10%) in which the cells should be cultured. This allows the cultures to be maintained at the normally optimum pH range of 6.9–7.4. The CO_2 is provided by a controlled atmosphere inside an incubator or alternatively in a sealed flask. The disadvantage of the bicarbonate–CO_2 buffer system is that the cultures may become alkaline very quickly when removed from the CO_2 incubator. To prevent this, the organic buffer Hepes (pH 7.0) may be added at a concentration up to ~25 mM. This allows the CO_2 level to be reduced to 2%.

Serum

Basal media used for the culture of mammalian cells are usually mixtures of a number of chemically defined components, such as vitamins, amino acids, salts, and sometimes protein supplements. Serum, which is a supernatant of clotted blood, is commonly used to provide animal cells with other components necessary for survival. These include attachment factors, micronutrients (trace elements, water-insoluble nutrients), growth factors (hormones, proteases), and protective elements (antitoxins, antioxidants, antiproteases). In most cases the growth requirements of the cell are not met until serum is added to the culture medium at a concentration of 5–10%.

Serum can be obtained from various animal sources—bovine or equine being the most common. One of the most effective supplements for cell growth is fetal calf serum because of its high content of embryonic growth factors.

SERUM-FREE MEDIA

The use of serum in culture media is associated with several problems including:

a. Batch to batch variation, which causes inconsistency in growth-promoting properties.
b. The high protein content, which can hinder product purification.
c. The risk of contaminants—viruses, mycoplasma, prions.
d. The risk of transmission of these contaminants to an end product used by humans, e.g., bovine spongiform encephalopathy (BSE) or mad cow disease.
e. Interfering with the effect of hormones or growth factors when studying their interaction with cells.
f. Limited availability of fetal calf serum, with periods of world shortages.
g. High cost of fetal calf serum, which can account for up to 85% of the overall cost of the medium when calculated for large-scale cultures.

Furthermore, serum-containing culture systems are becoming undesirable for large-scale processes in industry. The recent threat to human health caused by the undefined agents of BSE is likely to limit the continued use of bovine serum in culture processes used for the synthesis of health-care products such as viral vaccines.

For reduced serum, or serum-free formulations, various substitutes can be used. These include defined hormone cocktails such as the well-used HITES or ITES media, which contain hydrocortisone, insulin, transferrin, ethanolamine, and selenite. Alternatively, serum-free formulation may include growth factor extracts from endocrine glands such as epidermal or fibroblast growth factors. Every cell line is specific in its growth requirement and each serum-free formulation can only be applied to a limited number of cell types. Serum-free formulations have worked well for fast-growing tumorigenic or transformed cell lines, whereas their development for certain fastidious cell lines can be more difficult.

Since serum contains many undefined nutrients it has been difficult to reproduce the growth results associated with serum-supplemented media. Typical serum-free formulations have utilized a wide variety of components to mimic the action of serum. These have included purified proteins (animal or recombinant origin, see later), peptones, amino acids, inorganic salts, and animal cell or plant hydrolysates. Although undefined in nature, they can provide significant enhancements to cell growth and proliferation. However, the disadvantages are similar to that of serum supplementation, such as contamination and inconsistent cell growth from one batch to another.

Serum-free media formulations have been developed for commonly used cell lines [such as hybridoma and Chinese hamster ovary (CHO)] as well as for some anchorage-dependent cell lines. The difficulty in the creation of formulations for some cell lines has been their fastidiousness and it is only recently that an understanding of growth factors and hormones has allowed the development of effective serum-free formulations. These lines have included those involved in cancer research, immunology, and even cell transplantation. In addition, there has been success with human lymphocytes (145), smooth muscle cells (70), and umbilical cord vein endothelial cells (61). A few examples of serum-free formulations are shown in Table 3.

Serum-free media for certain cancer cell lines, such as melanocytes, have been developed. In most cases a keratinocyte serum-free formulation such as MCDB 153 has been used (62). However, these cells often require specific growth factors, such as melanocyte-stimulating hormone (alpha-MSH) (63) or dibutyryl cAMP (64), to induce proliferation. Otherwise the presence of normal keratinocytes or a keratinocyte-conditioned medium is required to promote melanocyte cell growth (62). Other cancer cell lines that have serum-free formulations in the literature include prostatic carcinomas (65), lung carcinoma (66–68), bladder carcinoma (protein-free, serum-free medium) (69).

The range of available serum-free media formulations and the demand for customized formulations are likely to increase as new cell lines are being established. Presently, there are still limited types of serum-free formulations available from commercial suppliers, and the selection still tends to be quite limited. Many claim to be "all-purpose" serum-free media to support a range of cell lines, but due to the variability of the growth factor requirement between cell lines, they remain quite limited in supporting cell growth. Further research in this area will result in the development of new and better formulations.

Chemically Defined Serum-Free Media

There is a clear distinction between serum-free media and chemically defined serum-free media. Chemically defined media require that all of the components must be identified and have their exact concentrations known. The supplements may be classified into several groups such as peptide growth factors, hormones, carrier proteins, hydrolysates, and attachment factors. The advantages of having chemically defined media are the same as those listed previously. It also allows researchers who are studying in the field of cell physiology (especially extracellular) and/or molecule–cell interactions to eliminate any variables that may arise due to the effects of unknown components in the medium.

The advantages of using media supplemented with defined growth factors and hormones include the following:

a. In many cases cells show enhanced growth characteristics in these defined media compared to serum-containing media. Some cells, notably differentiated types, cannot be maintained at all in serum-supplemented media and their maintenance in vitro is dependent on selected hormones and growth factors.
b. It is possible to select specific cell types from a mixed population of cells as may be obtained from a primary source. Such cell-specific selection requires careful manipulation of the medium composition.
c. Studies of the interaction of hormones or drugs on cultured cells are made possible by such media. The uncertain composition of serum and the binding effects of the associated proteins would limit such studies in serum-supplemented media.

Table 3 Examples of Serum-Free Formulations Designed for Various Cell Lines

Components	Molarity (M)				
	Hematopoietic progenitor cells (57)	MDCK (58)	Lung carcinoma (59)	CHO (60)	Human umbilical endothelial (61)
Inorganic salts					
Calcium chloride ($CaCl_2$)	1.49×10^{-3}	1.05×10^{-3}		1.05×10^{-3}	8.93×10^{-4}
Calcium nitrate ($Ca(NO_3)_2\cdot4H_2O$)		1.20×10^{-10}	4.24×10^{-4}	1.20×10^{-10}	
Cupric sulfate ($CuSO_4\cdot5H_2O$)		7.80×10^{-12}		7.80×10^{-12}	5.01×10^{-9}
Ferric nitrate ($Fe(NO_3)_3\,9H_2O$)		1.20×10^{-7}		1.20×10^{-7}	
Ferrous sulfate ($FeSO_4\,7H_2O$)	5.00×10^{-6}	1.50×10^{-6}		1.50×10^{-6}	1.50×10^{-6}
Potassium chloride (KCl)	4.44×10^{-3}	4.16×10^{-3}	5.30×10^{-3}	4.16×10^{-3}	3.71×10^{-3}
Potassium nitrate (KNO_3)	7.52×10^{-4}				3.76×10^{-7}
Magnesium chloride ($MgCl_2$ anhydrous)					3.01×10^{-4}
Magnesium sulfate ($MgSO_4$)	8.14×10^{-4}	4.07×10^{-4}	4.07×10^{-4}	4.07×10^{-4}	
Magnesium sulfate ($MgSO_4$ anhydrous)					2.41×10^{-3}
Mangenous sulfate ($MnSO_4\cdot H_2O$)	1.00×10^{-7}				1.46×10^{-6}
Ammonium molybdate ($(NH_4)_6Mn_7O_{24}.\,4H_2O$)	1.00×10^{-7}				
Nickelous chloride ($NiCl_2.\,6H_2O$)	5.00×10^{-8}				
Sodium chloride (NaCl)	7.76×10^{-2}	1.21×10^{-1}	1.03×10^{-1}	1.21×10^{-1}	1.04×10^{-1}
Sodium bicarbonate ($NaHCO_3$)	3.60×10^{-2}	0.02	2.38×10^{-2}	2.90×10^{-2}	2.50×10^{-2}
Sodium phosphate ($Na_2HPO_4\cdot H_2O$)		5.00×10^{-4}		5.00×10^{-4}	5.00×10^{-4}
Sodium phosphate ($NaH_2PO_4\cdot H_2O$)	9.06×10^{-4}	4.53×10^{-4}	5.63×10^{-3}	4.53×10^{-4}	4.53×10^{-4}
Stannous chloride ($SnCl_2$)	5.00×10^{-8}				
Sodium selenite ($Na_2SeO_3\cdot5H_2O$)	6.46×10^{-8}		3.00×10^{-8}	1.00×10^{-7}	4.79×10^{-8}
Ammonium metavanadate ($NaVO_3$)	5.00×10^{-7}				
Zinc sulfate ($ZnSO_4\cdot7H_2O$)		1.50×10^{-6}		1.50×10^{-6}	1.35×10^{-5}
Energy metabolism					

D-Glucose	2.50×10^{-2}	1.75×10^{-2}	1.11×10^{-2}	1.75×10^{-2}	1.75×10^{-2}
Sodium pyruvate	1.00×10^{-3}	5.00×10^{-4}		5.00×10^{-4}	5.00×10^{-4}
Nucleic acid derivatives					
2′-Deoxyadenosine	0.1 mg/mL				
2′-Deoxycytidine	0.1 mg/mL				
2′-Deoxyguanosine	0.1 mg/mL				
Hypoxanthine (Na)		1.50×10^{-5}		7.35×10^{-5}	
Thymidine		1.50×10^{-6}		4.13×10^{-5}	1.51×10^{-6}
Adenosine	0.1 mg/mL				
Cytidine	0.1 mg/mL				
Uridine	0.1 mg/mL				
Guanosine	0.1 mg/mL				
2′-Deoxythymidine	0.1 mg/mL				
Amino acids					
β-Alanine	2.81×10^{-4}				
L-Alanine	3.98×10^{-4}	5.00×10^{-5}		1.00×10^{-6}	2.52×10^{-4}
L-Arginine-HCl	1.89×10^{-4}	7.00×10^{-4}	1.10×10^{-3}	1.40×10^{-3}	1.04×10^{-3}
L-Asparagine	2.30×10^{-4}	5.00×10^{-5}	3.79×10^{-4}	1.06×10^{-4}	2.71×10^{-4}
L-Aspartic acid		5.00×10^{-5}	1.50×10^{-4}	1.03×10^{-4}	2.53×10^{-4}
L-Cystine-2HCl	3.81×10^{-4}	1.00×10^{-4}		2.00×10^{-4}	1.46×10^{-4}
L-Cysteine-HCl·H_2O		1.00×10^{-4}	2.06×10^{-4}		1.00×10^{-4}
L-Cystine	5.10×10^{-4}				
L-Glutamic acid	4.00×10^{-3}	5.00×10^{-5}	1.36×10^{-4}	9.14×10^{-5}	5.43×10^{-4}
L-Glutamine	2.00×10^{-3}	2.50×10^{-3}	2.05×10^{-3}	5.00×10^{-3}	9.00×10^{-3}
Glycine	4.00×10^{-2}	2.50×10^{-4}	1.33×10^{-4}	5.00×10^{-4}	2.50×10^{-4}
L-Histidine hydrochloride·H_2O		1.50×10^{-4}		1.50×10^{-4}	2.19×10^{-4}
L-Histidine			9.67×10^{-5}	4.00×10^{-4}	
L-Hydroxyproline	8.02×10^{-4}		1.53×10^{-4}		
L-Isoleucine	8.02×10^{-4}	4.16×10^{-4}	3.82×10^{-4}	8.32×10^{-4}	4.15×10^{-4}
L-Leucine	7.98×10^{-4}	4.51×10^{-4}	3.82×10^{-4}	1.35×10^{-3}	4.50×10^{-4}
L-Lysine-HCl	2.01×10^{-4}	4.99×10^{-4}	2.19×10^{-4}	1.50×10^{-3}	6.53×10^{-4}

(Continued)

Table 3 Examples of Serum-Free Formulations Designed for Various Cell Lines (*Continued*)

	Molarity (M)				
Components	Hematopoietic progenitor cells (57)	MDCK (58)	Lung carcinoma (59)	CHO (60)	Human umbilical endothelial (61)
L-Methionine	4.00×10^{-4}	1.16×10^{-4}	1.01×10^{-4}	2.30×10^{-4}	1.34×10^{-4}
L-Phenylalanine	3.48×10^{-4}	2.15×10^{-4}	9.09×10^{-5}	4.27×10^{-4}	4.69×10^{-4}
L-Proline	4.00×10^{-4}	1.50×10^{-4}	1.74×10^{-4}	4.45×10^{-4}	3.24×10^{-4}
L-Serine	7.80×10^{-5}	2.50×10^{-4}	2.86×10^{-4}	2.50×10^{-4}	7.26×10^{-4}
L-Threonine	7.80×10^{-5}	4.49×10^{-4}	1.68×10^{-4}	8.94×10^{-4}	4.49×10^{-4}
L-Tryptophan	4.62×10^{-4}	4.42×10^{-5}	2.45×10^{-5}	1.32×10^{-4}	4.42×10^{-5}
L-Tyrosine-2Na•$2H_2O$	8.03×10^{-4}	2.14×10^{-4}	1.10×10^{-4}	5.23×10^{-4}	2.49×10^{-4}
L-Valine		4.52×10^{-4}	1.71×10^{-4}	9.05×10^{-4}	5.54×10^{-4}
Vitamins					
Ascorbic acid	5.32×10^{-8}				1.00×10^{-3}
Biotin	8.30×10^{-6}	1.43×10^{-8}	8.00×10^{-6}	1.43×10^{-8}	4.15×10^{-8}
D-Calcium pantothenate	2.85×10^{-5}	4.60×10^{-6}	5.00×10^{-7}	4.60×10^{-6}	4.70×10^{-6}
Choline chloride	9.06×10^{-6}	6.41×10^{-5}	2.14×10^{-5}	6.41×10^{-5}	6.43×10^{-5}
Folic acid	4.00×10^{-5}	6.00×10^{-6}	2.20×10^{-6}	6.00×10^{-6}	6.00×10^{-6}
i-Inositol		7.00×10^{-5}	1.94×10^{-4}	7.00×10^{-5}	6.99×10^{-5}
Nicotinic acid (Niacin)	3.28×10^{-5}				
Nicotinamide		1.65×10^{-5}	8.10×10^{-6}	1.65×10^{-5}	1.65×10^{-5}
P-aminobenzoic acid	1.96×10^{-5}		7.20×10^{-6}		
Pyridoxal-HCl	1.06×10^{-6}		4.80×10^{-6}		9.82×10^{-6}
Pyridoxine hydrochloride		9.70×10^{-6}		9.70×10^{-6}	1.51×10^{-7}
Riboflavin	1.18×10^{-5}	6.00×10^{-7}	5.00×10^{-7}	6.00×10^{-7}	5.82×10^{-7}
Thiamine-HCl	9.59×10^{-9}	6.40×10^{-6}	2.90×10^{-6}	6.40×10^{-6}	6.43×10^{-6}
Vitamin B_{12}		5.00×10^{-7}	3.69×10^{-9}	5.00×10^{-7}	5.06×10^{-7}
Lipids and derivatives					
Cholesterol	2.00×10^{-6}				
i-Inositol			1.9×10^{-4}		

Linoleic acid (μg/mL)		1.5×10^{-4}		1.5×10^{-4}	42
Lipoic acid		5.1×10^{-4}		5.1×10^{-4}	105
Ethanolamine (mg/L)				6	
Prostaglandin E_1 (ng/mL)		25			
Phophatidylcholine (mg/L)				5	
BSA/oleic acid (mg/L)					200/1.6
Other compounds					
EGF (ng/mL)	10				10
Putrescine-2HCl		5.03×10^{-7}		5.03×10^{-7}	5.00×10^{-7}
Glutathione (reduced)			3.20×10^{-6}		
Hydrocortisone		5.00×10^{-8}	10	2.76×10^{-5}	2.76×10^{-6}
Insulin (mg/mL)		5	5	2	5
BSA	1% w/v				
2-Mercaptoethanol	5.00×10^{-5}				1.25×10^{-5}
Transferrin (mg/mL)	0.7	5	100	10	5
Stem cell factor (ng/mL)	10				
Interleukin-1β (ng/mL)	3				
Interleukin-3 (ng/mL)	100				
Interleukin-6 (ng/mL)	100				
Erythropoietin (U/mL)	1				
Triiodothronine		5.00×10^{-12}			
Estradiol (17-β)			1.00×10^{-8}		
bFGF (ng/mL)					10
Liver growth factor (gly-his-lys) (μg/mL)					50
VEGF (ng/mL)					5
N-Acetyl-L-cysteine					2.50×10^{-4}
Buffers/indicators					
Hepes	25	15			$1/25\times10^{-2}$
Phenol red	0.0346	2.04×10^{-5}	1.25×10^{-5}	2.04×10^{-5}	1.21×10^{-5}

Some examples of serum-free formulations that are not chemically defined are those that contain various serum substitutes, such as serum fractions or animal/plant hydrolysates. Although such serum-free media are more defined than serum-containing media, there remains the risk of introducing harmful contaminants such as viruses (71). Therefore, the only way to ensure a safe serum-free medium is to completely replace any animal-derived substances by non-animal-derived products, such as those of plant origin.

GROWTH FACTORS

This section includes information on some of the specific proteins used as supplements, which are often added to cell culture media in order to enhance cell growth. These components are supplements to basal media that will promote growth in the absence of serum or in reduced serum formulations.

Many polypeptides have been characterized as mitogenic in vitro for specific mammalian cell lines. This family includes fibroblast growth factor (FGF: aFGF and bFGF), insulin-like growth factor (IGF), epithelial growth factor (EGF), nerve growth factor (NGF), platelet-derived growth factor (PDGF), and transforming growth factor (TGF: α and β). These growth factors are active in the 1–10 ng/mL range.

Fibroblast Growth Factor

FGFs are pleiotropic growth factors that control cell proliferation, migration, and differentiation in a wide variety of cell types. They exist as two types: acidic FGF (p*I* 5.6) and basic FGF (p*I* > 9.0). Due to their affinity to heparin they are sometimes called heparin-binding growth factors, although only the activity of acidic FGF is altered by heparin.

Acidic FGF

This form of FGF carries 55% homology with basic FGF. Also known as endothelial growth factor, this is the less potent of the two mitogens. In the absence of heparin sulfate on the surface of target cells, or free heparin, FGF cannot exert its biological activity due to its instability and is easily damaged by proteolysis. For this reason it is often supplemented with heparin. In some cases synthetic dextrans have been used as an alternative to heparin (72).

Basic FGF

This is produced by most cell types investigated in vitro, and the FGF receptor is ubiquitous. Thus, bFGF can act as both an autocrine (stimulation of the same cell) and a paracrine growth factor (one cell stimulating an adjacent cell). Basic FGF stimulates cell proliferation of a wide variety of cell types through activation of protein kinase C (73).

It is a potent mitogen for many types of cultures derived from mesodermal cells, neuroectoderm, and several transformed cell lines (74). It also promotes the differentiation of adipocytes and ovarian ganulosa cells, and induces the synthesis of extracellular matrix (ECM) proteins such as collagen type IV (75).

Epithelial Growth Factor

EGF is a potent mitogen for a variety of cell lines including primary, mesenchymal, epidermal, and glial cells. Cells that are transformed tend to lose their requirement for EGF (76,77). The activity of EGF is linked to its activation of a transmembrane tyrosine kinase, which plays a role in cellular signaling and development. Many of its activities include the induction of gene expression, changes in ion flux, and mitogenesis (78). Its mitogenic activity is synergistic with that of other peptide growth factors such as IGF-1 and TGF (79,80).

Nerve Growth Factor

NGF is a well-characterized molecule required for the survival and differentiation of a variety of cell types both in the peripheral and in the central nervous system. It is not a potent mitogen, but it does induce the differentiation and enhances the survival of sympathetic neurons and PC12 phaeochromocytoma cells in culture (75). NGF initiates the majority of its biological effects by activation of the tyrosine kinase receptor TrkA. These signals are then propagated to other messengers involved during neuronal differentiation (81,82).

Transforming Growth Factor

Transforming Growth FactorAlpha

This factor bears 30% homology to EGF, and binds to EGF receptors with equal potency. Its actions are similar to those of EGF and it is a potent epithelial oncoprotein (83). Usually, a culture will not require EGF supplementation if TGFα is present. TGFα is synthesized as a large membrane-bound precursor, which is subsequently cleaved by proteases into the mature peptide. TGFα is found naturally in the embryonic kidney, adult brain, pituitary gland, skin, and placenta.

Transforming Growth Factor Beta

Three isoforms of TGFβ (β1, β2, and β3) exist in mammals. They play a critical role in growth regulation and development. All three growth factors are secreted by most cell types, generally in a latent form, requiring activation before they can exert biological activity. Activation of TGFβ requires an acidic environment or the presence of certain activators, such as plasmin or thrombospondin (85).

The highest concentrations of TGFβ1 are found in platelets and bone, with smaller amounts expressed by many cell types. It is closely conserved between species with $\approx$98% equivalency. Mature TGFβ1 in human, porcine, chicken, simian, and bovine species are identical, and differ from murine TGFβ1 in a single amino acid position.

The effects of TGFβ on proliferation and differentiation can be stimulatory or inhibitory. This is dependent on many factors, including cell type, growth conditions, the state of cell differentiation, and the presence of other growth factors. TGFβ promotes the growth of several mesenchymal cell lines in soft agar (e.g., fibroblasts) in combination with either TGFα or EGF. In contrast, TGFβ alone *inhibits* the growth of many cell lines in monolayer culture, which may result from its ability to stimulate the secretion of ECM proteins (collagens, fibronectin, and glycosaminoglycans) and protease inhibitors (86). TGFβ can inhibit monolayer growth in epithelial cells, endothelial cells (antagonized by bFGF), stem cells, and lymphocytes.

Insulin and IGFs

Insulin

Several of the mitogenic actions previously ascribed to insulin may be due to IGF contamination of the original preparations. Insulin is quite unstable at 37°C (particularly in media with high levels of cysteines) with >90% of its activity being destroyed over 1 hr (87). It is often added to media at relatively high concentrations and because insulin has a weak affinity for IGF-1 receptors, it may activate the same mitogenic responses as IGF-1 at these high doses. However, insulin is also added to most serum-free media for its ability to promote energy and anabolic metabolism (e.g., glucose uptake and oxidation, synthesis of glycogen, amino acid transport) (75). The ubiquitous insulin receptor is a dimer that has tyrosine kinase activity (88).

Insulin requires adequate zinc to exert its biological function and should be present in the culture medium.

Insulin-Like Growth Factors

These peptides are homologous to insulin and evolved by gene duplication (89). They were first discovered as the serum components that mediate the effects of growth hormone on cartilage—hence, their former name of somatomedins. IGF activity is regulated by a family of six IGF-binding proteins that circulate in the blood. They extend the life and modulate the activity of IGF on various tissues (90). Many of the metabolic actions of IGF overlap with that of insulin; however, they are more potent mitogens and synergize with FGF, EGF, and PDGF.

Insulin-Like Growth Factor 1. IGF-1 is secreted in low amounts by most tissues, and acts on a wide range of mesenchymal cells. Although IGF-1 binds to the insulin receptor with low affinity, its main actions are mediated by a specific IGF-1 receptor, which has intrinsic tyrosine kinase activity. This receptor also binds IGF-2 and, to a lesser extent, insulin (90).

Insulin-Like Growth Factor 2. Formerly known as multiplication-stimulating activity, this factor is produced by a more restricted range of tissues (e.g., fetal liver, muscle, skin, and adult brain). It exhibits insulin-like activity, but it is not as potent as IGF-1. Its receptor has lower affinity to insulin or IGF-1. This receptor is found on adipocytes and (in contrast to insulin and IGF-1 receptors) does not display tyrosine kinase activity (90).

Platelet-Derived Growth Factor

PDGF is a major mitogen for fibroblasts, smooth muscle cells, and other cells. It was originally identified as a component of whole-blood serum (40–60 ng/mL) that was subsequently isolated from human platelets.

PDGF consists of two structurally similar polypeptide chains (A and B), which combine to form homo- and heterodimers. The three-dimensional structure of PDGF-BB has been shown to have similarity to NGF, TGFβ, and vascular endothelial growth factor (VEGF)s, which have homologous amino acid sequences.

The PDGF isoforms promote a response by binding to two structurally similar binding sites of a protein tyrosine kinase receptor, designated α and β (91). Binding of PDGF can result in mitosis, chemotaxis, actin reorganization, or prevention of apoptosis in certain cell lines.

In serum both forms of PDGF bind to α_2-macroglobulin. PDGF synergizes with EGF and IGF-1 in stimulating normal Balb/C 3T3 fibroblasts to proliferate

(91). Its action is restricted to mesenchymal and neuroectoderm cells, which have specific PDGF receptors exhibiting tyrosine kinase activity (88).

Interleukin-6

Interleukin-6 (IL-6) is a cytokine that regulates many immune and hematopoietic systems. Its function is pleotropic and it exhibits many overlapping biological functions with other cytokines such as IL-11, leukemia inhibitory factor, oncastatin, ciliary neurotrophic factor, and cardiotrophin-1, as they share a common signal transducing receptor, gp130. It does not stimulate tyrosine kinase activity directly, but rather it stimulates other cytoplasmic tyrosine kinases and subsequent modification of transcription factors (92).

It is produced by many cell types including macrophages, fibroblasts, T-cells, and endothelial cells. Its effect upon some cell lines can be proliferative and mitogenic. Treatment of the PC 12 cell line with IL-6 causes differentiation and it can also protect them from cell death under serum deprivation (93,94). It is also mitogenic to other cell types such as stem cells and hybridomas (95). Similarly, IL-6 can stimulate B-cells to proliferate in vitro (96). Many of these effects can be synergistically enhanced with simultaneous treatment with other growth factors such as EGF, FGF, or NGF (97).

Not all responses to this cytokine are proliferative. Certain hybridomas increase their antibody production when treated with IL-6 (75). In addition, IL-6 synergizes with other ILs to enhance antibody production (92).

In cell culture, IL-6 can replace the use of feeder cells (such as peritoneal exudate cells, spleen cells, fibroblasts, or thymocytes) during cloning by limiting dilution and in the post-fusion stages of hybridoma culture.

Reconstitution and Storage of Growth Factors for Serum-Free Media

Many of the components of serum-free formulations are not particularly stable in a protein-free environment. These include recombinant proteins such as IGF, EGF, PDGF, FGF, TGF, and others. These compounds should be reconstituted in water with a protein carrier, such as bovine serum albumin (BSA), at a high concentration ($\sim$mg/mL). The solution should be kept frozen at $<-20^\circ$C depending on the instructions supplied with the product information sheet. The solution can then be thawed and introduced into the medium immediately before being used for cell culture, or even directly into the culture flask.

Many of the compounds may not require a protein carrier upon reconstitution (if animal components are not desirable in the medium). They can be dissolved in water, sometimes with a little acid at low concentration (10 μM). However, the absence of a carrier will reduce the stability of the recombinant protein and the compound may only be stable at $<-20^\circ$C for a few months.

LIPIDS

Lipids serve a variety of functions within the cell. They act as a structural component for cell membranes, in sensors for external signals, and as a source of metabolic energy. Serum itself contains lipids, but basal media formulations do not normally contain any lipid components. Certain lipids can be included as supplements to serum-free formulations and have been shown to enhance cell proliferation (98).

The types of lipids commonly found in serum include fatty acids, phospholipids, lecithin (phophatidylcholine), and cholesterol. Phospholipids are not only major components in cell membranes but also play an important role in signal transduction pathways regulating growth. When introduced extracellularly, phospholipids have shown growth stimulatory effects on many cell lines. Phosphatidic acid and lysophosphatidic are growth promoting for some anchorage-dependent cells, such as MDCK, mouse epithelial, and other kidney cell lines (58,99,100). Others, such as phosphatidylcholine, phosphatidylethanolamine, and phosphatidylinositol, stimulate the growth of human diploid fibroblasts in serum-free formulations (30). Ethanolamine is growth stimulatory for some hybridoma cell lines (101), and is often supplemented to serum-free formulations (102). Cholesterol has also been shown to be stimulatory for most mammalian cell lines and is a requirement for growth of certain cells, such as myelomas (103). Other specific fatty acids, such as oleic and linoleic, will enhance both the growth and the productivity of hybridomas grown in a serum-free medium over an extended number of passages (104,105).

CARRIER PROTEINS

These are important to promote the transport of essential nutrients or trace elements. A brief description of some of the carrier proteins follows.

Bovine Serum Albumin

BSA is commonly found in circulating plasma and plays an important role in cell culture. It is primarily used as a carrier of lipids, such as linoleic and oleic acids, since they are insoluble in aqueous solution. In addition to fatty acids it also has a high affinity for other compounds, such as various metals (Fe, Cu, and Ni). It also serves as a protective agent of toxic metabolites such as oxygen-free radicals and bilirubin (106). In addition, BSA is a transport protein for nitric oxide, which is important for neurotransmission (107).

In some cases BSA is undesirable for protein-free and serum-free formulations. In these cases alternative non-animal-derived compounds can be used as lipid carriers, such as cyclodextrin.

Transferrin

This is an important iron carrier protein and is essentially the source for all circulating plasma iron. The function of transferrin is to solubilize the iron, prevents free-iron toxicity, and facilitate the transport of iron into cells. Iron uptake occurs as an iron–transferrin complex and is internalized by receptor-mediated endocytosis (108).

For the purposes of culture media, human transferrin is more commonly used than bovine transferrin, due to its ability to deliver iron to a wider range of cell types (109).

ECM PROTEINS

A number of proteins are important to ensure the attachment of anchorage-dependent cells to their substratum. The attachment proteins (also referred to as ECM proteins) are present in serum and can also be secreted by the cells. Some adherent cells do not bind directly onto plastic and require ECM proteins for efficient plating and growth (110). Some of the commonly used ECM proteins in cell culture will be discussed.

Fibronectin

Fibronectin is a high-molecular weight multimeric glycoprotein found on the surface membrane of cells. There are two forms of fibronectin that are generally available. Plasma fibronectin has been isolated from blood plasma and plays a role in blood clotting, wound healing, and phagocytosis.

In cell cultures, fibronectin is important in cell-substratum adhesion of anchorage-dependent cells. As cells synthesize fibronectin, the need for media supplementation is dependent on the cell type. Transformed cells lose the ability to produce fibronectin and this may partly explain their loss of anchorage dependence. Many cell types show improved adherence and growth on fibronectin-coated surfaces (111). Furthermore, in the absence of serum, fibronectin has been shown to enhance the survival of primary fibroblasts and endothelial cells in culture by suppressing p53-regulated apoptosis via transduction signals through focal adhesion kinase (112).

In addition to adhesion, fibronectin plays an important role in stimulating many mitotic processes (such as the mitogen-activated protein kinase pathway) and has been shown to stimulate cell cycle progression in nontransformed cell lines, such as NIH-3T3 and BHK (113,114).

Laminin

Laminin is another ECM protein and is sometimes used as an alternative to fibronectin, especially for plating epithelial cells (115). It is similar in structure to fibronectin, but more predominant in differentiated tissues, and is one of the major components of the basal lamina. It also plays an important role in the regulation of neuronal migration, being a potent promoter of cell outgrowth in cultures, especially neural cell lines. In conjunction with a polylysine peptide, laminin has been shown to provide guidance for neuron attachment and axon outgrowth under serum-free conditions (116).

Pronectin

Pronectin is a recombinant peptide designed to replace fibronectin in serum-free media formulations. Similar in structure to fibronectin, it contains the same repeats of the RGD (Arginine-Glycine-Aspartate) cell attachment sequence. It has been used to promote attachment of various anchorage-dependent cells in both stationary and microcarrier cultures, such as the MDCK cell line (117).

CHOICE OF SUPPLEMENTS FOR SERUM-FREE MEDIA FORMULATIONS

Deciding which supplements are important for each cell line can be the most difficult task during development of a new serum-free medium. Rather than using classical approaches to development (e.g., Ham and Sato), using formulations that have been developed for similar cell lines can serve as a guide, or at least a starting point, in creating a new formulation. However, the growth factor requirements for any given cell line can vary dramatically from one another, so this approach does not always work.

There are many growth factors to choose from, and the effect of each may be cell line dependent. A brief guide, based on cell type, follows:

a. *Epithelial cells*: Growth factors that are often included are EGF, corticosteroids, and retinol. ECM proteins can also be benefi-cial, including fibronectin, laminin, and collagen, to promote cell adhesion.
b. *Mesenchymal cells*: Growth factors that are often included are acidic/basic FGF (basic being more potent than acidic), EGF (or TGFα), IGF/insulin (but IGF being more potent), PDGF (BB most potent), and sometimes TGFβ (but it is growth inhibitory for most cells). Other supplements can include transferrin, corticosteroids, ECM proteins such as fibronectin, and trace elements.
c. *Transformed cells*: These cells often have a lower requirement for growth factors. Supplements used have included IGF/insulin, carrier proteins BSA/fetuin, human transferrin, and trace elements.

Iron Salts as a Replacement for Transferrin

Transferrin is used in culture media as an iron transporter for animal cells. However, for protein-free media or for media that do not contain components of animal origin, transferrin can be replaced by simple ferric salts. This has been accomplished in many media formulations for various cell lines, such as hybridoma (102). Ferrous salts tend to oxidize quickly, however, so the medium should be used within a few weeks or they should be introduced just before the medium is to be used. Typical sources of iron in serum-free formulations have been ferrous sulfate or ferric citrate (119).

Undefined Serum Substitutes

In an attempt to further enhance the performance of serum-free formulations, various extracts of serum or plasma have been developed. These fractions may serve to replace the growth-promoting effects of whole serum. Although a serum-free medium containing these fractions is not chemically defined, it confers the advantage of lowering the protein content, which in turn can reduce the batch-to-batch variability. The only disadvantage is that the exact chemical composition is not known.

Cohn fractionation of blood is a method used to separate plasma fractions by sequential steps of ethanol precipitation. This method is used for the large-scale preparation of hemophilic factors from pooled human blood. Cohn fraction 4.1, which is a by-product of this process, has been found to possess growth-promoting properties that can substitute for whole serum in growth medium (120), and be used to promote the growth of MDCK cells.

Commercial sources of these substitutes are available and are designed as complete serum-free media for cell culture. These media formulations are usually tested on specific cell lines in comparison to serum-based media in growth-promoting ability. Some of the companies that supply these serum-substitutes are Invitrogen (Gibco), Hyclone, and Sigma.

Other types of substitutes have also been developed. Hydrolysates, which are enzymatic digests of proteins, are being widely used in cell culture. They are obtained from sources such as animal tissue, milk, and plants such as soy, wheat, or rice. The end product is a source of vitamins, lipids, minerals, and di- and tripeptides that supply a rich energy source of basic amino acids for the cell.

Hydrolysates are typically used as a supplement to a basal medium, although they can be used to replace it altogether. The types of cell lines that have been propagated with hydrolysates include CHO, hybridoma, BHK, Vero, lymphocytes, and others (121).

The advantages of using hydrolysates to replace serum include the following:

a. There is no excess protein, which can interfere with purification.
b. Any hormones that were introduced with serum are no longer present.
c. There is a significant reduction in the cost of the media.
d. Utilizing plant hydrolysates will decrease the risk of introducing animal-derived contaminants into the cell culture process.

Although serum-free formulations that contain these hydrolysates confer certain advantages, the problem of batch variation still remains. Furthermore, hydrolysates derived from animal tissue can still carry the risk of introducing contaminants, such as viruses or prions, into a cell culture process used in the production of biologicals (71). This can limit their use in the large-scale production of human therapeutics due to health regulatory concerns.

PRODUCTION OF BIOLOGICALLY ACTIVE SUBSTANCES BY SERUM-FREE CULTURES

Animal cell technology has allowed the production of a great diversity of biological substances including monoclonal antibodies, vaccines, hormones, and other regulatory molecules. The development of serum-free media has improved these processes by conferring many advantages as discussed previously.

The vaccine industry has seen the emergence of new processes for production utilizing serum-free culture systems. In many cases the traditional method of production has led to many problems including adverse allergic reactions (e.g., from chick eggs used to propagate the virus) to the transfer of biological contaminants from animal components used in the process. Furthermore, many of the more traditional processes were inefficient and time-consuming. The use of animal cells for the production of vaccines alleviated some of these problems by allowing large-scale production with minimal supervision and monitoring. However, the use of serum in the culture medium was still required, which posed the threat of transferring contaminants to the end product. The recent advancements in serum-free formulations have reduced the dependence on serum in industrial-scale processes. This has allowed the minimization of potential contamination by adventitious agents.

It has been demonstrated that serum-free cultures can have comparable productivity to their serum-supplemented counterparts and have been widely accepted in industry. Some examples of vaccines produced today under serum-free conditions are rabies, polio, and influenza (122–126). Other important viruses, such as HIV, adenovirus, and reovirus (to name a few) have also been produced under serum-free conditions (127–129).

Serum-free culture systems have also been used for the production of other biologically active substances. Cytokines, which include the lymphokines and interferons, are of particular interest to industry due to their involvement in the immune and inflammatory responses within the body. IL-2, for example, has been produced from BHK cells under serum-free conditions employing both stirred tank reactors and a hollow fiber system (130). The yields of the system ranged from 0.75 mg/L in the stirred tank system to 0.23 mg/L in the hollow fiber module, where the cell densities reached 3.0×10^7 cells/mL and 6.0×10^7 cells/mL, respectively. Another group discovered that IL-2 production between serum-containing and serum-free cultures resulted in identical titers (131).

The production of interferon has been achieved in a wide variety of cell lines under serum-free conditions. Gamma-interferon, for example, normally obtained from T- or B-lymphocytes, has been produced using CHO cells (132). In this case certain components of the serum-free medium were important for interferon production and affected the productivity of the culture system. They included BSA, sodium pyruvate, glutamate, methionine, proline, histidine, hydroxyproline, tyrosine, and phenylalanine. Similarly, the production of beta-interferon in Namalwa cells under serum-free conditions has been studied (133).

Other biologicals, such as tissue plasminogen activators (tPA), blood clotting factors, hormones, and polypeptide growth factors have been produced in vitro in serum-free cultures. Some examples of these processes include the production of tPA. In this serum-free system, cultures of fibroblasts on Cytodex-3 microcarriers gave yields comparable to that of 5% fetal bovine serum-containing media (134). Also, a serum-free culture of CHO cells in the production of recombinant human growth hormone allowed the elimination of a final purification step (135). The elimination of serum has allowed continuous nutrient optimization to improve the productivity (136).

The importance and wide-spread use of monoclonal antibodies in the pharmaceutical industry has led to the development of better serum-free formulations for large-scale production. They are becoming a standard for industry due to the high reproducibility of growth conditions, the ability to optimize nutrient requirements, and the reduction of possible contaminants. Furthermore, hybridoma cells grow well in suspension culture and are relatively easy to adapt to serum-free formulations compared to anchorage-dependent cell lines, which tend to be more fastidious. For these reasons they are commonly adapted to serum-free media for large-scale production systems.

In most cases the conversion to serum-free media has led to improvements for the biopharmaceutical industry, including increased productivity, efficiency, and cost-effectiveness.

PROTEIN-FREE MEDIA

Protein-free media are typically serum-free formulations in which the protein growth factors or other protein substitutes have been removed. These can be in the form of BSA, fetuin, or others such as transferrin. This can create problems in downstream product purification by necessitating an increase in purification steps. As a consequence, lower product yields or a decrease in purity can result. Therefore, protein-free media can greatly increase the productivity and efficiency of processes for producing biologicals.

The elimination of serum alone reduces the amount of protein in the medium. However, the removal of proteins such as transferrin, insulin, and fetuin from a serum-free formulation can drastically reduce the performance of the medium. To counter these effects other supplements can be used, such as animal tissue or plant hydrolysates. These compounds have been shown, in some cases, to increase the growth rate of cell lines in protein-free formulations compared to that of serum-containing media (121). Since these hydrolysates do not contain any protein they do not interfere with the purification procedures in downstream processes.

In many cases the yields obtained from protein-free cultures are higher than those of their serum-containing counterparts. For monoclonal antibody production an increase in productivity has been observed for many cell lines, including hybridoma, CHO, and BHK (137–139). Similar results have also been observed for the production of viruses, such as HIV (140,141).

STRATEGIES FOR THE DEVELOPMENT OF SERUM-FREE MEDIA

There are many analytical approaches for designing serum-free formulations. The objective is to identify growth-promoting components that may be present in serum and include them in a completely synthetic medium.

Ham's Approach

This method is a systematic approach to the development of serum-free media by the gradual reduction in the concentration of undefined supplements in the culture (142). It is a time-consuming but successful method to develop a single formulation for a given cell line.

For nutrients in the basal medium there is a wide concentration range over which the nutrient is not limiting. Starting with a serum-containing medium, a range can be established by adding the nutrient at concentrations of 0.1×, 1×, 10×, 100×, etc., of the original concentration. By comparing the growth response at each concentration, the optimum concentration range can then be determined through the construction of a growth response curve (Fig. 1). The optimum concentration for the nutrient is set near the center of the broad optimum plateau to reduce the likelihood that it will become growth limiting through the adjustments of other medium components. In the response curve shown in Fig. 1 the center of the optimum concentration is 10^{-5} M.

This procedure of optimization of the basal medium may reduce the dependence of serum supplementation to a low level (2%). The cells can adapt to reduced serum by the stimulation of anabolic processes. It may be necessary to allow the cells to grow for several generations at each step of serum reduction to allow the required adaptation to become established.

The ability of cells to grow in a serum content of <2% will vary and may depend on the availability of growth factors and hormones to replace those normally supplied by the serum. A dialysis step is useful to distinguish between the requirements for low- and high-molecular weight factors. The addition of selected components to cells growing at 50% maximal rates provides for easy methods of detecting growth enhancement.

Steps in Ham's approach to the design of serum-free media:

a. Use any means necessary to obtain growth of the cells of interest. This may include serum supplementation, the use of feeder layers or of conditioned media.
b. Select the combination of readily available media to obtain the maximum cell growth.
c. Replace all undefined supplements with dialyzed supplements. Add the undefined low-molecular weight supplements that were removed by dialysis.
d. Identify all the low-molecular weight supplements required for growth by a combination of analysis and trial additions.
e. Reduce the addition of dialyzed supplements to a level that supports less than maximal growth (for example, 50% of maximum). Thus, the undefined supplements are now rate limiting to cell growth.
f. Sequentially adjust the concentrations of all defined nutrient components of the basal media to experimentally determined optima.
g. Sequentially test defined low-molecular weight substances for growth enhancement.
h. At each stage of growth improvement, reduce the amount of undefined supplement so that it becomes rate limiting.
i. Repeat the procedures in steps (g) and (h) until no further reduction in the amount of dialyzed supplement is possible without affecting cell growth.

Sato's Approach

Sato and colleagues (27) have developed an approach based on the understanding of the interaction between cells and growth factors or hormones.

They argue that the survival of cells in culture is dependent on the ability of media components to perform the functions previously provided by the normal environment of the cell in vivo. Therefore, the design of defined media should be based on an attempt to reconstruct the extracellular environment involved in supporting cell growth in vivo. This environment involves a combination of hormones, growth factors, binding proteins, and attachment factors. It is argued that although serum supplements may support some cell growth, serum may be an unsuitable substitute for the in vivo environment in many cases. Serum contains substances that may never come in contact with some cells in vivo and the toxicity of high levels of serum has been reported in many instances. This includes reports of the presence in serum of inhibitors of neural differentiation (143).

Thus, the Sato approach involves the formulation of chemically defined media for optimization of cell growth from a list of isolated and purified growth-promoting substances. These formulations are not perceived as substitutes from serum supplementation but rather as an attempt to simulate the *in vivo* environment.

Top-Down vs. Bottom-Up Approaches to Serum-Free Media Development

There are two basic approaches to developing serum-free media formulations. The first (top-down) involves taking an existing formulation for a similar cell line, supplemented with serum, and selecting constituents that are stimulatory for growth. This is followed by the slow reduction of serum in the medium. The second approach (bottom-up) involves the selection of a basal medium, analyzing individual components for their effects on growth, and combining them to yield a serum-free formulation.

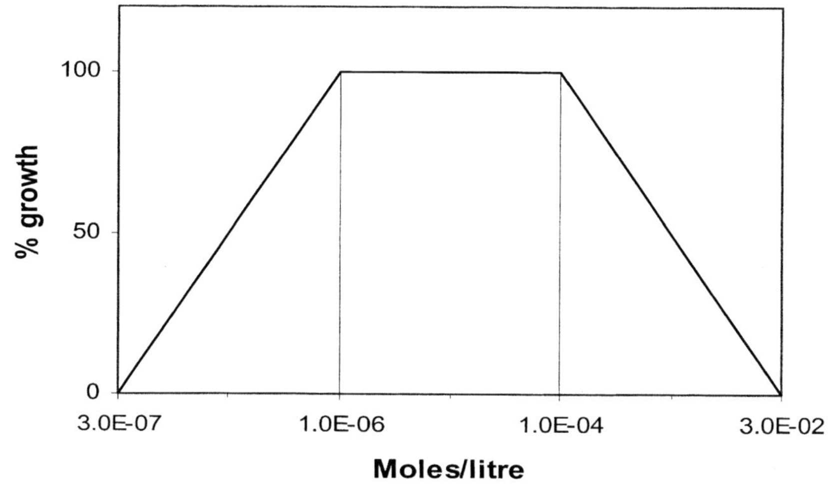

Figure 1 Idealized growth response curve for a single nutrient.

The top-down approach is often easier to pursue since a working serum-free formulation can often be developed more quickly. Cell lines that belong to the same group, such as epithelial or transformed, often require the same growth factors for growth. Therefore, a formulation that works for one epithelial cell line may work for another with minimal modifications to certain growth factors or hormones (such as EGF, FGF, etc.). For this reason serum-free formulations can be designed faster by this approach.

However, the drawback to the top-down approach is that many components in the formulation may be unnecessary, and often inhibitory for growth. This can often result in the "capping" of the optimal performance of the medium (i.e., the maximum growth may not be achieved) as improvements are hindered by the presence of unwanted compounds.

The bottom-up approach, although more labor-intensive and time-consuming, can lead to higher-quality media. Only the components that are required for growth are included in the formulation, allowing for greater control in optimizing the medium. Thus, media developed in this way tend to have higher growth rates and are more easily improved since inhibitory compounds are less likely to be present.

Plackett–Burman Statistical Approach to Serum-Free Media Development

Another method of serum-free media development involves the use of a factorial experimental design, modeled after the Plackett–Burman statistical approach. This statistical method enables the systematic study of complex components and identifies nutritionally important factors for cell proliferation. In addition, the specific interactions between components (whether beneficial or detrimental) can be observed since the combination of nutrients is more important than the individual components by themselves. A large number of components can be studied at once to determine which factors, or combination of factors, are important for a serum-free formulation. This approach is not limited to cell growth studies and can also be valuable in determining which components in the medium are important for productivity, such as antibody production.

The steps involved in the approach are outlined as follows:

a. Select the basal medium and components that are to be analyzed for their ability to enhance growth or productivity.
b. Establish two concentrations for each component: a "high" concentration and a "low" concentration.
c. Grow the cells in the selected basal medium (e.g., DMEM) with serum (5–10%) in a well-plate culture. Allow the cells to initiate growth and division (~24 hr).
d. Remove the serum-medium and replace with the same basal medium plus combinations of media components (no serum), e.g., DMEM+insulin (high/low concentration)+transferrin (high/low concentration).
e. Include a positive control (serum-supplemented medium) and a negative control (medium with no components).
f. Count the viable cells (via MTT assay or trypan-blue exclusion method) in each well after ~48 hr and/or assay for protein production (productivity).
g. Construct a matrix and determine which combinations of components, and at which concentration, are suitable for encouraging or sustaining cell growth (see Table 4).
h. Calculate the variances from all of the effects from the single factors and their interactions to determine which are significant. This will allow the determination of the best combination of components and thus the most suitable serum-free formulation.
i. Initiate an adaptation procedure for the cells to the new serum-free medium.

This stepwise approach has been utilized in many cases to design serum-free formulations. It has been attempted with the CHO cell line, where a serum-free

Table 4 Plackett–Burman Matrix

	Variables (components)			
Medium	A	B	C	D
1	+	+	+	+
2	+	+	+	−
3	+	+	−	−
4	+	−	−	−
5	−	+	+	+
6	−	+	+	−
7	−	+	−	−
8	−	−	+	+
9	−	−	+	−
10	−	−	−	+
11	+	+	−	+
12	+	−	+	−
13	−	+	−	+
14	+	−	+	+
15	+	−	−	+
16	−	−	−	−

Note: +/− represents high/low cocncentration of the component.

medium was optimized for human interferon-gamma production (132). A series of components, including many amino acids, BSA, and other growth factors, were studied. From this, glycine was discovered to affect the specific growth rate, as well as BSA, phenylalanine, and tyrosine. Other amino acids, such as methionine, proline, and histidine played an important role in interferon-gamma production. On the other hand, insulin, arginine, aspartate, and serine produced an inhibitory effect on both cell growth and interferon-gamma production. The end result was an optimized medium that yielded 45% higher productivity than previous formulations.

Similarly, using the same statistical approach, an improved medium for the production of erythropoietin from CHO cells was developed (144). This medium was based on an IMDM basal medium, supplemented with $Fe(NO_3)_3{\cdot}9H_2O$, $CuCl_2$,$\cdot ZnSO_4.7H_2O$, insulin, transferrin, and ethanolamine. Glutamate, serine, methionine, phosphatidylcholine, hydrocortisone, and pluronic F68 were identified as positive determinants for cell growth, and the final optimized medium yielded 79% higher productivities than serum-supplemented IMDM.

FUTURE PROSPECTS

The development of better serum-free formulations is ongoing in the scientific and industrial community. The push for the elimination of animal components in the production of biologicals is leading to considerable effort being placed on developing completely chemically defined serum-free formulations. This will probably lead to the elimination of hydrolysates, or other such undefined components in cell culture.

The many difficulties associated with serum-free development can make the whole task seem intimidating. The development of each formulation requires considerable investment of time. However, the advantages, such as overcoming regulatory hurdles, increasing the performance of a production system, or eliminating the variability of serum in studies of molecule–cell interactions, can make the investment worthwhile. A greater understanding of cell metabolism, hormone interaction, and nutrient requirements will allow better culture media formulations to be developed in the future.

REFERENCES

1. Morgan JG, Morton HJ, Parker RC. Proc Soc Exp Biol Med 1950; 73:1–8.
2. Eagle H. Science 1955; 122:501.
3. Dulbecco R, Freeman G. Virology 1959; 8:396–397.
4. Ham R. Clonal growth of mammalian cells in a chemically defined synthetic medium. Proc Natl Acad Sci 1965; 53:288.
5. Moore GE, Gerner RE, Franklin HA. Culture of normal human leukocytes. J Am Acad Sci 1967; 199:519–524.
6. Eagle H. Science 1959; 130:432.
7. Stoker MGP, MacPherson I. Virology 1961; 14:359.
8. Stoker MGP, MacPherson I. Nature 1964; 203:1355.
9. MacPherson L, Stoker MGP. Virology 1962; 16:147.
10. Yamane I, Matsuya Y, Jimbo K. An autoclavable powdered culture medium for mammalian cells. Proc Soc Exp Biol Med 1968; 127:335–336.
11. Stanners CP, Eliceiri GL, Green H. Two types of ribosome in mouse-hamster hybrid cells. Nat New Biol 1971; 230:52–54.
12. McCoy TA, Maxwell M, Kruse PF. Proc Soc Exp Biol Med 1959; 100:115.

13. Hsu TC, Kellogg DS. J Natl Cancer Inst 1960; 25:221.
14. Iwakata S, Grace JT. NY State J Med 1964; 64:2279.
15. Morton HJ. A survey of commercially available tissue culture media. In Vitro 1970; 5:89–108.
16. Park MS, Terasaki PI. Transplantation 1974; 18:520.
17. Moore GE, Kitamura H. Cell line derived from patient with myeloma. NY State J Med 1968; 68:2054–2060.
18. Fischer GA. Ann NY Acad Sci 1958; 76:673.
19. Fischer GA, Sartorelli AS. Methods Med Res 1964; 10:247.
20. Leibovitz A. Am J Hyg 1963; 78:173.
21. Trowell OA. Exp Cell Res 1959; 16:118.
22. Williams GM, Gunn JM. Long-term cell culture of adult rat liver epithelial cells. Exp Cell Res 1974; 89:139–142.
23. Biggers JD, Gwatkin RBL, Heyner SJ. Exp Cell Res 1961; 25:41.
24. Parker RC. NY Acad Sci (spec publ) 1957; 5:303.
25. Morgan JF, Morton HJ, Parker RC. Proc Soc Exp Biol Med 1950; 73:1.
26. Murakami H. In: Mizrahi A, ed. Advances in Biotechnological Processes. Vol. 11. New York: A. R. Liss, 1989:107.
27. Barnes D, Sato G. Serum-free cell culture: a unifying approach. Cell 1980; 22:649–655.
28. Iscove NN, Melchers F. Complete replacement of serum by albumin, transferrin, and soybean lipid in cultures of lipopolysaccharide-reactive B lymphocytes. J Exp Med 1978; 147:923–933.
29. Bettger WJ, Boyce ST, Walthall BJ, Ham RG. Rapid clonal growth and serial passage of human diploid fibroblasts in a lipid-enriched synthetic medium supplemented with epidermal growth factor, insulin, and dexamethasone. Proc Natl Acad Sci USA 1981; 78:5588–5592.
30. Ryan PA, Maher VM, McCormick JJ. Modification of MCDB 110 medium to support prolonged growth and consistent high cloning efficiency of diploid human fibroblasts. Exp Cell Res 1987; 172:318–328.
31. Boyce ST, Ham RG. Calcium-regulated differentiation of normal human epidermal keratinocytes in chemically defined clonal culture and serum-free serial culture. J Invest Dermatol 1983; 81:33s–40s.
32. McKeehan WL, Hamilton WG, Ham RG. Selenium is an essential trace nutrient for growth of WI-38 diploid human fibroblasts. Proc Natl Acad Sci USA 1976; 73: 2023–2027.
33. McKeehan WL, McKeehan KA, Hammond SL, Ham RG. Improved medium for clonal growth of human diploid fibroblasts at low concentrations of serum protein. In Vitro 1977; 13:399–416.
34. Morgan JF, Campbell ME, Morton HJ. Natl Cancer Inst 1955; 16:557.
35. Evans VJ, Bryant JC, Kerr HA, Schilling L. Exp Cell Res 1964; 36:439.
36. Evans VJ, Bryant JC, Fioramonti MC, McQuilkin WT, Sanford KK, Earle WR. Cancer Res 1956; 16:11.
37. Waymouth C. J Natl Cancer Inst 1959; 22:1003.
38. ISO. Water for Analytical Laboratory Use—Specification and Test Methods. International Organization for Standardization, 1987.
39. Neermann J, Wagner R. Comparative analysis of glucose and glutamine metabolism in transformed mammalian cell lines, insect and primary liver cells. J Cell Physiol 1996; 166:152–169.
40. Butler M, Christie A. Adaptation of mammalian cells to non-ammoniagenic media. Cytotechnology 1994; 15:87–94.
41. Butler M, Thilly WG. MDCK microcarrier cultured: seeding density effects and amino acid utilization. In Vitro 1982; 18:213–219.
42. Lambert K, Pirt SJ. The quantitative requirements of human diploid cells (strain MRC-5) for amino acids, vitamins and serum. J Cell Sci 1975; 17:397–411.

43. Roberts RS, Hsu HW, Lin KD, Yang TJ. Amino acid metabolism of myeloma cells in culture. J Cell Sci 1976; 21:609–615.
44. Arathoon WR, Telling RC. Uptake of amino acids and glucose by BHK 21 clone 13 suspension cells during cell growth. Dev Biol Stand 1981; 50:145–154.
45. Miller SWM, Blanch HW, Wilke CR. A kinetic analysis of hybridoma growth and metabolism in batch and continuous suspension culture: effect of nutrient concentration, dilution rate, and pH. Biotechnol Bioeng 2000; 67:853–871.
46. Linz M, Zeng AP, Wagner R, Deckwer WD. Stoichiometry, kinetics, and regulation of glucose and amino acid metabolism of a recombinant BHK cell line in batch and continuous cultures. Biotechnol Prog 1997; 13:453–463.
47. Holben DH, Smith AM. The diverse role of selenium within selenoproteins: a review. J Am Diet Assoc 1999; 99:836–843.
48. Mustacich D, Powis G. Thioredoxin reductase. Biochem J 2000; 346(Pt 1):1–8.
49. Seligman PA, Kovar J, Schleicher RB, Gelfand EW. Transferrin-independent iron uptake supports B lymphocyte growth. Blood 1991; 78:1526–1531.
50. Keenan J, Clynes M. Replacement of transferrin by simple iron compounds for MDCK cells grown and subcultured in serum-free medium. In Vitro Cell Dev Biol Anim 1996; 32:451–453.
51. Kan M, Yamane I. Effects of ferrous iron and transferrin on cell proliferation of human diploid fibroblasts in serum-free culture. In Vitro 1984; 20:89–94.
52. Wilmann M, Gautel M, Mayans O. Activation of calcium/calmodulin regulated kinases. Cell Mol Biol (Noisy-le-grand) 2000; 46:883–894.
53. Boisseau AM, Donatien P, Surleve-Bazeille JE, Amedee J, Harmand MF, Bezian JH, Maleville J, Taieb A. Production of epidermal sheets in a serum free culture system: a further appraisal of the role of extracellular calcium. J Dermatol Sci 1992; 3:111–120.
54. Daniels JT, Harris IR, Kearney JN, Ingham E. Calcium: a crucial consideration in serum-free keratinocyte culture. Exp Dermatol 1995; 4:183–191.
55. Praeger FC, Cristofalo VJ. The growth of WI-38 cells in a serum-free, growth factor-free, medium with elevated calcium concentrations. In Vitro Cell Dev Biol 1986; 22:355–359.
56. Peshwa MV, Kyung Y, McClure DB, Hu W. Cultivation of mammalian cells as aggregates in bioreactors: effect of calcium concentration on spatial distribution of viability. Biotechnol Bioeng 1993; 41:179–187.
57. Mobest D, Mertelsmann R, Henschler R. Serum-free ex vivo expansion of CD34 hematopoietic progenitor cells. Biotechnol Bioeng 1998; 60:341–347.
58. Bashir N, Kuhen K, Taub M. Phospholipids regulate growth and function of MDCK cells in hormonally defined serum free medium. In Vitro Cell Dev Biol 1992; 28A: 663–668.
59. Carney D, Bunn P, Gazdar A, Pagan J, Minna J. Selective growth in serum-free hormone-supplemented medium of tumor cells obtained by biopsy from patients with small cell carcinoma of the lung. Proc Natl Acad Sci USA 1981; 75:3185–3189.
60. Kim EJ, Kim NS, Lee GM. Development of a serum-free medium for the production of humanized antibody from Chinese hamster ovary cells using a statistical design. In Vitro Cell Dev Biol Anim 1998; 34:757–761.
61. Labitzke R, Friedl P. A serum-free medium formulation supporting growth of human umbilical cord vein endo-thelial cells in long-term cultivation. Cytotechnology 2001; 35:87–92.
62. Donatien P, Surleve-Bazeille JE, Thody AJ, Taieb A. Growth and differentiation of normal human melanocytes in a TPA-free, cholera toxin-free, low-serum medium and influence of keratinocytes. Arch Dermatol Res 1993; 285:385–392.
63. Hirobe T. Melanocyte stimulating hormone induces the differentiation of mouse epidermal melanocytes in serum-free culture. J Cell Physiol 1992; 152:337–345.

64. Furuya R, Akiu S, Naganuma M, Fukuda M, Hirobe T. The proliferation and differentiation of neonatal epidermal melanocytes in Fl hairless mice of HR-1 × HR/De in serum-free culture. J Dermatol 1998; 25:211–221.
65. Hedlund TE, Miller GJ. A serum-free defined medium capable of supporting growth of four established human prostatic carcinoma cell lines. Prostate 1994; 24:221–228.
66. Siegfried JM, Han YH, DeMichele MA, Hunt JD, Gaither AL, Cuttitta F. Production of gastrin-releasing peptide by a non-small cell lung carcinoma cell line adapted to serum-free and growth factor-free conditions. J Biol Chem 1994; 269:8596–8603.
67. Brower M, Carney DN, Oie HK, Gazdar AF, Minna JD. Growth of cell lines and clinical specimens of human non-small cell lung cancer in a serum-free defined medium. Cancer Res 1986; 46:798–806.
68. Masuda N, Fukuoka M, Takada M, Kudoh S, Kusunoki Y. Establishment and characterization of 20 human non-small cell lung cancer cell lines in a serum-free defined medium (ACL-4). Chest 1991; 100:429–438.
69. Ruck A, Jakobson E, Bjorkman S, Paulie S. Adaptation of human bladder carcinoma cell lines to serum-free growth. Evidence for autocrine growth, stimulation. Anticancer Res 1994; 14:55–60.
70. Dartsch PC, Weiss HD, Betz E. Human vascular smooth muscle cells in culture: growth characteristics and protein pattern by use of serum-free media supplements. Eur J Cell Biol 1990; 51:285–294.
71. Merten OW. Safety issues of animal products used in serum-free media. Dev Biol Stand 1999; 99:167–180.
72. Tardieu M, Gamby C, Avramoglou T, Jozefonvicz J, Barritault D. Derivatized dextrans mimic heparin as stabilizers, potentiators, and protectors of acidic or basic FGF. J Cell Physiol 1992; 150:194–203.
73. Shiokawa K, Asano M, Shiozaki C. Function, molecular structure and gene expression of fibroblast growth factor (FGF/HBGF). Nippon Rinsho 1992; 50:1893–1901.
74. Burgess WH, Maciag T. The heparin-binding (fibroblast) growth factor family of proteins. Annu Rev Biochem 1989; 58:575–606.
75. Jenkins N. Growth factors. In: Butler M, ed. Mammalian Cell Biotechnology. New York: Oxford University Press, 1991:39–55.
76. Gopas J, Ono M, Princler G, Smith MR, Tainsky MA, Siddiqui M, Wishniak O, Segal S, Kuwano M, Kung HF. EGF receptor activity and mitogenic response of Balb/3T3 cells expressing Ras and Myc oncogenes. EGF receptor activity in oncogene transformed cells. Cell Mol Biol 1992; 38:605–614.
77. Chua CC, Ladda RL. Protein kinase C and non-functional EGF receptor in K-ras transformed cells. Biochem Biophys Res Commun 1986; 135:435–444.
78. Haley JD. Regulation of epidermal growth factor receptor expression and activation: a brief review. Symp Soc Exp Biol 1990; 44:21–31.
79. Krane JF, Murphy DP, Carter DM, Krueger JG. Synergistic effects of epidermal growth factor (EGF) and insulin-like growth factor I/somatomedin C (IGF-I) on keratinocyte proliferation may be mediated by IGF-I transmodulation of the EGF receptor. J Invest Dermatol 1991; 96:419–424.
80. Simmons JG, Hoyt EC, Westwick JK, Brenner DA, Pucilowska JB, Lund PK. Insulin-like growth factor-I and epidermal growth factor interact to regulate growth and gene expression in IEC-6 intestinal epithelial cells. Mol Endocrinol 1995; 9:1157–1165.
81. Kaplan DR, Stephens RM. Neurotrophin signal transduction by the Trk receptor. J Neurobiol 1994; 25:1404–1417.
82. Yano H, Cong F, Birge RB, Goff SP, Chao MV. Association of the Abl tyrosine kinase with the Trk nerve growth factor receptor. J Neurosci Res 2000; 59:356–364.
83. Gangarosa LM, Dempsey PJ, Damstrup L, Barnard JA, Coffey RJ. Transforming growth factor-alpha. Baillieres Clin Gastroenterol 1996; 10:49–63.
84. Lawrence DA. Transforming growth factor-beta: a general review. Eur Cytokine Netw 1996; 7:363–374.

85. Moses HL, Coffey RJ Jr, Leof EB, Lyons RM, Keski-Oja J. Transforming growth factor beta regulation of cell proliferation. Cell Physiol Suppl 1987 (suppl):1–7.
86. Barnes D, Sato G. Methods for growth of cultured cells in serum-free medium. Anal Biochem 1980; 102:255–270.
87. Williams LT. Signal transduction by the platelet-derived growth factor receptor. Science 1989; 243:1564–1570.
88. Daughaday WH, Rotwein P. Insulin-like growth factors I and II. Peptide, messenger ribonucleic acid and gene structures, serum, and tissue concentrations. Endocr Rev 1989; 10:68–91.
89. Baxter RC. Insulin-like growth factor binding proteins in the human circulation: a review. Horm Res 1994; 42:140–144.
90. Heldin CH, Westermark B. Mechanism of action and in vivo role of platelet-derived growth factor. Physiol Rev 1999; 79:1283–1316.
91. Taga T, Kishimoto T. Gpl30 and the interleukin-6 family of cytokines. Annu Rev Immunol 1997; 15:797–819.
92. Kunioku H, Inoue K, Tomida M. Interleukin-6 protects rat PC 12 cells from serum deprivation or chemotherapeutic agents through the phosphatidylinositol 3-kinase and STAT3 pathways. Neurosci Lett 2001; 309:13–16.
93. Sterneck E, Kaplan DR, Johnson PF. Interleukin-6 induces expression of peripherin and cooperates with Trk receptor signaling to promote neuronal differentiation in PC12 cells. J Neurochem 1996; 67:1365–1374.
94. Sugasawara R. J Biotechnol 1988; 6:895.
95. Levy Y, Fermand JP, Brouet JC. Differential effects of low and high concentrations of interleukin 6 on human B cells. Eur J Immunol 1990; 20:2389–2393.
96. Wu YY, Bradshaw RA. Synergistic induction of neurite outgrowth by nerve growth factor or epidermal growth factor and interleukin-6 in PC 12 cells. J Biol Chem 1996; 271:13033–13039.
97. Glassy MC, Tharakan JP, Chau PC. Biotechnol Bioeng 1988; 32:1015.
98. Imagawa W, Bandyopadhyay GK, Wallace D, Nandi S. Phospholipids containing polyunsaturated fatty acyl groups are mitogenic for normal mouse mammary epithelial cells in serum-free primary cell culture. Proc Natl Acad Sci USA 1989; 86:4122–4126.
99. van Corven EJ, Groenink A, Jalink K, Eichholtz T, Moolenaar WH. Lysophosphatidate-induced cell proliferation: identification and dissection of signaling pathways mediated by G proteins. Cell 1989; 59:45–54.
100. Murakami H, Masui H, Sato GH, Sueoka N, Chow TP, Kano-Sueoka T. Growth of hybridoma cells in serum-free medium: ethanolamine is an essential component. Proc Natl Acad Sci USA 1982; 79:1158–1162.
101. Chen Z, Ke Y, Chen Y. A serum-free medium for hybridoma cell culture. Cytotechnology 1993; 1:169–174.
102. Sato JD, Kawamoto T, McClure DB, Sato GH. Cholesterol requirement of NS-1 mouse myeloma cells for growth in serum-free medium. Mol Biol Med 1984; 2:121–134.
103. Butler M, Huzel N. The effect of fatty acids on hybridoma cell growth and antibody productivity in serum-free cultures. J Biotechnol 1995; 39:165–173.
104. Butler M, Huzel N, Barnabe N. Unsaturated fatty acids enhance cell yields and perturb the energy metabolism of an antibody-secreting hybridoma. Biochem J 1997; 322: 615–623.
105. Emerson TE Jr. Unique features of albumin: a brief review. Crit Care Med 1989; 77:690–694.
106. Stamler JS, Jaraki O, Osborne J, Simon DI, Keaney J, Vita J, Singel D, Valeri CR, Loscalzo J. Nitric oxide circulates in mammalian plasma primarily as an S-nitroso adduct of serum albumin. Proc Natl Acad Sci USA 1992; 89:1614–1611.
107. Pollack S. Receptor-mediated iron uptake and intracellular iron transport. Am J Hematol 1992; 39:113–118.

108. Young SP, Garner C. Delivery of iron to human cells by bovine transferrin. Implications for the growth of human cells in vitro. Biochem J 1990; 265:587–591.
109. Kleinman HK, Luckenbill-Edds L, Cannon FW, Sephel GC. Use of extracellular matrix components for cell culture. Anal Biochem 1987; 166:1–13.
110. Orly J, Sato G. Fibronectin mediates cytokinesis and growth of rat follicular cells in serum-free medium. Cell 1979; 17:295–305.
111. Ilic D, Almeida EA, Schlaepfer DD, Dazin P, Aizawa S, Damsky CH. Extracellular matrix survival signals transduced by focal adhesion kinase suppress p53-mediated apoptosis. J Cell Biol 1998; 143:547–560.
112. Danen EH, Sonneveld P, Sonnenberg A, Yamada KM. Dual stimulation of Ras/mitogen-activated protein kinase and RhoA by cell adhesion to fibronectin supports growth factor-stimulated cell cycle progression. Cell Biol 2000; 151:1413–1422.
113. Hotchin NA, Kidd AG, Altroff H, Mardon HJ. Differential activation of focal adhesion kinase, Rho and Rac by the ninth and tenth FIII domains of fibronectin. J Cell Sci 1999; 112:2937–2946.
114. von der Mark K, Kuhl U. Laminin and its receptor. Biochim Biophys Acta 1985; 823:147–160.
115. Kam L, Shain W, Turner JN, Bizios R. Axonal outgrowth of hippocampal neurons on micro-scale networks of polylysine-conjugated laminin. Biomaterials 2001; 22: 1049–1054.
116. Varani J, Inman DR, Frigid SE, Hillegas WJ. Use of recombinant and synthetic peptides as attachment factors for cells on microcarriers. Cytotechnology 1993; 13:89–98.
117. Esty A. Receptor-specific serum-free cell attachment using a highly stable engineered protein polymer. Am Biotechnol Lab 1991; 9:44.
118. Kovar J. Growth-stimulating effect of ferric citrate on hybridoma cells: characterization and relation to transferrin function. Hybridoma 1988; 7:255–263.
119. MacLeod AJ. In: Butler M, ed. Mammalian Cell Biotechnology: A Practical Approach. Oxford: Oxford University Press, 1991:28–38.
120. Schlaeger EJ. The protein hydrolysate, Primatone RL, is a cost-effective multiple growth promoter of mammalian cell culture in serum-containing and serum-free media and displays anti-apoptosis properties. J Immunol Methods 1996; 194:191–199.
121. Kistner O, Barrett PN, Mundt W, Reiter M, Schober-Bendixen S, Eder G, Dorner F. A novel mammalian cell (vero) derived influenza virus vaccine: development, characterization and industrial scale production. Wien Klin Wochenschr 1999; 111:207–214.
122. Majoul S, Kharmachi H, Saadi M, Chouaib A, Kallel H. Adaption of vero cells to a serum-free medium for the production of rabies virus. In: Bernard A, et al, eds. Animal Cell Technology: Products from Cells, Cells as Products. Dordrecht: Kluwer Academic Publishers, 1999:467–469.
123. Montagnon BJ, Vincent-Falquet JC, Saluzzo JF. Experience with vero cells at Pasteur Merieux Connaught. Dev Biol Stand 1999; 98:137–140.
124. Voeten JT, Brands R, Palache AM, van Scharrenburg GJ, Rimmelzwaan GF, Osterhaus AD, Claas EC. Characterization of high-growth reassortant influenza A viruses generated in MDCK cells cultured in serum-free medium. Vaccine 1999; 17:1942–1950.
125. Perrin P, Madhusudana S, Gontier-Jallet C, Petres S, Tordo N, Merten OW. An experimental rabies vaccine produced with a new BHK-21 suspension cell culture process: use of serum-free medium and perfusion-reactor system. Vaccine 1995; 13:1244–1250.
126. Yahi N, Baghdiguian S, Fantini J. Production of a highly cytopathic HIV-1 isolate from a human mucosal epithelial cell line cultured on microcarrier beads in serum-free medium. In Vitro Cell Dev Biol Anim 1995; 31:62–66.
127. Gardner TA, Ko SC, Yang L, Cadwell JJ, Chung LW, Kao C. Serum-free recombinant production of adenovirus using a hollow fiber capillary system. Biotechniques 2001; 30:422–427.

128. Butler M, Burgener A, Patrick M, Berry M, Moffatt D, Huzel N, Barnabe N, Coombs K. Application of a serum-free medium for the growth of vsero cells and the production of reovirus. Biotechnol Prog 2000; 16:854–858.
129. Ryll T, Lucki-Lange M, Jager V, Wagner R. Production of recombinant human interleukin-2 with BHK cells in a hollow fibre and a stirred tank reactor with protein-free medium. J Biotechnol 1990; 14:377–392.
130. Valyakina TI, Vershinina NV, Gilyarevskii SD, Nesmeyanov VA. Production of interleukin 2 in serum-free medium. Mol Immunol 1984; 21:811–814.
131. Castro PM, Hayter P, Ison AP, Bull AT. Application of a statistical design to the optimization of culture medium for recombinant interferon-gamma production by Chinese hamster ovary cells. Appl Microbiol Biotechnol 1992; 55:84–90.
132. Miyaji H, Harada N, Mizukami T, Sato S, Fujiyoshi N, Itoh S. Efficient expression of human beta-interferon in Namalwa KJM-1 cells adapted to serum-free medium by a dhfr gene coamplification method. Cytotechnology 1990; 4:173–180.
133. Kim YN, Park YS, Kim HK, Jeon BC, Youn SE, Lee HY. Enhancement of the attachment on microcarriers and tPA production by fibroblast cells in a serum-free medium by the addition of the extracts of Centella asiatica. Cytotechnology 1993; 13:221–226.
134. Haldankar R, Kopchick JJ, Ridgway D. Stable production of a human growth hormone antagonist from CHO cells adapted to serum-free suspension culture. Biotechnol Prog 1999; 15:336–346.
135. Boedeker D. Production processes of licensed recombinant factor VIII preparations. Semin Thromb Hemost 2001; 27:385–394.
136. Stoll TS, Muhlethaler K, von Stockar U, Marison IW. Systematic improvement of a chemically-defined protein-free medium for hybridoma growth and monoclonal antibody production. J Biotechnol 1996; 45:111–123.
137. Voigt A, Zintl F. Hybridoma cell growth and anti-neuroblastoma monoclonal antibody production in spinner flasks using a protein-free medium with microcarriers. J Biotechnol 1999; 68:213–226.
138. Inoue Y, Lopez LB, Kawamoto S, Arita N, Teruya K, Seki K, Shoji M, Kamei M, Hashizume S, Shiozawa Y, Tachibana H, Ohashi H, Yasumoto K, Nagashima A, Nakahashi H, Suzuki T, Imai T, Nomoto K, Takenoyama M, Katakura Y, Shirahata S. Production of recombinant human monoclonal antibody using ras-amplified BHK-21 cells in a protein-free medium. Biosci Biotechnol Biochem 1996; 60:811–817.
139. Cinatl J Jr, Cinatl J, Rabenau H, Ebener U, Kornhuber B, Doerr HW. Increased HIV-1 production in chronically infected H9 cells grown in protein-free medium. Arch Virol 1992; 125:327–330.
140. Pinter C, Siccardi AG, Clivio A. Production of human immunodeficiency virus by chronically infected cells grown in protein-free medium. Cell Biol Int 1995; 19:507–515.
141. Ham RG. In: Fischer G, Wieser RJ, eds. Hormonally Defined Media. Berlin: Springer-Verlag, 1983:16–30.
142. Davis GE, Skaper SD, Manthorpe M, Moonen G, Varon S. Fetal calf serum-mediated inhibition of neurite growth from ciliary ganglion neurons in vitro. J Neurosci Res 1984; 12:29–39.
143. Lee GM, Kim EJ, Kim NS, Yoon SK, Ahn YH, Song JY. Development of a serum-free medium for the production of erythropoietin by suspension culture of recombinant Chinese hamster ovary cells using a statistical design. J Biotechnol 1999; 69:85–93.
144. Causey AL, Wooten RM, Clem LW, Bly JE. A serum-free medium for human primary T. lymphocyte culture. J Immunol Methods 1994; 175(1):115–121.

4

Cell Metabolism

Francesc Gódia and Jordi Joan Cairó
Department d'Enginyeria Química, ETSE,
Universitat Autònoma de Barcelona, Bellaterra, Barcelona, Spain

INTRODUCTION

Mammalian cell lines are used extensively for the production of therapeutic proteins in a variety of in vitro culture systems, and the characterization, understanding, and ultimately the redistribution of their basic metabolic behavior are needed to improve their performance and to design optimized production systems. The metabolic patterns of mammalian cells are altered substantially when the original homeostatic environment in a tissue of stressful pluricellular organisms is changed by the stressful in vitro culture system environment and the genetic alterations occurring during the immortalization process, leading to the establishment of a production cell line. Consequently, mammalian cell lines exhibit a highly deregulated metabolism when cultured in vitro. This is characterized by a very high and inefficient consumption of the main carbon, nitrogen, and energy sources, glucose and glutamine, and leads to the generation of waste metabolic end products, mainly lactate, ammonium, and some amino acids, such as alanine or proline. To understands this altered metabolism, one should consider the interaction of the different metabolic pathways in the cells. The main pathways reviewed in this chapter cannot be considered individually but as part of a more complex and flexible structure. In addition, the cellular compartmentalization of mammalian cells must also be considered in this analysis, particularly those processes occurring in the mitochondrion or in the cytosol. The central metabolic pathways considered comprise glycolysis, pentose phosphate pathway, tricarboxylic acid (TCA) cycle, oxidative phosphorylation, glutaminolysis, and the metabolism of other amino acids. They will be discussed in combination with the different transport mechanisms both from the external medium into the cells and from the cytosolic medium into the mitochondria, and, taking into account the interaction between them, by the malate shunt or the transamination (TA) pathways. A number of authors have reviewed different aspects of animal cell metabolism prior to this study, some very extensively (1–3).

In addition to the analysis of the metabolic pathways in animal cells and their interaction, this chapter also reviews the different possible strategies to mitigate the metabolic deregulation of cells in culture to obtain more physiologically and metabolically balanced patterns to allow the generation of more efficient processes. This can be pursued either by a redefinition of the process conditions (i.e., medium formulation, culture strategy), based on the knowledge of cell metabolism, or by a direct metabolic redistribution by means of cell engineering, to obtain a modified cell with a more balanced metabolic pattern.

CARBON AND ENERGY SOURCE METABOLISM

In this section, the analysis of the main pathways of animal cell metabolism is approached using as a starting point the most general situation encountered when in vitro culture is developed. Glucose and glutamine are the main compounds in the culture media, providing the required precursors for the biosynthetic pathways and energy to the cells. As widely reported, the metabolism of glucose and glutamine generates elevated quantities of lactate and ammonium, very often limiting the culture performance, demonstrating a clear inefficiency in the use of these compounds. It should be pointed out that this pattern is particularly acute in batch culture systems, which are most often used because of their simplicity, where cells are exposed to elevated glucose and glutamine levels in the medium, but this pattern is also found in fed-batch systems after a certain period of operation. This basic metabolic pattern is indeed the consequence and combination of different mechanisms: glucose and glutamine transport into the cells, glycolysis and glutaminolysis pathways, the TCA cycle, phosphorylative oxidation, generation of cellular energy, regeneration of reducing power, among others. It is also affected by cell metabolic flexibility, allowing the adjustment of certain pathways as a consequence of alteration in other pathways, and by cell compartmentalization, particularly reflected in the additional transport steps required for those reactions taking place in the mitochondria. Finally, it should be observed at this point that the cell metabolism will also change as a consequence of alterations in the physicochemical conditions of the culture and the changes in its composition. The understanding of these basic pathways, transport mechanisms, and how they interact provide the basis for cellular metabolism analysis to further design strategies to balance it or redistribute it, as discussed in the last sections of this chapter.

Glucose Transport

The cytoplasmatic membrane is impermeable to polar molecules such as glucose, and therefore the uptake of glucose by mammalian cells is achieved by means of transport proteins located in the plasmatic membrane. In the majority of mammalian cells the glucose transport mechanism facilitates saturable and bidirectional diffusion, driven by the concentration gradient across the cell membrane. Normally, since low amounts of free glucose are present in the cell cytoplasm due to rapid phosphorylation by hexokinases, the transport will take place in the uptake direction. Six different isoforms of glucose transporters have been isolated in different tissues. They have a similar size, between 492 and 524 amino acids, and they present identity in the amino acid sequence of 39–65%, and similarity between 50% and 76%. GLUT1 is the most abundant transporter in human erythrocytes; it is also abundant in brain, placenta, and epithelial tissues (4), although it is expressed at a certain level in all tissues. Thus, it has been suggested that GLUT1 could be, at least partially, the protein responsible for constitutive glucose transport. In addition to glucose, GLUT1 also allows the uptake of other hexoses and pentoses, although with low affinity (5). It evidences as well a clear asymmetry in the transport kinetics: The K_M value for the uptake is 1–2 mM, as the value for the glucose export is one order of magnitude higher. GLUT2 is a transporter with a lower affinity for glucose, 15–20 mM, and symmetric kinetics (6) found in different tissues but especially in the liver cells. It allows transport of glucose, galactose, mannose, and fructose as well (7). GLUT3 is the transporter more abundant in brain cells, and it is also expressed in kidney and placenta (8). The K_M value for this transporter is even lower than that for GLUT1, probably as a mechanism to ensure the uptake of glucose by brain cells since the glucose concentration is lower in the extracellular fluids than in the plasma. In fact, that the affinity of GLUT3 is the highest among all other glucose transporters (4). GLUT4 is the transporter more abundant in the adipic tissues and muscles, and it is responsible for the glucose uptake stimulated by insulin (9). GLUT5 is the more divergent of the isoforms, with a 40% identity from the rest, and has a higher affinity for fructose than for glucose (10). GLUT7 is the last isoform cloned from a cDNA library of rat liver (11). In addition to the facilitated diffusion transport, glucose uptake can also occur through the sodium symport process, which is based on the sodium concentration gradient across the cytoplasmatic membrane.

Glucose Metabolism

Once incorporated into the cytosol, glucose is further converted by two main pathways: glycolysis and the pentose phosphate cycle (PPC), as represented in Fig. 1. In the main pathway, glycolysis, glucose is converted to pyruvate, generating a number of intermediates for biosynthesis, two ATP molecules, and two NADH molecules. Glucose 6-phosphate is used for glycogen synthesis; alternatively it can be converted into 6-phosphogluconate and further to ribose 5-phosphate, through the PPC. The PPC allows the generation of NADPH reduction equivalents required for cell anabolism, and ribose 5-phosphate molecules for the synthesis of nucleic acids. Alternatively, ribose 5-phosphate can also be generated or transformed into fructose 6-phosphate or glyceraldehyde 3-phosphate, providing metabolic flexibility to the cell in order to balance the fluxes through these pathways. The flux through the PPC should be related to the nucleic acid requirements for DNA duplication or RNA transcription, and could probably be controlled by the cell cycle (2). Different authors have measured experimentally using labeled glucose, or estimated by indirect measurements and mass balancing, the percentage of the consumed glucose that enters the PPC. This amount is quite low in a number of cases, 4–8% (12–17), and increases to 14% for cultures performed at a low dissolved oxygen tension (18). Other studies of intracellular flux analysis report significantly higher values, in the range of 15–27% (19–21). Other biosynthesis precursors provided by the glycolysis pathways are: fructose 6-phosphate, required for the synthesis of amino sugar compounds; dihydroxyacetone-phosphate, which is further converted into glycerol 3-phosphate and used for lipid synthesis; and 3-phosphoglycerate, precursor for the synthesis of serine and glycine, used for purine and pyrimidine nucleotide synthesis.

Deregulated Glycolysis in Mammalian Cells

The rate of the glycolysis pathway in mammalian cells has been reported to be very high by different authors. In addition, this high glycolysis is linked to the generation of large amounts of lactate as an end product, even under fully aerobic conditions, indicating that a low percentage of pyruvate is incorporated into the TCA cycle for a

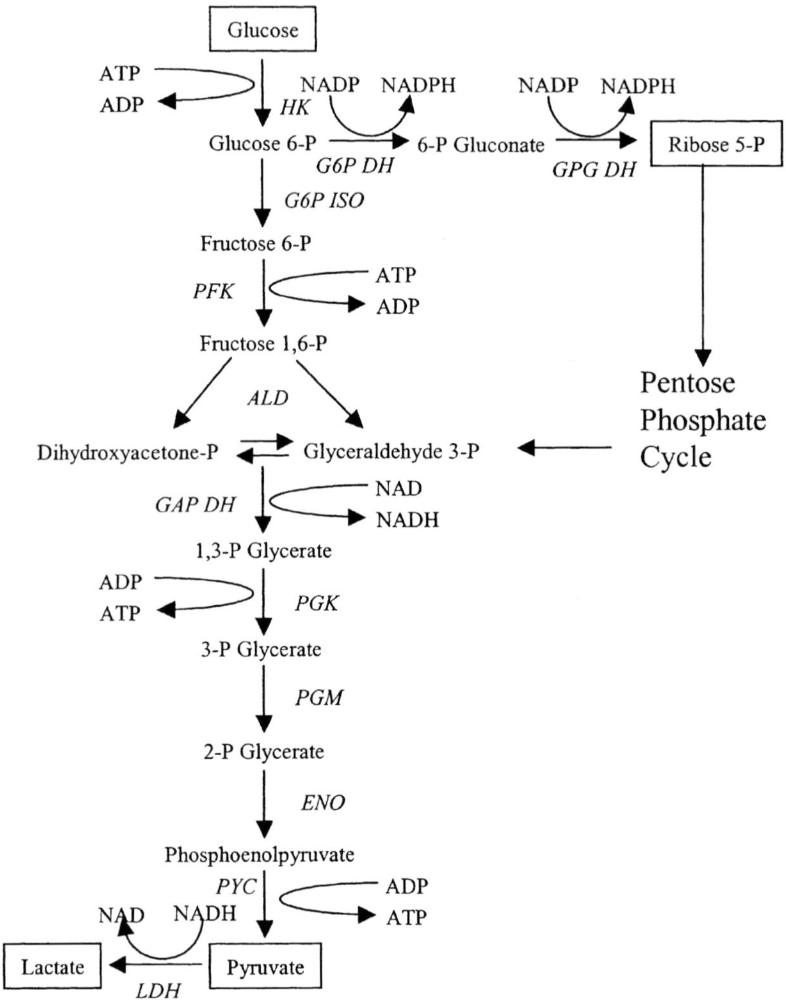

Figure 1 Overview of the glycolysis pathway and the PPC. The enzymes involved in the pathway are indicated in italics and the corresponding full names are as follows: AC, aconitase; AlaTA, alanine transaminase; aspTA, aspartate transaminase; ALD, aldolase; ATP, adenosine triphosphate; CS, citrate synthase; D, dilution rate in continuous cultures (h^{-1}); ENO, enolase; FU, fumarase; GAPDH, glyceraldehyde 3-phosphate dehydrogenase; GDH, glutamate dehydrogenase; GS, glutamine synthetase; G6PDH, glucose 6-phosphate dehydrogenase; G6PISO, glucose 6-phosphate isomerase; HK, hexokinase; ICDH, isocitrate dehydrogenase; α-KGDH, α-ketoglutarate dehydrogenase; LDH, lactate dehydrogenase; MDH, malate dehydrogenase; ME, malic enzyme; NADH, nicotine adenine dinucleotide; OUR, oxygen uptake rate; PAG, phosphate-activated glutaminase; PC, pyruvate carboxylase; PDH, pyruvate dehydrogenase; PDHC, pyruvate dehydrogenase complex; PEPCK, phosphoenolpyruvate carboxykinase; PFK, phosphofructokinase; 6PGDH, 6-phosphogluconate dehydrogenase; PGK, phosphoglycerate kinase; PGM, phosphoglycerate mutase; PPC, pentose phosphate cycle; PYC, pyruvate kinase; q_{glu}, specific consumption rate of glucose; q_{gln}, specific consumption rate of glutamine; q_{o_2} specific consumption rate of oxygen; SCoA S, succinyl coenzyme A synthetase; SDH, succinate dehydrogenase; TCA, tricarboxylic acid.

complete oxidation to CO_2 and H_2O, a pathway much more efficient regarding the energetic yield (36 ATP instead of 2 ATP when lactate is the end product). It has been demonstrated that the specific glucose consumption of different cell lines depends of the glucose concentration in the medium. This effect was already reported in early batch studies (22–24). The actual values for hybridoma cells can range from 7 nmol 10^{-9} cells per day at 5 mM to 24 nmol 10^{-9} cells per day at 16 mM in the culture medium (24). Continuous cultures in chemostat offer a more precise way to evaluate metabolic rates under specific conditions. Miller et al. (26) performed a series of step changes, increasing the glucose concentration in the feed medium. They showed that when glucose concentration increased from 0.1 to 8.4 mM, the specific consumption rate of glucose increased dramatically and reached a steady-state value of 2.9 nmol 10^{-9} cells per day, 67% higher than the value before the step change. Sanfeliu et al. (27) showed in a series of continuous runs decreasing stepwise the glucose feed to a hybridoma culture that when the glucose concentration was lower, it was more efficiently consumed, with a lower lactate generation (lactate being reduced from 10 to 2 mM). This has also been shown by other authors (27,28). They evaluated the minimum specific glucose uptake rate required to maintain cell metabolism and avoid the decrease of cell viability and density, 76.8 nmol 10^{-6} cells per hour. For Chinese hamster ovary (CHO) cells, Hayter et al. (30) evaluated the specific consumption rate of glucose under limiting conditions in chemostat cultures in the range of 118–123 nmol 10^{-6} cells per hour. This is considerably lower than the maximum value obtained by the same authors in batch experiments, 300 nmol 10^{-6} cells per hour, when cells are exposed to higher glucose concentrations (31). For a different CHO cell line, a series of continuous cultures at decreasing glucose concentration in the feed showed how the specific consumption rate decreased from 60 to 17.7 nmol 10^{-6} cells per hour, while the lactate-to-glucose yield decreased from 1.35 to 0.2, thus evidencing a much more efficient metabolism, with a much lower percentage of the consumed glucose transformed into lactate as end product (32). For baby hamster kidney (BHK) cells, Cruz et al. (33) observed the same trend by performing chemostat cultures at high and low glucose levels in the feed. For the high-glucose case, specific glucose consumption reached a value of 108 nmol 10^{-6} cells per hour. This value decreased to 13.2 nmol 10^{-6} cells per hour when low-glucose feeding was used. The intracellular flow distribution in this case reveals that for high-glucose feeding, 69% of the glucose consumed is converted into pyruvate, and most of it is further transformed into lactate, while there is no pyruvate entry into the TCA cycle. For the low-glucose feeding, 86% of the consumed glucose was converted to pyruvate, and 66% of pyruvate was channeled to the TCA cycle, as the percentage directed to lactate was dramatically decreased to 14%.

The analysis of the metabolic rates of glucose consumption and lactate formation leads to the conclusion that mammalian cells are highly deregulated and, in the presence of nonlimiting glucose concentrations, they would consume glucose at a rate much higher than what would be required to maintain cell growth. This phenomenon is still not well understood and has received various interpretations. Different authors (2,26,27) have pointed out that lactate generation from pyruvate may be used by the cells as a way to re-equilibrate their redox potential. Indeed, the high glycolysis rates would generate NADH, at high rates, which should be reduced to NAD in order to keep this pathway functional. Normally, NAD regeneration would take place inside the mitochondrion. However, the mitochondrial membrane is impermeable to NADH; the cells must use a shuttle system to allow NADH to be incorporated indirectly into the mitochondria and further oxidized through the respiratory chain. Some of these shuttle systems, such as the glycerol-phosphate, malate-aspartate, and malate-citrate, are described by Häggstrom (3) and have been identified in a number of cell lines. The hypothesis that remains is to what extent the rate of exchange in this shuttle system limits the rate of the cytosolic NADH generation. It has been reported that the flux through the malate-aspartate shuttle can be limited under certain conditions (34). Although the mechanisms of these shuttle systems have not been studied in detail, some precursor work showed that their efficiency could be low in some cases, compared to the glycolysis rate (35). In any case, this rate should be at least twice the elevated glycolysis rate. Therefore, the generation of lactate could be used by the cell as a means for NAD regeneration in the cytosol. Some other explanations have also been given (3): (a) direct competition by lactate dehydrogenase (LDH) , present at high levels in the cytosol; (b) too low concentrations of aspartate in the cytosol, which could limit the malate-aspartate shuttle activity; (c) accumulation of malate inside the mitochondria due to glutamine metabolism, which would make more difficult the influx of malate inside the mitochondria, as required by the malate-aspartate shuttle; (d) saturation of the respiratory system by NADH generated by the metabolism of glutamine inside the mitochondria; and (e) suppression of respiration by the low availability of ADP, which would be primarily consumed due to the fast glycolysis pathway. Finally, other authors (13,15,16) have attributed this abnormal lactate generation to a deficiency in some key enzymatic activities linking the glycolysis pathway with the TCA cycle, such as pyruvate dehydrogenase complex (PDHC), phosphoenolpyruvate carboxykinase (PEPCK), or pyruvate carboxylase (PC). In this situation the metabolism of pyruvate through the LDH enzyme would be the favored route.

Finally, it should also be mentioned here that this analysis is still not complete, since the fluctuations that the cells have at the level of glucose consumption and metabolism will be also linked to the consumption and metabolism of other medium compounds, namely amino acids such as glutamine. The interaction between the main metabolic pathways, glycolysis and glutaminolysis, through the TCA cycle, as a response to a given environmental condition is further discussed in the section on the flexibility of animal cell metabolism.

The Role of Mitochondria: The TCA Cycle, Respiration and ATP Generation

The TCA cycle (Fig. 2) is a major pathway in the central metabolism, having two main functions: It provides different compounds that are precursors for the cell anabolism and generates most of the metabolic energy. Regarding metabolism compartmentalization, this pathway takes place inside the mitochondria. It is also important to highlight its central role as an internal mechanism to balance the rate of glycolysis and glutaminolysis, as discussed later. The connection between TCA and glycolysis is made through pyruvate, which is first transported into the mitochondria and further transformed in acetyl coenzyme A (CoA). Acetyl CoA is also partially provided by the degradation of some amino acids: tyrosine, lysine, leucine, and isoleucine. As previously discussed, most of the pyruvate obtained in the metabolization of glucose is diverted to lactate, and therefore only a small percentage is incorporated into the mitochondria. However, some differences are observed in the reported results, probably due to the different cell lines and growth conditions employed by different authors. Some authors have reported very low values (13,15,16), where the flux from pyruvate to acetyl CoA, through PDHC, accounted for less than 1%

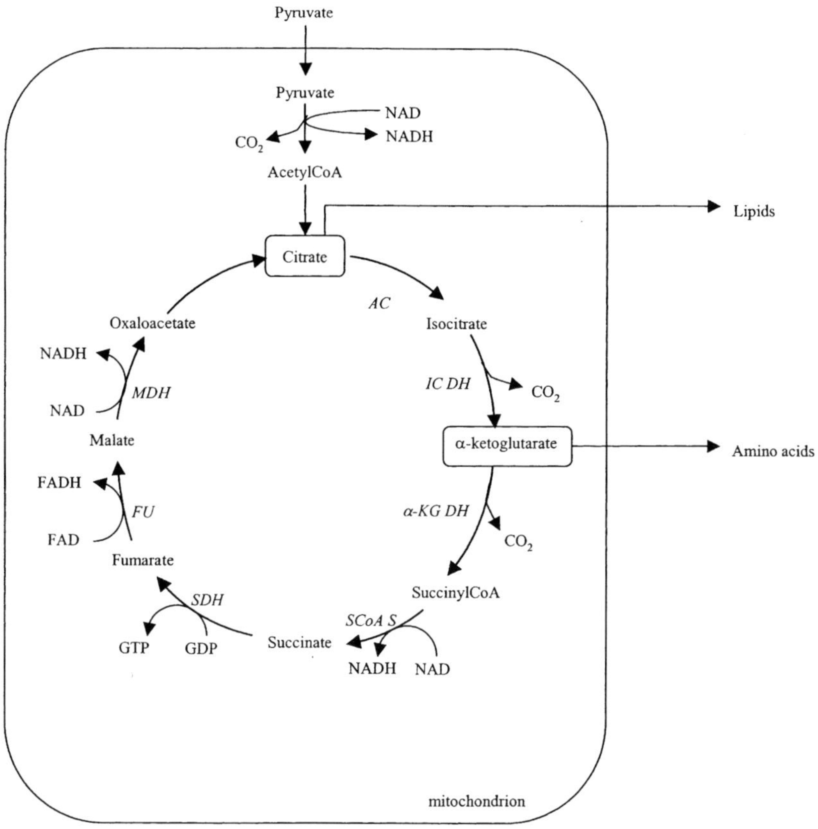

Figure 2 The TCA cycle. The enzymes involved in the pathway are indicated in italics and the corresponding full names are given in the list of abbreviations in Fig. 1.

of the rate of glucose consumption, compared with other reported values of 8–14% (17,21), and even up to 40% (14). Acetyl CoA enters the cycle reacting with oxaloacetate to form citrate. At this point, an important flux of citrate leaves the TCA cycle and is exported to the cytosol, where it contributes to lipid formation with concomitant production of pyruvate. Lipid formation is a highly demanding biosynthetic pathway in tumoral cell lines. The combination of a low incorporation of acetylCoA into the cycle and the outflow of citrate severely depletes the flux in the TCA cycle; this needs to be replenished in order to fulfill the energy generation function in the cell. This is done at the level of α-ketoglutarate. Glutamine is the precursor molecule for this, transformed to glutamate inside the mitochondria, and further to α-ketoglutarate. As a consequence, it is found that the flux from citrate to α-ketoglutarate can be as low as 10% of the flux through citrate synthase (12,14). This phenomenon is called the truncated citric acid cycle (36). The primary role of glutamine is to supply the necessary intermediate to keep the cycle operating, which reveals the importance of this compound as the main source of carbon and energy (37,38). Other possible anaplerotic reactions that could have a similar effect of TCA cycle replenishment at the level of the oxaloacetate molecule seem to be nonfunctional in mammalian cells, since no PEPCK or PC has been detected (13,15,16). Oxaloacetate is used for the biosynthesis of aspartate and asparagine. At the level of succinyl CoA there is a contribution from different amino acids such as isoleucine, valine, threonine, and methionine that can represent up to 10% of the flux from succynil CoA to succinate. At the level of succinate there is a new input, derived from the degradation of tyrosine and leucine, that can represent up to 7% of the flux from fumarate to malate (14,17). At the level of malate there is another important export flow from the cycle, used by the cells as an overflow mechanism to balance the flux through the cycle. Indeed, due to the high rate of glutamine uptake in most mammalian cells and the anaplerotic reaction of α-ketoglutarate incorporated into the cycle, the flux from succinate to malate is higher than the flux from oxaloacetate to citrate. Since the stoichiometry of the reaction between oxaloacetate and acetyl CoA is one to one, the cell must relieve any malate in excess from the cycle, in order to avoid accumulation of intermediates. Malate is therefore exported out of the TCA cycle, and further converted to pyruvate, as will be discussed in the next section.

The TCA cycle is directly linked to metabolic energy generation in the form of ATP, based in the respiratory system, located in the internal membranes of the mitochondria. The different molecules of NADH and FADH are oxidized, with oxygen as the final electron acceptor. Normally, a maximum of three ATP molecules per NADH and two ATP molecules per FADH can be reached, although some authors have suggested that mammalian cells can adjust their P/O ratio as a result of the culture conditions, adjusting to a more efficient metabolism when they are exposed to an oxygen limitation.

Glutamine Transport

Glutamine, as mentioned previously, is the second major component of cell culture media and is the source of most of the ammonium ions generated. It is important to mention that, under the normal conditions used for cell culture (37°C), glutamine is partially decomposed in a spontaneous way at a rate of 0.2–0.6 mM/day, forming pyrrolidine-carboxylic acid and ammonium ions (39). The rest of the glutamine is incorporated into the cell by means of the different amino acid transport systems, which are not specific for glutamine and also mediate the movement of other amino acids. Due to the critical role of glutamine in mammalian cell metabolism, different transporters have been identified, although their specific mechanisms and tissue-specific regulations still require further investigation. These transporters have been recently reviewed by Bode (40). In general, they can be classified into two main categories: Na^+ dependent and Na^+ independent. The first type utilizes the Na^+ electrochemical gradient, maintained by the Na^+/K^+-ATPase, to cotransport glutamine and Na^+ against their concentration gradient. The Na^+-dependent transporters identified include systems ACS (or B^0), $B^{0,+}$, y^+L, A, and N. The second type facilitates the selective movement of amino acids across the plasma membrane independent of the Na^+ gradient. The Na^+-independent transporters identified include systems L, $b^{0,+}$, and n. It is important to emphasize that glutamine is the source of intracellular glutamate and aspartate, but these charged amino acids cannot be transported into the cells from the culture at a rate that does not limit cell growth. Indeed, in proliferating cells, the activity of system $X_{Ag}-$, responsible specifically for the transport of glutamate and aspartate, is low, while that of the transport system A, responsible for the transport of glutamine, is high. Therefore, in rapidly proliferating cells, glutamine transport and further internal generation of glutamate and aspartate seem to be the way to obtain these amino acids. However, in some cases this different uptake rate has been used to obtain a low growth rate in cells and/or low ammonium generation in the medium, as further discussed in the section on redistribution of cell metabolism.

Another transport mechanism is the one around the mitochondria, since most of the glutamine metabolism is mitochondrial. The transport mechanism of glutamine into the mitochondria is still not well elucidated, although it does not appear to be rate limiting for glutamine metabolism (3). There is also no definitive understanding of the localization of the first enzyme of glutamine metabolism, phosphate-activated glutaminase (PAG), which could be located either in the mitochondrial intramembrane space or the mitochondrial matrix. This aspect was discussed in detail by Häggström (3), and the experimental evidence obtained so far indicated that a percentage of the glutamate-derived glutamine might be exported out of the mitochondria, while the rest could be used directly in the mitochondria either for its incorporation into the TCA or by TA reactions (discussed in the next section). The glutamate obtained by the deamidation of glutamine and released outside the mitochondria can be either used directly in the cytosol or transported back into the mitochondria. This can be done by two different transport mechanisms: the glutamate-aspartate exchanger and the proton symport system. The first of these mechanisms is directly related to the mitochondrial aspartate TA pathway for aspartate formation, otherwise its use would lead to a depletion of aspartate in the mitochondria of the cell (41).

Glutamine Metabolism

The first step in glutamine metabolism is its conversion to glutamate by PAG in the mitochondrial membrane, as already mentioned. In addition to the important role of glutamine as an energy precursor for the cell through TCA replenishment, glutamine and glutamate are two important precursor compounds for cell metabolism (42–44), the first necessary for the synthesis of pyridine, purine, amino sugars, NAD, and asparagine as a nitrogen source, and the second used in the synthesis of proline and ornithine. In the deamidation of glutamine by PAG and the subsequent deamination of glutamate, two ammonium ions are generated. From glutamate, two basic pathways may follow, glutamate dehydrogenase (GDH) or TA, as schematized in Fig. 3. In the GDH pathway, glutamate is deaminated to α-ketoglutarate, releasing the second ammonium group of the glutamine molecule as free ammonium. α-ketoglutarate is then incorporated into the TCA cycle and can they generate the subsequent intermediary metabolites. It is important to highlight the role of glutamine metabolism in the generation of oxaloacetate, since the anaplerotic reaction of TCA replenishment by means of PC reaction, connecting pyruvate to oxaloacetate, has been shown to be inactive in most mammalian cell lines (2). In addition, the GDH-metabolized glutamine can be basically channeled in three different ways, as shown in Fig. 4. It can be completely oxidized to CO_2, by means of the participation of the mitochondrial malic enzyme (ME), this possibility being the most efficient from energetic yield considerations (up to 27 ATP per glutamine consumed). Alternatively, and most frequent because of the TCA overflow generated by the rapid consumption of glutamine in most industrial cell lines, malate is exported out of the mitochondria to be further converted into pyruvate. This pathway, connecting the four carbon compounds of TCA and the three carbon compounds of glycolysis, is known as malate shunt, and has been evidenced in the most commonly used cell lines such as hybridomas, CHO, or BHK cells, using either direct or indirect measurements (12,14,17,33,45–47). In principle, this reaction could take place either

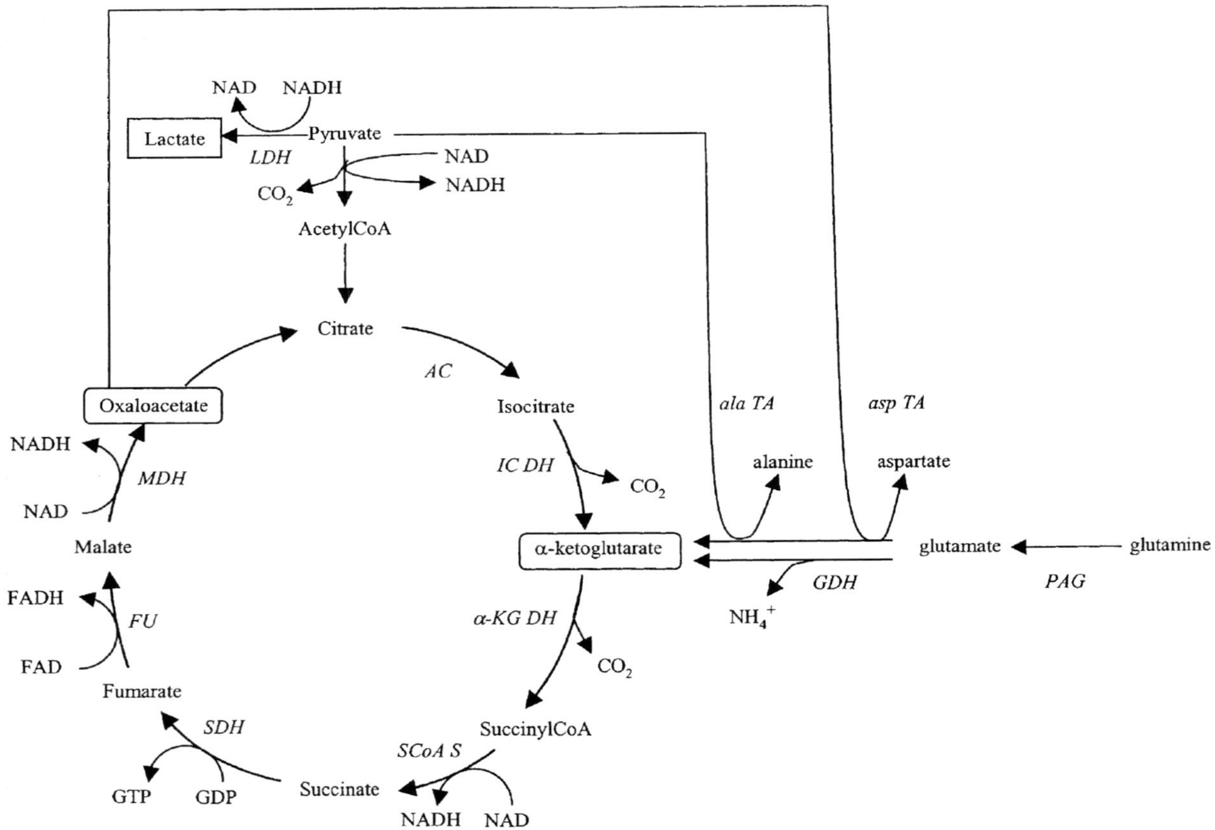

Figure 3 Overview of the main pathways involved in glutamine metabolism: glutaminolysis and transamination pathways. The enzymes involved in the pathway are indicated in italics and the corresponding full names are given in the list of abbreviations in Fig. 1

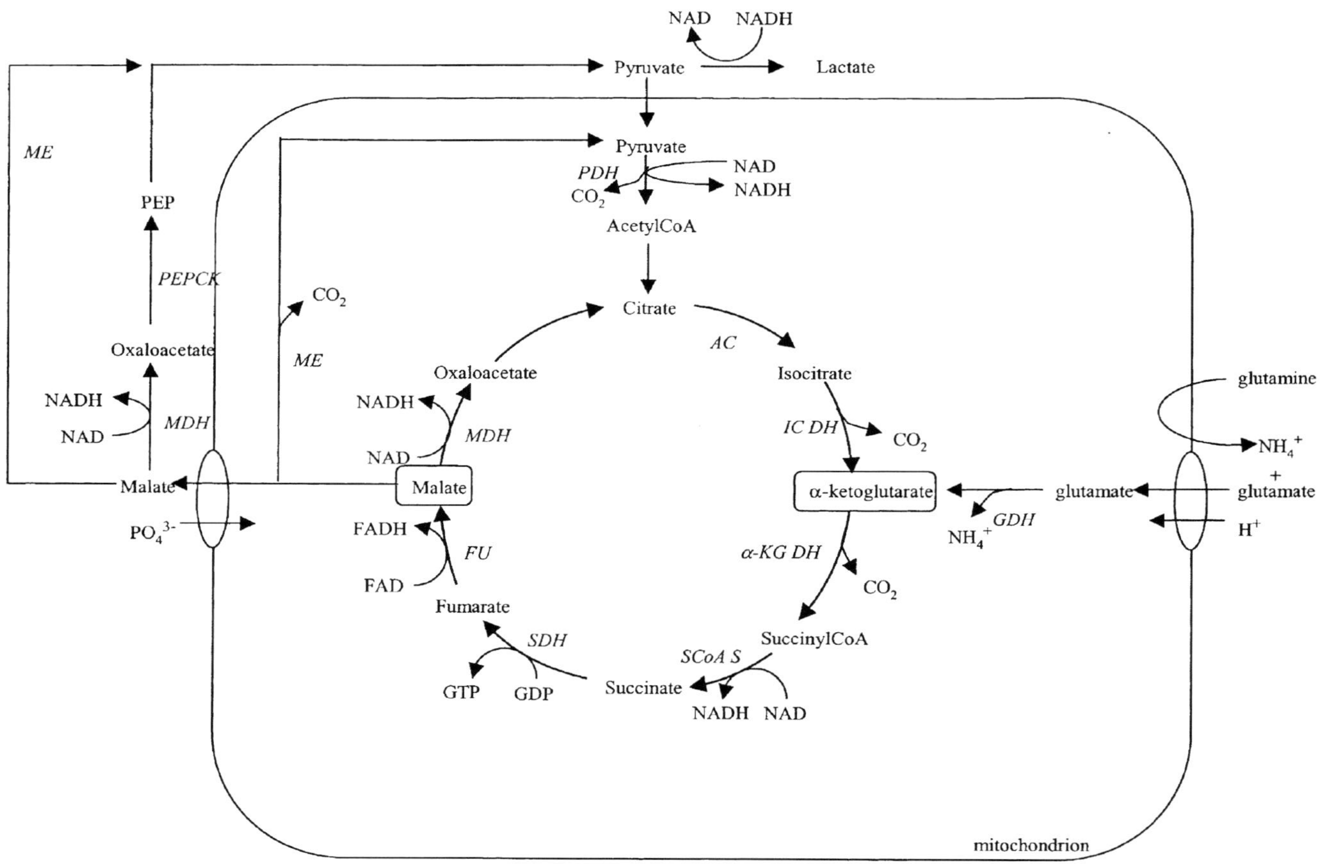

Figure 4 The glutaminolysis pathway and malate shunt, indicating alternative routes and main transport systems around the mitochondria. The enzymes involved in the pathway are indicated in italics and the corresponding full names are given in the list of abbreviations in Fig. 1.

through the cytosolic enzymes malate dehydrogenase (MDH) and PEPCK or through the cytosolic ME. Vriezen and van Dijken (45) have reported the detection of PEPCK in a myeloma cell line. However, various studies have shown the lack of PEPCK and also PC in hybridoma, CHO, and BHK cell lines (12,13,15,16). Bonarius et al. (21) have used ^{13}C-labeled compounds tracing and metabolite balance to characterize the metabolic pathway distribution of a hybridoma cell line showing that the PC activity was negligible, while the ME was active in the malate shunt, which had a rate of 8% of the glucose uptake rate.

In the TA pathway for glutamine metabolism (shown in Fig. 3), glutamate is also converted in α-ketoglurate, but the released ammonium ion is transferred in a stoichiometric proportion to oxaloacetate to form aspartate (in the so-called aspTA pathway), or to pyruvate to form alanine (in the so called alaTA pathway) as follows:

$$\text{Glu} + \text{Pyr} \rightarrow \text{Ala} + \alpha - \text{KG}$$
$$\text{Glu} + \text{Oxal} \rightarrow \text{Asp} + \alpha - \text{KG}$$

Because the molecules of oxaloacetate or pyruvate being transaminated are obtained from the glutamate metabolism through the TCA, these TA pathways have a loop structure. The TA pathways can be used by the cell as an internal mechanism to balance ammonium overflow due to the rapid glutamine consumption, as will be discussed in the next section. In addition, alanine and aspartate are important precursors, in particular for the biosynthesis of purine, pyrimidine, and asparagine. In Fig. 5 the two main possibilities for the localization of the TA pathways are considered: cytosolic or mitochondrial TA. The experimental evidence is still not conclusive enough on this aspect, as discussed in detail by Häggström (3). However, for aspartate formation, mitochondrial TA is considered the main active pathway, which could be explained by two facts: the equilibrium of the reaction between malate and oxaloacetate is favored toward malate formation and, in addition, this reaction would generate NADH in the cytosol, a compound already in excess due to the glycolysis activity. The aspTA pathway uses the glutamate-aspartate shuttle to incorporate glutamate into the mitochondria in an exclusive way, since other pathways would deplete aspartate in the mitochondria. In the case of alanine formation, both cytosolic and mitochondrial TAs may be active. For the cytosolic alanine formation, the α-ketoglutarate/malate exchange system in the mitochondrial membrane (48–50) plays a central role for the balance between cytolosic and mitochondrial reactions. In this exchange carrier, the α-ketoglutarate produced in the TA reaction is incorporated into the mitochondria and TCA, and in the exchange malate, is exported out of the mitochondria. In the cytosol, malate can be further converted to pyruvate by the cytosolic ME present in most cell lines, as previously discussed. Although a higher alaTA activity has been detected in the cytosol with respect to the mitochondria (45,51), this seems contradictory to the fact that ME is dependent on $NADP^+$, and therefore the malate shunt should be confined in the mitochondria (42,52). Thus, the specific localization of the enzymatic activities of the malate shunt remains to be clarified. The second possibility, alanine generation in the mitochondria, has been observed experimentally in different cell lines (53–56) when pyruvate is available. Lipid synthesis is also related to alanine generation in the mitochondria. In growing cells, acetyl CoA is required in the cytosol for the synthesis of cholesterol and fatty acids. Citrate, produced as an intermediate of TCA, is exported from the mitochondria through the citrate/malate exchanger, then it is converted in the cytosol to acetyl CoA for lipid synthesis and oxaloacetate, and

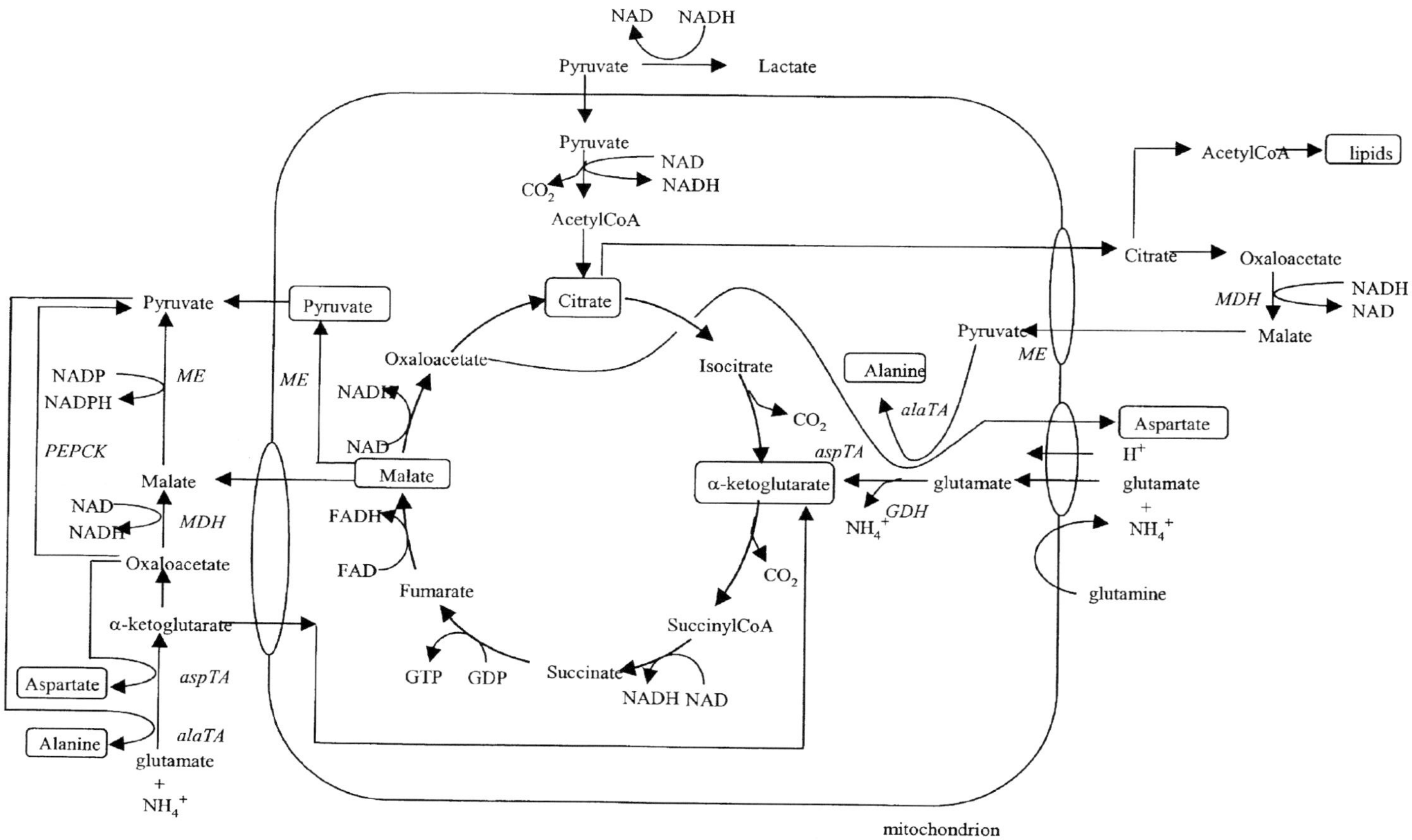

Figure 5 Illustration of the different glutamine metabolism possibilities, indicating the main transport systems around the mitochondria: cytosolic transamination pathways, mitochondrial transamination to aspartate, and interaction between glutamine metabolism and lipids cycle. The main enzymes involved in the pathway are indicated in italics and the corresponding full names are given in the list of abbreviations in Fig. 1.

further into malate, as shown in Fig. 5. Malate is incorporated back into the mitochondria through the citrate/malate exchanger, where it can be converted to pyruvate by the ME, and then transaminated to alanine by glutamate. In this case, the glutamate is either generated inside the mitochondria or incorporated through the proton symport system. In any case, it should be mentioned that part of the acetyl CoA requirements in the cell are also provided from glucose metabolism. These two pathways could occur in a simultaneous way in the cell. Indeed, Bonarius et al. (21) have recently described, using isotope labeling determinations and mass balancing, how the ME is participating simultaneously in the generation of extramitochondrial malate, both through the malate shunt and through the pyruvate/malate shuttle linked to the lipid synthesis.

As can be observed, the glutamine metabolism can be diverted to a number of different possible pathways that in turn share a number of intermediate metabolites and transport systems around the mitochondria. From the energy yield perspective, the GDH pathway is more efficient than TA (27 ATP instead of 9) (57), and usually is the favored pathway when glutamine metabolism must be accelerated in the cell, for example, as a consequence of glucose depletion (58). Also, glutamine and glucose metabolism are related to each other, as will be discussed specifically in the section on flexibility of animal cell metabolism. As a consequence of this metabolic flexibility, different cell lines will adjust the metabolic rates through GDH, alaTA, or aspTA pathways, depending on the specific culture conditions.

Deregulated Glutaminolysis in Mammalian Cells

Similar to glucose, glutamine is also consumed by most mammalian cell lines at high rates, normally associated with the general use of excess glutamine concentrations in the culture media. As is well known, this leads to the undesired accumulation of ammonium ions, in addition to an inefficient use of glutamine. The increase in glutamine consumption when glutamine concentration is raised, either as a pulse or a step change in a chemostat culture of a hybridoma cell line, was reported by Miller et al. (59), confirming early results seen in batch cultures (24,28,60,61). Vriezen et al. (62) performed a series of chemostat experiments with two different cell lines, a hybridoma and a myeloma, at glutamine concentrations in the range of 0.5–4 mM at a dilution rate of 0.03 per hour. For the limiting conditions, up to 2 mM glutamine in the feed, no residual glutamine was detected in the culture, and the specific consumption rate of glutamine was in the order of 20 nmol 10^{-6}cells per hour. This value increased to 36 nmol 10^{-6}cells per hour when excess glutamine was fed at a concentration of 4 mM without any further increase in cell concentration, and a parallel increase in the specific ammonium generation rate. Sanfeliu et al. (27) performed a series of continuous experiments with a hybridoma cell line, decreasing stepwise the glutamine concentration in the feed from 5 to 0.75 mM; they obtained a constant level of cell concentration with a gradual decrease of specific glutamine consumption, and a fourfold lower ammonium generation. They estimated a minimum specific requirement for glutamine of 30.72 nmol 10^{-6}cells per hour. Mancuso et al. (63) observed a significant reduction of specific glutamine consumption in a hybridoma cell line, when glutamine concentration in the medium was reduced from 2.4 to 1.2 mM in a continuous hollow fiber reactor. Under these conditions, they reached a glutamine concentration of 0.08 mM at the reactor outlet that was not limiting for energy metabolism. Similar results were obtained by Sharfstein (14). Cruz et al. (33) observed a similar pattern in continuous cultures of BHK cells: specific glutamine consumption was reduced from 33 to 16.8 nmol 10^{-6}cells per hour when the feed concentration was reduced from 0.52 to 0.14 mM.

The Flexibility of Animal Cell Metabolism: Interaction Between Glutamine and Glucose Metabolic Pathways

Up to this point the main characteristics of glucose and glutamine metabolism have been described, giving special emphasis to the apparent deregulation of the glucose and glutamine uptake rates, which takes place at higher rates than those strictly necessary for the cell functions when the cells are exposed to a nonlimiting concentration of these main nutrients. This analysis will now be completed by including account the interaction between glucose and glutamine metabolism, together with the metabolism of other amino acids, and the energy generation and redox balancing in the cell. Clearly, this interaction will depend greatly on the relative concentrations of glucose and glutamine (provided that the availability or physical condition of no other nutrients is limiting the culture performance), and can also be different for each particular cell line. For glucose, the general trend is that, above a minimum nonlimiting concentration in the range of 0.5–1 mM, glucose uptake will increase for increasing concentrations of glucose until a saturation maximum specific consumption value is reached. The increase in the specific consumption is accompanied by an increase in the lactate/glucose yield, which can change from values around 1–2 mol/mol or even higher, due to the contribution of lactate produced from glutamine, and a lower specific oxygen consumption due to a lower oxidative metabolism. Regarding the consequences of the different glucose uptake rates on glutamine metabolism, it seems that when both glucose and glutamine concentrations are above their limiting values, the specific consumption rates of glutamine and ammonium generation are not affected greatly by the changes in specific glucose consumption rates. Sanfeliu et al. (27) observed this phenomenon in a series of step change continuous experiments where glutamine in the feed is maintained at the nonlimiting concentration of 0.75 mM and glucose is changed in the range of 25–1.5 mM. In the glucose pulse experiment performed in a continuous culture of a hybridoma cell line, Miller et al. (26) observed how the glycolysis pathway reacted immediately to increasing glucose concentration. This change resulted in an increase in q_{glu} of 100–200%, together with an increase in lactate formation, and a decrease in oxygen consumption. However, glutamine consumption remained almost unchanged during the limited duration of the pulse, although lower amounts of ammonium and alanine were observed, evidencing an increased biosynthetic activity and leading to an increase in cell population after the step change.

On the other hand, when glucose levels are limiting and glutamine is present at nonlimiting concentrations, the cell metabolism adapts itself to the decrease in energy and biosynthesis molecules generation from glucose by increasing the glutamine consumption rate, with a concomitant increase in ammonium excretion and oxygen consumption. Also, it is observed that the glutamine metabolism is diverted at a higher percentage into the GDH pathway, at the expense of TA pathways [as determined experimentally using ^{15}N-labeled glutamine in cultures of myeloma and hybridoma cells under glucose limitation (58)], and more efficient use is taking place, with a lower lactate production to glucose consumption ratio. These trends are observed in a continuous culture of a hybridoma cell line performed under three different conditions: glutamine limitation (taken as a reference condition), low glucose, and low oxygen culture (45). The specific consumption rates of consumption/

production (in nmol 10^{-6} cells per hour) changed as follows: under low-glucose conditions, $q_{\rm glu}$ changed from 138 to 18, while $q_{\rm gln}$ increased from 20.8 to 41.4. Lactate production decreased from 163 to 10, ammonium generation increased from 6.31 to 33.9, as alanine generation decreased from 3.1 to 0.4. The ratio between the specific consumption rates of glucose and glutamine changed from 6.64 for the reference culture to 0.43 for the low-glucose conditions, again indicating the extent to which hybridoma cells could adapt their metabolism to the new conditions. The apparent yield of lactate to glucose changed as well from 1.18 to 0.57 mol/mol indicating that a lower amount of the metabolized glucose was diverted to lactate formation. Finally, some amino acids were produced instead of consumed (aspartic acid, glutamate, glycine, and threonine) evidencing this change in the cell metabolism. Interestingly, when the enzyme activities involved in the different pathways were measured, the observed expression levels were not significantly different among the situations studied, suggesting that the cells adapt their metabolic profile by means of the regulation of their enzymatic activities and not changing greatly the level of expression of the involved enzymes. Similar trends have been found in studies using other hybridoma cell lines (64), myeloma cells (65), and BHK cells (33). The results of Miller et al. (26), who obtained in a step change of glucose feeding to a hybridoma cell line from a limiting glucose situation, show that when glucose concentration in the feed is increased, the specific glutamine consumption and ammonium production decreases as the lactate-to-glucose yield increases with a lower oxygen consumption.

With respect to the interactions between glutamine and glucose consumption rates when glutamine concentration in the medium changes, the generally observed trend is that for nonlimiting concentrations, well above the $K_{\rm M}$ value for glutamine (found to be in the range of 0.09–0.15 mM) (61,66–68), glucose consumption remains constant at varying glutamine concentrations and consumption rates (27). When glutamine is the limiting substrate, glutamine and glucose consumption will follow a parallel profile, both increasing simultaneously at higher glutamine concentrations and decreasing at lower glutamine concentrations (59,62,69). This fact is also corroborated in many studies where glutamine feeding control strategies are used, keeping glutamine concentrations at a low level, for example, using fed-batch feeding. A decrease in glucose consumption rate is also observed when a decrease in glutamine consumption rate is achieved.

However, it has also been shown that these effects can be markedly different for different cell lines (62). The transition from limiting to nonlimiting glutamine concentrations can also give different patterns in the dynamics of the metabolism of these compounds. This is observed, for example, in two different step change experiments performed by Miller et al. (59) from ≈ 0 to 0.9 mM and from 0.2 to 3 mM glutamine concentration in the feed. In the first case, an increase in $q_{\rm glu}$, $q_{\rm gln}$, and $q_{\rm O_2}$, was observed. However, in the second case, the increase in $q_{\rm gln}$ was accompanied by a decrease in $q_{\rm glu}$ in the final steady state. Similar observations can also be drawn from the non-steady-state profiles of these experiments. In the first case, when glutamine is clearly limiting the cell metabolism, the immediate response to the increase of glutamine is a sharp increase of both glutamine- and glucose-specific consumption rates. However, in the second case, the specific consumption rate of glutamine is also clearly augmented after the change in the feed concentration, but the specific consumption rate for glucose remains constant. The dynamic studies performed by Mancuso et al. (63) and Sharfstein et al. (14) using labeled glucose also showed that the observed effects can depend on the rate at which the concentration is changed, as well on the magnitude of the perturbation introduced in the glutamine concentration. They observed that for a brief but important reduction in glutamine feed from 0.67 to $\approx$0 mM, the specific consumption rate of glutamine was drastically reduced, as the specific consumption rate for glucose was increased by 55%. On the other hand, a prolonged reduction in glutamine concentration from 0.3 to 0.08 mM (a concentration level that did not appear to be limiting in this case) provided an important reduction of the glutamine consumption rate, with almost unchanged specific glucose consumption rate.

One of the conclusions that can be derived from various studies performed on the complementarity of glucose and glutamine consumption and its dynamic nature is that cell metabolism will adapt to changing conditions, and that this will occur also as a function of the existing culture conditions at the time a given perturbation or change takes place. This fact is especially relevant in batch and fed-batch cultures, where the cells are exposed to different glucose/glutamine ratios and rates during changes in their concentrations. In this respect, the observation of the evidence of the multiplicity of steady states made by different authors in continuous culture experiments is very interesting (32,70,71). Indeed, it is observed that under the same experimental continuous culture conditions, different steady states can be obtained as a response/consequence of the previous culture conditions experienced by the cells. Europa et al. (70) started a continuous culture of a hybridoma cell line after two different scenarios of batch or fed-batch culture. They showed that the cells exposed to the fed-batch culture had shifted their metabolism to a much more efficient pattern, when compared to the cells grown in batch conditions: reduced glycolysis and glutaminolysis rates, lowered excretion of ammonium and lactate. This metabolic pattern was maintained in continuous cultivation, and the cells with improved metabolic profile were more efficient, reaching a concentration of 6.4×10^6 cells per milliliter in comparison with 1.9×10^6 cells per milliliter where continuous culture was started after a batch. Also, monoclonal antibody concentration was increased about fivefold. Follstad et al. (71) also reported similar observations in the course of the continuous culture of a hybridoma cell line, in which they shifted from an initial batch culture to a continuous culture ($D = 0.04$ per hour) in two different ways: a direct change to the desired dilution rate or a stepwise increase of the dilution rate to reach the final value. With the second strategy, double the cell concentration was reached, and the metabolic flux analysis revealed a more efficient metabolism in this case. Finally, Altamirano et al. (32) observed this phenomenon when studying the continuous culture of a CHO cell line at a constant dilution rate, with variations in the glucose concentration in the feeding medium. For the same 4.8 mM concentration in the feed, they obtained two different steady-state profiles, one with 1.5×10^6 cells per milliliter and the other with 2.43×10^6 cells per milliliter, the second being metabolically more efficient. The only difference between these situations was that one was obtained just after the initial batch culture, while the second, more efficient, was obtained after a gradual step change in the glucose concentration from 4.8 to 1.8 mM. This allowed the cells to develop a more efficient metabolic pattern, which they could maintain when the glucose concentration was brought back to the 4.8 mM concentration in the feed, and then allowed them to reach a higher concentration with the same feeding conditions.

The Metabolism of Other Amino Acids

Unlike prokaryotic cells, animal cells cannot synthesize the branched carbon chains found in branched-chain amino acids or the ring systems found in phenylalanine and the aromatic amino acids. Neither can they incorporate sulfur into covalently bonded structures. For this reason, amino acids are grouped as essential and nonessential. The nonessential amino acids are alanine, asparagines, aspartate, cysteine, glutamate, glutamine, glycine, proline, serine, and tyrosine. Glutamine, however, is required in high quantities due to its additional role as one of the main carbon and energy sources in the cell, as already discussed, and therefore needs to be supplemented in order to maintain cell metabolism and growth. The essential amino acids are arginine, histidine, isoleucine, leucine, lysine, methionine, phenylalanine, threonine, tryptophan, and valine. Arginine, cysteine, and tyrosine are considered generally nonessential since they are formed from the essential amino acids methionine and phenylalanine. However, the amounts of these amino acids must be appropriately balanced. If sufficient amounts of cysteine and tyrosine are present in the culture medium, the requirements for methionine and phenylalanine are markedly reduced. Conversely, if methionine and phenylalanine are present in only limited quantities, cysteine and tyrosine can become essential compounds in the medium formulation. It should be recognized that if the α-keto acids corresponding to the carbon skeletons of the essential amino acids are supplied in the culture, the aminotransferases in the cells will convert the keto acids to their respective amino acids, supplying their basic needs.

Amino acids must be supplied from the culture media in order to ensure the metabolic functions of mammalian cells and supply the required amounts in a balanced manner, as demonstrated by many authors (72–75), because the absence of any of these amino acids would trigger the onset of cell death (76,77).

When cultured in vitro, animal cells exhibit different profiles of amino acid consumption, which, although they may change for each particular cell line, exhibit some general trends. For most cases the amino acids can be divided into three groups (74,78,79). First, a number of them are consumed very rapidly: valine, isoluecine, leucine, lysine, and cysteine. Cysteine is the amino acid more rapidly consumed, as observed by different authors (72,75,79). Cysteine has an important role in protein conformation and its biological activity by binding peptide chains in disulfide bridges. The rapid consumption of branched-chain amino acids has been reported for various cell lines and media (78,80).

Second, there are some amino acids that remain constant or are consumed in very fair amounts: threonine, arginine, phenylalanine, serine, histidine, methionine, and glycine. Third, are those that are produced and released by the cells, under different culture conditions and cell physiological states: alanine, proline, asparagine, and glutamate. Alanine is produced in large amounts. Alanine is formed through TA reactions, as described previously, from pyruvate and glutamate formed in glutamine metabolism. Alanine release in important amounts during cell culture is related to a high glutamine metabolism. Up to 40% of the glutamine consumed can be diverted to alanine as end product. Proline is also synthesized directly from intracellular glutamate accumulation. Asparagine is formed by the amidation of aspartate, which in turn is formed by TA reactions involving glutamate and oxaloacetate, as described in the section on Glutamine Metabolism. Finally, glutamate accumulation in the culture medium is typically observed when cells stop their growth, and intracellular glutamate accumulation cannot be processed through other metabolic reactions.

The catabolism of the 20 different amino acids flows through a reduced number of intermediate compounds, and finally the carbonated skeleton of the molecules is incorporated into the TCA cycle, as indicated in Fig. 6. In this process, the amino

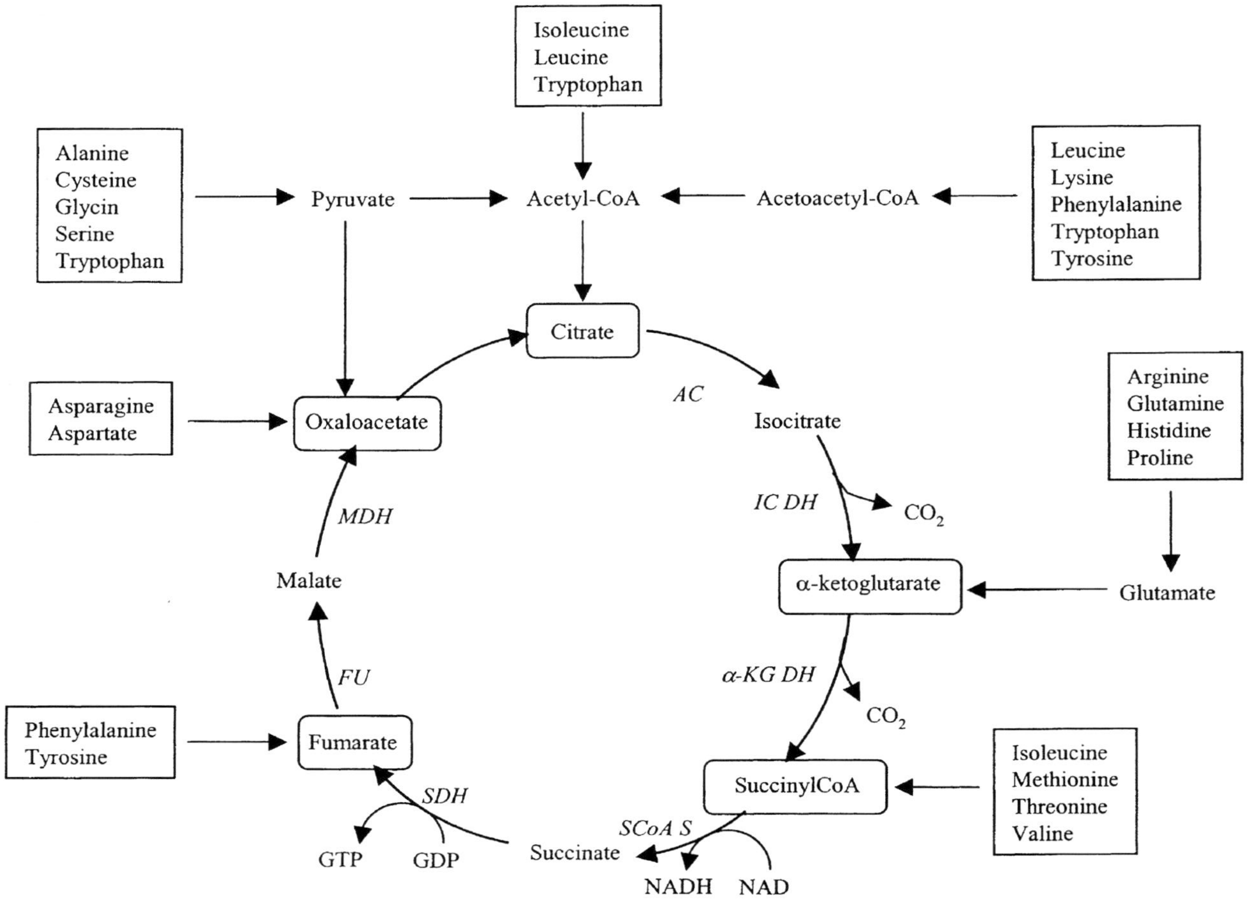

Figure 6 Interaction of the metabolism of the main amino acid compounds with the central metabolism of the cells.

groups of the different amino acid molecules are released by aminotransferases or transaminases. This group of enzymes is substrate specific, but has pyrodoxal phosphate as a prosthetic group. The acceptor for the released ammonium is always α-ketoglutarate, allowing the different amino groups to be funneled into L-glutamate. In turn, L-glutamate will be used by the cell for the different biosynthetic routes, such us protein synthesis, hormone metabolism, regulation of cell growth, production of metabolic energy, synthesis of purines and pyrimidines, nitrogen metabolism, and biosynthesis of urea. Alternatively, its excess will be converted into nitrogenated compounds excreted to the medium, such as alanine or aspartate.

With respect to amino acid transport into the cell, amino acids are transferred through membranes by specialized, energy-dependent, and passive transporters with overlapping substrate specificities. A number of Na^+-dependent amino acid transport systems with overlapping amino acid specificity have been described. In these transport systems, Na^+ and amino acids at high concentrations are cotransported down their concentration gradient to the interior of the cell. The ATP-dependent Na^+/K^+ pump exchanges the accumulated Na^+ for extracellular K^+, reducing intracellular Na^+ levels and maintaining the high extracellular Na^+ concentration required to drive this transport process. An extensive review of these transport systems has been reported by Castagna et al. (81).

THE ROLE OF OXYGEN AND CO_2 IN MAMMALIAN CELL METABOLISM

Oxygen is an important nutrient to support mammalian cell metabolism, since it is the final electron acceptor in the mitochondrial respiration chain, directly linked to the energy generation in the form of ATP molecules. In addition, the solubility of oxygen in normal culture media compositions is low, about 7 mg/L, and therefore oxygen supply to the cells may be a limiting factor, especially at high cell concentrations. The effect of oxygen level on cell metabolism has been studied by a number of authors (82–87), from very low concentrations to hyperoxic conditions.

Although different cell lines have different optimal values of percentage of oxygen saturation, some general trends are often observed: For a wide range of values of percentage of oxygen saturation, between 10% and 100%, cell metabolism is mostly unaffected, with optimal values in the range of 40–60%. When the oxygen level drops down from 10%, two trends can be observed: First, oxygen becomes the limiting substrate, cell growth is reduced, and the specific consumption rate of glucose and glutamine are also reduced; second, if the oxygen level drops very low (below 0.1–0.5%), cell metabolism is severely affected, the specific consumption rates of glucose and glutamine are very high, TCA and oxydative phosphorylation are inhibited, and lactate production is abnormally high. As anticipated, oxygen levels will have a direct effect on the cellular energetic metabolism and redox balance, as shown by Zupke et al. (88) when analyzing hybridoma growth and metabolic pattern at 0%, 1%, and 60% oxygen saturation. Low oxygen levels directly reduced the activity of the electron transport chain, and NADH accumulated intracellularly, forcing the cell to obtain energy from the anaerobic metabolism to lactate. However, in this situation the ATP generation is lower, and an acute change is observed with respect to the use of ATP for maintenance, 50% lower for the oxygen-limited growth when compared to 60% saturation level. This fact could be explained by a reduction of the effect of oxidative stress, or a lower use of ATP in futyle cycles. Regarding redox shuttle molecules in the cell, a higher NAD/NADH ratio was observed at increasing oxygen levels, both at the cytosol and at the mitochondria. This ratio was cell density dependent only for the cytosol pool, being markedly lower for a high cell concentration. On the other hand, and more unexpectedly, the NADP/NADPH ratio in the mitochondria was estimated to be lower at the high oxygen levels.

Different mammalian cell lines have been cultured under hyperoxic conditions (83,89), although the production of reactive oxygen species can induce cell death (90). Jan et al. (87) have studied the effect of oxygen concentration on the metabolism of a hybridoma cell line in continuous cultures at oxygen levels between 10% and 150%. They observed that the adaptation of cells up to 100% oxygen saturation was accompanied by the increased activity of various antioxidant enzymes, glutathione peroxidase, glutathione *S*-transferase, and superoxide dismutase, but there was a decrease in the activity of some of these enzymes at oxygen levels above 100%, suggesting an influence of the reactive oxygen radicals. The influence of the oxygen concentration was also observed on the ratio of the specific consumption rates of the two main compounds in the medium, glucose and glutamine. The ratio q_{glu}/q_{gln} increased from 2.54 to 4.17 when the percentage of oxygen saturation increased from 10% to 125%. Therefore, at higher oxygen levels glucose consumption is favored with respect to glutamine. This is accompanied by an increase of the lactate to glucose yield, from 1.35 to 1.66 mol/mol, and a lower percentage of glucose consumption being channeled into the TCA, an unexpected trend that could be explained by the effect of oxygen toxicity.

The oxygen uptake rate (OUR) of the cells in culture is a very relevant measurement that can be directly correlated to the physiological state of the culture; it is easy to measure on-line by an in-situ conventional oxygen probe, and can then be directly used to optimize nutrient feeding in bioreactors (91). The OUR is the direct product of the specific oxygen consumption rate and the concentration of viable (metabolically active) cells. Therefore, it is interesting to know if the specific oxygen consumption rate, q_{O_2} changes with the cell concentration, since in this case the OUR measurements would provide the concentration of viable cells, provided that q_{O_2} had been previously determined. The results in this matter are often contradictory; in some cases constant q_{O_2} over a wide range of oxygen concentrations has been found, while in some others an effect of cell concentration has been observed. These results are even harder to compare due to the different cell lines used, experimental determinations in batch or continuous cultures, and other measurement artifacts, as recently reviewed (91).

Carbon dioxide is one of the end products of mammalian cell metabolism, as it is produced in different metabolic steps. In some other reactions CO_2 can be used as a substrate, such as the carboxylation of pyruvate to oxaloacetic acid, although this reaction does not take place in many cell lines, as already discussed in previous sections. The relevance of CO_2 is also related to its generalized use in cell culture technology for pH regulation, in equilibrium with medium bicarbonate ($CO_2 + H_2O <=> HCO_3^- + H^+$). The influence of CO_2 generation by cell metabolism can be especially relevant in industrial applications, operating with systems optimized to reach high cell densities and oxygen supply. In such systems, the metabolically produced CO_2 will be built up, with an increase in CO_2 partial pressure, and can lead to reduced cell viability, decreased specific production rates, and polysialylation rates, while glycosylation seems not to be affected (92,93). Generally, pCO_2 levels in the range of 40–50 mmHg are considered optimal, and no deleterious effects are observed (94). However, pCO_2 levels in the range of 100–170 mmHg can be reached, especially in high-volume reactors (95). The influence of the bicarbonate buffer for pH control in this problem is important, and has been demonstrated by significant pCO_2 reduction when MOPS buffer and NaOH are used for pH control (96).

THE MAIN METABOLIC END PRODUCTS AND THEIR EFFECTS: AMMONIUM AND LACTATE

As previously mentioned, the metabolic deregulation of mammalian cells in culture, with very high glucose and glutamine consumption rates, leads to the accumulation of lactate and ammonium ions in the culture media, in the latter case also due to the spontaneous decomposition of glutamine under the usual culture conditions (39). In addition to the inefficient use of these two main compounds, with the possibility of reaching limiting concentrations due to their fast depletion, the effects of lactate and ammonium accumulation in the medium can be deleterious, either due directly to their effects or indirectly to the increase in medium osmolality. As a general pattern, it is well established that the direct inhibitory effects of ammonium are much more important than those of lactate.

Lactate generation by cell metabolism requires the use of pH control methods in order to avoid the direct negative effects of medium acidification on cell growth. In systems with pH control, the effects of lactate concentration on cell growth are only clearly observed for concentrations of 40 mM or higher (2,97). Therefore, lactate accumulation effects will be more important in systems like fed-batch cultures, where the extension of culture times is accompanied by an increased cumulative glucose consumption and then lactate accumulation. However, the observed effects of lactate on cell growth are primarily due to the increased medium osmolality, as the direct metabolic effects are observed later (97,98). It is also interesting to note the metabolic shift observed at the final stage of such fed-batch cultures, where lactate is being consumed instead of produced (99–101), and simultaneously alanine is consumed as well, inverting the normal metabolic profile toward its production. With respect to the influence of lactate on product formation, an inhibition effect has been reported for lactate concentrations above 20 mM (102), based on calculations of the total cell number. However, observations based on specific production rates revealed an increase of this magnitude at higher lactate concentrations, basically associated to the osmolality rather than to a direct lactate effect.

The effects of ammonium on cell metabolism are observed from much lower concentration levels, typically 2–4 mM, which are easier to reach in culture systems; for this reason they have been studied in more detail. The negative effects of ammonium ions on cell growth have been reported by a large number of authors using different cell lines (60,97,103–107) in batch cultures where different initial ammonium concentrations had been added. Interestingly, when CHO cells in a chemostat steady-state culture were exposed to increasing ammonium concentrations, no toxic effect could be observed up to relatively high levels (108). The same trend could be observed for hybridoma cells, which could be adapted to grow in ammonium concentrations previously considered toxic (109). Newland et al. (110) also demonstrated that they could grow cells to ammonium concentrations of 5 mM by controlled feeding, while batch growth at 5 mM initial added ammonium concentration was inhibited. One possible explanation for these observations may be related to the effects caused by ammonium ions in the NH_3/NH_4^+ equilibrium, the transport of the different ions through the cellular membranes (ammonia being more rapidly transported by diffusion than charged ammonium ions, which, on the other hand, can be transported directly by ion pumps in the plasma membrane), and the values of intra- and extracellular pH, as described in detail by Häggström (3). The effects of a sudden increase in the extracellular ammonium are first a rise in pH at the intracellular level, followed by acidification of the cytosol, reaching lower pH values for increasing concentrations of added ammonium (97,103,111). The reported results support the observation that these effects are more drastic when the addition is performed externally than when ammonium is accumulated slowly as a consequence of cell metabolism. In the latter case the cells have time to adjust their ion transport systems, the intracellular pH changes do not occur sharply, and the ammonium toxic effects are not observed until higher levels are reached. The activation of the different transport mechanisms in the plasma membranes associated to an increase in ammonium concentration is further supported by the generally described observation of increased glucose and glutamine consumption rates, associated to a higher maintenance energy demand (97,107,112). It is also observed in several cases that as the cells adapt to higher ammonium levels, they increase the percentage of glutamine metabolized through the GDH pathway, increasing the final output of alanine and decreasing the free ammonium ion generation from their metabolism, another possible explanation for the direct effect of the increase of activated sugars. High levels of ammonium have been shown to increase the intracellular amounts of UDP-GlcNAc and UDP-GalNAc (105), reaching levels toxic to the cells (2).

Regarding the influence of ammonium levels in the culture media on the obtained product, two aspects should be considered: quantity and quality. With respect to quantity, the results reported for those cases where specific production rates are evaluated seem to be cell dependent. Some authors have reported a decline in productivity with high ammonium concentrations (108), others have observed an increase in specific productivity (107), and still others have observed no influence (97). There is more consensus when examining the role of ammonium on the glycosylation pattern of the obtained product, and several authors have evidenced those effects (107,113–116).

Finally, it should also be considered that, although most attention has been given to the potential toxic effects of ammonium and lactate, there is experimental evidence that other substances may also accumulate as a result of the cellular metabolic activity, causing negative effects on cellular behavior (116).

REDISTRIBUTION OF CELL METABOLISM: TOWARD A MORE EFFICIENT CELL BEHAVIOR IN CULTURE

In the last part of this chapter, an overview is provided of the different possibilities for approaching a redistribution of mammalian cell metabolism to obtain more efficient cell behavior in culture. Basically, this means obtaining a culture profile without the severe deregulation of glucose and glutamine metabolism most often observed in conventional batch cultures (taken as the reference culture system), leading to limiting conditions in the culture (lactate or ammonium accumulation, media depletion or imbalance, high osmolality) and to an inefficient use of medium components. Avoiding or at least diminishing this situation should allow increased cell growth and most probably higher product concentrations and productivities. Several ways have been used to reach this general objective, which can be divided into three groups: optimization of the feeding strategy of glucose and glutamine, medium reformulation by substitution of glucose and/or glutamine, and metabolic engineering of the cells to obtain a modified metabolic profile. Indeed, these options should be considered as complementary rather than as opposing alternatives.

The evidence that high glucose and glutamine concentrations in the medium increase specific glucose and glutamine consumption rates to unnecessarily high levels has led to the definition of feeding strategies, in either fed-batch or perfusion systems, based on keeping low concentrations of these compounds in the culture medium and at the same time providing the necessary feed to keep cell metabolic activity unlimited. Various authors have demonstrated the validity of such an approach (118–123) (further discussed in other chapters of this book), which results in increased cell concentrations and productivities. It should be taken into account, though, that a precise application of this methodology, ensuring a minimal nonlimiting specific glucose/glutamine feeding at any point in a culture, should be based on a proper kinetic characterization of the cell line used, and on the necessary on-line instrumentation to follow correctly the cell growth dynamics and metabolic activity, allowing automatic adjustment of the feeding rates to the culture.

A second approach could be followed, on the basis of a replacement of glucose and/or glutamine by another compound, which should provide the cell with the anabolic precursors and energy for growth and maintenance, but which would be transported into or metabolized by the cell at a lower rate. This strategy should be easy to implement, since the rate-controlling step would be localized at the cell level substrate uptake, thus not requiring specific hardware to control extracellularly the substrate supply to the cells. Since the waste product of glutamine metabolism, ammonium, has the most harmful effects on most mammalian cells, its replacement is particularly interesting. The first possibility is the replacement of glutamine by peptide compounds containing glutamine, such as L-alanyl-L-glutamine and glycyl-L-glutamine (124,125), or by asparagine (104,108). One particular advantage of the use of peptide compounds is their thermal stability, allowing even thermal esterilization of the media, an important advantage from the process point of view. Glutamate has also been used as a glutamine substitute, with the inherent advantage of having less nitrogen atoms per molecule than glutamine, and therefore releasing less ammonium to the medium. However, glutamate adaptation is cell line dependent (126), and those cell lines lacking glutamine synthetase (GS) activity (for example, hybridoma cells), will not grow on glutamate (2) unless they are genetically modified, as will be discussed later. Also, combinations of these compounds can be formulated. Christie and Butler (127) reported the successful adaptation of a BHK cell line to a medium containing glucose (3 mM), asparagine (3 mM), and glutamine (0.5 mM); the last compound could also be replaced by peptide compounds ala-gln (1 mM) or gly-gln (2.5 mM), conferring thermal stability to the medium. This study also evidenced that the cells adapted to this medium progressively, as shown by the subculture results. During this process, the authors could follow the change in key enzymatic activities: GS increased significantly, considering that it was undetectable in a glutamine-based medium, PAG decreased significantly, alanine aminotransferase increases, in accordance with a higher TA activity and alanine secretion to the medium, while GDH levels did not correlate clearly with the substrate change. Regarding glucose, its substitution by alternative compounds that could be more slowly metabolized, generating a lower lactate concentration, has also been demonstrated. Wagner et al. (128) used a mixture of glucose and galactose for the culture of human tumor kidney cells. Reitzer et al. (37) showed that galactose and fructose can replace glucose in the culture of HeLa cells. Marquis et al. (129) substituted glucose by galactose in the culture of a lymphoblastoid cell line, with a marked decrease in lactate generation. Altamirano et al. (130) showed that for a CHO cell line a similar growth pattern is observed when glucose is replaced by mannose, while galactose and fructose are metabolized more slowly.

The third approach for metabolic redistribution is based on metabolic engineering techniques, either by incorporation of a nonexisting enzymatic activity, allowing for a new reaction to take place, or by modulation of an existing enzymatic reaction, for example, reducing the corresponding enzyme level to decrease the rate of a particular metabolic step, although primary metabolic pathways may be seen as highly regulated, rigid, and difficult to manipulate (131). This should be a possible approach, since, as discussed previously, primary metabolism of mammalian cells when grown in vitro occurs at much higher rates than necessary. Therefore, one could attain a reduction in the overall function of primary metabolism without a negative effect on cell growth or product formation. The advantage in this case is that the new self-regulated cell line does not need external feeding control systems, and would harbor an internal regulation system, operating independently of the fluctuations in concentrations of the main compounds in the culture medium. At the level of glycolysis, Paredes et al. (132) attempted a direct reduction of the glucose uptake rate of the cells by using an antisense RNA approach [the fundamentals on antisense RNA technologies can be reviewed in Ref. (133)] targeted to reduce the enzymatic levels of either the first step in the whole pathway, the glucose transport, or one step downstream of the pathway, the enzyme enolase. They obtained a 50% reduction of glucose-specific consumption rate in the first case and 22% in the second, although they also showed that the constructions were not stable after a limited number of subcultures. Chen et al. (134) directly focused on the problem of lactate accumulation, developing a genetically modified hybridoma deficient in LDH. This variant cell line, LDH-neo21, had a 50% reduction in the specific lactate production rate, when compared to the parental cell line from which it had been derived. Under the same conditions, using a batch culture with conventional glucose and glutamine concentrations, the new LDH-neo21 cells could reach a 30% higher final cell concentration, a decrease in glucose consumption by 25%, and a triplicated final concentration of the generated product. Irani et al. (135) tried a different approach, not just for the metabolic reaction being targeted, but also from the methodological point of view: they introduced a new gene activity instead of reducing an existing one. Their approach was to engineer the cells to compensate for the lack of enzyme activities connecting glycosylation to the TCA cycle, generally observed in many cell lines, as discussed before. Particularly, they introduced a yeast PC gene in a BHK cell line. In this way, pyruvate can be transformed to oxaloacetate in the cytosol and further to malate, by the cytosol MDH enzyme, which can be finally incorporated into the mitochondria and used to replenish TCA, compensating the low rate of incorporation of pyruvate into the TCA. In addition, it should be noted that the second reaction allows regeneration of NAD^+ in the cytoplasm, and will be an alternative to NAD regeneration via LDH and lactate generation.

Another possibility for cell engineering that has been extensively applied is the expression of glutamine synthetase as a way to control the rate of glutaminolysis. This would allow replacement of glutamine by glutamate in cells not expressing GS constitutively (136); this is particularly significant for hybridoma cell lines, very widely used in industrial processes, and has been applied by many authors (118,132,137). GS-modified cell lines can obtain glutamine from glutamate and, due to the lower uptake of glutamate and the low ammonium contents of this compound, a significantly low amount of ammonium ions is generated. The lower glutamate metabolism is accompanied also by a low glycolysis rate and lactate generation. Also, alanine secretion is almost nil due to the decrease in TA pathways as a consequence of the low ammonium generated. This last case of metabolic engineering of hybridoma cells offers a very good example of how the three approaches to redistribute cell metabolism can be combined to optimize cell culture processes: genetic modification of the cell line, redefinition of the culture medium, and definition of an optimized feeding strategy, based on on-line measurements. By this, excellent improvements in final cell concentration and monoclonal antibody concentration up to 2.7 g/L have been reported (100,137,138).

CONCLUSION

Mammalian cells have a very complex and flexible metabolism. Its understanding should be the basis for the design of modifications leading to the improvement of the output obtained from the culture of these cells in industrial processes. The metabolic pattern most often observed in in vitro culture of mammalian cells is highly deregulated, with uptake rates for the main compounds of their metabolism much higher that those strictly necessary. This fact severely limits the performance of the culture, as different adverse limitations can rapidly be met. Based on the knowledge of cell metabolism, different approaches can be followed, usually in a combined manner, to improve this performance: metabolic engineering, medium redefinition and balance, and optimized bioprocess engineering design based on metabolic requirements.

REFERENCES

1. Ozturk SS. Cell culture technology—an overview. In: Ozturk SS, ed. Cell Culture Technology. New York: Taylor and Francis, 2005:1.
2. Wagner R. Metabolic control of animal cell culture processes. In: Hauser H, Wagner R, eds. Mammalian Cell Biotechnology in Protein Production. Berlin: Walter de Gruyter, 1997:193.
3. Häggström L. Cell metabolism, animal. In: Spier R, ed. Encyclopedia of Cell Technology. New York: John Wiley & Sons, 2000:392.
4. Pessin JE, Bell GI. Mammalian facilitative glucose transporter family. Structure and molecular regulation. Annu Rev Physiol 1992; 54:911.
5. Carruthers A. Facilitated diffusion of glucose. Physiol Rev 1990; 70:1135.
6. Craik JD, Elliot KRF. Kinetics of 3-O-methyl-D-glucose transport in isolated rat hepatocytes. Biochem J 1979; 182:503.
7. Gould GW, Thomas HM, Jess TJ, Bell GI. Expression of human glucose transporters in *Xenopus oocytes*. Kinetic characterization and substrate specificities of the erythrocyte, liver, and brain isoforms. Biochemistry 1991; 30:5139.
8. Kayano T, Fukumoto H, Eddy RL, Fan YS, Byers MG, Shows TB, Bell GI. Evidence for a family of human glucose transporter-like proteins. Sequence and gene localization of a protein expressed in fetal skeletal muscle and other tissues. J Biol Chem 1988; 263:15245.
9. Kahn BB. Facilitative glucose transporters. Regulatory mechanisms and de-regulation in diabetes. J Clin Invest 1992; 89:1367.
10. Burant CF, Takeda J, Brot LE, Bell GI, Davidson NO. Fructose transporter in human spermatozoa and small intestine is GLUT5. J Biol Chem 1992; 267:14523.

11. Waddell ID, Zomerschoe AG, Voice MW, Burchell A. Cloning and expression of a hepatic microsomal glucose transport protein. Biochem J 1992; 286:173.
12. Mancuso A, Sharfstein ST, Tucker SN, Clark DS, Blanch HW. Examination of primary metabolic pathways in a murine hybridoma with carbon-13 nuclear magnetic resonance spectroscopy. Biotechnol Bioeng 1994; 44:563.
13. Fitzpatrick L, Jenkins HA, Butler M. Glucose and glutamine metabolism of a B-lymphocyte hybridoma grown in batch culture. Appl Biochem Biotechnol 1993; 43:93.
14. Sharfstein ST, Tucker SN, Mancuso A, Blanch HW, Clark DS. Quantitative in vivo nuclear magnetic resonance studies of hybridoma metabolism. Biotechnol Bioeng 1994; 43:1059.
15. Petch D, Butler M. Profile of energy metabolism in a murine hybridoma: glucose and glutamine utilization. J Cell Physiol 1994; 161:71.
16. Neerman J, Wagner R. Comparative analysis of glucose and glutamine metabolism in transformed mammalian cell lines, insect and primary liver cells. J Cell Physiol 1996; 166:152.
17. Paredes C, Sanfeliu A, Cardenas F, Cairó JJ, Gòdia F. Estimation of the intracellular fluxes for a hybridoma cell line by material balances. Enzyme Microb Technol 1998; 23:1.
18. Jan DCH, Petch DA, Huzel N, Butler M. The effect of dissolved oxygen on the metabolic profile of a murine hybridoma grown in serum-free medium in continuous culture. Biotechnol Bioeng 1997; 54:153.
19. Larrabee MG. The pentose cycle. Rigorous evaluation of limits to the flux from glucose using $^{14}CO_2$ data, with applications to peripherial ganglia of chicken embryos. J Biol Chem 1989; 264:15875.
20. Zupke C, Stephanopoulos G. Modelling of isotope distributions and intracellular fluxes in metabolic networks using atom mapping matrices. Biotechnol Prog 1994; 10:489.
21. Bonarius HPJ, Özemre A, Timmerarends B, Skrabal P, Tramper J, Schmid G, Heinzle E. Metabolix-flux analysis of continuously cultured hybridoma cells using $^{13}CO_2$ mass spectrometry in combination with ^{13}C-lactate nuclear magnetic resonance spectroscopy and metabolite balancing. Biotechnol Bioeng 2001; 74:528.
22. Renner ED, Plagemann PGW, Bernlohr RW. Permeation of glucose by simple and facilitated diffusion by Novikoff rat hepatoma cells in suspension culture and its relationship to glucose metabolism. J Biol Chem 1972; 247:5765.
23. Reitzer LJ, Wice BM, Kenell D. Evidence that glutamine, not sugar, is the major energy source for cultivated HeLa cells. J Biol Chem 1979; 254:2669.
24. Miller WM, Blanch HW, Wilke CR. A kinetic analysis of hybridoma growth and metabolism in batch and continuous suspension culture: effect of nutrient concentration, dilution rate and pH. Biotechnol Bioeng 1988; 32:947.
25. Pörtner R, Bohmann A, Lüdemann I, Märkl H. Estimation of specific glucose uptake rates in cultures of hybridoma cells. J Biotechnol 1994; 34:237.
26. Miller WM, Wilke CR, Blanch HW. Transient responses of hybridoma cells to nutrient additions in continuous culture: I. Glucose pulse and step changes. Biotechnol Bioeng 1989; 33:477.
27. Sanfeliu A, Paredes C, Cairó JJ, Gódia F. Identification of key patterns in the metabolism of hybridoma cells in culture. Enzyme Microb Technol 1997; 21:421.
28. Glacken MW, Fleischaker RJ, Sinskey AJ. Reduction of waste product excretion via nutrient control: possible strategies for maximizing product and cell yields on serum in cultures of mammalian cells. Biotechnol Bioeng 1986; 28:1376.
29. Hu WS, Dodge TC, Frame KK, Himes VB. Effect of glucose on the cultivation of mammalian cells. Dev Biol Stand 1987; 66:279.
30. Hayter PM, Curling EMA, Baines AJ, Jenkins N, Salmon I, Strange PG, Tong JM, Bull AT. Glucose-limited chemostat culture of Chinese hamster ovary cells producing recombinant interferon-γ. Biotechnol Bioeng 1992; 39:327.

31. Hayter PM, Curling EMA, Baines AJ, Jenkins N, Salmon I, Strange PG, Bull AT. Chinese hamster ovary cell growth and interferon production kinetics in stirred batch culture. Appl. Microbiol. Biotechnol 1991; 34:559.
32. Altamirano C, Illanes A, Casablancas A, Gámez X, Cairó JJ, Gòdia F. Analysis of CHO cells metabolic redistribution in a glutamate-based defined medium in continuous culture. Biotechnol Prog 2001; 17:1032.
33. Cruz HJ, Moreira JL, Carrondo MJT. Metabolic shifts by nutrient manipulation in continuous cultures of BHK cells. Biotechnol Bioeng 1999; 66:104.
34. Eigenbrodt E, Fister P, Reinacher M. New perspectives on carbohydrate metabolism in tumor cells. In: Baitner R, ed. Regulation of Carbohydrate Metabolism. Boca Raton: CRC Press, 1985:141.
35. Boxer GE, Devlin TM. Pathways of intracellular hydrogen transport. Science 1961; 134:1495.
36. Coleman PS, Lavietes BB. Membrane cholesterol, tumorigenesis and the biochemical phenotype of neoplasia. CRC Crit Rev Biochem 1981; 11:341.
37. Reitzer LJ, Wice BM, Kennell D. Evidence that glutamine, not sugar is the major source for cultured HeLa cells. J Biol Chem 1979; 254:2669.
38. Zielke HR, Zielke CL, Ozand PT. Glutamine: a major energy source for cultured mammalian cells. Fed Proc 1984; 43:121.
39. Trisch GL, Moore GE. Spontaneous decomposition of glutamine in cell culture media. Exp Cell Res 1962; 28:360.
40. Bode BP, Recent molecular advances in mammalian glutamine transport. J Nutr 2001; 131:2475S.
41. Schoolwerth AC, LaNoue KF. The role of microcompartmentation in the regulation of glutamate metabolism by rat kidney mitochondria. J Biol Chem 1980; 255:3403.
42. McKeehnan WL. Glutaminolysis in animal cells. In: Morgan J, ed. Carbohydrate Metabolism in Cultured Cells. New York: Plenum Press, 1986:111.
43. Prusiner S, Stadtman ER. The Enzymes of Glutamine Metabolism. New York: Academic Press, 1973.
44. Engstrom W, Zetterberg A. The relationship between purines, pyrimidines, nucleosides, and glutamine for fibroblast cell proliferation. J Biol Chem 1984; 218:607.
45. Vriezen N, van Dijken JP. Fluxes and enzyme activities in central metabolism of myeloma cells grown in chemostat culture. Biotechnol Bioeng 1998; 59:28.
46. Zupke C, Stephanopoulos G. Intracellular flux analysis in hybridomas using mass balances and in vitro ^{13}C NMR. Biotechnol Bioeng 1995; 45:292.
47. Bonarius HPJ, Hatzimanikatis V, Meesters KPH, de Gooijer CD, Schmidt G, Tramper J. Metabolic flux analysis of hybridoma cells in different culture media using mass balances. Biotechnol Bioeng 1996; 50:299.
48. Schoolwerth AC, LaNoue KF. Transport of metabolic substrates in renal mitochondria. Annu Rev Physiol 1985; 47:143.
49. Kilberg MS, Häussinger D. Mammalian Amino Acid Transport. Mechanisms and Control. New York: Plenum Press, 1992.
50. LaNoue KF, Schoolwerth AC. Metabolite transport in mitochondria. Annu Rev Biochem 1979; 48:871.
51. Masola B, Peters TJ, Evered DF. Transamination pathways influencing L-glutamine and L-glutamate oxidation by rat enterocyte mitochondria and the subcellular localization of L-alanine aminotransferase and L-aspartate aminotransferase. Biochim Biophys Acta 1985; 843:137.
52. McKeehan WL. Glycoslysis, glutaminolysis and cell proliferation. Cell Biol Int Rep 1982; 6:635.
53. Piva TJ, McEvoy-Bowe E. Oxidation of glutamine in HeLa cells: role and control of truncated TCA cycles in tumour mitochondria. J Cell Biochem 1998; 68:213.

54. Moreadith RW, Lehninger AL. The pathway of glutamate and glutamine oxidation by tumor cell mitochondria: role of mitochondrial $NAD(P)^+$-dependent malic enzyme. J Biol Chem 1984; 259:6215.
55. Evered DF, Masola B. The oxidation of glutamine and glutamate in relation to anion transport in enterocyte mitochondria. Biochem J 1984; 218:449.
56. Kovacevic Z, Brkljac O, Bajin K. Control and function of the transamination pathways of glutamine oxidation in tumour cells. Biochem J 1991; 273:271.
57. Häggström L. Energetics of glutaminolysis-a theoretical evaluation. In: Spier RE, Griffiths JB, Meignier B, eds. Production of Biologicals from Animal Cells in Culture. Oxford: Butterworth-Heinemann, 1991:79.
58. Martinelle K, Doverskog M, Jacobsson U, Chapman BE, Kuchel PW, Häggström L. Elevated glutamate dehydrogenase flux in glucose-deprived hybridoma and myeloma cells: evidence from $^{1}H/^{15}N$ NMR. Biotechnol Bioeng 1998; 60:508.
59. Miller WM, Wilke CR, Blanch HW. The transient responses of hybridoma cells to nutrient additions in continuous culture: II. Glutamine pulse and step changes. Biotechnol Bioeng 1989; 33:487.
60. Butler M, Spier RE. The effects of glutamine utilisation and ammonia production on the growth of BHK cells in microcarrier culture. J Biotechnol 1984; 1:187.
61. Jeong YH, Wang SS. Role of lutamine in hybridoma cell culture: effects on cell growth, antibody production, and cell metabolism. Enzyme Microb Technol 1995; 77:45.
62. Vriezen N, Romein B, Luyben KChAM, van Dijken JP. Effects of glutamine supply on growth and metabolism of mammalian cells in chemostat culture. Biotechnol Bioeng 1997; 54:272.
63. Mancuso A, Sharfstein ST, Fernandez EJ, Clark DS, Blanch HW. Effect of extracellular glutamine concentration on primary and secondary metabolism of a murine hybridoma: an in vivo ^{13}C nuclear magnetic resonance study. Biotechnol Bioeng 1998; 57:172.
64. Ljunggren J, Häggstrom L. Catabolic control of hybridoma cells by glucose and glutamine limited fed batch cultures. Biotechnol Bioeng 1994; 44:808.
65. Meijer JJ, van Dijken JP. Effects of glucose supply on myeloma growth and metabolism in chemostat culture. J Cell Physiol 1995; 162:191.
66. Glacken MW, Adema E, Sinskey AJ. Mathematical description of hybridoma culture kinetics: I. Initial metabolic rates. Biotechnol Bioeng 1988; 32:491.
67. Flickinger MC, Goebel NK, Bibila T, Boycejacinso S. Evidence for posttranscriptional stimulation of monoclonal antibody secretion by L-glutamine during slow glutamine growth. J Biotechnol 1992; 22:201.
68. Linz M, Zeng AP, Wagner R, Deckwer WD. Stoichiometry, kinetics and regulation of glucose and amino acid metabolism of recombinant BHK cell line in batch and continuous cultures. Biotechnol Prog 1997; 13:453.
69. Schmid G, Keller T. Monitoring hybridoma metabolism in continuous suspension culture at the intracellular level. Cytotechnology 1992; 9:217.
70. Europa AF, Gambhir A, Fu PC, Hu WS. Multiple steady states with distinct cellular metabolism in continuous culture of mammalian cells. Biotechnol Bioeng 2000; 67:25.
71. Follstad BD, Balcarcel RR, Stephanopoulos G, Wang DIC. Metabolic flux analysis of hybridoma continuous culture steady state multiplicity. Biotechnol Bioeng 1999; 63:675.
72. Lambert K, Pirt SJ. The quantitative requirements of human diploid cells (strain MRC-5) for amino acids, vitamins and serum. J Cell Sci 1975; 17:397.
73. Luan YT, Mutharasan R, Magee WE. Strategies to extend longevity of hybridomas in culture and promote yield of monoclonal antibodies. Biotechnol Lett 1987; 9:691.
74. Duval D, Demangel C, Munier-Jolain K, Miossec S, Geahel I. Factors controlling cell proliferation and antibody production in mouse hybridoma cells: I. Influence of the amino acid supply. Biotechnol Bioeng 1991; 38:561.
75. Jo EC, Park HJ, Park JM, Kim KH. Balanced nutrient fortification enables high density hybridoma cell culture in batch culture. Biotechnol Bioeng 1990; 36:717.

76. Simpson NH, Singh RP, Perani A, Goldenzon C, Al-Rubeai M. In hybridoma cell cultures, deprivation of any single amino acid leads to apoptotic death, which is suppressed by the expression of the bcl-2 gene. Biotechnol Bioeng 1998; 59:90.
77. Franêk K, Srámková K. Cell suicide in starving hybridoma culture: survival-signal effect of some amino acids. Cytotechnology 1997; 23:231.
78. Geaugey V, Duval D, Geahel I, Marc A, Engasser JM. Influence of amino acids on hybridoma cell viability and antibody secretion. Cytotechnology 1989; 2:119.
79. Sanfeliu A, Cairó JJ, Casas C, Solà C, Gòdia F. Analysis of nutritional factors and physical conditions affecting growth and monoclonal antibody production of the hybridoma KB-26.5 cell line. Biotechnol Prog 1996; 12:209.
80. Lazo PA. Amino acids and glucose utilization by different metabolic pathways in ascites-tumour cells. Eur J Biochem 1981; 117:19.
81. Castagna M, Shayakul Ch, Trotti D, Franca Sacchi V, Harvey WR, Hediger MA. Molecular characteristics of mammalian and insect amino acid transporters: implications for amino acid homeostasis. J Exp Biol 1997; 200:269.
82. Mizrahi A. Oxygen in human lymphoblastoid cell line cultures and effect of polymers in agitated and aerated cultures. Dev Biol Stand 1984; 55:93.
83. van der Valk P, Gille JJP, Oostra AB, Roubos EW, Sminia T, Joenje H. Characterization of an oxygen-tolerant cell line derived from Chinese hamster ovary. Cell Tissue Res 1985; 239:61.
84. Phillips HA, Scharer JM, Bols NC, Moo-Ypung M. Effects of oxygen on antibody productivity in hybridoma culture. Biotechnol Lett 1987; 9:745.
85. Miller WM, Wilke CR, Blanch HW. Effects of dissolved oxygen concentration on hybridoma growth and metabolism in continuous culture. J Cell Physiol 1987; 132:524.
86. Ozturk SS, Palsson BO. Effects of dissolved oxygen on hybridoma cell growth, metabolism and antibody production kinetics in continuous culture. Biotechnol Prog 1990; 6:437.
87. Jan DCH, Petch DA, Huzel N, Butler M. The effect of dissolved oxygen on the metabolic profile of a murine hybridoma grown in serum-free medium in continuous culture. Biotechnol Bioeng 1997; 54:153.
88. Zupke C, Sinskey AJ, Stephanopoulos G. Intracellular flux analysis applied to the effect of dissolved oxygen on hybridomas. Appl Microbiol Biotechnol 1995; 44:27.
89. Oiler AR, Buser CW, Tyo MA, Thilly WG. Growth of mammalian cells at high oxygen concentrations. J Cell Sci 1989; 14:43.
90. Cacciuttolo MA, Trinh L, Lumpkin JA, Rao G. Hiperoxya induces DNA damage in mammalian cells. Free Radic Biol Med 1993; 14:267.
91. Jorjani P, Ozturk SS. Effects of cell density and temperature on oxygen consumption rate for different mammalian cell lines. Biotechnol Bioeng 1999; 64:349.
92. Schmelzer AE, Miller WM, Dezengotita VM, Abston LR. Environmental effects on cell physiology and metabolism: response to elevated pCO2. In: Lindner-Olsson E, Chatzissavidou N, Lüllau E, eds. Animal Cell Technology: from Market to Target. Dordrecht: Kluwer Academic Publishers, 2001:121.
93. Zanghi JA, Schmeltzer AE, Mendoza TP, Knop RH, Miller WM. Bicarbonate concentration and osmolality are key determinants in the inhibition of CHO cell polysialylation under elevated pCO_2 or pH. Biotechnol Bioeng 1999; 65:182.
94. deZengotita VM, Kimura R, Miller WM. Effects of CO_2 and osmolality on hubridoma cells: growth, metabolism and monoclonal antibody production. Cytotechnology 1998; 28:213.
95. Gray DR, Chen S, Howarth W, Inlow D, Maiorella, BL. CO_2 in large-scale and high-density CHO cell perfusion culture. Cytotechnology 1996; 22:65.
96. Matanguihan R, Sajan E, Zachariou M, Olson C, Michaels J, Thrift J, Konstantinov K. Solution to the high dissolved CO_2 problem in high density perfusion culture of mammalian cells. In: Lindner-Olsson E, Chatzissavidou N, Lüllau E, eds. Animal Cell

Technology: from Market to Target. Dordrecht: Kluwer Academic Publishers, 2001:399.

97. Ozturk SS, Riley MR, Palsson BO. Effects of ammonia and lactate on hybridoma growth, metabolism, and antibody production. Biotechnol Bioeng 1992; 39:418.
98. Omasa T, Higashiyama K, Shioya S, Suga K. Effects of lactate concentration on hybridoma culture in lactate-controlled fed-batch operation. Biotechnol Bioeng 1992; 39:556.
99. Bibila TA, Ranucci CS, Glazomitsky K, Buckland BC, Aunins JG. Monoclonal antibody process development using medium concentrates. Biotechnol Prog 1994; 10:87.
100. Zhou W, Chen CC, Buckland BC, Aunins JG. Fed-batch culture of recombinant NSO myeloma cells with high monoclonal antibody production. Biotechnol Bioeng 1997; 55:783.
101. deZengotita VM, Miller WM, Aunins JG, Zhou W. Phosphate feeding improves high-cell-concentration NSO myeloma culture performance for monoclonal antibody production. Biotechnol Bioeng 2000; 69:566.
102. Glacken MW. Catabolic control of mammalian cell culture. Bio/Technology 1988; 6:1041.
103. McQueen A, Bailey JE. Effect of ammonium ion and extracellular pH on hybridoma cell metabolism and antibody production. Biotechnol Bioeng 1990; 35:1067.
104. Kurano N, Leist C, Messi F, Kurano S, Fiechter A. Growth behavior of Chinese hamster ovary cells in a compact loop reactor. 2. Effects of medium components and waste products. J Biotechnol 1990; 15:113.
105. Ryll T, Valley U, Wagner R. Biochemistry of growth inhibition by ammonium ions in mammalian cells. Biotechnol Bioeng 1994; 44:184.
106. Hassell T, Gleave S, Butler M. Growth inhibition in animal cell culture. The effect of lactate and ammonia. Appl Biochem Biotechnol 1991; 30:29.
107. Yang M, Butler M. Effects of ammonia on CHO cell growth, erythropoietin production, and glycosilation. Biotechnol Bioeng 2000; 68:370.
108. Hansen HA, Emborg C. Influence of ammonium on growth, metabolism and productivity of a continuous suspension Chinese hamster ovary cell culture. Biotechnol Prog 1994; 10:121.
109. Miller WM, Wilke CR, Blanch HW. Transient response of hybridoma cells to lactate and ammonia pulse and step changes in continuous culture. Bioprocess Eng 1988; 35:113.
110. Newland M, Kamal MN, Greenfield PF, Nielsen LK. Biotechnol Bioeng 1994; 43:434.
111. McQueen A, Bailey JE. Growth inhibition of hybridoma cells by ammonium ion: correlation with effects on intracellular pH. Bioprocess Eng 1991; 6:49.
112. Schneider YJ, Marison IW, Stockar U. The importance of ammonia in mammalian cell culture. J Biotechnol 1996; 46:161.
113. Andersen DC, Goochee CF. The effect of ammonia on the O-linked glycosylation of granulocyte colony-stimulating factor produced by Chinese hamster ovary cells. Biotechnol Bioeng 1995; 47:96.
114. Borys MC, Linzer DIH, Papoutsakis ET. Ammonia affects the glycosylation pattern of recombinant mouse placental lactogen-I by Chinese hamster ovary cells in a pH-dependent manner. Biotechnol Bioeng 1994; 43:505.
115. Thorens B, Vassalli P. Chloroquine and ammonium chloride prevent terminal glycosylation of immunoglobulins in plasma cells without affecting secretion. Nature 1986; 321:618.
116. Gawlitzek M, Valley U, Wagner R. Ammonium ion and glucosamine dependent increases of oligosaccharide complexity in recombinant glycoproteins secreted from cultivated BHK-21 cells. Biotechnol Bioeng 1998; 57:518.
117. Ronning OW, Schartum M, Winsnes A, Lindberg G. Growth limitation in hybridoma cell cultures: the role of inhibitory or toxic metabolites. Cytotechnology 1991; 7:15.
118. Bibila TA, Robinson DK. In pursuit of the optimal fed-batch process for monoclonal antibody production. Biotechnol Prog 1995; 11:1.

119. Zhou W, Chen CC, Buckland B, Aunins J. Fed-batch culture of recombinant NSO myeloma cells with high monoclonal antibody production. Biotechnol Bioeng 1997; 55:783.
120. Xie L, Wang DIC. High cell density and high monoclonal antibody production through medium design and rational control in a bioreactor. Biotechnol Bioeng 1996; 51:725.
121. Sauer PW, Burky JE, Wesson MC, Sternard LQ. A high yielding, generic fed-batch cell culture process for production of recombinant antibodies. Biotechnol Bioeng 2000; 67:585.
122. Konstantinov KB. Monitoring and control of the physiological state of cell cultures. Biotechnol Bioeng 1996; 52:271.
123. Ozturk SS. Engineering challenges in high density bioreactors. Cytotechnology 1999; 64:349.
124. Christie A, Butler M. Glutamine-based dipeptides are utilized in mammalian cell culture by extracellular hydrolysis catalyzed by a specific peptidase. J Biotechnol 1994; 37:277.
125. Holmlund AC, Chatzisavido N, Bell SL, Lindner-Olsson E. Growth and metabolism of recombinant CHO cell-lines in serum-free medium containing derivatives of glutamine. In: Spier RE, Griffiths JB, Macdonald C, eds. Animal Cell Technology: Developments, Processes and Products. Oxford: Butterworth-Heinemann, 1992:176.
126. Hassell T, Butler M. Adaptation to non-ammoniagenic medium and selective substrate feeding lead to enhanced yields in animal cell cultures. J Cell Sci 1990; 96:501.
127. Christie A, Butler M. The adaptation of BHK cells to a non-ammoniagenic glutamate-based culture medium. Biotechnol Bioeng 1999; 64:298.
128. Wagner A, Marc A, Engasser JM, Einsele A. Growth and metabolism of human tumor kidney cells on galactose and glucose. Cytotechnology 1991; 7:7.
129. Marquis CP, Barford JP, Harbour C, Fletcher A. Carbohydrate and amino acid metabolism during batch culture of a human lymphoblastoid cell line BTSN6. Cytotechnology 1996; 21:121.
130. Altamirano C, Paredes C, Cairò JJ, Godia F. Improvement of CHO cell culture medium formulation: simultaneous substitution of glucose and glutamine. Biotechnol Prog 2000; 16:69.
131. Stephanopoulos G, Valllino JJ. Network rigidity and metabolic engineering in metabolite overproduction. Science 1991; 252:1675.
132. Paredes C, Prats E, Cairò JJ, Azorin F, Cornudella LI, Gódia F. Modification of glucose and glutamine metabolism in hybridoma cells through metabolic engineering. Cytotechnology 1999; 30:85.
133. Scherczinger CA, Yates AA, Knecht DA. Variables affecting antisense RNA inhibition of gene expression. In: Baserga R and Denhart DT, eds. Antisense Strategies. New York: Academy of Sciences1992:660:45.
134. Chen K, Liu Q, Xie L, Sharp PA, Wang DIC. Engineering of a mammalian cell line for reduction of lactate formation and high monoclonal antibody production. Biotechnol Bioeng 2001; 72:55.
135. Irani N, Wirth M, van den Heuvel J, Wagner R. Improvement of the primary metabolism of cell cultures by introducing a new cytoplasmatic pyruvate carboxylase reaction. Biotechnol Bioeng 1999; 66:238.
136. Bell SL, Bebbington CR, Scott MF, Vardell JN, Spier RE, Bushell ME, Sanders PG. Genetic engineering of hybridoma glutamine metabolism. Enzyme Microb Technol 1995; 17:98.
137. Robinson DK, Chan CP, Yu I, Tsai PK, Tung J, Seamans TC, Lenny AB, Lee DK, Irwin J, Silberklang M. Characterization of a recombinant antibody produced in the course of a high yield fed-batch process. Biotechnol Bioeng 1994; 44:727.
138. Zhou W, Rehm J, Hu WSh. High viable cell concentration fed-batch cultures of hybridoma cells through on-line nutrient feeding. Biotechnol Bioeng 1995; 46:579.

5
Protein Glycosylation

Sarah W. Harcum
Department of Bioengineering, Clemson University, Clemson, South Carolina, U.S.A.

INTRODUCTION

There are many new human therapeutics that are glycosylated recombinant proteins. These recombinant glycoproteins require "correct" glycosylation for clinical efficacy. For this reason, uncontrolled changes in protein glycosylation are of great regulatory and industrial concern. Protein glycosylation is strongly influenced by many environmental factors, such as dissolved oxygen (DO), pH, carbon source, and temperature. In many instances, recombinant glycoproteins are highly expressed relative to native proteins, potentially exacerbating the influence of environmental factors that alter the expression or function of the enzymes controlling glycosylation.

Protein glycosylation occurs when a sugar group (oligosaccharide) is covalently bound to a protein. Oligosaccharide attachment can profoundly change the biological function of a protein. From a bioprocessing point of view, altering the oligosaccharide attachment can significantly change the biologic function/activity of the therapeutic protein. Unlike the biosyn-thesis of a polypeptide from mRNA, where the mRNA only codes for one polypeptide sequence, the attachment of an oligosaccharide to a protein is not template driven. Thus, numerous different oligosaccharide motifs can occur with a gradient of biological activity. Additionally, oligosaccharide structures can be located at many different positions on a protein, which adds to the complexity and variability of biological response. The glycosylation or nonglycosylation variability at a specific site in a protein is termed macroheterogeneity. Since oligosaccharides can branch and have stereochemistry, different oligosaccharides can be attached to the same site on a protein. This variation in the oligosaccharide group is termed microheterogeneity. The degree of macroheterogeneity and microheterogeneity for a glycoprotein is dependent on the attachment position in the protein, the protein conformation, the physiological status of the cell, the organism, the expression vector for recombinant glycoproteins, and the culture environment for in vitro cells. For recombinant glycoproteins, the most important *N*-glycan heterogeneity is due to the presence or absence of specific features in the oligosaccharide and the number of branches (1,2).

BIOLOGICAL FUNCTION AND THERAPEUTIC SIGNIFICANCE

The addition of an oligosaccharide to a protein can result in a diverse set of physiochemical and biological property changes. The oligosaccharides can impart structural changes to the protein as well as change specific recognition of the oligosaccharide by receptors. To determine the extent of these effects on biological function and physiochemical properties, each glycoprotein must be evaluated individually.

The biological activity of glycoproteins is highly variable and corresponds to the spectrum of glycoforms that are possible for a given protein. Thus, the U.S. Food and Drug Administration (FDA) will license a glycosylation pattern (fingerprint) as a part of a drug's specifications. As long as the manufacturer operates within a narrow set of process conditions, the drug will likely have the same lot-to-lot fingerprint. Being able to engineer host cell lines capable of producing recombinant proteins with greater fidelity under a wider range of process conditions has the potential for greatly improving quality control and increasing flexibility at the manufacturing level (3–5).

Rather than being entirely template driven, protein glycosylation is influenced by many environmental and genetic factors (6,7). Furthermore, host cell and expression vector selection for recombinant protein expression will greatly impact glycosylation patterns (6,8). In order to assure the efficient production of an efficacious therapeutic, it is therefore important to examine protein productivity and protein glycosylation together.

Physiochemical Changes

Physiochemical changes that can occur due to the addition or modification of an oligosaccharide include modified solubility, oligomerization, electric charge, mass, size, and viscosity. The addition of sialic acid not only increases the size and mass of the glycoprotein, but also changes its charge and solubility (9,10). Many of the physiochemical changes are due to the inherent properties of the sugar molecules. Commonly, the glycoprotein becomes more rigid due to the hydrogen bonding between the sugar residues, thus the size of the glycoprotein is altered. Also, the presence of attached oligosaccharides aids in protein folding for some glycoproteins (11). Other physiochemical properties of a protein that are altered due to the presence of attached oligosaccharides are stabilized protein conformation, increased thermal stability, and increased protease resistance (12).

Biological Changes

For the majority of enzymes, glycosylation does not affect biological activity; however, biological activity is not the same as therapeutic efficacy. Therapeutic efficacy is the combination of biological activity and clearance rate in the patient. A glycoprotein can have extremely high biological activity and zero efficacy if the glycoprotein is cleared from the body immediately. Conversely, a glycoprotein with relatively low biological activity can have high therapeutic efficacy if the clearance rate from the body is very low (9,10). Therapeutic glycoproteins that have higher efficacy due to the oligosaccharides include erythropoietin (EPO), tissue plasminogen activator (tPA), Pulmozyme (DNase), Factor VIII, and Factor IX.

The biological functions of the oligosaccharide on a glycoprotein are very diverse. The oligosaccharide can regulate the intracellular trafficking of the glycoprotein, regulate the localization of the glycoprotein, and determine the lifetime of the glycoprotein in circulation. Also, the oligosaccharides can modulate enzyme and hormone activity, act as cell receptors for lectins, antibodies, and toxins, participate in cell-to-cell interactions, and modify immunological properties. One example of modified hormone activity was observed for human chorionic gonadotropin (hCG): the deglycosylated hCG has activity in vitro, but no activity in vivo (9,13). Also, EPO has increased in vivo activity when the sialic acid content is higher (9).

In general, O-linked oligosaccharides are not as well understood; however, O-linked *N*-acetylglucosamine (O-GlcNAc) appears to modify proteins in a manner reciprocal to phosphorylation (6,14). All identified O-GlcNAc proteins are also phosphoproteins. The most important O-GlcNAc protein is probably RNA polymerase II. The control of gene expression in a cell requires the coordination of DNA enhancers and silencers, chromatin structures, transcription activators and repressors, mRNA stabilization, protein degradation, and post-translation modification. Yet, only one molecule, RNA polymerase II, is responsible for the expression (14).

Immunogenic Considerations

Protein glycosylation is a common posttranslational modification that contributes to the antigenic properties of a protein. These antigenic properties can be due to the carbohydrate unit itself or to modification of the peptide. The peptide–carbohydrate complex may elicit the immune response. The peptide may elicit the immune response due to protein conformation changes, thus exposing the antigenic peptide (15). The ABO blood groups are the first documented observation relating the glycosylation of a protein to an immune response (9). For therapeutic glycoproteins, an important manufacturing consideration is the immunogenicity of the glycoprotein. For example, the granulocyte-macrophage colony-stimulating factor (GM-CSF) produced in yeast has relatively little glycosylation compared to human GM-CSF. Yet, patients who received the yeast GM-CSF developed antibodies to native (human) GM-CSF. In particular, patients developed antibodies to a portion of the native protein that was normally covered by oligosaccharides (9).

It is also well documented that some monoclonal antibodies (MAbs) manufactured in mouse hybridoma cell lines and used as diagnostics and therapeutics can result in antibody production in the patient against the mouse MAbs. This immune response is thought to be partially mitigated by the sugar group, α1,3-galactose-β1,4-galactose-*N*-acetylglucosamine (Galα1,3-Galβ1,4-GlcNAc). Most mammalian-derived cells express Galα1,3-Galβ1,4-GlcNAc residues on membrane and secreted proteins; however, humans, apes, and Chinese hamster ovary (CHO) cells do not express this sugar group (6). Humans do not have the enzyme for this linkage, and therefore the body recognizes this oligosaccharide as foreign and mounts an immune response (16). Approximately 1% of the circulatory antibodies in a normal human are directed toward this particular sugar group, probably as a defense against enteric bacteria (3). The strong immune response in humans to the Galα1,3-Galβ1,4-GlcNAc-linked sugars is a major contributing factor preventing xenotransplantation (17).

STRUCTURES AND CONFORMATIONS OF OLIGOSACCHARIDES

There are two main types of glycosylation: N-linked oligosaccharides and O-linked oligosaccharides. N-linked oligosaccharides are found on asparagine (Asn) residues bonded through the R-group by an *N*-glycosidic bond. O-linked oligosaccharides are attached to the protein by an *O*-glycosidic bond at either a serine or a threonine R-group (8,18). The sugars that are commonly found in mammalian protein oligosaccharides are shown in Fig. 1. Glucose (Glc), galactose (Gal), mannose (Man), fucose (Fuc), and xylose (Xyl) are neutral sugars. GlcNAc and *N*-acetylgalactosamine (GalNAc) are amino sugars. *N*-acetylneuraminic acid (NeuAc) is an acidic sugar and is the mostly common sialic acid form found in adult humans. Since monosaccharides can form glycosidic bonds between any two hydroxyl groups, a glycosidic bond between two identical sugars can result in 11 unique disaccharides. A pentamer of identical sugars has 17,872 possible configurations compared to only one possible polypeptide configuration for a pentamer with identical amino acids. Fortunately

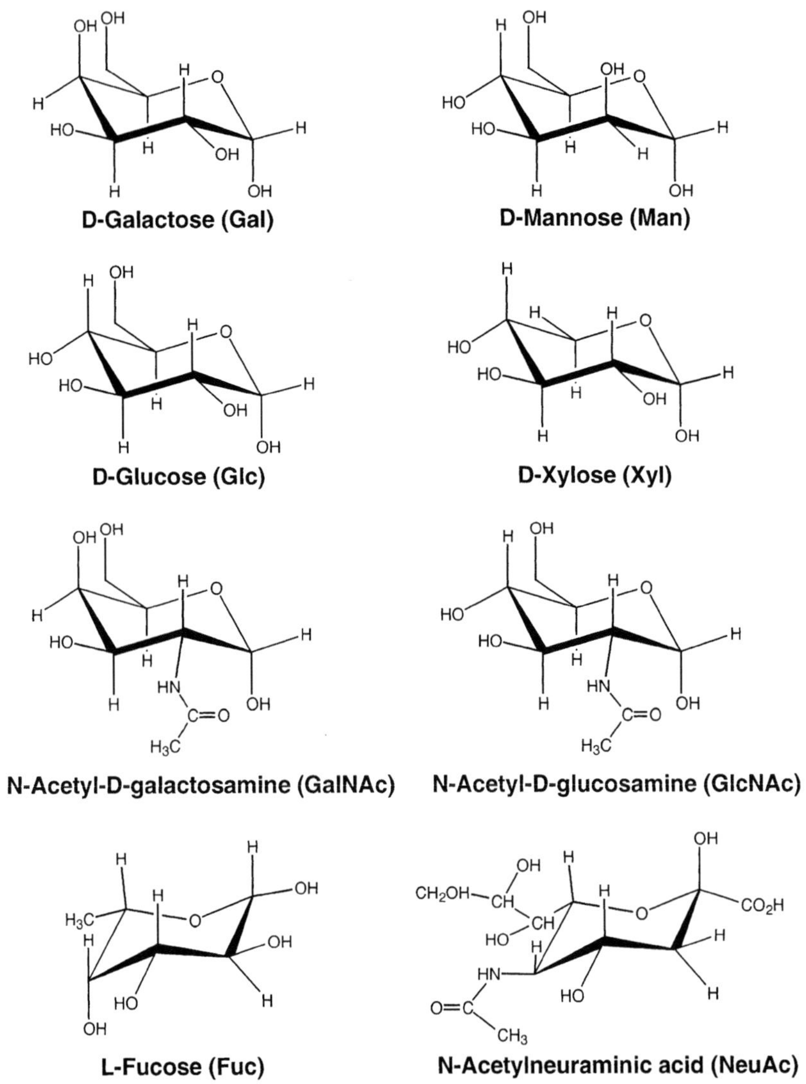

Figure 1 Common monosaccharides found in glycoproteins.

for glycobiologists, this extreme structural variety is not observed in nature (6). However, there still is a great deal of oligosaccharide structural variety in nature.

N-Linked Oligosaccharides

N-oligosaccharides or *N*-glycans only occupy asparagine (Asn) sites in a protein; however, not all asparagine sites in the protein will be glycosylated. The consensus sequence Asn-X-Ser/Thr (Ser: serine and Thr: threonine) is required, but does not guarantee glycosylation (8). Protein conformation also influences glycosylation. Hidden asparagine residues are less likely to be glycosylated (19), while asparagine sites near the N-terminal are more likely to be glycosylated.

N-linked oligosaccharide residues of a protein are all derived from a common 14-sugar-lipid complex. Five of the 14 core sugars are conserved. The conserved core *N*-oligosaccharides are $Man_3GlcNAc_2$. The core sugars are extensively modified by the addition and deletion of simple sugars following attachment to the protein (Fig. 2). These modifications are termed the trimming and branching reactions. The final N-linked oligosaccharide can have three distinct configurations: complex, high mannose, and hybrid. The complex *N*-oligosaccharides contain disaccharides like Galβ1,4-GlcNAc. The high-mannose *N*-oligosaccharides contain only mannoses at the ends of the oligosaccharide branches. Hybrid *N*-oligosaccharides have one complex branch or antennae, and one oligosaccharide mannose antennae branching off the core $Man_3GlcNAc_2$ group. Figure 2 shows the core *N*-oligosaccharides with the conserved pentasacchrides boxed. Also in Fig. 2, a generalized oligosaccharide is shown with common antennae as observed in mammals (6,20). For recombinant glycoproteins, the degree of branching can be important to efficacy. Additionally, the presence of a terminal sialic acid, the type of sialic acid present, the presence of a bisecting GlcNAc attached to the core mannose, the presence of terminal poly-*N*-acetyllactosamine, GlcNAc sulfation, and Man phosphorylation are important features of the *N*-glycan with respect to efficacy (1).

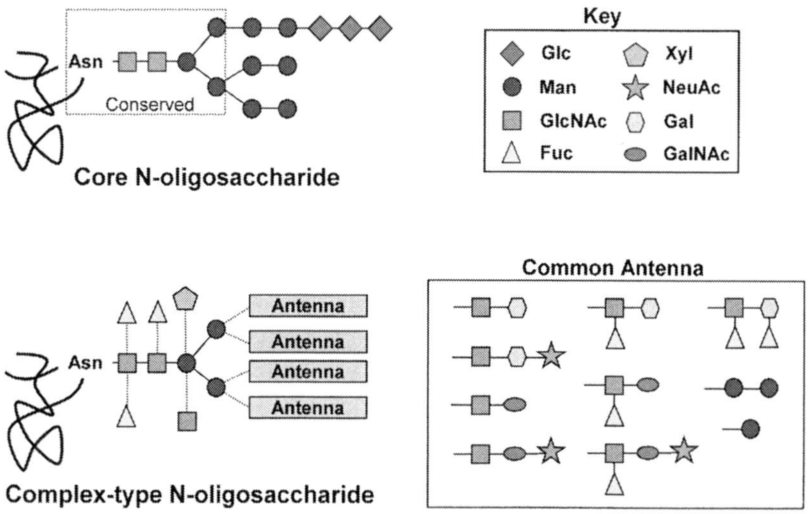

Figure 2 N-glycosylation structures. [Adapted from Refs. (1,20,21).] (*See color insert p. 1.*)

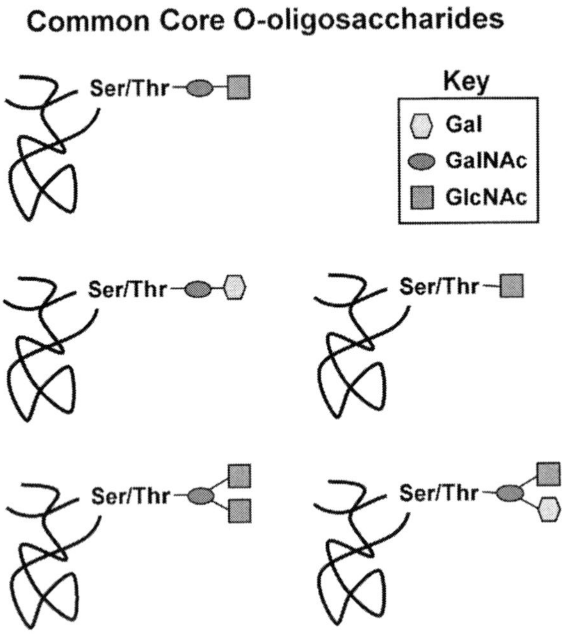

Figure 3 O-glycosylation structures. [Adapted from Refs. (6,20).] (*See color insert p. 2.*)

O-Linked Oligosaccharides

Protein O-glycosylation is more common in secreted and membrane-associated proteins (22). O-linked oligosaccharides are attached to the protein by an *O*-glycosidic bond at either a serine or a threonine R-group. No consensus sequence for O-glycosylation appears to exist, since site-directed mutagenesis does not influence site-specific glycosylation (18). O-glycosylated proteins are relatively resistant to proteases (23). *O*-oligosaccharides are typical smaller and less complex than *N*-oligosaccharides. The attached *O*-oligosaccharide can vary from 1 to 20 sugar residues. Unlike in *N*-oligosaccharides, the core structure is not conserved. There are at least seven different core structures, with the five most common shown in Fig. 3. *O*-oligosaccharides often contain GlcNAc, GalNAc, and galactose.

INTRACELLULAR BIOSYNTHESIS

Biosynthesis of *N*-Oligosaccharides

The N-glycosylation pathway contains three steps: the dolichol cycle and oligosaccharide transfer to the protein (Fig. 4), the trimming reactions where glucose and mannose residues are removed (Fig. 5), and the branching reactions that construct the oligosaccharide antennae (Fig. 5). The initial dolichol cycle reactions occur on the cytoplasmic face of the endoplasmic reticulum (ER). Sugars are added sequentially to dolichyl phosphate. When the oligosaccharide attached to the dolichyl phosphate contains two GlcNAc and five mannoses, the oligosaccharide group flips to the lumen side of the ER. Seven more sugars are added sequentially to the oligosaccharide from lipid precursors. Finally, oligosaccharyltransferase (OligoST) transfers the core oligosaccharide to the asparagine residue of a nascent, growing polypeptide chain. Next, trimming reactions remove three glucoses. Then, the terminal mannose residues can be removed by different ER mannosidases (MANs). The ER also contains glycosyltransferases than can reglucosylate glucose-free chains, which can cycle. Once the glycoprotein has folded and reached the Golgi complex, more mannose trimming can occur. Two more mannose residues can be trimmed from the oligosaccharide following the addition of a GlcNAc residue. Next, subsequent terminal glycosylation can include the addition of GlcNAc, galactose, sialic acid, and fucose. The number of branches in the oligosaccharide is variable. Additionally, the number and identity of the sugars present are variable. The ER glycoforms are less complex than the Golgi glycoforms; however, both types are not homogenous (11,24,25).

Core N-Oligosaccharide

The dolichol cycle and oligotransferase reactions are shown Fig. 4. The dolichyl cycle begins with the transfers of the *N*-acetylglucosamine phosphate residue of uracil diphosphate (UDP)–GlcNAc to dolichylphosphate (Dol-*P*) by the enzyme *N*-acetylglucosaminylphosphotransferase (UD-GlcNP) to produce *N*-acetylglucosamine dolichyl pyrophosphate (GlcNAc-*PP*-Dol). The reactions shown in Fig. 4 are accompanied by a diagram showing the sequential increase in the oligosaccharide-Dol-*P* entity. The second reaction, catalyzed by the enzyme *N*-acetylglucosaminyltransferase (GlcNAcT), produces $(GlcNAc)_2$-*PP*-Dol from UDP–GlcNAc and GlcNAc-*PP*-Dol. Mannosyltransferase I (ManT I) catalyzes the first mannose addition to $(GlcNAc)_2$-*PP*-Dol, and ManT II–V catalyze the addition of four more mannoses to produce $(Man)_5(GlcNAc)_2$-*PP*-Dol. Dolichylphospho-β-mannose

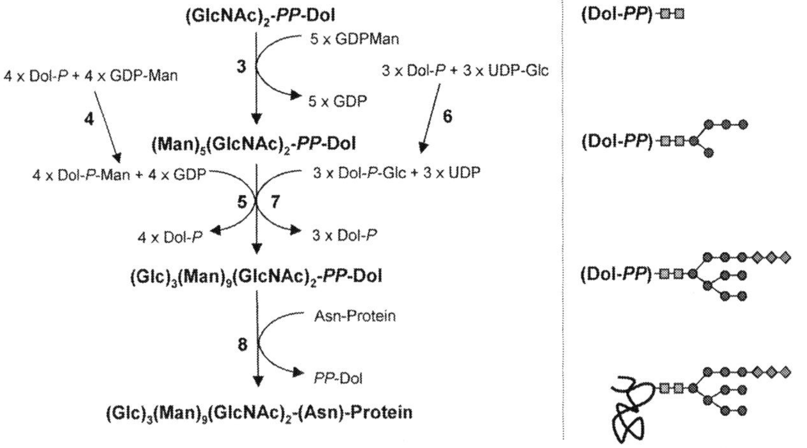

Figure 4 The dolichol cycle and oligosaccharide transfer reactions to form *N*-oligosaccharide core. [Adapted from Refs. (6,20).] (*See color insert p. 2.*)

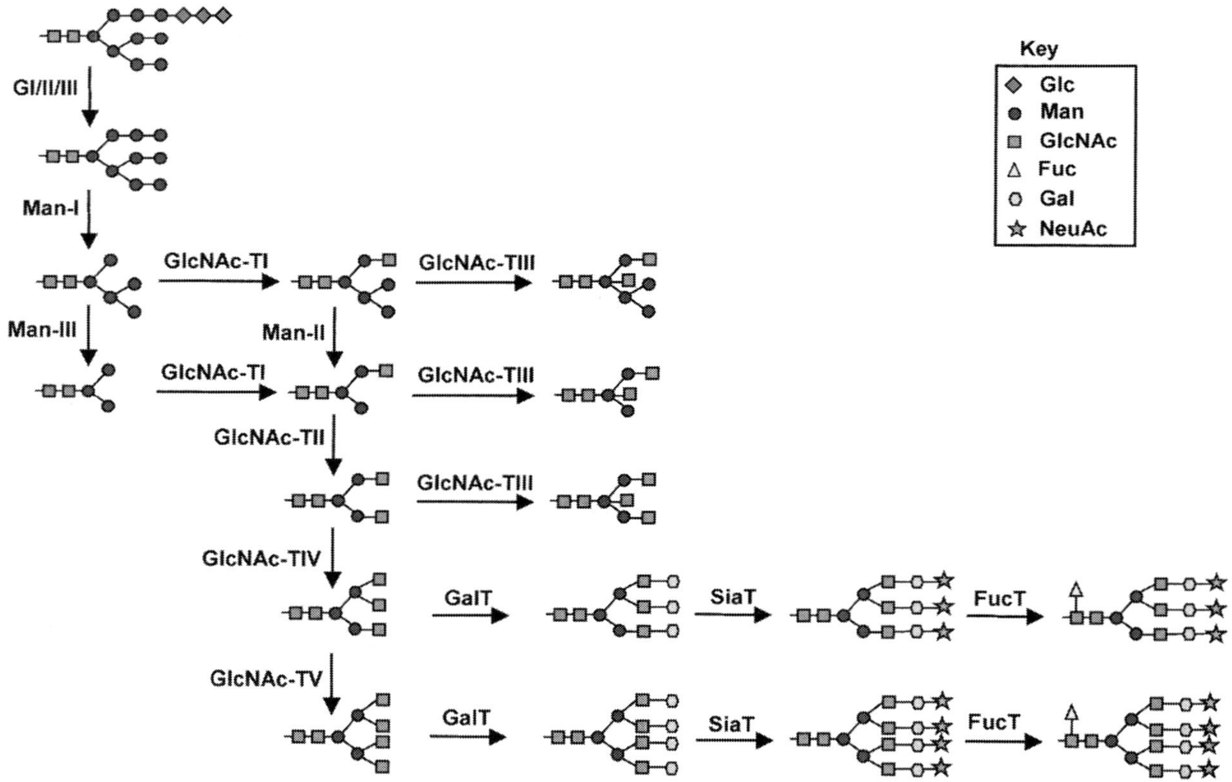

Figure 5 Trimming and branching reactions that modify the core *N*-oligosaccharides. [Adapted from Refs. (20,21).] (*See color insert p. 3.*)

synthetase (DPMS) transfers four mannoses from guanidine diphosphate-Man (GDP-Man) to dolichylphospho-β-mannose (Dol-*P*-Man). Dolichylphospho-β-glucose (Dol-*P*-Glc) is formed by the reaction of Dol-*P* with UDP–Glc catalyzed by the enzyme dolicylphospho-β-glucose synthetase (DPGS). Dol-*P*-Glc is the donor for the final three glucoses added to the $(Man)_5(GlcNAc)_2$-*PP*-Dol catalyzed by the enzymes glucosyltransferase I–III (GlcT I–III). ManT VI–IX catalyze the final four mannose additions. The ManT VI–IX and GlcT I–III reactions result in the formation of $(Glc)_3(Man)_9(GlcNAc)_2$-*PP*-Dol, the core *N*-oligosaccharide. OligoST transfers the $(Glc)_3(Man)_9(GlcNAc)_2$ residue from $(Glc)_3(Man)_9(GlcNAc)_2$-*PP*-Dol to the nascent protein at the exposed asparagine amino acid, resulting in the nascent glycoprotein. Macroheterogeneity results when not all of the asparagine residues have an attached core *N*-oligosaccharide. The complete $(Glc)_3(Man)_9$ $(GlcNAc)_2$ residue is required for further glycosylation processes (6,11,21,24).

Trimming and Branching Reactions of N-Oligosaccharides

Once the nascent glycoprotein is formed, the trimming and branching reactions specific to the species and cell type can occur in the Golgi. Trimming and branching reactions may or may not occur at a particular glycosylation site or in a particular cell type. In general, trimming reactions are catalyzed by two types of enzymes, glucosidases and endomannosidases. Figure 5 shows the common trimming and branching reactions that may occur. There are three glucosidases (GI, GII, and GIII), which remove the terminal glucose residues from the nascent core glycoprotein. Endomannosidases are able to remove one, two, or three glucose residues, plus a mannose residue, in one reaction. ER mannosidase and Golgi mannosidase I remove all four outer mannose residues. Branching reactions are very complex and may result in biantennary, triantennary, tetra-antennary, and penta-antennary oligosaccharides. Other common modifications of glycoproteins are galactosylation and sialylation. GlcNAc-TI is the key enzyme for the conversion of high-mannose type to hybrid- and complex-type N-linked oligosaccharides (Fig. 5). Other critical enzymes include GlcNAc-TII, GlcNAc-TIII, GlcNAc-TIV, GlcNAc-TV, GlcNAc-TVI, mannosidases II, II^X, and III, and α1,6-fucosyltransferase (FT). Galactosylation of N-linked oligosaccharides is catalyzed by a number of species-specific and cell-specific enzymes. The most common galactosyltransferases (GalTs) are β1,4-Gal-T and β1,3-Gal-T. Another common modification of glycoproteins is sialylation. Sialylation is very important for the efficacy of many highly glycosylated proteins, such as EPO and Factor VIII. Sialyltransferases (STs) catalyze the sialylation at the terminal end of the oligosaccharide. These are many different ST enzymes. Each type of ST catalyzes the addition of a particular sialic acid with a specific terminal linkage. In humans, the most common terminal sialic acid is NeuAc. The commonly observed sialic acid linkages in mammals are α2,3-, α2,6-, and α2,8-linked sialic acids (6). These complex trimming and branching reactions, occurring and not occurring, lead to N-linked microheterogeneity (11,21,24).

The key control points for N-glycosylation of recombinant proteins are (1) ER and Golgi mannosidase I and GlcNAc transferase I, which catalyzes the conversion of high-mannose glycans to complex-type glycans; (2) GlcNAc transferases II, IV, and V, which catalyze the degree of branching; (3) GlcNAc transferase III, which catalyzes the bisection of GlcNAc attached to the core mannose; (4) α2,3- and α2,6-STs, which catalyze the addition of terminal sialic acid; and (5) CMP–NeuAc–4-hydroxylase, which catalyzes the conversion of NeuAc to NeuGc (1).

Understanding the effects of bioprocess conditions on these controlling reactions could be used to improve product quality.

Biosynthesis of *O*-Oligosaccharides

O-glycosylation, unlike N-glycosylation, is not generated by the transfer of a lipid-sugar onto the protein, but is generated by the direct, sequential transfer of monosaccharides onto the protein. Mucin O-glycosylation, the most studied O-glycosylation reaction, is initiated by a monosaccharide (galactose) transfer to the nascent protein at a serine or threonine by a class of enzymes called UDP–GalNAc:polypeptide *N*-acetylgalactosaminyltransferases (ppGalNAc transferase). The oligosaccharide extension reactions are mediated by numerous enzymes. The more common enzymes are α2,3-, α2,6-sialylation transferases, β1,6-Gal-T, and β1,3-Gal-T. The final O-linked oligosaccharide is smaller and less complex that typical N-linked oligosaccharides. Typically, the *O*-oligosaccharide contains 1–20 sugar residues. Mammalian cytoskeleton proteins also include many nonmucin O-linked oligosaccharides. tPA, Factor IX, and other human clotting factors have O-linked oligosaccharides that contain O-fucosylated structures (9). An enzyme that catalyzes this reaction is α-O-FT (see Fig. 2 for the common fucose linkages). Another enzyme that has been isolated is xylosyltransferase, which transfers a xyloglycosyl side chain to clotting factors (see Fig. 2 for the common xylose linkages) (6). Due to the high variability of the O-glycosylation reactions, microheterogeneity is also observed for O-linked oligosaccharides (21,24).

GLYCOSYLATION POTENTIAL OF VARIOUS EXPRESSION SYSTEMS

The need for glycosylation of a recombinant protein is completely dependent on its effect on the efficacy (a function of biological activity and the clearance rate) of the protein in the patient. Bacteria typically do not glycosylate proteins, but can be utilized as hosts when efficacy is independent of glycosylation (for example, recombinant human insulin and human growth hormone). If the glycoform is critical for biological activity, the choice of the host organism becomes critical. All eukaryotic cells glycosylate proteins via common pathways; however, evolution has imparted many variations. Therefore, the selection of a host for a glycoprotein is completely dependent of the biological in vivo activity of the therapeutic (1).

Bacteria

Bacteria are generally considered to be unable to glycosylate proteins; however, eubacteria and archaebacteria are able to glycosylate surface proteins. The biosynthetic pathways are not well characterized, but the reactions appear to occur on the cell's outer membrane. In many respects the reactions are very similar to those of N-glycosylation in eukaryotic organisms. The oligosaccharide is preassembled with a dolichyl-linked intermediate and then transferred to the protein as a group. Some unfamiliar sugars are used (rhamnose, galactofuranose, and *N*-acetylmannosamine). Instead of UDP as the phosphate donor, thymine diphosphate may be used. Thus, at this time, recombinant proteins produced in bacteria are not glycosylated, but some of the mechanisms are present in prokaryotic organisms (6,26), and missing reactions could be added via metabolic engineering.

Yeast and Fungi

Yeast and fungi have been used to express numerous recombinant proteins that require posttranslational modification. These cells are easier to culture on a large-scale than mammalian cells due to the presence of cell walls, less complex media requirements, rapid growth, and a longer experience-base due to antibiotic fermentations and brewing. For glycoproteins, yeast and fungi typically add excessive mannose (termed hypermannosylation) groups, which can be antigenic (27,28). Yeast cells will also O-glycosylate different serine residues in a glycoprotein compared to humans (27). Despite these differences in glycosylation, there are two biopharmaceutical protein entities on the market that are produced in *Saccharomyces cerevisiae* (29), the hepatitis B vaccine [containing the hepatitis B surface antigen (HBsAg)] and platelet-derived growth factor. *Pichia pastoris,* a methylotrophic yeast strain, has been used recently to produce recombinant proteins. *P. pastoris* secretes the recombinant protein, without secreting many native proteins relative to *S. cerevisiae,* which improves purification efficiency. Although *P. pastoris* still hyperglycosylates the recombinant protein, the oligosaccharides are smaller (30–32). Thus, many researchers express potential recombinant glycoprotein therapeutics in *P. pastoris* (33).

Insect/Baculovirus Systems

Insect cells infected by baculovirus have the ability to synthesize large quantities of recombinant protein and have an established record for producing glycoproteins (34). The insect/baculovirus expression system is considered safer that mammalian cell culture, since baculoviruses do not infect humans. The N- and O-linked oligosaccharides produced in the insect/baculovirus system contain similar core structures as in other eukaryotic organisms, including humans. The *N*-oligosaccharides are usually the oligomannose type. A potential limitation for therapeutic glycoprotein production in insect cells is the lack of GalTs and STs. Also, sialic acid residues have not been detected in insect cells (34–36). The reducing-terminal GlcNAc residue has been observed to be α1,3-fucosylated (34), which is immunogenic in humans. Altmann et al. (34) stressed that there are many applications for recombinant protein that are not as stringent as human therapeutics, for example, veterinary therapeutics and diagnostics, for which the insect/baculovirus system maybe appropriate. Chen et al. (37) produced human glial cell line-derived neurotrophic factor (GDNF) in the baculovirus-infected *Trichoplusia ni* cells. The GDNF was glycosylated and biologically active in an animal model. Zhang et al. (38) noted that DO concentrations affect the final glycosylation state of a protein made in insect cells. The optimal glycosylation occurred at an intermediated DO; thus, glycosylation heterogeneity in insect cultures can be controlled. Also, there are a few researchers who are investigating the use of metabolic engineering to clone the missing enzymes into insect cells, such that more humanized proteins are produced, and could be potentially used as therapeutics (34,39).

Plants

Transgenic plants could provide the means to produce large quantities of recombinant protein using standard cultivation and harvesting practices. The drawbacks of transgenic plants are low accumulation in the cells, uncharacterized posttranslational events, and lack of downstream processing data. Many recombinant proteins have been expressed in plants. Therapeutically important recombinant proteins that have been expressed in transgenic plants include EPO, growth hormone, and hirudin. However, it does not appear that the glycosylation of these proteins was examined (40). Also, if sialic acid is required for efficacy, transgenic plants may not be well suited, since only buckwheat has been reported to produce sialic acid (10,41,42). Additionally, plants may produce antigenic sugars residues like α1,3-Fuc and Xylβ1,2-Man (42,43). Transgenic plants could provide an important alternative for the production of recombinant proteins, where glycosylation and sialylation are not required. More research is needed to fully characterize the capability of transgenic plants to glycosylate therapeutic proteins.

Transgenic Animals

Other potential sources of therapeutic glycoproteins are transgenic animals (44–48). Transgenic animal-derived therapeutic proteins have some potential cost savings over traditional mammalian cell culture. The raw materials and capital investment are lower for transgenic animals. The downstream purification costs are similar. The recombinant glycoprotein is expressed in the mammary gland and secreted into the milk. The glycoprotein has the potential to be very similar to the human protein. Fibrinogen (360 kDa) has been successfully produced in sheep at 5 g/L. The fibrinogen molecule has six (6) subunits and 29 intra- and interdisulfide bonds. The sheep-expressed protein was fully active (48). Alpha-1-antitrypsin (AAT) has also been expressed in a transgenic animal milk at 12 g/L. AAT is currently under clinical trials for cystic fibrosis treatment (48). However, the glycosylation pattern of the transgenically produced fibrinogen and AAT were not discussed. Bovine follicle-stimulating hormone (FSH) has been expressed in rabbit milk. FSH contains two subunits and requires extensive glycosylation and sialylation for activity. The rabbit produced FSH appeared glycosylated and sialylated on Western blot. More importantly, the rabbit FSH induced super ovulation in cows (47).

Mammalian Cell Culture

Species that are more closely related to humans may be better at producing glycoproteins that are similar to human glycoproteins. Rodent cell lines (the most common host cell lines) have been shown to have many differences in glycosylation compared to humans (8). For a very comprehensive list of recombinant protein hosts and their glycosylation characteristics, Jenkins et al. (8) provides a good summary. Some highlights will be presented here. Most mammalian-derived glycoproteins contain Galα1,3-Galβ1,4-GlcNAc residues; however, humans, apes, Old World monkeys, baby hamster kidney (BHK), and CHO cells do not express α1,3-GalT due to a frameshift mutation (6). This unusual sugar residue has the potential to be antigenic to humans (8). Therefore, for acute usage, this unusual sugar would not cause significant immune response problems; however, if the drug is to be administered chronically in large doses, the potential immune response needs to be considered. To date, MAbs produced in mouse hybridoma cell line have not had significant adverse effects due to this sugar residue.

As of January 2003, there were 17 biopharmaceutical glycoprotein entities (excluding MAbs) on the market (29,49–51). Two of these glycoproteins are produced in *S. cerevisiae,* 2 in BHK cells, 2 in unspecified mammalian hosts, 1 in a human cell line, and 12 in CHO cells (the number of hosts is greater than the number of glycoprotein entities, since some glycoproteins are manufactured by more than one company using different hosts). There are 22 licensed MAbs on the market as therapeutics or injectable diagnostics (29,49–51). MAbs are typically produced in mouse hybridomas, mouse–human hybridomas, or CHO cells. There are many more therapeutic glycoproteins in clinical trials, some of which are not produced in mammalian-derived cells or *S. cerevisiae.*

Mouse-Derived and Common Rodent-Derived Cells

Unlike humans, mouse hybridomas, mouse–human hybridomas, and C127 cells, these cells all express the sugar residue Galα1,3-Galβ1,4-GlcNAc. There are mixed reports regarding Galα1,3-Galβ1,4-GlcNAc expression in NS0 cells, and the rat Y0 myeloma cell line has been reported not to express Galα1,3-Galβ1,4-GlcNAc (8,52,53). Mouse and humans both express the enzymes α2,3-ST and α2,6-ST (8,52). However, mouse and mouse–human hybridomas express *N*-glycolylneuraminic acid (NeuGc) as the terminal sialic acid sugar more commonly than in humans. NeuGc is an oncofetal antigen in humans. Adult humans do not normally express NeuGc, and adult humans can tolerate this sugar in low levels ($< 1\%$ of the terminal sialic acid residues of a glycoprotein), but at higher levels ($>7\%$) NeuGc is known to elicit an anti-NeuGc response. Also, NeuGc containing glycoproteins have more rapid clearance rates in humans (8). For MAbs, where sialylation is low, no adverse events have been attributed to the anti-NeuGc. Baker et al. (53) reported that NS0 had variable sialylation, where 30% of the sialic acids were α1,3-linkages. It was also observed that the NS0-derived glycoprotein contained a higher NeuGc to NeuAc ratio than would be observed in adult humans.

CHO and BHK Cells

Like humans, CHO and BHK cells do not express the enzyme α1,3-GalT, due to a frameshift mutation. Therefore, recombinant glycoproteins made in these cells do not contain Galα1,3-Galβ1,4-GlcNAc. CHO-derived MAbs only contain Galα1,4-linked or GlcNAc as terminal sugars (54). Also, like humans, CHO and BHK only express NeuGc terminal sugars at very low levels. However, unlike humans, CHO and BHK lack a functional α2,6-ST gene, which results in all terminal sialic acids having only α2,3-linkages (8,55,56). Additionally, CHO cells will O-glycosylate different serine residues in a glycoprotein compared to humans and cannot sulfate GalNAc residues are commonly found in human hormones (27). Also, CHO lacks the α1,3-FT gene, which prevents Fucα1,3-GlcNAc (18). To help eliminate some of the differences between human and CHO/BHK protein glycosylation, both CHO and BHK have been genetically modified to express many of these missing genes, which has resulted in glycoproteins containing the human sugar linkages (8,16,52,55). Monaco (57) cloned a rat α2,6-ST into CHO and the α2,6-ST to α2,3-ST ratio approached the human ratio in some clones. Umaña et al. (58) overexpressed rat N-acetylglucosaminyltransferease III and V (GlcNAcT III, GlcNAcT V) in CHO cells using a tetracycline promoter. Severe growth inhibition occurred, which was attributed to a limited cellular capacity for glycosyltransferase expression. Weikert et al. (59) expressed β1,4-Gal-T and/or α2,3-ST in CHO cells, even though these genes are present in CHO cells. Improved galactosylation and sialy-lation were observed. Prati et al. (16) cloned CMP–sialic acid:Galβ1,3-GalNAc-α2,3-ST (ST3Gal I) gene in antisense into CHO cells to downregulate the corresponding enzyme activity. The human UDP–GlcNAc:Galβ1,3-GalNAc-*R*-β1,6-GlcNAcT (C2GnI) gene was also cloned into the same CHO cells. Increased O-glycosylation was observed. Thus, it is possible to redirect CHO cell metabolism to control glycosylation by metabolic engineering.

ENVIRONMENTAL EFFECTS ON RECOMBINANT THERAPEUTIC GLYCOPROTEINS

The interplay between cell growth conditions and the structure of therapeutic proteins produced through mammalian cell culture has been well documented (60–71). For example, Patel et al. (66) and Maiorella et al. (68) both observed that an MAb glycosylation pattern was strongly dependent on the culturing method. Maiorella et al. (68) observed that clearance rates were different for MAbs obtained from different culturing methods. Patel et al. (66) expressed concern regarding the traditional development protocols for therapeutic MAbs, where the initial clinical trials used MAbs harvested from ascites, since subsequent clinical trials would use material obtained from ascites. One criterion the FDA considers is bioequivency. If the clearance rates and efficacy are similar in vivo, then the material is acceptable. Problems arise in attempting to demonstrate bioequivency for glycoproteins produced in different systems, in part, due to sophisticated carbohydrate analysis techniques. However, understanding the effects of bioprocess conditions on glycosylation can be used to control the variation in glycosylation.

Nutrient Limitations and Feeding Strategies

Many literature studies have investigated glycosylation variations of recombinant proteins with respect to nutrient requirements including sugar feeding, nucleotide sugar feeding, oxygen levels, amino acid additions, and serum components (3,72–78). Serum is known to contain growth factors, which improve cell growth, and lipids, which improve shear resistance, but serum also contains waste products and proteases, which can be detrimental to the cell and glycoprotein products. Gawlitzek et al. (74) and Hayter et al. (76,77) both used glucose-limited chemostats to demonstrate that glycosylation was dependent on media components. Hayter et al. (76) examined the glycosylation pattern of interferon-γ (IFN-γ) produced by CHO cells at a constant dilution rate and at two different glucose concentrations. They were able to demonstrate that fully glycosylated IFN-γ occurred more readily when glucose was not limited. The authors concluded that this effect was to be due to the physiological state of the cells.

Hahn and Goochee (67) observed that transferrin, a glycoprotein, secreted by confluent and subconfluent cultures, contains different proportions of biantennary oligosaccharides, where the biantennary glycoprotein is more biologically active. Confluent cells produced more active transferrin than the subconfluent cultures. Hahn and Goochee (67) concluded that oligosaccharide synthesis is growth dependent, thus in standard batch cultures where the growth rate varies throughout the fermentation, the glycosylation pattern will vary.

Hooker et al. (70) examined the *N*-oligosaccharides of recombinant IFN-γ produced by CHO cells, where temperature, pH, and DO were held constant. The proportion of the biantennary structure decreased throughout the fermentation and the proportion of oligomannose and truncated species increased for both

N-glycosylation sites. This observation is very important for recombinant therapeutics, since the shift to oligomannose and truncated species typically increases the clearance rate, thus reducing the efficacy of the therapeutic. It was also suggested by the authors, that fed-batch fermentations might mitigate the deterioration of the glycoprotein (70). In a separate study, an MAb fed-batch fermentation, the MAb also had a decrease in the amount of complex glycoforms and an increase in the high-mannose and truncated glycoforms throughout the fermentation (69), which can detect and quantify microheterogeneity changes. The problem of high-mannose and truncated glycoforms is potentially a problem for any material produced in bioreactors by mammalian cells.

Gawlitzek et al. (74) examined the production of interleukin-2 (IL-2) under various nutrient limitations [glucose, fetal calf serum (FCS), glutamine, serine, aspartate, tryptophan, and oxygen]. They found the oligosaccharide profiles to be identical, regardless of the culture conditions. However, they also observed that the glycosylated to unglycosylated ratio varied depending on the culture conditions. Gawlitzek et al. (74) observed truncated IL-2, which is an indicator of proteolysis of the IL-2. Their experiments are critical to furthering the understanding of how glycosylation is influenced by culture conditions; however, none of the parameters tested were independent due to the culturing method. Each nutrient limitation lasted only 3 days and was immediately replaced by another nutrient limitation. Since cells need time to recover from stress, it was possible that the similar responses observed for oligosaccharide content were not steady-state responses to the nutrient limitations (stress), but that the effects of the early nutrient stress were still being detected at later time points. Gawlitzek et al. (75) investigated the role of the influence of FCS on glycosylation for suspension and microcarrier BHK-21 cultivated cells. They found that N-glycosylation and O-glycosylation of IL-2 were greater in the serum-free environment. Additionally, due to the recent outbreaks of mad cow's disease in Europe, numerous industrial cell culture-derived therapeutics, under clinical trial, are using serum- and protein-free media.

Jenkins et al. (3) varied the media lipid content for a batch CHO cell culture expressing IFN-γ to investigate the importance of lipids on N-glycosylation. Lipoprotein supplementation increased the amount of fully glycosylated IFN-γ compared to the control culture (3). Interestingly, if lipids are critical for N-glycosylation, then the cell culture conditions that alter the lipid content of cell membranes could also effect glycosylation, e.g., agitation, pH, and temperature. Castro et al. (78) stressed the importance of media composition after they observed that batch feeding did not prevent the decrease of fully glycosylated IFN-γ; however, the initial glutamine concentration did impact the glycosylation pattern. The authors also observed that the addition of serum to the media caused proteolysis of IFN-γ. Last, they found that the supplemental protein added to the media and the lipid composition of the media affected the glycosylation pattern (78).

Barnabe and Butler (79) investigated dual glucose and galactose feeding in hybridoma cells. It was observed that the corresponding intracellular nucleotide pool increased; however, no changes in glycosylation were observed. Hills et al. (63) fed NS0 cells expressing an IgG galactose, glucosamine, or CMP–sialic acid. The glucosamine feeding resulted in a 17-fold increase in the UDP–*N*-acetylhexose pools; however, the UDP–hexose pools decreased by 33%. The UDP–Glc to UDP–Gal ratio remained constant at 4:1. A 57% reduction in galactosylation was observed. The core Mans structure was unaffected. Galactose feeding resulted in a fivefold increase in the UDP–Gal pool, but no increase in β1,4-galactosylation. The CMP–sialic acid feeding did not result in increased sialylation, but it did result in a 44-fold increase in the CMP–sialic acid pool. Gu and Wang (80) fed *N*-acetylmannosamine to CHO cells expressing IFN-γ; however, no change in sialylation was observed. Gawlitzek et al. (81) observed that glucosamine feeding increased the UDP–GxxNAc pool (sum of the UDP–GlcNAc and UDP–GalNAc pools) for CHO producing TNFR–IgG (an immunoadhesin tumor necrosis factor–IgG fusion). However, the glucosamine feeding did not alter the glycosylation (81).

Baker et al. (53) demonstrated that the addition of glucosamine, uridine, and *N*-acetylmannosamine (ManNAc) significantly increased the intracellular UDP–HexNAc and CMP–sialic acid concentrations for both NS0 and CHO cells. However, these increased nucleotide sugar pools did not increase the level of sialylation. These additions did, however, improve the NeuAc to NeuGc ratio in NS0 cells to a more favorable (human) level. Altamirano et al. (72,73,82) examined the dual feeding of sugars to CHO cells expressing tPA to control growth, substrate consumption, and waster product buildup. Fructose and glucose feeding resulted in decreased growth rates, the substrate utilization, and toxic waste buildup.

Butyrate feeding has been observed to increase type I tPA production (83), which is less effective that type II tPA. Kim and Lee noted increased glycoprotein production in the presence of sodium butyrate, but also observed cell growth inhibition and increased apoptosis. Santell et al. (84) suggested that sodium butyrate increased glycosylation by increasing glycoprotein reutilization. Wang et al. (85) examined CHO cells expressing EPO in fluidized beds. They observed that the addition of glucose and sodium butyrate doubled the final EPO concentration without changing the glycosylation pattern.

Cruz et al. (86) produced a recombinant antibody-IL-2 fusion in BHK cells using protein-free media. Two different protein-free media were compared. Chemostat cultures were used and consistent productivity and glycosylation patterns were obtained. They also demonstrated that long-term cultivation is not adverse to product consistency. The higher titers obtained from the chemostats also increased the yield in the purification steps (86).

Media pH and Ammonium Concentration

It is well known that the ammonium ion content in the media alters intracellular pH and thereby affects cell growth and protein glycosylation (87). Miller et al. (88) conducted a comprehensive study of hybridomas in chemostats. The pH was varied and cell densities, viability, and antibody concentration were measured. Miller et al. (88) determined that pH was a very important parameter for successful cell culture. Borys et al. (89,90) investigated the effects of pH on mouse placental lactogen expressed in CHO cells. Their initial studies used tissue culture flasks (T-flasks), yet clearly demonstrated that the culture pH shifted the glycosylation pattern. Later experiments used cells grown on microcarriers; again, their results demonstrated that pH effected glycosylation. Next, the extracellular pH was held constant and the ammonium concentration was varied. Increased ammonium ion concentrations decreased intracellular pH, and resulted in less glycosylated proteins (90).

Valley et al. (91) noted that ammonia inhibited cell growth and influenced glycosylation, but were unsure of the mechanism. Valley et al. (91) used labeled NH_4Cl to detect N-labeled species. They observed that 60% of the UDP–GlcNAc pool was labeled, and 60–80% of the N-acetylated sugars in *N*-glycan structures were labeled. Therefore, they concluded that ammonium was used as a building block.

Gawlitzek et al. (81) observed that ammonium altered N-glycosylation in CHO expressing TNFR–IgG. Ammonium levels were increased from 1.0 to 15–mM, which resulted in a 40% decrease in terminal galactosylation and sialylation. Extracellular β-galactosidase and sialidase activity increased throughout the culture. The β-galactosidase activity was proportional to the initial ammonium level; however, the sialidase activity was independent of the initial ammonium level. Also, the incubation of TNFR-IgG with β-galactosidase and sialidase in vitro did not alter glycosylation. These results suggested that the effect of ammonium on glycosylation was biosynthetic and not degradative. mRNA levels and the enzyme activity of β1,4-Gal-T and α2,3-ST enzymes were observed in response to ammonium (13 mM). Both the mRNA and the enzyme activities were not observed to change. The β1,4-Gal-T and α2,3-ST enzymes were observed in this study to be very sensitive to pH in vitro. Therefore, Gawlitzek et al. (81) concluded that ammonium altered the carbohydrate biosynthesis of TNGR–IgG by a pH-mediated effect on glycosyltransferase activity. Other studies have shown that increased nucleotide sugar pools, do not increase glycosylation, as would be expected (61–63,79).

Yang and Butler (61,62) investigated the effects of ammonia on the glycosylation of EPO expressed in CHO cells. It was observed that increased ammonia concentration resulted in a reduction of the tetra-antennary structures and increased the tri- and biantennary structures. Also, the sialic acid content decreased with increased ammonia; however, the oligosaccharide sequence did not change. A later study by Yang and Butler (60) investigated ammonia and glucosamine levels on glycosylation. Both ammonia and glucosamine reduced the terminal sialylation, increased heterogeneity, and reduced the level of O-glycosylation. It was also observed that the UDP–*N*-acetylhexsamine pools increased (60).

Elevated CO_2 and Osmolality Levels

The effects of elevated CO_2 and osmolality on cell growth and production of tPA in CHO have been investigated (92,94). The effect of CO_2 on glycosylation is a very important bioprocess consideration, since media pH is commonly maintained via the HCO_3^-/CO_2 equilibrium and CO_2 is a metabolic by-product. It was observed that the growth rate of CHO was inhibited by CO_2 and higher osmolality. However, tPA productivity was not significantly affected by CO_2 and osmolality. Kimura and Miller (93) observed that in serum-free media, the amount of sialylation decreased slightly with elevated CO_2 and the monosaccharide content of the tPA was relatively constant. Schmelzer and Miller (94) grew CHO cells expressing tPA under hyperosmotic stress and elevated CO_2. The tPA sialylation was observed to change. They used amino acids (threonine, glycine, proline) to protect the cells from the hyperosmotic stress and elevated CO_2 levels. Some of the lost sialylation due to these stresses was recovered by the addition of these amino acids. Interestingly, the extracellular β-galactosidase levels were unchanged due to the hyperosmotic stress or elevated CO_2 levels. It has been suggested that tPA is relatively robust with respect to glycosylation. On the other hand, EPO is highly sensitive to culture conditions with respect to glycosylation and sialic acid content (95). EPO sialylation greatly influences the EPO clearance rate in the patient (2,96,97). Therefore, the effects of elevated CO_2 on glycosylation need to be considered for recombinant glycoproteins, individually.

DO Effects

The effect of the DO concentration on cell metabolism and growth has been investigated for numerous mammalian cell types (74,98–104). A few resear-chers have extended their investigations to include the influence of DO on N-glycosylation and the sialic acid content of glycoproteins (74,102–104). Interestingly, these authors observed that protein production increased with increased DO (102) and N-glycosylation was less sensitive to DO changes than to growth rate (74,101). Jan et al. (103) observed that even though cell viability, cell growth rate, and MAb productivity were not significantly influenced by DO changes (10–100% air saturation), the fluxes through metabolic pathways were significantly affected for a hybridoma cell line. Ozturk and Palsson (100) investigated cell growth, amino acid, carbohydrate, energy metabolism, and antibody production for a hybridoma cell line under oxygen stress. These authors observed that amino acid utilization was significantly increased in an oxygen-stressed environment. Although the glycosylation of the MAbs was not addressed, the authors' data indicate that glycosylation may have been altered, since the cell switched metabolism to adapt to the low DO.

Lin et al. (101) investigated the influence of DO on glycosylation of tPA in perfusion reactors. Although they observed significant chances in the growth rate, metabolism, and protein productivity, they did not observe significant changes in glycosylation. The methods they used to evaluate glycosylation were in vitro activity and sodium dodecyl sulfate polyacrylamide gel electrophoresis (SDS-PAGE), methods known to detect gross changes only. Additionally, the cells were cultivated in perfusion reactors that never reached steady state, in terms of cell density or media composition. The DO of the perfusion reactor was changed while the nutrient composition of the media was changing. Therefore, it is likely that any changes in glycosylation due to DO were masked by other nutrient influences on glycosylation.

One of the most complete studies on MAb glycosylation examined the influence of DO on glycosylation (104). Kunkel et al. (104) varied the DO from 10 to 100% air saturation in serum-free chemostats. Glycosylation was evaluated by fluorophore-assisted carbohydrate electrophoresis (FACE) and monosaccharide and oligosaccharide profiles were determined by high-pH anion-exchange chromatography with pulsed amperometric detection (HPAEC-PAD). The authors observed that galactosylation decreased as the DO decreased. This decrease in the galactose content of the oligosaccharide could be the result of disturbances in the ER. The authors also noted that cells at the lower DO concentrations had increased glycolytic flux and a decreased tricarboxylic acid cycle flux (104). The changes in the glycosylation of the MAb were directly related to changes in central metabolic pathway enzyme activity.

Temperature Effects

The literature regarding the affects of temperature on cell culture productivity and glycosylation is relatively limited. Research, thus far, has concluded that decreased growth temperatures have been associated with increased cell viability, decreased media consumption, and decreased oxygen consumption (64,65,105–107). With regard to cellular productivity, the effects of decre-ased culture temperature have been variable. Additionally, the effects of temperature on glycosylation have only been preliminarily investigated (64,108). Chuppa et al. (64) found that at decreased growth temperatures, the sialic acid content did not appear to change. Additionally, the authors observed that the total sugar content of the protein was lower at 37°C than at the lower culture temperatures. Kaufmann et al. (65) did not directly examine glycosylation changes; however, the authors observed changes in posttranslational modifications, specifically tyrosine phosphorylation. Also, Kaufmann et al. (65) used two-dimensional polyacrylamide gel electrophoresis (2D-PAGE) to examine the proteome. They observed 10 upregulated proteins at the lower culture temperature. Thus, they concluded that the molecular mechanisms underlying the change in protein expression are not well understood and need to be investigated further.

Effects of Extracellular Enzymes

Extracellular enzymes, namely proteases and glycosidases, are believed to play an important role in protein glycosylation (8,109). A properly glycosylated recombinant protein is thought to be more resistant to protease degradation (18). Since media protein components are also substrates for extracellular proteases, industry's switch to serum-free and protein-free medias could potentially increase the probability of proteolytic degradation of the recombinant glycoprotein.

There are five identified types of glycosidases: sialidases, β-galactosidases, β-hexosaminidases, mannosidases, and fucosidases (109). Glycosidases that are most important to recombinant protein production are: sialidases and β-galactosidases. Sialidases remove terminal sialic acids and β-galactosidases remove terminal galactoses. Both terminal sialic acid and galactose increase the half-life of a recombinant protein in circulation. For a glycosidase to cleave sugars from the glycoprotein, the glycosidase needs to be expressed in the cell, secreted into the media, have specificity to the glycoprotein, and be stable in the media. Therefore, culture conditions that minimize extracellular glycosidases will result in increased uniformity and stability of the recombinant glycoprotein (109).

Goochee and Gramer (110,111) characterized glycosidases from CHO, adenovirus-transformed human embryonic kidney (293), mouse myeloma (NS0), and a hybridoma cell line. The glycosidases isolated from the CHO cell supernatant were very stable over a wide range of pH and temperature. The glycosidases isolated from 293, NS0, and the hybridoma cell line supernatants were not very stable at neutral pH or 37°C. The glycosidase activity in CHO cells increased when the cells were damaged (112). Munzert et al. (113) reported both sialidase activity and loss of protein sialylation. Ferrari et al. (114) used antisense technology to reduce the sialidase expression in CHO producing DNase. The final product had a 20–30% increase in sialylation (114).

Kratje et al. (115) conducted a comprehensive study of CHO proteases. They determined that one of the main issues with serum-free cultivations is proteolytic attack and degradation of the recombinant protein. They observed that the cell-specific protease composition changed, depending on the culture environment. Protease activity was observed to spike after a media exchange. Normal protease levels returned after the cells had adapted to the fresh media. Yang and Butler (60) also noted increased protease activity in serum-free media, observed as truncated forms of EPO. They hypothesized that amino acid starvation might have caused the elevated protease activity. Satoh et al. (116) first reported two types of proteases in CHO, an endoprotease and an exopeptidase. The exopeptidase activity increased with culture time and correlated with the nonviable cell count. Endoprotease activity was observed to have cysteine protease characteristics and was secreted continuously by viable cells. No molecular weights were reported for these proteases. Froud et al. (117) observed extracellular proteases in CHO that were upregulated during amino acid starvation. Cartwright (118) stated that protease activity was directly due to amino acid starvation in CHO cells. Gawlitzek et al. (75) reported that eight amino acids were removed from IL-2 expressed in BHK cells. Goldman et al. (119) reported up to 10 amino acids removed from IFN-γ. Wang et al. (85) reported EPO degradation by proteases. EPO contained a protease-sensitive region that bisected the molecule. Therefore, protease and glycosidase activity should be assessed and monitored during the production of recombinant glycoproteins.

GLYCOSYLATION ANALYSIS

There are many analytical techniques available to determine and quantify protein glycosylation. These analytical techniques are essential in recombinant protein production to ensure product quality, lot-to-lot consistency, and product stability. Glycosylation analytical techniques need to be highly sensitive and able to detect and separate large molecules containing very small differences. Each analytical technique measures glycosylation based on different physical and chemical characteristics of the glycoprotein. Broadly, glycosylation patterns can be obtained via four methods: separating the glycoprotein by glycoforms, determining the glycoprotein monosaccharide content, cleaving oligosaccharides from the glycoprotein, and degrading the glycoprotein via peptidases (120). Separated glycopeptides can be analyzed by the same methods as intact glycoproteins. The structure and sugar linkages can be determined for the cleaved oligosaccharides. The analytical techniques that are commonly used to quantify glycoproteins are: electrophoresis, liquid chromatography (LC), mass spectrometry (MS), nuclear magnetic resonance (NMR), and exoglycosidase digestion. To fully characterize a glycoprotein to meet FDA requirements, at least two or more orthogonal analytical methods need to be used. The selection of the appropriate analytical techniques is determined by the amount sample required, cost per sample, desired resolution, and availability of equipment and technical expertise. All of these factors together must be considered in the selection of appropriate analytical techniques.

Material Analyzed and Expected Results

The sample preparation necessary to analyze glycoproteins depends on the analytical method selected; however, the glycoprotein usually needs to be affinity purified. Once a purified glycoprotein has been obtained, many of the analytical techniques are sensitive to salts. A desalting step is usually conducted prior to analysis. The whole or intact glycoprotein can be analyzed directly, or digested and the individual components analyzed. Figure 6 shows the breakdown of an intact glycoprotein for analysis into a glycopeptide mixture, individual glycopeptides, oligosaccharides, and monosaccharides. Each pool shown in Fig. 6 can be analyzed by a variety of methods. Glycopeptides are obtained by protease digestion of the glycoprotein. Trypsin is a commonly used protease (120). The glycopeptide mixture is separated into glycopeptide pools with the same peptide backbone. The individual glycopeptides can be analyzed for glycosylation or the oligosaccharides and monosaccharides can be analyzed after removal from the peptide. Oligosaccharides can be enzymatically or chemically cleaved from the glycoprotein or separated glycopeptides. Monosaccharides can be obtained directly from the glycoprotein or glycopeptide via acid

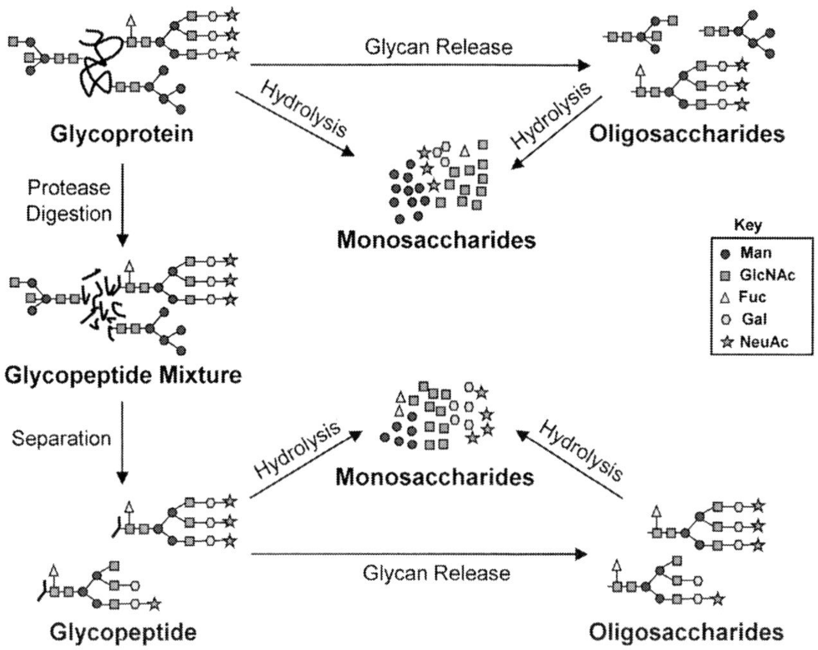

Figure 6 Fractionation methods to characterize glycoproteins. [Adapted from Refs. (20,54,121–132).] (*See color insert p. 3.*)

hydrolysis. Monosaccharides can also be obtained via acid hydrolysis of the oligosaccharides (120).

In this review, the type of information that can be obtained from analyzing the intact glycoprotein, glycopeptides, oligosaccharides, and monosaccharides will be discussed independently of the analytical techniques. Then, the analytical techniques will be discussed individually. Since no single technique can routinely provide complete characterization, how the various techniques are used together to obtain the glycosylation information for a glycoprotein will be discussed. Table 1 lists the analytical techniques commonly used to characterize the glycosylation of a glycoprotein. Details are presented regarding starting material, information obtained, sample size, and analysis difficulty.

Glycoprotein Analysis

Whole or intact glycoproteins can be analyzed based on size (molecular weight), charge, affinity, and oligosaccharide content. The presence of different oligosaccharides can change the molecular weight of the glycoprotein. The glycoprotein glycoforms can be separated from each other by size. Since sialic acid is a negatively charged sugar, and under some conditions other sugars will take a charge, the presence of different oligosaccharides can be used to separate the glycoforms by charge. The oligosaccharides attached to the glycoprotein also alter the shape of the glycoprotein. The glycoforms can be analyzed for affinity with lectins or antibodies against oligosaccharide groups. The level of resolution for the glycosylation analysis of a glycoprotein is relatively low. However, the analytical techniques used are also relatively easy, fast, and inexpensive. Analysis techniques used on the intact

Table 1 Analysis Techniques to Characterize the Glycosylation of a Glycoprotein

Technique	Glycoprotein	Glycopeptide	Oligosaccharide	Monosaccharide	Sample size	Analysis difficulty	Comments (Refs.)
Electrophoresis							
SDS-PAGE	Purity, degree of variability, and molecular weight	—	—	—	µg	Average	Coupled with exoglycosidase to determine oligosaccharide linkages (66,69,76,122)
IEF-PAGE	Glycoform profile and p*I*	Glycopeptide profile and p*I*	—	—	µg	Average	(68,89,90,126,133,134)
2D-PAGE	Glycoform profile, p*I*, and molecular weight	—	—	—	µg	Above average	(126,133)
Western blot	Oligosaccharide presence	—	—	—	ng	Average	Standardized lectins required (95,97,122,135)
FACE	—	—	Oligosaccharide mass and charge	Monosaccharide identification and mass	pmol	Average	Fluorophore required for detection (61,122,124,133,136–138)
Capillary (CE)	Glycoform profile	Glycopeptide profile	Oligosaccharide profile	—	fmol	Above average	Fluorophore improves oligosaccharide detection. Can be coupled with MS (119,127,128,133,134,139–141)
LC							
HPLC	—	Glycopeptide separation	Oligosaccharide separation, profile, sequencing	Monosaccharide separation, identification, and mass	µmol	Above average	Fluorophore required for oligo- and monosaccharide detection (68,74,129,137,130,142,143)
HPAEC-PAD	—	—	Oligosaccharide profile	Monosaccharide identification	pmol	Above average	(52,69,74,75,87,120,126,133,144)

				and mass			
MS							
FAB-MS	—	Glycopeptide mass	Oligosaccharide mass	—	nmol	High	(20,144)
ES-MS	—	Glycopeptide and oligosaccharide mass	Oligosaccharide mass and sequencing	—	fmol	High	Can be coupled with exoglycosidase to sequence oligosaccharides (20,122,125–127,145)
MALDI-MS and MALDI-TOF-MS	Initial location of oligosaccharide attachment	Glycopeptide and oligosaccharide mass	Oligosaccharide mass and sequencing	Monosaccharide anomericity	pmol	High	Can be coupled with exoglycosidase to determine linkages (20,54,122,131,146)
GC-MS	—	—	—	Monosaccharide identification and mass	nmol	High	(147,148)
LC–MS	—	Glycopeptide mass and profile	Oligosaccharide mass and profile	—	μg	High	(149,150)
NMR	—	—	Oligosaccharide linkages and sequencing	Monosaccharide linkage position and anomericity	mg	Very high	Nondestructive (13,52,122,144)
Exoglycosidase digestion	Oligosaccharide linkages and sequencing	Oligosaccharide linkages sequencing	Oligosaccharide linkages and sequencing	—	μg	Average	Can be coupled with SDS-PAGE, HPLC, CE, or MS (122,123,126,151)

Note: For common combinations, the type of information obtained, material analyzed, sample size, and analysis difficulty are presented.

glycoprotein are commonly used to assess the variability in fermentations, and can be used to verify lot-to-lot consistency for a therapeutic glycoprotein (133). Intact glycoprotein can by analyzed directly by methods such as SDS-PAGE, isoelectric focusing polyacrylamide gel electrophoresis (IEF-PAGE), 2D-PAGE, Western blots, capillary electrophoresis (CE), matrix-assisted laser desorption ionization mass spectrometry (MALDI-MS), and exoglycosidase digestion.

Glycopeptide Analysis

Glycopeptides obtained from the proteolytic digestion of a glycoprotein are usually separated into pools with a common peptide backbone. The glycopeptide/peptide profile may also be used to determine lot-to-lot consistency. For a glycoprotein with a known amino acid sequence, the peptide fragments can be predicted. Glycopeptide pools can be separated by charge. The charge is sensitive to the attached oligosaccharide, due to the relative size of the peptide and oligosaccharide. Different oligosaccharides attached to the same peptide can be identified. The glycopeptide fragment mass can be determined and the oligosaccharide mass calculated. The mass of the oligosaccharides attached at a specific asparagine is the most important information obtained from analysis of glycopeptides. However, since many sugars have similar masses, unambiguous oligosaccharide structures cannot always be resolved (121). The glycopeptides can by analyzed by methods such as IEF-PAGE, CE, high-performance liquid chromatography (HPLC), fast-atom bombardment mass spectrometry (FAB-MS), electrospray mass spectrometry (ES-MS), MALDI-MS, liquid chromatography mass spectrometry (LC–MS), and exoglycosidase digestion.

Oligosaccharide Analysis

Oligosaccharides can be obtained from glycoproteins and glycopeptides by either chemical or enzymatic cleavage (121,122,133). Hydrozinolysis is the most common method for cleaving N-and O-linked oligosaccharides nonselectively. Most oligosaccharides can be readily cleaved; however, reducing-end sugar moieties may be modified (18,120). Hydrozinolysis conditions need to be optimized to minimize the degradation of terminal GalNAc in O-linked oligosaccharides (122). *O*-glycosidic linkages can also be cleaved by β-elimination, with or without the reduction of the reducing sugar to an alditol (122). The enzyme *N*-glycosidaseF (PNGaseF) can be used to cleave *N*-oligosaccharides from the glycoprotein or glycopeptide at the *N*-glycosidic bond (52,75,120). Other endoglycosidases can be used to cleave specific bonds in the oligosaccharide or between the oligosaccharide and protein (121,123).

PNGaseF is the most commonly used enzymatic method for cleaving *N*-oligosaccharides. Briefly, in a PNGaseF digestion, the oligosaccharides are obtained from a purified glycoprotein or glycopeptide (20–200 μg) in non-Tris buffer and minimal salt. The glycoprotein or glycopeptide must be dried, by either a centrifugal vacuum evaporator or lyophilization. The glycoprotein or glycopeptide is dissolved in a phosphate buffer at pH 7.5. The sample is boiled for 5 min with SDS ($\approx$0.1% SDS) plus β-mercaptoethanol (50 mM), if needed. The detergent Nonidet P-40 is added to the sample to a final concentration of 0.75%. The enzyme (1 mU PNGaseF) is added to the sample. The digestion or cleavage reaction, at 37°C, takes at least 2 hr. Typically, the reaction is incubated overnight. After release of oligosaccharide, the protein is precipitated with three volumes of ethanol at −20°C. The sample is centrifuged for 5 min at 10,000 rpm. The supernatant is transferred to a clean tube.

The supernatant contains released oligosaccharides. The oligosaccharides can be stored at −70°C until further analysis (139).

Once the oligosaccharides have been liberated from either the glycoprotein or the glycopeptide (Fig. 6), an oligosaccharide profile can be obtained. Oligosaccharides can be analyzed by methods such as FACE, CE, HPLC, FAB-MS, ES-MS, MALDI-MS, LC–MS, NMR, and exoglycosidase digestion. Usually, the oligosaccharides from a protein are pooled for analysis, such that site-specific glycosylation is not determined. If the oligosaccharides are from a glycopeptide containing only one glycosylation site, then site-specific glycosylation can be determined. Oligosaccharides can be analyzed for mass, composition, linkages, and sequence information. Fluorophores are often used to label oligosaccharides to improve detection (124–126).

Monosaccharide Analysis

Monosaccharides can be obtained directly from an intact glycoprotein or glycopeptide pool by acid hydrolysis. Monosaccharides may also be obtained from oligosaccharides using a mild acid hydrolysis reaction (Fig. 6). The information about the monosaccharide composition will be the same for monosaccharides obtained directly from the glycoprotein and the oligosaccharide pool cleaved from that glycoprotein. The same is true for the glycopeptide/oligosaccharide pathway (Fig. 6). Many analytical techniques can determine the monosaccharide, the amount of each monosaccharide, and linkages. In high-resolution analysis techniques, the anomericity of the monosaccharides can also be determined.

Figure 1 shows the common monosaccharides found in protein oligosaccharides, where xylose is the least common. These sugars can be divided into three groups: neutral, amino, and acidic. Acid hydrolysis conditions have been optimized for the cleavage of each monosaccharide group, depending on the source (glycoprotein/glycopeptide or oligosaccharide) (124). For example, to hydrolyze neutral sugars from an intact glycoprotein, 50 μg of dried glycoprotein is incubated in 100 μL 2.0 M trifluoroacetic acid (TFA) for 5 hr at 100°C. After incubation, the samples are dried, and can be stored frozen at −70°C until analyzed. The incubation procedure for acidic sugar release is milder and for amino sugar, harsher. The incubation conditions for oligosaccharide hydrolysis are milder for each sugar type than the glycoprotein incubation conditions (124). The identity, composition, anomericity, and linkage information for the monosaccharides can be determined by FACE, HPLC, MALDI-MS, and NMR.

Electrophoresis Methods

Sodium Dodecyl Sulfate Polyacrylamide Gel Electrophoresis

Electrophoresis-based methods separate the material of interest by electric potential. For SDS-PAGE, the SDS coats a protein with a uniform negative charge, such that the proteins are separated by molecular weight when a voltage is applied. SDS-PAGE is a method accepted by FDA for determining the purity and quantity of therapeutic MAbs, and it is used extensively to compare material from different batches and to monitor quality during a batch (66,69). However, SDS-PAGE alone can only resolve gross changes in macroheterogeneity (18). For a small glycoprotein with few glycosylation sites, glycosylation variants can be detected and quantified, as demonstrated for IFN-γ under glucose-limited conditions (76). SDS-PAGE sensitivity has recently been improved by the commercial availability of various exoglycosidases. Exoglycosidases cleave specific oligosaccharide bonds on a glycoprotein, thus reducing the size of the protein. The spectrum of cleaved glycoproteins, resulting from numerous cleavage steps by individual glycosidase, is analyzed. The degree and type of glycosylation can be determined by comparing the cleaved protein to the original glycoprotein (122,123).

Isoelectric Focusing Polyacrylamide Gel Electrophoresis

Isoelectric focusing gels separate glycoproteins based on the p*I* of the glycoprotein. A pH gradient in the gel is created by ampholyte molecules. The glycoprotein migrates to the region of the gel where the p*I* of the glycoprotein matches the pH in the gel. At this point in the gel, the protein has a neutral charge and thus is immobile (126,133). This method is semiquantitative due to staining limitations. IEF-PAGE can be used to detect gross macroheterogeneity changes in the glycoprotein (18). Microheterogeneity changes are more difficult to detect due to similar p*I* values for most sugars. IEF-PAGE has been used to resolve multiple glycoforms for many MAbs due to macroheterogeneity, and thus can be used to compare glycosylation patterns resulting from different culturing methods (68,133). IEF-PAGE is used to verify lot-to-lot consistence for licensed MAbs. Additionally, IEF-PAGE has also been used to demonstrate differences in a glycoprotein produced in media at different pH values (89,90). IEF-PAGE has also been used to separate EPO glycoforms (134). The capability of IEF-PAGE to separate glycoforms by charge is important for therapeutic glycoproteins like EPO, where the sialic acid content correlates with efficacy. IEF-PAGE will separate glycoforms with different numbers of terminal sialic acid (133,134).

Two-Dimensional Polyacrylamide Gel Electrophoresis

2D-PAGE is a combination of SDS-PAGE and IEF-PAGE. The glycoprotein is separated by p*I* and molecular weight. Each glycoprotein is visualized as a series of spots at approximately the same size for the different glycoforms corresponding to different p*I* values. Since each glycoform is separated, 2D-PAGE can be coupled with MS to determine the glycoform associated with each spot. 2D-PAGE has become more reproducible lately due to the use of immobilized ampholytes (126). 2D-PAGE could be used to measure a recombinant protein during production for posttranslational changes (133).

Immunoblotting

A sensitive method for verifying glycosylation is to use immunolabels. Immunolabeling can be conducted with dot blots or Western blots. For immunoblotting (Western blotting), the glycoforms are separated by SDS-PAGE. This increases the sensitivity and decreases the amount of glycoprotein sample required. The separated glycoforms are transferred to a solid support, either polyvinlyidene difluoride or nitrocellulose. For dot blots, the glycoprotein is directly bound to the solid support. The hydroxyl groups of the monosaccharides are oxidized by mild periodate and derivatized to aldehyde groups with biotin or digoxigenin. A streptavidin–alkaline phosphatase conjugate can be used with biotin to produce color when an oligosaccharide group is present. Digoxgenin-labeled oligosaccharides are detected by an antidigoxgenin antibody–alkaline phosphatase conjugate. The combination of SDS-PAGE, exoglycosidase, and Western blot analysis techniques has been used successfully to characterize glycoforms of recombinant EPO (95,97,135).

Fluorophore-Assisted Carbohydrate Electrophoresis

Another electrophoresis-based technique is FACE. Oligosaccharides and monosaccharides can be analyzed by this method. FACE is relatively simple and inexpensive. The glycoprotein can be separated by SDS-PAGE from the other extracellular proteins to purify the glycoprotein. The glycoprotein of interest can be eluded from the gel. The glycoprotein can be digested to cleave the oligosaccharides or hydrolyzed to release the monosaccharides. The oligosaccharides or monosaccharides are labeled with 8-aminonaphthalene-1,3,6-trisulfonate (ANTS). ANTS imparts a negative charge to the labeled carbohydrates, thus the labeled oligosaccharides or monosaccharides are separated by mass on polyacrylamide gels for quantification (104,122,136). Other fluorophores can be used to label oligosaccharides and monosaccharides to be separated by electrophoresis. Two common fluorophores are 2-amino acridone and 2-aminoanthranilic acid. By varying the buffer, the oligosaccharides and monosaccharides can be separated by charge, which can be used to detect sialic acid substitutions (122). FACE allows for direct comparison of samples for quantification, since multiple samples can be run on a single gel. FACE is also compatible with glycosidase sequencing. Oligosaccharide can be cleaved by glycosidases. The oligosaccharides (cleaved and uncleaved) can be run on a single gel for easy comparison. FACE is also relatively tolerant of salts and detergents. This method can be used to monitor glycosylation changes during a fermentation due to bioprocess changes (133). Yang and Butler (61,62) used FACE to determine that increased ammonia concentrations resulted in a reduction in the number of sialic acid residues on EPO produced in CHO cells.

Capillary Electrophoresis

A newer technique for detecting glycoform heterogeneity is CE. This technique can separate glycoproteins, glycopeptides, or oligosaccharides. Very narrow capillaries are subjected to high voltage, which separate the glycoproteins, glycopeptides, or oligosaccharides by either molecular weight or charge. UV peak detection can be used to quantify the glycoproteins, glycopeptides, or oligosaccharides. CE is faster than SDS-PAGE or IEF-PAGE. In addition, smaller samples can be used compared to SDS-PAGE and IEF-PAGE (18). Typically, fused silica capillaries of 10–100 μm internal diameter are used. The effective lengths are 10–100 cm. Electric field strengths as high as 1000 V/cm are used. CE can be coupled to MS after desalting the eluted sample (127,133). Kakehi et al. (128) reviewed the application of CE to glycoproteins. Taverna et al. (133) provided a good review on the application of CE for recombinant glycoproteins and Kakehi et al. (140) reviewed the application of CE to bioprocess monitoring.

The analysis of glycoproteins by CE is possible due to the amide and aromatic amino acids in the glycoprotein adsorption of UV. Most glycoprotein separations are based on the number of sialic acids due to the inherent negative charge of sialic acid (127,128). Due to the complex interactions of the glycoprotein with the capillary wall, each glycoprotein separation needs to be optimized to improve resolution by changing the pH, buffer type, and organic modifiers (128). The major difficulty in the analysis of glycoproteins by CE is sample loss, which leads to decreased separation efficiency and poor resolution. Sample loss is due to the adhesion of the glycoprotein to the capillary wall. Strategies have been developed to minimize sample loss. One strategy is to lower the pH, which reduces the charge on the capillary wall, and thus the affinity of the glycoprotein to the capillary wall. Working below the p*I* of the glycoprotein has the same effect. A second strategy is to add components to the separation buffer that compete for the charged silanols on the capillary walls. The third strategy is to chemically modify the capillary wall (127). Once CE conditions are optimized for a particular glycoprotein, CE analysis is fast and reproducible. For example, Watson and Yao (134) were able to resolve EPO and GM-CSF by CE because of the different numbers of terminal sialic acids. Goldman et al. (119) used CE to rapidly determine the proportion of IFN-γ glycosylated during a 500 hr fermentation.

For glycopeptides, CE is able to separate tryptic digests of the glycoprotein into nonglycosylated and glycosylated species. The nonglycosylated peptides have shorter retention times. From the predicted peptide map, a one-to-one correspondence between the peaks and the predicted peptide entity can be determined. For example, tryptic digested EPO results in three glycopeptides, which produced 11 well-resolved peaks by CE. These glycopeptide peaks are due to the microheterogeneity of the oligosaccharides. Macroheterogeneity can also be determined from the relative amounts in the corresponding nonglycosylated peak (128). Thus, tryptic digests of glycoproteins could be used to monitor product quality during a fermentation, purification process, or stability study (127). Wu (141) used CE to monitor tPA glycopeptides, where sialic acid variants were identified.

Oligosaccharides can be detected directly by UV or after precolumn derivatization. Direct detection with UV is commonly conducted at 200 nm; however, recent research has shown that the resolution is better at 185 nm. The oligosaccharide profiles obtained by CE at 185 nm are similar to the oligosaccharide profiles obtained by HPAEC-PAD (128). To improve detection, the oligosaccharides can be labeled with trisodium 8-amino-1,3,6-pyrene trisulfonate (APTS), ANTS, or 2-aminopyridine. Sialic acid detection can be improved by labeling with 4,5-dintrocatechol-*O,O*-diactic acid (152). Laser-induced fluorescence can also be used to improve detection (139).

Liquid Chromatography

LC methods can distinguish individual oligosaccharides and quantify the individual oligosaccharides. There are two approaches for preparing the glycoprotein prior to the chromatography step. In one approach, the glycoprotein is cleaved into glycopeptides with a peptidase, and in the other, the oligosaccharides are cleaved from the glycoprotein. For the first method, the glycopeptides are separated by reverse-phase high-performance liquid chromatography (RPHPLC). Peptidases are selected such that only one oligosaccharide is associated with any peptide. This analysis technique requires that the amino acid sequence is known in order to predict the resulting peptides. Trypsin is the most common peptidase used due its broad specificity (120). Researchers who initially characterized recombinant β-trace protein used trypsin glycopeptides (52). The effect of different culture conditions on glycosylation have also been investigated using trypsin digested-glycopeptides to characterize the *N*-oligosaccharides (75).

Another very sensitive LC method uses Biogel 4 and radiolabeled glycoproteins that have been degraded by either hydrozinolysis or PNGaseF. Biogel 4 and hydrozinolysis were used to characterize the glycosylation of IgM obtained from cells with different physiological states: ascites, airlift fermenters, and human blood. The IgM obtained from the different culturing techniquess had unique glycosylation patterns (68).

LC coupled to MS (LC–MS) can provide detailed glycosylation by analyzing the glycopeptides or oligosaccharides (149,150). For glycopeptides, LC is used to separate the glycopeptides in a proteolytic digestion, and the glycosylation state at each site is determined by MS. Ohta et al. (149) were able to detect site-specific glycosylation differences in EPO obtained from three different sources. Additionally, LC–MS detected acetylation, sialylation, and sulfation differences in the three EPO sources. Oligosaccharide mapping by LC–MS can also determine carbohydrate heterogeneity. Kawasaki et al. (150) used LC–MS to compare the carbohydrate characteristic of EPO obtained from different sources.

HPLC Methods

HPLC methods are commonly used to characterize the glycosylation of a glycoprotein. Usually, the glycopeptides, oligosaccharides, or monosaccharides are analyzed. Anion-exchange HPLC separates charged groups, such as sialic acid; however, this method requires labeling and often the resolution is poor (129). Weak anion-exchange HPLC was used to demonstrate that the sialic acid content of EPO produced in CHO was reduced in the presence of elevated ammonia concentration (137). Labeled monosaccharide can be separated by RPHPLC with excellent resolution. Labeled oligosaccharides can also be detected by RPHPLC. One advantage of RPHPLC is that the buffers are salt-free, which is more compatible with MS systems. Typical labeling agents are anthranilic acid (ABA) for monosaccharides, phenylisothiocyanate (PITC) for amino sugar alcohols, and *o*-phenylene diamine (OPD) for sialic acids. ABA and OPD are detected by fluorescence, whereas PITC is detected by UV absorption. A monosaccharide sample is labeled separately with each dye to detect the different sugars. Typically, 5–50 μg of glycoprotein are hydrolyzed in 20% TFA at 100°C for 6–7 hr to obtain the monosaccharides (129,142).

Xu and Cacia (130) developed an RPHPLC method to monitor the manufacturing process development of TNK-tPA (a genetically engineered variant of tPA). The RPHPLC method was able to identity and quantify the tPA glycoforms. Wu (141) used RPHPLC to separate tPA glycopeptides and peptides prior to CE. Thus, macroheterogeneity and microheterogeneity were determined. Wan et al. (143) developed an HPLC–ES-MS method for the direct analysis of the terminal Gal content of a glycoprotein. Cell culture media were directly injected into the HPLC followed by the ES-MS. The results obtained were comparable with the CE methods used to analyze the purified glycoprotein.

High-pH Anion-Exchange Chromatography with Pulsed Amperometric Detection

Oligosaccharides and monosaccharides can be analyzed by HPAEC-PAD. Monosaccharides are obtained from the glycoprotein or glycopeptide by acid hydrolysis. Oligosaccharides are cleaved from the glycoprotein or a glycopeptide pool either enzymatically or chemically (18). The enzyme PNGaseF cleaves *N*-oligosaccharides from the glycoprotein or glycopeptide at the *N*-glycosidic bond (52,74,120). Hydrazinolysis can also be used to cleave both *N*- and *O*-oligosaccharides from the glycoprotein or glycopeptide (18,120). Monosaccharides may also be obtained by acid hydrolysis of the oligosaccharides.

Once the oligosaccharides have been liberated from either the glycoprotein or the glycopeptide, HPAEC-PAD can be used to rapidly and sensitively determine and quantify the glycosylation pattern of the glycoproteins, often termed the oligosaccharide profile (18,87,120,144). HPAEC-PAD exploits the electrochemical properties of carbohydrate; however, every sugar electrochemical response is different. To quantify oligosaccharides by HPAEC-PAD, an identical external standard must be used to calibrate the peak area. This method has been used detect both macroheterogeneity and microheterogeneity. Monosaccharides can be separated and quantified by HPAEC-PAD. Monosaccharide standards are required to quantify the monosaccharides for all anticipated sugars, since each monosaccharide peak height has a different proportionality based on mass.

Gawlitzek et al. (74,75) used HPAEC-PAD to demonstrate the effect of different culture conditions on the N-glycosylation of recombinant IL-2. Robinson et al. (69) and Patel et al. (66) both observed N-glycosylation pattern changes for an MAb production with different culture methods and/or conditions. Andersen et al. (87) demonstrated that HPAEC-PAD was also a sensitive method for determining O-glycosylation. They also were able to correlate increased ammonium ion concentration with a decrease in the proportion of α(2,6)-linked sialic acid, compared to terminal GalNAc for GM-CSF secreted by CHO cells (87). Kunkel et al. (104) demonstrated decreased the galactose content of the oligosaccharides by HPAEC-PAD.

Mass Spectrometry

MS can be used to determine the primary structure of a glycoprotein. Structural characteristics that can be obtained are branching, linkages, configuration, and identification of isomer sugars. MS requires that the sample be volatilized. Electron impact and chemical ionization are both used. Permethylation (chemical ionization) can be used to release the oligosaccharides by β-elimination. Soft ionization can be obtained by electron impact, which then allows for the direct ionization and desorption analysis of oligosaccharides, polypeptides, glycopeptides, and intact glycoproteins (20). Additionally, gas chromatography can be coupled to electron impact MS to determine composition and linkages. Biochemical methods, such as exoglycosidases, can be used to aid in the definition of terminal sequences (20). LC coupled to MS can also improve oligosaccharide analysis of glycoproteins and glycopeptides (54,149,150). Additionally, tandem MS–MS methods have been develop to further increase the sensitivity of MS techniques (121).

Fast-Atom Bombardment Mass Spectrometry

The first form of electron impact MS developed was FAB-MS developed in the 1970s. FAB-MS has the highest mass accuracy of the MS techniques and is sensitive enough to detect the mass change due to an asparagine being changed to an aspartate, a 1 g/mol change (18); however, the molecular weight maximum is only 15 kDa (1). For FAB-MS, the sample is dissolved in a matrix containing either glycerol or thioglycerol. An accelerated beam of atoms is fired at the sample. When the accelerated atoms or ions collide with the matrix, the sample is ejected into a high-vacuum source of the ion source. The sample molecules are ionized producing a positively charged species called quasimolecular ions (20). In a detailed study to determine the oligosaccharides of EPO produced in BHK-21 cells (EPO is commercially produced in CHO by Amgen), FAB-MS was used to identify and quantify 30 different oligosaccharides from three N-glycosylation and one O-glycosylation sites (144). FAB-MS is particular good for providing composition and sequence information.

Electrospray Mass Spectrometry

A less accurate and less expensive MS system is ES-MS. For ES-MS, the sample is contained in a stream of liquid that is injected into an atmospheric-pressure ion source of an MS by a capillary. The capillary is subjected to a high voltage that causes an aerosol of charged droplets to enter the MS. ES-MS is gentle, such that almost no ionization occurs and therefore very little fragmentation. Fragmentation of the glycoprotein can be induced in the ES-MS system by collisional activation, which is then detected by a second MS (also called ES-MS–MS). ES-MS can be directly coupled to a CE or HPLC system to separate the glycoforms, and then determine masses accurately to within 0.1% for relatively small molecules (<100 kDa) (1,20,145). Exoglycosidases can be used to enhance the structural detail for linkage positions and anomericity of the monosaccharides in the oligosaccharides (122).

Matrix-Assisted Laser Desorption Ionization Mass Spectrometry

MALDI-MS, also called laser desorption mass spectrometry (LD-MS), is less expensive than ES-MS. MALDI-MS was developed in the 1980s to analyzed large molecules (< 300 kDa) (1,131). Most MALDI-MS instruments use time-of-flight (TOF) to determine the mass to charge (m/z) ratio. MALDI-TOF-MS is 10–100 times more sensitive than FAB for underivatized oligosaccharides and intact glycoproteins, although FAB-MS is more sensitive for derivatized compounds (131,132). Also, MALDI-MS is more tolerant of salts that either FAB or ES. MALDI-MS has a higher mass range than ES-MS or FAB-MS; however, a glycoprotein with a lot of heterogeneity can result in a peak with a peak width of $\approx$8% (131).

For MALDI-TOF-MS, the sample is embedded in a low-molecular weight UV-absorbing matrix, and the sample is ionized by a pulsed laser. Very few fragments result, and thus MALDI-MS can be used for screening molecular ions for high throughput and high sensitivity (20,122). MALDI can be used to determine the molecular mass of an oligosaccharide, but not the exact linkages. For example, the number of Hex (Glc, Gal, Man, or Xyl) and HexNAc (GlcNAc or GalNAc) residues can be determined, but not exactly which Hex is present (122), Sialic acid is cleaved during the sample preparation, and therefore cannot be detected (122). Exoglycosidases can be used to obtain structural detail regarding the linkage position and anomericity of the monosaccharides in the oligosaccharides (122,131). Therefore, MALDI-MS is often coupled with other analytical techniques. For example, first the oligosaccharides are released from an acid denatured protease digested glycoprotein. The peptides are removed by RPHPLC and the oligosaccharides are desalted using size exclusion chromatography. The oligosaccharides are analyzed by HPAEC-PAD and the monosaccharide content determined by HPAEC-PAD, gas chromatography mass spectrometry (GC-MS), or FACE. Then, the oligosaccharides are analyzed by MALDI-MS (132). For example, the absence of GalNAc can be used to exclude O-glycosylation (146). Harvey (131) provides a comprehensive review of MALDI-MS applications and procedures. LD-MS was used to quantify the various *N*-oligosaccharide structures at the two N-glycosylation sites for recombinant IFN-γ produced by CHO cells, where temperature, pH, and DO were held constant. (70).

Gas Chromatography Mass Spectrometry

GC-MS can be used to identify and determine the composition of monosaccharides. Prior to analysis by GC-MS, the monosaccharide must be cleaved from the glycoprotein, glycopeptide, or oligosaccharide into its components. Acid hydrolysis, methanolysis, and solvolysis by hydrogen fluoride can be used to obtain free monosaccharides. Since most monosaccharides are nonvolatile, derivatization is required (147). Glycosidic bonds have different rates of cleavage. This property can be used to selectively cleave the neutral, amino, and acidic sugars as discussed previously for FACE monosaccharide analysis. GC-MS can also be used to determine the structure of an oligosaccharide up to 10 residues (148).

Each of the cleavage techniques has some limitation, which will be briefly discussed here. Merkle and Poppe (147) have provided a detailed review of GC-MS techniques and its limitations. Acid hydrolysis can result in the loss of the *N*-acetyl group from amino sugars. Re-N-acetylation is required before the sample is derivatized. Methanolysis does not cleave the GlcNAc attached to the asparagine efficiently, and when released the GlcNAc is not a methlyglycoside like the other cleaved monosaccharides. Solvolysis cleaves most glycosidic linkages without changing the *N*-acetyl groups. There is very little degradation of the released monosaccharides, so the monosaccharides maybe derivatized immediately. However, the HF used in solvolysis requires special handling and precautions (147). Derivatization of the monosaccharide results in alditol acetates or trimethylsilylation. The mass spectra fragmentation patterns are used to identify the monosaccharide. The peak heights are used to calculate the amount of each monosaccharide (147).

Nuclear Magnetic Resonance

NMR is a powerful tool for analyzing unknown configurations. Bonding can be determined and the conformation of an oligosaccharide precisely determined for an unknown species. This is very useful when many oligosaccharide configurations (microheterogeneity) can exist at a single site. NMR is also nondestructive, so samples are still available for further analysis. For example, in the β-trace protein, the Asn sites 25 and 97, depending on culture age, had up to 15 and 19 different oligosaccharide configurations, respectively, all of which were distinguished by NMR (52). NMR can determine the structure of the monomer, anomericity, the linkage between residues, and the nature and degree of substitutions (i.e., P, SO_4^-) (9,13,122). Nimtz et al. (144) were able to demonstrate that NMR provides structural detail, while HPAEC-PAD provides rapid sensitive analysis for batch comparisons. NMR is not a tool appropriate for production monitoring, since very large pure samples are required. However, NMR can be a critical analytical method for drug development in determining glycoprotein structures.

Exoglycosidase Digestion

Exoglycosidases are to carbohydrates, what restriction enzymes are to DNA. Exoglycosidases cleave very specifically the terminal monosaccharide on a glycoprotein, glycopeptide, or oligosaccharide. The exoglycosidase can be specific to a sugar or glycosidic linkage (153). Exoglycosidases can be used to determine the identity, and absolute and anomeric configuration, and with some enzymes, the linkages position of the oligosaccharide (126). Thus, oligosaccharides can be sequenced by sequentially removing sugars. The remaining oligosaccharide can be labeled with 7-amino-4-methycoumarin and separated by thin-layer chromatography or electrophoresis (153). For example, β-galactosidase from *Diplococcus pneumoniae* will recognize Galβ(1,4)-GlcNAc, but not Galβ(1,3)-GlcNAc (122). Exoglycosidases can be used with a colorimetric assay to monitor the Gal content of a recombinant protein for batch-to-batch consistency and to optimize the production pH, time, and temperature (151). There are well over 25 exoglycosidases commercially available that can be used to characterize a glycoprotein. O'Neill (123) provides a detailed review of exoglycosidases, including the history, characterization, and specificity.

CONCLUSIONS

Protein glycosylation is the most common posttranslational modification in eukaryotic cells. One-half to one percent of the genome appears to be involved in glycosylation, on the same order of magnitude as protein phosphorylation. There are two main types of glycosylation: N-linked oligosaccharides and O-linked oligosaccharides. All eukaryotic cells glycosylate proteins via common pathways; however, evolution has imparted many variations. The N-glycosylation pathway has three steps: the dolichol cycle and oligosaccharide transfer to the protein, the trimming reactions where glucose and mannose residues are removed, and the branching reactions that construct the oligosaccharide antennae. O-glycosylation, unlike N-glycosylation, is not generated by the transfer of a lipid-sugar onto the protein, but by the direct, sequential transfer of monosaccharides onto the protein. The glycosylation requirements of a recombinant glycoprotein are completely dependent on efficacy in the patient. Glycosylation is now recognized as a necessary component for many therapeutic enzymes, antibodies, and hormones. It is also recognized that many environmental factors affect glycosylation in cell culture. Recombinant glycoproteins can be produced in yeast, fungi, plants, insect cells, transgenic animals, and mammalian cells. However; to date, the only licensed therapeutic glycoproteins are produced by *S. cerevisiae* and mammalian cell culture.

Early therapeutic glycoprotein development was hindered by the lack of analytical techniques to measure glycosylation and an incomplete understanding of the cellular processes that controlled glycosylation. More ground is now being covered with respect to the cellular processes involved in glycosylation; however, there is still much work to be done. There are now many analytical techniques available to determine and quantify protein glycosylation. These analytical techniques are highly sensitive and able to detect and separate large molecules containing very small differences. The analytical techniques that are commonly used to quantify glycoproteins are electrophoresis, LC, MS, NMR, and exoglycosidase digestion. To fully characterize a glycoprotein, at least two or more orthogonal analytical techniques need to be used.

REFERENCES

1. James DC, Baker KN. Glycosylation of recombinant proteins. In: Drew SW, ed. Encyclopedia of Bioprocess Technology: Fermentation, Biocatalyst, and Bioseparation. New York: John Wiley & Sons, 1999:1336–1349.
2. Kronman C, Chitlaru T, Elhanany E, Velan B, Shafferman A. Hierarchy of post-translational modifications involved in the circulatory longevity of glycoproteins—demonstration

of concerted contributions of glycan sialylation and subunit assembly to the pharmacokinetic behavior of bovine acetylcholinesterase. J Biol Chem 2000; 275:29488–29502.
3. Jenkins N, Castro P, Menon S, Ison A, Bull A. Effect of lipid supplements on the production and glycosylation of recombinant interferon-γ expressed in CHO cells. Cytotechnology 1994; 15:209–215.
4. Geisow MJ. Glycoprotein glycans—roles and controls. Trends Biotechnol 1992; 10:333–335.
5. Liu DTY. Glycoprotein pharmaceuticals—scientific and regluatory considerations, and the United States Orphan Drug Act. Trends Biotechnol 1992; 10:114–120.
6. Bill RM, Revers L, Wilson IBH. Protein Glycosylation. Boston, MA: Kluwer Academic Publishers, 1998.
7. Goochee CF, Gramer MJ, Andersen DC, Bahr JB, Rasmussen JR. The oligosaccharides of glycoproteins: bioprocess factors affecting oligosaccharide structure and their effect on glycoprotein properties. Bio/Technology 1991; 9:1347–1355.
8. Jenkins N, Parekh RB, James DC. Getting the glycosylation right: implications for the biotechnology industry. Nat Biotechnol 1996; 14:975–981.
9. Lis H, Sharon N. Protein glycosylation: structural and functional aspects. Eur J Biochem 1993; 218:1–27.
10. Reuter G, Gabius HJ. Eukaryotic glycosylation: whim of nature or multipurpose tool? Cell Mol Life Sci 1999; 55:368–422.
11. Helenius A, Aebi M. Intracellular functions of N-linked glycans. Science 2001; 291:2364–2369.
12. Ashford DA, Alafi CD, Gamble VM, Mackay DJG, Rademacher TW, Williams PJ, Dwek RA, Barclay AN, Davis SJ, Somoza C, Ward HA, Williams AF. Site-specific glycosylation of recombinant rat and human soluble CD4 variants expressed in Chinese hamster ovary cells. J Biol Chem 1993; 268:3260–3267.
13. Lustbader JW, Birken S, Pollak S, Chait BT, Mirza UA, Ramnarain S, Canfield RE, Brown JM. Expression of human chorionic gonadotropin uniformly labeled with NMR isotopes in Chinese hamster ovary cells: An advance toward rapid determination of glycoprotein structures. J Biomol NMR 1996; 7:295–304.
14. Comer FL, Hart GW. O-GlcNAc and the control of gene expression. Biochim Biophys Acta 1999; 1473:161–171.
15. Lisowska E. The role of glycosylation in protein antigenic properties. Cell Mol Life Sci 2002; 59:445–455.
16. Prati EGP, Matasci M, Suter TB, Dinter A, Sburlati AR, Bailey JE. Engineering of coordinated up- and down-regulation of two glycosyltransferases of the O-glycosylation pathway in Chinese hamster ovary (CHO) cells. Biotechnol Bioeng 2000; 68:239–244.
17. Joziasse DH, Oriol R. Xenotransplantation: the importance of the Galαl, 3Gal epitope in hyperacute vascular rejection. Biochim Biophys Acta-Mol Basis Dis 1999; 1455: 403–418.
18. Jenkins N, Curling EMA. Glycosylation of recombinant proteins: problems and prospects. Enzyme Microb Technol 1994; 16:354–364.
19. Bulleid NJ, Bassel-Duby RS, Freedman RB, Sambrook JF, Gething M-JH. Cell-free synthesis of enzymically active tissue-type plasminogen activator. Biochem J 1992; 286:275–280.
20. Dell A, Morris HR. Glycoprotein structure determination by mass spectrometry. Science 2001; 291:2351–2356.
21. Dennis JW, Granovsky M, Warren CE. Protein glycosylation in development and disease. BioEssays 1999; 21:412–421.
22. Gerken TA, Owens CL, Pasumarthy M. Effects of O-glycosylation on protein structure. In: Arland T, Hotchkiss J, eds. Techniques in Glycobiology. New York: Marcel Dekker, 1997:247–269.
23. Jentoft N. Why are proteins O-glycosylated? Trends Biochem Sci 1990; 15:291–294.

24. Kaufman RJ. Secretion from animal cells. In: Drew SW, eds. Encyclopedia of Bioprocess Technology: Fermentation, Biocatalyst, and Bioseparation. New York: John Wiley & Sons, 1999:2366–2378.
25. Helenius J, Aebi M. Transmembrane movement of dolichol linked carbohydrates during N-glycoprotein biosynthesis in the endoplasmic reticulum 2002; 13:171–178.
26. Messner P. Bacterial glycoproteins. Glycoconjugate Journal 1997; 14:3–11.
27. Kalsner I, Schneider F-J, Geyer R, Ahorn H, Mauer-Fogy I. Comparison of the carbohydrate moieties of recombinant soluble Fc_{ε} receptor ($sFc_{\varepsilon}RII$/sCD23) expressed in *Saccaromyces cerevisiae* and Chinese hamster ovary cells. Different O-glycosylation sites are used by yeast and mammalian cells. Glycoconjug J 1992; 9:209–216.
28. Yang Y, Bergmann C, Benen J, Orlando R. Identification of the glycosylation site and glycan structures of recombinant endopolygalacturonase II by mass spectrometry. Rapid Commun Mass Spectrom 1997; 11:1257–1262.
29. Walsh G. Biopharmaceutical benchmarks. Nat Biotechnol 2000; 18:831–833.
30. Zhu A, Wang Z-K, Beavis R. Structural studies of α-N-acetylgalactosaminidase: effect of glycosylation on the level of expression, secretion efficiency, and enzyme activity. Arch Biochem Biophys 1998; 352:1–8.
31. Miele RG, Nilsen SL, Brito T, Bretthauer RK, Castellino FJ. Glycosylation properties of the *Pichia pastoris* expressed recombinant kringle-2 domain of tissue-type plasminogen activator. Biotechnol Appl Biochem 1997; 25:151–157.
32. Murphy KP, Gagne P, Pazmany C, Moody MD. Expression of human interleukin-17 in *Pichia pastoris*: purification and characterization. Protein Expr Purif 1998; 12:208–214.
33. Cereghino GPL, Cereghino JL, Ilgen C, Cregg JM. Production of recombinant proteins in fermenter cultures of the yeast *Pichia pastoris*. Current Opinion, in Biotechnology 2002; 13:329–332.
34. Altmann F, Staudacher E, Wilson IBH, Marz L. Insect cells as hosts for the expression of recombinant glycoproteins. Glycoconjug J 1999; 16:109–123.
35. Hooker AD, James DC. Monitoring recombinant glycoprotein heterogeneity. In: Jenkins N, ed. Animal Cell Biotechnology. Vol. 8. Totowa, NJ: Humana Press Inc., 1999:289–298.
36. Marchal I, Jarvis DL, Cacan R, Verbert A. Glycoproteins from insect cells: sialylated or not? Biol Chem 2001; 382:151–159.
37. Chen Z-Y, Sun J-X, Li J-H, He C, Lu C-L, Wu X-F. Preparation of recombinant human GDNF by baculovirus expression system and analysis of its biological activities. Biochem Biophys Res Commun 2000; 273:902–906.
38. Zhang FM, Murhammer DW, Linhardt RJ. Enzyme kinetics and glycan structural characterization of secreted alkaline phosphatase prepared using the baculovirus expression vector system. Appl Biochem Biotechnol 2002; 101:197–210.
39. Ailor E, Takahashi N, Tsukamoto Y, Masuda K, Rahman BA, Jarvis DL, Lee YC, Betenbaugh MJ. N-Glycan patterns of human transferrin produced in *Trichoplusia ni* insect cells: effects of mammalian galactosyltransferase. Glycobiology 2000; 10:837–847.
40. Kusnadi AR, Nikolov ZL, Howard JA. Production of recombinant proteins in transgenic plants: practical considerations. Biotechnol Bioeng 1997; 56:473–484.
41. Reuter G, Gabius H-J. Sialic acids: structure—analysis—metabolism—occurence—recognition. Biol Chem 1996; 377:325–342.
42. Ma JKC, Hein MB. Immunotherapeutic potential of antibodies produced in plants. Trends Biotechnol 1995; 13:522–527.
43. Bardor M, Faye L, Lerouge P. Analysis of the N-glycosylation of recombinant glycoproteins produced in transgenic plants. Trends Plant Sci 1999; 4:376–380.
44. Kemp PA, Jenkins N, Clark AJ, Freedman RB. The glycosylation of human recombinant alpha-1-antitrypsin expressed in transgenic mice. Biochem Soc Trans 1996; 24:S339–S339.

45. Ko JH, Lee CS, Kim KH, Pang MG, Koo JS, Fang NZ, Koo DB, Oh KB, Youn WS, Zheng GD, Park JS, Kim SJ, Han YM, Choi IY, Lim J, Shin ST, Jin SW, Lee KK, Yoo OJ. Production of biologically active human granulocyte colony stimulating factor in the milk of transgenic goat. Transgenic Res 2000; 9:215–222.
46. McClain DA, Alexander T, Cooksey RC, Considine RV. Hexosamines stimulate leptin production in transgenic mice. Endocrinology 2000; 141:1999–2002.
47. Coulibaly S, Besenfelder U, Miller I, Zinovieva N, Lassnig C, Kotler T, Jameson JL, Gemeiner M, Muller M, Brem G. Expression and characterization of functional recombinant bovine follicle-stimulating hormone (boFSH alpha/beta) produced in the milk of transgenic rabbits. Molecular Reproduction and Development 2002; 63:300–308.
48. Rohricht P. Transgenic protein production: part 2: process economics. BioPharm 1999:52–54.
49. www.fda.gov/appr2000. January 12, 2003.
50. www.fda.gov/appr2001. January 12, 2003.
51. www.fda.gov/appr2002. January 12, 2003.
52. Grabenhorst E, Hoffmann A, Nimtz M, Zettlmeissl G, Conradt HS. Construction of stable BHK-21 cells coexpressing human secretory glycoproteins human Gal(β1-4)GlcNAc-α2,6-sialyltransferase. Eur J Biochem 1995; 232:718–725.
53. Baker KN, Rendall MH, Hills AE, Hoare M, Freedman RB, James DC. Metabolic control of recombinant protein N-glycan processing in NS0 and CHO cells. Biotechnol Bioeng 2001; 73:188–202.
54. Sheeley DM, Merrill BM, Taylor LCE. Characterization of monoclonal antibody glycosylation: comparison of expression systems and identification of terminal α-linked Galactose. Anal Biochem 1997; 247:102–110.
55. Lamotte D, Buckberry L, Monaco L, Soria M, Jenkins N, Engasser J-M, Marc A. Na-butyrate increases the production and α2,6-sialylation of recombinant interferon-γ expressed by α2,6-sialyltransferase engineered CHO cells. Cytotechnology 1999; 29:55–64.
56. Grabenhorst E, Schlenke P, Pohl S, Nimtz M, Conradt HS. Genetic engineering of recombinant glycoproteins and the glycosylation pathway in mammalian host cells. Glycoconjug J 1999; 16:81–97.
57. Monaco L, Marc N, Eon-Duval A, Acerbis G, Distefano G, Lamotte D, Engasser J-M, Soria M, Jenkins N. Genetic engineering of α2,6-sialyltransferase in recombinant CHO cells and its effects on the sialylation of recombinant interferon-γ. Cytotechnology 1996; 22.
58. Umaña P, JeanMairet J, Bailey JE. Tetracycline-regulated overexpression of glycosyltransferases in Chinese hamster ovary cells. Biotechnol Bioeng 1999; 65:542–549.
59. Weikert S, Papac D, Briggs J, Cowfer D, Tom S, Gawlitzek M, Lofgren J, Mehta S, Chisholm V, Modi N, Eppler S, Carroll K, Chamow S, Peers D, Berman P, Krummen L. Engineering Chinese hamster ovary cells to maximize sialic acid content of recombinant glycoproteins. Nat Biotechnol 1999; 17:1116–1121.
60. Yang M, Butler M. Effects of ammonia and glucosamine on the heterogeneity of erythropoietin glycoforms. Biotechnol Prog 2002; 18:129–138.
61. Yang M, Butler M. Effect of ammonia on CHO cell growth, erythropoietin production and glycosylation. Biotechnol Bioeng 2000; 68:370–380.
62. Yang M, Butler M. Effect of ammonia on the glycosylation of human recombinant erythropoietin in culture. Biotechnol Prog 2000; 16:751–759.
63. Hills AE, Patel A, Boyd P, James DC. Metabolic control of recombinant monoclonal antibody N-glycosylation in GS-NSO cells. Biotechnol Bioeng 2001; 75:239–249.
64. Chuppa S, Tsai Y-S, Yoon S, Shackleford S, Rozales C, Bhat R, Tsay G, Matanguihan C, Konstantinov K, Naveh D. Fermentor temperature as a tool for control of high-density perfusion cultures of mammalian cells. Biotechnol Bioeng 1997; 55:328–338.

65. Kaufmann H, Mazur X, Fussenegger M, Bailey J. Influence of low temperature on productivity, proteome and protein phosphorylation of CHO cells. Biotechnol Bioeng 1999; 63:573–582.
66. Patel TP, Parekh RG, Poellering BJ, Prior CP. Different culture methods lead to differences in glycosylation of a murine IgG monoclonal antibody. Biochem J 1992; 285:839–845.
67. Hahn TJ, Goochee CF. Growth-associated glycosylation of transferrin secreted by HepG2 cells. J Biol Chem 1992; 267:23982–23987.
68. Maiorella BL, Winkelhake J, Young J, Moyer B, Bauer R, Hora M, Andya J, Thomson J, Patel T, Parekh R. Effect of culture conditions on IgM antibody structure, pharmacokinetics and activity. Biotechnology 1993; 11:387–392.
69. Robinson DK, Chan CP, Ip CY, Tsai PK, Tung J, Seamans TC, Lenny AB, Lee DK, Irwin J, Silberklang M. Characterization of a recombinant antibody produced in the course of a high yield fed-batch process. Biotechnol Bioeng 1994; 44:727–735.
70. Hooker AD, Goldman MH, Markham NH, James DC, Ison AP, Bull AT, Strange PG, Salmon I, Baines AJ, Jenkins N. N-Glycans of recombinant human interferon-γ change during batch culture of Chinese hamster ovary cells. Biotechnol Bioeng 1995; 48:639–648.
71. Hooker AD, Green NH, Baines AJ, Bull AT, Jenkins N, Strange PG, James DC. Constraints on the transport and glycosylation of recombinant IFN-γ in Chinese hamster ovary and insect cells. Biotechnol Bioeng 1999; 63:559–572.
72. Altamirano C, Illanes A, Casablancas A, Gamez X, Cairo JJ, Godia C. Analysis of CHO cells metabolic redistribution in a glutamate-based defined medium in continuous culture. Biotechnol Prog 2001; 17:1032–1041.
73. Altamirano C, Cairo JJ, Godia F. Decoupling cell growth and product formation in Chinese hamster ovary cells through metabolic control. Biotechnol Bioeng 2001; 76:351–360.
74. Gawlitzek M, Conradt HS, Wagner R. Effect of different cell culture conditions on the polypeptide integrity and N-glycosylation of a recombinant model glycoprotein. Biotechnol Bioeng 1995; 46:536–544.
75. Gawlitzek M, Valley U, Nimtz M, Wagner R, Conradt HS. Characterization of changes in the glycosylation pattern of recombinant proteins from BHK-21 cells due to different culture conditions. J Biotechnol 1995; 42:117–131.
76. Hayter PM, Curling EMA, Baines AJ, Jenkins N, Salmon I, Strange PG, Tong MM, Bull AT. Glucose-limited chemostat culture of Chinese hamster ovary cells producing recombinant human interferon-γ. Biotechnol Bioeng 1992; 39:327–335.
77. Hayter PM, Curling EMA, Gould ML, Baines AJ, Jenkins N, Salmon I, Strange PG, Bull AT. The effect of the dilution rate on CHO cell physiology and recombinant interferon-γ production in glucose-limited chemostat culture. Biotechnol Bioeng 1993; 42:1007–1085.
78. Castro PML, Ison AP, Hayter PM, Bull AT. The macroheterogeneity of recombinant human interferon-γ produced by Chinese hamster ovary cells is affected by the protein and lipid content of the culture medium. Biotechnol Appl Biochem 1995; 21:87–100.
79. Barnabe N, Butler M. The relationship between intracellular UDP-N-acetyl hexosamine nucleotide pool and monoclonal antibody production in a mouse hybridoma. J Biotechnol 1998; 60:67–80.
80. Gu X, Wang DIC. Improvement of interferon-γ sialylation in Chinese hamster ovary cell culture by feeding of N-acetylmannosamine. Biotechnol Bioeng 1998; 58:642–648.
81. Gawlitzek M, Ryll T, Lofgren J, Sliwkowski MB. Ammonium alters N-glycan structures of recombinant TNFR-IgG: degradative versus biosynthetic mechanisms. Biotechnol Bioeng 2000; 68:637–646.
82. Altamirano C, Paredes C, Cairo JJ, Godia F. Improvement of CHO cell culture medium formulation: simultaneous substitution of glucose and glutamine. Biotechnol Prog 2000; 16:69–75.

83. Andersen DC, Bridges T, Gawlitzek M, Hoy C. Multiple cell culture factors can affect the glycosylation of Asn-184 in CHO-produced tissue-type plasminogen activator. Biotechnol Bioeng 2000; 70:25–31.
84. Santell L, Ryll T, Etcheverry T, Santoris M, Dutina G, Wang A, Gunson J, Warner TG. Aberrant metabolic sialyation of recombinant proteins expressed in Chinese hamster ovary cells in high productivity cultures. Biochem Biophys Res Commun 1999; 258:132–137.
85. Wang MD, Yang M, Huzel N, Butler M. Erythropoietin production from CHO cells grown by continuous culture in a fluidized-bed bioreactor. Biotechnol Bioeng 2002; 77:194–203.
86. Cruz HJ, Conradt HS, Dunker R, Peixoto CM, Cunha AE, Thomaz M, Burger C, Dias EM, Clemente J, Moreira JL, Rieke E, Carrondo MJT. Process development of a recombinant antibody/interleukin-2 fusion protein expressed in protein-free medium by BHK cells. J Biotechnol 2002; 96:169–183.
87. Andersen DC, Goochee CF, Cooper G, Weitzhandler M. Monosaccharide and oligosaccharide analysis of isoelectric focusing-separated and blotted granulocyte colony-stimulating factor glycoforms using high-pH anion exchange chromatography with pulsed amperometric detection. Glycobiology 1994; 4:459–467.
88. Miller WM, Blanch HW, Wilke CR. A kinetic analysis of hybridoma growth and metabolism in batch and continuous suspension culture: effect of nutrient concentration, dilution rate, and pH. Biotechnology and Bioengineering 1988; 32:947–965.
89. Borys MC, Linzer DIH, Papoutsakis ET. Culture pH affects expression rates and glycosylation of recombinant mouse placental lactogen proteins by Chinese hamster ovary (CHO) cells. Bio/Technology 1993; 11:720–724.
90. Borys MC, Linzer DIH, Papoutsakis ET. Ammonia affects the glycosylation pattern of recombinant mouse placental lactogen-I by Chinese hamster ovary cells in a pH dependent manner. Biotechnol Bioeng 1994; 43:505–514.
91. Valley U, Nimtz M, Conradt HS, Wagner R. Incorporation of ammonium into intracellular UDP-activated N-acetylhexosamines and into carbohydrate structures in glycoproteins. Biotechnol Bioeng 1999; 64:401–417.
92. Kimura R, Miller WM. Effects of elevated pCO_2 and/or osmolality on the growth and recombinant tPA production of CHO cells. Biotechnol Bioeng 1996; 52:152–160.
93. Kimura R, Miller WM. Glycosylation of CHO-derived recombinant tPA produced under elevated pCO_2. Biotechnol Prog 1997; 13:311–317.
94. Schmelzer AE, Miller WM. Effects of osmoprotectant compounds on NCAM polysialylation under hyperosmotic stress and elevated pCO_2. Biotechnol Bioeng 2002; 77: 359–368.
95. Delorme E, Lorenzini T, Giffin J, Martin F, Jacobsen F, Boone T, Elliott S. Role of Glycosylation on the secretion and biological activity of erythropoietin. Biochemistry 1992; 31:9871–9876.
96. Takeuchi M, Inoue N, Strickland TW, Kubota M, Wada M, Shimizu R, Hoshi S, Kozutsumi H, Takasaki S, Kobata A. Relationship between sugar chain structure and biological activity of recombinant human erythropoietin produced in Chinese hamster ovary cells. Proc Natl Acad Sci USA 1989; 86:7819–7822.
97. Tsuda T, Kawanishi G, Ueda M, Masuda S, Sasaki R. The role of carbohydrate in recombinant human erythropoietin. Eur J Biochem 1990; 188:405–411.
98. Do KY, Fregien N, Pierce M, Cummings RD. Modification of glycoproteins by N-acetylglucosaminyltransferase V is greatly influenced by accessibility of the enzyme to oligosaccharide acceptors. J Biol Chem 1994; 269:23456–23464.
99. Rueckert RR, Mueller GC. Effect of oxygen tension on HeLa cell growth. Cancer Res 1960; 60:944–949.
100. Ozturk SS, Palsson BO. Effects of dissolved oxygen on hybridoma cell growth, metabolism, and antibody production kinetics in continuous culture. Biotechnol Prog 1990; 6:437–446.

101. Lin AA, Kimura R, Miller WM. Production of tPA in recombinant CHO cells under oxygen-limited conditions. Biotechnol Bioeng 1993; 42:339–350.
102. Chotigeat W, Watanapokasin Y, Mahler S, Gray PP. Role of enviromental conditions on the expression levels, glycoform pattern and levels of sialyltransferase for hFSH produced by recombinant CHO. Cytotechnology 1994; 15:217–221.
103. Jan DCH, Petch DA, Huzel N, Butler M. The effect of dissolved oxygen on the metabolic profile of a murine hybridoma grown in serum-free medium in continuous culture. Biotechnol Bioeng 1997; 54:153–164.
104. Kunkel JP, Jan DCH, Jamieson JC, Butler M. Dissolved oxygen concentration in serum-free continuous culture affects N-linked glycosylation of a monoclonal antibody. J Biotechnol 1998; 62:55–71.
105. Reuveny S, Velez D, Macmillan JD, Miller L. Factors affecting cell growth and monoclonal antibody production in stirred reactors. J Immunol Methods 1986; 86:53–59.
106. Furukawa K, Ohsuye K. Effect of culture temperature on a recombinant CHO cell line producing a C-terminal α-amidating enzyme. Cytotechnology 1998; 26:153–164.
107. Jorjani P, Ozturk SS. Effects of cell density and temperature on oxygen consumption rate for different mammalian cell lines. Biotechnol Bioeng 1999; 64:349–356.
108. Donaldson M, Wood HA, Kulakosku PC, Shuler ML. Glycosylation of a recombinant protein in the Tn5Bl-4 insect cell line: influence of ammonia, time of harvest, temperature, and dissolved oxygen. Biotechnol Bioeng 1999; 63:255–262.
109. Gramer MJ. Measurement of cell-culture gycosidase activity. In: Jenkins N, ed. Animal Cell Biotechnology. Vol. 8. Totowa, NJ: Humana Press, Inc., 1999:289–298.
110. Goochee CF, Gramer MJ. Glycosidase activities in Chinese hamster ovary cell lysate and cell culture supernatant. Biotechnol Prog 1993; 9:366–373.
111. Goochee CF, Gramer MJ. Glycosidase activities of the 293 and NS0 cell lines and of an antibody-producing hybridoma cell line. Biotechnol Bioeng 1994; 43:423–428.
112. Gramer MJ, Goochee C, Chock VY, Brousseau DT, Sllwkowski MB. Removal of sialic acid from a glycoprotein in CHO cell culture supernatant by action of an extracellular CHO cell sialidase. Biotechnology 1995; 13:692–698.
113. Munzert E, Muthing J, Buntemeyer H, Lehmann J. Sialidase activity in culture fluid of Chinese hamster ovary cells during batch culture and its effect on recombinant human antithrombin III integrity. Biotechnol Prog 1996; 12:559–563.
114. Ferrari J, Gunson J, Lofgren J, Krummen L, Warner TG. Chinese hamster ovary cells with constitutively expressed sialidase antisense RNA produce recombinant DNase in batch culture with increased sialic acid. Biotechnol Bioeng 1998; 60:589–595.
115. Kratje RB, Lind W, Wagner R. Evaluation of the proteolytic potential of in vitro cultivated hybridoma and recombinant mammalian cells. J Biotechnol 1994; 32:107–125.
116. Satoh M, Hosoi S, Sato S. Chinese hamster ovary cells continuously secrete a cysteine endopeptidase. In Vitro Cell Dev Biol Anim 1990; 26:1101–1104.
117. Froud SJ, Clements GJ, Doyle ME, Harris ELV, Lloyd C, Murray P, Preneta A, Stephens PE, Thompson S, Yarranton GT. Development of a process for the production of HIV 1 gpl20 from recombinant cell lines. In: Meignier B, ed. Production of Biological from Animal Cells in Culture. London: Butterworth-Heinemann, 1991:110–115.
118. Cartwright T. Adjusting cellular metabolism for optimum product yield. In: Animal Cells as Bioreactors: Cambridge Studies in Biotecnology. Vol. 11. Cambridge, U.K.: Cambridge University Press, 1994:96–111.
119. Goldman MH, James DC, Rendall M, Ison AP, Hoare M, Bull AT. Monitoring recombinant human interferon-γ N-glycosylation during perfused fluidized-bed and stirred-tank batch culture of CHO cells. Biotechnol Bioeng 1998; 60:596–607.
120. Parekh RB, Patel TP. Comparing the glycosylation patterns of recombinant glycoproteins. TIBTECH 1992; 10:276–280.
121. Merry T. Current techniques in protein glycosylation analysis: a guide to their application. Acta Biochim Pol 1999; 46:303–314.
122. Geyer H, Geyer R. Strategies for glycoconjugate analysis. Acta Anat 1998; 161:18–35.

123. O'Neill RA. Enzymatic release of oligosaccharides from glycoproteins for chromatographic and electrophoretic analysis. J Chromatogr 1996; 720:201–215.
124. Hu G-F. Fluorophore-assisted carbohydrate electropho-resis technology and applications. J Chromatogr 1995; 705:89–103.
125. Rudd PM, Dwek RA. Rapid, sensitive sequencing of oligosaccharides from glycoproteins. Curr Opin Biotechnol 1997; 8:488–497.
126. Packer NH, Harrison MJ. Glycobiology and proteomics: is mass spectrometry the Holy Grail? Electrophoresis 1998; 19:1872–1882.
127. Pantazaki A, Taverna M, Vidal-Madjar C. Recent advances in the capillary electrophoresis of glycoproteins. Anal Chim Acta 1999; 383:137–156.
128. Kakehi K, Honda S. Analysis of glycoproteins, glycopeptides and glycoprotein-derived oligosaccharides by high-performance capillary electrophoresis. J Chromatogr 1996; 720:377–393.
129. Higgins E, Bernasconi R. A comparison of oligosaccharide profiling methods. Arland T, In: Hotchkiss J, eds. Techniques in Glycobiology. New York: Marcel Dekker, 1997:431–442.
130. Xu Y, Cacia J. A reversed-phase HPLC assay for plasminogen activators. J Liq Chromatogr Relat Technol 2000; 23:1841–1850.
131. Harvey DJ. Matrix-assisted laser desorption/ionization mass spectrometry of carbohydrates. Mass Spectrom Rev 1999; 18:349–451.
132. Papac DI, Jones AJS, Basa LJ. Matrix-assisted laser desorption/ionization time-of-flight mass spectrometry of oligosaccharides separated by high pH anion-exchange chromatography. In: Hotchkiss AT, ed. Techniques in Glycobiology. New York: Marcel Dekker, 1997:33–65.
133. Taverna M, Tran NT, Merry T, Horvath W, Ferrier D. Electrophoretic methods for process monitoring and the quality assessment of recombinant glycoproteins. Electrophoresis 1998; 19:2572–2594.
134. Watson E, Yao F. Capillary electrophoretic separation of human recombinant erythropoietin (rhuEPO) glycoforms. Anal Biochem 1993; 210:389–393.
135. Takeuchi M, Taksaki S, Shimada M, Kobata A. Role of sugar chains in the in vitro biological activity of human erythropoietin produced in recombinant Chinese hamster ovary cells. J Biol Chem 1990; 265:12127–12130.
136. Friedman Y, Higgins EA. A method for monitoring the glycosylation of recombinant glycoproteins from conditioned medium, using fluorophore-assisted carbohydrate electrophoresis. Anal Biochem 1995; 228:221–225.
137. Yang M, Butler M. Enhanced erythropoietin heterogeneity in a CHO culture is caused by proteolytic degradation and can be eliminated by a high glutamine level. Cytotechnology 2000; 34:83–99.
138. Kunkel JP, Jan DCH, Butler M, Jamieson JC. Comparisons of the glycosylation of a monoclonal antibody produced under nominally identical cell culture conditions in two different bioreactors. Biotechnol Prog 2000; 16:462–470.
139. Guttman A. Capillary gel electrophoresis of 8-aminopyrene-3,6,8-trisulfonate-labeled oligosaccharides. In: Arland T, Hotchkiss J, eds. Techniques in Glycobiology. New York: Marcel Dekker, 1997:377–389.
140. Kakehi K, Kinoshita M, Nakano M. Analysis of glycoproteins and the oligosaccharides thereof by high-performance capillary electrophoresis—significance in regulatory studies on biopharmaceutical products. Biomed Chromatogr 2002; 16:103–115.
141. Wu SL. The use of sequential high-performance liquid chromatography and capillary zone electrophoresis to separate the glycosylated peptides from recombinant tissue plasminogen activator to a detailed level of microheterogeneity. Anal Biochem 1997; 253:85–97.
142. Anumula KR. Highly sensitive pre-column derivatization procedures for quantitative determination of monosaccharides, sialic acids, and amino sugar alcohols of glycoproteins

by reversed phase high-performance liquid chromatography. In: Arland T, Hotchkiss J, eds. Techniques in Glycobiology. New York: Marcel Dekker, 1997:349–357.

143. Wan HZ, Kaneshiro S, Frenz J, Cacia J. Rapid method for monitoring galactosylation levels during recombinant antibody production by electrospray mass spectrometry with selective-ion monitoring. J Chromatogr A 2001; 913:437–446.
144. Nimtz M, Martin W, Wray V, Kloppel K-D, Augustin J, Conradt HS. Structures of sialylated oligosaccharides of human erythropoietin expressed in recombinant BHK-21 cells. Eur J Biochem 1993; 213:39–56.
145. Geisow M. Electrospray ionization mass spectrometry—a powerful new analytical tool. Trends Biotechnol 1990; 8:301–303.
146. Rapp U, Resemann A, Mayer-Posner FJ, Schafer W, Feichtinger K. The fragmentation behavior of glycopeptides using the post-source decay technique and matrix-assisted laser desportion ionization mass spectrometry. In: Arland T, Hotchkiss J, eds. Techniques in Glycobiology. New York: Marcel Dekker, 1997:53–66.
147. Merkle RK, Poppe I. Carbohydrate composition analysis of glycoconjugates by gas-liquid chromatography/mass spectrometry. Methods Enzymol 1994; 230:1–15.
148. Thomsson KA, Karlsson NG, Karlsson H, Hansson GC. Analysis of permethylated glycoprotein oligosaccharide fractions by gas chromatography and gas chromatography-mass spectrometry. In: Arland T, Hotchkiss J, eds. Techniques in Glycobiology. New York: Marcel Dekker, 1997:335–347.
149. Ohta M, Kawasaki N, Itoh S, Hayakawa T. Usefulness of glycopeptide mapping by liquid chromatography/mass spectrometry in comparability assessment of glycoprotein products. Biologicals 2002; 30:235–244.
150. Kawasaki N, Ohta M, Itoh S, Hyuga M, Hyuga S, Hayakawa T. Usefulness of sugar mapping by liquid chromatography/mass spectrometry in comparability assessments of glycoprotein products. Biologicals 2002; 30:113–123.
151. Raju TS, Nayak N, Briggs J, O'Connor JV, Lerner L. A convenient microscale colormetric method for teminal galactose on immunoglobins. Biochem Biophys Res Commun 1999; 261:196–201.
152. Stefansson M, Novotny M. Resolution of oligosaccharides in capillary electrophoresis. In: Arland T, Hotchkiss J, eds. Techniques in Glycobiology. New York: Marcel Dekker, 1997:409–430.
153. Wong-Madden ST, Landry D, Gutherie EP. Discovery and Uses of Novel Glycosidases. In: Arland T, Hotchkiss J, eds. Techniques in Glycobiology. New York: Marcel Dekker, 1997:401–408.

6
Cell Culture Bioreactors

Christel Fenge and Elke Lüllau
AstraZeneca Biotech Laboratory, Södertälje, Sweden

INTRODUCTION

Historically, cell cultivation was first devised at the beginning of the 20th century (1,2). For many years, it was restricted to simple small-scale propagation systems applied to tissue culture for basic research. Simultaneously, large bioreactors were already in use for the production of commercially interesting secondary metabolites exploiting microbial fermentation. The development of industrial-scale cell culture bioreactors started in the mid-1950s in response to the need for mass cultivation techniques that were suitable for vaccine production as demanded by the massive vaccination programs launched at that time. Previously, roller bottles were used to culture primary cells at relatively limited scale for vaccine production. These initial cell culture bioreactors were specifically designed for adherent cells, examples are plate propagators and packed beds as reviewed in Refs. (3,4). The first commercially interesting suspension cell products [food-and-mouth disease vaccine produced in suspension culture of BHK cells and interferon produced in Namalva cells (5)] stimulated the adaptation of homogeneous bioreactor systems used for microbial culture to the requirements of the mechanically more sensitive animal cells. The advent of the monoclonal antibody era in the 1970s then gave rise to the development of a plethora of different bioreactors and culture systems suitable for suspension cell culture with special emphasis on increasing the product yield per unit volume through improved nutrient supply and waste product removal. These specialized systems include hollow fiber, fluidized bed reactors, and other different types of compartmentalized bioreactors based on cell immobilization and perfusion of fresh medium through the cell-containing compartment (6). The basic idea was to overcome the major limitations of cell cultivation, i.e., slow cell growth and low final cell densities, by providing an environment that allows the cells to continuously produce the product of interest at high levels. In parallel, a large number of cell retention devices for stirred tanks or airlift bioreactors were developed allowing for continuous exchange of medium in homogeneous systems. Recombinant DNA technology, which is the basis of modern biotechnology today, enabled the production of a series of protein therapeutics in mammalian cells in the 1980s. This novel opportunity had further impact on the development and optimization of bioreactors for suspension as well as anchorage-dependent cells in large scale. In this regard, a severe biomanufacturing bottleneck is forecasted for the coming decade due to the strong product pipeline for fully human monoclonal antibodies and ultra-large scale bioreactors beyond the $20\,m^2$ scale are discussed to meet this anticipated demand of several kilograms of protein per year (7). At the same time scaled-down systems are becoming more important to enable multiple parallel experimental approaches to be taken for cell line and process development. Since the mid-1990s, the available knowledge has also been applied to the design of bioreactors used for artificial organs (8–13) and systems for tissue and stem cell culture (14,15). Other more recent applications are the production of viral vectors for genetic vaccination or gene therapy (16–19).

This chapter aims to review the different cell culture bioreactor systems developed for suspension and anchorage-dependent cells during the last 50 years of industrial cell culture. Major emphasis is given to well-established bioreactors of broad applicability and an attempt is made to give a comprehensive overview of cell culture systems developed for specific applications that in many cases might not be commercially available. The primary role of a bioreactor is to provide containment with suitable conditions for cell growth and product formation. In principle, this can be achieved by mimicking the environment a cell is exposed to in a tissue, i.e., ensuring hemostasis by exclusion of microorganisms, thermostatization, continuous supply of oxygen, nutrients, and growth factors, and removal of CO_2 and waste products. Adherent cells will also require a surface for attachment. Therefore, one way to describe bioreactors for animal cells is to categorize them into either suitable for suspension or anchorage-dependent cell culture. Some systems may also be modified or adapted for the cultivation of both types. In general, bioreactors can be distinguished in homogeneous systems where the cells are uniformly distributed in the bioreactor and heterogeneous systems, which are characterized by a separation of cells and medium using different types of membranes. Homogeneous systems are well-mixed systems whereas in heterogeneous systems gradient formation can occur.

The description of the different bioreactors covers functional principles, engineering considerations, and scale-up. Examples of applications in terms of cell lines cultured, products recovered, or novel purposes other than production of proteins are given. Advantages and limitations of the different systems are discussed as well as general performance regarding cell densities and productivity. Typical bioreactor operation modes are described in general terms and the relative merits for different applications are discussed. Finally, an attempt is made to summarize technical, scientific, economic, and regulatory considerations related to the selection of a certain bioreactor system and their impact on the design.

BIOREACTORS FOR SUSPENSION CELL CULTURES

Small-Scale Culture Systems

Small-scale culture systems are characterized by a relatively simple design and low level of instrumentation and control. Traditionally, roller bottles and spinner flasks have been used for small-scale suspension culture although even T-flasks, Petri dishes, multiwell plates, and other stationary culture systems are applicable for suspension cell propagation in small-scale. Spinner flasks (Techne, Integra Biosciences, etc.) are available from 125 mL to 5 L working volume for operation in humidified CO_2 incubators. Larger flasks up to 36 L require access to warm rooms or can be operated with heating belts (Belco). They are generally made of glass but flasks made of plastics (NalgeNunc) are also available. The most primitive design consists of a flask with two side ports for inoculum and medium addition and a magnetically coupled stirrer either of a bar type on a central axis or a conical pendulum (20). The stirrer rate is maintained between 30 and 100 rpm, frequently as low as possible to prevent sedimentation of cells. The flasks are commonly kept in a humidified CO_2 incubator and gas transfer takes place via the headspace through slightly opened caps. Oxygen transfer coefficients in the rage of 0.1–4 h^{-1} are reported for spinners (21). A general drawback of this design is the very low oxygen transfer resulting in oxygen limitation already at relatively low cell densities. To maximize oxygen transfer in spinners the height to surface ratio should be kept low; i.e. the working volume should be minimized. To overcome this limitation but still keeping it a simple culture system a special floating stirrer has been suggested (20). Another solution to overcome gas transfer limitations is to submerge gas permeable membranes in the culture fluid [Biott Spinner (22)]. Heidemann et al. developed the so-called Superspinner (23) where microporous polypropylene membranes are fixed on a pendulum stirrer in a standard 1 L Schott glass bottle and the incubator air/CO_2 mixture is pumped through the membrane and the headspace using a simple air pump. This spinner flask set-up is also used in a recently developed small-scale screening system where pH and oxygen electrodes are introduced for monitoring and control (DASGIP) (24). Spinner flasks have been widely used for the cultivation of many different suspension cell lines (primary and transformed mammalian and insect cell lines) for a variety of applications within preclinical research, process development, and for inoculum expansion in production. In addition, shaker flasks are commonly used for the propagation of suspension cells, not only insect cells (25) but also hybridoma or CHO cells (26,27). It is also possible to culture suspension cells in sterile plastic tubes; e.g., centrifuge tubes, placed in a modified shaker apparatus or roller drums (New Brunswick Scientific and models from other manufacturers) inside a humidified CO_2 incubator. This set-up is commonly used for simple screening purposes in media development trials where numerous parallel experiments are undertaken. Microtiter plates are increasingly used for screening and optimization purposes in cell culture (28), in the area of high-throughput protein expression, and biopharmaceutical cell and process development. However, little knowledge is currently available on the fundamental characteristics affecting culture conditions in this system (29,30). Generally, scale-down techniques are becoming more important in industrial process development and optimization. A review by Palomares and Ramirez (31) summarizes the approach and provides examples and experimental configurations.

Another simple culture system is the WaveBioreactor™ (32) made for suspension cell culture of 100 mL up to 500 L volumes in disposable plastic bags (see Fig. 1). Oxygen is transferred through the gas permeable wall. Due to the rocking motion of the bag on a rocker base oxygen transfer is considerably improved compared to spinner culture. The system can be operated in CO_2 incubators or standalone in combination with a heater and a CO_2 control unit. Singh (32) describes the application of the system for monoclonal antibody production, adenovirus production with HEK 293 cells, and baculovirus production with insect cells. Due to the disposable nature of the bioreactor it offers clear advantages for cell therapy related to patient safety.

Stirred Tank Reactors

General

Stirred tank bioreactors (STR) are the most widely used bioreactor type to cultivate suspension cells, mainly due to the broad experience obtained in microbial

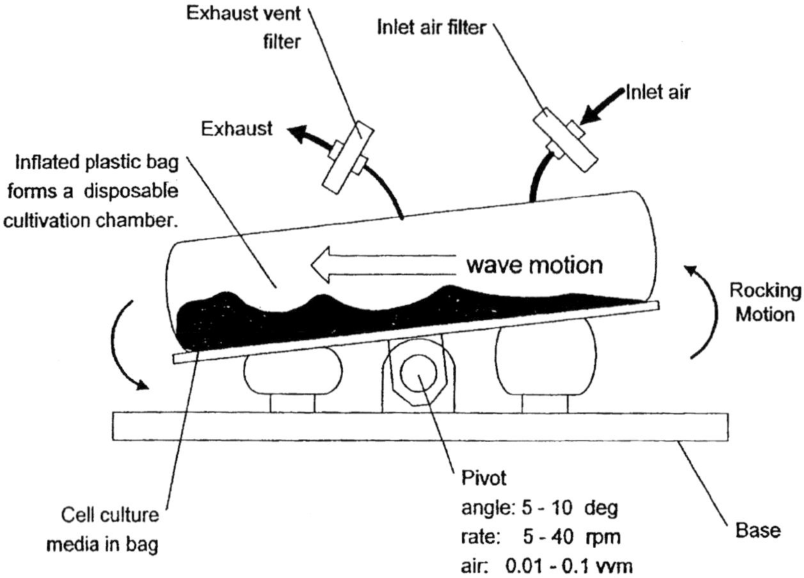

Figure 1 Schematic representation of the Wave™ bioreactor with wave-induced agitation. [Reprinted from Ref. (32), Copyright © 1999 with kind permission from Kluwer Academic Publishers.]

fermentation. They have successfully been used for the cultivation of a wide variety of suspension cells and cells adapted to growth in suspension such as hybridoma cells, CHO, BHK 21, HEK293, and others. Commercial applications are the production of monoclonal antibodies (33,34), recombinant proteins such as blood coagulation factor VIII, tPA, erythropoietin, and other proteins for replacement therapy (35,36), vaccines (37), growth factors, and interferon (5). Novel applications are in gene and cell therapy such as the expansion of hematopoietic cells (14) for reconstituting in vivo hematopoiesis in patients who have undergone intensive chemotherapy, the cultivation of human T-cells for immunotherapy (15), or for the production of virus vectors (17).

A schematic drawing of a typical bioreactor setup is given in Fig. 2. Major advantages are the relative ease of handling and the familiarity of technical personnel with this reactor type. Scale-up principles are better characterized compared to other bioreactor types and mechanical design principles for sterilization-in-place (SIP) and cleaning-in-place (CIP) are known to manufacturing engineers. Furthermore, regulatory agencies are experienced with products obtained from this bioreactor type in batch, fed-batch, and perfusion modes. Especially for contract manufacturers and pilot production facilities in the pharmaceutical industry the high flexibility in terms of applicable working volumes, suitability for different cell types, operation modes, and products is a clear economic advantage.

Large-scale stirred tanks are used widely in the chemical and biochemical industries up to scales of several thousand cubic meters. For animal cells maximum working volumes of 15000 L have been reported (5,38). In the past some pharmaceutical companies have used their own engineering departments for the construction of large bioreactors. In addition, a number of manufacturers provide stirred tank reactors from 0.5 L to 15 m^3. During the last 15 years, a consolidation of the number of

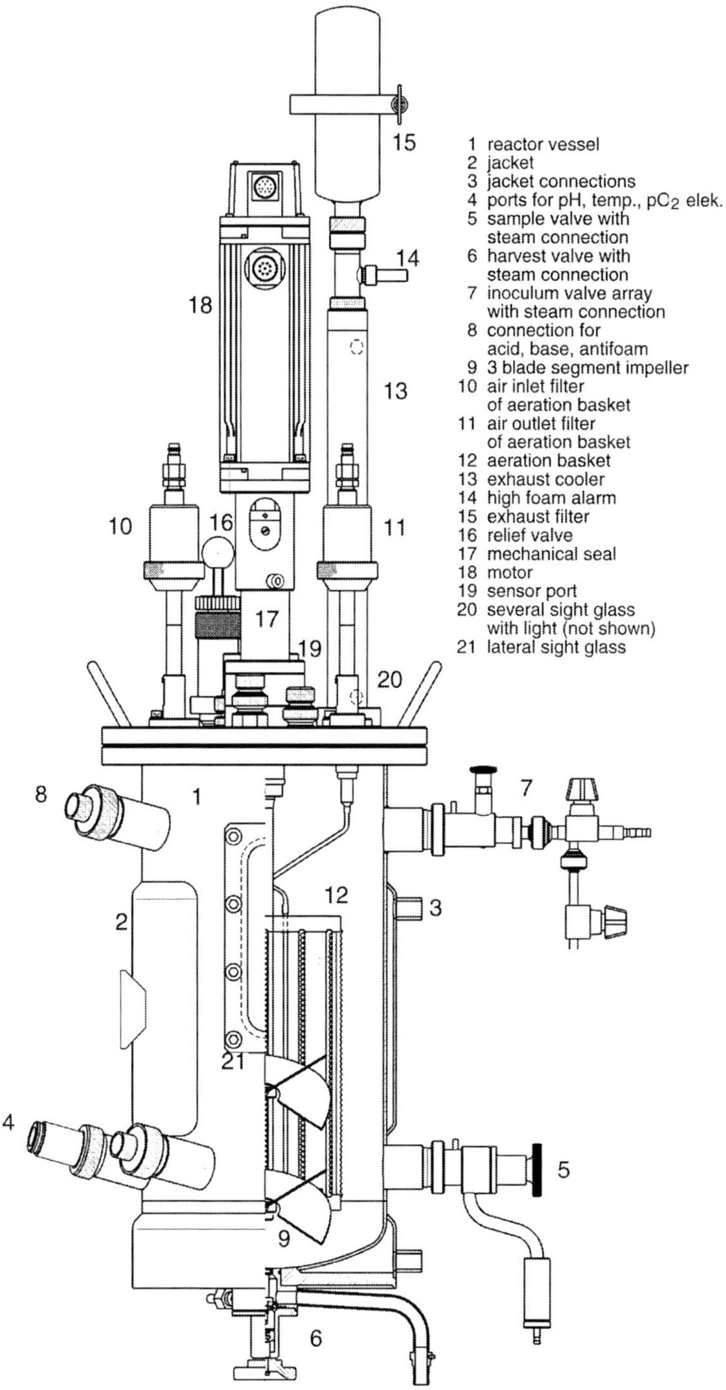

Figure 2 Schematic drawing of stirred tank bioreactor equipped with silicone tubing for bubble free oxygen transfer. 1 reactor vessel, 2 jacket, 3 jacket connections, 4 ports for pH, temperature, pO2 electrodes, 5 sample valve with steam connection, 6 harvest valve with steam connection, 7 inoculum valve array with steam connection, 8 connection for acid, base, antifoam, 9 three-blade segment impeller, 10 air inlet filter of aeration basket, 11 air outlet filter of aeration basket, 12 aeration basket, 13 exhaust cooler, 14 high foam alarm, 15 exhaust filter, 16 relief valve, 17 mechanical seal, 18 motor, 19 sensor port, 20 sight glass with light (not shown), 21 lateral sight glass. (Courtesy of B. Braun Biotech International.)

Table 1 International Suppliers of Stirred Tank Bioreactors

Name	Web address
Abec	http://www.abec.com
Applikon	http://www.applikon.com
B. Braun Biotech International	http://www.bbraunbiotech.com
Bioengineering	http://www.bioengineering.ch
Infors	http://www.infors.ch
New Brunswick Scientific	http://www.nbsc.com

bioreactor manufacturers has occurred. Major suppliers internationally active in this market segment at present are listed in Table 1.

General design criteria are derived from STRs for microbial culture (39,40) but modified to meet the requirements of the more sensitive animal cells. Specifically, the shear sensitivity of animal cells needs consideration regarding the design of impellers, use of baffles, aspect ratio, and oxygenation method.

Mixing

Agitation in stirred tanks aims for homogeneous suspension of the cells and the prevention of chemical (nutrients, waste products), physical (pH, oxygen, carbon dioxide), and thermal gradients in the vessel. Typically, large impellers (impeller diameter $\geq 0.5 \times$ vessel diameter) with an axial fluid flow characteristics such as marine-type impellers, large paddle impellers, or segmented impellers (see Fig. 3) are used to achieve nonturbulent bulk flow patterns at minimum shear rates (41–45). Turbine impellers as used in microbial bioreactors cause damage to many cell lines and sufficient mixing requires rather high stirrer speed. Pumping capacity can be maximized and mixing times minimized by the use of large agitators operated at low stirrer rates rather than small impellers at high stirrer rates (46). As the reactor size increases, especially with increasing height, it becomes necessary to install multiple impellers to prevent compartmentalization of the fluid. In such setups attention has to be paid to optimum positioning of impellers regarding mixing times and homogeneity at a

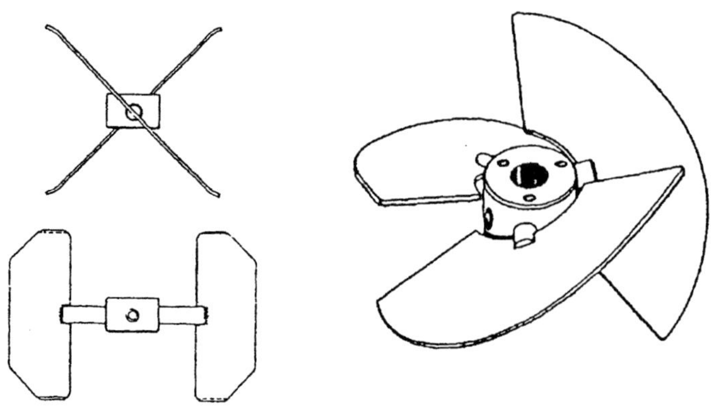

Figure 3 Schematic drawing of large paddle impeller (left) and three-blade segment impeller (right). [Reprinted from Ref. (42), Copyright © 1993, with kind permission from Kluwer Academic Publishers.]

given stirrer speed. Details regarding impeller choice, scale-up, and transitions between flow regimes are given in Ref. (47). Many other impeller types have been suggested in the past. Anchor mixers (41) provide superior mass transfer when using an aeration basket with silicon tubing wrapped around (48) due to its radial transport characteristics. Feder and Tolbert suggested a sail impeller that should provide low shear homogeneous mixing (49). Comparative studies using a shear-sensitive inorganic test system revealed, however, that lowest shear forces as determined by disruption of the flocks were obtained with a three-blade segment impeller followed by a large pitch bladed impeller (42). Different helical impeller types have been applied for insect and plant cell cultures (50,51). In combination with special baffles at the gas–liquid interface surface oxygen transfer rates of 4 up to 45 h^{-1} were reported for water and agar suspensions (51). The cell lift impeller (52) is characterized by a very specific design for gentle mixing by pumping the fluid upwards through an axial cylinder and gas transfer in a surrounding aeration cavity covered by a mesh screen (see Fig. 10). The annular cage impeller (53) is very similar in design but provides improved oxygen transfer. Even other mixer types such as the Vibromixer (54–56) and the vibrating mixer described by Monahan (57) are combined mixing and oxygen transfer devices. Blasey (56) reported on the advantageous application of the Vibromixer for keeping BHK21 cells in single cell suspension. A special tumbling membrane stirrer developed by Lehmann (1–100 L bioreactor scale) (58–60) provides gentle mixing and efficient gas transfer through hydrophobic polypropylene membranes. Despite the variety of impeller types described, marine impeller designs generating an upward fluid flow with a rather low radial dispersion are dominant for suspension culture in industrial bioreactor setups.

In the literature, a number of scale-up criteria valid for agitation rate and tolerable shear forces are discussed such as impeller tip speed (61), integrated shear factor (62), or Kolmogorov's theory of isotropic turbulence (63–65). For hybridoma cells a maximum tolerable tip speed of 1 m/sec is reported for laboratory scale (1–10 L) STRs (42,66). Other authors state tolerable tip speeds of up to 2 m/sec (46,67). A very important consideration for cell damage by agitation is vortex formation with bubble entrainment since this phenomenon is very likely to cause cell death (66). Other authors have investigated the effect of agitation rate on cell growth and product formation with and without sparging (68). They found that increasing stirrer speed alone did not cause cell damage but in combination with direct sparging a decrease in cell growth rate and maximum density was observed. The use of baffles is avoided in animal cell culture to minimize the generation of shear forces. If the agitation rate is too low dead zones are formed causing cell sedimentation and subsequent irreversible aggregation.

Aeration

Typical aspect ratios (height–diameter) for stirred tank reactors vary between 1:1 and 3:1. At low aspect ratios gas transfer via the headspace of the vessel is improved due to the relatively high surface to volume ratio. Stirred tank reactors used for microcarrier culture are frequently designed accordingly. However, higher aspect ratios offer advantages when direct sparging is used for oxygenation of the cell culture. Better dispersion and longer residence times of gas bubbles in the culture liquid are obtained resulting in improved gas transfer rates. Direct sparging of air or oxygen is performed with ringspargers (drilled holes > 0.5 mm diameter) similarly designed as those used in microbial fermenters or microspargers made of sintered stainless steel material (10–100 μM pore size). Especially in small bioreactors excessive sparging leads to considerable cell damage and death as the cells are disrupted when the bubbles burst at the gas–liquid interface. In larger bioreactors excessive sparging and foam formation can be minimized effectively by applying microsparging with pure oxygen at low flow rates with optimized pO_2 control parameters. The observation that shear stress through sparging is decreased as the reactor size increases is confirmed by Henzler's (48) theoretical treatment of sparging in different reactor scales.

The gas–liquid interfacial phenomena and their effects on growth, viability, and productivity of animal cells have been extensively studied by many authors (15,69–75). For a recent review see Ref. (76). Cell damage through bursting bubbles is even increased in serum-free or protein-free media due to the missing shear protection provided by serum or albumin. Therefore, surface active substances such as Pluronic F68, Polyvinylalcohol, Methocel, and other polymers are typically added to cell culture media (77,78). Furthermore, cells might be entrapped in stable foam layers generated by excessive aeration with microspargers (79). Foaming can be reduced by the addition of antifoam agents, although these agents may negatively interfere during downstream processing. Alternatively, the use of a hydrophobic net made of polysiloxane placed onto the liquid has been proposed (80). Despite the shear and foam related challenges, direct sparging is most frequently used in stirred tank reactors larger than 10 L due to the high oxygen transfer rates provided (48,81) and the simple scale-up, handling, in situ sterilization (SIP) and cleaning-in-place (CIP) of the equipment.

Another problem encountered with microsparging is carbon dioxide (CO_2) accumulation as a result of the low flow rates of pure oxygen (<0.01 vvm) applied to minimize cell disruption and foam formation. Elevated pCO_2 levels decrease the medium pH and may adversely affect productivity (82,83), cell growth, or glycosylation of protein products (84). Therefore, oxygen transfer and CO_2 stripping needs to be balanced carefully via optimized bubble size and airflow rate. Another possibility to remove excessive CO_2 is through intense ventilation of the headspace with air. Stripping can further be improved through a radial impeller directly below the liquid surface (48). Addition of HEPES (zwitterionic organic buffer) to increase the buffer capacity or sodium hydroxide or bicarbonate as corrective agents will, however, rather increase the pCO_2 and osmolarity (85) and may even lead to cell damage due to high local pH values (86). To overcome these problems an optimized direct sparging strategy taking into account the CO_2 accumulation was recently proposed by a research group at Bayer (87,88).

Bubble-free oxygen transfer methods were suggested for stirred tanks via thin, gas permeable silicone tubes (89,90) or microporous membranes made of PTFE (91,92) or polypropylene (59). These systems have significantly lower oxygen transfer capacities when compared to direct sparging. Scale-up of membrane aeration devices is very limited and the largest vessels constructed provide 100–300 L working volume (59,93).

Arranging many meters of tubing to maximize mass transfer and mixing is an engineering challenge. In particular, sedimentation of cells, cell aggregates, or microcarriers may occur. A comparison of oxygen transfer rates reported in the literature is given in Table 2. Oxygen consumption rates determined for different cell lines (CHO, BHK, hybridoma cells, insect cells, hematopoietic cells) are in the range of 0.02–0.6 μmol/10^6cells/hr (14,94–96). Based on these data it can be calculated that an oxygen transfer coefficient $k_{\mathrm{l}}a$ of 0.4–4 h^{-1} is necessary to supply sufficient

Table 2 Comparison of Oxygen Transfer Rates k_la for Different Aeration Systems Used in Stirred Tank Reactors

Aeration mode	k_la (1/hr)	Parameters affecting oxygen transfer	Operation parameter range	Comment	References
Monolayer	200 (roller) (0.53 μM O_2/cm^2/hr, T-flask)	Diffusion controlled	Roller bottle (200 rpm) T-flask	Limiting depth, 1.2 mm	21
Surface	0.3–1.6 (3.8 [81]) (4 [21])	Fluid flow pattern Surface area Stirrer speed	0.1–1 vvm	Only suitable for low cell densities and low O_2 demand	21,42,81
Sparging	0.02–0.1 vvm: 3–10 (0.8–15.5 [81])	Gas flow rate Bubble size (Stirrer rate minor effect on transfer)	0.005–0.1 vvm Ringsparger: 1 mm Microsparger: 10–100 μm pore size	Cell damage CO_2 accumulation pH drop	42,81,453
Caged sparging	0.02–0.1 vvm: 1–3	Spinning velocity Stirrer rate	0.01–1 vvm	Easy to scale up Suitable for microcarrier	42,453,454
Bubble free	0.1–1 vvm: 2–5 (3–6 [81])	Active surface area Stirrer rate Transmembrane pressure (Gas flow rate minor effect on transfer) Material	0.1–1 vvm Silicone Polypropylene, PTFE	No cell damage Scale-up difficult	42,59–61, 81,90,92

The represented values were obtained under typical cell culture conditions regarding stirrer rate, airflow, and temperature (measured in tap water or cell culture medium supplemented with serum). For details see the references.

oxygen to $\times 10^6$ cells per ml at a pO_2 of 30% air saturation. This rough estimation clearly indicates that at higher cell densities membrane oxygenation systems may not guarantee sufficient oxygen transfer even when using pure oxygen especially in a larger scale.

A detailed comparative discussion of oxygen transfer characteristics in stirred tanks using different oxygenation methods, i.e., surface, sparger, and membrane aeration, in connection with impeller design and medium additives such as Pluronic F 68 and albumin, is provided by Henzler (48), Aunins (97), and Moreira (81).

Perfusion

Suspension culture in STRs offers broad flexibility regarding operation mode. In the pharmaceutical industry typically batch, repeated batch and fed-batch processes are applied for production purposes. Continuous cultures such as in chemostat or cytostat mode (98,99) have been mainly used for scientific investigation and optimization of cell growth, metabolism, and productivity under different conditions (100–107). During the last 10 years perfusion culture, i.e., the continuous exchange of spent medium while retaining the cells in the bioreactor has become an established technology. Schematic bioreactor setups for internal and external perfusion devices are shown in Fig. 4. Perfusion culture has been used for production of monoclonal antibodies (108,109) and recombinant proteins (35,110–115). Large-scale operations ($\leq 1\,m^3$ bioreactor scale) are running at Bayer (35), Biovitrum (116), Chiron (84), and others. Devices for continuous harvest of cell-free supernatant from suspension cultures are based on either sedimentation or filtration as the separation principle. Critical parameters for the design of such devices are the size of suspension cells (10–30 μm), their low density [~5% greater than that of the medium (117)], and their

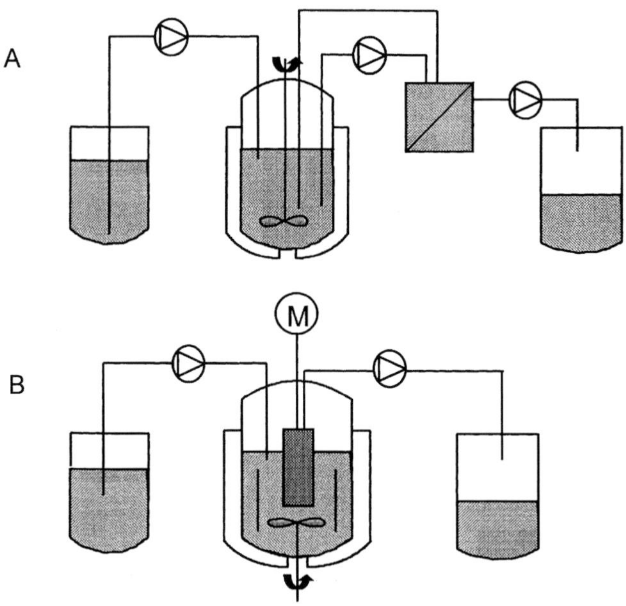

Figure 4 Schematic representation of perfusion operation of stirred tank bioreactors. (A) External cell retention unit recycling the cells back to the bioreactor. (B) Internal cell retention via a spinfilter.

shear sensitivity. For filtration-based separation devices, the content of cell debris and macromolecules in the culture supernatant, i.e., supplemented or cellular proteins and released DNA, frequently cause problems in operation longevity due to membrane clogging (118).

During the last 20 years a large number of different devices for operation in an external loop or satellite vessel of the bioreactor or inside the vessel have been described in the literature (see Table 3). However, only a limited number of systems are commercially available to date. Generally, external systems can easily be exchanged upon failure and offer the advantage of scale-up via application of parallel units. However, controlling critical culture parameters such as pO_2, pH, and temperature is difficult and the need to use pumps to recycle the cell suspension may cause mechanical damage to sensitive cells. The most commonly used devices for suspension cell perfusion culture are continuous centrifuges (119,120), different types of settlers (121–123), external cross-flow filtration devices (124,125), and spinfilters made of metal screens, either operated in the bioreactor (33,93) or an external loop (126).

In the following paragraph internal retention systems specific for stirred tanks will be described briefly. Microporous polypropylene membrane segments (0.2 μm pore size) of the above described membrane stirrer were hydrophilized with ethanol and applied for intermittent harvest of culture supernatant and feed of fresh medium (59,127,128) or even recycling of amino acid enriched harvest (129). A static perfusion module equipped with permanently hydrophilic microporous PTFE membrane was used for hybridoma and insect cell culture in 2–10 L stirred tanks (130). Rotating wire cages, so-called spinfilters, mounted on the stirrer axis or driven via an independent motor, have been used with single cell suspensions (nominal pore sizes in the range of 5–20 μm) (131,132) and aggregate forming cell lines (nominal pore sizes 40–120 μm) (133,134). Several authors described continuous perfusion cultivation of suspension cells over weeks up to months at retention rates >90% leading to maximum cell densities far beyond 1×10^7/mL (33,111,131–142). Especially at low pore sizes frequent clogging has been reported (33). At larger pore sizes noncomplete cell retention is obtained for single cells (typical cell diameter 10–20 μm). Therefore, the pore size has to be carefully balanced with the desired retention rate to avoid early termination of the culture. Factors affecting the retention degree of suspension cells, filtration performance, and scale-up have been investigated by several authors (93,141,143–146) but are still not completely understood. From the literature it can be deducted that pore size, surface velocity of the mesh, agitation rate, the fluid flow pattern in the stirred tank (132,141,147,148), and the type of applied mesh (142,149) are important parameters determining spinfilter performance. Due to their mechanical robustness scale-up has been shown in stirred tanks up to several hundred liters (93,137,140,145). Despite the vast literature it is rather challenging to successfully design and operate spinfilter bioreactors for suspension cells due to the empirical nature of the data gathered so far.

A variety of sedimentation devices for suspension cells suitable mainly for small-scale cell culture have been developed. Settlers permit only slow perfusion rates for single cell suspensions since mammalian cells settle at a rate of 2–10 cm/hr and retention is therefore often not complete. Several authors applied systems based on gravitational settling (121,150–153), and Tokashiki (154) succeeded to scale-up based on this principle to the 22 L scale and claimed a scale-up potential to some 200 L. Sato (155) and more recently Hosoi (156) perfused Namalva and hybridoma cells by applying centrifugal forces to facilitate settling in an enlarged hollow stirrer

Table 3 Perfusion Devices for Suspension Culture in Stirred Tank Bioreactors

Separation mode	Device	Reactor scale	Critical parameters	Comment	References
Internal settling	Inclined settling zones	<1–22 L	Settling surface area Harvest flow rate	Sensitive technology Narrow range of operation parameters for optimal cell retention Difficult to scale-up	121, 150–156, 399, 455
External settling	Vertical settlers Inclined settler with and without channels	1–25 L	Settling surface area Harvest flow rate Temperature	Sensitive technology Narrow range of operation parameters for optimal cell retention Difficult to scale-up	115, 122, 152, 158, 456–458
Acoustic aided settling	Acoustic settler	1 L to some 10 L	Harvest rate limited by cell concentration and input power	No fouling Narrow range of operation parameters for optimal cell retention Difficult to scale-up Selective retention of viable and dead cells obtainable	123, 160, 459–462
Centrifugation	Disc stack centrifuges (Westfalia) Centritech Cell Centrifuge (Kendro)	2–500 L	Centrifugal force Feed rate	No clogging or fouling Separation rate controlled by *g*-force and feed flow rate Separation of viable and dead cells achievable by adapting *g*-force Centritech cell centrifuge allows for maximum harvest rate of 2400 L/day (463)	119, 120, 159, 399, 403, 464, 465

Internal rotating filters (spinfilters)	Metal wire mesh cages Polymer meshes Ceramics, sintered glass	1–500 L	Pore size Spinning velocity Fluid flow characteristics	Operated on stirrer axis or separate axis Pore sizes 5 up to 20 μM even larger (53 μm, 120 μm) for aggregate forming cells, smaller pore sizes down to 1 μM for sintered porcelain	33, 93, 111, 131–138, 140–145 149, 466
External rotating filters	Metal wire mesh cages Polymer meshes Ceramics, sintered glass	4 L	Membrane material Spinning velocity Gap dimensions	Operated on a central axis Taylor vortex formation reduces fouling in certain setups	126, 138,467
Internal membrane filtration	Membrane stirrer Static perfusion module (B. Braun Biotech International)			Hydrophylized polypropylene (0.2 μm) Permanent hydrophilic PTFE (1–5 μm)	59, 127–130, 264, 468, 470
External membrane filtration	Hollow fiber modules Flat membrane modules Thin channel modules Stirred filtration Cells	1 L up to several hundred liters	Membrane material Flow rate Channel geometry	Micorporous membranes (02–5 μm) made of polypropylene, Nylon, PVDF, mixed cellulose ester (ME), polyether-sulfone (PES)	124, 125, 207, 396, 397, 471–477, 478

shaft. A more detailed description of these internal settling devices and their scale-up is given in Ref.(157). A major advantage is that failure due to clogging does not occur since separation is not carried out via a physical barrier and selective retention of viable cells can be achieved due to the twofold lower sedimentation velocity of nonviable cells (158). This observation is generally valid for sedimentation-based perfusion devices and has been shown also for external centrifuges applied to recycle the cells back to the bioreactor (159) and ultrasonic separation (160). Recent reviews on perfusion devices have be published by Tokashiki (157) and Woodside (161).

Airlift Reactors

General

In addition to STRs, airlift reactors are widely used for suspension cell culture. The largest published scale of airlifts applied for animal cell culture is $2\,m^3$, which is routinely used for monoclonal antibody production at LONZA (46,162). This relatively new bioreactor type has been scaled-up to $1500\,m^3$ for microbial fermentation and $17{,}000\,m^3$ for wastewater treatment (163) demonstrating its enormous scale-up potential. There are several publications on the use of this bioreactor type for suspension growth of mammalian (hybridoma, BHK, CHO, Namalva) (122,164–166) and insect cell lines (167). Even studies on microcarrier cell culture were reported (168,169) although it has generally been considered impractical due to microcarrier aggregation at the air–medium interface (70). Airlifts can be regarded as a type of bubble column since mixing is provided by the introduction of gas bubbles at the base of a tall column (170). In bubble columns the ascending bubbles cause random mixing. In airlifts, fluid circulation is obtained by mechanical separation of a channel for gas/liquid up-flow (riser) and a channel for down-flow (downcomer). These channels are connected at the top and the bottom of the column forming a closed loop. Fluid flow is driven by the density difference between the sparged fluid in the riser and the bubble-free fluid in the downcomer. Basically, two different geometric configurations can be distinguished: (a) external loop vessels where the circulation takes place through an external loop and (b) baffled vessels with either a cylindrical draft tube or a simple baffle resulting in an internal loop circulation (see Fig. 5). In airlifts mixing and oxygen transfer are generally coupled, i.e., the gas flow rate is controlled by the oxygen demand of the culture and determines the hydrodynamic conditions in the bioreactor. In order to decouple mixing and oxygen transfer, pH and pO_2 can be controlled by varying the composition of a carrier gas (air or nitrogen) regarding oxygen and CO_2 at a maintained total gas flow rate. Typical superficial gas velocities are in the range of 0.001–0.01 m/s (67) and corresponding k_la values of 0.7–20 h^{-1} were reported (46).

Major design considerations are geometrical configuration of the riser and downcomer, aspect ratio, sparger layout, positions for electrodes, sample ports, feed and base addition. The most widely used design for animal cell culture is a bubble column with a concentrical draft tube. Typical aspect ratios are in the range of 6:1 to 12:1 (67,162,171). Optimum mixing is obtained by keeping the cross-sectional area of the downcomer similar to that of the riser. It has also been shown that the cylindrical area below and above the draft tube is affecting mixing. A recent review of the design and scale-up of airlift bioreactors was given by Varley and Birch (46); a comprehensive description of the engineering principle can be found in Ref. (163).

The main advantage quoted for airlift over STRs beside its ease of scale-up is that no moving parts and mechanical seals are needed which improves the reliability

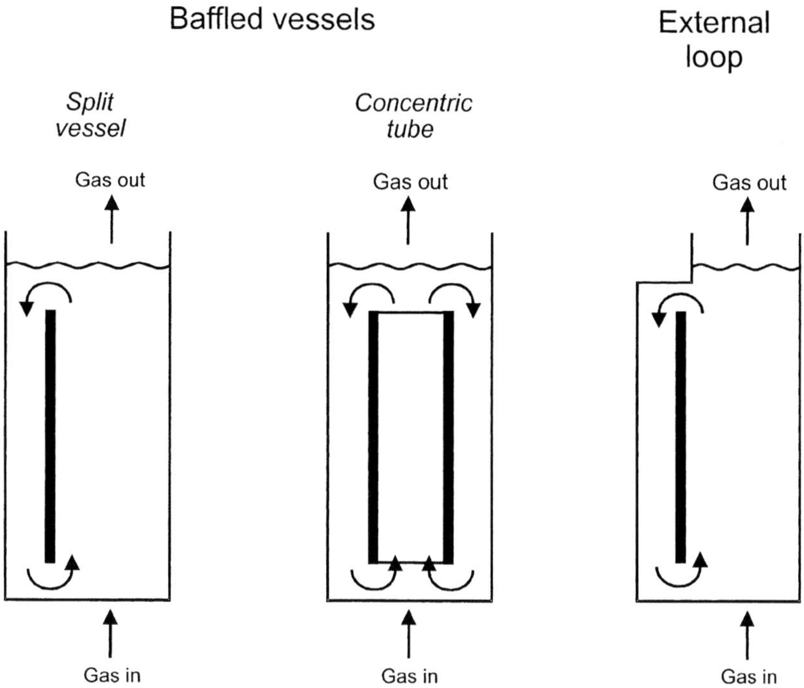

Figure 5 Airlift reactor configurations. [Reprinted from Ref. (170), Copyright © 1990, with permission from Elsevier Science Publishers.]

of sterile operation (162,170). Despite these advantages the airlift is not as widely used and modified for cell culture as the STR. This has historical reasons but might also be due to the limited flexibility in terms of working volumes and suitability for microcarrier culture. Another advantage frequently quoted is its gentler mixing action and suitability for shear-sensitive cells (170). However, similar considerations as described for direct sparging in stirred tank reactors are valid for cell damage and CO_2 accumulation (65).

A number of authors have studied the shear sensitivity of different cell lines such as hybridomas, BHK and insect cells in airlift and bubble column bioreactors and the effect of serum or surfactant concentration, sparger design, bubble size, and column height on cell survival (117,167,172–174). Although most reports on the use of airlift bioreactors are based on batch culture, it is generally possible to operate the system at high cell densities using external cell retention systems similar to those described for STRs. For example, Hülscher (122) used an external settler for the selective recycle of viable cells to an airlift loop reactor. Integrated perfusion devices on the basis of settling zones in external-loop airlifts are described in Ref. (175).

Modifications

A modification of the bubble column principle is the inclined plate bioreactor (176) designed to reduce the effect of bursting bubbles on cells. The inclination of the column leaves the bulk of the liquid bubble free while still providing effective liquid circulation. Another design modification to minimize cell damage due to sparging (74) is provided in the bubble bed bioreactor (177). In this stirred tank-bubble column hybrid bioreactor the residence time of bubbles was dramatically prolonged by floating the bubbles in a countercurrent flow produced by an impeller in the lower part of a central conical draft tube.

Other Reactor Types

Hollow Fiber Bioreactor

In contrast to the homogeneous bioreactor systems described so far, which support in the range of 10^7 cells/mL under perfusion conditions, the heterogeneous hollow fiber bioreactor (HFBR) is a high-intensity system where tissue-like cell densities ($>10^8\,mL^{-1}$) and structures are obtained. This bioreactor type was originally developed by Knazek in 1972 (178) starting from commercially available ultrafiltration modules containing cellulose acetate capillaries and further developed by adding gas permeable silicone polycarbonate capillaries for improved oxygen supply and CO_2 removal. The units are composed of a large number of semipermeable capillaries potted into a cylindrical housing. In many reports especially from research laboratories simple hemodialysis cartridges (8,58,179,180) have been applied. However, during the last 30 years a number of companies have commercially developed and automated systems for mammalian cell culture [some examples are: Technomouse™ (Integra Biosciences) (181), Cell-Pharm™ (Unisyn), Acusyst™ (Biovest International), Cellstasis™ (Genespan)].

The membranes are either of an ultrafiltration type (10–100 kDa) (58,178,180,182) or microporous (0.1–0.2 µm). Different membrane materials, for example, cellulose acetate (178), polypropylene (183), and polysulfone (0.2 µm) (184,185), have been used. In most applications the cells are grown in the extracapillary space (EC) while oxygen enriched medium is recirculated through the fibers (intracapillary space, IC) (179,186,187). A general setup is shown in Fig. 6. In many cases, 10 kDa ultrafiltration membranes are applied to enable the use of inexpensive basal medium where expensive large molecular weight growth factors are supplied only in the EC space (182). Cell secreted proteins are also retained in the EC space leading to a significant accumulation of the product in the EC space which substantially reduces downstream processing efforts. However, it was reported that inhibitory components might accumulate in the EC space, which negatively affect cell growth and productivity (188). Furthermore, proteases secreted by the cells or released from dead cells may cause product degradation during long-term operation. Due to product accumulation this bioreactor type is not suitable for toxic or feedback inhibited products. To improve nutrient supply and waste removal, protocols have been developed where fresh medium is supplied to a medium reservoir connected to the IC space after removal of spent medium (180). In other cases the medium reservoir is regularly replaced with fresh medium (185).

The basic concept of an axial flow hollow fiber bioreactor has been developed further over the years and a number of alternative designs have been described. An overview of cell culture systems and related operation modes and transport phenomena was given by Tharakan et al. (189), Piret and Cooney (190), and Brotherton and Chau (191). Uludag et al. (192) reviewed the technology in relation to its application in cell therapy and tissue transplantation.

In principle, three operation modes of hollow fiber systems can be distinguished: (a) open shell ultrafiltration, (b) closed shell ultrafiltration, and (c) crossflow

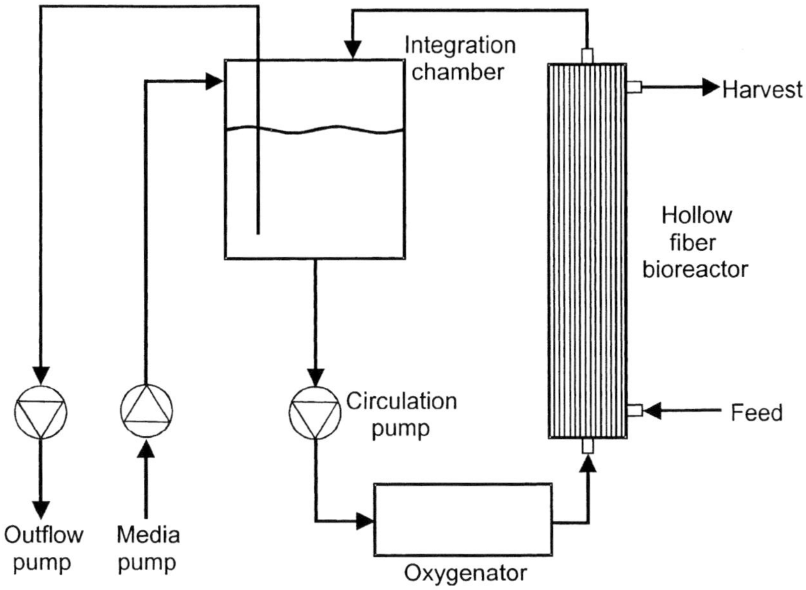

Figure 6 Schematic drawing of a hollow fiber cell culture bioreactor. The cells are kept in the extracapillary space while fresh oxygen enriched medium is recycled through the intracapillary space. [Reprinted from Ref. (182), with permission of Wiley-Liss, Inc., a subsidiary of John Wiley & Sons, Inc., Copyright ©1999 John Wiley & Sons.]

ultrafiltration (see Fig. 7). For a detailed description see Refs.(189,193). In the open shell ultrafiltration mode the transmembrane pressure is decreasing along the length of the fibers. In the closed shell ultrafiltration mode the transmembrane pressure is positive in the entrance half of the module and a back-flow of medium from the EC is obtained in the distal part. These pressure gradients along the axial flow bioreactor types (a, b) cause nutrient and oxygen gradients, which may result in uneven cell growth. These gradient phenomena are the reason for the limited scale-up

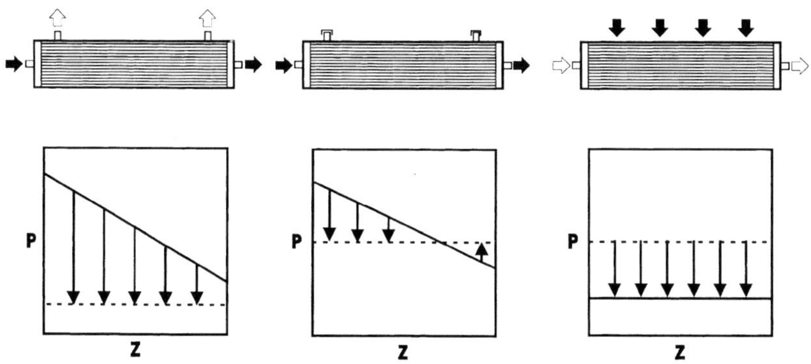

Figure 7 Three operation modes of hollow fiber systems (top) and their corresponding transmembrane pressure profile P (bottom). (Left) open shell ultrafiltration, (center) closed shell ultrafiltration, (right) crossflow ultrafiltration. [Reprinted from Ref. (193), with permission of Wiley-Liss, Inc., a subsidiary of John Wiley & Sons, Inc., Copyright ©1986 John Wiley & Sons.]

potential of these modules and a solution is to use multiple parallel units. A detailed review to model axial flow hollow fiber cell culture bioreactors is provided in Ref. (191).

Attempts to overcome the gradient-related problems have been made by modifying the operation of conventional HFBR. Manipulation of the bioreactor orientation (194) or the osmotic environment of the EC space (195) were suggested in addition to periodically reversing the flow direction in the IC space (194) or change the pressure in the EC space to generate a backflow of medium into the IC space (196). In order to improve oxygen supply to the EC space oxygen carriers were added to the medium that was recirculated through the IC space, which led to improved antibody production (197). Another approach is the development of alternative designs. In bioreactors with extra capillaries for oxygen supply the IC flow rate can be reduced and axial oxygen gradients are eliminated. By additional rotation of the culture chamber further improvement of the culture performance was claimed (Cellstasis™ product information, Genespan). Cima et al. (183) developed a radial flow hollow fiber reactor based on concentric microporous polypropylene fibers with a 200 μm annular space for cell growth. They investigated different operation modes resulting in a more uniform distribution of nutrients by forcing medium through the EC space. The modeling of an intercalated dead end hollow fiber bioreactor with arterial (inlet) and venous (outlet) fibers is presented in Ref. (198). An example for the crossflow operation mode is the flat-bed bioreactor developed by Ku et al. (199) where gradients are substantially reduced compared to the axial configurations. The fibers are used for oxygen supply and pH control. The cells were retained in the unit using microporous stainless-steel filters. However, in most commercially available systems modified axial flow configurations are used (Cell-Pharm™, Unisyn; Technomouse™, Integra Biosciences; Cellstat™, Genespan).

Hollow fiber bioreactors have mainly been used for the growth of suspension cells. However, using other membrane materials such as polypropylene (183) tissue-like cell densities of adherent cells were obtained and could be maintained over several weeks to months. Also coating of the fibers with polycationic materials such as polylysine (200) can be used to enable adherent cell growth. Many different types of mammalian cells have been cultured in hollow fiber bioreactors such as primary cells, tumor cells, and stable cell lines [BHK 21 and Vero cells as reviewed in Refs.(189,201)]. The most important application is the production of monoclonal antibodies (58,180,184,186,202) mainly for diagnostic and research purposes. Considerable attempts have been made to use the technology for the production of therapeutic antibodies, which resulted in the first registration of a hollow fiber produced drug, ProstaScint [Cytogen (203)] in 1996 (204). Other, more recent applications are the production of cells for cell and tissue therapy (205) and the use as artificial organs [liver (8,9), kidney (13), pancreas (10)].

Miscellaneous Dialysis Bioreactors

Several other bioreactors based on the dialysis principle were developed in the past.

Generally, dialysis culture offers the advantage of continuous supply of nutrients and removal of low molecular weight metabolic waste products while accumulating the product in the cell culture compartment. The stirred tank dialysis bioreactor setup eliminates the formation of gradients, the major disadvantage of the hollow fiber bioreactor. Compared to stirred-tank perfusion systems mass transfer is still diffusion controlled. Culturing hybridoma cells in simple dialysis bags

(206) placed in a spinner was further developed to setups where dialysis membranes (207,208) or dialysis modules (209) were submerged in STRs. In the latter systems the cells are cultured in suspension in the stirred vessels and medium from a reservoir is recycled through the dialysis membranes. Cell densities of more than $1 \times 10^7\,mL^{-1}$ and significantly increased product concentrations have been reported in an industrial application (209). Another modification of this principle is a two-compartment stirred tank vessel with cells on one side of the dialysis membrane and dialysis medium on the other side (210). Dialysis bioreactor technology was reviewed by Pörtner (211). Some historical examples of complex systems mainly for cell maintenance are the InVitron *Static Maintenance Reactor* (212) or the *Membroferm* bioreactor (137,213).

Microencapsulation Technique

A completely different technology for suspension cell culture based on immobilization and compartmentalization is microencapsulation. It was originally introduced by Lim and Sun (214) for the immobilization of mammalian cells used in bioartificial organ applications . This technique has been further developed and applied to monoclonal antibody production for commercial use (215,216). The cells are suspended in a solution of a naturally gelling polymer and the microspheres are generated subsequently. The microcapsules can be "cultured" in suspension in different bioreactor types. Various materials have been described for suspension cell entrapment in microcapsules; Ca-alginate (217), Na-alginate (215,216), agarose (218), cellulose sulphate (219). Additionally, composite gels combining the properties of alginate and agarose (220) or PEG and alginate (221) were developed. Collagen (222) and fibrin (67) were suggested for the cultivation of anchorage dependent cells. The diameters of the capsules range from 0.5 to 1 mm. Agarose beads have a lower mechanical strength compared to alginate (223,224). Alginate beads have been coated with polylysine to generate a semipermeable membrane with controllable molecular weight cut-off (225). Improvements of the mechanical strength were obtained using photosensitive polymers (226). The cell leakage of conventional alginate could be reduced using PEG–alginate composite beads (221). Even more materials and modifications are used for cell and gene therapy and tissue transplantation as reviewed by Uludag (192).

Microencapsulation of cells is a well-established technology in bacterial culture and especially in wastewater treatment. Much of the literature on the preparation of the microcapsules (227), their rheological and mechanical properties (224), as well as transport and mass transfer phenomena (228) is found there.

Typical advantages of microencapsulation are the shear protection of cells, the product concentration, and compartmentalization obtained with some microencapsulation methods [e.g., polylysin coated Na-alginate beads (216)] and the high cell densities in the particles in the range of $10^7\,mL^{-1}$ (220) to $10^8\,mL^{-1}$ (67). Another advantage of the microencapsulation technique often cited is the increased specific antibody production rate (220). However, a critical review of the literature shows that an increase in specific antibody productivity is cell line specific (229) and dependent on the cultivation conditions and applied parameters (230). Furthermore, serum-free (217) and even protein-free cultivation (220) could be established facilitated by the compartmentalization of cells and the bulk of the medium. However, a drawback of the technology is the diffusion-controlled transport of nutrients, waste products and—most critically—oxygen, which may lead to necrosis in the center of larger microcapsules. In case of product retention in the capsule this system is not suitable for proteolytically sensitive and feedback inhibited products.

Microcapsules have been used in different bioreactor types such as stirred tanks (215,216), airlifts (231), and fluidized beds (220) mainly in small-scale batch operation mode. Although, from a historical perspective, the development of this technology was stimulated by the need for efficient and economic production methods for monoclonal antibodies it cannot be considered as suitable for industrial-scale mammalian cell culture (67,157). The major application area of this technology today is cell and gene therapy and tissue transplantation (232–234).

BIOREACTORS FOR ANCHORAGE-DEPENDENT CELL CULTURES

Small-Scale Culture Systems

Adherent cell culture has a tradition going back to the beginning of the last century when tissue culture was first devised as a method to study the behavior of animal cells free of systemic variations. Basic techniques are described by Freshney (235) and an interesting historical review on the advances in tissue culture during the last century is given by Jensen (236). In anchorage-dependent monolayer culture, the achievable cell number is directly proportional to the available growth surface. Cell yields in the order of 10^5 cells/cm^2 are obtained for HeLa cells (235), CHO and HEK 293 cell lines (own data). Typical small-scale culture systems include Petri dishes and multiwell pates that are kept in a humidified CO_2 incubator. From a sterility point of view T-flasks available with and without vented caps are preferable (available surface area 25–225 cm^2, 500 cm^2 triple layer (some suppliers are NalgeNunc, Corning Costar, Integra Biosciences, Falcon). Culture flasks used to be made of glass. However, nowadays, disposable plastic ware is standard. Coating the plastic surface with L-lysine or other substances (235) may improve attachment of less anchorage dependent cells such as HEK 293 cells. Especially in roller bottle culture this helps preventing the formation of suspended aggregates. One of the simplest systems for scaling up monolayer cultures is the Cell Factory^{TM} (NalgeNunc) providing culture surface areas of 600–24 000 cm^2 and larger (100 000 cm^2) via multiple interconnected plastic layers (235). Gas transfer is obtained via diffusion and a critical liquid depth of a few millimeters should not be exceeded. Spier (21) reported oxygen transfer rates of 0.53 μM/cm^2/hr for monolayer aeration.

Especially for commercial vaccine production roller bottles have been applied for over 40 years (37,236) (surface area 850 cm^2, 1500 cm^2, and other sizes; available from Falcon, Corning, etc.). Roller bottles require a specific apparatus for constant rotation. Major advantages of this traditional "large-scale" culture method are (a) the increased surface area, (b) the constant mixing preventing gradients, (c) very high oxygen transfer rates [200 h^{-1} (21)], and (d) the flexibility in number of units applied for a certain task. Handling large numbers of multiple culture units for industrial production of vaccines or recombinant protein is very tedious and prone to contamination. Therefore, successful attempts have been made toward automation using laboratory robots such as the Cellmate^{TM} (The Automation Partnership), which is able to handle roller bottles and T-flasks (237).

The CellCube^{TM} system (Corning Costar, available surface area 8500—85,000 cm^2) combines the advantage of a large surface area typical for multilayer systems with good oxygen transfer and mixing capabilities. Medium is continuously pumped through the system which can be oxygen and nutrient enriched in an external loop. This system has been used at Merck for the development of Hepatitis A vaccine (238) and by other authors for virus propagation (239).

Although the described methods are feasible for large-scale production there are considerable drawbacks such as (a) high costs for labor, equipment, and consumables, (b) considerable risk of contamination, and (c) relatively poor opportunity to control culture parameters at optimum set-points. This led to the development of scaleable bioreactor systems for anchorage-dependent cell culture. The most prominent examples are microcarrier culture in stirred tanks (240), fluidized (241), or packed (242) beds or cell culture in solid bed bioreactors (243). These bioreactors are described in more detail Chapter 11. The initial impetus for the development of such techniques was provided by the mass vaccination campaigns of the 1950s against viral diseases and later by the advent of recombinant DNA technology used for the production of drugs in animal cells.

Microcarrier Culture in Stirred Tank Reactors

The basic idea of microcarrier culture was to develop a unit operation system with similar scale-up and environmental control potential as suspension cell culture in stirred tanks. Van Wezel (240) introduced the use of dextran carriers for the growth of human fibroblast like cells in 1967. A historical review of the development of microcarrier technology is presented in Ref. (244). In principle, this technique comprises the cultivation of anchorage dependent cells on small solid particles suspended in growth medium. Under proper conditions the cells attach and spread on the carrier surface and grow to a confluent monolayer. A series of different types have been introduced and attention has been paid to the chemical and physical properties of the carriers such as size, density, and surface charge. General requirements are nontoxicity, good adhesion properties, and a buoyant density of 1.03–1.04 g/mL to be easily kept in suspension. Furthermore, uniform size distribution, suitability for microscopic monitoring of cell growth, autoclavability, and a high batch-to-batch consistency are important quality features for industrial application of the technology. Typically, the carrriers have diameters of around 100–200 μm and are made of dextran, plastic, gelatin, glass, or cellulose (244,245). Even the use of flat microcarriers in STRs has been reported (246). For good cell attachment a number of physicochemical factors are critical; a positive or negative surface charge, the charge density, the presence of cations in the medium. Very often the surface is coated or modified in a specific way to improve cell attachment. In this regard, essential proteins of the extracellular matrix such as collagen, proteoglycans, fibronectin, laminin, elastin, and chondronectin affect cell adhesion positively (247). A comparative study of cell growth and product formation for 13 different carrier types using a recombinant CHO cell line producing monocyte-colony inhibition factor showed clear differences in cell growth and specific productivity for the different carrier types (248).

In 1986, porous microcarriers made of gelatin (249) were introduced and have been used in many applications since than. Due to the porous structure an increased surface area in comparison to smooth beads is obtained, which enables to achieve an improved maximum cell density and productivity. This carrier type is also suitable for the propagation of suspension cells that can be entrapped in the pores. Furthermore, it was claimed that shear sensitive cells are protected inside the beads (250). An extensive reference list on this carrier type is found on the supplier's homepage (Cultisphere™, Percell Biolytica). Other materials were used to produce porous carriers for application in STR such as silicone (251) and cellulose [Cytopore™, Amersham

Biosciences; Cellsnow™, Kirin Brewery Co. Ltd. (252); Cytocell™, Asahi Chemical Industry (22)]. Probably the commercially most important microcarrier types for stirred tank culture are Cytodex™ (Amersham Biosciences) and Cultisphere™ (Percell Biolytica) carriers. A comprehensive overview on different microcarrier types was given by Butler (245).

This technology has primarily been used for the large-scale production of vaccines (37,253–258) and interferon up to the several thousand liters scale (259,260) and also for the production of recombinant therapeutic proteins (244,245,252,261). For these applications, harvesting and separation of the cells from the carriers is not necessary except during volume expansion when cells need to be detached to inoculate a larger bioreactor. Attempts made to use bead-to-bead cell transfer have never been very successful despite for certain cell lines under specific conditions (262,263). Therefore, in general, methods are used where the cells are detached enzymatically or by using buffer solutions containing chelating agents such as EDTA. In case of the gelatin beads enzymatic treatment leads to the complete dissolution of the carriers, which is facilitating subsequent manipulations such as cell harvest and inoculation. Other applications of stirred tank microcarrier culture are the production of proteins and cells used in drug discovery and research (264,265).

For adherent cell culture in stirred tanks suggested microcarrier densities range from 1 g/L under batch up to 10 g/L (266) under fed-batch or perfusion conditions providing cell densities of $106\,mL^{-1}$ to more than $10^7\,mL^{-1}$. In contrast to suspension cell culture in STRs retention of immobilized cells on the relatively large carriers is easily and reliably achieved with spinfilters (pore sizes of $> 75\,\mu m$) (42,136) or even simpler means such as stagnant zones (267). Problems observed during perfusion of single cell suspension such as filter fouling or incomplete cell retention have rarely been reported.

Attention has to be paid to mixing and aeration of microcarrier cultures in stirred tank reactors. Especially cells grown on smooth carriers are sensitive to overagitation. Growth arrest and even detachment of cells has been observed above critical stirrer rates (42,268,269). The stirring regimen is even more important during inoculation where intermittent profiles have been suggested to improve attachment (270). Several researchers investigated hydrodynamic effects on cells in agitated carrier cultures (64,65,271,272). Three potential mechanisms of cell damage were postulated: bead collision with the impeller or other stationary surfaces, bead-to-bead collision, and bead interaction with turbulent fluid eddies of the same size as the beads. Gregoriades (272) found that cell damage of cells attached to microcarriers was observed at energy dissipation levels orders of magnitude lower than those values reported to cause damage to suspended animal cells. However, hydrodynamic stress can be controlled at tolerable rates providing a well-mixed culture by choosing large marine-type impellers that generate a strong axial flow at low stirrer rates (42).

Another problem observed during scale-up is oxygen supply. Generally, bubble-free aeration methods as described earlier in this chapter are preferable for carrier culture since microcarriers tend to be entrapped in foam layers generated by direct sparging and cell damage has been observed (168). Through the optimization of direct sparging (control algorithms, gas mixture, pore size of the sparger, gas flow rates) foam formation can be minimized and this oxygenation method becomes applicable for carrier culture especially in larger scale. In contrast to freely suspended cells, it has been shown that cells attached to microcarriers are not damaged by bubble rupture at gas-medium interfaces. Instead it was suggested that cell damage is caused by bubble-attached cells being removed from the bead as a consequence of the hydrodynamic drag force of the rising bubble (273). In the same study different surfactants were tested for their suitability to protect the carrier attached cells but none gave satisfactory results.

In conclusion, microcarrier culture in stirred tank reactors is a well-established, scaleable technology for anchorage-dependent cells offering a series of advantages such as high growth surface-to-culture volume ratio, simple cell separation, perfusion, and direct monitoring and control. Homogeneous microcarrier culture was also reported in airlift bioreactors (168,169) although it is not as established as stirred tank cultivation.

Fluidized Bed Bioreactors

The fluidized bed technology was introduced for animal cell culture in the 1980s by the company Verax Corporation (274,275). The cells were grown on porous microspheres, made of collagen and weighted with a noncytotoxic steel to achieve a specific gravity of > 1.6, so that the carriers remained suspended in the high-velocity ($\sim$70 cm/min) upward-fluid flow of culture medium. The microspheres (500 µm diameter) had a sponge-like structure of interconnected pores and channels with a diameter in the order of 20–40 µm allowing the cells to enter easily and populate the interior of the carrier (276). The fluidized bed was contained in a column type bioreactor that was connected to an external recirculation loop. A gas exchanger (hollow fiber cartridge), pO_2, pH, and temperature sensors, heating elements, and a circulation pump that controlled the expansion of the fluidized bed were located in the loop. Dissolved oxygen levels were monitored at the inlet and outlet of the gas exchanger in order to measure the oxygen transfer rate of the reactor and to control the oxygen flow (277). Typically, oxygen transfer rates approached 10 mmol/L/hr. Carbon dioxide could also be supplied to the gas exchanger for the purpose of pH control. Should the cell culture medium become to acidic addition of base kept the pH at a given set point. This type of bioreactor, now no longer on the market, was available from 0.4 L (research scale) to 24 L expanded bed volume size for production of proteins for human use. The design principles of the bioreactor as well as examples for cultivation of hybridoma and recombinant CHO cells at different scales are comprehensively described in Refs.(274,278,279). The general concept of the Verax-system inspired many research groups to investigate and optimize the fluidized bed technology for animal cell culture (Fig. 8).

One focus of research was the microcarrier itself. The growth of cells on different matrix materials, such as glass and polyethylene, was investigated. Porous Siran™ carriers made of borosilicate glass (Schottwerke, Mainz, Germany) were shown to be a cost-effective alternative to the weighted Verax microspheres (280–282). Chemical modifications of borosilicate Siran™ glass resulted in similar cell densities per milliliter packed bed volume as reported for the Verax microspheres (283,284). In contrast to stirred tank systems, a significantly higher specific gravity of >1.5 of the carriers is typically chosen to achieve homogenous fluidization in a fluidized bed bioreactor.

A major disadvantage of the fluidized beds was the progressive depletion of oxygen from the bottom to the top of the expanded bed. This problem of an oxygen gradient along the axis of the tubular reactor could be overcome by integration of a membrane oxygenation module directly into the fluidized bed (285,286). Alternatively, the integration of an in-line gasification tube module developed at the

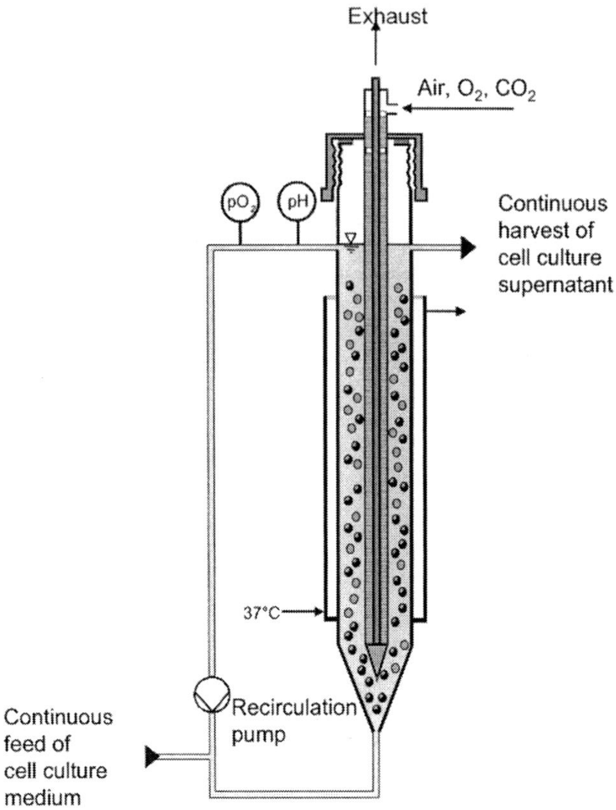

Figure 8 Schematic drawing of the fluidized bed bioreactor system with in-line aeration module for bubble and gradient free oxygen supply. (Courtesy of Papaspyrou Biotechnology.)

Research Center Jülich was suggested (287). This latter fluidized bed bioreactor with improved oxygenation is commercially available form B. Braun Biotech International (Melsungen, Germany). A schematic drawing of this bioreactor is presented in Fig. 9. This bioreactor type is available with settled carrier volumes ranging from 0.02 to 0.5 L. Typically, it is operated with gelatinized porous Siran^{TM} microcarriers that are available in different diameters. This system is used in the biotech industry for large-scale production of proteins for research and medical applications (288).

An alternative fluidized bed system based on internal recirculation of the culture medium was developed at the Institute of Applied Microbiology (289,290) in cooperation with Vogelbusch GmbH, Vienna, Austria (Cytopilot^{TM}). This system is available from laboratory scale to production scale (Vogelbusch GmbH). The fundamental differences between a conventional fluidized bed system and the Cytopilot^{TM} are illustrated in Fig. 9.

The Cytopilot^{TM} bioreactor comprises lower and upper cylindrical chambers. A draft tube in the fermentor replaces the external recycle loop. The lower chamber houses an axial flow impeller that provides a fluid flow to expand the bed. Oxygen is homogeneously microsparged into the downcomer close to the impeller and uniformly distributed in the upper chamber by a designed liquid flow and a gas distribution plate at the bottom entrance of the fluidized bed. The bed expands or contracts

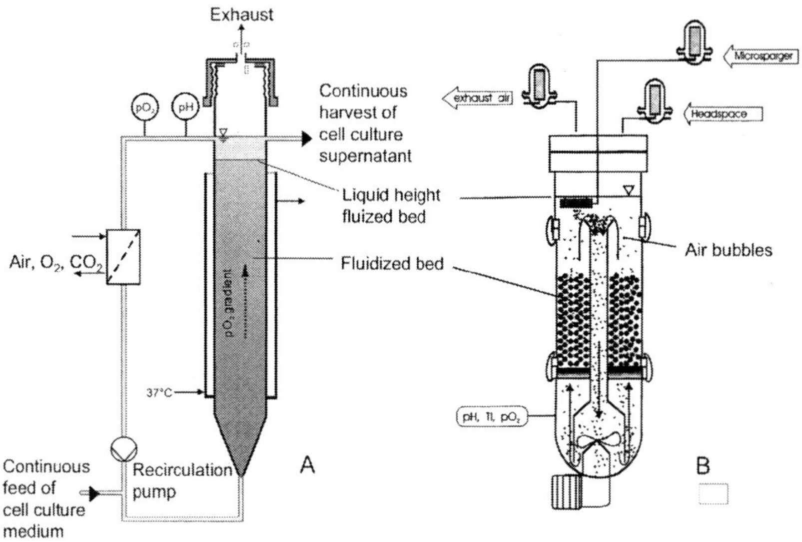

Figure 9 (A) Schematic drawing of a conventional fluidized-bed reactor for animal cell technology. (B) Schematic drawing of the modular fluidized bed bioreactor design Cytopilot^{TM} developed by the Institute of Applied Microbiology, Vienna. (Courtesy of Blüml, G.)

as a function of the stirrer speed that creates the necessary hydrodynamic pressure to lift the settled microcarriers. The lower chamber is additionally equipped with a heating circuit (double water jacket), sampling and harvest ports, pH and pO_2 sensors.

Bluml and colleagues developed a microcarrier type that consists of polyethylene weighted with chalk and silicates (291). The buoyant density of this carrier type is dependent on the ratio of these three compounds providing various densities suitable for use in stirred tank, fluidized bed or packed bed systems (291,292). This former "IAM-carrier" is now commercialized as Cytoline^{TM} (Amersham Biosciences) and is very often used in combination with the Cytopilot^{TM}.

The application of different types of microcarriers is described elsewhere in this book and has been reviewed previously (3,241). Verax microspheres, Siran^{TM} (Schottwerke) and Cytoline^{TM} (Amersham Biosciences) carriers have been successfully evaluated for protein production with anchorage dependent cell lines and suspension cells such as hybridomas. Very often "laboratory home-made" bioreactor designs were used for all kinds of comparative studies as summarized in Table 4.

Reliable determination of the cell density in immobilized cultures used to be a draw-back when using Siran^{TM} or Cytoline^{TM} carriers, since both carrier types cannot be solubilized enzymatically as is possible with proteinaceous materials such as collagen. This problem of growth monitoring has been overcome recently by the use of dielectric spectroscopy (284,293).

Packed Bed Bioreactors

Fixed bed or packed beds are an alternative bioreactor technology to fluidized beds for the immobilization of cells. Both terms are used as synonyms for immobilization systems that have been deduced from large-scale fixed bed reactors ori-

Table 4 Applications of Fluidized Bed Bioreactor Systems

Reactor type	Microcarrier type	Microcarrier properties: average diameter (mm); pore diameter (μm)	Cell line	Cell density (10^6/mL settled bed volume)	Application	References
Verax	Verax microspheres	0.5; 20–40	CHO	300–400	Protein expression	197, 275
			Fibroblast	300–400	Protein expression	275
			Hybridoma	100–200	Protein expression	275, 376
			Hybridoma	20	Protein expression	402
			MDBK	100–200	Research	479
			Vero	ND	Virus production	480
Laboratory home-made	Calcium alginate	ND	Hybridoma	39	Protein expression	220, 481
Laboratory home-made	Polyscience™	105–210	BHK	20[a]	Research	482
Laboratory home-made	Siran™	0.4–0.7; <120	BHK	70	Protein expression	483
		0.4–1.0; <120	BHK	40[a]	Research	482
		0.5–0.7; <120	BHK	300–400	Protein expression	282, 484
			Hybridoma	42	Protein expression	281
		0.1–1.0; <120	CHO, C-127	40–50	Research	485
		0.6; <120	CHO	10	Research	486
			GPK	ND	Research	242, 426, 484
Laboratory home-made	Siran™ modified	0.4–0.7; <120	Hybridoma	20	Protein expression	286
			Hybridoma	100–200	Other	284
			Hybridoma	ND	Pilot scale	487
			Hybridoma	50	Research	488
			CHO	100–300	Research	283
	Soda glass modified	1.0–2.0; 60–300	CHO	10–50	Research	283
			C-127	26	Protein expression	489
Laboratory home-made	Verax microspheres	0.5; 20–40	C-127	21	Protein expression	489
Cytopilot™	Cytoline™	0.4–1.1; 10–400	CHO	30–200	Protein expression	490–493
			HEK293	ND	Research	494
	Polyethylene	1.5–2.0; 100–300	CHO	50–300	Research	292, 495
		1200–1500; ND	CHO	120	Pilot scale	289
	Polypore™	2.0–2.5; 10–400	Hybridoma	ND	Research	496
Bioengineering (modified)	Cell aggregates	0.5	CHO	200–300	Research	497

[a] 10^6 cells/mL expanded bed volume; ND: not determined.

ginally designed for wastewater treatment. Initial work with mammalian cells started in the 1950s and aimed for tissue-like growth of cells in small-scale packed beds of cellophane (294), Raschig rings (295), and glass spirals (296). However, glass beads became the most widely used matrix. In 1976, Spier and Whiteside reported the production of foot and mouth disease vaccine in a packed bed system that could be scaled up to a unit process containing 100 L of medium (297,298). Glass-bead systems were considered for the production of interferon (299), herpes simplex virus (300), tissue plasminogen activator, and acetylcholinesterase (301).

The original design of packed beds was very simple and required two connected vessels; one holding the bed material, a second containing the culture medium and instrumentation for maintaining temperature, pH, and dissolved oxygen (4). The packed or fixed bed systems, initially operated with solid glass beads, did not achieve widespread use, mainly because they had a low surface area per unit volume and they were not suitable for suspension cells. These disadvantages were overcome by replacement of solid glass beads through porous glass spheres (280,302,303). Porous carriers were soon considered to have a greater potential for scale-up regarding both achievable cell density and increased surface-to-volume ratio (3,260).

Typically, packed bed systems are composed of a cylindrical bioreactor chamber filled with porous glass beads, a gas-exchanger (hollow fiber module or a stirred tank with air/oxygen sparging), a medium reservoir, and a pump that circulates medium between the bioreactor and the reservoir (301,302). As an alternative to glass carriers, other macroporous materials have been placed into bioreactors to support cell growth (251,304). The packed bed culture systems are designed for the immobilization of anchorage dependent or suspension cells providing high cell densities in the macroporous matrixes for the production of secretory proteins and lytic viruses. Different applications are listed in Table 5. Some of the major advantages of packed or fixed bed reactors are the low surface shear rates, the high cell density, and productivity achievable and that no particle–particle abrasion is obtained. Reported disadvantages are the blockage of the pores due to high cell densities, poor oxygen transfer and the risk for medium channeling in the bed (242). Porous Siran™ carriers of 3–5 mm diameter were shown to give highest cell yields. This is due to the open bed structure (~1 cm^2 channel cross-sectional area) and reduced channel blockage caused by increases in biomass, uneven distribution of the inoculum and media channeling within the bed (280,302,305,306).

Alternative matrices such as nonwoven fibrous polyester disks were evaluated (Fibra-cel™, New Brunswick Scientific) and packed into column-type reactors (307–310) or in a basket mounted into a modified stirred tank reactor (311–313). A schematic drawing of the CelliGen Plus™ bioreactor (New Brunswick Scientific) is shown in Fig. 10. These two alternative setups were used for the production of recombinant proteins and viruses. A similar fermentor concept has been developed by Meredos GmbH, Bovenden, Germany.

Packed or fixed bed perfusion bioreactors have been investigated for other applications than protein production. Research teams evaluated collagen coated reticulated polyvinyl formal resin (PVF) as a matrix for primary hepatocyte cultures for use as a bioartifical liver (12,314,315). Other researchers investigated the ex vivo expansion of hematopoietic cells in porous glass carriers for bone marrow transplantation or gene therapy applications (316). Packed bed bioreactors mimicking lung and liver function were characterized as potential model systems for pharmacokinetic analysis of new drugs (317).

Table 5 Applications of Packed Bed Bioreactor Systems

Microcarrier type	Microcarrier properties: average diameter (mm); pore diameter (μm)	Cell line	Cell density (10^6/mL settled bed volume)	Application	Reactor type	References
Siran™	3.0–5.0	Hybridoma		Research	External loop with STR	242, 305, 498
	60–300	CHO	20	Research	reservoir, laboratory	280, 302, 305
		GPK		Protein expression	Home-made	302
		Transfectoma		Protein expression		499
		JHAN		Virus production		500
		MDCK		Virus production		255
		U937		Virus production		500
		Hybridoma		Research	Radial-flow fixed bed	501
		Hybridoma		Protein expression	Meredos, modified STR	502
		CHO		Protein expression	Laboratory home-made	503
		U937		Virus production	Applikon STR modified	504
		JHAN		Research		504
	Various, 1.5;	BHK	10–11	Research		505
	< 120	Hybridoma	10	Protein expression	Laboratory home-made dialysis/radial-flow fixed bed	305, 323
Siran™ modified	0.4–0.7; < 120	Bone marrow cells	1–2	Cell therapy	Laboratory home-made	506
		Hematopoietic cells		Cell therapy		316
Siran™	2 × 2	CHO	10	Research	Laboratory home-made	507
Raschig rings		C-127	10			507
Static mixer	Laboratory scale	MRC-5		Virus production	Laboratory home-made	328
					External loop with reservoir	508
Polyurethane foam (PFU)	5 × 5 × 5	HEK 293		Research	Celligen™ with external loop	509
	10 × 10 × 1; 250	HEK 293	30–68	Protein expression	External loop, laboratory home-made	486

	0.8 × 37 × 37 (plates) or 1 × 47 (discs)	Vero	50–110	Research		304, 510, 511
		CHO	24–48			
Polyvinyl formal resin	2 × 2 × 2	Hepatocyte	8–12	Research	External loop, laboratory home-made	12, 314
Glass fiber		CHO	68	Protein expression	Annular packed bed, laboratory home-made	325
	50–120	MRC-5		Virus production	Laboratory home-made	326
		Murine bone marrow		Research	Laboratory home-made	327
Fibra-cel™		HEK 293		Protein expression	CelliGen Plus™ with internal retention device	312
		BHK		Virus production		313
		CHO		Research		512
		Hybridoma		Research		512
		Fibroblast	< 100	Protein expression		513
		Hybridoma		Protein expression	CelliGen™ with external loop	311
		Sf21 insect cells	6	Research		307, 309
Cytodex 3		H4IIE (rat hepatoma)		Other	External loop, laboratory home-made	317
		L2				
Ceramic particles		AtT20 (rat pituitary)	510	Research	Laboratory home-made	53, 514
Ceramic foam		CHO	10	Research	Laboratory home-made	507
		C-127				507
Cellsnow		CHO	29	Other	External loop, Applikon STR	515
		Hybridoma		Research	Laboratory home-made	501

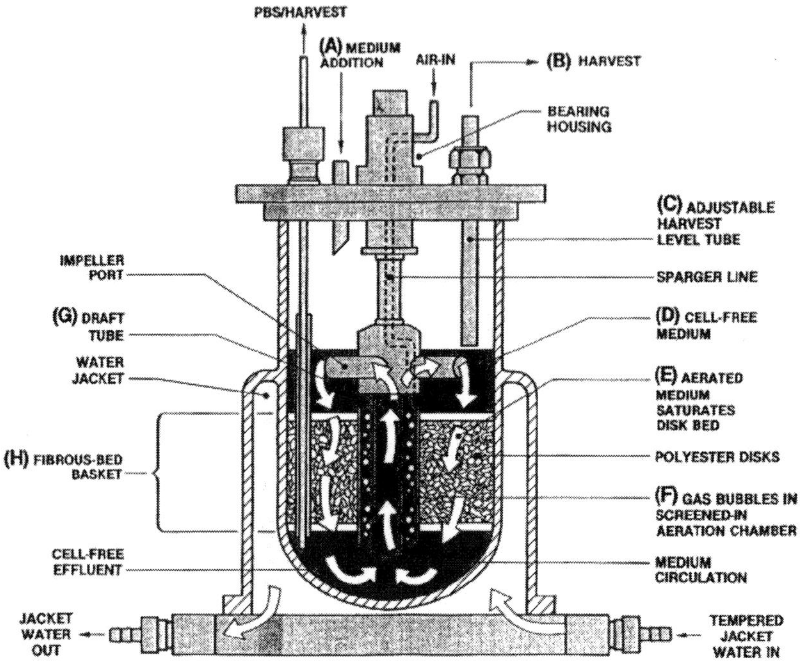

Figure 10 Schematic drawing of the Celligen Plus^{TM} bioreactor with cell-lift impeller. (Courtesy of New Brunswick Scientific.)

Other Reactor Types

The Opticell system, an automatically controlled system for anchorage-dependent and suspension cell culture, was developed in the 1980s. Anchorage-dependent cells were grown on a nonmodified ceramic matrix (35,243,318,319) whereas suspension cells required a modified ceramic matrix (320). The scalability of this system is proportional to the amount of surface provided and is limited in contrast to stirred microcarrier systems, fluidized bed and packed bed bioreactors (321).

A dialysis bioreactor with a radial-flow fixed bed for animal cell culture was derived form a dialysis membrane bioreactor used for microbial fermentation. This bioreactor consisted of two chambers, which were separated by a cylindrical membrane. The inner chamber was equipped with a centrifugal pump and packed with porous glass spheres and the centrifugal pump induced the radial flow of the medium. Aeration was done in the outer chamber that also served as medium reservoir (211,303,322–324). The fiber bed bioreactor was designed for anchorage dependent cell culture and was composed of a concentric-cylinder airlift reactor, in which the annulus is a packed bed of glass fibers. In this bioreactor, oxygen-containing gas is sparged into the inner draft tube. Bubble-free medium flows down through the fiber bed in the outer cylinder providing convective oxygen and nutrient transfer to the cells (325,326). This bioreactor concept has been applied to large-scale production of murine bone marrow cells (327). Design and scale-up of a bioreactor based on cell attachment and growth on Koch-Sulzer static mixing elements was reported for viral vaccine production (277,328).

BIOREACTOR OPERATION MODES

The following operation modes for cell culture bioreactors have been described in the literature; batch, fed-batch, continuous culture without cell retention (chemostat or cytostat), and continuous culture with cell retention (perfusion). During perfusion culture sometimes a cell bleed is applied or obtained due to noncomplete cell retention. The kinetics that is observed during this type of continuous culture is a combination of the two types of continuous mode, i.e., perfusion and chemostat. The different modes are discussed in more detail in this chapter. For a quick overview, the typical operation schemes and associated kinetics are shown in Fig. 11; the mass balance equations for ideal systems (i.e., excluding cell death, cell lysis, accumulation of product, product degradation, or nutrient degradation) are given in Table 6. Common for the different operation modes is that the temperature, the oxygen partial pressure, and the pH of the culture are controlled. This means continuous supply of oxygen and removal of carbon dioxide. A comparison of the different culture modes is attempted in Table 7.

Batch and Repeated Batch Cultures

In batch culture the bioreactor is charged with cells and medium, no further medium is added or withdrawn during the process and at the end of the culture the whole reactor content is harvested at once. The cells grow exponentially while nutrients are consumed and metabolic waste products are accumulated until the maximum cell density for the given medium is reached due to nutrient limitation or waste product inhibition (see Fig. 11). Historically, ammonium, lactate, and CO_2 have been considered as inhibitory metabolic waste products (82,83,266,329). However, there are reports in the literature on small inhibitory proteins (330,331) suggesting a key role in growth control (99). The length of

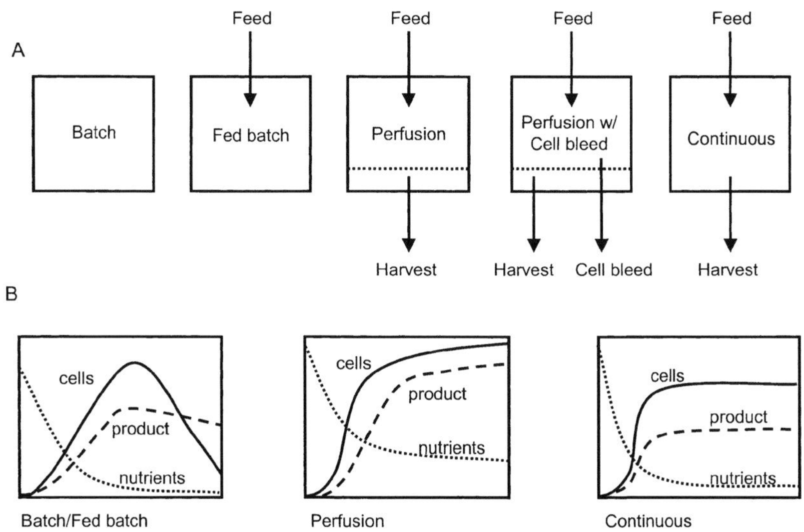

Figure 11 Operation principle (A) and kinetics (B) of cell growth, nutrient consumption, and product formation during batch, fed-batch, perfusion, and continuous operations. Shown are the concentrations of nutrients, product, and viable cells versus process time.

Table 6 Mass Balance Equations for Ideal Systems (No Cell Death, Cell Lysis, Accumulation of Product, Product Degradation)

	Growth	Substrate consumption	Product formation
Batch	$dC/dt = \mu C$	$dS_j/dt = q_{Si}C$	$dP_j/dt = q_{Pj}C$
Fed-batch	$dCV_R/dt = \mu C V_R$	$dS_i V_R/dt = q_{Si} C V_R$	$dP_j V_R/dt = q_{Pj} C V_R$
Continuous w/o cell retention	$dC/dt = \mu C - DC$	$dS_i/dt = -q_{Si}C + DS_i$ $D = F/V_R$	$dP_j/dt = q_{Pj}C - DP_j$
Continuous w/cell retention (perfusion)	$dC/dt = \mu C$	$dS_j/dt = -q_{Si}C + D(F_i - S_i)$	$dP_j/dt = q_{Pj}C - DP_j$
Continuous with partial cell retention (perfusion with cell bleed)	$dC/dt = \mu C - D_c C$ $Dc = Fc/VR$	$dSi/dt = -q_{Si}C + D(F_i - S_j)$ $D = F/V_R$	$dP_j/dt = q_{Pj}C - DP_j$

C: cell number, t: time, S_i: substrate concentration in the reactor, F_i: substrate concentration in feed, P_j: product concentration, D: medium exchange rate, D_c: cell bleed rate, F: feed flow rate, F_c: flow rate of cell bleed, V_R: reactor volume, μ: specific growth rate, qs_i: specific substrate consumption, qp_j: specific production rate.

Table 7 Comparison of Different Bioreactor Operation Modes

	Batch	Fed-batch	Chemostat	Perfusion
Cell density (mL^{-1})	10^6	10^6–$\leq 10^7$	10^5–10^6	10^7–$< 10^8$
Product concentration in arbitrary units	1	2–10	1	5–10[a] 5–100[b]
Culture duration	Days	Weeks	Months	Months
Space-time yields	Low	Moderate	Low	High
Scale-up	+++	+++	++	+ to ++[c]
Reliability	+++	++	++	+ to ++[c]

[a]Without product accumulation due to semipermeable membranes.
[b]With product accumulation.
[c]Scalability and reliability dependent on perfusion system.

the subsequent stationary phase is dependent on the cell type and finally the cells are dying during the declining phase. Sometimes a short lag phase after inoculation is observed which has been attributed to conditioning of the medium or the age of the preculture (332). In standard cell culture media cell densities in the range of 1–$3 \times 10^6\,mL^{-1}$ are reached with continuous cell lines. Significantly higher densities of $\geq 10^7\,mL^{-1}$ (333) have been reported for hybridoma cells using fortified media. However, this may not be generally achievable due to growth inhibition by metabolic waste products (334). Product formation can be growth associated, i.e., the specific productivity increases with increasing growth rate and ceases when the culture is entering stationary phase, or nongrowth associated, i.e., product formation increases with decreasing growth rate or is only observed during stationary phase (see Fig. 12). Several authors published data analyzing product formation in relation to the culture phase (329,335–337). The product formation kinetics has tremendous impact on the time of harvest during batch culture and the choice of cultivation mode in general (338).

Batch cultivation is simple and reliable and therefore the method of choice for many industrial applications (339). However, there has been an on-going debate on the pros and cons of continuous versus batch culture (340). Due to its simplicity it is also the most frequently applied mode for investigations of medium and culture parameters on growth, product formation and metabolism in process development, optimization, and scale-up (329,335,341–344). Therefore, most of the data available for commonly applied cell lines are derived from batch culture.

Repeated batch culture is very similar to batch culture with the exception that a fraction of the cell suspension is left in the bioreactor when the culture is harvested and fresh medium is added for a new batch growth cycle. For the success of this culture mode a high viability of the remaining cell suspension is important and optimization of the time when to initiate a new batch is necessary.

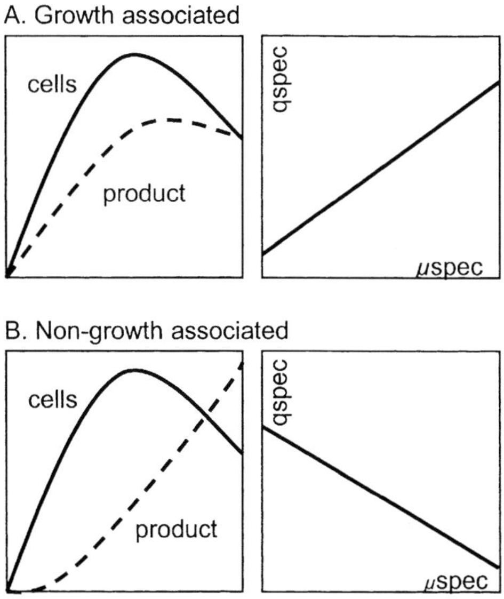

Figure 12 Basic types of product formation kinetics during batch operation. Abbreviations: μspec: specific growth rate, qspec: specific production rate.

Fed-Batch Culture

During fed-batch culture a continuous or intermittent feed of nutrients is applied to prolong cell growth and product formation in comparison to simple batch operation (see Fig. 11). Since no cell suspension or supernatant is withdrawn the bioreactor volume increases until the whole content is harvested at once. In fed-batch operation, in general, higher product concentrations are obtained due to the prolonged accumulation of product. Controlled feeding of key nutrients is often performed and the nutrient concentration in the culture is maintained at low levels (334,345–351), aiming to minimize the formation of toxic metabolites (329,352), e.g., lactate and ammonia to improve productivity (348,353–355). These approaches mainly involve development of nutrient feeding solutions and feeding rate control strategies to avoid nutrient depletion and to reduce lactate and ammonia formation by controlling glucose and glutamine concentrations at low levels.

This cultivation mode has been applied widely to hybridoma culture (see above references) and insect cell culture (356–364). Less reports are found on CHO fed-batch culture (365–368), however, it is a well-established technology for large-scale production of recombinant protein and antibody medicines. It has been applied to a range of cell lines such as HEK 293 (369), BHK (370), FS-4 (266). Using fed-batch mode the culture duration is typically prolonged from a few days to weeks. The cell density and productivity increase is dependent on the cell type. In insect cell fed-batch culture densities in the range of $5 \times 10^7\,mL^{-1}$ have been reported (364), which is equivalent to a 10-fold increase over batch. In CHO fed-batch cultures cell densities in the range of $5 \times 10^6\,mL^{-1}$ have been reported (365). In hybridoma culture the density could be increased to around $3–17 \times 10^6\,mL^{-1}$ (371) and productivity up to 10-fold compared to simple batch operation (334,348,372). Using fed-batch culture in combination with nutrient-fortified medium even higher cell densities ($> 10^7\,mL^{-1}$) were reported (373). For a review of fed-batch culture in monoclonal antibody production, see Refs. (351,353). The mathematical treatment was described by Glacken (374). In conclusion, fed-batch culture mode has been applied to improve productivity over conventional batch mode. Although it is a more sophisticated strategy with regard to process control it is still equipment-wise sharing the advantages of the simple and robust batch culture technique (see Table 7).

Continuous Culture

Continuous culture is characterized by a continuous flow of fresh medium into the bioreactor and removal of cell suspension from the culture system at the same rate, keeping the reactor volume at a constant level (see Fig. 11). An important criterion is that the reactor content is well mixed to ensure that the concentrations of cells, product and metabolites in the reactor are equal to those in the effluent stream. When cell growth is limited by a single nutrient, e.g., glucose, glutamine or oxygen, continuous culture is also called chemostat culture. For microbial systems, typically the cell density is constant over a wide range of dilution rates and only affected by a concentration change of the limiting nutrient in agreement with a saturation-type growth kinetics as described by the Monod model. Published data on continuous culture of animal cells differ from the expected simple Monod model in many cases suggesting that more complex equations govern the system (99,375). A linear decrease of the viable cell density with increasing dilution rates has been observed (375) or a relatively stable cell density at low dilution rates, which decreased toward higher dilution rates (100,104) as well as the opposite (376). Other researchers found data in accordance with a simple Monod model (377).

Microbial systems can be kept at constant cell densities (turbidostat) under nonsubstrate limited conditions by controlling the dilution rate. A similar approach was applied to run a cytostat (98) where almost arbitrary cell densities were maintained at similar dilution rates using identical feed medium. Despite the deviating behavior of continuous cultures, steady states regarding cell density and growth rate, product and metabolite concentrations, metabolic and production rates are obtained both in chemostat and cytostat operation after at least three to five times the residence time of the medium in the system. Typically, dilution rates applied are in the range of growth rates observed during batch culture, i.e., 0.3–1.4 day^{-1} and should not exceed the maximum growth rate to avoid wash-out of cells. Cell densities and product concentrations are in the same order as determined during batch culture. In contrast to batch culture, where transient changes of most parameters are observed, the continuous mode offers the advantage to study the effect of different culture variables under steady state environmental conditions.

Growth and metabolism have been characterized over a wide range of dilution rates for hybridoma cells (100–102,375,376,378–381), CHO cells (382,383), and BHK cells (384,385). Metabolic shifts toward a more efficient energy metabolism were described at very low nutrient concentrations for hybridoma (386) and BHK cells (387). Europa et al. (385) observed multiple steady states at similar dilution rates depending on the history of the culture allowing it to adjust to more desirable conditions for antibody expression.

Continuous culture has been used to investigate the effect of different culture conditions on a variety of cellular responses. Examples are the glycosylation pattern of BHK (387), CHO (367), and NS0 cells (388), cell death in hybridoma culture (330), Bcl-2 overexpression and viability (389), the effect of dissolved oxygen on cell density, energy metabolism and antibody production (390), shear sensitivity, cell cycle distribution and productivity of hybridoma cells (377), the effect of pCO_2 (83). Heidemann determined the K_s-values (Monod constants) of essential amino acids for CHO and hybridoma cell lines in continuous culture (107). Schmid et al. (391) employed the evolutionary potential of continuous culture to adapt hybridoma cells to increased shear forces.

Although continuous culture has proven to be a powerful research tool it is not very suitable for the commercial production of therapeutic proteins. There is a risk of rather unstable expression after a certain period in culture (185,392–394). The selection of low producer cell clones due to their frequently increased growth rate (395) is a major disadvantage besides the low cell densities and productivities obtained in continuous culture (see Table 7).

Perfusion Culture

Another type of continuous culture is perfusion culture where the cells are retained in or recycled back to the bioreactor while fresh medium is supplied and cell-free supernatant continuously removed at the same rate. Suspension cells may be retained in or recycled back to the stirred tank or airlift bioreactors using filtration, centrifugation or sedimentation techniques (see Table 3). Other methods are immurement in hollow fiber bioreactors, flat membrane reactors and dialysis reactors, entrapment in porous matrices cultured in fixed and fluidized beds or encapsulation in gelling polymers. Methods applied for anchorage dependent cells include microcarrier culture in stirred tanks equipped with filters or settlers, cultivation on porous microspheres or other growth substrates in fluidized and packed beds, culture in ceramic matrices and plate bioreactors. The details of these techniques have been described in previous sections of this review.

During perfusion operation the cell density increases steadily according to the chosen medium perfusion rate D until nutrient or oxygen limitation or waste product inhibition occurs and a quasi steady-state with respect to cell, metabolite and product concentration is reached (396). Typically, cell densities in the range of 5×10^6 to $5 \times 10^7\,mL^{-1}$ have been reported for homogeneous suspension cultures and even up to $5 \times 10^8\,mL^{-1}$ for hollow fiber systems (397). Thus, the engineering challenge regarding oxygen and medium supply is considerable (398). Advantages of this technique are the high cell densities and product concentrations obtained leading to increased space-time yields when compared to batch, chemostat and even fed-batch operation (33,150,152,161,209) (397,399–401). It is difficult to make a general statement on the achievable productivity increase since it is dependent on a series of parameters related to the cell, the medium and the applied system. Typically, a productivity increase in the range of 5–10-fold is found in the literature (109).

Although perfusion is more complex than batch culture and therefore considered to be more prone to contamination it has become an established technique for large scale commercial production of recombinant proteins (35,110,116,209) (402,403). Under certain circumstances, e.g., when growth associated product formation is observed or when product formation is dependent on the cell specific perfusion rate (398), it is better to control the growth rate and cell density by applying a cell bleed (130,404). Another advantage is that the accumulation of dead cells and debris and associated negative effects such as release of proteolytic enzymes or DNA are prevented which is otherwise observed in membrane based cell retention systems.

Growth, metabolism, and product formation during perfusion with a controlled cell bleed has been investigated by a series of researchers (103,104,405,406). Frequently, a natural cell bleed is obtained in packed and fluidized bed cultures or during spinfilter, settler or centrifuge based perfusion of suspension cells (93,132,152,159). The productivity of hybridoma perfusion systems was further increased by the application of fortified or nutrient enriched media feeds (407) or by a separately controlled substrate feed (109) to minimize toxic metabolite production.

SELECTION AND DESIGN OF BIOREACTORS

General

There is no single, superior bioreactor suitable for all applications. The bioreactors or cultivation systems describe offer different advantages and their applicability is limited for various reasons. Therefore, some considerable thought has to be given to the selection of a specific bioreactor type and its design. Literature precedents may provide some guidance, however, they should be reviewed carefully and alternatives should be investigated. It is a time consuming and expensive task to undertake an experimental evaluation to compare the relative performance of competing culture systems for a specific application. A range of technical, biological and regulatory issues govern the decision on whether inherent characteristics or complexities would make the system risky for a commercial purpose. In many cases, especially in a nonindustrial setting the choice of reactor system is mainly influenced by the familiarity of the operator with the technology, the availability of space and services; and the investment and operational costs. For industrial applications, homogeneous culture of suspension cells, cell aggregates or adherent cells on microcarriers in STRs can be considered as the predominant technology due to their scale-up potential, flexibility and environmental control features. Some guidelines for mechanical design, materials selection and finishes especially for STRs are given in Refs. (408,409) and summarized in Table 8.

Figure 13 tries to summarize the different issues that need to be considered for selection of a suitable bioreactor type and its intrinsic design specification. The grouping is somewhat arbitrary since one aspect may belong to different categories and parameters are related. It is recommended to start off with a clear definition of the purpose of the bioreactor and potential future changes to the initial framework of conditions. The next step in the process of bioreactor selection and design is to compile a requirement specification list for the application under consideration taking into account the users specific circumstances. Such a document could simply follow the different issues listed in Fig. 13 and the level of detail will be governed by the expertise of the team involved with the task.

Important general issues independent of the reactor type are the availability and quality of service. The need for service can vary with the circumstances. A large company with many bioreactors of the same type probably might have a competent internal technical support. The initial availability and quality of steam, pressurized air, process gases, electricity, cooling water and drain limits the choice of systems or at least impact strongly on the design. Therefore, hollow fiber bioreactors are frequently found in standard laboratory facilities since these systems are designed accordingly. However, most bioreactor types are available in designs appropriate to limited media and electricity supplies.

Table 8 Recommendations for the Mechanical Design of Stirred Tank Cell Culture Bioreactors

Materials	Stainless steel 316 L (1.4404) or other high alloy grades for improved chloride corrosion resistance
	Borosilicate glass for vessels or sight glasses
	Silicone rubber and EPDM (ethylene propylene diene monomer) for gaskets
	PTFE (polytetrafluoroethylene) for electrode sleeves, support bearings
	Ceramics and graphite for mechanical seals
Finishes	Electropolished surface, $R_a < 0.4\,\mu m$ (R_a: arithmetic mean roughness)
	Avoid internal threads
Welding	Automated welding methods such as orbital welding
Aspect ratio	1:1 to 3:1, typically 2:1
Impeller	Axial flow impellers such as marine type, large pitch bladed, segment impeller
	Impeller diameter ≥ 0.5 vessel diameter
Sparger	Ringsparger (pore size ~ 1 mm), microsparger (pore size 5–10 μm) or bubble-free via silicone tubings

[Modified from Ref. (408).]

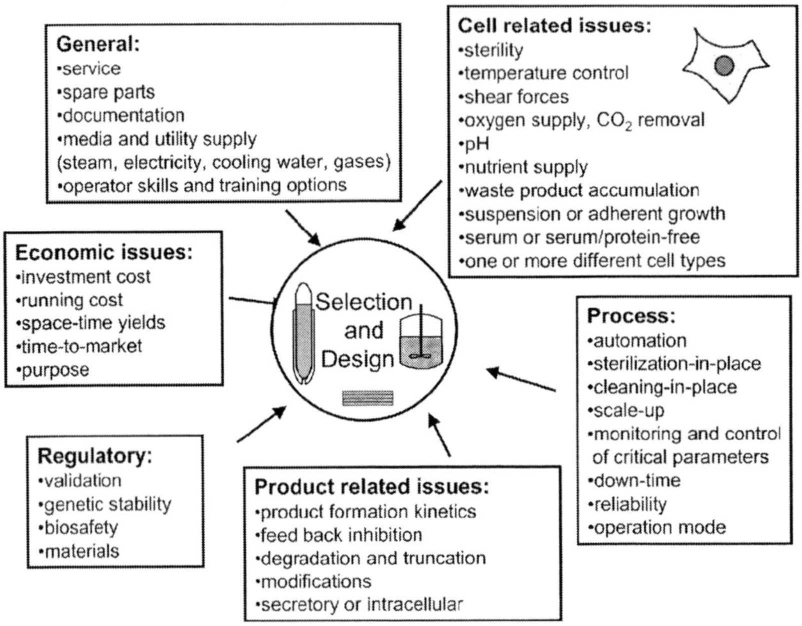

Figure 13 Factors affecting the design and selection of bioreactors.

Sterile Operation

The most important requirement for cell culture bioreactors is long-term sterile operation due to the slow growth of animal cells (maximum population doubling times in the order of a day are typical). Appropriate sterile design is very important for continuous processes, which may be conducted for periods of months to up to half a year (35,275,410). Small-scale equipment made of glass is generally autoclaved whereas larger bioreactors made of stainless steel are commonly in-situ steam sterilized. Especially, the design of seals, valves, pumps and transfer lines and the sterilization procedures and sequences have to be considered. Double mechanical seals are used for rotating shafts or alternatively magnetically coupled systems. In the sterile, cell and product containing part of the bioreactor, membrane valves are applied as well as peristaltic pumps, which are also appropriate for cell recycle. For successful sterilization complete drainage of condensate and removal of residual air has to be ensured to avoid cold spots and air pockets. If presterilized units are used, coupling of reservoirs and transfer of medium and inoculum needs special attention. In particular, the degree of complexity of a bioreactor or even bioprocess plant has a large impact on a careful sterile design. Interesting contributions on bioreactor sterility and sterilization are found in (411–415).

Economic Issues

Even though many people feel that economic issues should not determine the selection of a certain bioreactor system it may be the ultimate criterion for the decision. Here, short-term issues such as the investment cost have to be balanced carefully to rather long-term issues such as running costs, achievable space-time-yields and if the bioreactor shall be used for different purposes. Other long-term considerations are time-to-market for a biopharmaceutical or time-to-patient for a medical application, which are affected by the time needed for validation and the probability of process related difficulties during clinical tests.

Process-Related Issues

For industrial applications sterilization-in-place and cleaning-in-place, the degree of process automation, appropriate process control and monitoring of critical culture parameters as well as compatibility to up- and downstream operations are important issues for the choice of a reactor system. Furthermore, downtime and reliability of the bioreactor and process are other critical factors. These parameters are mainly dependent on the complexity of the operations, the operator skills and training. In a commercial context, scalability of the culture system has to be emphasized in particular due to the frequently enormous production capacities required to secure supply to the patients. Yearly production capacities of ten to hundred of kilograms were estimated for therapeutic monoclonal antibodies depending on the dose (348,416). Other examples demanding large production capacities are recombinant therapeutic proteins such as tPA, EPO, blood factor VIII (36), and vaccines used in broad vaccination programs (37).

Regulatory Issues

For the production of biopharmaceuticals regulatory guidelines affecting the design and operation of bioreactors have to be considered. Issuing authorities are the Food and Drug Authority in the United States (417), the EMEA in Europe (418), and national regulatory authorities. The international conference of harmonization (419) aims to provide one single guiding document, which is independent of the country. These guidelines cover equipment construction, installation and operational qualification and finally process validation where in a number of consecutive runs the consistency of the process has to be demonstrated via a reproducible quantity and quality of the product. However, they cannot be read like a technical instruction manual, instead an assessment of the specific application has to be made. Concerning equipment construction the 21 Code of Federal Regulations, parts 210 and 211 states that "Equipment shall be constructed so that surfaces that contact components, in-process materials, or drug products shall not be reactive, additive, or absorptive so as to alter the safety, identity, strengths, quality, or purity of the drug product beyond the official or other established requirements" (420–422). Relevant technical guidance documents on pharmaceutical engineering are provided by the "Society for Pharmaceutical Engineering" (423).

Process safety regulations govern the contamination control of the product with impurities such as residual host cell protein, DNA, microbes and adventitious viruses. In multipurpose facilities, cross-contamination caused from previous production campaigns of other substances have to be considered. A major focus concerning the mechanical design of a bioreactor or plant in this regard is on sterility and cleaning, i.e., removal of contaminants as proven during validation. More recently, the use of disposable materials and bioreactors to circumvent contamination problems and minimize validation costs has been discussed.

Other important aspects related to regulatory requirements are product degradation, truncation, and modification, which may be affected by the choice of the culture system and the process parameters. An example is the potentially increased product degradation in high cell density systems with product accumulation due to the increased concentration of proteases and the eventual release of intracellular proteases due to cell death and subsequent lysis. Proteolytic degradation of complex proteins has been observed in cell cultures of BHK cells known for their high proteolytic activity (424–426). Additionally, the genetic stability of a production cell line will impact on the choice of the process mode and culture conditions and therefore indirectly affects the selection of a bioreactor type. This is especially important for continuous operation when growing cultures in perfusion and chemostat mode. One clear advantage in this respect are growth arrested continuous cultures, which can be obtained in high cell density systems without cell bleed or via changing the culture temperature (427), adding chemicals (428,429), or using genetically engineered cell lines (430–432).

Biosafety Issues

Biosafety assessment of the cell line and process will have impact on the containment design of the culture to control the risk to health or environment posed by the genetically modified organism. Biosafety regulations or guidelines are issued by the National Institute of Health (433) in the United States, the World Health Organisation WHO (434), the European Commission (435,436), and different national authorities. These regulations are relevant for the design of seals, valves, air vents, sampling ports, addition ports, and other parts of a bioreactor where a potential risk for release of hazardous organisms can be assumed. No particular design solutions are prescribed since the focus is on performance of the containment design in order to allow for technical improvement. Detailed technical information valid for the EC is found in the CEN (European Committee for Standardization) standards (437) which are a technical complement to the legislation. Some containment design considerations for bioreactors are described in (438).

Cell Line and Product-Related Issues

Probably better covered in the scientific literature are design and selection considerations related to the cell type and product class. A basic demand is appropriate temperature control. This needs special attention in the case of cells passing through external loops where control of the environmental conditions can be poor dependent on the design and the residence time of the fluid in the loop. Commonly, mammalian cells are grown at 37°C and insect cells at 27°C. However, the beneficial effect of a decreased culture temperature has been shown for various mammalian cells (328,427,439,440). Other temperature control profiles have been suggested to control growth and productivity (441). It is known that membrane fluidity of mammalian cells is temperature dependent and therefore a temperature effect on shear sensitivity cannot be excluded (442,443). All this emphasizes the importance of a reliable and precise control of culture temperature.

Parameters affecting the shear sensitivity of animal cells include the concentration of serum, proteins, and surfactants. Although the shear tolerance is cell line dependent (66,71,444) and encapsulation in gel-forming polymers or entrapment in macroporous carriers may protect the cells or reduce the accessibility (250) it clearly affects the design and operation conditions of stirred tank, airlift, fluidized, and packed bed bioreactors. The generated shear rates depend on the design and operation parameters of impellers and pumps and in particular the sparging conditions as outlined in different sections of this chapter. Bubble-free aeration is gentler, however scale-up is difficult regardless the system used. Due to the intense research and development related to shear force generation, operation conditions and bioreactor configurations have been identified that are suitable for large-scale gentle cell cultivation in stirred tank, airlift and fluidized bed bioreactors. In parallel, other systems have been developed where the cells are not exposed to mechanical forces, e.g., hollow fiber bioreactors (189), the static maintenance bioreactor (212), and the membrane bioreactor (213). The quantification of the shear sensitivity of different cell lines under various process conditions (66,443) in combination with practical experience showed clearly that animal cells are not at all that sensitive to mechanical forces as originally assumed (48).

The minimization of mechanical forces is not the only design criterion for oxygenation systems. CO_2 removal and sufficient oxygen transfer are other important criteria for the selection of a bioreactor system. Controlling the oxygen partial pressure in cell cultures is of utmost importance for cell growth, productivity and metabolism as shown by many authors (382,445,446). In general, oxygen supply is a major engineering challenge in heterogeneous system, e.g., fluidized bed, packed bed and solid bed bioreactors, and high cell density systems, e.g., hollow fiber systems, perfusion and fed-batch bioreactors (398). The oxygen transfer characteristics and limitations of the different systems are reviewed in the individual sections describing the reactor type. Different technical solutions for improved oxygen supply are also discussed. Related to oxygen supply is also the control of an appropriate culture pH. Generally, mammalian cell culture media are buffered using the bicarbonate buffer system and the culture pH is affected by CO_2 and lactate formation. If base addition is used to compensate for acid formation a concurrent increasing osmolality is observed. It has been shown that the culture pH, CO_2, and an increased osmolality may affect productivity and glycosylation (82–85). Therefore, for sensitive cell lines a culture system should be chosen allowing for pH monitoring and control via addition of base or ventilation of the culture system avoiding major osmolality changes. Beside temperature, pH, and pO_2 control, the supply of nutrients and removal of waste products affect cell cultures. Significantly improved product yields are typically obtained from continuous perfusion and fed-batch processes in comparison to simple batch mode.

Product feedback inhibition needs consideration in high cell density culture since it will prohibit the application of high cell density systems where the product is accumulated in the cell containing compartment (e.g., hollow fiber systems, microencapsulation, membrane filtration in combination with fouling or fed-batch systems) (447). The product localization, i.e., secreted into the medium, membrane bound or intracellular, and the nature of the product, i.e., produced by the cells or the cell itself, is affecting bioreactor design and selection. For the production of secretory products continuous perfusion operation of fluidized and packed bed bioreactors provide an interesting alternative to conventional stirred tank technology. If the cells need to be harvested for further processing easy accessibility and quick, safe and reliable harvest procedures are advantageous. The product formation kinetics dictates the optimum operation mode and therefore indirectly the selection of a bioreactor type, since not all reactors are suitable for a given mode.

The cell type, i.e., anchorage dependent or suspension, has significant impact on the design and selection of an appropriate bioreactor. Although a series of bioreactors can be modified to be suitable for the cultivation of both cell types (e.g., stirred tank, airlift, fluidized bed with or without different types of solid or macroporous carriers) there are applications that require a certain bioreactor type since significant improvements in productivity are obtained, or the cell line cannot be grown or do not show the required functionality in a "multipurpose" culture system. A prominent example is artificial organs where a certain three dimensional architecture is required for an organ-like behavior (448). Especially in the case of a single purpose bioreactor one would decide for the most appropriate type and design for the specific application, i.e., yielding optimum productivity and product quality. However, true multipurpose design is required when different cell lines are cultured for the production of various intracellular, membrane bound or secretory products, e.g., in drug discovery using recombinant proteins for structure determination or as tools for high throughput screening of novel drugs. Other applications that benefit from multipurpose design are process development and pilot production for different classes of biopharmaceuticals.

Although a series of growth and productivity promoting serum-free or even protein-free media have been developed during the last 15 years (449) not all processes are conducted at low protein concentrations. This is due to the requirement of a particular cell line for a certain protein concentration and/or availability of growth factor(s) or simply the reluctance to adapt the cell line to low protein serum-free conditions. As the viability decreases cellular proteins and DNA are released in the culture supernatant due to cell lysis. The increasing concentration of proteins and DNA are causing membrane fouling and reduced performance of membrane based cell retention systems (118,161,450,451). Therefore, the use of centrifuges or settlers is frequently preferred. Another important aspect using low protein serum-free media is the increased chemical reactivity of such media especially in the presence of high chloride concentrations degrading certain stainless steel qualities. In conclusion, the choice of a bioreactor system and its design is very individual and needs careful consideration of the specific demands of the application.

CONCLUSION

A wide range of reactor types has been suggested in the literature for animal cell culture including stirred tanks, airlift reactors, fluidized bed reactors, packed bed and solid bed reactors, hollow fiber and membrane reactors. The number of different types is even increased for small-scale simple systems such as spinner, roller, and shake flasks. Typical operation modes are batch, fed-batch, and perfusion although the suitability is strongly dependent on the application, the cell line, and the product. For certain bioreactor systems a broad applicability for different cell lines and purposes has been demonstrated. However, there is no single cell culture system that is universally suitable for all applications in cell culture technology. For commercial production of recombinant proteins, monoclonal antibodies, or vaccines, homogenous culture of suspension cells, cell aggregates, or adherent cells immobilized on microcarriers in SRTs operated in batch, fed-batch, or perfusion mode can be considered as the predominant technology due to its maturity, scale-up potential, flexibility, and environmental control. The specialist reactor designs are experiencing a renaissance at present within the field of artificial organs and tissue culture for replacement therapy (452). At the same time, conditions for using the different suggested reactor designs for the production of viral vectors for genetic vaccination or gene therapy are being investigated (16–19). Apart from these novel applications bioreactor development has reached a high degree of maturation and the focus of cell culture technology has moved to biological issues and novel applications of cell and tissue culture.

ACKNOWLEDGMENTS

The authors would like to express their gratitude to Ian Taylor for critical reading of this manuscript, to Jack Kidane, Yannick Wurm, and Eva Eriksson for helping with the set-up of the literature database, and to AstraZeneca and especially Anders Karlström for economic support and encouragement. Special thanks go to our families for their love, support, and patience, which allowed completion of this manuscript.

REFERENCES

1. Harrison RG. Observations on the living developing nerve fiber. Proc Soc Exp Biol Med 1907; 4:140–143.
2. Carrel A. On the permanent life of tissues outside the organism. J Exp Med 1912; 15:516–528.
3. Griffiths JB. Advances in animal cell immobilization technology. In: Spier RE, Griffiths JB, eds. Animal Cell Biotechnology. Vol. 4. New York: Academic Press, , 1990:149–166.
4. Spier RE, Kadouri A. The evolution of processes for the commercial exploitation of anchorage-dependent animal cells. Enzyme Microb Technol 1997; 21:2–8.
5. Finter NB. Animal cell culture: the problems and rewards. In: Spier RE, Griffiths JB, Meignier B, eds. Production of Biologicals from Animal Cells in Culture. Oxford: Butterworth-Heinemann, 1991:3–12.
6. Scheirer W. Animal cell immobilization. In: Durant G, Bobichon L, Florent J, Eds. 8th International Biotechnology Symposium. Paris: Avignon 1988:216–227.
7. Sinclair A. Biomanufacturing capacity: current and future requirements. J Commercial Biotechnol 2001; 8(1):43–50.
8. Nyberg SC, Shatford RA, Peshwa MV, White JG, Cerra F, Hu WS. Evaluation of a hepatocyte-entrapment hollow fiber bioreactor: a potential bioartificial liver. Biotechnol Bioeng 1993; 41:194–203.
9. Hu WS, Friend JR, Wu FJ, Sielaff T, Peshwa MV, Lazar A, Nyberg SL, Remmel RP, Cerra FB. Development of a bioartificial liver employing xenogenic hepatocytes. Cytotechnology 1997; 23(1):29–38.
10. Maki T, Otsu I, O'Neil JJ, Dunleavy K, Mullon CJP, Solomon BA, Monaco AP. Treatment of diabetes by xenogenic islets without immunosuppression: use of a vascularized bioartificial pancreas. Diabetes 1996; 45:342–347.
11. Pörtner R, Rössing S, Stange J, Fassnacht D. Fixed-bed reactors for animal cell culture cultivation: an approach to artificial organs. In: Merten O-W, Perrin B, Griffiths B, eds. New Developments and New Applications in Animal Cell Technology. Dordrecht: Kluwer Academic Publishers, 1998:657–659.
12. Yanagi K, Moyoshi H, Ohshima N. Improvement of metabolic performance of hepatocytes cultured in vitro in a packed-bed reactor for use as a bioartificial liver. J ASAIO 1998:M436–M440.
13. Moussy Y. Bioartificial kidney. I. Theoretical analysis of convective flow in hollow fiber modules: application to a bioartificial hemofilter. Biotechnol Bioeng 2000; 68(2):142–152.
14. Collins PC, Nielsen LK, Patel SD, Papoutsakis ET, Miller WM. Characterization of hematopoietic cell expansion, oxygen uptake, and glycolysis in a controlled, stirred-tank bioreactor system. Biotechnol Prog 1998; 14(3):466–472.

15. Carswell KS, Papoutsakis ET. Culture of human T cells in stirred bioreactors for cellular immunotherapy applications: shear, proliferation, and the IL-2 receptor. Biotechnol Bioeng 2000; 68(3):328–338.
16. Boyd JE, Borland L, Fisher CE. Scale-up and optimisation issues in gene therapy viral vector production. In: Merten O-W, Perrin B, Griffiths B, eds. New Developments and New Applications in Animal Cell Technology. Dordrecht: Kluwer Academic Publishers, 1998:509–513.
17. Ostrove JM, Iyer P, Marshall J, Vacante D. Comparison of manufacturing techniques for adenovirus production. In: Merten O-W, Perrin B, Griffiths B, eds. New Developments and New Applications in Animal Cell Technology. Dordrecht: Kluwer Academic Publishers, 1998:515–521.
18. Iyer P, Ostrove JM, Vacante D. Comparison of manufacturing techniques for adenovirus production. Cytotechnology 1999; 30(1/3):169–172.
19. Merten OW, Cruz PE, Rochette C, Geny-Fiamma C, Bouquet C, Goncalves D, Danos O, Carrondo MJT. Comparison of different bioreactor systems for the production of high titer retroviral vectors. Biotechnol Prog 2001; 17(2):326–335.
20. De Bruyne NA. The design of bench-scale reactors. In: Spier RE, Griffiths JB, eds. Animal Cell Biotechnology. Vol. 3. New York: Academic Press, 1988:141–176.
21. Spier RE, Griffiths B. An examination of the data and concepts germane to the oxygenation of cultured animal cells. Dev Biol Stand 1984; 55:81–92.
22. Shirokaze J, Nogawa M, Ogura R. High density culture macroporous microcarrier. In: Spier RE, Griffiths JB, Berthold W, eds. Animal Cell Technology: Products of Today, Prospects for Tomorrow. Oxford: Butterworth-Heinemann, 1994:261–263.
23. Heidemann R, Riese U, Lutkemeyer D, Buntemeyer H, Lehmann J. The Super-Spinner: a low cost animal cell culture bioreactor for the CO_2 incubator. Cytotechnology 1994; 14(1):1–9.
24. Selander M, Hassel J, Dartsch Ch, Fenge C, Lüllau E. Characterisation and evaluation of a new screening system for animal cell culture. In: Lindner-Olsson E, Lüllau E, Chatzisavido N, eds. Animal Cell Technology: From Target to Market. Proceedings of the 17th ESACT Meeting 2001. Dordrecht: Kluwer Academic Publishers, 2002:341–343.
25. Goosen MFA. Insect cell cultivation techniques for the production of high-valued products. Can J Chem Eng 1991; 69:450–456.
26. Martinelle K, Doverskog M, Jacobsson U, Chapman BE, Kuchel PW, Häggström L. Elevated glutamate dehydrogenase flux in glucose-deprived hybridoma and myeloma cells: evidence from $^{1}H/^{15}N$ NMR. Biotechnol Bioeng 1998; 60(4):508–517.
27. Cowger NL, O'Connor KC, Hammond TG, Lacks DJ, Navar GL. Characterization of bimodal cell death of insect cells in a rotating-wall vessel and shaker flask. Biotechnol Bioeng 1999; 64(1):14–26.
28. Hunt L, Batard P, Jordan M, Wurm F. Fluorescent proteins in animal cells for process development: optimization of sodium butyrate treatment as an example. Biotechnol Bioeng 2002; 77(5):528–537.
29. Weiss S, John GT, Klimant I, Heinzle E. Modeling of mixing in 96-well microplates observed with fluorescence indicators. Biotechnol Prog 2002; 18:821–830.
30. Hermann R, Lehmann M, Büchs J. Characterization of gas-liquid mass transfer phenomena in microtiter plates. Biotechnol Bioeng 2003; 81(2):178–186.
31. Palomares LA, Ramirez OT. Bioreactor scale-down. In: Spier RE, ed. Encyclopedia of Cell Technology. Vol. 1. New York: Wiley , 2000:174–183.
32. Singh V. Disposable bioreactor for cell culture using wave-induced agitation. Cytotechnology 1999; 30(1/3):149–158.
33. Reuveny S, Velez D, Miller L, MacMillan JD. Comparison of cell propagation methods for their effect on monoclonal antibody yield in fermentors. J Immunol Methods 1986; 86:61–69.
34. Schürch U, Cryz SJ Jr, Lang AB. Human hybridoma producing antibodies to pseudomonas aeruginosa: scale up and optimization of IgM production in stirred bioreactors.

In: Spier RE, Griffiths IB, MacDonald C, eds. Animal Cell Technology: Developments, Process and Products. Oxford: Butterworth-Heinemann, 1992:336–341.
35. Bödeker BGD, Newcomb R, Yuan P, Braufman A, Kelsey W. Production of recombinant factor VIII from perfusion culture: I. Large-scale fermentation. In: Spier RE, Griffiths IB, MacDonald C, eds. Animal Cell Technology: Products of Today, Prospects for Tomorrow. Oxford: Butterworth-Heinemann, 1994:580–583.
36. Griffiths B. Animal cell products, overview. In: Spier RE, ed. Encyclopedia of Cell Technology. Vol. 1. New York: Wiley, 2000:70–76.
37. Aunins J. Viral vaccine production in cell culture. In: Spier RE, ed. Encyclopedia of Cell Technology. Vol. 2. New York: Wiley, 2000:1182–1217.
38. Werner RG. Potential and efficiency in the biotechnical process. Pharm Technol Europe 1994; 20–28.
39. Henzler H-J. Verfahrenstechnische Auslegungsunterlagen fur Rührbehälter als Fermenter. Chem-Ing-Tech 1982; 54(5):461–476.
40. Reuss M. Stirred tank bioreactors. In: Asenjo JA, Merchuk JC, eds. Bioreactor System Design. New York: Marcel Dekker, 1995:207–255.
41. Prokop A, Rosenberg MZ. Bioreactor for mammalian cell culture. In: Fiechter A, ed. Vertebrate Cell Culture II and Enzyme Technology. Advances in Biochemical Engineering/Biotechnology. Vol. 39. Berlin: Springer-Verlag, 1989:29–71.
42. Fenge C, Klein C, Heuer C, Siegel U, Fraune E. Agitation, aeration and perfusion modules for cell culture bioreactors. Cytotechnology 1993; 11:233–244.
43. Nienow AW, Langheinrich C, Stevenson NC, Emery AN, Clayton TM, Slater NKH. Homogenisation and oxygen transfer rates in large agitated and sparged animal cell bioreactors: Some implications for growth and production. Cytotechnology 1996; 22:87–94.
44. Palomares LA, Ramirez OT. Bioreactor scale-up. In: Spier RE, ed. Encyclopedia of Cell Technology. New York: Wiley, 2000:182–201.
45. Doran PM. Bioreactors, stirred tank. Spier RE, ed. Encyclopedia of Cell Technology. Vol. 1. New York: Wiley, 2000:249–278.
46. Varley J, Birch J. Reactor design for large scale suspension animal cell culture. Cytotechnology 1999; 29(3):177–205.
47. Nienow A. Gas dispersion performance in fermenter operation. Chem Eng Prog 1990; 86(2):61–71.
48. Henzler HJ, Kauling DJ. Oxygenation of cell cultures. Bioprocess Eng 1993; 9:61–75.
49. Feder J, Tolbert WR. The large-scale cultivation of mammalian cells. Sci Am 1983; 248:24–31.
50. Kamen AA, Tom RL, Caron AW, Chavarie C, Massie B, Archambault J. Culture of insect cells in a helical ribbon impeller bioreactor. Biotechnol Bioeng 1991; 38:619–628.
51. Jolicoeur M, Chavarie C, Carreau PJ, Archambault J. Development of helical-ribbon impeller bioreactor for high-density plant cell suspension culture. Biotechnol Bioeng 1992; 39:511–521.
52. Johnson M, André G, Chavarie C, Archambault J. Oxygen transfer rates in a mammalian cell culture bioreactor equipped with a cell-lift impeller. Biotechnol Bioeng 1990; 35:43–49.
53. Shi Y, Ryu DDY, Park SH. Performance of mammalian cell culture bioreactor with a new impeller design. Biotechnol Bioeng 1992; 40:260–270.
54. Girard HC, Okay G, Kilvilcim Y. Use of the vibrofermenter for multiplication of BHK cells in suspension and for replication of FMD virus. Bull Off Int Epiz 1973; 79:805–822.
55. Thalmann E, Berg GJ, Kopp TJ. Kinetics of BHK-21 cells in a bioreactor with a new aeration system through a mesh (CHEMCELL). Dechema Biotechnology Conferences 2. VCH Verlagsgesellschaft, 1988:3–11.

56. Blasey HD, Brethon B, Hovius R, Vogel H, Tairi AP, Lundström K, Rey L, Bernard AR. Large scale transient 5-HT3 receptor production with the Semliki Forest Virus Expression System. Cytotechnology 2000; 32:199–208.
57. Monahan PB, Holtzapple MT. Oxygen transfer in a pulse bioreactor. Biotechnol Bioeng 1993; 42:724–728.
58. Jäger J, Büntemeyer H, Lehmann J. Increasing the rate of production of monoclonal antibodies by combining a stirred reactor with a hollow fiber perfusion system. In: Neijssel OM, van der Meer RR, Lüyben KChAM, eds. Proceedings of the 4th European Congress on Biotechnology. Amsterdam: Elsevier Scientific Publishers, 1987:581–584.
59. Lehmann J, Vorlop J, Büntemeyer H. Bubble-free reactors and their development for contiuous culture with cell recycle. In: Spier RE, Griffiths JB, eds. Animal Cell Biotechnology. Vol. 3. New York Academic Press, 1988:221–237.
60. Vorlop J, Lehmann J. Scale-up of bioreactors for fermentation of mammalian cell cultures, with special reference to oxygen supply and microcarrier mixing. Chem-Ing-Tech 1988; 11:171–178.
61. Midler M Jr, Finn RK. A model system for evaluating shear in the design of stirred fermentors. Biotechnol Bioeng 1966; 8:71–84.
62. Hu WS. Quantitative and Mechanistic Analysis of Mammalian Cell Cultivation on Microcarriers. Ph.D. Thesis, Massachusetts Institute of Technology, Cambridge, MA, 1983.
63. Kolmogorov ANCR. Dokl Acad Sci USSR 1941; 30:301–305.
64. Cherry RS, Papoutsakis ET. Hydrodynamic effects on cells in agitated tissue culture reactors. Bioprocess Eng 1986; 1:29–41.
65. Cherry RS, Papoutsakis E. Understanding and controlling fluid-mechanical injury of animal cells in bioreactors. In: Spier RE, Griffiths JB, eds. Animal Cell Biotechnology. Vol. 4. New York: Academic Press, 1990:71–121.
66. Kunas KT, Papoutsakis ET. Damage mechanisms of suspended animal cells in agitated bioreactors with and without bubble entrainment. Biotechnol Bioeng 1990; 36:476–483.
67. Griffiths JB. Overview of cell culture systems and their scale-up. In: Spier RE, Griffiths JB, eds. Animal Cell Biotechnology. Vol. 3. New York: Academic Press, 1988:179–220.
68. Oh SKW, Nienow AW, Al-Rubeai M, Emery AN. The effects of agitation intensity with and without continuous sparging on the growth and antibody production of hybridoma cells. J Biotechnol 1989; 12:45–62.
69. zMurhammer DW, Goochee CF. Sparged animal cell bioreactors: mechanism of cell damage and Pluronic F-68 protection. Biotechnol Prog 1990; 6(5):391–397.
70. Papoutsakis ET. Fluid-mechanical damage of animal cells in bioreactors. Trends Biotechnol 1991; 9:427–437.
71. Merchuk JC. Shear effects on suspended cells. Adv Biochem Eng/Biotechnol 1991; 44:65–95.
72. Jöbses I, Martens D, Tramper J. Lethal events during gas sparging in animal cell culture. Biotechnol Bioeng 1991; 37:484–490.
73. Zhang S, Handa-Corrigan A, Spier RE. Foaming and media surfactant effects on the cultivation of animal cells in stirred and sparged bioreactors. J Biotechnol 1992; 25(3):289–306.
74. Jordan M, Sucker H, Widmer F, Eppenberger HM. Cell–bubble interaction during aeration are strongly influenced by surfactants in the medium and can be minimized in the newly developed bubble bed reactor. In: Spier RE, Griffiths JB, Berthold W, eds. Animal Cell Technology: Products of Today, Prospects for Tomorrow. Oxford: Butterworth- Heinemann, 1994:302–304.
75. Dey D, Emery AN. Problems in predicting cell damage from bubble bursting. Biotechnol Bioeng 1999; 65(2):240–245.
76. Chisti Y. Animal-cell damage in sparged bioreactors. Trends Biotechnol 2000; 18:320–432.

77. Papoutsakis ET. Media additives for protecting freely suspended animal cells against agitation and aeration damage. Trends Biotechnol 1991; 9:316–324.
78. Michaels JD, Nowak JE, Mallik AK, Koczo K, Wasan DT, Papoutsakis ET. Analysis of cell-to-bubble attachment in sparged bioreactors in the presence of cell-protecting additives. Biotechnol Bioeng 1995; 47:407–419.
79. Chisti Y. Animal cell culture in stirred bioreactors: observations on scale-up. Process Biochem 1993; 28:511–517.
80. Ishida M, Haga R, Nishimura N, Matuzaki HNR. High cell density suspension culture of mammalian anchorage independent cells: oxygen transfer by gas sparging and defoaming with a hydrophobic net. Cytotechnology 1990; 4:215–225.
81. Moreira JL, Cruz PE, Santana PC, Feliciano AS, Lehmann J, Carrondo MJT. Influence of power input and aeration method on mass transfer in a laboratory animal cell culture vessel. J Chem Tech Biotechnol 1995; 62:118–131.
82. Kimura R, Miller WM. Effects of elevated pCO_2 and/or osmolality on the growth and recombinant tPA production of CHO cells. Biotechnol Bioeng 1996; 52:152–160.
83. deZengotita VM, Kimura R, Miller WM. Effects of CO_2 and osmolality on hybridoma cells: growth, metabolism and monoclonal antibody production. Cytotechnology 1998; 28(1–3):213–227.
84. Gray DR, Chen S, Howarth W, Inlow D, Maiorella BL. CO_2 in large-scale and high-density CHO cell perfusion culture. Cytotechnology 1996; 22(1):65–78.
85. Schmelzer AE, deZengotita VM, Miller WM. Considerations for osmolality measurement under elevated pCO_2: comparison of vapor pressure and freezing point osmometry. Biotechnol Bioeng 2000; 67(2):189–196.
86. Langheinrich C, Nienow AW. Control of pH in large-scale, free suspension animal cell bioreactors: alkali addition and pH excursions. Biotechnol Bioeng 1999; 66(3):171–179.
87. Qi H, Jovanoic G, Micheals J, Konstantinov K. The art and science of micro-sparging in high-density perfusion cultures of animal cells. In: Lindner-Olsson E, Lüllau E, Chatzisavido N, eds. Animal Cell Technology: From Target to Market. Proceedings of the 17th ESACT Meeting 2001. Dordrecht: Kluwer Academic Publishers, 2002:412–415.
88. Mantanguihan R, Sajan E, Zachariou M, Olson C, Micheals J, Thrift J, Konstantinov K. Solution to the high dissolved CO_2 problem in high-density perfusion culture of mammalian cells. In: Lindner-Olsson E. Lüllau E, Chatzisavido N, eds. Animal Cell Technology: From Target to Market. Proceedings of the 17th ESACT Meeting 2001. Dordrecht: Kluwer Academic Publishers, 2002:399–402.
89. Miltenburger H. Züchtung von tierischen Zellen in Suspensions- und Monolayerkulturen in FermentationsgefaBen. DE 2940 446 Al, Apr 9, 1981.
90. Kuhlmann W. Optimization of a membrane oxygenation system for cell culture in stirred tank reactors. Dev Biol Stand 1987; 66:263–268.
91. Mano T, Kimura T, Iijima S, Takashiki K, Takeuchi H, Kobayashi T. Comparison of oxygen supply methods for cultures of shear-stress sensitive organisms including animal cell culture. J Chem Tech Biotechnol 1990; 47:259–271.
92. Schneider M, Reymod F, Marison IW, von Stockar U. Bubble-free oxygenation by means of hydrophobic porous membranes. Enzyme Microb Technol 1995; 17:839–847.
93. Iding K, Lütkemeyer D, Fraune E, Gerlach K, Lehmann J. Influence of alterations in culture conditions and changes in perfusion parameters on the retention performance of a 20 μm spinfilter during a perfusion cultivation of a recombinant CHO cell line in pilot scale. Cytotechnology 2000; 34:141–150.
94. Ruffieux PA, von Stockar U, Marison IW. Measurement of volumetric (OUR) and determination of specific (qO_2) oxygen uptake rates in animal cell cultures. J Biotechnol 1998; 63(2):85–95.
95. Ducommun P, Ruffieux P, Furter M, Marison I, von Stockar U. A new method for on-line measurement of the volumetric oxygen uptake rate in membrane aerated animal cell cultures. J Biotechnol 2000; 78(2):139–147.

96. Chico E, Jager V. Perfusion culture of baculovirus-infected BTI-Tn-5Bl-4 insect cells: a method to restore cell-specific-trace glycoprotein productivity at high cell density. Biotechnol Bioeng 2000; 70(5):574–586.
97. Aunins JG, Henzler HJ. Aeration in cell culture bioreactors. In: Stephanopoulos G, ed. Bioprocessing; Biotechnology. Vol. 3. VCH: Weinheim, Germany, 1993:219–281.
98. Fazekas de St, Groth S. Automated production of monclonal antibodies in a cytostat. J Immunol Methods 1983; 57:121–136.
99. Zeng AP, Deckwer WD, Hu WS. Determinants and rate laws for growth and death of hybridoma cells in continuous culture. Biotechnol Bioeng 1998; 57(6):642–654.
100. Miller WM, Blanch HW, Wilke DR. A kinetic analysis of hybridoma growth and metabolism in batch and continuous suspension culture: Effect of nutrient concentration, dilution rate, and pH. Biotechnol Bioeng 1988; 32:947–965.
101. Miller WM, Wilke CR, Blanch HW. Transient responses of hybridoma cells to nutrient additions in continuous culture: I. Glucose pulse and step changes. Biotechnol Bioeng 1989; 33:477–486.
102. Miller WM, Wilke CR, Blanch HW. Transient responses of hybridoma cells to nutrient additions in continuous culture: II. Glutamine pulse and step changes. Biotechnol Bioeng 1989; 33:487–499.
103. Robinson DK, Widmer J, Memmert K. Effect of specific growth rates on productivity in continuous open and partial cell retention animal cell bioreactors. J Biotechnol 1992; 22(1–2):41–50.
104. Hiller GW, Clark DS, Blanch HW. Cell retention-chemostat studies of hybridoma cells—analysis of hybridoma growth and metabolism in continuous suspension culture on serum-free medium. Biotechnol Bioeng 1993; 42:185–195.
105. Figueroa C, Rice C, Mered M, Naveh D. Continuous fermentation consumption kinetics for basal media design. In: Beuvery EC, Griffiths JB, Zeijlemaker WP, eds. Animal Cell Technology: Developments Towards the 21st Century. Dordrecht: Kluwer Academic Publishers, 1995:169–174.
106. Meijer JJ, Romein B, Dijken JP, Luyben KCAM. Steady-state analysis of hybridoma in chemostat culture. In: Spier RE, Griffiths JB, Berthold W, eds. Animal Cell Technology: Products of Today, Prospects for Tomorrow. Oxford: Butterworth-Heinemann, 1994:539–541.
107. Heidemann R, Lutkemeyer D, Lehmann J. Determination of K_s-values (Monod constants) for essential amino acids using CHO and hybridoma cells in batch and chemostat cultures. In: Beuvery EC, Griffiths JB, Zeijlemaker WP, eds. Animal Cell Technology: Developments Towards the 21st Century. Dordrecht: Kluwer Academic Publishers, 1995:735–739.
108. Pinton H, Rabaud JN, Engasser JM, Marc A. Optimization of OKT3 hybridoma cultures in a perfusion stirred reactor. In: Spier RE, Griffiths IB, Meignier B, eds. Production of Biologicals from Animal Cells in Culture. Oxford: Butterworth-Heinemann, 1991:486–488.
109. Yang JD, Angelillo Y, Chaudhry M, Goldenberg C, Goldenberg DM. Achievement of high cell density and high antibody productivity by a controlled-fed perfusion bioreactor process. Biotechnol Bioeng 2000; 69(1):74–82.
110. Tolbert WR, In: Srigley WR, Prior CP. Perfusion culture systems for large-scale pharmaceutical production. In: Spier RE, Griffiths JB, eds. Animal Cell Biotechnology. Vol. 3. Orlando: Academic Press, 1988:373–393.
111. Avgerinos GC, Drapeau D, Socolow JS, Mao I, Hsiao K, Broeze RJ. Spin filter perfusion system for high density cell culture: production of recombinant urinary type plasminogen activation in CHO cells. Biotechnology 1990; 8:54–58.
112. Li S, Roder B, Wirth M. Fermentation of ATIII producing BHK cells in a double membrane perfusion bioreactor. In: Spier RE, Griffiths JB, Meignier B, eds. Production of Biologicals from Animal Cells in Culture. Oxford: Butterworth-Heinemann, 1991:445–450.

113. Jager V, Kobold A, Kohne C, Deutschmann SM, Grabenhorst E. High density insect cell culture for the production of recombinant proteins with baculovirus expression system. In: Spier RE, Griffiths JB, Berthold W, eds. Animal Cell Technology: Products of Today, Prospects for Tomorrow. Oxford: Butterworth-Heinemann, 1994:207–211.
114. Horwath BE. Facility design and validation considerations for continuous cell culture processes. In: Beuvery EC, Griffiths JB, Zeijlemaker WP, eds. Animal Cell Technology: Developments Towards the 21st Century. Dordrecht: Kluwer Academic Publishers, 1995:553–559.
115. Lamotte D, Syraczek I, Marc A. Cell-settler perfusion system for the production and glycosylation of human interferon by clumped cells. In: Merten O-W, Perrin B, Griffiths B, eds. New Developments and New Applications in Animal Cell Technology. Dordrecht: Kluwer Academic Publishers, 1998:395–397.
116. Bachinger T, Riese U, Eriksson R, Mandenius MF. Monitoring cellular state transitions in a production-scale CHO-cell process using an electronic nose. J Biotechnol 2000; 76(1):61–71.
117. Hiilscher M. Auslegung von Airlift-Schlaufenreaktoren fur die Kultivierung von Tierzellen im Hinblick auf Zellschddigung. Fortschrittsberichte VDI Reihe 3, VDI, 1990.
118. Esclade LRJ, Carrel S, Peringer P. Influence of the screen material on the fouling of spin filters. Biotechnol Bioeng 1991; 38:159–168.
119. Gronvik K-O, Frieberg EL, Malmstrom U. Centritech Cell –a new separation device for mammalian cells. In: Spier RE, Grififths JB, Stephenne J, Crooy PJ, eds. Advances in Animal Cell Biology and Technology for Bioprocesses. Guildford: Butterworth, 1989:428–433.
120. Hamamoto K, Ishimaru K, Tokashiki M. Perfusion culture of hybridoma cells using a centrifuge to separate cells from culture mixture. J Ferment Bioeng 1989; 67:190–194.
121. Tokashiki M, Arai T. High density culture of mouse-human hybridoma using perfusion, apparatus with multi-settling zones to separate cells from the culture medium. Cytotechnology 1989; 2:5–8.
122. Hiilscher M, Scheibler U, Onken U. Selective recycle of viable animal cells by coupling of airlift reactor and cell settler. Biotechnol Bioeng 1992; 39:442–446.
123. Trampler F, Sonderhoff SA, Pui PWS, Kilburn DG, Piret JM. Acoustic cell filter for high density perfusion culture of hybridoma cells. Biotechnology 1994; 12:281–284.
124. Velez D, Miller LO, Macmillan DJ. Use of tangential flow filtration in perfusion propagation of hybridoma cells for production of monoclonal antibodies. Biotechnol Bioeng 1989; 33:938–940.
125. Goergen JL, Marc A, Engasser JM, Rabaud JN, Pierry G, Geaugey V, Geahel I, Hache J. Development of a new membrane reactor for large scale mammalian cell culture. In: Spier RE, Griffiths JB, Berthold W, eds. Animal Cell Technology: Products of Today, Prospects for Tomorrow. Oxford: Butterworth-Heinemann, 1994:287–292.
126. Fraune E, Meichsner EFS, Kamal MN. A new spinfilter design. In: Carrondo MJT, Griffiths JB, Moreira JLP, eds. Animal Cell Technology: From Vaccine to Genetic Medicine. Dordrecht: Kluwer Academic Publishers, 1997:283–287.
127. Schiitz C, Jager V, Dreisel AJ, Wagner R. Cultivation of insect cell lines stirred membrane reactors. In: Spier RE, Griffiths JB, Meignier B, eds. Production of Biologicals from Animal Cells in Culture. Oxford: Butterworth-Heinemann, 1991:460–466.
128. Biintemeyer H, Lehmann J. Cultivation strategy and bioreactor system for optimal medium re-use and detoxification during continuous perfusion processes. In: Spier RE, Griffiths JB, MacDonald C, eds. Animal Cell Technology: Developments, Process and Products. Oxford: Butterworth-Heinemann, 1992:269–275.
129. Kempken R, Biintemeyer H, Lehmann J. Long term application of medium recycling for economic antibody production. In: Spier RE, Griffiths JB, MacDonald C, eds. Animal Cell Technology: Developments, Process and Products. Oxford: Butterworth-Heinemann, 1992:264–268.

130. Fenge C, Fraune E, Schiigerl K. Perfusion bioreactor performance at different cell bleed rates. In: Spier RE, Griffiths JB, MacDonald C, eds. Animal Cell Technology: Developments, Process and Products. Oxford: Butterworth-Heinemann, 1992:365–372.
131. Himmelfarb P, Thayer PS, Martin HE. Spin filter culture: the propagation of mammalian cells in suspension. Science 1969; 164:555–557.
132. Siegel U, Fenge C, Fraune E. Spin filter for continuous perfusion of suspension cells. In: Spier RE, Griffiths JB, MacDonald C, eds. Animal Cell Technology: Developments, Process and Products. Oxford: Butterworth-Heinemann, 1992:434–436.
133. Varecka R, Scheirer W. Use of a rotating wire cage for retention of animal cells in a perfusion fermentor. Dev Biol Stand 1987; 66:269–272.
134. Fenge C, Buzsaky F, Fraune E, Lindner-Olsson E. Evaluation of spin filter during perfusion culture of recombinant CHO cells. In: Spier RE, Griffiths JB, MacDonald C, eds. Animal Cell Technology: Developments, Process and Products. Oxford: Butterworth-Heinemann, 1992:429–433.
135. Thayer PS. Spin filter device for suspension cultures. In: Kruse PF, Patters MK, eds. Tissue Culture: Methods and Applications. New York: Academic Press, 1973:345–351.
136. Van Wezel AL, van der Velden-de Groot CAM, Haan HH, van den Heuvel N, Schasfoort R. Large scale animal cell cultivation for production of cellular biologicals. Dev Biol Stand 1985; 60:229–236.
137. Scheirer W. High-density growth of animal cell within cell retention fermenters equipped with membranes. In: Spier RE, Griffiths JB, eds. Animal Cell Biotechnology. Vol. 3. Orlando: Academic Press, 1988:263–281.
138. Tolbert WR, Prior CP. Perfusion culture. In: Miller AO, ed. Advanced Research on Animal Cell Technology. Dordrecht: Kluwer Academic Publishers, 1989:119–137.
139. Favre E, Peringer P, Wolden C, Kyung YS, Hu WS. Retention of animal cells in a rotating wire-cage bioreactor: a mechanistic study. In: Murakami H, ed. Trends in Animal Cell Culture Technology. Weinheim: VCH, 1990:127–132.
140. Bayeler W, Thaler T, Clements R. Some aspects of scaling-up animal cell processes. In: Murakami H, ed. Trends in Animal Cell Culture Technology. Weinheim: VCH: 1990:225–233.
141. Hu WS, Scholz MT, Favre E, Seamans TC. A new look at animal cell bioreactor development. In: Spier RE, Griffiths JB, Meignier B, eds. Production of Biologicals from Animal Cells in Culture. Oxford: Butterworth-Heinemann, 1991:370–380.
142. Sumeghy Z. Improved cell retention system based on the rotating sieve technique. In: Spier RE, Griffiths JB, MacDonald C, eds. Animal Cell Technology: Developments, Process and Products. Oxford: Butterworth-Heinemann, 1992:437–447.
143. Yabannavar VM, Singh V, Connelly NV. Mammalian cell retention in a spinfilter perfusion bioreactor. Biotechnol Bioeng 1992; 40:925–933.
144. Favre E, Thaler T. An engineering analysis of rotating sieves for hybridoma cellretention in stirred tank bioreactors. Cytotechnology 1992; 9:11–19.
145. Deo YM, Mahadevan MD, Fuchs R. Practical considerations in operation and scale-up of spin-filter based bioreactors for monoclonal antibody production. Biotechnol Prog 1996; 12:57–64.
146. Trocha M, von Rohr R, Stimeghy Z. Contribution to the understanding of retention of animal cells in spinfilters. In: Carrondo MJT, Griffiths IB, Moreira JLP, eds. Animal Cell Technology: From Vaccine to Genetic Medicine. Dordrecht: Kluwer Academic Publishers, 1997:405–409.
147. Jem KJ, Fateen S, Michaels J. Mixing phenomena in industrial bioreactors with perfusion spin filters. In: Spier RE, Griffiths IB, Berthold W, eds. Animal Cell Technology: Products of Today, Prospects for Tomorrow. Oxford: Butterworth-Heinemann, 1994:392–396.
148. Phiss R, Doyen D, van der Pol L, Romette J-L. Progress in animal cell perfusion technology with superspin. In: Spier RE, Griffiths IB, Berthold W, eds. Animal Cell Tech-

nology: Products of Today, Prospects for Tomorrow. Oxford: Butterworth-Heinemann, 1994:218–221.
149. Jan DCH, Emery AN, Al-Rubeai M. Optimization of spin-filter performance in the intensive culture of suspended cells. In: Spier RE, Griffiths JB, MacDonald C, eds. Animal Cell Technology: Developments, Process and Products. Oxford: Butterworth-Heinemann, 1992:448–451.
150. Kitano K, Shintani Y, Ichimori Y, Tsukamoto K, Sasai S, Kida M. Production of human monoclonal antibodies by heterohybridomas. Appl Microbiol Biotechnol 1986; 24:282–286.
151. Takazawa Y, Tokashiki M, Hamamoto K, Murakami H. High cell density perfusion of hybridoma cells recycling high molecular weight components. Cytotechnology 1988; 1:171–178.
152. Bart BC, Davis RH, Kompala DS. Inclined sedimentation for selective retention of viable hybridomas in a continuous suspension bioreactor. Biotechnol Prog 1990; 6:458–464.
153. Knaack C, Andre G, Chavarie C. Conical bioreactor with internal lamella settler for perfusion culture of suspension cells. In: Spier RE, Griffiths JB, Berthold W, eds. Animal Cell Technology: Products of Today, Prospects for Tomorrow. Oxford: Butterworth-Heinemann, 1994:230–233.
154. Tokashiki M, Arai T. High density culture of hybridoma cells using a perfusion culture apparatus with multi-settling zones. In: Spier RE, Griffiths JB, Meignier B, eds. Production of Biologicals from Animal Cells in Culture. Oxford: Butterworth-Heinemann, 1991:467–469.
155. Sato S, Kawamura K, Fuyiyoshi N. Animal cell cultivation for production of biological substances with a novel perfusion culture apparatus. J Tissue Culture Methods 1983; 8(4):167–171.
156. Hosoi S, Miyaji H, Satoh M, Kurimoto T, Mihara A, Fujiyoshi N, Itoh S, Sato S. Optimization of cell culture conditions for production of biologically active proteins. Cytotechnology 1991; 5:17–34.
157. Tokashiki M, Yokoyama S. Bioreactors designed for animal cells. In: Hauser HJ, Wagner R, eds. Mammalian Cell Biotechnology in Protein Production. de Gruyter, 1997:279–317.
158. Searles JA, Todd P, Kompala DS. Perfusion culture of suspended CHO cells employing inclined sedimentation. In: Spier RE, Griffiths JB, Berthold W, eds. Animal Cell Technology: Products of Today, Prospects for Tomorrow. Oxford: Butterworth-Heinemann, 1994:240–242.
159. Bjorling T, Malmstrom U. Bleeding out dead cells from fermentation systems. Spier RE, Griffiths JB, Meignier B, eds. Production of Biologicals from Animal Cells in Culture. Oxford: Butterworth-Heinemann, 1991:394–399.
160. Gaida Th, Doblhoff O, Strutzenberger K, Katinger HBW, Groschl B, Benes E. Scale-up of ultrasonic resonance field cell separation devices used in cell technology. In: Beuvery EC, Griffiths JB, Zeijlemaker WP, eds. Animal Cell Technology: Developments Towards the 21st Century. Dordrecht: Kluwer Academic Publishers, 1995:699–703.
161. Woodside SM, Bowen BD, Piret JM. Mammalian cell retention devices for stirred perfusion bioreactors. Cytotechnology 1998; 28:163–175.
162. Birch JR, Arathoon R. Suspension culture of mammalian cells. Bioprocess Technol 1990; 10:251–270.
163. Chisti MY, Moo-Young M. Airlift reactors: characteristics, applications and design considerations. Chem Eng Commun 1987; 60:195–242.
164. Rhodes M, Birch J. Large scale production of proteins from mammalian cells. Biotechnology 1988; 6:518–523.
165. Konopitzky K, Kanzler O, Winhab K. Monoclonal antibody production using an airlift fermenter system consisting of a continuous seed fermenter and a fed batch production

fermenter. In: Spier RE, Griffiths JB, Meignier B, eds. Production of Biologicals from Animal Cells in Culture. Oxford: Butterworth-Heinemann, 1991:390–393.
166. Boraston SJFRC, Marshall CM, Gardiner SAM. The effect of high gas sparge rates in airlift fermenter culture on hybridoma cell growth and antibody production in low protein medium. In: Spier RE, Griffiths JB, MacDonald C, eds. Animal Cell Technology: Developments, Process and Products. Oxford: Butterworth-Heinemann, 1992:218–220.
167. Tramper J, Smit D, Straatman J, Vlak JM. Bubble-column design for growth of fragile insect cells. Bioprocess Eng 1988; 3:37–41.
168. Martens DE, Nollen EAA, Hardeveld M, Groot CAMV, Gooijer CD, Beuvery EC, Tramper J. Death rate in a small air-lift loop reactor of Vero cells grown on solid microcarrier and in macroporous microcarriers. Cytotechnology 1997; 23(1/3):61–75.
169. Grima EM, Chisti Y, Mooyoung M. Characterization of shear rates in airliftbioreactors for animal cell culture. J Biotechnol 1997; 54(3):195–210.
170. Merchuk JC. Why use air-lift bioreactors? Trends Biotechnol 1990; 8:66–71.
171. Bliem R, Konopitzky K, Katinger H. Industrial animal cell reactor systems: aspects of selection and evaluation. In: Fiechter A, ed. Advances in Biochemical Engineering/Biotechnology. Vol. 44. Berlin Springer, 1991:1–25.
172. Handa A, Emery AN, Spier R. On the evaluation of gas–liquid interfacial effects on hybridoma viability in bubble column bioreactors. Dev Biol Stand 1989; 66:241–252.
173. Handa-Corrigan A, Emery AN, Spier RE. Effect of gas–liquid interfaces on the growth of suspended mammalian cells: mechanisms of cell damage by bubbles. Enzyme Microb Technol 1989; 11:230–235.
174. Huelscher M, Onken U. Mechanical damage of suspended animal cells in directly sparged bioreactors. In: Spier RE, Griffiths JB, MacDonald C, eds. Animal Cell Technology: Developments, Process and Products. Oxford: Butterworth-Heinemann, 1992:195–202.
175. Su WW. Bioreactors, perfusion. In: Spier RE, ed. Encyclopedia of Cell Technology. Vol. 1. New York: Wiley, 2000:230–242.
176. Lu GZ, Thompson BG, Suresh MR, Gray MR. Cultivation of hybridoma cells in an inclined bioreactor. Biotechnol Bioeng 1995; 45:176–186.
177. Sucker HG, Jordan M, Eppenberger HM, Wimer F. Bubble bed reactor: a reactordesign to minimize the damage of bubble aeration on animal cells. Biotechnol Bioeng 1994; 44:1246–1254.
178. Knazek RA, Gullino PM, Kohler PO, Dedrick RL. Cell culture in artificial capillaries: an approach to tissue growth in vitro. Science 1972; 178:65–67.
179. Labecki M, Bowden B, Piret J. Two-dimensional analysis of protein transport in the extracapillary space of hollow-fibre bioreactors. Chem Eng Sci 1996; 51(17):4197–4213.
180. Dowd JE, Weber I, Rodriguez B, Piret JM, Kwok KE. Predictive control of hollow-fiber bioreactors for the production of monoclonal antibodies. Biotechnol Bioeng 1999; 63(4):484–492.
181. Nagel A, Effenberger E, Koch S, Liibbe L, Marx U. Human cancer and primary cell culture in the new hybrid bioreactor system Technomouse. In: Spier RE, Griffiths JB, Berthold W, eds. Animal Cell Technology: Products of Today, Prospects for Tomorrow. Oxford: Butterworth-Heinemann, 1994:296–301.
182. Gramer MJ, Poesch DM, Conroy MJ, Hammer BE. Effect of harvesting protocol on performance of a hollow fiber bioreactor. Biotechnol Bioeng 1999; 65(3):334–340.
183. Cima LG, Blanch HW, Wilke CR. A theoretical and experimental evaluation of a novel radial-flow hollow fiber reactor for mammalian cell culture. Bioprocess Eng 1990; 5:19–30.
184. Altshuler GL, Sowek JA, Belfort G. IgG production in microporous hollow fiberbioreactors. In: Neijssel OM, van der Meer RR, Luyben KChAM, eds. Proceedings of the 4th European Congress on Biotechnology. Amsterdam: Elsevier Science Publishers, 1987:127–129.

185. Heath C, Dilwith R, Belfort G. Methods for increasing monoclonal antibody production in suspension and entrapped cell cultures: biochemical and flow cytometric analysis as a function of medium serum content. J Biotechnol 1990; 15:71–90.
186. Dhainaut F, Meterreau JL, Mas MP, Potentini C, Mignot G. Scale-up of human IgGl production by a lymphoblastoid cell line in hollow fiber system. In: Spier RE, Griffiths JB, MacDonald C, eds. Animal Cell Technology: Developments, Process and Products. Oxford: Butterworth-Heinemann, 1992:527–529.
187. Goffe RA, Shi JY, Nguyen AKC. Medium additives for high performance cell culture in hollow fiber bioreactors. In: Beuvery EC, Griffiths JB, Zeijlemaker WP, eds. Animal Cell Technology: Developments Towards the 21st Century. Dordrecht: Kluwer Academic Publishers, 1995:187–191.
188. Kidwell WR. Filtering out inhibition. Biotechnology 1989; 7.
189. Tharakan JP, Gallagher SL, Chau PC. Hollow-fiber bioreactors for mammalian cell culture. Adv Biotechnol Processes 1988; 7:153–184.
190. Piret JM, Cooney CL. Immobilized mammalian cell cultivation in hollow fiber bioreactors. Biotechnol Adv 1990; 8:763–783.
191. Brotherton JD, Chau PC. Modeling of axial-flow hollow fiber cell culture bioreactors. Biotechnol Prog 1996; 12(5):575–590.
192. Uludag H, De Vos P, Tresco PA. Technology of mammalian cell encapsulation. Adv Drug Deliv Rev 2000; 42(1–2):29–64.
193. Tharakan JP, Chau PC. Operation and pressure distribution of immobilized cell hollow fiber bioreactors. Biotechnol Bioeng 1986; 28:1064–1071.
194. Piret JM, Cooney CL. Mammalian cell and protein distributions in ultrafiltration hollow fiber bioreactors. Biotechnol Bioeng 1990; 36:902–910.
195. Taylor DG, Piret JM, Bowen BD. Protein polarization in isotropic membrane hollow fiber bioreactors. AIChE J 1994; 40:321–333.
196. Hirschel MD, Gruenberg MX. An automated hollow fiber system for the large scale manufacture of mammalian cell secreted product. In: Lydersen BK, ed. Large Scale Cell Culture Technology. London: Macmillan, 1987:115–144.
197. Shi Y, Sardonini CA, Goffe RA. The use of oxygen carriers for increasing the production of monoclonal antibodies from hollow fibre bioreactors. Res Immunol 1998; 149(6):576–587.
198. Brotherton JD, Chau PC. Modeling analysis of an intercalated-spiral alternate-dead-ended hollow fiber bioreactor for mammalian cell cultures. Biotechnol Bioeng 1990; 35:375–394.
199. Ku K, Kuo MJ, Delente J, Wildi BS, Feder J. Development of a hollow-fiber system for large-scale culture of mammalian cell's. Biotechnol Bioeng 1981; 23:79–95.
200. Tyo MA, Bubulian BJ, Menken BZ, Murphy T. Large-scale mammalian cell culture utilizing Acusyst technology. In: Spier RE, Griffiths IB, eds. Animal Cell Biotechnology. Vol. 3. Orlando: Academic Press, 1988:357–371.
201. Heath C, Belford G. Immobilization of suspended mammalian cells: analysis of hollow fiber and microcapsule bioreactors. Adv Biochem Eng/Biotechnol 1987; 34:1–31.
202. Marx U, Tanzmann H, Jaeger V, Kiessig ST, von Baehr R. Simultaneous cultivation of two hybridoma cell lines in a hollow fiber bioreactor. In: Spier RE, Griffiths JB, MacDonald C, eds. Animal Cell Technology: Developments, Process and Products. Oxford: Butterworth-Heinemann, 1992:534–537.
203. Cytogen: http://www.cytogen.com/prosta_mstr.html.
204. ProstaScint: http://www.fda.gov/cber/products/capcytl0289 6.htm.
205. Lamers CHJ, Griend RJ, Gratama JW, Bolhuis RLH. Activation and expansion of human cytotoxic T lymphocytes in hollow fiber bioreactors. In: Spier RE, Griffiths JB, Berthold W, eds. Animal Cell Technology: Products of Today, Prospects for Tomorrow. Oxford: Butterworth-Heinemann, 1994:735–737.

206. Adamson SR, Fitzpatrick SL, Behie LA, Gaucher GM, Lesser BH. In vitro production of high titre monoclonal antibody by hybridoma cells in dialysis culture. Biotechnol Lett 1983; 5(9):573–578.
207. Hagedorn J, Kargi F. Coiled tube membrane bioreactor for cultivation of hybridoma-cells producing monoclonal antibodies. Enzyme Microb Technol 1990; 12:824–829.
208. Kleuser B, Kocher HP, Mernmert K, Stenz M, Straubhaar B, Zurini M. Configuration of a simple dialysis membrane bioreactor suitable for large-scale application. In: Bernard A, Griffiths B, Noe W, Wurm F, eds. Animal Cell Technology: Products from Cells, Cells as Products. Dordrecht: Kluwer Academic Publishers, 1999:303–305.
209. Comer MJ, Kearms MJ, Wahl I, Muster M, Lorenz T, Szperalski B, Koch S, Behrendt U, Brunner H. Industrial production of monoclonal antibodies and therapeutic proteins by dialysis fermentation. Cytotechnology 1990; 3:295–299.
210. Markl H, Lechner M, Gotz F. A new dialysis fermenter for the production of high concentrations of extracellular enzymes. J Ferment Bioeng 1990; 69:244–249.
211. Portner R, Markl H. Dialysis cultures. Appl Microbiol Biotechnol 1998; 50:403–414.
212. Tolbert WR, White PJ, Feder J. Perfusion culture systems for production of mammalian cell biomolecules. In: Feder J, Tolbert WR, eds. Large-scale Mammalian Cell Culture. Orlando: Academic Press, 1985:1–18.
213. Klement G, Scheirer W, Katinger HWD. Construction of a large scale membrane reactor system with different compartments for cells, medium and product. Dev Biol Stand 1985; 66:221–226.
214. Lim F, Sun AM. Microencapsulated islets as bioartificial endocrine pancreas. Science 1980; 210(21):908–910.
215. Rupp RG. Use of cellular microencapsulation in large-scale production of monoclonalantibodies. In: Feder J, Tolbert WR, eds. Large-Scale Mammalian Cell Culture. Orlando: Academic Press, 1985:19–38.
216. Duff RG. Microencapsulation technology: a novel method for monoclonal antibody production. Trends Biotechnol 1985; 3:167–170.
217. Lee GM, Palsson BO. Immobilization can improve the stability of hybridoma antibody productivity in serum-free media. Biotechnol Bioeng 1990; 36:1049–1055.
218. Nilsson K, Scheirer W, Katinger HWD, Mosbach K. Production of monoclonal antibodies by agarose-entrapped hybridoma cells. Methods Enzymol 1986; 121:352–361.
219. Merten OW, Dautzenberg H, Palfi GE. A new method for the encapsulation of mammalian cells. Cytotechnology 1991; 7:121–130.
220. Shen BQ, Greenfield PF, Reid S. Hybridoma cells in a protein-free medium within a composite gel perfusion bioreactor. Cytotechnology 1994; 16(1):51–58.
221. Seifert DB, Phillips JA. Porous alginate-poly (ethylene glycol) entrapment system for the cultivation of mammalian cells. Biotechnol Prog 1997; 13(5):569–576.
222. Ray NG, Vournakis JN, Runstadler PW, Tung AS, Venkatasubramanian K. Physiology of hybridoma and recombinant Chinese hamster ovary cells immobilized in collagen matrix. In: De Bont JM, Visser J, Mattiasson B, Tramper J, eds. Physiology of Immobilized Cells. Amsterdam: Elsevier, 1990:679–711.
223. Nilsson K. Methods for immobilizing animal cells. Trends Biotechnol 1987; 5:73–78.
224. Vogelsang C, Wijffels RH, Ostgaard K. Rheological properties and mechanical stability of new gel-entrapment systems applied in bioreactors. Biotechnol Bioeng 2000; 70(3):247–253.
225. King GA, Kaugulis AJ, Faulkner P, Goosen MFA. Alginate-polylysine microcapsules of controlled membrane molecular weight cutoff for mammalian cell culture engineering. Biotechnol Prog 1987; 3:231–340.
226. Lu MZ, Lan HL, Wang FF, Chang SJ, Wang YJ. Cell encapsulation with alginate and -phenoxycinnamylidene-acetylatedpoly(allylamine). Biotechnol Bioeng 2000; 70(5):479–483.

227. Serp D, Cantana E, Heinzen C, Stockar UV, Marison IW. Characterization of an encapsulation device for the production of monodisperse alginate beads for cell immobilization. Biotechnol Bioeng 2000; 70(1):41–53.
228. Martinson A, Storro L, Skjak-Braek G. Alginate as immobilization material: III. Diffusional properties. Biotechnol Bioeng 1992; 39:186–194.
229. Lee GM, Kim SJ, Palsson BO. Enhanced specific antibody productivity of calcium alginate-entrapped hybridoma is cell line-specific. Cytotechnology 1994; 16(1):1–15.
230. Lee GM, Chuck AS, Palsson BO. Cell culture conditions determine the enhancement of specific monoclonal antibody productivity of calcium alginate-entrapped S3H5/$_{\gamma}$2aA2 hybridoma cells. Biotechnol Bioeng 1993; 41:330–340.
231. Kwong C-W, Hsieh J-H, Syu M-L, Chou S-S. Anti-hepatitis B surface antigenmonoclonal antibody IgM production in suspension and immobilized cell bioreactors. Biotechnol Lett 1989; II(6):377–382.
232. Tai IT, Sun AM. Microencapsulation of recombinant cells: a new delivery system for gene therapy. FASEB J 1993; 7(11):1061–1069.
233. Basic D, Vacek I, Sun AM. Microencapsulation and transplantation of genetically engineered cells: a new approach to somatic gene therapy. Artificial Cells, Blood Substitutes, & Immobilization Biotechnology 1996; 24(3):219–255.
234. Chang PL, Van Raamsdonk JM, Hortelano G, Barsoum SC, MacDonald NC, Stockley TL. The in vivo delivery of heterologous proteins by microencapsulated recombinant cells. Trends Biotechnol 1999; 17(2):78–83.
235. Freshney IR. Culture of Animal Cells: A Manual of Basic Techniques. New York: Willey-Liss, 1994.
236. Jensen MD. Production of anchorage-dependent cells-problems and their possible solutions. Biotechnol Bioeng 1981; XXIII:2703–2716.
237. Archer R, Wood L. Production tissue culture by robots. In: Spier RE, Griffiths JB, MacDonald C, eds. Animal Cell Technology: Developments, Processes and Products. Oxford: Butterworth-Heinemann, 1992:403–408.
238. Aunins JG, Bibila TA, Gatchalian S, Hunt GR, Junker BH, Lewis JA, Seifert DB, Licari P, Ramasubramanyan K, Ranucci CS, Seamans TC, Zhou W, Waterbury W, Buckland BC. Reactor development for the hepatitis a vaccine VAQTA. In: Carrondo MJT, Griffiths B, Moreira JLP, eds. Animal Cell Technology: From Vaccine to Genetic Medicine. Dordrecht: Kluwer Academic Publishers, 1997:175–183.
239. Beeksma LA, Kompier R. Cell growth and virus propagation in the Costar Cellcube system. In: Beuvery EC, Griffiths JB, Zeijlemaker WP, eds. Animal Cell Technology: Developments Towards the 21st Century. Dordrecht: Kluwer Academic Publishers, 1995:661–663.
240. Van Wezel AL. Growth of cell-strains and primary cells on micro-carriers in homogeneous culture. Nature 1967; 216:64–66.
241. Lundgren B, Bluml G. Microcarriers in cell culture production. In: Subramanian G, ed. Bioseparation and Bioprocessing. Vol. 2. Weinheim: Wiley-VCH, 1998:165–222.
242. Looby D, Griffiths B. Immobilization of animal cells in porous carrier culture. Trends Biotechnol 1990; 8:204–209.
243. Berg GJ, Bodeker BGD. Employing a ceramic matrix for the immobilization of mammalian cells in culture. In: Spier RE, Griffiths JB, eds. Animal Cell Biotechnology. Vol. 3. Orlando: Academic Press, 1988:321–335.
244. Groot CAMV. Microcarrier technology, present status and perspective. Cytotechnology 1995; 18:51–56.
245. Butler M. A comparative review of microcarriers available for growth of anchorage-dependent animal cells. In: Spier RE, Griffiths JB, eds. Animal Cell Biotechnology. Vol. 3. Orlando: Academic Press, 1988:3:283–303.
246. Gatot C, Trampler F, Wanderpepen M-P, Harfield J, Oudshoorn A, Johansson A, Nielsen V, Miller AOA. Use of a new microcarrier with two-dimensional geometry for the culture of anchorage-dependent cells in sonoperfused, continuously stirred tank bior-

eactors. In: Merten O-W, Perrin B, Griffiths B, eds. New Developments and New Applications in Animal Cell Technology. Dordrecht: Kluwer Academic Publishers, 1998:377–380.
247. Groot CAMV. Microcarrier technology, present status and perspective. In: Beuvery EC, Griffiths JB, Zeijlemaker WP, eds. Animal Cell Technology: Developments Towards the 21st Century. Dordrecht: Kluwer Academic Publishers, 1995:899–905.
248. Kong D, Gentz R, Zhang J. Cell growth and protein formation on various microcarriers. Cytotechnology 1999; 29(2):151–158.
249. Nilsson K, Buzsaky F, Mosbach K. Growth of anchorage dependent cells on macroporous microcarrier. Biotechnology 1986; 4:989–990.
250. Almgren J, Nilsson C, Petersson AC, Nilsson K. Cultispher-macroporous gelatin microcarrier-new applications. In: Spier RE, Griffiths JB, Meignier B, eds. Production of Biologicals from Animal Cells in Culture. Oxford: Butterworth-Heinemann, 1991:434–438.
251. Looby D, Racher AJ, Fuller JP, Griffiths B. Evaluation of porous silicone carrier (Immobasil G) for animal cell culture. In: Beuvery EC, Griffiths JB, Zeijlemaker WP, eds. Animal Cell Technology: Developments Towards the 21st Century. Dordrecht: Kluwer Academic Publishers, 1995:783–786.
252. Nomura N, Matsumura M, Kataoka H, Saijo K, Ohno T. Production of human growth hormone by protein-free cultivation of anchorage-dependent cells with porous cellulose carrier. In: Beuvery EC, Griffiths JB, Zeijlemaker WP, eds. Animal Cell Technology: Developments Towards the 21st Century. Dordrecht: Kluwer Academic Publishers, 1995:817–821.
253. Long RJ. The production and regulation of veterinary virus vaccines. In: Spier RE, Griffiths JB, eds. Animal Cell Biotechnology. Orlando: Academic Press, 1994; 6:281–313.
254. Chun BH, Woo GJ, Cheung HS, Moon HM. Production of live attenuated varicella-zoster virus in human embryonic lung cells microcarrier. In: Carrondo MJT, Griffiths JB, Moreira JLP, eds. Animal Cell Technology: From Vaccine to Genetic Medicine. Dordrecht: Kluwer Academic Publishers, 1997:185–189.
255. Looby D, Tree J, Talukder A, Hayes K, House H, Stacey G. The production of influenza virus by immobilised MDCK cells. In: Merten O-W, Perrin B, Griffiths B, eds. New Developments and New Applications in Animal Cell Technology. Dordrecht: Kluwer Academic Publishers, 1998:551–554.
256. Berry JM, Barnabe N, Coombs KM, Butler M. Production of reovirus type-1 and type-3 from Vero cells grown on solid and macroporous microcarriers. Biotechnol Bioeng 1999; 62(1):12–19.
257. Moran E. A microcarrier-based cell culture process for the production of a bovine respiratory syncytial virus vaccine. Cytotechnology 1999; 29(2):135–149.
258. Merten OW, Kallel H, Manuguerra JC, Tardy-Panit M, Crainic R, Delpeyroux F, Werf SV, Parrin P. The new medium MDSS2N, free any animal protein supports cell growth and production of various viruses. Cytotechnology 1999; 30(1/3):191–201.
259. Montagnon B, Vincent-Falquet JC, Fanget B. Thousand litre scale microcarrier culture of Vero cells for killed polio virus vaccine; promising results. Dev Biol Stand 1984; 55:37–42.
260. Griffiths B. Animal cells—the breakthrough to a dominant technology. Cytotechnology 1990; 3:109–116.
261. Smiley AL, Hu W-S, Wang DIC. Production of human immune interferon by recombinant mammalian cells cultivated on microcarriers. Biotechnol Bioeng 1989; 33:1182–1190.
262. Kamiya K, Yanagida K, Shirokaze J. Subculture method for large scale cell culture using macroporous microcarrier. In: Beuvery EC, Griffiths IB, Zeijlemaker WP, eds. Animal Cell Technology: Developments Towards the 21st Century. Dordrecht: Kluwer Academic Publishers, 1995:759–763.

263. Wang Y, Ouyang F. Bead-to-bead transfer of Vero cells in microcarrier culture. Cytotechnology 1999; 31(3):221–224.
264. Miithing J, Duvar S, Nerger S, Biintemeyer H, Lehmann J. Microcarrier cultivation of bovine aortic endothelial cells in spinner vessels and a membrane stirred bioreactor. Cytotechnology 1995; 18(3):193–206.
265. Fenge C, Jansson L, Froberg T, Danielsson A, Jonsson M, Sygowski L, Moore C, Snyder D, Liillau E, Wood M. Process development for functional membrane receptor production in mammalian cells. In: Lindner-Olsson E, Liillau E, Chatzisavido N, eds. Animal Cell Technology: From Target to Market. Proceedings of the 17th ESACT Meeting 2001. Dordrecht: Kluwer Academic Publishers, 2002:441–445.
266. Glacken MW, Fleischaker RJ, Sinskey AJ. Reduction of waste product excretion via nutrient control: possible strategies for maximizing product and cell yields on serum in cultures of mammalian cells. Biotechnol Bioeng 1986; 28:1376–1389.
267. Butler M, Imamura T, Thomas J, Thilly WG. High yields from microcarrier cultures by medium diffusion. J Cell Sci 1983; 61:351–363.
268. Croughan MS, Wang DIC. Growth and death in overagitated microcarrier cell culture. Biotechnol Bioeng 1989; 33:731–744.
269. Alves PM, Moreira JL, Rodrigues JM, Aunins JG, Carrondo MJT. Two-dimensional versus three-dimensional culture systems: effects on growth and productivity of BHK cells. Biotechnol Bioeng 1996; 52(3):429–432.
270. Ng YC, Berry JM, Butler M. Optimization of physical parameters for cell attachment and growth on macroporous microcarriers. Biotechnol Bioeng 1996; 50(6):627–635.
271. Croughan MS, Hamel JF, Wang DIC. Hydrodynamic effects on animal cells grown in microcarrier cultures. Biotechnol Bioeng 2000; 67(6):841–852.
272. Gregoriades N, Clay J, Ma N, Koelling K, Chalmers JJ. Cell damage of microcarrier cultures as a function of local energy dissipation created by a rapid extensional flow. Biotechnol Bioeng 2000; 69(2):171–182.
273. Bauer P, Hancock L, Rathman J, Chalmers JJ. Cell-microcarrier adhesion to gas–liquid interfaces and foam. Biotechnol Prog 2000; 16(1):125–132.
274. Dean RC Jr, Karkare SB, Ray NG, Runstadler PW Jr, Venkatasubramanian K. Large-scale culture of hybridoma and mammalian cells in fluidized bed bioreactors. Ann NY Acad Sci 1987; 506:129–146.
275. Runstadler PW, Cernek SR. Large-scale fluidized-bed, immobilized cultivation of animal cells at high densities. In: Spier RE, Griffiths JB, eds. Animal Cell Biotechnology. Vol. 3. Orlando: Academic Press, 1988:305–320.
276. Vournakis JN, Runstadler PW. Optimization of the microenvironment for mammalian cell culture in flexible collagen microspheres in a fluidized-bed bioreactor. In: Ho CH, Wang DIC, eds. Animal Cell Bioreactors. Oxford: Butterworth, 1991:305–326.
277. Paul EDL. Design and scale-up of an anchorage-dependent mammalian cell bioreactor. Ann NY Acad Sci 1988;642–649.
278. Ray NG, Tung AS, Hayman EG, Vournakis JN, Runstadler PWJ. Continuous cell cultures in fluidized-bed bioreactors. Cultivation of hybridomas and recombmant Chinese hamster ovary cells immobilized in collagen microspheres. Ann NY Acad Sci 1990; 589:443–457.
279. Ray NG, Tung AS, Runstadler PW, Vournakis IN. Enhanced productivity of hybridoma and recombinant CHO cell cultures by Pluronic F-68 and other medium components, and by increased perfusion rates, in fluidized-bed bioreactors. In: Spier RE, Griffiths JB, Meignier B, eds. Production of Biologicals from Animal Cells in Culture. Oxford: Butterworth-Heinemann, 1991:502–511.
280. Looby D, Griffiths JB. Immobilisation of animal cells in fixed and fluidized porous glass sphere reactors. In: Spier RE, Griffiths JB, Stephenne J, Crooy PJ, eds. Advances in Animal Cell Biology and Technology for Bioprocesses. Guildford: Butterworths, 1989:336–344.

281. Kratje R, Jager V, Wagner R. Comparison of the production efficiency of mammalian cells grown in a fluidized bed and in a stirred tank reactor. In: Spier RE, Griffiths JB, Meignier B, eds. Production of Biologicals from Animal Cells in Culture. Oxford: Butterworth-Heinemann, 1991:528–532.
282. Kratje R, Jager V, Wagner R. Evaluation of the production of recombinant human interleukin-2 in a fluidized bed bioreactor. Biotechnol Bioeng 1992; 39:233–242.
283. Ltillau E, Biselli M, Wandrey C. Growth and metabolism of CHO-cells in porous glass carriers. In: Spier RE, Griffiths JB, Berthold W, eds. Animal Cell Technology: Products of Today, Prospects for Tomorrow. Oxford: Butterworth-Heinemann, 1994:252–255.
284. Noll TBM, Wandrey C. On-line biomass monitoring of immobilised hybridoma cells by dielectrical measurements. In: Carrondo MJT, Griffiths JB, Moreira JLP, eds. Animal Cell Technology: from Vaccine to Genetic Medicine. Dordrecht: Kluwer Academic Publishers, 1997:289–294.
285. Hambach B, Biselli M, Runstadler PW, Wandrey C. Development of a reactor-integrated aeration system for cultivation of animal cells in fluidized beds. In: Spier RE, Griffiths JB, MacDonald C, eds. Animal Cell Technology: Developments, Process and Products. Oxford: Butterworth-Heinemann, 1992:381–385.
286. Born C, Biselli M, Wandrey C. Production of monoclonal antibodies in a pilot scale fluidized bed bioreactor. In: Beuvery EC, Griffiths JB, Zeijlemaker WP, eds. Animal Cell Technology: Developments Towards the 21st Century. Dordrecht: Kluwer Academic Publishers, 1995:683–686.
287. Biselli M, Born C, Schroder B, Wandrey C. Gasification-tube module and reactor for cell cultivation. US 5,601,757, February 11, 1997.
288. Papaspyrou, M. Personal communication. 2001 June.
289. Reiter M, Bluml G, Gaida T, Zach N, Unterluggauer F, Doblhoff-Dier O, Noe M, Plail R, Huss S, Katinger H. Modular integrated fluidized bed bioreactor technology. Biotechnology 1991; 9(11):1100–1102.
290. Reiter M, Zach N, Gaida T, Bluml G, Doblhhoff-Dier O, Unterluggauer F, Katinger H. Oxygenation in fluidized bed bioreactors using the microsparging technique. In: Spier RE, Griffiths JB, MacDonald C, eds. Animal Cell Technology: Developments, Process and Products. Oxford: Butterworth-Heinemann, 1992:386–392.
291. Bluml G. Entwicklung und Etablierung eines macroporosen Carrier aus Polyethylen zur Immobilisierung von (tierischen) Zellen. Ph.D. Thesis. Instituted for Applied Microbiology, University of Agriculture, Vienna, Austria, 1994.
292. Bluml G, Reiter M, Zach N, Gaida T, Schmatz C. Development of a new type of macroporous carrier. In: Spier RE, Griffiths JB, MacDonald C, eds. Animal Cell Technology: Developments, Process and Products. Oxford: Butterworth-Heinemann, 1992:501–504.
293. Ducommun P, Bolzonella I, Rhiel M, Pugeaud P, von Stockar U, Marison IW. On-line determination of animal cell concentration. Biotechnol Bioeng 2001; 72(5):515–522.
294. Earle RW, Evans VJ, Schilling EX. Extension of cellophane substrate procedure to growth of in vitro cultures of large areas. J Natl Cancer Inst 1950; 10:943–967.
295. Earle WR, Evans VJ, Stanford KK, Shannon JE, Waltz HK. Influence of glass and cellophane substrates on proliferation of strain L cells in tissue culture. J Natl Cancer Inst 1951; 12:563–567.
296. Earle WR, Schilling EL, Shannon JE. Growth of animal tissue cells on three dimensional substrates. J Natl Cancer Inst 1951; 12:179–193.
297. Spier RE, Whiteside JP. The production of foot-and-mouth disease virus from BHK 21 C13 cells grown on the surface of glass spheres. Biotechnol Bioeng 1976l; 18:649–657.
298. Whiteside JP, Whiting B, Spier RE. Development of a methodology for the production of FMD virus from BHK 21 Vol. 3. C13 monolayer cells grown in a 100-L glass sphere propagator. Dev Biol Stand 1979; 42:113–119.
299. Burbidge C. The mass culture of human diploid fibroblast in packed beds of glass beads. Dev Biol Stand 1980; 46:169–172.

300. Griffiths JB, Thornton B, McEntee I. The development and use of microcarriers and glass sphere culture techniques for the production of herpes simplex virus. Dev Biol Stand 1982; 50:103–110.
301. Brown PC, Figueroa C, Costello MAC, Oakley R, Maciukas SM. Protein production from mammalian cells grown on glass beads. In: Spier RE, Griffiths JB, eds. Animal Cell Biotechnology. Vol. 3. Orlando: Academic Press, 1988:251–262.
302. Looby D, Griffiths JB. Fixed bed porous glass sphere (Porosphere) bioreactors for animal cell culture. Cytotechnology 1988; 1:339–346.
303. Bohmann A, Portner R, Schmieding J, Kasche V, Markl H. The membrane dialysis bioreactor with integrated radial-flow fixed bed – a new approach for continuous cultivation of animal cells. Cytotechnology 1992; 9(1–3):51–57.
304. Lazar A, Reuveny S, Muzrahi M, Avtalion M, Whiteside JP, Spier RE. Production of biologicals by animal cells immobilized in polyurethan foam matrix. In: Spier RE, Griffiths JB, eds. Modern Approaches to Animal Cell Technology. Oxford: Butterworth, 1987:437–448.
305. Looby D, Racher A, Griffiths JB, Dowsett AB. The immobilisation of animal cells in fixed and fluidized porous sphere reactors. In: De Bont JAM, Visser J, Mattiasson B, Tramper J, eds. Physiology of Immobilised Cells. Amsterdam: Elsevier Science Publishers, 1990:255–246.
306. Racher AJ, Looby D, Griffiths JB. Use of lactate dehydrogenase release to assess changes in culture viability. Cytotechnology 1990; 3(3):301–307.
307. Kompier R, Kislev N, Segal I, Kadouri A. Use of a stationary bed reactor and serum-free medium for the production of recombinant proteins in insect cells. Enzyme Microb Technol 1991; 13:822–827.
308. Wang G, Zhang W, Freedman D, Eppstein L, Kadouri A. Continuous production of monoclonal antibodies in Celligen packed bed reactor using Fibra-Cel carrier. In: Spier RE, Griffiths JB, MacDonald C, eds. Animal Cell Technology: Developments, Process and Products. Oxford: Butterworth-Heinemann, 1992:460–464.
309. Kadouri A. Cultivation of anchorage-dependent mammalian cells and production of various metabolites. Colloids Surfaces B: Biointerfaces 1994; 2:265–272.
310. Kang SH, Kim BG, Lee GM. Justification of continuous packed-bed reactor for retroviral vector production from amphotropic 0CRIP murine producer cell. Cytotechnology 2000; 34(1/2):151–158.
311. Wang G, Zhang W, Jacklin C, Freedman D, Eppstein L, Kadouri A. Modified CelliGen-packed bed bioreactors for hybridoma cell cultures. Cytotechnology 1992; 9:41–49.
312. Kaufman JB, Wang G, Zhang W, Valle MA, Shiloach J. Continuous production and recovery of recombinant Ca^{2+} binding receptor from HEK 293 cells using perfusion through a packed bed bioreactor. Cytotechnology 2000; 33(1):3–11.
313. Gumtisderelioglu M, Aslankaraoglu E, Giirhan SI. Rabies virus production in non-woven polyester fabric (NWPF) packed-bed reactors. Biotechnol Appl Biochem 2001; 33:167–172.
314. Yanagi K, Miyoshi H, Fukuda H, Ohshima N. A packed-bed reactor utilizing porous resin enables high density culture of hepatocytes. Appl Microbiol Biotechnol 1992; 37:316–320.
315. Miyoshi H, Yanagi K, Fukuda H, Ohshima N. Long-term performance of albumin secretion of hepatocytes cultures in a packed-bed reactor utilising porous resin. Artificial Organs 1996; 20(7):803–807.
316. Meissner P, Werner P, Schronder B, Herfurth C, Wandrey C, Biselli M. Expansion of human hematopoietic progenitor cells in a fixed bed bioreactor. In: Merten O-W, Perrin B, Griffiths B, eds. New Developments and New Applications in Animal Cell Technology. Dordrecht: Kluwer Academic Publishers, 1998:635–636.
317. Ghanem A, Shuler ML. Characterization of a perfusion reactor utilizing mammalian cells on microcarrier beads. Biotechnol Prog 2000; 16(3):471–479.

318. Lydersen BK, Pugh GG, Paris MS, Sharma BP, Noll LA. Ceramic matrix for large scale animal cell culture. Biotechnology 1985; 3:63–67.
319. Bodeker BGD, Hewlett G, Nerger D, Schlumberger HD. Verfahren und Vorrichtung zur Kultivierung immobilisierter oder adharierter Saugerzellen sowie dessen Verwendung bei der Herstellung von Proteinen. EP 0270905, June 15, 1988.
320. Bodeker BGD, Hiibner GE, Hewlett G, Schlumberger HD. Production of human monoclonal antibodies from immobilized cells in the Opticell culture system. Dev Biol Stand 1987; 66:473–479.
321. Werner RG, Merk W, Walz F. Fermentation with immobilized cell cultures. Arzneimittelforschung 1988; 38(2):320–325.
322. Lechner M, Markl H, Gotz F. Lipase production of *Streptococcus carnosus* in a dialysis fermentor. Appl Microbiol Biotechnol 1988; 28:345–349.
323. Kurosawa H, Markl H, Niebuhr-Redder C, Matsumura M. Dialysis bioreactor with radial-flow fixed bed for animal cell culture. J Ferment Bioeng 1991; 72(1):41–45.
324. Bohmann A, Portner R, Markl H. Performance of a membrane dialysis bioreactor with a radical-flow fixed bed for the cultivation of a hybridoma cell line. Appl Microbiol Biotechnol 1995; 43:772–780.
325. Chiou TW, Murakami S, Wang DIC, Wu WT. A fiber bed bioreactor for anchorage dependent animal cell culture: Part 1. Bioreactor design and operations. Biotechnol Bioeng 1991; 37:756–761.
326. Junker BH, Chiou T, Wang DIC, Buckland BC. Cultivation of virus antigen in fibroblast cells using a glass fiber bed reactor. Biotechnol Bioeng 1993; 42:635–642.
327. Highfill JG, Haley SD, Kompala DS. Large-scale production of murine bone marrow cells in an airlift packed bed bioreactor. Biotechnol Bioeng 1996; 50(5):514–520.
328. Junker BH, Seamans TC, Ramasubramanyan K, Aunins J, Paul E, Buckland BC. Cultivation of attenuated hepatitis A virus antigen in a titanium static mixer reactor. Biotechnol Bioeng 1994; 44:1315–1324.
329. Glacken MW, Adema E, Sinskey AJ. Mathematical descriptions of hybridoma culture kinetics: I. initial metabolic rates. Biotechnol Bioeng 1988; 32:491–506.
330. Lee YK, Yap PK, Teoh AP. Correlation between steady-state cell concentration and cell death of hybridoma cultures in chemostat. Biotechnol Bioeng 1995; 45:18–26.
331. Biintemeyer H, Siwiofa S, Lehmann J. Inhibitors of cell growth: accumulation and concentration. In: Carrondo MJT, Griffiths JB, Moreira JLP, eds. Animal Cell Technology: From Vaccine to Genetic Medicine. Dordrecht: Kluwer Academic Publishers, 1997:651–655.
332. Goergen J-L, Marc A, Engasser J-M. Kinetics and simulation of animal cell processes. In: Hauser HJ, Wagner R, eds. Mammalian Cell Biotechnology in Protein Production. de Gruyter, 1997:345–371.
333. Jo E-C, Kim D-L, Moon HM. Step-fortifications of nutrients in mammalian cell culture. Biotechnol Bioeng 1993; 42:1218–1228.
334. Xie L, Wang DIC. Fed-batch cultivation of animal cells using different medium design concepts and feeding strategies. Biotechnol Bioeng 1994; 43:1175–1189.
335. Merten O-W, Keller H, Cabanie L, Leno M, Hardefelt M. Batch production and secretion kinetics of hybridomas: pulse-chase experiments. Cytotechnology 1990; 4:77–89.
336. Philips PJ, Marquis CP, Barford JP, Harbour C. Analysis of some batch and continuous kinetic data of hybridoma cells. Cytotechnology 1991; 6:189–195.
337. Bushell ME, Bell SL, Scott MF, Snell K, Spier RE, Wardel JN, Sanders PG. A three phase pattern in growth, monoclonal antibody production, and metabolite exchange in a hybridoma bioreactor culture. Biotechnol Bioeng 1993; 42:133–139.
338. Merten O-W, Palfi GE, Element G, Steindl F. Specific kinetic patterns of production of monoclonal antibodies in batch cultures and consequences on fermentation processes. In: Spier RE, Griffiths JB, eds. Modern Approaches to Animal Cell Technology. Oxford: Butterworth, 1987:381–396.

339. Noe W, Bux R, Werner RG, Berthold W. Aspects of continuous culture versus batch mode for animal cells in industrial processes. In: Spier RE, Griffiths JB, MacDonald C, eds. Animal Cell Technology: Developments, Process and Products. Oxford: Butterworth-Heinemann, 1992:354–360.
340. Kadouri A, Spier RE. Some myths and messages concerning the batch and continuous culture of animal cells. Cytotechnology 1997; 24(2):89–98.
341. Dalili M, Sayles GD, Ollis D. Glutamine-limited batch hybridoma growth and antibody production: experiment and model. Biotechnol Bioeng 1990; 36:74–82.
342. Leelavatcharamas V, Emery AN, Al-Rubeal M. The effects of initial serum concentration, pH and agitation rate on cell aggregation during batch culture of Chinese hamster ovary (CHO) cells. In: Beuvery EC, Griffiths JB, Zeijlemaker WP, eds. Animal Cell Technology: Developments Towards the 21st Century. Oxford: Kluwer Academic Publishers, 1995:217–221.
343. Doverskog M, Bertram E, Ljunggren J, Ohman L, Sennerstam R, Haggstrom L. Cell cycle progression in serum-free cultures of Sf9 insect cells: modulation by conditioned medium factors and implications for proliferation and productivity. Biotechnol Prog 2000; 16(5):837–846.
344. Zeiser A, Elias CB, Voyer R, Jardin B, Kamen AA. On-line monitoring of physiological parameters of insect cell cultures during the growth and infection process. Biotechnol Prog 2000; 16(5):803–808.
345. Agrawal P, Koshy G, Ramseier M. An algorithm for operating a fed-batch fermentor at optimum specific-growth rate. Biotechnol Bioeng 1989; 33:115–125.
346. Ljunggren J, Haggstrom L. Catabolic control of hybridoma cells by glucose and glutamine limited fed batch cultures. Biotechnol Bioeng 1994; 44:808–818.
347. Romein B, van Londen P, Meijer JJ, Hellinga C, van Dijken JP, Luyben, KChAM. Substrate level control in fed-batch hybridoma cultures. In: Spier RE, Griffiths JB, Berthold W, eds. Animal Cell Technology: Products of Today, Prospects for Tomorrow. Oxford: Butterworth-Heinemann, 1994:434–436.
348. Robinson DK, Chan CP, Ip CY, Tsai PK, Tung I, Seamans TC, Lenny AB, Lee DK, Irwin J, Silberklang M. Characterization of recombinant antibody produced in the course of a high yield fed-batch process. Biotechnol Bioeng 1994; 44:727–735.
349. Noe W, Schorn P, Bux R, Berthold W. Fed-batch strategies for mammalian cell culture. In: Spier RE, Griffiths JB, Berthold W, eds. Animal Cell Technology: Products of Today, Prospects for Tomorrow. Oxford: Butterworth-Heinemann, 1994:413–418.
350. Dardenne M, Cherlet M, Engasser JM, Marc A. Kineteic studies of fed-batch hybridoma cultures: Effect of various feeding compositions and flow rates. In: Spier RE, Griffiths JB, Berthold W, eds. Animal Cell Technology: Products of Today, Prospects for Tomorrow. Oxford: Butterworth-Heinemann, 1994:542–544.
351. Xie L, Wang DIC. Integrated approaches to the design of media and feeding strategies for fed-batch cultures of animal cells. Trends Biotechnol 1997; 15:109–113.
352. Martinelle K, Westlund A, Haggstrom L. Ammonium ion transport—a cause of cell death. Cytotechnology 1996; 22:251–254.
353. Bibila TA, Robinson DK. In pursuit of the optimal fed-batch process for monoclonal antibody production. Biotechnol Prog 1995; 11:1–13.
354. Zhou W, Chen C-C, Buckland B, Aunins J. Fed-batch culture of recombinant NSO myeloma cells with high monoclonal antibody production. Biotechnol Bioeng 1997; 55:783–792.
355. Sauer PW, Burky JE, Wesson MC, Sternard HD, Qu L. A high-yielding, generic fed-batch cell culture process for production of recombinant antibodies. Biotechnol Bioeng 2000; 67(5):585–597.
356. Nguyen B, Jarnagin K, Williams S, Chan H, Barnett J. Fed-batch culture of insect cells: a method to increase the yield of recombinant human nerve growth factor (rhNGF) in the baculovirus expression system. J Biotechnol 1993; 31(2):205–217.

357. Haggstrom L, Ljunggren J, Ohman L. Metabolic engineering of animal cells. Ann NY Acad Sci 1996; 782:40–52.
358. Doverskog M, Ljunggren J, Ohman L, Haggstrom L. Physiology of cultured animal cells. J Biotechnol 1997; 59(1–2):103–115.
359. Bedard C, Perret S, Kamen AA. Fed-batch culture of Sf-9 cells supports 3×10^7 cells per ml and improves baculovirus-expressed recombinant protein yields. Biotechnol Lett 1997; 19(7):629–632.
360. Riley MR, Rhiel M, Zhou XJ, Arnold MA, Murhammer DW. Simultaneous measurement of glucose and glutamine in insect cell culture media by near infrared spectroscopy. Biotechnol Bioeng 1997; 55(1):11–15.
361. Chan LCL, Greenfield PF, Reid S. Optimizating fed-batch production of recombinant proteins using the baculovirus expression vector system. Biotechnol Bioeng 1998; 59(2):178–188.
362. Mendonca RZ, Palomares LA, Ramirez OT. An insight into insect cell metabolism through selective nutrient manipulation. J Biotechnol 1999; 72(1–2):61–75.
363. Wang SL, Bentley WE, Liu CC, Wang MY. The use of glucose to regulate pH values of culture media and increase the production of baculo virus (BmNPV) and foreign protein (HBsAg). Process Biochem 1999; 34(3):295–301.
364. Elias CB, Zeiser A, Bedard C, Kamen AA. Enhanced growth of Sf-9 cells to a maximum density of 5.2×10^7 cells per mL and production of beta-galactosidase at high cell density by fed batch culture. Biotechnol Bioeng 2000; 68(4):381–388.
365. Xie L, Nyberg G, Gu X, Li H, Mollborn F, Wang DIC. Gamma-interferon production and quality in stoichiometric fed-batch cultures of Chinese hamster ovary (CHO) cells under serum-free conditions. Biotechnol Bioeng 1997; 56(5):577–582.
366. Liibben H, Kretzmer G. Diauxic cell behavior enables detoxification of CHO cell culture medium during fed batch cultivation. In: Merten O-W, Perrin B, Griffiths B, eds. New Developments and New Applications in Animal Cell Technology. Dordrecht: Kluwer Academic Publishers, 1998:267–271.
367. Nyberg GB, Balcarcel RR, Follstad BD, Stephanopoulos G, Wang DIC. Metabolic effects on recombinant interferon-y glycosylation in continuous culture of Chinese hamster ovary cells. Biotechnol Bioeng 1998; 62(3):336–347.
368. Andersen DC, Bridges T, Gawlitzek M, Hoy C. Multiple cell culture factors can affect the glycosylation of Ans-184 in CHO-produced tissue-type plasminogen activator. Biotechnol Bioeng 2000; 70(1):25–31.
369. Kyung Y-S, Hu W-S. Enhanced productivity of protein C by recombinant human cells in automated fed-batch cultures. Cytotechnology 1995; 17(2):109–115.
370. Lenas P, Kitade T, Watanabe H, Honda H, Kobayashi T. Adaptive fuzzy control of nutrients concentration in fed-batch culture of mammalian cells. Cytotechnology 1997; 25(1/3):9–15.
371. Xie L, Wang DIC. High cell density and high monoclonal antibody production through medium design and rational control in a bioreactor. Biotechnol Bioeng 1996; 51:725–729.
372. Zhou W, Rehm J, Europa A, Hu W-S. Alteration of mammalian cell metabolism by dynamic nutrient feeding. Cytotechnology 1997; 24:99–108.
373. Jo EC, Park HJ, Kim D, Moon HM. Repeated fed-batch culture of hybridoma cells in nutrient-fortified high-density medium. Biotechnol Bioeng 1993; 42:1229–1237.
374. Glacken MW, Huang C, Sinskey AJ. Mathematical descriptions of hybridoma culture kinetics. III. Simulation of fed-batch bioreactors. J Biotechnol 1989; 10:39–66.
375. Frame KK, Hu WS. Kinetic study of hybridoma cell growth in continuous culture. I. A model for non-producing cells. Biotechnol Bioeng 1991; 37:55–64.
376. Ray NG, Karkare SB, Runstadler PW. Cultivation of hybridoma cells in continuous cultures: kinetics of growth and products formation. Biotechnol Bioeng 1989; 33:724–730.

377. Martens DE, Gooijer CD, Groot CAMV, Beuvery EC. Effect of dilution rate on growth, productivity, cell cycle and size, and shear sensitivity of a hybridoma cell in a continuous culture. Biotechnol Bioeng 1993; 41:429–439.
378. Hiller GW, Aeschlimann AD, Clark DS, Blanch HW. A kinetic analysis of hybridoma growth and metabolism in continuous suspension culture on serum-free medium. Biotechnol Bioeng 1991; 38:733–741.
379. Sanfeliu A, Paredes C, Cairo JJ, Godia F. Analysis of glucose and glutamine metabolism of hybridoma cells by continuous culture experiments. In: Carrondo MJT, Griffiths JB, Moreira JLP, eds. Animal Cell Technology: From Vaccine to Genetic Medicine. Dordrecht: Kluwer Academic Publishers, 1997:785–789.
380. Cherlet M, Marc A. Intracellular pH monitoring as a tool for the study of hybridoma cell behavior in batch and continuous bioreactor cultures. Biotechnol Prog 1998; 14(4):626–638.
381. Vriezen N, Vandijken JP. Fluxes and enzyme activities in central metabolism of myeloma cells grown in chemostat culture. Biotechnol Bioeng 1998; 59(1):28–39.
382. Heidemann R, Liitkemeyer D, Bfintemeyer H, Lehmann J. Effects of dissolved oxygen levels and the role of extra- and intracellular amino acid concentrations upon the metabolism of mammalian cell lines during batch and continuous cultures. Cytotechnology 1998; 26(3):185–197.
383. Nyberg GB, Balcarcel RR, Follstad BD, Stephanopoulos GI, Wang DIC. Metabolism of peptide amino-acids by Chinese hamster ovary cells grown in a complex medium. Biotechnol Bioeng 1999; 62(3):324–335.
384. Cruz HJ, Moreira JL, Carrondo MJT. Metabolic shifts by nutrient manipulation in continuous cultures of BHK cells. Biotechnol Bioeng 1999; 66(2):104–113.
385. Europa AF, Gambhir A, Fu PC, Hu WS. Multiple steady states with distinct cellular metabolism in continuous culture of mammalian cells. Biotechnol Bioeng 2000; 67(1):25–34.
386. Follstad BD, Balcarcel RR, Stephanopoulos G, Wang DIC. Metabolic flux analysis of hybridoma continuous culture steady state multiplicity. Biotechnol Bioeng 1999; 63(6):675–683.
387. Cruz HJ, Peixoto CM, Nimtz M, Alves PM, Dias EM, Moreira JL, Carrondo MJT. Metabolic shifts do not influence the glycosylation patterns of a recombinant fusion protein expressed in BHK cells. Biotechnol Bioeng 2000; 69(2):129–139.
388. Baker KN, Rendal MH, Hills AE, Hoare M, Freedman RB, James DC. Metabolic control of recombinant protein iV-glycan processing in NSO and CHO cells. Biotechnol Bioeng 2001; 73(3):188–202.
389. Simpson NH, Singh RP, Emery AN, Al-Rubeai M. Bcl-2 over-expression reduces growth rate and prolongs Gl phase in continuous chemostat cultures of hybridoma cells. Biotechnol Bioeng 1999; 64(2):174–186.
390. Jan DCH, Petch DA, Huzel N, Butler M. The effect of dissolved oxygen on the metabolic profile of a murine hybridoma grown in serum-free medium in continuous culture. Biotechnol Bioeng 1997; 54:153–164.
391. Schmid G, Huber F, Kerschbaumer R. Adaptation of hybridoma cells to hydrodynamic stress under continuous culture conditions. In: Spier RE, Griffiths JB, MacDonald C, eds. Animal Cell Technology: Developments, Process and Products. Oxford: Butterworth-Heinemann, 1992:203–205.
392. Ozturk SS, Palsson BO. Loss of antibody productivity durig long-term cultivation of a hybridoma cell line in low serum and serum free media. Hybridoma 1990; 9:167–175.
393. Frame KK, Hu WS. The loss of antibody productivity in continuous cultures of hybridoma cells. Biotechnol Bioeng 1990; 35:469–476.
394. Barnes LM, Bentley CM, Dickson AJ. Characterization of the stability of recombinant protein production in the GS-NSO expression system. Biotechnol Bioeng 2001; 73(4):261–270.

395. Morrison CJ, McMaster WR, Piret JM. Differential stability of proteolytically active and inactive recombinant metalloproteinase in Chinese hamster ovary cells. Biotechnol Bioeng 1997; 53(6):594–600.
396. Mercille S, Johnson M, Lanthier S, Kamen AA, Massie B. Understanding factors that limit the productivity of suspension-based perfusion cultures operated at high medium renewal rates. Biotechnol Bioeng 2000; 67(4):435–450.
397. De la Broise D, Noiseux M, Massie B, Lemieux R. Hybridoma perfusion systems: a comparison study. Biotechnol Bioeng 1992; 40:25–32.
398. Ozturk SS. Engineering challenges in high density cell culture systems. Cytotechnology 1996; 22(1/3):3–16.
399. Tokashiki M, Arai T, Hamamoto K, Ishimaru K. High density culture of hybridoma cells using a perfusion culture vessel with external centrifuge. Cytotechnology 1990; 3:239–244.
400. Hu WS, Peshwa MV. Animal cell bioreactors-recent advances and challenges to scale-up. Can J Chem Eng 1991; 69:409–420.
401. Runstadler PW. Perfusion versus batch culture: which is the preferred process? In: Spier RE, Griffiths JB, MacDonald C, eds Animal Cell Technology: Developments, Process and Products. Oxford: Butterworth-Heinemann, 1992:347–353.
402. Runstadler PW, Tung AS, Ray NG. Fluidized-bed bioreactor perfusion culture for the production of therapeutic proteins. In: Spier RE, Griffiths JB, MacDonald C, eds. Animal Cell Technology: Developments, Process and Products. Oxford: Butterworth-Heinemann, 1992:495–500.
403. Bjorling T, Dudel U, Fenge C. Evaluation of a cell separator in large scale perfusion culture. In: Beuvery EC, Griffiths JB, Zeijlemaker WP, eds. Animal Cell Technology: Developments Towards the 21st Century. Dordrecht: Kluwer Academic Publishers, 1995:671–675.
404. Zeng AP, Deckwer WD. Model simulation and analysis of perfusion culture of mammalian cells at high cell density. Biotechnol Prog 1999; 15(3):373–382.
405. Schmid G, Wilke CR, Blanch HW. Continuous hybridoma suspension cultures with and without cell retention: kinetics of growth, and product formation. J Biotechnol 1992; 22:31–40.
406. Banik GG, Heath CA. Partial and total cell retention in a filtration-based homogeneous perfusion reactor. Biotechnol Bioeng 1995; 11:548–588.
407. Banik GG, Heath CA. High-density hybridoma perfusion culture, limitation vs. inhibition. Appl Biochem Biotechnol 1996; 61:211–229.
408. Chisti Y. Build better industrial bioreactors. Chem Eng Prog 1992; January:55–58.
409. Gray DR. Bioreactor operations—preparation, sterilization, charging, culture initiation and harvesting. In: Spier RE, ed. Encyclopedia of Cell Technology. Vol. 1. New York: Wiley, 2000:138–174.
410. Wolf MX, Hirschel MD. Growing TIL and LAK cells in hollow fiber bioreactors. In: Spier RE, Griffiths JB, eds. Animal Cell Biotechnology. Vol. 6. Orlando: Academic Press, 1994:237–258.
411. Threlfall G, Garland SG. Equipment sterilisation. In: Spier RE, Griffiths JB, eds. Animal Cell Biotechnology. Orlando: Academic Press, 1985.
412. Harper GJ. Air sterilisation. In: Spier RE, Griffiths JB, eds. Animal Cell Biotechnology. Vol. 1. Orlando: Academic Press, 1985.
413. Chisti Y. Assure bioreactor sterility. Chem Eng Prog 1992; 80–85.
414. Baseman HJ. SIP/CIP validation. Pharm Eng 1992; 12(2):37–46.
415. Young JH. Sterilization of various diameter dead-ended tubes. Biotechnol Bioeng 1993; 42:125–132.
416. Chadd HE, Chamow SM. Therapeutic antibody expression technology. Curr Opin Biotechnol 2001; 12:188–194.
417. FDA: http://www.fda.gov/.
418. EMEA: http://www.emea.eu.int/.

419. ICH: http://www.ifpma.org/ichl.html.
420. Current good manufacturing practice for finished pharmaceuticals, 21; C. F. R. 211. http://www.access.gpo.gov/nara /cfr/waisidx_01/21 cfr211_01.html.
421. Current good manufacturing practice in manufacturing, processing, packing, or holding of drugs, general, 21; C. F. R. 210.http://www.access.gpo.gov/nara/cfr/waisidx_01/21 cfr2 10_01.html.
422. CDER: http://www.fda.gov/cder.
423. ISPE: http://www.ispe.org/guides.htm.
424. Adamson R. Design and operation of a recombinant mammalian cell manufacturing process for rFVIII. Ann Hematol 1994; 68:9–14.
425. Teige M, Weidemann R, Kretzmer G. Problems with serum-free production of antithrombin III regarding proteolytic activity and product quality. J Biotechnol 1994; 34:101–105.
426. Kratje RB, Reiman A, Hammer I, Wagner R. Cultivation of recombinant baby hamster kidney cells in a fluidized bed bioreactor with porous borosilicate glass. Biotechnol Prog 1994; 10:410–420.
427. Chuppa S, Tsai YS, Yoon S, Shackleford S, Rozales C, Bhat R, Tsay G, Matanguihan C, Konstantinov K, Naveh D. Fermentor temperature as a tool for control of high-density perfusion cultures of mammalian cells. Biotechnol Bioeng 1997; 55(2):328–338.
428. Palermo UP, DeGraaf ME, Marotti KR, Rehberg E. Post LE Production of analytical quantities of recombinant proteins in Chinese hamster ovary cells using sodium butyrate to elevate gene expression. J Biotechnol 1991; 19:35–48.
429. Oh SKW, Vig P, Chua F, Teo WK, Yap MGS. Substantial overproduction of antibodies by applying osmotic pressure and sodium butyrate. Biotechnol Bioeng 1993; 42:601–610.
430. Fussenegger M, Mazur X, Bailey JE. A novel cytostatic process enhances the productivity of Chinese hamster ovary cells. Biotechnol Bioeng 1997; 55(6):927–939.
431. Fussenegger M, Schlatter S, Datwyler D, Mazur X, Bailey JE. Controlled proliferation by multigene metabolic engineering enhances the productivity of Chinese hamster ovary cells. Nat Biotechnol 1998; 16(5):468–472.
432. Mazur X, Fussenegger M, Renner WA, Bailey JE. Higher productivity of growth-arrested Chinese hamster ovary cells expressing the cyclin-dependent kinase inhibitor p27. Biotechnol Prog 1998; 14:705–713.
433. NIH Classification of Human Etiologic Agents on the Basis of Hazard—NIH Guidelines on Recombinant DNA Molecules: http://www4.od.nih. gov/oba/RAC/ guidelines/appendix_b. htm.
434. WHO Laboratory Biosafety Manual: http://www.who.int/ dsa/cat98/lab8.htm.
435. Council Directive of 26 October 1998 amending Directive 90/219/EEC on the contained use of genetically modified micro-organisms, Directive 98/8 I/EC OJ L 330 5 December 1998, pp. 13–31. http://biosafety.ihe.be/GB/Dk.Eur.GB/ Cont. Use/98_81 /98_8 l_TC.html.
436. Council directive of 12 October 1993 amending directive 90/679/EEC on the protection of workers from risks related to exposure to biological agents at work, Directive 93/88/EEC OJ L 268 29 October 1993, p. 71. http:// www.biosafety.be/GB/ Dir.Eur.GB/Other/93_88/TC.html.
437. CEN: list and status of TC233 standards: http:// biosafety.ihe.be/CEN/ CEN_List.html.
438. Miller SR, Bergman D. Biocontainment design considerations for biopharmaceutical facilities. J Indust Microbiol 1993; 11:223–234.
439. Reuveny S, Velez D, Macmillan J, Miller L. Factors affecting cell growth and monoclonal antibody production in stirred reactors. J Immunol Methods 1986; 86:53–59.
440. Rossler B, Liibben H, Kretzmer G. Temperature: a simple parameter for process optimization in fed-batch cultures of recombinant Chinese hamster ovary cells. Enzyme Microb Technol 1996; 18:423–427.

441. Jenkins N, Hovey A. Temperature control of growth and productivity in mutant Chinese hamster ovary cells synthesizing a recombinant protein. Biotechnol Bioeng 1993; 42:1029–1036.
442. Ramirez OT, Mutharasan R. The role of the plasma membrane fluidity on the shear sensitivity of hybridomas grown under hydrodynamic stress. Biotechnol Bioeng 1990; 36:911–920.
443. Ludwig A, Tomeczkowski J, Kretzmer G. Influence of the temperature on the shear stress sensitivity of adherent BHK 21 cells. Appl Microbiol Biotechnol 1992; 38:323–327.
444. Shi Y, Ryu DDY, Ballica R. Rheological properties of mammalian cell culture suspensions: hybridoma and HeLa cell lines. Biotechnol Bioeng 1993; 41:745–754.
445. Almgren A, Nilsson K, Nilsson L. Dissolved oxygen- the critical parameter at cell growth and expression of rec. protein in porous microcarrier system. Spier RE, Griffiths JB, MacDonald C, eds. Animal Cell Technology: Developments, Process and Products. Oxford: Butterworth-Heinemann, 1992:212–214.
446. Lim HS, Chang KH, Kim JH. Effect of oxygen partial pressure on production of animal virus (VSV). Cytotechnology 1999; 31(3):265–270.
447. Stachura ME, Lapp CA, Tyler JM, Lee YS. Medium flow rate modulates autocrine-paracrine feedback of GH and PRL release by perfused GH3 cells. In Vitro Cell Dev Biol 1990; 26(5):482–492.
448. Bader A, Boker K, Knop E, Friihauf N, Oldhafer K, Ringe B, Pichlmayr R, Sewing K-F. Reconstruction of liver tissue in-vitro geometry of characteristic flat bed-, hollow fiber-, and spouted bed bioreactors with reference to the in vivo liver. Artificial Organs 1995; 19:941–950.
449. Hesse F, Wagner R. Developments and improvements in the manufacturing of human therapeutics with mammalian cell cultures. Trends Biotechnol 2000; 18(4):173–180.
450. Bowen WR, Gan Q. Properties of microfiltration membranes: flux loss during constant pressure permeation of bovine serum albumin. Biotechnol Bioeng 1991; 38:688–696.
451. Mercille S, Johnson M, Lemieux R, Massie B. Filtration-based perfusion of hybridoma cultures in protein-free medium: reduction of membrane fouling by medium supplementation with DNase I. Biotechnol Bioeng 1994; 43:833–846.
452. McAdams TA, Miller WM, Papoutsakis ET. Hematopoietic cell culture therapies (Part I): cell culture considerations. Trends Biotechnol 1996; 14(9):341–349.
453. Thaler T, Vanak J. Oxygen transfer characteristics in cell culture fermenters: direct sparging–membrane oxygenation–bubble-free aeration through a rotating sieve. In: Spier RE, Griffiths JB, Meignier B, eds. Production of Biologicals from Animal Cells in Culture. Oxford: Butterworth-Heinemann, 1991:451–456.
454. Oosterhuis NMG, Thalman G, Berg GJ, Kopp TJ. Growth of mammalian cells using a new bubblefree aeration and perfusion systems: the Chemcell. Med Fac Landbouww Rijksuniv Gent 1988; 53(4a):1655–1662.
455. Takazawa Y, Tokashiki M. Production of human-mouse chimeric antibody by high cell density perfusion culture. Cytotechnology 1989; 2:95–101.
456. Stevens J, Eickel S, Onken U. Lamellar clarifier- a new device for animal cell retention in perfusion culture systems. In: Spier RE, Griffiths JB, Berthold W, eds. Animal Cell Technology: Products of Today, Prospects for Tomorrow. Oxford: Butterworth-Heinemann, 1994:234–239.
457. Thompson KJ, Wilson JS. A compact gravitational settling device for cell retention. In: Spier RE, Griffiths JB, Berthold W, eds. Animal Cell Technology: Products of Today, Prospects for Tomorrow. Oxford: Butterworth-Heinemann, 1994:227–229.
458. Wen ZY, Teng XW, Chen F. A novel perfusion system for animal cell cultures by two step sequential sedimentation. J Biotechnol 2000; 79(1):1–11.
459. Doblhoff-Dier O, Gaida T, Katinger H, Burger W, Groeschl M, Benes E. A novel ultrasonic resonance field device for the retention of animal cells. Biotechnol Prog 1994; 10:428–432.

460. Woodside SM, Bowen BD, Piret JM. Experimental and theoretical analysis of cell separations by ultrasonic forces. In: Carrondo MJT, Griffiths JB, Moreira JLP, eds. Animal Cell Technology: From Vaccine to Genetic Medicine. Dordrecht: Kluwer Academic Publishers, 1997:251–256.
461. Merten OW. Constructive improvement of the ultrasonic separation device ADI 1015. Cytotechnology 2000; 34(1):175–179.
462. Ryll T, Dutina G, Reyes A, Gunson J, Krummen L, Etcheverry T. Performance of small-scale CHO perfusion cultures using an acoustic cell filtration device for cell retention: characterization of separation efficiency and impact of perfusion on product quality. Biotechnol Bioeng 2000; 69(4):440–449.
463. Kendro. Product information Centritech Cell®. 2000.
464. Jager V. A novel perfusion system for the large-scale cultivation of animal cells based on a continuous flow centrifuge. In: Spier RE, Griffiths JB, MacDonald C, eds. Animal Cell Technology: Developments, Processes and Products. Oxford: Butterworth-Heinemann, 1992:397–402.
465. Kamen AA. Use of the Centritech lab centrifuge for perfusion culture of hybridoma cells in protein-free medium. Biotechnol Prog 1996; 12(6):855–864.
466. Yabannavar VM, Singh V, Connelly NV. Scaleup of spinfilter perfusion bioreactor for mammalian cell retention. Biotechnol Bioeng 1994; 43:159–164.
467. Rebsamen E, Goldinger W, Scheirer W, Merten O-W, Palfi GE. Use of a dynamic filtration method for separation of animal cells. Dev Biol Stand 1987; 66:273–277.
468. Fraune E, Fenge C, Kuhlmann W, Broly H. Development of perfusion bioreactors for high density cultures. In: White M, Reuveny S, Shaferman A, eds. Biologicals from Recombinant Microorganisms and Animal Cells—Production and Recovery. Weinheim: VCH, 1990:159–164.
469. Duvar S, Muthing J, Mohr FL, Lehmann J. Scale up cultivation of primary human umbilical vein endothelial cells on macrocarriers from spinner vessels to bioreactor fermentation. Cytotechnology 1996; 21(1):61–72.
470. Werner A, Duvar S, Miithing I, Btintemeyer H, Ltinsdorf H, Strauss M, Lehmann J. Cultivation of immortalized human hepatocytes HepZ on macroporous CultiSpher G microcarriers. Biotechnol Bioeng 2000; 68(1):59–70.
471. Martin N, Brennan A, DeNome L, Shaevitz J. High productivity in mammalian cell culture. Biotechnology 1987; 5:838–840.
472. Kloppinger M, Freitag G, Fraune E, Miltenburger HG. Flow cytometric process monitoring of hybridoma cells in batch and perfusion culture. In: Spier RE, Grififths JB, Stephenne J, Crooy PJ, eds. Advances in Animal Cell Biology and Technology for Bioprocesses. Oxford: Butterworth, 1989:125–128.
473. Goebel NK, Kuehn R, Flickinger C. Methods for determination of growth-rate-dependent changes in hybridoma volume, shape and surface structure during continuous recycle. Cytotechnology 1990; 4:45–57.
474. Winzeler HB. Membran-Filtration mit hoher Trennleistung und minimalem Energiebedarf. Chimia 1990; 44:288–291.
475. Pinton H, Silva AL, Goergen JL, Marc A, Engasser JM, Rabaud JN, Pierry G. Control of the maximal cell density in a membrane perfusion reactor. In: Spier RE, Griffiths JB, Berthold W, eds. Animal Cell Technology: Products of Today, Prospects for Tomorrow. Oxford: Butterworth-Heinemann, 1994:470–475.
476. Goergen JL, da Silva AL, Marc A, Engasser JM. Fast propagation of hybridoma cells in a forced-flow membrane perfusion reactor. In: Beuvery EC, Griffiths JB, Zeijlemaker WP, eds. Animal Cell Technology: Developments Towards the 21st Century. Dordrecht: Kluwer Academic Publishers, 1995:705–710.
477. Zumbusch Pv, Portner R, Brun B, Brunner G, Markl H. On the use of electric fields to prevent membrane fouling of process integrated cross flow filtration at hybridoma cell culture. In: Carrondo MJT, Griffiths JB, Moreira JLP, eds. Animal Cell Technology:

From Vaccine to Genetic Medicine. Oxford: Kluwer Academic Publishers, 1997:215–220.
478. Vogel JH, Kroner KH. Controlled shear filtration: A novel technique for animal cell separation. Biotechnol Bioeng 1999; 63(6):663–674.
479. Marique T, Paul T, Blankaert D, Guerra IT, Werenne J. Collagenase production by a cell line in collagen microspheres in a fluidized-bed reactor and its modelization. In: Spier RE, Griffiths JB, Berthold W, eds. Animal Cell Technology: Products of Today, Prospects for Tomorrow. Oxford: Butterworth-Heinemann, 1994:245–247.
480. Marique TH, Malarme D, Stragier P, Werenne J. On-line monitoring of growth and viral infection of VERO cells in collagen microspheres in a fluidized-bed bioreactor. In: Spier RE, Griffiths IB, Berthold W, eds. Animal Cell Technology: Products of Today, Prospects for Tomorrow. Oxford: Butterworth-Heinemann, 1994:369–371.
481. Shen BQ, Greenfield PF, Reid S. Calcium alginate immobilized hybridomas grown using a fluidized-bed perfusion system with a protein-free medium. Cytotechnology 1994; 14:109–117.
482. Keller J, Dunn IJP, Heinzle E. Improved performance of the fluidized-bed reactor for the cultivation of animal cells. In: Spier RE, Griffiths IB, Meignier B, eds. Production of Biologicals from Animal Cells in Culture. Oxford: Butterworth-Heinemann, 1991:513–515.
483. Kratje RB, Wagner R. Evaluation of production of recombinant human interleukin-2 in fluidized bed bioreactor. Biotechnol Bioeng 1992; 39:233–242.
484. Kratje R, Wagner R. The production of recombinant human interleukin-2 in a fluidized bed bireactor. In: Spier RE, Griffiths JB, MacDonald C, eds. Animal Cell Technology: Developments, Processes and Products. Oxford: Butterworth-Heinemann, 1992:488–494.
485. Preissmann A, Wiesmann R, Buchholz R, Werner RG, Noe W. Investigations on oxygen limitations of adherent cells growing on macroporous microcarriers. Cytotechnology 1997; 24(2):121–134.
486. Kawakubo Y, Matsushita T, Funatsu K. Tissue plasminogen activator (t-PA) production by a high density culture of weakly adherent human embryonic kidney cells using a polyurethan-foam packed-bed culture system. Appl Microbiol Biotechnol 1994; 41:413–418.
487. Heine H, Spies M, Wandrey C. Effects of ammonium and lactate during continuous hybridoma fermentations in a fluidized-bed bioreactor. In: Merten O-W, Perrin B, Griffiths B, eds. New Developments and New Applications in Animal Cell Technology. Dordercht: Kluwer Academic Publishers, 1998:281–283.
488. Heine H, Biselli M, Wandrey C. High cell density cultivation of hybridoma cells: spin filter vs immobilised culture. In: Bernard A, Griffiths B, Noe W, Wurm F, eds. Animal Cell Technology: Products from Cells-Cells as Products. Oxford: Butterworth-Heinemann, 1999:83–85.
489. Reuschenbach P. Production of t-PA by C127 cells cultivated on different porous microcarrier in a perfused fluidized bed reactor. Spier RE, Griffiths JB, MacDonald C, eds. Animal Cell Technology: Developments, Processes and Products. Oxford: Butterworth-Heinemann, 1992:511–513.
490. Kong D, Cardak S, Chen M, Gentz R, Zhang J. High cell density and productivity culture of Chinese hamster ovary cells a fluidized bed bioreactor. Cytotechnology 1999; 29(3):215–220.
491. Bluml G, Gaida TH, Zach N, Assadian A, Schmatz C, Katinger H. Fluidized bed technology: Influence of fluidization velocity on nutrient consumption and product expression. In: Merten OW, Perrin B, Griffiths B, eds. New Developments and New Applications in Animal Cell Technology. Dordrecht: Kluwer Academic Publishers, 1998:385–387.
492. Wang MD, Yang M, Huzel N, Butler M. Erythropoetin production from CHO cells grown in a fluidized-bed bioreactor with macroporous beads. In: Bernard A, Griffiths

B, Noe W, Wurm F, eds. Animal Cell Technology: Products from Cells-Cells as Products; 1999:307–309.
493. Muller D, Unterluggauer F, Kreismayr G, Schmatz C, Assadian A, Wiederkum S, Preis S, Vorauer-Uhl K, Doblhoff-Dier O, Katinger H. Continuous perfused fluidized bed technology, increased productivities and product concentration. In: Bernard A, Griffiths B, Noe W, Wurm F, eds. Animal Cell Technology: Products from Cells, Cells as Products. Oxford: Butterworth-Heinemann, 1999:319–321.
494. Valle MA, Kaufman J, Bentley WE, Shiloach J. Evaluation of porous microcarriers in fluidized bed reactors for protein production by HEK 293 cells. In: Merten O-W, Perrin B, Griffiths B, eds. New Developments and New Applications in Animal Cell Technology. Dordrecht: Kluwer Academic Publishers, 1998:381–384.
495. Unterluggauer F, Doblhoff O, Reiter M, Bluml G, Zach N, Krai G, Schmatz G, Katinge H. Scale-up of fluidized bed reactor operated with porous glass carriers. In: Spier RE, Griffiths JB, MacDonald C, eds. Animal Cell Technology: Developments, Processes and Products. Oxford: Butterworth-Heinemann, 1992:505–507.
496. Loibner AP, Zach N, Doblhoff-Dier O, Reiter M, Bayer M, Katinger H. On-line glucose control of animal cell cultures in fluidized bed. In: Spier RE, Griffiths JB, Berthold W, eds. Animal Cell Technology: Products of Today, Prospects for Tomorrow. Oxford: Butterworth-Heinemann, 1994:372–375.
497. Reiter R, Bliiml G, Gaida T, Zach N, Schmatz C, Borth N, Hohenwarter O, Katinger H. High density aggregate culture of recombinant cho cells in fluidized bed. In: Spier RE, Griffiths JB, MacDonald C, eds. Animal Cell Technology: Developments, Process and Products. Oxford: Butterworth-Heinemann:421–423.
498. Racher AJ, Looby D, Griffiths JB. Metabolite parameter modulating mab production in a fixed bed, glass sphere bioreactor. In: Spier RE, Griffiths JB, MacDonald C, eds. Animal Cell Technology: Developments, Process and Products. Oxford: Butterworth-Heinemann, 1992:455–559.
499. Ludemann I, Portner R, Schaefer C, Schick K, Sramkova K, Reher K, Neumaier M, Franek F, Markl, H. Improvement of the culture stability of non-anchorage—dependent animal cells grown in serum-free media through immobilization. Cytotechnology 1996; 19(2):111–124.
500. Clarke JB, Griffiths JB. Bioreactors for production of HIV— a comparative evaluation of flask, suspension and fixed bed cultures. In: Spier RE, Griffiths JB, Meignier B, eds. Production of Biologicals from Animal Cells in Culture. Oxford: Butterworth-Heinemann, 1991:519–522.
501. Portner R, Bohmann A, Markl H. Application of dialysis techniques for effective production of biologicals with immobilized animal cells. In: Beuvery EC, Griffiths JB, Zeijlemaker WP, eds. Animal Cell Technology: Developments Towards the 21st Century. Dordrecht: Kluwer Academic Publishers, 1995:823–827.
502. Fassnacht D, Portner R. Experimental and theoretical considerations on oxygen supply for animal cell growth in fixed-bed reactors. J Biotechnol 1999; 72(3):169–184.
503. Jannin J, Gaspard C, Zaot C, Fraue T, Dhainaut F, Mignot, G. Comparison of two systems based on porous beads for the production of recombinant von Willebrand factor by CHO cells. In: Spier RE, Griffiths JB, Meignier B, eds. Production of Biologicals from Animal Cells in Culture. Oxford: Butterworth-Heinemann, 1991:525–527.
504. Clarke JB, Griffiths JB. The effect of bioreactor configuration on production of HIV and cell-virus interaction. Cytotechnology 1990; 4(2):145–153.
505. Racher AJ, Griffiths JB. Development of a bioreactor process for recombinant cell lines. In: Spier RE, Griffiths JB, Berthold W, eds. Animal Cell Technology: Products of Today, Prospects for Tomorrow. Oxford: Butterworth-Heinemann, 1994:248–251.
506. Schroder B, Herfurth C, Meilkier P, Link H, Wandrey C. Differentiation of bone marrow cells immobilized on microcarriers in a fluidized bed reactor. In: Carrondo MJT, Griffiths JB, Moreira JLP, eds. Animal Cell Technology: From Vaccine to Genetic Medicine. Dordrecht: Kluwer Academic Publishers, 1997:589–593.

507. Preissmann A, Bux R, Schorn P, Noe W. Comparative study for the propagation of anchorage-dependent cells using different forms macroporous carriers. In: Beuvery EC, Griffiths JB, Zeijlemaker WP, eds. Animal Cell Technology: Developments Towards the 21st Century. Dordrecht: Kluwer Academic Publishers, 1995:841–845.
508. Grabner R, Paul E. Use of motionless mixer as cell culture propagator. US 4,296,204, October 20, 1981.
509. Lazar A, Reuveny S, Kronman C, Velan B, Shafferman A. Evaluation of anchorage-dependent cell propagation systems for production of human acetylcholinesterase by recombinant 293 cells. Cytotechnology 1993; 13(2):115–123.
510. Matsushita T, Hidaka H, Kamihata K, Kawakubo Y, Funatsu K. High density culture of anchorage-dependent animal cells by polyurethane foam packed-bed culture systems. Appl Microbiol Biotechnol 1991; 35:159–164.
511. Lazar A. Immobilisation of animal cells in fixed bed reactors. Biotech Adv 1991; 9:411–424.
512. Treat JC, Sandelin KK, Beiderman BC, Kirschner RJ. Low-and high-producing animal cells cultures in perfused, packed beds of non-woven fabric. In: Carrondo MJT, Griffiths JB, Moreira JLP, eds. Animal Cell Technology: From Vaccine to Genetic Medicine. Dordrecht: Kluwer Academic Publishers, 1997:313–318.
513. Kadouri A, Wang G, Zhang W, Jacklin C, Freedman D. Design and performance of perfusion based packed bed bioreactor for the production of secreted protein. In: Animal Cell Technology: from Vaccine to Genetic Medicine. Kluwer Academic Publishers, 1997:307–312.
514. Park S, Stephanopoulos G. Packed bed bioreactor with porous ceramic beads for animal cell culture. Biotechnol Bioeng 1993; 41:25–34.
515. Thewall PE, Brindle KM. Analysis of CHO-K1 cell growth in a fixed bed bioreactor using magnetic resonance spectroscopy and imaging. Cytotechnology 1999; 30:121–132.

7

Aeration, Mixing and Hydrodynamics in Bioreactors

Ningning Ma, Mike Mollet, and Jeffrey J. Chalmers
Department of Chemical and Biomolecular Engineering, The Ohio State University, Columbus, Ohio, U.S.A.

INTRODUCTION

Over the last 20 years, there has been a significant increase in the number of large-scale animal cell culture processes for commercial, therapeutic products. A number of these processes are conducted in bioreactors on the order of 10,000 L; yet, no single universal methodology on how to design, operate , and scale-up bioreactors exists. Conceptually, the optimal scale-up (or scale-down) methodology would provide an optimum, constant "microenvironment" surrounding each cell that is scale independent. This constant "microenvironment" would not only include the concentration of all nutrients and byproducts, but also all hydrodynamic forces acting on the cells. This lack of methodology is the result of the complexity of introducing/removing gas-phase nutrients/byproducts into/from the culture and the turbulent nature of the hydrodynamics. The gas-phase challenges are exacerbated by the low solubility of oxygen in the liquid phase, which requires an active oxygen introduction system (either gas sparging or the use of membranes) in culture sizes typically greater than 100 L. Equally challenging is the perceived, and in some cases actual, sensitivity of the cells to hydrodynamic forces generated in bioreactors to facilitate the mixing of nutrients and the suspension of the cells.

Agitated tanks dominate the design of animal cell bioreactors, although there are reports of relatively large scale 2000 L airlift bioreactors (1). One reason is the vast practical knowledge accumulated in the fermentation and chemical engineering fields over the last century. A second reason is the relative ease of design, operation, and scale-up. However, unlike agitated tanks used for chemical mixing and/or reaction, animal cells in some cases are sensitive to hydrodynamic forces. This sensitivity varies according to cell lines, whether the cells are in suspension or attached to microcarriers, and potentially other factors.

Except for blood cells, all other animal cells are attached to surfaces in their native state. Several cell lines, such as Chinese hamster ovary (CHO) cells, hybridoma cells, and insect cells, have been adapted to grow in suspension. Microcarriers provide a way to culture surface anchored cells in high concentration. However, as will be discussed later, microcarrier cultures present unique challenges since experimental and semiexperimental studies indicated that cells attached to microcarriers are significantly more sensitive to hydrodynamic forces than freely suspended cells.

The actual mechanisms by which cells are damaged as a result of hydrodynamic forces are still not well understood. This paucity of knowledge is a result of our inability to understand turbulent flow at a sufficiently small scale, the interfacial phenomena in the cell–air–medium system, the complexity of cellular mechanical properties, and the cells' physiological responses to physical strain. However, significant progress has been made in this understanding. This chapter will attempt to summarize our current understanding of aeration, mixing, and the relationship of the hydrodynamic forces caused by these processes to animal cells.

AERATION FOR CELL CULTURE BIOREACTORS

Oxygen is a key nutrient in cellular metabolism. While low oxygen concentration affects cell growth and yield, excessive oxygen tensions can also be toxic to the cells (2,3). Glacken et al. (4) attributed the toxicity to the generation of superoxide radicals that may affect cells directly or indirectly by oxidizing medium components. The oxygen requirement of animal cells varies according to different cell lines and culture conditions. For common cell types, the typical oxygen uptake rate (OUR) is between 3×10^{-10} and 2×10^{-8} mg/cell hr (5). The classical and typically used aeration method is sparging, which is bubbling gas directly into the culture medium for liquid volumes $>$ 10–100 L. Other methods are usually problematic, although they are able to outperform sparging in some aspects. Even though the oxygen demands of animal cell cultures are several orders of magnitude lower than microbial cultures, oxygenation is usually the primary challenge in cell culture scale-up. The importance of aeration increases with the escalation of bioreactor size and cell concentration.

Basic Mass Transfer Concepts

The oxygen transfer rate is determined by the mass transfer resistance and the driving force:

$$\text{OTR} = K_{g}a(P_{O_2} - P^*_{O_2}) \tag{7.1}$$

or

$$\text{OTR} = K_{l}a(C^*_{O_2} - C_{O_2}) \tag{7.2}$$

where OTR represents the oxygen transfer rate per unit reactor volume, $1/K_{g}a$ and $1/K_{l}a$ are the transfer resistances, and $(P_{O_2} - P^*_{O_2})$ and $(C^*_{O_2} - C_{O_2})$ are the corresponding driving forces. More specifically, P_{O_2} and $P^*_{O_2}$ are the partial oxygen pressure in the gas phase and the equilibrium partial oxygen pressure in the liquid phase, respectively, C_{O_2} and $C^*_{O_2}$ are the dissolved oxygen concentration and the saturating oxygen concentration, respectively, K_g and K_l are the corresponding oxygen transfer coefficients, and a is the interfacial area per bioreactor volume.

Equations (7.1) and (7.2) are interchangeable via Henry's law constant,

$$k_l = HK_g \tag{7.3}$$

and

$$P^*_{O_2} = HC_{O_2} \tag{7.4}$$

A simple analysis of Eqs. (7.1) and (7.2) indicates that increasing the oxygen transfer coefficients, the interfacial area, and/or the driving force can boost the OTR. Increasing the vessel pressure is a very effective and economical way to obtain a higher driving force. Large-scale bioreactors are typically rated to withstand 40 psi because of the need of "in place" sterilization (6). Operating a bioreactor at 40 psi would increase the driving force by a factor of 2–2.5 relative to the driving force at atmospheric pressure. However, a further increase of pressure would require a reinforced vessel, and the associated increase in carbon dioxide concentration in the culture may result in inhibitory physiological effects. Another relatively simple step to increase the OTR is using pure oxygen. Inspection of Eqs. (7.1) and (7.2) indicates that using pure oxygen can significantly increase the value of $P^*_{O_2}$ and $C^*_{O_2}$ thereby increasing the OTR. However, as will be discussed later, the sparging of gas in the vessel also affects the CO_2 removal rate from the system; consequently, any increase in OTR must be balanced by the negative effect it has on the removal of CO_2 from the system.

Surface Aeration

The easiest way to provide nutrient and byproduct exchange between a gas phase and the culture medium is through surface aeration, which, as the name implies, is the transfer of gas to the culture through the headspace in the bioreactor (7). Simple surface aeration is sufficient for low-density culture in small vessels, such as T-flasks, roller bottles, and bench-scale vessels, where a relatively large ratio of air–medium interface to volume exists. The actual culture volume at which surface aeration is insufficient depends on a number of factors; volumes up to 100 L have been effectively operated with surface aeration alone. Surface impellers and subsurface impellers facilitate the bulk oxygen diffusion (8,9). Aunins et al. (9) increased the OTR by a factor of 50% by moving the impeller within 0.13 times the impeller diameter of the liquid surface (9). Hu et al. (7) reported a fourfold increase of oxygen supply by using a surface impeller in addition to the normal impeller. Even if other aeration methods are utilized, flushing air or pure oxygen through the headspace compartment is an additional means of increasing the overall OTR (10).

Perfluorocarbons to Supply Oxygen

Perfluorocarbons are hydrocarbons whose hydrogen atoms are partly or totally replaced by fluorine. The solubility of oxygen in perfluorocarbons is 10–20 times higher than that in pure water (11). The perfluorochemicals are hydrophobic and immiscible with water-based media; therefore, the addition of perfluorocarbons yields a four-phase mixture in sparged bioreactors. Perfluorocarbons, the discrete phase, improve the oxygen transfer speed by working as gas vectors that transfer oxygen and carbon dioxide between air and the medium. The benefits of using perfluorocarbons in cell culture systems are summarized by Lowe et al. (12) as follows:

1. chemically and biologically inert;
2. ease of sterilization (e.g., by autoclave);
3. recoverable and recyclable;

4. high respiratory gas solubility;
5. scavengers of gaseous cellular products;
6. provide a two-phase (perfluorocarbon-liquid and aqueous medium) interface and physical support for cell division (in stationary cell culture).

Several studies have been conducted with animal and insect cell cultures in which the medium was supplemented with perfluorocarbon emulsions (13–16). These studies found that the exponential growth phase was prolonged, along with a more than sixfold increase in cell concentration, which was consistent with theoretical predictions based on the enhanced oxygen transfer. However, this method introduces additional cost and problems in downstream processing and has not been reported for industrial use (1).

Membrane Aeration

Membrane aeration replaces the need for gas sparging by providing a large interfacial area for oxygen diffusion. Most membranes could be categorized either as a microporous type or as a diffusion type. For the first type of membranes, the medium is in direct contact with air in the micropores within the membrane. The position of the air–liquid interface is held stationary via both the hydrophobic force and pressure. For the diffusional type of membrane, oxygen diffuses first from the gas phase into the highly oxygen soluble membrane, normally silicon rubber, and then into the culture medium. Of all the membrane module geometries possible, the hollow fiber modules have received the most attention. Aunins and Henzler (5) summarized the properties of a number of commercial membranes.

Without introducing a moving gas–medium interface into the liquid phase, membrane aeration is able to avoid physical damage to relatively fragile insect and animal cells. This method can also provide a high oxygen transfer rate. Aunins et al. (17) and Moreira et al. (18) compared the performance of three different aeration methods: surface, membrane, and sparging. The results obtained by Moreira et al. (18) with a laboratory-scale vessel showed that membrane aeration could give large OTRs, comparable to sparging. However, it requires higher gas pressure and flow rates.

Membrane aeration is appropriate in small- to intermediate-scale bioreactors (10–500 L), which have large oxygen demand and are processing flow sensitive cells (5). However, it is not widely used because of the complexity of the process, the limited design data, and the difficulties in maintenance. Proteins and cell debris can deposit on the pore wall, potentially altering its hydrophobic property. It also increases the difficulty in cleaning, especially at larger scales (19).

Sparging

Sparging gas directly into the medium is the most simple and commonly used "large-scale" way to provide oxygen in bioreactors (19,20). In fact, agitated, sparged tank, bubble column, and airlift reactor account for most of bioreactors used in large-scale cell culturing. The oxygen transfer speed in sparged bioreactors is determined by (1) design parameters, such as sparger size and type, (2) operation parameters, such as superficial gas velocity and agitation intensity, and (3) material properties of gas and culture, such as densities, viscosities, and surface tension. Normally, smaller bubbles give larger interface and hence higher OTRs at the same volume speed (21). Bubble coalescence is an important factor because of its effects on bubble size. However, most bioreactors are operated at low gas flow rate and low agitation rate based on the precieved flow sensitivity of cells. In this situation, the coalescence is not severe and the bubble size far from the sparger is almost equal to the bubble size at the sparger (1).

There are several correlations for the oxygen transfer coefficient. Several published correlations in noncommercial size fermentors share the form

$$K_g a = a_1 (P_g/V_1)^{\alpha_1} (U_s)^{\beta_1} \tag{7.5}$$

where P_g/V_1 is the power input per unit volume of gas dispersion, U_s is the superficial gas velocity, and a_1, α_1, and β_1 are constants independent of scale and impeller type (22–25). While care should be taken in the use of any correlation, Eq. (7.5) allows one to observe the most important groups of variables that affect the oxygen transfer coefficient (6).

CO_2 Accumulation and Removal

The role of CO_2 in cell culture is complex. First of all, the possible solute species of carbon dioxide can take the form of CO_2, HCO_3^-, H_2CO_3, or CO_3^{2-}. Since sodium bicarbonate is typically used as the pH buffer, the variation of CO_2 concentration may affect the pH of the medium through the following relationship:

$$\begin{aligned}[CO_2]_T &= [CO_2] + [H_2CO_3] + [HCO_3^-] + [CO_3^{2-}] \\ &= [CO_2]\left(1 + \frac{K_1}{[H^+]} + \frac{K_1 K_2}{[H^+]^2}\right)\end{aligned} \tag{7.6}$$

where the subscript T denotes total CO_2 concentration, and K_1 and K_2 are equilibrium constants (26). Secondly, CO_2 is both a nutrient and a potentially inhibitory byproduct in typical cell cultures. Therefore, both CO_2 accumulation and excessive removal should be avoided. Finally, the addition of either air or O_2 to the system can, and usually does, remove CO_2 from the system. This balance between O_2 addition and CO_2 removal leads to more complexity with respect to gas exchange.

Drapeau (27) studied the effect of increasing dissolved CO_2 on recombinant CHO cells. A 3-day batch study showed that the cellular growth rate and specific productivity decreased dramatically with increased pCO_2 while the dissolved oxygen, pH, as well as other variables were held constant. Gray et al. (28) also observed this inhibitory effect in a system in which pure oxygen microbubbles were added to the culture. Even though the microbubbles were able to meet the oxygen requirement for up to 2.1×10^7 CHO cells per mL, poor CO_2 removal restricted the cell density to only 2.4×10^6 cells per mL. Significant accumulation of dissolved CO_2 was also reported with silicon diffusion aerators where the mass transfer coefficient of oxygen was much higher than that of CO_2 (29).

In contrast to the problem of CO_2 build-up due to the low level of gas–medium exchange, the opposite effect can happen in bubble columns or airlift bioreactors. In these systems, one primary purpose of gas sparging is for mixing; hence, rates of sparging significantly exceed the rate needed for gas exchange. Consequently, the CO_2 concentration can drop to inhibitory low levels. Birch et al. (30) observed that the pCO_2 dropped from 0.08 to 0.008 bar in a large-scale airlift bioreactor with a constant sparging rate. The reduced pCO_2 causes a higher pH and hinders prime metabolism reactions as well as the synthesis of purines and pyrimidines (5).

Detrimental Effects of Sparging

The observation of cell damage caused by introducing bubbles into culture goes back to the beginning of suspended animal cell culture (31,32). However, it was the work of Kunas and Papoutsakis (33) and Oh et al. (34) that revealed the importance of cell–bubble interactions at gas–medium interfaces as opposed to pure agitation with respect to physical cell damages. Before their studies, the rule of thumb was that the hydrodynamic forces resulting from the action of impellers were more damaging. Both authors reported that as long as air was not introduced into the vessel, hybridoma cells could withstand impeller speeds >450 rpm, which is significantly higher than that typically used in bioreactors. Continuing this study, Kunas and Papoutsakis (35) built a bioreactor that had no air–medium interface at the top of the vessel to prevent bubble entrainment from the central vortex. With this system, no cell damage was observed at impeller speeds up to 600 rpm.

Potential cell damage areas in sparged vessels include the bubble generation region, the bubble rising region, and the bubble disengagement region. When the bioreactor is agitated and the sparger lies under the impeller, the impeller region is also a possible damage area where bubbles coalesce and breakup occurs.

Several researches suggested that cell damage took place in the bubble disengagement region at the surface of the culture (36–40). Through a microscope–video system, Handa et al. (36) observed rapid cell movements when the bubble film drained, and violent turbulent oscillations and surface deformations of cells when surrounding bubbles ruptured. Using a higher-resolution video system and specially designed columns, Bavarian et al. (37) presented photographs of insect cells attached to rising bubbles, trapped in the foam layer, and attached to the bubble film at the gas–medium interface.

When a bubble sitting at the gas–medium interface ruptures, a highly energetic event occurs, resulting in a rapid upward and downward jet of fluid (it is this upward jet of fluid that produces the "fizz" in carbonated soft drinks). To determine if this rupture process and the subsequent upward jet is sufficient to kill cells, Trinh et al. (40) developed a device in which a large number of single bubbles could be ruptured in a short time. Cell damage in the bulk liquid was measured and fluid in the upward jet was also collected for microscopic analysis. Statistically significant data indicated that on average $\sim 10^3$ cells were killed per 2 mm bubble rupture. In addition, all of the cells collected from the upward jet were killed and the concentration of cells in the upward jet was twice that in the bulk suspension.

Based on these experimental observations, Boulton and Blake (41) and Garcia-Briones et al. (42) conducted computer simulations of the bubble rupture process and obtained similar results. The simulations indicated that the rupture of small air bubbles in pure water generated intensive energy dissipation, which was three to four orders of magnitude higher than what is typically created in bioreactors solely due to agitation (42,43). The level of the energy dissipation rate increases rapidly with decrease in bubble size (41,42).

Several semiempirical correlations are available to predict the specific death rates of suspended cells in sparged vessels. Tramper et al. (44) suggested that cell damage followed first-order death rate kinetics and that there was a hypothetical killing volume associated with each air bubble. According to this argument, the first-order death rate constant, K_d, of a sparged culture can be represented by

$$K_d = 4FX'/(\pi^2 D^2 H) \tag{7.7}$$

where F is the airflow rate into the vessel, X' is the specific hypothetical killing volume, i.e., the hypothetical killing volume divided by the air bubble volume, D is the column diameter, and H is the height of the column. This relationship has been verified with different cell types, bubble sizes, and H/D ratios (45,46). Literature values of X' vary between 2×10^{-3} and 2×10^{-2}. Trinh et al. (40) suggested that this hypothetical killing volume was a thin layer of liquid surrounding the bubble. This is consistent with both experimental (47,48) and simulation results (42), which indicated that the fluid in the upward jet originated from a thin film surrounding the original bubble and the hydrodynamic forces that the fluid experiences in this thin film is exceedingly high.

A second correlation was proposed by Wang et al., which related the local specific death rate, p, to the local specific bubble interfacial area, a, and the local bubble breakup/bursting frequency:

$$p = \left(\frac{k_2 s}{K}\right) a \tag{7.8}$$

where k_2 is the intrinsic cell inactivation rate constant, s is the equivalent thickness of the inactivation region around a deformed bubble, and K is the Michaelis–Menton saturation constant for cells absorbed into an inactivation zone around a bubble. Although neither correlation provides a mechanistic explanation for the cell damage, they provide the following insights: (1) cell damage is proportional to the bubble–medium interfacial area and (2) cell damage is unrelated to the vessel diameter and height.

Cell Interactions with Gas–Medium Interfaces

Parker and Barsom (49) reported that the chemical composition of the surface microlayer, which is the thin layer of liquid in the air–liquid interface, was very different from that of the bulk water. Hydrophobic materials, including cellular components , tend to accumulate in this layer. As stated previously, Bavarian et al. (37) and Chalmers and Bavarian (38) observed the attachment of insect cells to various gas–medium interfaces. Several representative photographs are shown in Fig. 1. In further studies, Garcia-Briones and Chalmers (50) observed the attachment of cells to the bubble film that separated the air in the bubble from the air above the gas–medium interface. Figure 2a, illustrates the large number of cells on the film. However, when Pluronic F-68 (PF-68) was present in the medium, almost no cells were observed attached to this film (Fig. 2b).

Thermodynamics of Cell Adhesion to Bubbles

Dahlback et al. (51) demonstrated that the tendency of bacteria to attach to the air–water interface was proportional to the hydrophobicity of the cell surface. Similarly, Chattopadhyay et al. (52) suggested that the cell–bubble attachment of insect cells was also the result of hydrophilic–hydrophobic interactions. Continuing this study,

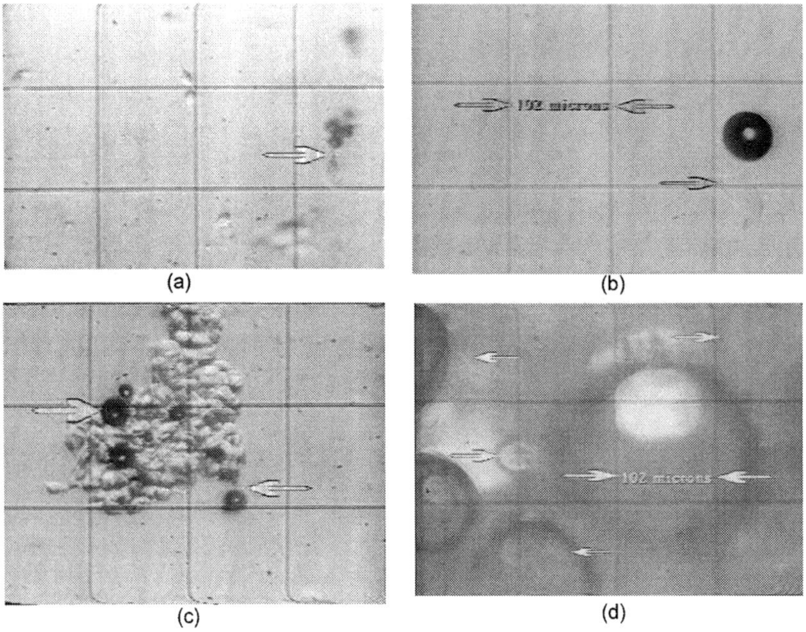

Figure 1 Microscopic images of insect cells attached to rising bubbles (a,b), a clump of cells rising with bubbles (c), and cells trapped in the foam layer (d).

Chattopadhyay et al. (53) tried to explain the cell–bubble attachment from a thermodynamic point of view.

During the attachment of a cell to a gas–liquid interface, a region of the cell–liquid interface and the gas–liquid interface are replaced by a cell–gas interface . The variation of free energy for this process can be represented as

$$\Delta F^{adh} = \gamma_{cv} - \gamma_{lv} - \gamma_{cl} \tag{7.9}$$

where ΔF^{adh} is the change in free energy and γ_{cv},γ_{lv},γ_{cl} are surface tensions of the interfaces of cell–gas, liquid–gas, and cell–liquid, respectively. For a thermodynamically feasible process, the change of free energy must be negative. Chattopadhyay et al. (53) examined the surface tensions and free energy changes for five different cell lines in different media. γ_{cl} values were found much smaller than the values of γ_{cv} and γ_{lv}. Consequently, the governing factor determining whether ΔF^{adh} is negative or positive is the value of γ_{cv} relative to γ_{lv}. Consistent with this model, most protective additives, such as PF-68 and Methocel E-50, are able to lower the value of γ_{lv} below that of γ_{cv} , thereby creating a condition in which ΔF^{adh} is positive (53,54).

Protective Additives to Prevent Damage and Proposed Mechanism

Over the years a number of compounds have been found to be able to prevent cell damage when added to the culture. These compounds include: hydroxyethyl starch, derivatives of cellulose, serum, tryptosephosphate, polyvinyl alcohol (PVA), bovine serum albumin, polyethylene glycols (PEGs), pluronics and dextran (5,52,55).

A number of mechanisms have been suggested to explain the observed protective effects. In general, these mechanisms can be grouped into three major categories:

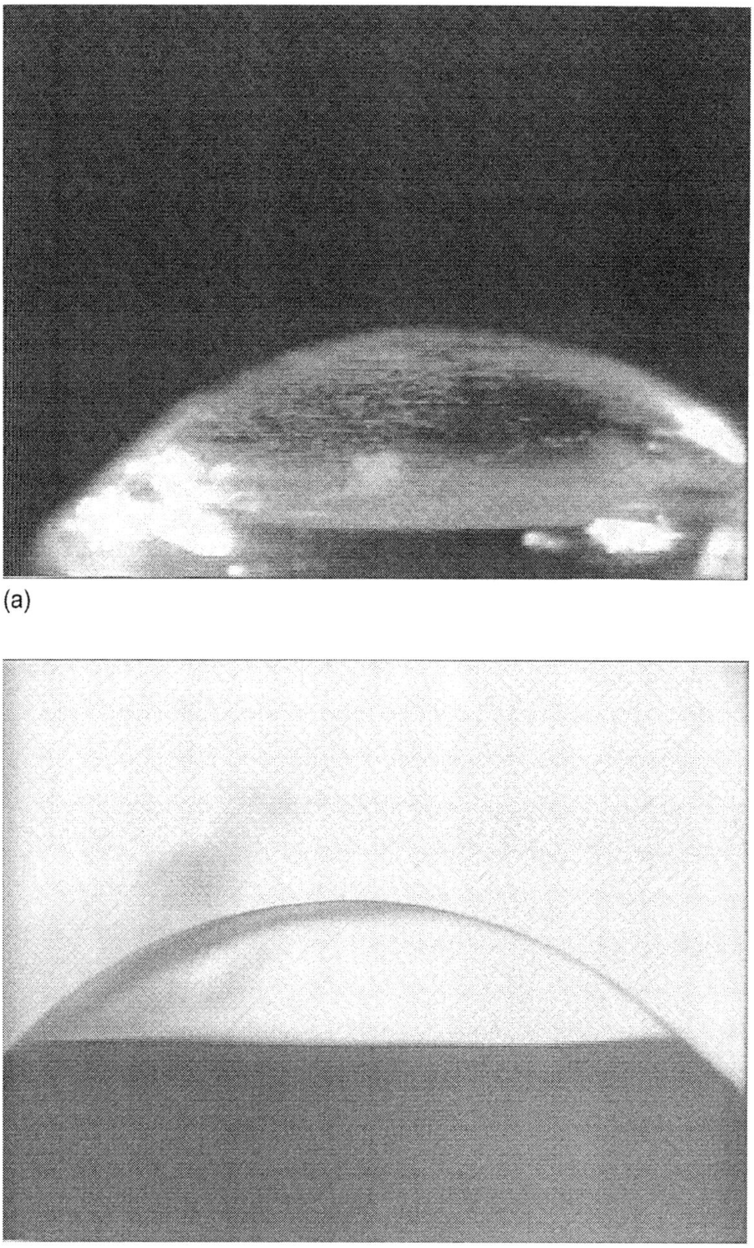

(a)

(b)

Figure 2 Microscopic image of cells attached to the bubble film (a), and the same system containing Pluronic F-68 indicating no cells attached to the bubble film (b).

(1) nutritive effects, (2) the suppression of bubble–cell attachments, and (3) the physical strengthening of the cell membranes. Because experimental results demonstrated that the time required to achieve measurable protective effects after the addition of the supplements is short, it has been argued that the major mechanism of protection is not nutritive (33,56,57).

Various studies indicate that cell–air bubble adhesions are inhibited when protective additives are present. Mizrahi (58) and Jordan et al. (59) attributed the protective effects to the surface active nature of the additives. Chattopadhyay et al. (52) and Micheals et al. (54) compared the effectiveness of several additives and found that the most effective compound could decrease the air–medium interfacial tension rapidly and significantly. In addition, both groups suggested that the lowering of the dynamic surface tension was more important than lowering the equilibrium surface tension. As stated in the previous section, Chattopadhyay et al. (52,53) used a thermodynamic analysis to explain these observations.

Besides the alteration of bubble film properties, another possible explanation for the reducing of cell–bubble adhesion is a surfactant–cell membrane interaction. Wu (60,61) found that the hydrophobicity of two types of insect cells decreased when PF-68, methylcellulose, or PEG was present and suggested that the weakened hydrophobicity reduced the cell–bubble bonding. The protective effect of PF-68 in the nonaerated cultures was also noticed by several research groups. Goldblum et al. (56) suggested the adsorption of PF-68 to the cell membrane made cells stronger in withstanding strains. Ramirez and Mutharasan (62) observed that PF-68 decreased the plasma membrane fluidity significantly; implying that a more rigid membrane increased the cells resistance to shear forces. Zhang et al. (21) found the membrane bursting tension of hybridoma cells increased with PF-68 or serum present, but it was recognized that the strengthened membrane alone was not enough for the protection observed in bioreactors.

Despite these studies, a mechanistic understanding of the protective roles of the additives is far from being elucidated. This can be attributed to our inability to understand cell damage mechanisms and the complex interfacial phenomena within bioreactors. As a result, the choice of the protective additives and the concentration is mainly based on empirical experience (63).

Pluronic F-68 and Antifoam

PF-68, a nonionic surfactant, is widely used to protect animal and insect cells from sparging and agitation-related injuries. Because of its essential role in cell cultures, this section will specifically discuss the properties and issues surrounding PF-68 and foaming. PF-68 is one of a family of block copolymer of poly(oxypropylene) and poly(oxyethylene) with an average molecular weight of 8400 Da (64). Unlike many other surfactants, PF-68 does not have a distinct critical micelle concentration, which renders its interfacial properties more complex than the already very complex surfactant–fluid systems. The protective ability of PF-68 increases initially with the concentration, but levels off after $0.5\,kg/m^3$(0.05%, or 0.5 g/L). Typically, the reported concentration adopted is $0.5–3\,kg/m^3$(65). The interaction of PF-68 with cell membrane and bubble–liquid surface decreases the mass transfer rate, especially the OTR (64). In addition, Murhammer and Goochee (66) noticed that Pluronic inhibited the trypan blue uptake into insect cells, which is consistent with speculation that PF-68 interacts with cell membranes. Finally, some initial reports indicate that high concentrations can be toxic (1,67).

The addition of PF-68 or other surface-active compounds can enhances the stability of foam at the gas–medium interface. While a stable foam layer might be desirable if it allows cells attached to bubbles or trapped in the fluid surrounding bubbles in the foam to drain back into the culture medium, excessive foaming can be a serious operating problem in the cell culturing. Foaming process can lead to loss of cells and product, reduction of working volume, and increased risk of contamination (68).

Both mechanical and chemical methods have been studied in foam control. Rotational impellers and stirrers positioned above the liquid surface can suppress the foam accumulation via the shear forces. The mechanical defoaming methods do not introduce any supplements into the culture. But they have large power consumption rates, difficulties in operation, and insufficient breaking efficiencies. Chemical methods usually use hydrophobic oils, insoluble hydrophobic particles, or mixtures of both as the antifoam agents. A number of mechanisms have been proposed to explain the action of these antifoam agents including: (1) the building up of hydrophobic bridges between two film surfaces, (2) the replacement of the adsorbed surfactant on the interface, and (3) the lateral stress subjected to the film liquid during oil spreading on the film surface (68,69). Synergistic effect exists for the mixed type oil–particle antifoamers, which are found very effective at low concentrations (10–1000 ppm) (69). The effect of antifoams on mass transfer is complex. At low concentration, the mass transfer efficiency is decreased dramatically; however, at high concentration, the efficiency has been reported to be increased (70–72). Antifoam increases the bubble coalescence and the resulting large bubbles may cause vigorous vibrations in airlift bioreactors (73).

Sparger Design

Proper selection of a sparger is important for increasing aeration and minimizing cell damage and foaming. Christi (74) considered two kinds of spargers for the scale-up of a bioreactor and found the larger multihole sparger was preferable over a porous metal sparger (74). This preference was the result of foaming problems associated with the porous metal, presumably due to the smaller bubbles that this sparger created. Orton and Wang (39) suggested macrospargers producing bubbles in the range of 6–8 mm diameter were suitable for minimizing foaming and simultaneously reducing bubble-related cell death. However, microhole spargers provide significantly higher gas–medium interfaces and thereby higher oxygen mass transfer coefficient for the same gas flow rate, therefore, effectively, reducing the overall superficial gas flow rate (28).

Studies of sparger designs on cell viability are scarce. Murhammer and Goochee (75) found that when the air velocity leaving the gas orifice was high, indicated by a high pressure drop, net cell growth was inhibited. They suggested the higher level of fluid turbulence aroused by the high pressure drop brought about cell damage. Gary et al. (28) developed a model to determine the best sparger hole size that can meet the requirements of aeration and CO_2 stripping at the same time.

MIXING AND SHEAR STRESS

Mixing is required in suspended cell cultures to keep the suspension homogeneous and to improve the mass and heat transfer rates. In most bioreactors, mixing is achieved by impeller agitation and gas sparging. Mixing within a bioreactor can be divided into macromixing and micromixing, although there is no distinct boundary defining the two. Macromixing generally refers to the overall flow patterns in the whole vessel and is usually characterized by the global concepts: i.e., power addition per total volume, type of turbine, mixing times, etc. Some commonly used methodologies in measuring the mixing time were summarized by Nagata (76). Micromixing is generally focused on the intensity of mixing in localized regions and, most fundamentally, on the length scale of molecular level, i.e., the mixing of different types of molecules, and the smallest eddy level. Characterization of micromixing includes the Kolmogorov microscale and more recently the local intensity of energy dissipation (77).

The agitation intensity in cell cultures is typically much lower than that in microbial cultures because of the assumed higher level of mammalian and insect cells fragility and significantly lower OUR. Quantitatively, this agitation intensity difference is ~100-fold, i.e., a global average of $10\,W/m^3$ (1×10^2 ergs/cm^3 s) in animal cell culture vessels versus $1000\ W/m^3$ (1×10^4 ergs/cm^3 s) in microbial fermentation vessels (1,44). The commonly used lower level of agitation is generally thought not sufficient for mixing in many cases. This was dramatically illustrated by Ozturk (29). In his study, local cell lysis was observed as result of base addition in the form of concentrated drops added at the top of fluid in the bioreactor. The resulting cell debris formed a "snow ball" like structure surrounding the base inlet and a drop in the cell density was detected. Consistent with this experimental observation, Reuss et al. (78) presented computer simulations indicating that insufficient mixing conditions could result in localized regions of pH as high as 9; despite that the overall pH value was controlled at 7.2. In addition to the heterogeneous pH distribution, Nienow et al. (79) stressed that significant spatial gradient in dO_2 can also exist. The uneven distribution of nutrient, waste, product, and pH becomes more severe when vessel scale increases (1).

Overview of the Hydrodynamics in Bioreactors

Extensive work has been conducted to understand the fluid dynamics in both bioreactors and, in general, stirred vessels. Nevertheless, a precise characterization of the turbulent flow is still out of reach. Theoretical solutions are restricted by the closure problem, where the available equations are less than the unknown variables. Direct measurements are also limited because of the incapability of current techniques to give a high resolution, three-dimensional, instantaneous velocity or pressure field. Two of the most commonly used methods, the laser Doppler anemometer (80–83) and the constant temperature anemometry (84,85), are restricted to measure a single point as a function of time. A newly developed particle tracking velocimetry method enables simultaneously full field velocity measurement. However, the resolution of the velocity map is only moderate. Hence, velocity vectors acquired at different time points need to be summed up and the resulting data are time averaged (86).

No matter how random the flow is in the agitated vessels, predictable macroscale flow patterns exist. Different macrocirculation structures aroused by axial- or radial-type impellers are well known. Chapman et al. (87–89) reviewed the bulk flow patterns in sparged and stirred tanks for different impellers: disk turbines, pitched blade turbines pumping up or down, marine impellers, and pitched blade turbines with disks. Besides the bulk circulation, the passage of blades superimposes a periodical velocity to the surrounding fluid and after each blade, there exists an intense trailing vortex emanating from the tip of the blade (80,82,90–92).

Turbulence can be characterized as an array of eddies with different sizes. The mechanical energy delivered to the fluid through the blades cascades from larger to smaller eddies until it researches the smallest ones where mechanical energy is converted to thermal energy. The smallest eddy is usually defined by the Kolmogorov microscale. It should be noted that the energy is not cascaded strictly layer by layer. It has been suggested that the energy contained in one eddy can be transferred to eddies much smaller than it although the closest one, in size, receives the largest share. Similarly, heat is not generated exclusively by eddies in the Kolmogorov mircoscale level.

Most of the hydrodynamic studies conducted with the agitated tanks are not specifically for bioreactors. However, despite the relatively lower power input that has historically been used in cell culture, these studies can be applied to the design of animal cell bioreactors (93).

Cell Damage Caused by Flow Forces

It is generally believed that animal and insect cells are sensitive to hydrodynamic forces due to their relatively larger size and lack of cell wall. Both lethal and sublethal effects of hydrodynamic forces on cells are well documented (94–98). The first reports of suspended animal cell damage in agitated bioreactors came with the attempts to cultivate mammalian cells at scales larger than a couple of liters (99–101). In a pioneering work, Augenstein et al. (102) studied the effect of shear stresses on cell suspension of two strains of mammalian tissue cells, HeLa S3 and mouse L929, by passing the cultures through capillary tubes of different diameters. They found the cell death rate could be correlated to either the average wall shear stress or the power dissipation. At stresses between 10 and 200 N/m^2 cell death was observed. In another pioneering study, Midler and Finn (103) observed that the damage of shear-sensitive protozoa cells in a uniform shear device showed a two-phase behavior: a rapid primary damage followed by a slow decline of viable cells. This result indicates that cell damage is not only related to the magnitude of the hydrodynamic forces, but also to the exposure time. Peterson et al. (104) studied the effect of mode of cell growth and culture age on the tolerance of shear stress. The cells cultured in spinner flasks were found much less sensitive to shear than the cells grown in T-flasks. In addition, cells from either the lag or the stationary phases of batch cultures were more sensitive to mechanical damage than cells from the exponential growth phase.

Viscometers are the prevalent devices to quantify the cell sensibility to hydrodynamic forces. Studies using various types of viscometers were reviewed extensively by Joshi et al. (94). The advantages of viscometers are that the flow fields are uniform and well defined and that the shear rate and exposure time can be controlled accurately. However, various researchers have suggested that normal forces arising from extensional flow, which are typical in turbulent flow bioreactors, are more damaging to cells than purely shear flow at the same level of energy dissipation (105–108). At the same time, the hydrodynamic forces in a mixed vessel are not constant, but can range over several orders of magnitude with only a small region of the vessel exhibiting high levels (43). Consequently, more relevant studies are those in which the high hydrodynamic forces are abrupt and rapid. McQueen et al. (107) and Gregoriades et al. (108) pumped cell suspensions through transient, abrupt contraction flow devices. Gregoriades et al. quantified the local energy dissipation in the contraction region and found that the damage of CHO cells attached to microcarriers became significant when the maximum energy dissipation exceeded 10^4 $ergs/cm^3$ s. This magnitude of energy dissipation is consistent with the experimentally determined values of local energy dissipation that has been reported in spinner vessels to damage cells attached to micocarriers (92).

It is generally accepted that cells attached to microcarriers are more susceptible to hydrodynamic forces than the suspended cells. This increased susceptibility is believed to be the result of the attached cells not able to rotate freely; rotation relaxes the stress and torque the cells are subjected to by the moving fluid (105,106). Using a similar flow device and the same cell line, Ma et al. quantitatively compared the damage of freely suspended CHO cells to the results of Gregoriades (109). The freely suspended CHO cells were observed to be able to withstand energy dissipation rates over three orders of magnitude higher than CHO cells attached to the 200 μm microcarriers.

Apart from the death and lysis due to the detrimental flow environment, sublethal level of shear stress may elicit various physiological responses, such as modified metabolism pathways and gene expressions. Under intense agitations, glucose consumption and mitochondrial activity were found to be increased (110,111). Studies of McDowell and Papoutsakis (112) indicated that step changes in the rotation speed from 80 to 300 or 400 rpm resulted in altered metabolism, mRNA levels, and surface receptor content of a human promyelocytic leukemia cell line. Descriptions of further sublethal affects can be found in Al-Rubeai et al. (113) and Goswami et al. (114).

Quantification of Cell Damages

A number of parameters have been suggested to quantify the hydrodynamic forces that cells experience. Garcia and Chalmers (105) suggested that the ideal parameters "cannot come from the geometry of the system producing the flow, but from 'intrinsic characteristics' of the flow itself" and they should represent the local environment, not the average one. Based on these criteria, impeller speed, impeller tip speed, and integrated shear stress are not appropriate since they are system (vessel) dependent.

Laminar shear stress can be accurately measured in a number of viscometers and parallel plate devices and is able to characterize the local flow. However, while it is an effective quantitative parameter for laminar, two-dimensional flow, the characterization of time dependent, three-dimensional turbulent flow in the bioreactors requires a second-order shear stress tensor consisting of six distinct shear stress vectors. Hence, it is difficult, if not impossible, to extend the results acquired in well-defined flow devices to bioreactors. Meanwhile, as stated earlier, the normal forces produced in the turbulent flow bioreactors are more damaging than the pure laminar shear forces where the flow is half extensional and half rotational (105–108).

Energy dissipation has been widely used to quantify local mixing performance (115). The concept that energy dissipation can be related to cell damage was first suggested by Bluestein and Mockros with respect to hemolysis of red blood cells (116). The energy dissipation rate is a scalar and consists of contributions from all the shear stress components in the three-dimensional flows:

$$\varepsilon = \mu[\nabla U + \nabla U^{\mathrm{T}}] : (\nabla U) \tag{7.10}$$

where μ is the viscosity and (∇U) is the velocity gradient tensor. The Kolmogorov microscale, which is a well-known parameter suggested for use in bioreactor scale-up, is also based on the local energy dissipation (117,118). The challenge in using energy dissipation as a universal parameter is the difficulty in measuring its local distribution and value. Compounding this difficulty is Taylor's study of drop disruption, which found that while it is possible to disrupt a drop of oil in water in both extensional and pure shear flow, the magnitude of the energy dissipation rate at which the breakup occurred was higher for the pure shear flow (119). Taylor attributed this difference to the rotation of the droplet in the pure shear flow, which relaxed the stress the drops experienced. Motivated by this discussion, Garcia and Chalmers (105) proposed a parameter called "amount of rotation," which represents the ratio of the time scale of liquid deformation to the time scale of rotation, to supplement energy dissipation rate in describing the damaging forces.

The Distribution of Energy Dissipation in Typical Bioreactors

In the bioreactor agitated with impeller(s), the power introduced through the shaft is not dissipated evenly to the whole fluid volume. Instead, various studies indicated that 30–70% of the total power input was dissipated in a small region around the impeller that accounted for only 10% of the total vessel volume (83). In addition, the distribution between the high and low values of local energy dissipation in the vessel can vary by two to three orders of magnitude (43,83). Because of this heterogeneity in power dissipation, the maximum energy dissipation rate as well as the bulk average should be considered when designing and operating bioreactors.

Operation/Scale-Up of Bioreactors with Respect to Mixing

The scale-up and optimum operation of a bioreactor is as much an art as a science. The optimum agitated culture should provide adequate mass transfer without zones of physiological or physical damages to cells. Unfortunately, this scenario cannot be predicted or designed from purely first principles. Consequently, a number of empirical, semifundamental or correlational scale-up arguments have been proposed and used.

Rules of Thumb

Chalmers (63) and Tramper et al. (120) described some of the common rules of thumb for scale-up bioreactors, which can be summarized as

1. anchorage-dependent cells are more "shear sensitive" than suspended cells;
2. sparging would kill all the suspended cells in a short time if there is no "shear-protective" additives added;
3. marine impellers produce less shear damage to cells than Rushton-type impellers;
4. sparging is the simplest and the most straightforward method for aeration;
5. "shear sensitivity" is cell type and sometimes cell clone specific;
6. the minimum velocity in the bulk phase should be at least twice the terminal settling speed of the cells;
7. tip speed of about 2 m/s is needed to ensure sufficient homogeneity.

Scale-Up based on Mass Transfer and Mixing

One method to scale up bioreactors is to focus primarily on maintaining proper rates of mass transfer. This method has been advocated since it became apparent after a number of studies that many of the typically used animal cell lines are relatively tough and can withstand significant agitation. Specifically, Kunus and Papoutsakis (121) demonstrated that hybridoma cells were able to withstand up to 700 rpm without damage as long as gas bubbles, either sparged or entrained, did not rupture at the gas–liquid interface. This specific bioreactor had a volume of ~2 L, a diameter of 11.8 cm, and used a pitched blade impeller of 7 cm diameter. In a complementary study, Oh et al. (122) also reported that hybridoma cells were able to withstand up to 450 rpm in a 2-L bioreactor. One word of caution is in order, however. While a number of commonly used animal cells are relatively tough, this is not necessarily true for every cell line, or even different clones of the same cell line. While not published, numerous informal reports exist of clonal variations in a cell's resistance to hydrodynamic forces and that the toughest one is usually selected.

Scale-up is often restricted by oxygen delivery and carbon dioxide stripping, especially when high cell density is desired (123). Intensified sparging allows higher cell concentration. However, Ozturk (29) found that cell damage and foaming problems limited the sparging rate under 0.1 vvm. Longer mixing time brings local high oxygen and carbon dioxide gradients, which are known to be deleterious to the cells. Tremper et al. (120) suggested that the time required by the cells to move through various areas of different conditions should be smaller than their biological response time. The mixing time, Θ_{m}, in agitated vessels without aeration could be approximated by the following equations (124,125):

$$C_{\mathrm{m}} = 5.3(\mathrm{T/D})^2 \quad \text{for H/T} = 1 \tag{7.11}$$

$$C_{\mathrm{m}} = 3.3(\mathrm{T/D})^{2.43} \quad \text{for H/T} > 1 \tag{7.12}$$

where

$$C_{\mathrm{m}} = \Theta_m N Po^{1/3} \tag{7.13}$$

Nienow et al. (79) rearranged Eq. (7.11) to give:

$$\Theta_{\mathrm{m}} = 5.9(\bar{\varepsilon}_T)^{-1/3}(D/T)^{-1/3}T^{2/3} \tag{7.14}$$

This equation indicates that mixing time can be kept constant during scale-up by increasing the volume average power input ($\bar{\varepsilon}_T$). However, during scale-up, the volume average energy input normally could not be kept as high as that in the small scale (126). As a result, the mixing time increases even at a significant speed due to the contributions from both the enlarged tank size (*T*) and lowered power input per volume ($\bar{\varepsilon}_T$). Equation (7.14) is proposed to be universal to both axial and radial flow impellers. A comparison of two types of impellers, one Rushton turbine and one A310 hydrofoil, with similar D/T ratio indicated that the same mean power input gave similar mixing times (79).

Scale-Up Based on Mechanical Cell Damage

A number of correlation-based parameters have been suggested and used in the scale-up of animal cell bioreactor. These parameters include impeller tip speed, bulk mean power input, impeller Reynolds number, and tip speed (34,127,128). However, as Kossen acknowledged, it is not possible to hold more than one of these parameters constant at the same time during scale-up (129). Sinskey et al. (130) proposed to use the "integrated shear factor" (ISF), which was defined by dividing the impeller tip speed with the distance between the impeller tip and the bioreactor wall. While working with cells adhered to microcarriers, Sinskey et al. observed a steep cell density drop when the ISF exceeded 60 s^{-1}.

The next level of sophistication is the semifundamentally based method of scale-up, popularly called the Kolmogorov eddy-length model (106,113,114,131). The Kolmogorov microscale, obtained by the classical dimensional analysis of turbulence by Kolmogorov, denotes the smallest eddy size in the flow field at which the inertial forces are balanced by the viscous forces (132,133). Even though true isotropic flow does not exist in agitated tanks, local isotropy can be assumed in fully turbulent areas where the smallest eddies, which dissipate the majority of the mechanical energy, lose their original orientation in the cascade process and become random. As indicated by Eq. (7.15), the Kolmogrov microscale, η, is only a function of the turbulence intensity, represented by the local energy dissipation rate, ε, and the liquid property, represented by the kinematic viscosity, ν. Thus, it has no direct relationship with system geometry and large-scale flow patterns:

$$\eta = \left(\frac{\nu^3}{\varepsilon}\right)^{1/4} \tag{7.15}$$

The Kolmogorov eddy-length model proposes that when η reaches the size of a microcarrier (~200 μm) with cells attached or the size of a freely suspended cell (~10 μm), cell damage takes place. Good predictions were reported for cells on microcarriers in small-scale vessels, but difficulties were encountered when the model is applied to larger vessels or cells in suspension (134). This limitation arises from the fully turbulent requirement, the local isotropic turbulence assumption, and most probably the difficulty in determining the highest local energy dissipation level (92,95). In the original Kolmogorov eddy-length model, it was assumed that the energy added through the impeller was dissipated mainly in an annular region around the impeller. However, as was discussed previously, the highest levels of energy dissipation are located in much smaller volumes that appear to be associated with the trailing vortices emanating from the impellers. By focusing on the trailing vortices, Aloi and Cherry (135) reached a better agreement between the bioreactor results and the better-defined capillary flow results.

The Selection of Impeller(s)

Nagata (76) compared the discharge efficiencies of various impellers where the efficiency is defined as the ratio of the power number, Np, versus discharging flow number, Nq. Because the discharge flows are the direct cause of mixing, impellers with lower Np/Nq ratios are superior in producing better bulk mixing and less shear force. The results indicated that the preferable impellers were those having retreated blades, large width, pitched blades, and small diameter compared to the vessel diameter. Nienow (126) correlated the mixing directly with the average energy dissipation rate. The result revealed that in single impeller agitated systems with $H/T = 1$, mixing time is unrelated to the impeller type, but to the mean energy dissipation and the impeller and vessel diameters (Eq. (7.14)). The diameter of the impellers in turbulent media is normally between one-fourth to one-half of the tank diameter.

A retrofitting analysis based on Eq. (7.14) shows that the impellers with a larger diameter but with lower bulk energy dissipation may decrease the cell damage without impairing the overall mixing. In addition, impellers that yield less severe local energy dissipation at the same bulk average energy dissipation level have advantages in reducing alleged cell damages. In this sense, hydrofoils are superior to pitched blade turbines (PBT) and PBTs are superior to RTs. Axial flow impellers with aerodynamically shaped blades and large projected area have recently received more attention because large project area produces less shear force because of the more uniform energy dissipation (126,136).

Multiple-impeller systems are common for high H/T ratio vessels. For the same power input, multiple-impeller systems require lower impeller speed, which results in a lower intensity of local energy dissipation. However, the circulations aroused by the impellers tend to be independent and several compartments of flow can be formed. This reduces the overall mixing because the bulk mixing is determined by the much weaker interstage mass exchange speed. Increasing the impeller speed, decreasing the impeller distance, and utilizing axial flow impellers were found to be helpful in cutting the mixing time (137,138). Rushton turbine impellers are commonly used as the bottom impeller in order to disperse bubbles. Combining the Rushton turbine with axial flow impeller(s) were superior to multiple-Rushton turbines with respect to bulk blending. Mishra and Joshi (138) reported that the disk turbine (DT)–pitched blade turbine downstream (PTD) combination gave higher fluid transport at the same level of energy consumption as the DT–DT combination or a single DT. Arjunwadkar et al. (139) compared nine dual-impeller combinations comprising of DT, PTD, and pitched blade turbine upflow (PTU). The DT–PTD combination was revealed to be the optimum due to its efficient gas hold up and lower energy dissipation level. Bittorf and Kresta (140) found that the active volume of mean circulation around axial impellers corresponded to a height equivalent of two-thirds of the tank diameter. This indicates that even though the axial flow impellers give rise to strong axial directional flows, the spacing between impellers should be lower than $2/3T$ to retain strong interstage mass transfer.

REFERENCES

1. Varley J, Birch J. Reactor design for large scale suspension animal cell culture. Cytotechnology 1999; 29:177–205.
2. Kilburn DG, Webb FC. The culvation of animal cells at controlled agitated bioreactors with and without bubble entrainment, Biotechnol Bioeng 1968; 10:801–814.
3. Meilhoc E, Wittrup KD, Bailey JE. Influence of dissolved oxygen concentration on growth, mitochondrial function, and antibody production of hybridoma cells in batch culture. Bioprocess Eng 1990; 5:263–274.
4. Glacken MW, Adema E, Sinskey AJ. Mathematical description of hybridoma culture kinetics: II. The relationship between thiol chemistry and the degradation of serum activity, Biotechnol Bioeng 1989; 33:440–450.
5. Aunins JG, Henzler HJ. Aeration in cell culture bioreactors. In: Stephanopoulos G, ed. Biotechnology Vol. 3: Bioprocessing. Wienheim: VCH 1993.
6. Charles M, Wilson J. Fermentor design. In: Lydersen BK, D'Elia NA, Nelson KL, eds. Bioprocessing Engineering: Systems, Equipment and Facilities. New York: John Wiley & Sons, 1994.
7. Hu WS, Meier J, Wang DIC. Use of surface aerator to improve oxygen transfer in cell culture. Biotechnol Bioeng 1986; 28:122–125.

8. Zlokarnik M. Scale up of surface aerators for waste water treatment. Adv Biochem Eng 1979; 11:157–180.
9. Aunins JG, Woodson BA, Hale TK, Wang DIC. Effects of paddle impeller geometry on power input and mass transfer in small-scale animal cell culture vessels. Biotechnol Bioeng 1989; 34:1127–1132.
10. Backer MP, Metzger LS, Slaber PL, Nevitt KL, Boder GB. Large scale production of monoclonal antibodies in suspension culture. Biotechnol Bioeng 1988; 32:993–1000.
11. Riess JG, LeBlanc M. Solubility and transport phenomena in perfluorochemicals relevant to blood substitution and other biomedical applications. Pure Appl Chem 1982; 54:2383–2406.
12. Lowe KC, Davey MR, Power JB. Perfluorochemicals: their applications and benefits to cell culture. Trends Biotechnol 1998; 16:272–277.
13. Hamamoto K, Tokashiki Y, Ichikawa Y, Murakami H. High cell density culture of hybridoma using perfuorocarbon to supply oxygen. Agric Biol Chem 1987; 51:3415–3416.
14. Cho MH, Wang SS. Enhancement of oxygen-transfer in hybridoma cell-culture by using a perfluorocarbon as an oxygen carrier. Biotechnol Lett 1988; 10:855–860.
15. Ju L-K, Armiger WB. Use of perfluorocarbon emulsion in cell-culture. BioTechniques 1992; 12:258–263.
16. Gotoh T, Mochizuki G, Kikuchi K. Perfluorocarbon-mediated aeration applied to recombinant protein production by virus-infected insect cells. Biochem Eng J 2001; 7:69–78.
17. Aunins JG. Aeration in cell culture bioreactors. In: Rehm HJ, Reed G, Klein J, Winter J, eds. Biotechnology: A Multi-volume Comprehensive Treatise. New York: John Wiley & Sons, 2000.
18. Moreira JL, Cruz PE, Santana PC, Feliciano AS. Influence of power input and aeration method on mass transfer in a laboratory animal cell culture vessel. J Chem Tech Biotechnol 1995; 62:118–131.
19. Chalmers JJ. Cells and bubbles in sparged bioreactors. Cytotechnology 1994; 15:311–320.
20. Smith GL. Large-scale animal cell culture. In: Lydersen BK, D'Elia NA, Nelson KL, eds. Bioprocessing Engineering: Systems, Equipment and Facilities. New York: John Wiley & Sons, 1994.
21. Zhang S, Handa-Corrigan A, Spier RE. Oxygen transfer properties of bubbles in animal cell culture media. Biotechnol Bioeng 1992; 40:252–259.
22. Van't Riet K. Review of measuring methods and results in non-viscous gas–liquid mass transfer in stirred vessel. Ind Eng Chem Process Des Dev 1979; 128:357–364.
23. Kawase Y, Halard B, Moo Young M. Liquid phase mass transfer coefficients in bioreactors. Biotechnol Bioeng 1992; 39:1133–1140.
24. Vasconcelos JMT, Aleves SS. Direct dynamic kla measurement in viscous fermentation broths: the residual gas holdup problem. Chem Eng J 1991; 47:B35–B44.
25. Blanch HW. Aeration. In: Perlman D, ed. Annual Reports in Fermentations Vol. 3. New York: Academic press 1979.
26. Blanch HW, Clark DS. Biochemical Engineering. New York: Marcel Dekker Inc., 1996.
27. Drapeau D. SIM Annual Meeting, Orlando, Florida, 1990.
28. Gary DR, Chen S, Howarth W, Inlow D, Maiorella RL. CO_2 in large-scale and high-density CHO cell perfusion culture. Cytotechnology 1996; 22:65–78.
29. Ozturk SS. Engineering challenges in high density cell culture systems. Cytotechnology 1996; 22:3–16.
30. Brich JR, Thompson PW, Boraston R, Oliver S, Lambert K. The large-scale production of monoclonal antibodies in airlift fermenters. In: Webb C, Mavituna F, eds. Plant and Animal Cells: Process Possibilities. Chichester: Ellis Horwood, 1987.
31. Swim HE, Parker RF. Effect of Pluronic F-68 on growth of fibroblasts in suspension on rotary shakers. Proc Soc Biol Med 1960; 103:252–254.

32. Ruyan WS, Gyer RP. Growth of L cell suspension on a Warburg apparatus. Proc Soc Biol Med 1963; 103:252–254.
33. Kunas KT, Papousakis ET. Damage mechanisms of suspended animal cells in agitated bioreactors with and without bubble entrainment. Biotechnol Bioeng 1990; 36:476–483.
34. Oh SKW, Nienow AW, Al-Rubeai M, Emary AN. The effect of agitation intensity with and without continuous sparging on the growth and antibody production of hybridoma cells. J Biotechnol 1989; 12:45–62.
35. Kunas KT, Papousakis ET. The portective effect of serum against hydrodynamics damage of hybridoma cells in agitated and surface aerated bioreactors. J Biotechnol 1990; 15:57–70.
36. Handa A, Emary AN, Spier PE. On the evaluation of gas-liquid interfacial effects on hybridoma viability in bubble column bioreactors. Dev Biol Stand 1987; 66:241–253.
37. Bavarian F, Fan LS, Chalmers JJ. Microscopic visualization of insect cell-bubble interactions I: Rising bubbles, air–medium interface, and the foam layer. Biotechnol Prog 1991; 7:140–150.
38. Chalmers JJ, Bavarian F. Microscopic visualization of insect cell-bubble interactions II: The bubble film and bubble rupture. Biotechnol Prog 1991; 7:151–158.
39. Orton D, Wang DIC. Fluorescent visualization of cell death in bubble aerated bioreactors. Cell Culture Engineering III, Engineering Foundation, February 2–7, 1991.
40. Trinh K, Garcia-Briones MA, Hink FH, Chalmers JJ. Quantification of damage to suspended insect cells as a result of bubble rupture. Biotechnol Bioeng 1994; 43:37–45.
41. Boulton GM, Blake JR. Gas bubble bursting at a free surface. J Fluid Mech 1993; 154:437–466.
42. Garcia-Briones MA, Brodkey RS, Chalmers JJ. Computer simulations of the rupture of a gas bubble at a gas–liquid interface and its implications in animal cell damage. Chem Eng Sci 1994; 49:2301–2320.
43. Mollet M, Ma N, Zhao Y, Brodkey RS, Taticek K, Chalmers JJ. Bioprocess Equipment: Characterization of Energy Dissipation Rate and its potential to damage cells. Biotechnol. Progess 2004; 20:1437–1448.
44. Tramper J, Smith JD, Sreaatman J, Valk JM. Bubble-column design for growth of fragile insect cells. Bioprocess Eng 1988; 3:37–41.
45. Jobses I, Martens D, Tramper J. Lethal events during gas sparging in animal cell culture. Biotechnol Bioeng 1991; 37:476–483.
46. Martens DE, de Gooijer CD, Beuvery EC, Tramper J. Effect of serum concentration on hybridoma viable cell density and production of monoclonal antibodies in CSTRs and on shear sensitivity in air-lift loop reactors. Biotechnol Bioeng 1992; 39:891–897.
47. MacIntyre F. Bubbles: a boundary-layer "microtome" for micron-thick sample of a liquid surface. J Phys Chem 1968; 72:589–592.
48. MacIntyre F. Flow patterns in breaking bubbles. J Geophys Res 1972; 77:5211–5228.
49. Parker B, Barsom G. Biological and chemical significance of surface microlayers in aquatic ecosystems. BioScience 1970; 20:87–93.
50. Garcia-Briones MA, Chalmers JJ. Cell–bubble interactions: mechanisms of suspended cell damage. Ann NY Acad Sci 1992; 665:219–229.
51. Dahlback B, Hermansson M, Kjelleberg S, Norkans B. The hydrophobicity of bacteria—an important factor in their initial adhesion at the air–water interface. Arch Microbiol 1981; 128:267–270.
52. Chattopadhyay D, Rathman J, Chalmers JJ. The protective effect of specific medium additives with respect to bubble rupture. Biotechnol Bioeng 1995; 45:473–480.
53. Chattopadhyay D, Rathman J, Chalmers JJ. A thermodynamic explanation for cell adhesion to gas–liquid interfaces. Biotechnol Bioeng 1995; 48:649–658.
54. Michaels JD, Nowak JE, Mallik AK, Koczo K, Wasan DT, Papoutsakis ET. Interfacial properties of cell culture media with cell-protecting additives. Biotechnol Bioeng 1995; 47:420–430.

55. Papoutsakis ET. Fluid-mechanical damage of animal cells in bioreactors. Trends Biotechnol 1991; 9:427–437.
56. Goldblum S, Bae Y-K, Hink WF, Chalmers JJ. Protective effect of methylcellulose and other polymers on insect cells subjected to laminar shear stress. Biotechnol Prog 1990; 6:383–390.
57. Michaels JD, Petersen JF, McIntire LV, Papoutsakis ET. Protection mechanisms of freely suspended cells (CRL 8018) from fluid-mechanical injury. Viscometer and bioreactor studies using serum, Pluronic F68 and polyethylene glycol. Biotechnol Bioeng 1991; 38:169–180.
58. Mizrahi A. Production of human interferons—an overview. Dev Biol Stand 1983; 55:93–102.
59. Jordan M, Sucker H, Einsele A, Widmer F, Eppenberger HM. Interactions of serum and Pluronic F-68 on the physical properties of the bubble surface. Biotechnol Bioeng 1994; 43:446–454.
60. Wu J. Insights into protective effects of medium additives on animal cells under fluid stresses: the hydrophobic interactions. Cytotechnolology 1996; 22:103–109.
61. Wu J, Ruan Q, Lam HYP. Effect of surface-active medium additives on insect cell surface hydrophobicity relating to cell protection against bubble damage. Enzyme Microbiol Technol 1997; 21:341–348.
62. Ramirez OT, Mutharasan R. The role of the plasma membrane fluidity on the shear sensitivity of hybridomas grown under hydrodynamic stress. Biotechnol Bioeng 1990; 36:911–920.
63. Chalmers JJ. Animal cell culture, effects of agitation and aeration on cell adaptation. In: Spier R, Griffiths JB, Scragg AH, eds. Encyclopedia of Cell Technology. New York: John Wiley & Sons, 2000.
64. Murhammer DW, Goochee CF. Scaleup of insect cell cultures: protective effects of Pluronic F-68. BioTechnology 1988; 6:1411–1418.
65. Chisti Y. Animal-cell damage in sparging bioreactors. Trends Biotechnol 2000; 18:420–432.
66. Elibol M, Mavituna F. A remedy to oxygen limitation problem in antibiotic production: addition of perfluorocarbon. Biochem Eng J 1999; 3:1–7.
67. Patricia B. The Effect of Surfactant Type and Concentration on Animal Cell Culture. M.S. Thesis, Ohio State University, 1996.
68. Vardar-Sukan F. Foaming: consequences, prevention and destruction. Biotechnol Adv 1998; 16:913–948.
69. Pugh RJ. Foaming, foam films, antifoaming and defoaming. Adv Colloid Interface 1996; 64:67–142.
70. Kawase Y, Moo-Young M. The effect of antifoam agents on mass transfer in bioreactors. Bioprocess Eng 1990; 5:169–173.
71. Liu H, Chiung WC, Wang YC. Effect of lard oil, olive oil and castor oil on oxygen transfer in an agitated fermenter. Biotechnol Tech 1994; 8:17–20.
72. Morao A, Maia CI, Fonseca MMR, Vasconcelos JMT, Alves SS. Effect of antifoam addition on gas–liquid mass transfer in stirred fermenters. Process Eng 1999; 20:165–172.
73. Al-Masry WA. Effects of antifoam and scale-up on operation of bioreactors. Chem Eng Proc 1999; 38:197–201.
74. Christi Y. Animal cell culture in stirred bioreactors: observations on scale up. Bioprocess Eng 1993; 9:191–196.
75. Goochee CF, Murhammer DW. Sparged animal cell bioreactors: mechanism of cell damage and pluronic F-68 protection. Biotechnol Prog 1990; 6:391–397.
76. Nagata S. Mixing: Principles and Applications. Tokyo: Kodansha, 1975.
77. Zhou G, Kresta SM. Impact of tank geometry on the maximum turbulence energy dissipation rate for impellers. AIChE J 1996; 42:2476–2490.

78. Reuss M, Schmalzriedt JM. Structured modelling of bioreactors. In: Galindo E, Ramirez, eds. Advances in Bioprocess Engineering. Dordrecht: Kluwer Academic Publishers, 1994.
79. Nienow AW, Langheinrich C, Stevenson NC, Emery AN, Clayton TM, Slater KH. Homogenisation and oxygen transfer rates in large agitated and sparged animal cell bioreactors: some implications for growth and production. Cytotechnology 1996; 22:87–94.
80. Yianneskis M, Popiolek Z, Whitelaw JH. An experimental study of the steady and unsteady flow characteristics of stirred reactors. J Fluid Mech 1987; 175:537–555.
81. Stoots CM, Calabrese RV. Mean velocity field relative to a Rushton turbine blade. AIChE J 1995; 41:1–11.
82. Wu H, Patterson GK. Laser-Doppler measurements of turbulent flow parameters in a stirred mixer. Chem Eng Sci 1989; 44:2207–2221.
83. Zhou G, Kresta SM. Distribution of energy dissipation between convective and turbulent flow for three frequently used impellers. Trans IChemE 1996; 74:379–389.
84. Rao MA, Brodkey RS. Continuous flow stirred tank turbulence parameters in the impeller stream. Chem Eng Sci 1972; 27:137–156.
85. Wernersson ES, Tragardh C. Scale-up of Rushton turbine-agitated tanks. Chem Eng Sci 1999; 54:4245–4256.
86. Zhao Y, Brodkey RS. Averaged and time-resolved, full-field (three-dimensional), measurements of unsteady opposed jets. Can J Chem Eng 1998; 76:536–545.
87. Chapman CM, Nienow AW, Cooke M, Middleton JC. Particle gas–liquid mixing in stirred vessels. 2. Gas–liquid mixing. Chem Eng Res Des 1983; 61:82–95.
88. Chapman CM, Nienow AW, Cooke M, Middleton JC. Particle gas–liquid mixing in stirred vessels. 3. 3 phase mixing. Chem Eng Res Des 1983; 61:167–181.
89. Chapman CM, Nienow AW, Cooke M, Middleton JC. Particle gas–liquid mixing in stirred vessels. 4. Mass transfer and final conclusions. Chem Eng Res Des 1983; 61:182–185.
90. Van't Riet K, Smith JM. The trailing vortex system produced by Rushton turbine agitators. Chem Eng Sci 1975; 30:1093–1105.
91. Tatterson GB, Yuan HS, Brodkey RS. Stereoscopic visualization of the flows for pitched blade turbines. Chem Eng Sci 1980; 35:1369–1375.
92. Venkat RV, Stock LR, Chalmers JJ. Study of hydrodynamics in microcarrier culture spinner vessels: a particle tracking velocimetry approach. Biotechnol Bioeng 1996; 49:456–466.
93. Zhang ZB, Thomas CR. Micromanipulation—a new approach to study animal cell damage in bioreactors. Genet Eng Biotechnol 1993; 13:19–29.
94. Joshi JB, Elias CB, Patole MS. Role of hydrodynamic shear in the cultivation of animal, plant and microbial cells. Chem Eng J 1996; 62:121–141.
95. Thomas CR, Zhang Z. The effect of hydrodynamics on biological materials. In: Galindo E, Ramirez OT, eds. Advances on Bioprocess Engineering II. Dordrecht: Kluwer Academic, 1998:137–170.
96. van der Pol L, Tramper J. Shear sensitivity of animal cells from a culture-medium perspective. Trends Biotechnol 1998; 16:323–328.
97. Chisti Y. Shear sensitivity. In: Flickinger MC, Drew SW, eds. Encyclopedia of Bioprocess Technology: Fermentation, Biocatalysis, and Bioseparation, Vol. 5. New York: John Wiley & Sons, 1999:2379–2406.
98. Chalmers JJ. Shear sensitivity of insect cells. Cytotechnology 1996; 20:163–171.
99. Bryant JC, Schilling EL, Farle WR, Evans VJ. Massive fluid suspension cultures of certain mammalian tissue cells. III. Exploratory supplementation experiments with chemically defined medium NCTC 109. J Nat Cancer Inst 1960; 24:859–871.
100. Runyan WS, Geyer RP. Growth of L cell suspensions on a Warburg apparatus. Proc Soc Exp Biol Med 1963; 112:1027–1030.

101. Telling RC, Elswoth R. Submerged culture of hamster kidney cells in a stainless steel vessel. Biotechnol Bioeng 1965; 7:417–434.
102. Augenstein DC, Sinskey AJ, Wang DIC. Effect of shear on the death of two strains of mammalian tissue cells. Biotechnol Bioeng 1971; 13:409–418.
103. Midler M Jr, Finn RK. A model system for evaluating shear in the design of stirred fermentors. Biotechnol Bioeng 1966; 8:71–84.
104. Peterson HJ, McIntire LV, Papoutsakis BT. Shear sensitivity of cultured hybridoma cells (CRL-8018) depends on mode of growth, culture age and metabolite concentration. J Biotechnol 1988; 7:229–246.
105. Garcia MA, Chalmers JJ. Flow parameters associated with hydrodynamic cell injury. Biotechnol Bioeng 1994; 44:1089–1098.
106. Croughan MS, Wang DI. Growth and death in overagitated microcarrier cell cultures. Biotechnol Bioeng 1989; 33:731–744.
107. McQueen A, Meilhoc E, Bailey JE. Flow effects on the viability and lysis of suspended mammalian cells. Biotechnol Lett 1987; 9:831–836.
108. Gregoriades N, Clay J, Ma N, Koelling K, Chalmers JJ. Cell damage of microcarrier cultures as a function of local energy dissipation created by a rapid extensional flow. Biotechnol Bioeng 2000; 69:171–182.
109. Ma N, Koelling K, Chalmers JJ. The fabrication and use of a transient contraction flow device to quantify the sensitivity of mammalian and insect cells to hydrodynamic forces. Biotechnol Bioeng 2002; 80:428–437.
110. Frangos JA, Eskin SG, McIntire LV, Ives CL. Flow effects on prostacyclin production by cultured human-endothelial cells. Science 1985; 227:1477–1479.
111. Al-Rubeai M, Oh SKW, Musaheb R, Emery AN. Modified cellular metabolism in hybridomas subjected to hydrodynamic and other stresses. Biotechnol Lett 1990; 12:323–328.
112. McDowell CL, Papoutsakis ET. Increased agitation intensity increases CD13 receptor surface content and mRNA levels, and alters the metabolism of HL6O cells cultured in stirred tank bioreactors. Biotechnol Bioeng 1998; 60:239–250.
113. Al-Rubeai M, Singh RP, Goldman MH, Emery AN. Death mechanisms of animal cells in conditions of intensive agitation. Biotechnol Bioeng 1995; 45:463–472.
114. Goswami J, Sinskey AJ, Steller H, Stephanopoulos GN, Wang DIC. Apoptosis in batch cultures of Chinese hamster ovary cells. Biotechnol Bioeng 1999; 62:632–640.
115. Kresta S. Turbulence in stirred tanks: anisotropic, approximate, and applied. Can J Chem Eng 1998; 76:563–575.
116. Bluestein M, Mockros LF. Hemolysis effects of energy dissipation in flowing blood. Med Biol Eng 1969; 7:1–6.
117. Croughan MS, Hamel JF, Wang DIC. Hydrodynamic effects on animal cells grown in microcarrier cultures. Biotechnol Bioeng 1987; 29:130–141.
118. Cherry RS, Papoutsakis ET. Hydrodynamic effects on cells in agitated tissue culture reactors. Bioprocess Eng 1986; 1:81–89.
119. Taylor GI. The formation of emulsions in definable fields of flow. Proc R Soc Lond A 1934; 146:501–523.
120. Tremper J, Vlak JM, Gooijer CD. Scale up aspects of sparged insect-cell bioreactors. Cytotechnology 1996; 20:221–229.
121. Kunus KT, Papoutsakis ET. Damage mechanisms of suspended animal cells in agitated bioreactors with and without bubble entrainment. Biotechnol Bioeng 1990; 36:476–483.
122. Oh SKW, Nienow AW, Al-Rubeai M, Emary AN. The effect of agitation intensity with and without continuous sparging on the growth and antibody production of hybridoma cells. J Biotechnol 1989; 12:45–62.
123. Henzler HJ, Kauling DJ. Oxygenation of cell cultures. Bioprocess Eng 1993; 9:61–75.
124. Ruszkowski S. A rational method for measuring blending performance, and comparison of different impeller types. Proc 8th Euro Conf on Mixing, I. Chem E Symp Ser 1994; 136:283–292.

125. Cooke M, Middleton JC, Bush J. Mixing and mass transfer in filamentous fermentations. In: King R, ed. 2nd Int Conf on Bioreactor Fluid Dynamics. Amsterdam: Elsevier Applied Science, 1988.
126. Nienow AW. Gas–liquid mixing studies: a comparison of Rushton turbines with some modern impellers. Che Eng Res Des 1996; 74A:417–423.
127. Dodge TC, Hu WS. Growth of hybridoma cells under different agitation conditions. Biotechnol Lett 1986; 8:683–686.
128. Lee GM, Huard TK, Kaminski MS, Palsson BO. Effect of mechanical agitation on hybridoma cell growth. Biotechnol Lett 1988; 10:625–628.
129. Kossen NWF. Scale-up. In: Galindo E, Ramirez OT, eds. Advances in Bioprocess Engineering. Dordrecht: Kluwer Academic Publishers, 1994:53–65.
130. Sinskey AJ, Fleischaker RJ, Tyo MA, Giard DJ, Wang DIC. Production of cell derived products: virus and interferon. Ann NY Acad Sci 1981; 369:47–59.
131. Cherry RS, Papoutsakis ET. Physical mechanism of cell damage in microcarrier cell culture bioreactors. Biotechnol Bioeng 1988; 32:1001–1014.
132. Kolmogrov A. The local structure of turbulence in incompressible viscous fluid for very large Reynolds number. C R Acad Sci USSR 1941; 30:301–305.
133. Kolmogrov A. Dissipation of energy in the locally isotropic turbulence. C R Acad Sci USSR 1941; 32:16–18.
134. Cherry RS, Kwon KY. Transient shear stresses on a suspension cell in turbulence. Biotechnol Bioeng 1990; 36:563–571.
135. Aloi LE, Cherry RS. Cellular response to agitation characterized by energy dissipation at the impeller tip. Chem Eng Sci 1996; 51:1523–1529.
136. Fort I, Ambros F, Medek J. Axial impeller with aerodynamically shaped blades of large projected area. Presentation given by Ivan Fort at OSU, 2000.
137. Breman BB, Beenackers AACM, Bouma MJ, Van der Werf MH. Axial liquid mixing and a gas–liquid multi-stage agitated contactor (MAC). Che Eng Res Des 1996; 74A:669–678.
138. Mishra VP, Joshi JB. Flow generated by a disc turbine. 4. Multiple-impellers. Chem Eng Res Des 1994; 73:657–668.
139. Arjunwadkar SJ, Sarvanan K, Kulkarni PR, Pandit AB. Gas–liquid mass transfer in dual impeller bioreactor. Biochem Eng J 1998; 1:99–106.
140. Bittorf KJ, Kresta SM. Active volume of mean circulation for stirred tanks agitated with axial impellers. Chem Eng Sci 2000; 55:1325–1335.

8
Instrumentation and Process Control

Mark Riley
Agricultural and Biosystems Engineering, The University of Arizona, Tucson, Arizona, U.S.A.

INTRODUCTION

Cultivation of animal cells in an environment optimal for manufacture of desired products requires monitoring and control of a substantial number of physical and chemical parameters. Physical parameters include temperature, fluid flow, and agitation rates. Chemical parameters include the dissolved oxygen (DO) concentration, pH, cell density, and the concentrations of nutrients, cofactors, growth factors, desired products, and wastes. Alterations in the levels of these factors can impact the productivity of a cultivation. Neglecting any of these can potentially introduce substantial variations in the quantity or quality of the desired product. Development of instrumentation to monitor these parameters and to provide aggressive process control schemes has been the subject of substantial research in recent years and now forms a core competency of cell culture technology. This chapter provides an overview of instrumentation applied to monitor cell cultures and describes the application of instrumentation to process control.

Tremendous progress has recently been made in characterizing the influence of the cellular environment on physiology, particularly in identifying and quantifying factors that adversely affect cell growth, viability, and productivity (1). Achieving high cell concentrations and high viability continue to be common themes in bioprocess engineering research. To reach these goals requires analytical process monitoring schemes and suitable control methodologies to best utilize this information. Use of online measurements to monitor bioprocesses for identifying the physiological state of cells and controlling environmental conditions to maintain the cells in the optimal physiological state has become substantially more prevalent in the past 10 years. Novel control strategies such as expert systems and artificial neural networks (2) have been introduced and applied in bioprocess monitoring and control. While tremendous progress has been achieved in recent years, substantial challenges remain for the full integration of process monitoring with process control. Specific research areas of critical importance for online monitoring and control include: production of genetically modified cultures, production of monoclonal antibodies (MAbs) by cell lines (where control of key nutrients is essential to maintain cell viability and productivity), and development of control strategies using adaptive control or neural networks. Vaidyanathan and coworkers (3) provide a critical assessment of recent developments in measurement technologies with an evaluation of their potential application in industrial production or process development.

The growth of cell cultures is regulated by a complex interaction between the external physical and chemical conditions of the liquid environment of the bioreactor and the internal biochemical processes of the cells. The former conditions are under the direct control of the bioprocess engineer whereas the latter can be modified through genetic manipulations. The challenge, therefore, is to regulate the environment so that the "ideal" conditions are obtained for cell growth or production. For example, hybridoma cells can be highly sensitive to the concentration of available nutrients including glucose. In a batch culture, either excess glucose or a shortage of glucose can diminish the amount of antibody obtained, as demonstrated in the model predictions of Fig. 1. This simple model is based on Michaelis–Menten relations to describe the uptake of metabolites, and Monod kinetics with inhibition to describe cell growth. Hybridomas have a clear optimum in glucose requirements to maximize productivity. Unfortunately, identification of optimal bioprocess parameters can be difficult as optima typically are dependent on the values of many other process parameters.

Most traditional efforts in bioprocess monitoring apply well-established techniques to quantify physical factors such as temperature, pressure, agitation, power input, and flow rates and to quantify chemical factors such as pH or oxygen and carbon dioxide in the exit gas. More recently techniques for determining redox potential, DO, and nutrient concentrations have become more prevalent. The area in greatest need for instrumentation development is for biological factors: cell density, cell viability, nutrient concentrations (glucose and glutamine), waste concentrations (lactate and ammonia), and desired product concentrations (4). Direct measurement schemes are being developed but in general have so far seen only limited adoption

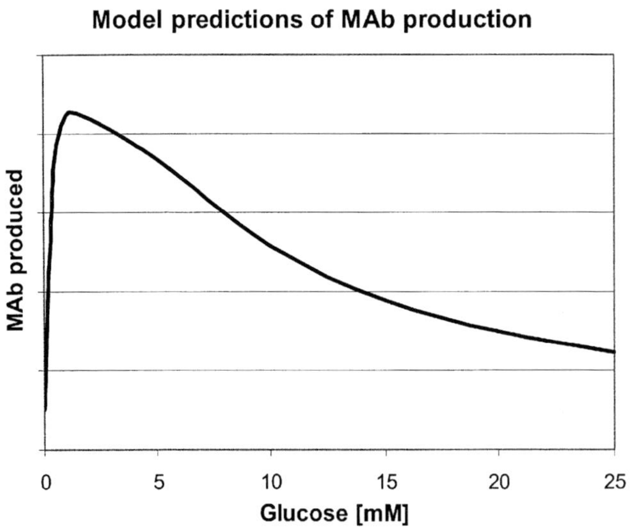

Figure 1 Impact of varying glucose concentration on monoclonal antibody (MAb) produced by a batch hybridoma cell culture. Model predictions based on cell growth, death, and metabolism accounting for variations of glucose concentrations. Note that a fairly narrow range of glucose concentrations around 1 mM yield maximal MAb production.

partly due to perceived low reliability. Many of these techniques will be described in this chapter.

A complete process control scheme involves: (a) measurement of a response variable, (b) comparison of this variable to set points, and (c) activation of a control scheme. An ideal scheme for monitoring a bioreactor must include an analytical device (a sensor), an appropriate sampling methodology, and a control system to utilize the information gained (5). The measurements should be rapid and noninvasive; invasive or direct sampling methods run the risk of introducing contamination. A bioreactor with a high degree of monitoring and control systems is presented in Fig. 2. Note the interconnectivity of monitoring and process control components.

Monitoring is an important exercise in the operation of bioprocesses and is a subject that has attracted intense research activity in recent years (3,5). In order to achieve optimal production or conversion, the factors that influence the performance of a bioprocess should be measured, preferably online, so that the process can be monitored in real-time and appropriate control or remedial action implemented. Such measurement can be achieved either in situ or ex situ, with in situ approaches desirable. In cases where direct analytical information is not available, inferential approaches can be adopted.

Analysis of the cell environment can be performed either "online" or "offline". Here we define "offline" as a measurement that is performed substantially outside of the cell cultivation and often requiring direct intervention by an operator. "Online" analyses then include measurements performed inside a bioreactor (in situ) or outside of the bioreactor, but connected directly to the bioreactor interior. Historically, online analyses have included measurements of pH, DO concentrations, agitation rates, and temperature. Measurements of other chemical parameters such as the cell density, cell viability, and nutrient and product concentrations were performed

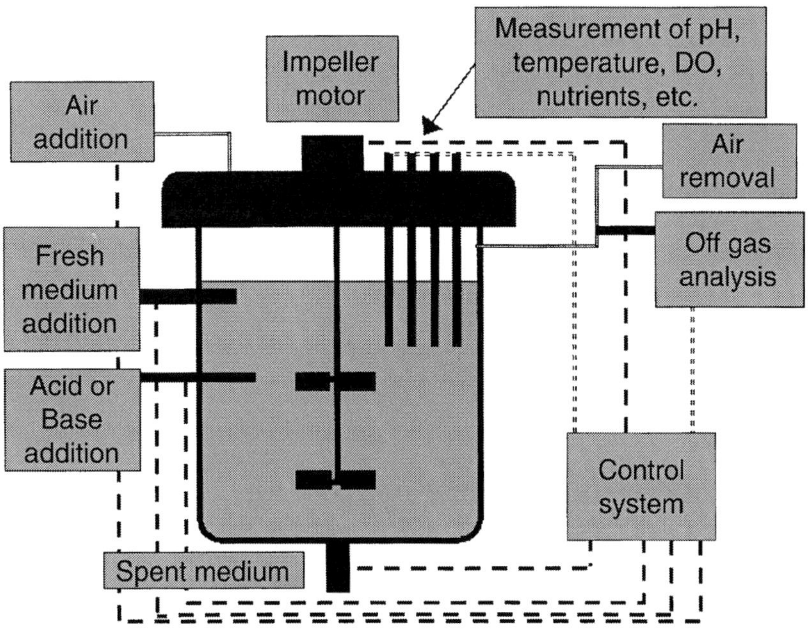

Figure 2 A highly monitored and controlled bioreactor for animal cell culture. Note that all monitoring and process actions follow from decisions made by the control system.

offline. Development of flow injection analyzers, biosensor probes, and noninvasive techniques has opened new possibilities for performing more analyses in the preferable online manner.

The function of a sensor is to supply information on the state of the biological process with the ultimate goal of translating this physical or chemical effect into an electrical signal that can be amplified, recorded, and analyzed. For implementation of a control scheme, it is critical that this information reflect the current or future state of the process. Therefore, process sensing must be performed rapidly and preferably with minimal operator intervention.

A bioprocess is significantly different from a standard synthetic chemical reaction in many ways that impact the required instrumentation. For a comparison, Fig. 3 presents schematics of a chemical sensor and a biochemical sensor. Biological reactions tend to be stable compared with many chemical processes and most major variations in a biological reaction occur over long time scales (hours to days) in the absence of equipment failures. However, if the cultivated cells have been subjected to an acute stress, they may not be able to recover and the desired product yield may not be obtained.

Another difference between sensors for a chemical process compared with those for a biological process lies in the complex milieu in which cells grow. The culture media contains proteins, amino acids, sugars, cell debris, and salts. Together these compounds tend to coat or foul surfaces of sensing elements. The methods used to develop sensing technologies for a bioprocess also can depend on whether the cell cultures are anchorage independent (or suspension) or are anchorage dependent, requiring a surface upon which to grow (Fig. 4). As anchorage-dependent cells strongly attach to many surfaces, they tend to foul sensing elements much more readily than do anchorage-independent cells. Appropriate, nonsticky, materials that

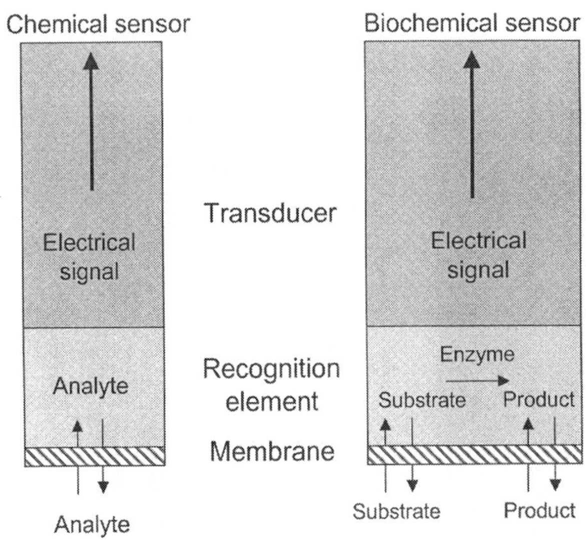

Figure 3 Chemical and biochemical sensors. Both contain recognition elements, transducers, and usually are covered by membranes. Note that the biochemical sensor relies upon a biological reaction (enzymatic or some binding event) to convert the substrate into another component that can more easily be quantified. Chemical sensors do not require such conversions for the analytes to be readily quantified.

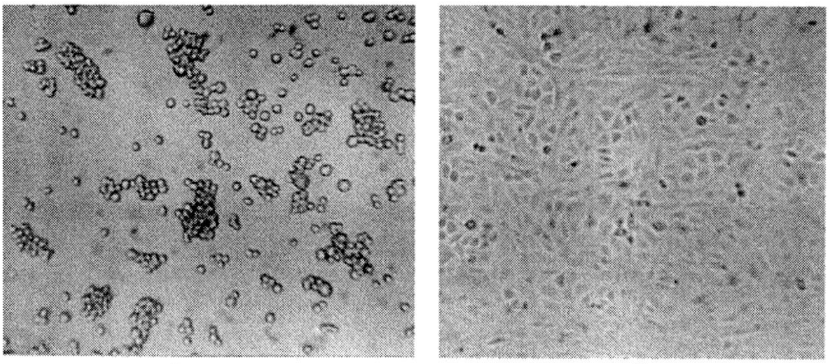

A B

Figure 4 (A) Photomicrograph of rat alveolar macrophages, an anchorage-independent cell type. (B) Photomicrograph of rat type II epithelial cells, an anchorage-dependent cell type.

discourage cell attachment must be selected for biosensor components, although some degree of fouling is likely for any surface placed into a bioreactor environment. Figure 5 displays an example of the interconnectivity between cells attached to a surface. Attachment of one cell frequently encourages the attachment of further cells.

Due to the biological nature of the contents of cell culture reactors, sensors must meet several strict requirements:

1. The most important factor is that the presence of the sensor must not increase the risk of culture contamination either directly or indirectly. This is critical due to the fairly long times required to establish productive

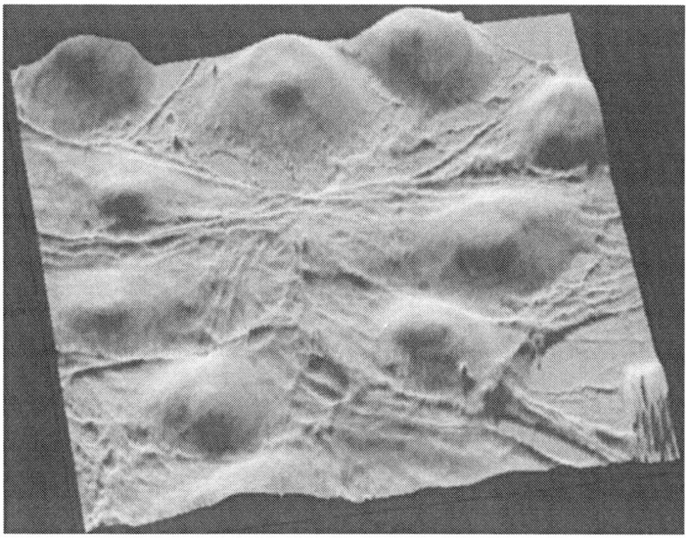

Figure 5 Atomic force microscopic image of epithelial cells on a glass surface. Cells are strongly attached to the surface and to each other through multiple connections. (Image provided by Christophe Juncker and Joseph Simmons, University of Arizona.) Length of the field is 100 μm; height of cells is 2.72 μm.

culture systems owing to the slow growth rates of animal cells. This problem is addressed in a number of ways. The sensor can be employed without contacting the interior of the bioreactor; the sensor can be located downstream of the bioreactor; or the sensor can be sterilized together with the bioreactor (in which case the sensor must be able to withstand harsh sterilization conditions).

2. To minimize the need to remove and replace the sensor, it must have a robust and reliable response with long times between failures. Replacement of sensor elements can often lead to loss of process control or to introduction of contaminants.
3. The sensor must be insensitive to the complex and potentially harsh environment of a bioreactor. In many cases, multiple phases are present (liquid, gas, biomass, and suspended bubbles). Additionally, the presence of proteins from serum, cell debris, or desired products must not influence the measurement.
4. The sensor must be highly specific to the desired variable to be quantified. Cell culture media contains many sugars, amino acids, and small molecules. The presence of structurally similar compounds should not confound measurement.

Sizeable differences exist between the use of monitoring and control methodologies applied in an academic setting (where capital costs dominate decisions) and those in an industrial environment (where time and productivity can be more critical issues). Olsson and coworkers (6) have analyzed the most frequently used bioprocess monitoring components and analytical techniques, with an emphasis on these differences.

MONITORING AND CONTROL OF CELL ENVIRONMENT

Temperature

Temperature is one of the most critical variables to be monitored for the maintenance of healthy cell cultures and so instrumentation and control strategies are well developed and highly accurate. An accuracy of $\pm 0.5^{\circ}C$ or less is typically considered adequate for cell culture monitoring, while variations on temperature of 1–1.5°C can in many cases lead to an unacceptable level of production (7).

The most commonly applied temperature measurement devices for biological processes are resistance temperature devices (RTDs). RTDs have high accuracy and reproducibility with moderately high cost and response time constants of several seconds. RTDs quantify solution temperature based on the changing resistance of a metal conductor with temperature. Platinum is ideal due to its high linearity and broad applicable temperature range and hence is the most commonly applied material (4).

Thermocouples are also used to quantify temperature in bioprocesses. Thermocouples are lower in cost and more rugged than RTDs; however, temperature measurement is not as accurate or as stable as that obtained from RTDs. For these reasons, thermocouples are usually employed for temperature measurements of processes that are less sensitive to temperature fluctuations or for physically rough environments.

Regardless of the type of temperature measurement employed, this sensing element should be physically separated from the bioreactor interior. For small bioreactors, the thermosensing device is often placed in a deep well that typically runs from the bioreactor head plate downward. This arrangement shields the sensor from corrosion and direct impingement from material flow and allows easy removal and maintenance of the device without the need to shut down the bioreactor. It is critical that both the thermowell and the temperature sensing device be adequately shielded from exterior (i.e., outside of the bioreactor) environmental temperature fluctuations. The thermowell must not be so long that it impacts agitation of the culture medium or be subjected to high shear that may lead to mechanical fatigue.

For use in control schemes, the signal from the sensing device is first amplified, linearized, and then transmitted to a controller, where it is compared to the set-point value. Control of bioreactor temperature is often achieved by regulating the temperature or flow rate of water in an external water jacket or in internal heating or cooling coils. Differences in the temperature of this heat transfer fluid and the culture medium alter the rate of heat transfer.

Simple on–off control usually is adequate for laboratory-scale bioreactors, but better control can be achieved by use of PI (proportional-integral) or PID (proportional-integral-derivative) control schemes (8). A two-way heating and cooling action using PID control can provide $\pm 0.2°C$ control on small bioreactors (4). For larger bioreactors, full PID control using both steam and cooling water are required to provide adequate control. Internal coils including hollow baffles or external heat exchangers may affect heat exchange. Alterations in the agitation rate may also impact solution temperatures.

Flow Measurements

The flow of materials into and out of a bioreactor must be accurately measured for fed batch or continuous flow bioreactors. Numerous types of flow metering devices for sterile or nonsterile solutions are commercially available with the most common being rotameters. These are flow-through devices in which a fluid enters the meter, its force acts upon a float, the float rises within a sight glass with calibrated gradations, and the position of the float can be read by an operator (7). While rotatmeters are simple to install and to apply, they are not amenable to automation.

Magnetic meters are ideal for monitoring of sterile process fluids, as long as the fluid has some electrical conductivity (7). The flow of a conducting fluid through a magnetic field generates a voltage that is a linear function of velocity. This voltage is an ideal output for data logging.

Venturi and orifice meters are also commonly used to monitor flow into and out of bioreactors. Unfortunately, they provide low accuracy and are often difficult to clean and sterilize. Therefore, their use has been somewhat limited in bioreactors.

Dissolved Oxygen

Animal cell cultures require oxygen for the production of energy from organic carbon sources. Oxygen can be a limiting species for animal cell cultivations due to its low solubility in water and its high rate of consumption by the cells. The solubility of oxygen in culture media is $\sim 6.6\,\mu g/mL$ at 37°C (9). Note that this refers to the saturation concentration of oxygen in culture media in contact with a 5% CO_2–95% air gas mixture (standard for cell cultivations) (10). The specific oxygen consumption rate can be cell line dependent. For example, at 37°C, these rates can be 0.31 pmol/cell hr for Chinese hamster ovary (CHO) cells, 0.30 for BHK cells, and

0.22 for hybridoma cells (11), although specific values will depend on a number of factors including the cell growth rate, carbon source, etc.

Due to variations in cell concentrations and the high consumption rate, the amount of oxygen dissolved in the liquid culture medium is in a state of dynamic equilibrium. At a constant temperature, the amount of DO follows Henry's law:

$$C_L = HC_G$$

where C_L is the concentration of DO in the culture media, C_G is the concentration of oxygen in the vapor phase above the media, and H is Henry's law constant. Henry's law constant varies with temperature and composition of the medium.

The most commonly applied means for quantifying DO is the Clark-type electrode, developed in the 1950s by Leland Clark. These consist of an electrode covered by a membrane that is selectively permeable to oxygen. Oxygen diffuses across the membrane and is reduced at a noble metal cathode that is negatively polarized with respect to a reference anode. Polarographic electrodes require that an external voltage be applied for negative polarization and oxygen reduction; galvanic electrodes rely on a voltage generated from the use of electrons from lead, zinc, or cadmium anode (12).

Dissolved oxygen is most commonly quantified amperometrically, by the reduction of oxygen at the cathode and the formation of silver chloride at the anode. An electrolyte solution connects the anode and cathode, and with a polarizing voltage. Oxygen from a fluid diffuses across a gas permeable membrane and reacts with two water molecules and incorporates four electrons to yield four hydroxyl ions.

A polarographic electrode (Clark-type) usually contains a platinum cathode, a silver/silver chloride anode, and an electrolyte such as potassium chloride. When a voltage of −0.6 to −0.8 V is applied to the anode, the following half-cell reactions result:

$$\text{Cathode: } O_2 + 2H_2O + 4e^- \rightarrow 4OH^-$$

$$\text{Anode: } Ag + KCl \rightarrow AgCl + K^+ + e^-$$

$$\text{Net reaction: } O_2 + 2H_2O + 4Ag + 4KCL \rightarrow 4OH^- + 4AgCl + 4K^+$$

The reduction of oxygen produces a voltage dependent current that is directly proportional to the oxygen activity in solution. The rate of oxygen diffusion through the gas permeable membrane is typically the limiting step (12).

A galvanic (potentiometric) electrode with a silver cathode and a lead anode has the following half-cell reactions:

$$\text{Cathode: } O_2 + 2H_2O + 4e^- \rightarrow 4OH^-$$

$$\text{Anode: } Pb \rightarrow Pb^{2+} + 2e^-$$

$$\text{Net reaction: } O_2 + 2H_2O + 2Pb \rightarrow 2Pb(OH)_2$$

By carefully selecting the anode material, the cathode will be charged at −0.6 to −0.8 V with respect to the anode (12). Galvanic electrodes also have a slow step of oxygen diffusion and so yield a linear relation between current output and DO concentration. One difficulty with amperometric DO probes is that they tend to have a high rate of failure.

A spectrophotometric method can be used for determination of DO concentrations in suboxic water (less than 1.0 mg/L DO concentration). The method is based on a Rhodazine-DTM colorimetric technique adapted by White and coworkers (13), which minimizes atmospheric interaction with the water sampled. This technique is sensitive to 0.2 μmol/L (0.006 mg/L)—an order of magnitude lower than the amperometric method and much higher than generally required for cell culture work. The technique was originally developed for quantifying oxygen in environmental applications such as ground water, lakes, and reservoirs.

Oxygen concentrations in bioreactor media have also been quantified using fluorescence quenching of suitable dyes (14). Blue light (450 nm) is produced following the excitation of a ZnS:Ag phosphor from beta particles released by a self-powered radioluminescent (RL) light source (Pm-147), which excites a ruthenium complex immobilized in a membrane (14). Analytical information is acquired by measuring the magnitude of oxygen-induced fluorescence quenching of the ruthenium complex. This oxygen sensor shows good stability and reversibility, with detection limits of 0.25 Torr in gaseous samples and 0.028 ppm in aqueous samples. This sensor was recently applied to monitor the oxygen content of fibroblast cell cultures sent on the space shuttle mission STS-93 (15).

Gaseous measurement of oxygen can be applied by taking advantage of the paramagnetic nature of oxygen. A dumbbell-shaped, glass rotor is suspended by a torsion spring in a non uniform magnetic field. Oxygen preferentially accumulates in this field and displaces the nonmagnetic rotor. This rotation is opposed by the torsion spring and occurs in proportion to the oxygen tension (16). The rotation of the rotor can then be read against a calibrated scale. The partial pressure of oxygen in the exit stream is quantified; however, water vapor must be removed from the gas phase in order to minimize sensor drift (17).

Carbon Dioxide

The concentration of gases in the vapor phase leaving a bioreactor can be indicative of the cellular metabolism and so is a commonly used measure of activity. Carbon dioxide (CO_2) and oxygen are the primary components of interest. At high viable cell concentrations in large-scale mammalian cell culture processes, dissolved CO_2 can accumulate due to low removal rates at reduced surface-to-volume ratios (18). High concentrations of dissolved CO_2 can inhibit cell growth and metabolism and can impact the glycosylation of some desired protein products. Bachinger and coworkers (19) used a gas sensor array to detect bacterial contamination in CHO cell cultures. Measurement of head space composition provides a means to detect contamination earlier than conventional methods.

CO_2 in the gas phase is commonly quantified using relatively inexpensive infrared (IR) analyzers and are available with many incubators and bioreactors. These sensors are generally stable as long as the ambient temperature is maintained constant and the vapor environment contains a consistent level of moisture. Both of these factors are controlled to maintain cell viability and so CO_2 sensor failure for these reasons is rare.

Early optical schemes to quantify CO_2 were based on a "wet covered" type sensor, in which a pH-sensitive dye in an aqueous buffer is covered by a gas permeable, ion impermeable, membrane (20). Eventually, measurement approaches evolved to "solid-water droplet" type sensors and then onto "solid" sensors. Mills and Eaton

(20) have reviewed the use of optical sensors for measurement of carbon dioxide, particularly through use of luminescence-based sensors.

Luminescence resonance energy transfer has been used as an alternative optical means to quantify CO_2 (21). The basic principle is radiationless energy transfer from a ruthenium complex as luminescent donor to thymol blue (a common pH indicator) as acceptor. These are both embedded in a hydrophobic matrix. In the presence of CO_2 thymol blue is protonated and changes its color from blue to yellow resulting in a decrease in the rate of energy transfer and consequently an increase in decay time. Other dyes can also be used as acceptors and interfaced with a fiber-optic luminescent sensors (22).

Some novel schemes have been developed to quantify CO_2 using IR absorption. For example, dissolved CO_2 in mammalian cell cultures has been quantified using in situ fiber-optic sensors (18). This group applied an YSI 8500 sensor that had a response time of 6 min with a sensitivity of 0.5% and high linearity. These measurements were not affected by culture pH or the presence of varying concentrations of cellular metabolites. Chang and coworkers (23) applied a fluorescence lifetime-based sensing film to quantify CO_2 in cell cultures. Uttamlal and Walt (24) developed a fiber-optic CO_2 sensor using a pH-sensitive dye in a bicarbonate buffer for online monitoring of bioreactors. Zhang and coworkers (25) have developed stabilization methods to remove the effect of temperature variations on IR CO_2 measurements. Recent developments in other fiber-optical chemical sensors have been review by Wolfbeis (26).

Gas chromatography (GC) can also be applied to monitor exhaust gases including CO_2 and others volatile components. Ethanol, acetaldehyde, and carboxylic acids can be quantified and used to evaluate the cellular metabolic rate. A limitation of applying GC for bioreactor monitoring is that the method requires intermittent injections, often spaced 15 min apart, to monitor transients (17). Most metabolic changes occur on a longer time scale, but the process is not readily amenable to some control schemes.

Mass spectrometry (MS) can be employed to quantify several components simultaneously with a rapid response time (less than a minute) and high sensitivity (detecting concentrations less than 10 μM). MS is gaining in acceptance for process monitoring, although cost of the instrumentation can still be substantial compared with other monitoring schemes. MS instruments are often multiplexed to monitor multiple bioreactors at a time.

pH

Next to temperature, pH is considered the most critical parameter to monitor in a biological process based on the degree of cellular damage that may ensue upon loss of control. Variations of pH of 0.1 units away from optimal can have a substantial impact on cell growth and productivity for certain cell types. While most cell culture media provide substantial buffering of pH, cellular metabolism invariably decreases pH due to the secretion of lactic acid and CO_2 as metabolic byproducts. Hydrogen ions can have a substantial impact on cell metabolism and growth by altering rates of substrate uptake and product release. Typical animal cell culture media begins at a pH of 7.4, but can quickly drop to below 7 within hours, depending on cellular productivity and culture buffering capacity. Difficulties in controlling pH can be related to the logarithmic relationship between pH and the hydrogen ion concentration.

pH-measuring devices are based on the formation of a galvanic cell with two metal conductors connected by an electrolyte. The metal conductors each form a half-cell, one that serves as the sensing element, while the other serves as the reference. A typical glass pH electrode includes a thin-walled glass membrane of 0.2–0.5 mm (which makes the electrode very delicate) enclosing an internal standard solution, usually composed of aqueous HCl (27). Into this solution is placed a rod of silver covered with silver chloride which acts as an internal reference. The key to the selectivity of the electrode is the glass membrane. The glass chosen consists of chemically bonded Na_2O and SiO_2 and is low in Al_2O_3 and B_2O_3 (27). The surface layers of the glass consist of fixed silicate groups associated with sodium ions. When this electrode is first placed in water, the sodium ions exchange with the solvated protons in the water thus forming a hydrated gel layer 50–5000 Å thick (4). This layer exists on both sides of the membrane and is essential for the operation of the electrode.

The use of the pH-responsive glass is restricted to the bulb-shaped element at the end of the electrode, thus limiting the response to only a small sample area. Figure 6 presents a schematic of the surface of a glass pH probe placed in solutions of varying acidity. A constant potential is maintained at the inner surface of the glass membrane by filling the bulb with a buffered solution of stable and accurate hydrogen ion activity. An $Ag/AgCl_2$ electrode is generally used as the electrical lead out from this system. As the pH of the sample varies, this causes a change in potential on the outer surface of the membrane. To measure this a reference electrode is necessary to complete the circuit. The combined electrode is constructed as an integral part of the electrode assembly and consists of an $Ag/AgCl_2$ electrode in KCl electrolyte saturated with $AgCl_2$ (4).

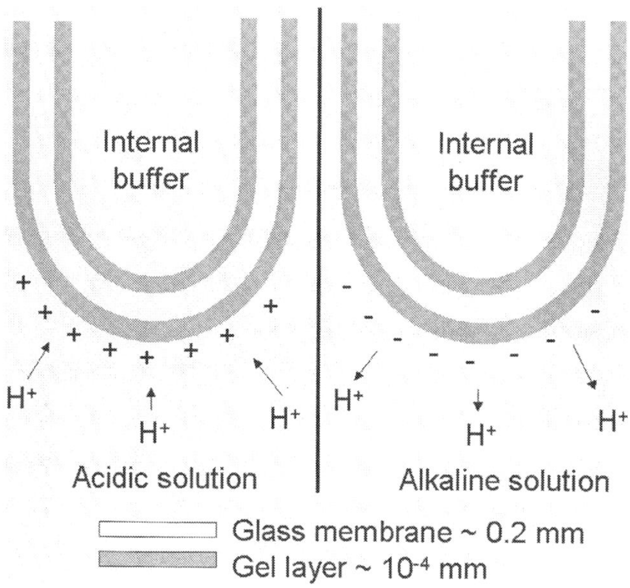

Figure 6 Schematic representation of the tip of a glass pH electrode in two solutions. The external gel layer carries either a positive or negative charge based on the acidity of the surrounding solution.

When the electrode is placed in a solution of unknown pH, the activity of the H^+ ions in the test solution is likely different than that of the H^+ ions in the hydrated layer which sets up a potential difference between the solution and the surface of the membrane. The magnitude of this boundary potential will be determined by the difference in activities. A similar boundary potential will exist at the inner glass wall. The potential can be related to the activity by the Nernst equation:

$$E(\text{cell}) = E^* + 0.0591 \log \text{a}\,(\text{H}^+)$$

where $a(\text{H}^+)$ is the activity of H^+ and E^* includes the standard electrode potential of the glass electrode, the potential of the reference electrode, and liquid junction potentials. pH is then defined as:

$$\text{pH} = -\log \text{a}(\text{H}^+)$$

Note that the degree of hydration of the outer surface will also change if the electrode is allowed to dry out. It is important that the electrode be stored either in an electrolyte solution (e.g., 4 M KCl) or wrapped with a damp cloth.

The glass electrode generally exhibits a Nernstian response over most of its working range; however, at extremes of pH, the behavior becomes non-Nernstian (27). Alkaline errors are due to the response of the glass membrane to other ions and so lead to pH measurements that are lower than their true value. These can occur at pHs around 9–10. Acid errors occur at the opposite end of the pH scale and are due to the impact of variations in the activity of water in solution. The net effect is for the sensor to report an acid pH that is lower than the true value.

An alternative pH sensing approach employs a diaphragm that links an electrolyte solution and the sample to be measured. However, sulfur-containing compounds such as the amino acid cysteine can cause these pH probes to fail due to the formation of silver sulfide deposits on the diaphragm surface.

Several important points in the application of pH sensors emerge from the Nernst equation (7):

1. The millivolt output is positive below a pH of 7 and negative above a pH of 7.
2. The temperature effect at a pH of 7 is 0; however, at pHs away from this point, the effect increases linearly with pH.

Ideally, pH calibration should involve measurement at three points (4,7 and 10) so as to best correct for differences between the Nernst equation and experimental conditions.

Once a probe is calibrated, there are a number of problems that can lead to erroneous or unreliable readings. These include slow response, drift, and reduced sensitivity. Inside of a bioreactor these problems are often the result of (7):

1. The electrode surface has been coated by proteinatious material from serum, cell debris, or cellular products. Often this will result in a slow measurement response with pH readings higher or lower than anticipated.
2. The glass surface has been etched or dried. Overly high pH readings and difficulty in calibrating over an extended range may result.
3. Probe terminals are shorted. The probe may report a constant pH of 7 regardless of the true value.

Control of pH is achieved through pulses of acid or base (usually only base is required). A proportional controller can be applied to respond to pHs lower than a set point with addition of a pulse of base (often KOH, NaOH, $NaHCO_2$, or Na_2CO_3). The frequency of measurement and pulse addition and duration of pulse addition can be modified to effect varying levels of control. For pilot plant bioreactors, control of pH at ±0.03 pH units is obtainable using pumps operating on a time-proportional basis (4). Most media applied for animal cell cultivation have substantial buffering capacity around 7.4 providing some degree of safety around the pH desired by cultures.

pH has also been quantified by optical means (28). For example, the diffuse reflectance of a conductive polymer such as PoAnis/TSA and cellulose acetate fixed on the end of a fiber-optic probe which allows measurements in the pH range from 4.9 to 10.5 with a precision of ±0.01 pH and a response time of 5 min for a full pH variation from acid to base (28). Unfortunately, the sensor response can be a function of the ionic strength of the solution and on the supporting electrolyte used to adjust the ionic strength. Lin (29) reviewed developments of optical and fiber-optic pH sensors in the 1990s including methods for immobilization of pH indicators. Optical and fiber-optic pH sensors are reviewed, including pH sensors based on conductive polymers, imaging fibers, microparticles and nanospheres, as well as micrometer and submicrometer fiber-optic pH sensors, distributed fiber-optic pH sensors, pH sensors for high acidity and alkalinity, pH sensors with broad dynamic range and linear response, and CO_2 and NH_3 sensors based on pH indicators.

While major efforts in cell culture systems are focused on increasing the viable cell density, such increases in cell numbers will also result in an increased oxygen demand and will require better mixing approaches (30). Additionally, the accompanying increase in CO_2 production and accumulation and the resulting reduction in pH are also important implications for process engineering. Such pH reduction is typically controlled by the addition of sodium carbonate, but with current bioreactor operating schemes, pH excursions away from pH set points can be found in bioreactor regions in close proximity to the location of base addition due to poor local liquid mixing (30). Enhanced mixing schemes are required to homogenize the bioreactor media without damaging cell function (31,32).

Metabolites and Products

Animal cell culture technology has advanced due to increased understanding of cellular metabolism including the effects of product inhibition. Glucose and glutamine are the primary sources of energy and building blocks for animal cells; ammonia and lactate are the primary metabolic byproducts that may influence cellular productivity. As little as 4 mM ammonia can reduce the cell growth rate of hybridoma cells by 50% and accelerate metabolic rates in an undesired fashion (33). Cultivation of hybridoma cells under tightly controlled glucose and glutamine concentrations can increase the cellular production of MAbs (34). Nutrients and products have been quantified through a variety of methods including electrochemiluminescence, optical sensing, enzymatic conversions, chromatography, and nephelometry (35).

While monitoring provides important information on the bioprocess, oftentimes, predictive information is also required. This can be achieved through use of biological models. Unstructured models are empirically based on experimental observations. Structured models are theoretically based on knowledge of the individual cellular processes. Miller and coworkers (36) developed an unstructured model for hybridoma growth, death, and MAb production based on measurements of the viable cell density and dilution rate. Dalili and coworkers (37) developed a model to characterize the influence of glutamine on hybridoma growth and MAb production. In agreement with experimental results, Dalili's model predicted that the maximum viable cell concentration increased with glutamine levels in the range of 0.5–2.0 mM. Several groups have developed structured models to characterize the interplay of both nutrients and metabolic byproducts on cell behavior (38–41). These models focus on the individual metabolic rates and so provide information which is of most use for development of bioreactor control schemes to optimize MAb production.

Glucose and lactate concentrations have been monitored online using a commercially available analyzer (Model 2700, Yellow Springs instruments, Yellow Springs, OH) during batch and perfusion hybridoma cell cultures (42). A schematic of this sensing element is presented in Fig. 7. Cell-free samples from the reactor were obtained using a 0.45 μm hollow fiber filtering system placed in a circulation loop. A process control strategy was developed to control the concentrations of glucose and lactate in a perfusion reactor by adjusting the feed rate to maintain desired set points. During exponential growth the control algorithm successfully adjusted the perfusion rate while maintaining constant glucose and lactate concentrations. Glucose consumption and lactate accumulation rates were used for the estimation of viable cell density in the reactor (42).

Recent evidence suggests that there may be a substantial need for monitoring a greater number of components in bioreactors for animal cells than are currently taken into account in such models or are measured online. Simpson and coworkers (43) have demonstrated that depletion of any single amino acid in a hybridoma cell cultivation can lead to apoptotic cell death including both essential and nonessential

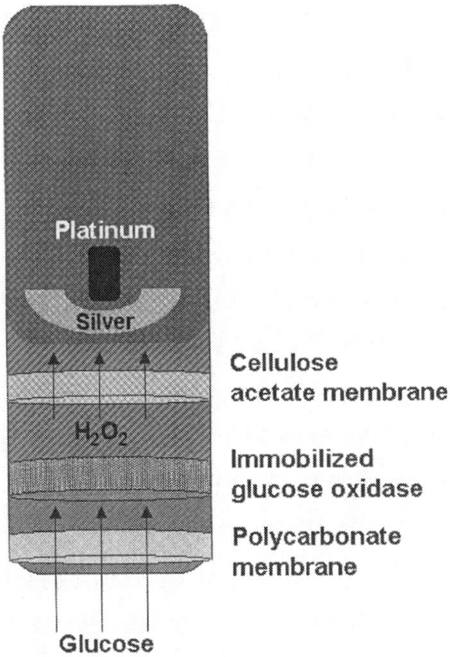

Figure 7 Schematic representation of the sensing mechanism of the YSI glucose analyzer. Glucose diffuses across the polycarbonate membrane, reacts with the immobilized glucose oxidase to generate hydrogen peroxide, which diffuses across a cellulose acetate membrane, and is detected by a platinum–silver electrode. (Adapted from the YSI 2700 analyzer manual.)

amino acids. Based on this result, monitoring the glucose and glutamine concentration alone may not suffice for maintaining a healthy culture. Likely all or nearly all of the high-level components in a cultivation must be tracked. Many standard culture media for animal cells contain 17 components present at concentrations of 0.5 mM or greater (44). Currently, reliable methods are not available to rapidly quantitate the concentrations of all these chemical species in a manner amenable to the implementation of intelligent control schemes. Metabolic engineering of animal cells provide the potential to improve productivity of cell cultures in generating complex proteins with correct post-translational modifications; however, for these approaches to be truly optimized, the complex requirements for the growth and production must be integrated with an understanding of the cellular environment (45). Mulchandani and Bassi (5) have recommended that a specific area of importance for the development of online process improvement is the monitoring and control of nutrient concentrations for the maintenance of cell viability and production.

Once a monitoring and/or modeling scheme is developed for an application, the gathered information must be put to use so as to optimize the culture productivity. However, traditional control theory does not take into account the inner metabolic processes that govern the cell behavior such as the interplay between multiple metabolic pathways. Konstantinov (46) presented a methodology for the design of systems capable of performing advanced monitoring and control functions based on the physiological state of a cell population. The physiological state should incorporate the specific rate of metabolism, metabolic ratios for product generation, and the specific growth rate. The goal of the control system then becomes to maintain the physiological state of the cell as close as possible to a predetermined trajectory of maximum efficiency.

Enzymatic Methods

Accurate medium concentration information is required to implement intelligent bioreactor control schemes. For many years, mammalian cell metabolism was determined by removing samples from a bioreactor and analyzed offline by analytical techniques such as high-performance liquid chromatography (HPLC). Online chromatographic analyses were eventually applied to speed the characterization process, but measurement time can be substantial and the periodic removal of samples creates concerns for bioreactor contamination not to mention the undesired loss of some amount of the generated product. The advent of enzymatic sensors has reduced the time required for analysis of bioreactor media.

A biosensor is an analytical device that combines the specificity of a biological sensing element for the analyte of interest with a transducer to produce a signal proportional to the target analyte concentration (5). Enzymes have been employed in a wide range of sensing schemes to quantify cellular nutrients, wastes, and products. Most include enzymes from the classes of hydrolases, lyases, and oxidoreductases (47). Enzymatic biosensor probes provide rapid measurements from within a bioreactor, thus bypassing some of the aforementioned problems. Such enzymatic biosensors can be difficult to sterilize and can be unstable when used over long operation periods because of enzymatic degradation, and, hence, frequent recalibrations are required to ensure accurate measurements. Enzymatic systems also require a separate sensing element for each analyte, complicating measurement of multiple species.

Measurements using enzymatic biosensors are most often performed offline due to the need to isolate the enzyme from the bioreactor environment. Typically in an enzymatic sensing scheme, the analyte to be quantified reacts specifically with an enzyme and generates a product that can be readily detected. In some cases, multiple reaction steps are required. For example, glucose can be quantified by the following reaction scheme:

$$\text{Glucose} + O_2 \rightarrow \text{gluconic acid} + H_2O_2$$

$$\text{Fluorescein} + H_2O_2 \rightarrow \text{fluorescein}^*(\text{fluorescent}) + 2H_2O$$

The first reaction is catalyzed by glucose oxidase, the second by horseradish peroxidase (48) and is a frequently applied scheme for glucose quantification. The amount of fluorescence detected will correlate to the amount of glucose present, as long as excess oxygen is available. This scheme could also be used to quantify oxygen if sufficient glucose were available. Alternatively, H_2O_2 could be quantified amperometrically versus an Ag/AgCl reference electrode. Many other similar enzymatic methods have been applied for glucose measurement in biological samples.

Glutamine can be quantified by the following reaction scheme:

$$\text{Glutamine} + O_2 \rightarrow \text{glutamate} + NH_3$$

$$\text{Glutamine} + O_2 \rightarrow \alpha\text{-ketoglutarate} + NH_3 + H_2O_2$$

The first reaction is catalyzed by glutaminase; the second is catalyzed by glutamate oxidase. Such an approach provides multiple options for the ultimate detection of glutamine. H_2O_2 could be measured amperometrically or could be converted by the horseradish peroxidase reaction shown above. The liberated NH_3 could be quantified using an ammonia detection scheme.

When applying enzymes as the biological recognition element in a biosensor, both operational stability and long-term stability must be taken into account. If the enzymes are to be immobilized to a surface (often a requirement for application in which the sensing element is to be separated from the culture medium), the immobilization technique will have a great impact on sensor stability. Chemical methods such as covalent cross-linking to a support material typically yield long stability, but at a reduced enzymatic activity. In some cases, the activity of an immobilized enzyme can be reduced by as much as a factor of 20 compared with a soluble preparation (49). Physical methods such as entrapment of the enzyme within a membrane or gel material provide greater activity, but at the expense of overall stability of the enzyme and of the sensor as a whole. Culture media proteins, cells, and cell debris can readily foul membranes.

Flow Injection Analysis

Flow injection analysis (FIA) is based on the introduction of a liquid sample into a moving aqueous carrier stream. The injected sample forms a pulse which is transported toward a detector which continuously records the sample absorbance, electrical potential, fluorescence, or other physical parameter that changes as a result of the sample material (50). Typical output is a broad peak because the sample disperses along the carrier stream. Detection methods include amperometric, potentiometric, fluorescence, chemiluminescence, UV/Vis absorbance, or turbidometric (51). Advantages of FIA include a reduced risk of contamination, ease of recalibration or replacement of sensing elements, small sample requirements, and short response time. Such methods provide rapid analyses with measurements requiring as little as 20 sec. A disadvantage is that each sensor can monitor only one analyte and that sensitivity and selectivity are often not as high as required for implementation of control strategies (52). However, multiple sensing elements can be integrated to provide concentration information on a wide range of components.

An FIA system typically includes a selector valve to switch injection flow between a sample to be analyzed and calibration standards; a multichannel pump to deliver sample, reagents, and calibration solutions; an injection valve with a fixed-volume loop (usually 20–50 μL); and a detector (5). If multiple analytes are to be quantified, the sample flow is often split into different channels, each of which is used to analyze for one component. Isolation of these channels can simplify analysis by providing physical isolation of each sensing array, thus reducing interference or cross-talk between analysis steps. FIA can be difficult to apply in automatic control systems as frequent sampling of culture medium will reduce some of the natural advantages of this technique.

Examples of components in bioreactor media that have been quantified by FIA include ammonia (53–55), glucose (13,56–59), glutamine (13,56), lactate (57,59,60), and antibodies (61). Quantification of glucose can proceed through amperometric measurement in which immobilized glucose oxidase produces hydrogen peroxide which can be quantified through anodic oxidation (56). Quantification of glutamine can be based on measurement of ammonia ions produced in a flow-through enzyme reactor containing immobilized glutaminase enzyme, followed by subsequent downstream potentiometric detection by an ion-selective membrane electrode (56).

White and coworkers (13) applied three amperometric FIA biosensors based on rhodinized carbon electrodes to quantify glucose, glutamine, and glutamate in parallel in mammalian cell perfusion cultures. The geometry of an FIA system (in which the sensors are not directly incorporated into the bioreactor) permits easy replacement of sensing elements. The inclusion of frequent recalibrations of the sensors maintained accuracy of measurements.

A solid-state, diffuse reflectance-based fiber-optic sensor has been used to quantify ammonia by employing immobilizing chlorophenol red, a weak acid chromophoric indicator dye, in a microporous polypropylene membrane (62). A schematic flow injection scheme is presented in Fig. 8. This sensor uses flow-injection to carry a 10-μL aliquot of the sample across the treated membrane. Ammonia in the sample diffuses through the air-filled pores within the membrane structure before

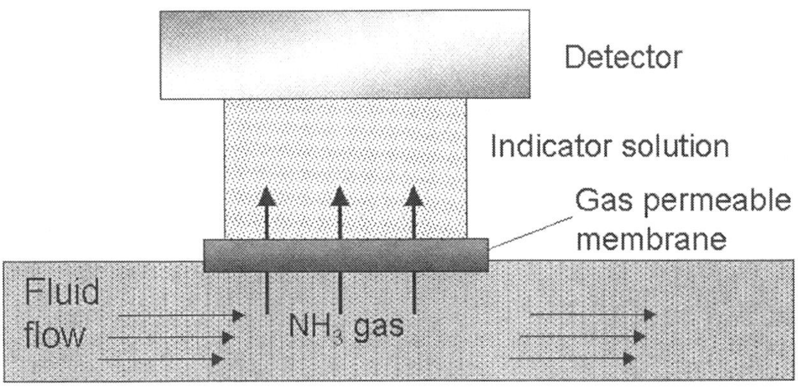

Figure 8 Schematic of an ammonia sensor applied in a flow scheme. NH_3, and not NH_4^+, can traverse the gas-permeable membrane, interact with a suitable indicator solution that may change color, or fluoresce in response to NH_3 that is monitored by the light detector.

reacting with the indicator dye. A reversible acid–base reaction between ammonia and chlorophenol red results in a measurable change in the reflectance at 560 nm. Response characteristics include a peak response within 20 sec, a limit of detection of 0.2 ± 0.1 mM ammonia, and a dynamic range of up to 60 mM ammonia.

A flow-injection sensor using an expanded microbed as the enzyme reactor has been developed to quantify glucose and lactate in bioreactors (63). The expanded bed reactor is capable of handling a mobile phase containing suspended matter like cells and cell debris, thus eliminating the need to pretreat the sample to remove particulate matter. Glucose oxidase and lactate oxidase were immobilized and used to quantify glucose and lactate, respectively, with an assay time of 6 min (63).

The FIA system has also been used as a two-step immunoassay-based determination of human prolactin (hPRL) concentration along with its degree of glycosylation (64) in which the antibodies are immobilized on the surface of a carrier. The results of the two-step FIA method were found to agree with those obtained by the standard methods while requiring minimal analysis time (10 min) and elimination of operator intervention.

Schugerl (51) provides an excellent review of recent developments in monitoring product formation processes using FIA, spectroscopic methods, and biosensors. This review also covers advanced control of indirectly evaluated process variables by means of state estimation/observer, with the use of structured and hybrid models, expert systems and pattern recognition for process optimization. Christensen (65) also provides a detailed review on the use of FIA and sequential injection analysis (SIA) for monitoring biological processes.

Spectroscopic Methods

The aforementioned sensing methods are destructive and require that samples be removed from the reactor prior to analysis. The methods also typically quantify a small number of analytes, following the axiom of one sensor for one analyte. Efficient bioprocess control requires real-time monitoring of biomass, substrates, intermediates, and nutrients and is performed preferably without destroying or removing bioreactor media. Optical sensors have the advantages that they can be performed quickly, require no sample preparation, and can be noninvasive. Currently, optical sensors are available to quantify DO and dissolved carbon dioxide (66); however, methods are currently in development to quantify a variety of nutrients and wastes in addition to quantifying biomass.

Spectroscopy using visible, UV, or near-infrared (NIR) light provide attractive alternatives to monitor the concentration of nutrients and wastes in cell culture media. The approach of applying spectroscopy to quantify components in bioreactors has evolved from development of portable glucose sensors designed for use by diabetics (67). Spectroscopic measurements can be noninvasive, nondestructive, rapid, require no sample preparation, and used to quantify multiple chemical species simultaneously. Spectroscopic measurements are based on introducing a beam of light to a sample, collecting transmitted or reflected light, and correlating the amount of light absorbed to the composition. Any nonsymmetric chemical species may be quantified by identification of its characteristic absorbance features in the NIR. Conversely, symmetric species can readily be quantified using Raman spectroscopy. Many of the critical metabolites, such as glucose, glutamine, ammonia, and lactate have distinct spectral features in the IR, thus suggesting that they may be easily quantified (Fig. 9). Figure 10 presents IR spectra of animal cell culture media,

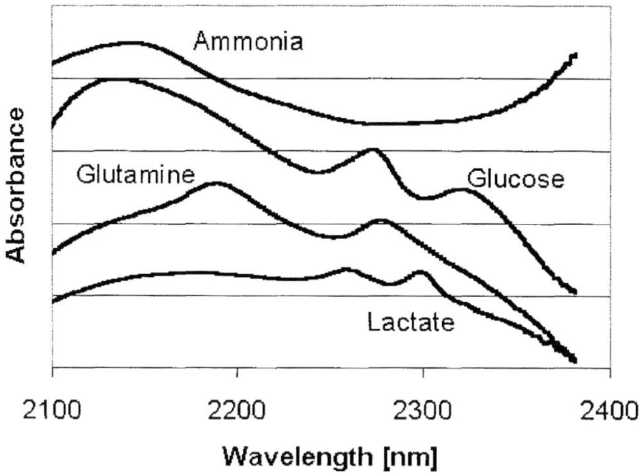

Figure 9 NIR spectra of ammonia, glucose, glutamine, and lactate in a physiologically balanced salt solution. Note that each component has a distinct spectral feature that may be used for identification and quantification.

demonstrating that in a complex solution, identification of the metabolite features can be difficult.

Each of these measurement types provides specific advantages. FT-NIR provides the greatest light penetration depth and so can be used for thicker samples so that they provide a higher degree of light scattering. FT-mid-IR (MIR) provides information that is more easily discernible as being specific for certain analytes as

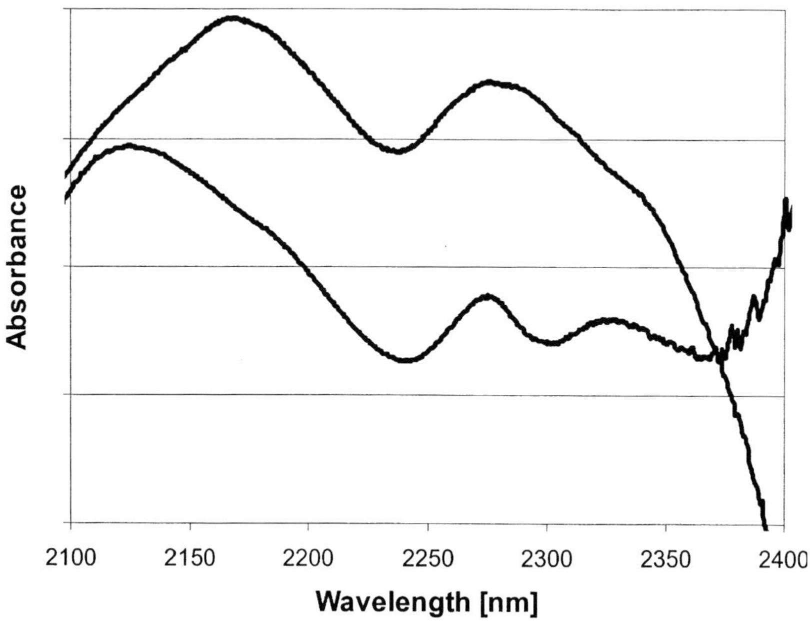

Figure 10 Infrared spectra of animal cell culture media. The bottom spectrum is the commercially available DMEM; the top spectrum is DMEM +10% horse serum.

these wavelengths are closer to the fundamental IR absorptions. FT-Raman is advantageous when the interference due to water is to be minimized, such as when the water content is variable. Sivakesava and coworkers (68) compared these three methods for monitoring a lactic acid fermentation and found that the MIR region was better suited for quantifying glucose, lactic acid, and biomass. Rhiel (69) has also reported on the merits of the MIR region over the NIR. They applied MIR spectroscopy to quantify glucose and lactate in CHO cell cultures and were able to make accurate measurements over the course of 60 days without the need to recalibrate.

The amount and frequencies of light absorbed by the sample can be correlated to the type and concentration of chemical species present in the sample. The resultant absorbance spectra contain complex spectral features that require the use of chemometric methods such as partial least-squares (PLS) regression analysis to extract the analyte concentration information. These analysis methods are routinely practiced in the agricultural and petrochemical industries.

In cell cultures, NIR spectroscopy (NIRS) has been used to quantify the concentrations of glucose and maltose (70); glucose and glutamine (71); glutamine and asparagine (72); glucose, glutamine, ammonia, lactate, and glutamate (73); and glucose, sucrose, and fructose (74). Spectroscopic methods have recently been applied to monitor biological reactors. For yeast cell fermentations, the concentrations of ethanol (75); fructose, glycerol, glucose, and ethanol (76); and yeast cell density (77) have been quantified. In bacterial fermentations, the concentrations of lactate, glucose, and biomass (78); acetate, ammonia, and the cell density (79); NH_4OH (80); and exopolysaccharide, lactic acid, and lactose concentrations (81) have been quantified. In the past 5 years, use of spectroscopic techniques has expanded substantially. Pollard and coworkers (82) used attenuated total reflectance spectroscopy in a pilot plant to quantify sugars, phosphate, and proline for a fungal culture. Attenuated total reflectance is most useful for evaluating liquid samples with a high reflectivity. Unfortunately, many suitable crystals used for this method also permit strong attachment by proteins and cellular debris, thus rapidly fouling the sensing element.

Most of the aforementioned studies applied spectroscopic methods to quantitate the composition of yeast and bacterial cell growth media that have relatively few components compared with mammalian cell growth media. Often, the primary carbon source is present at levels of 100 mM or more in bacterial growth media (76). In contrast, performing such measurements on animal cell cultures is more difficult due to the greater complexity of animal cell culture media and due to the lower concentrations of many nutrients. For example, RPMI 1640 media used for animal cell culture contains 11 mM glucose, and also has 20 amino acids, in addition to vitamins, inorganic salts, and various amounts of undefined serum. There have been fewer published works on the development of NIR spectroscopic techniques for mammalian cell cultivation. Yano and Harata (83) quantified glucose, glutamine, ammonia, and lactate in mammalian cell cultivations; similarly, McShane and Cote (84) quantified glucose, lactate, and ammonia in cell culture media; Jung,(85) quantified glucose using a fiber-optic measurement. Recently, as many as 19 components have been quantified simultaneously in animal cell culture media (86). The presence of horse serum had no apparent effect on measurement accuracy, as long as serum was present in the calibration sample sets.

The large number of chemical species present in mammalian cell growth media produces NIR spectra with multiple, overlapping absorbance features. PLS regression analysis is often applied to quantitatively correlate spectral features such as absorbance peak shapes and heights to the concentration of a specific analyte present in the sample (87). PLS regression methods require that analyte concentrations in the calibration samples must be random and uncorrelated. This requirement of uncorrelated species concentrations complicates the construction of calibration models for bioreactors.

For monitoring biological processes, the requirements on calibration samples typically are satisfied through use of 70–100 calibration samples with independent concentrations of each component (88–92). Developing samples with independent concentrations of each component, collecting spectra, and analyzing the calibration can demand many weeks of laboratory work. If any substantial modification is made to the bioprocess (pH, temperature, absence or presence of new chemical constituents), this calibration may no longer be applicable, thus requiring a lengthy recalibration process. Such difficulties have so far limited the application of NIR spectroscopy for many biological processes.

Calibration of spectroscopic techniques presents a sizeable difficulty. If a single bioreactor run is used to generate samples for calibration, unwanted correlations in the analyte concentrations will arise. If such a calibration model were applied to a second bioreactor run that has slightly different initial conditions, the spectroscopic measurements are likely not to be accurate (87). Samples collected from batch cultures typically contain highly correlated metabolite concentrations as a result of metabolic relations. For example, the concentrations of glucose and lactate are often anticorrelated, as are glutamine and ammonia. Calibrations based on such samples can only be reliably used to predict concentrations in new samples if a similar correlation structure was present; unusual variations can lead to highly inaccurate measurements (93). This problem can be avoided if calibration samples are drawn from multiple bioreactors. Fayolle and coworkers (76) operated four calibration fermentations, while Hall and coworkers (79) ran five calibration fermentations. The use of multiple runs does not guarantee that species concentrations will be uncorrelated because nutrient levels consistently diminish and wastes accumulate and these rates are intrinsically linked through the cell metabolism. The use of synthetic calibration samples (which are produced through addition of known amounts of each component) simplifies the removal of correlations by producing a large set of samples with uncorrelated analyte concentrations (90,91). Unfortunately, this approach requires prohibitive preparation time. This preparation time depends on the number of varying components, but can be from 4 weeks to several months.

A recognized problem for spectroscopic measurements arises when a working calibration is applied to a similar, but not identical system. Common changes that create difficulties in transferring calibrations include varying background composition, temperature alterations, and path length modifications. For bioreactors, a change in the type of culture medium could be expected to affect spectroscopic measurements. Standardization methods have been developed to reduce the time required to generate transferable calibrations for new conditions, by applying samples intermediate or common to the two measurement conditions.

Adaptive calibration schemes have also been applied. These methods often involve spiking samples with additional amounts of the analytes to be quantified (94). In this approach, a limited number of calibration samples are collected, typically 5–10, from bioreactors operated under normal conditions over time. These samples are each subdivided into 10–20 additional samples. To each of these is added a known amount of the analytes to be quantified. The initial large volume samples provide spectral information on uncontrolled variability including cell debris, low concentration wastes, and the breakdown of serum. The spiked material provides a greater variety of analyte concentrations and can increase the calibration range. Note that this approach can only be applied to increase the analyte levels. Rhiel (95) applied a similar adaptive calibration procedure to quantify glucose, lactate, glutamine, and ammonia in cultures of PC-3 cells and obtained very low measurement errors.

An alternative and more rigorous approach to construct spectroscopic calibration models for bioreactor monitoring is to produce synthetic samples with each of the compounds present in the culture media, taking into account changing concentrations over the course of the cultivation. For example, at early times, glucose and glutamine concentrations would be high, whereas lactate and ammonia would be very low. This would be reflected in the synthetic samples produced. Typically, 80–100 samples are required to generate a reliable calibration, so this can be a very time-consuming procedure, particularly when taking into account the low-level amino acids and carbohydrates that typically play a minor role in cell metabolism. This approach has been successfully used to quantify 19 components simultaneously (86); however, calibration development time was several months.

Solution pH, temperature, and ionic strength are important experimental parameters that can affect the spectroscopic measurements. Each of these parameters can alter the position and intensity of absorbance bands in the NIR. Small band shifts can be observed due to changes in the hydrogen bonding nature of the sample caused by changes in the pH, temperature, and ionic strength. Changes in temperature by a few degrees may alter the position of the water absorbance features normally located around 1923 and 2632 nm. The pH, temperature, and ionic strength of the samples are adjusted so as to not be affected by addition of analytes and maintained at levels observed in the bioreactor. If such variations become unavoidable, digital Fourier filtering may be employed to reduce these effects. Such methods have been shown to compensate for significant alterations in absorbance features brought about by temperature changes of four degrees (96). The Fourier filtering step in spectral preprocessing effectively discriminates between broad baseline features and the relatively narrow analyte bands, and, hence, enhances the analyte information relative to the baseline variation.

Vaidyanathan and coworkers (97) analyzed the sensitivity of spectroscopic calibrations to variations in analyte concentrations and to background variations. They found that simple models, such as uni- or bivariate linear regression models or PLS models using no more than four factors, performed reasonably well when subjected to variations. However, models based on weak absorbance features were vulnerable to changes in the background matrix. This study highlights the need for development of robust models, particularly without over-modeling the available calibration data. Over-modeling results from using too few calibration samples or from applying too many calibration factors. An over-modeled calibration can provide unreliable results when challenged with even small variations in the sample material.

The challenge, therefore, for applying NIRS lies in developing a calibration that provides accurate concentration measurement to which a substantial amount of confidence can be applied without the calibration being overly specific so that it can be applied to reaction schemes with small variations. A recently published study brings into question the validity of some calibration schemes for measurement of glucose from NIR spectra collected in a time-dependent manner (98). This group developed a "phantom" glucose data set by purposely omitting glucose in a number of samples that were collected consecutively. Within the data processing step, arbitrary and nonzero glucose values were assigned to successive phantom spectra, and multivariate calibration models generated for glucose based on PLS regression. The result was that PLS could be "tricked" into predicting the presence of glucose in samples devoid of glucose. Chance temporal correlations between assigned glucose concentrations and some uncontrolled experimental parameter are responsible for this apparent model functionality. This study presents sizeable evidence that one must take extreme care in developing calibrations that do not contain correlations between components, both quantifiable and nonquantifiable.

One of the practical constraints that may limit the application of NIRS to monitor biological processes is the limit of detection of the technique. Many of the aforementioned results provide measurement errors in the neighborhood of 0.12–1.2 mM. There is no straightforward method for converting the standard error of prediction (SEP), normally provided as the comparable metric for PLS analysis, to a limit of detection (LOD). A reasonable approximation is that the LOD is three times that of the SEP (analogous to the use of three times the measurement standard deviation as a predictor of the LOD in other measurement schemes). Therefore, a 0.5 mM SEP would yield a 1.5 mM LOD. This and other information has led some researchers to believe that 1 mM is the practical limit of detection for compounds with a molar absorbtivity similar to that of glucose. Ideal cultivation conditions for hybridoma cells typical require maintaining concentrations of nutrients and wastes at concentrations as low as 1 mM each.

This evaluation of the LOD of NIRS is lacking in that the quantified SEP is highly dependent on the concentration range evaluated. For example, measurement of glucose over a concentration range of 0–30 mM yields an SEP of 0.49 mM, while measurement of glucose over a range from 0 to 1 mM yields an SEP of 0.12 mM. Clearly, the SEP does not scale directly with the concentration range, but increases at a rate less than an increase in the concentration range. Measurements of glucose and glutamine may be reliably performed at concentrations as low as 0.2 mM (92), the approximate level required for adequate control.

Sensing in Microbioreactors

In recent years, microscale approaches have been developed to provide information on cell cultures maintained in highly scaled down processes (99). Instead of performing process development on the 5–500 L scale, cultures are grown in 0.5–5 mL volume reactors. Working at such a small scale permits analysis of many cultures in parallel and in high throughput, thus providing a relatively inexpensive and fairly rapid means to evaluate a large number of process conditions simultaneously. This philosophy of high throughput has been adopted from the genomic and proteomic screening fields. Challenges in this area lay in development of suitable reactor systems, in development of analytical techniques for analyzing small sample volumes (often as little as nanoliters of material), and in analyzing scale up of such a magnitude.

Several groups have explored microbioreactors. Kostov (100) developed 2 mL microbioreactors to evaluate cultivation conditions of *Escherichia coli* fermentations. They monitored pH, DO, and optical density using light emitting diodes, photodetectors, and an oxygen-sensing device. Their approach is simple and inexpensive and can be applied to many other types of culture systems. Their results corresponded closely to those obtained in a 1 L fermentation. Riley and coworkers (101) have applied light emitting diodes and photodetectors as biological sensing elements to track the metabolism of lung cell cultures grown in plastic spectrophotometric cuvettes. A schematic of this reactor and sensing components is presented in Fig. 11. This approach is used to track the impact of environmental and biological toxins on culture metabolism, ultimately correlated to other metrics of biological toxicity. This device is small, inexpensive, and portable, thus providing a means to evaluate the safety of airborne, soluble, and particulate materials.

A significant challenge in working with microbioreactors lies in quantifying the desired products. Oftentimes, ELISAs (enzyme-linked immunosorbent assays) are used to quantify products of cell cultivations that rely on the highly specific and strong binding of a monoclonal antibody to their antigens. Typically, ELISAs require a minimum of 4–5 hr to perform and require frequent intervention by a technician or by expensive robotics. These assays can be scaled down and automated on the surface of a microchip, particularly on etched glass surfaces.

A number of groups have recently developed microchip biosensors for quantifying small compounds (102–104). Similar methods can be used to sequence DNA (105), although capillary immunoassays have also been developed (106). Quantification of as little as 2100 molecules in a single sample has been reported (107). Management of thermal gradients is a prime concern (99). These approaches employ microfabrication techniques, including photolithography and chemical etching, to generate a series of channels on the surface of a glass or silica substrate. Each channel is independently wired to permit capillary electrophoretic separation that isolates molecules based on the charge to mass ratio of each. Such fabrication techniques permit tremendous flexibility in the size of channels and on the flow rates of fluids. Each design layout may be replicated with a high degree of consistency. Sample volumes may be as small as 50 nL, which require separation times of ~30 sec (102). Detection methods vary, but are often based on laser-induced fluorescence. Sensitivity and the ability to detect low levels of cytokine secretion are difficulties with the current technologies. Figure 12 presents a photomicrograph of an etched glass surface employed for detection of cytokines produced by cell cultures. The separation between lanes is ~10 μm. This device is produced using chrome deposition on glass, followed by deposition of a photoresist that is exposed to UV light through a mask and rinsed away. The pattern is etched into the glass with a buffered oxide etch of hydrofluoric acid for 2 min

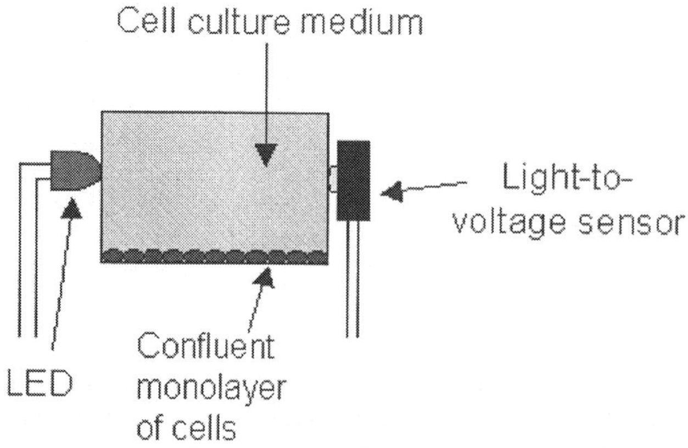

Figure 11 Schematic of a microbioreactor with LED-based sensing of cell metabolism.

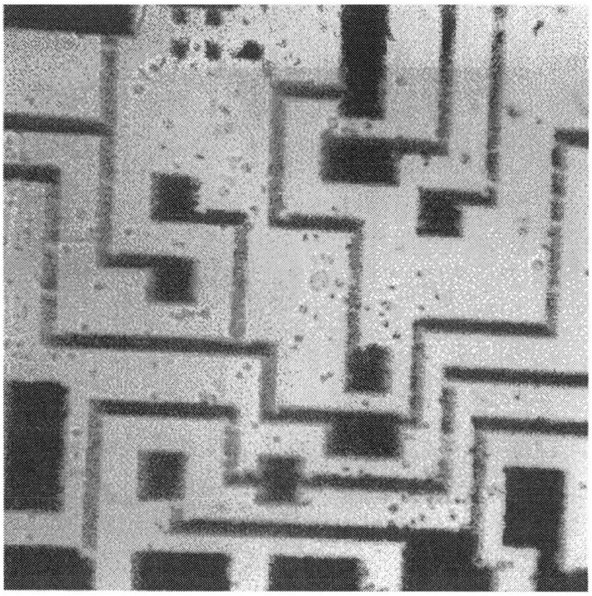

Figure 12 Photomicograph of an etched glass device developed with channels for the separation and quantification of cytokines secreted by cell cultures. Epithelial cells have been seeded within several of these channels. (Device and image produced by Jennifer Wipf, Victor Wells, and Mark Riley, University of Arizona.)

followed by removal of any remaining chrome. The result is a glass surface etched with the pattern of the mask.

On-chip measurements have been used to separate antibodies from unbound antigen in a competitive-binding assay in which either an antibody or antigen is labeled with a fluorescent tag (102,103,106). The typical operation involves adding a sample with an unknown concentration of an antigen to a solution containing an antibody reactive to this antigen and a known concentration of fluorescently labeled antigen. Any antigen present in the sample will displace an equal amount of labeled antigen from the antibody. Capillary electrophoresis separates the free antigen (both labeled and unlabelled) from the antibody–antigen complex based on the charge-to-mass ratio. Laser-induced fluorescence is used to quantify the labeled antigen whether it be bound or free, thus resulting in an electropherogram that can be related back to the concentration of antigen in the original sample. The entire procedure can be rapid (as little as 30 sec), requires minute sample volumes (around 50 nL), and has low detection limits (around 1 ng/mL) (108). A limitation in using these techniques is the requirement of high electrical field strengths on the order of 800 V/cm (103). This high field strength would disrupt cellular processes and so these assays must be performed in a cell-free environment.

Electronic Noses

Electronic noses are also being applied to monitor the composition of the head space or gas exit streams from bioreactors. Bachinger and coworkers (109) measured the off-gas composition from perfusion cultivations of a CHO cell line producing recombinant proteins. In general, this approach involves the use of multiple gas sensors that provide information on multiple analytes at one time. Often, neural networks are employed to analyze the data to search for patterns in the data. For example, the CO_2 evolution rate may be strongly correlated to the production of CH_4 or other volatile compounds. This correlation and the proportion of specific compounds produced can be specific for certain types of organisms and can be predictive of the energy state of the culture.

Cell Density

Maximizing the cell density through increases of the growth rate and cell yield is often one of the hallmarks of a successful mammalian cell culture process. Ideally, cell density measurements should be accomplished inside the bioreactor without the need to remove a cell sample and perform microscope hematocytometer counts. Several methods have been developed in recent years to address this.

The concentration or cell density present in a bioreactor can be quantified through a number of direct or indirect means. Indirect methods assume some constant value of the cellular metabolic rate per viable cell and thus quantify the consumption of certain analytes such as glucose or oxygen, by the production of lactate, or reduction of the pH. Such approaches provide only minimal accuracy as the cellular metabolism can change over the course of a bioreactor run.

The oxygen uptake rate (110), glucose uptake rate, and the rate of waste generation have been applied. A major limitation of these methods is the reliance on the assumption of constant cellular metabolic rates. Time-independent metabolic models have been used to circumvent this difficulty by correlating the specific metabolic rate to the limiting substrate concentration, providing a continuous means to determine the specific uptake rate from online measurement of the limiting substrate. An advantage of this approach is that the method can be applied to determine online cell concentration in both freely suspended and immobilized cell cultures.

Direct methods of cell quantification are significantly more accurate and more robust. These can include spectroscopic measurements, turbidity measurements, and optical density measurements. Unfortunately, these measures do not provide quantification of the number or percentage of viable cells, but rather return total cell numbers. Cells do strongly absorb light at 260 nm, but, this wavelength is also strongly absorbed by proteins. IR measurements of cell density have been applied with good success (111). A variety of optical density probes have been evaluated for use in online monitoring of cell densities in bioreactors (111).

Online optical cell density probes have been used to continuously monitor the cell densities in mammalian cell bioreactors and to achieve advanced bioreactor controls. Backscattering probes were found to produce the most linear response with respect to the concentration of hybridoma cells whereas transmission probes had a lower resolution (112). Fouling of the probe surface due to attachment of cells is always of concern. Wu and coworkers (112) found little fouling after 2 weeks of operation, but the presence of attached debris became apparent after 3 weeks. The impact of fouling on these measurements could be removed. A laser turbidity sensor has been used to monitor online the cell concentration in batch hybridoma cultivation with reasonably good correlation between offline cell counts and the sensor signal (113). Marose and coworkers (114) recently reviewed developments in optical-density probes, in situ microscopy, optical biosensors, and fiber-optic sensors for bioprocess monitoring.

Dielectric permittivity and electrical impendence spectroscopy have been used to quantify viable biomass for mammalian, yeast, and bacteria cells (115–120) with measurements being sensitive to changes in cell volume and to changes in the viable cell population.

Cell density, or biomass content, can also be quantified through spectroscopic methods. Sivakesava and coworkers (121) compared FT-MIR spectroscopy and FT-Raman spectroscopy techniques to quantify *Saccharomyces cerevisiae* during ethanol fermentation. They report that FT-MIR was more successful because the Raman scattering of the cultures was weak, thus leading to lower useful signals.

ASEPTIC SAMPLING

As animal cell cultures require strict maintenance of sterility, all sensing and measurement must be undertaken within this constraint. Practically, this requires either that an aseptic sampling method be designed or that the sensors be sterilizable by steam of chemical methods. Long-term stability of the sensor helps to minimize the need for frequent replacement and sterilization. In general, it is preferable that the monitoring of a bioreactor be performed in situ, without the need to remove samples. Typically, such measurements can be performed with less operator intervention, do not consume culture medium or cells, and do not increase the risk for introducing contaminants. However, monitoring schemes for many desired process variables cannot be quantified in situ, thus requiring aseptic sample removal. Examples of ex situ measurements include FIA, HPLC, and GC analyses performed at a location separate from the bioreactor environment (122).

Aseptic sampling from liquid systems includes nonmembrane filters, dialysis membranes, spin filters, and ultrafiltration membranes (123–125). Rotating membrane samplers have been shown to provide better filtration performances than stationary samplers due primarily to a lower propensity to clog with cells or cell debris (126). Once collected from a bioreactor, it is preferable that the sample be transported directly to a sensing device. If allowed to incubate outside of a bioreactor for any significant time it may permit the growth of unwanted microorganisms that would skew measurements of pH, nutrients, or wastes.

An ideal situation, if a sample must be removed from the bioreactor, is to employ an autosampler, possibly consisting of a peristaltic pump, an interval timer, and a fraction collector (123,124). The autosampler repetitively removes samples of user-determined size at repetitive time intervals and injects them into the sensing device. Such an approach requires less operator intervention, thus permitting a greater degree of automatic control of the process. However, such automatic sampling often does increase the chance of introducing culture contamination and for small-scale processes the volume of material removed for measurement can be substantial compared to the total bioreactor volume. Due to these limitations, it is recommended that autosampling be applied preferentially for bioreactors for which the frequent (i.e., hourly) measurement of culture parameters can lead to a direct increase in process performance.

Acoustic filtration has also been used to retain cells within a bioreactor while still permitting frequent sampling or use of high medium perfusion rates (127). High media flow rates or high cell concentrations can significantly reduce the separation performance. The relative merits and limitations of other methods to retain cells and permit bioreactor sampling have been reviewed by Woodside and coworkers (128).

ESTIMATION OF RATES AND METABOLIC RATIOS FROM ONLINE MEASUREMENTS

Online measurements can provide a number of valuable metrics of the health and productivity of a culture.

Oxygen Uptake Rate

The oxygen uptake rate (OUR) is a good indicator of cellular activity, and even under some conditions a good indicator of the number of viable cells. Hassan and coworkers (53) report a comparison of the viable cell counts, packed cell volume, intracellular nucleotide ratios, cell cycle analysis, online OURs, and optical density for the prediction of the end of exponential growth to optimize transfer times during scale-up of CHO cell cultures. Viable cell concentration, packed cell volume, and relative abundance of cells in S-phase were not very reliable at determining the end of exponential growth during the process; however, online determination of OUR and offline determination of intracellular nucleotide ratios (U-ratio) were very sensitive to changes in growth rate, enabling clear determination of the end of exponential growth within a short time. Optical density showed an inflection along with OUR and U-ratio but was less sensitive in determining the end of exponential growth (53).

Methods to quantify the OUR uptake rate and the oxygen transfer rate (OTR) begin with a mass balance on oxygen within a bioreactor. This can be written as the accumulation being equal to the rate of transfer into the culture minus that which is consumed:

$$\frac{dC}{dt} = k_{\mathrm{L}}a(C^* - C) - \mathrm{OUR} \times \mathrm{X}$$

where $k_{\mathrm{L}}a$ is the mass transfer coefficient, C is the concentration of oxygen in solution, C^* is the equilibrium solubility of oxygen (oxygen saturation), and X is the cell density (cells/L). The units on OUR are typically (g $O_2/10^6$cells-time). The oxygen transfer rate is

$$\mathrm{OTR} = k_{\mathrm{L}}a(C^* - C)$$

The three most common methods to evaluate the delivery of oxygen to a culture involve different manipulations or boundary conditions of the mass balance. The accuracy of the OUR measurements and the required analytical devices are quite different from method to method (129). Note that it is advisable that control of DO concentrations be combined with pH control to avoid accumulation of CO_2 and to avoid stripping of volatile compounds from the medium.

Static Method

In the static method, the consumption of oxygen is assumed to be negligible as non-respiring (or dead) cells are employed, thus simplifying the mass balance (12). Oxygen is removed from the bioreactor and then resupplied. This method is also called static gassing out. Specifically, a culture of cells is inoculated into the bioreactor, allowed to grow to some typical density, and then are administered either a metabolic blocking agent or a toxin such as azide. Oxygen is removed from the head space by purging with nitrogen. Aeration is turned on at the typical gas flow rate and the impeller is operated at a typical level. The increase in the oxygen concentration is followed until oxygen saturation (C^*) is reached. The change of the DO concentration in time is described by:

$$\ln\left(1-\frac{C}{C^*}\right) = -k_{\mathrm{L}}a \times t$$

The slope of a plot of $\ln(1-C/C^*)$ versus time gives a slope of $-k_{\mathrm{L}}a$. Care must be taken to ensure that oxygen is not consumed within the bioreactor during measurement otherwise $k_{\mathrm{L}}a$ will be underestimated. The static method provides a relatively easy and rapid means for evaluating the potential delivery of oxygen to a culture.

Dynamic Method

In the dynamic method, also called dynamic gassing out, aeration to an active culture is briefly turned off and the unsteady-state mass balance of oxygen tracked (130). It is critical that the oxygen concentration not drop to such a low level that the culture is negatively impacted. Often this is defined as C_{crit} equal to the oxygen concentration for which the growth rate is 99% of its maximal level (12).

Figure 13 presents a schematic description of the dynamic method. Initially, the bioreactor contains a healthy and active culture. At some time, t_0, the aeration is turned off and the concentration of oxygen decreases due to that consumed by the cells. A short time later (much before the oxygen level reaches C_{crit}), the aeration is turned back on. It is very important that C not approach C_{crit} so that the rate of oxygen uptake is independent of the oxygen concentration. The DO concentration will steadily rise until it reaches the steady-state level (C_{s}). Prior to reaching the steady-state, both the DO concentration and the time are recorded twice (C_1 and

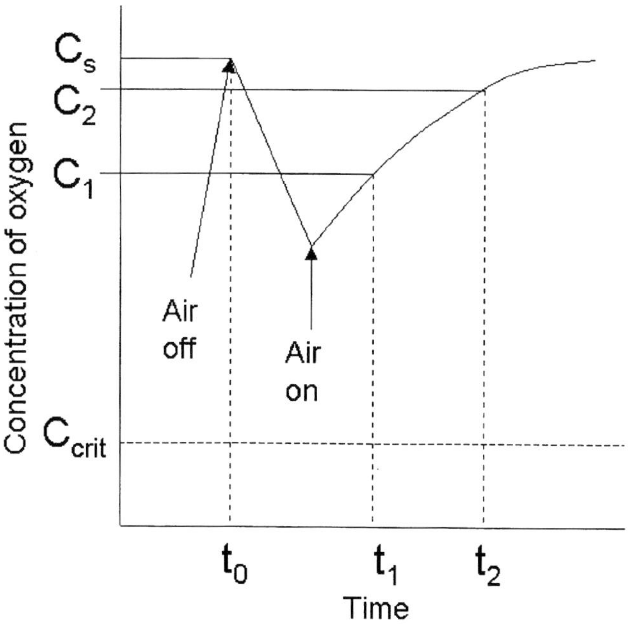

Figure 13 Dynamic method for oxygen transfer measurement as described in the text.

t_1; C_2 and t_2). Analysis begins with the oxygen mass balance evaluated at times representing the steady-state and the unsteady-state conditions. We require expressions for both the OUR and the OTR.

The OUR can be evaluated based on the steady-state condition which yields,

$$\text{OUR} \times X = k_{\text{L}}a(C^* - C_s)$$

where C_S reflects the steady-state DO concentration. This relation can be solved for $k_{\text{L}}a$ and substituted into the equation above to yield,

$$\frac{\text{d}c}{\text{d}t} = k_{\text{L}}a(C_s - C)$$

If $k_{\text{L}}a$ is constant over time, we can integrate the above relation, and using the boundary conditions of $C = C_1$ at $t = t_1$ and $C = C_2$ at $t = t_2$ we obtain,

$$k_{\text{L}}a = \frac{\ln\left(\frac{C_s - C_1}{C_s - C_2}\right)}{t_2 - t_1}$$

The OUR can then be calculated from the mass transfer coefficient, the cell density, the oxygen solubility and the steady-state oxygen concentration.

The dynamic method is widely used as it can be applied without the need to significantly alter the operation of a bioreactor. Errors can arise due to slow-responding DO electrodes and due to the comparatively small oxygen uptake rate of many animal cell cultures. Measurement time should be kept significantly shorter than the doubling time of the cells so as to avoid errors introduced by changing cell densities. Singh (131) described a computational method for the determination of oxygen uptake from the dynamic response of DO probe. For a hybridoma cell culture, the OUR for exponential growth in a standard serum-containing culture medium was 0.15 mM $O_2/10^9$cell-hr (131).

Oxygen Balance Method

A third method for quantifying oxygen transfer to a bioreactor is the oxygen balance method, which is based on an evaluation of the gas–liquid mass transfer (17). This approach requires measurement of the oxygen concentration of gas streams into and out of the bioreactor. A steady-state mass balance will yield the rate of oxygen transfer into the medium,

$$N_{\text{A}} = \frac{1}{RV}\left[\left(\frac{F_{\text{g}}P_{\text{A}}}{T}\right)_{\text{in}} - \left(\frac{F_{\text{g}}P_{\text{A}}}{T}\right)_{\text{out}}\right]$$

where N_{A} is the rate of mass transfer into the medium, R is the gas constant, V is the volume of liquid medium, F_{g} is the volumetric gas flow rate, T is the temperature, P_{A} is the oxygen partial pressure in the gas phase, and "in" and "out" refer to gas streams entering or leaving the bioreactor (130).

A primary difficulty in obtaining accurate measurements with this approach is that the partial pressure of oxygen in the streams entering or leaving the bioreactor will differ by only a small amount and thus need to be measured very accurately. The bioreactor must be operating at a steady state. Advantages of this method are that the mass transfer properties may be obtained from a single measurement and normal bioreactor operation need not be disturbed. Note that the oxygen concentration in the medium (C_s) will also necessarily be quantified so as to determine the mass transfer coefficient from:

$$k_La = \frac{N_A}{(C^* - C_s)}$$

Proper mixing and an even distribution of oxygen throughout the medium is required for an accurate measurement of C_s and hence for reliable calculations of k_La. It is advisable to measure the DO concentration at multiple locations within the bioreactor to ensure an accurate value for C_s (11). Good mixing within the bioreactor environment is critical.

Sulfite Oxygen

This method is based on the oxidation of sodium sulfite to sulfate by oxygen in the presence of a catalyst such as a divalent cation (12).This is a relatively fast reaction compared to oxygen transfer, and so is limited by the OTR. Unreacted sulfite can be removed from the bioreactor and quantified using a back titration with an iodine–starch reaction for quantification. The sulfite oxidation method yields higher k_La values than other methods and is influenced by pH and operating conditions in a difficult to characterize manner. For these reasons, this method is not highly recommended.

Measurement of OUR can be difficult due to the very low specific consumption rate ($\sim 2 \times 10^{13}$ mol cell/hr), the sensitivity of the cells to variations in DO concentration, and challenges in providing variable oxygen levels without damaging the cells (129). This delivery issue has been resolved by using a gas-permeable membrane to provide the required amounts of oxygen delivery directly into the culture which avoids foaming of increased liquid shear stresses due to gas bubbles in solution (132). Unfortunately, this approach is not easily scaled up. A mass balance on oxygen is applied to quantify the OUR. For CHO cells during batch and continuous culture OUR values of 2.85×10^{-13} and 2.54×10^{-13} mol O_2/cell hr, respectively (132). Changes in the OUR can be used to track transitions in the culture behavior including the end of exponential growth.

Fed-batch cultures were implemented to study the metabolism of HEK-293 cells (133).Glucose, measured every 30 mm by a FIA biosensor system, was maintained at 1 mM throughout the culture using an adaptive nonlinear controller based on minimal process modeling. Maintaining a low glucose concentration, thus decreasing the rate of glycolysis, significantly reduced lactate production.

The rates of glucose and glutamine uptake as well as the lactate and ammonia production were compared to those obtained in batch mode with an initial glucose concentration of 21 mM. Basically, three phases were observed in both culture modes. The metabolic shift from the first to the second phase was characterized by a significant reduction in glucose consumption and lactate production while maximum growth rate was maintained. The specific respiration rate appeared unchanged during the first two phases, suggesting that no change occurred in the oxidative pathway capacity. In the third phase, cell growth became slower very likely due to glutamine limitation.

A system for measuring the OUR based on an oxygen balance on the liquid phase was developed (132). A gas-permeable membrane would provide the required quantity of oxygen into the culture, while avoiding problems of foaming or shear stress generally linked to sparging. This aeration system allowed maintenance of a known and constant k_La value through cultures up to 400 hr. OUR was measured online for a CHO cell line was determined during batch growth and continuous culture as, respectively, equal to 2.85×10^{-13} and 2.54×10^{-13} mol O_2/cell hr.

Respiratory Quotient

Oxygen is a key substrate in animal cell metabolism and its consumption and the production of CO_2 are thus parameters of great interest for bioprocess monitoring and control. The respiratory quotient (RQ) for a culture is defined as the molar ratio of CO_2 formed to O_2 consumed (17). This parameter is often used to describe the relative efficiency of the cellular metabolism for a given set of process conditions. Until recent years, RQ had infrequently been used to evaluate animal cell metabolism primarily due to difficulties in measuring the CO_2 evolution rate (134). The oxygen uptake rate has traditionally been more readily quantified accurately.

An important complication in the determination of the CO_2 evolution rate is that standard culture media is buffered by bicarbonate, and so the CO_2 balance is affected by accumulation and therefore the RQ cannot directly be calculated from gas-phase measurements (135). A significant portion of the gas-phase CO_2 was found to evolve from the culture medium (134). The introduction or loss of CO_2 from the culture medium during preparation and sterilization must be taken into account. Modeling of liquid- and gas-phase mass balances and data filtering methods have been used to account for this CO_2 reservoir (134,135).

RQ values for mammalian cells grown under near optimal conditions are generally 1.0 ± 0.05 (134). In general, an increase in RQ has been shown to correlate with an increase in cellular productivity and so RQ could serve as a means for estimating the physiological state of the cells.

RQ values for Sf-9 insect cells have been quantified using online gas-phase O_2 measurements and IR CO_2 measurements (136). Linear relationships between viable cell densities and both oxygen uptake rate and carbon dioxide evolution rate were obtained in exponentially growing cultures. The extent of the increase in CER following infection and the time postinfection at which maximum CER was attained were negatively correlated with the multiplicity of infection at multiplicities below the level required to infect all the cells in a culture. The relative permittivity, carbon dioxide evolution rate, and the cell volume profiles were closely matched during the growth phase of cells when grown in a batch or fed-batch culture (137). The relationship became more complex when the cultures were either in stationary phase, or in the postinfection phase.

Metabolic Rates (Glucose Consumption/Lactate Production)

An important question in bioprocess modeling is how many rate equations must be specified to fully account for all relevant biological processes (138). The number of rate equations for which practical information can be obtained is constrained by the yield equations, which represent the balances of reducing power, energy in the form of ATP, and the various elements involved in cell metabolism. These balances are derived from a simplified picture that divides metabolism into catabolic, anabolic, respiratory, and product formation pathways. In most practical situations, a limited number of metabolic ratios are monitored and used to implement process decisions. The number of rate measurements available versus the number needed defines the state estimation problem in bioprocess control.

Stoichiometric ratios of lactate yield from glucose (Lac/Glc) and ammonium yield from glutamine (Amm/Gln) are often applied to monitor the health and productivity of a cell culture. Zeng (139) evaluated several other stoichiometric ratios: ammonium yield from the total consumption of amino acids (NH_4^+/TAA), consumption of total amino acids to glutamine (TAA/Gln), essential amino acids to glutamine (EAA/G1n), glutamine to glucose (Gln/Glc), and oxygen to glucose (OUR/Glc). Several commonly employed animal cell lines (hybridoma, BHK, and CHO) demonstrated similar patterns of variation of stoichiometry. In continuous culture, Lac/Glc and Gln/Glc are primarily determined by the residual glucose concentration while TAA/Gln and EAA/Gln correlate well with the residual glutamine concentration (139). Ammonium formation not only is a function of glutamine concentration but also is affected by the consumption of other amino acids, particularly at low residual glutamine concentrations. NH_4^+/TAA turned out to be a more suitable parameter to describe the ammonium formation. Thus, these stoichiometric ratios could be used as a means to predict and eventually control the concentrations of nutrients that are otherwise difficult to determine online.

Metabolic flux analysis is a useful tool for unraveling relationships between metabolism and cell function. Material balancing can provide estimates of major metabolic pathway fluxes, provided all significant metabolite uptake and production rates are measured (140). Many serum-free media formulations contain small amounts of yeast extracts and plant or animal tissue hydrolysates consisting of metabolizable materials that are not easily identified or quantified. For chemostat steady states of CHO cultures grown in a steady-state chemostat culture with hydrolysate-supplemented medium, consistent flux analyses were obtained only when amino acids liberated from protein supplements were taken into account (140).

Since cell growth is accompanied by an enthalpy change, heat dissipation quantified by calorimetry can serve as an index for the cellular metabolic rate (141). Such measurements need to be performed in conjunction with the viable cell concentration measurements by an approach such as quantification of the dielectric. The ratio of the two signals gives scalar heat flux (142). Comparison of heat flux with glucose and glutamine fluxes indicated that the former most accurately reflected decreased metabolic activity. The set of stoichiometric coefficients in the reaction were related through the extent of reaction to overall metabolic activity. This approach can be used for medium optimization based on enrichment of amino-acids to improve cell growth while decreasing catabolic fluxes.

APPLICATION OF ONLINE MONITORING OF THE CELL ENVIRONMENT

Obtaining satisfactory performance from a cell culture requires maintenance of operating conditions at certain set points. Due to unpredictable process upsets, such as variable pumping rates, temperature fluctuations, or altered cell productivity, process parameters invariably differ from one run to the next. In some cases, such fluctuations can be predicted through previous experience with the process or through mathematical modeling of the process. Proper use of instrumentation requires a process control algorithm to determine the requisite response to process offsets.

For example, animal cells cultivated in batch systems rapidly exhaust the supply of glucose and glutamine and generate ammonia and lactate as metabolic byproducts. Several approaches have been used to extend the supply of nutrients, increase culture longevity, and maximize the final product concentration. Continuous and perfusion culture systems replace fresh medium into the reactor while removing used or spent medium, thus avoiding metabolite limitation and product inhibition. Such methods also dilute the products generated, which extends the purification steps required to obtain a concentrated product. Fortified media with abnormally high glucose and glutamine concentrations have been used in batch cultivations; however, such methods often do not substantially improve the cell density or production as the levels of ammonia and lactate increase, thus inhibiting cell productivity (143).

An ideal approach for maintaining extended productivity is to add nutrients to the bioreactor only as needed by the cells. Highly concentrated solutions of glucose, glutamine, and amino acids added slowly to a bioreactor have been shown to improve product concentrations by two- to fourfold over standard batch conditions (143–146). The composition of medium supplements should be based on the stoichiometric demands of the cells as determined by analysis of the spent culture medium. Such measurements should be performed rapidly so that the medium supplement matches the current cellular requirements.

Bioreactor control has been an active area of research and has attracted more attention in recent years. This is due to the new developments in related areas that can be exploited to overcome the inherent difficulties in bioreactor control. Beginning with conventional regulatory control of operating variables such as temperature, pH and DO concentration, research in bioreactor control has undergone significant changes including neural network based approaches. Rani and Rao (147) provide a summary of recent developments in the control of batch, fed-batch, and continuous bioreactors.

Effective control requires that the characteristic times of the physical system and the control system be less than the characteristic times of the biological system. In most cases, the limiting physical characteristics are heat and oxygen mass transfer within the culture medium. These can be manipulated through changes in the power supplied to the impeller, in the impeller design, and in the bioreactor size and geometry. For most animal cell cultures, the cells maintain a doubling time on the order of 16–24 h, and typically the cell response to a process alteration is much slower than that observed with cultured microbes. The response time of the sensors themselves also need to be taken into account when developing a monitoring and control scheme.

A typical monitoring and control scheme includes the following steps (148):

1. Observe or measure one or more process variables.
2. Compare the observed variable to a desired set point.
3. Decide what course of action is to be taken.
4. Act on the decision.

Important factors for the development of a monitoring and control scheme for a bioprocess include (148):

1. The rate at which a change can be effected in a process is dependent on the size of the system.
2. Sensors for many environmental variables that may impact the process are not available or may not be practical for every system.
3. Some changes can be made in only one direction. For example, one can readily increase the concentration of glucose in a bioreactor, but reducing glucose is only feasible by diluting the entire culture or by waiting for the cells to take action.

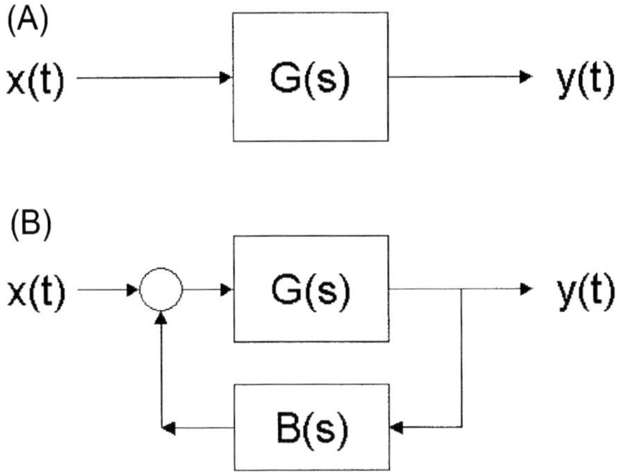

Figure 14 Schematic of (A) open loop control, (B) closed loop control. Open loop control is significantly easier to implement, but requires substantial knowledge of a process. Closed loop requires frequent monitoring of the process. x(t) represents a process input, G(s) is the process, and y(t) is the process output. B(s) represents the monitoring mechanism that feeds into a comparator and decision support component.

4. Many process changes that may appear to be critical for basic scientific investigation may have little impact at the production level.
5. The high degree of uniformity that may be achieved in a small laboratory scale bioreactor usually cannot be reproduced in large production vessels.

The type of control action may be either a closed loop, or feedback control, in which a correction is enacted based on observable characteristics of the process, or may follow an open loop control in which process modifications are made at predetermined times and levels. Open loop control is significantly more difficult to implement correctly for many biological processes and requires either an accurate model of the process or substantial experience in running the process. Open loop control is sometimes called preprogrammed control in that process modifications are determined prior to start of the process. For this approach to work well, the process must be highly characterized and relatively invariant from one run to the next. For example, an operator may know from experience with an individual process that over 24 hr the cell density will double, thus necessitating an increase in the impeller rotation to increase oxygen transfer. Open loop control is easy to implement and is often applied in laboratory situations; however, it is not practical for most industrial-scale processes or for processes with substantial variability. Closed loop control involves modifying one more or process parameters based on feedback obtained from sensors monitoring the process. A schematic of these processes is presented in Fig. 14.

Fed-Batch Systems

Mammalian cells have the ability to proliferate under a variety of nutrient environments by utilizing different combinations of nutrients, especially glucose and the amino acids. Under conditions often used in in vitro cultivation, the cells consume nutrients in great excess of what is needed for construction of biomass and desired products. They also produce large amounts of metabolic wastes with lactate, ammonia, and some nonessential amino acids such as alanine as the most prominent ones. By controlling glucose and glutamine at low levels, cellular metabolism can be altered and can result in reduced glucose and glutamine consumption as well as in reduced metabolite formation.

A research group at the University of Minnesota has made tremendous strides in applying aggressive control to improve MAb production. Using a fed-batch reactor to manipulate glucose at a low level (as compared to a typical batch culture), cell metabolism was altered to a state with substantially reduced lactate production (149). The culture was then switched to a continuous mode and allowed to reach a steady state. At this steady state, the concentrations of cells and MAb were substantially higher than a control culture that was initiated from a batch culture without first altering cellular metabolism. The lactate and other metabolite concentrations were also substantially reduced as compared to the control culture. This newly observed steady state was achieved at the same dilution rate and feed medium as the control culture. The paths leading to the two steady states, however, were different, and so these results demonstrate steady-state multiplicity. At this new steady state, not only was glucose metabolism altered, but the metabolism of amino acids was altered as well. The amino acid metabolism in the new steady state was more balanced, and the excretion of nonessential amino acids and ammonia was substantially lower. This approach of reaching a more desirable steady state with higher concentrations of cells and product opens a new avenue for high-density- and high-productivity-cell culture.

This and similar approaches target the ideal bioreactor operation in which metabolites are added to the bioreactor only as needed by the cells as determined by an analysis of the culture medium and the stoichiometric demands of the cells. This is determined either through use of an online monitoring scheme, or more commonly, through offline analyses of culture medium constituents. Measurements should be performed rapidly so that the medium supplement matches the current cellular requirements. In most cases, decisions to add individual nutrients or medium supplements are done ad hoc by a bioprocess engineer, although such process decisions are likely to be automated in the near future.

The addition of highly concentrated solutions of glucose, glutamine, and amino acids can substantially improve product concentrations. Increases in product concentrations were observed by two- to fourfold over standard batch conditions (143–145). Oh and coworkers (150) developed an interactive system for controlled feeding of glucose and glutamine to hybridoma cells so as to maximize antibody production.

A more sophisticated control scheme applies concentrated solutions of key nutrient components fed periodically to cell cultures using a simple feeding control strategy based on the integral of viable cell concentrations over time (151). This approach assumed constant specific nutrient consumption rates. Through effective nutritional control, both cell growth phase and culture lifetime were prolonged significantly. These studies suggest close relationships among nutrient depletion, cell metabolism transition, and cell death.

Fed-batch cultures of HEK-293 cells were monitored using an FIA glucose measurement and controlled at a concentration of 1 mM using an adaptive nonlinear controller based on minimal process modeling (133). Lactate production was significantly reduced by maintaining a low glucose concentration, thus decreasing the rate of glycolysis.

A number of schemes have been applied to minimize the instantaneous concentration of inhibiting metabolites present in a culture so as to increase the yield of product in fed-batch bioreactors. Schwabe and coworkers (152) applied a balanced supply of substrates from concentrated feed solutions based on offline estimated oxygen and glucose uptake rates. While implementation of this control scheme increased product concentration threefold, inhibition by ammonia could not be avoided. To address this difficulty, a dialysis membrane was used to remove ammonia and in which a concentrated medium was fed to the cells and inhibitory metabolites removed into a buffer solution. This improved approach led to a 10-fold increase of the product concentration compared with batch cultures (152).

Perfusion Systems

Continuous and perfusion culture systems replace fresh medium to the reactor and remove spent medium, thus avoiding metabolite limitation and product inhibition. Such methods also dilute products, thus extending purification steps required to obtain a concentrated product. The key parameter in perfusion culture is the rate of medium replacement (*D*) and this is often based on the required rate for feeding a single nutrient such as glucose or glutamine. Increases in this dilution rate can lead to an increase in the viable cell density, X_v, and to increases in the cell growth rate (153). Such an approach has the added advantage of improving the amount of product generated without the need for increasing the bioreactor scale, provided that both viable cell yield per perfusion rate and specific cellular productivity remain constant at higher *D*.

Significant reductions in cellular productivity have been observed at high dilution rates due to the high gas sparging rate needed to meet the oxygen demand (153). Apparently, increases in hydrodynamic shear stress imparted to the culture via intensification of gas sparging resulted in a gradual increase in specific glucose consumption and lactate production rates, while no variations were observed in glutamine-consumption rates. As a result, while glutamine was the sole limiting-nutrient under nonsparging conditions, both glutamine and glucose became limiting under sparging conditions (153).

Alternatively, the dilution rate does not need to be a constant, but can be manipulated based on the feedback obtained through online measurements of nutrients, wastes, cell density, or cell viability. Controlled feeding of nutrient supplements to a perfused hybridoma cell culture was used to enhance monoclonal antibody productivity (154). This controlled-fed perfusion approach significantly increased the volumetric antibody productivity by nearly twofold over the perfusion process, and surpassed fed-batch and batch processes by almost 10-fold. This increase in productivity can be attributed to both increased cell density as well as reduced product dilution.

Perfusion cultures of CHO cells have been performed using an acoustic filtration system to retain the cells within the bioreactor (127). High packed cell volumes could be maintained without additional bleeding off of cells. Perfusion of up to 50 days was performed without loss of performance of the acoustic filter. Using the protease-sensitive product rhesus thrombopoietin, cultivation in perfusion mode drastically reduced proteolysis when compared to a batch culture without addition of protease inhibitors such as leupeptin (127).

If high perfusion rates are to be employed, cells are often immobilized within a rigid support material in order to avoid cell washout while still permitting removal of wastes. A disadvantage of this system is that the cell density cannot be directly monitored. One group has been able to obtain high cell density of immobilized hybridomas in a bubble-column bioreactor filled with hollow glass cylinders (155).

The parameters monitored during the cultivation were pH, temperature, DO, glucose, lactate, and monoclonal antibody. The glucose uptake rate was used to estimate the cell concentration along the time using offline measurement techniques.

Feed-Forward Control Using Mathematical Relations

Prior to designing a control strategy for a biological system, one typically constructs a simplified process model that is used to provide insight into the process dynamics. This model is based on conservation of mass and energy which may be approximated by linear, time-dependent equations with the variables that are to be controlled taken as the dependent variables. The variables that are to be adjusted so as to keep the controlled variable at its set point are the manipulated variables. For example, a mass balance on glucose in a continuous bioreactor may be as follows:

$$V\frac{\mathrm{d}C(t)}{\mathrm{d}t} = F[C_i(t) - C(t)] - \frac{FX}{Y_{X/S}}$$

where V is the volume of the bioreactor, $C(t)$ is the time-dependent concentration of glucose, $C_i(t)$ is the concentration of glucose in the feed, F is the volumetric flow rate, X is the cell density, and $Y_{x/s}$ is the yield of cells from glucose. The flow rate and feed glucose concentrations are variables which are manipulated so as to maintain the control variable, the glucose concentration in the bioreactor, at its set point.

A difficulty in applying traditional reactor control schemes developed from the chemical process industry is that even for mature bioprocesses, fundamental understanding of cell growth and metabolism remains quite limited. We are far from developing biological control models based on first principles; rather, most models remain highly empirical. The field of metabolic engineering has great promise in bridging this gap, but application of such methods directly to process control remains a challenge addressed by a small number of investigators (156,157). The problem in developing fundamental process control models is exacerbated by the highly nonlinear behavior of cell culture, the highly interactive nature of critical process variables, and the batch or fed-batch nature of most commercial processes (148).

Bioreactor control schemes are often based on one of two approaches (146). The first relies on a biological model that characterizes the cellular nutritional requirements. The control scheme attempts to match these target requirements based on the known input of cells and metabolites to the bioreactor, the expected cell growth, and the expected decrease in nutrients through metabolism and spontaneous degradation. Glacken and coworkers (34) showed that maintenance of glucose and glutamine concentrations based on a metabolic model improved monoclonal antibody production 10-fold over batch cultivations. This approach is limited by the need for kinetic models that accurately describe cell growth and metabolism. The second control approach requires frequent measurements of the composition of the growth medium which are used to adjust the supply of metabolites added to the bioreactor to minimize the production of wastes. This approach is limited primarily by the availability of rapid measurement schemes that can quantify multiple species.

Conventional controller designs employ Laplace transforms and so require the use of linear equations (146). Biological processes are inherently nonlinear and have a multimodal character (158,159). Linear approximations are often applied over short time frames and for limited operating conditions. Most proposed methods fail to find the global optima for control of such processes (159). Even if a globally optimal control protocol can be designed for a specific process, in many instances the implementation of the control scheme presents an even more significant challenge than manual control.

Novel Strategies for Manipulating Cellular Activity

In recent years, the development of advanced systems for bioprocess monitoring and control has become an area of intensive research. Along with traditional techniques, there are several new approaches that are increasingly being applied to bioprocess operations. A detailed description of control strategies is beyond the scope of this chapter. Several examples of these novel approaches are summarized below.

1. Expert systems—Expert systems utilize an extensive database of knowledge about a specific process and how it has performed historically. This has been used to develop process control and bioreactor state prediction based on this extensive knowledge (160).
2. Genetic algorithms—Roubos and coworkers (161) applied an evolutionary program, based on a genetic algorithm to calculate optimal control policies for bioreactors. This genetic algorithm (GA) is used as a nonlinear optimizer in combination with simulation software and constraint handling procedures. A GA searches parameter space by manipulating, cross-breading, and translocating "genes" which represent individual parameter settings.
3. Stochastic optimization—Stochastic optimization has been proposed as a reliable alternative to designing a control scheme (162). These stochastic algorithms are used to successfully solve case studies taken from the recent literature. The advantages of these alternative techniques include ease of implementation, global convergence, and good computational efficiency.
4. Ant colony algorithm—The ant colony algorithm, which mimics the cooperative search behavior of ants in real life, has been employed for the dynamic optimization of fed-batch bioreactors (163). The algorithm rapidly converges to optimal feed rate profiles, which maximize the overall production of the desired product and the profits in a computationally efficient and robust manner. The optimal profiles evolved are easy to implement in plant operation.

Other more traditional predictive control schemes have been applied to determine the required medium feed rates for perfusion bioreactors (164). Adaptive software routines were developed to estimate the current and predict the future glucose uptake and lactate production of hybridoma cell cultures in hollow fiber systems. The current and future glucose uptake rates were used to select the perfusion feed rate in a designed response to deviations from the set point values. The use of the predictive controller routine decreased the glucose and lactate concentration variances up to sevenfold, and antibody yields increased by 10–43% (164).

The dynamic open-loop control of bioprocesses has been studied by a number of groups (165). Development of suitable control schemes can be performed by using a control vector parameterization concept, which makes use of second-order sensitivities to obtain exact gradients for the objective function of the underlying dynamic process model. Extensions of this method can result in efficient methodologies for solving general dynamic optimization problems, even for high levels of control discretization as typical of most bioprocesses.

Wang and Cheng (166) claim to have developed an efficient method for simultaneously determining optimal feeding rates and operation parameters for cell cultures based on a finite dimensional optimization using the control parameterization technique. The optimal production rate obtained by the simultaneous optimization approach could be significantly improved with comparison to a simplified optimization problem.

One research group has hypothesized that inaccuracies in process optimization can be traced back to limitations in the process models applied (167). This group formulated a control algorithm based on cell mass and lactate production and calculated the process–model mismatch at each sampling time. These deviations were used to reoptimize the substrate concentrations throughout the length of the cultivation. The cell mass produced using dynamic optimization was compared to the cell mass produced for a nonoptimized case, and for a one-time optimization at the beginning of the batch. A single offline optimization of substrate concentration at the start of the batch increased the yield of cell mass by 27% over a nonoptimized fermentation, whereas frequent reoptimization increased yield of cell mass per batch by 44% compared to the single offline optimization. Monoclonal antibody productivities also tracked with cell densities, thus suggesting that frequent revision of control parameters can improve performance of bioreactors (167).

An issue that is being recognized as of tremendous concern but is only recently addressed is the role process changes have in altering not just the concentration of the desired product but also the functionality and activity of the product. Moran and coworkers (168) developed an approach to validate ranges of control parameters for a cell culture process producing a monoclonal antibody. Specifically, the structure and functional activity of a monoclonal antibody produced at the numerical boundaries of fed-batch culture control parameters were examined with the goal of assuring that antibody produced under varying culture conditions was of consistent quality. All antibody preparations were identical to each other and to the current antibody reference standard or control. Glycosylation analysis of certain samples from the study demonstrated that the distribution of glycoforms of the antibody was not affected by the varying process control conditions of the fed-batch cultures.

Process efficiency can be strongly influenced by the cellular state that should be monitored, interpreted, and controlled. In most control systems the cellular state is not explicitly considered, rather control schemes are based on the implicit assumptions that the growth environment will directly map to the process efficiency. This limitation is well realized (57) and explicit monitoring and control of cellular physiology are considered to be among the most challenging tasks of modern bioprocess engineering. For example, the onset of widespread apoptotic (gene-directed) cell death may be difficult to predict but may have a sizeable impact on culture productivity. The physiological state of a culture can be composed of several process variables including specific metabolic rates, metabolic rate ratios, and others (46). The real-time monitoring of many of these is possible using commercial sensors and has been described above. This approach reduces the goal of a control scheme to maintaining the physiological state of the cell as close as possible to an "ideal" trajectory, providing maximum efficiency for the culture as a whole. For a detailed description of this approach see Konstantinov (46).

SUMMARY

The application of cell culture technology to the production of high-value products has benefited greatly from advances in process monitoring and control. Spectroscopic methods that provide noninvasive information on the cell environment and on cell physiology are likely to have substantial impacts on process development. The methods of IR spectroscopy and Raman spectroscopy in addition to the new approach of near-field scanning optical microscopy have the potential to provide subcellular information that may be of tremendous use to bioprocess development. Variations in the forces and chemical components encountered by an individual cell within a bioreactor will also be the target of future monitoring and control methodologies. While many new technologies have been developed over the past 5–10 years, integration into existing processes has been slow. Advances in the near future will likely evolve from innovative designs that incorporate monitoring of the culture environment from the cellular level to the entire process.

REFERENCES

1. Hu WS, Aunins JG. Large-scale mammalian cell culture. Curr Opin Biotechnol 1997; 8:148–153.
2. Ritzka A, Ulber R, Hitzmann B, Scheper T. Control of bioprocesses with neural networks. Chemie Ingenieur Technik 2002; 74:149–152.
3. Vaidyanathan S, Macaloney G, Vaughn J, McNeil, B, Harvey, LM.Monitoring of submerged bioprocesses. Crit Rev Biotechnol 1999; 19:277–316.
4. Scragg AH. Bioreactors in Biotechnology. New York: Ellis Horwood, 1991.
5. Mulchandani A, Bassi AS. Principles and applications of biosensors for bioprocess monitoring and control. Crit Rev Biotechnol 1995; 15:105–124.
6. Olsson L, Schulze U, Nielsen J. On-line bioprocess monitoring—an academic discipline or an industrial tool? Trends Anal Chem 1998; 17:88–95.
7. Hartnett T. Instrumentation and control of bioprocesses. In: Lydersen BK, D'Elia NA, Nelson KL, eds. Bioprocess Engineering: Systems, Equipment, and Facilities. New York: Wiley & Sons, 1994.
8. Seborg DE, Edgar TF, Mellichamp DA. Process Dynamics and Control. New York: Wiley and Sons, 1989.
9. Fleischaker RJ, Sinskey AJ. Oxygen-demand and supply in cell culture. Eur J Appl Microbiol 1981; 12:193–197.
10. Riley MR, Muzzio FJ, Reyes SC. Effect of oxygen limitations on monoclonal antibody production by immobilized hybridoma cells. Biotechnol Prog 1997; 13:301–310.
11. Jorjani P, Ozturk SS. Effects of cell density and temperature on oxygen consumption rate for different mammalian cell lines. Biotechnol Bioeng 1999; 64:349–356.
12. Hesselink PGM. Operational Modes of Bioreactors. Chapter 2. Oxford: Butterworth-Heinemann, 1992.
13. White SF, Turner APF, Bilitewski U, Bradley J, Schmid RD. On-line monitoring of glucose, glutamate, and glutamine during mammalian cell cultivations. Biosens Bioelectron 1995; 10:543–551.
14. Chuang H, Arnold MA. Radioluminescent light source for optical oxygen sensors. Anal Chem 1997; 69:1899–1903.
15. Reece JS, Miller MJ, Arnold MA, Waterhouse C, Delaplaine T, Cohn L, Cannon T. Continuous oxygen monitoring of mammalian cell growth on space shuttle mission STS-93 with a novel radioluminescent oxygen sensor. Appl Biochem Biotechnol 2003; 104:1–22.
16. Mukhopadhyay SN, Das DK. Oxygen Responses, Reactivities, and Measurements in Biosystems. Boca Raton, FL: CRC Press, 1994.

17. Bailey JE, Ollis DF. Biochemical Engineering Fundamentals. 2nd ed. New York: McGraw Hill, 1986.
18. Pattison RN, Swamy J, Mendenhall B, Hwang C, Frohlich BT. Measurement and control of dissolved carbon dioxide in mammalian cell culture processes using an in situ fiber optic chemical sensor. Biotechnol Prog 2000; 16:769–774.
19. Bachinger T, Riese U, Eriksson RK, Mandenius CF. Gas sensor arrays for early detection of infection in mammalian cell culture. Biosens Bioelectron 2002; 17(5):395–403.
20. Mills A, Eaton K. Optical sensors for carbon dioxide: an overview of sensing strategies past and present. Quim Anal 2000; 19:75–86.
21. Neurauter G, Klimant I, Wolfbeis OS. Microsecond lifetime-based optical carbon dioxide sensor using luminescence resonance energy transfer. Anal Chim Acta 1999; 382: 67–75.
22. Marazuela MD, Moreno-Bondi MC, Orellana G. Luminescence lifetime quenching of a ruthenium(II) polypyridyl dye for optical sensing of carbon dioxide. Appl Spectrosc 1998; 52:1314–1320.
23. Chang Q, Randers-Eichhorn L, Lakowicz JR, Rao G. Steam-sterilizable, fluorescence lifetime-based sensing film for dissolved carbon dioxide. Biotechnol Prog 1998; 14:326–331.
24. Uttamlal M, Walt DR. A fiberoptic carbon-dioxide sensor for fermentation monitoring. Biotechnology 1995; 13:597–601.
25. Zhang GJ, Lui JF, Yuan M. Novel carbon dioxide gas sensor based on infrared absorption. Opt Eng 2000; 39:2235–2240.
26. Wolfbeis OS. Fiber optic chemical sensors and biosensors. Anal Chem 2000; 72: 81R–89R.
27. Evans A. Potentiometry and Ion Selective Electrodes. London: Wiley, 1987.
28. Sotomayor MDT, DePaoli MA, deOliveira WA. Fiber-optic pH sensor based on poly(*o*-methoxyaniline). Anal Chim Acta 1997; 353:275–280.
29. Lin J. Recent development and applications of optical and fiber-optic pH sensors. Trends Anal Chem 2000; 19:541–552.
30. Langheinrich C, Nienow AW. Control of pH in large-scale, free suspension animal cell bioreactors: alkali addition and pH excursions. Biotechnol Bioeng 1999: 66:171–179.
31. Lamberto DJ, Alvarez MM, Muzzio FJ. Experimental and computational investigation of the laminar flow structure in a stirred tank. Chem Eng Sci 1999; 54:919–930.
32. Mavros P. Flow visualization in stirred vessels—a review of experimental techniques. Chem Eng Res Des 2001; 79:113–127.
33. Ozturk SS, Riley MR, Palsson BO. Effects of ammonia and lactate on hybridoma growth, metabolism, and antibody production. Biotechnol Bioeng 1992; 39:418–431.
34. Glacken MW, Huang C, Sinskey AJ. Mathematical descriptions of hybridoma culture kinetics. III. Simulation of fed-batch bioreactors. J Biotechnol 1989; 10:39–66.
35. Baker KN, Rendall MH, Patel A, Boyd P, Hoare M, Freedman RB, James DC. Rapid monitoring of recombinant protein products: a comparison of current technologies. Trends Biotechnol 2002; 20(4):149–156.
36. Miller WM, Wilke CR, Blanch HW. Transient responses of hybridoma cells to nutrient additions in continuous culture: I. Glucose pulse and step changes. Biotechnol Bioeng 1989; 33:477–486.
37. Dalili M, Sayles GD, Ollis DF. Glutamine-limited batch hybridoma growth and antibody production: experiment and model. Biotechnol Bioeng 1990; 36:74–82.
38. Bibila TA, Flickinger MC. A model of interorganelle monoclonal antibody transport and secretion in mouse hybridoma cells. Biotechnol Bioeng 1991; 38:767–780.
39. Bonarius HPJ, Hatzimanikatis V, Meesters KPH, de Gooijer CD, Schmid G, Tramper J. Metabolic flux analysis of hybridoma cells in different culture media using mass balances. Biotechnol Bioeng 1996; 50:299–318.
40. Zeng AP. Mathematical modeling and analysis of monoclonal antibody production by hybridoma cells. Biotechnol Bioeng 1996; 50:238–247.

41. Suzuki E, Ollis DF. Enhanced antibody production at slowed growth rates: experimental demonstration and a simple structured model. Biotechnol Prog 1990; 6:231–236.
42. Ozturk SS, Thrift JC, Blackie JD, Naveh D. Real-time monitoring and control of glucose and lactate concentrations in a mammalian cell perfusion reactor. Biotechnol Bioeng 1997; 53:372–378.
43. Simpson NH, Singh RP, Perani A, Goldenzon C, Al-Rubeai M. In hybridoma cultures, deprivation of any single amino acid leads to apoptotic death, which is suppressed by the expression of the bcl-2 gene. Biotechnol Bioeng 1998; 59:90–98.
44. Anonymous, Sigma Chemical Catalog, St. Louis, MO, 2004.
45. Godia F, Cairo JJ. Metabolic engineering of animal cells. Bioprocess Biosyst Eng 2002; 24:289–298.
46. Konstantinov KB. Monitoring and control of the physiological state of cell cultures. Biotechnol Bioeng 1996; 52:271–289.
47. Guilbault GG. Analytical Uses of Immobilized Enzymes. New York: Marcel Dekker, 1984.
48. Valcarcel MJ, DeCastro M. Flow Injection Analysis Principles and Applications. New York; Wiley and Sons, 1987:329–357.
49. Okeson CD, Riley MR. Glutamine replenishment and ammonia removal in hybridoma cell cultures via immobilized glutamine synthetase. Biochem Eng J 2001; 9(2):125–133.
50. Ruzicka J, Hansen EH. Flow Injection Analysis. New York: Wiley and Sons, 1981: 6–20.
51. Schugerl K. Progress in monitoring, modeling and control of bioprocesses during the last 20 years. J Biotechnol 2001; 85:149–173.
52. Schugerl K, Hitzmann B, Jurgens H, Kullick T, Ulber R, Weigal B. Challenges in integrating biosensors and FIA for on-line monitoring and control. Trends Biotechnol 1996; 14:21–31.
53. Hassan SSM, Marei SA, Badr IH, Arida HA. Novel solid-state ammonium ion potentiometric sensor based on zirconium titanium phosphate ion exchanger. Anal Chim Acta 2001; 427:21–28.
54. Stoll TS, Ruffieux PA, vonStockar U, Marison IW. Development of an on-line control system for the cultivation of animal cells in a hollow-fiber reactor using flow injection analysis and a visual programming language. J Biotechnol 1996; 51:37–48.
55. Aminot A, Kerouel R, Birot D. A flow injection-fluorometric method for the determination of ammonium in fresh and saline waters with a view to in situ analyses. Water Res 2001; 35:1777–1785.
56. Meyerhoff ME, Trojanowicz M, Palsson BO. Simultaneous enzymatic/electrochemical determination of glucose and L-glutamine in hybridoma media by flow injection analysis. Biotechnol Bioeng 1993; 41:961.
57. Palmisano F, Rizzi R, Centonze D, Zambonin PG. Simultaneous monitoring of glucose and lactate by an interference and cross-talk free dual electrode amperometric biosensor based on electropolymerized thin films. Biosens Bioelectron 2000; 15:531–539.
58. Yabuki S, Mizutani F, Hirata Y. Glucose-sensing electrode based on glucose oxidase-attached polyion complex membrane containing peroxidase and ferrocene. Electroanalysis 2001; 13:380–383.
59. Kumar MA, Thakur MS, Senthuran A, Senthuran V, Karanth NG, Hatti-Kaul R, Mattiasson B. An automated flow injection analysis system for on-line monitoring of glucose and L-lactate during lactic acid fermentation in a recycle bioreactor. World J Microbiol Biotechnol 2001; 17:23–29.
60. Barrado E, Vela MH, Lapa RAS, Lima JLFC. An FIA-enzymatic method for the determination of L(+) lactate: application in parenteral solution analysis. Quim Anal 2000; 19:241–246.
61. Reinecke M, Scheper T. Fast online flow injection analysis system for IgG monitoring in bioprocesses. J Biotechnol 1997; 59:145–153.

62. Spear SK, Rhiel M, Murhammer DW, Arnold MA. Ammonia measurements in mammalian cell culture media with a diffuse reflectance-based fiberoptic ammonia sensor. Appl Biochem Biotechnol 1998; 75:175–186.
63. Nandakumar MP, Lali AM, Mattiasson B. On-line monitoring of glucose and/or lactate in a fermentation process using an expanded micro-bed flow injection analyzer. Bioseparation 1998; 8:229–235.
64. Reinecke M, Stephanopoulos G. Flow injection analysis for simultaneous quantification of prolactin concentration and glycosylation macroheterogeneity in cell culture samples. Cytotechnology 2000; 34:237–242.
65. Christensen LH, Marcher J, Schulze U, Carlsen M, Min RW, Nielsen J, Villadsen J. Semi-on-line analysis for fast and precise monitoring of bioreaction processes. Biotechnol Bioeng 1996; 52:237–247.
66. Harms P, Kostov Y, Rao G. Bioprocess monitoring. Curr Opin Biotechnol 2002; 13(2):124–127.
67. Cote GL. Noninvasive and minimally-invasive optical monitoring technologies. J Nutr 2001; 131(5):1596S–1604S.
68. Sivakesava S, Irudayaraj J, Ali D. Simultaneous determination of multiple components in lactic acid fermentation using FT-MIR, NIR, and FT-Raman spectroscopic techniques. Process Biochem 2001; 37(4):371–378.
69. Rhiel M, Ducommun P, Bolzonella I, Marison I, von Stockar U. Real-time in situ monitoring of freely suspended and immobilized cell cultures based on mid-infrared spectroscopic measurements. Biotechnol Bioeng 2002; 77(2):174–185.
70. Bellon-Maurel V, Vallat C, Goffinet D. Quantitative analysis for individual sugars during starch hydrolysis by FT-IT/ATR spectrometry. Part II: Influence of external factors and wavelength parameters. Appl Spectrosc 1995; 49:563–568.
71. Chung H, Arnold MA, Rhiel M, Murhammer DW. Simultaneous measurement of glucose and glutamine in aqueous solutions by near infrared spectroscopy. App Bioche Biotechno 1995; 50:109–125.
72. Zhou X, Chung H, Arnold MA, Rhiel M, Murhammer DW. Selective measurement for glutamine and asparagine in aqueous media by near-infrared spectroscopy. In: Rogers KR, Mulchandani A, Zhou W, eds. Biosensor and Chemical Sensor Technology: Process Monitoring and Control. Chapter 12. American Chemical Society Symposium Series 613. Washington, DC: ACS Books, 1995:116–132.
73. Chung H, Arnold MA, Rhiel M, Murhammer DW. Simultaneous measurements of glucose, glutamine, ammonia, lactate, and glutamate in aqueous solutions by near-infrared spectroscopy. Appl Spectrosc 1996; 50:270–276.
74. Cadet F, Robert C, Offmann B. Simultaneous determination of sugars by multivariate analysis applied to mid-infrared spectra of biological samples. Appl Spectrosc 1997; 51:369–375.
75. Cavinato AG, Mayes DM, Ge Z, Callis JB. Noninvasive method for monitoring ethanol in fermentation processes using fiber-optic near-infrared spectroscopy. Anal Chem 1990; 62:1977–1982.
76. Fayolle P, Picque D, Perret B, Latrille E, Corrieu G. Determination of major compounds of alcoholic fermentation by middle-infrared spectroscopy: study of temperature effects and calibration methods. Appl Spectrosc 1996; 50:1325–1330.
77. Ge Z, Cavinato AG, Callis JB. Noninvasive spectroscopy for monitoring cell density in a fermentation process. Anal Chem 1994; 66:1354–1362.
78. Vaccari G, Dosi E, Campi AL, Gonzales-Vara yRA, Matteuzzi D, Mantovani G. A near-infrared spectroscopy technique for the control of fermentation processes: an application to lactic acid fermentation. Biotechnol Bioeng 1994; 43:913–917.
79. Hall JW, McNeil B, Rollins MJ, Draper I, Thompson BG, Macaloney G. Near-infrared spectroscopic determination of acetate, ammonium, biomass, and glycerol in an industrial *Escherichia coli* fermentation. Appl Spectrosc 1996; 50:102–108.

80. Macaloney G, Draper I, Preston J, Anderson KB, Rollins MJ, Thompson BG, Hall JW, McNeil B. At-line controls and fault analysis in an industrial high cell density *Escherichia coli* fermentation, using NIR spectroscopy. Food Bioproduct Testing 1996; 74:212–220.
81. Macedo MG, Laporte MF, Lacroix C. Quantification of exopolysaccharide, lactic acid, and lactose concentrations in culture broth by near-infrared spectroscopy. J Agric Food Chem 2002; 50(7):1774–1779.
82. Pollard DJ, Buccino R, Connors NC, Kirschner TF, Olewinski RC, Saini K, Salmon PM. Real-time analyte monitoring of a fungal fermentation, at pilot scale, using in situ mid-infrared spectroscopy. Bioprocess Biosyst Eng 2001; 24(1):13–24.
83. Yano T, Harata M. Prediction of the concentration of several constituents in a mouse-mouse hybridoma culture by near infrared spectroscopy. J Ferment Bioeng 1994; 77:659–662.
84. McShane MJ, Cote GL. Near-infrared spectroscopy for determination of glucose, lactate, and ammonia in cell culture media. Appl Spectrosc 1998; 52:1073–1078.
85. Jung BJ, Lee S, Yang IH, Good T, Cote GL. Automated on-line noninvasive optical glucose monitoring in a cell culture system. Appl Spectrosc 2002; 56:51–57.
86. Riley MR, Crider HM, Nite ME, Garcia RA, Woo J, Wegge RM. Simultaneous measurement of 19 components in animal cell culture media by near infrared spectroscopy. Biotechnol Prog 2001; 17(2):376–378.
87. Haaland DM, Thomas EV. Partial least-squares methods for spectral analyses. 1. Relation to other quantitative calibration methods and the extraction of qualitative information. Anal Chem 1988; 60:1193–1202.
88. Riley MR, Rhiel M, Zhou X, Arnold MA, Murhammer DW. Simultaneous monitoring of glucose and glutamine in insect cell cultures by NIR spectroscopy. Biotechnol Bioeng 1997; 55(1):11–15.
89. Riley MR, Arnold MA, Murhammer DW. Matrix enhanced calibration procedure for multivariate calibration models with near-infrared spectra. Appl Spectrosc 1998; 52(10):1339–1347.
90. Riley MR, Arnold MA, Murhammer DW. Effect of sample complexity on quantification of analytes in aqueous samples by near infrared spectroscopy. Appl Spectrosc 2000; 54(2):255–261.
91. Riley MR, Okeson CD, Frazier BL. Rapid calibration development for near infrared spectroscopic monitoring of mammalian cell cultivations. Biotechnol Prog 1999; 15(6):1133–1141.
92. Riley MR, Crider HM. The effect of analyte concentration range on measurement errors obtained by NIR spectroscopy. Talanta 2000; 52:473–484.
93. Rhiel MH, Amrhein MI, Marison IW, von Stockar U. The influence of correlated calibration samples on the prediction performance of multivariate models based on mid-infrared spectra of animal cell cultures. Anal Chem 2002; 74(20):5227–5236.
94. Riley MR, Arnold MA, Murhammer DW, Walls EL, DelaCruz N. Adaptive calibration scheme for quantification of metabolites in insect cell media by near infrared spectroscopy. Biotechnol Prog 1998; 14:527–533.
95. Rhiel M, Cohen MB, Murhammer DW, Arnold MA. Nondestructive near-infrared spectroscopic measurement of multiple analytes in undiluted samples of serum-based cell culture media. Biotechnol Bioeng 2002; 77(1):73–82.
96. Hazen KH, Arnold MA, Small GW. Temperature-insensitive near-infrared spectroscopic measurement of glucose in aqueous solutions. Appl Spectrosc 1994; 48:477–483.
97. Vaidyanathan S, Macaloney G, Harvey LM, McNeil B. Assessment of the structure and predictive ability of models developed for monitoring key analytes in a submerged fungal bioprocess using near-infrared spectroscopy. Appl Spectrosc 2001; 55(4): 444–453.
98. Arnold MA, Burmeister JJ, Small GW. Phantom glucose calibration models from simulated noninvasive human near-infrared spectra. Anal Chem 1998; 70:1773–1781.

99. Haswell SJ, Skelton V. Chemical and biochemical microbioreactors. Trends Anal Chem 2000; 19:389–396.
100. Kostov Y, Harms P, Randers-Eichorn L, Rao G. Low-cost microbioreactor for high-throughput bioprocessing. Biotechnol Bioeng 2001; 72:346–353.
101. Riley MR, Jordan KA, Cox ML. Development of a cell-based biosensor to evaluate the toxicity of inhaled materials. Biochemical Engineering Journal, Volume 19, Issue 2, 95–99 (2004).
102. Schultz NM, Huang L, Kennedy RT. Capillary electrophoresis-based immunoassay to determine insulin content and insulin secretion from single Islets of Langerhans. Anal Chem 1995; 67:924–929.
103. Koutny LB, Schmalzing D, Taylor TA, Fuchs M. Microchip electrophoretic immunoassay for serum cortisol. Anal Chem 1996; 68:18–22.
104. Schmalzing D, Koutny LB, Taylor TA, Nashabeth W, Fuchs M. Immunoassay for thyroxine in serum using capillary electrophoresis and micromachined devices. J Chromatogr 1997; 697:175–180.
105. Mitnik L, Novotny M, Felten C, Buonocore S, Koutny L, Schmalzing D. Recent advances in DNA sequencing by capillary and microdevice electrophoresis. Electrophoresis 2001; 22(19):4104–4117.
106. Schmalzing D, Buonocore S, Piggee C. Capillary electrophoresis-based immunoassays. Electrophoresis 2000; 21(18):3919–3930.
107. Taylor J, Picelli G, Harrison DJ. An evaluation of the detection limits possible for competitive capillary electrophoretic immunoassays. Electrophoresis 2001; 22(17): 3699–3708.
108. Shimura K, Hoshino M, Kamiya K, Katoh K, Hisada S, Matsumoto H, Kasai K. Immunoassay of serum alpha (1)-antitrypsin by affinity-probe capillary isoelectric focusing using a fluorescence-labeled recombinant antibody fragment. Electrophoresis 2002; 23(6):909–917.
109. Bachinger T, Riese U, Eriksson RK, Mandenius CF. Electronic nose for estimation of product concentration in mammalian cell cultivation. Bioprocess Eng 2000; 23:637–642.
110. Ducommun P, Ruffieux PA, Furter MP, Marison I, von Stockar U. A new method for on-line measurement of the volumetric oxygen uptake rate in membrane aerated animal cell cultures. J Biotechnol 2000; 78:139–147.
111. Junker BH, Reddy J, Gbewonyo K, Greasham R. Online and in-situ monitoring technology for cell-density measurement in microbial and animal cell cultures. Bioprocess Eng 1994; 10:195–207.
112. Wu P, Ozturk SS, Blackie JD, Thrift JC, Figueroa C, Daveh D. Evaluation of and application of optical cell density probes in mammalian-cell bioreactors. Biotechnol Bioeng 1995; 45:495–502.
113. Konstantinov KB, Pambayun R, Matanguihan R, Yoshida T, Purusich CM, Hu WS. On-line monitoring of hybridoma cell growth using a laser turbidity sensor. Biotechnol and Bioeng 1992; 40:1337–1342.
114. Marose S, Lindemann C, Ulber R, Scheper T. Optical sensor systems for bioprocess monitoring. Trends Biotechnol 1999; 17:30–34.
115. Davey CL, Davey HM, Kell DB. On the dielectric properties of cell suspensions at high-volume fractions. Bioelectrochem Bioenerg 1992; 28:319–340.
116. Bragos R, Gamez X, Cairo J, Riu PJ, Godia F. Biomass monitoring using impedance spectroscopy, electrical bioimpedance methods: applications to medicine and biotechnology. Ann NY Acad Sci 1999; 873:299–305.
117. Siano SA. Biomass measurement by inductive permittivity. Biotechnol Bioeng 1997; 55:289–304.
118. Morita S, Umakoshi H, Kuboi R. Characterization and on-line monitoring of cell disruption and lysis using dielectric measurement. J Biosci Bioeng 1999; 88:78–84.

119. Zeiser A, Bedard C, Voyer R, Jardin B, Tom R, Kamen AA. On-line monitoring of the progress of infection in Sf-9 insect cell cultures using relative permittivity measurements. Biotechnol Bioeng 1999; 63:122–126.
120. Yardley YE, Kell DB, Barrett J, Davey CL. On-line, real-time measurements of cellular biomass using dielectric spectroscopy. Biotechnol Genet Eng Rev 2000; 17:3–35.
121. Sivakesava S, Irudayaraj J, Demirci A. Monitoring a bioprocess for ethanol production using FT-MIR and FT-Raman spectroscopy. J Ind Microbiol Biotechnol 2001; 26(4):185–190.
122. Danielsson B. Fermentation monitoring. Curr Opin Biotechnol 1991; 2:17–22.
123. Webb C, Dean JF, Marshal RD. Automatic aseptic sampling of fermentation broth. Biotechnology 1990; 8:926–928.
124. Page WJ, Rudy B. An autosampler for fermentation. Biotechniques 1993; 14:228.
125. vandeMerbel NC, Lingeman H, Brinkman UAT. Sampling and analytical strategies in on-line bioprocess monitoring and control. J Chromatogr A 1996; 725:13–27.
126. Picque D, Corrieu G. Performances of aseptic sampling devices for online monitoring of fermentation processes,. Biotechnology and Bioengineering 1992; 40:919–924.
127. Ryll T, Dutina G, Reyes A, Gunson J, Krummen L, Etcheverry T. Performance of small-scale CHO perfusion cultures using an acoustic cell filtration device for cell retention: characterization of separation efficiency and impact of perfusion on product quality. Biotechnol Bioeng 2000; 69(4):440–449.
128. Woodside SM, Bowen BD, Piret JM. Mammalian cell retention devices for stirred perfusion bioreactors. Cytotechnology 1998; 28:163–175.
129. Ruffieux PA, von Stockar U, Marison IW. Measurement of volumetric (OUR) and determination of specific (qO(2)) oxygen uptake rates in animal cell cultures. J Biotechnol 1998; 63:85–95.
130. Doran P. Bioprocess Engineering Principles. London: Academic Press, 1997.
131. Singh V. On-line measurement of oxygen uptake in cell culture using the dynamic method. Biotechnol Bioeng 1996; 52:443–448.
132. Ducommun P, Bolzonella I, Rhiel M, Pugeaud P, von Stockar U, Marison IW. On-line determination of animal cell concentration. Biotechnol Bioeng 2001; 72:515–522.
133. Siegwart P, Cote J, Male K, Luong JHT, Perrier M, Kamen A. Adaptive control at low glucose concentration of HEK-293 cell serum-free cultures. Biotechnol Prog 1999; 15:608–616.
134. Bonarius HPJ, Degooider CD, Tramper J, Schmid G. Determination of the respiration quotient in mammalian cell culture in bicarbonate buffered media. Biotechnol Bioeng 1995; 45:524–535.
135. Neeleman R, van den End EJ, van Boxtel AJB. Estimation of the respiration quotient in a bicarbonate buffered batch cell cultivation. J Biotechnol 2000; 80:85–94.
136. Kamen AA, Bedard C, Tom R, Perret S, Jardin B. On-line monitoring of respiration in recombinant-baculovirus infected and uninfected insect cell bioreactor cultures. Biotechnol Bioeng 1996; 50:36–48.
137. Zeiser A, Elias CB, Voyer R, Jardin B, Kamen AA. On-line monitoring of physiological parameters of insect cell cultures during the growth and infection process. Biotechnol Prog 2000; 16:803–808.
138. Andrews GF. The yield equations in the modeling and control of bioprocesses. Biotechnol Bioeng 1993; 42:549–556.
139. Zeng AP, Hu WS, Deckwer WD. Variation of stoichiometric ratios and their correlation for monitoring and control of animal cell cultures. Biotechnol Prog 1998; 14:434–441.
140. Nyberg GB, Balcarcel RR, Follstad BD, Stephanopoulos G, Wang DIC. Metabolism of peptide amino acids by Chinese hamster ovary cells grown in a complex medium. Biotechnol Bioeng 1999; 62:324–335.

141. Guan YH, Kemp RB. On-line heat flux measurements improve the culture medium for the growth and productivity of genetically engineered CHO cells. Cytotechnology 1999; 30:107–120.
142. Kemp RB, Guan YH. The application of heat flux measurements to improve the growth of mammalian cells in culture. Thermochim Acta 2000; 349:23–30.
143. Xie L, Wang DIC. Fed-batch cultivation of animal cells using different medium design concepts and feeding strategies. Biotechnol Bioeng 1994; 43:1175–1189.
144. Jo EC, Park HJ, Kim DI, Moon HM. Repeated fed-batch culture of hybridoma cells in nutrient-fortified high-density medium. Biotechnol Bioeng 1993; 42:1229–1237.
145. Bibila TA, Ranucci CS, Glazomitsky K, Buckland BC, Aunins JG. Monoclonal antibody process development using medium concentrates. Biotechnol Prog 1994; 10:87–96.
146. Bibila TA, Robinson DK. In pursuit of the optimal fed-batch process for monoclonal antibody production. Biotechnol Prog 1995; 11:1–13.
147. Rani KY, Rao VSR. Control of fermentors—a review. Bioprocess Eng 1999; 21:77–88.
148. Reilly MT, Charles M. The use of on-line sensors in bioprocess control. In: Twork JV, Yacynych AM. Sensors in Bioprocess Control. New York, NY: Marcel Dekker, 1990.
149. Europa AF, Gambhir A, Fu PC, Hu WS. Multiple steady states with distinct cellular metabolism in continuous culture of mammalian cells. Biotechnol Bioeng 2000; 67:25–34.
150. Oh GS, Izuishi T, Inoue T, Hu WS, Yoshida T. Interactive dual control of glucose and glutamine feeding in hybridoma cultivation. J Ferment Bioeng 1996; 81:329–336.
151. Zhou WC, Mulchandani A. Recent advances in bioprocess monitoring and control. Biosens Chem Sens Technol ACS Symp Ser 1995; 613:88–98.
152. Schwabe JO, Portner R, Markl H. Improving an on-line feeding strategy for fed-batch cultures of hybridoma cells by dialysis and 'Nutrient-Split'-feeding. Bioprocess Eng 1999; 20:475–484.
153. Mercille S, Johnson M, Lanthier S, Kamen AA, Massie B. Understanding factors that limit the productivity of suspension-based perfusion cultures operated at high medium renewal rates. Biotechnol Bioeng 2000; 67:435–450.
154. Yang JD, Angelillo Y, Chaudhry M, Goldenberg C, Goldenberg DM. Achievement of high cell density and high antibody productivity by a controlled-fed perfusion bioreactor process. Biotechnol Bioeng 2000; 69:74–82.
155. Rodrigues MTA, Vilaca PR, Garbuio A, Takagi M, Barbosa S, Leo P, Laignier NS, Silva AAP, Moro AM. Glucose uptake rate as a tool to estimate hybridoma growth in a packed bed bioreactor. Bioprocess Eng 1999; 21:543–546.
156. Koffas M, Roberge C, Lee K, Stephanopoulos G. Metabolic engineering. Annu Rev Biomed Eng 1999; 1:535–557.
157. Shanks JV, Stephanopoulos G. Bridging the gap between gene and product. Curr Opin Biotechnol 2000; 11:169–170.
158. Bree MA, Dhurjati P, Geoghegan RF Jr, Robnett B. Kinetic modelling of hybridoma cell growth and immunoglobulin production in a large-scale suspension culture. Biotechnol Bioeng 1988; 32:1067–1072.
159. Banga JR, Alonso AA, Singh RP. Stochastic dynamic optimization of batch and semicontinuous bioprocesses. Biotechnol Prog 1997; 13:326–335.
160. Konstantinov KB, Zhou WC, Golini F, Hu WS. Expert-systems in the control of animal cell culture processes—potentials, functions, and perspectives. Cytotechnology 1994; 14:233–246.
161. Roubos JA, van Straten G, van Boxtel AJB. An evolutionary strategy for fed-batch bioreactor optimization; concepts and performance. J Biotechnol 1999; 67:173–187.
162. Carrasco EF, Banga JR. Dynamic optimization of batch reactors using adaptive stochastic algorithms. Ind Eng Chem Res 1997; 36:2252–2261.
163. Jayaraman VK, Kulkarni BD, Gupta K, Rajesh J, Kusumaker HS. Dynamic optimization of fed-batch bioreactors using the ant algorithm. Biotechnol Prog 2001; 17:81–88.

164. Dowd JE, Weber I, Rodriguez B, Piret JM, Kwok KE. Predictive control of hollow-fiber bioreactors for the production of monoclonal antibodies. Biotechnol Bioeng 1999; 63:484–492.
165. Balsa-Canto E, Banga JR, Alonso AA, Vassiliadis VS. Efficient optimal control of bio-processes using second-order information. Ind Eng Chem Res 2000; 39:4287–4295.
166. Wang FS, Cheng WM. Simultaneous optimization of feeding rate and operation parameters for fed-batch fermentation processes. Biotechnol Prog 1999; 15:949–952.
167. Dhir S, Morrow KJ, Rhinehart RR, Wiesner T. Dynamic optimization of hybridoma growth in a fed-batch bioreactor. Biotechnol Bioeng 2000; 67:197–205.
168. Moran EB, McGowan ST, McGuire JM, Frankland JE, Oyebade IA, Waller W, Archer LC, Morris LO, Pandya J, Nathan SR, Smith L, Cadette ML, Michalowski JT. A systematic approach to the validation of process control parameters for monoclonal antibody production in fed-batch culture of a murine myeloma. Biotechnol Bioeng 2000; 69:242–255.

9
Cell Culture Kinetics and Modeling

An-Ping Zeng
GBF National Research Institute for Biotechnology, Division of Biochemical Engineering, Braunschweig, Germany

Jing-xiu Bi
Institute of Process Engineering, National Laboratory of Biochemical Engineering, Chinese Academy of Sciences, Beijing, P.R. China

INTRODUCTION

Kinetics of cell growth and metabolism is desired for a better understanding of cell physiology and for the optimization and control of animal cell cultures. Consequently, it has been the objective of a large number of investigations in the past (1–22). A large number of factors were reported to affect the kinetics of growth, death, and product formation of animal cells in bioreactors. Despite these efforts, uncertainties and even controversies often exist in the literature regarding the relative importance of these factors in different cell lines (23–26). This may be mainly due to the lack of quantitative knowledge of the significance of these factors and particularly of the range of concentrations or values in which these factors are important. A mathematical description of the interactions of different factors in kinetic models can be very useful to solve this problem because they make it possible to separate and identify the effects of different factors. It can help better in interpreting and understanding experimental results obtained under different conditions. Models can also be used to check different hypotheses regarding regulation mechanisms. In addition, they can be useful as a guide tool for developing model-based process control and monitoring strategies.

For these reasons, considerable efforts have been made in the past to develop mathematical models to describe the growth, death, and metabolism of animal cell culture (2–5,8–10,23–38). Some of the kinetic models were reviewed or compared by Tziampazis and Sambanis (39), Pörther and Schäfer (40), and Goergen et al. (41). Whereas structured models may be attractive due to their general applicability, they suffer from the shortcomings of unknown or complicated mechanisms of regulation of cell growth and death. In addition, they contain a large number of model parameters. The identification and estimation of these parameters are often difficult. Therefore, most of the present models for animal cell culture are unstructured models.

This chapter first gives an introduction to the kinetic characterization of cell culture in batch and continuous operations. This is followed by a brief description of major factors affecting the kinetics of cell culture and their quantitative analysis. The last part of this chapter describes typical models for kinetic analysis and simulation. Future research needs in cell culture kinetics are then discussed. No effort is made in this chapter to cover the kinetics and modeling of adherent cells and insect cells. This chapter is also not intended to be a complete survey of kinetics of animal cell culture. Rather, it emphasizes the quantification of cell growth and metabolism and the derivation of rate equations and mathematical models for kinetic analysis and simulations.

KINETIC CHARACTERIZATION OF CELL CULTURE

Kinetics of Cells in Batch Culture

Kinetics of animal cell culture has been extensively studied in batch culture (2,3,15,18,19,22). Advances of batch culture include simple operation and process monitoring, short cultivation time, and relatively stable kinetic characteristics of cell lines if recombinant cells are used.

Time Profile of the Concentration of Cells, Nutrients, and Metabolites

Figure 1 shows typical time profiles of concentrations of cells, nutrients, and metabolites of a recombinant BHK-21 cell line in batch culture (22). Also shown in Fig. 1 are the effects of initial concentration of glucose and glutamine on the growth kinetics of this cell line. Both the viable cell number and the specific growth rate (see following text for definition) reach a maximum during the cultivation and decline afterwards in all cultures (Fig. 1A,B and E–F). This decline in growth is not due to the exhaustion of any macronutrient components (e.g., glucose and amino acids) or the inhibition of metabolic byproducts (e.g., lactate and ammonium) because the decline of μ in most cultures occurs in the early stage of cultivation. Normally, lactate concentration increases until glucose is depleted and decreases thereafter (Fig. 1C and G). The reconsumption of lactate is particularly obvious in cultures with low initial glucose concentration (Fig. 1C). Ammonium formation is closely correlated with the consumption of glutamine (Fig. 1D and H).

The initial concentrations of glucose and glutamine can significantly affect the time profile of batch culture. However, for most cell culture media with initial glucose and glutamine concentrations above a certain level (about 1 mM in the case of BHK cells), these macronutrients have no significant effect on the maximum viable cell concentration and the specific cell growth rate (22).

For cell growth in batch culture, the inoculum age and size can have significant influences on the kinetic behavior. Martial et al. (42) showed that increasing the age of inoculum resulted in a longer lag phase, a reduced maximal cell concentration, and a lower maximum growth rate of hybridoma cells seeded with the same initial cell density. For a number of cell lines it has been demonstrated that the initial cell density significantly affects growth in batch culture and a critical inoculum size exists (43,44). Relatively less is known about the effects of cell density on the metabolism and product formation of batch culture. Ozturk and Palsson (45) found no significant effects of initial cell density (10^2 to 10^5 cells/mL) on the nutrient uptake rate and monoclonal antibody (MAb) formation rate of a hybridoma batch culture. This seems to disagree with some of the results from continuous cultures.

Intracellular Content and Excreted Protein Component

The intracellular content of many intermediary metabolites, proteins, and nucleic acid may vary during batch culture and is particularly useful for kinetic

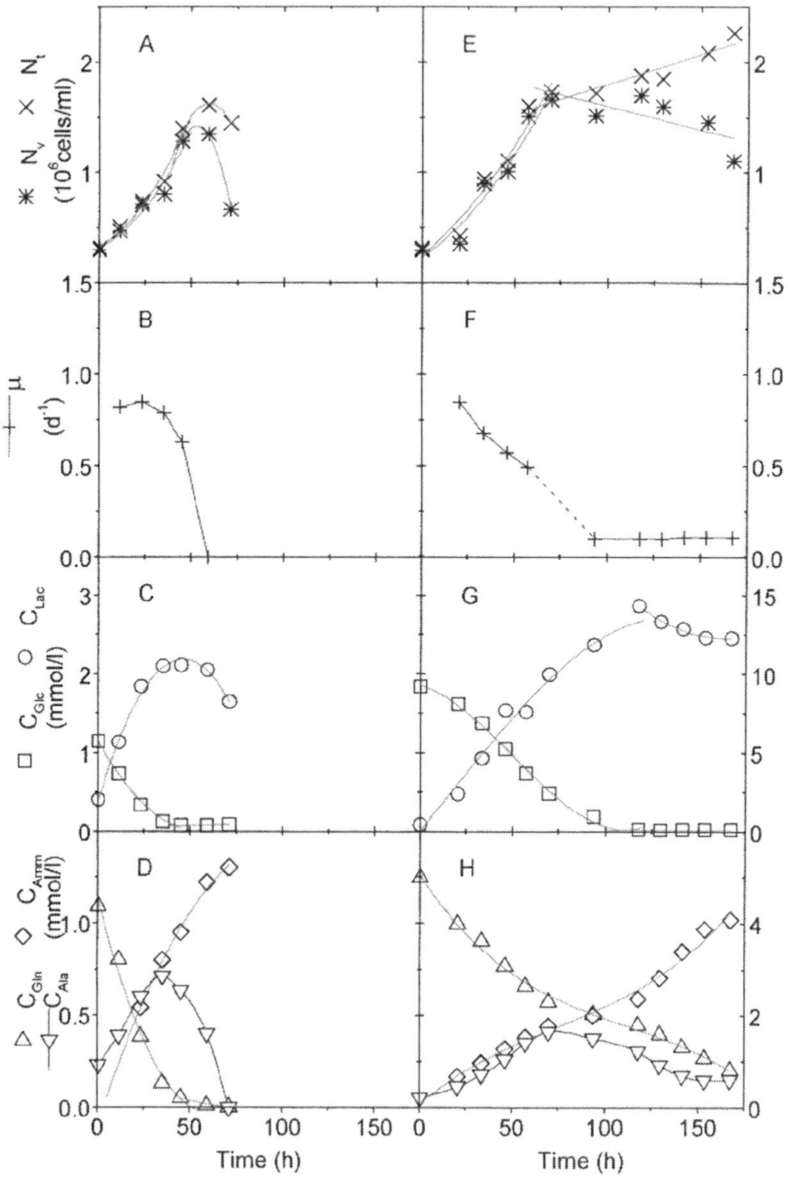

Figure 1 Time courses of viable and total cell number (N_v and N_t), specific mitotic rate (μ), and concentrations of glucose (C_{Glc}), glutamine (C_{Gln}), lactate (C_{Lac}), ammonium (C_{Amm}), and alanine (C_{Ala}) in batch cultures of a recombinant BHK-21 cell line with different initial concentrations of glucose and glutamine. Batch culture I (A, B, C, D): initial glucose = 1 mmol/L; initial glutamine = 1 mmol/L. Batch culture II (E, F, G, H): initial glucose = 10 mmol/L; initial glutamine = 5 mmol/L [From Linz et al. (22)].

characterization of growth and metabolism. The intracellular content may be studied either by cell disruption, extraction, and subsequent analysis of the different substances, or by in situ flow cytometry or NMR analysis. For example, Ryll and Wagner (46) studied the variation of intracellular ribonucleotide pools during "perfused" batch cultivations of hybridoma and BHK cell lines. These authors found

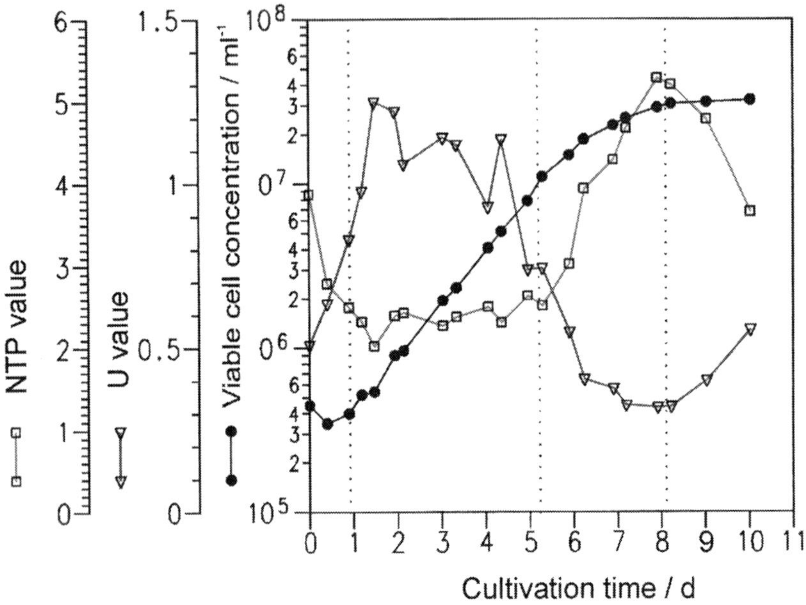

Figure 2 Change of NTP and U values during the growth cycle of a "perfused" batch culture of a hybridoma cell line (46).

that three cell specific regularities, i.e., the NTP ratio (NTP = (ATP + GTP)/(UTP + CTP)), the uridine (U) ratio (U = UTP/(UTP−GNAc)), and the combined ratio NTP/U, are particularly useful for characterizing the cell growth. As shown in Fig. 2 for a hybridoma cell line, the NTP value reduces rapidly during the adaptation (lag) phase and remains at a nearly constant and low value during the exponential growth phase. During the phase of reduced growth rate it dramatically increases. The U-value changes in an opposite direction to the NTP value. The intracellular nucleotide pools also sensitively respond to the presence of growth inhibitor such as ammonium. Studies of different cell lines in batch, chemostat, and perfusion cultures suggest that the content of nucleotide pools is cell line specific. For a given cell line the intracellular nucleotide ratios can be used as a sensitive parameter for process monitoring and control (46–48).

The variation of the U-value as shown in Fig. 2 is primarily due to the change of the content of UDP-N-acetylhexosamines (UDPGNAc pool, primarily UDPG1c-NAc and UDPGalNAc). Rapidly growing cells have a low UDPGNAc pool that rapidly increases when growth slows down. Since UDPGlcNAc and UDPGalNAc, as activated sugars, are precursors for the synthesis of oligosaccharides and present in the O- and N-glycanes of glycoproteins, they can influence the expression of carbohydrates in glycoproteins (49). Since the UDPGNAc pool significantly changes in batch culture, the glycosylation of excreted proteins can also vary during a batch culture (50). In particular, ammonium ion concentration can strongly affect the UDPG-NAc pool and hence the glycosylation of excreted proteins (50,51). For a recombinant Chinese hamster ovary (CHO) cell line producing an immunoadhesin tumor necrosis factor-immunoglobulin it was shown that as ammonium increased from 1 to 15 mM, a concomitant decrease of up to 40% in terminal galactosylation and sialylation of the product occurred (50). The proportion of the nonglycosylated form of excreted protein seems to generally increase during batch culture (52). In addition to the changes of concentration of nutrients and metabolites, variations of glycosylation of secreted proteins can also be caused by modifications in the intracellular protein maturation processes or to the extracellular action of proteases or glycosidases.

Variation of Cell Morphology

Cell morphology (cell size and cell surface structure) significantly changes during batch culture (53–55). The cell size usually increases during the exponential growth phase but decreases during the stationary or decline phase. In the late phase of a batch culture, the cell surface (membrane) undergoes significant structural changes resulting in the disappearance of microvilli and the appearance of blebs. The cell size change is related to cell cycle and population change (56). The variation of cell morphology is correlated with the growth and death of cells. The increase of cell size in the growth phase is due to active cell proliferation with a high portion of cells in the S-phase of the cell cycle. On the other hand, the decline of cell size and changes of cell membrane toward the end of a batch culture can be explained by the mechanisms of cell death (54,57).

Mammalian cells principally die by two mechanisms: necrosis and apoptosis (also called programmed cell death). These two types of cell death display distinctive differences in their morphological and biochemical features (Fig. 3). Morphologically, necrotic cell death is characterized by a progressive hydration of the cytoplasma (swelling) that is followed by membrane and organelle disruption, leakage of lysosomes into cytoplasm, nuclear disintegration, and finally complete disruption of the cell. The morphology of apoptotic cell death markedly contrasts with that of necrosis (Fig. 3). Important features of apoptosis include condensation and fragmentation of chromatin, loss of volume (cell shrinkage), and modification of the cytoskeleton. Following these events the plasma membrane convolutes and the cell eventually fragments to form membrane-bound vesicles (apoptotic bodies) that contain intact cytoplasmic organelles and, usually, nuclear fragments. In the body or under in vivo conditions in tissues apoptotic bodies are recognized and engulfed by phagocytic or nearby cells. The recognition of apoptotic bodies has been suggested to be due to changes of the membrane including occurrence of the signal molecule phosphatidylserine on the membrane surface. A number of investigations have shown that in the physical and chemical environments of a bioreactor most production cell lines including hybridoma, myeloma, CHO, and baby hamster kidney (BHK) cells primarily die by apoptosis (57–63).

Determination of Cell Population Heterogeneity for Kinetic Studies

For the study of cell culture kinetics it is important to quantify the different types of cells and to follow their variations during cultivation. The determination of cell population heterogeneity can be conventionally based on either morphological or biochemical features. For example, based on the properties of cell membrane and the structural morphology of cell nuclei the method of acridine orange and ethidium bromide staining can be used to distinguish four types of cells, namely viable nonapoptotic (VNA), viable apoptotic (VA), nonviable apoptotic (NVA), and nonviable nonapoptotic cells (NVNA) (i.e., necrotic dead cells) (Fig. 3). As shown in Fig. 4, the relative numbers of both living and dead apoptotic cells are low at the beginning of the batch culture. Whereas the number of living apoptotic cells remains at a low level (maximum 2–5% of the total cells) throughout the cultivation, the number of

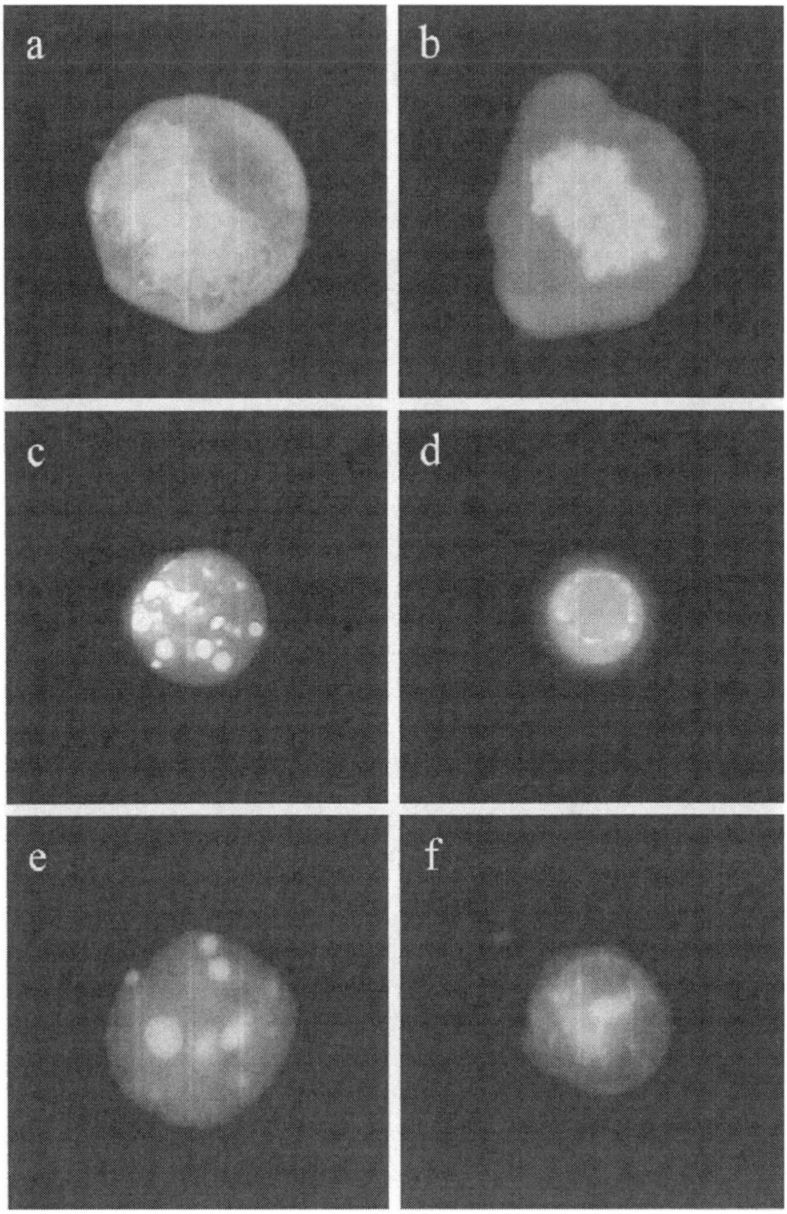

Figure 3 Acridin orange/ethidium bromide staining of hybridoma cells, taken with 1000 times amplification. Living cells, green; dead cell, orange: (a) living nonapoptotic cell, (b) living nonapoptotic cell during division, (c) living apoptotic cell (clear DNA condensation and nuclear fragmentation), (d) living apoptotic cell (ring form condensation of DNA near the nuclear membrane), (e) dead apoptotic cell, (f) dead nonapoptotic (necrotic) cell. [From Ref. (57).] (*See color insert p. 4.*)

apoptotic death cells increases continuously and accounts for more than 97% of the total dead cells at the end of the cultivation. These results demonstrate that, first, apoptosis is the prevailing death mechanism of this cell line and, second, the kinetics of transition of living apoptotic to dead cells is much faster than the kinetics of apoptosis induction (described subsequently).

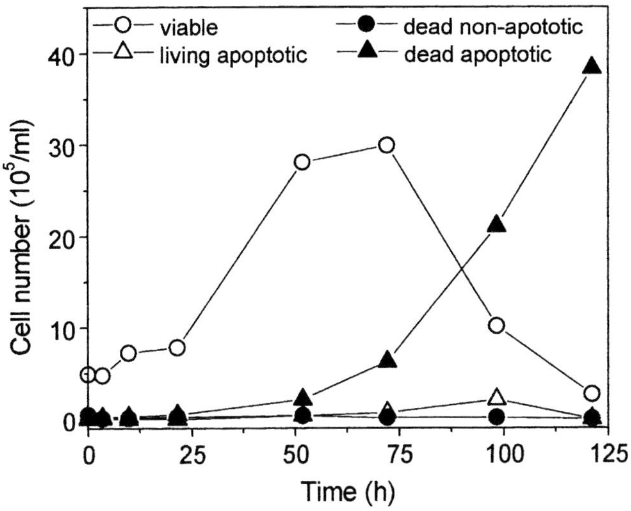

Figure 4 Typical time profile of cell populations in a batch culture of the hybridoma cell line HyGPD YK-1–1 (N_t = total cell number, N_v = viable cell number, N_d = dead cell number, Vit = vitality) as determined by acridine orange/ethidium bromide staining (VNA = viable nonapoptotic, VA = viable apoptotic, NVA = nonviable apoptotic, NVNA = nonviable nonapoptotic cells).

Flow cytometry is also a very useful tool for studying cell population heterogeneity. By staining cells with proper fluorophores, such as propidium iodide and acridine orange, flow cytometry can be used to quantify the progress of cells through the cell cycle (53,54). During batch culture, the proportion of cells in the G1, S, and G2/M phases significantly changes. In the exponential growth phase, most of the cells are in the S and G2 phases, whereas during the stationary and decline phases the majority of cells are in the G1-phase. Flow cytometry can be also used to determine the populations of living, apoptotic, and dead cells by combination with propidium iodide and annexin staining (54).

In contrast to the large body of knowledge concerning the molecular mechanisms, biochemical features, and physiological consequences of apoptosis, relatively little is known about the kinetics and dynamics of cell death (for both apoptosis and necrosis). This may be mainly due to two reasons. First, apoptotic cell death is a relatively fast process. In vivo, it may take only a few hours from the early stage of induction (evanesce to recognition) to the end stages of degradation and phagocytosis (64). Second, reliable methods for quantification of apoptosis have only recently become available, particularly for suspension culture (65,66). Therefore, most of the studies dealing with cell death have been descriptive or qualitative. To better understand and intervene in the apoptotic process a quantitative and systematic knowledge of cell death is important. Linz (57) reported on the kinetics of apoptosis in three mammalian cell lines (hybridoma, BHK, and CHO) and in the human carcinoma cell line HeLa in normal batch cultures and cultures under different stress conditions. Similarly, as shown in Fig. 4 for the hybridoma cell culture, apoptosis is the prevailing death mechanism of all these cell lines and the kinetics of transition of living apoptotic to dead cells is much faster than the kinetics of apoptosis induction.

Estimation of Specific Rates of Cell Growth, Death, and Metabolism

The specific mitotic rate as defined in Eq. (9.1) is normally used to calculate the specific rate of cell growth μ (hr^{-1} or day^{-1}) in batch culture:

$$\mu = \frac{dN_t}{dt} \cdot \frac{1}{N_v} \tag{9.1}$$

Similarly, Eq. (9.2) is often used to calculate the specific death rate (k_d) of cells:

$$k_d = \frac{dN_d}{dt} \cdot \frac{1}{N_v} \tag{9.2}$$

In Eqs. (9.1) and (9.2), N_t, N_v, and N_d are the total cell number, viable cell number, and dead cell number per unit culture volume, respectively. N_t, N_v, and N_d can be determined by counting the viable and dead cells using the trypan blue exclusion method with a hemocytometer. N_t can be also determined by counting the cell nuclei using the so-called CASY Counter after disruption of the cells. This method is particularly suitable for cultures with aggregate formation.

To better quantify the kinetics of apoptotic and necrosis cell death processes, Linz et al. (56) introduced the terms apoptosis induction rate (k_{ap}^{in}) apoptotic death rate (k_{ap}^{death}) and necrotic death rate (k_{nec}^{death}):

$$k_{ap}^{in} = \left(\frac{dN_{NVA}}{dt} + \frac{dN_{VA}}{dt}\right) \times \frac{1}{N_{VNA}} \tag{9.3}$$

$$k_{ap}^{death} = \left(\frac{dN_{NVA}}{dt}\right) \times \frac{1}{N_{VA}} \tag{9.4}$$

$$k_{nec}^{death} = \left(\frac{dN_{NVNA}}{dt}\right) \times \frac{1}{N_{VNA}} \tag{9.5}$$

where N_{VNA}, N_{VA}, N_{NVA}, and N_{NVNA} are the number of viable nonapoptotic, viable apoptotic, nonviable apoptotic, and nonviable nonapoptotic cells per unit volume of culture, respectively.

The specific rates of consumption of major nutrients (e.g., glucose and glutamine) and formation of metabolites (e.g., lactate and ammonium) in batch culture can be calculated as:

$$q_{Glc} = -\frac{dC_{Glc}}{dt}\frac{1}{N_v} \tag{9.6}$$

$$q_{Lac} = \frac{dC_{Lac}}{dt}\frac{1}{N_v} \tag{9.7}$$

$$q_{Gln} = \left[-\frac{dC_{Gln}}{dt} - kC_{Gln}\right]\frac{1}{N_v} \tag{9.8}$$

$$q_{Amm} = \left[\frac{dC_{Amm}}{dt} - kC_{Gln}\right]\frac{1}{N_v} \tag{9.9}$$

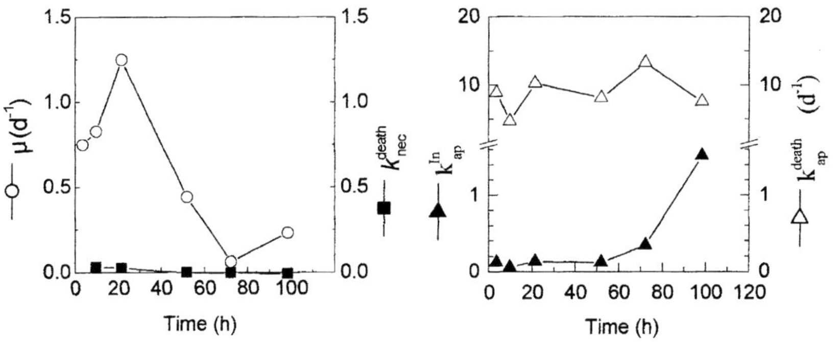

Figure 5 Specific rates of cell growth, apoptosis induction and death (apoptosis and necrosis) of hybridoma cells in a batch culture (time profile of cell population shown in Fig. 4). [From Ref. (57).]

In Eqs. (9.8) and (9.9), k is a rate constant accounting for the spontaneous decomposition of glutamine as a first order reaction and having a value of $4.675 \times 10^{-3}\,\mathrm{hr}^{-1}$ at 37°C (67,68).

To calculate the specific rates it is recommended to first fit the experimental data to appropriate polynomial functions by the least-squares method. For these purposes, experimental data with relatively short time intervals are required. The time course of some variables sometimes has to be divided into several sections to find functions describing the data satisfactorily. The derivatives with respect to time of these functions are then used to calculate the specific rates of cell growth, death, and metabolism.

Figure 5 depicts the time courses of μ, ($k_{\mathrm{ap}}^{\mathrm{in}}$), ($k_{\mathrm{ap}}^{\mathrm{death}}$), and ($k_{\mathrm{nec}}^{\mathrm{death}}$) for the batch culture shown in Fig. 4. It is interesting to note that whereas ($k_{\mathrm{ap}}^{\mathrm{death}}$) is more or less constant throughout the cultivation, ($k_{\mathrm{ap}}^{\mathrm{in}}$) is quite low during the growth phase and significantly increases during the death phase. In general, ($k_{\mathrm{ap}}^{\mathrm{death}}$) is much higher than ($k_{\mathrm{ap}}^{\mathrm{in}}$) (particularly during the growth phase) and μ. The fast kinetics of apoptotic cell death in comparison to the induction explains why only a very low population of living apoptotic cells can be experimentally determined. Because of this fast kinetics of apoptotic cell death and because cells of some cultures undergo a fast secondary necrosis after apoptotic death, these cultures may not display the so-called typical hallmarks of apoptotic cell death such as DNA fragmentation and hence are sometimes considered not to die by apoptosis. A careful and fast measurement of many parameters is important for a reliable kinetic characterization of cell death.

The specific rates of metabolism of the batch culture follow a similar trend as the specific cell growth rate (Fig. 5). This gives the impression that cell growth was limited by the availability of macronutrients. As shown in Fig. 1, however, the decline of specific cell growth rate cannot be overcome by simply increasing the nutrient concentration, indicating that other factors are involved (26). The relationships between the specific rates of growth and metabolism will be discussed in the following section.

Results from kinetic study of batch cultures suggest that the specific consumption rate of glucose and specific formation rate of lactate are closely related. This also seems to apply to the relationships between the specific consumption rate of glutamine and the specific formation rate of ammonium. Zeng et al. (69) analyzed the stoichiometric ratios of a number of cell lines under diverse experimental conditions including batch cultures. The results are summarized in the section of continuous culture.

Cell Kinetics in Continuous Culture

Advantages and Application of Continuous Culture

The advantages and potentials of continuous culture for cell culture technology can be viewed from two general aspects (70). From an application point of view, these include high volumetric productivity, savings in labor and energy costs (e.g., inoculum preparation, reactor cleaning, and sterilization), uniform product quality, better automation and process control, and the use of more efficient and economic methods of medium preparation and downstream processing. If cell recycling (perfusion culture) is applied, the concentrations of cells and product can be significantly improved. As a research tool, continuous culture provides well-defined cultivation conditions for genetic, biochemical and physiological characterizations of cells. It allows an independent variation of growth parameters, enabling reliable kinetic studies of cell growth and metabolism for process optimization. The transition behavior of a continuous culture upon shift-up or shift-down of a variable is also a powerful tool to study the regulation of growth and metabolism. In addition, continuous culture is a useful means for selecting strains and/or subclones with improved growth characteristics and productivity. On the other hand, the mutation and instability of some producing strains due to the selection pressure in continuous culture is a major obstacle for the industrial application of this cultivation technique, particularly in the case of genetically modified cells (71).

Continuous operation is used for the production of monoclonal antibodies and recombinant proteins by animal cell culture. The major applications of continuous culture are however still found in fundamental studies and process optimization in laboratory scale. Animal cell cultures have been quantitatively studied in depth with the help of continuous culture (8,9,13,14,22,72,73).

Time Profile and Steady State of Continuous Culture

Figure 6 shows a typical time profile of total cell number (N_t) determined by counting cell nuclei, the cell viability, and the concentrations of glucose, lactate, glutamine, and ammonium of two series of continuous cultures at a constant dilution rate but with varied concentrations of glucose and glutamine in medium (22).

Steady states (with the exception of cultures Cl-2 in Fig. 6A and C2-2 in Fig. 6B) with varied residual concentrations of nutrients and hence varied growth limitation were approached at relatively constant specific growth rates. The steady states had altered consumption rates of nutrients and formation rates of products. In the experiment series C1, N_t increased up to 5×10^6 cells/mL with increasing glutamine in the feed (C_{Gln}^{in}) at low concentration range but leveled off above a glutamine concentration of 3.5 mM. The residual glucose concentration (C_{Glc}) was very low (between 0.20 and 0.27 mM) throughout the cultivation. The residual glutamine concentration (C_{Gln}) was also very low (between 0.04 and 0.16 mM) at low C_{Gln}^{in} (between 1 and 3.5 mM). Under these conditions, growth may be partly limited by glucose and/or glutamine. A further increase of C_{Gln}^{in} from 3.5 to 5 mM led to an increase of C_{Gln} from 0.14 to 0.31 mM. Since there was no increase of cell number in this range of C_{Gln}^{in} it can be assumed that growth was not limited by the availability of glutamine under these conditions. The increased consumption of glutamine was accompanied by a corresponding increase of ammonium production. Lactate concentration was below 1 mM throughout the cultivation. In the culture series C2, N_t increased with increasing glucose in the feed (C_{Gln}^{in}) at low concentration range

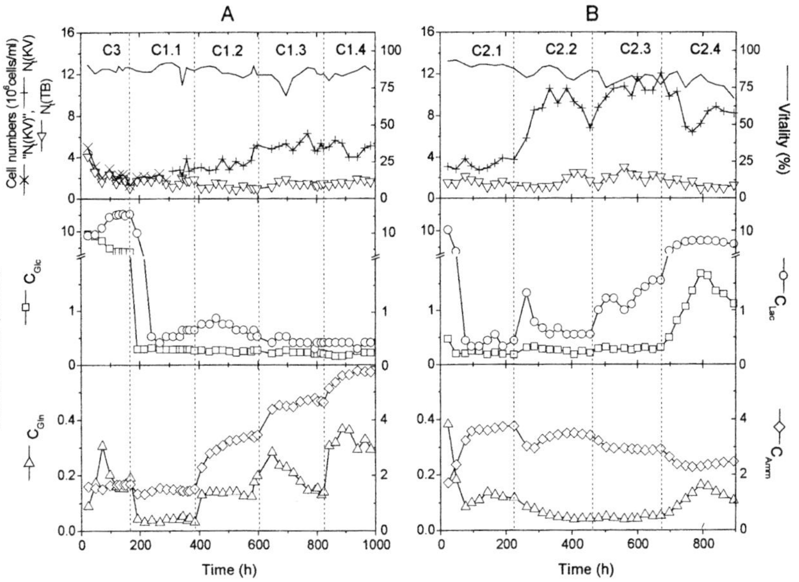

Figure 6 Total cells numbers determined by counting cells and cell nuclei (N_t), viability, and concentrations of glucose, lactate, glutamine, and ammonium of continuous culture of a BHK-21 cell line at a dilution rate of 1.48 day^{-1}. (A) Culture series 1 had a constant glucose concentration (2 mmol/L) but increasing glutamine concentrations (1–5 mmol/L) in the feed. For the first 165 hr (C3) the culture had an glucose concentration of 20 mmol/L and a glutamine concentration of 3 mmol/L in the feed. (B) Culture series 2 had a constant glutamine concentration(2.5 mmol/L) but increasing glucose concentrations (1–10 mmol/L) in the feed (22). [From Ref. (22).]

as well (Fig. 6B). It reached a maximum value of 11×10^6 cells/mL at an initial glucose concentration of 6 mM and decreased afterwards with further increases of $C_{\text{Glc}}^{\text{in}}$. The residual glucose concentration was in a similar range to that of culture C1 (0.18–0.29 mM) at $C_{\text{Glc}}^{\text{in}}$ below 10 mM. At $C_{\text{Glc}}^{\text{in}} = 10$ mM, C_{GlC} increased to 1.36 mM and was obviously not growth limiting. Throughout the cultivation, C_{Gln} was in the limitation range (between 0.04 and 0.13 mM) as found in culture series C1. Culture series C2 was accompanied with an increased production of lactate. In contrast, the ammonium concentration decreased.

Estimation of Specific Rates of Cell Metabolism in Continuous Cultures

Steady-state data from continuous culture can be used to calculate the specific rates of cell metabolism under clearly defined physiological conditions. Under steady-state conditions, the specific rates of cell growth, death, nutrient consumption, and product formation of a conventional continuous culture can be calculated as follows:

$$\mu = D\frac{N_t}{N_v} \tag{9.10}$$

$$k_d = \mu - D \tag{9.11}$$

$$q_{\text{Glc}} = (C_{\text{Glc}}^{\text{in}} - C_{\text{Glc}})\frac{D}{N_v} \tag{9.12}$$

$$q_{\mathrm{Lac}} = (C_{\mathrm{Lac}} - C_{\mathrm{Lac}}^{\mathrm{in}})\frac{D}{N_{\mathrm{v}}} \tag{9.13}$$

$$q_{\mathrm{Gln}} = [(C_{\mathrm{Gln}}^{\mathrm{in}} - C_{\mathrm{Gln}}) \cdot D - k \cdot C_{\mathrm{Gln}}]\frac{1}{N_{\mathrm{v}}} \tag{9.14}$$

$$q_{\mathrm{Amm}} = [(C_{\mathrm{Amm}} - C_{\mathrm{Amm}}^{\mathrm{in}}) \cdot D - k \cdot C_{\mathrm{Gln}}]\frac{1}{N_{\mathrm{v}}} \tag{9.15}$$

where D is the dilution rate (hr^{-1} or day^{-1}); C_j^{in} and C_j are the concentrations of substance j in the feed and in the culture, respectively.

In a perfusion culture (cell recycle) with complete cell separation (i.e., by means of membrane filtration) a cell-free permeation stream is withdrawn at a flow rate of F_{p}. A stream of culture with cells at the same concentration as in the reactor is also withdrawn at a flow rate of F_{B}. The perfusion rate (P) of the culture is an important process variable:

$$P = \frac{F_{\mathrm{P}} + F_{\mathrm{B}}}{V} \tag{9.16}$$

where V is the reactor volume of the culture. In this case, the cell bleed rate (D_{B}) is used for the calculation of cell growth and death rates:

$$D_{\mathrm{B}} = \frac{F_{\mathrm{B}}}{V} \tag{9.17}$$

$$\mu = D_{\mathrm{B}}\frac{N_{\mathrm{t}}}{N_{\mathrm{v}}} \tag{9.18}$$

$$k_{\mathrm{d}} = \mu - D_{\mathrm{B}} \tag{9.19}$$

Thus, the perfusion rate P should be used instead of the dilution rate D in Eqs. (9.12)–(9.15). In practice, cell retention devices such as centrifuges and spin filters may not completely separate cells from culture broth. In such a case, the above equations need further modifications. The term retention ratio (r) may be used to describe the ratio of cell number in the permeate to the total cell number in culture broth:

$$r = \frac{N_{\mathrm{P}}}{N_{\mathrm{t}}} \tag{9.20}$$

where N_{P} is the cell number per unit permeate volume. A modified form of Eq. (9.17) is needed to describe the real or effective cell bleed rate (D_{B}^{*}):

$$D_{\mathrm{B}}^{*} = \frac{F_{\mathrm{B}} + F_{\mathrm{P}} \cdot r}{V} = D_{\mathrm{B}}(1 - r) + rP \tag{9.21}$$

For high cell density perfusion culture cell lysis may considerably contribute to the death of cells (41,74). In such a case, caution should be taken in estimating the total cell number N_{t} in the above equations. The measurement of release in the culture medium of the cytoplasmic enzyme lactate dehydrogenase has been suggested as a means to estimate the total cell number (74).

INFLUENCES OF ENVIRONMENTAL AND PHYSIOLOGICAL CONDITIONS AND RATE EQUATIONS

A large number of factors have been reported in the literature to affect the kinetics of growth, death, metabolism, and product formation of animal cells in bioreactors. These include environmental physicochemical parameters, composition and levels of nutrients, growth factors and the formation of metabolites, and some not yet identified autocrine factors. In this section, only some of these factors are addressed in view of kinetic analysis and optimization of cell culture. More general information about the effects of some of the factors and cell physiology can be found in this book.

Effect of Nutrients and their Consumption Rates

Glucose and glutamine are the two major nutrients for animal cell culture that can strongly affect the kinetics. Their effects on the kinetics of cell culture have been shown in Figs. 1 and 6 for the batch and continuous cultures of BHK cells, respectively. Table 1 summarizes the steady-state specific rates of cell growth, nutrient consumption, and product formation for the continuous culture under varied initial glucose and glutamine concentrations in medium. The specific growth rates of the continuous cultures are relatively constant (1.0–1.2 day^{-1}).

Despite a relatively constant growth rate, the cultures exhibit strongly varied specific rates of nutrient consumption and metabolite formation. As found for several other cell lines in batch and continuous cultures (75), q_{Glc} strongly depends on the residual glucose concentration. For the BHK cell line it begins to significantly increase at $C_{Glc} > 0.2\,mM$. Similarly, q_{Gln} is mainly affected by the residual glutamine concentration and begins to increase notably at $C_{Gln} > 0.1\,mM$ except for the culture C2-4, which has an obvious glucose excess. q_{Lac} is somewhat less sensitive to glucose compared to q_{Glc}. The former begins to increase significantly at $C_{Glc} > 1\,mM$. q_{Amm} is similarly sensitive to glutamine as q_{Gln}.

Obviously, the linear maintenance model previously used to describe the nutrient consumption of animal cells (4) cannot be generally applied. Based on the observation that the nutrient consumption rate is not only dependent on cell growth rate

Table 1 Total Cell Number, Residual Concentrations of Glucose and Glutamine and Specific Rates of Cell Growth and Metabolism at Steady or "Quasi-Steady" States in Continuous Cultures Shown in Fig. 6

Continuous culture #	C1-1	C1-2	C1-3	C1-4	C2-1	C2-2	C2-3	C2-4
N_t(max.) (10^6 cells/mL)	3.3	4.2	4.9	5.0	3.8	8.9	11.1	8.5
C_{Glc} (mmol/L)	0.27	0.26	0.20	0.22	0.18	0.22	0.29	1.36
C_{Gln} (mmol/L)	0.04	0.16	0.14	0.31	0.12	0.04	0.05	0.13
μ (day^{-1})	1.01	1.01	1.03	0.98	1.01	1.08	1.10	1.20
q_{Glc} (mmol/10^9 cells hr)	0.022	0.018	0.016	0.015	0.010	0.020	0.024	0.050
q_{Lac} (mmol/10^9 cells hr)	0.009	0.006	0.004	0.004	0.004	0.003	0.006	0.038
q_{Gln} (mmol/10^9 cells hr)	0.011	0.024	0.028	0.042	0.025	0.014	0.011	0.014
q_{NH3} (mmol/10^9 cells hr)	0.019	0.034	0.041	0.047	0.041	0.018	0.012	0.014

Cell number determined by trypan blue exclusion method.
[From Ref. (22).]

but also strongly on the nutrient concentration (Table 1), Zeng and Deckwer (75) proposed the following kinetic expressions for the rates of consumption of glucose and glutamine of animal cells:

$$q_{\mathrm{Glc}} = m_{\mathrm{Glc}} + \frac{\mu}{Y_{\mathrm{Glc}}^{\max}} + \Delta q_{\mathrm{Glc}}^{\mathrm{Glc(max)}} \frac{C_{\mathrm{Glc}} - C_{\mathrm{Glc}}^{*}}{C_{\mathrm{Glc}} - C_{\mathrm{Glc}}^{*} + K_{\mathrm{Glc}}^{\mathrm{Glc}}} + \Delta q_{\mathrm{Gk}}^{\mathrm{Gln(max)}} \frac{C_{\mathrm{Gln}} - C_{\mathrm{Gln}}^{*}}{C_{\mathrm{Gln}} - C_{\mathrm{Gln}}^{*} + K_{\mathrm{Glc}}^{\mathrm{Gln}}} \tag{9.22}$$

$$q_{\mathrm{Gln}} = m_{\mathrm{Gln}} + \frac{\mu}{Y_{\mathrm{Gln}}^{\max}} + \Delta q_{\mathrm{Gln}}^{\mathrm{Glc(max)}} \frac{C_{\mathrm{Gln}} - C_{\mathrm{Gln}}^{*}}{C_{\mathrm{Gln}} - C_{\mathrm{Gln}}^{*} + K_{\mathrm{Gln}}^{\mathrm{Glc}}} + \Delta q_{\mathrm{Gln}}^{\mathrm{Gln(max)}} \frac{C_{\mathrm{Gln}} - C_{\mathrm{Gln}}^{*}}{C_{\mathrm{Gln}} - C_{\mathrm{Gln}}^{*} + K_{\mathrm{Gln}}^{\mathrm{Gln}}} \tag{9.23}$$

for $C_{\mathrm{Glc}} \geq C_{\mathrm{Glc}}^{*}$ and $C_{\mathrm{Gln}} \geq C_{\mathrm{Gln}}^{*}$, Eqs. (9.22) and (9.23) express the substrate consumption rates as a sum of the substrate consumption rate under substrate limitation and two additional consumption rates owing to the excess of glucose and glutamine, respectively. Similarly to the substrate consumption model of microbial cells under substrate-sufficient conditions (23), Eqs. (9.22) and (9.23) may be called "excess kinetics" of animal cells. m_{Glc} and m_{Gln} are the maintenance requirements for glucose and glutamine that are often negligible in conventional batch and continuous cultures. C_{Glc}^{*} and C_{Gln}^{*} are defined as the concentrations of glucose and glutamine under dual limitations. They are often very small compared to concentrations of glucose (C_{Glc}) and glutamine (C_{Gln}) in the culture and can be neglected. $Y_{\mathrm{Glc}}^{\max}$, $Y_{\mathrm{Gln}}^{\max}$, $\Delta q_{\mathrm{Glc}}^{\mathrm{Glc(max)}}$, $\Delta q_{\mathrm{Glc}}^{\mathrm{Gln(max)}}$, $\Delta q_{\mathrm{Gln}}^{\mathrm{Gln(max)}}$, $\Delta q_{\mathrm{Gln}}^{\mathrm{Glc(max)}}$, $K_{\mathrm{Glc}}^{\mathrm{Glc}}$, $K_{\mathrm{Glc}}^{\mathrm{Gln}}$, $K_{\mathrm{Gln}}^{\mathrm{Gln}}$ and $K_{\mathrm{Gln}}^{\mathrm{Glc}}$ are constants with different physiological meanings (75).

Equations (9.22) and (9.23) satisfactorily described experimental data from batch, fed-batch, and continuous cultures of several animal cell lines. For the continuous culture shown in Fig. 6 (data summarized in Table 1) the following rate equations were established (22):

$$q_{\mathrm{Glc}} = \frac{\mu}{3.19} + 0.25 \frac{C_{\mathrm{Glc}} - 0.06}{C_{\mathrm{Glc}} + 7.74} - 0.017 \frac{C_{\mathrm{Gln}} - 0.07}{C_{\mathrm{Gln}} + 0.59} \quad (\mathrm{mmol}/10^9 \text{ cells h}) \tag{9.24}$$

$$q_{\mathrm{Gln}} = \frac{\mu}{3.19} + 0.044 \frac{C_{\mathrm{Gln}} - 0.05}{C_{\mathrm{Gln}} + 0.22} \quad (\mathrm{mmol}/10^9 \text{ cells h}) \tag{9.25}$$

Using the rate equations described above the interaction and regulation of glucose and glutamine utilization of animal cells were quantitatively analyzed (75). The results indicated that, whereas q_{Glc} is affected by glutamine, q_{Gln} appears to be not or less significantly affected by glucose. This is in accordance with the flux and enzyme analyses of metabolism of different animal cell lines reported by Neermann and Wagner (76). The authors found that the flux from pyruvate to acetyl-CoA is absent or very weak due to the lack of the enzyme pyruvate dehydrogenase. The previously reported effect of glucose on glutaminolysis in literature seems to be attributed to the indirect influence of glucose on cell growth and residual glutamine concentration. On the other hand, it is shown that glutamine can have positive or negative influences on the glycolysis, depending on the nature of growth limitation. The kinetic analysis also indicated that the utilization rate of glucose and glutamine is mainly affected by the residual concentrations of the respective compounds and less by the growth rate. Similar kinetic expressions as Eqs. (9.22) and (9.23) may also be used for the uptake or consumption of other major nutrients.

Effect of Metabolic Byproducts and their Formation Rates

Ammonium and lactate are two major byproducts in mammalian cell cultures. Lactate excretion is due to incomplete oxidation of glucose in the glycolysis pathway; pyruvate (the endproduct of this pathway) is transformed into lactate to maintain the oxidative state of the cell. The main source of ammonium formation is amino acid metabolism, particularly glutamine which serves as a protein constituent and also as the main energy source (77–79). At low concentration of glutamine, the consumption of other amino acids, especially essential amino acids, can significantly contribute to the formation of ammonium (69).

Ammonium and lactate are often considered to be the dominating factors inhibiting cell growth and indeed have been shown to be toxic above certain concentrations (15,80). For example, ammonium level >2 mM or lactate level >20 mM were reported to inhibit cell growth and MAb production of hybridoma and CHO cells (2,5,15,81). The toxic action of lactate is probably due to the effects of pH and osmolarity of culture medium at relatively high concentration (4,15). An elevated ammonium concentration may reduce metabolic efficiency by forcing excretions of potentially valuable intermediate metabolites, such as alanine, to achieve ammonium detoxification (82,83). High concentrations of lactate and ammonium up to the critical levels can occur at the end of batch culture or in high cell density cultures. Therefore, removing or reducing lactate and ammonium formation has often been suggested as an important goal of process optimization. However, as shown by Zeng et al. (26) and discussed in section "Rate laws of cell growth and death," the inhibitory effects of ammonium and lactate may be overestimated or even misinterpreted in many cases.

The formation rate of ammonium and lactate is mainly determined by the nutrient concentrations and growth rate. In animal cell culture there are clear stoichiometric relationships between the consumption of nutrients and formation of metabolites (69,84). Rate equations for the formation of the two major metabolites lactate and ammonium were therefore proposed which have a similar structure like Eqs. (9.22) and (9.23) (34):

$$q_{\mathrm{Lac}} = m_{\mathrm{Lac}} + \frac{\mu}{Y^{\mathrm{m}}_{\mathrm{Lac}}} + \Delta q^{\mathrm{m}}_{\mathrm{Lac}} \frac{C_{\mathrm{Gln}}}{C_{\mathrm{Gln}} + K^{\mathrm{Glc}}_{\mathrm{Lac}}} \tag{9.26}$$

$$q_{\mathrm{NH_3}} = m_{\mathrm{NH_3}} + \frac{\mu}{Y^{\mathrm{m}}_{\mathrm{NH_3}}} + \Delta q^{\mathrm{m}}_{\mathrm{NH_3}} \frac{C_{\mathrm{Gln}}}{C_{\mathrm{Gln}} + K^{\mathrm{Gln}}_{\mathrm{NH_3}}} \tag{9.27}$$

The parameters $m_{\mathrm{Lac}}, m_{\mathrm{NH_3}}, Y^{\mathrm{m}}_{\mathrm{Lac}}, Y^{\mathrm{m}}_{\mathrm{NH_3}}, \Delta q^{\mathrm{m}}_{\mathrm{Lac}}, \Delta q^{\mathrm{m}}_{\mathrm{NH_3}}, K^{\mathrm{Glc}}_{\mathrm{Lac}}$ and $K^{\mathrm{Gln}}_{\mathrm{NH_3}}$ are constants having similar physiological meanings as those of Eqs. (9.22) and (9.23).

From kinetic analysis of nutrient consumption and byproduct formation it can be concluded that for an effective utilization of nutrients and for a reduction of toxic byproducts the glucose and glutamine levels should be controlled at low levels (e.g., ca. 0.1–0.5 mM depending on cell lines). The critical nutrient levels may also depend on cell density under certain conditions. For example, Ljunggren and Häggström (78) investigated glucose and glutamine-limited fed-batch cultures of hybridoma cells and found that glucose and glutamine concentrations below 1 mM did not limit cell growth. Glucose limitation alone did not reduce ammonium formation, in comparison to a reduced ammonium release by about 50 and 80% caused by glutamine limitation and dual glucose and glutamine limitation, respectively (78). The metabolism of glucose and amino acids of a recombinant BHK cell line at low levels of nutrient concentration (glucose concentration <5 mM and glutamine concentration <1 mM) was quantitatively characterized by Linz et al. (22). It was demonstrated that the uptake rates of glucose and glutamine are markedly reduced at low levels of glucose and glutamine, resulting in a more efficient energy metabolism and biosynthesis, and a reduced formation rate of lactate and ammonium.

Effect of Temperature

Cultivation temperature can significantly influence the growth and metabolism of animal cell cultures (85–89). Animal cell cultures are normally carried out at physiological temperature (37°C). Temperature stress (heat shock) was known to adversely affect many functions in animal cells [for recent reviews see Refs. (90,91)]. The most common cellular responses to heat shock include changes in gene expression, resulting in the synthesis of specific stress proteins, i.e., heat-shock proteins (HSPs). HSPs are a group of proteins that are highly conserved in all organisms from bacteria to mammals, many of which act as molecular chaperones for protein folding.

Recent studies focusing on the response of cells to decreased temperatures led to the discovery of cold inducible proteins. However, unlike HSPs, cold-shock proteins are not conserved among all species [for review see Ref.(92)]. Lowering the culture temperature generally suppresses cell growth but its effects on cellular productivity are variable among different cell lines and expression systems (85,93–96). It has been reported that a temperature shift from 37 to 30–33°C can lead to a reduced growth rate, prolong the total generation time, and increase the viability of cultured mammalian cells (89). Cold-induced growth arrest was found to be G1-phase specific in CHO batch cultures (87). The prolonged culture viability at lower temperatures is considered to be a result of delayed onset of apoptosis that causes a rapid decrease in the percentage of cells in S-phase in CHO batch culture (87). CHO cells engineered to synthesize secreted alkaline phosphatase (SEAP) were also characterized by shifting the cultivation temperature from 37 to 30°C (89,87). This temperature shift resulted in a growth arrest mainly in the G1-phase of the cell cycle and a concomitant increase of specific productivity. However, the effects of low temperature on the protein production rate of mammalian cells seem to depend on cell lines and protein products (85). For example, Sureshkumar and Mutharasan (96) reported that hybridoma cells cultivated at low temperatures had a reduced specific productivity of MAb. However, Kaufmann et al. (89) found an enhanced SEAP productivity in CHO cells grown at 30°C. An up to 1.7-fold higher specific productivity and prolonged culture viability resulted in an overall 3.4-fold higher product yield in low-temperature cultivations compared to standard cultivations at 37°C. Although several reports suggest that controlled proliferation increases the productivity of mammalian cells, its influence on product quality should be investigated to better evaluate the potential of this strategy. The majority of pharmaceutically important proteins produced with mammalian cells require post-translational modifications for full therapeutic efficacy. Glycosylation is a particularly critical parameter for product quality because oligosaccharide structures can influence the solubility, stability, bioactivity, immunogenicity, and pharmacokinetics of a pharmaceutical protein (98). In this connection, it may be stated that cultivations at lower temperatures are favorable because of the more consistent quality (glycosylation and molecule fragmentation) and improved molecule integrity of protein products (99).

Culture temperature also affects the metabolic rates of animal cells (100). Low temperatures normally reduce cellular metabolic activities as reflected by the decreased specific rates of glucose uptake, lactate production, glutamine uptake, ammonium production, oxygen uptake, and CO_2 evolution. The decreased oxygen demand at lower temperatures makes it possible to support a higher cell concentration in a bioreactor (86). Furthermore, lower temperatures also result in a significant decrease in the $NaHCO_3$ addition for pH control.

Ludwig et al. (101) studied the influence of temperature on the shear sensitivity of adherent BHK-21 cells grown at temperatures between 28 and 39°C. It was found that decreasing the temperature lowered the growth rate and increased the shear resistance in BHK cells. Cell morphology also changed at low temperatures. At 28°C the cells tented mostly to be more spherical or triangular, as opposed to a confluent monolayer at 37°C. This phenomenon was explained by an increased rigidity of the lipid bilayer of cell membrane.

Weidemann et al. (85) compared batch and repeated-batch cultures of BHK-21 cells for the production of AT III at 33 and 37°C. At these temperatures, the specific growth rate of the cells was 0.50 and 0.62 days^{-1}, while the specific glucose uptake rate was 0.45 and 0.58 ng/cell/day, respectively. A higher product titer was reached at 33°C, a temperature at which the medium demand can be rationalized due to reduced nutrient consumption.

Jorjani and Ozturk (88) quantitatively studied the effect of temperature on three different mammalian cell lines (BHK, murine hybridoma, and CHO). In all cases, the specific oxygen uptake rate (OUR) qO_2 decreased by about 10% for one degree of reduced temperature. The effect of temperature was observed to be exponential and can be well described by Arrhenius' equation:

$$qO_2 = qO_2^0 \cdot e^{-E/RT} \tag{9.28}$$

where E is the activation energy, R is the ideal gas constant, and T is the absolute temperature in degrees Kelvin. E was found to be similar for different cell lines (between 80 and 90 kJ/mol), indicating a similar mechanism for the effect of temperature on oxygen consumption. The effects of temperature on cell growth, nutrient consumption, and formation of metabolites and a protein product in high-density perfusion cultures were also quantitatively studied for the temperature range 34–37°C. Whereas the nutrient consumption and formation of lactate and ammonium decreased with decreasing temperature in a nearly linear manner, the formation of the protein product increased. Kinetic expressions for these observed effects have not been established yet. It may be possible that Arrhenius equation can also be applied to describe the temperature effects on primary cell metabolism.

In conclusion, decreasing the culture temperature from 37 to about 30–34.0°C appears to be advantageous for many animal cell cultures, especially for high-density perfusion cultures of animal cells.

Effect of pH

pH affects the metabolism, growth, and protein production of animal cell culture in various ways. The effects of pH on cell metabolism and growth have been examined in cell lines of HeLa (102), lymphoblastoid (103), hybridoma (4,104), and HL60 (105). The optimum pH for growth varies with cell lines. It is worth mentioning that a pH excursion as small as 0.2 pH units can profoundly influence the growth, metabolism, and productivity of animal cells in some cases (106,107). Problems in product quality and uniformity may also arise if the culture pH is not properly controlled. In fact, an effective pH control is not always obtainable in cell culture, especially in large-scale bioreactors, since spatial heterogeneities in the culture system may exist due to imperfect mixing or cell density effect (108) that results in a pH gradient. For example, Akatov et al. (108) showed that in T-flask cells of Chinese hamster fibroblasts had a local pH of 6.5, although the pH of the bulk liquid remained at 7.6.

Osman et al. (109,110) investigated the effect of pH shifts on cell growth and productivity in batch culture of a GS-NS0 mouse myeloma cell line. They reported that pH shifts above 0.2 units caused a transient increase in apoptotic cell death. However, cultures shifted to pH values between 7.0 and 8.0 continued to grow and the apoptotic fraction returned to the initial levels. Cultures shifted to pH values above pH 8.0 and below pH 7.0 did not recover, resulting in cell death. After the pH shift, a maximum specific growth rate was observed over the range of pH 7.3–7.5 and the maximum viable cell number was observed at pH 7.3. A maximum volumetric antibody production, resulting from increased culture longevity, was found at pH 7.0. It was also observed that glucose consumption rate increased with increasing pH. Exposure of cells to a pH value >8.5 for more than 10 min caused a decrease in the proportion of viable cells and induced a lag phase in cell growth.

pH can influence the structure of proteins either directly or indirectly by influencing the cellular glycosylation pathways. For example, the extracellular pH value was found to affect the glycoform distribution of IgG from hybridomas (111) and mPL-I in CHO cells (112). Glycoforms of mPL-I with similar molecular sizes were expressed between pH values of 7.2 and 8.0. However, a decreased glycosylation of mPL-I occurred at both lower pH ($pH < 6.9$) and higher pH ($pH > 8.22$).

The intracellular pH (pHi) plays a decisive role in mediating the pH effects mentioned earlier. It can be conveniently measured by using flow cytometry with carvoxy-SNARF-1 (113). Ishaque and Al-Rubeai (113) determined the relationships among pHi, apoptosis, and cell cycle in hybridoma cells. It is reported that temporal changes in the distribution of proliferative capacity (S-phase), metabolic activity (pHi), and cell death population dynamics can be effectively and reliably determined using this approach. It is shown, for example, that intracellular acidification precedes the occurrence of apoptosis during batch culture, suggesting that the decrease in pHi can be used as an indicator of cellular deterioration and cell death. pHi has also been reported to affect the localization of hexokinase, resulting in an increase in glycolysis with increasing pHi (114).

Dissolved Oxygen and Oxygen Uptake Rate

Dissolved oxygen concentration (DO or pO_2) is an important variable for mammalian cell culture. Oxygen is essential for the efficient generation of ATP and must be continually supplied to cultured cells due to the low solubility of oxygen in culture medium (7.8 mg/L in water at 25°C). Elevated oxygen concentration (usually above 100% of air saturation), or hyperoxia, has a negative effect on cell growth (115–117). Hyperoxia can damage cellular macromolecules such as DNA (118), proteins (119–121), and lipids (121). This can, in turn, damage the cell membrane and induce cell death (123). The toxicity of oxygen is associated with the production of intracellular reactive oxygen species (ROS) (123) that damage a variety of cellular components or impair their functions.

Dissolved oxygen also has profound effects on cellular metabolism. A frequently observed trend is that once DO drops below a critical value (typically between 1 and 10% of air saturation) the oxygen consumption rate decreases and the specific glucose consumption and lactate production rates increase. The optimum DO value for antibody production differs from that for cell growth (124,125). Miller et al. (125) reported an increase in the total and viable cell number when DO was decreased to a critical value of 0.5% air saturation with continuous culture of a hybridoma cell line, while the optimum DO for antibody production was 50% air saturation. Also Ozturk and Palsson (126) investigated the effects of DO on hybridoma cell in a continuous stirred tank bioreactor where the DO value was varied between 0 and 100% of air saturation. Cell growth was inhibited at both high and low DO values. Cell viability was higher at low DO. Interestingly, they showed that the consumption rates of glucose, glutamine, and oxygen remained relatively constant at DO values above 1% of air saturation. But when growth became oxygen-limited (DO 1–0%), the consumption rates of glucose and glutamine increased by 2–3-folds and a corresponding increase of formation of lactate and ammonium was observed. It was concluded in this study that the antibody concentration was the highest at DO = 35% of air saturation, while the specific antibody production rate was insensitive to DO.

While a low level of oxygen concentration (e.g., below 10% of air saturation) is normally beneficial for keeping a high cell viability and thus for extending the production phase, it may become limiting for cellular metabolism, leading to conditions of hypoxia. Under these conditions, certain cellular functions such as mitochondrial activity can be impaired. The mitochondrial activity has been reported to strongly depend on the oxygen level in the range of 2–10% of air saturation for different cell lines (125–128). The declined activity of mitochondria can reduce the specific secretion rate of proteins and also the volumetric accumulation of product in most cases. Thus, although hypoxia is considered beneficial in delaying cell death and minimizing chromosomal damage (129), these effects may not be as critical as ensuring an appreciate delivery of oxygen to cells in a bioreactor to avoid starvation or low product yield. Animal cells used for protein production should therefore be cultivated within an optimal range of DO. The optimum oxygen concentration for cell growth varies with cell type and has been reported to be in the range of 10–50% of air saturation (115,125,126). For the formation of antibodies by hybridomas a DO range between 20 and 70% of air saturation has been reported to be optimal for the volumetric productivity.

Under optimal oxygen supply the specific oxygen consumption rate of most cultured animal cells ranges from 0.15 to 0.8×10^{-9} mmol O^2/cell/hr (Table 2). The volumetric oxygen demand of 0.1–1.0 mmol O^2/L/hr at a typical cell concentration of 2×10^6 cells/mL is much lower than the oxygen demand (13–199 mmol O^2/L/hr) observed for mold and yeast cultures (139). However, unlike microorganisms, animal cells do not have cell wall and cannot be stirred vigorously in culture.

Table 2 Specific Oxygen Uptake Rate (qO_2) Reported for Different Cell Lines

qO_2 (10–12 mol/cells/h)	Cell line	References
0.15–0.36	KS1/4 (hybriboma)	130
0.21–0.25	NB1 (hybridoma)	131
0.234 ± 0.014	Cla (hybridoma)	132
0.19–0.4	AB20143.2 (hybridoma)	125, 133
0.023–0.087	167.4G5.3 (hybridoma)	126
0.33–0.37	HB-32 (hybridoma)	53
0.2	HFN7.1 (hybridoma)	134
0.46 ± 0.05	MAK (hybridoma)	135
0.23–0.42	X-D (hybridoma)	12
0.219–0.406	NS0 (myeloma)	136
0.199	CHO	137
0.5/0.8	CHO	138

[From Refs. (144,145).]

Although hybridomas have been successfully grown in airlift reactors (131), it has been shown that mammalian cells can be damaged by gas bubbles used to increase the gas–liquid interfacial area in sparged reactors (140,141). An increase in oxygen demand will result from the use of perfusion cultures or fed-batch cultures that can increase the cell concentration by 10–50 times (73,135). With the development of high cell density culture aeration strategies that can meet the high oxygen demand while efficiently removing carbon dioxide and minimizing bubble damage to the cells have become an important issue (142).

The rate of oxygen uptake can be estimated from the consumption rates of nutrients. Stoichiometrially, the specific oxygen consumption rate (qO_2) of mammalian cells can be written as (69):

$$qO_2 = \alpha\left(q_{\mathrm{Glc}} - \frac{1}{2}q_{\mathrm{Lac}}\right) + \beta q_{\mathrm{TAA}} \tag{9.29}$$

where q_{TAA} is the specific consumption rate of total amino acids. α and β are stoichiometric coefficients representing the consumption of oxygen due to oxidation of glucose and amino acid in the energy metabolism. Originally, α is assumed to be 6 mol/mol based on the assumption that glucose that is not converted into lactate is completely oxidized in the tricarboxylic acid cycle (135). An examination of experimental data from three hybridoma cell cultures (5,13,143) reveals that this assumption leads to a unrealistic negative value for β. Obviously, glucose that is not converted to lactate is not completely oxidized (76). α should be therefore estimated from experimental data. α appears to be not cell line specific for hybridomas culture, whereas the value of β should be individually estimated for each culture (69). The values of α and β estimated for three cultures are summarized in Table 3. Equation (9.29) described the experimental data satisfactorily. It is interesting to note in Table 3 that α is only about half of the theoretical value assumed for complete oxidation of glucose. The two cultures, which were grown on media with serum and supplements rich in amino acids, respectively, have also almost the same β value. The culture that was grown on a serum-free medium has a much lower β value, suggesting that a greater portion of the amino acids is used for

Table 3 Stoichiometric Coefficients for Specific Oxygen Consumption Rate [Eq. (9.35)]

Culture	(α mol/mol)	(β mol/mol)	Reference
Hybridoma AB2-143.2 (medium with serum)	2.89	1.64	133
Hybridoma X-D (medium without serum)	2.89	0.76	13
Murine hybridoma[a]	2.89	1.69	143

[a] Growth on medium with or without Primatone RL, an enzymatic hydrolysate of animal tissue. In the medium without Primotone RL amino acids equivalent in composition and concentration to those of Primotone RL were added.

biosynthesis in this case. Under conditions of glutamine excess, Eq. (9.29) can be reduced to:

$$
\begin{aligned}
qO_2 &= \alpha\left(q_{\text{Glc}} - \frac{1}{2}q_{\text{Lac}}\right) + \beta\frac{q_{\text{TAA}}}{q_{\text{Gln}}}q_{\text{Gln}} \\
&= \alpha\left(q_{\text{Glc}} - \frac{1}{2}q_{\text{Lac}}\right) + \beta^{*}q_{\text{Gln}}
\end{aligned}
\tag{9.30}
$$

where β^{*} is a constant that is about 1.6 times as high as β.

A useful parameter for online control of animal cell culture is the ratio of consumption rates of oxygen (OUR) to glucose (Q_{Glc}). The ratio OUR/Glc was used to control the feeding rate of glucose and other nutrient components so as to keep the glucose concentration at a relatively low level (135). However, the latter changes during the cultivation. This variation of the stoichiometric ratio can be described by a modified form of Eq. (9.30):

$$
\frac{\text{OUR}}{Q_{\text{Glc}}} = \frac{qO_2}{q_{\text{Glc}}} = \alpha\left(1 - \frac{1}{2}\frac{q_{\text{Lac}}}{q_{\text{Glc}}}\right) + \beta\frac{q_{\text{TAA}}}{q_{\text{Gln}}}\frac{q_{\text{Gln}}}{q_{\text{Glc}}}
\tag{9.31}
$$

The rate equations presented above for the product formation and consumption of nutrients may be directly used to predict the ratio of oxygen and glucose consumptions as functions of residual concentrations of glucose and glutamine and the cell growth rate. The ratio of consumption of total amino acids to glutamine can be estimated as:

$$
\frac{q_{\text{TAA}}}{q_{\text{Gln}}} = \text{TAA}_{\text{min}}e^{A/C_{\text{Gln}}}
\tag{9.32}
$$

where TAA_{min} is the minimum ratio of consumption of total amino acids (including glutamine) to glutamine and typically has a value about 1.6 mol/mol for most cell cultures (69). A is a constant and has a typical value of 0.01–0.02. Similarly, ratio equations for $q_{\text{Lac}}/q_{\text{Glc}}$ and $q_{\text{Gln}}/q_{\text{Glc}}$ have been derived that can be used to convert Eq. (9.31) into the following simplified functional form:

$$
\begin{aligned}
\frac{\text{OUR}}{Q_{\text{Glc}}} = \frac{qO_2}{q_{\text{Glc}}} &= \alpha\left(1 - \frac{\text{LG}_{\text{max}}}{2}\cdot\frac{C_{\text{Glc}}}{C_{\text{Glc}} + K_{\text{LG}}}\right) \\
&\quad + \beta\cdot\text{TAA}_{\text{min}}\cdot e^{A/C_{\text{Gln}}}\cdot\text{GG}_{\text{min}}\cdot e^{B/C_{\text{Glc}}}
\end{aligned}
\tag{9.33}
$$

In Eq. (9.33), LG_{max} is the maximum lactate yield from glucose, having a value of about 1.5–1.7 mol/mol for most cell lines; K_{LG} is a saturation constant reflecting the sensitivity of lactate formation to the residual concentration of glucose, and has values between 0.1 and 0.3 mmol/L for most cell lines; GG_{min} is the minimum ratio of consumptions of glutamine to glucose. It typically has values between 0.2 and 0.4 mol/mol. B is a constant that depends on the cell lines. For hybridoma cells B has a typical value of 0.03–0.05 mM. Equations (9.31) and (9.33) can be used to predict the upper and low limits for OUR/Glc and the influence of nutrient availability.

Xiu et al. (145) presented a method to estimate the rates of oxygen uptake and carbon dioxide evolution of animal cells based on material and energy balances. For this purpose, lumped compositions, molecular weight, and reductance degree of cellular proteins, MAb, biomass, and amino acid consumption (excluding glutamine and alanine) are used, which were found to be relatively constant for different hybridoma cell lines and can be used as regularities. The calculated rates of oxygen uptake and carbon dioxide evolution agreed well with experimental values of several different cultures reported in the literature. The method of Xiu et al. (145) gives comparable results as calculated with Eq. (9.36) or on the basis of a detailed metabolic reaction network.

Dissolved Carbon Dioxide

Carbon dioxide is produced via catabolic reactions and is required for the synthesis of pyrimidines, purines, and fatty acids in animal cells. CO_2 produced by cells enters the culture broth in a dissolved form. In aqueous solution, CO_2 forms carbonic acid, which dissociates to bicarbonate and carbonate ions. The reactions involved can be written as:

$$CO_2(\text{gas}) \Leftrightarrow CO_2(\text{aq.}) + H_2O \Leftrightarrow H_2CO_3 \Leftrightarrow H^+ + HCO_3^- \Leftrightarrow 2H^+ + CO_3^{2-} \quad (9.34)$$

The dissociation constants of H_2CO_3 (K_1) and HCO_3^- (K_2) can be written as:

$$K_1 = \frac{C_{HCO_3^-} C_{H^+}}{C_{CO_2} + C_{H_2CO_3}} \quad (9.35)$$

$$K_2 = \frac{C_{CO_3^{2-}} C_{H^+}}{C_{HCO_3^-}} \quad (9.36)$$

At 37°C, $pK_1 = 6.02$ and $pK_2 = 10.3$ according to Arrua et al. (146).

If the concentration of CO_2 dissolved is in equilibrium with the gas phase, the average concentration of CO_2 ($[C_{CO_2}]^*$) throughout the liquid can be written as:

$$[CO_2]^* = H_{co_2} P \chi_{co_2} \quad (9.37)$$

where H_{co_2} is Henry's constant for CO_2 [2.99×10^{-2} mM/mmHg at 37°C; Spérandio and Paul (147)], P the total pressure of gas phase, and χ_{co_2} the mol fraction of CO_2 in exit gas. $[C_{CO_2}]^*$ in Eq. (9.37) can be taken as the sum of C_{CO_2} (aq.) and $C_{H_2CO_3}$. Thus, from Eqs. (9.35) and (9.36) and according to the definition of pH,

$$[HCO_3^-] = \frac{[C_{CO_2}]^* K_1 10^{pH}}{[H^+]^{St}} \quad (9.38)$$

$$[CO_3^{2-}] = \frac{[C_{CO_2}]^* K_1 K_2 10^{2pH}}{([H^+]^{St})^2} \quad (9.39)$$

$[H^+]^{St}$ is the standard concentration of H^+, having the value of 1 mol/L. The concentration of CO_2 dissolved in the culture broths is in general higher than that calculated by assuming CO_2 in the broth to be in equilibrium with CO_2 in the gas phase, since a pressure gradient is required for CO_2 transfer from the liquid phase to the gas phase. The extent of this difference is dependent on the evolution rate of CO_2 and the mass transfer coefficient of the cultivation systems and can be calculated according to Zeng (148).

At pH values (7–7.2) normally used for animal cell cultures the decomposition of HCO_3^- to CO_2^{2-} can be ignored. By further assuming an effective equilibrium between CO_2 (aq.) and HCO_3^-, Eq. (9.34) can be simplified as:

$$\mathrm{CO_2}(gas) \longleftrightarrow \mathrm{CO_2(aq.)} + H_2O \longleftrightarrow H^+ + HCO_3^- \tag{9.40}$$

The equilibrium bicarbonate concentration ($[HCO_3^-]$; mM) at 37°C can be related to the partial pressure of CO_2 [$P_{\chi_{CO_2}}$ in Eq. (9.37), expressed here as pCO_2 in mmHg] in the gas phase and the medium pH via the following equation [147]:

$$\log[\mathrm{HCO_3}] = \mathrm{pH} + \log[p\mathrm{CO_2}] - 7.543 \tag{9.41}$$

Thus, $[HCO_3^-]$ will increase with increasing pCO_2 and/or pH. It should be emphasized again that real concentration of $[HCO_3^-]$ in a culture system is higher than that calculated by the above equation and depends on the mass transfer of the system.

The increase in $[HCO_3^-]$ under elevated pCO_2, as well as the increase in cation concentration due the addition of a base such as NaOH to control pH, results in a concomitant increase in medium osmolality. High osmolality by itself may have detrimental effects on cell growth and metabolism (see section "Effect of pH"). It is therefore important to dissect the effects of dissolved CO_2 and osmolality (149). Many results reported in literature concerning the effects of dissolved CO_2 are in fact combined effects of CO_2 and osmolality. Caution should also be taken in comparing the effects of dissolved CO_2 in different culture systems due to the fact that the dissolved concentration of CO_2 and the corresponding concentration of $[HCO_3^-]$ can change significantly depending on the mass transfer characteristics of the culture system and the physiochemical properties of the culture medium (148). For example, CO_2 accumulation is high in cultures oxygenated with a low flow rate and small bubbles of pure O_2 where pCO_2 was predicted to reach 150–200 mmHg (137,150,151). pCO_2 values in this range have been reported for a 200-L culture sparged with small amounts of O_2 (142), 1800–2500-L production bioreactors (151,152), and a high cell density perfusion bioreactor (153). The physiological pCO_2 range for proper growth of animal cells is 31–54 mmHg (154).

Carbon dioxide partial pressure in the range of 120–200 mmHg has been shown to inhibit growth and recombinant protein production and protein glycosylation (137,142,152,155). For example, Drapeau et al. (152) showed that the growth rate and the cell-specific M-CSF production of CHO cells were both 40% lower at a pCO_2 of 165 mmHg than at 53 mmHg. With another CHO cell culture, Gray et al. (137) found dose-dependent decreases in the cell density, viability, and specific production rate—such that the total productivity of a recombinant viral antigen in a 10-L perfusion bioreactor decreased by 69% at $pCO_2 = 148$ mmHg. Decreases in specific productivity of BHK-21 cells have been observed beginning at 50–80 mmHg (153). It has also been reported that recombinant protein production by infected Sf-9 insect cells is markedly delayed under 115 mmHg pCO_2 (156). Early cell death has been observed in NS/0 myeloma cell cultures with a final pCO_2 of 120 mmHg (142).

Due to the formation of bicarbonate, an increase in pCO_2 at a constant pH will result in a proportional increase in osmolality. deZengotital et al. (157) demonstrated that the growth rate of a hybridoma AB2-143.2 cell line in well-plate culture decreased with increasing pCO_2, with a 45% decrease at 195 mmHg pCO_2 under a partial osmolality compensation (to 361 mOsm/kg). Inhibition was more extensive without osmolality compensation, with a 63% decrease in growth rate at 195 mmHg pCO_2 and 415 mOsm/kg. Also, the death rate of the hybridoma cells increased with increasing pCO_2, with 31- and 64-fold increases at 250 mmHg pCO_2 for osmolality at 401 and 469 mOsm/kg, respectively. The specific glucose consumption and lactate production rates were 40–50% lower at 140 mmHg pCO_2. However, there was little further inhibition of glycolysis at higher pCO_2. But the specific antibody production rate was not significantly affected by pCO_2 or osmolality within the range tested. Interestingly, quite different results were obtained with the same hybridoma cell line in continuous culture. For example, the death rate k_d decreased only slightly at 140 mmHg in continuous culture, while in well-plate and batch cultures k_d was more than twice as great as that for the control at 140 mmHg. The authors explained this difference by the high residual nutrients and low byproduct levels in the continuous culture. It is also possible that the different concentrations of dissolved CO_2 and thus HCO_3^- in the different culture systems also contributed to the different observations.

Osmolality and Salt

Osmolality is one of the most important physical factors in mammalian cell cultures. Most cell culture media are designed to have an osmolality in the range of 270–330 mOsm/kg, which is known to be acceptable for most cells (158). Osmolality has been shown to affect both cell growth and protein production. The impact of osmolality on cell growth is cell-type specific, with CHO cells exhibiting less growth inhibition than hybridoma cells exposed to similar medium osmolality (150,157–160). Osmolality compensation by decreasing the concentration of NaCl in the basal medium has been found to partially mitigate the inhibitory effect of elevated pCO_2 on cell growth and protein production (150,157) and on neural cell adhesion molecule polysialylation by CHO cells (161). Additionally, HCO^{3-} free perfusion medium has been used to greatly decrease the build-up of pCO_2 in perfusion culture (137).

As mentioned earlier, dissolved CO_2 often interferes or even masks the effects of osmolality. Recently, deZengotita et al. (149) performed a thorough study to decouple the effects of pCO_2 and osmolality on the growth and metabolism of hybridoma cell by using low-salt basal media and corresponding compensation of osmolality. Under control conditions (40 mmHg; 320 mOsm/kg), cell growth and metabolism was similar in DMEM: F12 with 2% fetal bovine serum and serum-free HB GRO. In both media, pCO_2 and osmolality made dose-dependent contributions to the inhibition of hybridoma cell growth and synergized to more extensively inhibit growth when combined. Elevated osmolality was associated with increased apoptosis.

Specific antibody production also increased with osmolality although not with pCO_2. On the other hand, osmolality had little effect on glycolysis while elevated pCO_2 (with or without osmolality compensation) inhibited glycolysis in a dose-dependent fashion in both media. Mammalian cell lines derived from different organs have very distinct functions and therefore very different enzymatic and metabolic patterns. They may also have different responses to osmotic stress. For example, AB2-143.2 cells increase the specific formation rate of MAb (q_{MAb}) in response to gradual osmotic stress in continuous culture. In contrast, IND1 cells decrease q_{MAb} during gradual osmotic stress. Although the two cell lines differ in antibody production, they have very similar intracellular antibody content profiles (104,160). Both cell lines show: (1) a constant antibody content during the exponential phase in control culture with a decrease as cells enter the stationary phase; (2) maintenance of exponential-phase antibody content into the stationary phase after batch osmotic shock; and (3) no change in antibody content in response to gradual osmotic stress.

Cell culture longevity and thus product formation in fed-batch culture of hybridomas is often limited by elevated medium osmolality caused by repeated nutrient feeding. The use of hypoosmolar medium can overcome some of the problems cause by osmolality. As shown by Ryu and Lee (162), the use of hypoosmolar medium (223 mOsm/kg) as an initial medium in fed-batch culture can delay the onset of severe cell death of hybridoma cells, resulting in improved cell longevity and a substantial increase in the final antibody concentration.

Rate Laws of Cell Growth and Death

In most unstructured models, the specific growth rate μ is often expressed as a function of the glucose and/or glutamine concentration (8,9). In some of the models inhibition terms for lactate and ammonia are also included. Similar models are also used for the death rate (k_d). Pörtner and Schäfer (40) presented a survey of the unstructured models for cell growth and death rates. They also made a quantitative comparison between selected unstructured growth models and experimental data. It was noticed that a very limited data set covering a relatively narrow range of experimental conditions has often been used for model set-up. This practice of model formulation can lead to incorrect conclusions and model parameters that are not realistic. For example, inhibition constants estimated for lactate and ammonia considerably differ from those experimentally separately determined; saturation constants estimated for glucose ranges from as low as 0.03 mM to as high as 1 mM for similar hybridoma cell lines.

Zeng et al. (26) analyzed experimental data from six hybridoma cell lines grown on serum-containing and serum-free media in both normal continuous and perfusion cultures with respect to the significance of nutrients and products in determining the growth and death rates of cells and with respect to their mathematical description. It was interesting to find that for many continuous cultures there seems to exist direct correlations between μ and certain culture variables (Fig. 7A). For example, if one plots μ versus the residual glutamine concentration of a continuous hybridoma culture studied by Hiller et al. (12), an apparent strong dependency of μ on C_{Gln} is observed. The growth rate curve appears to follow a Monod-type saturation kinetic that would give an apparent saturation constant as low as 0.01–0.05 mM for glutamine. Hiller et al. (13) studied the same cell line in perfusion culture with glucose and glutamine concentrations in medium different from those used in the normal continuous culture. Interestingly, the growth rate of the perfusion culture apparently shows similar correlation with C_{Gln}, following again a kind of saturation kinetic with a saturation constant as high as 1.5–2 mM and a minimum C_{Gln} of about 1.2 mM for growth (Fig. 7A). Similar apparent dependencies can also be demonstrated for more cultures in the literature (4,8,21,36,40,72). It is worth mentioning that the continuous culture of Frame and Hu (8) was clearly shown to be

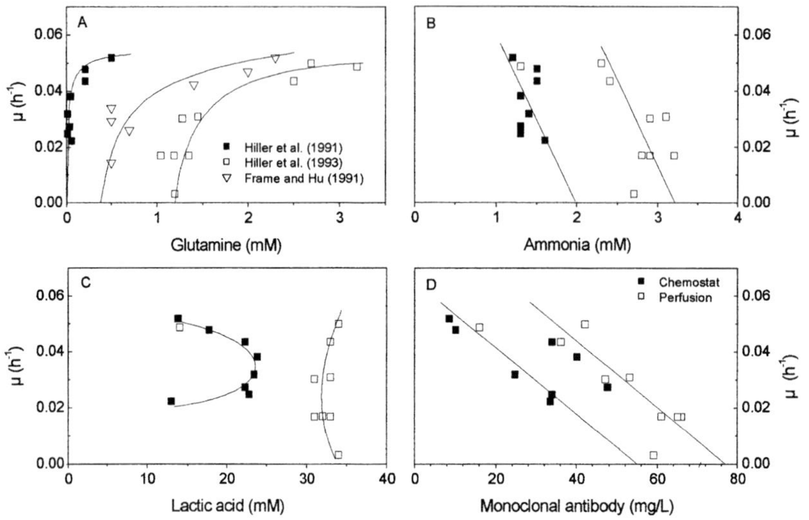

Figure 7 Apparent correlations between the specific growth rate (μ) and culture variables in two normal continuous cultures and a perfusion culture: (A) residual concentration of glutamine; (B) culture concentration of ammonia; (C) concentration of lactic acid; and (D) monoclonal antibody. [From Ref. (26).]

glucose-limited for some of the steady states. The possibility of a glutamine limitation was experimentally excluded. The data of Frame and Hu (8) are also depicted in Fig. 7A for comparison. An apparent K_m value of about 1 mM would be estimated from these data if the culture were assumed to be glutamine-limited. The medium used for the latter culture had an intermediate concentration of glutamine (3.61 mM) compared to 2.5 mM for the chemostat culture of Hiller et al. (12) and 6 mM for the perfusion culture. It is thus expected that the apparent saturation curve for the cell line of Hiller et al. (13) would move toward the left side if lower initial glutamine concentrations were used in the medium for perfusion. By carrying out a series of continuous cultures at varied initial glutamine concentration any apparent K_m values for glutamine between 0.01 and 2.5 mM would be obtained for one and the same cell line under similar environmental conditions. Obviously, an assumption of glutamine limitation for these cultures would be very implausible. It should be emphasized, however, that such a controversial conjecture is only obvious if one compares experimental data obtained by independent changes of culture conditions such as the medium composition and the ratio of D_B to D. In conclusion, the apparent strong dependency of μ on C_{Gln} in Fig. 7 may not necessarily mean a glutamine limitation. It can be just coincidental. In analogy, the apparent correlation of μ with glucose for some of the cultures studied in literature may not necessarily mean a glucose limitation as well.

Ammonium and lactic acid are often considered to be the dominating inhibitors limiting growth as they are the main byproducts of the substrate metabolism and have been shown to be toxic to cell growth above certain concentrations (15,80). Plotting μ versus ammonium and lactate concentrations for the cultures of Hiller et al. (12,13) give pictures of Fig. 7B and C. As in the case of glutamine a certain degree of correlation is observed between μ and the ammonium concentration. Again, the correlation depends strongly on the operation mode and obviously on the initial concentration of glutamine in the medium. Figure 7C reveals that there is no clear correlation between μ and the lactate concentration. It can be stated that for most cell cultures lactate and ammonium cannot be the dominant factors determining growth rate. The possibility of growth inhibition by the product MAb was also examined (Fig. 7D). In a similar way as argued for ammonium, MAb produced appears not to be a dominant inhibitor in the lines examined.

Three conclusions can be drawn regarding the significance of nutrient limitation and product inhibition and the modeling of cell growth rate (26). First, in all the cultures examined neither the macronutrients such as glucose, glutamine, and other amino acids nor the three products lactate, ammonium, and MAb are clearly dominant factors affecting the growth rate. Second, the apparent correlation of μ with some of these variables can be simply coincidental due to stoichiometric and/or kinetic interdependencies of the variables if they are not independently varied. Finally, none of the unstructured growth models existing in the literature can be generally applied, especially if they are intended to cover a relatively wide range of experimental conditions. Similar analysis and conclusions can be made with respect to the significance of nutrients and products in affecting the cell death rate.

For seven continuous cultures it was then found that μ almost linearly correlates with the ratio of the viable cell concentration (N_V) to the dilution (or perfusion) rate (D) irrespective of the operation mode, as typically shown in Fig. 8. Similarly, the specific death rate (k_d) of all the cultures is a function of the ratio of the total cell concentration (N_t) to the dilution (perfusion) rate (Fig. 9). For most of the cultures k_d also linearly correlates with μ if the effect of N_t/D is taken into account.

Based on these observations and other experimental evidence that suggests the formation of a not yet identified critical factor or autoinhibitor by the cells the following simple rate equation was derived for the specific growth rate of hybridoma cells (26):

$$\mu = \mu_{\max}\left(1 - \frac{C_I}{C_I^*}\right)\frac{C_{Glc}}{C_{Glc}+K_{Glc}}\frac{C_{Gln}}{C_{Gln}+K_{Gln}}\frac{K_{Lac}}{C_{Lac}+K_{Lac}} \times \frac{K_{NH_3}}{C_{NH_3}+K_{NH_3}} \tag{9.42a}$$

where C_I is the concentration of the autoinhibitor, C_I^* is the maximum concentration of the autoinhibitor above which cells cease to grow. Assuming a specific formation rate q_I for this inhibitor by viable cells, Eq. 9.42a(a) can be transformed to:

$$\mu = \mu_{\max}\left(1 - \alpha\frac{N_v}{D}\right)\frac{C_{Glc}}{G_{Glc}+K_{Glc}}\frac{C_{Gln}}{C_{Gln}+K_{Gln}}\frac{K_{Lac}}{C_{Lac}+K_{Lac}} \times \frac{K_{NH_3}}{C_{NH_3}+K_{NH_3}} \tag{9.42b}$$

for continuous culture, with the constant

$$\alpha = \frac{q_I}{C_I^*} \tag{9.43}$$

Similarly, the following rate equation has been proposed for the specific death rate in continuous culture:

$$k_d = (\beta_0 + \beta\mu)\frac{N_t}{D} \tag{9.44}$$

where β_0 and β are constants.

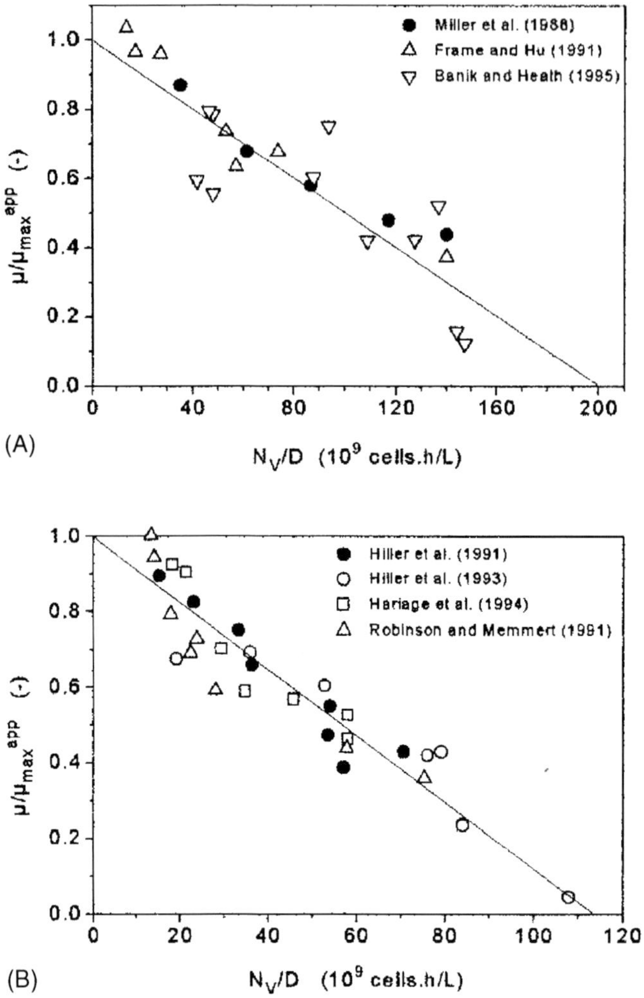

Figure 8 Relationship between μ/μ^{app} and the reciprocal of the specific dilution (perfusion) rate N_V/D for cells grown on (A) serum-containing media and (B) serum-free media. [From Ref. (26).]

In most cases, Eq. (9.42) can be simplified as:

$$\mu = \mu_{max}^{app}\left(1 - \alpha\frac{N_v}{D}\right) \tag{9.45}$$

where $\mu_{max}^{app} = f$ (nutrients, lactate, ammonia, etc.) $\approx$ constant is an apparent maximum specific growth rate that includes the influences of other inhibitors such as ammonium and lactate. The physiological meaning of α can be easily understood by setting $\mu = 0$ in Eq. (9.45). Under these conditions,

$$\alpha = \left(\frac{D}{N_v}\right)_{min} \tag{9.46}$$

According to Eq. (9.46), α has the unit "liter per unit biomass per hour" and can be defined as the minimum specific dilution (or perfusion) rate (based on viable

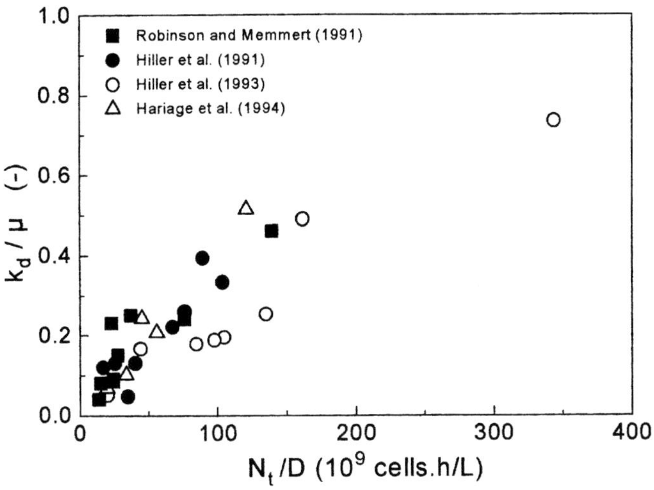

Figure 9 Correlations between k_d/μ and N_t/D for hybridoma cells grown in different continuous cultures with serum-free media. [From Ref. (26).]

cells) of a continuous culture to keep a non-negative cell growth rate. For perfusion culture the perfusion rate P should replace the dilution rate D in Eqs. (9.42)–(9.45).

Equations (9.44) and (9.45) describe the experimental data of seven continuous cultures reasonably well. To calculate the growth rate of a batch culture the original form of Eq. (9.42a) should be used. Equation (9.44) should be modified for batch culture. It is worth mentioning that irrespective of the cell lines, cells grown on serum-containing media have almost the same α value that is distinctively different from that of cells grown on serum-free media (Fig. 8). This indicates that under the current cultivation conditions the formation rate of the autoinhibitor(s) or the sensitivity of cell growth and death to the autoinhibitor(s) is mainly affected by the medium composition; α can be used as a quantitative parameter for comparison of medium performance.

The positive relationship between μ and k_d as envisaged by Eq. (9.44) is in strict contrast with the prevailing view in the literature that k_d is inversely related to μ (3,10,40). The apparent negative correlation between k_d and μ found for some normal continuous cultures appears to be coincidental due to the stoichiometric and kinetic interdependencies of culture variables as discussed earlier for the correlations of μ with concentrations of nutrients and products. In fact, the inverse relationship between k_d and μ is often not valid for perfusion cultures. The apparent inverse relationship between k_d and μ in a normal continuous culture can be understood with the help of Eq. (9.44). At low dilution rates (and thus low values of μ), the ratio $\mu(N_t/D)$ is normally higher than that at high dilution rates. Hence, k_d is higher at lower growth rates. But this is by no means an intrinsic relationship. This is also supported by the experimental results of Vomastek and Franek (59). They studied the kinetics of the development of apoptosis in mouse B-cell hybridoma cultures grown on both protein-free medium and serum-containing medium. The B-cell hybridoma apoptosis was found to be associated with cell proliferation and metabolic activity.

Product Formation: Monoclonal Antibody

Biologically, the rate of synthesis and secretion of MAbs depends on the rate of peptide chain synthesis, chain assembly, interorganelle transport, intravesicular degradation, and release from the cell membrane (30,163). The genetic make-up of a particular cell line may dictate these events intracellularly. However, the rate of these individual steps and the overall rate of MAb production can also be greatly affected by cell growth and culture environmental conditions. These include growth rate, cell division cycle and viability, components of culture medium or broth, cultivation conditions, reactor type and operation mode, and hydrodynamic stress (18,20,163–169). It is not the purpose of this chapter to assess all of these factors. Instead, only those that are adjustable for a given cultivation system (i.e., for given reactor type, operation mode, and hydrodynamic parameters) and relevant for kinetic modeling are mentioned.

The effects of cell cycle, growth rate, and viability have received the most attention in literature. Both synthesis and secretion are shown to be maximum in the G1/early S phases (166,170). It seems to be generally accepted that MAb is mainly produced by secretion of viable cells (29,164,171), though some experimental results showed that high death rate of cells can apparently cause increase in the specific MAb production rate (q_{MAb}) (166). In many cell lines q_{MAb} was found to be higher at lower growth rate (8,165,172–174). These results are consistent with the finding that a prolongation of the G1-phase and an increase in death rate increased q_{MAb} of different cell lines (172). In addition to the effect of cell cycle it is understood that an elevated growth rate may make a significant portion of metabolic energy unavailable for the synthesis of antibody, leading thus to a decreased MAb synthesis rate.

In most studies the G1-phase arrest of cell cycle and the reduction of growth rate were achieved by starvation of cells for an essential nutrient or energy source or by the addition of DNA synthesis inhibitors such as thymidine and hydroxyurea, or genotoxic agents such as adriamycin (165,172,175). These approaches are not adequate to achieve a prolonged production phase with high productivity since they interfere with cell viability and/or disturb metabolic processes necessary for protein synthesis. Regulation of cell growth rate and cell cycle through the use of fed-batch culture with careful nutrient feeding was shown to be feasible with a murine myeloma cell line (176). Recently, the use of conditional overexpression of tumor suppressor genes such as p21, p27, or p53 (cytostatic) to regulate cell growth rate has received increasing interest (177–180). The MAb productivity of myeloma and CHO cell lines could be improved by about 3–4-folds through cytostatic gene overexpression and cell growth arrest at the G1-phase of the cell cycle. Preliminary study into the cytostatic function and its relationship to productivity demonstrated that the arrest of CHO cells by p21 overexpression correlates with the cellular response to accumulation of mitochondrial mass, ribosomal protein S6, and intracellular protein uncoupling cell growth (180).

By combining the cell cycle theory with the estimated number of MAb-coded mRNA molecules per cell and by considering the relative time length of the individual phases, Suzuki and Ollis (165) developed a structured model that predicted an enhanced MAb production at low growth rate. In contrast to this theoretical analysis and most of the experimental reports, Robinson and Memmert (11) showed with a transfactant of a myeloma (SP2/O) cell line grown in a serum-free medium that q_{MAb} increased linearly with μ. Borth et al. (181) also showed with a human–mouse heterohybridoma (3D6-LC4) that q_{MAb} increased with μ. It was also reported that MAb production is not growth associated in some cell lines (181,182). Ray et al. (7) reported on an optimal μ for q_{MAb}. It is not clear if these differences are due to the inherent characteristics of different cell lines or to some extent due to the very different experimental conditions that may significantly affect MAb formation as well (24).

A major concern about the use of continuous culture for MAb production is the possible loss of culture productivity. It has been reported for several cell lines that hybridomas may dramatically lose their MAb productivity soon after revival from liquid nitrogen storage (thawing). This loss of productivity is not reversible. It may last for a few weeks or months for cells to reach a stable MAb productivity in continuous culture depending on the cell line and/or culture conditions. The mechanisms for the loss of MAb productivity in hybridomas are not well understood. In a culture of mouse–mouse hybridoma cells the loss of MAb was shown to be mainly due to the occurrence of a nonproducing subpopulation of cells (9). Other mechanisms may involve the loss of heavy chain and/or light chain production and the loss of the chromosomes containing the gene loci for the antibody chains similar to those for myelomas. Whatever the mechanisms are the available literature data suggest that it is important to consider culture stability when studying the effects of cell growth and other environmental conditions on MAb production, particularly when comparing kinetic data of different cell lines or experimental data obtained in batch culture.

In a comprehensive study of hybridoma growth and product formation using a statistic experimental method, Gaetner and Dhurjati (18) showed that the concentrations of base medium, glucose, glutamine, serum, lactate, and ammonium all have significant effects. q_{MAb} increased with increasing concentrations of base medium and serum but decreases with lactate and ammonium. The negative effects of lactate and ammonium at high concentrations have also been confirmed for other cell lines (168,183). Concerning the effect of glucose concentration it was shown to inversely affect the formation of MAb (168,183). This negative glucose effect seems to be consistent with the finding that the immunoglobulin heavy chain binding protein, which is involved in antibody assembly, is identical to the 78 kD glucose-regulated protein (184). The latter is induced by glucose starvation. In contrast, high glutamine concentration in the culture was found to stimulate MAb formation (163,169,182,183,185). Other factors affecting MAb production include temperature, pH, dissolved oxygen, vitamin, amino acids, osmolarity, and probably cell density. Whereas some of these factors (such as pH and pO_2) are constant or do not change very much during a cultivation and therefore may not be important for kinetic modeling, changes of other factors or the depletion of an essential nutrient may strongly alter the product formation pattern. It is thus important to identify possible stoichiometric and/or kinetic limitations when modeling the product formation.

Several structured models have been proposed to describe the intracellular synthesis and transport of MAbs (29,30,165,186). In contrast, few works dealt with the quantitative description of the influences of extracellular culture conditions. A review of these models was given by Tziampazis and Sambanis (39). In most of the models, merely variables associated with cell growth such as the specific growth rate, the death rate, and cell cycle, are considered. Some authors considered the effects of other factors such as serum, glutamine, lactate, and ammonium on q_{MAb} in mathematical terms (82,182,187). These expressions were, however, all derived for very specific experimental conditions and normally consider only one factor. Zeng (24) made an attempt to mathematically describe the effects of culture conditions on MAb production by hybridoma cells, with particular emphasis on cultures under unsteady-state conditions. The following general rate equation that takes account of productivity loss during long-term cultivation, cell proliferation, and the effects of nutrients and toxic products was proposed:

$$q_{\mathrm{MAb}} = (\alpha + \beta k_{\mathrm{d}}) \prod_i \frac{C_i}{C_i + K_i} \prod_j \frac{K_j}{C_j - C_j^* + K_i} (B + e^{-A\Delta t}) \tag{9.47}$$

In Eq. (9.47), k_{d} is the specific death rate of cells. C_i and C_j are the concentration (or intensity) of a stimulating factor i and the concentration (or intensity) of an inhibiting factor j, respectively. α, β, K_i, and K_j are constants. The term $(B + e^{-A\Delta t})$ describes the loss of MAb productivity of cells during long-term cultivation.

For a reliable assessment of effects of different factors and for a comparison of kinetic data on MAb production it is important to consider possible loss of antibody productivity, the time dependence of which can be modeled by an exponential function plus a constant term $(B + e^{-A\Delta t})$. The latter gives an indication of stability of MAb productivity of a given cell line under the experimental conditions. Among the parameters related to cell proliferation the cell death rate appeared to be useful for modeling MAb production under both steady-state and unsteady-state conditions. Model analysis of the so-called cell density effect suggested that it can be attributed to the varying availability of nutrients. Among others, the relative concentration of glucose and glutamine may be decisive for the production of antibody under certain conditions. Furthermore, the specific formation rate of antibody in perfusion culture was found to be strongly affected by the perfusion rate, indicating that antibody production is limited by component(s) of the medium not yet identified (24).

Cell Density Effect

As mentioned earlier, cell density has been reported to be an important parameter for the kinetics of batch culture (44,45). For a continuous perfusion culture at relatively high density, Banik and Heath (72,73) showed that the specific rates of glucose consumption, glutamine consumption, and lactate production decreased significantly with increasing cell density. Similar effects of cell density were also reported by other authors (16,188). In the work of Banik and Heath (72,73), the specific antibody production was also a strong function of cell density, increasing as cell density increased independent of growth rate. Similar results were reported for a mouse–human hybridoma grown in serum-containing medium (189). However, constant q_{MAb} values were obtained with the same cell line grown in serum-free medium over a density range of 3.5×10^5 to 7.6×10^6 mL^{-1} viable cells. Other investigators found decreased q_{MAb} values with increasing cell density (190).

Density-dependent apoptosis has been reported for several cell lines (191–193). Normally, this kind of apoptosis was observed under low cell density. It is thought to be due to the lack or reduction of survival signals provoked by direct cell-to-cell contact or soluble growth factors (autocrine factors) produced by the cells. High cell density can also induce apoptosis which may have an important physiological role in vivo such as for the prevention of tissue hyperoxia and maintenance of appropriate tissue volume.

The cell density effects on cell growth and death are mathematically described in the rate Equations (9.44) and (9.45) Little is known about the effect(s) of cell density on the metabolism and MAb production. In the literature mechanisms modulated by the so-called humoral factors (43) and autocrine growth factors (194) were proposed to describe cell density dependent growth. In most of the experimental studies, however, no factors other than cell density itself and/or physical environments could be identified as the cause(s) for the observed effects (25,44,72,73,195). In the work of Zeng (25), the rate equations mentioned earlier (9.22), (9.23), (9.26), (9.27), and (9.47) were used to examine the significance of possible cell density effect(s) on two hybridoma cultures grown in perfusion bioreactors. By replacing the nutrient concentration C_s with a "specific concentration of nutrient c_s"

$$c_S = \frac{C_S}{N_V} \tag{9.48}$$

Equations (9.22), (9.23), (9.26), (9.27), and (9.47) satisfactorily describe the consumption rates of glucose and glutamine and the formation rates of lactate, ammonium, and MAb over a viable cell density range of 10^6–$10^7\,\text{mL}^{-1}$. A density effect can be generally assumed for these cultures. However, the effect is only significant for the rates of glucose consumption and lactate formation of one cell line under the experimental conditions. The observed variations of metabolism and antibody production rate at different cell density appear to be attributed to the varying availability of nutrients. Among others, the relative concentration of glucose and glutamine may be decisive for the uptake of glucose and glutamine and for the production of lactate, ammonium, and antibody under certain conditions.

MODELS FOR SIMULATION OF CELL CULTURE

Experimental data from kinetic studies in batch and continuous cultures build the basis for deriving quantitative relationships (rate equations) that correlate the rates of cell growth, death and metabolism with culture and environmental conditions. This set of rate equations represents a kinetic model of the cellular process. Combined with balance or differential equations governing the system, these rate equations can be used to simulate the time courses of concentrations of cells, nutrients, and metabolites in the culture under different conditions. Kinetic models for biological processes can be generally classified into four categories (98,196). In nonsegregated and unstructured models the cell population is treated as "one-component solute" or as an "average cell." Cell growth is considered to be "balanced." In segregated models the heterogeneity of cell population is considered and in structured models the intracellular structure and processes of cells are taken into account.

Unstructured Models

The simplest and frequently used kinetic models for cell culture are nonsegregated and unstructured models. Although the formulation of this kind of models is of empiric and phenomenological nature, they are usually based on fundamental observations of biological processes and can be used to qualitatively and quantitatively describe many important features of cell culture such as the dependency of growth rate on a limiting nutrient or an inhibiting substance. Because of the simple form and limited model parameters unstructured models are particularly suitable for process control, optimization, and scale-up of industrial processes. A major drawback of unstructured models is that they cannot be used for prediction and are generally not suitable for dynamic simulation.

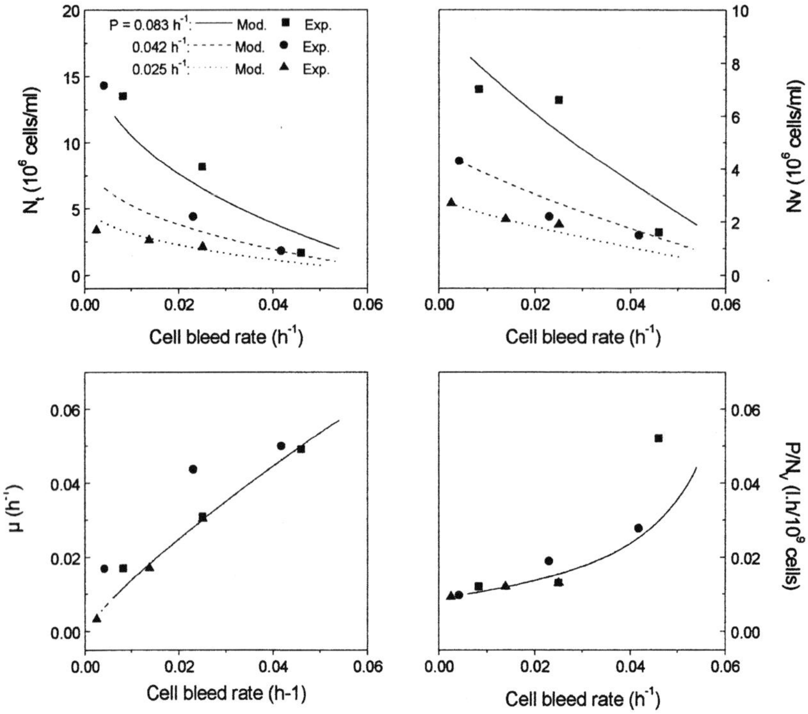

Figure 10 Comparisons of model simulations and experimental results of total and viable cell concentrations, growth and specific perfusion rate of a perfusion culture with the cell line X-D at different perfusion and cell bleed rates. Experimental data: points. Simulations: solid lines for $P = 0.083\,h^{-1}$, dotted lines for $P = 0.042\,h^{-1}$; dashed lines for $P = 0.025\,h^{-1}$. [From Ref. (197).]

The rate equations presented earlier were used to simulate the behavior of perfusion cultures at high cell density (197). As an example, Figs. 10 and 11 show comparisons of model simulations and experimental results of a perfusion culture. Model

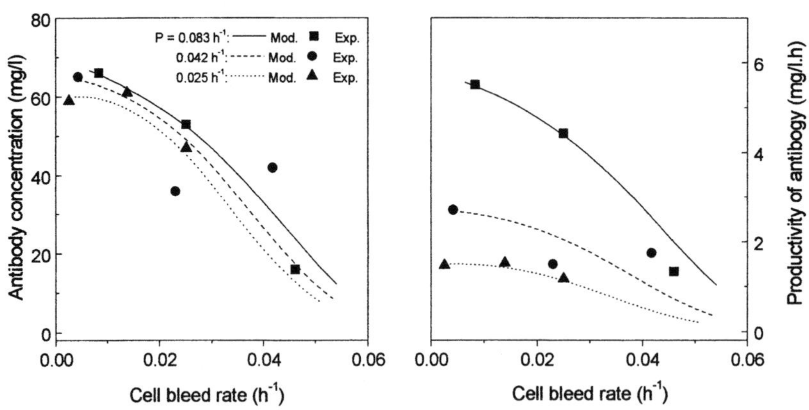

Figure 11 Comparisons of model simulations and experimental results of concentration and productivity of monoclonal antibody of a perfusion culture with the cell line X-D at different perfusion and cell bleed rates. Symbols as in Fig. 10. [From Ref. (13).]

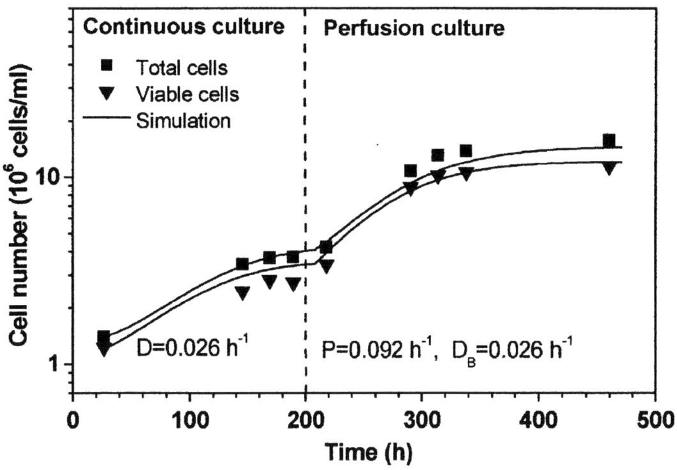

Figure 12 Simulation of time course of growth of hybridoma cells in conventional continuous and perfusion cultures.

simulations of steady-state behavior are in good agreement with experimental results under varied perfusion and cell bleed rates except for cultures with very low viability. The rate equations can be also used to simulate the time courses of batch and continuous cultures under non-steady states. Figure 12 shows, for example, the kinetic simulation of a culture that was first operated as a conventional continuous culture and then switched to perfusion. The effect of perfusion operation was described satisfactorily. It should be mentioned that under certain dynamic conditions such as sudden changes of nutrient supply the model may fail to describe the pitfall of cellular stress responses. The simulation capacities of the model appear to be mainly limited by the applicability of the cell growth and death rate expressions under these conditions. In general, unstructured kinetic models are severely limited for simulation of dynamic behavior of cell culture. To this end, structured models are preferred.

Structured Models

Structured or mechanistic models are based on knowledge of intracellular structure and bioreactions and their regulation mechanisms. One of the important advantages is their prediction ability. Once the model is experimentally verified, it can be in principle extrapolated for conditions outside the experimental range. It should be, however, mentioned that even the simplest living cell is so highly complicated system that any mathematical description is only an approximation of the real biological processes inside the cell. The underlying mechanisms used for model formulation are often not those that explain the biological processes on molecular basis but rather working hypothesis. Caution should thus be taken in interpreting the simulation results, especially for extrapolation. The experimental verification of a model does not necessarily guaranty the correctness of the hypothesis.

In the literature a number of structured or mechanistic models have been proposed for animal cell culture (27–30,32,35,37,198). Recently, cybernetic approach has also been used to model the kinetic behavior of cell culture, especially with respect to a possible occurrence of multiplicity in continuous culture (38). In the following, only two examples of these models are briefly introduced.

"Chemically or Metabolically" Structured Models

In the chemically or metabolically structured models the biomass is divided into several chemical components (pools) such as cell membrane (lipids), protein, and nuclei acids. The stoichiometrical relationships (balance equations) and rate equations for these components build the basis for the modeling of the whole cell. Chemically or metabolically structured models can be generally formulated as follows (196):

$$\frac{dc_j}{dt} = \sum_i^n r_{ij} - \mu c_j \tag{9.49}$$

$$\mu = \widehat{V} \sum_i^n \sum_j^m r_{ij} \tag{9.50}$$

where i stands for the n conversion reactions; j for the m components; c_j for the intracellular concentration of the component j; r_{ij} for the formation or consumption rate (based on intracellular volume) of the component j in the ith conversion processes including transport through the cell membrane; $\widehat{V}$ the intracellular volume per mass unit of biomass. The key and at the same time the difficulty in applying this modeling approach is the formulation of reliable kinetics rate expressions for r_{tj}.

The above model formulation was first practically used by Domach et al. (199) and Domach and Shuler (200) for the development of the so-called "single-cell model" for the growth of *Escherichia coli* on glucose. Batt and Kompala (27) applied this modeling approach and developed a structure framework for simulating continuous cell culture. The cell is considered to be composed of four pools: amino acids, nucleotides, proteins, and lipids. The rate equations for these pools are mainly saturation type kinetic expressions. Interactions among the pools and the extracellular environment occur via nutrient consumption, metabolite formation, and transport of nutrients and metabolites.

A more promising structured model for cell culture is the extension of the above mentioned "single-cell model" (200) for CHO cells by Wu et al. (32). In the single-cell model for CHO, cells are considered to be composed of 18 component pools. The model can accommodate additional mechanistic details by further subdivision of any of the pools. Stoichiometry of pseudochemical reactions relating the mass or information flow of the pools is considered. The rate equations are also generally of saturation type with a feedback control term and have altogether more than 100 parameters. The effects of glucose, glutamine, other amino acids, and the cellular metabolic byproducts on cellular metabolism are considered in these rate equations. This model describes experimental data of CHO cells under both steady-state and transient conditions quite well. Wu et al. (19) further extended this model to study the regulation of intracellular pH. It provides a direct mechanistic link between intracellular pH control and cellular metabolism.

Cell Cycle Model

In cell cycle models the biochemical features and kinetics of cells passing through the different phases of a cell cycle are considered. The cell cycle is generally considered to consist of four essential phases: the gap 1 phase (G1-phase), the DNA synthesizing

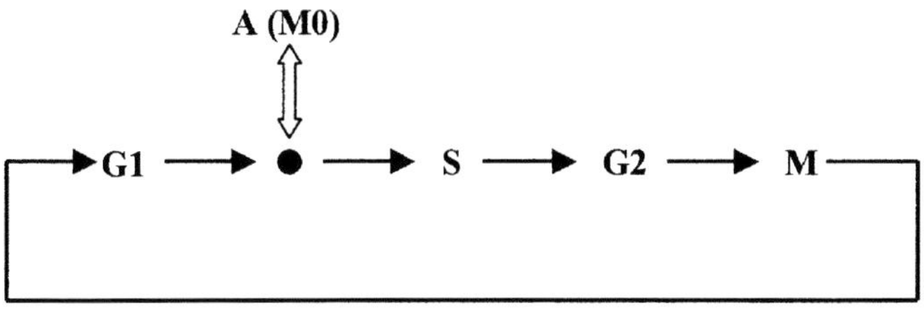

Figure 13 The successive phases of cell cycle.

phase (S-phase), the gap 2 phase (G2-phase), and the mitotic phase (M-phase) as shown in Fig. 13 (28). After M-phase, which consists of nuclear division (mitosis) and cytoplasmic division (cytokinesis), the daughter cell enters a new cycle. A new cell cycle starts with the G1-phase, in which the cells, whose biosynthetic activities have been greatly slowed during mitosis, resume a high rate of biosynthesis. The S-phase begins when DNA synthesis starts and ends when DNA content of nucleus has doubled and chromosomes have replicated. The cell then enters the G2-phase, which ends when mitosis starts. Between the G1 and S phases a cell may stop dividing unless signaled to continue through the whole cycle. The cells stopping at late G1 is called an arrested cell and symbolized by "A" in Fig. 13. The arrested cells are also called to be in the G0 phase. Park and Ryu (201) measured the population distribution of hybridoma cells during perfusion culture by flow-cytometric analysis. The results showed that, during the reduced growth period of perfusion culture, the fraction of cells in the S-phase decreased, and the fraction of cells in the G1/G0-phase increased with decreasing growth rate. The fraction of cells in the G2/M-phase was relatively constant during the whole period of perfusion culture. Ramirez and Mutharasan (44) determined the cell cycle and mean cell volume distribution in batch culture by cytometric analysis. The mean cell volume closely correlated with the fraction of cells in the S-phase. On single cell basis, a cell generally increases in volume and mass from the beginning of the G1-phase through the S- and G2-phase until the end of the G2-phase.

Besides the morphological and biochemical differences cells at different cell cycle phases have remarkably different kinetic properties, particularly with respect to MAb productivity of hybridoma cells. Several investigations have shown that the maximum antibody synthesis rate occurs in the late G1-phase or early S-phase of hybridoma cells and appears to correlate with the faction of arrested cells (28,44,170,201). The morphological and biochemical differences of distinguished cell population during a cell cycle are the basis of cell cycle models (28,31). For example, the specific antibody production rate q_{MAb} can be represented as a summation of the fraction of cells in each of the five phases multiplied by the antibody production rate of a cell in that phase:

$$q_{\mathrm{MAb}} = k_{\mathrm{G1}} f_{\mathrm{G1}} + k_{\mathrm{S}} f_{\mathrm{S}} + k_{\mathrm{G2}} f_{\mathrm{G2}} + k_{\mathrm{M}} f_{\mathrm{M}} + k_{\mathrm{A}} f_{\mathrm{A}} \tag{9.51}$$

where k_{G1}, k_{S}, k_{G2}, and k_{M} are the averaged antibody production rates in the G1-, S-, G2-, and M-phase, respectively; k_{A} is the antibody production rate of a cell arrested in the G1-phase; f_{G1}, f_{S}, f_{G2}, and f_{M} are the fractions of cells traveling the G1-, S-, G2-, and M-phase, respectively; and f_{A} is the fraction of cells arrested in the G1-phase.

The frequency of cells located at time t $(0 \leq t \leq t_C)$ measured from the beginning of the G1-phase is proportional to $2^{1-t/t_C}$, where the total cell cycle time t_C is the sum of t_{G1}, t_S, t_{G2}, and t_M. t_{G1} includes the time of a cell arrested in the G1-phase and normally increases as the specific growth rate decreases, while t_S, t_{G2}, and t_M are approximately constant independent of the growth rate. Thus, t_C also increases as μ decreases.

The fractions of cells traversing the individual phases can be calculated as:

$$f_{G1} = (1 = f_A) \int_0^{x_{G1}} 2^{1-x} \frac{dx}{A} \tag{9.52}$$

$$f_S = (1 = f_A) \int_{x_{G1}}^{x_S} 2^{1-x} \frac{dx}{A} \tag{9.53}$$

$$f_{G2} = (1 = f_A) \int_{x_S}^{x_{G2}} 2^{1-x} \frac{dx}{A} \tag{9.54}$$

$$f_M = (1 = f_A) \int_{x_{G2}}^{1} 2^{1-x} \frac{dx}{A} \tag{9.55}$$

The fraction f_A is related to the specific growth rate μ and the cell cycle time t_C as follows:

$$f_A = 1 - \mu \cdot t_C / \ln 2 \tag{9.56}$$

In a similar way, specific rates for growth and death, nutrient consumption, and formation of other metabolites may be related to the cell cycle. For example, in the work of Linardos et al. (31) the death rate of the hybridoma is assumed proportional to the fraction of cells arrested in the G0-phase.

An underlying assumption in cell cycle models is that the rates of biosynthesis and degradation are constant in each cycle phase; the culture conditions affect only the viable cell number and the distribution of cells among the phases. This assumption may be only valid for certain cell systems under particular conditions.

Application of Models for Kinetic Analysis

As demonstrated earlier, mathematic models (rate equations) are useful for kinetic analysis of effects and interactions of different factors on cell growth and metabolism. They are particularly helpful for obtaining quantitative knowledge about the significance of these factors and their range of concentrations or values in which they are important. Models are also valuable for kinetic analysis of the overall performance and for the control of cell culture. This is briefly illustrated in the following for the kinetic analysis of perfusion culture (197). As evidenced in Figs. 10 and 11, kinetic analysis of simulations and experimental results indicates that in perfusion cultures with a complete cell separation cell bleed rate is a key parameter that strongly affects all the process variables whereas the perfusion rate mainly affects the total and viable cell concentrations and the volumetric productivity of MAb. Growth rate, viability, and specific perfusion rate of cells are only a function of the cell bleed rate. This also applies to cultures with partial cell separation in the permeate if the effective cell bleed rate [Eq.(9.21)] is used. From the kinetic analysis it is clear that the (effective) cell bleed rate of a perfusion culture should be carefully chosen and controlled separately from the perfusion rate. In general, a low cell bleed rate that warrants reasonable cell viability is desirable for the production of antibodies. Furthermore, model simulations indicate the existence of an optimum initial glucose concentration in the feed. For several cell lines considered, the initial glucose concentration used in normal cell culture media is obviously too high. The initial glutamine concentration can also be reduced to a certain extent without significantly impairing the growth and antibody production but considerably reducing the ammonium concentration. The mathematical model can be used to predict these optimum conditions and may also be used for process design.

In this connection, it is worth mentioning that the specific perfusion rate based on viable cells (P/N_v) was proposed as a key control parameter for process scale-up and product consistency of perfusion cultures (202). The rationale behind this strategy can be understood in view of the linear relationship between μ and N_v/P [Eqs.(9.45) and Fig. 8] under conditions of no obvious limitation of nutrients as discussed. In Fig. 10, the relationship between P/N_v and D_B is examined for varied perfusion rates. Surprisingly, P/N_v turned out to be almost only a function of D_B. The simple relationship between P/N_v and D_B underlines that P/N_v is a useful control parameter since it determines D_B which in turn fixes the cell viability, cell growth, and death rates. However, the so-called "high-gain" range for a P/N_v control algorithm corresponds to high cell bleed rates or high perfusion rate in systems with partial cell separation. From a practical point of view, the effective cell bleed rate seems to be a more direct and more favorable control parameter than P/N_v, at least for systems with a complete cell separation. In these systems, the desired cell bleed rate may be chosen based on considerations of cell concentration and physiology (e.g., viability, growth and death rates) and final product concentration (Figs. 10 and 11). As mentioned earlier, a cell bleed rate as low as possible should be chosen for achieving high concentrations of cells and product. At such low values, D_B is quite sensitive to P/N_v. Thus, a small fluctuation in P/N_v may lead to relatively large changes of D_B and other process variables. This may be of disadvantage for process stability and consistency. Furthermore, an online measurement of viable cell concentration is not an easy task and is subjected to fluctuations. On the other hand, the control of D_B is simple and reliable. An online measurement of viable cell concentration is not needed. Once the D_B value is chosen the desired concentrations of cells and product can be achieved by varying the perfusion rate.

In general, simulations of cell culture, in particular detailed, structured, and dynamic simulations have a wide range of applications such as in planning and design of experiments, testing and development of theories and hypothesis concerning metabolism, data reconciliation, and inference of difficult to measure variables and advanced process optimization and control (35).

CONCLUDING REMARKS

Our knowledge about the kinetics of cell culture has been considerably augmented in recent years. Both unstructured and structured models are available to simulate the kinetic behavior of several cell lines in "balanced" batch growth or steady states of continuous cultures. These models are useful for quantitatively analyzing the influences of different culture and reactor operating conditions and for process control and optimization. However, difficulties are still encountered in simulating the dynamic behavior of cell culture under strongly disturbed or unexpected stress conditions, particularly in high cell density fed-batch and continuous cultures.

The prediction capacities of the models for these cultures are generally poor. To combat this situation, more quantitative knowledge on intracellular processes and structured models based on in vivo fluxes and intracellular parameters are needed. In particular, more research is needed to understand and to mathematically describe the kinetics and dynamics of the cell death machinery. It has been generally demonstrated that most commercially important mammalian cell lines primarily die by apoptosis in the physical and chemical environments of bioreactors. In this respect, the rapid development of cell and molecular biology concerning the genetics, biochemistry, and physiology of apoptosis gives a good chance to develop more comprehensive structured kinetic model for the growth and death of cell culture.

Till now, the intrinsic kinetic relationships among growth (cell cycle), death, product formation, and quality are not well understood. In view of the importance of glycoproteins as pharmaceutical products, the kinetics of glycosylation of proteins deserves more attention. Due to the diversity of glycosylation patterns and the many factors involved in the post-translational modifications this represents a challenge for kinetic modeling.

Another challenge for model development is the incorporation of genomic and proteomic data into kinetic analysis. Presently, there is a flood of genomic and proteomic data produced for many different cells. However, they normally only give static or structural information on the genes and proteins possibly involved in the cellular processes. The mechanisms and kinetic control of the functionality of the many genes and proteins under cell culture conditions remain to be explored. Research into these aspects is not only important for a more predictable and targeted improvement and control of animal cell culture but will have impact in many other areas of life science. In fact, understanding genome-wide kinetic behavior of cells is one of the central themes of postgenomic research. Cell culture as a clearly defined and reproducible biological system may be a good model system for this endeavor.

REFERENCES

1. Hu W-S, Dodge TC, Frame KK, Himes VB. Effect of glucose on the cultivation of mammalian cells. Dev Biol Stand 1987; 66:279–290.
2. Glacken MW, Adema E, Sinskey AJ. Mathematical description of hybridoma culture kinetics. I. Initial metabolic rates. Biotechnol Bioeng 1988; 32:491–506.
3. Glacken MW, Huang C, Sinskey AJ. Mathematical description of hybridoma culture kinetics. III. Simulation of fed-batch bioreactors. J Biotechnol 1989; 10:39–66.
4. Miller WM, Blanch HW, Wilke CR. A kinetic analysis of hybridoma growth and metabolism in batch and continuous suspension culture: effect of nutrient concentration, dilution rate, and pH. Biotechnol Bioeng 1988; 32:947–965.
5. Miller WM, Wilke CR, Blanch HW. Transient responses of hybridoma cells to nutrient additions in continuous culture: I. Glucose pulse and step changes. Biotechnol Bioeng 1989; 33:477–486.
6. Miller WM, Wilke CR, Blanch HW. Transient responses of hybridoma cells to nutrient additions in continuous culture: II. Glutamine pulse and step changes. Biotechnol Bioeng 1989; 33:487–499.
7. Ray NG, Karkare SB, Runstadler PW. Cultivation of hybridoma cells in continuous cultures: kinetics of growth and product formation. Biotechnol Bioeng 1989; 33: 724–730.
8. Frame KK, Hu WS. Kinetic study of hybridoma cell growth in continuous culture. I. A model for non-producing cells. Biotechnol Bioeng 1991; 37:55–64.

9. Frame KK, Hu WS. Kinetic study of hybridoma cell growth in continuous culture. II. Behavior of producers and comparison to non-producers. Biotechnol Bioeng 1991; 38:1020–1028.
10. Linardos TI, Kalogerakis N, Behie LA. The effect of specific growth rate and death rate on monoclonal antibody production in hybridoma chemostat cultures. Can J Chem Eng 1991; 69:429–438.
11. Robinson DK, Memmert K. Kinetics of recombinant immunoglobulin production by mammalian cells in continuous culture. Biotechnol Bioeng 1991; 38:972–976.
12. Hiller GW, Aeschlimann AD, Clark DS, Blanch HW. A kinetic analysis of hybridoma growth and metabolism in continuous suspension culture on serum-free medium. Biotechnol Bioeng 1991; 38:733–741.
13. Hiller GW, Clark DS, Blanch HW. Cell retention-chemostat studies of hybridoma cells—analysis of hybridoma growth and metabolism in continuous suspension culture on serum-free medium. Biotechnol Bioeng 1993; 42:185–195.
14. Hiller GW, Clark DS, Blanch HW. Transient responses of hybridoma cells in continuous culture to step changes in amino acid and vitamin concentrations. Biotechnol Bioeng 1994; 44:303–321.
15. Ozturk SS, Riley MR, Palsson BO. Effects of ammonia and lactate on hybridoma growth, metabolism, and antibody production. Biotechnol Bioeng 1992; 39:418–431.
16. Shirai Y, Hashimoto K, Yamaji H, Kawahara H. Growth kinetics of hybridoma cells in high cell density culture. J Ferment Bioeng 1992; 73:159–165.
17. Schmid G, Wilke CR, Blanch HW. Continuous hybridoma suspension cultures with and without cell retention: kinetics of growth metabolism and product formation. J Biotechnol 1992; 22:31–40.
18. Gaertner JG, Dhurjati P. Fractional factorial study of hybridoma behavior. 1. Kinetics of growth and antibody production. Biotechnol Prog 1993; 9:298–308.
19. Gaertner JG, Dhurjati P. Fractional factorial study of hybridoma behavior. 2. Kinetics of nutrient uptake and waste production. Biotechnol Prog 1993; 9:309–316.
20. Kurokawa H, Ogawa T, Kamihira M, Park YS, Iijima S, Kobayashi T. Kinetic study of hybridoma metabolism and antibody production in continuous culture using serum-free medium. J Ferment Bioeng 1993; 76:128–133.
21. Harigae M, Matsmura M, Kataoka H. Kinetic study on HBs-MAb production in continuous cultivation. J Biotechnol 1994; 34:227–235.
22. Linz M, Zeng A-P, Wagner R, Deckwer W-D. Stoichiometry, kinetics, and regulation of glucose and amino acid metabolism of a recombinant BHK cell line in batch and continuous cultures. Biotechnol Prog 1997; 13:453–463.
23. Zeng A-P, Deckwer W-D. Mathematical modeling and analysis of glucose and glutamine utilization and regulation in cultures of continuous mammalian cells. Biotechnol Bioeng 1995; 47:334–346.
24. Zeng A-P. Analysis and mathematical modeling of monoclonal antibody production by hybridoma cells. Biotechnol Bioeng 1996; 50:238–247.
25. Zeng A-P. Quantitative assessment of effect of cell density on metabolism and monoclonal antibody production of hybridoma cells. J Biotechnol 1996; 45:243–251.
26. Zeng A-P, Deckwer W-D, Hu W-S. Determinants and rate laws of growth and death of animal cells in continuous cultures. Biotechnol Bioeng 1998; 57:642–654.
27. Batt BC, Kompala DS. A structured kinetic modeling framework for the dynamics of hybridoma growth and monoclonal antibody production in continuous suspension cultures. Biotechnol Bioeng 1989; 34:515–531.
28. Suzuki E, Ollis DF. Cell cycle model for antibody production kinetics. Biotechnol Bioeng 1989; 34:1398–1402.
29. Bibila TA, Flickinger MC. A structured model for monoclonal antibody synthesis in exponentially growing and stationary phase hybridoma cells. Biotechnol Bioeng 1991; 37:210–226.

30. Bibila TA, Flickinger MC. A model of interorganelle monoclonal antibody transport and secretion in mouse hybridoma cells. Biotechnol Bioeng 1991; 38:767–780.
31. Linardos TI, Kalogerakis N, Behie LA. Cell cycle model for growth rate and death rate in continuous suspension hybridoma cultures. Biotechnol Bioeng 1992; 40:359–368.
32. Wu P, Ray NG, Shuler ML. A single-cell model for CHO cells. Ann NY Acad Sci 1992; 665:152–186.
33. Chotteau V, Bastin G. A systematic approach for modeling animal cell cultures: methodology and experimental results. Cytotechnology 1994; 16:196–196.
34. Zeng A-P. A kinetic model for product formation of microbial and mammalian cells. Biotechnol Bioeng 1995; 46:314–324.
35. Sanderson CS, Phillips PJ, Barford JP. Structured modelling of animal cells. Cytotechnology 1996; 21:149–153.
36. Bakker WAM, Schaefer T, Beeftink HH, Tramper J, De-Gooijer CF. Hybridomas in a bioreactor cascade: modeling and determination of growth and death kinetics. Cytotechnology 1996; 21:263–277.
37. Jang JD, Sanderson CS, Chan LCL, Barford JP, Reid S. Structured modeling of recombinant protein production in batch and fed-batch culture of baculovirus-infected insect cells. Cytotechnology 2000; 34:71–82.
38. Guardia MJ, Gambhir A, Europa AF, Ramkrishna D, Hu W-S. Cybernetic modeling and regulation of metabolic pathways in multiple steady states of hybridoma cells. Biotechnol Prog 2000; 16:847–853.
39. Tziampazis E, Sambanis A. Modelling of cell culture processes. Cytotechnology 1994; 14:191–204.
40. Pörtner R, Schafer T. Modeling of hybridoma cell growth and metabolism—a comparison of selected models and data. J Biotechnol 1996; 49:119–135.
41. Goergen JL, Marc A, Engasser JM. Kinetics and simulation of animal cell processes. Hause H, Wagner R, eds. Mammalian Cell Biotechnology in Protein Production. Berlin: De Gruyter, 1997:345–371.
42. Martial A, Marc A, Engasser JM. Influence of inoculum age on hybridoma culture kinetics. Cytotechnology 1991; 5:165–171.
43. Sand T, Condi R, Rosenberg A. Metabolic crowding effect in suspension of cultured lymphcytes. Blood 1977; 50:337–386.
44. Hu W-S, Meier J, Wang DIC. A mechanistic analysis of the inoculum requirement for the cultivation of mammalian cells on microcarriers. Biotechnol Bioeng 1985; 27:585–595.
45. Ozturk SS, Palsson BO. Effect of initial cell density on hybridoma growth, metabolism, and monoclonal antibody production. J Biotechnol 1990; 16:259–278.
46. Ryll T, Wagner R. Intracellular ribonucleotide pools as a tool for monitoring the physiological state of in vitro cultivated mammalian cells during production processes. Biotechnol Bioeng 1992; 40:934–946.
47. Barnabé N, Butler M. The effect of glucose and glutamine on the intracellular nucleotide pool and oxygen uptake rate of a murine hybridoma. Cytotechnology 2000; 34:47–57.
48. Schoenherr I, Stapp T, Ryll T. A comparison of different methods to determine the end of exponential growth in CHO cell cultures for optimization of scale-up. Biotechnol Prog 2000; 16:815–821.
49. Gawlitzek M, Valley U, Nimtz M, Wagner R, Conradt HS. Characterization of changes in the glycosylation pattern of recombinant proteins from BHK21 cells due to different culture conditions. J Biotechnology 1995; 42:117–131.
50. Gawlitzek M, Ryll T, Lofgren J, Sliwkowski MB. Ammonium alters N-glycan structures of recombinant TNFR-IgG: degradative versus biosynthetic mechanisms. Biotechnol Bioeng 2000; 68:637–646.
51. Gawlitzek M, Valley U, Wagner R. Ammonium ion and glucosamine dependent increases of oligosaccharide complexity in recombinant glycoproteins secreted from cultivated BHK21 cells. Biotechnol Bioeng 1998; 57:518–528.

52. Curling EMA, Hayter PM, Baines AJ, Bull AT, Gull K, Strange PG, Jenkins N. Recombinant human interferon-γ: differences in glycosylation and proteolytic processing lead to heterogeneity in batch culture. Biochem J 1990; 272:333–337.
53. Ramirez OT, Mutharasan R. Cell cycle- and growth phase-dependent variations in size distribution, antibody productivity, and oxygen demand in hybridoma cultures. Biotechnol Bioeng 1990; 36:839–843.
54. Al-Rubeai M. Apoptosis and cell culture technology. Adv Biochem Biotechnol 1997; 59:225–249.
55. Lloyd DR, Holmes P, Jackson LP, Emery AN, Al-Rubeai M. Relationship between cell size, cell cycle and specific recombinant protein productivity. Cytotechnology 2000; 34:59–70.
56. Al-Rubeai M, Emery AN, Chalder S. Flow cytometric study of cultured mammalian cells. J Biotechnol 1991; 19:67–81.
57. Linz M. Qualitative und quantitative Charakterisierung des Zelltodes in Säugerzellinien—Induktion und Kinetik von Apoptose. Dissertation, Technical University of Braunschweig, Germany, 2000.
58. Franek F, Vomastek T, Dolnikova J. Fragmented DNA and apoptotic bodies document the programmed way of cell death in hybridoma cultures. Cytotechnology 1992; 9:117–123.
59. Vomastek T, Franek F. Kinetics of development of spontaneous apoptosis in B cell hybridoma cultures. Immunol Lett 1993; 35:19–24.
60. Singh RP, Al-Rubeai M, Gregory CD, Emery AN. Cell death in bioreactors: a role of apoptosis. Biotechnol Bioeng 1994; 44:720–726.
61. Mercille S, Massie B. Induction of apoptosis in nutrient-deprived cultures of hybridoma and myeloma cells. Biotechnol Bioeng 1994; 44:1140–1154.
62. Mercille S, Massie B. Induction of apoptosis in oxygen-deprived cultures of hybridoma cells. Cytotechnology 1994; 15:117–128.
63. Al-Rubeai M, Singh RP, Goldman MH, Emery AN. Death mechanisms of animal cells in conditions of intensive agitation. Biotechnol Bioeng 1995; 45:463–472.
64. Vermes I, Haanen C, Steffens-Nakken H, Reutelingsperger C. A novel assay for apoptosis: flow cytometric detection of phosphatidylserine expression on early apoptotic cells using fluorescein labeled annexin V. J Immunol Meth 1995; 184:39–51.
65. Darzynkiewicz Z, Bedner E, Traganos F. Critical aspects in the analysis of apoptosis and necrosis. Human Cell 1998; 11:3–12.
66. Poirier J. Apoptosis Techniques and Protocols. New Jersey: Humana Press, 1997.
67. Ozturk SS, Palsson BO. Chemical decomposition of glutamine in cell culture media: effect of media type, pH, and serum concentration. Biotechnol Prog 1990; 6:121–128.
68. Tritsch GL, Moore GE. Spontaneous decomposition of glutamine in cell culture media. Exp Cell Res 1962; 28:360–364.
69. Zeng A-P, Hu W-S, Deckwer W-D. Variation of stoichiometric ratios and their correlation for monitoring and control of animal cell cultures. Biotechnol Prog 1998; 14: 434–441.
70. Zeng A-P. Continuous culture. Demain AL, Solomon NAManual of Industrial Microbiology and Biotechnology. 2nd ed. Washington, DC: American Society for Microbiology, 1999:151–164.
71. Werner RG, Walz F, Noé W, Konrad A. Safety and economic aspects of continuous mammalian cell culture. J Biotechnol 1992; 22:51–68.
72. Banik GG, Heath CA. An investigation of cell density effects on hybridoma metabolism in a homogeneous perfusion reactor. Bioprocess Eng 1994; 11:229–237.
73. Banik GG, Heath CA. Hybridoma growth and antibody production as a function of cell density and specific growth rate in perfusion culture. Biotechnol Bioeng 1995; 48:289–300.

74. Wagner A, Marc A, Engasser JM, Einsele A. The use of lactate dehydrogenase release kinetics for the evaluation of death and growth of mammalian cells in perfusion reactor. Biotechnol Bioeng 1992; 39:320–326.
75. Zeng A-P, Deckwer W-D. A kinetic model for substrate and energy consumption of microbial growth under substrate-sufficient conditions. Biotechnol Prog 1995; 11:71–79.
76. Neermann J, Wagner R. Comparative analysis of glucose and glutamine metabolism in transformed mammalian cell lines, insect and primary liver cells. J Cell Physiol 1996; 166:152–169.
77. Jenkins HA, Bulter M, Dickson AJ. Characterization of glutamine metabolism in two related murine hybridomas. J Biotechnol 1992; 23:167–182.
78. Ljunggren J, Häggström L. Glutamine limited fed-batch culture reduce the overflow metabolism of amino acids in myloma cells. Cytotechnology 1992; 8:45–56.
79. Sharfstein ST, Tucker SN, Manuco A, Blanch HW, Clark DS. Quantitative in-vivo nuclear magnetic resonance studies of hybridoma metabolism. Biotechnol Bioeng 1994; 43:1059–1074.
80. Schneider M, Marison IW, von Stockar U. The importance of ammonia in mammalian cell culture. J Biotechnol 1996; 46:161–185.
81. Andersen DC, Goochee CF. The effect of ammonia on the O-linked glycosylation of granulocyte colony-stimulating factor produced by Chinese hamster ovary cells. Biotechnol Bioeng 1995; 47:96–105.
82. Ryll T, Valley U, Wagner R. Biochemistry of growth inhibition by ammonia ions in mammalian cells. Biotechnol Bioeng 1994; 40:934–946.
83. Chang YHD, Grodzinsky AJ, Wang DIC. In situ removal of ammonia and lactate through electrical means for hybridoma cells. Biotechnol Bioeng 1995; 47:308–318.
84. Xie L, Wang DIC. Stoichiometric analysis of animal cell growth and its application in medium design. Biotechnol Bioeng 1994; 43:1164–1174.
85. Weidemann R, Ludwig A, Kretzmer G. Low temperature cultivation—a step towards process optimization. Cytotechnology 1994; 15:111–116.
86. Chuppa S, Tsai YS, Yoon S, Shackleford S, Rozales C, Bhat R, Tsay G, Matanguihan C, Konstantinov K, Naveh D. Fermentor temperature as a tool for control of high-density perfusion cultures of mammalian cells. Biotechnol Bioeng 1997; 55:328–338.
87. Moore A, Mercer J, Dutina G, Donahue CJ, Bauer KD, Mather JP, Etcheverry T, Ryll T. Effects of temperature shift on cell cycle, apoptosis and nucleotide pools in CHO cell batch cultures. Cytotechnology 1997; 23:47–54.
88. Jorjani P, Ozturk S-S. Effects of cell density and temperature on oxygen consumption rates for different mammalian cell lines. Biotechnol Bioeng 1999; 64:349–356.
89. Kaufmann H, Mazur X, Fussenegger M, Bailey JE. Influence if low temperature on productivity, proteome and protein phosphorylation of CHO cells. Biotechnol Bioeng 1999; 63:573–582.
90. Hightower LE, Hendershot LM. Molecular chaperones and the heat shock response at Cold Spring Harbor. Cell Stress Chaperones 1997; 2:1–11.
91. Sonna LA, Fujita J, Gaffin SL, Lilly CM. Effects of heat and cold stress on mammalian gene expression (invited review). J Appl Physiol 2002; 92:1725–1742.
92. Thieringer HA, Jones PG, Inouye M. Cold shock and adaptation. Bioassays 1998; 20:49–57.
93. Bloemkolk JW, Gray MR, Merchang F, Mosmann TR. Effect of temperature on hybridoma cell cycle and MAb production. Biotechnol Bioeng 1992; 40:427–431.
94. Giard DJ, Fleischaker RJ, Fabrikant M. Effect of temperature on the production of human fibroblast interferon. Proc Soc Exp Biol Med 1982; 170:155–159.
95. Reuveny S, Kim YP, Kemp CW, Shiloach J. Effect of temperature and oxygen on cell growth and recombinant protein production in insect cell cultures. Appl Microbiol Biotechnol 1993; 38:619–623.

96. Sureshkumar GK, Mutharasan R. The influence of temperature on a mouse-mouse hybridoma growth and monoclonal antibody production. Biotechnol Bioeng 1991; 37:292–295.
97. Kaufmann H, Mazur X, Marone R, Bailey JE, Fussenegger M. Comparative analysis of two controlled proliferation strategies regarding product quality, influence on tetracycline-regulated gene expression, and productivity. Biotechnol Bioeng 2001; 72: 592–602.
98. Bailey JE. Mathematical modeling and analysis in biochemical engineering: past accomplishments and future opportunities. Biotechnol Prog 1998; 14:8–20.
99. Wainwright LJ, Field MC. Quality control of glycosyl phosphatidylinositol anchor attachment in mammalian cells: a biochemical study. Biochem J 1997; 321:655–664.
100. Ducommun P, Ruffieux PA, Kadouri A, von Stockar U, Marison IW. Monitoring of temperature effects on animal cell metabolism in a packed bed process. Biotechnol Bioeng 2002; 7:838–842.
101. Ludwig A, Tomeczkowski J, Kretzmer G. Influence of the temperature on the shear stress sensitivity of adherent BHK21 cells. Appl Microbiol Biotechnol 1992; 38: 323–327.
102. Barton ME. Effect of pH on the growth cycle of HeLa cells in batch suspension culture without oxygen control. Biotechnol Bioeng 1971; 13:471–492.
103. Birch JR, Edwards DJ. The effect of pH on the growth and carbohydrate metabolism of a lymphoblastoid cell line. Dev Biol Stand 1980; 46:59–63.
104. Reddy S, Miller WM. Effects of abrupt and gradual osmotic stress on antibody production and content in hybridoma cells that differ in production kinetics. Biotechnol Prog 1994; 10:165–173.
105. McDowell CL, Papoutsakis ET. Serum increases the CD 13 receptor expression, reduces the transduction of fluid-mechanical forces, and alters the metabolism of HL60 cells cultured in agitated bioreactors. Biotechnol Bioeng 1998; 60:259–268.
106. Schmid G, Blanch HW, Wilke CR. Hybridoma growth, metabolism and product formulation in HEPES-buffered medium: II. Effect of pH. Biotechnol Lett 1990; 12:633–638.
107. Wayte J, Boraston R, Bland H, Varley J, Brown M. pH: effects on growth productivity of cell lines producing monoclonal antibodies: control in large scale fermenters. Cytotechnology 1997; 22:87–94.
108. Akatov VS, Lezhnev EI, Vexler AM, Kublik LN. Low pH value of pericellular medium as a factor limiting cell proliferation in dense cultures. Exp Cell Res 1985; 160:421–428.
109. Osman JJ, Birch J, Varley J. The response of GS-NSO myeloma cells to pH shifts and pH perturbations. Biotechnol Bioeng 2001; 75:63–73.
110. Osman JJ, Birch J, Varley J. The response of GS-NSO myeloma cells to single and multiple pH perturbations. Biotechnol Bioeng 2002; 79:398–407.
111. Rothman RJ, Warren L, Vliegenthart JF, Hard KJ. Clonal analysis of the glycosylation of immunoglobulin G secreted by murine hybridomas. Biochemistry 1989; 28: 1377–1384.
112. Borys MC, Linzer DIH, Papoutsakis M. Culture pH affects expression rates and glycosylation of recombinant mouse placental lactogen proteins by Chinese hamster ovary (CHO) cells. BioTechnology 1993; 11:720–724.
113. Ishaque A, Al-Rubeai M. Use of intracellular pH and annexin-V flow cytometric assays to monitor apoptosis and its suppression by bcl-2 over-expression in hybridoma cell culture. J Immunol Methods 1998; 221:43–57.
114. Miccoli L, Oudard S, Sureau F, Poirson B, Dutrillaux B, Poupon M. Intracellular pH governs the subcellular distribution of hexokinase in a glioma cell line. Biochem J 1996; 313:957–962.
115. Shu CH, Liao CC. Optimization of L-phenylalanine production of *Corynebacterium glutamicum* under product feedback inhibition by elevated oxygen transfer rate. Biotechnol Bioeng 2002; 77:131–141.

116. Fassnacht D, Portner R. Experimental and theoretical considerations on oxygen supply for animal cell growth in fixed-bed reactors. J Biotechnol 1999; 72:169–184.
117. Mercille S, Johnson M, Lanthier S, Kamen AA, Massie B. Understanding factors that limit the productivity of suspension-based perfusion cultures operated at high medium renewal rates. Biotechnol Bioeng 2000; 67:435–450.
118. Joenje H, Oostra AB. Effect of oxygen tension on chromosomal aberrations in Fanconi anaemia. Hum Genet 1983; 65:99–101.
119. Rancourt RC, Hayes DD, Chess PR, Keng PC, O'Reilly MA. Growth arrest in Gl protects against oxygen-induced DNA damage and cell death. J Cell Physiol 2000; 193:26–36.
120. Fehrenbach E, Northoff H. Free radicals, exercise, apoptosis, and heat shock proteins. Exerc, Immunol Rev 2001; 7:66–89.
121. Cochrane CG, Schraufstatter IU, Hyslop PA, Jackson JH. Oxy-radicals. Cerutti PA, Frodpvocj I, McCord, eds. Molecular Biology and Pathology. New York: Liss, 1988:125.
122. Shigenaga MK, Hagen TM, Ames BN. Oxidative damage and mitochondrial decay in aging. Proc Acad Sci 1994; 91:10771–10778.
123. Chen YC, Tsai SH, Lin-Shiau SY, Lin JK. Elevation of apoptotic potential by anoxia hyperoxia shift in NIH3T3 cells. Mol Cell Biochem 1999; 197:147–159.
124. Mizrahi A. Oxygen in human lymphoblastoid cell line cultures and effect of polymers in agitated and aerated cultures. In: Proceedings of the 5th Meeting of the ESACT, Copenhagen, Denmark. Basel: Karger, 1982:93–102.
125. Miller WM, Wilke CR, Blanch HW. Effect of dissolved oxygen concentration on hybridoma growth and metabolism in continuous culture. J Cell Physiol 1987; 132:524–530.
126. Ozturk SS, Palsson BO. Effects of dissolved oxygen on hybridoma cell growth, metabolism, and antibody production kinetics in continuous culture. Biotechnol Prog 1990; 6:437–446.
127. Orsi A, Beltran B, Clementi E, Hallen K, Feelisch M, Moncada S. Continuous exposure to high concentrations of nitric oxide leads to persistent inhibition of oxygen consumption by J774 cells as well as extraction of oxygen by the extracellular medium. Biochem J 2000; 346(Pt 2):407–412.
128. Lin AA, Miller WM. Modulation of glutathione level in CHO cells. Effects of oxygen concentration and prior exposure to hypoxia. Ann NY Acad Sci 1992; 665:117–126.
129. Packer L, Fuehr K. Low oxygen concentration extends the lifespan of cultured human diploid cells. Nature 1977; 267:423–425.
130. Backer MP, Metzger LS, Slaber PL, Nevitt KL, Boder GB. Large-scale production of monoclonal antibodies in suspension culture. Biotechnol Bioeng 1988; 32:993–1000.
131. Boraston R, Thompson PW, Garland S, Birch JR. Growth and oxygen requirement of antibody-producing mouse hybridoma cells in suspension culture. Dev Biol Stand 1983; 55:103–111.
132. Dorresteijn RC, Degooijer CD, Tramper J, Beuvery EC. A method for simultaneous determination of solubility and transfer coefficient of oxygen in aqueous media using off-gas mass spectrometry. Biotechnol Bioeng 1994; 43:149–154.
133. Miller WM, Wilke CR, Blanch HW. Transient responses of hybridoma metabolism to changes in the oxygen supply rate in continuous culture. Bioprocess Eng 1988; 3: 103–113.
134. Eyer K, Oeggerli A, Heinzle E. On-line gas analysis in animal cell cultivation: II. Methods for oxygen uptake rate estimation and its application to controlled feeding of glutamine. Biotechnol Bioeng 1995; 45:54–62.
135. Zhou W, Hu W-S. On-line characterization of a hybridoma cell culture process. Biotechnol Bioeng 1994; 44:170–177.
136. Yoon S, Konstantinov KB. Continuous, real-time monitoring of the oxygen-uptake rate (OUR) in animal-cell bioreactors. Biotechnol Bioeng 1994; 44:983–990.
137. Gray DR, Chen S, Howarth W, Inlow D, Maiorella L. CO_2 in large-scale and high-density CHO cell perfusion culture. Cytotechnology 1996; 22:65–78.

138. Lovrecz G, Gray P. Use of on-line gas analysis to monitor recombinant mammalian cell cultures. Cytotechnology 1994; 14:167–175.
139. Bailey JE, Ollis DF. Biochemical Engineering Fundamentals. New York: McGraw-Hill, 1977.
140. Kilburn DG, Webb FC. The cultivation of animal cells at controlled oxygen dissolved oxygen partial pressure. Biotechnol Bioeng 1968; 10:801–814.
141. Meier SJ, Hatton TA, Wang DIC. Cell death from bursting bubbles: role of cell attachment to rising bubbles in sparged reactors. Biotechnol Bioeng 1999; 62:468–478.
142. Aunins J, Henzler HJ. Aeration in cell culture bioreactors. Stephanopoulos G, Rehm HJ, Reed G, Puhler A, Stadler PJW, eds. Biotechnology: A Multi-Volume Comprehensive Treatise. Bioprocessing. Vol. 3. Weinheim, Germany: VCH Verlag, 1993:223–281.
143. Bonarius HPJ, Hatzimanikatis V, Meesters KPH, de-Gooijer CD, Schmid G, Tramper J. Metabolic flux analysis of hybridoma cells in different culture media using mass balances. Biotechnol Bioeng 1996; 50:299–318.
144. Ruffieux PA, von Stockar U, Marison IW. Measurement of volumetric (OUR) and determination of specific (qO_2) oxygen uptake rates in animal cell cultures. J Biotechnol 1998; 63:85–95.
145. Xiu Z-L, Deckwer W-D, Zeng A-P. Estimation of rate of oxygen uptake and carbon dioxide evolution of animal cell culture using material and energy balances. Cytotechnology 1999; 29:159–166.
146. Arrua LA, McCoy BJ, Smith JM. Gas–liquid mass transfer in stirred tanks. AIChE J 1990; 36:1768–1772.
147. Spérandio M, Paul E. Determination of carbon dioxide evolution rate using on-line gas analysis during dynamic degradation experiments. Biotechnol Bioeng 1997; 53:243–252.
148. Zeng A-P. Effect of CO_2 absorption on the measurement of CO_2 evolution rate in aerobic and anaerobic continuous cultures. Appl Microbiol Biotechnol 1995; 42:688–691.
149. deZengotita VM, Schmelzer AE, Miller WM. Characterization of hybridoma cell responses to elevated pCO_2 and osmolality: intracellular pH, cell size, apoptosis, and metabolism. Biotechnol Bioeng 2002; 77:369–380.
150. Kimura R, Miller WM. Glycosylation of CHO-derived recombinant tPA produced under elevated pCO_2. Biotechnol Prog 1997; 73:311–317.
151. Zupke C, Green J. Modeling of CO_2 concentration in small scale and large scale bioreactors. Proceedings of Cell Engineering VI, San Diego, CA, 1998.
152. Drapeau D, Luan Y-Y, Whiteford JC, Lavin DP, Adamson SR. Cell culture scale-up in stirred tank reactors. Proceedings of the Annual Meeting of the Society of Industrial Microbiology, Orlando, FL, 1990.
153. Taticek R, Petersen S, Konstantinov K, Naveh D. Effect of dissolved carbon dioxide and bicarbonate on mammalian cell metabolism and recombinant protein productivity in high density perfusion culture. Proceedings of Cell Culture Engineering VI, San Diego, CA, 1998.
154. Schmelzer AE, Miller WM. Effects of osmoprotectant compounds on NCAM polysialylation under hyperosmotic stress and elevated pCO_2. Biotechnol Bioeng 2002; 77: 359–368.
155. Grampp GE, Blumen TK, Kelley K, Derby P, Sleemen LA, Hettwer D. Environmental control of sialic acid composition in glycoproteins secreted by mammalian cells. Proceedings of Cell Culture Engineering IV, San Diego, CA, 1994.
156. Garnier A, Voyer R, Tom R, Perret S, Jardin B, Kamen A. Dissolved carbon dioxide accumulation in a large scale and high density production of TGFβ receptor with baculovirus infected Sf-9 cells. Cytotechnology 1996; 22:53–63.
157. deZengotita VM, Kimura R, Miller WM. Effects of CO_2 and osmolality on hybridoma cells: growth, metabolism and monoclonal antibody production. Cytotechnology 1998; 28:213–227.
158. Freshney RI. In: Alan R, ed. Culture of Animal Cells. 2nd ed. New York: Liss, 1987.

159. Kurano N, Leist C, Messi F, Kurano S, Fiechter A. Growth behavior of Chinese hamster ovary cells in a compact loop bioreactor: 1. Effects of physical and chemical environments. J Biotechnol 1990; 15:101–111.
160. Ozturk SS, Palsson BO. Effect of medium osmolality on hybridoma growth, metabolism, and antibody production. Biotechnol Bioeng 1991; 37:989–993.
161. Zanghi JA, Schmelzer AE, Mendoza TP, Knop RH, Miller WM. Bicarbonate concentration and osmolality are key determinants in the inhibition of CHO cell polysialylation under elevated pCO_2 or pH. Biotechnol Bioeng 1999; 65:182–191.
162. Ryu JS, Lee G. Application of hypoosmolar medium to fed-batch culture of hybridoma cells for improvement of culture longevity. Biotechnol Bioeng 1999; 62:120–123.
163. Flickinger MC, Goebel NK, Bibila T, Boyce-Jacino S. Evidence for posttranscriptional stimulation of monoclonal antibody secretion by l-glutamine during slow hybridoma growth. J Biotechnol 1992; 22:201–226.
164. Reuveny S, Velez D, Macmillan JD, Miller L. Factors affecting cell growth and monoclonal antibody production in stirred reactors. J Immunol Methods 1986; 86:53–59.
165. Suzuki E, Ollis DF. Enhanced antibody production at slowed growth rates: experimental demonstration and a simple structured model. Biotechnol Prog 1990; 6:231–236.
166. Al-Rubeai M, Emery AN. Mechanisms and kinetics of monoclonal antibody synthesis and secretion in synchronous and asynchronous hybridoma cell cultures. J Biotechnol 1990; 16:67–76.
167. Duval D, Demangel C, Munier-Jolain K, Miossec S, Geahel I. Factors controlling cell proliferation and antibody production in mouse hybridoma cells: I. Influence of the amino acid supply. Biotechnol Bioeng 1991; 38:561–570.
168. Racher AJ, Looby D, Griffiths JB. Influence of ammonium ion and glucose on mAb production in suspension and fixed bed hybridoma cultures. J Biotechnol 1993; 29:145–156.
169. Chua FK, Yap MGS, Oh SKW. Hyper-stimulation of monoclonal antibody production by high osmolarity stress in eRDF medium. J Biotechnol 1994; 37:265–275.
170. Kromenaker SJ, Srienc F. Cell cycle kinetics of accumulation of heavy and light chain immunoglobulin proteins in a mouse hybridoma cell line. Cytotechnology 1994; 14: 205–218.
171. Renard JM, Spagnoli R, Mazier C, Salles MF, Mandine E. Evidence that monoclonal antibody production kinetics is related to the integral of the viable cells curve in batch systems. Biotechnol Lett 1988; 10:91–96.
172. Al-Rubeai M, Emery AN, Chalder S, Jan DC. Specific monoclonal antibody productivity and the cell cycle—comparisons of batch, continuous and perfusion cultures. Cytotechnology 1992; 9:85–97.
173. Flickinger MC, Goebel NK, Bohn MA. Determination of specific monoclonal antibody secretion rate during very slow hybridoma growth. Bioprocess Eng 1990; 5:155–164.
174. Hayter PM, Kirkby NF, Spier RE. Relationship between hybridoma growth and monoclonal antibody production. Enzyme Microb Technol 1992; 14:454–461.
175. Mercille S, Massie B. Apoptosis-resistant NS/0-E1B-19K myeloma cells exhibit increased viability and chB43 monoclonal antibody productivity using cell cycle modulators. Cytotechnology 1998; 28:189–203.
176. Dempsey J, Ruddock S, Osborne M, Ridley A, Sturt S, Field R. Improved fermentation processes for NS0 cell lines expressing human antibodies and glutamine synthetase. Biotechnol Prog 2003; 19:175–178.
177. Fussenegger M, Schlatter S, Datwyler D, Mazur X, Bailey JE. Controlled proliferation by multigene metabolic engineering enhances the productivity of Chinese hamster ovary cells. Nat Biotechnol 1998; 16:468–472.
178. Watanabe S, Shuttleworth J, Al-Rubeai M. Regulation of cell cycle and productivity in NSO cells by the over-expression of p21CIPl. Biotechnol Bioeng 2002; 77:1–7.

179. Ibarra N, Watanabe B, Bi JX, Shuttleworth J, Al-Rubeai M. Modulation of cell cycle for enhancement of antibody productivity in perfusion culture of NSO cells. Biotechnol Prog 2003; 19:224–228.
180. Bi JX, Shuttleworth J, Al-Rubeai M. Uncoupling of cell growth and proliferation results in enhancement of productivity in $p21^{CIP1}$-arrested CHO cells. Biotechnol Bioeng 2004; 85:721–749.
181. Borth N, Heider R, Assadian A, Katinger H. Growth and production kinetics of human X mouse and mouse hybridoma cells at reduced temperature and serum content. J Biotechnol 1992; 25:319–331.
182. Dalili M, Sayles GD, Ollis D. Glutamine-limited batch hybridoma growth and antibody production: experiment and model. Biotechnol Bioeng 1990; 36:74–82.
183. Shaughnessy TS, Kargi F. Growth and product inhibition kinetics of T-cell hybridomas producing lymphokines in batch and continuous culture. Enzyme Microb Technol 1990; 12:669–684.
184. Bole DG, Hendershot LM, Kearney JF. Posttranslational association of immunoglobulin heavy chain binding protein with nascent heavy chains in nonsecreting and secreting hybridomas. J Cell Biol 1986; 102:1558–1566.
185. Omasa T, Ishimoto M, Higashiyama K-L, Shioya S, Suga KI. The enhancement of specific antibody production rate in glucose- and glutamine-controlled fed-batch culture. Cytotechnology 1992; 8:75–84.
186. Sambanis A, Lodish HF, Stephanopoulos G. A model of secretory protein trafficking in recombinant AtT-20 cells. Biotechnol Bioeng 1991; 38:280–295.
187. Kurokawa H, Ogawa T, Kamihira M, Park YS, Iijima S, Kobayashi T. Kinetic study of hybridoma metabolism and antibody production in continuous culture using serum-free medium. J Ferment Bioeng 1993; 76:128–133.
188. Murdin AD, Thorpe JS, Groves DJ, Spier RE. Growth and metabolism of hybridomas immobilized in packed beds: comparison with static and suspension cultures. Enzyme Microb Technol 1989; 11:341–346.
189. Takazawa Y, Tokashiki M, Murakami H, Yamada K, Omura H. High-density culture of mouse-human hybridoma in serum-free defined medium. Biotechnol Bioeng 1988; 31:168–172.
190. Seamans TC, Hu W-H. Kinetics of growth and antibody production by a hybridoma cell line in a perfusion culture. J Ferment Bioeng 1990; 70:241–245.
191. Ishizaki Y, Burne J, Raff MC. Autocrine signals enable chondrocytes to survive in culture. J Cell Biol 1994; 126:1069–1077.
192. Mathieu C, Jozan S, Mazars P, Come MG, Moisand A, Valette A. Density-dependent induction of apoptosis by transforming growth factor beta-1 in a human ovarian carcinoma cell line. Exp Cell Res 1995; 216:13–20.
193. Saeki K, Yuo A, Kato M, Miyazona K, Yazaki Y, Takaku F. Cell density dependent apoptosis in HL-60 cells, which is mediated by an unknown soluble factor, is inhibited by transforming growth factor β1 and overexpression of bcl-2. J Biol Chem 1997; 272:20003–20010.
194. Lauffenburger D, Cozens C. Regulation of mammalian cell growth by autocrine growth factors: Analysis of consequences for inoculum cell density effects. Biotechnol Bioeng 1989; 33:1365–1378.
195. Wohlpart D, Kirwan D, Gainer J. Effects of cell density and glucose and glutamine levels on the respiration rates of hybridoma cells. Biotechnol Bioeng 1990; 36:630–635.
196. Fredrickson AG. Formulation of structured growth models. Biotechnol Bioeng 1976; 18:1481–1486.
197. Zeng A-P, Deckwer W-D. Model simulation and analysis of perfusion cultures of animal cells at high cell density. Biotechnol Prog 1999; 15:373–382.
198. Wu P, Ray NG, Shuler ML. A computer model for intracellular pH regulation in Chinese hamster ovary cells. Biotechnol Prog 1993; 9:374–384.

199. Domach MM, Leung SK, Cahn RE, Cocks GG, Shuler ML. Computer model for glucose-limited growth of a single cell of *Escherichia coli* B/rA. Biotechnol Bioeng 1984; 26:203–216.
200. Domach MM, Shuler ML. A finite model for an asynchronous culture of *E. coli*. Biotechnol Bioeng 1984; 26:877–884.
201. Park SH, Ryu DDY. Cell cycle kinetics and monoclonal antibody productivity of hybridoma cells during perfusion culture. Biotechnol Bioeng 1994; 44:361–367.
202. Ozturk SS. Engineering challenges in high density cell culture system. Cytotechnology 1996; 22:3–16.

10

Fed-Batch Cultivation of Mammalian Cells for the Production of Recombinant Proteins

Liangzhi Xie and Weichang Zhou
Merck Research Laboratories, Merck & Co., Inc., Sumneytown Pike, West Point, Pennsylvania, U.S.A.

Fed-batch cultivation of mammalian cells is routinely used for industrial production of biologicals as a simple means of process intensification to increase cell concentration and to extend culture lifetime for a high product concentration and volumetric productivity. The operation is simple and sufficiently flexible to be implemented in existing facilities without major equipment modifications. However, its performance is strongly dependent upon the effectiveness of the nutrient feeding strategy to prevent nutrient depletion/accumulation and byproduct accumulation, so that critical nutritional and biochemical parameters including osmolality and carbon dioxide concentration can be maintained within levels suitable to promote cell growth or minimize cell death for product expression.

In this chapter, recent developments of efficient nutrient feeding strategies to achieve highly productive fed-batch cultures are reviewed. The two important strategic aspects, the design of concentrated nutrient feeding solutions through understanding of cellular nutritional requirements and metabolisms, and the control of nutrient feeding rates based on critical process parameters such as cell concentration, oxygen uptake rate (OUR), and nutrient concentrations, are discussed in detail. Examples of fed-batch cultures using industrial cell lines of choice including hybridoma, murine myeloma SP2/0 and NS0, and Chinese hamster ovary (CHO) cells, in which a viable cell concentration of more than 1×10^{10} cells/L and a protein production of approximately 3 g/L were achieved, are given to illustrate the capabilities and potential of such feeding strategies and cultures.

INTRODUCTION

The cultivation of mammalian cells has been a routine means for industrial production of biologicals including viral vaccines and vectors, human therapeutic proteins and monoclonal antibodies (MAb) (1–4). A detailed list of approved biotechnology drugs in 2004 can be found at http://www.bio.org/speeches/pubs/er/approveddrugs.asp. Production of attenuated live viruses as viral vaccines and vectors generally involves in vitro cultivation of mammalian cells and virus propagation steps. While very diversified bioreactors including disposable T-flasks, roller bottles, Nunc cell factories, Costar cubes and stirred tank microcarrier suspension bioreactors, are used to support growth of attachment-dependent cells, process operations are generally quite simple. Cells are mostly cultivated in a batch mode to a targeted cell concentration for virus propagation. Thereafter, culture media are exchanged to provide sufficient nutrients and/or to remove inhibitory metabolites for virus propagation prior to infection. Occasionally, perfusion mode is used for the propagation of nonlytic viruses, such as hepatitis A. Fed-batch operation is seldom used (1).

On the other hand, cultivation of mammalian cells for industrial production of human therapeutic proteins such as interferons, interleukins, growth hormones, blood growth or clotting factors, and MAb occurs mainly in the standard stirred-tank suspension bioreactors, which are operated in various operation modes: batch, fed-batch, perfusion, and continuous (2,4). For products licensed prior to the 1990s, batch operation is the dominant mode. Thereafter, either perfusion or fed-batch operations are pursued to meet higher standards for process economics and product quality necessary for commercialization. These operations provide the means for process intensification by increasing cell concentration and/or minimizing cell death, in order to extend culture lifetime for a high product concentration and volumetric productivity. As a result, lower-cost bulk manufacturing becomes feasible, thus enabling the industry to meet the challenge of providing high dosages, which are often required for therapeutic efficacy. This is particularly true for MAb. For example, a dose of 15 mg/kg of body weight is needed for an anti-respiratory syncytial virus monoclonal antibody (Synagis™, MedImmune, Inc.) expressed in murine myeloma NS0 cells. Other MAb, such as anti-tumor necrosis factor alpha monoclonal antibody for the treatment of active Crohn's (inflammatory bowel) disease (Remicade™, Centocor, Inc.) and anti-HER (human epidermal growth factor receptor) 2 monoclonal antibody for the treatment of metastatic breast cancer (Herceptin®, Genentech, Inc.), which are both expressed in CHO cells, require relatively high doses as well. Such requirements have been the driving forces for the development of even larger bioreactors and more intensive cell culture processing research since the 1990s. Efficient perfusion and fed-batch culture strategies have been systematically developed in parallel with intensive cell line and culture medium development to achieve a lower manufacturing cost of goods and higher product purities.

Fed-batch cultivation has become a platform technology for large-scale manufacturing of therapeutic proteins and MAb. It is simple to operate and sufficiently flexible to be implemented, even in the existing facilities which were previously designed for batch operations. No major modification of bioreactors and other equipment is required. In addition, it allows one to obtain a high product concentration and a high yield of product on medium for reduction of bulk manufacturing costs. Fed-batch operation can be viewed as a variation of the simple batch operation. Cells are first grown up to a time point in batch mode. Thereafter, concentrated solutions of single or multiple nutrient components are added to provide nutrients, while minimizing volume increase or culture dilution. The outcome is strongly dependent upon the effectiveness of the nutrient feeding strategy implemented. In general, salt-free nutrient concentrates are fed at rates to prevent nutrient depletion/accumulation, so that the nutritional and biochemical parameters including osmolality, carbon dioxide and other by-product concentrations can be maintained within levels suitable to promote cell growth or minimize cell death for product expression. In order to achieve this objective, an effective nutrient feeding strategy, which involves design of nutrient feeding solutions and control of the feeding rate, needs to be developed and implemented. Typically, concentrated solutions without inorganic salt components are designed based on stoichiometric nutritional analysis through understanding of cellular nutritional requirements and metabolisms. Their feeding rates are then controlled based on critical process parameters measured either off-line or on-line, such as cell concentration, OUR, and nutrient consumption.

In addition to the objectives of maximizing viable cell concentration over an extended culture lifetime for enhanced protein production, fed-batch operation can be very effectively used to manipulate cellular metabolism by controlling major carbon and energy sources such as glucose and glutamine at lower levels. As a result, lactate and ammonia production is reduced, and nutrients are utilized very efficiently.

In this chapter, cellular nutritional requirements and metabolisms are described for designing nutrient feeding solutions, followed by a summary of off-line and on-line measurements of process parameters, which are used for dynamic control of nutrient feeding rates. Examples of fed-batch cultures of industrial cell lines including hybridoma, murine myeloma SP2/0 and NS0, and CHO cells are then reviewed to illustrate applications of different nutrient feeding strategies to enable high performance cultures with maximum viable cell concentrations of more than 1×10^{10} cells/L and final product protein concentrations on the order of 3 g/L as well as the objective of cellular metabolism alterations in order to reduce lactate and ammonia production for efficient nutrient utilization.

CELL METABOLISM

Mammalian cells require a complex nutrient environment for growth and survival in vitro. Typical cell culture media are composed of carbon sources such as glucose, amino acids, vitamins, inorganic salts, buffers and various other components including growth factors. Some of these components, such as inorganic salts, are generally abundant, while others may limit cell growth. Most components may be consumed, while others, such as trace metals, are required for biological functions with little consumption. Understanding of the functions of medium components greatly contributed to the successful development of serum-free media by studying serum-replacement components. This has been extensively reported (5–10) and will not be covered in this chapter. Instead, the metabolism of glucose, amino acids, vitamins, and phosphate, which is relevant to nutrient feeding solution development, is reviewed in detail.

Glucose

Glucose plays many crucial roles in mammalian cell metabolism (Fig. 1). It is the primary source for energy production. Intermediates generated from its metabolic pathways also provide key precursors for cell mass synthesis. Several important

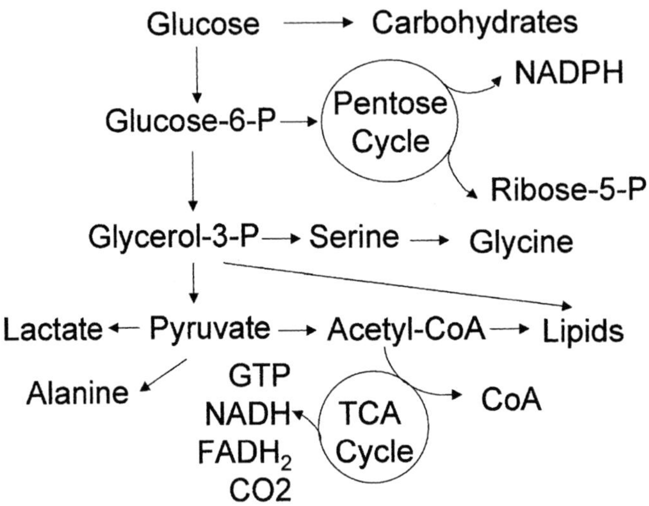

Figure 1 Schematic of simplified glucose metabolic pathways.

metabolic pathways, such as the glycolytic pathway, the pentose cycle, and the TCA cycle, are involved in the catabolism of glucose. The glycolytic pathway generates ATP and key intermediates for the syntheses of cellular components such as lipids in cell membranes and sugar chains attached to glycoproteins (11). The pentose cycle generates reducing power (NADPH) for lipid synthesis as well as ribose for nucleotide synthesis (12). The TCA cycle converts glucose carbon skeletons further into the end-product CO_2.

Lactate is a major by-product from incomplete glucose oxidation in the glycolytic pathway in mammalian cell culture. When glucose is converted into lactate, only two ATP molecules are generated in comparison with the maximum of 28 ATP molecules from complete oxidation in the TCA cycle, assuming the mechanistic P/O ratios of 2.5 for NADH and 1.5 for $FADH_2$ (13,14). This radical difference in energy production efficiency between the two different glucose pathways results in dramatic variations in glucose consumption and lactate production, depending on the percentage of glucose consumed in each of the two pathways.

Glucose metabolism is strongly influenced by environmental conditions, such as temperature, pH, and glucose concentration. Both glucose consumption and lactate production decrease significantly at lower temperature in cultures of hybridomas (15), recombinant hamster cells (16), and BHK cells (17). The effects of pH on glucose metabolism have been studied extensively as well. A high culture pH usually leads to high glucose consumption and lactate production as demonstrated in cultures of human liver HLM cells (18), hybridomas (19–21), and HeLa cells (22).

Most importantly, the consumption rate of glucose is dependent upon its concentration in culture. In HeLa cell cultures, a glucose concentration of about 2 mM is sufficient to sustain optimal cell growth (23). Further increase in the glucose concentration fails to increase the cell growth rate, but leads to a significant increase in lactate production. In human diploid fibroblast cell cultures, the minimum glucose concentration required for normal cell growth is even lower (24,25). With daily addition of glucose while maintaining a glucose concentration of 80 μM, the cell growth rate is equivalent to that in cultures with an initial glucose concentration of 5 mM. Although the growth rate is identical, the total glucose consumption at 80 mM glucose is only 7% of that at 5 mM, suggesting a significantly improved glucose utilization efficiency at a reduced concentration. In cultures of rat hepatoma cells, it is observed that the glucose-6-phosphate formed at low glucose concentrations (5–50 μM) is exclusively used by the cells for macromolecular synthesis and ATP production in the TCA cycle (26). The rate of CO2 production (from the TCA cycle) and glucose incorporation into macromolecules reaches a maximum at a 1-mM glucose concentration in culture, and the excessive glucose-6-phosphate formed at a higher glucose concentration is converted into lactate. Not surprisingly, lactate production rate is strongly dependent on glucose concentration in cultures of chick embryo fibroblasts (27). A linear relationship between the specific glucose consumption rate and the specific lactate production rate has also been observed in various cell culture systems (28,29).

Lactate is primarily derived from glucose, although glutamine can be a contributor as well, especially under glucose limited conditions (25,30). Experiments with radiolabeled compounds reveal that over 90% of glucose is transformed into lactate by ascite-tumor cells (31). In cultures where the glucose concentration is not intentionally controlled at a low level, the molar ratio of lactate production to glucose consumption usually ranges from 1 to 2, an indication that more than 50% of the glucose consumption is converted into lactate (32–35). The ratio increases with the residual glucose concentration in culture (29,36). The cause for the high glycolytic activity observed in cell culture is not well understood to date. It is hypothesized that excessive glycolysis is the result of a high influx of glucose across the cytoplasmic membrane due to a high extracellular glucose concentration and the absence or ineffectuality of membrane permeability control (37). Glycolysis is assumed to be a defense mechanism against the high intracellular glucose level. This is consistent with the fact that lactate is rarely produced in vivo (except under anaerobic conditions) where glucose concentration is tightly controlled by insulin. It has been observed that transformed and tumor cells are consistently associated with an elevated rate of glycolysis (lactate production from glucose) and there is a strong correlation with the degree of malignancy (38). For example, viral transformation of chick embryo fibroblasts reduces the apparent glucose K_m of glycolysis by over 10-fold, from 1.5 mM to 0.1 mM glucose, which results in an enhanced glycolytic activity at low glucose concentrations (27). Apparently, the molar ratio of lactate production to glucose consumption can differ significantly among different cell lines even under similar culture conditions.

Accumulation of lactate to a certain concentration in culture inhibits cell growth. The change in extracellular pH and osmolality as a result of lactate production is at least partially responsible for cell growth inhibition (39,40). The conversion of 1 mol of glucose to 2 mol of lactate is coupled with the generation of 2 mol of protons. These have to be neutralized by base addition when pH is controlled. Both the one extra mole of lactate produced per mole of glucose and the addition of a concentrated base solution for neutralization contribute to an osmolality increase. In an effort to understand the toxicity of lactate itself, the effect of lactate addition to culture on the initial hybridoma cell growth rate has been extensively studied by Glacken and coworkers (41). Through statistically designed experiments, the effects of lactate and other factors on growth rate were incorporated into a cell growth rate model. The addition of 40 mM sodium lactate has no considerable effect on the cell growth rate or the maximum cell concentration, while the addition of 80 mM causes a significant reduction in both the cell growth rate and the maximum cell concentration (42). The inhibitory effects of lactate and ammonia on cell growth are interactive as shown by Hassell and coworkers (43) when the lactate concentration exceeds a certain level. With proper pH control, an osmolality increase is responsible for the inhibitory effect of lactate in a lactate-controlled fed-batch culture (44). Evidence from many different cases indicates that lactate itself has little negative influence on cell growth (39,42,45).

The minimum concentration of glucose required to achieve a reasonable cell growth rate is typically below 2 mM (24,32,46), but lactate production increases significantly at a glucose concentration >2 mM. It is therefore possible to minimize lactate production without compromising cell growth by controlling the glucose concentration below 2 mM. This strategy has been successfully demonstrated in numerous fed-batch cultures (33,47–53).

Glutamine

Glutamine is a unique amino acid in cell metabolism and hence is discussed separately. A schematic diagram for glutamine metabolism is shown in Fig. 2. Like other amino acids, glutamine is incorporated into polypeptides for protein synthesis. Glutamine is unstable and decomposes into ammonia and pyrrolidone-carboxylic acid through first-order nonenzymatic kinetics. The amide group is essential for nucleotide synthesis, and it cannot be replaced by ammonia or the amino group. The amide group is also a donor for asparagine synthesis from aspartate. In cell culture, a significant percentage of glutamine is converted into glutamate and ammonia via glutaminase. The amino group in glutamate is then converted into proline, transaminated for the production of alanine, aspartate and serine/glycine via aminotransferases, or removed as ammonia via glutamate dehydrogenase. The glutamine carbon skeleton enters the TCA cycle as α-ketoglutarate. Under normal culture conditions, glutamine can contribute to a significant percentage (20–30%) of total energy production through the oxidation in the TCA cycle (54). Under a glucose limited condition, glutamine is consumed as the major energy source for the growth of human fibroblasts (25).

Ammonia is a major by-product from glutamine metabolism. It can be derived from the glutamine amide group via nonenzymatic degradation and glutaminase. It can also be derived from the glutamine amino group via glutamine dehydrogenase. The nonenzymatic glutamine degradation decreases at a reduced glutamine

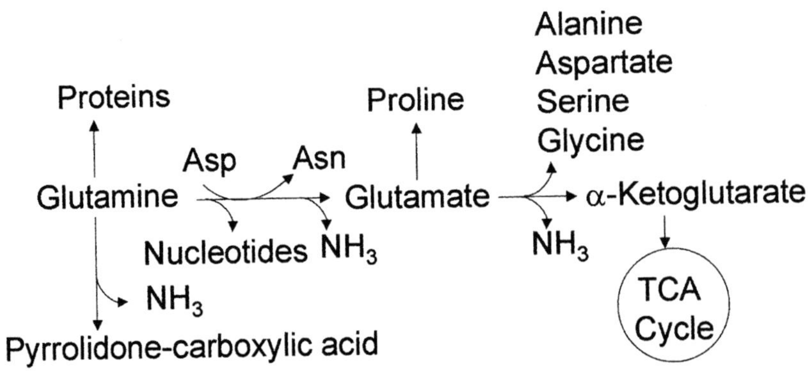

Figure 2 Schematic of glutamine metabolism.

concentration owing to the first-order kinetics, and its degradation rate is affected by pH (55), temperature (56), and serum concentration (57,58). The apparent glutamine K_m for glutaminase is in the range of 2–5 mM (59–61), while the glutamine K_m for cell growth is in the range of 0.15–0.53 mM (62,63). Glutamate generated in the mitochondria due to high glutaminase activities is also deaminated by glutamate dehydrogenase. In cultures of rat kidney cells using ^{15}N-labeled amino acids, about 50% of the total ammonia production is derived from the glutamine amide group and about 30% from the glutamine amino group (64). More recent studies by Street et al. (65) in cultures of HeLa and CHO cells using $^{1}H/^{15}N$-NMR techniques confirm that ammonia is predominately derived from the amide group of glutamine via glutaminase.

Glutamine concentration has little effect on cell growth rate until it is reduced to an extremely low level due to the low glutamine K_m required for cell growth, but glutamine concentration has a direct impact on the ammonia production rate. This is because the production of ammonia from glutamine increases with an increased glutamine consumption rate (29,35,47,51,52,66,67) that increases with glutamine concentration (44,47,49).

It is well documented that accumulation of ammonia in culture is toxic to mammalian cells (68). The ammonia level in human blood is usually maintained below 0.05 mM (69), while levels as high as 10 mM have been observed in tumor animal cell cultures (70,71). It appears that the ammonia level which retards cell growth depends on the cell line and culture conditions. Doyle and Butler (72) studied the effects of culture pH on the toxicity of ammonia in a murine hybridoma culture and proposed that the toxic species is ammonia instead of the ammonium ion. With the addition of NH_4Cl, Glacken et al. (47) observed no significant inhibition on cell growth at 4 mM, but a 50% reduction in growth rate was observed at 7 mM. Hassell et al. (43) studied the effects of ammonium on several cell lines. At a concentration of 2 mM exogenously added ammonium, four cell lines showed no significant inhibitory effects; two cell lines showed a 50% growth rate reduction; and two other cell lines showed a reduction of >75% in cell growth rate. In addition, ammonia affects protein glycosylation patterns. Borys et al. (73) showed that ammonia concentrations in the range of 3–9 mM inhibit the N-linked glycosylation in a pH-dependent fashion in a CHO cell culture. Similar inhibitory effects on the O-linked glycosylation were observed in a CHO culture by Andersen and Goochee (74). Newland et al. (75) reported that a gradual addition of ammonia to a batch culture up to 10 mM had no inhibitory effects while the spiking of 2 mM ammonium at the inoculation resulted in substantial growth inhibition. This is supported by the results obtained in cultures fed ammonium chloride where no growth inhibition was observed up to 10 mM ammonia (76,77). Further evidence was obtained in a hybridoma culture where cells grew normally under 12.5 mM of exogenous ammonia after more than 100 hr of adaptation. These data suggest that some cells may be able to adapt to a high ammonia level.

Reduction in ammonia accumulation is one of the successful approaches to achieve a high cell concentration, elongate the culture lifetime, and increase the product concentration and culture productivity. One of the strategies then is to take advantage of the significant differences in the glutamine K_m values for glutaminase and for cell growth. The cell growth rate can be maintained slightly below its maximum at a reasonably low glutamine concentration but the ammonia production from glutamine catalyzed by glutaminase can be significantly reduced. This strategy has been implemented successfully in fed-batch cultures with several mammalian cell lines (33,35,47,49,51,52,66,67). Another strategy is to replace glutamine with a glutamine-based dipeptide, which is more thermally stable (8,78). As a result, nonenzymatic glutamine degradation is largely eliminated. The demand for exogenous glutamine can be completely eliminated by either genetic modification or clone-selection for intrinsic expression of glutamine synthetase (GS) (49,79–82). Glutamine synthetase expression allows for glutamine synthesis from glutamate and ammonia, and hence eliminates ammonia production during growth and the need for exogenous glutamine at the same time. Methods have also been developed to remove ammonium through filter paper and an agarose-filled salt bridge (83,84).

Amino Acids

Thirteen amino acids including glutamine are essential for cell survival in vitro for most animal cell lines (85), while only eight (not including glutamine) are indispensable in vivo (69). The seven so called nonessential amino acids (NAAs) can be synthesized by the cells mainly from glutamine. Amino acid metabolism in culture has been investigated extensively (86–95). A correlation between the relative consumption of essential amino acids (EAAs) and amino acid composition of protein produced is observed except for valine, isoleucine, and leucine (89). This indicates that the major role of the essential amino acids is to provide precursors for protein synthesis. The consumption rates of valine, isoleucine, and leucine are affected by their concentrations in culture (96), presumably due to consumption for energy production in the TCA cycle and/or for lipid synthesis. However, catabolism of amino acids is not desirable because ammonia is one of the end products. When the concentrations of amino acids are controlled at low levels in fed-batch hybridoma cultures, the amino acid consumption correlates with the stoichiometric requirement for protein and nucleotide synthesis (35), suggesting minimum amino acid oxidation. On the other hand, several amino acids, especially isoleucine, are oxidized in the TCA cycle in batch cultures without nutritional control, contributing to ammonia accumulation (97).

Variations in the production of NAAs have been observed among different cell lines (98) and under different conditions for the same cell line (99). Alanine is often secreted into culture by tumor cell lines (33,98). Although provision of NAAs has no significant effect on cell growth rate (100), eliminating NAAs from culture medium results in an increased intracellular glutamine level that leads to an increased glutamine consumption and ammonia production (94). Inclusion of NAAs in a stoichiometrically balanced nutrient feeding solution reduces the demand for glutamine, NAA synthesis, and ammonia production (101). However, it is essential to maintain the concentrations of glutamine, glucose, and other essential amino acids at very low levels in order to force the cells to uptake NAAs.

Vitamins

Ten vitamins are essential for cell growth in vitro. Many important enzymes and cofactors are synthesized from vitamins (11). Nicotinamide is the substrate for the synthesis of NAD, NADH, NADP, and NADPH, while riboflavin provides precursor for the synthesis of FAD and $FADH_2$. The key coenzyme, CoA, in the pyruvate decarboxylation is synthesized from pantothenate. Pyridoxal phosphate participates in many key metabolic reactions such as transamination in the nonessential amino acid synthesis and amino acid decarboxylation. Thiamine pyrophosphate is a key coenzyme in the pentose phosphate pathway and the decarboxylation of pyruvate and α-ketoglutarate. Biotin is the CO_2 carrier in carboxylation reactions. Choline and inositol are required for lipid synthesis. Choline is also one of the major donors for the one carbon metabolism. Because of the low requirement and lack of routine analytic methods for vitamins, quantification of their demands for cell growth is not routinely performed (102). The vitamin composition in culture media is thus usually determined empirically (85). In a continuous hybridoma cell culture, step increases in vitamin B_1 and B_{12} concentrations improve cell growth and antibody production (96).

Phosphate

Phosphate is essential for life, as it is required for energy transfer within cells and for the synthesis of nucleic acids and phospholipids (103). It is present in nucleotides and important coenzymes, such as nicotinamide adenine dinucleotide (NAD), derived from nucleotides (104). It plays an essential role in biological energy transfer as ATP is the major form of chemical energy currency in all cells. A significant fraction of the cell membrane lipids is phospholipids (103). Therefore, availability of phosphorus, present as part of phosphate, can limit the total cell number achieved.

DESIGN OF CULTURE MEDIA AND FEEDING SOLUTIONS

Design of culture media and nutrient feeding solutions is principally based on stoichiometric nutritional analysis through understanding of cellular nutritional requirements and metabolism. To avoid a large osmolality increase and bioreactor dilution due to nutrient feeding, salt-free nutrient concentrates are generally developed as feeding solutions.

Stoichiometric Equation Governing Cell Growth

If one ignores the complex biological reactions on going in living cells and treats a cell culture bioreactor as a black box, one can use a simple equation (10.1) to describe the cell culture process: consumption of nutrients equals generation of cell mass, products, and by-products. The stoichiometric coefficients, defined as the amount of nutrient consumption or product formation divided by the total number of cells produced, can be calculated from measurements of net consumption of nutrients and net accumulation of cells, protein product, and by-products.

$$\begin{aligned} &A_1[\text{Glucose}] + A_2[\text{Glutamine}] + \cdots + A_i[\text{Amino Acid}]_i \\ &\quad + \cdots = [\text{Cell}] + B_1[\text{Lactate}] + B_2[\text{Ammonium}] + \cdots \end{aligned} \tag{10.1}$$

The stoichiometric coefficients measured from a culture can be employed to balance the nutrient requirements for cell growth by formulating culture media and feeding solutions according to the stoichiometric ratios. Use of a culture medium and a feeding solution which is stoichiometrically balanced may allow one to simplify the culture medium and/or feeding solution as one component. Therefore, measurement and control of one nutrient in the feed solution gives automatic control of the other nutrients. This approach has been successfully applied in fed-batch cultures of hybridomas (53,105,106) and NS0 cells (81).

However, the stoichiometric coefficients measured under one culture condition can be dramatically different from another. The consumption rate and utilization efficiency of glucose and glutamine vary significantly under different culture environments. This is because there are many alternative pathways to achieve the same metabolic functions. For example, ATP can be produced from glucose through complete oxidation into CO_2 and H_2O in the TCA cycle or through partial oxidation into lactate in the glycolytic pathway. However, the efficiency of these two alternative pathways differs dramatically. The exact cellular mechanism of controlling each metabolic pathway in mammalian cell culture is not yet fully understood, but it has been observed that nutrient utilization efficiency decreases when there are excessive amounts of nutrients available.

Table 1 shows the stoichiometric coefficients measured from four hybridoma cultures: one batch culture in a 2-L bioreactor, two fed-batch cultures in T-flasks, and one fed-batch culture in a 2-L bioreactor. In the batch culture, the nutrient concentrations are high, especially in the early phase of the culture. In the 2-L fed-batch culture, the nutrient concentrations are kept at low levels throughout the culture by continuous addition of an optimized feeding solution. The two T-flask cultures are fed manually and periodically, hence the nutrient concentrations are maintained lower than the batch culture but higher than the 2-L fed-batch. The stoichiometric coefficients measured under these conditions vary significantly, especially for glucose,

Table 1 Stoichiometric Coefficients Measured from Hybridoma Cultures Under Different Conditions

	Unit	Batch	FB 1[a]	FB 2[a]	FB 3[a]
Consumption				2	
Glc	pmol/cell	6.9	3.7	3.7	2.2
Gln	pmol/cell	1.4	0.86	1.0	0.72
Ile	pmol/cell	0.16	0.08	0.09	0.10
Leu	pmol/cell	0.18	0.13	0.16	0.19
Lys	pmol/cell	0.11	0.15	0.15	0.14
Met	pmol/cell	0.04	0.04	0.04	0.05
Phe	pmol/cell	0.06	0.05	0.06	0.06
Thr+Arg	pmol/cell	0.18	0.23	0.20	0.20
Tyr	pmol/cell	0.05	0.05	0.06	0.05
Val+Trp	pmol/cell	0.14	0.11	0.15	0.15
Ala	pmol/cell	−0.59	−0.30	−0.20	−0.17
Asn	pmol/cell	0.02	0.06	−0.01	0.09
Asp	pmol/cell	0.03	0.18	0.07	0.13
Glu	pmol/cell	−0.17	−0.04	0.00	−0.08
Gly	pmol/cell	−0.01	0.13	0.04	0.13
Pro	pmol/cell	0.00	0.00	0.00	0.00
Ser	pmol/cell	0.04	0.08	0.06	0.03
Production					
Cells	cell/cell	1	1	1	1
Antibody	pg/cell	18	20	40	39
Lac	pmol/cell	11	3.6	3.4	0.40
NH3	pmol/cell	0.79	0.46	0.43	0.22

[a]Fed-batch cultures conducted in T-flasks (1 and 2) with periodical manual feeding (52) and bioreactor (3) with continuous automatic feeding (35).
[Data calculated from Ref. (97).]

glutamine, alanine, lactate, and ammonium, as shown in Table 1. The control of the concentrations of glucose, glutamine, and amino acids in the fed-batch cultures leads to a dramatic improvement in nutrient utilization efficiencies, when compared with the uncontrolled batch culture. Different nutrient utilization efficiencies at different nutritional environments are also observed more recently in continuous hybridoma cultures, where multiple steady-states with distinct cellular metabolism exist (107).

Recognizing the variable efficiencies of nutrient utilization at different nutritional environment in mammalian cell cultures, Xie and Wang (51,108) controlled the nutritional environment to achieve maximum nutrient utilization efficiency and minimum by-product accumulation by maintaining the nutrient concentrations at low but sufficient levels to support a reasonable cell growth rate. It is obviously a challenge to control the concentrations of more than 30 nutrients (glucose, amino acids, vitamins, phosphate, etc.) at low levels without running into depletion/accumulation of one or several components, and it is not practical to use a feed-back control strategy for all of the nutrients individually. However, this problem can be solved if the stoichiometric coefficients in Eq. (10.1) are known. For example, in a simple chemical reaction $A + B + 2C = D + E$, the concentrations of all three reactants in a reactor can be simultaneously maintained constant through the control of a single parameter, the feeding rate of a mixture of A, B, and C formulated at a molar ratio of 1:1:2. The feeding rate can be controlled either by on-line monitoring of a single reactant or product, or by calculation of known reaction kinetics. Similarly, the nutritional environment can be controlled with a single nutrient feeding stream if the stoichiometric Eq. (10.1) can be determined. Obviously, the stoichiometric coefficients are not constant due to the multiple alternative metabolic pathways involved in the complex cellular metabolic reaction network, and hence there is no definitive solution. This is similar to a complex chemical process with multiple side reactions and by-products.

Theoretically, for a given cell line, there exists a set of minimum stoichiometric coefficients for nutrients and by-products. Under ideal conditions, the nutrient utilization efficiencies reach their maxima and by-product formation is minimized. These minimum stoichiometric coefficients can be determined from material and energy balances under the conditions where the most efficient pathways for essential biological functions are utilized and alternative pathways that provide nonessential biological functions, such as synthesis of nonessential amino acids, are eliminated. Such an ideal system can be characterized as the following:

1. Glucose is utilized solely for essential biological functions: energy production, reducing power generation, carbohydrate and lipid syntheses, and sugars for nucleotide synthesis.
2. Glutamine is utilized solely for essential biological functions: protein and nucleotide synthesis.
3. Energy production is derived from complete oxidation of glucose in the TCA cycle.
4. Reducing power is generated from the pentose cycle to meet the demands for lipid synthesis and no excessive reducing power is generated.
5. No lactate is produced.
6. No nonessential amino acids are synthesized.
7. Amino acids (except for glutamine) are utilized for protein synthesis only.

Under the above conditions, the minimum stoichiometric coefficients for all the 20 amino acids and glucose can be calculated from the cell composition and energy requirements. For example, with the known composition of amino acids for the cellular protein mixture and the total amount of cellular protein per cell, one can calculate the requirement of each individual amino acid needed to produce a cell, with the exception of glutamine that is also required for nucleotide synthesis. The coefficients for vitamins are less variable and are estimated based on the measured consumption rates reported in the literature (102). The model has been described in detail by Xie and Wang (51,52,97,108).

In reality, the nutrient utilization efficiencies are not at the theoretical maxima (with the minimum stoichiometric coefficients) at all times, especially for glucose and glutamine. Even when the concentrations of glucose and glutamine are controlled at very low levels, production of lactate, ammonium, and nonessential amino acids cannot be completely eliminated. Therefore, some relaxation from the ideal model on lactate, nonessential amino acids, and ammonium production is needed. The degree of relaxation can be determined empirically from historical cultures, especially from glucose and glutamine-limited cultures. Over- or under-relaxation may result in suboptimal process performance such as reduced nutrient utilization efficiency, increased by-product formation, lower than expected cell growth rate, and/or accumulation of some nutrients. To reach the most efficient nutrient utilization efficiencies in practical applications, one may need to optimize the degree of relaxation in a stepwise fashion (i.e., iteration).

Cell Composition

Cell composition is needed to calculate the stoichiometric coefficients for glucose and amino acids (51,52,97,108). However, measuring cell composition accurately is not a trivial task. The percentage (dry cell basis) of macromolecules (proteins, lipids, carbohydrates, DNA, and RNA) varies from cell line to cell line as shown in Table 2. However, for rough estimates, using cell compositions from a different cell line should not generate gross errors in medium formulations. The dry cell weight also varies significantly from cell line to cell line but such a measurement can be conducted (52). The dry cell weights of a few example cell lines are: 250 pg/cell for hybridoma CRL 1606 cell line (52); 302 pg/cell for 293 SF cell line (109); and 400 pg/cell for a CHO cell line (112).

The mean amino acid composition of the cellular proteins does not differ significantly from one cell line to another and from mammalian cells to *Escherichia coli* as shown in Table 3. This is probably because cellular proteins represent a pool of a

Table 2 Dry Cell Compositions (wt%)

Macromolecule	Hybridoma[a]	293 5–19S[b]	293[c]	*E. coli*[d]
Proteins	72.9	64.1	73.0	70
Lipids	13.5	24.7	10.0	10
Carbohydrates	3.5	N/A	N/A	5
DNA	1.4	4.0	N/A	5
RNA	3.8	N/A	N/A	10

[a] From Xie and Wang (51)
[b] From Nadeau et al. (109).
[c] From Nadeau et al. (110).
[d] From Lehninger (111).

Table 3 Molar Percentages of Amino Acid Composition of Cellular Proteins

Amino acids	207 Protein average[a]	Hybridoma[b]	293[c]	*E. coli*[d]
Ala	9.0	8.3	5.5	9.6
Arg	4.7	5.9	8.4	5.5
Asn	4.4	4.4	4.5	4.5
Asp	5.5	4.7	4.9	4.5
Cys	2.8	2.8	3.1	1.7
Glu	6.2	6.2	7.5	4.9
Gln	3.9	5.0	6.8	4.9
Gly	7.5	8.5	4.8	4.9
His	2.1	2.2	3.0	1.8
Ile	4.6	4.2	3.3	5.4
Leu	7.5	8.1	7.9	8.4
Lys	7.0	6.8	10	6.4
Met	1.7	2.2	0.06	2.9
Phe	3.5	3.2	3.9	3.5
Pro	4.6	5.3	4.8	4.1
Ser	7.1	6.9	5.1	4.0
Thr	6.0	5.7	5.9	4.7
Trp	1.1	1.1	1.1	1.1
Tyr	3.5	2.6	5.0	2.6
Val	6.9	6.0	4.5	7.9

[a] Calculated from 207 random proteins with know amino acid sequence (113).
[b] Measured from hydrolyzed total cellular proteins of hybridoma cells (51).
[c] Measured from hydrolyzed total cellular proteins of 293 cells (110).
[d] Measured from hydrolyzed total cellular proteins of *E. coli* (114).

large number of individual proteins. These data suggest that the requirements for amino acids in protein synthesis are similar among different cell lines, which makes it feasible to formulate culture media and feeding solutions with a generic composition of amino acids. In cases where the specific protein production rate is as high as 50 pg/cell-day, the amount of the protein product can be a significant percentage of the total cellular proteins, and the amino acid sequence of this individual protein could be different from the average of the cellular proteins. In this case, the amino acid composition of the protein product is needed to obtain accurate results. In most cases, however, the mean amino acid composition can be used without significant errors due to low protein production and/or similarity of the amino acid composition to the mean.

Stoichiometric Coefficients for Vitamins, Metal Ions, and Phosphate

In addition to glucose and amino acids, mammalian cells also consume significant amounts of vitamins, metal ions, and phosphorus in the form of phosphate. The stoichiometric coefficients for vitamins can be estimated from the measured consumption rate in a batch culture from the literature (102). The stoichiometric coefficients for metal ions can be easily calculated from the measured metal content of cells. As most of the inorganic salts are present in excess in the basal medium to maintain a physiological osmolality and to provide appropriate trans-membrane differentials, there is no need to include them in the feed solution except for iron and some trace metals (35). Phosphate is a major component of cell mass but it has been largely neglected by most investigators in the past. In typical basal culture media, the phosphate concentration is in the range of 0.5–1 mM, which is more than sufficient in batch cultures with a relatively low maximum cell concentration. Phosphate requirements for cell synthesis are dependent on the cell line. A phosphate requirement for synthesis of cell mass was determined to be 0.2 mmol/10^9 cells for mouse fibroblast cell clone 929-L (115). This is significantly higher than for NS0 and hybridoma cells. In a fed-batch culture of NS0 cells without phosphate feeding, a phosphate requirement of 0.18 mmol/10^9 cells was estimated by dividing the available phosphate by the total cell concentration achieved (81). Somewhat lower values have been calculated in the same way for hybridoma fed-batch cultures [0.068 mmol/10^9 cells (53) and 0.12 mmol/10^9 cells (97)]. Clearly, feeding strategy optimization efforts without phosphate feeding may extend the culture lifetime, but cannot further increase the total cell concentration. The benefits of phosphate feeding have been demonstrated by Xie and Wang (35) who assumed a phosphate yield of 0.084 mmol/10^9 cells for development of feeding solutions in a hybridoma fed-batch culture, and more recently in an NS0 myeloma fed-batch culture by deZengotita et al. (116).

MONITORING OF CRITICAL PROCESS PARAMETERS

The key to implementing any kind of control strategy is to reliably measure off-line and/or on-line process parameters, which are directly controlled or used to control inferred parameters. Use of in situ biosensors in bioreactors would be ideal, but is not essential (117). As in situ biosensors are still limited to measurement of temperature, pH, dissolved oxygen (DO) concentration, CO_2, and culture turbidity by optical density (OD) probes for estimation of cell concentration, off-line and ex situ on-line measurement techniques are techniques of choice for measuring concentrations of cells, nutrients and metabolites (118–124). Metabolic activities such as OUR, can be determined either on-line or off-line. As sensors for measuring temperature, pH and DO have been extensively described elsewhere (119–124), this chapter will mainly review the methods for determination of concentrations of cells, nutrients and metabolites as well as metabolic activities.

Determination of Cell Concentrations

Off-line determination of dry cell weight, as is commonly used for microbial cultivation, is not sufficiently accurate for following cell growth kinetics and control purposes in mammalian cell culture due to relatively low cell concentration. Instead, off-line numeric counting of viable and dead cells per unit of volume using a hemocytometer has been widely used. For on-line determination of cell growth kinetics, various methods including indirect determination based on metabolic activity (125), in situ OD probes (126–130) and microscopy systems (131–133) have been employed. While OD probes measure culture turbidity, which is then correlated with total cell concentration (126–130), the in situ microscope takes images from mammalian cell bioreactors, which are analyzed for cell number, cell size, and cell morphology (131,132).

The approach of using indirect measurements such as nutrient consumption and metabolite production to determine cell concentrations is not straightforward and may result in incorrect prediction. OD measurements are direct, simple, and noninvasive. They rely on the principle of light transmission or back scattering using different light sources including laser lights with a precise wavelength for culture turbidity. Through an off-line calibration, the culture turbidity is converted to total cell concentration on-line. It is not possible to differentiate viable and dead cell concentrations and to obtain microscopic information such as cell size and cell shape based on OD measurements. However, their utility cannot be underestimated, in particular for actively growing cultures with a high and consistent cell viability. For example, a number of OD probes have been successfully used to determine total cell concentrations on-line (128–130) and to estimate the physiological state of growing cells by determining specific cell growth rate and specific metabolic activities in combination with other on-line measurements such as OUR (129). Nutrient feeding rates can also be controlled in fed-batch and continuous perfusion cultures based on OD measurements (126,127,130).

The drawbacks of the OD probes can be overcome by use of an in situ microscope, which is directly installed in a bioreactor port (131,132). In situ images from mammalian cell bioreactors can be taken for analysis of cell number, cell size, and cell morphology. Differentiation of viable and nonviable cells can be achieved by brightness of cells with or without staining by trypan blue or by cell size (133).

Determination of Concentrations of Nutrients and Metabolites

On-line measurements of nutrient and metabolite concentrations provide means for direct feedback control of the measured parameters at desired levels. Due to the lack of in situ sensors, these parameters are measured outside the bioreactor by analyzers such as high pressure liquid chromatography (HPLC) and/or sensors coupled to flow injection analysis (FIA) (122,124). For this purpose, on-line aseptic sampling devices are required to provide continuous sampling. Selection of a simple and reliable sampling device is the key to the implementation of external on-line analysis. These devices, generally based on sterile membrane filtration (122,134,135), can be installed in situ or ex situ to provide 1–2 mL/min of sample for external analyzers. Such devices work generally very well for small molecular weight nutrients and metabolites such as glucose, lactate and amino acids. Potential operational problems may exist when using either in situ or ex situ membrane devices for large molecular weight compounds such as recombinant proteins. With long-term usage, the membrane may foul and large molecular weight compounds may not completely pass through these membrane devices. In situ sampling devices are much preferred over ex situ, as ex situ devices have several disadvantages including contamination risks associated with potential failure. Ex situ devices require a pump to maintain a high circulation rate across the filter membrane and the cells may be subjected to excessive mechanical shear and may suffer from oxygen or nutrient limitation which may alter their metabolic activities since they are outside the bioreactor.

Applications of FIA have been widely reported for on-line analysis of numerous low molecular weight compounds in the culture broth, such as glucose, lactate, glutamine, and phosphate (122,136–145). Proteins can also be assayed using either turbidimetric immunoassay or a heterogeneous assay with a working principle similar to affinity chromatography (135,146,147). In an FIA system, samples to be analyzed are injected into a moving nonsegmented carrier stream of buffers or reagents. The reaction between the sample and the carrier stream in the manifold, as the sample is transported toward a detector for continuous recording, results in a change in either absorbance, pH, DO, conductivity or other physical parameters. As a result, a typical FIA output is a peak because of the dispersion of the injected sample zone. The peak height or the area under the peak can be related to the concentration of the measured component based on prior calibrations. An FIA system consists of five essential components: (a) a selector valve used to switch flow between samples to be analyzed, calibration standards or wash/purge solutions; (b) a multi-channel peristaltic pump module; (c) an injection valve used to load a specific volume of samples or standards (20–50 μL) into a carrier stream at programmed intervals, (d) a manifold where reactions take place, and (e) a detector/transduction system with a recorder or computer. Typically as many as 120 samples can be analyzed within an hour, which is more than sufficiently frequent in mammalian cell culture since the residence time of a sample is < 30 sec. In addition to efforts for establishment of new FIA systems with expanded applications (148), further refinement has been achieved by including on-line calibration based on an internal standard (149) and improving the robustness, reliability and automation (147,150).

Culture broth samples also can be assayed by HPLC off-line or on-line for glucose, amino acids and other compounds (49,122,151,152). The results may be used for feedback control (49,152).

Determination of Metabolic Activities

Metabolic activities can be determined using concentration profiles of cells, nutrients and metabolites. They also can be inferred by determining other process parameters, such as base addition for pH control, OUR, carbon dioxide evolution rate (CER). Measurement of base addition for pH control can be related to lactate production and correlated with glucose consumption. OUR can be determined in different ways in mammalian cell culture.

Establishment of gas and liquid phase balances through the analysis of exhaust gas composition can be used for estimation of both OUR and CER (153–159). Often it is insensitive and complicated due to large gas hold up in the head space, use of bicarbonate-buffered medium and low oxygen consumption rates. These problems can be resolved by using optimized aeration methods where surface aeration with air can be combined with direct sparging with an oxygen-enriched gas. The problem of CO_2 release from bicarbonate in the medium in the calculation of CER can be solved through the use of models to calculate the portion of CO_2 released from bicarbonate. As a result, on-line gas analysis has been successfully performed using both infrared CO_2 and paramagnetic O_2 analyzers (154,155) or a mass spectrometer (156–159) for OUR and CER estimation.

Several simple and reliable alternative techniques to on-line gas analysis have found applications for on-line OUR determination in mammalian cell culture. They include so-called dynamic (129,160), stationary (161) and recirculation (162) methods. In the dynamic method, a DO concentration profile is dynamically created by changing DO set-point. The DO concentration is increased to a higher set point by intensified oxygenation through use of an oxygen enriched gas and/or an increase agitation rate. Subsequently, oxygen supply is interrupted and nitrogen gas is used to flush the O_2 from the bioreactor head space. As a result, the DO concentration decreases due to oxygen removal by nitrogen flush and oxygen consumption by the culture. The time profile of the DO decrease is used to calculate the OUR. This method is very sensitive and can accurately determine OUR even at a very low cell concentration. The measurement frequency can be as high as several times within a hour when necessary. The DO varies within a pre-determined range. Such a variation of DO could affect cell growth and product formation. This is however not an issue for most mammalian cells, as minimal DO effects are observed between 20% and 80% of air saturation.

The stationary method allows the determination of OUR continuously. In this method, the DO concentration is controlled at a constant level by varying the O_2 composition in the gas phase via variation of the gas flow rate of O_2, CO_2, and N_2. OUR is calculated through an O_2 balance in the liquid phase. This method is however not very sensitive, especially in batch cultures, when OUR is relatively small. The data quality relies strongly on the quality of DO control tuning. In both the dynamic and stationary methods, K_La values for oxygen transfer are either predetermined and assumed to be constant or can be adjusted at different culture conditions.

In the recirculation method, the culture broth is circulated through an external recycle loop at a certain flow rate. The difference between the DO concentrations in the bioreactor and the external loop is proportional to the pumping rate and the OUR of the culture. Thus, the pumping rate needs to be optimized to provide a relatively large difference between the two DO concentrations without causing an O_2 limitation for the cells in the loop. In addition, accurate determination and characterization of the external loop volume and residence time distribution are needed.

DYNAMIC NUTRIENT FEEDING STRATEGIES

Once feeding solutions are designed based on stoichiometric nutritional analysis through understanding of cellular nutritional requirements and metabolism, the control of the feeding rate becomes the key to ensure that nutrients are maintained within desired ranges. Feed-back and/or feed-forward control concepts are usually used to dynamically adjust the feeding rates based on the consumption of an indicator nutrient, which can be directly measured and/or inferred by an off-line or on-line measurement. Advanced control concepts, such as adaptive control strategies using mathematical models, expert systems and artificial neural networks, have been widely applied for real-time fault diagnosis, estimation of nonmeasurable parameters, on-line prediction and state estimation in microbial fermentation processes. They may also find applications in mammalian cell culture (117,163), despite the complexity of the culture medium and cellular metabolism.

Control of Nutrient Feeding Rates Based on Direct Measurements

When an indicator nutrient such as glucose or glutamine is measured on-line, the nutrient feeding rate can be adjusted based on its consumption. In a fed-batch hybridoma culture, Kurokawa et al. (49) controlled glucose and glutamine concentrations at low levels to reduce the production of inhibitory metabolites, lactate and ammonium, through controlled feeding of concentrated glucose and glutamine solutions. A higher cell concentration was obtained with the controlled feeding strategy than in batch culture. For feedback control, glucose and glutamine concentrations were measured on-line using an HPLC. This approach can also be used to control the feeding of a solution of multiple nutrients with a stoichiometrically balanced composition. By controlling the indicator nutrient at a desired level through addition of the feeding solution, other nutrients are added at the same time and their levels controlled.

Control of Nutrient Feeding Rates Based on Indirect Measurements

When the indicator nutrient is not measured directly, a correlation between the measured parameters and the consumed nutrients needs to be established first. For example, through establishment of a correlation between base addition for pH control and lactate production/glucose consumption, the amount of glucose consumed, i.e., the nutrient feeding solution to be added, can be inferred from the amount of base added for pH control (164). The same is true when using viable cell concentration and OUR measurements as indicators of nutrient consumption (81,53,116,165). Based on the measured viable cell concentration, the amount to be fed can be estimated from the integral of viable cell concentration over time (IVC) assuming a constant specific growth rate, specific uptake rates, or cell yields over time. The feeding volume of a particular nutrient solution(s) can be calculated using Eq. (10.2):

$$V_i = \frac{q_i V_c}{C_i} \int_{t_n}^{t_{n+1}} x_v \, dt \tag{10.2}$$

where V_i is the volume of the concentration of solution i required at time t_n, q_i is the specific consumption rate of nutrient i, C_i is the concentration of i in the feeding solution, V_c is the reactor volume at time n, and x_v is the viable cell concentration. To project the viable cell concentration for time t_{n+1} when the next feeding would take place, a predetermined specific growth rate of μ 1/hr can be used in the cell growth phase (Eq. 10.3).

$$(x_v)_{n+1} = (x_v)_n \exp(\mu \, (t_{n+1} - t_n)) \tag{10.3}$$

During the stationary and death phases, the same viable cell concentration can be assumed at time t_n and t_{n+1}.

Similarly, the stoichiometrically balanced feeding solution can be added on-line based on OUR measurements, once a direct correlation between the specific OUR and specific nutrient uptake including amino acid consumption rates is established. Nutrients can be fed on-line when OUR measurements take place according to Eq. (10.4):

$$V_{\text{AAfeed}(n)} = \frac{V_c \times OUR \times \Delta t}{\alpha C_{\text{feed}}} \tag{10.4}$$

where V_{AA} feed (n) is the volume of nutrient feeding solution at time t_n, V_c is the reactor volume at time t_n, Δt is the interval of OUR measurement, and C_{feed} is the concentration of a representative nutrient, α is the stoichiometric feeding coefficient (μmol oxygen/μmol representative nutrient).

EXAMPLES OF FED-BATCH CULTURES

Hybridoma, murine myeloma SP2/0 and NS0, and CHO cells are widely cultivated in fed-batch operation for production of recombinant proteins and MAb. Efficient nutrient feeding strategies have been developed to achieve highly productive fed-batch culture, in which a viable cell concentration of more than 1×10^{10} cells/L and a protein production of approximately 3 g/L were obtained, in the following sections, selected fed-batch examples of these cell lines are given to demonstrate the fed-batch culture capacities and potential.

Hybridoma Cells

Numerous publications on fed-batch cultures of hybridoma cells exist, even though their use for commercial manufacturing of licensed therapeutic proteins has not been wide spread. Measurements of base addition for pH control (164), glucose and glutamine (49), viable cell concentration and OUR (53,81,165) were used to control feeding rates of optimized nutrient feeding concentrates.

Hybridoma cells generally consume a lot of glucose and glutamine, and produce high levels of lactate and ammonia, which may be inhibitory to cell growth. As described above, lactate and ammonium production can be significantly reduced by controlling glucose and glutamine concentrations at low levels through controlled feeding of concentrates (49). This feeding strategy leads to a higher cell concentration when compared with a batch culture. However, fed-batch culture capacity can be significantly improved by also feeding other nutrients in addition to glucose and glutamine at the same time.

To improve fed-batch culture performance, a salt-free nutrient concentrate was developed and fed according to on-line measurement of OUR and its stoichiometric relation to glucose consumption rate (53). As the feeding solution contained multiple nutrients with a stoichiometrically balanced composition, these nutrients were controlled simultaneously through feeding. Glucose and other essential nutrient (e.g., amino acids) concentrations were maintained at low, but not limited levels. This resulted in a sixfold increase in viable cell concentration when compared with a batch culture (12 vs. 2×10^9 viable cells/L) (Fig. 3). In addition, a high cell viability of >90% was achieved throughout the growth phase. It is interesting to note that once the viable cells reached the maximum concentration, no further increase of the total cells was observed. Despite the high cell concentrations achieved in this fed-batch culture, monoclonal antibody production was relatively low at a maximum of less than 60 mg/L due to a very low cell-specific productivity. Clearly, cell metabolism alterations were observed continuously throughout the culture. By controlling the glucose concentration at a low level, lactate production was significantly reduced. In fact, little lactate was produced after 100 hr despite a large amount of glucose consumed, while oxygen consumption was increased (Fig. 4a). Such cell metabolism alterations from high lactate production and low oxygen consumption to low lactate production and high oxygen consumption can be better illustrated when the molar ratios of oxygen consumption and lactate production to glucose consumption are estimated throughout the culture (Fig. 4b). With a decreased lactate production derived from glucose consumed, glucose consumed was increasingly oxidized, reaching the theoretical maximum molar ratio of 6 between oxygen and glucose consumption. The glucose carbon balance was established with glucose consumption largely converted to lactate and completely oxidized to CO_2 in the TCA cycle (Fig. 4b) (53). Ammonia and alanine levels were also reduced when the concentrations of amino acids including glutamine were controlled at low levels (165). It was also clearly demonstrated that cell metabolism can be manipulated by dynamic nutrient feeding control. Even though the viable cell concentration could be maintained for a long period of time, the maximum total and viable cell concentrations remained almost unchanged in these fed-batch cultures (165). These results suggest that some other

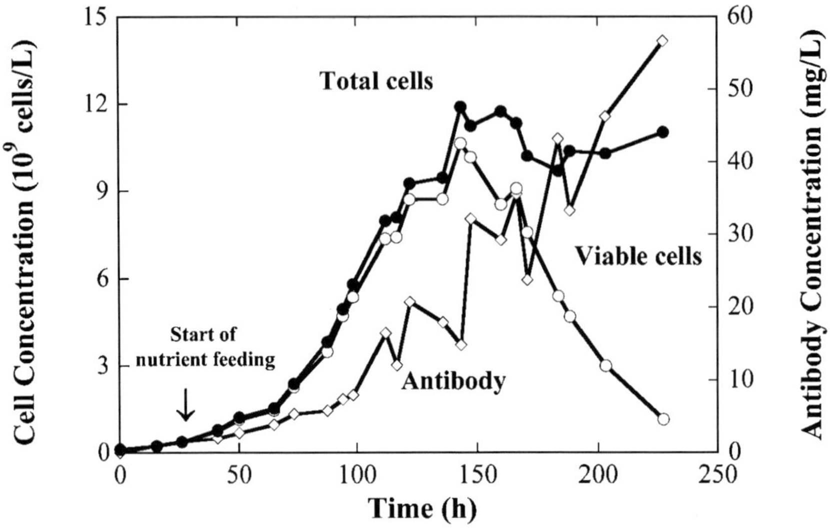

Figure 3 Growth kinetics and monoclonal antibody production of a fed-batch hybridoma culture. Viable cell concentration (O), total cell concentration (•), monoclonal antibody concentration (♦).

nutrients that were not included in the feeding solution were limiting factors for further cell growth improvement. Availability of phosphate was likely limiting for additional cell growth, as a phosphate consumption of 0.068 mmol/10^9 cells was estimated by dividing the phosphate amount available in the initial medium by the total cell concentration achieved.

Phosphate limitation seems to be eliminated in a hybridoma fed-batch culture by using a stoichiometric feeding strategy when dialyzed serum, phosphate and trace salts were also fed (35,51,52). The total phosphate requirement per cell was assumed to be twice of the phosphate requirement for DNA and RNA synthesis (0.084 mmol/10^9 cells) (35). In an initial un-optimized fed-batch culture, multiple nutrients, including glucose, glutamine, and all essential amino acids, were controlled at relatively constant and low levels throughout the entire fed-batch culture (Fig. 5a,b) (52). This type of stable nutritional environment approaches the steady-state nutrient environment achieved in typical continuous cultures. The feeding solution and feeding strategy were subsequently further optimized, which led to improved process performance (35). Bioreactor culture was conducted with a specially designed initial medium that contained low nutrient concentrations just sufficient to support reasonable initial cell growth rates but insufficient to sustain cell growth. Continuous feeding of an optimized and stoichiometrically balanced nutrient solution was controlled to maintain the nutritional environment constant, especially glucose and glutamine, whose concentrations were kept low throughout the 550-hr culture except for a brief spike during the stationary phase at the late stage of the culture. As a result, lactate and ammonia production rates were drastically reduced as shown in Fig. 6(a,b) (35). The average specific lactate and ammonia production rates were reduced by 83- and six-fold, respectively, when compared with a simple batch culture. In fact, only 3.4% of the total glucose consumption was converted into lactate, suggesting glucose metabolism approached the theoretical maximum efficiency under these controlled conditions as confirmed by material

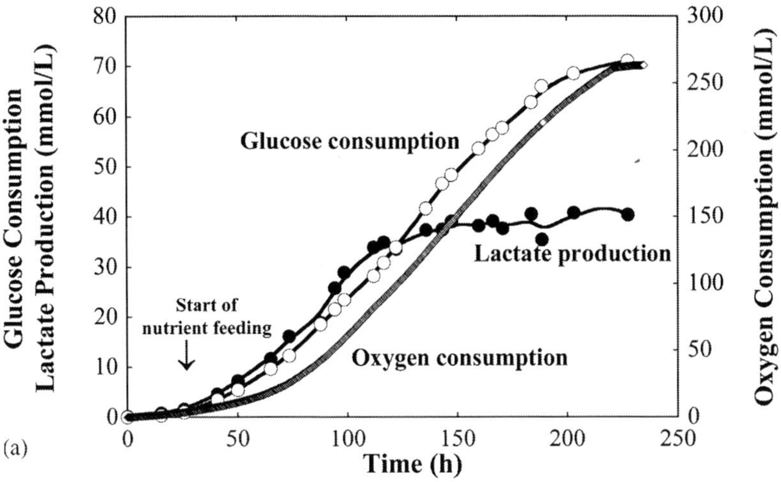

(a)

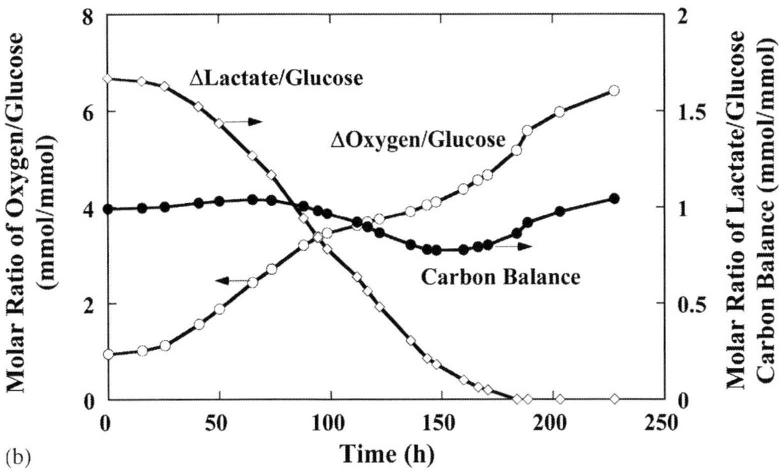

(b)

Figure 4 Continuous cell metabolism alterations in a fed-batch hybridoma culture. (a) Time course of cumulative glucose (O) and oxygen consumption (♦), and lactate production (•). (b) Carbon balance (•) and differential ratio of lactate production to glucose consumption (♦), and oxygen consumption to glucose consumption (O).

and energy balances (54,101). The control of the culture environment avoided nutrient depletion or accumulation and delayed the onset of cell growth inhibition from lactate and ammonia accumulation. The culture lifetime was extended from 120 hr in batch to above 550 hr in fed-batch with maximum concentrations of 1.7×10^{10} cells/L for viable cells and 5.0×10^{10} cells/L for total cells (35). Accompanying the improved cell growth and culture life time for a very high integral of viable cell concentration over time of 4.0×10^{12} cells hr/L, the final product concentration was increased from about 90 mg/L in a batch culture to about 2.4 g/L in a fed-batch culture using optimized and stoichiometrically balanced feeding solutions (Fig. 7) (35,97). It should be noted that the cell line employed was not genetically modified nor the clone selected for enhanced cell specific productivity, and hence the improvement in product concentration and productivity was solely due to the process optimization.

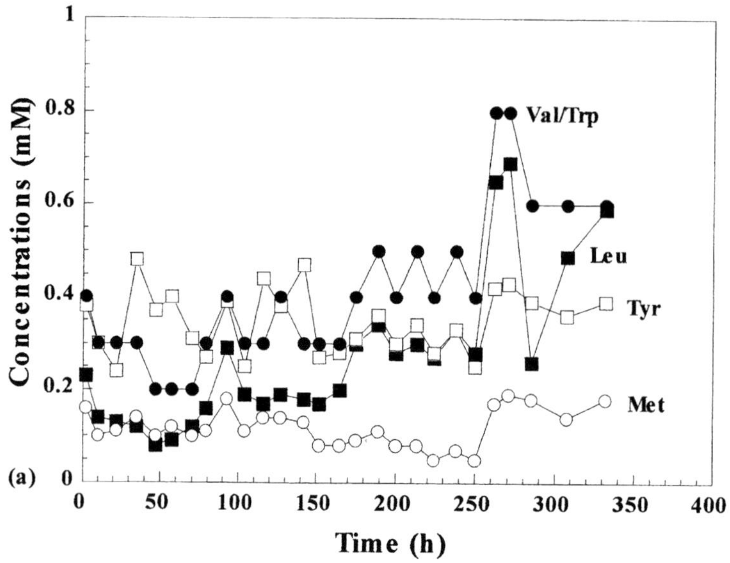

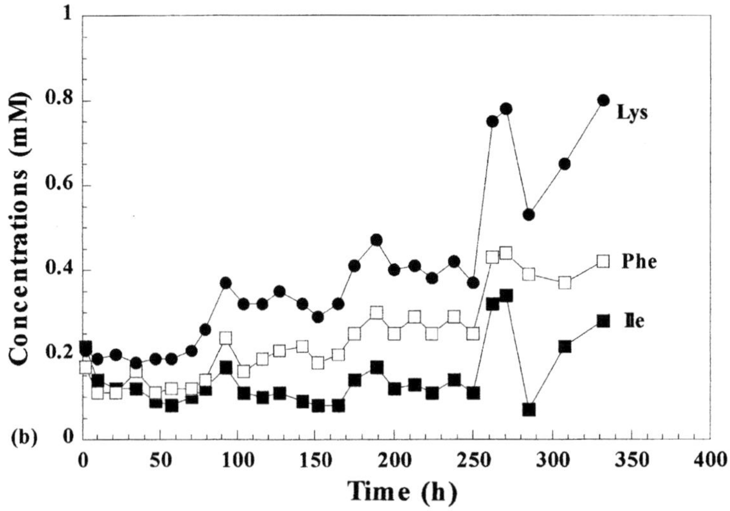

Figure 5 Control of the concentrations of multiple nutrients (glucose, glutamine, and other essential amino acids) in a fed-batch hybridoma culture through automatic feeding of a stoichiometrically balanced nutrient solution. (a) Met, Tyr, Leu, Val, and Trp. (b) Ile, Phe, and Lys.

Murine Myeloma SP2/0 and NS0 Cells

Murine myeloma SP2/0 and NS0 cell line are two principal mammalian cell lines for industrial production of recombinant proteins and MAb (28,81,166–168). Two commercial therapeutic MAb, Synagis^{TM} (Medlmmune, Inc.) and Zenapax® for prevention of kidney transplant rejection (Hoffmann-La Roche, Inc.), along with a number of monoclonal antibody product candidates, are expressed in these murine myeloma cell lines.

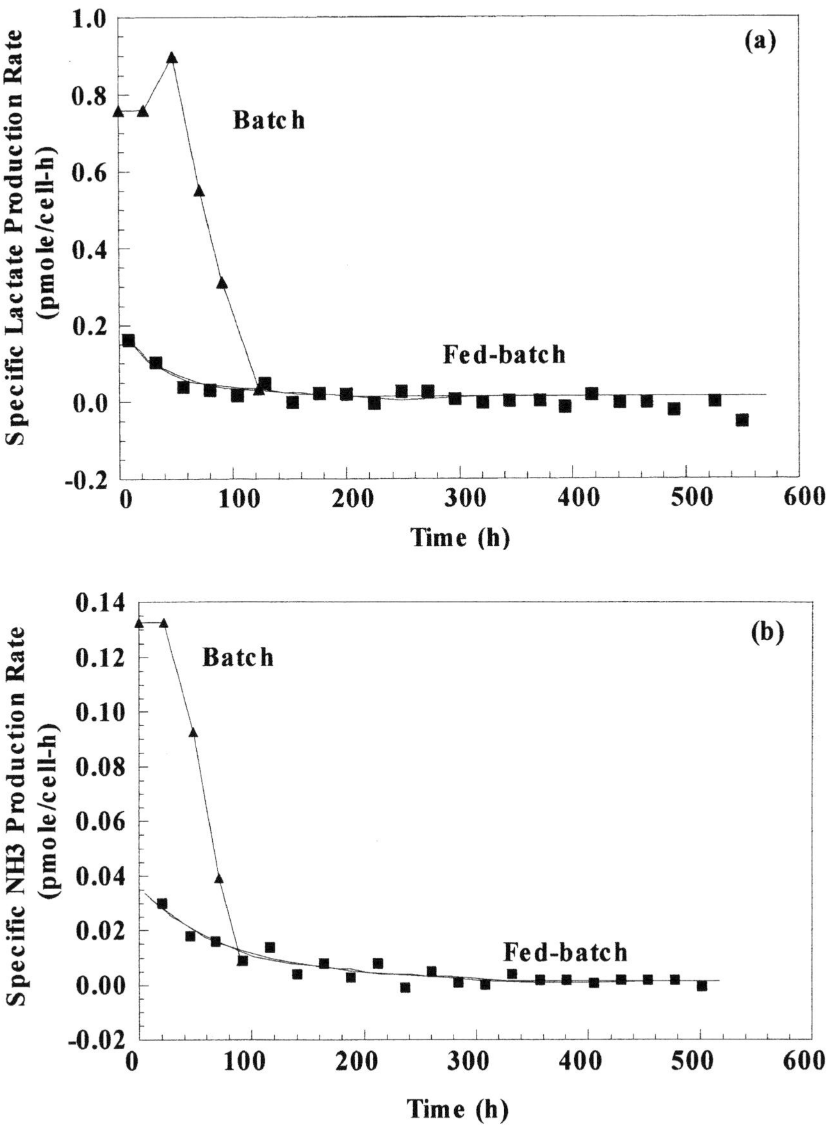

Figure 6 Reduction of the specific lactate and ammonia production rates through stoichiometrically balanced nutrient feeding in a fed-batch hybridoma culture. (a) Specific lactate production rate; (b) specific ammonia production rate.

Sauer et al. (28) developed a generic simple fed-batch protocol for different murine myeloma SP2/0 cell lines producing humanized MAb, and this was successfully scaled up to 750 L. A concentrated feeding solution primarily composed of amino acids, vitamins, and glucose was fed periodically with a predetermined post-feed target glucose concentration as an indicator nutrient. Due to simple and infrequent nutrient additions, glucose concentration was maintained at a fairly high level throughout the cultivation. Not surprisingly, lactate accumulated to a very high level (~110 mM). Nevertheless, monoclonal antibody concentrations were increased up to 7.6-fold when compared with batch culture.

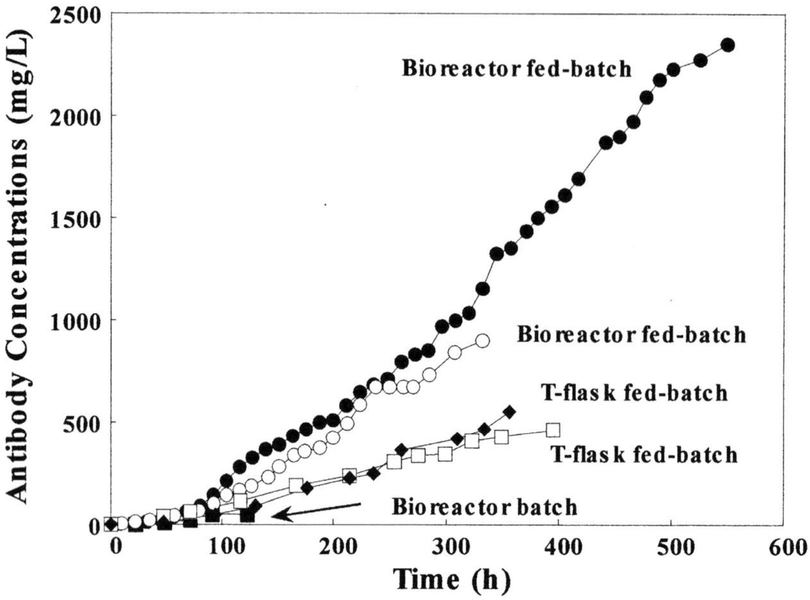

Figure 7 Enhancement of monoclonal antibody production through optimization of a balanced nutrient feeding in several fed-batch hybridoma cultures when compared with a batch culture.

NS0 myeloma cells transfected with a GS gene are frequently used to express recombinant proteins (80,166,169–171). GS expression allows the cells to synthesize glutamine from glutamate and ammonia thus refraining them from the need for exogenous glutamine, and lessening ammonia accumulation (80). These NS0 cells have different nutritional requirements and metabolic characteristics from wild-type NS0 cells (172). They consume significantly more glutamate, asparagine and other amino acids, less glucose and produce less lactate in culture compared to wild-type myeloma cells and hybridoma cells (164,172). Neither lactate nor ammonia accumulates in culture to a level inhibitory to cell growth or causing necrotic cell death (169,173). Most NS0 cells are cholesterol auxotrophs and require exogenous cholesterol for optimal growth (174), even though they can be adapted to cholesterol-independence (175,176). Cholesterol has a very low solubility in aqueous solution and is generally delivered using soybean lecithin emulsions or complex protein solutions in serum-free media. Supplementation of these complex protein solutions to high levels may produce inhibitory effects on the cell growth (177), so overfeeding of these solutions should be avoided. Furthermore, it has been shown that like many hybridoma cells, both wild-type and recombinant NS0 myeloma cells in culture die primarily by apoptosis, which can be induced by deprivation of essential nutrients such as glucose and amino acids (173,178–182). Nutrient feeding can prevent or delay the onset of apoptosis and suppress the extent of apoptosis.

To develop highly productive fed-batch cultures of an amplified NS0 cell line for production of MAb, important nutrients were maintained at their initial concentrations throughout the cultivation to achieve a high integral of viable cell concentration over time (IVC). Concentrated solutions of key nutrient components such as cholesterol and insulin were fed periodically using a feeding control strategy based on measurements of viable cell concentration and OUR. Feeding amounts were adjusted daily based on the IVC and/or OUR and assumed constant specific nutrient consumption rates or yields to maintain concentrations of the key nutrient components around their initial levels. Through effective nutritional control, both cell growth phase and culture lifetime were prolonged significantly resulting in a maximal viable cell concentration of 6.6×10^9 cells/L, as compared to $< 1.5 \times 10^9$ cells/L in a batch culture, and a final IVC of 1.6×10^{12} cells hr/L at 672 hr (Fig. 8) (53). OUR, which correlated well with viable cell concentration, reached a maximum of 1.0 mmol/L/hr (Fig. 8). The MAb concentration increased in parallel to the IVC and reached a final value of more than 2.7 g/L (Fig. 9), significantly higher than 0.2–0.3 g/L generally obtained in batch cultures. Cellular metabolism shifts were repeatedly observed. Accompanying the culture phase transition from the exponential growth to the stationary phase, lactate, which was produced in the exponential growth phase, was consumed. The time point at which this metabolism shift occurred corresponded to a rapid decrease of OUR which was likely caused by nutrient depletion. This transition coincided with the onset of ammonia, glutamate and glutamine accumulation. With removal of the nutrient depletion by increasing the daily nutrient feeding amount, OUR recovered and viable cell concentration increased, while cell metabolism shifted again. Instead of consumption, lactate was produced again. These results suggest close relationships among nutrient depletion, cell metabolism transitions, and cell death.

Subsequent to the above batches, phosphorus depletion was identified in fed-batch GS-NS0 cultures. Experiments using a less productive GS-NS0 clone generally yielded a maximum total cell concentration of $5–7 \times 10^9$ cells/L without phosphate feeding, which is consistent with the maximum total cell concentration supported by the phosphate available in the medium. By also feeding a phosphate solution to eliminate phosphorus depletion, the cell growth phase was prolonged significantly, resulting in a total cell concentration of ca. 17×10^9 cells/L, which is much greater

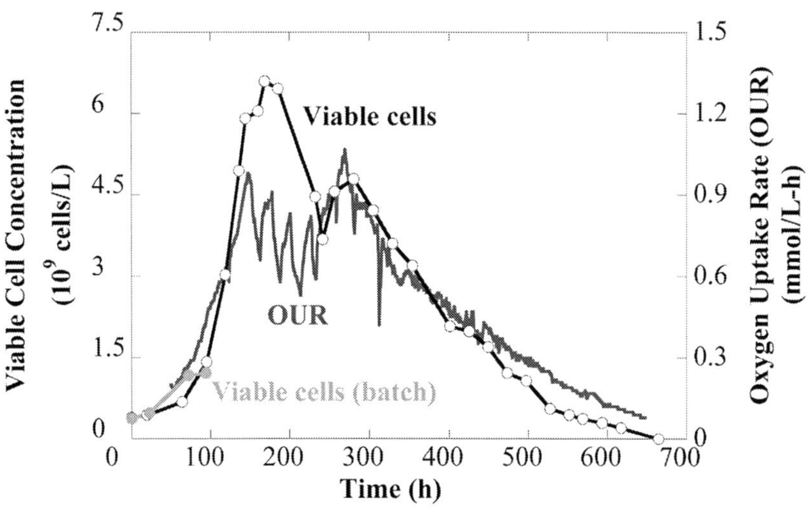

Figure 8 Growth kinetics of a fed-batch NSO culture when compared with a batch culture. Viable cell concentration (fed-batch) (O), viable cell concentration (batch) (•), on-line measured oxygen uptake rate (solid line).

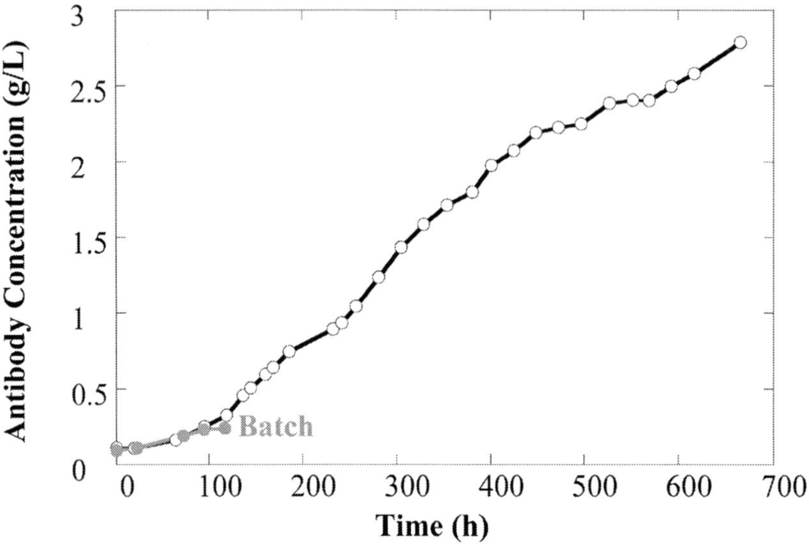

Figure 9 Monoclonal antibody production in a fed-batch NSO culture (O) compared with a batch culture (•).

than ca. 7×10^9 cells/L without phosphate feeding. The maximum viable cell concentration reached about 10×10^9 cells/L, twice as high as that without phosphate feeding (Fig. 10). Apoptosis was also delayed and suppressed with phosphate feeding. A nonapoptotic viable cell population of 6.5×10^9 cells/L, as compared to 3×10^9 cells/L without phosphate feeding, was obtained and successfully maintained for about 70 hr. These results are consistent with the knowledge that phosphorus is an essential part of many cell components including phospholipids, DNA, and

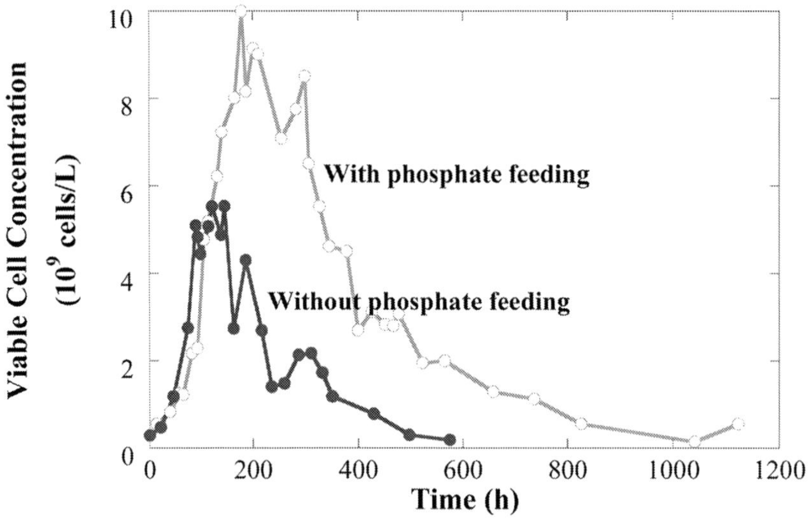

Figure 10 Growth kinetics of fed-batch NSO cultures with or without phosphate feeding. Viable cell concentration (with phosphate feeding) (O), viable cell concentration (without phosphate feeding) (•).

RNA. As a result of phosphate feeding, phosphate depletion was avoided (Fig. 12) and a much higher integral of viable cell concentration over time of 3.0×10^{12} cell hr/L was achieved, resulting in a correspondingly higher MAb titer of ca. 1.3 g/L (Fig. 11). This was a significant improvement over 5×10^9 viable cells/L, a final IVC of 1.1×10^{12} cell hr/L at 575 hr, and a final MAb concentration of 0.5 g/L without phosphate feeding (Fig. 11) (81). It was also noted that phosphate feeding delayed the cell metabolism shift from lactate production to lactate consumption typically observed in GS–NS0 cultures. The results highlight the importance of phosphate feeding in high cell concentration NS0 cultures. By further data analysis, a direct correlation between the total cell concentration and amount of phosphorus consumed was obtained (Fig. 12). The phosphorus requirement was not affected by phosphate feeding, and was determined to be ca. 6 mg or 0.18 mmol/10^9 cells (Fig. 12).

Such highly productive fed-batch GS–NS0 cultures have also been successfully developed for industrial production of MAb. Tsao (167) reported successful development of a simple GS–NS0 fed-batch culture expressing an Mab against RSV (Synagis®). By increasing the maximum viable cell concentration to ca. 5×10^9 cells/L, as compared to ca. 1.5×10^9 cells/L in its first-generation fed-batch process, the MAb concentration was increased to ca. 2.8 g/L after 3 weeks, as compared to ca. 0.8 g/L. This fed-batch process was developed at small scales (<10 L), successfully scaled up, and demonstrated at 2000 L scale with consistent performance.

CHO Cells

CHO cells are used for manufacturing of a number of licensed therapeutic proteins and MAb. They include tissue plasminogen activator, erythropoietin, granulocyte colony stimulating factor, factors VIII and IX, deoxyribonuclease I, glucocerebrosidase, beta-interferon, MAb against GPIIb/IIIa, CD20, tumor necrosis factor alpha, tumor necrosis factor receptor, and HER2. CHO cells are the most popular indus-

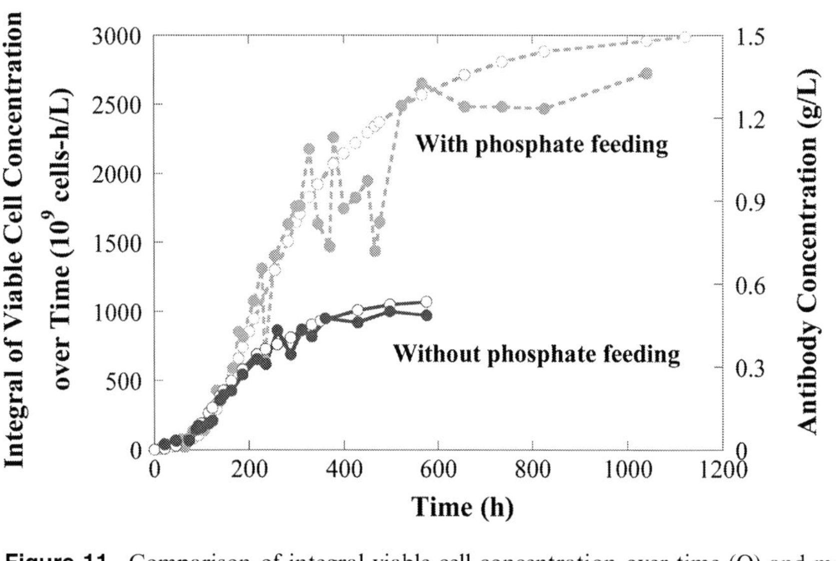

Figure 11 Comparison of integral viable cell concentration over time (O) and monoclonal antibody production (•) in NSO fed-batch cultures with (dotted lines) or without phosphate feeding.

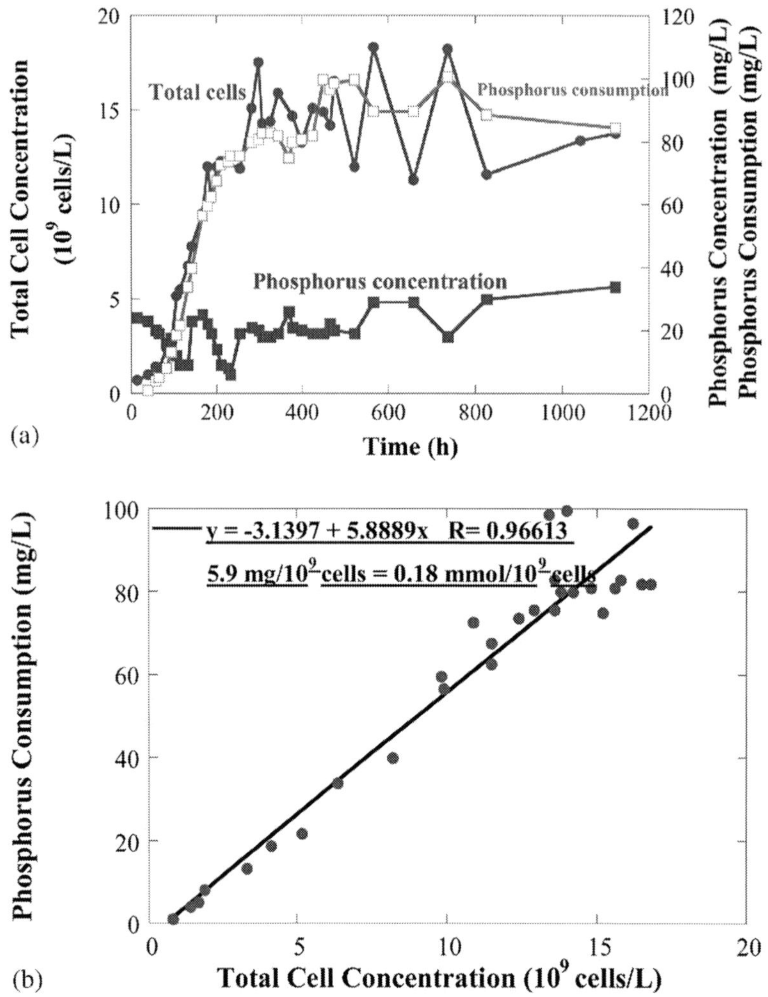

Figure 12 Estimation of phosphorus requirement for synthesis of NSO cells in a fed-batch culture. (a) Time course of total cell concentration (•), phosphorus concentration (■), phosphorus consumption (□). (b) Correlation between phosphorus consumption and NSO cells obtained in culture.

trial cells, and are also being used to express a number of product candidates. In general, sequences encoding for therapeutic proteins and MAb are incorporated into a vector also coding for dihydrofolate reductase for the purpose of selection and gene amplification. Despite their numerous applications, publications on CHO processes are very limited. Only sketchy information is available on the CHO fed-batch culture operations (183–185). Both Chang et al. (186) and Seewoester et al. (187) reported the BASF Bioresearch Corporation's (now Abbott) efforts on developing a generic high-yield fed-batch process for large-scale manufacturing of human MAb including one anti-TNF. Through use of an effective nutrient feeding strategy, fed-batch cultures with an MAb production in the order of 3 g/L were developed and scaled up to 3000 L.

The control of the nutritional environment in fed-batch cultures through the stoichiometrically balanced nutrient feeding was also shown to improve product

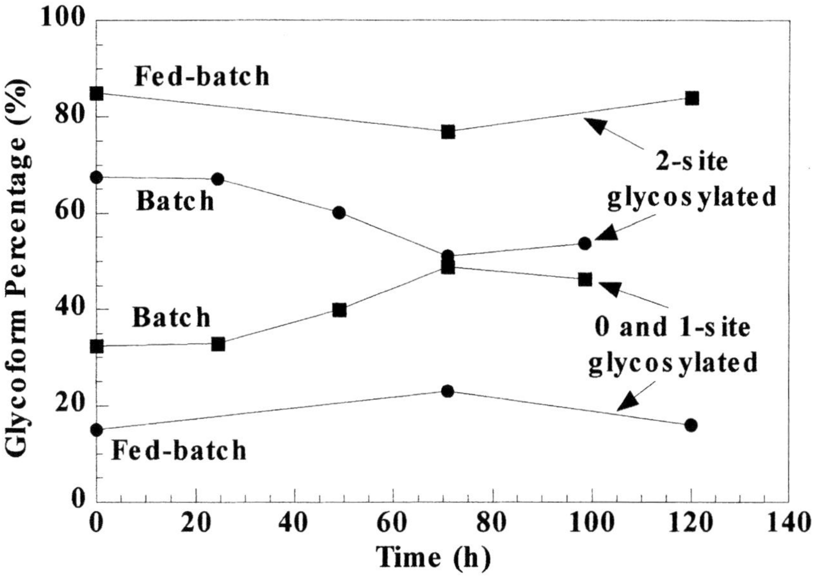

Figure 13 Improved protein glycosylation site-occupancy in a fed-batch CHO culture producing γ-interferon when compared with a batch culture.

quality. Xie and coworkers (112) were able to apply the stoichiometric model developed for hybridoma cells directly to a CHO culture producing gamma-interferon. The only modification to the model was the dry cell weight which was measured directly from a batch CHO culture. All other data, such as cell composition and amino acid percentage in cellular protein were used directly from measurements obtained from hybridoma cells. As a result, it took < 3 months from the initiation of the study to the completion of the fed-batch optimization. When compared with a batch culture, final product concentration was increased by 4.5-fold. Furthermore, the glycosylation efficiency of the gamma-interferon was improved as shown in Fig. 13(a,b) (112). The percentage of gamma-interferon with two sites glycosylated was maintained at a high level throughout the fed-batch culture, in comparison with a gradual decrease in the two site glycosylated form and an increase in one site glycosylated gamma-interferon in the uncontrolled batch culture.

CONCLUSION

Fed-batch cultivation of mammalian cells including hybridoma, murine myeloma SP2/0 and NS0, CHO has been established as a platform technology for the industrial production of recombinant therapeutic proteins and MAb. It is a simple means for productivity improvement to reduce bulk manufacturing costs and improve manufacturing capacities. This operation is also very effective for manipulating cellular metabolism and product quality. The key for development of a high performance fed-batch culture is to develop and implement an effective nutrient feeding strategy. This includes rational design of concentrated nutrient feeding solutions through understanding of cellular nutritional requirements and metabolisms. In addition, effective control of nutrient feeding rates based on critical process parameters such as cell concentration, OUR, and nutrient concentrations is important to maintain a desired nutritional environment for improved cell growth, minimized cell death, maximized nutrient utilization efficiencies, and formation of high-quality products.

REFERENCES

1. Aunins JG. Viral vaccine production in cell culture. In: Spier RE, ed. Encyclopedia of Cell Technology. Vol. 2. New York: Wiley & Sons, 2000:1182–1217.
2. Lubiniecki AS. Historical reflections on cell culture engineering. Cytotechnology 1998; 28:139–145.
3. Griffiths B. Animal cell products, overview. In: Spier RE, ed. Encyclopedia of Cell Technology. Vol. 1. New York: Wiley & Sons, 2000:70–76.
4. Chu L, Robinson DK. Industrial choices for protein production by large-scale cell culture. Curr Opin Biotechnol 2001; 12:180–187.
5. Barnes D, Sato G. Methods for growth of cultured cells in serum-free medium. Anal Biochem 1980; 102:255–270.
6. Brown RL, Griffith RL, Neubauer RH, Rabin H. Development of a serum-free medium which supports the long-term growth of human and nonhuman primate lymphoid cells. J Cell Physiol 1983; 115:191–198.
7. Hamilton WG, Ham RG. Clonal growth of Chinese hamster cells in protein-free media. In Vitro 1977; 13:537–547.
8. Jayme DW. Nutrient optimization for high density biological production applications. Cytotechnology 1991; 5:15–30.
9. Kano-Sueoka T, Errick JE. Roles of phosphoethanoamine, ethanolamine, and prolactin in mammary cell growth. In: Surbasku DA, Pardee AB, Salo G, eds. Growth of Cells in Hormonally Defined Media. Cold Spring Harbor Conferences on Cell Proliferation, 1982:221–230.
10. Kawamoto T, Sato JD, Le A, McClure DB, Sato GH. Development of a serum-free medium for growth of NS-1 mouse myeloma cells and its application to the isolation of NS-1 hybridomas. Anal Biochem 1983; 130:445–453.
11. Stryer L. Biochemistry. 3rd ed. New York: W.H. Freeman and Company, 1988.
12. Reitzer LJ, Wice BM, Kennell D. The pentose cycle: control and essential function in HeLa cell nucleic acid synthesis. J Biol Chem 1980; 255:5616–5626.
13. Hinkle PC, Kumar MA, Resetar A, Harris DL. Mechanistic stoichiometry of mitochondrial oxidative phosphorylation. Biochemistry 1991; 30:3576–3582.
14. Stoner CD. Determination of the $P/2e^-$ stoichiometries at the individual coupling sites in mitochondrial oxidative phosphorylation. J Biol Chem 1987; 262:10445–10453.
15. Sureshkumar GK, Mutharasan R. The influence of temperature on a mouse–mouse hybridoma growth and monoclonal antibody production. Biotechnol Bioeng 1991; 37:292–295.
16. Chuppa S, Tsai Y-S, Yoon S, Shackleford S, Rozales C, Bhat R, Tsay G, Matanguihan C, Konstantinov K, Naveh D. Fermentor temperature as a tool for control of high-density perfusion cultures of mammalian cells. Biotechnol Bioeng 1997; 55:328–338.
17. Weidemann R, Ludwig A, Kretzmer G. Low temperature cultivation—a step towards process optimization. Cytotechnology 1994; 15:111–116.
18. Paul J. Carbohydrate and energy metabolism. In: Willmer EN, ed. Cells and Tissues in Culture. New York: Academic Press, 1965:239–276.
19. Miller W, Blanch H, Wilke C. A kinetics analysis of hybridoma growth and metabolism in batch and continuous suspension culture: effect of nutrient concentration, dilution rate, and pH. Biotechnol Bioeng 1988; 32:947–965.
20. Ozturk SS, Palsson BO. Growth, metabolic, and antibody production kinetics of hybridoma cell culture: 2. Effects of serum concentration, dissolved oxygen concentration, and medium pH in a batch reactor. Biotechnol Prog 1991; 7:481–494.

21. Schmid G, Blanch HW, Wilke CR. Hybridoma growth, metabolism, and product formation in Hepes-buffered medium: II. Effect of pH. Biotechnol Lett 1990; 12:633–638.
22. Barton ME. Effect of pH on the growth cycle of HeLa cells in batch suspension culture without oxygen control. Biotechnol Bioeng 1971; 13:471–492.
23. Eagle H, Barran S, Levy M, Schulze HO. The utilization of carbohydrates by human cell cultures. J Biol Chem 1958; 233:551–558.
24. Zielke HR, Ozand PT, Tyson J, Sevdalian DA, Cornblath M. Growth of human diploid fibroblasts in the absence of glucose utilization. Proc Natl Acad Sci USA 1976; 73: 4110–4114.
25. Zielke HR, Zielke CL, Ozand PT. Glutamine: a major energy source for cultured mammalian cells. Feder Proc 1984; 43:121–125.
26. Renner ED, Plagemann PGW, Bernlohr RW. Permeation of glucose by simple and facilitated diffusion by Novikoff rat hepatoma cells in suspension culture and its relationship to glucose metabolism. J Biol Chem 1972; 247:5765–5776.
27. Fagan JB, Racker E. Determinants of glycolytic rate in normal and transformed chick embryo fibroblasts. Cancer Res 1978; 38:749–758.
28. Sauer PW, Burky JE, Wesson MC, Sternard HD, Qu L. A high-yielding, generic fed-batch cell culture process for production of recombinant antibodies. Biotechnol Bioeng 2000; 67:585–597.
29. Zeng A-P, Hu WS, Deckwer W. Variation of stoichiometric ratios and their correlation for monitoring and control of animal cell cultures. Biotechnol Prog 1998; 14:434–441.
30. Zielke HR, Sumbilla CM, Sevdalian DA, Hawkins RL, Ozand PT. Lactate: a major product of glutamine metabolism by human diploid fibroblasts. J Cell Physiol 1980; 104:433–441.
31. Lazo PA. Amino acids and glucose utilization by different metabolic pathways. Eur J Biochem 1981; 117:19–25.
32. Cruz HJ, Moreira JL, Carrondo MJT. Metabolically optimised BHK cell fed-batch cultures. J Biotechnol 2000; 80:109–118.
33. Ljunggren J, Haggstrom L. Catabolic control of hybridoma cells by glucose and glutamine limited fed batch cultures. Biotechnol Bioeng 1994; 44:808–818.
34. Miller WM, Wilke CR, Blanch HW. Transient responses of hybridoma cells to nutrient additions in continuous culture: 1. Glucose pulse and step changes. Biotechnol Bioeng 1989; 33:477–486.
35. Xie L, Wang DIC. High cell density and high monoclonal antibody production through medium design and rational control in bioreactor. Biotechnol Bioeng 1996; 51:725–729.
36. Cruz HJ, Moreira JL, Carrondo MJT. Metabolic shifts by nutrient manipulation in continuous cultures of BHK cells. Biotechnol Bioeng 1999; 66:104–113.
37. Graff S, Moser H, Kastner O, Graff AM, Tannenbaum M. The significance of glycolysis. J Natl Cancer Inst 1965; 34:511–519.
38. Burk DM, Woods M, Hunter J. On the significance of glycolysis for cancer growth with special reference to Morris rat heptoma. J Natl Cancer Inst 1967; 38:839–863.
39. Omasa T, Higashiyama K, Shioya S, Suga K. Effects of lactate concentration on hybridoma culture in lactate-controlled fed-batch operation. Biotechnol Bioeng 1992; 39:556–564.
40. Patel SD, Papoutsakis ET, Winter JN, Miller WM. The lactate issue revisited: novel feeding protocols to examine inhibition of cell proliferation and glucose metabolism in hemotopoietic cell cultures. Biotechnol Prog 2000; 16:885–892.
41. Glacken MW, Adema E, Sinskey AJ. Mathematical descriptions of hybridoma culture kinetics: I. Initial metabolic rates. Biotechnol Bioeng 1988; 32:491–506.
42. Chang YHD. Augmentation of Mass Transfer Through Electrical Means and Nutrient Enrichment for Suspension and Entrapment Cell Cultures. Ph.D. thesis, MIT, Cambridge, MA, 1994.
43. Hassell T, Gleave S, Butler M. Growth inhibition in animal cell culture the effect of lactate and ammonia. Appl Biochem Biotechnol 1991; 30:29–41.

44. Omasa T, Ishimoto M, Higashiyama K, Shioya S, Suga K. The enhancement of specific antibody production rate in glucose- and glutamine-controlled fed-batch culture. Cytotechnology 1992; 5:75–84.
45. Kurano N, Leist C, Messi F, Kurano S, Fiechter A. Growth behavior of Chinese hamster ovary cells in a compact loop bioreactor: 2. Effects of medium components and waste products. J Biotechnol 1990; 15:113–128.
46. Hayter PM, Curling EMA, Gould ML, Baines AJ, Jenkins N, Simon I, Strange PG, Bull AT. The effect of the dilution rate on CHO cell physiology and recombinant interferon-D production in glucose-limited chemostat culture. Biotechnol Bioeng 1993; 42:1077–1085.
47. Glacken MW, Fleischaker RJ, Sinskey AJ. Reduction of waste product excretion via nutrient control: possible strategies for maximizing product and cell yields on serum in cultures of mammalian cells. Biotechnol Bioeng 1986; 28:1376–1389.
48. Hu WS, Wang DIC. Mammalian cell culture technology: a review from an engineering perspective. In: Thilly WG, ed. Mammalian Cell Technology. London: Butterworths, 1986:167–198.
49. Kurokawa H, Park YS, Lijima S, Kobayashi T. Growth characteristics in fed-batch culture of hybridoma cells with control of glucose and glutamine concentrations. Biotechnol Bioeng 1994; 44:95–103.
50. Siegwart P, Côté J, Male K, Luong JHT, Perrier M, Kamen A. Adaptive control at low glucose concentration of HEK-293 cell serum-free cultures. Biotechnol Prog 1999; 15:608–616.
51. Xie L, Wang DIC. Fed-batch cultivation of animal cells using different medium design concepts and feeding strategies. Biotechnol Bioeng 1994; 43:1175–1189.
52. Xie L, Wang DIC. Stoichiometric analysis of animal cell growth and its application of medium design. Biotechnol Bioeng 1994; 43:1164–1174.
53. Zhou W, Rehm J, Hu W-S. High viable cell concentration fed-batch cultures of hybridoma cells through on-line nutrient feeding. Biotechnol Bioeng 1995; 46:579–587.
54. Xie L, Wang DIC. Energy metabolism and ATP balance in animal cell cultivation using a stoichiometrically based reaction network. Biotechnol Bioeng 1996; 52:591–601.
55. Gilbert JB, Price VE, Greenstein JP. Effect of anions on the non-enzymatic desamidation of glutamine. J Biol Chem 1949; 180:209–218.
56. Tritsch GL, Moore GE. Spontaneous decomposition of glutamine in cell culture media. Exp Cell Res 1962; 28:360–364.
57. Lin A, Agrawal P. Glutamine decomposition in DMEM: effect of pH and serum concentration. Biotechnol Lett 1988; 10:695–698.
58. Ozturk SS, Palsson BO. Chemical decomposition of glutamine in cell culture media: effect of media type, pH, serum concentration. Biotechnol Prog 1990; 6:121–128.
59. Goldstein L, Schooler JM. Regulation of ammonia production in the rat kidney. In: Weber G, ed. Advances in Enzyme Regulation. New York: Pergamon Press, 1967:71–89.
60. Katunuma N, Katsunuma T, Towatari T, Tomino I. Regulatory mechanisms of glutamine catabolism. In: Prusiner S, et al., eds. The Enzymes of Glutamine Metabolism. New York: Academic Press, 1973:227–258.
61. Kovacevic Z. The pathway of glutamine and glutamate oxidation in isolated mitochondria from mammalian cells. Biochem J 1971; 125:757–763.
62. Glacken MW, Huang C, Sinskey AJ. Mathematical descriptions of hybridoma culture kinetics. III. Simulation of fed-batch bioreactors. J Biotechnol 1989; 10:39–66.
63. McKeehan WL. Glycolysis, glutaminolysis and cell proliferation. Cell Biol Int Rep 1982; 6:635–650.
64. Nissim I, States B, Yudkoff M, Segal S. Characterization of amino acid metabolism by cultured rat kidney cells: study with ^{15}N. Am J Physiol 1987; 22:F1243–F1252.
65. Street JC, Delort AM, Braddock PSH, Brindle KMA. $^{1}H/^{15}N$ NMR study of nitrogen metabolism in cultured mammalian cells. Biochem J 1993; 291:485–492.

66. Ljunggren J, Haggstrom L. Glutamine limited fed-batch culture reduces ammonium ion production in animal cells. Biotechnol Lett 1990; 12:705–710.
67. Ljunggren J, Haggstrom L. Glutamine limited fed-batch culture reduces the overflow metabolism of amino acids in myeloma cells. Cytotechnology 1992; 8:45–56.
68. Ozturk SS. Ammonia toxicity, animal cells. In: Flickinger MC, Drew SW, eds. The Encyclopedia of Bioprocess Technology: Fermentation, Biocatalysis, and Bioseparation. New York: John Wiley & Sons, Inc., 1999:121–136.
69. Bender DA. Amino Acid Metabolism. 2nd ed. New York: John Wiley & Sons, 1985.
70. Glacken MW. Development of Mathematical Descriptions of Mammalian Cell Culture Kinetics for the Optimization of Fed-Batch Bioreactors. Ph.D. thesis, MIT, Cambridge, MA, 1987.
71. Jo EC, Park HJ, Park JM, Kim KH. Balanced nutrient fortification enables high-density hybridoma cell culture in batch culture. Biotechnol Bioeng 1990; 36:717–722.
72. Doyle A, Butler M. The effect of pH on the toxicity of ammonia to a murine hybridoma. J Biotechnol 1990; 15:91–100.
73. Borys MC, Linzer DIH, Papoutsakis ET. Ammonia affects the glycosylation patterns of recombinant mouse placental lactogen-I by Chinese hamster ovary cells in a pH-dependent manner. Biotechnol Bioeng 1994; 43:505–514.
74. Andersen DC, Goochee CF. The effect of ammonia on the O-linked glycosylation of granulocyte colony-stimulated factor produced by Chinese hamster ovary cells. Biotechnol Bioeng 1995; 47:96–105.
75. Newland M, Kamal MN, Greenfield PF, Nielsen LK. Ammonia inhibition of hybridomas propagated in batch, fed-batch, and continuous culture. Biotechnol Bioeng 1994; 43:434–438.
76. Bedard A, Tom R, Kamen A. Growth, nutrient consumption, and end-product accumulation in SF-9 and BTI-EAA insect cell cultures: insights into growth limitation and metabolism. Biotechnol Prog 1993; 9:615–624.
77. Hansen HA, Emborg C. Influence of ammonium on growth, metabolism, and productivity of a continuous suspension Chinese ovary cell culture. Biotechnol Prog 1994; 10:121–124.
78. Christie A, Butler M. Glutamine-based dipeptides are utilized in mammalian cell culture by extracellular hydrolysis catalyzed by a specific peptidase. J Biotechnol 1994; 37:277–290.
79. Altamirano A, Paredes C, Cairo JJ, Godia F. Improvement of CHO cell culture medium formulation: simultaneous substitution of glucose and glutamine. Biotechnol Prog 2000; 16:69–75.
80. Bebbington CR, Renner G, Thomson S, King D, Abrams D, Yarranton GT. High-level expression of a recombinant antibody from myeloma cells using a glutamine synthetase gene as an amplifiable selectable marker. Bio/Technology, 1992; 10:169–175.
81. Zhou W, Chen C, Buckland B, Aunins J. Fed-batch culture of recombinant NS0 myeloma cells with high monoclonal antibody production. Biotechnol Bioeng 1997; 55:783–792.
82. Christie A, Butler M. The adaptation of BHK cells to a non-ammoniagenic glutamate-based culture medium. Biotechnol Bioeng 1998; 64:298–309.
83. Chang YHD, Grodzinsky AJ, Wang DIC. Nutrient enrichment and in-situ removal of ammonium and lactate through electrical means for suspension hybridoma cultures. Biotechnol Bioeng 1995; 47:319–326.
84. Chang YHD, Grodzinsky AJ, Wang DIC. In-situ removal of ammonium and lactate from suspension hybridoma cultures through electrical means. Biotechnol Bioeng 1995; 47:308–318.
85. Eagle H. Nutrition needs of mammalian cells in tissue culture. Science 1955; 122: 501–504.
86. Chung RA, Brown RW, Beachm LH, Huang IL. Amino acid metabolism of different mammalian cell lines. Can J Biochem 1966; 44:1145–1157.

87. Eagle H. Amino acid metabolism in mammalian cell cultures. Science 1959; 130: 432–437.
88. Griffiths JB. The quantitative utilization of amino acids and glucose and contact inhibition of growth in cultures of the human diploid cell, WI-38. J Cell Sci 1970; 6:739–749.
89. Kruse PF, Miedema E, Carter HC. Amino acid utilizations and protein synthesis at various proliferation rates, population densities, and protein contents of perfused animal cell and tissue cultures. Biochemistry 1967; 6:949–955.
90. Lucy JA, Rinaldini LM. The amino acid metabolism of differentiating skeletal myoblasts in vitro. Exp Cell Res 1959; 17:385–398.
91. McCarty K. Selective utilization of amino acids by mammalian cell cultures. Exp Cell Res 1962; 27:230–240.
92. Pasieka AE, Morton HJ, Morgan JF. The metabolism of animal tissues cultivated in vitro II. Amino acid metabolism of chick embryonic kidney, chick embryonic liver, and monkey kidney cortex cultures. Can J Biochem Physiol 1958; 55:171–184.
93. Pasieka AE, Morton HJ, Morgan JF. The metabolism of animal tissues cultivated in vitro III. Amino acid metabolism of strain L cells in completely synthetic media. Can J Biochem Physiol 1958; 36:771–782.
94. Patterson MKJ. Uptake and utilization of amino acids by cell cultures. In: Rotbblat GH, Gristofalo VJ, eds. Growth, Nutrient, and Metabolism of Cells in Culture. New York: Academic Press, 1972:171–209.
95. Sinclair R, Leslie I. Amino acid and glucose uptake in relation to protein synthesis in cells growing in tissue culture. Biochim Biophys Acta 1959; 32:58–68.
96. Hiller GW, Clark DS, Blanch HW. Transient responses of hybridoma cells in continuous culture to step changes in amino acid vitamin concentrations. Biotechnol Bioeng 1994; 44:303–321.
97. Xie L. Stoichiometric Medium Design and Nutritional Control in Fed-Batch Cultivation of Animal Cells. Ph.D. thesis, MIT, Cambridge, MA, 1997.
98. Duval A, Demangel C, Munier-Jolain K, Miossec S, Geahel I. Factors controlling cell proliferation and antibody production in mouse hybridoma cells: 1. Influence of the amino acid supply. Biotechnol Bioeng 1991; 38:561–570.
99. Miller WM, Wilke CR, Blanch HW. The transient responses of hybridoma cells to nutrient additions in continuous culture: II. Glutamine pulse and step changes. Biotechnol Bioeng 1989; 33:487–499.
100. Stoner GD, Merchant DJ. Amino acid utilization by L-M strain mouse cells in a chemically defined medium. In Vitro 1972; 7:330–343.
101. Xie L, Wang DIC. Material balance studies on animal cell metabolism using a stoichiometrically based reaction network. Biotechnol Bioeng 1996; 52:579–590.
102. Lambert K, Pirt SJ. The quantitative requirements of human diploid cells (strain MRC-5) for amino acids, vitamins, and serum. J Cell Sci 1975; 17:397–411.
103. Mathews CK, van Holde KE. Biochemistry. Redwood City, CA: Benjamin/Cummings Publishing Company, Inc., 1990.
104. Bailey JE, Ollis DF. Biochemical Engineering Fundamentals. 2nd ed. New York: McGraw-Hill, 1986.
105. Bushell ME, Bell SL, Scott MF, Spier RE, Wardell JN, Sanders PG. Enhancement of monoclonal antibody yield by hybridoma fed-batch culture, resulting in extended maintenance of viable cell population. Biotechnol Bioeng 1994; 44:1099–1106.
106. Fike R, Kubiak J, Price P, Jayme D. Feeding strategies for enhanced hybridoma productivity: automated concentrate supplementation. BioPharm 1993; 6:49–54.
107. Europa AF, Gambhir A, Fu PC, Hu WS. Multiple steady states with distinct cellular metabolism in continuous culture of mammalian cells. Biotechnol Bioeng 2000; 67:25–34.
108. Xie L, Wang DIC. Applications of improved stoichiometric model in medium design and fed-batch cultivation of animal cells in bioreactor. Cytotechnology 1994; 15:17–29.

109. Nadeau I, Jacob D, Perrier M, Kamen A. 293SF metabolic flux analysis during cell growth and infection with an adenoviral vector. Biotechnol Prog 2000; 16:872–884.
110. Nadeau I, Sabatie J, Koehl M, Perrier M, Kamen A. Human 293 cell metabolism in low glutamine-supplied culture: interpretation of metabolic changes through metabolic flux analysis. Metab Eng 2000; 2:277–292.
111. Lehninger AL. Bioenergetics. 2nd ed. Palo Alto, CA: W.A. Benjamin, 1965.
112. Xie L, Nyberg G, Gu X, Li H, Mollborn F, Wang DIC. Gamma-interferon production and quality in stoichiometric fed-batch cultures of Chinese hamster ovary (CHO) cells under serum-free conditions. Biotechnol Bioeng 1997; 56:577–582.
113. Klapper MH. The independent distribution of amino acid near neighbor pairs into polypeptides. Biochem Biophys Res Commun 1977; 78:1018–1024.
114. Neidardt FC. Chemical composition of *Escherichia coli*. In: Neidardt FC, ed. *Escherichia coli* and *Salmonella typhimurium*: Cellular and Molecular Biology. Washington, DC: American Society for Microbiology, 1987:156–167.
115. Higuchi K. An improved chemically defined culture medium for strain L mouse cells based on growth responses to graded levels of nutrients including iron and zinc ions. J Cell Physiol 1970; 75:65–72.
116. deZengotita VM, Miller WM, Aunins JG, Zhou W. Phosphate feeding improves high-cell-concentration NS0 myeloma culture performance for monoclonal antibody production. Biotechnol Bioeng 2000; 69:566–576.
117. Hu WS, Piret JM. Mammalian cell culture process. Curr Opin Biotechnol 1992; 3:110–114.
118. Zhou W, Mulchandani A. Recent advances in bioprocess monitoring and control. In: Rogers KR, Mulchandani A, Zhou W, eds. Biosensor and Chemical Sensor Technology: Process Monitoring and Control. ACS Symposium Series 613. Washington, DC: American Chemical Society, 1995:88–98.
119. Glacken MW. Bioreactor control and optimization. In: Ho CS, Wang, DIC, eds. Animal Cell Bioreactors. Stoneham: Butterworth-Heinemann, 1991.
120. Loch G, Sonnleitner B, Fiechter A. On-line measurements in biotechnology: Techniques. J Biotechnol 1992; 25:23–53.
121. Schügerl K. Common instruments for process analysis and control. In: Schügerl, K, eds. Biotechnology. Vol. 4. New York: VCH, 1991:5–25.
122. Schügerl K. On-line analysis of broth. In: Schügerl K, ed. Biotechnology. Vol. 4. New York: VCH, 1991:149–180.
123. Scheirer W, Merten OW. Instrumentation of animal cell culture reactors. In: Ho CS, Wang DIC, eds. Animal Cell Bioreactors. Stoneham: Butterworth-Heinemann, 1991:405–443.
124. Kretzmer G. On-line analysis in animal cell culture. In: Spier RE, ed. Encyclopedia of Cell Technology. Vol. 2. New York: Wiley & Sons, 2000:968–978.
125. Ducommun P, Bolzonella I, Rhiel M, Pugeaud P, von Stockar U, Marison IW. On-line determination of animal cell concentration. Biotechnol Bioeng 2001; 72:515–522.
126. Yamane T, Hibino W, Ishihara K, Kadotani Y, Kominami M. Fed-batch culture automated by use of continuously measured cell concentration and cell volume. Biotechnol Bioeng 1992; 39:550–555.
127. Yamane T. Application of an on-line turbidimeter for the automation of fed-batch cultures. Biotechnol Prog 1993; 9:81–85.
128. Konstantinov KB, Pambayun R, Matanguihan R, Yoshida T, Perusich CM, Hu WS. On-line monitoring of hybridoma cell growth using a laser turbidity sensor. Biotechnol Bioeng 1992; 40:1337–1342.
129. Zhou W, Hu WS. On-line characterization of a hybridoma cell culture process. Biotechnol Bioeng 1994; 44:170–177.
130. Wu P, Ozturk SS, Blackie JD, Thrift JC, Figueroa C, Daveh D. Evaluation and applications of optical cell density probes in mammalian cell bioreactors. Biotechnol Bioeng 1995; 45:495–502.

131. Scheper T, Staerk E, Hitzmann B, Boehl D, Lindemann C. 2-D Spectrofluorometry for bioprocess monitoring and control. Oral presentation at Biochemical Engineering XII, Rohnert Park, CA, 2001.
132. Bittner C, Wehnert G, Scheper T. In situ microscopy for on-line determination of biomass. Biotechnol Bioeng 1998; 60:24–35.
133. Maruhashi F, Murakami S, Baba K. Automated monitoring of cell concentration and viability using an image analysis system. Cytotechnology 1994; 15:281–289.
134. Schügerl K. Aseptic sampling. In: Christiansen C, Munck L, Villadsen J, eds. Proceedings of the 5th European Congress on Biotechnology. Vol. 2. Copenhagen: Munksgaard, 1990:1188–1193.
135. Fenge C, Fraune E, Freitag R, Scheper T, Schügerl K. On-line monitoring of monoclonal antibody formation in high density perfusion culture using FIA. Cytotechnol 1991; 6:55–63.
136. Garn M, Gisin M, Thommen C. A flow injection analysis system for fermentation monitoring and control. Biotechnol Bioeng 1989; 34:423–428.
137. Renneberg R, Trott-Kriegeskorte G, Lietz M, Jaeger V, Pawlowa M, Kaiser G, Wollenberger U, Schubert F, Wagner R, et al. Enzyme sensor-FIA-system for on-line monitoring of glucose, lactate and glutamine in animal cell cultures. J Biotechnol 1991; 21:173–85.
138. Cattaneo MV, Male KB, Luong JHT. Chemiluminescence fiber-optic biosensor system for the determination of glutamine in mammalian cell cultures. Biosensors Bioelectron 1992; 7:569–574.
139. Dremel BAA, Li SY, Schmid RD. On-line determination of glucose and lactate concentrations in animal cell culture based on fiber optic detection of oxygen in flow-injection analysis. Biosensors Bioelectron 1992; 7:133–139.
140. Male KB, Luong JHT et al. Determination of urinary glucose by a flow injection analysis amperometric biosensor and ion-exchange chromatography. Appl Biochem Biotechnol 1992; 37:243–254.
141. Blum LJ. Chemiluminescent flow injection analysis of glucose in drinks with a bienzyme fiberoptic biosensor. Enzyme Microb Technol 1993; 15:407–411.
142. Cattaneo MV, Luong JHT. On-line chemiluminescent assay using FIA and fiber optics for urinary and blood glucose. Enzyme Microb Technol 1993; 15:424–428.
143. Cattaneo MV, Luong JHT. Monitoring glutamine in animal cell cultures using a chemiluminescent fiber optic biosensor. Biotechnol Bioeng 1993; 41:659–665.
144. Male KB, Luong JHT, Tom R, Mercille S. Novel FIA amperometric biosensor system for the determination of glutamine in cell culture system. Enzyme Microb Technol 1993; 15:26–32.
145. Meyerhoff EM, Trojanowicz KM, Palsson OB. Simultaneous enzymatic/electrochemical determination of glucose and L-glutamine in hybridoma media by flow-injection analysis. Biotechnol Bioeng 1993; 41:964–969.
146. Schulze B, Middendorf C, Reinecke M, Scheper T. Automated immunoanalysis systems for monitoring mammalian cell cultivation processes. Cytotechnology 1994; 15:259–269.
147. Hitzmann B, Schulze B, Reinecke M, Scheper T. The automation of two immune-FIA-systems using the flexible software system CAFCA. In: Rogers KR, Mulchandani A, Zhou W, eds. Biosensor and Chemical Sensor Technology: Process Monitoring and Control. ACS Symposium Series 613. Washington, DC: American Chemical Society, 1995:165–175.
148. Nandakumar MP, Karlsson EN, Mattiasson B. Integrated flow-injection processing for on-line quantification of plasmid DNA during cultivation of *E. coli*. Biotechnol Bioeng 2001; 73:406–411.
149. Kyröläinen M, Hakanson H, Mattiasson B. On-line calibration of a computerized biosensor system for continuous measurements of glucose and lactate. Biotechnol Bioeng 1995; 45:122–128.

150. Hitzmann B, Gomersall R, Brandt J, van Putten A. An expert system for the supervision of a multi channel flow injection analysis system. In: Rogers KR, Mulchandani A, Zhou W, eds. Biosensor and Chemical Sensor Technology: Process Monitoring and Control. ACS Symposium Series 613. Washington, DC: American Chemical Society, 1995:133–143.
151. Buentemeyer H. Off-line analysis in animal cell culture, methods. In: Spier RE, ed. Encyclopedia of Cell Technology. Vol. 2. New York: Wiley & Sons, 2000:945–959.
152. Holzhauer-Rieger K, Zhou W, Schügerl K. On-line high-performance liquid chromatography for determination of cephalosporin C and by-products in complex fermentation broth. J Chromatogr 1990; 499:609–615.
153. Heinzle E. Present and potential applications of mass spectrometry for bioprocess research and control. J Biotechnol 1992; 25:81–114.
154. Lovrecz G, Gray P. Use of on-line gas analysis to monitor recombinant mammalian cell cultures. Cytotechnology 1994; 14:167–175.
155. Bonarius HPJ, de Gooijer CD, Tramper J, Schmid G. Determination of the respiration quotient in mammalian cell culture in bicarbonate buffered media. Biotechnol Bioeng 1995; 45:524–535.
156. Backer MP, Metzger LS, Slaber PL, Nevitt KL, Boder GB. Large-scale production of monoclonal antibodies in suspension culture. Biotechnol Bioeng 1988; 32:993–1000.
157. Behrendt U, Koch S, Gooch DD, Steegmans U, Comer MJ. Mass spectrometry: a tool for on-line monitoring of animal cell cultures. Cytotechnology 1994; 14:157–165.
158. Oeggerli A, Eyer K, Heinzle E. On-line gas analysis in animal cell cultivation: I. Control of dissolved oxygen and pH. Biotechnol Bioeng 1995; 45:42–53.
159. Eyer K, Oeggerli A, Heinzle E. On-line gas analysis in animal cell cultivation: II. Methods for oxygen uptake rate estimation and its application to controlled feeding of glutamine. Biotechnol Bioeng 1995; 45:54–62.
160. Fleischaker RJ, Sinskey AJ. Oxygen demand and supply in cell culture. Eur J Appl Microbiol Biotechnol 1981; 12:193–197.
161. Ramirez OT, Mutharasen R. Cell cycle- and growth phase-dependent variations in size distribution, antibody productivity, and oxygen demand in hybridoma cultures. Biotechnol Bioeng 1990; 36:839–848.
162. Kilburn DG. In: Bulter M, ed. Mammalian Cell Biotechnology. Oxford: Oxford University Press, 1991.
163. Konstantinov KB, Zhou W, Golini F, Hu WS. Expert systems in the control of animal cell culture processes: potentials, functions, and perspectives. Cytotechnology 1994; 14:233–246.
164. Zhou W, Hu WS. Optimization of fed-batch mammalian cell culture processes. In: Galindo E, Ramirez OT, eds. Advances in Bioprocess Engineering. Proceedings of the International Symposium on Bioprocess Engineering. Cuernavaca, Mexico, June 20–24. Dordrecht: Kluwer, 1994.
165. Zhou W, Rehm J, Europa A, Hu WS. Alteration of mammalian cell metabolism by dynamic nutrient feeding. Cytotechnology 1997; 24:99–108.
166. Zhou W, Bibila TB, Glazomitsky K, Montalvo J, Chan CP, DiStefano DJ, Munshi S, Robinson DK, Buckland BC, Aunins JG. Large scale production of recombinant mouse and rat growth hormone by fed-batch GS-NS0 cell cultures. Cytotechnology 1996; 22:239–250.
167. Tsao E. Practical approaches for bioreactor yield improvement. Oral presentation at the 2001 Waterside Monoclonal Conference, Norfolk, VA, 2001.
168. Moran EB, McGowan ST, McGuire JM, Frankland JE, Oyebade IA, Waller W, Archer LC, Morris LO, Pandya J, Nathan SR, Smith L, Cadette ML, Michalowski JT. A systematic approach to the validation of process control parameters for monoclonal antibody production in fed-batch culture of a murine myeloma. Biotechnol Bioeng 2000; 69:242–255.
169. Bibila TA, Ranucci CS, Glazomitsky K, Buckland BC, Aunins JG. Monoclonal antibody process development using medium concentrates. Biotechnol Prog 1994; 10:87–96.

170. Bibila TA, Robinson DK. In pursuit of the optimal fed-batch process for monoclonal antibody production. Biotechnol Prog 1995; 11:1–13.
171. Robinson DK, DiStefano DJ, Gould SL, Cuca G, Seamans TC, Benincasa D, Munshi S, Chan CP, Lee DK, Stafford-Hollis J, Hollis G, Jain D, Ramasubramanyan K, Mark GE, Silberklang M. Production of engineered antibodies in myeloma and hybridoma cells–enhancements in gene expression and process design. In: Wang H, Imanaka T, eds. Antibody Engineering. American Chemical Society Symposium Series, 604. Washington, DC: American Chemical Society, 1995:1–14.
172. Gould SL, DiStefano DJ, Cuca G, Robinson DK, Silberklang M. Major metabolic changes accompany transfection and selection for high level expression of recombinant genes. In Vitro Cell Dev Biol 1992; 28:162A.
173. Mercille S, Massie B. Induction of apoptosis in nutrient-deprived cultures of hybridoma and myeloma cells. Biotechnol Bioeng 1994; 44:1140–1154.
174. Sato DJ, Cao HT, Kayada Y, Cabot MC, Sato HH, Okamoto T, Welsh CJ. Effects of proximate cholesterol precursors and steroid hormones on mouse myeloma growth in serum-free medium. In Vitro Cell Dev Biol 1988; 24:1223–1228.
175. Birch JR, Boraston RC, Metcalfe H, Brown ME, Bebbington CR, Field PR. Selecting and designing cell lines for improved physiological characteristics. Cytotechnology 1994; 15:11–16.
176. Keen MJ, Steward TW. Adaptation of cholesterol-requiring NS0 mouse myeloma cells to high density growth in a fully defined protein-free and cholesterol-free medium. Cytotechnology 1995; 17:203–211.
177. Gould SL, DiStefano DJ, Silberklang M, Robinson DK. Serum-free animal cell culture medium development using an automated multi-well growth monitoring device. Presented at the Congress on Cell and Tissue Culture, San Diego, CA, 1993.
178. DiStefano DJ, Mark GE, Robinson DK. Feeding of nutrients delays apoptotic death in fed-batch cultures of NS0 myeloma cells. Biotechnol Lett 1996; 18:1067–1072.
179. Franek K. Starvation-induced programmed cell death of hybridoma cells: Prevention by amino acid mixtures. Biotechnol Bioeng 1995; 45:86–90.
180. Franek F, Chladkova-Sramkova K. Apoptosis and nutrition: involvement of amino acid transport system in repression of hybridoma cell death. Cytotechnology 1995; 18:113–117.
181. Itoh Y, Ueda H, Suzuki E. Overexpression of bcl-2, apoptosis suppressing gene: prolonged viable culture period of hybridoma and enhanced antibody production. Biotechnol Bioeng 1995; 48:118–122.
182. Murray K, Ang CE, Gull K, Hickman JA, Dickinson AJ. NS0 Myeloma cell death: influence of bcl-2 overexpression. Biotechnol Bioeng 1996; 51:298–304.
183. Andersen DC, Bridges T, Gawlitzek M, Hoy C. Multiple cell culture factors can affect the glycosylation of Asn-184 in CHO-produced tissue-type plasminogen activator. Biotechnol Bioeng 2000; 70:25–31.
184. Ryll T, Dutina G, Reyes A, Gunson J, Krummen L, Etcheverry T. Performance of small-scale CHO perfusion cultures using an acoustic cell filtration device for cell retention: characterization of separation efficiency and impact of perfusion on product quality. Biotechnol Bioeng 2000; 69:440–449.
185. Gawlitzek M, Ryll T, Lofgren J, Sliwkowski MB. Ammonium alters N-glycan structures of recombinant TNFR-IgG: degradative versus biosynthetic mechanisms. Biotechnol Bioeng 2000; 68:637–646.
186. Chang DYH, Seewoester T, Turner BG, Moesta P. Development of a technology platform for production of human monoclonal antibodies in recombinant CHO cell culture. Oral presentation at Cell Culture Engineering VII, Santa Fe, NM, 2000.
187. Seewoester T, Chang DYH, Turner BG, Moesta P. Development of a generic high-yield fed-batch process for large-scale GMP manufacturing of human monoclonal antibodies. Oral presentation at 219th American Chemical Society National Meeting, San Francisco, CA, 2001.

11
Optimization of High Cell Density Perfusion Bioreactors

Dhinakar S. Kompala
University of Colorado, Boulder, Colorado, U.S.A.

Sadettin S. Ozturk
Centocor Inc., Malvern, Pennsylvania, U.S.A.

INTRODUCTION

This chapter discusses the benefits and the techniques of high cell density perfusion bioreactors for culturing mammalian cells. Perfusion refers to the continuous inflow of nutrient medium coupled with the outflow or harvesting of spent medium, while the cells are fully or partially retained within the bioreactor. In contrast to the chemostat bioreactor, in which the outflow stream contains cells at the same concentration as in the bioreactor, the outflow stream of a perfusion bioreactor is either completely or substantially depleted of cells. The concept of perfusion was applied as early as 1912 to keep small pieces of tissue alive for extended periods of observation under a microscope (1,2). A historical account on the development of perfusion bioreactors for mammalian cell cultivation is given in an earlier review (3). While many of the perfusion techniques discussed earlier pertain to attachment cultures and are still applicable, this chapter focuses more on perfusion techniques for suspension bioreactor cultures, as these homogeneous cultures have become increasingly popular in the ensuing decade (4).

Case for Perfusion Bioreactor Cultures

In addition to batch and fed-batch bioreactor cultures, which are traditionally preferred for the production of the therapeutic proteins in recombinant bacterial cells, chemostat and high cell density perfusion bioreactor cultures are becoming increasingly acceptable for culturing of mammalian cells due to a several key advantages. First, continuous addition of fresh nutrient medium and removal of spent medium provide an easy method of removing the metabolic by-products, ammonia and lactate, which are toxic to cells if allowed to accumulate as in batch or fed-batch cultures (5,6). Second, the facile secretion of recombinant proteins from mammalian cells (compared to intracellular hoarding of the heterologous proteins by bacteria) eliminates the need for cell harvesting at the end of a batch or fed-batch process, and cell disruption to obtain the secreted protein product. Third, due to the long doubling times involved in the growth of mammalian cells, it is more efficient to retain the viable and productive cells inside the bioreactor rather than growing the cells repeatedly in batch cultures. Fourth, even partial retention of viable and productive mammalian cells inside the bioreactor results in higher cell density compared to the low cell density batch or chemostat cultures. As the protein synthesis rate in the bioreactor is directly proportional to the number of viable and productive cells (7), the higher cell density in perfusion bioreactors results directly in a higher protein production rate. Next, at the low cell *growth* rates prevalent in the high cell density perfusion cultures, many hybridoma cell lines exhibit higher specific (per cell) production rates (6,8), resulting in much higher protein production rates. Finally, the shorter residence time of the product proteins inside the perfusion bioreactor is advantageous in minimizing their exposure to proteases, sialidases and other degradative enzymes (9) that are released into the culture medium from the accumulating dead cells in the fed-batch bioreactors.

As a note of caution, monoclonal antibody production in long-term continuous bioreactor cultures of hybridoma cells has sometimes been found to be unstable (8,10,11). However, this gradual instability of antibody production in the slower growing hybridoma cultures is in marked contrast to the rapid instabilities observed in heterologous protein expression from recombinant bacterial cultures (12). With the slower dynamics of protein production loss in mammalian cells, perfusion bioreactor cultures may still be sufficiently stable over several weeks or months of continuous operation and result in much higher protein production over the repeated batch cultures in the similar sized bioreactors. Thus, perfusion bioreactor cultures represent a superior bioreactor operation for large-scale production of therapeutic proteins from mammalian cells, if the instability of the specific protein synthesis rate is not a serious issue.

Industrial Application of Perfusion Bioreactors

As mentioned in Chapter 1 of this book, there are a number of products in the market derived from cell culture technology. A greater number of products are in development and it is very likely that some of these will be approved for commercial production. The demand for cell culture–derived products is high in most cases and the manufacturing cost is significant. Some of the antibodies, for instance, are administered at high doses and a production capacity of 500–1000 kg/year is required to meet the demand. There is no question that the cell culture processes must be optimized with respect to overall yield and productivity to meet these demands. As discussed, one of the key advantages of perfusion bioreactors is the higher protein production rate. Due to high cell densities achievable in a perfusion bioreactor, the volumetric production rate can be 4–10 times higher as compared to a fed-batch bioreactor. This higher production rate allows the cell culture process to utilize compact bioreactors, which are easier to operate and control. Product quality is another key advantage for perfusion bioreactors especially when the product is labile and prone to degradation in the bioreactor. Even though the batch or fed-batch operation is widely accepted by many companies, these key advantages resulted in perfusion processes for many products (13–15).

Table 1 presents marketed cell culture products from perfusion processes. The first industrial scale perfusion process was developed by Centocor in the late 1980s

Table 1 Pharmaceutical Products in the Market Produced from Perfusion Systems

Product	Company	Year	Sales ($M/year)[a]
Centoxin	Centocor	1991	n/a
Cerezyme	Genzyme	1994	740
ReoPro	Centocor	1994	400
Gonal-F	Serono	1997	600
Remicade	Centocor	1998	1,500
ReFacto	Wyeth	2000	224
Kogenate-FS	Bayer	2000	424
Zigris	Eli Lilly	2001	160

[a] These data are for 2003. [From Ref. (73).]

for the production of Centoxin®, an antibody for the treatment of sepsis. In 1991, the drug was approved for use in Europe. Centoxin failed the clinical trials in the United States and was discontinued. Centocor later developed the process and obtained approval for ReoPro® in 1994 using perfusion technology. The same year Genzyme obtained approval for Cerezyme® produced in a perfusion process. Other perfusion culture products like Gonal-F® from Serano, Remicade® from Centocor, ReFacto® from Wyeth, Kogenate-FS® from Bayer, and Zigris® from Eli-Lily reached the market in quick succession. As seen in Table 1, the products from the perfusion process have a sizable market. The sales of perfusion-based drugs are significant; they are multimillion dollar drugs, some of them blockbusters. Remicade® is a monoclonal antibody produced at a rate of several hundred kilograms a year with sales reaching $1.5 billion/year.

CELL RETENTION SYSTEMS

The key feature in operating a perfusion bioreactor is substantial or full retention of cells inside the bioreactor. The methods for retaining cells inside the bioreactor are primarily determined by whether the cells are growing attached to surfaces or growing in either single cell suspension or cell aggregates. While most mammalian cells historically were grown attached to a surface or a matrix, the difficulties in scaling up these surfaces for the production of large amounts of therapeutic proteins have encouraged efforts to adapt many industrial host cell lines to grow as suspended cells. With the increasing success of adaptation of these cells to grow in suspension, the previously developed methods of perfusion cultures for attached cells are now falling out of favor. Nevertheless, for any cells that are difficult to adapt for growth in suspension, the traditional techniques for perfusion cultures are summarized below. A more complete discussion of these older techniques may be found in previous reviews (3,16).

Cell Retention Systems for Heterogeneous Cultures

The different heterogeneous systems summarized below were developed for retaining mammalian cells inside the bioreactor through methods of cell attachment to surfaces and entrapment of cells in different devices. The heterogeneity of these systems arises from the segregation of cells into defined compartments within the bioreactors, i.e., the cells are not uniformly distributed throughout the culture volume in the bioreactor.

Fixed (Immobilized) Bed

Various matrices (e.g., glass beads, sponges, fibers, etc.) are typically packed in smaller vessels and inoculated with attachment-prone cells for the production of viruses (16–18) and proteins (19–22). Glass beads are the most popular matrices and they range in size from 3 to 5 mm (23). These perfusion bioreactor vessels have been scaled up to 100 L, but are usually much smaller (24). While lytic viral production is typically accomplished in batch processes, Brown et al. (20) have used glass beads as the attachment matrix for attachment-prone mammalian cells in perfusion bioreactors of volume up to 60 L for periods ranging from 4 months to over 1 year. Scale-up of bioreactors is typically limited by the weight-carrying capacity of glass beads, the uneven inoculation of cells onto the matrix in a deep bed, and poor oxygen transfer.

Ceramic Matrix Immobilization (Opticell)

The Opticell system using a ceramic matrix with either a nonporous or porous surface has been used for the perfusion culture of both suspension and attached cells (25–28). The porous surfaces can entrap suspension cells and provide a larger surface area for the attached cells, while nonporous smooth surfaces are suitable for harvesting cells. The system uses a computer to control the common bioreactor parameters, such as pH, oxygen, and perfusion rates. Inoculation of the complete matrix requires a four-step process to ensure that all the channels of the cylindrical ceramic matrix are loaded with cells. Scale-up of these ceramic cartridges is limited by the gradients in nutrient concentration along the length of the cylinder.

Hollow Fiber Reactors

Hollow fiber cartridge systems have been quite successful as small-scale perfusion bioreactors for the production of monoclonal antibodies from hybridoma cultures (29–31). Ultrafiltration capillary fibers enhance the antibody concentrations along with the high density of hybridoma cells in the extracapillary space, while nutrients and waste products are easily exchanged with lumen. (More details of the materials used in these cartridges are discussed in another chapter of this book.) Scale-up of these bioreactors has been hampered by the pressure drop as well as spatial variations in nutrient and waste product concentrations along the length of the reactor (32). Several other flow patterns have also been explored to overcome some of these problems to different degrees of partial success (33,34).

Microencapsulation of Mammalian Cells

Encapsulation of living tissues inside polymeric semipermeable microspheres has been developed for potential biomedical applications (35) and easily adapted for mammalian cells as well (36). The porosity of the polymer layer can be controlled to allow the permeation of smaller nutrient and waste metabolite molecules while retaining the larger protein product inside the microcapsules (37). The encapsulation technology is based on polyionic binding of poly-L-lysine, poly-L-orinithine, poly-L-glutamate or other basic polymers with gelled calcium alginate bead containing the mammalian cells, followed by liquefaction of the capsule interior. Another polymeric material used in microencapsulation of mammalian cells is chitosan biopolymer, which coacervates onto the gelled alginate beads (38). Size of the gelled beads can be controlled in the range of 100–500 μm by using different syringe sizes as well as application of an electrostatic field (39). A claimed advantage of concentrating the protein products inside the microcapsules is countered by the possible degradation of the protein by the proteases excreted by the dead cells accumulating inside the microcapsules. Recovery of the accumulated product from the capsule interior presents another complex operation.

Macroporous Matrix, Fluidized Systems

Microcarrier culture systems, discussed more extensively in Chapter 6, can be easily used in perfusion bioreactors, as it is very easy to separate the microcarrier beads from the harvest stream. The microcarrier beads introduced by Van Wezel et al. (40) provide a surface matrix for the attachment and growth of attachment-dependent cells. These beads are suspended in a homogeneous nutrient environment through gentle agitation. Porous microcarrier beads increase the surface available for cell attachment and growth. Two porous beads that have been used in perfusion cultures of attachment-dependent mammalian cells are the Verax collagen beads (41–43) and the gelatin beads (42). With the ease of adapting many industrial host mammalian cells to growth in suspension cultures, these processes are no longer in vogue in comparison with the methods of cell retention in homogeneous systems discussed below.

Cell Retention Systems for Homogeneous Cultures

In homogeneous systems, mammalian cells are grown in suspension cultures and the cells are distributed uniformly throughout the bioreactor culture volume. A major advantage of using a suspension culture is its scalability (4) and the ability to monitor and control results due to the relatively uniform environment. While suspension cultures are naturally used for attachment of independent cells like hybridoma and myeloma cells, the increasing success of the suspension adaptation protocols for other mammalian cells used in the industry (e.g., CHO and BHK) makes it possible to use the scalable suspension bioreactors for these cells as well. Following the previously established paradigms of recombinant microbial cultures and mammalian viral production systems, batch and fed-batch processes are also commonly used for the large-scale production of therapeutic proteins from mammalian cell cultures. However, in light of the given earlier, it is gradually becoming commonplace to consider the continuous perfusion cultures for the production of therapeutic proteins if the cells are sufficiently stable.

The key difference between a chemostat and a perfusion bioreactor is the substantial or full retention of the freely suspended cells inside perfusion bioreactors. Higher cell densities in the perfusion bioreactor are achieved through partial or full retention of suspended cells from the harvest or outlet stream back to the bioreactor. There are several methods for retaining cells inside bioreactors or recycling the cells back to the bioreactor. These methods are discussed qualitatively in this chapter, with an emphasis on optimization and scale-up potential of these retention systems for industrial scale bioreactors. A more quantitative discussion of these cell retention systems may be found in the previous review by Woodside et al. (14).

Spin Filters

The earliest cell retention device used for mammalian cell cultures is the spin filter, developed by Himmelfarb et al. in 1969 (44) and successfully scaled up to larger (40 L) bioreactors by Tolbert et al. (45). A spin filter is a cylindrical membrane or stainless steel mesh of defined pore size, usually attached to the impeller shaft of the bioreactor, as shown schematically in Figure 1. The cell-free culture medium passes through the filter and is pumped out of the bioreactor, while the cells are retained inside the bioreactor. With successful cell retention by the spin filter, higher cell densities (typically an order of magnitude higher) are achieved easily in early spin-filter perfusion bioreactors (6,45–49), resulting in a similar increase in bioreactor volumetric productivity. Contrary to a chemostat culture, the perfusion culture with a spin-filter is an unsteady state process, with the total and viable cell concentrations not reaching a true steady state unless a cell bleed stream is removed from the bioreactor.

The perfusion culture duration is significantly longer than that of a fed-batch culture process, however clogging of the spin filter leads to the eventual termination of the perfusion bioreactor. The operating life of the spin filter perfusion bioreactor is mainly a function of the effective pore size of the filter and is also affected slightly by the filter materials and rotational speed. With a 5 μm pore size, the spin filter for a hybridoma perfusion culture, clogged within 7 days (6). With a larger 15 μm pore stainless steel mesh spin-filter, Deo et al. (46) operated a 500 L continuous perfusion bioreactor for 30 days with the mesh clogging increasing with time. Yabannavar et al. (47) increased the pore size further to 25 μm–significantly larger than the mean size of a single mammalian cell; they and found a significant reduction in cell retention efficiency. With the larger 50 μm pore size filter, there was no detectable retention of single cells (47).

Avgerinos et al. (48) examined the effect of filter materials on fouling in perfusion cultures of a CHO cell line grown initially on microcarrier beads of diameters ranging from 115 to 195 μm and gradually forming aggregates between 200 and 600 μm in diameter. Stainless steel mesh filters ranging in pore size from 44 to 105 μm became plugged within 11–21 days, while fouling was not a significant problem for up to 54 days with a hydrophobic polymer (ethylene-tetrafluoroethylene) membrane. Esclade et al. (46,50) suggested that the high surface charge density of metals

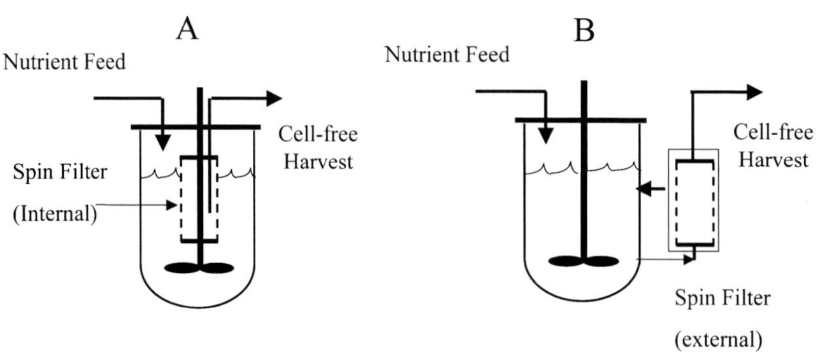

Figure 1 Spin filter for retention of mammalian cells for a high density perfusion bioreactor. Cell-free harvest medium is withdrawn from inside the cylindrical spin filter and the cells are retained inside the bioreactor. The spin filter can be used as an internal (A) and as an external (B) device.

compared to a polyamide membrane might be responsible for the observed differences in fouling of these materials over a week-long perfusion run. However, durability, autoclavability, and reusability of the stainless steel mesh filters make them more attractive than the polymeric membranes.

Deo et al. (46) have further identified rotational velocity as another key parameter affecting the possible maximum perfusion rate before filter clogging. The rate of filter fouling increases with the perfusion rate and cell concentration both. Even with these limitations, spin filter technology has been scaled up successfully to a perfusion bioreactor volume of 500 L in the reported literature (46). Because replacement of internal spin filters is impractical, spin-filters should be designed with the appropriate choice of pore size and materials, and operated at an optimal rotational speed to prolong the high cell density culture operation before filter clogging eventually forces the termination of perfusion bioreactor.

External Filtration

Considering the difficulties in removing the clogged spin filter from the bioreactor without terminating the perfusion culture, it may be advantageous to move the cell retention function to an external filtration module. Cross-flow microfiltration is the preferred method for continuously removing the cell free harvest on the permeate side while returning the concentrated cell suspension to the bioreactor, as shown schematically in Figure 2. Both flat-plate and hollow-fiber cartridges have been used traditionally as the filtration devices for this external filtration step (51–55). Here again, clogging of the filter with time becomes a serious issue. However, it is possible to replace the external filter cartridge when clogging reduces permeate flow significantly, typically between 5 and 7 days.

Majority of reported studies used microporous membranes with 0.2–0.65 μm pore size, with a few studies also using 5 and 10 μm pore membranes. While small pores are expected to clog more quickly than larger pore filters, the decreased cell retention efficiency (<70%) of the 10 μm pore membrane (56) places an upper limit on the pore size that can be used. Permeate flux is another key parameter affecting

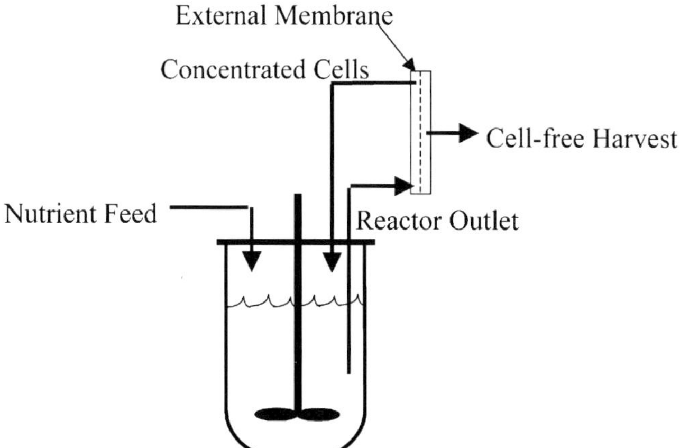

Figure 2 External filtration for retention of mammalian cells. Cell-free harvest is removed from the permeate side of the filtration device, while the concentrated cells are returned to the bioreactor.

rate of fouling (52), with faster fluxes drastically reducing the time of clog-free operation. Increasing the flow rate of cell suspension through the filtration device reduces the membrane fouling rate; however, higher flow rates result in higher shear rates at the membrane surface. High shear rates have been shown to cause significant cell damage and a drop in cell viability (52).

Some recent efforts are underway to reduce fouling through Taylor vortices in an external vortex flow filter (57,58). These efforts have extended the culture duration to 45 days with 0.8 μm pore vortex flow filter (57) and 60 days with a 10 μm steel mesh vortex flow filter (58). However, all the published results using external filter cartridges were conducted in the smaller laboratory scale perfusion bioreactors (less than 4 L reactor volume), highlighting the difficulties encountered in scaling up the two-dimensional filtration surface area necessary for retaining cells and recycling them back to the bioreactor. Persistent problems with filter clogging have been avoided successfully by the three cell retention methods discussed next.

Alternating Tangential Flow (ATF) System

Filter fouling and clogging problems with the filtration-based cell retention system may be reduced by ATF through the filter (59). In the ATF system developed by Refine Technology (Edison, NJ, U.S.A.), a 0.2 μm hollow fiber cartridge is connected to the bioreactor by a single port. Cell suspension is alternatively pumped into the filtration module and back to the bioreactor by a fast diaphragm pump (Fig. 3). The pump is partitioned into two chambers by a flexible, medical-grade silicone diaphragm (Fig. 4). The controller cycles filtered air to and from one of the pump chambers and a positive or negative pressure gradient is produced relative to the bioreactor. The unit is attached to the bioreactor using a single port and a cell free harvest is pulled out from the permeate of the hollow fiber cartridge. The system can be scaled up based on the surface area of the hollow fiber cartridge. Based on the

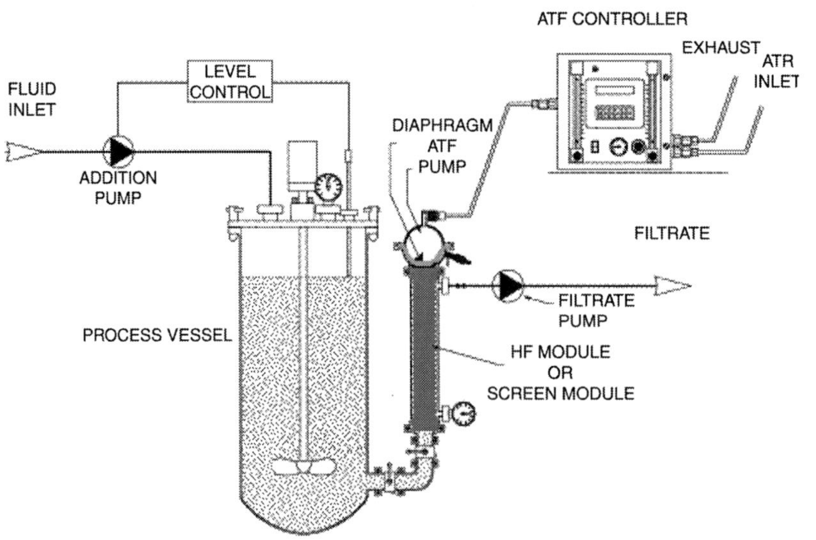

Figure 3 Alternating tangential filtration (ATF) system for cell retention. A diaphragm pump and control system generate alternating tangential flow through the hollow fibers. (Courtesy of Refine Technology.)

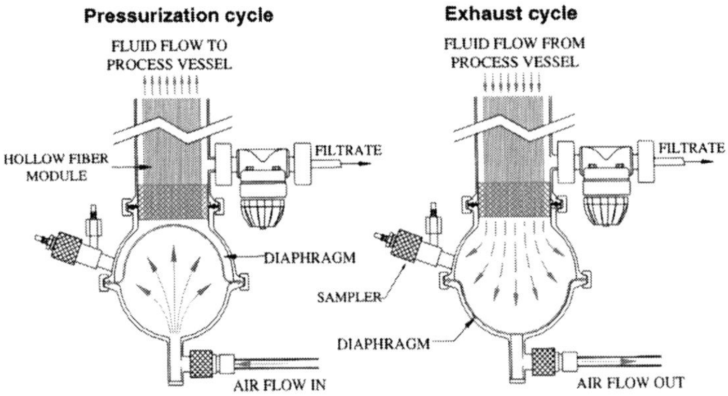

Figure 4 The action of a diaphragm pump in the ATF system is driven by pressure difference generated across a silicone diaphragm. The pressurization and exhaust cycles generate flow across the fibers resulting in a scrubbing effect, which minimizes the fouling of fibers. (Courtesy of Refine Technology.)

filter sizes available in the market it is possible to run the unit at more than 1000 L/day perfusion rates. The ATF system is designed for suspension cells but a slight modification can make it work with attachment-dependent cells. In anchorage dependent, microcarrier-based cultures, the modular nature of the ATF system allows replacement of the hollow fiber with a screen module (SM). This converts the ATF system to a highly effective perfusion device for microcarrier-based cultures. The screen module retains the microcarriers in the system as the microcarrier free medium is removed continuously for perfusion.

Cell Sedimentation

Vertical Sedimentation. Gravitational sedimentation of cells in quiescent liquid volume has been used to clarify the culture broth in a separate cell settling chamber, to remove the clarified liquid as the harvest stream, and to return the settled cells to the bioreactor (60–63). The settling chambers uses gravity to settle and separate the cells (Fig. 5). While settling of the cells can be carried out in any geometry, a conical shape offers some advantages for vertical sedimentation. Due to increased diameter in the cone, the linear liquid velocity becomes lower than the settling velocity of the cells and the cells settle. Kitano et al. (60) demonstrated the utility of a vertical sedimentation device placed above the bioreactor, with a single pipe transporting cells from bioreactor to the settler and the settled cells from the settler to the bioreactor. Such devices have also been used by other researchers (61,62) to achieve high cell densities in small-scale perfusion bioreactors. Hülscher et al. (63) improved the piping of the settler, by separating the inlet to the settler and the settled cell recycle to the bioreactor, and demonstrated a complete recycle of viable cells and selective removal of only dead cells in the settler harvest stream. However, these vertical sedimentation devices have a narrow range of upward flow rates, which can be used to clarify the cells from the harvest stream. Hence, the scale-up of the vertical sedimentation devices to larger scale bioreactors has not been successfully demonstrated. The size of these devices at high perfusion rates needs to be huge to effectively separate mammalian cells in suspension with low settling velocities. The cone settlers can be used at a large scale for microcarrier, based bioreactor systems however. In this case, settling is used to keep the microcarriers in the bioreactor during perfusion.

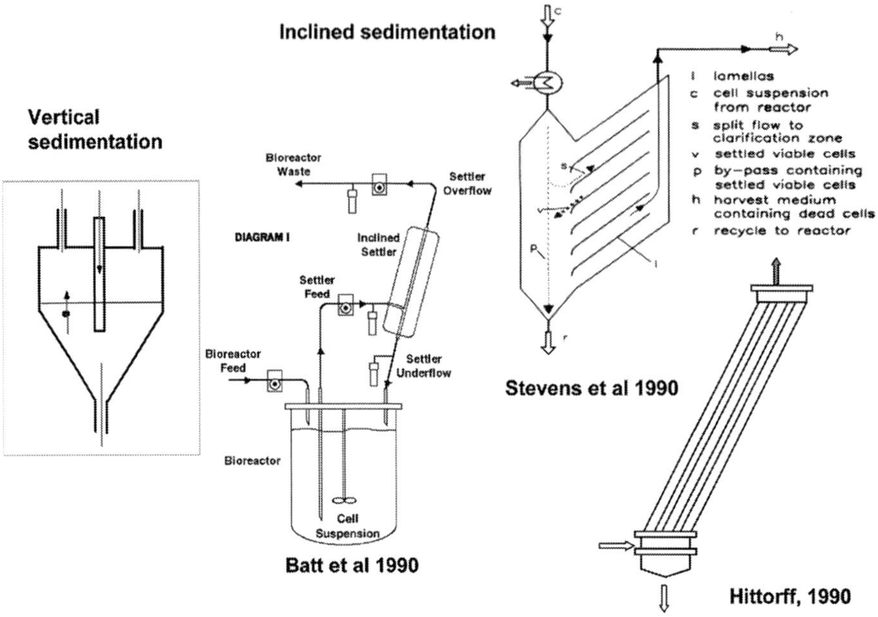

Figure 5 Sedimentation based cell retention systems. When the linear velocity of the fluid is less than the effective settling velocity of cells, cells can be separated and returned back to the bioreactor.

Inclined Sedimentation. Inclined sedimentation overcomes the limitations of vertical sedimentation in terms of efficiency and scalability. In these systems, cell suspension is forced to flow up between a pair of inclined plates. The cells settle through a short vertical distance between the plates and move down the surface of the plates to be returned back to the bioreactor. The settling efficiency in inclined sedimentation systems is enhanced due to the Boycott effect (64). In addition, the cell retention capacity of inclined sedimentation devices is easily increased by scaling up the settling area by increasing the width and length, and by stacking a number of plates over each other. The systems can be made very compact by decreasing the distance between the plates. Although inclined sedimentation is an established method for enhanced particle separation in other industries, its application to cell culture technology is relatively new. Early results on its application to perfusion bioreactors were presented in the late 1980s and these systems had been fully developed for commercial use in the next decade.

Batt et al. (65) first demonstrated the use of an inclined sedimentation device as a *selective* cell retention device to achieve high cell densities in a perfusion bioreactor. Larger live hybridoma cells settling on the bottom plate of the settler roll down this plane and are returned to the bioreactor, while the smaller dead cells and cell debris do not completely settle and are removed in the settler overflow stream. Because of the continuous removal of dead cells and cell debris, this perfusion bioreactor can be operated at a steady state for longer periods without any buildup of dead cells or cell debris inside the bioreactor. Due to the selective recycle of live cells from the settler to the bioreactor, high cell densities (over 10 million cells/mL) are easily achieved in the perfusion bioreactor. With the reduced vertical sedimentation path in the inclined settlers, much higher perfusion flow rates can be used in these devices, overcoming the limitations of the vertical sedimentation devices. The inclined settlers have been scaled up to 40 L (66) and to 500 L perfusion bioreactors (67) through stacking up of multiple inclined settlers, which are known as the lamellar settler (68,69). At these production scales, the inclined sedimentation devices have exhibited greater than 95% cell retention efficiency for over 20 days of high cell density perfusion culture.

Searles et al. (70,71) have adapted the inclined settler, as shown schematically in Figure 5, for the perfusion culture of attachment-prone Chinese hamster ovary cells. By increasing the settler inlet flow and liquid recycle to the bioreactor independent of the harvest or perfusion rate, the cells that settled in the device are continuously swept back into the bioreactor. Searles et al. have shown that more than 90% of the live cells entering the settler are returned to the bioreactor at the same residence time as the liquid residence time in the settler. The small fraction of live cells (between 5 to 10%) that enter and form a sediment layer in the settler are easily dislodged and resuspended into the liquid flowing down to the bioreactor by bubbling sterile headspace gas into the settler as first demonstrated in industrial scale perfusion bioreactors (P.C. Brown, Chiron, personal communication). Searles et al. have also investigated other modifications such as cooling the settler to 4°C to slow down the metabolism of live cells in the settler and release any attached cells, as well as vibrating the settler to dislodge any cells attached on the settler. These strategies have been successfully implemented in industrial scale continuous perfusion bioreactors (72).

Tyo et al. (73) described a laboratory scale perfusion system based on a conical lamellar settler. This system was incorporated into a spinner flask to grow cells to high cell densities. Knaack et al. (74) extended this concept to design a conical bioreactor with internal settling zone. The bioreactor was shaped as a cone and conical plates were placed on the top to retain the cells in the bioreactor. The design of a lamella settler can be simplified by the use of plates rather than cones. Several investigators presented successful operation of inclined plate settlers (66,68,69,72,75). These plate settlers are compact in size can be operated at high perfusion rates. Figure 5 presents some of the inclined plate settler designs. In these systems cell suspension flows up between the plates in laminar flow conditions. While they are in transit, the cells settle on the plates and return back to the bioreactor (15). Typically, the settlers are angled at 30° from vertical and the plates are separated by 0.5–2 cm distance. The dimensions of the plates are typically 5–15 cm in width and 50–100 cm in length. The number of plates in a settler can vary between 5 and 20 plates. The efficiency of a settler for cell separation can be enhanced by cooling the cells to room temperature or lower, and by applying an intermittent vibration to the plates (69). Figure 6 shows an industrial scale inclined plate settler currently available for commercial use (72).

Thus, inclined settlers have been scaled up to production scale perfusion bioreactors to continuously remove dead cells and cell debris and to selectively recycle the viable and productive cells back to bioreactors. While this retention device has no moving parts to damage the cells and is easy scalable, one drawback expressed by the industrial practitioners is that the deceptively simple device is bound by the inflexible performance characteristics governed by the different sedimentation rates of larger live and smaller dead cells.

Centrifugation

Another approach to circumventing the problem of filter clogging is the use of continuous cell centrifuges with rotating mechanical seals (76,77). While this cell

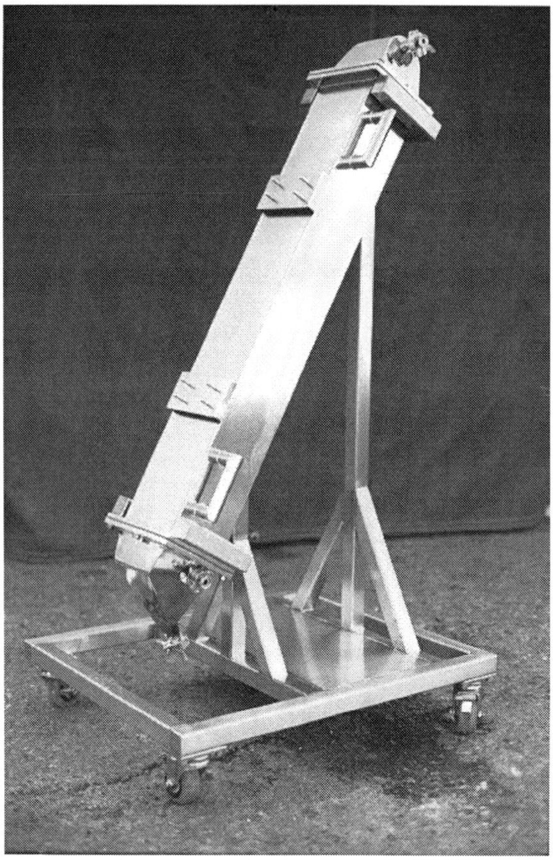

Figure 6 Inclined plate settler used at industrial scale cultivation of mammalian cells. (Courtesy of Biotechnology Solutions, California, U.S.A.)

retention method has been shown to retain cells at 100% efficiency in 12- and 15-day perfusion cultures (78), the effect of repeated centrifugation on cell viability remains a concern. The potential failure of the rotating seals (76) over a long perfusion run led to the development of the Centritech laboratory centrifuge with an ingenious tubing connection to enable continuous centrifuging (79). Johnson et al. (80) have investigated this centrifuge with a small-scale perfusion culture of hybridoma cells, and raised concerns of shear stress and nutrient deprivation in the pelleted cells. Intermittent operation of the centrifuge was suggested as a possible method to overcome these concerns. In addition, serious issues remain concerning the durability of the centrifuge components, such as the air barrier, the centrifuge insert, and the supernatant tubing. Thus, while trying to eliminate the problems of clogging in the simpler filtration devices, the centrifuges introduce the additional problems of more complex mechanical devices and their effects on cell viability. Figure 7 shows centrifugation systems that can be used for perfusion. The use of centrifugation was demonstrated for large-scale production of biologicals from mammalian cell culture (76,81–83). The perfusion rates were reported to reach 3600 L/day in these systems and the cell retention efficiency was more than 90%.

Centrifugation cell separation also resulted in interesting bioreactor designs for the culture of mammalian cells. A centrifugal bioreactor is essentially a continuous

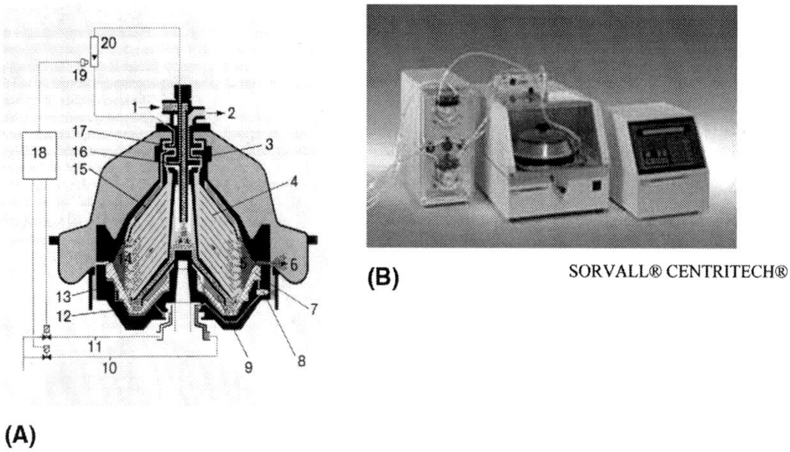

Figure 7 Centrifugation-based cell separation systems: (A) Continuous centrifuge from Westfalia. (Courtesy of Westfalia.) (B) The Sorvall/Centritech system. (Courtesy of Sorvall/Centritech.)

centrifugation device where the cells are retained in the bioreactor by centrifugation and perfused continually by the incoming medium stream (79,84). This system is very compact and has potential for large-scale culture. However, it needs to be scaled-up and issues related to aeration and mixing must be addressed.

Ultrasonic Separation

Ultrasonic separation is based on acoustic aggregation of suspended cells in the standing wave field due to differences in density and compressibility between the cells and the medium, and the enhanced cell sedimentation upon the removal of the field (85). The ultrasonic resonator, shown schematically in Figure 8, is not susceptible to fouling or mechanical failure, as it does not have a physical barrier or moving parts. Trampler et al. (86) have utilized this cell retention device on a 1 L bioreactor in a month-long perfusion culture of hybridoma cells with no detectable impact on cell viability. Cell retention at higher flow rates from a 5 L bioreactor requires a higher power input than the ultrasonic wave generator (87). More recently, this ultrasonic separation device has been scaled up to retain cells from the harvest stream of a 100 L perfusion bioreactor (88) by housing four of the ultrasonic units operating at even higher power in cooling water. The higher power requirement (90 W) and multiplexing design of the larger device highlight potential problems in further scaling up this cell retention device due to the nonuniformity of the force generated in larger standing wave fields and the removal of heat generated within this device.

Hydrocyclones

Hydrocyclones are used to separate solids from suspension using centrifugal forces generated by a flow field (Fig. 9). The suspension enters to the hydrocyclone tangentially to the wall. The solids are concentrated and removed in the underflow and the clarified liquid is removed from overflow. The system is very attractive to perfusion bioreactors because of its small volume and its high separation efficiency. However, operation of hydrocyclone requires high flow rates and a relatively high pressure drop. The use of a hydrocyclone for cell separation was demonstrated first for yeast

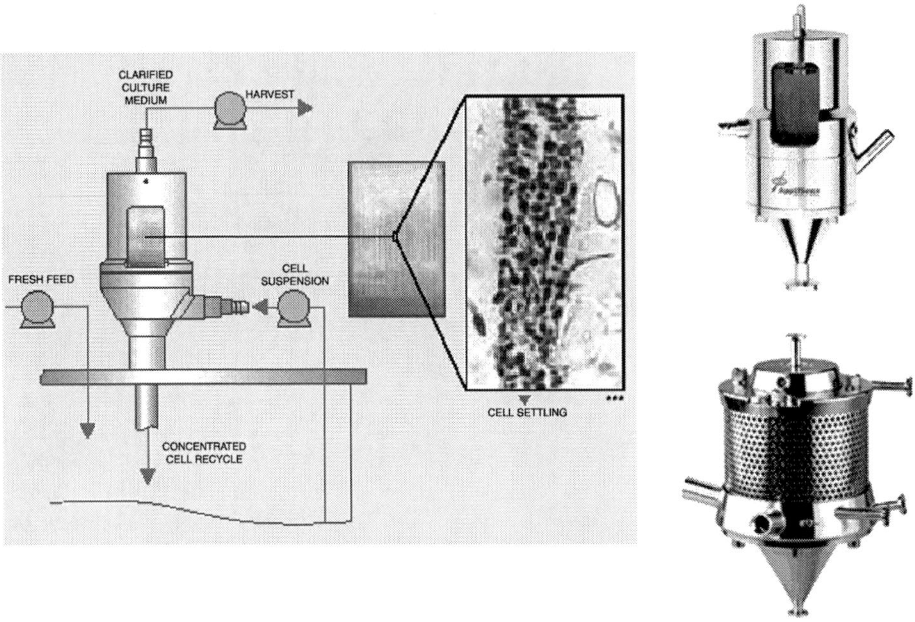

Figure 8 Ultrasonic cell separation device. The standing wave field in the resonator chamber aggregates and retains cells from the effluent stream. Periodic shut-down of the power to the resonator and effluent pump allows the aggregated cells to settle out of the resonator and into the bioreactor. (Courtesy of Attlikon.)

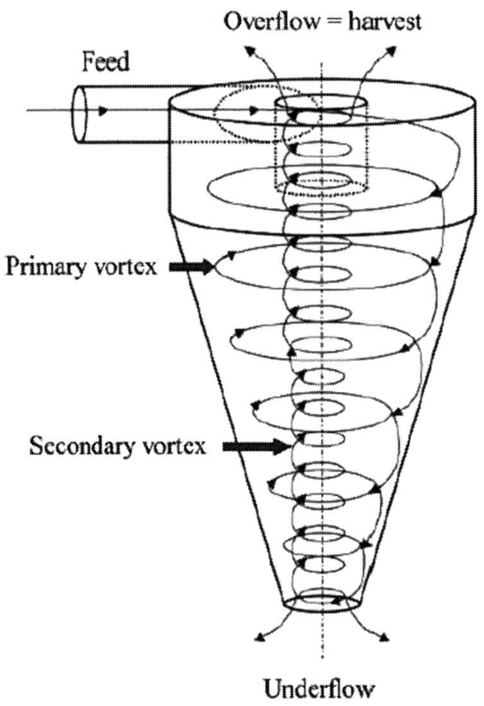

Figure 9 Cell separation using hydrocyclones. [From Ref. (91).]

fermentation (89). A recent work by Jockwer et al. (90) reported successful use of a hydrocyclone for CHO cell perfusion culture. A small unit with 3 cm diameter could be operated at 500 L/day perfusion rates with cell separation efficiency of more than 70%. The damage to the cells at high flow rates was minimal. The drop in viability was only 5%. The system could be operated at higher separation efficiencies (>90%); however, the drop in viability was higher (11%).

Scale-Up Potential of Cell Retention Systems for Industrial Application

It is clear that there are many options for cell retention and, depending on the application, one system may be a better choice than others. The simplicity in operation, robustness, scalability, and process economics are the factors that help determine the applicability of the system for use in industrial settings.

Although a lot of effort has been devoted to their development, cell retention systems for heterogeneous cultures such as fixed beds, hollow fiber systems, micro encapsulation, and immobilized cell systems have proved to be unsuitable for large-scale production. The main reasons for the absence of any encapsulation or immobilization methods in large-scale production processes are the difficulties in scale-up, intensive labor, and the cost involved in cell immobilization. Similarly, the complexity of the operation limits the acceptance of cell attachment systems such as fiber beds, ceramic cartridges, and microporous carriers for commercial production. Most of the bioreactor systems such as OptiCell and the Verax fluidized bed reactor have been abandoned. Only a limited number of technologies such as hollow fibers find application for small quantity production of antibodies for diagnostic purposes.

The cell retention systems for homogeneous systems have a better acceptance record for industrial applications. These systems use internal or external cell retention devices to separate the cells and they are easier to operate and easier to scale-up. The capacity of the cell retention system is measured in terms of cell density, the degree of cell retention, the duration for the operation, and perfusion flow rate. In order to discuss the limits in their performance, perfusion systems can be categorized into two categories: (a) filtration-based systems and (b) open perfusion systems.

Filtration based systems include spin filters, external filtration, and ATF. Since they are filtration based, the degree of cell retention can be very high and if one excludes the spin filter system, the retention is 100% and does not change with the flow rate. Figure 10 presents the performance of filtration-based cell retention systems. Under limiting conditions the number of cells achievable in the bioreactor will be dependent on the medium composition and the medium exchange rate (13). Medium exchange rate and medium depth [nutrient concentration of medium (72,91,92)] both result in an increase in cell density. The limitation of filtration based cell retention system exhibits itself in run length. These systems eventually clog and either the run must be terminated (for an internal cell separation device) or the device needs to be replaced (for an external cell separation device). The time to clog can vary between 10 and 100 days. A number of factors such as cell density, viability, perfusion rate, filter material, and pore size are involved in determining the time to clog.

Open perfusion systems include gravitational setters, centrifuges, ultrasonic separation devices, and hydrocyclones. The performance charts for these systems are presented in Figure 11. As in the case of filtration based systems, the maximum cell density attainable in these systems increases with the perfusion rate and with the medium depth. These open systems, by definition, do not clog so they can be operated

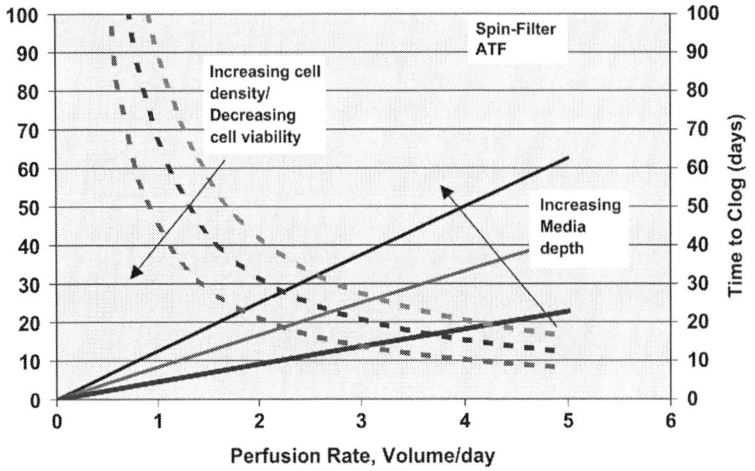

Figure 10 Performance chart for filtration based cell retention systems. Cell density (*solid lines*) and time to clog (*dashed lines*) plotted against perfusion rate.

indefinitely. The operating limit for these systems is exhibited in the degree of cell retention. Unlike filtration based systems, the degree of cell retention is reduced at high perfusion rates. This reduction is dependent on such factors as cell diameter, aggregation, and other process parameters. The open perfusion systems require a high degree of cell retention, typically more than 90%. The loss of cells in open perfusion systems should be less than the growth rate of the cells for a stable operation. The degree of tolerable cell loss or the minimum degree of cell retention limits the maximum perfusion rate. This maximum perfusion rate then limits the maximum number of cells attainable in the bioreactor.

Table 2 summarizes the scale-up potential of different cell retention systems. Cell entrapment systems can achieve very high cell densities (>50 million cells/mL) and they can be operated for a long time. The 3-D microcarrier system developed by

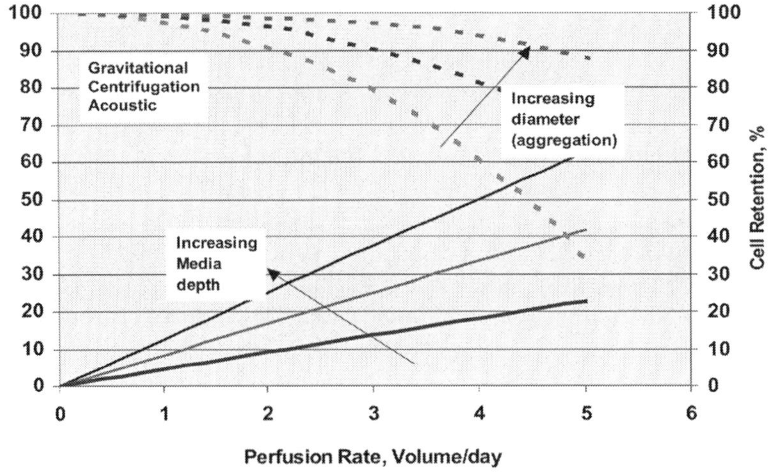

Figure 11 Performance chart for open cell retention systems. Cell density (*solid lines*) and cell retention (*dashed lines*) plotted against perfusion rate.

Table 2 Scale-up Potential of Different Cell Retention Systems[a]

Cell retention method		Cell density (MM/mL)	Perfusion rate (L/Day)	Culture duration (days)
Cell entrapment				
3-D microcarrier	Internal	100	600	200
Ceramic cartridge	Internal	50	200	90
Fiber-bed	Internal	50	100	90
Spin-filter	External, internal	15	1000	60
ATF	External	20	1000	60
Inclined settlers	External, internal	20	2000	200
Centrifugation	External	20	3600	90
Acoustic separator	External, internal	20	200	60
Hydrocyclones	External	20	500	20

[a] The values are gathered from published reports and from the authors' own experience (72).

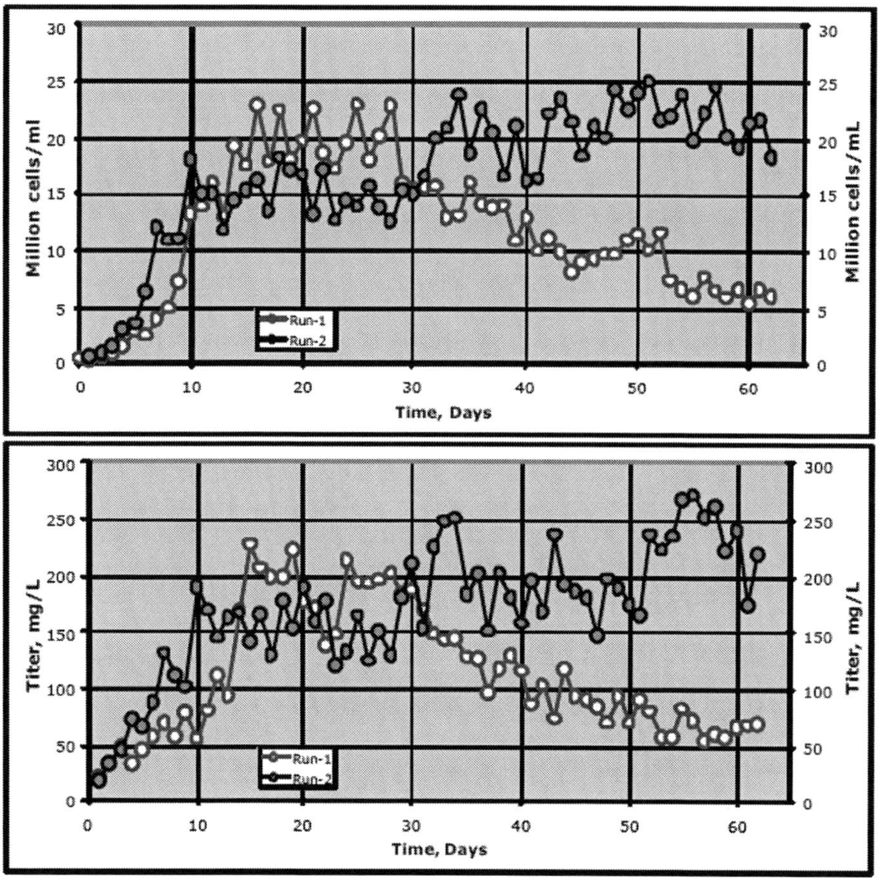

Figure 12 Consistency and reproducibility issues for long-term perfusion culture. Small differences in bioreactor operation can accumulate over time resulting in a big difference in performance. [From Ref. (107).] (*See color insert p. 5.*)

Verax could operate for more than 6 months at perfusion rates close to 600 L/day (92,93). However, this system, along with other cell entrapment systems, has been abandoned due to complexities in its operation. The operation of acoustic separation and hydrocyclones has been limited to small-scale bioreactors and more work needs to be done to utilize their full potential. A 200 L/day acoustic separation (BioSep) system is available and a 1000 L/day unit is being developed. Currently there are three systems that can be used at industrial scale. Alternating tangential filters (ATF), "gravitational" (replace with "inclined settlers"), and centrifuges can reach cell densities more than 20 million cells/mL, can be operated for a long time (>60 days) and at high perfusion rates (>1000 L/day). These systems are being optimized gradually and higher performance can be expected in the future in terms of cell density and achievable perfusion rates (Fig. 12).

CONTROL AND OPERATION OF PERFUSION BIOREACTORS

Consistency and Reproducibility of Perfusion Cultures

Compared to batch or fed-batch systems, perfusion cultures generally result in a more uniform environment for the cells. Thus, better consistency and reproducibility is expected from a perfusion bioreactor. However, as discussed widely in the literature (13,72), this is not always the case and effective control strategies should be in place to obtain consistency.

The perfusion systems operate for a long time and if there are slight differences in operation, the results can vary from one bioreactor to another. Small differences and minor incidents can accumulate over time and result in completely different outcomes. Figure 13 presents data on cell density and product titer from two different large-scale cultures using an open perfusion system. The bioreactors were initiated using the same inoculum and operated under the same conditions. A few incidents including a problem in the feed in one of the bioreactors resulted in completely different performance over a 60-day run. Both cell densities and the product titer in bioreactor Run1 were lower for much of the time. This example demonstrates the need for a close monitoring and control of perfusion bioreactors.

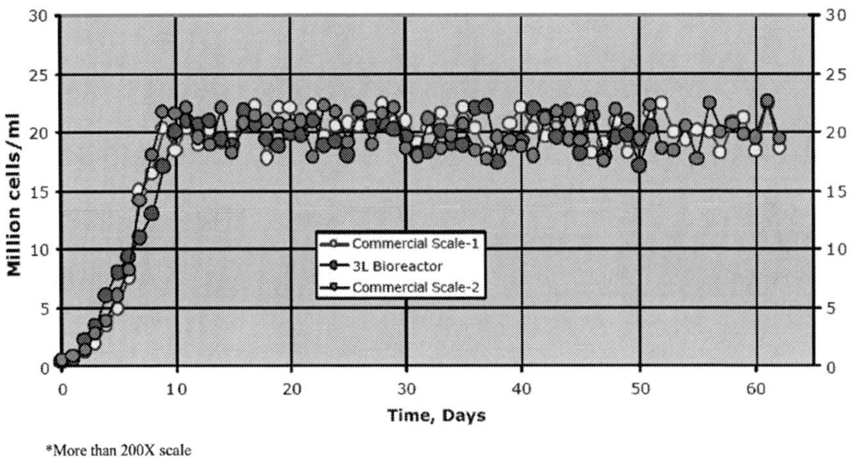

Figure 13 Scalability of perfusion bioreactors operated by CSPR control. [From Ref. (107).] (*See color insert p. 5.*)

Off-Line and On-Line Monitoring of Perfusion Bioreactors

Physical parameters (such as pH, dissolved oxygen, and temperature) in a perfusion bioreactor are typically monitored on-line and controlled in real time. Determination of cell density, viability, metabolite, and product concentration is performed using off-line sampling. The information obtained from these measurements is used to assess the culture performance and to make adjustments to the bioreactor if necessary. The frequency of sampling and making adjustments depends on the application. While a daily sampling is sufficient for most applications, more frequent and almost real-time monitoring and control are warranted for some processes (94–96).

On-line monitoring of perfusion cultures was demonstrated for cell density (13,94,96), oxygen consumption rate (97,98), metabolite concentrations (95), and for product (99). These measurements can be used to calculate cellular activities such as cell growth rates, cell specific metabolic, and production rates. In addition, these measurements can allow real-time control of perfusion bioreactors (13).

Operation of Perfusion Bioreactors by Dynamic Perfusion Rate Adjustments

In addition to physical parameters such as pH, dissolved oxygen, and temperature, perfusion rate and cell density are the important parameters to be monitored with perfusion bioreactors. Typically, the perfusion bioreactors start with a batch growth phase with no feeding. During that period, cells are allowed to grow to cell densities of 1–2 million cells/mL. Then the perfusion operation starts with continuous harvesting and feeding. The perfusion rate typically refers to the harvest flow rate, which is manually set to the desired value. A weight control for the bioreactor activates the feed pump so that a constant volume in the bioreactor can be maintained (91). Alternately a level control can be achieved by pumping out culture volume above a predetermined level in a variable bleed stream (100).

The perfusion rate in the bioreactor must be adjusted to deliver sufficient nutrients to the cells. As the cell density increases in the bioreactor, the perfusion rate must be increased. The dynamic adjustment of the perfusion rate allows cells to grow and produce without nutrient depletion. There are several methods for perfusion rate adjustments and they are described in detail in the following sections.

Control of Perfusion Rate Using Cell Density Measurements

Cell density is the most important measurement used for perfusion rate adjustments. Depending on how the cell density measurements are conducted, perfusion rates can be adjusted daily or in real time. Several on-line probes have been developed for the estimation of cell density (96). Some of these cell density probes are robust and reliable, and can be used for automatic control of perfusion rates (13,101). These cell density probes can also be used to control the cell density at a desired set point by removing excess cells from the bioreactor.

Control of Perfusion Rate Using Oxygen Consumption

Oxygen consumption or oxygen demand to the bioreactor is an indicator for cell density. These variables can then be used to adjust the perfusion rate. Oxygen consumption rates can be obtained using several techniques (13,98,101,102). Most of these techniques result in continuous on-line measurements. The perfusion rate can be adjusted proportionally to the oxygen consumption rate. The use of the oxygen consumption rate for perfusion rate control has been demonstrated successfully in laboratory and pilot plant scale bioreactors (97,98,102).

If the oxygen consumption rate is difficult to obtain, the gas flow rate to the bioreactor can also be used as an indicator for cell density. Thus, the perfusion rate can be adjusted based on the oxygen flow rate to the bioreactor (103).

Control of Perfusion Rate Using Metabolite Measurements

On-line or off-line measurement of metabolites allows determination of metabolic rates of the cells through a mass balance in the system. This metabolic rate information can then be used to adjust the perfusion rate to the bioreactor (43). Ozturk et al. (95) described the method of controlling glucose and lactate by dynamically adjusting the perfusion rate using a simple feed-forward algorithm.

Cell Specific Perfusion Rate Control

Ozturk et al. (13,101) introduced the concept of cell specific perfusion rate (CSPR) for the design, optimization, and control of perfusion bioreactors using mass balances for substrate and products. The steady state mass balance equations for the substrates and products can be written as:

$$D \cdot (S_o - S) = q_s \cdot X_v \quad \text{(for substrates)}$$
$$D \cdot (P - P_o) = q_p \cdot X_v \quad \text{(for products)}$$

where c is the dilution rate; X_v is viable cell density; S, P are substrate and product concentrations in the reactor, respectively; S_o, P_o, are their concentrations in the feed; and q_s, q_p are the rates of substrate utilization and product formation.

Using CSPR, which is defined as $\text{CSPR} = D/X_v$, transforms these mass balance equations to:

$$S = (S_o - q_s)/\text{CSPR} \quad \text{(for substrates)}$$
$$P = P_o + q_p/\text{CSPR} \quad \text{(for products)}$$

Therefore, if the CSPR is maintained as a constant, and the cellular activities do not change with time and cell density, then the CSPR control allows operation of high density reactors at constant medium composition, thus allowing consistent and optimal production.

The implementation of CSPR control to perfusion bioreactors was successfully demonstrated in pilot size bioreactors (13). A highly automated bioreactor control system monitors the cell density using optical density probe, determines the oxygen consumption rate in real time, and combines these measurements to estimate viable cell density and cellular activities. The system then controls the cell density and perfusion rate at set-points. A bioreactor with CSPR control could be run in an auto-pilot mode, make the necessary adjustments in response to the perturbations, and maintain a constant and uniform environment to the cells.

OPTIMIZATION OF PERFUSION BIOREACTORS

Design and Operation of Perfusion Bioreactors at High Cell Densities

The bioreactors used for perfusion are not very different from those used for batch/fed-batch cultures, except that they are more compact in size and are connected to a cell retention device. The production rate in a 1000 L perfusion bioreactor can reach or even exceed the output of a 10,000 L fed-batch bioreactor.

Cell Retention

The design and optimization of cell retention devices is key for the development of a perfusion process. The device should be robust, reliable, and scalable. There is no question that the addition of a cell retention device introduces complexity to the operation and that is why the industry is reluctant to accept perfusion processes for cell culture applications. In-house expertise is needed to design, operate, and scale-up these systems. As presented in Table 2, there are already a variety of cell retention technologies that can be used at industrial scale. These technologies can be acquired easily and with some extra development and fine-tuning, these systems can be customized for a specific application. While the current systems can be developed further to accommodate higher perfusion rates, new technologies may emerge in the field of cell retention. Even though these developments are likely to be gradual, they will have a major impact on the perfusion bioreactors.

Aeration

High cell densities achieved in perfusion bioreactors pose unique challenges with respect to aeration, mixing, and process control. Use of sparging can deliver sufficient oxygen at high densities. However, foaming and cell lysis due to sparging require a careful sparger design and aeration strategies. Aeration also serves as a means to remove CO_2 from the bioreactors. Accumulation of CO_2 due to the differences in mass transfer coefficients can be a problem, as illustrated in the literature. This problem needs to be addressed especially in large-scale bioreactors. In addition to increased base addition, CO_2 accumulation can affect the cell growth and productivity and product quality (13,67,104).

Mixing and Homogenization

Issues with mixing are elevated in perfusion bioreactors because of high cell densities. The mixing times in cell culture reactors can be as high as several minutes due to gentler agitation (105). At higher cell densities, viscosity in the bioreactor increases, thereby decreasing mixing efficiency. Poor mixing in bioreactors can result in gradients in pH and dissolved oxygen as discussed elsewhere (13).

In bioreactor mixing is used to suspend the cells, to disperse the bubbles for aeration, and to homogenize the added base. In poor mixing conditions, base addition, if not properly done, can cause localized cell lysis and impact the culture negatively. When time is not allowed for the homogenization of the base, hot spots with high pH are generated. Poor mixing also promotes heterogeneity of the culture and imposes control problems. Depending on the mixing and the eddy size distribution in the reactor, cells can aggregate to different sizes. Aggregation can induce segregation of the cells, with bigger ones migrating to the low agitation zones.

Process Control

While a batch or fed-batch process can be operated with minimal automation, a high-level process control scheme is desirable for controlling the perfusion bioreactors. Perfusion bioreactors are more dynamic in nature because of two reasons: (a) the cell densities change over a wide range during the operation, and (b) the system attains a high metabolic state due to the high cell densities involved. The bioreactors need to be monitored closely and adjustments need to be performed in a timely fashion. As mentioned before, several on-line monitoring and real-time control strategies have been developed for perfusion bioreactors. Some of these systems are highly automated and require complicated implementation strategies.

Scalability of Perfusion Bioreactors

Even though the perfusion bioreactors are more difficult to design and operate compared to batch or fed-batch counterparts, the scalability is rather straightforward. The success of scalability is mostly determined by the cell retention system and the control strategy used. Some of the cell retention systems such as spin filters are more difficult to scale-up. On the other hand, inclined plate settlers can be scaled-up easily using geometric similarities. A reliable control strategy, such as CSPR control, ensures the cellular environment to be similar at different scales and secures the scalability. Figure 13 presents scale-up data for a perfusion process where the process was scaled-up from 3 L bioreactors to 1000 L. As can be seen, the process is scaled-up successfully from laboratory to commercial scale.

New Directions in Perfusion Cultures: Medium Enrichment

The performance of a perfusion bioreactor is directly related to the cell concentration. The nutritional composition of the medium used, or the medium depth, is one of the key variables determining the maximum number of cells in the bioreactor. The other variable is the medium exchange rate in the bioreactor. Most of the current medium formulations can support about 10 million cells/mL (calculated based on cell yields on different nutrients) at growth conditions with an exchange rate of one volume per day. If the same medium is used for perfusion, the medium has to be exchanged 10 volumes per day to maintain a cell density of 100 million cells/mL. This magnitude of medium exchange, or perfusion rate, can impact the perfusion process significantly. First of all the cell retention system should still be functional at these high perfusion rates. Second, there will be an issue with medium utilization and cost. Third, the product from the bioreactor will be diluted at high perfusion rates.

Requirement for high medium exchange and relatively low product concentration from perfusion bioreactors have been the major disadvantages for perfusion processes. To deal with these issues several strategies have been developed.

A simple method for minimizing the medium utilization is to operate the bioreactor at low perfusion rates. If the cell density in the bioreactor can be maintained at low perfusion rates, and if the product is stable and the productivity of the cells is not impacted, the bioreactor can be operated at lower perfusion rates. This results in higher product concentration in the bioreactor without impacting the overall productivity. Figure 14 summarizes data obtained from a perfusion bioreactor for monoclonal antibody production. Initially the bioreactor is operated at a one volume per day exchange rate and a product concentration of 250 mg/L is obtained. A drop in perfusion rate from one volume per day to 0.5 volumes per day roughly doubles the concentration of product. Finally, a drop to 0.3 volumes per day results in product concentrations around 650 mg/L. While the results from Figure 14 are encouraging and a higher product concentration can be obtained, it may be noted that the volumetric productivity in this case has decreased by 20%.

A more effective solution for minimization of medium utilization and enhancing product concentration lies in the development of special concentrated or fortified

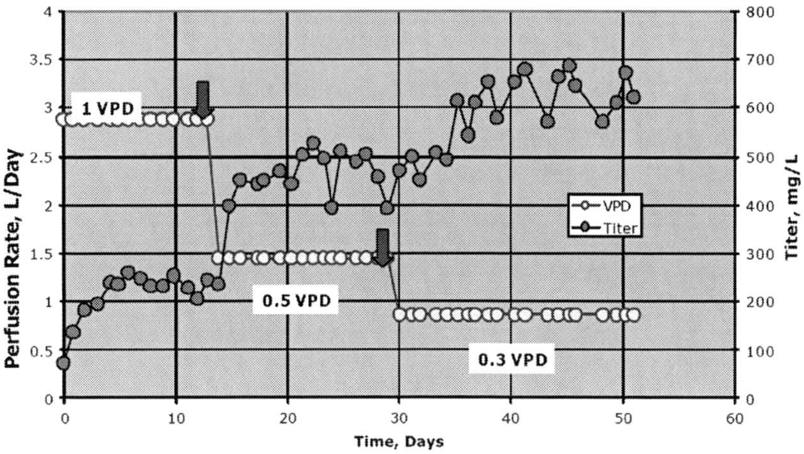

Figure 14 Decrease in perfusion rate can be used as a tool to increase product concentration from perfusion bioreactors. This approach does not offer any benefit in volumetric productivity. In most cases, the volumetric productivity is negatively impacted. [From Ref. (107).] (*See color insert p. 6.*)

medium for perfusion reactors (43,106). Research in the early 1990s demonstrated the use of 2- or 3-fold concentrated mediums for perfusion cultures. When the medium is concentrated and formulated at optimal osmolarity, the necessary perfusion rates are reduced by several fold (13,43,106). The use of enriched medium for perfusion has been shown to be successful. Accumulation of knowledge in cell metabolism and medium formulation over the years resulted in further developments of medium fortification.

The approach for medium enrichment is similar to fed-batch process development. If one analyzes the cell metabolism and the nutritional requirements of the cells, certain components can be added to the feed medium to reach higher cell

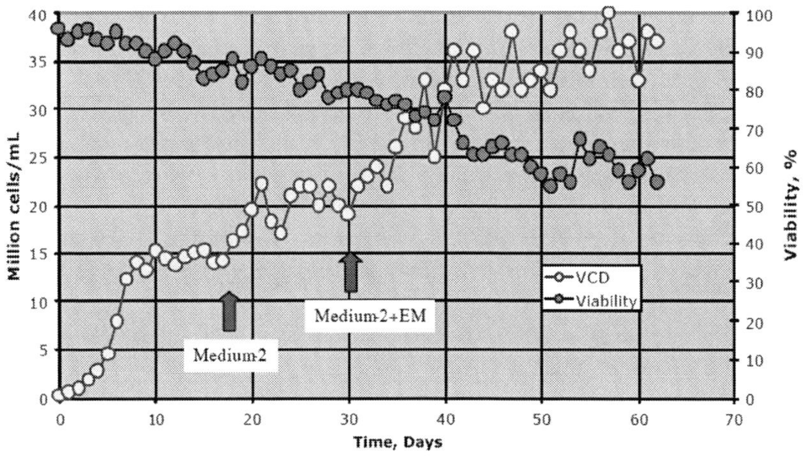

Figure 15 The use of enriched medium in a perfusion bioreactor. Switching to the enriched perfusion medium more than doubles cell density in the bioreactor [From Ref. (107).] (*See color insert p. 6.*)

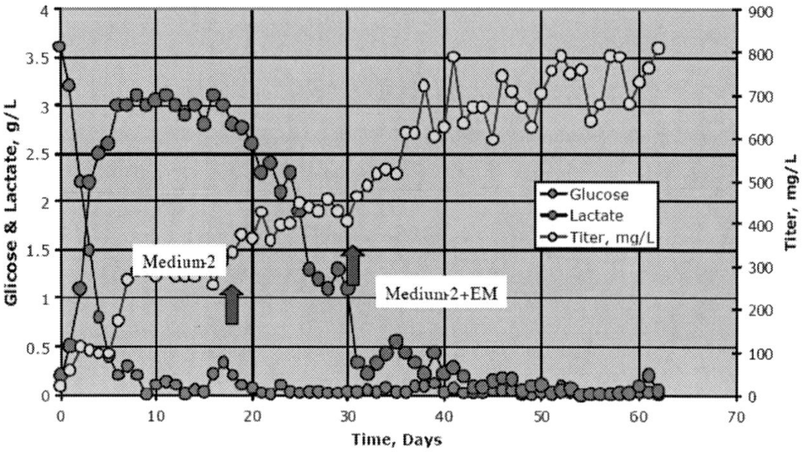

Figure 16 The use of enriched medium in a perfusion bioreactor. The product concentration reaches to 750 mg/L in enriched medium. After switching to the enriched medium a shift in lactate metabolism is observed. The lactate production drops to zero in enriched medium. [From Ref. (107).] (*See color insert p. 7.*)

densities. Instead of adding these components to the bioreactor as a feeding solution in the fed-batch case, these components can be added to the feed medium to the perfusion bioreactor resulting in an enriched medium (73,107).

Figure 15 presents the use of enriched medium for a CHO perfusion bioreactor for monoclonal antibody production. The perfusion bioreactor starts with a base medium (Medium-1) and a cell density of about 15 million cells/mL is obtained. The medium is switched on Day 17 to Medium-2 (a richer medium), and to Medium-2 supplemented with enriched medium (EM) on Day 30. Due to medium enrichment during the course of this run, the viable cell density reaches 35 million cells/mL. The viability in the bioreactor remains more than 50% during the run (107). Figure 16 summarizes the data on metabolites and product concentration for the same perfusion bioreactor. The product concentration increases by more than 100% and reaches 750 mg/L in enriched medium. The data in Figure 16 demonstrate an interesting feature of enriched medium. A drop in glucose and a corresponding increase in lactate concentration are obtained in regular medium (Medium-1). This metabolic profile is very typical for CHO cells in culture. After the switch to enriched medium, a significant drop in lactate levels is observed. In fact the lactate concentration in the perfusion bioreactor reaches zero after the switch to enriched medium. These data show an efficient utilization of glucose in enriched medium.

These results demonstrate that there is great potential for the development and utilization of enriched medium for perfusion cultures. These developments are parallel to the feeding strategies for fed-batch processes. Understanding of cell metabolism and a careful formulation of enriched medium will definitely open new frontiers for perfusion bioreactors (107).

CONCLUSIONS

A number of mammalian cell retention methods are now available for achieving high cell density in a perfusion culture, with several of these methods scaled up successfully to industrial scale bioreactors. Culture productivity in high density perfusion cultures can be as high as that of fed-batch cultures while the culture duration can be extended significantly over that of the fed-batch cultures. The high medium exchange rate of the perfusion culture not only provides the necessary nutrient supply to the high concentration of cells in a bioreactor, but also removes the secreted products quickly from the bioreactor, thereby minimizing their exposure to degradative enzymes that accumulate in a fed-batch reactor. Consequently, the size of a perfusion bioreactor can be much smaller than that used for batch and fed-batch culture processes for the same amount of desired production. A large number of industrial perfusion processes are already approved by the regulatory agencies for production of therapeutic processes; previous concerns about getting such processes approved are no longer an issue. If a given mammalian cell line is sufficiently stable in its protein production, then perfusion cultures represent a superior bioreactor operating strategy for maximizing production of the protein.

ACKNOWLEDGMENTS

The first author (DSK) expresses his gratitude to the National Science Foundation (grants BCS-8857719 and BES-9817249) for the support of his research on perfusion bioreactors.

REFERENCES

1. Burrows MT. A method of furnishing a continuous supply of new medium to a tissue culture in vitro. Anatomical Rev 1912; 6:141–144.
2. Patterson MKJ. Perfusion and mass culture systems. TCA Man 1975; 1:243–249.
3. Griffiths B. Perfusion systems for cell cultivation. In: Lubiniecki AS, ed. Large Scale Mammalian Cell Culture Technology. New York: Marcel Dekker, Inc., 1990:217–250.
4. Birch JR, Arathoon R. Suspension culture of mammalian cells. In: Lubiniecki AS, ed. Large Scale Mammalian Cell Culture Technology. New York: Marcel Dekker, Inc., 1990:251–270.
5. Reuveny S, Velez D, Miller L, Macmillan JD. Factors affecting cell growth and monoclonal antibody production in stirred reactors. J Immunol Methods 1986; 86:53–59.
6. Reuveny S, Velez D, Miller L, MacMillan JD. Comparison of cell propagation methods for their effect on monoclonal antibody yield in fermentors. J Immunol Methods 1986; 86:61–69.
7. Luan YT, Mutharasan R, Magee WE. Strategies to extend longevity of hybridomas in culture and promote yield of monoclonal-antibodies. Biotechnol Lett 1987; 9:691–696.
8. Miller WM, Blanch HW, Wilke DR. A kinetic analysis of hybridoma growth and metabolism in batch and continuous suspension culture: effect of nutrient concentration, dilution rate, and pH. Biotechnol Bioeng 1988; 32:947–965.
9. Gramer MJ, Goochee CJ. Glycosidase activities in Chinese-Hamster ovary cell lysate and cell-culture supernatant. Biotechnology Progress 1993; 9:366–373.
10. Frame KK, Hu WS. The loss of anitbody productivity in continuous cultures of hybridoma cells. Biotechnol Bioeng 1990; 35:469–476.
11. Ozturk SS, Palsson BO. Loss of antibody productivity during long-term cultivation of a hybridoma cell line in low serum and serum free media. Hybridoma 1990; 9:167–175.
12. Bentley WE, Kompala DS. Stability in continuous cultures of recombinant bacteria. Biotechnol Lett 1990; 12:329–334.

13. Ozturk SS. Engineering challenges in high density cell culture systems. Cytotechnology 1996; 22:3–16.
14. Woodside SM, Bowen BD, Piret JM. Mammalian cell retention devices for stirred perfusion bioreactors. Cytotechnology 1998; 28:163–175.
15. Voisard D, Meuwly F, Ruffieux P-A, Beaer G, Kadourin A. Potential of cell retention techniques for large-scale high-density perfusion culture of suspended mammalian cells. Biotechnol Bioeng 2003; 82:751–765.
16. Spier RE, Whiteside JP. The production of foot-and-mouth disease virus from BHK 21 C13 cells grown on the surface of glass spheres. Biotechnol Bioeng 1976; 18:649–657.
17. Griffiths JB, Thornton B, McEntee I. The development and use of microcarriers and glass sphere culture techniques for the production of Herpes simplex virus. Dev Bio Stand 1982; 50:103–110.
18. Kang SH, Kim BG, Lee GM. Justification of continuous packed bed reactor for retroviral vector production from amphotropic OCRIP muring producer cell. Cytotechnology 2000; 34:151–158.
19. Burbidge C. The mass culture of human diploid fibroblast in packed beds of glass beads. Dev Bio Stand 1980; 46:169–172.
20. Brown PC, Figueroa C, Costello MAC, Oakley R, Maciukas SM. Protein production from mammalian cells grown on glass beads. In: Spier RE, Griffiths JB, eds. Animal Cell Biotechnology. Vol. 3. London: Academic Press, 1988:251–262.
21. Wang G, Zhang W, Freedman D, Eppstein L, Kadouri A. Continuous production of monoclonal antibodies in Celligen packed bed reactor using Fibra-Cel carrier. In: Spier RE, Griffiths JB, MacDonald C, eds. Animal Cell Technology: Developments, Process and Products. Butterworth-Heinemann, 1992:460–464.
22. Kaufman JB, Wang G, Zhang W, Valle MA, Shiloach J. Continuous production and recovery of recombinant Ca^{2+} binding receptor from HEK 293 cells using perfusion through a packed bed bioreactor. Cytotechnology 2000; 33:3–11.
23. Griffiths JB, Looby D, Racher AJ. Maximisation of perfusion systems and process comparison with batch-type cultures. Maximisation of perfusion cultures. Cytotechnology 1992; 9:3–9.
24. Whiteside JP, Spier RE. The scale-up from 0.1 to 100 litre of a unit process system based on 3 mm diameter glass spheres for the production of four strains of FMDV from BHK monolayer cells. Biotechnol Bioeng 1981; 23:551–565.
25. Pugh GG, Berg GJ. Two ceramic matrices for the long-term growth of adherent or suspension cells. In: Moody GW, Baker PB, eds. Bioreactors and Biotransformations. London: Elsevier, 1983:121–131.
26. Lydersen BK, Pugh GG, Paris MS, Sharma BP, Noll LA. Ceramic matrix for large scale animal cell culture. Biotechnology 1985; 3:63–67.
27. Berg GJ, Bödeker BGD. Employing a ceramic matrix for the immobilization of mammlian cells in culture. In: Spier RE, Griffiths JB, eds. Animal Cell Biotechnology. Vol. 3. Academic Press, 1988:321–335.
28. Bödeker BGD, Newcomb R, Yuan P, Braufman A, Kelsey W. Production of recombinant factor VIII from perfusion culture: I Large-scale fermentation. Animal Cell Technology: Products of Today, Prospects for Tomorrow, 1994.
29. Tharakan JP, Gallagher SL, Chau PC. Hollow-fiber bioreactors for mammalian cell culture. Adv Biotechnol Processes 1988; 7:153–184.
30. Piret JM, Cooney CL. Immobilized mammalian cell cultivation in hollow fiber bioreactors. Biotechnol Adv 1990; 8:763–783.
31. Hopkinson J. Hollow fiber cell culture systems for economical cell-product manufacturing. Biotechnology 1985; 3:225–230.
32. Piret JM, Devens DA, Cooney CL. Nutrient and metabolite gradients in mammalian hollow-fiber bioreactors. Can J Chem Eng 1991; 69:421–428.
33. van Brunt J. Immobilized animal cells: the gentle way to productivity. Biotechnology 1986; 4:505–510.

34. Tharakan JP, Chau PC. A radial flow hollow fiber reactor for the large-scale culture of mammalian cells. Biotechnol Bioeng 1989; 28:329–342.
35. Lim F, Sun AM. Microencapsulated islets as bioartificial endocrine pancreas. Science 1980; 210:908–910.
36. Rupp RG. Use of cellular microencapsulation in large-scale production of monoclonal antibodies. In: Feder J, Tolbert WR, eds. Large-Scale Mammalian Cell Culture. Orlando: Academic Press, 1985:19–38.
37. King GA, Kaugulis AJ, Faulkner P, Goosen MFA. Alginate-polylysine microcapsules of controlled membrane molecular weight cutoff for mammalian cell culture engineering. Biotechnol Prog 1987; 3:231–340.
38. Daly MM, Knorr D. Chitosan-alginate complex coacervate capsules; effects of calcium chloride, plasticizers, and polyelectrolytes on mechanical stability. Biotechnol Prog 1988; 4:76.
39. Poncelet D, de Smet BP, Beaulieu C, Neufeld RJ. Scale-up of Gel Bead and Microcapsule Production in Cell Immobilization. Boca Raton: CRC Press, 1993.
40. Van Wezel AL, van der Velden-de Groot CAM, Haan HH, van den Heuvel N, Schasfoort R. Large scale animal cell cultivation for production of cellular biologicals. In: Karger S., ed. Developmental Biology Stand, Vol. 60. 1985:229–236.
41. Karkare SB, Phillips PG, Burke DH, Dean RC. Continuous Production of Monoclonal Antibodies by Chemostatic and Immobilized Hybridoma Culture. Orlando: Academic Press, 1985.
42. Nilsson K, Buzsaky F, Mosbach K. Growth of anchorage dependent cells on macroporous microcarrier. Biotechnology 1986; 4:989–990.
43. Runstadler PW, Ozturk SS, Ray NG. An evaluation of hybridoma cell-specific productivity: perfused immobilized, continuous suspension, and batch suspension Cultures. In: Murakami H, ed. Animal Cell Technology: Basic Applied Aspects. Dordrecht, The Netherlands: Kluwer Academic Publishers, 1992.
44. Himmelfarb P, Thayer PS, Martin HE. Spin filter culture: the propagation of mammalian cells in suspension. Science 1969; 164:555–557.
45. Tolbert WR, Feder J, Kimes RS. Large scale rotating filter perfusion system for high density growth of mammalian suspension cultures. In Vitro 1981; 17:885–890.
46. Deo YM, Mahadevan MD, Fuchs R. Practical considerations in operation and scale-up of spin-filter based bioreactors for monoclonal antibody production. Biotechnol Prog 1996; 12:57–64.
47. Yabannavar VM, Singh V, Connelly NV. Scaleup of spinfilter perfusion bioreactor for mammalian cell retention. Biotechnol Bioeng 1994; 43:159–164.
48. Avgerinos GC, Drapeau D, Socolow JS, Mao J, Hsiao K, Broeze RJ. Spin filter perfusion system for high density cell culture: production of recombinant urinary type plasminogen activation in CHO cells. Biotechnology 1990; 8:54–58.
49. Yabannavar VM, Singh V, Connelly NV. Mammalian cell retention in a spinfilter perfusion bioreactor. Biotechnol Bioeng 1992; 40:925–933.
50. Esclade LRJ, Carrel S, Peringer P. Influence of the screen material on the fouling of spin filters. Biotechnol Bioeng 1991; 38:159–168.
51. Velez D, Miller L, Macmillan DJ. Use of tangential flow filtration in perfusion propagation of hybridoma cells for production of monoclonal antibodies. Biotechnol Bioeng 1989; 33:938–940.
52. Maiorella B, Dorin G, Carion A, Harano D. Crossflow microfiltration of animal cells. Biotechnol Bioeng 1991; 37:121–126.
53. De la Broise D, Noiseux M, Lemieux R. Long-term perfusion culture of hybridoma: a "grow or die" cell cycle system. Biotechnol Bioeng 1991; 38:781–787.
54. Shi Y, Ryu DDY, Park SH. Monoclonal antibody productivity and the metabolic pattern of perfusion cultures under varying oxygen tensions. Biotechnol Bioeng 1993; 42:430–439.

55. S. Zhang Z, Handa-Corrigan A, Spier RE. A comparison of oxygenation methods for high-density perfusion cultures of animal cells. Biotechnol Bioeng 1993; 41:685–692.
56. De la Broise D, Noiseux M, Massie B, Lemieux R. Hybridoma perfusion systems: a comparison study. Biotechnol Bioeng 1992; 40:25–32.
57. Mercille S, Johnson M, Lemieux R, Massie B. Filtration-based perfusion of hybridoma cultures in protein-free medium: reduction of membrane fouling by medium supplementation with DNase I. Biotechnol Bioeng 1994; 43:833–846.
58. Roth G, Smith CE, Schoof GM, Montgomery JLA, Horwitz JI. Using an external vortex flow filtration device for perfusion cell culture. BioPharm 1997; 30–35.
59. Furey J. Continuous cell culture using the ATF system. Genetic Eng News 2000; 20: 52–53.
60. Kitano K, Shintani Y, Ichimori Y, Tsukamoto K, Sasai S, Kida M. Production of human monoclonal antibodies by heterohybridomas. Appl Microbiol Biotechnol 1986; 24:282–286.
61. Takazawa Y, Tokashiki M. High cell density perfusion culture of mouse–human hybridoma. Appl Microbiol Biotechnol 1989; 32:280–284.
62. Shitani Y, Kohno Y-I, Sawada H, Kitano K. Comparison of culture methods for human–human hybridomas secreting anti-HBsAg human monoclonal antibodies. Cytotechnology 1991; 6:197–208.
63. Hülscher M, Scheibler U, Onken U. Selective recycle of viable animal cells by coupling of airlift reactor and cell settler. Biotechnol Bioeng 1992; 39:442–446.
64. Boycott AE. Sedimentation of blood corpuscles. Nature 1920; 104:532–534.
65. Batt BC, Davis RH, Kompala DS. Inclined sedimentation for selective retention of viable hybridomas in a continuous suspension bioreactor. Biotechnol Prog 1990; 6:458–464.
66. Brown PC, Lane JC, Wininger MT, Chow R. On-line removal of cells from continuous suspension culture. In: Spier RE, Griffiths JB, Meignier B, eds. Production of Biologicals from Animal Cells in Culture. Butterworth-Heinemann, 1991:416–420.
67. Gray DR, Chen S, Howarth W, Inlow D, Maiorella BL. CO_2 in large-scale and high-density CHO cell perfusion culture. Cytotechnology 1996; 22:65–78.
68. Stevens J, Eickel S, Onken U. Lamellar clarifier- a new device for animal cell retention in perfusion culture systems. In: Spier RE, Griffiths JB, Berthold W, eds. Animal Cell Technology: Products of Today, Prospects for Tomorrow. Butterworth-Heinemann, 1994:234–239.
69. Thompson KJ, Wilson JS. A compact gravitational settling device for cell retention. In: Spier RE, Griffiths JB, Berthold W, eds. Animal Cell Technology: Products of Today, Prospects for Tomorrow. Butterworth-Heinemann, 1994:227–229.
70. Searles JA, Todd P, Kompala DS. Viable cell recycle with an inclined settler in the perfusion culture of suspended recombinant Chinese hamster ovary cells. Biotechnol Prog 1994; 10:198–206.
71. Searles JA, Todd P, Kompala DS. Perfusion culture of suspended CHO cells employing inclined sedimentation. In: Spier RE, Griffiths JB, Berthold W, eds. Animal Cell Technology: Products of Today, Prospects for Tomorrow. Butterworth-Heinemann, 1994:240–242.
72. Ozturk SS. Challenges and opportunities for high density perfusion cultures, Cancun, Mexico, 2004. Paper presented in Cell Culture Engineering, Cancun, Mexico, 2004.
73. Tyo M, Spier, Raymond E. Dense cultures of animal cells at the industrial scale. Enzyme Microb Technol 1987; 9:514–520.
74. Knaack C, Andre G, Chavarie C. Conical bioreactor with internal lamella settler for perfusion culture of suspension cells. Animal Cell Technology: Products of Today, Prospects for Tomorrow, 1994.
75. Hittorff M. Ruckhaltung von Partikeln in Kontinuierlich Durchstromten Reactoren. Diploarbeit, Universitat Dartmund, 1990.

76. Tokashiki M, Arai T, Hamamoto K, Ishimaru K. High density culture of hybridoma cells using a perfusion culture vessel with external centrifuge. Cytotechnology 1990; 3:239–244.
77. Jäger V. A novel perfusion system for the large-scale cultivation of animal cells based on a continuous flow centrifuge. In: Spier RE, Griffiths JB, MacDonald C, eds. Animal Cell Technology: Developments, Processes & Products. Butterworth-Heinemann, 1992:397–402.
78. Jäger J. High density perfusion culture of animal cells using a novel continuous flow centrifuge. Murakami H, ed. Animal Cell Technology: Basic and Applied Aspects. Dordrecht, The Netherlands: Kluwer Academic Publishers, 1992:209–216.
79. van Wie BJ, Broouns TM, Elliot ML, Davis WC. A novel continuous centrifugal bioreactor for high-density cultivation of mammalian and microbial cells. Biotechnol Bioeng 1991; 38:1190–1202.
80. Johnson M, Lanthier S, Massie B, lefebvre G, Kamen AA. Use of the centritech lab centrifuge for perfusion culture of hybridoma cells in protein-free medium. Biotechnol Prog 1996; 12:855–864.
81. Björling T, Dudel U, Fenge C. Evaluation of a cell separator in large-scale perfusion culture. In: Beuvery EC, Griffiths JB, Zeijlemaker WP, eds. Animal Cell Technology: Developments Towards the 21st Century. Dordrecht, The Netherlands: Kluwer Academic Publishers, 1995:671–675.
82. Chatzisavido N, Björling T, Fenge C, Boork S, Lindner-Olsson E, Apelman S. K K, ed. A Continuous Cell Centrifuge for Lab Scale Perfusion Processes of Mammalian Cells. Dordrecht, The Netherlands: Kluwer, 1994:463–468.
83. Takamatsu H, Hamamoto K, Ishimaru K, Yokoyama SMT. Large-scale perfusion culture process for suspended mammalian cells that uses a centrifuge with multiple settling zones. Appl Microbiol Biotechnol 1996; 45.
84. Wrotnowski C. Major trends in technologies for cell culture. Gen Eng News 2000; 20: 8–58.
85. Kilburn DG, Clarke DJ, Coakley WT, Bardsley DW. Enhanced sedimentation of mammalian cells following acoustic aggregation. Biotechnol Bioeng 1989; 34:559–562.
86. Trampler F, Sonderhoff SA, Pui PWS, Kilburn DG, Piret JM. Acoustic cell filter for high density perfusion culture of hybridoma cells. Biotechnology 1994; 12:281–284.
87. Doblhoff-Dier O, Gaida T, Katinger H, Burger W, Groeschl M, Benes E. A novel ultrasonic resonance field device for the retention of animal cells. Biotechnol Prog 1994; 10:428–432.
88. Gorenflo VM, Piret JM. Optimization of large-scale ultrasonic filter performance of a 100 L CHO cell perfusion bioreactor, Poster II-47. In: Engineering Foundation Conference on Cell Culture Engineering VIII Snowmass, Colorado, 2002.
89. Yuan H, Rickwood D, Smyth IC, Thew MT. An investigation into the possible use of hydrocyclones for the removal of yeast from beer. Bioseparation 1996; 6:159–163.
90. Jockwer A, Medronho RA, Wagner R, Anspach FB, Deckwer W-D. The use of hydrocyclones for mammalian cell retention in perfusion bioreactors. In: Linder-Olsson E, ed. Animal Cell Technology: From Targets to Market. Dordrecht, The Netherlands: Kluwer Academic Publishers, 2001.
91. Ozturk SS. Scale-up and optimization of high density cell culture bioreactors. In: Ramirez EGaOT, ed. Advances in Bioprocess Engineering. Dordrecht, The Netherlands: Kluwer Academic Press, 1994.
92. Runstadler PW, Ozturk SS, Ray NG. An evaluation of hybridoma cell-specific productivity: perfused immobilized, continuous suspension, and batch suspension cultures. In: Murakami H, ed. Animal Cell Technology: Basic Applied Aspects. The Netherlands: Kluwer Academic Publishers, 1992.
93. Runstadler PW, Tung AS, Ray NG. Fluidized-bed bioreactor perfusion culture for the production of therapeutic proteins. In: Spier RE, Griffiths JB, MacDonald C, eds.

Animal Cell Technology: Developments, Process and Products. Butterworth-Heinemann, 1992:495–500.
94. Ozturk SS, Blackie J, Wu P, Taticek R, Matanguihan C, Tsai Y, Yoon S, Thrift J, Naveh D. On-Line Monitoring for Consistent and Optimal Production of Biologicals from Mammalian Cell Cultures. Dordrecht, The Netherlands: Kluwer Academic Publishers, 1996.
95. Ozturk SS, Thrift JC, Blackie JD, Naveh D. Real-time monitoring and control of glucose and lactate concentrations in a mammalian cell perfusion reactor. Biotechnol Bioeng 1997; 53:372–378.
96. Wu P, Ozturk SS, Blackie JD, Thrift JC, Figueroa C, Naveh D. Evaluation and applications of optical cell density probes in mammalian cell bioreactors. Biotechnol. Bioeng 1995; 42:312–323.
97. Ozturk SS, Blackie J, Thrift J, Naveh D. Perfusion rate control based on on-line monitoring of oxygen consumption rates in mammalian cell fermentation. In ACS National Meeting, San Diego, CA, 1995.
98. Kyung Y.-S, Peshwa MV, Gryte DM, Hu W.-S. High density culture of mammalian cells with dynamic perfusion based on on-line oxygen uptake rate measurements. Cytotechnology 1994; 14:183–190.
99. Ozturk SS, Blackie J, Thrift J, Naveh D. Real-time monitoring of mammalian cell fermentation for monoclonal antibodies using a computer controlled immunoDetection (BioCad/RPM) System. Biotechnol Bioeng 1996; 48:95–101.
100. Lipscomb ML, Mowry MC, Kompala DS. Production of a secreted glycoprotein reporter from an inducible promoter system in a perfusion bioreactor. Biotechnol Prog 2004; 20:1402–1407.
101. Ducommun P, Ruffieux P, Furter M, Marison I, von Stockar U. A new method for on-line measurement of the volumetric oxygen uptake rate in membrane aerated animal cell cultures. J Biotechnol 2000; 78:139–147.
102. Ruffieux PA, von Stockar U, Marison IW. Measurement of volumetric (OUR) and determination of specific (qO_2) oxygen uptake rates in animal cell cultures. J Biotechnol 1998; 63:85–95.
103. Jorjani P, Ozturk SS. Effect of Temperature and Cell Density on the Respiration Rate of Different Cell Lines. Biotechnol Bioeng 1999; 349–356.
104. Buntemeyer H, Marzahl R, Lehmann J. A direct computer control concept for mammalian cell fermentation processes. Cytotechnology 1994; 15:271–279.
105. Kimura R, Miller WM. Effects of elevated pCO_2 and/or osmolality on the growth and recombinant tPA production of CHO cells. Biotechnol Bioeng 1996; 52:152–160.
106. Michaels JD, Nowak JE, Mallik AK, Koczo K, Wasan DT, Papoutsakis ET. Analysis of cell-to-bubble attachment in sparged bioreactors in the presence of cell-protecting additives. Biotechnol Bioeng 1995; 47:407–419.
107. Ozturk SS. Development of Highly Effective Cell Culture Perfusion Bioreactors for Commercial Production of Biologicals: Past, Present, and Future (Invited Lecture). In 3rd Recombinant Protein Production Meeting, Tavira, Algarve, Portugal, 2004.

12
Cell Separation and Product Capture

Thomas Seewoester
Amgen, Thousand Oaks, California, U.S.A.

INTRODUCTION

This chapter focuses on the different approaches for separating mammalian cells from cell culture medium containing recombinant proteins of therapeutic interest and their further primary isolation. Today the most common products from mammalian cells are secreted proteins. The isolation of the target protein requires in most cases the initial removal of particles of different physical and chemical properties such as cells and cell debris (1,2). This clarification step describes the transition from product generation to product isolation (3). The unit operation has to perform a proper solid–liquid phase separation of the cell suspension in order to enable further product purification, as the subsequent steps in general require feed streams with low particle occurrence. Upon the removal of larger particles such as cells and debris the clarified cell supernatant undergoes a primary purification step. The main objective of this unit operation is to remove proteins and other molecules with very different physical and chemical character and capture the target protein in a preferably significant reduced liquid volume. If both objectives, capture and volume reduction, cannot be accomplished in one step, an ultra- or diafiltration is performed upon initial capture in order to reduce the liquid volume and transfer the target protein into appropriate buffer conditions for the subsequent fine purification process step. The following paragraphs will give an overview of different methods and established techniques used for the removal of cells and cell debris and the subsequent primary purification of the target protein. Special emphasis is put on techniques developed and applied to clinical and commercial manufacturing of recombinant proteins and patents related to those operations.

REMOVAL OF CELLS AND CELL DEBRIS

The majority of commercially manufactured recombinant products are secreted proteins and the separation of cells or cell debris from the product of interest is an essential process step (4). However, some applications such as virus production schemes require the harvest of intact cells, as the product remains in most cases intracellular during its expression. In other cases the protein might be associated with the cell membrane like for membrane-anchored receptors. Given the fragility and high value of cell-derived products, cell separation must be accomplished with minimum cell damage and loss of product activity. Choosing the optimal clarification process can be challenging (5) even for microbial-derived products (6). Although the cell mass in mammalian cell systems is lower than in yeast or microbial-based production processes, a considerable amount of solids has to be removed when operating at a commercial manufacturing scale. The following paragraphs will present in more detail the most common techniques used for separating mammalian cells.

Depth Filtration

Depth filtration is the preferred normal-flow filtration method for removal of cells as the anisotropic character of the filter unit allows a significant higher solid holding capacity than isotropic membranes used for sterile filtration of liquids. Some filters used for clarification combine anisotropic character, large area and capacity with a second layer of isotropic character, which allows sterile filtration quality of the filtrate (7–9). Depth filters are no absolute filters. Their highly porous structure allows only a filter rating over a wide range of particle sizes and it enables gas to pass through the matrix. Further details on the manufacturing and structure of depth filters (Fig. 1) can be found in Singhvi et al. (10) and several patents from Cook et al. (11–13). In contrast to other cell separation methods, depth filtration is applied only if the product of interest is secreted and can pass the filtration membrane, as cells cannot be recovered from the filter matrix. Depth filtration systems generally consist of a series of filter cartridges (Fig. 2) with decreasing pore size rating, thus protecting and extending the life of the following filter cartridge. As depth filters are not considered sterile filters, the final filter of a depth filtration cascade is usually an absolute 0.2 or 0.1 μm filter. Choosing the correct type and combination of filter cartridge for each application is challenging, but ultimately will lead to an optimized filtration train and reduced overall filtration costs. Depth filters contain diatomaceous earth and are in most cases positively charged. Therefore, separation across depth filtration membranes is based on size exclusion and adsorption. This has to be considered for the correct filter sizing as cells, contaminants, impurities or even product can adsorb to the filter matrix (14,15). Filter sizing for depth filters should be undertaken using the volume end point (V_{max}), pressure end point (P_{max}) and turbidity end point

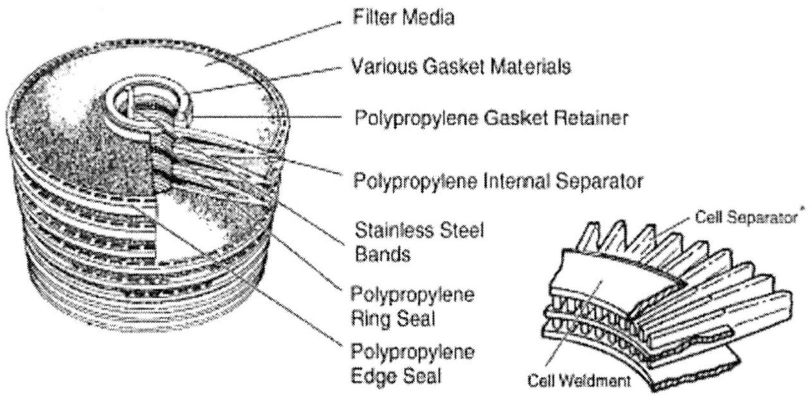

Figure 1 Structure of depth filter. (Courtesy of CUNO Corporation.)

Figure 2 Depth filter and housings. (Courtesy of CUNO Corporation.)

(T_{max}) methods (16). Small-scale studies on CHO cells expressing an IgG antibody showed no loss in product concentration and total quantity upon depth filtration clarification (10). However, when depth filtration is used in a larger manufacturing scale, a reduced product concentration and some loss of material should be expected due to hold up volumes in the equipment and remaining water from the initial and necessary water rinse of the depth filter. A step yield of about 95% should be expected for process volumes of more than 1000 L. More information on large-scale filtration cartridges can be found in recent patents from Millipore (17,18). Depth filtration used for cell removal has several advantages when compared with other methods such as tangential flow filtration (TFF) or centrifugation. In particular, the ease of use, the low initial cost of disposable filter units and the ease of validation allow a rapid and robust development of the clarification step. However, once the manufacturing scale reaches liquid volumes of more than 1000–3000 L, the costs of the disposable depth filter units could become inhibiting. At this scale TFF or centrifugation will become the preferred way for cell removal and these techniques might replace an initial depth filtration process.

Tangential Flow Filtration

Membranes have always been an integral part of biotechnology processes. TFF or cross-flow microfiltration competes with centrifugation, depth filtration and expanded-bed chromatography for the initial harvest of therapeutic products from mammalian, yeast, and bacterial cell cultures (19). In a cross-flow microfiltration the feed stream is applied in a tangential flow across a separation membrane in order to reduce fouling and clogging of the membrane. The filter unit is incorporated into a

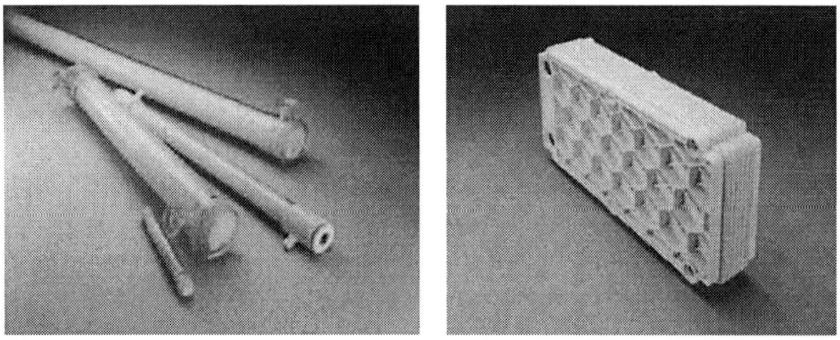

Figure 3 Different configurations of tangential flow filtration modules. (Courtesy of Millipore Corporation.)

loop system and the feed stream passes the membrane several times. Filtration generally occurs at low pressure with high permeation fluxes across the membrane. As a result, the concentration of particles that cannot pass the membrane increases over time. The filtrate is particle-free and in most cases of 0.2 μm filtered quality thus allowing a subsequent chromatography step without any further particle removal. TFF units are available in various configurations such as hollow fiber, spiral wound, flat sheet, tubular, and capillary (Figs. 3 and 4). A comparison of configuration characteristics and a generic approach to choose the best-suited configuration can be found in a review article from Belfort et al. (20). The most common configuration for cell removal is probably the hollow fiber. Hollow fiber cartridges with larger inner diameter prevent the unit from clogging with cells, however the filtration area per unit decreases with increasing lumen diameter. Robert van Reis, one of the leading authors in this field reported on systems designed for harvest volumes of 12,500 L. The re-circulation rate in the feed stream loop operates at 33,000 L/hr with

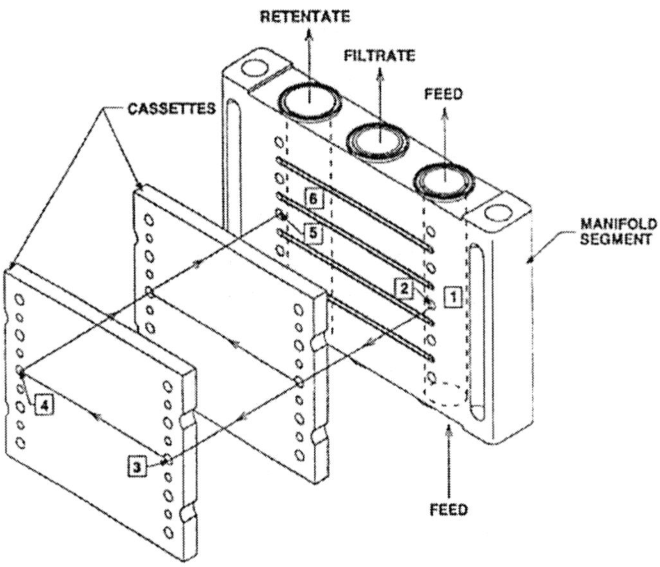

Figure 4 Schematic of plate and frame TFF unit showing liquid flow paths. [From Ref. (127).]

a filtration rate of 4,800 L/hr. As the filtration process equals a cell concentration generating a cell sludge, the step yield can be improved by washing the cell sludge with additional buffer. Theoretical step yields of more than 99% can be achieved with two diavolumes resulting in a moderate dilution of the feed stream by 7% (21). Although TFF applies low shear stress to intact cells within the filtration unit, the circulation pump in the feed stream loop becomes a critical and often limiting equipment unit. However, turbulent flow reaching Reynolds numbers of up to 71,000 is reported without causing cell damage (22). The study concludes that if cell lysis is of no concern filtration can be conducted at high shear rates and in turbulent flow using conditions like those typically applied for bacterial cell concentration. But if cell damage needs to be avoided the tangential flow rate should be limited to an average shear rate of <3000/sec, and transmembrane pressure must be limited to avoid cell deformation into the filter pores.

Tangential flow filtration has several advantages such as potentially 100% cell recovery, a simple product-washing step and a good biological containment. The temperature control is easy and the filtration capacity can be increased rapidly by adding additional modules. The disadvantages on the other hand range from time-dependent permeate flux and long retentate residence times to limited control over separation performance. In some cases the costs for membrane cleaning and replacement could become an additional disadvantage. In order to address some of those disadvantages several improvements were made in recent years. Luque et al. (23) and Gehlert et al. (24) reported on improved filtration performance using helically coiled hollow fiber modules. Chang et al. (25) describe proper selection of operating conditions such as shear rate, flux, membrane pore size, and membrane chemistry. Kwon et al. determined the critical flux at pore sizes ranging from 0.46 to 11.9 μm (26) and Li et al. assessed depolarization models (27). Other authors investigated the mechanism of membrane fouling (28,29), impact of fine particles (30) or developed optimization diagrams (31). Backpulsing has been found as an effective way to reduce membrane fouling (32) and charged membranes (33) or the addition of filtration aids to the feed stream (34) can further improve the clarification performance. Some of the recent patents in this field may serve as a valuable additional literature source (35–51). Modifications of the traditional TFF principle use different rotating filter configurations as reported by Kroner et al. (52), Parnham et al. (53) and Miller et al. (54,55). Vogel and Kroner (56) describe a system with a rotating conical-shaped rotor above a stationary filtration membrane to apply a hydrodynamic lift to cells, which minimizes concentration polarization. Besides the application of cross-flow microfiltration toward cell removal, TFF is widely established for ultrafiltration processes. Observations and improvements made on ultrafiltration processes are described later in this chapter, but can be adapted in many cases to mammalian cell removal processes as well.

Sedimentation

Sedimentation utilizes the density differences between cells or particles and their surrounding medium. Under normal gravitational force the settling velocity of cells or particles in cell culture medium is fairly low. This prevents sedimentation to act as an effective way to remove cells rapidly during a harvest process. Exceptions are processes, which require adherent cells cultivated on microcarriers. The density and mass increase of a confluent microcarrier results in a significantly higher settling velocity and allows for a reasonably short settling and harvest time even at a commercial scale of 2000 L. Gravitational settlers are used only in perfusion processes (57) where a gentle but robust cell separation and retention is required. Various external settling devices are described (58,59), but an external inclined settler is the most common settling device in commercial manufacturing. External settling devices are usually cooled to decrease cell metabolism, as the cell retention time in the settling loop can be several hours. The slow operation of settlers is circumvented by increased settling area, thus allowing even perfusion rates of 10 volumes per volume and day. Some novel types of compact settlers have been designed, which allow for a separation between viable and dead cells. These systems take advantage of the higher settling velocity of viable cells over nonviable cells, which shrink upon death (60,61). An interesting variation is the sedimentation field-flow fractionation (62) and the sedimentation in a specially designed elongated chamber (63) both applied to blood cells. Overall, the sedimentation technique is most appropriate for selective removal of nonviable cells in longlasting continuous cultivation systems due to the robust nature of the technique and simple equipment. For rapid and quantitative removal of viable and nonviable cells as well as cell debris the settling velocity has to be increased by applying higher gravitational forces to the feed stream. This is the main principle of centrifugation.

Centrifugation

The solid concentrate produced by centrifugation differs from that produced by filtration. At best centrifugation produces a cell paste, but often it yields only a concentrated suspension. Filtration in contrast produces a relatively dry cake thus minimal product loss, which is the major advantage. However, many biological feeds that can be centrifuged cannot be effectively or economically filtered, so that centrifugation is often a very attractive alternative. The five basic types of centrifuges are tubular bowls, disk stack type, basket type, decanter, and disk decanter.

Tubular bowl centrifuges provide very high centrifugal forces (64) resulting in a good dewatering performance. The bowl can be cooled easily, a real advantage in protein work. The liquid feed stream is applied through the bottom and clarified liquid is removed from the top. Solids such as cells deposit on the bowl's wall as a thick paste until the limited solid holding capacity of the bowl is reached. Then the bowl must be dismantled and cleaned, which might be a significant disadvantage for large-scale operations. Tubular bowl centrifuges are usually used to continually separate two liquid streams of different density when the solid content is small (64). Wang et al. (65) reported on minimal decrease in cell viability using a tubular centrifuge generating *g*-forces of 19,000 *g*.

Disk stack centrifuges (Fig. 5) on the other hand offer continuous operation within a compact volume and the variability in the conical disk stacks allows for large sedimentation areas combined with flexibility depending on the feed streams. The feed usually enters at the top, is accelerated very quickly (66) and the clarified liquid flows out at an annular slid near the feed. Disk stack types can be distinguished between the methods of solid discharge. Solids are either removed intermittently, as in the tubular bowl, involving complete stop of the separation, or solids can be discharged continuously or semi continuously, out of orfices on the side of the centrifuge. The disk-nozzle type allows a true continuous discharge of solids through nozzles in the underflow and liquid through gravity in the underflow. The desludger type on the other hand periodically closes off solid withdrawal, thus causing more compaction and dewatering than the disk-nozzle type. Disk centrifuges have a very

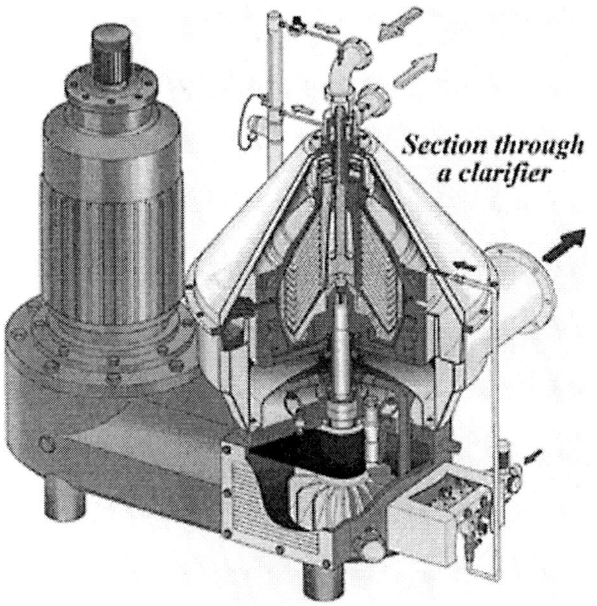

Figure 5 Cross section through a continuous disk stack centrifuge. (Courtesy of Westfalia Separator.)

high liquid throughput, but if the solids are continuously discharged the liquid content of the discharge is higher than a tubular bowl design, a potential disadvantage if good dewatering performance is required. Continuous disk stack centrifuges are used in perfusion processes (67) or lysis-free cell separation (68). Takagi et al. (69) found sustained productivity of a tPA-expressing CHO cell when very low *g*-forces of 67 *g* were applied. Scale-down models are described in Mannweiler et al. (70).

The third type of centrifuge is the basket type, which is essentially a combination of a tubular centrifuge and a filter. It consists of a rapidly rotating perforated basket on which the filter cake accumulates. Liquid passes under centrifugal forces through the cake and perforation and leaves the centrifuge. This type of centrifuge is very efficient for washing accumulated solids such as cells.

The fourth centrifuge type is the decanter (71), which is also used in truly continuous processes. A decanter can handle a large range of feed concentrations resulting in a dry solid cake. A disadvantage is the decanter's inability to obtain good overflow quality, which means solids are carried into the liquid phase, thus resulting in a less clarified liquid phase (64). To obtain a dry solid and a clear liquid, a combination of decanter followed by a disk centrifuge is used, which lead to a fifth type of centrifuge, the disk decanter.

The disk decanter is a combination of decanter and disk centrifuge in one single unit. The feed stream first passes through a disk stack and the ejected sludge enters the outer decanter zone where the sludge is further separated to form a dry cake. Although the material strength limitations do not allow for high rotational speeds and therefore only for a limited range of *g*-forces, the disk decanter can handle large and variable feed concentrations with very good liquid clarity in the overflow (64).

In addition to the described five main centrifuge types several other types were developed for a special application. A widely used centrifuge for perfusion processes is the Centritech centrifuge from Kendro (72–77). The separation chamber consists of a sterile bag stretched over a bowl rotor. Solids accumulate in the lower zone of the bag, whereas the liquid remains in the upper zone. Both zones can be pneumatically separated and the solids and liquids are removed through different outlets of the bag. Sheeler (78) describes a modified fixed angle rotor for continuous-flow centrifugation and various patents focus on improving whole blood separation (79) using different bag configurations (80–92), separated compartments within the centrifuge bowl (93,94), or modifying agents (95).

The above described cell separation techniques ranging from depth filtration over TFF to sedimentation and centrifugation are all utilized in today's commercial manufacturing processes for recombinant proteins. TFF and centrifugation dominate the commercial scale, where depth filtration can be seen in addition at the clinical manufacturing scale. It is difficult, if not impossible, to describe one of the techniques as the best. Every method has its unique advantages and disadvantages. Choosing the one over the other primarily depends on physical factors such as the cell culture process, shear sensitivity of the host cell, desired clarification grade, or liquid volume to be processed. It often depends also on economical factors such as capital budget, time constraints for delivery, installation and validation, or multi-product considerations. Sometimes a capital budget or ease of operation initially drives a decision in one direction, and later during scale-up for commercial lot sizes another cell separation technique is preferred and needs to be implemented. This change can have a significant impact on the final product quality, which can result in additional work and necessary improvements of the fine purification steps to meet established product quality specifications. A potential change of the cell separation step should be considered and discussed early on in each project.

PRODUCT CAPTURE AND PRIMARY PURIFICATION

Product separation usually begins with the separation of biomass from liquid as described in the previous paragraphs. In many cases, the desired product is in the liquid phase and the separated supernatant will undergo directly further purification steps. The biomass is discarded or sold as by-product.

In some cases, especially in microbial fermentations, the product of interest is not secreted and remains intracellular. Releasing the trapped material involves rupturing the cell wall. Different chemical and mechanical methods can be applied to break the cell wall and release the product. The most common methods are osmotic shock, enzyme digestion, solubilization, lipid dissolution, alkali treatment, homogenization, grinding, ultrasonication, and milling. Mechanical techniques are preferred for large-scale applications. As the focus of this chapter is on products from mammalian cells, which are almost all secreted products, the reader interested in microbial fermentations is referred to the literature mentioned at the end of this chapter.

The two objectives of the primary purification step are first to capture the product out of the clarified feed stream with minimal additional binding of by-products or contaminants. The second objective is to significantly reduce the liquid volume for the following fine purification steps, because smaller liquid volumes allow for smaller purification equipment and reduced run times, a major constraint in large-scale operations. To reach both objectives different product capture techniques and methods are used. They range from very economic but less specific methods such as precipitation and extraction to more product-specific but sometimes expensive methods such as affinity chromatography or ion exchange chromatography followed by ultra- and diafiltration. The next few paragraphs will highlight the basic principles and recent developments in those areas.

Precipitation

Protein precipitation describes a process in which the protein of interest is separated from the liquid by adding a reagent to the solution, which in turn forms an insoluble aggregate with the protein of interests. The insoluble precipitate can be then recovered by centrifugation (96). Upon recovery, the initially added reagent is removed again. It is the intention to recover the protein in either an unchanged molecular form or one, which can be readily returned to that form (97). A widely used method for precipitation of proteins is salting-out by salts such as ammonium or sodium sulfate. The salt competes with the protein for water molecules. At high salt concentrations the protein cannot bind enough water molecules to stay in solution and it precipitates out. Although ammonium sulfate is fairly cheap, its corrosive character and difficulty to handle and dispose are sometimes disadvantages for large-scale applications (98). Another method is the isoelectric precipitation (99,100) where the pH of the solution is adjusted to the isoelectric point of the protein of interest. The protein has a net charge of zero and a substantially reduced solubility. The addition of a weak polar solvent such as ethanol or isopropyl alcohol (101) to an aqueous solution of a protein reduces the effective dielectric constant of the solution and the protein solubility decreases considerably. A fourth method is the precipitation by nonionic polymers such as dextrans and polyethylene glycols. The polymer excludes the protein from part of the solution and reduces the effective amount of water available for their solvation. Ionic polyelectrolytes (102) such as alginate or carboxymethylcellulose (103) act similar to flocculating agents (104) with some salting-out and molecular exclusion action, but they have a high probability of causing structural changes to a protein (Fig. 6). Metal ions are used for precipitation because of their high precipitation capability especially at very dilute protein solutions. Metal salts when applied at low concentrations can be removed with subsequent ion exchange resins or through chelating agents. Heat induced precipitation is applied for large-scale plasmid purification (105). Irwine and Tipton (106,107) and Gupta (108) reported on a new method using selective agents for affinity precipitation thus overcoming the unspecific nature of protein precipitation. All described precipitation methods focus on creating a well-balanced, reversible, biochemical environment in which the protein of interest has a very low solubility. It is important to evaluate carefully the impact of physical parameters such as pH (Fig. 7), ion strength or temperature and engineering parameters (109) such as mixing time, heat distribution, or scalability on the overall process performance when precipitation is considered as protein capture step.

Extraction

Extraction takes advantage of the partitioning of a solute between two liquid phases. The protein is more soluble in one of two liquid phases and the protein concentration increases in the liquid phase of higher solubility as a result of depletion from the other liquid phase (110). In many cases one phase is water and the second phase is an organic solvent, which is added to the process to initiate the extraction process. Common organic solvents are butanol, amyl acetate, or polyethylene glycol.

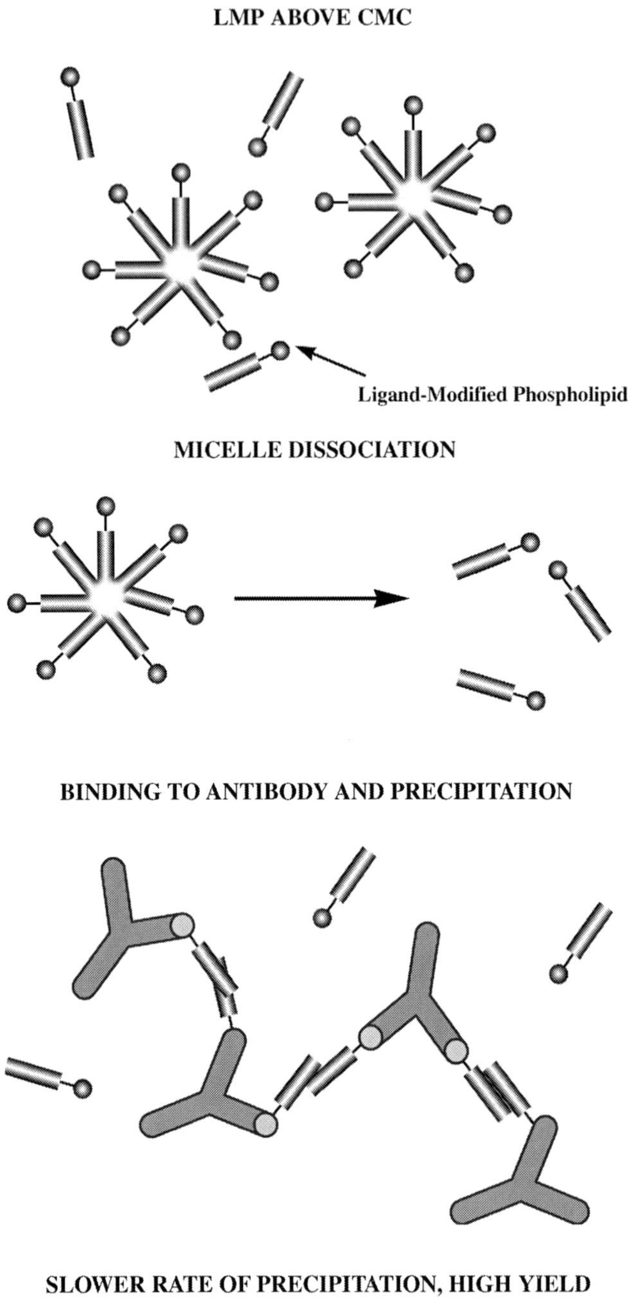

Figure 6 Example of precipitation using ligand-modified phospholipids. [From Ref. (102).]

However, the low specificity of the method when applied as an initial purification step to a complex protein mixture and the problems related to processing large volumes of organic solvents makes extraction not to a preferred choice for manufacturing processes of cell culture-derived proteins. However, extraction offers a very simple method to transfer a protein out of a very diluted aqueous solution into an organic phase, thus reducing the liquid volume and increasing the protein concentra-

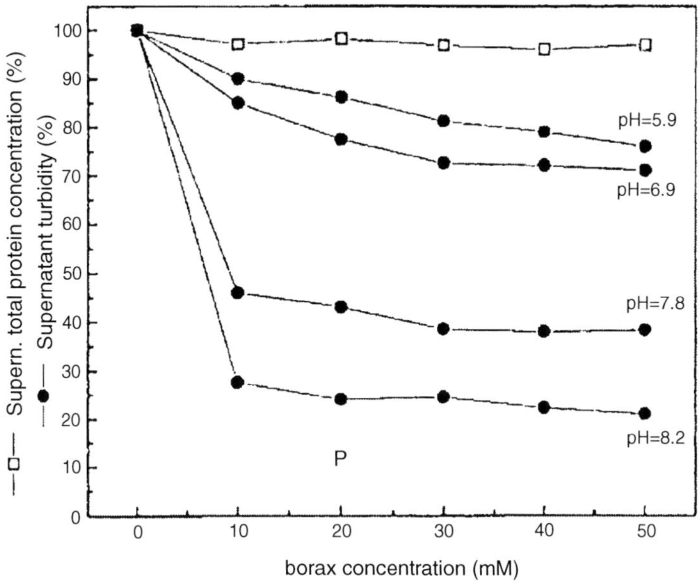

Figure 7 Effect of pH on precipitation by addition of borax. [From Ref. (104).]

tion. The method is used successfully during the initial purification of recombinant proteins derived from transgenic goats (111) and transgenic corn (112) as the starting mixture is less complex than cell culture-derived feed streams (Fig. 8). The extraction process can be optimized by using counterions, which binds to the protein of interest, by altering the pH of the aqueous solution or by mixing at low shear rates (113). Recent developments focused on improving the specificity of the extraction method and elimination of the organic solvent by implementing affinity-based reversed micelles (114,115) composed of unbound Cibacron blue (116) in two aqueous liquid phases. These developments address the major concerns the industry has for extraction with organic solvents and they could lead to significant changes in the way commercial product capture is performed today. Performing extraction in a two-phase aqueous system with high protein specificity would allow a direct combination with other purification techniques, which are also aqueous-based systems.

Ion Exchange Chromatography

Ion exchange chromatography is the most used method to capture a protein out of an aqueous solution. Chapter 13 describes the ion exchange chromatography in greater detail and therefore this paragraph will explain the principle only very briefly. Depending on the p*I* of the protein, either an anion- or a cation-resin is used to bind the protein of interest (117). Other proteins or contaminants that cannot bind to the resin under the biochemical conditions will pass through the column and are discarded. The column is rinsed with buffer until all nonbound molecules have been removed and the bound protein is eluted with a second buffer. Due to a complex feed stream other proteins or contaminants might bind to the column as well and optimization of the loading and elution conditions is very critical. In order to improve the specificity during the elution step buffer gradients are usually very shallow or low pH or salt steps are applied. Very often the protein is already 90% pure, but it is more

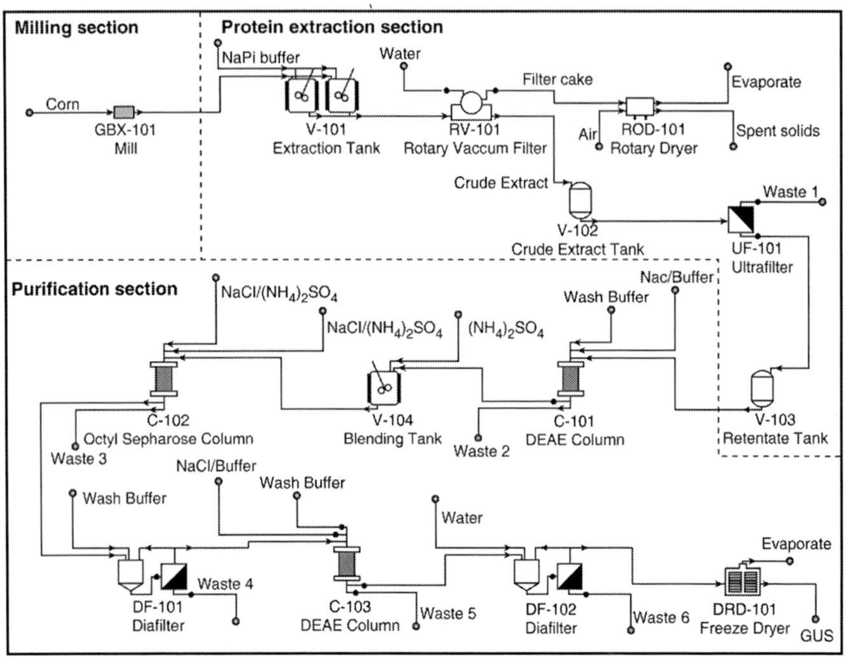

Figure 8 Process flow diagram of purification of recombinant proteinfrom transgenic corn using extraction as first process step. [From Ref. (112).]

diluted than before in an increased total liquid volume. Then the second objective of any product capture process, a significant volume reduction, can only be accomplished by adding an ultrafiltration step.

Ultrafiltration describes a TFF process (118–120). Water can pass the filtration membrane, but the protein of interest cannot, thus increasing the protein concentration and reducing the liquid volume (Fig. 9). Choosing the correct membrane grade (121) for a commercial scale is not only dependent upon the molecular weight of the protein, but also upon shear sensitivity, membrane fouling (122), costs and time constraints. van Reis et al. (123) describe a 400-fold scale-up of an ultrafiltration process at Genentech and large-scale considerations. Several factors influence the performance of an ultrafiltration process and they can be applied easily to any TFF. Burns and Zydney (124) shed light on the effect of the solution pH and Meireles et al. (125) report on effects of operating conditions on membrane fouling. A constant concentration of fully retained protein near the filtration membrane can further improve the filtration performance and product yield (126) as well as charged or modified membranes can help to retain the target protein (127,128). The obtained protein concentrate is either frozen and stored until the fine purification resumes or it is directly purified further. However, often the following purification step requires a lower ionic strength or a complete different buffer for optimal performance. Therefore, it is very common to perform a diafiltration (129) following the ultrafiltration step as both processes use the same equipment. In a diafiltration process the removed liquid volume is constantly replaced with another buffer solution. The total volume remains constant. Other methods such as size exclusion chromatography or counter current dialysis are also applicable for diafiltration, but the TFF is the preferred alternative for commercial manufacturing (130).

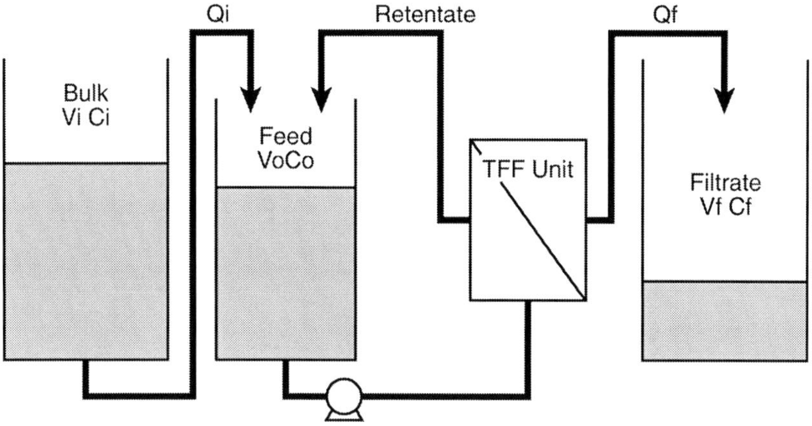

Figure 9 Diagram of tangential flow filtration used for ultra filtration step. [From Ref. (130).]

Affinity Chromatography

In order to further improve the capture step toward product specificity and volume reduction, affinity chromatography became very popular over the last 10 years (131–137). Affinity chromatography refers to the use of immobilized natural ligands (138), which specifically interact with the desired protein (139). The ligand is immobilized on a suitable resin and packed into a column. The protein-containing liquid is passed over the column under conditions under which the protein binds to the ligand. Other proteins or molecules cannot bind to the ligand, pass through, and are discarded. The elution condition, usually a low pH buffer, allows a complete elution of the target protein. Affinity separation is highly specific for the material that binds initially and therefore elution conditions usually do not have to be optimized and tweaked as much as for the previously described ion exchange chromatography, which is less specific for binding properties and requires more optimization of the elution conditions. Very well-defined ligands for the purification of monoclonal antibodies are protein A and protein G (Fig. 10), which can be bound to different resins showing similar purification yields, but differences in capacity and pressure drop (140). A significant disadvantage for large-scale applications of any chromatography step is the limiting flow rate than can be applied to a packed resin. This is especially critical for the initial capture step where large liquid volumes are processed. Two different methods were developed to overcome this problem especially for large-scale applications (141–143).

The expanded bed chromatography allows the affinity resin or any other ion exchange resin to expand in the column up to 20-fold of the packed volume. During the loading process the resin is floating in the liquid stream. Upon loading the resin is mechanically compressed, and the protein is eluted into a small volume. Expanded bed chromatography is applied to various production scenarios such as *Escherichia coli* (144,145), yeast (146,147), or mammalian cells expressing monoclonal antibodies (148–153), interleukin-2 (154), interleukin-2 receptor (155), or interferon (156). Palsson et al. (157) reported on an improved resin that allows for flow rates as high as 3000 cm/hr.

The second alternative to enable rapid processing of larger liquid volumes is an affinity membrane in a cross-flow configuration (158). Affinity membranes combine specific adsorption with filtration. Again, protein A and protein G as well as antibo-

(a)

	Poros 50	Poros LP	Prosep	Sepharose	Stramline
Pressure drop (Pa·h·cm^{-2})	22.1	12.4	2.1	7.6	0.7
Saturation capacity (g/L)	25	24	26	38	29
Dynamic capacity (g/L)	17.5 ± 0.1	14.8 ± 0.3	13.0 ± 0.3	10.9 ± 0.3	7.5 ± 0.1
Purified antibody					
Yield (%)	104 ± 1	106 ± 3	103 ± 2	100 ± 2	105 ± 4
Antibody concentration (g/L)	7.1 ± 0.1	6.4 ± 0.4	5.2 ± 0.2	3.8 ± 0.2	3.2 ± 0.1
DNA (ng/mg)	41 ± 3	48 ± 3	40 ± 4	29 ± 2	98 ± 18
Host-cell proteins (mg/g)	2.5 ± 0.2	2.7 ± 0.7	3.7 ± 0.2	4.9 ± 1.2	6.1 ± 20
Protein A (ng/mg)	4.6 ± 0.5	7.7 ± 0.4	3.1 ± 0.5	5.7 ± 1.7	6.0 ± 1.7
Areal production rate					
R_w (g·h^{-1} cm^{-2})	0.17	0.26	0.38	0.20	–
U_l (cm·h^{-1})	690	1020	1490	720	–
L_c (cm)	10.9	11.3	34.3	28.4	–
U_c (cm·h^{-1})	860	1470	1500	750	–
Q_s (g·L^{-1})	16.7	10.5	14.6	20.6	–
Volumetric production rate					
R_w (g·h^{-1} L^{-1})	17	23	23	13	–
U_l (cm·h^{-1})	700	1000	1000	750	–
L_c (cm)	10.3	11.3	11.1	10.6	–
U_c (cm·h^{-1})	910	1470	1500	740	–
Q_s (g·L^{-1})	16.4	10.6	9.5	5.9	–

(b)

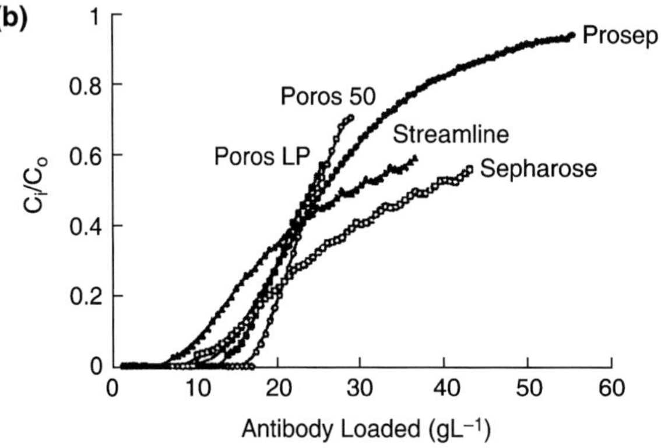

Figure 10 Comparison of various resins used for affinity chromatography. (a) Comparative performance values. (b) Breakthrough curves. [From Refs. (140,149).]

dies are preferred product-specific ligands that can be bound to a large variety of membrane matrices (159). Unfortunately, the cleaning of membranes with protein ligands is difficult, as the ligand has to be still active upon cleaning and sanitization of the capture unit. This led to the development of alternative specific ligands that are less sensitive to the cleaning processes in commercial manufacturing. Zeng (160) reviews several membrane types cross-linked with different specific ligands and Tejeda et al. (161) describe a design optimization based on the Thomas kinetic model. Affinity membranes with expensive protein-based ligands are not common in large-scale manufacturing as potential fouling of the membrane combined with the limited or insufficient cleaning regiments would make this process step uneconomical.

CONCLUSIONS

The primary purification of secreted proteins consists of a cell separation step and a product capture step. Different methods are available for both steps and choosing the right one can be challenging and difficult. After starting with initial biochemical and physical considerations to find a suitable method it is also very important to consider scalability, validation efforts, economical feasibility, and regulatory concerns early on in any project. Further on, the primary purification needs to be integrated into the cell culture as well as the fine purification process to ensure a streamlined and efficient manufacturing process with minimal steps and product losses. This chapter tried to highlight process steps that are common in large-scale manufacturing of biological therapeutics. The focus was on recently published articles, reviews, and patents. For further and also more comprehensive information, the interested reader should also consult books devoted to this topic such as "Bioseparations" from Belter et al. (162), or a chapter from Seely et al. (163). Stanbury et al. (164) give an excellent overview of clarification and capture steps related to microbial systems. The books from Scott and Hughes (165), Wang (166) and Rickwood et al. (167) focus on the individual cell separation steps, whereas Sofer and Hagal (168) give a comprehensive overview on process chromatography. Finally "Biopharmaceutical process validation" from Sofer and Zabriskie (169) describes constraints and hurdles that have to be considered for the necessary validation of a clarification and capture process.

REFERENCES

1. Hanisch W. Cell harvesting. In: McGregor WC, ed. Membrane Separations in Biotechnology. Marcel Dekker, 1986:61–87.
2. Lee S. The primary stages of protein recovery. J Biotechnol 1989; 11:103–118.
3. Berthold W, Kempken R. Interaction of cell culture with downstream purification: a case study. Cytotechnology 1994; 15:229–242.
4. Mammalian cell culture clarification. Millipore application note AN1511EN00.
5. Mackay D, Salusbury T. Choosing between centrifugation and crossflow microfiltration. Chemical Engineer 1988; 447:45–50.
6. Manzoni M, Bergomi S, Rollini M, Cavazzoni V, Felisaz C, Marinoni C. Extracellular K5 polysaccharide antigen recovery performed by different cell separation technologies. Biotechnol Lett 2000; 22:759–766.
7. Wang IF, Ditter JF, Morris R (USF Filtration and Separations Group). Microfiltration membranes having high pore density and mixed isotropic and anisotropic structure. Patent US5906742, 1999.
8. Kersten J, Mathias JM. (Baxter Travenol), Low volume, large area filter for IV or blood filtration. Patent US4517090, 1985.
9. Tanabe K (Nippon Roki Corporation). Multi-layer filter cartridge. Patent US5215661, 1993.
10. Singhvi R, Schorr C, O'Hara C, Xie L, Wang DIC. Clarification of animal cell culture process fluids using depth microfiltration. Bio Pharm April 1996:35–41.
11. Cook N, Gutman RG, Weight KR, Buttery RA, Grimes M (Pall Corporation). Depth filter media. Patent EP0475708 Bl, 1992.
12. Cook NJ, Weight KR, Gutman RG, Buttery RA, Grimes M (Pall Corporation). Depth filter media. Patent US5468382, 1995.
13. Cook NJD, Weight KR, Gutman, RG, Buttery RA (Pall Corporation). Depth filter media. Patent US5279731, 1994.
14. Savage M, Hinskens W. Use of structured depth filters for virus removal. Patent EP0798003A2, 1997.
15. Charlton HR, Relton JM, Slater NKH. Characterization of a generic monoclonal antibody harvesting system for adsorption of DNA by depth filters and various membranes. Bioseparations 1999; 8:281–291 (DF4).
16. Filter sizing methods. Millipore Application Note AN1512EN00.

17. Stankowski R, Hunt S, Hubbard JD (Millipore Corporation). Process and filter for filtering a slurry. Patent WO01/19490A1, 2001.
18. Proulx S, Perivolotis G, Straeffer G, Gagne GA, King K, Stankowski RJ (Millipore Corporation). Filtration cartridge and process for filtering a slurry. Patent WO00/32290, 2000.
19. van Reis R, Zydney A. Membrane separations in biotechnology. Curr Opin Biotechnol 2001; 12:208–211.
20. Belfort G, Davis RH, Zydney AL. The behavior of suspensions and macromolecular solutions in crossflow microfiltration. J Membr Sci 1994; 96:1–58.
21. van Reis R, Leonard LC, Hsu CC, Builder SE. Industrial scale harvest of proteins from mammalian cell culture by tangential flow filtration. Biotechnol Bioeng 1991; 38:413–422.
22. Maiorella B, Dorin G, Carion A, Harano D. Crossflow microfiltration of animal cells. Biotechnol Bioeng 1991; 37:121–126.
23. Luque S, Mallubhotla H, Gehlert G, Kuriyel R, Dzengeleski S, Pearl S, Belfort G. A new coiled hollow-fiber module design for enhanced microfiltration performance in biotechnology. Biotechnol Bioeng 1999; 65:247–257.
24. Gehlert G, Luque S, Belfort G. Comparison of ultra- and microfiltration in the presence and absence of secondary flow with polysaccharides, proteins, and yeast suspensions. Biotechnol Prog 1998; 14:931–942.
25. Chang DJ, Hsu FC, Hwang SJ. Steady state permeate flux of cross flow microfiltration. J Membr Sci 1995; 98:97–106.
26. Kwon DY, Vigneswaran S, Fane AG, Ben Aim R. Experimental determination of critical flux in cross-flow microfiltration. Sep Purif Technol 2000; 19:169–181.
27. Li H, Fane AG, Coster HGL, Vigneswaran S. An assessment of depolarisation models of crossflow microfiltration by direct observation through the membrane. J. Membrane Science 2000; 172:135–147.
28. Maa Y, Hsu CC. Investigation on fouling mechanisms for recombinant human growth hormone sterile filtration. J Pharm Sci 1998; 87:808–812.
29. Ho C, Zydney AL. Effect of membrane morphology on the initial rate of protein fouling during microfiltration. J Membr Sci 1999; 155:261–275.
30. Tanaka T, Kamimura R, Itoh K, Nakanishi K, Matsuno R. Factors affecting the performance of crossflow filtration of yeast cell suspension. Biotechnol Bioeng 1993; 41:617–624.
31. van Reis R, Saksena S. Optimization diagram for membrane separations. J Membr Sci 1997; 129:19–29.
32. Meacle FM, Aunins A, Thornton R, Lee A. Optimization of the membrane purification of a polysaccharide–protein conjugate vaccine using backpulsing. J Membr Sci 1999; 161:171–184.
33. van Reis R, Brake JM, Charkoudian J, Burns DB, Zydney AL. High-performance tangential flow filtration using charged membranes. J Membr Sci 1999; 159:133–142.
34. Mateus M, Cabral JMS. Recovery of 6-alpha-methylprednisolone from biotransformation medium by tangential flow filtration. Bioprocess Eng 1989; 4:169–174.
35. Hauk DI, Tanny GB (Gelman Instrument Company). Fluid purification system. Patent US4244820, 1981.
36. Nodes R, Hoffman J (Sartorius). Flat filter element for the filtration of fluids. Patent US4678578, 1987.
37. Lenk RP, Durning AG, Klimchak RJ, Portnoff J, Tomsho ML (Liposome Company). Method for size separation of particles. Patent WO89/00846, 1989.
38. Marinaccio PJ, Repetti RV (Cuno). Cross-flow filtration. Patent US4888115, 1989.
39. Pierrard P, Bernard M, Largeteau D. Flat filtering element with a membrane forming a lamellar filtration cell and a tangential flow pressure filter including stacks of such elements. Patent US4816150, 1989.

40. VanReis RD (Genentech). Improved tangential flow filtration process and apparatus. Patent WO92/04970, 1992.
41. VanReis RD (Genentech). Tangential flow filtration process and apparatus. Patent US5256294, 1993.
42. Moller JK (Dow Danmark A/S). Membrane filtration unit and cassette, and method of making membrane filtration unit. Patent EP0529682 Bl, 1993.
43. van Reis RD (Genentech). Tangential flow filtration process and apparatus. Patent US5490937, 1996.
44. Lutz H (Millipore). Membrane filtration with optimized performance. Patent WO96/34679, 1996.
45. Pearl SR, Proulx AG, Hillier B (Millipore). Manifold apparatus for tangential flow filtration apparatus. Patent WO96/39245, 1996.
46. Chahine GL (Dyna Flow). Microfiltration system with swirling flow around filter medium. Patent US5500134, 1996.
47. Lutz H (Millipore). Membrane filtration with optimized addition of second liquid to maximize flux. Patent US5597486, 1997.
48. Sklar E, Graham PL, Larson J (Gelman Sciences). Selective separation filter device. Patent US5601711, 1997.
49. VanReis R (Genentech). Hollow fiber co-flow filtration device. Patent WO99/07458, 1999.
50. van Reis RD (Genentech). Tangential-flow filtration system. Patent US6054051, 2000.
51. van Reis RD (Genentech). Tangential-flow filtration system. Patent US6221249B1, 2001.
52. Kroner KH, Nissinen V, Ziegler H. Improved dynamic filtration of microbial suspensions. Bio/Technology 1987; 5:921–926.
53. Parnham CS, Davis RH. Protein recovery from cell debris using rotary and tangential crossflow filtration. Biotechnol Bioeng 1995; 47:155–164.
54. Miller JD, Fendya TJ, Ryan JE, Hurwitz MF, Musto EM, Randhan H, Vogelman H. Dynamic filter system. Patent US6106713, 2000.
55. Miller JD, Geibel SA, Hurwitz MF, Fendya TJ, Alex T. Dynamic filter system. Patent US006117322A, 2000.
56. Vogel JH, Kroner KH. Controlled shear filtration: a novel technique for animal cell separation. Biotechnol Bioeng 1999; 63:663–674.
57. Thompson KJ, Wilson JS. A compact gravitational settling device for cell retention. In: Spier RE, Griffiths JB, Berthold W, eds. Animal Cell Technology: Products of Today, Prospects for Tomorrow. Oxford: Butterworth-Heinemann Ltd, 1994:227–229.
58. Searles JA, Todd P, Kompala DS. Perfusion culture of suspended CHO cells employing inclined sedimentation. In: Spier RE, Griffiths JB, Berthold W, eds. Animal Cell Technology: Products of Today, Prospects for Tomorrow. Oxford: Butterworth-Heinemann Ltd, 1994:240–242.
59. Stevens J, Eickel S, Onken U. Lamellar clarifier – a new device for animal cell retention in perfusion culture systems. In: Spier RE, Griffiths JB, Berthold W, eds. Animal Cell Technology: Products of Today, Prospects for Tomorrow. Oxford: Butterworth-Heinemann Ltd, 1994:234–239.
60. Huelscher M, Scheibler U, Onken U. Selective recycle of viable animal cells by coupling of airlift reactor and cell settler. Biotechnol Bioeng 1992; 39:442–446.
61. Tokashiki M, Arai T. High density culture of mouse-human hybridoma cells using a perfusion culture apparatus with multi-settling zones to separate cells from the culture medium. Cytotechnology 1989; 2:5–8.
62. Jiang Y, Myers MN, Giddings JC. Separation behavior of blood cells in sedimentation field-flow fractionation. J Liq Chrom Rel Technol 1999; 22:1213–1234.
63. SITAR. Device and method for the separation of human or animal cells of different densities from cellular dispersions which contain them. Patent WO99/23471, 1999.

64. Mahar JT. Scale-up and validation of sedimentation centrifuges, part I: scale-up. BioPharm September 1993:42–51.
65. Wang DIC, Sinskey TJ, Gerner RE, de Filippi RP. Effect of centrifugation on the viability of Burkitt lymphoma cells. Biotechnol Bioeng 1968; 10:641–649.
66. Zuccato R (Frau). Centrifugal separator with rotating seals on the fixed upper head. Patent US4936821, 1990.
67. Jaeger V. A novel perfusion system for the large-scale cultivation of animal cells based on a continuous flow centrifuge. In: Spier RE, Griffiths JB, Berthold W, eds. Animal Cell Technology: Products of Today, Prospects for Tomorrow. Oxford: Butterworth-Heinemann Ltd, 1992:397–402.
68. Tebbe H, Luetkemeyer D, Guderman F, Heidemann R, Lehmann J. Lysis-free separation of hybridoma cells by continuous disc stack centrifugation. Cytotechnology 1996; 22:119–127.
69. Takagi M, Ilisa M, Yoshida T. Selective retention of active cells employing low centrifugal force at the medium change during suspension culture of Chinese hamster ovary cells producing tPA. J Biosci Bioeng 2000; 89:340–344.
70. Mannweiler K, Hoare M. The scale-down of an industrial disc stack centrifuge. Bioprocess Eng 1992; 8:19–25.
71. Larsson LA (Noxon AB). Decantation centrifuge with peripheral washing nozzles. Patent US6193643B1, 2001.
72. Engler R. High efficiency batch cell harvesting with centritech. Kendro Application Brief S00207.
73. Bracco C. Centritech perfusion system for increased expansion of cancer cells. Kendro Application Brief S00283.
74. Bracco C, Kemp C. Comparison of two perfusion systems: the centritech Labll and a hollowfiber system for the production of monoclonal antibody. Kendro Application Brief S00238.
75. Engler R, Bracco C, Soepangkat P. Centritech perfusion system for anchorage dependent microcarriers. Kendro Application Brief S00208.
76. Engler R, Kemp CW. Serum free production of monoclonal antibody in a perfusion system. Kendro Application Brief S00166.
77. Johnson M, Lanthier S, Massie B, Lefebvre G, Kamen AA. Use of centritech lab centrifuge for perfusion culture of hybridoma cells in protein-free medium. Biotechnol Prog 1996; 12:855–864.
78. Sheeler P. Continuous-flow centrifugation using SS-34/KSB and TZ-28/GK rotors. Kendro Application Brief S00110.
79. Kambic H, Nose Y. Spin doctors: new innovations for centrifugal apheresis. Ther Apher 1997; 1:284–305.
80. Johansson AS, Hoegman CF. Method of separating blood components. Patent US4608178, 1986.
81. Wada S, Kuhlemann B (Miles Laboratories). Container for fine separation of blood and blood components. Patent US4857190, 1989.
82. Dennehey TM, West JC (Baxter International). Centrifugal fluid processing system and method. Patent US5078671, 1992.
83. Spielberg R (Miles). Blood filtering system. Patent US5092996, 1992.
84. Pall DB, Gsell TC, Matkovich VI, Bormann T (Pall Corporation). System and method for processing biological fluid. Patent US5217627, 1993.
85. Lake WC, Giesler R, VanEpps D, Chapman JR, Martinson JA, Ellis DR, Aono F, Bischof DF (Baxter International). Continuous centrifugation process for the separation of biologic components from heterogeneous cell populations. Patent WO94/27698, 1994.
86. Biesel W, Geibel J, Brass H (Fresenius). Method and apparatus for the separation of blood into its components. Patent US5281342, 1994.

87. Pall DB, Gsell TC, Matkovich VI, Bormann T (Pall Corporation). System for processing biological fluid. Patent US5587070, 1996.
88. Pall DB, Gsell TC, Matkovich VI, Bormann T (Pall Corporation). System and method for processing biological fluid. Patent US5616254, 1997.
89. Lake WC, Giesler R, Van Epps D, Chapman JR, Martinson JA, Ellis DR, Aono F, Bischof DF (Baxter International). Continuous centrifugation process for the separation of biological components from heterogeneous cell populations. Patent US5641622, 1997.
90. Thomas BS (Becton Dickinson and Company). Centrifugal hematology disposable. Patent US6153148, 2000.
91. Hoegberg N, Haellgren E, Pihlstedt P (Sangistech AB). Ring bag for the centrifugation of blood components. Patent USD440300S, 2001.
92. Trese MT, Williams GA, Hartzer M, Dailey WA (Nuvue Technologies). Method for purification of a blood component. Patent US6207066B1, 2001.
93. Schoendorfer DW, Berthe CE. Continuous centrifugation system and method for directly deriving intermediate density material from a suspension. Patents US4944883, 1990, and US5053127, 1991.
94. Holm NE (Bristol-Myers Squibb Company). Centrifuge with annular filter. Patents US5733446, 1998, and US5795489, 1998.
95. Dorn AR (Sepratech Corporation). Reagent for cell separation. Patent WO87/06233, 1987.
96. Garcia FAP. Protein precipitation. In: Kennedy JF, Cabral JMS, eds. Recovery Processes for Biological Material. Wiley, 1994:355–367.
97. Chang ST, Tobler SA, Fernandez EJ. Unfolding and conformational distributions during protein precipitation. Biotechnol Prog 2001; 17:583–585.
98. Bell DJ, Hoare M, Dunhill P. The formation of protein precipitates and their centrifugal recovery. In: Fichter A, ed. Advances in Biochemical Engineering/Biotechnology. Springer1983:26:1–72.
99. Zaman F, Kusnadi AR, Glatz CE. Strategies for recombinant protein recovery from canolaby precipitation. Biotechnol Prog 1999; 15:488–492.
100. Thiering R, Hofland G, Foster N, Witkamp GJ, van de Wielen L. Carbon dioxide-induced soybean protein precipitation: protein fractionation, particle aggregation and continuous operation. Biotechnol Prog 2001; 17:513–521.
101. Galindo E, Albiter V. High-yield recovery of xanthan by precipitation with isopropyl alcohol in a stirred tank. Biotechnol Prog 1996; 12:540–547.
102. Kilpatrick PK, Lisi JF, Carbonell RG. Selective precipitation of antibody with ligand-modified phospholipids: effect of lipid chain length. Biotechnol Prog 1997; 13:446–452.
103. Lali A, Balan S, John R, D'Souza F. Carboxymethyl cellulose as a new hetero-bifunctional ligand carrier for affinity precipitation of proteins. Bioseparations 1998; 7:195–205.
104. Tsoka S, Ciniawskyj OC, Thomas ORT, Titchener-Hooker NJ, Hoare M. Selective flocculation and precipitation for the improvement of virus-like particle recovery from yeast homogenate. Biotechnol Prog 2000; 16:661–667.
105. Lee AL, Sagar S (Merck & Co.). Method for large scale plasmid purification. Patent US6197553B1, 2001.
106. Irwine JA, Tipton KF. Affinity precipitation methods. Methods Mol Biol 1996; 59: 217–238.
107. Irwine JA, Tipton KF. Affinity precipitation: a novel approach to protein purification. Essays Biochem 1995; 29:137–156.
108. Gupta MN, Mattiasson B. Affinity precipitation. In: Mukherjea RN, ed. Downstream Processes in Biotechnology. Tata McGraw-Hill, 1992:43–53.
109. Stavrinides S, Shamlou PA, Hoare M. Effects of engineering parameters on the precipitation, recovery and purification of proteins. In: Shamlou PA, ed. Processes for solid-liquid suspensions. Oxford: Butterworth-Heinemann Ltd, 1994:118–157.

110. Gailliot FP. Initial extraction and product capture. Methods Biotechnol 1998; 4:53–89.
111. Kutzko JP, Hayes ML, Sherman LT (Genzyme Transgenics Corporation). Purification of biologically active peptides from milk. Patent WO97/42835, 1997.
112. Evangilista RL, Kusnadi AR, Howard JA, Nikolov ZL. Process and economics evaluation of the extraction and purification of recombinant beta-glucuronidase from transgenie corn. Biotechnol Prog 1998; 14:607–614.
113. Bleier JE, Kim EHJ, Dong Chen X. Assembly and mixing of two-stage aqueous two-phase protein extractions. Biotechnol Prog 2001; 17:697–702.
114. Pires MJ, Aires-Barros MR, Cabral JMS. Liquid–liquid extraction of proteins with reversed micelles. Biotechnol Prog 1996; 12:290–301.
115. Choe J, VanderNoot VA, Linhardt RJ, Dordick JS. Parameters affecting the efficiency of affinity-based reversed micellar extraction and separation (ARMES) in glycoprotein purification. Biotechnol Prog 1997; 13:440–445.
116. Zhang T, Liu H, Chen J. Affinity extraction of BSA with reversed micellar system composed of unbound Cibacron blue. Biotechnol Prog 1999; 15:1078–1082.
117. Grandics P, Szathmary S. Chromatography media and their uses. Patent US5466377, 1995.
118. Dosmar M, Brose D. Crossflow ultrafiltration,. Meltzer TH, Jornitz MW, eds. Filtration in the Biopharmaceutical Industry. Dekker, 1998:493–532.
119. Schratter P. Purification ad concentration by ultrafiltration. Methods Mol Biol 1996; 59:115–134.
120. Cheryan M. Ultrafiltration in bioprocessing. In: Mukherjea RN, ed. Downstream Processes in Biotechnology. Tata McGraw-Hill, 1994:75–97.
121. Ghose TK. Ultrafiltration membranes in biotechnology. Hind Antibiot Bull 1995; 36:173–190.
122. Rodgers VGJ. Protein adsorption on ultrafiltration membrane surfaces and effects on transport. Surfactant Sci Ser 1999; 79:225–246.
123. van Reis R, Goodrich EM, Yson CL, Frautschy LN, Dzengeleski S, Lutz H. Linear scale ultrafiltration. Biotechnol Bioeng 1997; 55:737–746.
124. Burns DB, Zydney AL. Effect of solution pH on protein transport through ultrafiltration membranes. Biotechnol Bioeng 1999; 64:27–37.
125. Meireles M, Aimar P, Sanchez V. Albumin denaturation during ultrafiltration: Effects of operating conditions and consequences on membrane fouling. Biotechnol Bioeng 1991; 38:528–534.
126. van Reis R, Goodrich EM, Yson CL, Frautschy LN, Whiteley R, Zydney AL. Constant Cwall ultrafiltration process control. J Membr Sci 1997; 130:123–140.
127. Koboyashi T. Molecular size exclusion by charged ultrafiltration membranes. Recent Res Dev Pur Appl Chem 1999; 1:85–112.
128. Kaul R, Mattiasson B. Affinity ultrafiltration for protein purification. Ngo TT, ed. Molecular Interactions in Bioseparations. Kluwer Academic Publishers, 1994:487–498.
129. Dutre B, Traegardh G. Macrosolute-microsolute separation by ultra-filtration: a review of diafiltration processes and applications. Desalination 1994; 95:227–267.
130. Kurnik RT, Yu AW, Blamk GS, Burton AR, Smith D, Athalye AM, van Reis R. Buffer exchange using size exclusion chromatography, countercurrent dialysis, and tangential flow filtration: models, development, and industrial application. Biotechnol Bioeng 1995; 45:149–157.
131. Olson WP (ed.). Affinity chromatography. In: Separation Technology. Interpharm Press 1996:303–351.
132. Wilchek M, Chaiken I. An overview of affinity chromatography. Methods Mol Biol 2000; 147:1–6.
133. Wilchek M, Miron T. Thirty years of affinity chromatography. React Funct Polym 1999; 41:263–268.
134. Lowe CR. Affinity chromatography and related techniques: perspectives and Trends. Adv Mol Cell Biol 1997; 15B:513–522.

135. Clonis YD. Process affinity chromatography. Bioprocess Technol 1991; 9:401–445.
136. Burton SJ. Affinity chromatography. In: Verrall MS, ed. Downstream Processes. Wiley, 1996:193–207.
137. Cutler P. Affinity chromatography. Methods Mol Biol 1996; 59:157–168.
138. Parikh I, Cuatrecasas P. Affinity chromatography: an overview. Ngo TT, ed. Molecular Interactions in Bioseparations. Kluwer Academic Publishers, 1994:3–13.
139. West I, Goldring O. Lectin affinity chromatography. Methods Mol Biol 1996; 59:177–185.
140. Fahrner RL, Blank, GS, Zapata GA. Expanded bed protein A affinity chromatography of a recombinant humanized monoclonal antibody: process development, operation, and comparison with a packed bed method. J Biotechnol 1999; 75:273–280.
141. Harakas NK. Biospecific affinity chromatography. Bioprocess Technol 1994; 18:259–316.
142. Jones K. A review of biotechnology and large scale affinity chromatography. Chromatographia 1992; 32:469–480.
143. Goward CR. Affinity chromatography and its application in large-scale separations. In: Subramanian G, ed. Process Scale in Liquid Chromatography. VCH, 1995:193–214.
144. Chaiken I. Affinity chromatography in biology and biotechnology. Probing macromolecular interactions using immobilized ligands. Chromatogr Sci Ser 1994; 63:219–227.
145. Ujam LB, Clemmitt RH, Chase HA. Cell separation by expanded bed adsorption: use of ion exchange chromatography for the separation of *E. coli* and *S. cerevesiae*. Bioprocess Eng 2000; 23:245–250.
146. Clemmitt RH, Chase HA. Immobilised metal affinity chromatography of beta-galactosidase from unclarified *Escherichia coli* homogenates using expanded bed adsorption. J Chromatogr 2000; 874:27–43.
147. Thommes J, Halfar M, Gieren H, Curvers S, Takors R, Brunschier R, Kula MR. Human chymotrypsinogen B production from *Pichia pastoris* by integrated development of fermentation and downstream processing. Part 2. Protein recovery. Biotechnol Prog 2001; 17:503–512.
148. Raymond F, Rolland D, Gauthier M, Jolivet M. Purification of a recombinant protein expressed in yeast: optimization of analytical and preparative chromatography. J Chromatogr 1998; 706:113–121.
149. Fahrner RL, Whitney DH, Vanderlaan M, Blank GS. Performance comparison of protein A affinity-chromatography sorbents for purifying recombinant monoclonal antibodies. Biotechnol Appl Biochem 1999; 30:121–128.
150. Luetkemeyer D, Ameskamp N, Tebbe H, Wittler J, Lehmann J. Estimation of cell damage in bench- and pilot scale affinity expanded bed chromatography for the purification of monoclonal antibodies. Biotechnol Bioeng 1999; 65:114–119.
151. Feuser J, Halfar M, Luetkemeyer D, Ameskamp N, Kula MR, Thommes J. Interaction of mammalian cell culture broth with adsorbents in expanded bed adsorption of monoclonal antibodies. Process Biochem 1999; 34:159–165.
152. Kent UM. Purification of antibodies using affinity chromatography. Methods Mol Biol 1999; 115:23–28.
153. Hill CR. Immunoaffinity purification with monoclonal antibodies. In: Birch JR, Lennox ES, eds. Monoclonal Antibodies. Wiley-Liss, 1995:121–136.
154. Bailon P, Smart JE, Weber DV (Hoffmann-La Roche). Purification of interleukin-2 by receptor-affinity chromatography. Patent US5162503, 1992.
155. Cerretti DP, Cosman DJ, Dower SK, March CJ, Urdal DL, Larsen AD (Immunex Corporation). Purification of the IL-2 receptor. Patent US5317087, 1994..
156. Borg H (Bionative AB). Interferon purification process. Patent US5391713, 1995.
157. Palsson E, Gustavsson PE, Larsson PO. Pellicular expanded bed matrix suitable for high flow rates. J Chromatogr 2000; 878:17–25.
158. Klein E. Affinity membranes: a 10-year review. J Membr Sci 2000; 179:1–27.

159. Castilho LR, Deckwer WD, Anspach FB. Influence of matrix activation and polymer coating on the purification of human IgG with protein A affinity membranes. J Membr Sci 2000; 172:269–277.
160. Zeng X. Membrane chromatography: preparation and applications to protein separation. Biotechnol Prog 1999; 15:1003–1019.
161. Tejeda A, Ortega J, Magana L, Guzman R. Optimal design of affinity membrane chromatographic columns. J Chromatogr A 1999; 830:293–300.
162. Belter PA, Cussler EL, Hu W.-S. Bioseparations – Downstream Processing for Biotechnology. John Wiley, 1988.
163. Seely RJ, Tomusiak ME, Kuhn RH. Recovery operations. In: Sofer G, Zabriskie DW, eds. Biopharmaceutical Process Validation. Dekker, 2000:129–142.
164. Stanbury PF, Whitaker A, Hall SJ. Principles of fermentation technology. Butterworth–Heinemann, 1995.
165. Scott K, Hughes R. Industrial membrane separation technology. Blackie Academic and Professional, 1996.
166. Wang WK. Membrane separations in biotechnology. Marcel Dekker, 2001.
167. Rickwood D, Ford T, Steensgaard J. Centrifugation. John Wiley, 1994.
168. Sofer G, Hagal L. Handbook of Process Chromatography. Academic Press, 1997.
169. Sofer G, Zabriskie DW. Biopharmaceutical Process Validation. Marcel Dekker, 2000.

13
Downstream Processing

Ron Bates
Bristol-Myers Squibb, Syracuse, New York, U.S.A.

INTRODUCTION

The production and recovery of recombinant proteins, although common, is by no means simple, and requires skills from many disciplines including molecular biology, biochemistry, genetics, and engineering. Since Watson and Crick's discovery of DNA (1), a lot of research has focused on the upstream side of recombinant protein production, and numerous well-established vectors are available along with documented procedures to create the desired expression system. Protein production involves manipulating the cells to generate the desired protein, but just getting the cells to produce the amino acid sequence of interest does not guarantee a usable protein. The protein may not fold correctly, it may not have its activity, or it may be degraded either within the cell, in the culture medium, or during the recovery operations. The culture medium where the product is produced, and hence from where the product must be removed, can be a very complex mixture of polysaccharides, salts, proteins, amino acids, lipids, proteases, nucleic acids, steroids, vitamins, growth factors, whole cells, prions (if animal-derived supplements are used), and adventitious agents. Proteins are recovered from complex mixtures, and often the recovery and purification of that protein may not be economically feasible. It has been estimated that the expense attributed to the purification of a recombinant product can approach 60% (2) to 80% (3) of the total production cost. The successful production of a biological product must involve teamwork between the upstream and downstream engineers, since every component added to boost protein production must eventually be removed.

The terminology in this chapter includes numerous downstream processing terms, and several of the key terms are defined here. The terms adsorbent, particles, stationary phase, and resin all refer to the packing material in a chromatography column. Solute refers to a component to be recovered, separated, or purified. The terms solute, protein, biomolecule, and component are used throughout this chapter interchangeably. Throughput is defined as the amount of product processed per unit time; thus a higher throughput refers to a larger amount of material processed in the same unit of time. Productivity is throughput per unit of measure, like milligrams of product per gram of stationary phase per day, or mg/g/day. Recovery is the amount of material obtained from the feed, and a high recovery means that very little product is lost during processing.

Downstream processing in cell culture technology is defined as a series of unit operations resulting in a purified therapeutic protein, and this chapter focuses on the purification of therapeutic proteins produced in genetically engineered cell cultures. It is assumed throughout this chapter that the desired protein is secreted into the cell culture medium, and only separation techniques directly used to purify and recover proteins are discussed. Identification techniques such as Western blotting or SDS-PAGE are not included in this chapter.

PURITY REQUIREMENTS FOR BIOLOGICAL PURIFICATION PROCESSES

Most therapeutic proteins are parenteral (i.e., injected or infused directly into the body) and thus require extremely high purity, and they must have a statistically low chance to carry harmful substances such as viruses, nucleic acids (or fragments of DNA/RNA), pyrogens (or endotoxins), or immunogenic proteins. Isoforms of proteins, i.e., different glycosylation patterns, and dimers, may also need to be removed during purification, but these will have to be addressed on a case-by-case basis.

The high purity must be achieved with a high product recovery in a reasonable amount of time to make the process economically acceptable. Thus, the number and complexity of the unit operations should be kept to a minimum. Unfortunately, there are no general purification schemes since all therapeutic proteins and their production medium are intrinsically different. There are, however, common steps and unit operations that should provide the reader with a good starting point. To remove harmful agents, such as virus particles, endotoxins, and other impurities it is good practice to employ several orthogonal purification techniques. For example, a purification process could involve three unit operations—ion exchange, affinity, and size exclusion—each having an orthogonal basis for purification.

POTENTIAL PRODUCT CONTAMINANTS DERIVED FROM ANIMAL CELL CULTURE PROCESSES

Viral Contamination

The presence of viral contamination in biotechnology and especially in cell culture processing is a major concern. As is addressed elsewhere in this book, viruses can be introduced into the process by numerous routes, including but not limited to the following (4–6):

- source of cell lines from infected animals
- viral establishment of the cell line
- use of contaminated reagents or equipment
- improper handling of the cell line.

There are three common, complimentary methods applied to reduce the risk of viral contamination in a protein production process (4–6):

- selection and use of virus-free cell lines and raw materials

- ensuring an adequate clearance achieved by downstream processing
- meticulous testing after appropriate steps to show absence of contaminating virus.

The selection of a tested (for the absence of viruses) cell line, and the use of raw materials shown to be free of viruses are essential in the production of therapeutic products by cell culture technology. Even with a "virus-free" cell line, the absence of viral contamination is not guaranteed, and the downstream processing must be shown to have a sufficient viral clearance. Several techniques are known to have good viral clearances, such as chromatography, but verification of viral clearance is necessary and should include viral spikes and challenge studies utilizing appropriate viral agents, e.g., generic as well as specific types of virus particles. Furthermore, meticulous testing for the absence of infectious viruses must be conducted at appropriate stages during the recovery process.

Endotoxins

Endotoxins (also called pyrogens) are lipopolysaccharides consisting of three completely different sections. There is a nonpolar lipid region, often called Lipid A, an oligosaccharide core and a heteropolysaccharide region, often called O-Antigen (7). Endotoxins are found in the cell membrane of gram-negative bacteria, and the release of endotoxins occurs during cell division, cell growth, and cell death. Unfortunately, endotoxins can be found just about everywhere, so removing them from a process, even a nonbacterial cell culture, can be quite problematic. The European Pharmacopoeia reports the threshold level of endotoxins for intravenous use at 5 endotoxin units (EU) per kilogram of body weight per hour (8), where an EU represents the biological activity of the endotoxin. This requirement does not mean that ultra-low endotoxin levels are needed in every case. For drugs only requiring low doses, the endotoxin level can be much higher than for an antibody, for example, with a dose of a gram or more.

Ideally, in cell culture supernatants the endotoxin level should be zero, but this is usually not the case. Endotoxins are found everywhere, including the process water used in the cell culture. Endotoxin-free water can be used, but it is quite expensive and not always available. Petsch and Anspach (7) provided a table showing the endotoxin concentrations in common sources. Tap water has an endotoxin concentration of 1–20 EU/mL, and murine IgGI cell culture filtrates have concentrations <100 EU/mL. This can be compared to a high-density *Escherichia coli* culture with a concentration after homogenization of $>2 \times 10^6$ EU/mL.

Due to the high likelihood of endotoxins in the process fluid, endotoxin removal should be built into the downstream processing. The standard protein purification protocol often includes an ion-exchange step, a hydrophobic interaction step, and some sort of size-based separation. These three steps frequently provide an adequate endotoxin clearance, but the product should be tested for the presence of endotoxins. Bischoff et al. lowered a culture's endotoxin level from 1.2×10^7 to <0.22 EU/mL using a three-step process: ultrafiltration, anion-exchange, and immobilized metal affinity chromatography (IMAC) (9).

The selective removal of endotoxins from protein solutions is extremely variable. Adsorptive chromatographic methods are frequently used, and, for most basic proteins, anion-exchange chromatography is effective. The endotoxins, with their highly negatively charged oligosaccharide core, bind strongly to the positively charged stationary phase, and the proteins are either not bound or eluted with weak mobile phases (10–13). Reduction in endotoxin levels by three to four orders of magnitude are commonly seen. Process streams containing acidic proteins should not be processed through an anion-exchange step with the hopes of reducing endotoxin levels. In this situation, hydrophobic adsorption could be utilized, exploiting the lipid region of the endotoxins. Affinity resins, such as Protein A, also provide a good endotoxin clearance for both basic and acidic proteins.

In addition to nonspecific adsorbers, numerous endotoxin-specific adsorbents are commercially available. These so-called detoxy gels include but are not limited to Polymyxin B, histamine and histidine, and polycationic ligands (such as poly ethyleneimine). Petsch and Anspach provide an excellent review article on endotoxin removal, complete with structures of the endotoxin-specific ligands (7).

Nonadsorptive methods are also used to remove endotoxins, but with less success. Ultrafiltration is used to remove endotoxins from water streams, but it is not very effective for protein solutions, due to the large relative sizes of both the endotoxins and the desired protein. Li and Luo (14) used a salt to aggregate endotoxins for easy removal by ultrafiltration, and two-phase aqueous extraction with detergents is occasionally used for pyrogen removal (15–17).

Cell-Related Impurities

While making the desired protein, cells need to produce a variety of components to survive and replicate, including nucleic acids, other host cell proteins, proteases, lipids, etc. All of these need to be removed from the desired protein. Ideally, whole cells are removed from the product-containing medium without cell rupture, thus limiting the quantities of cell-related impurities, but whole-cell removal does not guarantee the absence of cell debris. Additionally, even when the intact cells are removed, it is still necessary to select the unit operations to ensure that any trace quantities of cell-related impurities are removed.

GENERAL PRINCIPLES FOR THE SELECTION AND SEQUENCE OF DOWNSTREAM PROCESSING STEPS

The process goal is to recover a protein with an acceptable impurity profile in sufficient quantities, and each step in the process must be chosen for a specific reason. For example, a size-based technique could be used as a final step to remove a low-level dimer impurity. This size-based operation would not, however, be very useful at removing a deamidated version of the desired protein. The process should therefore be designed for the earlier steps to remove the deamidated form. Ultimately, the entire purification scheme should be optimized, not necessarily each step.

Downstream processing of biomolecules from recombinant sources can be loosely broken down into three generic zones: clarification and initial purification, intermediate processing, and polishing, with one or two unit operations per stage. Figure 1 shows a typical flow sheet for a protein recovery process highlighting the three stages and commonly used techniques for each stage. It is possible to combine or overlap the zones, but all three stages are often needed to provide adequate purification and clearance of viral particles and endotoxins. Some proteins have been recovered in one stage, but this is rare, and it would be difficult to convince the safety and analytical departments of the robustness of the process for adventitious agent removal.

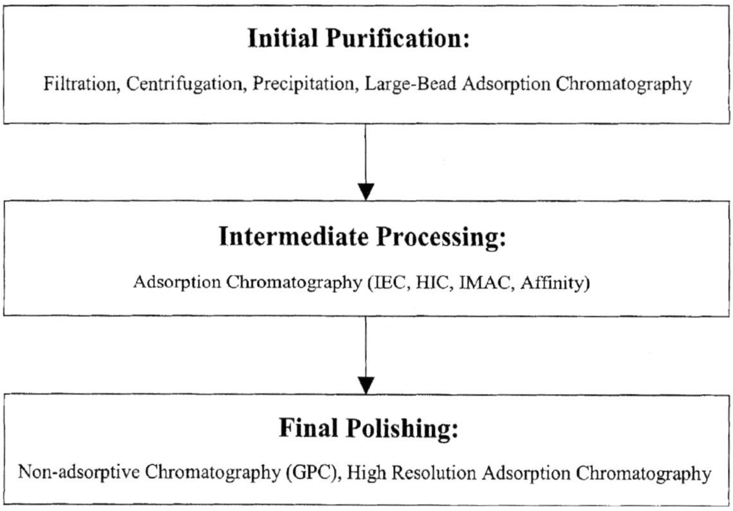

Figure 1 Overview of typical protein purification process.

The single most used purification technique in the downstream processing of therapeutic proteins is chromatography. It is employed throughout the downstream processes, but, to meet the specific requirements of each zone, different sized chromatographic resins with different efficiencies, selectivities and flow restrictions are used at each stage of the purification (18), as is described below.

Occasionally, the purification process can be set up such that minimal processing is needed between chromatography columns. For example, ion-exchange steps often leave the product-containing effluent in a high-salt solution, and hydrophobic interaction columns use high-salt solutions as feed. Therefore, the effluent from an ion-exchange column can occasionally be added to a hydrophobic interaction chromatography (HIC) column with minimal preconditioning (often just the addition of more salt). The reverse is possible as well, i.e., the low-salt effluent from a hydrophobic column can often be directly loaded onto an ion-exchange column.

Clarification and Initial Purification

In the initial product conditioning, the main goal is to reduce the working volume of the process stream and to remove cell debris and some harmful contaminants. The resolving strength of unit operations in the first stage is not as important as speed and recovery. Typical unit operations include filtration, chromatography, precipitation, and centrifugation. For example, a centrifugation and/or filtration step can be used to separate whole cells from the product-containing supernatant (called clarification). Subsequently, the product-containing material can be applied directly onto a chromatographic column for a crude purification, thus removing proteases and other harmful components as well as reducing the volume of the feedstock. For monoclonal antibodies, this first column is often a Protein A column (see section on Recovery of Monoclonal Antibodies later in this chapter). Ion-exchange and hydrophobic interaction columns also work well during the initial purification stage.

The initial purification performs crude separations while reducing the volume of feedstock, often large volumes of highly viscous, cell culture broth. Flow through packed beds, like chromatographic columns, is dependent on the size of the particles packed into the column, the viscosity of the fluid traveling through the bed and the length of the packed bed. The pressure drop across a packed bed is inversely proportional to the square of the particle diameter and directly proportional to the viscosity of the solution. Chromatographic resins and/or column hardware have a maximum allowable pressure rating, typically around 100 psi. Realizing that one of the main goals of the initial purification stage is to remove harmful degradants as quickly as possible and to capture the most product in the shortest length of time, the maximum attainable flow rate needs to be used. Since the viscosity of the cell culture fluid is fixed, an increase in the flow rate can only be realized with an increase in particle diameter. Therefore, to process large volumes of broth quickly, large diameter particles must be used. The so-called big beads (100–300 μm diameter) allow the processing of large volumes of viscous solutions without losing flow due to pressure limitations. The purity and recovery expected from the initial product conditioning are on the order of 70% and 85%, respectively; after clarification, typically only one purification step, utilizing unit operations that have high loading capabilities with high to moderate recoveries, is employed before the intermediate processing stage.

Intermediate Processing

The intermediate processing stage removes most of the remaining contaminants, such as virus particles, nucleic acids, endotoxins, and other host cell proteins, that were not removed in the initial purification stage. Typically, two or more orthogonal processes are used to ensure adequate impurity removal and adventitious agent clearance. Unit operations with a large capacity, high recovery and good resolving power, such as adsorption chromatography, endotoxin gels, and membrane processing are commonly employed in the intermediate processing stage. Chromatographic processes in the intermediate purification stage often utilize 30–100 μm particles to achieve high recovery and purity over one or two complimentary steps. The chromatographic processes involve the use of many types of adsorptive resins including ion exchange, hydrophobic interaction, and affinity-based. The effluent from one column often requires additional processing, such as buffer exchange, and/or a concentration step, before the next purification step. More information on chromatographic theory and practice is given below.

Polishing

The final polishing stage removes impurities closely related to the desired protein, such as deamidated isoforms and aggregates, and often involves only one unit operation having both very high resolution and extremely high recovery (typically >95%). Chromatographic resins used for polishing are frequently high-performance particles (10–30 μm diameter) that offer good resolution but at reduced flow rates and loading capacity. In general, the feedstocks for a final polishing step are very clean, often with only one or two minor impurities, and small diameter particles, with very high efficiencies, are required to remove these impurities. Unfortunately, these highly efficient particles also limit the processing capability (i.e., productivity) of the chromatographic step. Only small volume feedstocks are used in the final purification steps, and thus a concentration step is often required before the column. Most polishing steps involve loading a relatively small amount of feed onto a column with a very high recovery of product. Final processing resins are frequently size-based, such as gel permeation chromatography (GPC), but ion exchange, hydrophobic interaction, and affinity-based resins are also used with good success.

Combination of Initial and Intermediate Processing

The number of unit operations should be minimized to maximize the recovery of the desired protein. One way to accomplish this is to combine recovery steps. Recently, expanded bed adsorption (EBA), a technique that attempts to combine the initial and intermediate processing steps, has gained widespread use, especially on the laboratory scale. EBA uses a fluidized bed of adsorbent particles with a large intraparticle void volume to allow large particles, like whole cells, to pass unadsorbed, while secreted molecules, like proteins, are adsorbed onto the resin. Once the feedstock is passed over the column, the direction of flow is switched, resulting in a typical packed bed column. Thus, unclarified cell culture broth can be used as feedstock resulting in fairly pure protein in only one step. EBA has been successfully applied to unclarified cell culture broths without any detectable cell damage (19) and is explained in more detail below.

Global Process Requirements

The goal of downstream processing is the recovery of active protein in sufficient quantity at adequate purity. To meet this goal, several unit operations need to be developed, optimized, and implemented, but it is not necessary to achieve the highest possible resolution from each step, especially if it is at the cost of recovery of product or productivity. The requirements of the process should be achieved globally. The goal of each step should focus on the strength of that step, and it should also consider the strengths of the other steps in the process. For example, if a troublesome impurity eluted closely to the desired product in an ion-exchange step, the ion-exchange step could be developed to remove that impurity. Unfortunately, adjusting the conditions to remove that impurity may negatively affect the throughput of the step, and it could also affect the overall yield by requiring the step to discard the portion of the column effluent that contains the impurity. Another approach would be to determine if a different unit operation would remove that impurity more efficiently: in the previous example, if the impurity is a dimer it may be easily removed by a GPC column. During the development phase of the process, the strengths of each processing step should be maximized to produce the most productive process while still maintaining recovery and purity.

INITIAL PRODUCT CONDITIONING

The following section delineates the strengths and weaknesses of methods used for clarifying cell culture broths, for removing proteases and other harmful contaminants, and for the reduction in the working volume of the process stream. Common initial steps are membrane processes, like filtration, and centrifugation to remove whole cells and cell debris. After cell removal, crude chromatography, extraction, and precipitation are popular secondary steps to eliminate contaminants from the desired protein.

Membrane Processing

Membranes are an integral part of most protein purification schemes. A standard first step in the recovery of secreted proteins is to simply pass the cell-containing culture fluid over a membrane, where the permeate flows through the membrane and the cells are retained on the membrane surface (20). Additionally, the concentration or buffer exchange of process streams is frequently accomplished using a membrane unit. Membranes can be chemically inert, i.e., nonadsorptive, or contain functional groups to adsorb solutes. Nonadsorptive membranes separate solutes on the basis of size; a liquid (with or without particulates) is forced through a porous membrane. The small molecules (molecules smaller than the diameter of the pores) pass through the membrane, while the larger molecules are retained (21). Typically, only crude separations, large molecules from small molecules, are achieved using nonadsorptive membranes, but this crude separation can offer a very good recovery of the desired product.

There are two main classes of nonadsorptive membrane processes in the pharmaceutical industry, microfiltration (MF) and ultrafiltration (UF), although nanofiltration and reverse osmosis (RO) are also used to a lesser extent. MF is defined as the clarification or removal of microparticulates from a liquid stream. These microparticulates can be cells, cellular debris, aggregates, precipitates, bacteria or any other nondissolved substance in a cell culture medium. MF membranes are classified by the diameter of the smallest particle that is retained by the membrane, typically on the order of 0.1–10 μm. UF is the separation of dissolved components based on molecular size, such as the recovery of peptides from a cell culture supernatant, removal of protein dimers or elimination of virus particles from a process stream. Membranes used for UF are classified by molecular weight cut-off (MWCO), which is the molecular weight in Daltons (Da) of the smallest molecule that is retained by the membrane. Typical UF membranes range from 1000 to 500,000 MWCO. Nanofiltration is similar to UF but with MWCO ranging from 100 to 1000 Da. Reverse osmosis membranes are capable of separating salts from aqueous solutions and are classified by the percent rejection of a salt (typically sodium chloride).

Microfiltration is usually operated in a dead-end configuration where the solution is forced through a membrane situated perpendicular to the direction of flow. Dead-end filtration units are commonly used in applications where large differences between solutes exist, such as the removal of whole cells from a culture medium. Unfortunately, the buildup of retained material on the membrane surface results in a reduction of the fluid flow passing through the membrane, and this buildup requires frequent cleaning or replacement of the membrane.

Ultrafiltration processes are operated in either dead-end or cross flow (also known as tangential flow) configuration. Cross-flow filtration (CFF) utilizes a membrane situated parallel to the direction of flow, and thus only a portion of the feed flows through the membrane. The feed is kept under pressure and forced to flow along the surface of the membrane, preventing a large cake from forming on the surface of the membrane and allowing much longer operation lifetimes than dead-end filtration. The permeate side of the membrane in CFF is either kept at ambient pressure or under vacuum to promote flow through the membrane. The stream that does not pass through the membrane is called the retentate and exits the unit at a higher solute concentration. The concentration of the retentate is determined by the ratio of permeate to retentate flow. This ratio is controlled by altering the operating conditions, e.g., MWCO, feed pressure, feed flow rate, permeate pressure. The most common applications of CFF in the downstream processing of proteins are the concentration, buffer exchange and desalting of protein streams, such as after a chromatography column.

The difference between the feed pressure and the permeate pressure of membranes is called the transmembrane pressure (TMP). The TMP is used to monitor the filtration step (both in dead-end and CFF). As the TMP rises to a predetermined level, the operation needs to be stopped, and the membrane needs to be either cleaned or replaced.

There are various types and sizes of membrane units available commercially, such as tubular, hollow fiber, and flat sheet (cassette type, spiral wound units, etc.) (21), and they can be manufactured with functional groups, much the same as chromatography resins but with the opportunity for better productivity. Derivatized membranes utilize adsorption to selectively bind solutes from a process stream. Adsorptive membranes offer better separation efficiencies, but they are more difficult to develop and often get replaced by a chromatographic column. Membrane suppliers produce large lines of adsorptive membranes including ion exchange, hydrophobic, affinity, endotoxin, and viral-specific functional groups. There are numerous reports of researchers using derivatized membranes, such as Lee et al.'s work with an affinity membrane to purify enzymes (22), and most derivatized membranes are used similarly to chromatography resins, explained in more detail below.

Another benefit of membrane processing is the ease of scale-up. The process is scaled directly from the laboratory based on the cross-sectional area, assuming the same geometry at both scales. The accompanying flow rates are also scaled based upon the ratio of the cross-sectional areas, i.e., a five-fold increase in cross-sectional area results in a five-fold increase in flow rate, throughput, and membrane area. The pressure drop and process time should be the same in both the laboratory and manufacturing facility.

Centrifugation

Centrifugation is used to separate or concentrate materials suspended in a liquid medium, based on sedimentation rates in an increased gravitational field. Two particles of different masses will settle in a tube at different rates in response to gravity. In the downstream processing of biomolecules, the particles are usually cells, subcellular organelles, viruses, and large molecules such as protein and nucleic acid aggregates. Typically, a centrifugation step in cell culture processing results in a solid pellet (which is discarded) and a clear supernatant (containing the product).

The increased gravitational field is generated in the centrifuge by the revolution of a rotor around a fixed point. The centrifugal force (measured and reported as multiples of the force of gravity) generated by this rotation is used to increase the settling rate of biomolecules, and it is proportional to the rotation rate of the rotor (in rpm) and the distance between the rotor center and the centrifuge tube. Therefore, a given centrifuge may use multiple rotor sizes to produce different separation conditions. Each centrifuge has either a specific graph (a nomograph) or a table relating rotation rate (rpm) to centrifugal force for each size of rotor it accepts. The centrifugal force is also dependent upon the design of the rotor itself, e.g., fixed angle or swinging bucket.

Centrifugation is frequently used as a first step to separate whole cells from the supernatant. It is also used to separate protein aggregates from the supernatant following a precipitation step. Centrifugation is a well-studied and well-used unit operation (23) with a host of references on its use in bioprocessing (24–35, for example).

The types of rotors used in centrifuges are fixed angles, swinging buckets, continuous flow, or zonal, depending upon whether the sample is held at a given angle to the rotation plane, is allowed to swing out on a pivot and into the plane of rotation, designed with inlet and outlet ports for separation of large volumes, or a combination of these. The details of even basic centrifuges are too extensive to cover in this chapter. Fixed angle rotors are the work-horse elements of a cell culture laboratory, and the most common fixed angle is a rotor holding eight centrifuge tubes at an angle of 34° from the vertical. In a fixed angle rotor, the materials are forced against the side of the centrifuge tube and then slide down the wall of the tube.

Precipitation

Historically, precipitation was the most common initial purification step, and it was commonly used for fermentations where the product was not secreted. Recently, especially with animal cell cultures, the trend has been to replace traditional precipitation steps with filtration or chromatographic steps, especially in large-scale processes. A brief description of the theory and practice of precipitation, as well as several references, are given below.

The interactions between water and the polar side-chains of amino acids allow proteins to become hydrated and thus soluble in aqueous solutions. When these interactions are prevented, proteins interact with each other and form aggregates that eventually fall out of solution. The purposeful disruption of the water–protein interaction is called precipitation. There are two main classes of precipitation used in the recovery of biologies: solvent and salt (or polymer), and either class is used in two main facets in the biotechnology industry. The first is to remove components that are drastically different from the desired protein (such as proteases or nucleic acids) and prepare crude cell extracts for the next step, i.e., chromatography. The second is to precipitate all the material out of solution to change buffer or provide a storage medium.

Occasionally, once the water–protein interaction has been sufficiently disrupted, the protein aggregates will grow with a density far greater than water and fall directly out of solution, but it is more common, especially in a laboratory setting, to follow the precipitation with centrifugation.

It is also worth noting that precipitation is an equilibrium-driven interaction—the solubility of the protein is reduced but not eliminated, and thus not all of the protein is removed from solution. As an example, consider two solutions of the same protein at 10 and 5 mg/mL under identical conditions. Further assume that the solubility of the protein is reduced to 0.1 mg/mL upon addition of salt. It is obvious that the more concentrated solution will have a greater recovery of protein, due to a larger percentage of protein remaining in solution in the less concentrated sample. Thus, the concentration of a dilute feedstock is often employed directly before a precipitation step to increase the yield.

The solubility of proteins depends on, among other things, the salt concentration and the pH of the solution. At low salt concentrations, the presence of salt stabilizes the various charged groups on the protein, thus attracting the protein into the solution and enhancing its solubility. As the salt concentration is increased, a point is reached where addition of more salt results in less water available to solubilize the protein. The protein eventually starts to precipitate out of solution when sufficient water molecules are no longer available. These two phenomena are referred to as salting-in and salting-out, respectively. Salting-out of proteins is the most common form of protein precipitation.

Another major factor in the precipitation of proteins is the solution pH. A general rule of thumb for pH effects is proteins are much less soluble at their isoelectric point (pI) than at other pHs. Using organic modifiers or polymers such as polyethylene glycol (PEG) often further reduces the solubility. Protein stability is much more sensitive to changes in pH, however, than to either a change in the organic content of the solution or changes in the ionic strength of the solution. Very slight changes in pH can have drastic effects on protein stability.

Using salts to precipitate proteins out of solution is accomplished in one of the following two methods. The first is to simply add the salt, in solid form, to the feedstock. It is important to note that adding salt to a solution increases the total volume of solution, which must be taken into account if a precise salt concentration is required. The second method involves making up a concentrated salt solution, at the pH and temperature of the feedstock, and adding the saturated solution directly to the feedstock. In the laboratory setting either method is trivial to perform, but on a large scale it is often not possible to use a concentrated solution, since even a 50% saturated solution requires a two-fold increase in the volume of the working stream. In addition, the solubility of proteins in high-salt solutions is typically greater at lower temperatures. Thus a clear solution in a cold box at 2 M ammonium sulfate could become turbid or even show precipitates upon warming to room temperature.

The ability of anions to cause precipitation typically follows the Hofmeister series, i.e., sulfate stronger than acetate which is stronger than chloride, etc.; additionally, monovalent cations are better than divalent cations. Ammonium sulfate is the most commonly used salt due to its high solubility and low cost. The solubility of ammonium sulfate in water is 0.53 kg L^{-1} or 4.05 M at 20°C, and it is common to see precipitation steps using a 50% ammonium sulfate solution. The recent literature contains numerous examples using ammonium sulfate and other salts to precipitate proteins. Nagashima et al. used a 30–50% saturated ammonium sulfate solution to precipitate glucuronate:baicalein 7-o-glucuronosyltransferase with a 94% recovery (36). Menhard and Zenk used ammonium sulfate (a 0–70% saturated solution gradient) as an initial conditioning step in the purification of acetal coenzyme A (37). Ammonium sulfate precipitation offers several additional benefits over traditional precipitation methods, such as increased stability of most proteins and reduction in proteolysis and bacterial growth (21). In fact, ammonium sulfate precipitates are often the preferred long-term storage method for proteins. Numerous other researchers have used ammonium sulfate in one capacity or another in the recovery of biologies (38–41).

Solvent precipitation involves increasing the organic component of a protein solution until the desired protein falls out of solution. The organic solvent must be miscible with the protein solution, nonreactive with the desired protein, and effective at causing aggregation of the desired protein. The most common solvents are acetone and ethanol, although other alcohols, ethers, and ketones can be used. Aggregation of protein molecules during solvent precipitation is likely caused by electrostatic interactions between the charged groups of individual protein molecules. As the solution becomes more nonpolar, the polar groups on neighboring protein molecules interact, causing aggregation and eventually precipitation. Unfortunately, too much organic solvent often causes denaturation, so the optimal percent solvent should be chosen very carefully.

Contrary to salt or pH precipitation, solvent precipitation allows for operation at temperatures below the freezing point of water, since all miscible organic solvents lower the freezing point of the solution to well below 0°C. Typical solvent precipitation conditions (using either acetone or ethanol) are 20% solvent v/v on an additive basis (e.g., 20 mL of solvent to 80 mL of aqueous solution) with an aqueous solution having an ionic strength between 0.05 and 0.2 M, operated at 0°C. Generally, precipitation with solvents is greater at lower temperatures, and since proteins are more stable at lower temperatures, this benefit of solvent precipitation can be exploited.

The use of high molecular weight, water soluble, organic polymers is also commonly used to precipitate proteins from solution (21). Solutions of polyethylene glycol (PEG), up to 20% w/v, are frequently used, often resulting in the precipitation of all the protein mass in solution (42). Polyelectrolytes, which are organic polymers with multiple charged sites, are also used to precipitate proteins (43,44). The use of organic polymers in precipitation is very similar to the use of ammonium sulfate. Affinity ligands, either covalently bound to soluble polymers or in free solution, are used to selectively precipitate proteins from solution (45–49). In affinity precipitation, the ligands exploit the protein's specific functionality to form conjugates and precipitate out of solution. The interaction between the protein and the affinity ligand is very similar to the interaction used in affinity chromatography, and, in fact, many of the same ligands are used. Unfortunately, organic polymers are more difficult to remove than either salt or organic solvent from the precipitated proteins.

Aqueous Two-Phase Extraction

Another common initial purification step is liquid–liquid extraction, where solutes partition between two liquid phases. To form two distinct phases, the two liquids must be immiscible, such as hexane and water. Traditional extraction processes use an organic solvent and an aqueous solution to form two distinct phases. Unfortunately, proteins often denature in organic solvents, and thus extraction with an organic solvent is not a viable solution. An alternative is to use two immiscible aqueous phases.

Aqueous two-phase systems form when two solutions containing either two hydrophilic polymers or a polymer and a salt, in the appropriate amounts, are mixed. Typical polymers used in aqueous two-phase systems are PEG and dextran. Ammonium sulfate and sodium sulfate are common salts for the generation of aqueous two-phase systems. Fractionation of solutes between the two phases can be adjusted by slight alterations in the concentration, composition, or temperature of either phase. Aqueous two-phase extraction (ATPE) exploits the subtle differences of the two phases to selectively partition proteins between the two phases, or it can partition proteins away from contaminants like DNA.

The behavior of proteins in the aqueous phases is not well understood, but it is known that altering the composition and/or the concentration of aqueous solutions effects either the hydrophobicity or the ionic nature of that phase. Walter and coworkers offer two good references providing detailed descriptions of different polymer/polymer and polymer/salt combinations, temperature and solvent effects on partitioning, etc. (50,51). In addition to these references, numerous current uses of ATPE can be found for many different proteins (22,52–63) and amino acids (64). Diamond and Hsu report that ATPE is an efficient, economically viable initial purification step (65). ATPE can offer an alternative to other initial processing steps while still meeting the purity and recovery requirements, and ATPE can be operated in a continuous fashion, thus improving productivity over traditional batch operations.

In addition to the standard polymers, researchers also add specific ligands to polymer chains, e.g., affinity ligands, to enhance the partitioning in ATPE (21). Dyes are also chemically attached to PEG to alter the protein partitioning (66,67). Thermoseparating polymers are used as well. Thermoseparating polymers are polymers that when combined with water and heated above a critical temperature, called the cloud point, form two phases. The polymers are easily recycled by cooling the liquid below the cloud point (61–63,68) thus forming one homogenous phase.

Batch Adsorption

Researchers also use activated porous particles, such as chromatography resins, to capture the desired component by adsorbing the solute onto the resin in batch mode. A typical batch adsorption step involves the addition of the resin to the feedstock solution in a buffer that promotes adsorption of the desired solute. The resin, with the desired component, is then filtered to remove the initial buffer. The filtered resin is combined with a much smaller volume of a buffer that is unfavorable to adsorption. The same type of filtration is again used with the filtrate containing the desired component. The choice of buffer conditions to be used should be optimized to initially only bind the desired solute and close analogs, and the desorption buffer should be strong enough to allow for a large reduction in the overall process volume with a good recovery of product.

Any adsorptive particle can be used in batch mode, such as ion-exchange, hydrophobic, or affinity resins. Gagnon reports the use of anion-exchange chromatographic resins as batch adsorbers to remove host cell membranes, lipids, DNA, and pH indicator dyes from the desired protein (69). The adsorbent in this case is added directly to the harvest fluid, stirred gently for a sufficient length of time, and filtered through a 0.2 μm membrane. In a laboratory setting, this filtration is done in a dead-end configuration (often a stirred-cell) using vacuum to pull the supernatant through the membrane. EBA combines batch adsorption with traditional chromatography to both capture and purify the desired protein, as explained in more detail below.

Product Concentration

The concentration of product streams is very important in protein purification, especially before nonadsorptive purification steps like size-exclusion chromatography (SEC). The most common concentration step in the downstream processing of proteins is ultrafiltration operated in CFF mode; a membrane is chosen such that the desired solutes and large biomolecules are retained, but the liquids and smaller molecules pass through the membrane, thus permitting reduction until the desired process volume is obtained. Other processes such as batch adsorption, dialysis, adsorption chromatography and precipitation also offer significant concentration opportunities at the capture stage, but UF is far and away the easiest and most commonly used.

Although used less than UF, ammonium sulfate precipitation is also used as a concentration step. A saturated ammonium sulfate solution can be added to a post-column effluent, for example, precipitating out all the protein in the solution. After centrifugation, the resulting pellets are redissolved in the appropriate volume of solution required for the next step. Although dialysis and ultrafiltration are more common, ammonium sulfate concentration does have benefits, especially when stability of the desired protein is a major issue.

Buffer Exchange

The removal of salts and the exchange of buffers is typically accomplished using either UF, dialysis, or GPC. In UF, the desired buffer is added to the protein solution at a rate matching the volume removal rate of the permeate stream.

Dialysis involves using a piece of tubing or a bag deliberately manufactured to allow small molecular weight solutes to pass unrestricted while prohibiting larger molecular weight (>15 kDa for example) solutes from passing through the material. A typical dialysis step involves loading the concentrated sample into a dialysis bag and placing that bag into a vessel filled with the desired buffer under mild agitation. Sufficient time is allowed for equilibrium to be reached and the process is repeated. The equilibrium nature of dialysis necessitates at least one repetition of the process. For example, if the buffer volume is 50-fold greater than the sample volume, the maximum dilution of the original sample is 51-fold.

Dialysis is a time-intensive unit operation. Typically, a well-mixed sample can reach 90% of equilibrium in 2–3 hr (21); however, the remaining 10% can take appreciably longer. The simple operation and equipment needed to perform a dialysis step makes it a very popular laboratory technique, but it is less useful in large-scale operations.

Gel permeation chromatography also exchanges buffers by size-based operations. In GPC, the proteins are sterically prohibited from entering the pores of the column packing, and thus they elute much faster than smaller ions. A buffer exchange, GPC step involves the following: equilibrating a column packed with the adsorptive stationary phase with the desired buffer, load the sample onto the column, and immediately after the sample has been loaded, flow the desired buffer through the column. The desired solute should elute (in the desired buffer) much faster than the original buffer/salt. A rule of thumb for GPC is that the volume of feed loaded onto the column should be <5% of the total column volume to ensure adequate distance between the back edge of the solute band and the front edge of the original buffer band.

Gel permeation chromatography is typically much faster than dialysis and is used more frequently on large scale. More information on GPC is given below.

CONVENTIONAL CHROMATOGRAPHIC METHODS

General Principles of Chromatography

Chromatography is one of the oldest and most widely used purification methods, with Tswett (70) generally being credited with its invention. Since Tswett's early work separating plant dyes, much has changed in the "state of the art." In the 1950s column chromatography emerged, drastically increasing the speed and efficiency of chromatography, but the column packings used in those first columns were structurally soft, thus prohibiting the use of high flow rates. By the early 1960s, rigid column packings and better solvent delivery systems were available, and high-performance liquid chromatography (HPLC) was born. HPLC is used today in every recovery and purification laboratory in the biotechnology industry.

Chromatography encompasses a wide range of separation modes that can be categorized loosely as either: nonadsorptive (e.g., size-exclusion) or adsorptive (e.g., ion-exchange). The adsorptive modes of chromatography include (but are not limited to) ion-exchange, hydrophobic interaction, reversed-phase, affinity, and immunoaffinity. In this chapter, chromatography refers to the batch-wise loading of feed onto a column packed with particles to separate moieties into distinct zones (i.e., peaks) that are kept apart by zones containing only mobile phase (i.e., "empty zones").

Chromatographic theory revolves upon the amount of time solutes spend interacting with the stationary phase. The most common way to represent this interaction is called the capacity factor, k', defined as:

$$k' = \frac{\text{time spent in absorbent}}{\text{time spent in liquid phase}} = \frac{t_r - t_0}{t_0} \tag{13.1}$$

where t_r is the retention time of the solute, and t_0 is the retention time of an unadsorbed solute. Figure 2 shows an ideal chromatogram with two components, and this figure, hopefully, provides a graphical explanation of the terms used above and below.

For components to be separated on a chromatography column, the time the individual solutes spend in the adsorbent must be sufficiently different. Typically, larger differences in k' values lead to better separations. The separation between two peaks is called the separation factor (α) and is defined as the ratio of the capacity factors of each component:

$$\alpha = \frac{k'_2}{k'_1} \tag{13.2}$$

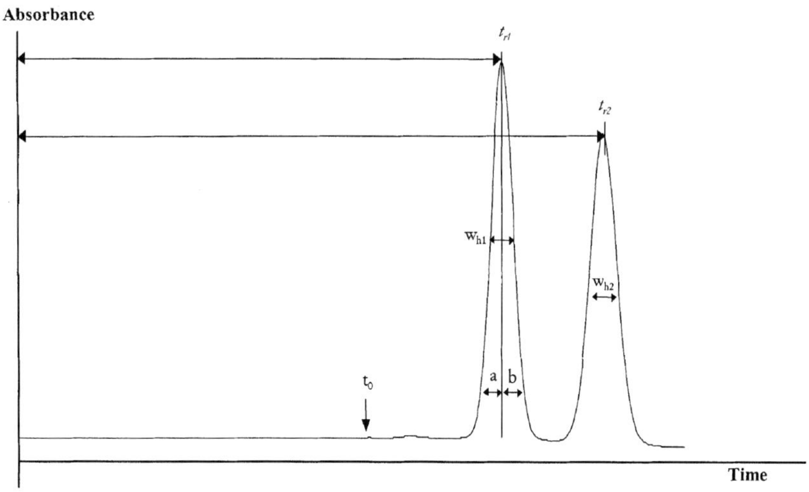

Figure 2 Generic elution chromatogram. A standard two-peak separation separated by a small empty zone. The y-axis is the response from the detector and the x-axis is the time. The variables are further explained in the text with t_r referring to retention time of each component and w_h referring to the width of each peak.

It is worthwhile to note that a large α does not guarantee a good separation. The retention times used to calculate k' are the time when the peak apex exits the column, and there is no factor in α to account for the shape of a peak. Thus, a situation can arise where a large difference in retention times exist, but where the widths of the peaks are sufficiently large and overlap, preventing pure product from being recovered.

Extremely large peak widths can result from an inefficient column or from poorly chosen elution conditions resulting in nonspecific binding or gradual desorption. Large peak widths can be wide Gaussian-shaped peaks, or they can be the result of tailing or fronting. Generally, the larger diameter resins have wider peaks and lower column efficiencies.

The efficiency of a column is typically reported as either the number of theoretical plates, N (analogous to the number of plates for a distillation column) or the height equivalent to a theoretical plate (*HETP*). Both N and *HETP* are related to the width of the protein band exiting the column. This can be clearly seen from the following relations:

$$N = 5.54\left(\frac{t_r}{w_h}\right)^2 \tag{13.3}$$

$$\text{HETP} = \frac{L}{N} = \frac{L}{5.54}\left(\frac{w_h}{t_r}\right)^2 \tag{13.4}$$

where w_h is the peak width at half peak height and L is the column length. Smaller plate heights (HETP values) are analogous to higher efficiencies. The efficiency of a column is often reported as the number of plates per meter of bed length, and efficiencies for a 90 μm packed bed are on the order of 5000 plates/m. Smaller (and hence more efficient) particles should have considerably more plates per meter.

Another commonly reported term is the resolution (R_s), which is often used as a measure of the relative separation between two peaks, expressed as:

$$R_s = \left(\frac{1}{4}\right)\frac{(\alpha - 1)}{\alpha}\sqrt{N}\frac{k'_{av}}{(1 + k'_{av})} \tag{13.5}$$

where, k'_{av} refers to the average of the capacity factors. An R_s of 1.5 correlates to complete resolution (also called baseline resolution) between peaks, and "empty zones" exist between the peaks. As is intuitively obvious, the higher the R_s, the better the separation between peaks. It must be realized that the above development assumes Gaussian peak shapes. For non-Gaussian peak shapes, the asymmetry of the peak should also be examined. The asymmetry factor (A_s) is defined by:

$$A_s = \frac{b}{a} \tag{13.6}$$

where b is the second half peak width measured from the apex of the peak to the point on the peak at 10% peak height, and a is the first half peak width also measured at 10% peak height to the apex of the peak, as shown in Fig. 2. Acceptable asymmetry factors range from 0.8 to 1.2. Peaks that have long fronts or tails are undesirable, since even if extremely asymmetric peaks are well resolved, the purification is achieved at the cost of throughput and productivity.

The brief theoretical development given above can be used to determine how well a packed chromatography column will separate the desired protein from the other components. It can also be used to compare separations using different conditions. The efficiency of the column, coupled with the resolution between the desired protein and the closest troublesome impurity, can be used to screen columns for applicability. Once columns providing the necessary resolving ability are determined, the amount of protein that the column can separate per batch needs to be examined. Obviously, to maximize productivity, the most protein that can be loaded onto the column while still meeting the required purity needs to be determined.

The ability of a chromatographic resin to adsorb a protein is called the protein uptake capacity or the binding capacity. Resin manufactures report protein binding capacities at specific conditions, and it is important to note that the binding capacity can be very dependent on the specific conditions. In addition, the model proteins used by the manufacturers do not have the same binding capacity as the desired protein, but, in a pinch, they can be used for comparison. There are two types of binding capacities reported by manufacturers and in the literature: batch and dynamic. Batch (or static) capacities are determined using batch adsorption experiments. Dynamic binding capacities involve measuring the breakthrough of proteins through a packed column under given conditions at specific flow rates. The capacity, whether it is dynamic or batch, depends on the properties of the adsorbent, the properties of the protein and the experimental conditions. The column capacity is especially important during scale-up, where the size of the column is an important consideration.

The column hardware and packing technique are as important to the success of the chromatographic step as the choice of the adsorbent and the experimental conditions. Each resin manufacturer supplies a suggested packing method and packing solvent. After packing the column, the bed should be visually inspected for channeling, and an efficiency test should be run. The efficiency test is run with a nonadsorbing substance (like acetone for an ion-exchange resin) or a compound of known characteristics (like the protein of interest or a test compound suggested by the manufacturer). From the solute peak, the efficiency of the column can be calculated and compared to manufacturer specifications.

Productivity Concerns

Productivity refers to the amount of material that can be processed per equipment size per unit of time (usually either $kg/L_{stationary\ phase}/year$ or $g/m^2/day$). Throughput is defined as the amount of product processed per unit time; thus a higher throughput refers to a larger amount of material processed in the same unit of time. For process-scale chromatography, the economics must also be favorable. The costs of a chromatography step involve the cost of the resin, buffers, hardware, utilities, labor, etc., and all of these costs are directly influenced by both the productivity and the throughput of the step. Higher productivities allow for smaller sizes of equipment to process the same amount of material and thus lower capital costs. Another way to interpret this is that more product produced per day relates to a smaller number of days needed to operate the unit operation, but if bigger units are used to shorten the campaign length, the increased capital costs could outweigh the reduction in campaign length. The throughput and productivity of a chromatographic step are dependent on many factors including the size and type of resin used, the buffering species, the flow rate, the separation efficiency, the length of the

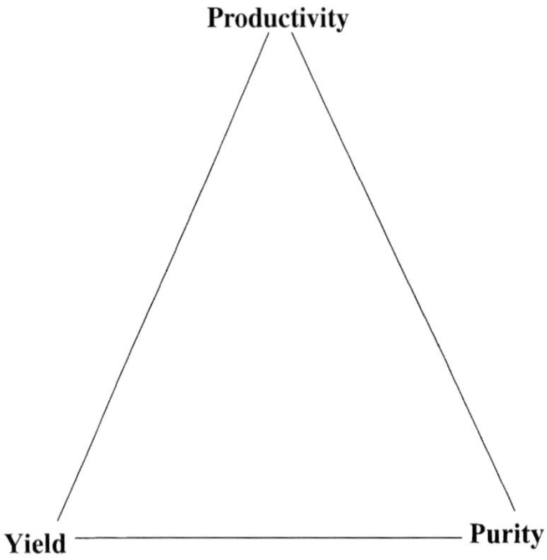

Figure 3 Productivity triangle. The maximization of the productivity of a unit operation is dependent on two main constraints, yield and purity.

column, the capacity of the resin, etc. Unfortunately, it is not possible to maximize all of these factors. The productivity triangle shown in Fig. 3 displays the tradeoff that must be made in preparative chromatography. To maximize productivity, either the yield or the purity can be optimized, but not both. For example, consider optimizing (hence maximizing productivity) an initial purification step where yield is much more important than purity. High capacity, large diameter resins should be used to increase flow rate through the packed bed, but large diameter resins are typically not very efficient, and thus, the purity of the product will be low. The separation resulting from this step will be crude at best, but the feedstock will be processed quickly; this is especially important in the initial purification, where proteases could be present in the feedstock.

The productivity of each step needs to be examined to maximize overall process throughputs. It is also important to realize that optimization of each step does not always transfer to the optimal overall process, and hence the global goal must also be taken into account during the development of each step, as discussed above.

Adsorptive vs. Nonadsorptive Chromatography

Chromatography is the use of resins to selectively retard the flow of solutes across a packed bed. The resins used in chromatography are grossly classified as either adsorptive or nonadsorptive. Solutes physically bind, or adsorb, onto the surface of the stationary phase in adsorptive chromatography. For example, ionic interactions between a charged solute and the oppositely charged stationary phase cause adsorption in ion-exchange chromatography (IEC). A typical adsorptive chromatography cycle involves loading the feedstock onto the column, washing the less strongly bound impurities from the column, eluting the desired component with a stronger eluent, removing the more strongly adsorbed solutes with a harsh mobile phase, and regenerating the stationary phase for the next cycle. However, care must be used when selecting the conditions for adsorption to ensure that the protein of interest does not bind too strongly and require harsh, possibly denaturing, elution conditions. In any adsorptive chromatographic separation, it is desirable to optimize the conditions to allow for the weakest binding of the solute to the stationary phase. Strongly bound solutes require large volumes of mobile phase or harsher mobile phases to desorb. The binding, washing and eluting conditions are chosen in adsorption chromatography to suit the situation. Adsorptive chromatographic methods provide many options: the solute of interest can be strongly bound to the stationary phase, while the impurities are either not bound or weakly bound; the desired protein can be weakly bound to the stationary phase, while impurities are either not bound or very strongly bound; or the product can be unretained by the column while the impurities are adsorbed by the stationary phase.

The use of nonadsorptive resins in chromatography mostly refers to GPC, which is also known as SEC and gel filtration chromatography. In GPC, the separation is based on size: larger molecules are restricted from the smaller pores of the resin and thus elute faster (and in less volume) than the smaller molecules that have access to a larger volume of the stationary phase. There is no adsorption and thus no concentration of the solute bands. A concentration step is frequently required before this chromatographic method. Furthermore, as the solute band disperses (due to diffusion) while it passes through the stationary phase, the volume of the feed must be less than the elution volume of the desired component, to allow for complete separation. One useful benefit of the inert nature of the stationary phase in GPC is the use of mild operating conditions, often simply the eluent from the previous step, can be used.

Gradient vs. Isocratic Elution

Isocratic elution refers to the case where the solutes are eluted off the column using the same mobile phase as the loading buffer. Isocratic elution is very effective at separating solutes that have similar binding affinities. If the solutes encompass a large range of affinities, then isocratic elution becomes less attractive due to the retention time and volume of mobile phase needed to elute the strongly retained components as compared to the weakly retained components. One instance where isocratic elution is frequently used is where the desired solute is not retained by the column, but the impurities that need to be removed are. The feed can be continuously loaded onto the column until the active binding sites on the column are occupied by the impurities. The column can then be washed to remove the impurities and equilibrated with the starting buffer to begin another purification. Isocratic elution is also used when the solutes are weakly bound to the adsorbent. Another benefit of isocratic elution is that since the loading and elution buffers are the same, no equilibration between cycles is needed, and in some applications, where equilibration steps are long, this is extremely beneficial. It is important to realize that the longer the solutes are on the column in isocratic elution, the wider the peak gets.

Gradient elution can be either continuous or step-wise. In a continuous gradient, the strength of the mobile phase is continuously increased to selectively desorb solutes. The shape of the gradient can be anywhere from convex to linear to concave, although linear is the most common. Continuous gradient elution allows for the separation of solutes having drastically different binding affinities, in a relatively short time and small volume. Unfortunately, before the next cycle can be started, the column needs to be equilibrated with the loading buffer, thus lengthening the effective cycle time. Step gradients involve the instantaneous change in mobile phase composition (e.g., 10–40 mM NaCl) allowing for the separation of solutes having vastly different binding characteristics in a relatively short amount of time. Step gradients are commonly used in systems purifying more than one component. Another common example of step gradient elution is where the impurities are not bound, but the desired solute is. Here, the feed is loaded onto the column until the stationary phase is saturated with the product. Next, a wash step with the loading buffer can be used to remove any unbound or weakly bound components still in the stationary phase matrix, and finally, the desired solute can be eluted with a step change to a stronger eluent. The adsorbent will then need to be equilibrated before the next cycle. In large-scale chromatographic processes, isocratic or step gradients are preferred due to inherent complexities associated with generating a reproducible, continuous gradient.

Types of Resin Selectivities

Ion-Exchange Chromatography

The charged nature of proteins is exploited in IEC to differentiate between individual proteins, as well as between nucleic acids and proteins. IEC involves the reversible adsorption of charged molecules to an oppositely charged functional group immobilized on a particle. Under certain conditions, some biomolecules bind irreversibly to ion-exchange resins, and it is recommended that another form of chromatography be used for this situation. The desorption of the desired solute from the stationary phase (referred to as the elution of the solute) should occur under nondenaturing conditions. IEC is the most commonly used form of chromatography to separate proteins due to its simplicity, nondenaturing operating conditions, and relatively low cost (71).

All proteins have a backbone of amino acids, which may or may not be charged. In a sense, proteins can be thought of as large polyions, with the charge dependent on the pH of the solution. Of the 20 naturally occurring amino acids, lysine, arginine, and histidine (below pH 7) are positively charged, and asparatate, glutamate, and cysteine (above pH 8) are negatively charged.

Additionally, at neutral pH, the N-terminal amines are positively charged, and the C-terminal carboxylic acids are negatively charged. The remaining amino acids do not contribute to the overall net charge on the protein. They do, however, affect the binding of the protein to various chromatographic resins, due to nonspecific interactions. It is also worthwhile to note that the charge on these amino acids is affected by both the neighboring amino acids and the pH of the local environment. The point at which the number of positive charges and negative charges are equal (i.e., where the protein is electrostatically neutral) is called the pI of the protein. The pI is useful in determining which type of ion-exchange resin to use. This will be explained in more detail below.

An IEC step, like most adsorptive chromatography steps, involves four main stages: equilibration of the stationary phase, adsorption of the sample or solute onto the stationary phase, desorption of solute, and regeneration of the stationary phase in preparation for another run. Occasionally, a wash or cleaning step is used either to flush solutes less strongly bound to the matrix before the elution of the desired protein or after the elution of the desired protein to remove strongly bound components. Sanitization might also be needed, depending on the process.

The first stage of an ion-exchange operation is the equilibration of the charged stationary phase with a buffer at a specific ionic strength and pH, thus associating the charged functional groups with simple counter-ions (typically either sodium or chloride). The equilibrium conditions should be chosen such that the adsorption of the desired compound is favorable but not too strong or irreversible.

The second stage is the loading of the feed onto the stationary phase in either the equilibration buffer or another buffer favorable to adsorption. During this stage, the solute molecules either displace the adsorbed counter-ions and bind reversibly to the stationary phase or they travel through the column unretained. A more detailed discussion on the pros and cons of using the stationary phase to bind the desired solutes or the impurities is given elsewhere in this chapter.

The third stage involves changing the conditions of the mobile phase to be unfavorable for ionic interactions, by either increasing the ionic strength or adjusting the pH. This can be done via a constantly changing gradient or via one or more step changes. On a large scale, step changes or completely isocratic operation are preferred.

The fourth stage in an IEC step is to regenerate the resin and prepare it for the next cycle. Regeneration typically involves a very high-salt buffer to remove all bound components and fully associate the functional groups with counter-ions. The nature of the regeneration buffer is dictated by the feed and by the process. Regeneration can involve extremely high pH solutions, like 1 N NaOH for complex mixtures.

The type of ion-exchange is classified by the functional group attached to the stationary phase. Anion-exchange chromatography involves a positively charged stationary phase (or column matrix) with negatively charged counter-ions. Negatively charged solutes (also called anions) exchange positions with the counter-ions and adsorb onto the stationary phase until either displaced by a compound with a higher affinity or until the charge is altered by changes in the mobile phase pH. Cation-exchange chromatography works in much the same way, but with a negatively charged matrix and positively charged counter-ions.

Anion- and cation-exchange chromatography are further subdivided into two classes, strong or weak—e.g., strong anion-exchange chromatography—based on whether or not the functional group carries a constant or pH-dependent charge. Resins classified as weak have functional groups with nominal pK_as, thus allowing for a variable charge over a pH range. Table 1 lists the types of exchanger and common functional groups attached to the stationary phase. Levison et al. present physical data on 70 commercially available ion-exchange resins (72). The authors list experimentally derived data on swelling, ion-exchange capacity, protein binding capacity, flow rate/pressure drop, and overall performance.

The type of ion-exchange resin to be used in a chromatographic step depends on three important factors: scale (or scale-up), specific requirements of the step (including purity, capacity, speed, and resolution), and the pI of the protein of interest. If a chromatographic step is to be scaled-up, the resin needs to be available in sufficient quantities. A great method developed in the laboratory is worthless if the resin is not available in large enough quantities. Additionally, the resin available in bulk should have the same characteristics as the analytical-size column; no *surprises* should occur upon scale-up. The best way to verify this is to pack an analytical column from the same lot of material to be used on large scale and confirm that the performance is the same.

When choosing resins, the specific requirements of the step should also be taken into account. For example, if the step is to be used as a capture step, then throughput and yield are more important than resolution and purity; in turn a large diameter resin should be used. Alternatively, if the chromatographic step is needed as

Table 1 Types of Ion-exchange Chromatography

Type of exchanger	Functional group	Common name
Strong anion-exchanger	Quaternary amine (*N*,*N*,*N*-trimethyl-2 hydroxy-propylamine)	Q
Weak anion-exchanger	Tertiary amine (diethylaminoethyl)	DEAE
Strong cation-exchanger	Sulfonic acid (sulfoxyethyl or sulfoxymethyl)	S
Weak cation-exchanger	Carboxylic acid (carboxymethyl)	CM

a final polishing step, high resolution is the most important criteria, and a high-efficiency small diameter resin should be used. Most ion-exchange resin manufacturers offer a full line of stationary phases from high performance, small beads to the so-called "big beads."

The choice of the functional group attached to the particle is dependent on the pI of the protein, the pH range over which the protein is stable, and the types of impurities present. At a pH below its pI, a protein has a net positive charge (more positively charged amino acids than negatively charged ones) and will adsorb onto a negatively charged stationary phase, i.e., a cation-exchange resin. Conversely, at a pH above its pI, a protein has a net negative charge and will adsorb onto an anion-exchange resin. It should also be noted that although net charge is a good first approximation, it does not always provide the necessary information. It has been postulated by many researchers that protein adsorption occurs via charged patches on the protein, and these charged patches do not necessarily reflect the overall net charge. Although very important, this discussion is beyond the scope of this chapter.

Righetti and Caravaggio (73,74) and Malamud and Drysdale (75) present tables containing the isoelectric points of many common proteins. Unfortunately, most cell culture researchers are working with proteins that are not going to be listed in these tables. In these situations, a protein titration curve can be generated via batch experiments at different pH values or electrophoretically. Modern electrophoresis equipment allows for fast, simple isoelectric point determination using isoelectric focusing (IEF). In addition to pI, the pH range over which the protein is stable is also important. If the protein is stable over a region above its pI, an anion-exchange resin should be used; conversely, if it is stable below its pI, a cation-exchanger should be used. For proteins that are stable above and below their pI, either type of ion-exchange resin can be used. Simple batch adsorption experiments with several resins at various pH values provide enough information to determine a starting point for column development work. Batch adsorption experiments (with several resins) could also be used to determine situations where the protein of interest is bound to the stationary phase and the contaminants are not. Adsorption experiments are an effective way to quickly screen a host of resins under numerous conditions. Two model proteins, lysosyme and bovine serum albumin (BSA), are commonly used to determine the type of ion-exchange resin that is needed. Lysosyme is an extremely basic protein (pI $>$ 10) and is not retained on an anion-exchanger under normal conditions. It is, however, retained quite well on cation-exchangers. BSA (pI = 4.7) provides the opposite situation; BSA is retained strongly on an anion-exchanger at neutral pH, but is unretained at pH 4.5 or less (76). It is, however, retained on a cation-exchange resin at pH values < 4.5.

The choice of buffer conditions, both equilibration and elution, are quite product-specific, but the following can be used as a general guideline. For an anion-exchanger, the starting pH should be as high as possible, taking into account product stability, with an ionic strength that is as high as possible to allow the product protein to bind but as many impurities as possible to flow through the column unretained. Typically, the equilibration buffer should be 0.5–1 pH unit higher than the pI with a salt level of 50 mM. The buffering species used in the equilibration buffer should not interact chemically with the product and should provide an adequate buffering capacity at the desired pH. More information about the choice of buffering species is provided below. The elution buffer is usually at the same pH as the equilibration buffer with a higher concentration of salt, typically 0.5–1.0 M. Gradient elution involves gradually blending the elution buffer into the equilibration buffer until either 100% elution buffer is being used or the desired protein has eluted. A cation-exchanger would use a presaturation buffer that is 0.5–1 pH unit below the pI of the protein and the same 50 mM salt concentration; the elution condition would again be at the same pH and 0.5–1.0 M salt.

For both anion- and cation-exchangers, a third buffer—often called a cleaning buffer—is employed after elution to remove strongly bound components. The cleaning buffer typically has a much higher salt level, often as high as 4 M, and occasionally is at pH extremes, such as pH 13 or pH 1.

In certain instances, the buffer conditions can be manipulated to allow either the product to flow through the column unretained and the impurities to be bound to the adsorber, or the product to be bound and the impurities to be unretained. This is especially useful to remove nucleic acids (which are negatively charged) from a basic protein on an anion-exchange column. Care must be taken when using chromatography resins to bind impurities since not all impurities are likely to bind to the resin. If used early in the purification process, flow-through columns are extremely productive, but it is generally accepted that a binding step (i.e., where the desired solute is bound to a resin) is needed downstream of a flow-through chromatographic step.

The choice of buffering species is very important and should not be taken lightly. The most important criteria include using buffering species that provide buffering capacity at the desired pH, do not interact chemically with the protein and are nontoxic. Interaction of the buffering species with the adsorbent is the subject of some debate. If the buffering species is such that it is oppositely charged than the adsorber, then when the buffering species is charged, it will compete with the protein for sites on the adsorber. Fortunately, the capacity of the adsorber for small ions (like charged buffering species) is usually much greater than that for proteins, which minimizes the competitive adsorption effects due to the ability of small ions to bind to many more adsorption sites. To be safe, the buffering species should form ions of the same charge as the adsorber and thus not compete for binding sites. There are numerous published tables and charts providing pK_as of different buffering species (77–80), but many of these species are hazardous and/or toxic. Good et al. (81) published a list of biologically favorable buffering species, the so-called "Good's Buffers," that are commonly used in the downstream processing of proteins. Lastly, some situations require volatile buffers, although this is less common; examples of these are formic acid, acetic acid, ammonium acetate, and ammonium carbonate.

IEC, like all adsorptive chromatographic methods, can be operated using either isocratic or gradient elutions, although most isocratic elutions actually involve the use of step gradients, i.e., the instantaneous change from one buffer to another. Additionally, it is worth noting that the traditional sense of gradient elution involves the precolumn mixing of two streams to generate a gradually changing mobile phase. This gradient can be of ionic strength or pH. The use of salt (or ionic strength) gradients is explained above.

In the late 1970s and early 1980s Sluyterman and co-workers developed a technique, termed chromatofocusing, utilizing internally generated pH gradients (82–88). The technique utilizes a column with a buffering capacity, like DEAE, and an eluent containing a low ionic strength, synthetic polymer mixture containing thousands of individual buffering species with a broad spectrum of pK_as and thus a very linear titration curve and an even buffering capacity over a large pH range. Their method involved equilibrating a weak-base ion-exchange column with a buffer containing a simple basic buffering species (same charge as the adsorbent) at a high pH. The protein sample is adsorbed onto the adsorbent, and then the focusing buffer (titrated to a lower pH with acetic acid, for example) is pumped through the packed column. A linear pH gradient is generated internally (without precolumn mixing) by the titration of the adsorbent by the polyampholyte buffer. In use, these polyampholyte buffers have many limitations, such as their high cost and their known propensity to associate with proteins, thereby restricting the widespread acceptance of chromatofocusing and its use on a preparative scale.

Chromatofocusing typically results in very high resolution and is frequently used as a final polishing step. Nagashima et al. used chromatofocusing as a final polishing step, even though they used a weak-base ion-exchange step in the intermediate processing step (36). Sluyterman and Wijdenes (84) experimented with using simple buffering species to internally generate continuous pH gradients, but concluded that the polyampholyte buffers produced better pH gradients. Recently, investigators have developed techniques involving retained, internally generated pH gradients using simple buffering species to avoid the complications associated with polyampholyte buffers (89–93). Frey and co-workers developed a chromatofocusing technique using pH step gradients (containing one or multiple pH steps) (89,90). Their procedure involved equilibrating the column at one pH with one counterion (acetate, for example) and eluting at a lower pH with a second counter-ion (chloride, for example). Bates and co-workers expanded the earlier work using simple buffering species to generate linear pH gradients to separate proteins (91–93).

Hydrophobic Chromatography

The separation of biological components by hydrophobic interactions was first suggested by Tiselius in the late 1940s (94), and since then, reversed phase chromatography (RPC) has been widely used for the analysis of a variety of substances (95). Hydrophobic interactions are the binding of hydrophobic moieties in the presence of water (96).

The term reversed phase arises from operating a column in a *reverse mode* to what was traditionally done (called normal phase). In normal phase chromatography, the solute partitions between a polar stationary phase, often underivatized silica, and a nonpolar, organic mobile phase. In RPC, long alkyl chains are bonded to the silica backbone creating a hydrophobic, nonpolar stationary phase. The solutes partition between the hydrophobic stationary phase and the polar mobile phase (e.g., water). The strength of the mobile phase is increased by adding an organic modifier (like acetonitrile or methanol) to the water, thus decreasing the surface tension of the stationary phase, and desorbing the solute. RPC is typically operated using a continuous, linearly increasing gradient of organic modifier, i.e., from 5% to 95% acetonitrile, generated by precolumn mixing of the two "pure" solutions.

Reversed phase chromatography is usually performed with small particles (e.g., 10 μm diameter) and is a very efficient step offering the potential for high resolution. Highly polar proteins are eluted under mild conditions in RPC, and in these instances RPC is used for low-loading, high-resolution steps, i.e., final polishing steps. Comuzzi et al. used RPC after a heparin affinity column to purify milligram quantities of RANTES (regulated on activation, normal T cell expressed and secreted) (97). The authors concentrated the effluent (by ultrafiltration) from the affinity chromatography step and loaded it onto a C-4 column. The RANTES was eluted using a 60-min, linear acetonitrile gradient with an approximate 80% yield. Other researchers have also used RPC in the purification of proteins, but RPC is not as common as a similar technique, HIC, in the downstream processing of proteins.

Hydrophobic interaction chromatography (HIC) and RPC are similar techniques, separating solutes based upon interactions between surface hydrophobic patches on the protein and the hydrophobic ligands on the surface of the stationary phase. RPC typically uses long alkyl chains having 4, 8, 12 or 18 carbon atoms, while HIC adsorbents contain short alkyl chains, 8 carbon atoms or less, or phenyl groups. Additionally, the capacities of HIC columns are much less than the capacities of RPC columns, due to the degree of substitution of the functional groups. In contrast to the high organic content of RPC mobile phases, HIC utilizes aqueous mobile phases that are less likely to denature proteins; the solutes are adsorbed on a mildly hydrophobic stationary phase with a high-salt aqueous mobile phase and eluted with a low-salt aqueous mobile phase. The high-salt concentration of the loading buffer increases the surface tension of the mobile phase, thus favoring adsorption onto the stationary phase (98). This "hydrophobic effect" is defined as an unfavorable interaction between a nonpolar moiety and water (99). The "salting out" of proteins enhances binding in HIC, and hence the precipitation-promoting salts (PO_4^{3-}, SO_4^{2-}, NH_4^+, K^+, etc.) from the Hofmeister lyotropic salt series (100) are preferred, but not required, in HIC. After adsorption, the solutes are desorbed off the stationary phase using a mobile phase containing a lower concentration of salts, at the same pH. Horváth and colleagues have published a series of articles related to binding in HIC revolving around the solvophobic theory (97,101–104). Nagashima et al. used a HIC step consisting of a phenyl functional group as an intermediate purification step to purify an enzyme (36). Numerous other groups have reported the use of HIC in the recovery of proteins (105–112).

In HIC, pH effects are worth considering, but care should be taken to not denature the protein of interest. Several pH values should be tested within the stability range of the protein; this can be accomplished in a batch adsorption mode before column studies are initiated. Frequently, increasing the pH (and thus deprotonating amino acids) will decrease the hydrophobic interactions.

Temperature effects have also been shown to be important in HIC (102,103), but it must be realized that temperature control on a large scale is not as easy as in the laboratory.

The development of a HIC step should involve examining several salt concentrations, pHs, and ligand types. The loading of the feed onto the hydrophobic resin should be performed under the lowest possible salt concentration where the desired protein binds, but conditions where precipitation occurs should be avoided. During method development, the screening via batch adsorption experiments should focus on initial pH and the choice of salt. A common starting point for method development is 1.0 M ammonium sulfate at a pH where the protein is stable. The ligand attached to the stationary phase is also important, and several different ligands should be examined. Once the resin and loading conditions are set, the type of elution needs to be determined, either isocratic or gradient. A gradient elution would involve simply decreasing the salt concentration of the mobile phase by precolumn mixing of the high salt buffer with a low or no salt buffer until the desired protein elutes. Gradients are typically easier to develop, since a gradient with a fairly shallow slope works most of the time. Isocratic elution offers the possibility for better resolution between components, as described above, but they frequently require a larger volume of mobile phase to elute the desired component. Determining the optimal isocratic conditions is also a time-consuming operation, often involving numerous iterations.

Thiophilic Chromatography

Thiophilic chromatography was first described by Porath and colleagues (113–115) in their efforts to purify immunoglobulins from serum and since has been used by other researchers to purify proteins from serum, ascites fluid, hybridomas, and prostate-specific antigen proteins (112–126). Thiophilic chromatography goes by many names such as T-Gel, thiophilic interaction chromatography (TIC), and thiophilic affinity chromatography. Thiophilic interactions result from having a nucleophilic molecule three atoms away from the sulfone group (21), such as mercaptoethanol bound to divinyl sulfone-activated agarose. Other novel ligands are also used (116, for example).

Thiophilic interactions between the sulfone–thioether ligand and active sites on the protein are very specific, promoted by water structure-forming salts (like sodium sulfate) (113,118,119), and adsorption is enhanced by high-salt concentrations. Similar to HIC, loading occurs using high concentrations of sodium, potassium or ammonium sulfate, and elution is accomplished by decreasing the salt concentration either via continuous or step gradients. Berna et al. compare and contrast the adsorption of proteins from serum in TIC and HIC; they concluded that both TIC and HIC are salt-promoted, but the salt promotes different types of weak molecular interactions in each mode (117). Sulfate salts have been shown to increase binding, and chloride salts have been shown to decrease binding (21).

Affinity and Pseudo-affinity Chromatography

Affinity adsorbers can be loosely classified into two categories: highly specific adsorbers and group adsorbers (21). Highly specific adsorbers use the actual enzyme or substrate or an analog as the ligand, and those ligands are immobilized onto the stationary phase. Group adsorbers use ligands that bind to a wide range of proteins that have similar structures or functions, such as using immobilized Protein A to purify antibodies. Highly specific immunoaffinity resins are either available commercially or easily produced using activated particles that can be loaded with a ligand of choice from resin manufacturers (127, for example), but the specifics involved with choosing and loading of the ligand are beyond the scope of this chapter. For the remainder of this chapter, affinity chromatography will be used to refer to group adsorbers. A list of common affinity resins and literature references are given in Table 2.

The two most common affinity resins for antibody purification are Protein A and G. Both resins utilize immobilized bacterial proteins that bind specifically to the Fc region of IgG antibodies (128). Protein A is more specific to mammalian IgGs (like fully humanized IgGs of subclasses 1, 2 and 4), while Protein G shows greater specificity towards mouse or rat IgG (128). On either Protein A or Protein G, the loading of the feedstock is done at or near physiological pH and ionic strength, e.g., 10 mM sodium phosphate, 150 mM NaCl, pH 8.0. After loading the antibody-containing feedstock onto the affinity resin, the resin is then washed with 5–10 column volumes of buffer under loading conditions. The elution of the antibody off the column requires a strong elution buffer (e.g., 1.0 M acetic acid, pH 3) and the antibody-containing effluent should be neutralized, dialyzed or diafiltered as soon as possible, to limit the degradation of the antibody in the strong elution buffer. Examples of the use of several resins are given below.

Pseudo-affinity adsorbers, like immobilized metal affinity resins, bind proteins through interactions common to many groups of proteins. IMAC separates proteins based on their affinity for metal ions. Two amino acids, histidine and cysteine, form complexes with transition metals at neutral pHs. Typically, the affinity of a protein for an IMAC column is proportional to the number of surface histidine residues. This is often exploited by adding a poly-histidine fusion to recombinant proteins to simplify the purification. Fusion and tagged proteins are beyond the scope of this chapter, but the reader is encouraged to review the literature on using fusion proteins to simplify recovery using IMAC columns (129–135), affinity tags (e.g., green fluor-

Table 2 Examples of Affinity Chromatography Resins

Ligand	Specificity	References
2′5′ ADP	Enzymes which have $NADP^+$ as cofactor	21
5′ AMP	Enzymes which have NAD^+ as cofactor, and ATP-dependent kinases	21
Blue (Cibacron Blue)	Broad range of enzymes which have nucleotide cofactors, serum albumin, etc.	21,179,180
Bezhydroxamic acid (BHA)	Peroxidase	181–184
Red (Procion Red)	Broad range of enzymes which have nucleotide cofactors, etc.	21
Protein A	Fc region of IgG and related molecules	174,185,186
Protein G	Similar to Protein A, but different affinities for IgGs from different species	187
HCIC	Similar antibody binding affinities to Protein A, but with a different ligand	137–139
IMAC	Proteins containing surface histidines	121,162

escent protein) (38,136–139), or other fusions (140–144). The most common metal ions used in IMAC are Zn^{2+}, Cu^{2+}, Ni^{2+} and Fe^{2+}.

The development of a purification step should begin with several analytical experiments (either in column or batch adsorption mode) to determine the best metal ion to use. Typically, copper offers stronger binding, and zinc is often weaker. Loading of the protein-containing feed solution onto the IMAC column should be done at relatively high ionic strength (0.5–1.0 M NaCl) with a slightly alkaline pH (typical buffers include tris acetate at 50 mM or sodium phosphate at 25–50 mM). After loading the protein, the column should be washed with 5–10 column volumes of the loading buffer to remove weakly bound contaminants. The elution buffer is at the same conditions used for loading but with an increased amount of a substance with a strong affinity for the immobilized metal ion, e.g., ammonium chloride (150 mM) or imidazole (50 mM). EDTA can also be used to elute proteins from an IMAC column, but it must be realized that EDTA will strip the metal–protein complex from the column, thus offering no further resolution. EDTA is frequently used to clean all strongly bound protein from the column matrix. As with all adsorptive methods, isocratic or gradient elution can be used, and it is up to the reader to decide which is best for the situation at hand. Lastly, the issue of metals leaching off the column is worth considering, especially for large-scale operations.

Recently another affinity resin has gathered much attention, especially in the recovery of antibodies. Hydrophobic charge induction chromatography (HCIC) utilizes a ligand that is antibody-selective and allows for elution under much gentler conditions. HCIC was first described by Burton and Harding (145) using a pH-dependent, dual-mode ligand (4-mercaptoethylpyridine, see Fig. 4) that favor hydrophobic binding under loading conditions (physiological pH) and electrostatic repulsion under elution conditions. A typical HCIC step involves equilibrating the column with a 50 mM tris buffer at pH 8 (phosphate-buffered saline, pH 7, is also used) and loading the sample onto the column. The loaded column is then washed with the loading buffer before an instantaneous step change to the elution buffer (50 mM sodium acetate, pH 4.0) rapidly desorbs the antibodies. The pooled antibody containing fractions are then immediately dialyzed (or buffer exchange) to a more suitable buffer. Schwartz et al. (146) compare HCIC to traditional Protein A purification for several systems. They reported recovery values in the 92–98% range with purities >95%, which were at least as good as those realized from a Protein A recovery system. A four-log reduction in viral load was also seen in the HCIC step, and

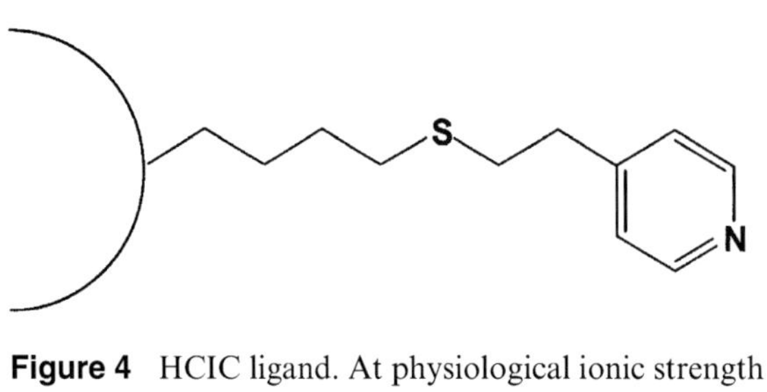

Figure 4 HCIC ligand. At physiological ionic strength and pH values, the aromatic pyridine ring is uncharged and hydrophobic. The 4-MEP is known to bind strongly and selectively to IgG subclasses from most species. When the pH of the solution is dropped below the pK_a of the nitrogen (4.8), the ligand has a positive charge; under such conditions, most antibodies carry a positive charge, and therefore electrostatic repulsion occurs.

significant DNA reduction was realized. Additionally, Guerrier et al. (147) used HCIC to purify milligram quantities of whole $hIgG_1$ and antibody fragments.

Gel Permeation Chromatography

Gel Permeation chromatography has been used extensively in the purification of proteins, nucleic acids, and other biological macromolecules since its inception (148,149). Gel chromatography (also called GPC, gel filtration chromatography or SEC) is a mild purification technique that separates molecules based on size, using an inert (noninteracting) matrix with a carefully controlled pore structure. The pore structure of the gel matrix is such that smaller molecules are able to diffuse into more of the matrix than larger molecules, and hence the larger molecules elute first from the column followed by the other solutes, in order of decreasing size. It is important to note that the solute "size" reported from a gel chromatography assay is dependent on the shape of the molecule. Two proteins of the same molecular weight can have different elution characteristics depending on their shape, e.g., one is a rod-like molecule and one is spherical in shape. Most commercially available "calibration" standard samples assume a spherical shape.

Gel permeation chromatography is a very mild purification technique that allows operation at conditions suited for the stability of the desired protein, and it can be operated in the presence of cofactors, at high or low pH, at 37°C or 4°C, with or without detergents, etc. It is also a suitable technique for both preparative practices and analytical techniques. Preparative gel permeation chromatography is frequently used as a polishing step where low capacities are acceptable and the feed stock is relatively clean (150). Dimers and degradation products are the most common impurities removed from the product in gel chromatography steps. Additionally, GPC is frequently used as a buffer-exchange step.

It is generally believed that gel filtration separates solutes based on steric hindrances, and that gel filtration can be used to determine the molecular weight of components (151–153). To determine the molecular weight of an unknown sample, a set of standards of known molecular weight is used to determine a calibration curve (also called a selectivity curve). The set of standards would contain similarly shaped molecules with molecular weights greater and less than the expected molecular weight of the desired protein. A typical calibration curve involves plotting the logarithm of the molecular weight vs. the average distribution coefficient ($K_{\mathrm{d,av}}$).

$$K_{\mathrm{d,av}} = \frac{V_{\mathrm{e}} - V_{\mathrm{o}}}{V_{\mathrm{t}} - V_{\mathrm{o}}} \tag{13.7}$$

where V_{e} is defined as the volume of mobile phase needed to elute the solute from the column, V_{o} is defined as the volume needed to elute a solute that does not enter the pores of the resin, and V_{t} is the total volume of the column. The volume of mobile phase needed to elute the solute from the column can be calculated by multiplying the retention time and the flow rate. The subsequent plotting of the distribution coefficients and the log of the molecular weights of the standards results in a pseudo-linear graph from which the molecular weight of an unknown sample can be determined from its elution volume. Again, it is imperative to realize that the shape of the molecule is very important and can skew the results drastically. Obviously, the relative molecular weights of compounds are less important than purity and recovery when GPC is used preparatively.

NONCONVENTIONAL CHROMATOGRAPHIC METHODS

Displacement Chromatography

Displacement chromatography is a type of elution chromatography where the more strongly bound solutes push (or displace) the less retained solutes from the stationary phase. The different moieties are separated into adjacent zones of highly pure individual components. In contrast to standard elution chromatography, no "empty zones" exist between solute zones. The zones exit the column in order of increasing binding strength in a so-called "displacement train," with the last moiety to exit, having the strongest binding strength, being the displacer.

The displacers can be ampholytes (154,155) indigenous proteins (156), other proteins (157,158), or small molecules (159–161). The presence of a displacer, and hence a displacement train, allows the solutes to be loaded at levels significantly higher than normal elution chromatography, without the detrimental peak tailing that is seen in overloaded elution chromatography. Displacement can occur with any adsorption media, and numerous examples can be found in the literature for displacement in IEC (162), IMAC (163), HIC (164), hydroxyapatite (165), and affinity-based chromatography (166). The choice of displacer is dependent on both the type of resin used and the protein to be purified, and there are numerous alternatives for each type of resin.

The choice of displacer is obviously a crucial factor in the utility of displacement chromatography for the desired solute. Displacers should be chosen such that they are nontoxic, stable, inert with respect to the components in the feed, soluble in the mobile phase, easily removed from the desired solute and have a stronger affinity for the stationary phase than any of the components in the feed. Another factor to consider is the concentration of the displacer. Too little displacer results in broad zones of insufficient purity, and too much displacer results in insufficient volume available to form a displacement train and hence little to no separation between components.

A typical purification cycle consists of presaturating a column with a buffer that is favorable to adsorption. Next, the solute is loaded onto the column followed directly by the loading of a solution containing the displacer. The column is then subjected to a buffer that is unfavorable to adsorption. This elution buffer allows the formation of the "displacement train" and collection of product. The last step involves regenerating the column in preparation for the next cycle.

Expanded Bed Chromatography

As mentioned above, EBA or expanded bed chromatography allows the loading of unclarified feedstocks directly onto a recovery column. Expanded bed chromatography combines the capture and intermediate purification steps by simultaneous utilization of the same resin and equipment for two separate steps. The first part of EBA is the generation of a fluidized bed by an upward flow. The second step is a downward flow resulting in a traditional adsorption chromatographic step. As with any fluidized bed, it is imperative that the flow rate is greater than the minimum fluidization velocity to ensure adequate mixing. A fluidized bed has much greater space between the particles (called intraparticle void volume) than a traditional packed bed, allowing the use of viscous feed solutions at high flow rates with minimal pressure drop. A typical EBA step would involve the loading of a viscous feed solution in such a manner to create a fluidized bed where the cell debris and other nonadsorbed material flow through the bed. This is followed by a wash step to remove debris and unbound components from the resin. After completion of the loading and washing steps, the direction of flow is reversed and a standard elution chromatography step is run. The type of elution depends on the nature of the separation and the type of functional group attached to the bead. Many researchers have reported the use of EBA to purify proteins from bacteria, yeast and mammalian cultures (167–175). Garke et al. used EBA with cation exchange resin as an initial step in the purification of recombinant human basic fibroblast growth factor (173). Feuser et al. discussed the use of EBA for the purification of monoclonal antibodies (MAbs) and report that hybridoma cells interact with an expanded bed of cation exchange resin but do not interact significantly with Protein A adsorbent (19). Fahrner et al. reported that expanded bed Protein A chromatography had similar purification results as typical fixed bed chromatography (174). The removal of whole cells from product-containing supernatant is a common use of EBA, but if the cells rupture, proteases, nucleic acids and other undesirable moieties will be released. Fortunately, several researchers have shown that EBA is a mild technique, even for shear sensitive hybridoma cells (175), and thus cell rupture is usually not an issue.

DESIGN OF PURIFICATION PROTOCOLS

Although, there are no generic protein purification procedures, several guidelines will assist in the method development process. To ensure adequate purity and adventitious agent clearance, two to three orthogonal purification techniques should be used in the recovery process. It is important to maximize the strengths of each unit operation and not to demand too much from each step. The design of the purification protocol should utilize techniques that are adequate for that stage of purification; for example, a size exclusion chromatographic method should not be used as a primary recovery step; instead an UF step should be used if size-based retention is desired. Lastly, the entire purification protocol must be chosen such that the protein is not denatured or rendered inactive.

Recovery of Non-MAb Proteins

A generic protocol utilizing the techniques and guidelines explained above for the purification of most non-MAb proteins would include the following steps:

- Initial purification—microfiltration to remove the whole, intact cells, followed by either a low resolution chromatography step or an ammonium sulfate precipitation to crash most proteins out of solution.
- Intermediate—HIC and IEC (order interchangeable) to remove impurities with different hydrophobic and ionic characteristics, respectively, as well as to drastically reduce the DNA/RNA and viral load.
- Polishing–HIC or IEC using very high-resolution particles followed by GPC to remove any dimers.

In between unit operations, buffer exchange and concentration steps will occasionally be needed. Of course, this "generic" protocol does not work in all instances, and most certainly it is not the most productive or efficient process. Several simple modifications to the protocol can be quickly made based upon known attributes of the protein. For example, if surface histidines are present, an IMAC column can replace either the HIC or the IEC step in the intermediate stage. Lastly, nucleic

Table 3 Examples of Protein Purification

Protein	Primary recovery	Intermediate purification	Final polishing	References
UDP-glucuronate	Precipitation	AEC HIC	GPC CF	36
MT5-MMP	Centrifugation	AEC	IAC	188
2-hydroxy-isoflavanone dehydratase	HAC Precipitation	AEC HIC AEC	CF GPC	189
4-hydroxy-benzoate glucosyl-transferase	Precipitation GPC	AEC GPC HAC Affinity chromatography	CF AEC	180
UBGAT	Precipitation GPC	HIC GPC	CF	112

AEC, anion-exchange chromatography; CF, chromatofocusing; HAC, hydroxyapatite chromatography; IAC, immunoaffinity chromatography; GPC, gel permeation chromatography.

acid and virus-specific affinity resins can be used in the protocol to ensure adequate clearance. Table 3 presents several purification schemes used by other researchers.

Recovery of Monoclonal Antibodies

The recovery of monoclonal antibodies has been covered in some detail above in various sections, and the references provided in those sections contain many purification protocols. An article by Iyer et al. (176) provides information on many different issues that arise during the development of an antibody purification process. Most MAb purification schemes employ similar chromatographic techniques, e.g., Protein A, IEC, HIC, and affinity, used in various orders depending on the impurities present. Other researchers have developed one-step antibody purification methods using novel affinity or immunoaffinity ligands (127, for example). As provided in the previous section for non-MAb proteins, below is a quick, generic protocol for the purification of MAbs:

- Initial purification—centrifugation, depth filtration or microfiltration to remove whole, intact cells, followed by big-bead Protein A, HIC or batch adsorption or chromatography.
- Intermediate processing—Protein A to bind all the antibodies, followed by IEC to purify based on ionic characterization.
- Polishing—GPC or high-performance cation-exchange chromatography to remove dimers or high-performance IEC (like chromatofocusing) to remove close analogs based on charge.

Again, the buffer exchange and concentration steps have been omitted, but they will most likely be needed between the techniques. If EBA is used, the microfiltration step can be eliminated. In addition, endotoxin-specific columns (so-called detoxy gels) are frequently utilized in-between purification zones to ensure adequate endotoxin removal.

The current industrial trend is to avoid using Protein A, thus avoiding potential leachable or extractable issues with the column matrix. In these non-Protein A processes, the protocol is similar to that used for non-MAb proteins discussed above. Yazaki et al. (177) discuss the purification of antibody fragments using a nonaffinity-based purification process. The authors utilized batch adsorption of an anion-exchange resin followed by vacuum filtration through a 0.2μm membrane to clarify the harvest supernatants. The clarified broth was then passed through two complimentary chromatographic columns (with dialysis before and after each column) to produce sufficient quantities of minibody and diabody fragments for both preclinical and early clinical supplies. Directly following clarification and dialysis, the broth was applied to a ceramic hydroxyapatite (calcium-hydroxide-phosphate functional groups) column at low salt and near biological conditions. After sufficient washing (until the baseline—measured at 260 nm—returned to zero), a linear gradient of K_2HPO_4 was used to elute the desired components off the hydroxyapatite resin. The fractions containing product were then dialyzed and adsorbed onto a strong anion-exchange column. The column was then washed until the baseline returned to zero and eluted with an increasing linear salt (sodium chloride) gradient. The product-containing fractions were then concentrated and sterile filtered. The authors also examined the use of a hydrophobic interaction column instead of the hydroxyapatite column. The product was eluted off the column using a decreasing salt gradient, but formation of a minibody dimer was noticed. The authors reported recoveries of between 40% and 81% for each chromatographic step. The purified product was lastly filtered through a 0.04 μm filter to ensure viral removal.

Additionally, researchers at the University of Minnesota reported using a novel, modified zirconia resin with a step gradient in salt to replace Protein A in

Table 4 Examples of Monoclonal Antibody Purification

Protein	Primary recovery	Intermediate purification	Final polishing	References
MAb24	Ultrafiltration	Protein A	GPC	186
Polyhistidine-tagged antigen	Centrifugation	IMAC	GPC AEC	121
BsMAbs	Centrifugation Ultrafiltration	Protein A	HAC	190
Polyhistidine-tagged firefly luciferase	Centrifugation EBA (IMAC)			191
Peroxidase-tagged antibodies	Precipitation	Affinity chromatography		181
IL-18	Anti-porcine IL-18 affinity chromatography			127

AEC, anion-exchange chromatography; HAC, hydroxyapatite chromatography; IMAC, immobilized metal affinity chromatography; EBA, expanded bed adsorption; GPC, gel permeation chromatography.

the purification of MAbs (178). Numerous other examples are available in the literature, some delineated in Table 4.

SCALE-UP STRATEGIES FOR CHROMATOGRAPHIC PURIFICATION OF BIOLOGICS

Although gram quantities of protein can frequently be produced using bench-top equipment, the need for larger quantities is occasionally required. In the biotechnology industry, this is akin to the advancement of a potential biological candidate. To produce more material, either several batches can be run in parallel or the laboratory scale procedure can be scaled up. The procedure for scaling up, if done properly, is quite simple, but before the scale-up can be done, several factors must be considered. The same resins (or packing material) used in the laboratory should be used on the preparative scale to provide the same separation results. Ensuring that the packing material used in the laboratory is available in sufficient quantities is very important for obvious reasons. Additionally, the cost, continuity and consistency of the bulk resin must be thoroughly examined. The cost of the resin is not always a major part of the recovery cost, especially if the resin can be reused over numerous cycles, but the continuity of the supply of resin and the batch-to-batch consistency are vital to the success of a large campaign. If the resin is to be part of a FDA-filed process, an audit of the manufacturer is advisable, examining factors such as quality assurance protocols, batch sizes, and quality control data. The ability of a resin to be cleaned-in-place (CIP) is very important if resin reuse is planned. The CIP conditions often involve harsh solutions (such as 1 M NaOH), and the chemical stability of the packing material must be confirmed under these conditions. Laboratory studies should include a column lifetime study, where complete cycles (loading, eluting, CIP, regeneration) are continuously repeated until column degradation is noticed. Occasionally, these studies are conducted at elevated temperatures to extrapolate past the number of cycles tested, but these elevated temperature studies should be interpreted carefully. They frequently can provide misleading results. The most accurate method would involve operating the stability testing under the same conditions of the preparative separation.

The actual scale-up is a relatively straightforward task involving several simple steps, listed below:

- The amount of protein to be recovered is needed
- The maximum protein loading at laboratory scale is required
- The effective resin capacity needs to be determined
- The size of the preparative column needs to be calculated.

Before the preparative method can be determined, the mass of protein that is going to be needed at preparative scale must be identified. The mass of protein needed depends on its intended use (preclinical, phase I, phase II, etc.), the size and potency of the bioreactor, and whether or not multiple lots of material will be pooled prior to chromatographic purification. Once the total amount of protein needed is known, the maximum amount that can be loaded onto the laboratory column must be determined. The easiest way to determine the maximum loading is to incrementally increase the amount of feed applied to the column until the purity and/or yield of the desired protein becomes unacceptable. The increase of protein load on the column can be accomplished in two ways, either volume or concentration overloading. If the step involves adsorption chromatography, and the feed is loaded in a buffer that promotes adsorption of the solutes onto the stationary phase, then the amount of solute added can be increased by simply increasing the volume of the feed solution at the same concentration loaded onto the column. An alternative is to increase the concentration of the feed in order to add the same feed volume but more mass of protein. Adding the same volume of solution is critical in nonadsorptive chromatographic steps.

Once the maximum loading is known, the effective column capacity can be calculated. The maximum mass of protein loaded onto the column ($m_{p,max}$) divided by the total column volume (V_c) yields the effective column capacity, CAP_{effect}:

$$\mathrm{CAP}_{\mathrm{effect}} = \frac{m_{\mathrm{p,max}}}{V_{\mathrm{c}}} \tag{13.8}$$

Having the capacity of the packing material, combined with knowing the mass of protein that is needed per cycle, the volume of the preparative column can be calculated. The quotient of the mass of protein required (m_{prep}) by the effective column capacity yields the preparative volume of resin needed to purify m_{prep} of protein:

$$V_{\mathrm{prep}} = \frac{m_{\mathrm{prep}}}{\mathrm{CAP}_{\mathrm{effect}}} \tag{13.9}$$

From the volume of resin needed it is trivial to calculate the dimensions of the preparative chromatography column. Although scale-up based solely on column volumes are routinely performed (i.e., the dimensions of the column are not important) keeping the length of the preparative and laboratory column the same is the safest situation. The length of column is directly related to the efficiency of the column, and it is also directly related to the pressure drop across the column (i.e., longer columns have higher pressure drops, all other factors being equal). Therefore, whenever possible it is best to scale-up based upon inner diameter, keeping the length of the column fixed. Thus, once the volume of the preparative column is known, the required cross-sectional area (and hence the diameter) of the preparative column can be calculated from the length of the analytical column. For example, if 10 L of resin are required and the length of the laboratory column used was 20 cm, the inner diameter (ID) of the preparative column should be 25.2 cm, as seen below:

$$\mathrm{ID} = \sqrt{\frac{4V_{\mathrm{prep}}}{\pi L}} = \sqrt{\frac{4 \times 10\ \mathrm{L} \times (1000\,\mathrm{cm}^3/\mathrm{L})}{\pi \times 20\,\mathrm{cm}}} = 25.2\,\mathrm{cm} \tag{13.10}$$

Therefore, in this example, the preparative column would be 20 cm × 25.2 cm ID.

It is also very important to remember that as the purification progresses to large pilot or production scale, the weight of the bed becomes significant. In these instances, extra care should be taken to keep the length of the column short. Most laboratory columns are <30 cm long, and if the scale-up is performed based on inner diameter, this does not become an issue.

REFERENCES

1. Watson J, Crick F. A structure for Deoxyribose Nucleic Acid. Nature 1953; 171:737.
2. Howard J, Fox S. Genet. Eng News 1994; 14:8.

3. Regnier FE. High performance liquid chromatography of biopolymers. Science 1983; 222:245.
4. ICH Secretariat Guidelines, "Q5A: Viral Safety Evaluation of Biotechnology Products Derived from Cell Lines of Human or Animal Origin", Geneva, Switzerland, 1997.
5. ICH Secretariat Guidelines, "Q6B: Specifications: Test Procedures and Acceptance Criteria for Biotechnological/Biological Products", Geneva, Switzerland, 1999.
6. US Food and Drug Administration, Federal Register, Volume 63, Number 185 (1998) 51074.
7. Petsch D, Anspach R. Endotoxin removal from protein solutions. J Biotechnol 2000; 76:97.
8. European Pharmacopoeia Commission, European Pharmacopoeia 3rd ed., Strasbourg, France, 1997.
9. Bischoff R, Speck D, Lepage P, Delatre L, Leoux K, Brown S, Roitsch C. Purification and biochemical characterization of recombinant α_1-antitrypsin variants expressed in Escherichia coli. Biochem 1991; 30:3464.
10. Gerba C, Hou K. Endotoxin removal by charge-modified filters. Appl Environ Microbiol 1985; 50:1375.
11. Hou K, Zaniewski R. Depyrogenation by endotoxin removal with positively charged depth filter cartridge. J Parenter Sci Technol 1990; 44:204.
12. Hou K, Zaniewski R. Endotoxin removal by anion-exchange polymeric matrix. Biotechnol Bioengr 1990; 12:315.
13. Neidhardt E, Luther M, Recny M. Rapid, two-step purification process for the preparation of pyrogen-free murine immunoglobulin G1 monoclonal antibodies. J Chromatogr 1992; 590:255.
14. Li L, Luo R. Use of Ca^{2+} to re-aggregate lipopolysaccharide (LPS) in hemoglobin solutions and the subsequent removal of endotoxins by ultrafiltration. Biotechnol Tech 1998; 12:119.
15. Bordier C. Phase separation of integral membrane proteins in Triton X-l 14 solution. J Biol Chem 1981; 256:1604.
16. Aida Y, Pabst M. Removal of endotoxin from protein solutions by phase separation using triton X-114. J Immunol Methods 1990; 132:191.
17. Liu S, Tobias R, McClure S, Styba G, Shi Q, Jackowski G. Removal of endotoxin from recombinant protein preparations. Clin Biochem 1997; 30:455.
18. Sofer G. Preparative chromatographic separations in pharmaceutical, diagnostic, and biotechnology industries: current and future trends. J Chromatogr A 1995; 707:23.
19. Feuser J, Halfar M, Lütkemeyer D, Ameskamp N, Kula M, Thömmes J. Interaction of mammalian cell culture broth with adsorbents in expanded bed adsorption of monoclonal antibodies. Process Biochem 1999; 34:159.
20. van Reis R, Leonard L, Hsu C, Builder S. Industrial scale harvest of protein from mammalian cell culture by tangential flow filtration. Biotechnol Bioeng 1991; 38:413.
21. Scopes R. Protein Purification Principles and Practices. 3rd Ed., New York: Springer, 1994.
22. Lee C, Wang N, Ju Y. Purification of a large protein using ion-exchange membranes. Separation Science and Technol 1995; 30:509.
23. Miller S, Ambler C, Bennett R, Dahlstrom D, Darji J, Emmett R, Gray J, Gurnham C, Jacobs L, Klepper R, Michalson A, Oldshue J, Silverblatt C, Smith J, Todd D. in Perry's Chemical Engineers Handbook, 6th Edition, Perry R, Green D, Maloney J, ed., McGraw-Hill Book Company, New York, 1984, Chapter 19, 19-89–19-103.
24. Fischer R, Liao Y, Drossard J. Affinity-purification of a TMV-specific recombinant full-size antibody from a transgenic tobacco suspension culture. J Immunol Methods 1999; 226:1.
25. Hunt A, Cropp C, Chang G. A recombinant particulate antigen of Japanese encephalitis virus produced in stably-transformed cells is an effective noninfectious antigen and subunit immunogen. J Virological Methods 2001; 97:133.

26. Yan T, Zhou M, Fu M, Wang Y, Yu R, Li J. Inhibition of egg hatching success and larvae survival of the scallop, Chlamys farreri, associated with exposure to cells and cell fragments of the dinoflagellate Alexandrium tamarense. Toxicon 2001; 39:1239.
27. Otoguro M, Hayakawa M, Yamazaki T, Iimura Y. An integrated method for the enrichment and selective isolation of Actinokineospora spp. in soil and plant litter. J Appl Microbiol 2001; 91:118.
28. Ebner S, Neyer S, Hofer S, Nussbaumer W, Romani N, Heufler C. Generation of large numbers of human dendritic cells from whole blood passaged through leukocyte removal filters: an alternative to standard buffy coats. J Immunol Methods 2001; 252:93.
29. Timmins J, Scianimanico S, Schoehn G, Weissenhorn W. Vesicular release of Ebola virus matrix protein VP40. Virology 2001; 283:1.
30. Yuksel S, Thompson K, Ellis A, Adams A. Purification of Piscirickettsia salmonis and associated phage particles. Diseases of Aquatic Organisms 2001; 44:231.
31. Li S, Hearn M. Isolation of thecal cells: an assessment of purity and steroidogenic potential. J Biochem Biophys Methods 2000; 45:169.
32. Thornton D, Gray T, Nettesheim P, Howard M, Koo J, Sheehan J. Characterization of mucins from cultured normal human tracheobronchial epithelial cells. American J Physiology Lung Cellular and Molecular Physiology 2000; 278:L1118.
33. Maurin M, Raoult D. Isolation in endothelial cell cultures of chlamydia trachomatis LGV (Serovar L2) from a lymph node of a patient with suspected cat scratch disease. J Clinical Microbiology 2000; 38:2062.
34. Hernandez D, Townsend L, Uzieblo M, Haan M, Callahan R, Bendick P, Glover J. The American Surgeon 2000; 66:355; discussion 359.
35. Ito Y. Centrifugal precipitation chromatography: principle, apparatus, and optimization of key parameters for protein fractionation by ammonium sulfate precipitation. Anal Biochem 2000; 277:143.
36. Nagashima S, Hirotani M, Yoshikawa T. Purification and characterization of UDP-glucuronate: baicalein 7-O-glucuronosyltransferase from Scutellaria baicalensis Georgi. cell suspension cultures. Phytochemistry 2000; 53:533.
37. Menhard B, Zenk M. Purification and characterization of acetyl coenzye A: 10-hydroxytaxane O-acetyltransferase from cell suspension cultures of Taxus chinensis. Phytochemistry 1999; 50:763.
38. Husereau DR, Suresh MR. A general affinity method to purify peroxidase-tagged antibodies. J Immunol Methods 2001; 249:33.
39. Darling J, Sullivan W Jr, Carter D, Ullman B, Roos D. Recombinant expression, purification, and characterization of Tooplasma gondii adenosine kinase. Molecular and Biochemical Parasitology 1999; 103:15.
40. Hakamatsuka T, Mori K, Ishida S, Ebizuka Y, Sankawa U. Purification of 2- hydroxyisoflavanone dehydratase from the cell cultures of Pueraria lobata. Phytochemistry 1998; 49:497.
41. Li S, Wang Z, Heide L. Purification of UDP-glucose: 4-hydroxybenzoate glucosyltransferase from cell cultures of Lithospermum erythrorhizon. Phytochemistry 1997; 46:27.
42. Thrahs S, Otto J, Deits T. Effect of divalent ions on protein precipitation with polyethylene glycol: mechanism of action and applications. Protein Expression Purif 1991; 2:83.
43. Sternberg M, Hershberger D. Separation of proteins with polyacrylic acids. Biochim Biophys Acta 1974; 342:195.
44. Curling J. Methods of Plasma Protein Fractionation. New York: Academic Press, 1980.
45. Morris J, Hoffman A, Fisher R. Affinity precipitation of proteins by polyligands. Biotechnol Bioeng 1993; 41:991.
46. Flygare S, Griffin T, Larsson P, Mosbach L. Affinity precipitation of dehydrogenases. Anal Biochem 1983; 133:409.
47. Hayet M, Vijayalakshmi M. Affinity precipitation of proteins using bis-dyes. J Chromatogr 1986; 376:157.

48. Morris J, Fisher R. Complications encountered using cibacron blue F3G-A as a ligand for affinity precipitation of lactate dehydrogenase. Biotechnol Bioeng 1990; 36:737.
49. Pearson J, Burion S, Lowe C. Affinity precipitation of lactate dehydrogenase with a triazine dye derivative: Selective precipitation of rabbit muscle lactate dehydrogenase with a procion blue H-B analog. Anal Biochem 1986; 158:382.
50. Walter H, Brooks DE, Fisher D. Partitioning in Aqueous Two Phase System. New York: Academic Press, 1985.
51. Berthold W, Walter J. Protein purification: Aspects of processes for pharmaceutical products. Biologicals 1994; 22:135.
52. Gaikaiwari R, Shendye A, Kulkarni N, Mala N. Biotechnol Appl Biochem 1996; 23:237.
53. Anderson E, Johansson A, Hägerdal GB. α-amylase production in aqueous two-phase systems with Bacillus subtilis. Enzyme Microb Technol 1985; 7:333.
54. Tanuja S, Srinivas N, Raghavarao K, Gowthaman M. Aqueous two phase extraction for downstream processing of amyloglucosidase. Process Biochem 1997; 32:635.
55. Srinivas N, Rashimi K, Raghavarao K. Extraction and purification of a plant peroxidase by aqueous two-phase extraction coupled with gel filtration. Process Biochem 1999; 35:43.
56. Guan Y, Lilley T, Treffry T, Zhou C, Wilkinson P. Use of aqueous two-phase systems in the purification of human interferon-αl from recombinant Escherichia coli. Enzyme and Microbial Technology 1996; 19:446.
57. Raghavarao K, Rastogi N, Gowthaman M, Karanth N. Aqueous two phase extraction for downstream processing of enzymes/proteins. Adv Appl Microbiol 1995; 41:97.
58. Suzuki M, Kamihira M, Shiraishi T, Takeuchi H, Kobayashi T. Affinity partitioning of protein a using a magnetic aqueous two-phase system. J Fermentation Bioengin 1995; 80:78.
59. Pan I, Li Y. Rapid process for purification of an extracellular β-xylosidase by aqueous two-phase extraction. J Chromatogr B 2001; 754:179.
60. Lu M, Albertsson P, Johansson G, Tjerneld F. Use of aqueous two-phase systems in sample preparation for polymerase chain reaction-based detection of microorganisms. J Chromatogr B 1996; 680:65.
61. Persson J, Nyström L, Ageland H, Tjerneld F. Purification of recombinant apolipoprotein A-l_{Milano} expressed in Escherichia coli using aqueous two-phase extraction followed by temperature-induced phase separation. J Chromatogr A 1998; 718:97.
62. Persson J, Nyström L, Ageland H, Tjerneld F. Purification of recombinant and human apolipoprotein A-l using surfactant micelles in aqueous two-phase systems: Recycling of thermoseparating polymer and surfactant with temperature-induced phase separation. Biotechnol Bioeng 1999; 65:371.
63. Persson J, Johansson H, Tjerneld F. Purification of protein and recycling of polymers in a new aqueous two-phase system using two thermoseparating polymers. J Chromatogr A 1999; 864:31.
64. Li M, Zhu Z, Mei L. Partitioning of amino acids by aqueous two-phase system combined with temperature-induced phase formation. Biotechnol Prog 1997; 13:105.
65. Diamond A, Hsu J. Aqueous two-phase systems for biomolecule separation. Adv Biochem Eng/Biotechnol 1992; 47:89.
66. Kopperschläger G, Lorenz G, Usbeck E. Application of affinity partitioning in an aqueous two-phase system to the investigation of triazine dye—enzyme interactions. J Chromatogr 1983; 259:97.
67. Miller W, Bünemann H, Schuetz H, Eigel A. Affinity Chromatography and Related Techniques. Gribnau T, Visser J, Nivard R, eds., Amsterdam: Elsevier Scientific, 1982:437.
68. Alred P, Tjerneld F, Kozlowski A, Harris J. Bioseparation 1992; 2:363.
69. Gagnon P. Purification Tools for Monoclonal Antibodies. Tucson, AZ: Validated Biosystems Press.
70. Tswett M. Ber Deut Botan Ges 1906; 24:234.

71. Bonnerjera J, Oh S, Hoare M, Dunhill P. Bio/Technology 1986; 4:954.
72. Levison P, Mumford C, Streater M, Brandt-Nielsen A, Pathirana N, Badger S. Performance comparison of low-pressure ion-exchange chromatography media for protein separation. J Chromatogr A 1997; 760:151.
73. Righetti P, Caravaggio T. Isoelectric points and molecular weights of proteins: A table. J Chromatogr 1976; 127:1.
74. Righetti P, Tudor G. Isoelectric points and molecular weights of proteins: A new table. J Chromatogr 1981; 220:115.
75. Malamud D, Drysdale J. Isoelectric points of proteins: A table. Anal Biochem 1978; 86:620.
76. Strong J, Frey D. Experimental and numerical studies of the chromatofocusing of dilute proteins using retained pH gradients formed on a strong-base anion-exchange column. J Chromatogr A 1997; 769:129.
77. Catalog. Sigma-Aldrich Fine Chemicals Company, St. Louis, MO, 2001.
78. Budavari S, ed. Merck Index, 12th Edition, 1996.
79. Melander W, Stoveken J, Horvath C. Mobile phase effects in reversed-phase chromatography. II. Acidic amine phosphate buffers as eluents. J Chromatogr 1979; 185:111.
80. Albert A, Serjent E. The Determination of Ionization Constants. London: Chapman and Hall, Ltd, 1971:82.
81. Good N, Winget G, Winter W, Izawa S, Singh R. Hydrogen ion buffers for biological research. Biochem 1966; 5:467.
82. Sluyterman L, Elgersma O. Chromatofocusing: Isoelectric focusing on ion-exchange columns: I. General Principles. J Chromatogr 1978; 150:17.
83. Sluyterman L, Wijdenes J. Chromatofocusing: Isoelectric focusing on ion-exchange columns: II. Experimental verification. J Chromatogr 1978; 150:31.
84. Sluyterman L, Wijdenes J. Chromatofocusing: III. The properties of a deae-agarose anion exchanger and its suitability for protein separations. J Chromatogr 1981; 206:429.
85. Sluyterman L, Wijdenes J. Chromato focusing: IV. Properties of an agarose polyethyleneimine ion exchanger and its suitability for protein separations. J Chromatogr 1981; 206:441.
86. The Chromatofocusing Handbook: Principles and Methods, Pharmacia Biotech, Snits & Design AB, Sweden, 1981.
87. Hearn M, Lyttle D. Buffer-focusing chromatography using multicomponent electrolyte elution systems. J Chromatogr 1981; 218:483.
88. Sluyterman L, Koojstra C. Ten years of chromatofocusing: a discussion. J Chromatogr 1989; 470:317.
89. Frey D, Barnes A, Strong J. Numerical studies of multicomponent chromatography employing pH gradients. AIChE J 1995; 41:1171.
90. Strong J, Frey D. Experimental and numerical studies of the chromatofocusing of dilute proteins using retained pH gradients formed on a strong-base anion-exchange column. J Chromatogr A 1997; 769:129.
91. Bates R, Frey D. Quasi-linear pH gradients for chromatofocusing using simple buffer mixtures: local equilibrium theory and experimental verification. J Chromatogr A 1998; 814:43.
92. Bates R, Kang X, Frey D. High-performance chromatofocusing using linear and concave pH gradients formed with simple buffer mixtures: I. Effect of buffer composition on the gradient shape. J Chromatogr A 2000; 890:25.
93. Kang X, Bates R, Frey D. High-performance chromatofocusing using linear and concave pH gradients formed with simple buffer mixtures: II. Separation of proteins. J Chromatogr A 2000; 890:37.
94. Tiselius A. Chromatographic analysis, Discussions of the Faraday Society. London: Hazell, Watson and Winey Ltd, 1949:447.

95. Horváth C, Melander W. Liquid-chromatography with hydrocarbonaceous bonded phases–theory and practice of reversed phase chromatography. J Chromatographic Science 1977; 15:393.
96. Ben-Nairn A. Hydrophobic Interactions. New York: Plenum Press, 1980.
97. Comuzzi B, Arcelloni C, Polo S, Nardese V, Lusso P, Paroni R. Multi-step purification strategy for RANTES wild-type and mutated analogues expressed in a baculovirus system. J Chromatogr B 2000; 737:47.
98. Melander W, Corrandini D, Horváth C. Salt-mediated retention of proteins in hydrophobic-interaction chromatography: Application of solvophobic theory. J Chromatogr 1984; 317:67.
99. Tanford C. The hydrophobic effect and the organization of living matter. Science 1978; 200:1012.
100. Hofmeister F. Zur Lehre von der Wirkung der Sake. II. Arch Exp Pathol Pharmakol 1888; 24:247.
101. Melander W, Horváth C. Salt effects on hydrophobic interactions in precipitation and chromatography of proteins: An interpretation of the lyotropic series. Arch Biochem Biophys 1977; 183:200.
102. Haidacher D, Vailaya A, Horváth C. Temperature effects in hydrophobic interaction chromatography. Proc Natl Acad Sci USA 1996; 93:2290.
103. Vailaya A, Horváth C. Retention thermodynamics in hydrophobic interaction chromatography. Ind Eng Chem Res 1996; 35:2964.
104. Roettger B, Myers J, Ladisch M, Regnier F. in Protein Purification: From Molecular Mechanisms to Large-Scale Processes. Ladisch M, Willson R, Painton C, Builder S, eds. ACS Symposium Series. Washington, DC: American Chemical Society1990:427:80.
105. Hofstee B. Hydrophobic affinity chromatography of proteins. Anal Biochem 1973; 52:430.
106. Fausnaugh J, Regnier F. Solute and mobile phase contributions to retention in hydrophobic interaction chromatography of proteins. J Chromatogr 1986; 359:131.
107. Pavlu B, Johansson U, Nyhlén C, Wichman A. Rapid purification of monoclonal antibodies by high-performance liquid chromatography. J Chromatogr 1986; 359:449.
108. Alpert A. Hydrophobic interaction chromatography of peptides as an alternative to reversed-phase chromatography. J Chromatogr 1988; 444:269.
109. Okamura H, Tsutsui H, Komatsu T, Yutsudo M, Haruka A, Tanimoto T, Torigoe K, Okura T, Nukada Y, Hattori K, Akita K, Namba M, Tanabe F, Konishi K, Fukuda S, Kurimoto M. Cloning of a new cytokine that induces IFN-γ production by T cells. Nature 1995; 378:88.
110. Ushio S, Namba M, Lkura T, Hattori K, Nukada Y, Akita K, Tanabe F, Konishi K, Micallef M, Fujii M, Torigoe K, Tanimoto T, Fukuda S, Ikeda M, Okamura H, Kurimoto M. Cloning of the cDNA for human IFN-gamma-inducmg factor, expression in Escherichia coli, and studies on the biologic activities of the protein. J Immunol 1996; 156:4274.
111. Hakamatsuka T, Mori K, Ishida S, Ebizuka Y, Sankawa U. Purification of 2-Hydroxyisoflavanone Dehydratase from the Cell Cultures of Pueraria Lobata in Honour of Professor G. H. Neil Towers 75th Birthday. Phytochemistry 1998; 49:497.
112. Nagashima S, Hirotani M, Yoshikawa T. Purification and characterization of UDP-glucuronate: baicalein 7-O-glucuronosyltransferase from Scutellaria baicalensis Georgi. cell suspension cultures. Phytochemistry 2000; 53:533.
113. Hutchens T, Porath J. Thiophilic adsorption-a comparison of model protein behavior. Biochem 1987; 26:7199.
114. Porath J, Maisano F, Belew M. Thiophilic adsorption–a new method for protein fractionation. Fed Eur Biochem Soc (FEBS) Lett 1985; 185:306.
115. Porath J. Salt-promoted adsorption: Recent developments. J Chromatogr 1986; 376:331.

116. Scholz G, Vieweg S, Leistner S, Seissler J, Scherbaum W, Huse K. A simplified procedure for the isolation of immunoglobulins from human serum using a novel type of thiophilic gel at low salt concentration. J Immunol Methods 1998; 219:109.
117. Berna P, Berna N, Porath J, Oscarsson S. Comparison of the protein adsorption selectivity of salt-promoted agarose-base adsorbents: Hydrophilic, thiophilic and electron donor-acceptor adsorbents. J Chromatogr A 1998; 800:151.
118. Bridonneau P, Lederer F. Behaviour of human immunoglobulin G subclasses on thiophilic gels: comparison with hydrophobic interaction chromatography. J Chromatogr B 1993; 616:197.
119. Hutchens T, Porath J. Thiophilic adsorption of immunoglobulins – analysis of conditions optimal for selective immobilization and purification. Anal Biochem 1986; 159:217.
120. Kreutz F, Wishart D, Suresh M. Efficient bispecific monoclonal antibody purification using gradient thiophilic affinity chromatography. J Chromatogr A 1998; 714:161.
121. Nopper B, Kohen F, Wilchek M. A thiophilic adsorbent for the one-step high-performance liquid chromatography purification of monoclonal antibodies. Anal Biochem 1989; 180:66.
122. Belew M, Juntti N, Larsson A, Porath J. A one-step purification method for monoclonal antibodies based on salt-promoted adsorption chromatography on a 'thiophilic' adsorbent. J Immunol Methods 1987; 102:173.
123. Lihme A, Heegaard P. Thiophilic adsorption chromatography: The separation of serum proteins. Anal Biochem 1991; 192:64.
124. Oscarsson S, Porath J. Covalent chromatography and salt-promoted thiophilic adsorption. Anal Biochem 1989; 176:330.
125. Knudsen K, Hansen M, Henriksen L, Anderson B, Lihme A. Sulfone-aromatic ligands for thiophilic adsorption chromatography: Purification of human and mouse immunoglobulins. Anal Biochem 1992; 201:170.
126. Chandha K, Kawiaski E, Sulkowski E. Thiophilic interaction chromatography of prostate-specific antigen. J Chromatogr B 2001; 754:521.
127. Affinity Chromatography: Principles and Methods, Amersham Pharmacia Biotech, 2001.
128. Kessler S. Rapid isolation of antigens from cells with a staphylococcal protein A-antibody adsorbent: parameters of the interaction of antibody-antigen complexes with protein A. J Immunol 1975; 115:1617.
129. Colangeli R, Heijbel A, Williams A, Manca C, Chan J, Lyashchenko K, Gennaro M. Three-step purification of lipopolysaccharied-free, polyhistidine-tagged recombinant antigens of Mycobacterium tuberculosis. J Chromatogr B 1998; 714:223.
130. Schmiedl A, Breitling F, Winter C, Queitsch I, Dübel S. Effects of unpaired cysteines on yield, solubility and activity of different recombinant antibody constructs expressed in E. coli. J Immunol Methods 2000; 242:101.
131. Lehr R, Elefante L, Kikly K, O'Brien S, Kirkpatrick R. A modified metal-ion affinity chromatography procedure for the purification of histidine-tagged recombinant proteins expressed in Drosophila S2 cells. Protein Expression and Purification 2000; 19:362.
132. Zhang C, Reslewic S, Glatz C. Suitability of immobilized metal affinity chromatography for protein purification from canola. Biotechnol Bioengin 2000; 68:52.
133. Chen H, Luo S, Chen K, Li C. Affinity purification of Schistosoma japonicum glutathione-S-transferase and its site-directed mutants with glutathione affinity chromatography and immobilized metal affinity chromatography. J Chromatogr A 1999; 852:151.
134. Sloane R, Ward J, O'Brien S, Thomas O, Dunnill P. Expression and purification of a recombinant metal-binding T4 lysozyme fusion protein. J Biotechnol 1996; 49:231.

135. Muneta Y, Shimoji Y, Yokomizo Y, Mori Y. Production of monoclonal antibodies to porcine interleukin-18 and their use for immunoaffinity purification of recombinant porcine interleukin-18. J Immunol Methods 2000; 236:99.
136. Cha H, Dalai N, Vakharia V, Bentley W. Expression of human interleukin-2 as a fusion with green fluorescent protein in suspended Sf-9 insect cells. J Biotechnol 1999; 69:9.
137. Cha H, Dalai N, Pham M, Bentley W. Purification of human interleukin-2 fusion protein produced in insect larvae is facilitated by fusion with green fluorescent protein and metal affinity ligand. Biotechnol Progress 1999; 15:283.
138. Cha H, Dalai N, Pham M,Vakharia V, Rao G, Bentley W. Insect larval expression process is optimized by generating fusions with green fluorescent protein. Biotechnol Bioengin 1999; 65:316.
139. Andersson C, Wikmna M, Lövgren-Bengtsson K, Lundén A, Ståhl S. In vivo and in vitro lipidation of recombinant immunogens for direct iscom incorporation. J Immunol Methods 2001; 255:135.
140. D'Alatri L, Di Massimo A, Anastasi A, Pacilli A, Novelli S, Saccinto M, De Santis R, Mele A, Parente D. Production and characterization of a recombinant single-chain anti ErbB2-clavin immunotoxin. Anticancer Res 1998; 18:3369.
141. Miranda A, Benítez J, Acevedo B, Domínguez M, Villareal A, Palenzuela D, Rodriguez L, Rodriguez J, Núñez Z, Gavilondo J. A comparison of VDRL and immunoassays developed with a recombinant TmpA antigen in the screening of antibodies to Treponema pallidum. Serodiagnosis and Immunotherapy in Infectious Disease 1997; 8:149.
142. Ericson M, Mooney N, Charron D. Prokaryotic production of the human T cell receptor Vß2 chain and binding to toxic-shock syndrome toxin-1. Gene 1996; 168:257.
143. Setterquist R, Smith G, Oakley T, Youn-Hyung L, Fox G. Sequence, overproduction and purification of Vibrio proteolyticus ribosomal protein L 18 for in vitro and in vivo studies. Gene 1996; 183:237.
144. Negro A, Onisto M, Masiero L, Garbisa S. Synthesis and refolding of human TIMP-2 from E. coli, with specific activity for MMP-2. FEBS Letters 1995; 360:52.
145. Burton S, Harding D. Hydrophobic charge induction chromatography: salt independent protein adsorption and facile elution with aqueous buffers. J Chromatogr A 1998; 814:71.
146. Schwartz W, Judd D, Wysocki M, Guerrier L, Birck-Wilson E, Boschetti E. Comparison of hydrophobic charge induction chromatography with affinity chromatography on protein A for harvest and purification of antibodies. J Chromatogr A 2001; 908:251.
147. Guerrier L, Flayeux I, Boschetti E. A dual-mode approach to the selective separation of antibodies and their fragments. J Chromatogr B 2001; 755:37.
148. Porath J, Flodin P. Gel filtration: a method for desalting and group separation. Nature 1959; 183:1657.
149. Janson J. On the history of the development of Sephadex®. Chromatographia 1987; 23:361.
150. Andersson T, Hagel L. Some properties and applications of Superose 6B. Anal Biochem 1984; 141:461.
151. Hagel L, Lundström H, Andersson T, Lindblom H. Properties, in theory and practice, of novel gel filtration media for standard liquid chromatography. J Chromatogr 1989; 476:329.
152. Hagel L, Hartmann A, Lund K. Analytical gel filtration of dextran for study of the glomerular barrier function. J Chromatogr 1993; 641:63.
153. Janson J, Ryden L, eds. Protein Purification: Principles, High Resolution Methods, and Applications. New York: VCH Publishers, Inc., 1989:63.
154. Leaback D, Robinson H. Ampholyte displacement chromatography - a new technique for the separation of proteins illustrated by the resolution of beta-N-acetyl-D-hexosaminidase isoenzymes unresolvable by isoelectric focusing or conventional ion-exchange chromatography. Biochem Biophys Res Commun 1975; 67:248.

155. Edmond P, Pagé M. An approach to ampholyte-displacement chromatography. J Chromatog 1990; 200:57.
156. Narahari C, Strong J, Frey D. Displacement chromatography of proteins using a self-sharpening pH front formed by adsorbed buffering species as the displacer. J Chromatogr A 1998; 825:115.
157. Skidmore G, Chase H. Two-component protein adsorption to the cation exchanger S Sepharose® FF. J Chromatog 1990; 505:329.
158. Felinger A, Guiochon G. Comparison of maximum production rates and optimum operating/design parameters in overloaded elution and displacement chromatography. Biotech Bioeng 1993; 41:134.
159. Jungbauer A, Uhl K, Tauer C, Steindl F, Buchacher A, Schönhofer W, Unterluggauer F. Displacement effects in large scale chromatography. Biotech Bioeng 1992; 39:579.
160. Cramer S. Displacement chromatography. Nature 1991; 351:251–252.
161. Luo Q, Andrade J. Phytic acid as an efficient low-molecular mass displacer for anion-exchange displacement chromatography of proteins. J Chromatogr B 2000; 741:23.
162. Jayaraman G, Li Y, Moore J, Cramer S. Ion-exchange displacement chromatography of proteins: Dendritic polymers as novel displacers. J Chromatogr A 1995; 702:143.
163. Ivanov A, Galaev I, Kazakov S, Mattiasson B. Thermosensitive copolymers of N-vinylimidazole as displacers of proteins in immobilized metal affinity chromatography. J Chromatogr A 2001; 907:115.
164. Shukla A, Sunasara K, Ropp R, Cramer S. Hydrophobic displacement chromatography of proteins. Biotech Bioeng 2000; 68:672.
165. Kasper B, Breier J, Vogt S, Freitag R. Protein displacement chromatography on hydroxy- and fluoroapatite columns. Bioseparation 1996; 6:247.
166. Manseth E, Skjervold P, Flengsrud R. Sample displacement chromatography of Atlantic Salmon (Salmo salar) thrombin. Journal of Biochemical and Biophysical Methods 2004; 60:39.
167. Hansson M, Stahl S, Hjorth R, Uhlen M, Moks T. Single-step recovery of a secreted recombinant protein by expanded bed adsorption. Bio/Technology 1994; 12:285.
168. Zurek H, Kubis E, Keup P, Hoerlein D, Beunink H, Thömmes J, Kula M, Hollenberg C, Gellissen G. Production of two aprotinin variants in Hansenula polymorpha. Process Biochem 1996; 31:679.
169. Barnfield-Frej A, Hjorth R, Hammarström A. Pilot scale recovery of recombinant annexin V from unclarified escherichia coli homogenate using expanded bed adsorption. Biotech Bioeng 1994; 44:922.
170. Chang Y, Chase H. Ion exchange purification of G6PDH from unclarified yeast cell homogenates using expanded bed adsorption. Biotech Bioeng 1996; 49:204.
171. Thömmes J, Halfar M, Lenz S, Kula M. Purification of monoclonal antibodies from whole hybridoma fermentation broth by fluidized bed adsorption. Biotech Bioeng 1995; 45:205.
172. Batt B, Yabannavar V, Singh V. Expanded bed adsorption process for protein recovery from whole mammalian cell culture broth. Bioseparation 1995; 5:41.
173. Garke G, Dechwer W, Anspach F. Preparative two-step purification of recombinant human basicfibroblast growth factor from high cell-density cultivation of Eschericia coli. J Chromatogr B 2000; 737:25.
174. Fahrner R, Blank G, Zapata A. Real-time control of antibody loading during Protein A affinity chromatography using an on-line assay. J Biotech 1999; 75:273.
175. Lütkemeyer D, Ameskamp N, Tebbe H, Wittier J, Lehmann J. Estimation of cell damage in bench- and pilot-scale affinity expanded-bed chromatography for the purification of monoclonal antibodies. Biotech/Bioeng 1999; 65:114.
176. Iyer H, Henderson F, Cunningham E, Webb J, Hanson J, Bork C, Conley L. Considerations during development of a Protein A-based antibody purification process. BioPharm 2002; 15:14.

177. Yazaki P, Shively L, Clark C, Cheung C, Le W, Szpikowska B, Shively J, Raubitschek A, Wu A. Mammalian expression and hollowfiber bioreactor production of recombinant anti-CEA diabody and minibody for clinical applications. J Immunol Methods 2001; 253:195.
178. Clausen A, Subramanian A, Carr P. Purification of monoclonal antibodies from cell culture supernatants using a modified zirconia based cation-exchange support. J Chromatogr A 1999; 831:63.
179. Xu D, Leveugle B, Kreutz F, Suresh MR. Mimetic ligand-based affinity purification of immune complexes and immunoconjugates. J Chromatogr B 1998; 706:217.
180. Li S, Wang Z, Heidi L. Purification of UDP-glucose: 4-hydroxybenzoate glucosyltransferase from cell cultures of Lithospermum erythrorhizon. Phytochemistry 1997; 46:27.
181. Husereau D, Suresh M. A general affinity method to purify peroxidase-tagged antibodies. J Immunol Methods 2001; 249:33.
182. Veitch N, Williams R. The use of methyl-substituted benzhydroxamic acids as structural probes of peroxidase substrate binding. Eur J Biochem 1995; 229:629.
183. Henriksen A, Schuller D, Meno K, Welinder K, Smith A, Gajhede M. Structural interactions between horseradish peroxidase C and the substrate benzhydroxamic acid determined by X-ray crystallography. Biochem 1998; 37:8054.
184. de Ropp J, Mandal P, La Mar G. Solution 1H NMR investigation of the heme cavity and substrate binding site in cyanide-inhibited horseradish peroxidase. Biochem 1999; 38:1077.
185. Feuser J, Walter J, Kula M, Thömmes J. Cell/adsorbent interactions in expanded bed adsorption of proteins. Bioseparation 1999; 8:99.
186. Fischer R, Liao Y, Drossard J. Affinity-purification of a TMV-specific recombinant full-size antibody from a transgenic tobacco suspension culture. J Immunol Methods 1999; 226:1.
187. Aybay C, Imir T. Development of a rapid, single-step procedure using Protein G affinity chromatography to deplete fetal calf serum of its IgG and to isolate murine IgGl monoclonal antibodies from supernatants of hybridoma cells. J Immunol Methods 2000; 233:77.
188. Wang X, Yi J, Lei J, Pei D. Expression, purification and characterization of recombinant mouse MT5-MMP protein products. FEBS Lett 1999; 462:261.
189. Hakamatsuka T, Mori K, Ishida S, Ebizuka Y, Sankawa U. Purification of 2-hydroxyisoflavanone dehydratase from the cell cultures of Pueraria lobata in honour of Professor G. H. Neil Towers 75th birthday. Phytochemisty 1998; 49:497–505.
190. Ford C, Osborne P, Mathew A, Rego B. Affinity purification of novel bispecific antibodies recognising carcinoembryonic antigen and doxorubicin. J Chromatogr B 2001; 754:427.
191. Michel P, Torkkeli T, Karp M, Oker-Blom C. Expression and purification of polyhistidine-tagged firefly luciferase in insect cells – a potential alternative for process scale-up. J Biotechnology 2001; 85:49.

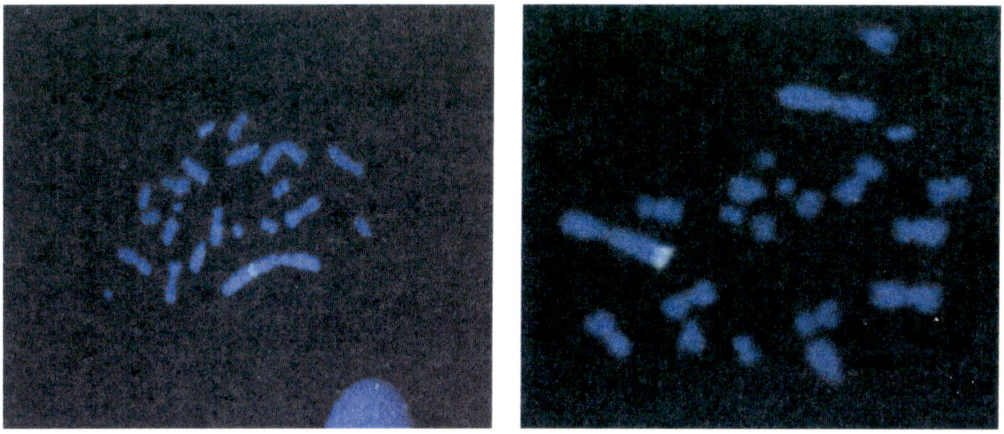

Figure 2.6 Fluorescent in situ hybridization of recombinant sequences integrated into the chromosomes of CHO cells. Photos are of two different cell lines resulting from random integration of the plasmid DNA during a single transfection event.

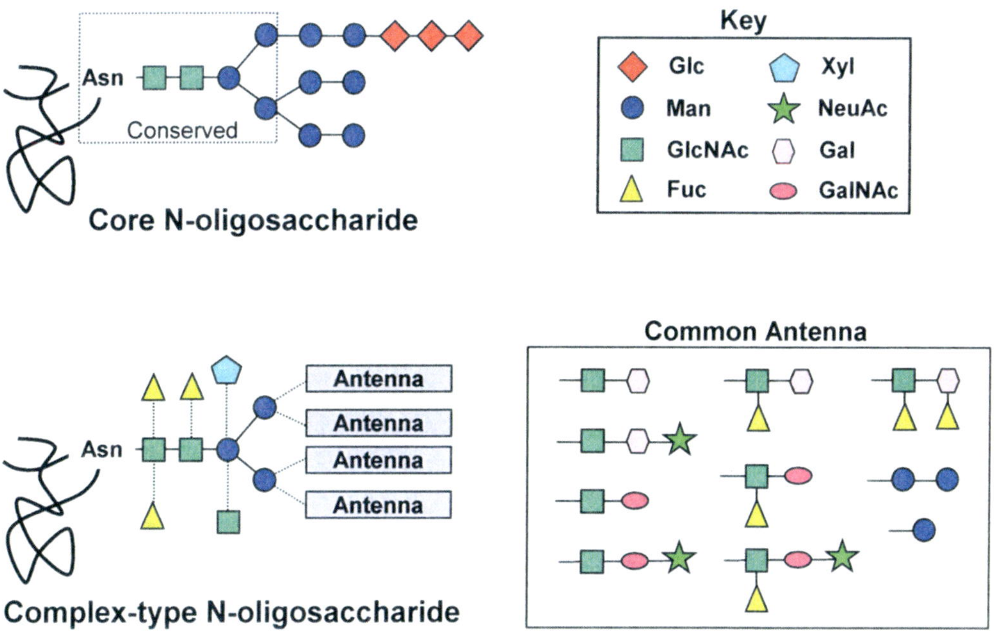

Figure 5.2 N-glycosylation structures. [Adapted from Refs. (1,20,21).]

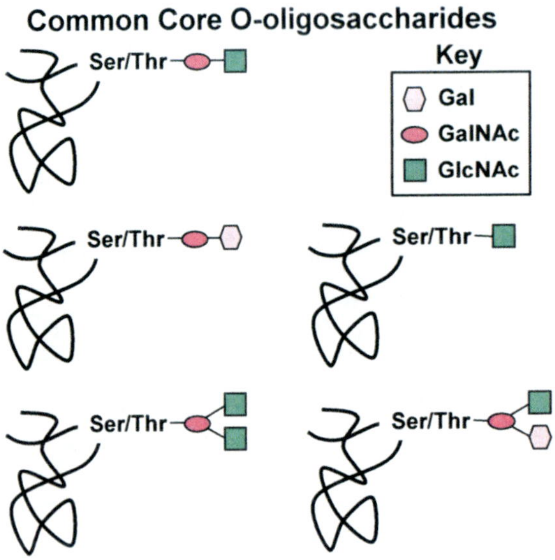

Figure 5.3 O-glycosylation structures. [Adapted from Refs. (6,20).]

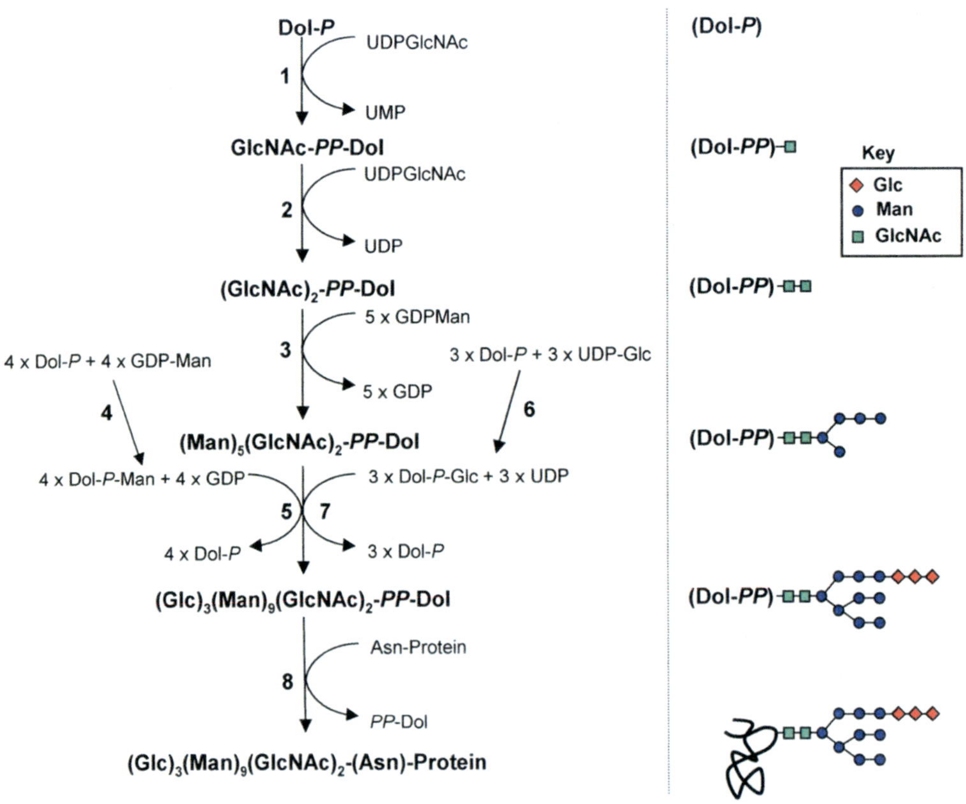

Figure 5.4 The dolichol cycle and oligosaccharide transfer reactions to form *N*-oligosaccharide core. [Adapted from Refs. (6,20).]

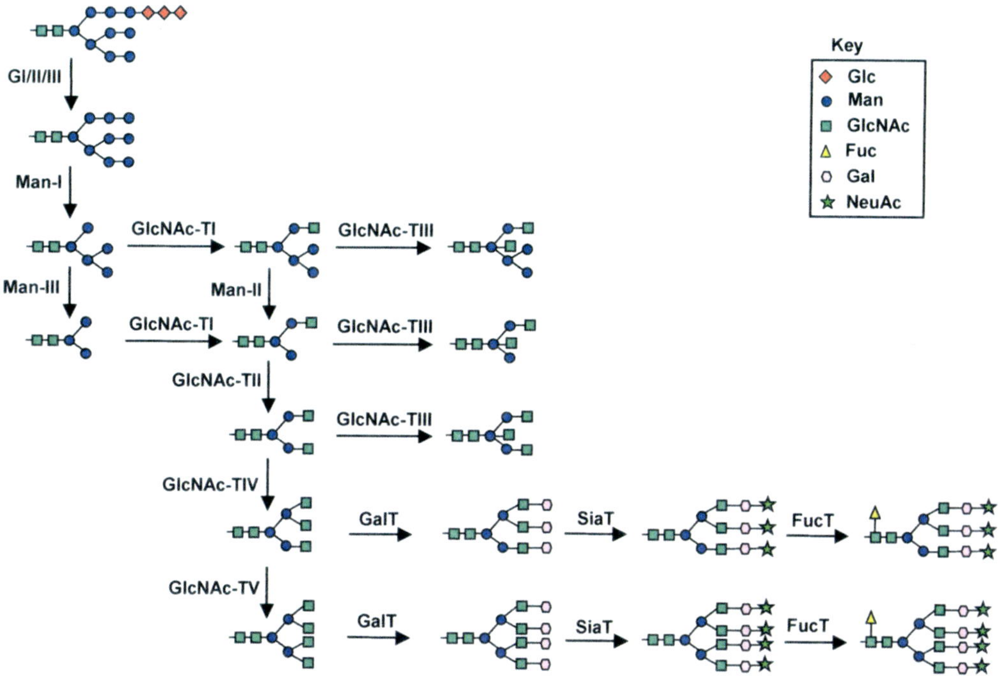

Figure 5.5 Trimming and branching reactions that modify the core *N*-oligosaccharides. [Adapted from Refs. (20,21).]

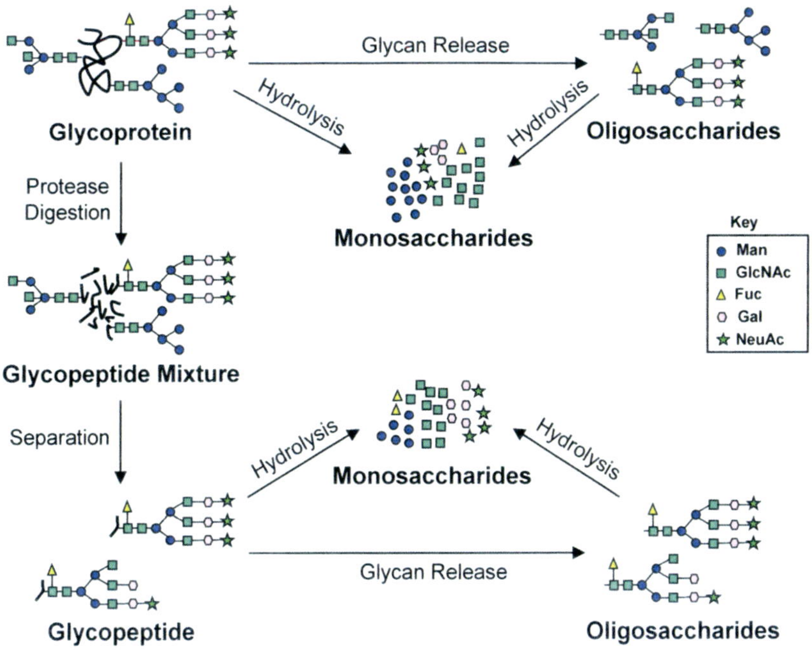

Figure 5.6 Fractionation methods to characterize glycoproteins. [Adapted from Refs. (20,54,121–132).]

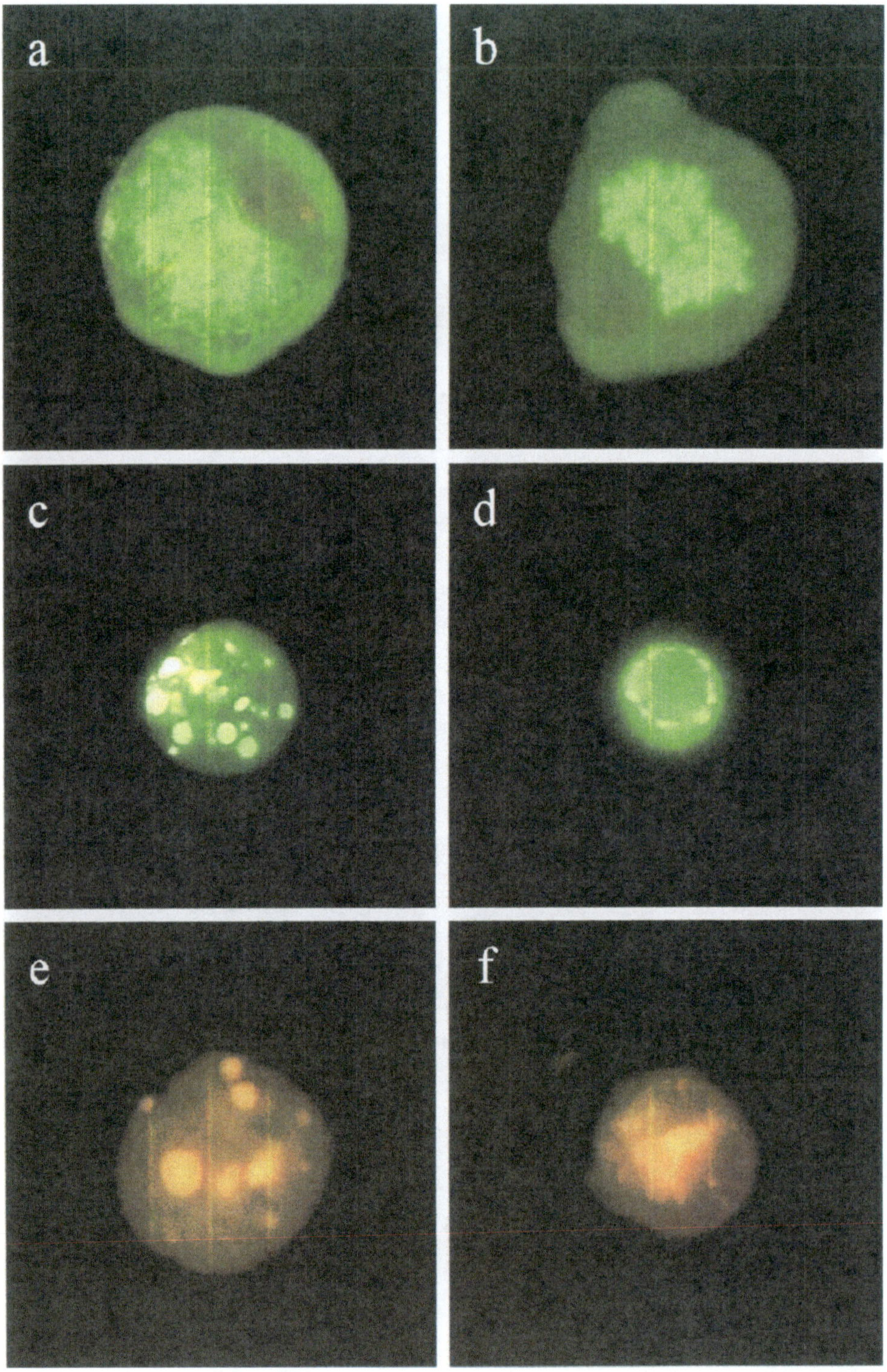

Figure 9.3 Acridin orange/ethidium bromide staining of hybridoma cells, taken with 1000 times amplification. Living cells, green; dead cell, orange: (a) living nonapoptotic cell, (b) living nonapoptotic cell during division, (c) living apoptotic cell (clear DNA condensation and nuclear fragmentation), (d) living apoptotic cell (ring form condensation of DNA near the nuclear membrane), (e) dead apoptotic cell, (f) dead nonapoptotic (necrotic) cell. [From Ref. (57).]

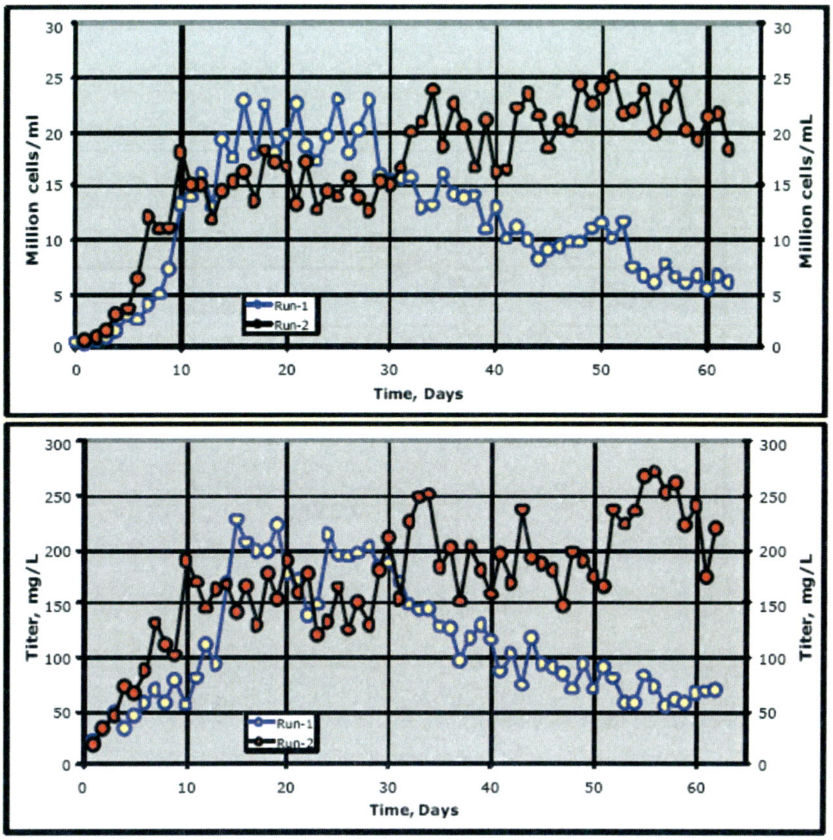

Figure 11.12 Consistency and reproducibility issues for long-term perfusion culture. Small differences in bioreactor operation can accumulate over time resulting in a big difference in performance. [From Ref. (107).]

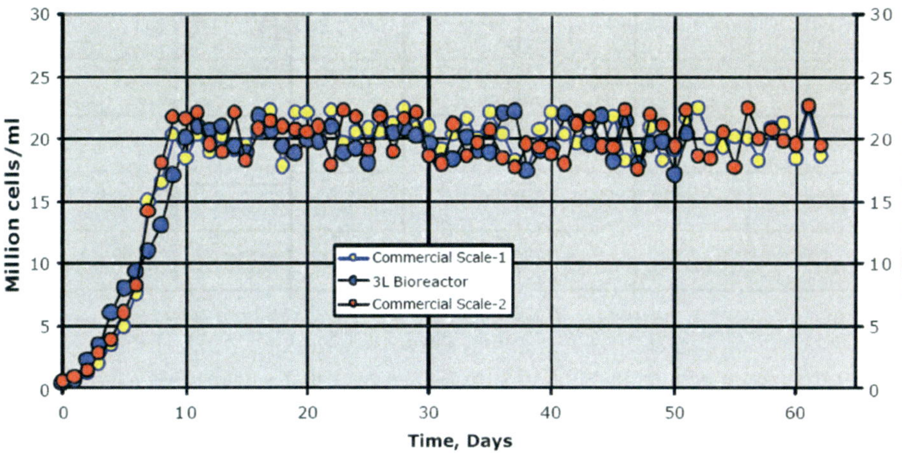

*More than 200X scale

Figure 11.13 Scalability of perfusion bioreactors operated by CSPR control. [From Ref. (107).]

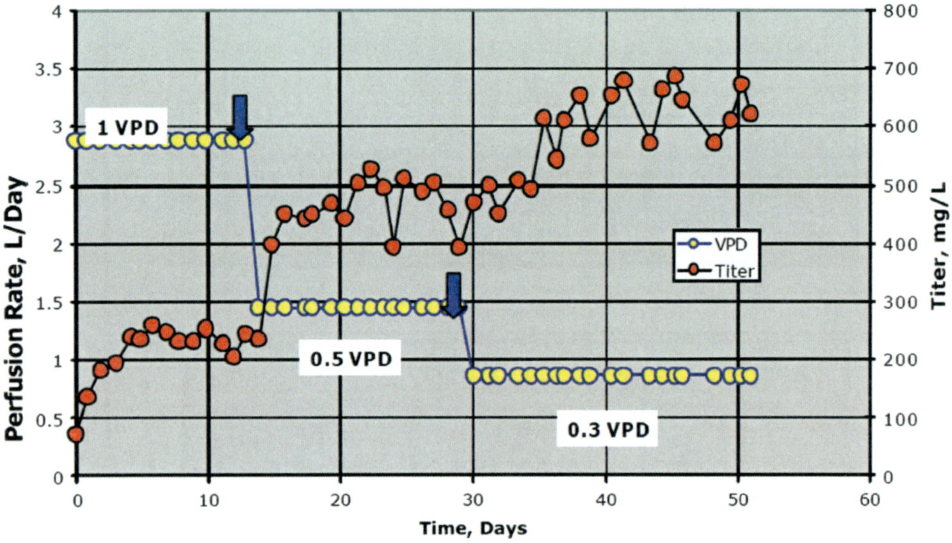

Figure 11.14 Decrease in perfusion rate can be used as a tool to increase product concentration from perfusion bioreactors. This approach does not offer any benefit in volumetric productivity. In most cases, the volumetric productivity is negatively impacted. [From Ref. (107).]

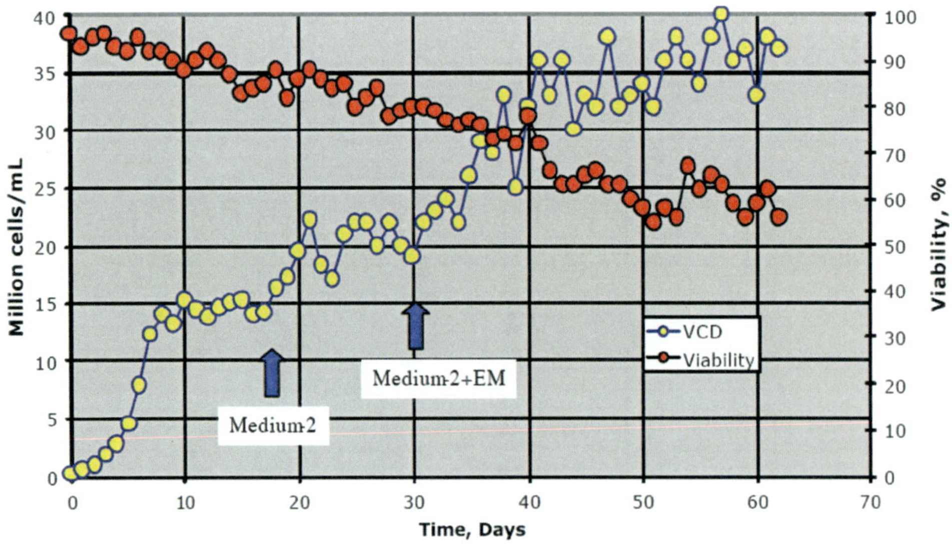

Figure 11.15 The use of enriched medium in a perfusion bioreactor. Switching to the enriched perfusion medium more than doubles cell density in the bioreactor. [From Ref. (107).]

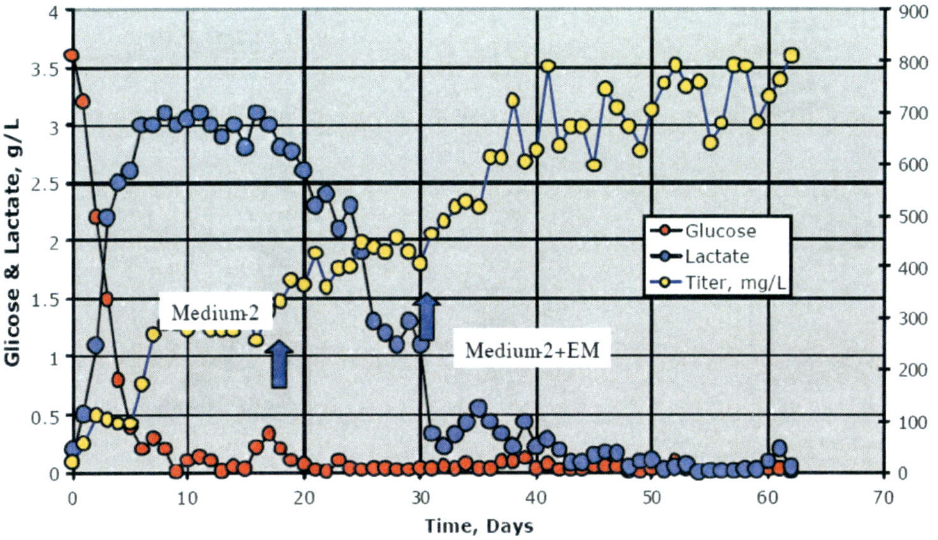

Figure 11.16 The use of enriched medium in a perfusion bioreactor. The product concentration reaches to 750 mg/L in enriched medium. After switching to the enriched medium a shift in lactate metabolism is observed. The lactate production drops to zero in enriched medium. [From Ref. (107).]

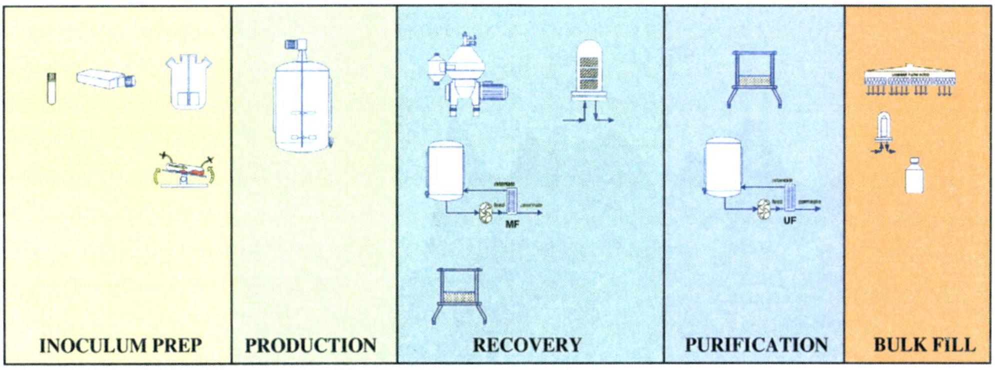

Figure 16.2 Basic steps in bulk manufacturing of a cell culture based therapeutic.

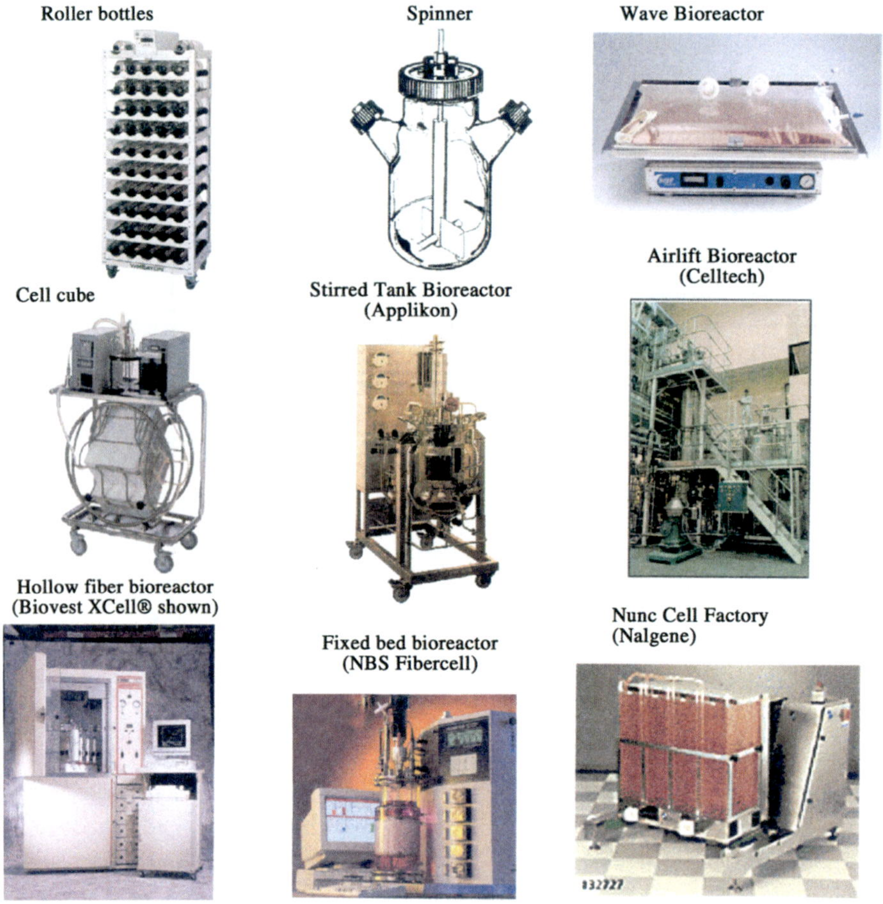

Figure 16.3 Representative cell culture systems successfully used in GMP licensed production systems.

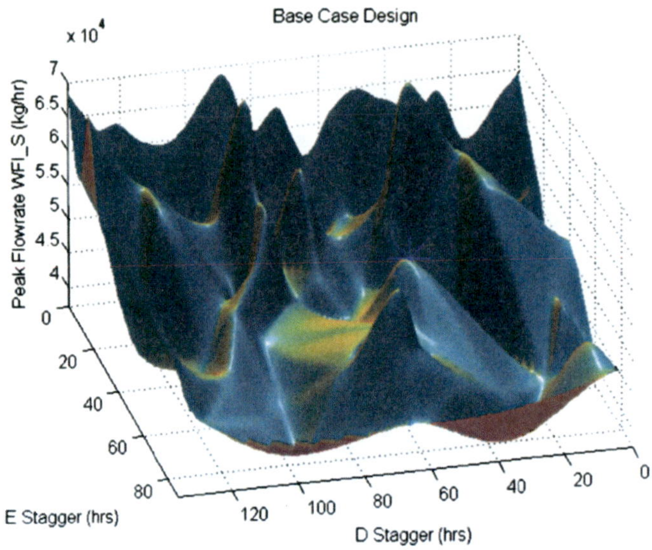

Figure 16.6 Optimization of processes must be careful to select robust operating range solutions. This is necessary to avoid an unstable points where, for example, the delay of an operation could inadvertently result in a cascading of problems excessive demands, or unavailable equipment. (Response surface plot courtesy of CRB Consulting Engineers, Inc.)

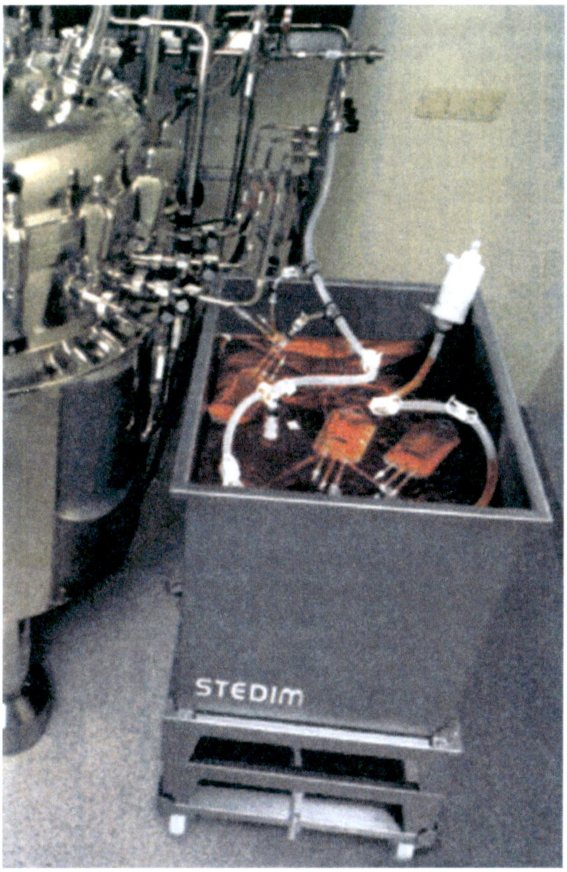

Figure 16.7 Stainless steel pallet tank used for holding media and buffer bags (photos courtesy of Stedim).

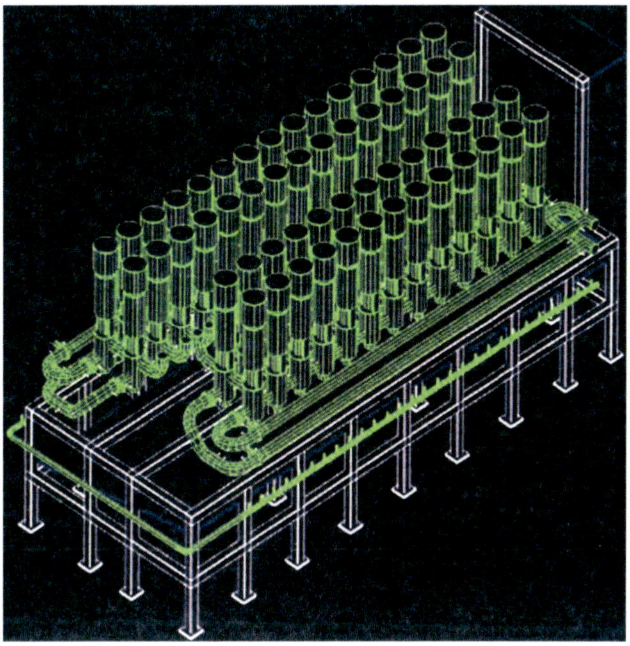

Figure 16.9 3D design renderings of equivalent transfer panel and mix-proof valve arrays. In this case, when bids were received, the capital cost of the transfer panel and the valve array were nearly identical. (Courtesy of CRB Consulting Engineers.)

Figure 16.13 Walkable ceiling interstitial space. The air handling units are on the grating floor above. The walkable ceiling allows easy access to HVAC control instruments and dampers, as well as allowing for the servicing of lights from outside the clean envelope. (Courtesy Clestra Cleanrooms Inc.)

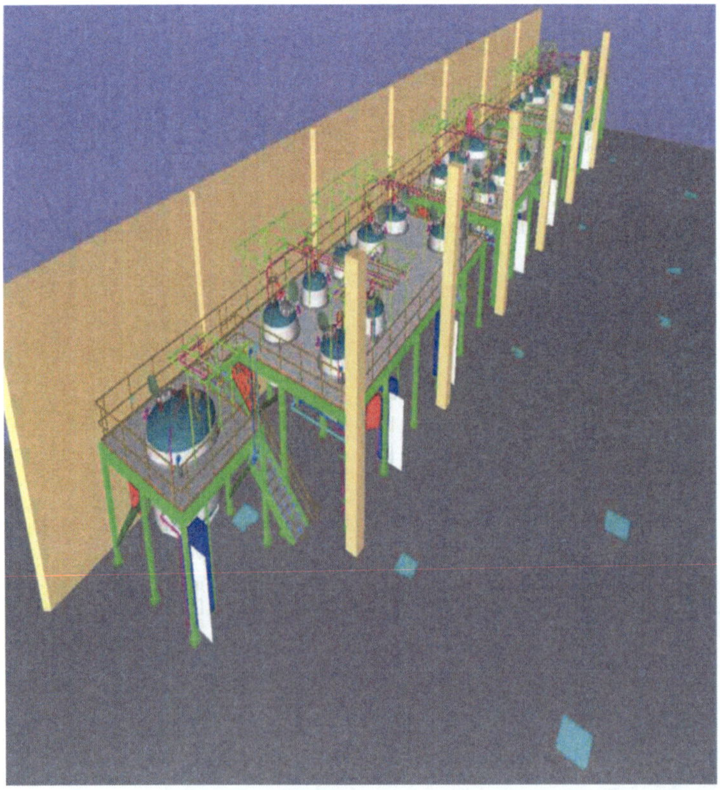

Figure 16.17 Rendering of buffer hold module for Biogen-IDEC's NIMO facility in Oceanside, CA. (Courtesy of Biogen Idec.)

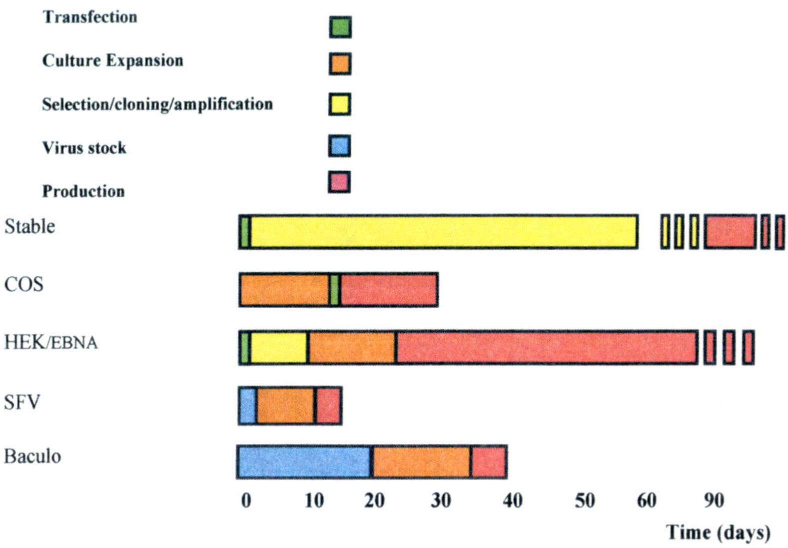

Figure 17.1 Transient expression timings. The average times for the various step operations have been estimated for each system. Time 0 corresponds to the availability of the expression vector (whether virus or nucleic acid). Repetitive operations are indicated by leaving some empty time between each of them. Baculovirus technology is used for comparative purposes.

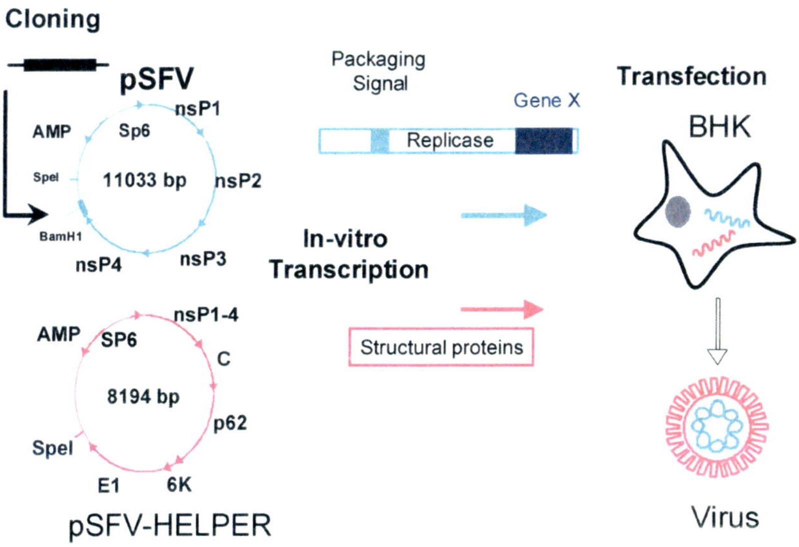

Figure 17.4 Semliki Forest virus expression system. In vivo packaging of recombinant SFV RNA transcripts into infectious particles. After co-transfection of RNA transcipts from the expression and the helper vector, the replicase, coded on the recombinant vector, drives RNA replication and transcription for both plasmids. Structural genes from the helper are translated into proteins. While the capsid protein stays cytosolic and binds RNA to form the nucleocapsid, the membrane proteins undergo post-translational modifications. El and p62 are dimerized in the ER and are transported to the plasma membrane where virus budding occurs. Since the viral packaging signal is only found on the recombinant RNA, selective packaging of the pSFVdhfr occurs and as a result yields nonreplicative infectious SFV particles. [Adapted from Ref. (39).]

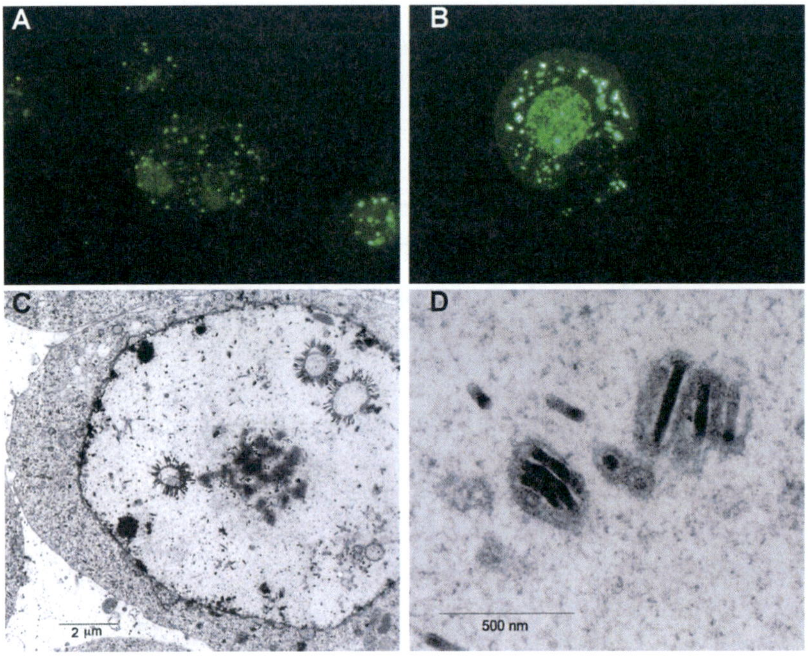

Figure 18.3 Micrographs of infected insect cells. Epifluorescence micrographs: (A) Sf9 cells at 72 hpi (× 560); (B) Tn5Bl-4 cells at 72 hpi (× 560). Cells in A and B are expressing a fusion protein containing GFP and VP2 of rotavirus. (Baculovirus provided by Prof. J. Cohen, Institut National de la Recherche Agronomique (INRA), France.) (C) Transmission electron microscopy of Sf9 cells infected with a recombinant AcMNPV. (D) Enlarged section of the nucleus. Note the large nucleus in C and enveloped multiple virions in D.

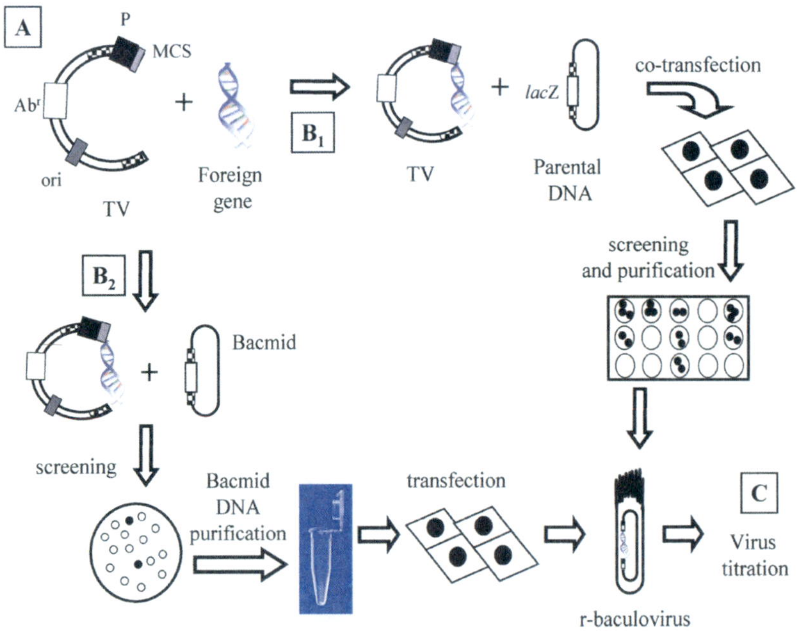

Figure 18.4 Main steps for obtaining an r-baculovirus. (A) The transfer vector (TV) is depicted with its main elements (ori, bacterial replication origin; Ab^r, antibiotic resistance marker; P, promoter; MCS, multiple-cloning site; patterned areas, homologous recombination sites). Depending on the transfer vector used, two routes can be chosen: (B1) Traditional method in which recombination occurs in the insect cells; (B2) Bacmid-transposition method.

14

Formulation, Filling and Packaging

M. E. M. Cromwell
Early Stage Formulation, Genentech, Inc., South SanFrancisco, California, U.S.A.

INTRODUCTION

There are many biotechnology products approved by the Food and Drug Administration (FDA) for either the detection or treatment of human diseases, as presented in Chapter 1. More than 350 biotechnology-produced drugs and vaccines are currently being tested in clinical trials, with hundreds more in earlier stages of development (1). The approved products treat a wide variety of conditions and diseases, including hemophilia, multiple sclerosis, AIDS-related illnesses, growth failure, infertility, cancer, diabetes, hepatitis, anemia, Crohn's disease, diabetic ulcers, prevention of transplant rejection, stroke, and acute myocardial infarctions. The successful treatment of a disease state requires that the drug be delivered in an active form over a specific timeframe to the location in the body where it is needed. This necessitates development of a suitable formulation and drug delivery system that ensures the stability of the active compound, the delivery of the drug to the site of action, and its presence at the site of action over a desirable timeframe.

In the development of a formulation, the degradation of the protein is assessed under several conditions to determine under which conditions the active compound is most stable. Typically, variations in pH, ionic strength, buffer components, tonicifiers, and surfactants are investigated. Stressed conditions such as exposure to elevated temperatures, freezing, or harsh lighting are used to purposefully damage the protein. These conditions may not reflect the actual storage conditions, but can give insight into the mechanisms by which the protein may degrade.

The goal of a successful formulation is to minimize the degradation of the protein during storage as a formulated bulk (drug substance) and over the shelf life of the final drug product. Storage conditions for the purified drug substance must be determined to ensure minimal changes to the protein before the material is manufactured as the final drug product. It is in the manufacturer's best interest to have as much stability as possible for the drug substance for two reasons: (a) the ability to build adequate inventory of the drug substance frees the manufacturer to produce the bulks at will instead of trying to time the manufacture of the drug substance to meet the needs of the market and (b) degradation of the protein while stored as the drug substance minimizes the shelf life of the drug product. Typically, at least 2 years of shelf life is required to ensure suitable time for the manufacture, testing, and distribution of the drug product. Any degradation observed during storage of the drug substance would adversely impact the expiration dating for the drug product. The selection of the form of the drug product (e.g., liquid or solid state) and the final storage conditions are made with the goal of achieving the most cost-efficient and stable product possible.

In addition to developing a stable product, it is also necessary to determine the most desirable form for the administration and marketing of the product. Drug products administered in a hospital setting have different requirements for development compared to those administered by a patient (or parent of a patient) at home. Consideration must also be given to the formulation/delivery system for any competing products that are on the market. For example, if the competition sells a drug that is administered orally, it would not be prudent to sell a drug that must be administered by injection on a daily basis unless it provided superior efficacy or safety benefits. Patients would not have the motivation to switch to a more painful and less convenient route of administration otherwise! Thus, input from the field regarding the user preferences and competitive products must be considered throughout product development.

This chapter will discuss the strategy used for development of a therapeutic formulation for clinical trials that will eventually lead to a marketed product. What questions does the formulation scientist need to answer to develop a suitable drug product? First, the degradation and inactivation mechanisms for the protein are determined. After all, it is those reactions that the successful formulation scientist is attempting to prevent! "How can the degradation be minimized?" Next, consideration is given to the state and composition of the formulation. "Will it be a liquid product? Is it better to have a lyophilized product that can be stored at ambient temperature rather than a liquid product that must be stored under refrigerated conditions?" With the decision regarding state, the excipients that will comprise the formulation are selected. "What should the pH of the formulation be? Is a bulking agent necessary to form a good lyophilized cake? Does the formulation need to be isotonic? With which excipients is the formulation most stable?" Linked to both the selection of state and excipients is a decision regarding the route of administration and a prototype of the final product. "Will the drug be injected? Is sustained delivery desired? What container and closure system will be used? Are delivery devices available to enhance the value of the product? How will the drug product be packaged?" Once the selection of delivery route, excipients, and physical state is made, studies to examine the stability of the product under recommended and stressed storage conditions must be performed to ensure product quality and safety. "Does the product require refrigeration? Should the bulk be stored frozen? Is there sufficient stability for a viable product?" These studies utilize analytical methodologies to evaluate the stability of the protein. "What does each of the assays tell the scientist about the degradation of the protein? Are the assays stability indicating?" It is through this process that an active protein drug may be formulated into a viable drug product.

DEGRADATION/INACTIVATION

Several routes of degradation may be exhibited by proteins. In some cases, denaturation of the protein may cause its inactivation. Some of the potential sites of degradation may be anticipated from the primary structure of the protein, such as the high probability of deamidation of Asn when followed by Gly (2). Other sources of instability are only discovered during the course of studies in which the stability of the protein is assessed.

These instabilities fall into two general classes: *physical instability*, in which the protein changes its tertiary or quaternary structure, and *chemical instability*, in which a chemical reaction causes a change in one or more of the amino acids in the protein. In the development of a protein formulation, the amount of degradation products of either type that are formed over the shelf life of the drug product must be minimized. The denatured protein may have altered activity, pharmacokinetics, or safety. There is concern that the denatured protein may cause an immunogenic response when administered to a patient, even if no response is observed with the native protein. For this reason, it is crucial for the formulation scientist to develop a drug product that minimizes any changes to the protein over its shelf life.

Physical Instability

Physical instability is caused by *aggregation* or *surface denaturation* of the protein. Soluble aggregates result from proteins that self-associate into discrete units such as dimers or trimers through ionic interactions, hydrophobic interactions, or from a change in disulfide bonding of the native protein. Precipitation, which is the formation of *insoluble aggregates*, usually results from nonspecific protein interactions although in some proteins it may result from the formation of extremely large, ordered aggregates. Some disease states are the result of the formation of insoluble protein aggregates *in vivo*. Amyloidosis (AL) and light chain deposition disease (LCDD) are caused by aggregates of immunoglobulin light chain fragments either as ordered, fibrillar aggregates (AL) or as amorphous aggregates (LCDD) (3–6). Alzheimer's disease has been associated with the presence of amyloid fibrils that form plaques (7–9).

In some cases, it is desirable to have a pharmaceutical protein in an aggregated state, because it is the bioactive form of the protein. An example of this is surfactant protein B (SP-B), a pulmonary surfactant protein necessary for normal lung function in neonatal infants (10,11). The protein exists exclusively as a homodimer in which the monomers are linked by a disulfide bond. In studies investigating efficacy of the SP-B monomer compared to the dimer in transgenic mice, it was found that although the surfactant action was preserved in the monomeric form of the protein, altered lung hysteresis was noted (12). The authors concluded that SP-B dimerization is required for optimal lung function.

Aggregation of a protein may also be desired when the aggregate is a more stable form of the protein. For example, insulin is formulated in the presence of Zn^{2+} that coordinates insulin dimers to form an ordered hexameric form of the protein. Zn^{2+} added to the formulation has been shown to increase the physical stability of an insulin solution (13).

Most often, however, aggregation is not desired. The presence of a non-native aggregate is cause of concern to biopharmaceutical scientists, because this aggregate may have altered activity, clearance, and toxicity compared to the native protein. Aggregates of ribonuclease A were found to be less active (14). Covalent aggregates of insulin resulted in the appearance of antibodies to the protein in the blood of insulin-using diabetic patients (15).

Due to the potential effects on safety and activity, it is necessary to minimize the aggregate content in the final formulation. In the formulation screen for the recombinant humanized monoclonal antibody to vascular endothelial growth factor (rhuMAb VEGF), a reversible self-association of the protein was observed whose levels were found to be dependent on protein concentration, pH, and ionic strength of the formulation (16). It would be possible to minimize the aggregate content in the formulation by controlling these variables.

In addition to aggregation, *surface denaturation* causes physical instability of the protein. Surface denaturation occurs when the protein interacts with the container surface or when it comes into contact with an interface (e.g., air–liquid interface). At these interfaces the protein may partially unfold, leading to a non-native structure. This unfolding can result in aggregation, precipitation, or an adsorptive loss of the protein to the container surface. At relatively high protein concentrations (>10 mg/mL), adsorptive losses are rarely observed because the amount of protein lost to the surface is very small compared to the total amount of protein in solution. At low protein concentrations (<1 mg/mL) the adsorptive losses are often significant and must be considered during the selection of a formulation and a container/closure system for the product. The protein interaction with the container surface is usually nonspecific and is caused by hydrophobic interactions.

One study investigating protein adsorption examined the stability of Factor VIII, a protein used to treat hemophilia B, in polyvinylchloride (PVC) mini-bags (17). These bags are commonly used in clinical settings for administration of drugs to patients because of the ease of administration. After 48 hr of storage at room temperature in these PVC mini-bags, the activity of Factor VIII dropped to 2% of the expected concentration of 2 IU/mL or 44% of the expected concentration of 10 IU/mL (17). These losses in activity were determined to be due to protein adsorption onto the PVC surface. Patients treated with Factor VIII that had been stored in PVC bags would have received significantly lower doses of the protein than expected.

For liquid formulations, shaking the formulation increases the air/liquid interface in the formulation and often leads to protein denaturation. Several proteins are susceptible to denaturation by shaking, including human growth hormone (hGH) (18,19) and recombinant factor XIII (20), both of which formed insoluble aggregates after shaking.

For lyophilized formulations, it is often critical to add a surfactant to the formulation to minimize aggregation. Denaturation of the protein may occur during the freezing or dehydration portion of the process. Interactions of the protein at the solid/air, liquid/air, or water/ice interfaces may result in aggregation, and it is these interactions that the addition of surfactant to the formulation may minimize (21).

Chemical Instability

Chemical degradation of proteins generally involves several common reactions in the protein. Some information regarding the chemical reactivity may be deduced from the primary sequence of the protein. Powell et al. (22) compiled hydropathy and flexibility information for 71 proteins and found that the hydropathy value coupled with known "hotspot" sequences predicted degradation with high certainty for deamidation and fragmentation. *Deamidation* primarily occurs through the hydrolysis of Asn or Gln residues, often from the formation of an intermediate cyclic imide. Oxidation may occur through several different mechanisms, including reactions with free radicals and metal ions. Disulfide exchange results from either the reduction or beta-elimination of existing disulfide bonds that, in turn, form new bonds or from the creation of new disulfide bonds from free thiol groups. Isomerization may result from several pathways. For example, proline isomerization may be caused by peptidyl prolyl isomerase that exists in trace amounts in the purified bulk. Fragmentation may result from enzymatic cleavage or from hydrolysis of the peptide bond. Each of these sources of chemical instability is detailed below.

Deamidation and Succinimide Formation

Deamidation of proteins at Asn or Gln residues proceeds through one of two pathways. Direct hydrolysis of the amide bond occurs under both basic and acidic conditions and results in cleavage of the peptide bond and thus the formation of fragments. This reaction will be discussed further under Fragmentation. The second route for deamidation in proteins occurs through the formation of a succinimide intermediate. This pathway involves a unimolecular interaction in which a deprotonated amide nitrogen undergoes nucleophilic attack on a carbonyl side chain of Asn to form the succinimide. Loss of an ammonia molecule from this structure results in the formation of a carboxylic acid. Scheme 1 shows the deamidation reaction for Asn. Asn residues typically are much more susceptible to deamidation compared to glutamine (23). Both the succinimide and the deamidated forms are considered to be degradation products. The susceptibility of Asn to deamidation has been studied by Robinson and Rudd (2) who found that an Asn followed in sequence by Gly deamidated at a much faster rate compared to an Asn adjacent to bulkier residues. This is presumably because the smaller amino acid allowed greater flexibility in the peptide chain, facilitating the formation of the cyclic imide. Because water is a reactant in this mechanism, the deamidation reactions occur much more readily in solution than in solid phase. Deamidation is one of the key routes of degradation for insulin (24) and DNase (25).

Succinimide formation may also result from the degradation of Asp or Glu (26). This reaction may occur under either basic or acidic conditions. As mentioned previously, the presence of succinimide is considered a degradation product of the

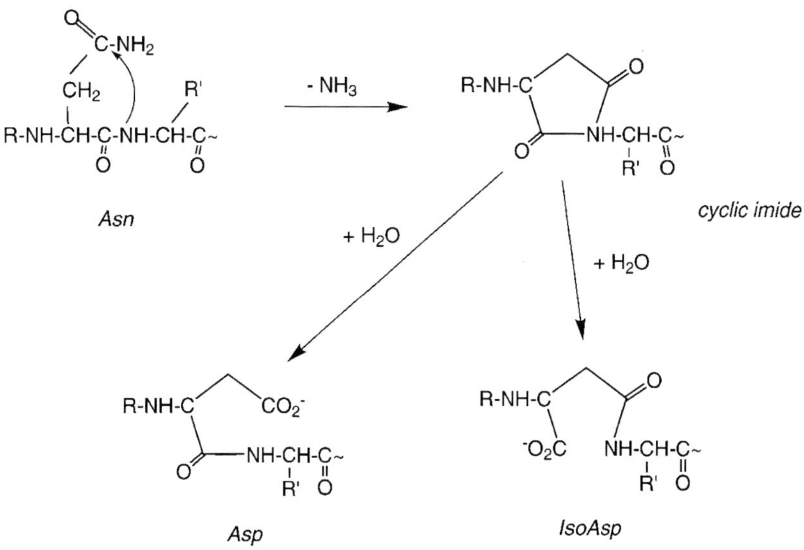

Scheme 1

protein. Conversion of Asp to the succinimide product was observed to be optimal at pH 4–5 in basic fibroblast growth factor (bFGF) (27).

Oxidation

Oxidation of proteins and peptides is extremely common, and there are several reviews that cover this topic in depth (23,28–30). Many amino acids are susceptible to oxidation. Methionine, cysteine, histidine, tryptophan, and tyrosine all undergo oxidation under specific conditions. Oxidation may be photolytic, metal-catalyzed, or through the presence of reactive oxygen species in the system that have been introduced as contaminants during the manufacturing process (e.g., peroxide). Even trace amounts of an oxidizing agent may wreak havoc with a protein, because the oxidizing agent serves as the initiator for a propagation reaction. Photolysis occurs when light causes the formation of reactive oxygen species that in turn attack the protein. Molecular oxygen may be activated by converting it to an excited singlet state [$^1O_2$] or by reducing it to species such as the superoxide radical ($^{\bullet}O_2^-$), hydrogen peroxide (H_2O_2), or hydroxyl radical ($^{\bullet}OH$) (29).

Metal ions lead to protein oxidation either by reacting directly with an amino acid to form a radical or by generating reactive oxygen species in solution. Amino acids that chelate transition metal ions (e.g., histidine) are most prone to metal-catalyzed oxidation, because the complex generates the reactive oxygen species which lead to oxidation (31). The proximity of the nascent reactive oxygen species to the chelated amino acid makes that amino acid most susceptible to attack.

Several commonly used solvents and/or excipients may contain trace levels of reactive oxygen species which could cause oxidation in proteins (30). Bleach, a commonly used cleaning agent, may leave trace levels of hypochlorite on storage containers and processing equipment. Polysorbate 20 and polysorbate 80 are widely used excipients in protein formulations that help prevent aggregation and surface denaturation. During the manufacture of the polysorbates, a bleaching step is sometimes used that results in the presence of trace levels of alkylhydroperoxide and hydrogen peroxide in the final polysorbate product (32). The polysorbates generate peroxides on storage (32). Oxidizing agents have been introduced as contaminants with mannitol (33).

To examine the effects of oxidation on proteins, one must understand which amino acids are most susceptible to oxidation and what the resulting products of oxidation are. Methionine is one of the most readily oxidized amino acid in proteins. Under mild oxidation conditions, the Met-sulfoxide is formed. This is the primary oxidation mechanism under acidic conditions (34). The reaction to form Met-sulfoxide is reversible chemically or enzymatically. If oxidation occurs under more harsh conditions then the sulfone is formed. The reactions for the formation of these two products are displayed in Scheme 2.

Histidine may undergo either photo- or metal-catalyzed oxidation. Photolytic oxidation results in the formation of 2-oxohistidine, also known as 2-oxoimidazoline, as shown in Scheme 3. Because His is an effective chelating agent, it is highly sensitive to metal-catalyzed oxidation. Additional oxidation products may be observed through metal-catalyzed oxidation of His, including the formation of Asp or Asn, but the mechanism for the formation of these products is not understood (29).

Cysteine has a free thiol group that may oxidize to form a disulfide bond (Cystine) as shown in the reaction displayed in Scheme 4. This reaction is favored at higher pHs, where the thiol is deprotonated (35). The oxidation of Cys may be spontaneous (resulting from the presence of O_2) or metal-catalyzed. Spontaneous

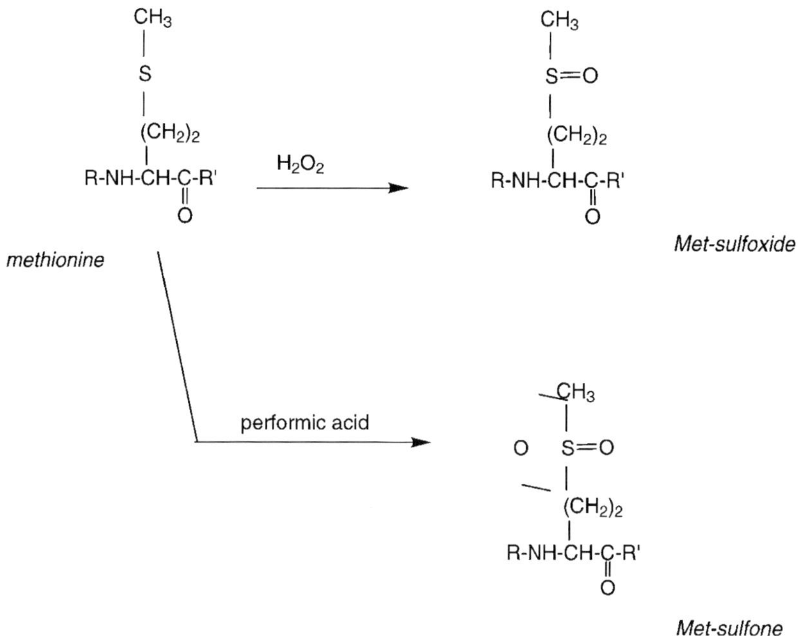

Scheme 2

oxidation of free thiols in bFGF was promoted in the presence of heparin and led to aggregation of the protein (36).

Tryptophan oxidizes primarily via reactive oxygen species, although it may occur as a result of photoxidation in the presence of dyes (28). Trp and Met are the only amino acids that are capable of being oxidized below pH 4. The primary oxidation products of Trp are *N*-formylkynurenine and kynurenine (37,38), shown in Scheme 5, with many other products observed including Gly and Ala (39). As with His, the detailed mechanism for understanding the oxidation of Trp is poorly understood.

Tyrosine is oxidized via a photolytic mechanism to produce 3,4-dihydroxyphenylalanine and bityrosine as displayed in Scheme 6 (40,41). Since bityrosine formation may be intermolecular, the product of oxidation is often the formation of aggregates.

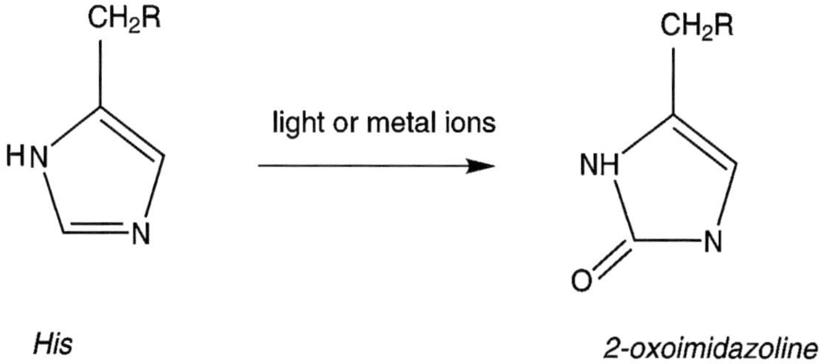

Scheme 3

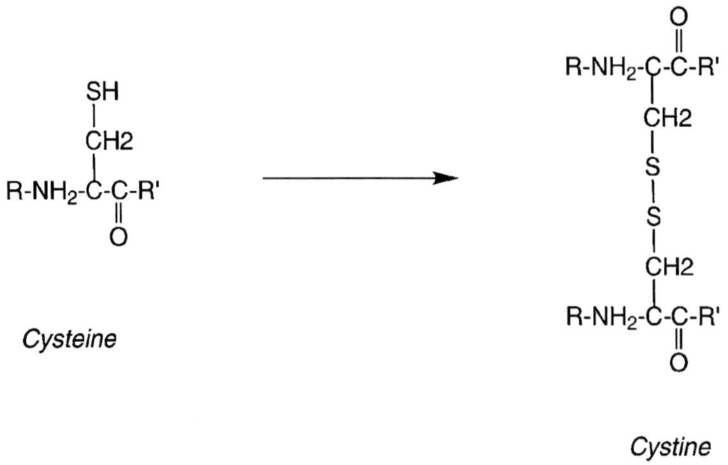

Scheme 4

Because protein oxidation may occur during at any stage of the manufacture of the product, it is necessary to assess the susceptibility of the protein to oxidation through several mechanisms. Hydrogen peroxide, *tert*-butyl hydroperoxide (TBHP), and light are often used to promote oxidation in protein samples. Metal ions such as Cu^{2+} and Fe^{3+} have been added to protein formulations to purposefully assess whether a protein is susceptible to metal-catalyzed oxidation.

The oxidation of recombinant interferon-gamma (IFN-γ) and recombinant tissue-plasminogen activator (t-PA) was investigated by Keck (42). Two of five Met in IFN-γ and three of five Met in t-PA were oxidized to the sulfoxide state with TBHP. In each case, the Met was located on the surface of the protein. No other amino acids were oxidized in this experiment. Additionally, oxidation of IFN-γ by H_2O_2 resulted in the conversion of all five Met to the Met-sulfoxide, again with no other oxidation

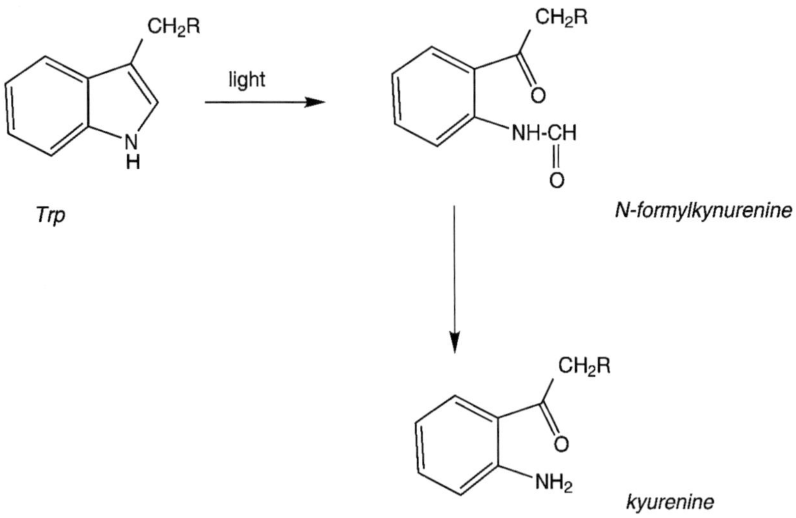

Scheme 5

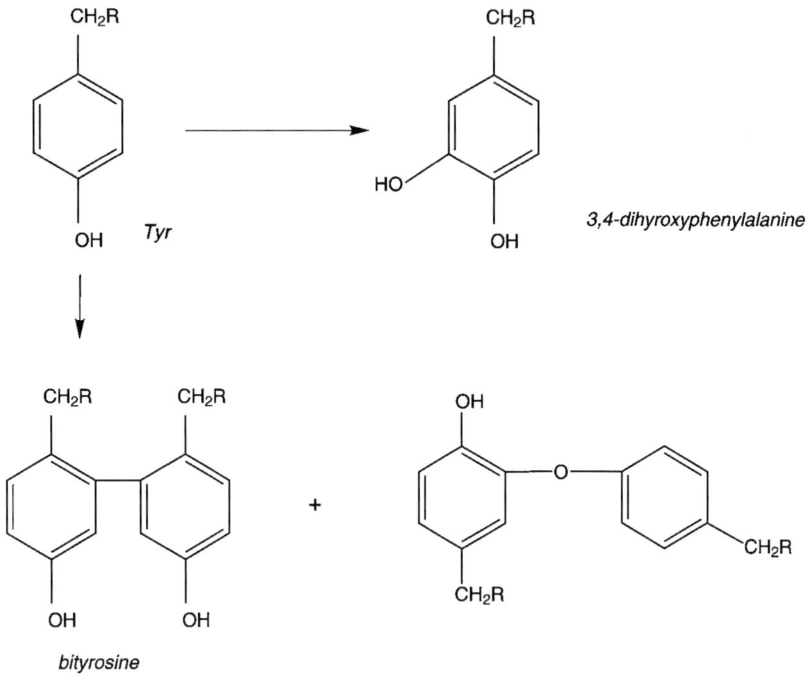

Scheme 6

products observed. DalleDone (43) probed the oxidation of actin with TBHP and found that while none of the Met oxidized, one of five Cys did undergo modification which caused a marked decrease in the rate of actin polymerization. Oxidation of human epidermal growth factor 1–48 (hEGFl-48) was achieved using light exposure (44). In this protein, the lone Met converts to Met-sulfoxide on exposure to light when the protein solution is stored in glass. The rate of oxidation was observed to be greatest in colorless glass. Storage in amber glass afforded some protection, while storage in foil-wrapped glass showed the slowest rate of oxidation. Metal-catalyzed photooxidation of human growth hormone (hGH) was found to specifically oxidize His 21, one of three residues involved in a cation-binding site (45). Other potential sites of oxidation, including three Met and two additional His, were found to be unaffected by photooxidation of hGH.

Oxidation of relaxin has been extensively studied and serves as a useful example of how oxidation may be achieved through different mechanisms. Relaxin is a two chain, disulfide-linked hormone with a molecular weight of ~6 kDa. Early formulation screens by Cipolla and Shire (46) demonstrated that oxidation of relaxin was enhanced by light or in the presence of methylcellulose, a carbohydrate used to make a topical formulation. Under these conditions, Met 4 and Met 25 on the B chain formed sulfoxides. Oxidation of relaxin by hydrogen peroxide gave similar results, with only the sulfoxide form of the two Met observed (47). No other amino acids displayed oxidation. Li et al. (48) demonstrated that metal-catalyzed oxidation achieved by a combination of ascorbate/cupric chloride/oxygen resulted in modification of His 12 on the A chain in addition to the previously mentioned Met, as well as physical instability of the protein at pH >6.0. The physical instability was postulated to be an indirect result of the His modification, and this hypothesis was supported in later studies comparing the protein to the porcine version which does not contain His 12 (49). These examples demonstrate that oxidation products of proteins depend highly on the mechanism used to achieve oxidation.

Disulfide Exchange

Disulfide exchange occurs when a disulfide bond undergoes β-elimination to form free thiols, and the free thiols reoxidize with incorrect pairings. When disulfide exchange occurs, a change in secondary, tertiary, or quaternary structure may be observed. Lyophilized insulin was shown to undergo β-elimination followed by formation of mixed disulfides when stored at high temperature and/or high humidity (50). β-elimination has been proposed as the mechanism for degradation of recombinant human macrophage colony-stimulating factor (rM-CSF) stored under alkaline conditions (51).

Isomerization/Racemization

All amino acids except Gly are susceptible to isomerization or racemization. Enzymes such as peptidyl prolyl isomerases exist solely to isomerize specific amino acids. The cyclic imide pathway of degradation for both Asn and Asp can result in a racemic mixture of products. Isomerization has been observed in Asp residues of bFGF (27) and Asn residues of insulin (52).

Fragmentation

Fragmentation results from multiple pathways including deamidation, Asp-Pro cleavage, and protease activity. Direct hydrolysis of Asn-containing peptide bond could occur causing fragmentation and deamidation. Cleavage after Asnl0l in aging α-crystalline was determined to occur through a deamidation pathway (53). Asp, because of the electronegative carbon center in the carbonyl group, is particularly susceptible to cleavage of the peptide bond when followed by Pro in the protein sequence (28). Recombinant human interleukin 11 (54) and rM-CSF (51) both degrade primarily via cleavage at Asp-Pro sites in acidic solution. Finally, fragmentation may occur due to protease activity. While the final protein bulk obtained from manufacturing is highly pure, it is sometimes possible for trace amounts of proteases or other enzymes to copurify with the protein of interest. The quantities of the proteases may be too small to observe analytically; however, even a trace amount of a protease can result in significant fragmentation in the protein during storage. Fragmentation at locations on a protein other than those described above may be the only evidence for the presence of a protease in the formulation.

Glycation

Glycation occurs when a sugar molecule chemically reacts with one of the amino acid side chains to form a carbohydrate adduct on the protein. This occurs via a Maillard reaction in which the carbonyl of a reducing sugar undergoes a condensation reaction with the amino group on Arg, Lys, or N-terminal amino acid. The product of this reaction has a Schiff's base. This reaction is accelerated in the solid state and is favored when the formulation pH is less than neutral (55,56). Amadori rearrangement of the Schiff's base product may lead to browning of the solution.

Glycation has been observed in a lyophilized formulation of relaxin in which a glucose adduct was identified (48). Formulation of recombinant human DNase with lactose in a spray-dried state resulted in the addition of lactose molecules to five of the six Lys in the protein (57). A similar reaction was observed in a lyophilized formulation of human growth hormone in lactose (58). Due to the tendency of reducing sugars to undergo the Maillard reaction with proteins, they are not the carbohydrates of choice for use in formulations.

Summary

In the development of a formulation, the potential degradation mechanisms for the protein must be considered. From the primary structure, one may anticipate some of the potential degradation pathways for a given protein (e.g., deamidation at Asn-Gly; fragmentation at Asp-Pro). Other types of degradation (e.g., surface denaturation) cannot be predicted as easily and must be determined experimentally. With the information from an assessment of the primary structure of the protein, the formulation scientist now begins the task of developing a suitable system for storage of the protein.

FORMULATION DEVELOPMENT

The first stage of formulation development is called preformulation. During the preformulation stage, short studies are conducted to assess the relative types and rates of degradation observed as a function of pH, protein concentration, and temperature. It is from these early studies that the formulation scientist first gains information regarding which specific degradation mechanisms may be important to the protein of interest.

With the results of the preformulation studies, the formulation scientist can begin to address some of the specific questions pertaining to the development of the final formulation. The *state* of the dosage form is selected. The inactive ingredients in the formulation, *excipients,* which promote stability of the final product, are screened. Finally, supportive *accelerated stability studies* are conducted to aid in the selection of the final formulation.

State

The purified bulk that is obtained from the manufacturing process of proteins typically arrives as a liquid solution. A decision is made by the formulation scientist regarding the state of the final drug formulation (i.e., liquid or solid state) to be produced from the liquid bulk. The selection of state for the final product involves balancing several criteria, including cost, stability, route of administration, dose, and the intended storage conditions for the final product.

Many different routes are employed to achieve solid-state formulations including crystallization, lyophilization, spray-drying, and spray-freeze-drying. *Crystallization* is the only one of these techniques to form a purposefully ordered solid. Formulations of crystallized glucose oxidase and crystallized lipase have been reported to be more stable than their amorphous counterparts (59), presumably because the crystal packing structure does not allow reactions to occur that require flexibility of the peptide chain. However, Pikal and Rigsbee (60) noted that amorphous insulin exhibited greater stability compared to the crystalline form of the protein. The authors postulated the decreased stability of the crystalline form was due to local configurational differences around Asp21, the primary site of degradation (60). The investigation of crystallized protein formulations is expected to be a hot topic in the next decade. Because little information is available regarding actual examples of crystalline protein formulation production and stability, the remaining portion of this section will focus on amorphous solid-state formulations obtained by one of the drying processes.

The most common method to achieve a solid-state formulation is through *lyophilization,* or freeze-drying, in which the formulation is frozen and the bulk water is removed by sublimation. The resulting cake is composed of the protein, any nonvolatile excipients, and a small amount of residual water tightly associated with the protein. Excipients such as sucrose, mannitol, or glycine are often added to formulations to be lyophilized to serve as bulking agents and to protect the protein during the freezing and drying processes. There is a wealth of literature available that examines the lyophilization process and its effects on protein integrity (61,62).

Spray-drying (SD) uses atomization to form microdispersed droplets. The water in these droplets quickly evaporates when passed through a stream of hot gas, resulting in the formation of a fine powder of microparticles containing protein and excipients. Although the protein solution passes quickly through the hot gas stream and evaporation provides some cooling, the potential for thermal degradation of the protein is a concern using this technique. This technique was successfully applied to the preparation of powders of hGH and t-PA (63).

Spray-freeze-drying (SFD) is a combination of the two previously described drying techniques in which the microdispersed liquid particles are generated through the jet nozzle in the absence of heat, collected and frozen in liquid nitrogen before sublimation occurs. The frozen particles are then lyophilized. SFD is used in place of SD when the protein cannot withstand the temperatures in the SD process. Preparation of SFD powders of DNase and an anti-IgE monoclonal antibody were found to be superior to those prepared by SD (64).

How does one decide which state to pursue for a dosage form? Production of a liquid formulation is cheaper than any of the solid-state forms because it takes fewer steps to manufacture, thus requiring a shorter time. At most, a formulated liquid bulk may need to be diluted before a liquid fill is performed, although often the bulk is filled directly into the final container. For a lyophilized product, the filled containers must undergo the freeze-drying process which can take upwards of one week. For spray-dried formulations, the bulk must be further processed to achieve the powder formulation that then must be filled into the appropriate final containers. SFD formulations have the added cost of both powder production and lyophilization prior to filling. These added manufacturing steps result in a higher production cost for the solid state. In addition, losses in protein yield are observed for SD and SFD processes. Thus, the cost in terms of time, money, and sometimes product yield for the liquid formulation is smaller than that for solid-state formulations.

If the cost is higher, why would one pursue a solid-state formulation? One critical issue is stability. Generally, solid-state formulations degrade much more slowly than liquid state formulations because water plays a key role in several of the degradation mechanisms (e.g., deamidation) for proteins. The decreased rate of degradation in a solid-state formulation potentially enables storage of the drug product at higher temperatures than would be allowed for liquid products which are generally stored under refrigerated conditions. It should be noted, however, that some mechanisms of degradation are accelerated in the solid state (e.g., glycation) and simply switching to a solid-state formulation does not ensure that no degradation will be observed.

The ability to store a drug product under nonrefrigerated conditions is particularly critical for products, such as vaccines, intended for export to third world areas. Although sometimes inconvenient, it is easy to find refrigerators for storage of drugs in the United States and other industrialized countries. In developing countries, access to refrigeration may be quite limited. It is critical that drugs manufactured for use in developing areas have sufficient stability under local ambient conditions to enable their use. Solid-state formulations may allow one to achieve that goal.

The need for a multidose formulation may also dictate the use of a solid-state formulation. As the name implies, multidose products are intended to provide the patient with a product that contains several doses of the therapeutic within one container. Multidose formulations contain preservatives to kill any bacteria and prevent mold growth that may result from repeated entry into the drug product. Phenol and benzyl alcohol are two widely used preservatives in protein-based parenteral pharmaceuticals. Frequently, the addition of a preservative to the formulation compromises the long-term stability of the drug product, typically because the protein becomes physically unstable and/or exhibits oxidation. If the formulation scientist can obtain sufficient short term stability (e.g., 2 weeks) for a formulation containing a preservative, then the use of a solid-state product may enable production of a multiuse formulation. In this case, the preservative is NOT added to the liquid bulk used to prepare the solid state. Instead, when the solid-state formulation is reconstituted prior to use, a preservative is included in the water for reconstitution. Thus, the final product to be used is a multidose formulation that will experience only short-term exposure to the preservative.

The desired protein concentration may also dictate the use of a solid-state formulation. At large scale, protein solutions are concentrated using ultrafiltration. During ultrafiltration, the protein concentration at the membrane can become several fold higher than the concentration of the bulk solution. The result of this high protein concentration is limited diffusion of the solvent through the protein layer which in turn limits the bulk protein concentration that may be achieved using this process. Depending on the protein, the maximum concentration that may be reached using ultrafiltration varies between 40 and 200 mg/mL. This concentration limitation is very important when one is developing a formulation intended for administration subcutaneously (SC) or intramuscularly (IM). Both of these routes of administration require that a small volume be given per injection. For proteins that require high concentrations for efficacy, the concentration achievable through ultrafiltration may be insufficient to meet the dose necessary for SC or IM administration.

To work around this concentration limitation, formulation scientists often use a solid-state formulation as a means to achieving a higher protein concentration for the final drug product. By reconstituting the solid formulation with less water than was used to initially formulate the drug, one gets an increased protein concentration in the reconstituted product as depicted in Fig. 1. It is important for the formulation scientist to remember that in addition to the protein, the excipients in the formulation are also increased in concentration when the solid formulation is reconstituted to a lower final volume. Care should be taken to maintain a suitable final product osmolality for injection.

The ability to achieve a higher concentration on reconstitution of the solid-state product also may allow the formulation scientist to side-step potential

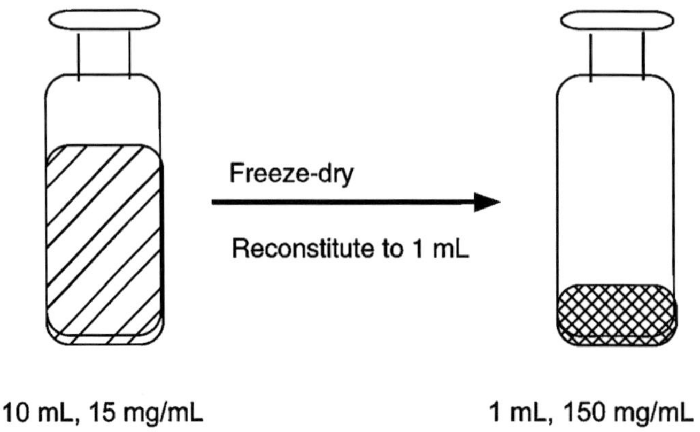

Figure 1 Using reconstitution to achieve high protein concentrations. The vial on the left has been filled with 10 mL of a 15 mg/mL aqueous protein solution. After lyophilization, this vial contains 150 mg of the protein plus any nonvolatile excipients. Reconstitution with 1 mL of water would yield a final formulation that contains 150 mg/mL protein plus nonvolatile excipients at concentrations 10-fold higher than originally filled into the vial. If the formulation scientist plans to use this method for achieving a high concentration formulation, it is necessary to consider the final concentration of excipients.

stability-limiting aggregation that may result in a liquid formulation of high concentration protein.

Excipient Selection

Excipients are the inactive ingredients added to stabilize the drug substance and the final drug product. The excipients may serve many purposes including maintaining solution pH, adjusting solution osmolality, antioxidant, preservative, bulking agent, and minimizing surface denaturation. While it might be tempting to add a little bit of each type of additive to the formulation, the desire is to add only what is necessary for stabilization of the drug. This is because the added excipients may interact with the protein or with each other, thus compromising the desired activity. Also, an increased complexity of the formulation due to the number of excipients used also increases the risk of errors in the manufacture of the formulation which can result in product failure.

Formulation scientists tend to only use excipients that have been previously incorporated in other formulations. The reason for this is very clear: the demonstrated safety of the excipients in humans. For a new excipient to be used for a particular route of administration, the FDA requires that safety of that excipient is demonstrated in a human clinical trial. This may be an additional trial to that used to study the safety of the active drug product, and therefore it is a significant added expense to the development of the drug product. If an excipient that has been used previously accomplishes the same goal in a formulation as one that has not been previously tested, it is prudent for the formulation scientist to proceed with the previously tested excipient.

For a list of previously used excipients, one is often referred to the GRAS list, additives that are Generally Regarded As Safe by the FDA. Caution should be used when consulting this list as it refers to compounds that have been administered orally and specifically to food additives. The safety of excipients administered by other delivery routes may be quite different than those administered orally. The FDA publishes the *Inactive Ingredients Guide*, which lists all inactive ingredients in approved drugs with the route of administration, the number of drugs approved containing the ingredient, and the range of concentrations that are used (65). The Physician's Desk Reference (PDR) includes formulation information for specific prescription and over-the-counter drugs (66). Information from the PDR has been compiled by Powell and coworkers (67) to provide the formulation scientist easier access to this information.

Buffer and pH

The first variable typically specified for a formulation will be the pH. Several factors will aid the selection of pH: the stability of the protein, solubility of the protein, and acceptability of the pH for the route of administration. The pH at which the drug undergoes minimal degradation is usually the preferred pH of the formulation. However, proteins reach minimal solubility when the solution pH approaches the isoeletric point (p*I*) of the protein. At the p*I*, the net charge on the protein is minimized and as the solution pH reaches the p*I*, precipitation is often observed. Generally one must be greater than one pH unit away from the p*I* to achieve meaningful solubility. The final consideration is the pH that will be acceptable for administration. For drugs that are administered via the IM or SC routes, and especially to the lungs and eyes, it is desirable to be as close as possible to the tissue fluid pH, pH 7.4. Administration of a drug at a pH that differs significantly from neutral could result in pain on administration, possibly due to cellular damage at the site of administration (68,69).

Using the desired pH and the list of buffers that have been previously used for drugs, the formulation scientist quickly narrows the buffers available for testing that meet both criteria. A listing of some of the commonly used buffers for injectables with their corresponding pK_as is displayed in Table 1. It is notable that there are actually very few choices at any pH. This list is typically decreased further if one is planning to formulate a solid-state product, because some of the buffering components (e.g., acetic acid) are volatile. A reconstituted formulation that originally contained sodium acetate/acetic acid would no longer have a buffering species or pH

Table 1 Approximate p*K*a for Commonly Used Buffers for Parenteral Administration

Buffer salt	p*K*a at 25°C
Acetate	4.76
Citrate	3.13
	4.76
	6.40
Glycine	2.35
	9.78
Histidine	6.04
Phosphate	2.15
	6.82
	12.38
Succinate	4.21
	5.64
Tris	8.30

control. Tris and other amine buffers are known to have a strong temperature dependency on pH. For this reason, analysis of protein stability as a function of pH under elevated temperature conditions is often difficult to interpret because the pH varies with the temperature.

The concentration of the buffering species to be used also needs to be assessed. As with excipients in general, the rule of thumb is to use only the concentration of buffering species necessary to maintain pH or solubility of the formulation. One of the reasons to minimize the concentration of buffer components in the formulation is to prevent pain on administration. Products that are formulated at pHs that are not equivalent to that of the human tissue fluids are somewhat painful when administered by SC, IM, topically to open wounds, or in the eye. The use of a buffer that maintains the pH of the formulation delays the equilibration of the administered drug's pH to that of the body. Thus, higher concentrations of the buffering species prolong this equilibration and result in pain on use. As anyone can attest who has been on the receiving end of administration of a painful formulation, the development of a drug product should minimize the potential for pain due to the excipients added. A second reason to minimize the buffer concentration is to limit the extent of buffer-catalyzed reactions. It has been observed with the peptide gonadorelin that an increase in the concentration of phosphate buffer led to an increase in the degradation rate of the peptide, and this increase in degradation rate was independent of the ionic strength of the buffer (70).

A preformulation screen performed for hybrid (BDBB) interferon-α (IFN-α) provides an interesting study in the balance of factors that must be considered on selection of pH and buffering species used for a formulation (71). The authors compared the degradation of IFN-α in liquid solutions ranging in pH from 1.0 to 7.6 using citrate, acetate, glycine-HCl, or phosphate as the buffering species. The results of this study are displayed in Fig. 2, in which the relative rate of degradation is plotted versus pH. Formulations prepared with glycine-HCl at pH 5.0 and 6.0 did not allow sufficient solubility of the protein, presumably due to the proximity of the bulk pH to the p*I* of the protein (5.5). Formulations in acetate (pH 3.4–5.0) and glycine-HCl (pH 1.0, 2.0) were found to degrade the fastest as assessed by reversed-phase high-performance liquid chromatography, a technique used to monitor oxidation and fragmentation for this protein. Citrate formulations ranging in pH from 2.0 to 4.5 showed the next highest rates of degradation. Glycine-HCl (pH 3.0–4.5) and phosphate (pH 7.6) formulations displayed the slowest rates of degradation. These results demonstrate that for IFN-α (a) formulation at low pH (e.g., 4.0) and near neutral pH (7.6) provide the most stability according to this method and (b) glycine-HCl is the preferred buffering species at pH 4.5 compared to acetate and citrate. No difference in the rate of degradation was observed when the concentration of phosphate in the pH 7.6 formulation was changed two-fold.

While the relative rates of degradation for the pH 7.6 and pH 4.0 formulations of IFN-α were almost equivalent, the mechanisms of degradation were found to be very different. The pH 7.6 formulations exhibited higher levels of aggregation with concurrent decreased solubility as compared to the pH 4.0 samples. Oxidation was observed at pH 7.6, while none was observed at pH 4.0. Lower molecular weight species were formed at pH 4.0. Thus, the next stage of formulation development for this protein for the pH 7.6 solution would require the use of excipients to minimize aggregation and oxidation, while stabilization of the pH 4.0 solution would require determination of the mechanism for fragmentation and the steps that could be taken to minimize degradation by this route.

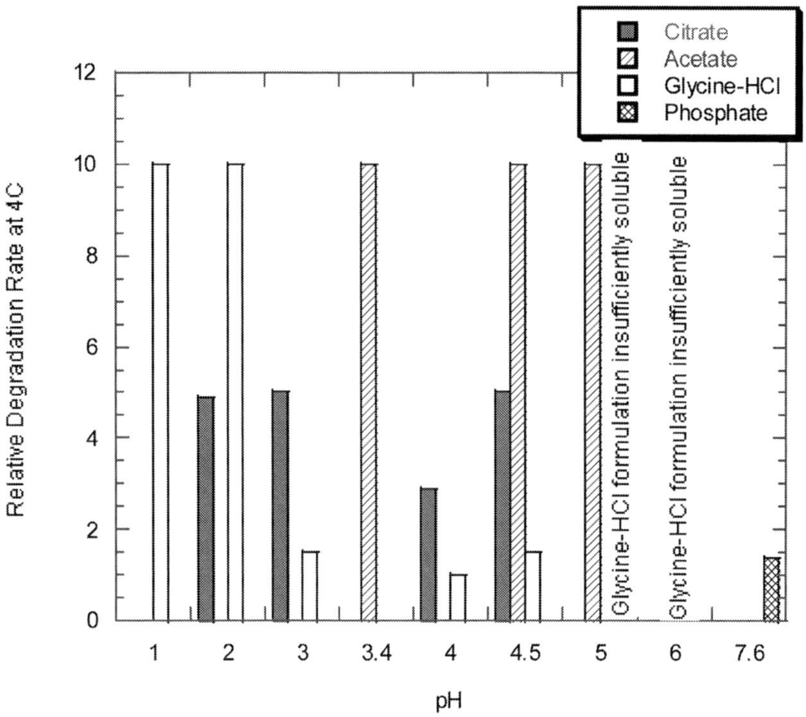

Figure 2 The relative rates of degradation for hybrid (BDBB) interferon-α formulations (70). The maximum rate of degradation assigned was 10. Glycine-HCl formulations at pH 5 and pH 6 were insufficiently soluble to examine degradation rates. Phosphate was only used in the pH 7.6 formulation.

Salts, Sugars, and Polyols

After selection of pH and buffer, the osmolality of the solution is adjusted using salts, sugars, or polyols. The osmolality of blood is ~290 mmol/kg (72–74), and this is typically the target osmolality used for formulations that will be administered SC, IM, or to the eye. Solutions that are more than two- or three-fold hypoosmolar or hyperosmolar may result in cell lysis, an undesirable side effect of dose administration (75). Solutions that are administered IV do not necessarily have to be isoosmolar, because the dilution into the large volume of blood quickly normalizes the osmolality of the drug.

For liquid formulations, the choice of either using a salt or a carbohydrate to adjust the osmolality of the solution is made by the impact on protein stability. Sodium chloride is one of the most commonly used salts in formulation of both traditional pharmaceutics as well as biological pharmaceutics. It is extremely safe, well-tolerated, and inexpensive. However, the presence of sodium chloride in a formulation of rhuMAb HER2 was found to increase oxidation when the formulation was stored in stainless steel containers, presumably because the sodium chloride promoted corrosion of the stainless steel (76).

Sugars and polyols have often been used in formulations, as well, particularly those that are solid state. Commonly used carbohydrates include mannitol, sucrose, and trehalose. As previously described, the formulation scientist is strongly advised to steer clear of incorporation of reducing sugars (e.g., glucose, lactose) in formulations to avoid glycation and browning of the solution.

In lyophilized, SD, and SFD formulations, carbohydrates are employed as both bulking agents and as lyoprotectants. A bulking agent is necessary for these solid-state formulations because the physical amount of protein is very small. The addition of a bulking agent makes the sample amount more adequate for handling and generally improves the appearance of the final product. More importantly, the carbohydrate additives usually impart physical stability to the protein during the drying process. Specific ratios of carbohydrate to protein were found to be important to the stabilization of the protein structure in the solid state for two recombinant humanized antibodies (77,78) and for recombinant human interleukin-1 (79).

Surfactants

Surfactants are added to formulations to minimize denaturation of the protein at interfaces, typically liquid/air, solid/air, and liquid/container interfaces. Surfactants bind to hydrophobic areas of proteins (80). By minimizing the accessibility of the hydrophobic contacts in a protein solution, the surfactant reduces the protein–protein interactions that lead to aggregation (81). Furthermore, surfactants compete with the protein for binding at hydrophobic surfaces and are added to minimize protein losses due to surface denaturation (26). The concentration of surfactant required in these formulations generally is generally above the critical micelle concentration (CMC) for the particular surfactant. Polysorbate 20, polysorbate 80, and pluronic F68 are commonly used surfactants in protein therapeutics.

Several studies have demonstrated the need for surfactants in protein formulations. A formulation of recombinant factor VIII SQ required either polysorbate 80 or polysorbate 20 to prevent losses due to surface adsorption (82). Another study examining the stability of factor VIII SQ found that polysorbate 80 prevented losses of protein activity upon filtration and freeze–thaw (83). Polysorbate 20 was added to a formulation of recombinant factor XIII to stabilize the protein against both agitation and freeze–thaw-induced aggregation (20). Polysorbate 80 was added to a freeze-dried formulation of recombinant hemoglobin and found to protect the protein from aggregation resulting from the freeze–thaw process, although long-term stability of the protein against aggregation was not achieved (81). Pluronic F68, Brij 35, and polysorbate 80 all prevented aggregation of recombinant human growth hormone that was induced by vortexing the solution (18).

Antioxidants

Antioxidants are incorporated into formulations in which oxidation is a degradation mechanism for the protein. Typically, the antioxidant is a compound that is highly susceptible to oxidation and serves to scavenge any oxidative species in the solution before the oxidants can attack the protein. Addition of Met and thiosulfate to a formulation of rhuMAb HER2 effectively inhibited oxidation of the protein (76). In cases where metal-catalyzed oxidation is observed, the incorporation of a chelating agent such as ethylenediaminetetraacetic acid (EDTA) may reduce the rate of oxidation of the protein by effectively scavenging free metal ions in solution before they have a chance to oxidize the protein. An alternative approach to reducing metal-catalyzed oxidation is to include Zn^{2+} in the formulation to bind to residues which may be susceptible to oxidation. Because Zn^{2+} does not promote oxidation, it protects the amino acid to which it chelates.

Preservatives

The next class of excipients that may be added to formulations are preservatives. These usually are not added to enhance stability to the formulation, but rather to give flexibility in the use of the drug product. With single use formulations (those lacking a preservative), there is a requirement that each drug-filled container be entered only once. This is a safety precaution, because exposure of the drug product in the container/closure system to air may result in the introduction of bacteria to the drug product. This is not a problem if the drug is used immediately, because there is not sufficient time for the bacteria to colonize. However, on storage it is possible for a significant number of bacteria to grow, particularly in formulations that are produced at neutral pH and that contain carbon and nitrogen sources on which the bacteria may feed. It would be unwise to inject a patient with a drug full of bacteria!

For this reason, a drug product that is to be used multiple times (multidose) must contain a preservative to prevent bacterial growth. A list of preservatives that have been used in pharmaceutical formulations is shown in Table 2. However, most of these are not usually compatible with protein formulations. Some, such as the parabens, are not active in the presence of nonionic surfactants, excipients that are typically required in protein formulations (84). Others may not be acceptable for a particular route of administration. Benzalkonium chloride, a commonly used preservative in topical formulations, causes ototoxicity when applied to the ear (85). As with buffering species, the list of preservatives available to the formulation scientist quickly narrows to just a few compounds including benzyl alcohol, phenol, *m*-cresol, and benzethonium chloride.

In screening preservatives for use in formulations, it is necessary to determine what levels of preservatives are efficacious at preventing bacterial growth in the particular formulation. Preservative challenge tests are conducted in which the protein formulation is spiked with different organisms such as *P. aeroginosa.* The levels of each organism are monitored over time to determine if the preservative acted to kill or prevent growth of the particular organism. The United States Pharmacopeia (USP) and the European Pharmacopeia (EP) have prescribed levels of efficacy that a preserved formulation must pass to be used (86,87). Because the efficacy of the preservative is dependent on the other excipients in the formulation, the formulation scientist must be careful to assess the final formulation conditions in the preservative challenge test. Even a change in protein concentration can affect the efficacy of the preservative.

Proteins are generally not exceedingly stable in the presence of preservatives. Preservatives are typically small hydrophobic compounds that may interact with hydrophobic regions of the protein, leading to a disruption in protein structure. Lam and coworkers (88) found that addition of benzyl alcohol to a formulation of IFN-ϒ resulted in a loss of tertiary structure as determined by CD spectroscopy. Aggregation of interleukin-1 receptor was found to be the predominant pathway for degradation in the presence of phenol, *m*-cresol, and benzyl alcohol (89).

Other

There are numerous other excipients that may be required for use in formulating proteins. Amino acids such as His, Arg, and Gly have been used as bulking and solubilizing agents. Metal ions such as Ca^{2+} and Zn^{2+} are sometimes added for maintaining structure or activity of the protein. Other additives may be required to meet the needs of a specific route of administration, such as the necessity of including

Table 2 Commonly Used Preservatives and Some of the Major Incompatibilities of the Preservative that are of Particular Interest to a Formulator of Protein Therapeutics

Preservative	Route of administration	Incompatibilities
Benzalkonium chloride	IM, inhalation, nasal, opthalmic, otic, topical	Citrate, methylcellulose, surfactants, some plastics, some rubbers
Benzethonium chloride	IM, IV, opthalmic, otic	Anionic surfactants
Benzoic acid	IM, IV, irrigation, oral, rectal, topical, vaginal	Alkalis, heavy metals, kaolin
Benzyl alcohol	Injections, oral, topical, vaginal	Oxidizing agents, strong acids, nonionic surfactants, methylcellulose, some plastics
Bronopol	Topical	Sulfydryl compounds, aluminum
Butylparaben	Injections, oral, rectal, topical	Nonionic surfactants, some plastics
Cetrimide	Topical, opthalmic	Anionic and nonionic surfactants, metals
Chlorhexidine	Topical, opthalmic	Anionic materials, Ca^{2+}, Mg^{2+}, viscous materials
Chlorobutanol	IM, IV, SC, inhalation, nasal,otic, optalmic, topical	Some plastics, rubber, carboxymethylcellulose, polysorbate 80
Chlorocresol	Topical	Some plastics, rubber, nonionic surfactants, methylcellulose
Cresol	IM, intradermal, SC, topical	Nonionic surfactants
Ethylparaben	Oral, topical	Nonionic surfactants, some plastics, silica
Imidurea	Topical	None listed
Methylparaben	IM, IV, SC, opthalmic, oral, otic, rectal, topical, vaginal	Nonionic surfactants, sorbital, some plastics
Phenol	Injections	Acts as a reducing agent, albumin, gelatin
Phenoxyethanol	Topical	Nonionic surfactants, PVC, cellulose derivatives
Phenylethyl alcohol	Nasal, ophthalmic, otic	Oxidizing agents, proteins, polysorbates
Phenylmercuric acetate/borate	Ophthalmic	Halides, amino acids, some plastics, rubber
Phenylmercuric nitrate	IM, ophthalmic, topical	Halides, aluminum, amino acids, some plastics, rubber
Propylparaben	IM, IV, SC, inhalation, ophthalmic, oral, otic,	Nonionic surfactants, some plastics

(Continued)

Table 2 Commonly Used Preservatives and Some of the Major Incompatibilities of the Preservative that are of Particular Interest to a Formulator of Protein Therapeutics (*Continued*)

Preservative	Route of administration	Incompatibilities
	rectal, topical, vaginal	
Sodium benzoate	Dental, IM, IV, oral, rectal, topical	Quaternary compounds, gelatin, calcium salts, nonionic surfactants
Sodium propionate	Oral	
Sorbic acid	Oral, topical	Bases, oxidizing agents, reducing agents, nonionic surfactants, plastics, heavy metal salts
Thimerosal	IM, IV, SC, ophthalmic, otic, topical	Sodium chloride solutions, proteins, some plastics, rubber

[From Ref. (84)]

a gelling agent for topical formulations. To minimize losses due to protein adsorption, additives may be incorporated in the formulation to prevent or reduce surface denaturation of the active protein. Previously unpublished dilution studies were performed using transforming growth factor-beta1 (TGF-beta1) to determine the lowest concentration feasible at which the protein could be reasonably formulated. TGF-beta1 is a potent cytokine, and very small amounts are required for biological action. As shown in Fig. 3, in the absence of an additive to the formulation the recovery of TGF-beta1 from either a glass container or from a polypropylene tube was ~80–90% at 10 μg/mL. At 10 ng/mL, the recovery dropped to 20–30% of the expected concentration. In the presence of a 0.5% (w/v) 100 bloom gelatin solution, the recovery increased to 100% at concentrations of 100 ng/mL or greater, and almost 70% at 10 ng/mL. Thus, to achieve consistent protein dosages, gelatin addition to the formulation was required to minimize the losses of the TGF-beta1 to the container surfaces used for storage of the drug and for analysis of the formulation. In the last 10 years, there has been great concern regarding the use of animal-derived excipients in pharmaceutical formulations. There is a possibility of contamination of an excipient by either an undetected virus (e.g., hepatitis C, HIV) or prion that may cause transmissible spongiform encephalitis. For this reason, formulators have moved away from plasma-derived additives (e.g., human serum albumin, gelatin) and animal-sourced excipients (e.g., polysorbate 80) to either recombinant or vegetable-derived sources for these excipients. For example, the lyophilized formulation of recombinant factor VIII SQ was modified to remove human albumin (present to minimize surface adsorption and to function as a bulking agent and stabilizer) and to replace it with a combination of polysorbate 80, histidine, sucrose, and sodium chloride to achieve a stable product that would not have the potential of containing human viral particles (83).

Accelerated Stability Studies

Once the formulation scientist has narrowed the list of excipients to test in a formulation, an accelerated stability study is initiated. In this study, potential formulations

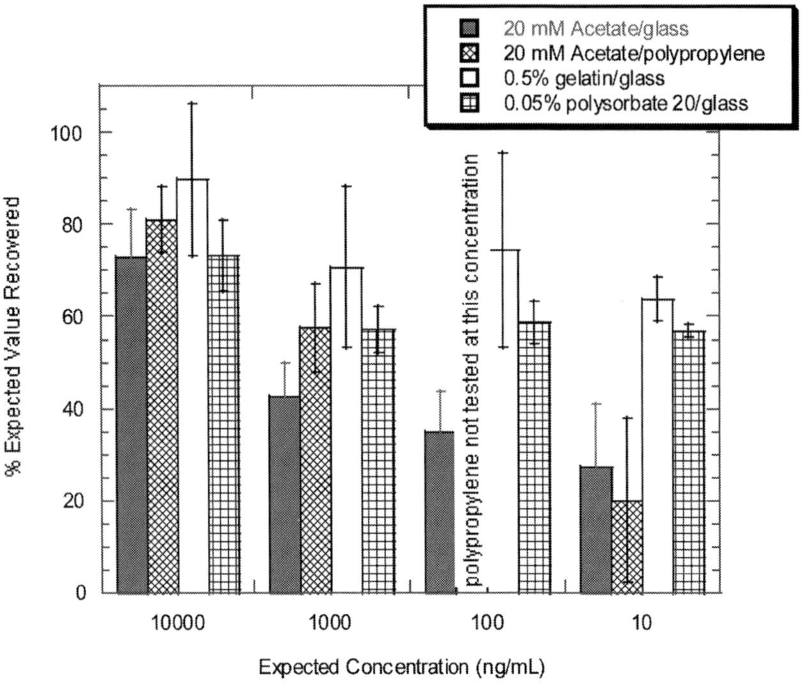

Figure 3 Effect of additives and container type on the recovery of TGF-beta1 after dilution. Protein concentrations were determined by ELISA. The data for sodium acetate in glass represent the average of five experiments. The data for 0.5% gelatin in glass represent the average of three experiments. The other samples were tested in only one experiment. Two samples were tested in each experiment for each condition. Error bars indicate the standard deviation.

are exposed to conditions such as elevated temperature, harsh light, mechanical stress, and freezing to assess which formulation provides the most stability to the active protein.

Accelerated temperature studies are conducted to assist the formulation scientist in the selection of the optimal formulation for a particular protein. Degradation is often accelerated at higher temperatures, allowing a faster assessment regarding formulation parameters. The temperatures selected for use in the accelerated temperature studies in part depend on the T_m, melting temperature, for the protein. Generally, protein formulations are assessed under refrigerated temperatures, at room temperature, and at least one higher temperature at which protein degradation is forced. Care must be taken to avoid temperatures that are too high; degradation at extremely high temperatures (e.g., 60°C) may not be representative of that which would occur under normal storage conditions. Using an Arrhenius plot of the data collected from several temperatures, it may be possible to estimate the shelf-life for a product under the recommended storage conditions.

In addition to exposure to elevated temperatures, an assessment is made of the protein degradation under exposure to freezing conditions. Protein solutions may be stored in the bulk form at −20 to −40°C for long-term storage. Data from several proteins, including hGH (90) and hemoglobin (81) show that the freezing process may result in aggregation or denaturation. This damage may be due to formation of ice, to a change in pH caused by freezing, or to cold denaturation of the protein (21). As ice forms in the solution, the remaining components of the formulation are concentrated into the liquid phase. The resulting concentrate may be conducive to protein denaturation. Furthermore, a shift in pH of the solution during freezing may occur when one of the buffer salts (e.g., disodium phosphate) has limited solubility in this concentrated solution.

Exposure to harsh lighting is used to support handling of the drug product during the filling, finishing, and inspection of the protein. Excessive degradation on exposure to light may necessitate the use of colored glass to limit the amount of light exposure for vialed proteins. The previously mentioned study of hEGF1-48 demonstrated that the rate of photooxidation decreased when the product was stored in amber glass compared to the rate observed on storage in colorless glass (44). However, use of colored glass or other types of packaging made to limit light exposure of the product makes visual inspection of the product more difficult.

The susceptibility of the protein to degradation as a result of mechanical stress is conducted to support the manufacture and distribution of the protein. Protein solutions are often shaken mildly for 24–48 hr to provide information on mechanical stress-induced degradation. With shaking, the protein formulation's exposure to the air/liquid and liquid/container interfaces is maximized. Under these conditions, protein solutions may exhibit physical instability or oxidation. Stirring may also provide information on degradation induced by mechanical stress on a protein. Unlike shaking, stirring a protein solution generally does not significantly change the exposure of the protein solution to the interfaces. Rather, stirring results in an increase in the shear stress experienced by proteins. Physical instability is the main route of degradation from shear stress.

Summary

Information from preformulation and formulation studies allows the formulation scientist to determine the state and composition of the drug substance and drug product. Using accelerated stress conditions such as exposure to elevated temperature, harsh lighting, freezing, and shaking helps the scientist to elucidate the likely degradation products for the protein of interest. Data from these studies are instrumental in determining storage conditions for the drug substance and drug product. Still to be considered, however, are the route of administration and the use of a delivery device for the drug product. These impact the final selection of the formulation and will be covered in the next section.

DRUG DELIVERY

A drug product can only be successful if it is delivered in a timely manner to the site of action in a way that will be amenable to the patient and in a way to ensure product quality. Different *routes of administration* may be used to achieve either systemic or local delivery of the protein. *Devices* such as needle-free injectors and nebulizers may be used to deliver the protein and to enhance patient compliance with use of the drug. Both the route of administration and the decision to use a device are optimally determined early in clinical development of the protein so that there is plenty of clinical experience with the final product. Some considerations regarding the selection of route and device are given below.

Route of Administration

One of the key pieces to development of a successful drug product is the ability to deliver the drug to the site of action with minimal discomfort or inconvenience to the patient. For small molecule therapeutics, there is a wide range of options available for drug administration. Delivery via injection (IV, IM, and SC), oral, nasal, ocular, transmucosal (buccal, vaginal, rectal), and transdermal routes is possible with small molecule drugs. However, the size of proteins and the complexity of their structures severely limit the routes of administration available to proteins.

Injection

Parenteral injections of proteins often provide the fastest route of development for protein-based formulations (91). For systemic delivery of proteins, intravenous administration is considered to be the most efficient approach. Because of the efficiency of IV administration, the bioavailability of drugs is determined by comparing the blood levels achieved with the route of interest to that obtained via IV administration. For drugs administered through a route that has poor bioavailability, larger doses must be given to achieve the same serum levels of the drug compared to those achieved by IV administration. This additional amount of drug required would then lead to higher manufacturing costs associated with each dose.

Administration by injection is not without its drawbacks: it is considered to be painful, inconvenient, and invasive. For proteins in competition with traditional therapeutics that can be delivered via noninvasive methods, this is considered to present a significant marketing disadvantage. For SC and IM delivery, there is a volume limitation for injection. Typically, a maximum of 1.2 mL may be injected to a single SC site (~3 mL for IM), with larger volume doses necessitating the use of additional injection sites. For this reason, formulations intended for SC or IM delivery usually contain higher protein concentrations than those intended for IV delivery. The achievable protein concentration in the formulation may prove to limit the ability to deliver proteins by SC or IM routes.

Oral

Oral delivery of proteins has been the lofty goal of many formulation scientists and are the topic of several review articles (92–97). It is a traditional route of delivery for small molecules. However, the ability to deliver proteins orally presents several challenges to the biotechnological formulation scientist. First and foremost, the purpose of the gastrointestinal tract is to aid in the digestion and absorption of dietary proteins. The low pH and high enzymatic content of the gastric fluid make the environment highly unfavorable for protein therapeutics. This environment could be circumvented by designing a formulation with an enteric coating that allows the protein to escape unscathed from the stomach into the small intestine, where the pH is closer to neutral and the enzymatic content is lower. However, the permeability of intact proteins through the stomach and intestinal membranes is extremely low. These are membranes that by design allow passage of small molecules and amino acids from the gut into the bloodstream. Protein therapeutics, in general, are too large and too hydrophilic to pass through these membranes. The low permeability, which may be extremely variable, translates to low bioavailability. For a successful oral formulation to be developed for protein-based pharmaceutics, it would most likely require the addition of a permeation enhancer to facilitate absorption of the drug. Only proteins requiring a low dose for efficacy with a large therapeutic window are viable candidates for oral delivery.

Pulmonary

While routine oral delivery of proteins has not been realized, some protein formulations have been developed for pulmonary delivery. Pulmonary delivery can result in either parenteral or local administration of the drug and, like oral delivery, is considered noninvasive. As with other routes of delivery, the size of the protein may limit its ability to be delivered systemically via the pulmonary route of administration. Pulmozyme®, a DNase based formulation approved for the treatment of cystic fibrosis (CF), is delivered to the lungs by a nebulizer to clear blockage of the airways in the CF patient (25). A formulation for insulin to be administered by inhalation for systemic delivery of the protein has been developed (98,99).

Topical

Topical delivery of small molecule drugs has been used to deliver therapeutics transdermally and transmucosally. Again, due to the size of the protein it is often not feasible to deliver enough protein to achieve systemic levels of the drug (100). It is possible to deliver protein-based therapeutics locally by a topical formulation. Regranex®, a platelet-derived growth factor based topical formulation, has been approved by the FDA for the treatment of diabetic ulcers (101).

Delivery Devices

Traditional parenteral delivery of biotechonological products has been accomplished through the use of a syringe equipped with a needle. Using this, it is possible to deliver proteins IV, IM, or SC. Devices have been developed to make this method of administration either less painful or more user friendly. Robertson et al. (102) have written a review of available devices for delivery of insulin. Devices have been manufactured to shield the needle to prevent accidental needle-sticks. Prefilled cartridges coupled with a pen type device reduce the need for the patient/nurse to withdraw drug into the syringe, thereby reducing the steps necessary for delivering the drug. Needle-free injectors have been developed that use a propellant to drive the drug solution through the dermis to deliver protein SC or IM (102). This type of device would not be suitable for use with protein formulations that are susceptible to denaturation by shear stress.

Implantable pumps have been developed to administer drugs continuously (103,104), and there has been renewed interest in these devices for delivery of insulin (105). A small palm-sized pump is surgically placed subcutaneously in the abdomen of the patient with a catheter extending from the pump to the desired site of administration. The battery-operated pump is controlled externally with a device that adjusts delivery rate and volume. The pump may be refilled using a long needle. When the battery dies, the pump must be replaced. A clinical trial comparing delivery of insulin from an implantable pump to SC injection demonstrated several advantages of the pump, including lower levels of antibodies to insulin, improved lipid metabolism, and no weight gain (106). Another study demonstrated reduced glycemic variability and hypoglycemic incidences with pump therapy (107).

There are several formulation concerns related to delivery of proteins from an implantable pump. Because this pump is placed within the body, it requires a formulation that is stable at body temperature, 37°C, over the length of time the pump will deliver the protein until it is refilled. Motion of the body results in agitation of the solution contained within the pump, and therefore it is necessary to ensure that the formulation is stable to shaking. The high surface-to-volume ratio of the catheter used for delivery may result in losses of protein due to surface adsorption. Each of these potential pitfalls may be overcome with the appropriate selection of excipients for most proteins.

Depot delivery systems offer an alternative for continuous delivery of protein therapeutics. Cleland et al. (78) have written a review that covers the current technologies for depot delivery. With depot delivery, a formulation is injected or implanted that provides sustained release of the therapeutic. The delivery may be for either local or systemic administration of the drug. Typically, the formulation contains a biodegradable matrix such as hyaluronate or poly(lactide-co-glycolide) (PLG), which entraps the therapeutic drug and slowly degrades in the presence of enzymes. During degradation of the gel or polymer, the drug incorporated in the matrix is released. Careful selection of the matrix may yield delivery times that release drug over several days or weeks. As with the implantable pump, depot formulations require sufficient stability at body temperature to ensure delivery of active protein. Nutropin Depot™ is a formulation of hGH in PLG microspheres that has been shown in clinical trials to release active hGH for 1 month after injection (108).

For pulmonary delivery of proteins, there are three broad categories of devices available: nebulizers, metered dose inhalers (MDI), and dry particle inhalers (DPI). Each of these must meet specific criteria for successful use, such as generation of appropriately sized particles for inhalation, chemical and physical compatability of the device with the drug formulation, and consistency of dose delivery (109). Jet nebulizers use a pump to pressurize a liquid solution held in a reservoir through a nozzle, forcing the creation of tiny droplets that are deposited in the lung airways when the droplet stream is inhaled. The efficiency of delivery with this device is typically low, because there is a large holdup volume in the reservoir. Because the larger droplets that are formed do not leave the reservoir, the protein may experience several trips through the jet. The shear stress that the protein encounters during these repeated trips may lead to protein denaturation. Additionally, milliliters of dosing solution are required per administration, and the length of time required for drug administration is not convenient (15–20 min). Ultrasonic nebulizers, which generate the droplet stream by ultrasound, require much smaller volumes (25–50 dL) and delivery times (110). However, the small volume necessitates a high concentration formulation that is able to withstand the ultrasonic exposure without degrading.

MDIs have not been used extensively with protein therapeutics. With an MDI, a small amount of drug is aerosolized using a propellant, typically a hydrofluoroalkane. Poor delivery efficiency and reproducibility, coupled with concern regarding the stability of proteins in the presence of the hydrofluoroalkanes have led to a lack of interest in pursuing the use of this type of device for pulmonary delivery of proteins (110).

Dry powder inhalers that are currently marketed use the patient's inspiration to move the powder from the device to the lungs, although powered DPIs are also under development (110). Fine powders of the protein formulation are stored either in a large reservoir in the DPI (for multidose administration) or in individual blister packs (for single-dose administration). As the patient inhales, the DPI is activated to release powder into the air stream. For this method of delivery to be successful, the protein formulation must retain the fine, dry powder consistency necessary for delivery. Exposure to high humidity may cause the particles to clump which may lead to clogging of the device and a decrease in delivery efficiency. For the multidose formulation, a preservative is required in the powdered formulation.

Summary

Selection of the appropriate route of administration and delivery device is critical for the commercial success of a drug product. While injections are the most efficient delivery method for proteins, they are not always the most suitable from the patient's perspective. Few routes of administration (IV, IM, SC, pulmonary, topical for local delivery) have been successful to date with protein therapeutics due to the size and complexity of the protein structure. Consideration of the bioavailability via a given route must be made when determining the dose required. Use of a delivery device such as an implantable pump, needle-free injector, or a dry-powder inhaler may yield a product with a commercial advantage over a competitor's product.

STABILITY STUDIES

Stability studies of the drug substance and drug product are required to support expiration dating by the FDA and other regulatory agencies in submissions for product approval. If a device is being used to administer the drug, studies must also be conducted to demonstrate the compatibility of the formulation with the device. These stability studies are typically conducted in a Good Manufacturing Practice environment using established protocols and analytical methods. The International Conference on Harmonisation (ICH) has recommended for adoption by the regulatory bodies of the US, Europe, and Japan a guideline for conducting stability testing for biotechnological products (111). The guidelines for biotechnological products are different from those used for traditional small molecule drugs due to the complexity of the protein structure. Three areas that are covered in this guideline are *selection of batches, the stability indicating profile*, and *storage conditions.* Information from this guideline is summarized below.

Selection of Batches

The selection of batches refers to the specific drug substance and drug product batches that will be tested to support expiration dating. At the time of regulatory submission for approval (e.g., BLA), a minimum of 6 months of stability data is needed on at least three batches of drug substance and final drug product for which the manufacture, container, and storage are representative of the manufacturing scale of production. The quality and process used to manufacture each must be representative of material used in preclinical and clinical trials. Expiration dating is based upon real time/real temperature storage data.

Stability-Indicating Profile

The stability-indicating profile for a biotechnological product generally comprises information from a battery of assays, not from a single stability-indicating assay. To support the expiration dating, the stability of the drug product and drug substance must be assessed by methods that have been validated and are detailed in a protocol specific to that product. This protocol includes the testing intervals and the specifications that the product must meet. Some of the methodologies that could be used for this purpose are described in a later portion of this chapter.

Storage Conditions

The storage conditions for biotechnological products need to be clearly defined because protein stability is generally temperature dependent. Unlike small molecule drugs, accelerated temperatures studies are not used to determine expiration dating for biotechnological products. However, accelerated temperature studies may be used to provide supporting data and to determine the pathway(s) of degradation for the protein. Humidity control is generally not a concern for biotechnological products because most are packaged in containers that protect against humidity. For those products that are not, stability data gathered at different humidities must be provided. Studies should be conducted to support shipping and handling of the product. These may include exposure to light and to a wider range of temperatures than would be tested for recommended storage temperatures. Interactions, or lack thereof, of the product with the container/closure must be assessed. For products that are further prepared for administration in the field (e.g., reconstitution of a lyophilized product followed by dilution into an IV bag containing saline) data must be provided that supports expiration dating and recommended storage conditions of the diluted product that will be administered to a patient.

Summary

In the application to the FDA for product approval, the section describing drug product and drug substance stability information is one of the critical sections related to manufacturing as it will define the expiration dating for the drug substance and the drug product. Relatively short expiration dating for either the drug substance or drug product would result in increased frequency of manufacturing and a higher level of complexity in product distribution.

METHODOLOGY FOR ASSESSING PROTEIN STABILITY

As stated throughout this chapter, one goal of the formulation scientist is to understand and minimize degradation of the protein in the formulation during storage. Additionally, the FDA and other regulatory agencies require that the purity and potency of pharmaceuticals are monitored during the shelf life of the products. Achieving these requirements involves using a combination of analytical techniques such as chromatography, electrophoresis, and spectroscopy among others. Because proteins are capable of denaturing via several mechanisms, it is necessary to use more than one technique to demonstrate stability. Jones has written an in depth review of analytical methodologies used to assess protein stability (112). Some of the more commonly used methodologies are described briefly below.

Liquid Chromatography

High-performance liquid chromatography, HPLC, is the workhorse for many of the analytical technologies used to assess protein stability. HPLC relies on protein separation by interaction of the protein with a resin, called the stationary phase.

The protein is removed (eluted) from the resin using a solution called the mobile phase. Differences in the interactions between protein variants (e.g., aggregates, deamidated products) and the stationary phase are exploited to achieve separation. The length of time that the protein variants are retained by the resin, the retention time, is determined to gain understanding of the nature of the variants. There are four basic types of chromatographic separations used by formulation scientists: size exclusion chromatography (SEC), ion exchange chromatography (IEC), hydrophobic interaction chromatography (HIC), and reversed phase chromatography (RPC).

In SEC, also referred to as gel filtration or gel permeation chromatography, the protein variants are separated according to differences in hydrodynamic radius. The stationary phase for SEC employs porous microparticles while the mobile phase is typically an aqueous salt solution. When the protein and its variants enter the SEC column containing the stationary phase, they are greeted with a tortuous path to reach the end of the column. The smaller variants (e.g., fragments) have access to a greater number of paths through the porous material that results in a larger accessible volume for these variants and therefore a longer retention time on the column. Larger variants (e.g., aggregates) have fewer paths (detours) available, and therefore elute first from the SEC column. The smallest particles are those that elute with the longest retention time. Molecular weight is usually estimated by a comparison of the retention time of a variant to those of molecular weight standards. However, because the retention time is determined by the hydrodynamic radius rather than molecular weight alone, it is often important to confirm estimated molecular weights using a second technique such as static light scattering or mass spectrometry.

One advantage that SEC has for the formulation scientist is the ability to study protein aggregation /fragmentation in the environment of different formulations. Also, it is possible to add denaturants such as SDS to the mobile phase to gain a greater understanding of the sources of molecular weight heterogeneity.

IEC exploits charge differences between protein variants to achieve separation. The resin used in the IEC stationary phase is either for cation exchange or anion exchange. For proteins with basic p*I*s, cation exchange chromatography is typically used. The protein is injected onto the stationary phase where it interacts with the resin. Using a gradient of salt and/or pH, the protein is eluted from the column when the pH or ionic strength of the mobile phase is sufficient to overcome the protein's ionic interaction with the column. The products of deamidation are typically resolved using IEC chromatography.

With HIC, the protein variants are separated by differences in hydrophobicity. In this technique, the stationary phase is comprised of a resin with a hydrophobic backbone. The protein is injected onto this column under high ionic strength conditions. This promotes hydrophobic interactions. A mobile phase gradient is used in which the ionic strength decreases with time. Less hydrophobic variants elute with an earlier retention time than those with higher hydrophobicity. Products of oxidation may often be resolved by HIC.

RPC is used to separate protein variants by hydrophobicity as well. The stationary phase is also comprised of particles containing hydrophobic backbones. However, separation is achieved using a gradient that employs an increasing content of an organic solvent such as acetonitrile or methanol. Trifluoroacetic acid (TFA) is typically added to the mobile phase to improve the interactions of the protein with the resin through ion pairing with charged residues on the protein. Unlike the other chromatographies described, this technique often results in denaturation of the protein during chromatography due to exposure of the protein to both low pH

(TFA) and the organic solvent. As with HIC, products of oxidation may be separated by RPC. RPC generally offers much better resolution than HIC, but has limitations on the size of proteins that may be separated. With all of the HPLC techniques, it is often possible to gain more information by modifying the protein prior to injection either by reduction or through mild enzymatic treatment. By purposefully fragmenting the protein, it may be possible to identify the site of degradation or to at least increase the resolution achieved in the chromatographic separation. Peptide maps generated from complete enzymatic digests, together with mass spectrometry, have proven extremely useful for determining sites of degradation in many proteins, including rhM-CSF and relaxin (47,51).

Electrophoresis

In electrophoresis, protein variants are separated due to differences in their mobility in the presence of an electrical field. Traditionally, electrophoresis has been performed using a stationary slab gel. The most common electrophoretic technique is SDS-PAGE, sodium dodecyl sulfate-polyacrylamide gel electrophoresis. In this technique, a polyacrylamide gel provides a sieving matrix for the protein solution. The protein is mixed with an SDS solution prior to loading onto the gel. By coating the protein with SDS, the charge to mass ratio becomes uniform between proteins. This allows proteins to be separated solely on molecular weight in the presence of the electrical field. The protein/SDS solution is loaded onto the gel which is continually bathed in an SDS-containing buffer. A voltage is applied across the gel, and the protein is mobilized. Smaller proteins travel the gel faster than larger ones. Protein visualization occurs through staining of the gel, usually with either established silver staining or Coomassie staining techniques. Molecular weights are estimated by comparison of mobilization distance to molecular weight standards.

Native PAGE is conducted in the same manner as SDS-PAGE, with the exception that SDS is added neither to the protein solution nor to the bathing solution surrounding the gel during electrophoresis. In the absence of SDS, protein separation results from a combination of the charge and mass of the protein. For this reason, analysis of a native gel is more difficult than that of traditional SDS-PAGE. However, native gels can provide information on protein interactions under very specific conditions (buffer, pH) that could not be studied using SDS-PAGE.

Isoelectric focusing (IEF) separates proteins on a gel by exploiting differences in the isoelectric point (p*I*) of the protein variants. At the p*I*, the sum of all of the charges on the protein is zero. Degradation that results in a change in the p*I* (e.g., deamidation, fragmentation) may be detected using IEF. Separation by IEF occurs in the presence of ampholytes that maintain a pH gradient in the presence of an electrical field. The protein variants migrate on the gel until they reach the pH in which they are neutral. The p*I* of a variant is determined by comparison to the location of p*I* standards.

During the 1990s technological advances were achieved in the field of capillary electrophoresis (CE), resulting in its emergence as one of the primary tools in assessing protein degradation. In CE, an electric field is applied across a capillary containing a separation medium and the protein solution to achieve separation of the protein variants. Once the separation occurs, the solution in the capillary is mobilized and passes in front of a detector to assess the separation. UV and fluorescence detectors are commonly used to assess protein separations. CE has the ability to separate proteins based on size (cf. SEC, SDS-PAGE), charge (cf. IEC), p*I* (cf.

IEF), and hydrophobicity (cf. HIC, RPC). The amount of protein required for CE analysis is substantially smaller than that required for traditional HPLC or gel electrophoretic techniques. This makes CE attractive when only small amounts of protein are available. It is not feasible, however, to use CE as a separation technique when one wishes to collect the fractionated proteins for use in another assay because the amount of protein is too small.

The separation medium used in the capillary dictates the type of separation that will be observed. In capillary zone electrophoresis (CZE), an aqueous buffered solution at a specific pH is used. Protein variants are separated according to their charge to mass ratio at that specific pH. In capillary isoelectric focusing (CIEF), ampholytes are used in the separation medium. Proteins are separated by p*I* in CIEF. Micellar electrokinetic capillary electrophoresis (MEKC) employs micelle forming surfactant solutions to achieve separations that resemble RPC. Various sieving matrices may be used in the capillary in conjunction with SDS to achieve separations based on molecular weight.

Spectroscopy

Spectroscopy measures the interactions of the protein chromophores with light. Information regarding the concentration and conformation of the protein may be obtained through different types of spectroscopy.

Using ultraviolet/visible (UV/Vis) absorption spectroscopy, it is possible to measure the protein concentration using Beer's law: $A = \varepsilon l c$, where A is the measured absorbance of a solution, ε is the absorptivity of the protein, l is the pathlength of the cell used to determine the absorbance, and c is the protein concentration. Proteins typically exhibit two strong, broad absorption bands in the UV/Vis part of the spectrum. The first and most intense band is centered at 214 nm and arises from absorption of light by the peptide backbone. The second absorption band is typically found at ~280 nm. This band arises from absorbance from the aromatic side chains of Trp, Tyr, and Phe. Disulfide bonds may exhibit weak absorption in this range as well.

During stability studies, the concentration of the protein is monitored to ensure that there is no loss due to physical instability of the protein. Most typical routes of chemical degradation do not result in a change of absorbance. A change in color of the solution may occur as a result of degradation of some excipients, (notably His) or in the presence of a reducing sugar.

The UV/Vis absorption spectrum has the capability of providing important information regarding the stability of the formulation by using the spectrum to ascertain the level of light scattering that the solution exhibits. The intensity of light scattering by particulates depends on the fourth power of the frequency of light and on the size of the particulates. As one progresses to bluer (shorter) wavelengths, the intensity of light scattering dramatically increases. The formulation scientist can exploit this to monitor formulations for low levels of particulates, particularly before the particulates are large enough to be observed by visual inspection. Dynamic laser light scattering may also be employed for this purpose, although the equipment for this is usually more costly than the average spectrophotometer.

Circular dichroism (CD) is often used by the formulation scientist to monitor the secondary and tertiary structures of the protein of interest. CD results from the preferential absorption of left vs. right circularly polarized light. In the far UV region of the spectrum (190–250 nm), the CD absorption of proteins arises from peptide bonds. Each type of secondary structural element (e.g., alpha helix, beta sheet) has a unique CD spectrum. Using algorithms which compare the spectrum of the protein of interest to libraries of spectra from proteins with known secondary structural elements allows the formulation scientist to estimate the contribution of each secondary structure element to the total structure of the protein. This ability may aid the formulation scientist in the selection of a suitable formulation for a protein. A combination of excipients which results in an increase in the content of random coiled structure within the protein would not provide a suitable formulation for long term stability of the protein.

Signals in the near UV region (250–400 nm) of the CD spectrum of a protein predominantly arise from absorption of the aromatic side chains of Trp, Tyr, and Phe. The dichroism in this region is highly dependent on the tertiary structure of the protein. For this reason, the near UV CD spectrum is often used to monitor changes in the native environment of proteins that result from instability. For example, Lam (88) used the near UV CD spectrum of IFN-γ to determine there was an incompatability of the protein with a combination of excipients (succinate, benzyl alcohol) which resulted in a loss of structure.

The secondary structure of proteins may also be assessed using vibrational spectroscopy. Fourier transformed infrared (FTIR) and Raman spectroscopy both provide information on the secondary structure of proteins. The bulk of the literature using vibrational spectroscopy to study protein structure has involved the use of FTIR. Water produces vibrational bands that interfere with the bands associated with proteins. For this reason, most of the FTIR literature focuses on the use of this technique to assess structure in the solid state or in the presence of non-aqueous environments. Recently, differential FTIR has been used in which a water background is subtracted from the FTIR spectrum. This workaround is limited to solutions containing relatively high protein concentrations.

Raman spectroscopy offers the advantage of being able to collect spectra in aqueous environments, because the vibrational frequencies arising from water are almost invisible in the Raman spectrum. However, this technique has not been used as extensively as FTIR because the technique was more time consuming and the equipment more costly. The recent development of FT-Raman systems has allowed this technique to be used more readily than in the past.

Visual Inspection

In addition to the various physical techniques, a simple visual inspection of the formulation is conducted to determine if there are changes to the formulation. The human eye is extremely sensitive to many changes observed in protein formulations, and for this reason visual inspection is a valuable tool in the bag of techniques used to assess stability. For liquid formulations, the appearance of precipitates or a change in color of the formulation signifies trouble. For lyophilized formulations, the visual appearance of the lyophilized cake is an important characteristic of the formulation. Collapse or discoloration of the cake could indicate a compromised formulation.

Potency/Activity Assays

The most important assessment during stability studies is analysis of the protein by an activity assay. Degradation of a protein that results in compromised activity is of great concern because this affects the potency of the drug product. Expiration dating of the drug product is made to minimize the loss of potency.

Potency of drug candidates is typically assessed *in vitro*. An appropriate potency assay must be biologically relevant to the clinical indication for which the drug is targeted. For example, the potency of a growth factor whose action is believed to induce proliferation of epithelial cells *in vivo* would need to be assessed by an in vitro epithelial cell-based proliferation assay.

Thermal Techniques

Thermal techniques such as isothermal calorimetry and differential scanning calorimetry have been used in formulation screens to predict the formulation with the greatest stability based on the assumption that excipients that increase the T_m of the protein will stabilize the molecule at the recommended storage temperature (19). For example, a screen of preservatives performed during formulation development for interleukin-1 receptor found that the T_m for the formulation correlated with the extent of aggregation observed on storage at 37°C (89). Such thermal analysis may be useful to rank order formulations.

Other

Other methodologies may be employed for specific drug products. For microencapsulated, SD, and SFD formulations an assessment of the particle size may be critical. Lyophilized, SD, and SFD formulations must be monitored for moisture content. Gel-based formulations typically require monitoring the viscosity of the semisolid formulation.

Summary

The use of appropriate analytical techniques during formulation development aids the formulation scientist in determining the degradation products and the rate of formation of those products. Since proteins are complex structures, a combination of techniques must be used to ensure that the formulation scientist is able to assess potential degradation of he protein. This information is critical for the successful development of a therapeutic formulation.

FILLING, FINISHING AND PACKAGING

Filling, finishing, and packaging are the processes by which the drug substance is turned into the final product and are performed according to current Good Manufacturing Practices (cGMPs). These steps in the manufacturing process are tightly regulated to ensure patient safety. A thorough review of the cGMPs related to filling, finishing, and labeling of drug products is given by Willig and Stoker (113). Only the highlights of the requirements are presented below.

Filling

The FDA "Guidelines for Sterile Drug Products Produced by Asceptic Processing" describes the critical points that must be considered when filling the final product (65). Biotechnological products cannot be terminally sterilized, because the heat or radiation used for sterilization of the final product damages the protein. Therefore, the filling operation is performed in an aseptic environment to minimize the risk of introduction of bacteria to the product. At this stage in manufacture, the presence of bacteria in the product of that could present a serious safety risk for patients that use the drug.

To minimize the potential for introducing microbes and other foreign particles to the final product, several precautions are used in the filling area (114). All components used for the filling operation, including the components for the final package (e.g. vial, stopper, blisterpack) and filling equipment, are presterilized. For a liquid drug substance, a sterile filtration is performed into a presterilized filling vessel to ensure that the drug substance is bacteria-free. Any area in which the sterilized product or the open container/closure systems are exposed to the environment must meet special requirements. Class 100 rooms are used in which the air is HEP A-filtered at the point of delivery and contains less than 100 particles sized $\sim$0.5 μm/ft^3.The airflow is unidirectional in this room. The microbial load should be no higher than 0.1 CFU per cubic foot of air. Operators working in class 100 rooms are required to be completely gowned in sterile garb. They must minimize movements and talking within the room so as not to disturb the airflow. For lyophilized products, the vials are filled with the liquid drug substance and partially stoppered before being placed into a lyophilizer contained within the class 100 area. Validation of the filling environment and operations is accomplished by performing media fills periodically and testing the sterility of the media-filled units. At least 3000 units should be filled to be able to detect a contamination rate of one in one thousand units with a 95% confidence interval.

Filling and packaging by blow/fill/seal technology has gained wider acceptance for use in the biotechnology industry in the past decade (115–118). The blow/fill/seal process is one continual, integrated operation in which the container is formed by "blowing" a molten plastic such as low density polyethylene into a mold, filled using a nozzle provided with a class 100 HEP A air-flow, and then sealed by molding of the still molten plastic at the neck. Due to the highly automated procedure used, minimal personnel are required for the filling operation. This reduces the sources of potential viable and non-viable particulate contamination in the final product.

Finishing and Packaging

Once the protein formulation has been filled into the final container, several packaging and finishing steps are performed before the product may be shipped for use. For vialed formulations, the final vial must be capped to ensure the stopper does not inadvertently come off of the vial. This protects potential leakage from the vial or the introduction of contaminants to the vial. The product is inspected and representative samples are tested to ensure consistency, safety, and potency. The final container is labeled with product information such as the name of the drug, the manufacturer, the amount of the drug in the container, the lot number, and the expiration date. Once the labeling has occurred, a second inspection is performed on a representative sample of units to ensure the correct label has been used. Outer packaging is selected (e.g., a carton) in which all of the components for the final product are assembled for sale. This packaging may be as simple as one vial in a carton or as elaborate as one containing a lyophilized product packaged with a vial of diluent and a syringe to enable easy access to the tools necessary for reconstitution of the product. This package is appropriately labeled to include the name of the drug, the manufacturer, the amount of drug per unit, the number of units, and the expiration date among other information. Included in the final package is a "package insert," a very detailed description of the composition of the drug product, its intended use, and any side effects noted with use of the drug. Once packaged, the product is ready for distribution.

FUTURE PROSPECTS

With the development of biotechnology-derived products for clinical indications that are not life-threatening, there will be an increased drive to move away from drugs that must be given by injection, particularly by IV administration. The ability to formulate proteins at high concentrations will allow more to be administered subcutaneously. Crystalline forms of proteins are now being investigated as potential formulations and with the possibility that these could be given as depots to allow extended delivery of the protein. Devices to increase the convenience of delivery will continue to be attractive for protein therapeutics. Development of formulations for pulmonary delivery should increase significantly within the next decade. Pulmonary delivery is non-invasive, and the advent of formulations that may be administered by dry-powder inhalers make this route extremely attractive to patients.

REFERENCES

1. Editors' and Reporters' Guide to Biotechnology: Approved Biotechnology Drugs 1999. Washington, DC: The Biotechnology Industry Organization, 1999.
2. Robinson AB, Rudd CJ. Deamidation of glutaminyl and asparaginyl residues in peptides and proteins. In: Horecker BL, Stadman ER, eds. Current Topics in Cellular Regulations. Vol.8. New York: Academic Press, 1974:8:247–295.
3. Ganeval D, Noel LH, Preud'home JL, Droz D, Grunfeld JP. Light-chain deposition disease: its relation with AL-type amyloidosis. Kidney Int 1984; 26(1):1–9.
4. Buxbaum JN, Chuba JV, Hellman GC, Solomon A, Gallo GR. Monoclonal immunoglobulin deposition disease: light chain and light and heavy chain deposition diseases and their relation to light chain amyloidosis. Clinical features, immunopathology, and molecular analysis. Ann Intern Med 1990; 112(6):455–464.
5. Buxbaum J. Mechanisms of disease: monoclonal immunoglobulin deposition. Amyloidosis, light chain deposition disease, and light and heavy chain deposition disease. Hematol Oncol Clin North Am 1992; 6(2):323–346.
6. Cogne M, Silvain C, Khamlichi AA, Preud'homme JL. Structurally abnormal immunoglobulins in human immunoproliferative disorders. Blood 1992; 79(90):2181–2195.
7. Pike CJ, Burdick D, Walencewicz AJ, Glabe CG, Cotman CW. Neurodegeneration induced by beta-amyloid peptides in vitro: the role of peptide assembly state. J Neurosci 1993; 13:1676–1687.
8. Lorenzo A, Yankner BA. Amyloid fibril toxicity in Alzheimer's disease and diabetes. Ann N Y Acad Sci 1996; 777:89–95.
9. Geula C, Mesulam MM, Saroff DM, Wu CK. Relationship between plaques, tangles, and loss of cortical cholinergic fibers in Alzheimer disease. J Neuropathol Exp Neurol 1998; 57(1):63–75.
10. Thompson MW. Surfactant protein B deficiency: insights into surfactant function through clinical surfactant protein deficiency. Am J Med Sci 2001; 321(1):26–32.
11. Weaver TE, Conkright JJ. Function of surfactant proteins B and C. Annu Rev Physiol 2001; 63:555–578.
12. Beck DC, Ikegami M, Na C-L, Zaltash S, Johansson J, Whitsett J, Weaver TE. The role of homodimers in surfactant protein B function in vivo. J Biol Chem 2000; 275(5): 3365–3370.

13. Brange J, Havelund S, Hommel E, Sorensen E, Kuhl C. Neutral insulin solutions physically stabilized by addition of Zn^{2+}. Diabet Med 1986; 3(6):532–536.
14. Townsend MW, DeLuca PP. Nature of aggregates formed during storage of freeze-dried ribonuclease A. J Pharm Sci 1991; 80(1):63–66.
15. Robbins DC, Cooper SM, Fineberg SE, Mead PM. Antibodies to covalent aggregates of insulin in blood of insulin-using diabetic patients. Diabetes 1987; 36(7):838–841.
16. Moore JMR, Patapoff TW, Cromwell MEM. Kinetics and thermodynamics of dimmer formation and dissociation for a recombinant humanized monoclonal antibody to vascular endothelial growth factor. Biochemistry 1999; 38(42):13960–13967.
17. McLeod AG, Walker IR, Zheng S, Hayward CP. Loss of factor VIII activity during storage in PVC containers due to adsorption. Haemophilia 2000; 6(2):89–92.
18. Katakam M, Bell LN, Banga AK. Effect of surfactants on the physical stability of recombinant human growth hormone. J Pharm Sci 1995; 84(6):713–716.
19. Bam NB, Cleland JL, Yang J, Manning MC, Carpenter JF, Kelley RF, Randolph TW. Tween protects recombinant human growth hormone against agitation-induced damage via hydrophobic interactions. J Pharm Sci 1998; 87(12):1554–1559.
20. Kreilgaard L, Jones LS, Randolph TW, Frokjaer S, Flink JM, Manning MC, Carpenter JF. Effect of tween 20 on freeze-thawing- and agitation-induced aggregation of recombinant human factor XIII. J Pharm Sci 1998; 87(12):1597–1603.
21. Carpenter JF, Izutsu K-L, Randolph TW. Freezing- and drying-induced perturbations of protein structures and mechanisms of protein protection by stabilizing additives. In: Rey L, May JC, eds. Freeze-Drying/Lyophilizatino of Pharmaceutical and Biological Products. New York: Marcel Dekker, Inc., 1999.
22. Powell MF et al. A compendium and hydropathy/flexibility analysis of common reactive sites in proteins: reactivity at Asn, Asp, Gin, and Met motifs in neutral pH solution. Pharm Biotechnol 1996; 9:1–140.
23. Cleland JL, Powell MF, Shire SJ. The development of stable protein formulations: a close look at protein aggregation, deamidation, and oxidation. Crit Rev Ther Drug Carrier Syst 1993; 10(4):307–377.
24. Pikal MJ, Rigsbee DR. The stability of insulin in crystalline and amorphous solids: observation of greater stability for the amorphous form. Pharm Res 1997; 14(10):1379–1387.
25. Shire SJ. Stability characterization and formulation development of recombinant human deoxyribonuclease I [Pulmozyme (Dornase alpha)]. In: Pearlman R, Wang J, eds. Formulation, Characterization, and Stability of Protein Drugs. New York: Plenum Press, 1996.
26. Wang Y-CJ, Hanson MA. Parenteral formulations of proteins and peptides: stability and stabilizers. J Parent Sci Technol 1988; 42(2S):S3–S26.
27. Shahrokh Z, Eberlein G, Buckley D, Paranandi MV, Aswad DW, Stratton P, Mischak R, Wang YJ. Major degradation products of basic fibroblast growth factor: detection of succinimide and iso-aspartate in place of aspartate. Pharm Res 1994; 11(7):936–944.
28. Manning MC, Patel K, Borchardt RT. Stability of protein pharmaceuticals. Pharm Res 1989; 6(11):903–918.
29. Li S, Shoeneich C, Borchardt RT. Chemical instability of protein pharmaceuticals: mechanisms of oxidation and strategies for stabilization. Biotechnol Bioeng 1995; 48:590–500.
30. Nguyen TH. Oxidation degradation of protein pharmaceuticals. In: Cleland JL, Langer R, eds. Formulation and Delivery of Proteins and Peptides. ACS Symposium Series 567. Washington, DC: American Chemical Society, 1994:59–71.
31. Stadtman ER. Oxidation of free amino acids and amino acid residues in proteins by radiolysis and by metal-catalyzed reactions. Annu Rev Biochem 1993; 62:797–821.
32. Donbrow M, Azaz E, Pillersdorf A. Autoxidation of polysorbates. J Pharm Sci 1978; 67(12):1676–1681.

33. Dubost DC, Kaufman JM, Zimmerman JA, Bogusky MJ, Coddington AB, Pitzenberger SM. Characterization of a solid state reaction product from a lyophilized formulation of a cyclic heptapeptide. A novel example of an excipient-induced oxidation. Pharm Res 1996; 13(12):1811–1814.
34. Brot N, Weissbach H. Biochemistry and physiological role of methionine sulfoxide residues in proteins. Arch Biochem Biophys 1983; 223(1):271–281.
35. Philipson L. On the oxidation of 2,3-dimercaptopropanol. Biochem Biophys Acta 1962; 56:375–366.
36. Shahrokh Z, Sluzky V, Stratton PR, Eberlein GA, Wang YJ. Disulfide-linked oligomerization of basic fibroblast growth factors: effect of sulfated compounds. In: Cleland JL, Langer R, eds. Formulation and Delivery of Proteins and Peptides. ACS Symposium Series 567. Washington, DC: American Chemical Society, 1994:85–99.
37. Adam W, Ahrweiler M, Sauter M, Schmiedeskamp B. Oxidation of indoles by singlet oxygen and dimethyldioxirane: isolation of indole dioxanes and epoxides by stabilization through nitrogen acylation. Tetrahedron Lett 1993; 34:5247–5250.
38. Adam W, Ahrweiler M, Peters K, Schmiedeskamp B. Oxidation of N-acylindoles by dimethyldioxirane and singlet oxygen: substituent effects on thermally persistent indole epoxides and dioxetanes. J Org Chem 1994; 59:2733–2739.
39. Holt LA, Milligan B, Rivett DE, Stewart FH. The photodecomposition of tryptophan peptides. Biochim Biophys Acta 1977; 499(1):131–138.
40. Creed D. The photophysics and photochemistry of the near-UV absorbing amino acids—II. Tyrosine and its simple derivatives. Photochem Photobiol 1984; 39:563–575.
41. Giulivi C, Davies KJA. Dityrosine: a marker for oxidatively modified proteins and selective proteolysis. In: Packer L, ed. Methods in Enzymology. Vol.233. New York: Academic Press, 1994:363–371.
42. Keck RG. The use of *t*-butyl hydroperoxide as a probe for methionine oxidation in proteins. Anal Biochem 1996; 236:56–62.
43. DalleDonne I, Milzani A, Colombo R. The *tert*-butyl hydroperoxide-induced oxidation of actin Cys-374 is coupled with structural changes in distant regions of the protein. Biochemistry 1999; 38:12471–12480.
44. Senderoff RI, Wootton SC, Boctor AM, Chen TM, Giordani AB, Julian TN, Radebaugh GW. Aqueous stability of human epidermal growth factor 1–48. Pharm Res 1994; 11(12):1712–1720.
45. Chang SH, Teshima GM, Milby T, Gillece-Castro B, Canova-Davis E. Metal-catalyzed photooxidation of histidine in human growth hormone. Anal Biochem 1997; 244:221–227.
46. Cipolla DC, Shire SJ. Techniques in Protein Chemistry II. New York: Academic Press, 1991:543–555.
47. Nguyen TH, Burnier J, Meng W. The kinetics of relaxin oxidation by hydrogen peroxide. Pharm Res 1993; 10(11):1563–1571.
48. Li S, Patapoff TW, Overcashier D, Hsu C, Nguyen TH, Borchardt RT. Effects of reducing sugars on the chemical stability of human relaxin in the lyophilized state. J Pharm Sci 1996; 95(9):873–877.
49. Khossravi M, Shire SJ, Borchardt RT. Evidence for the involvement of histidine A(12) in the aggregation and precipitation of human relaxin induced by metal-catalyzed oxidation. Biochemistry 2000; 39:5876–5885.
50. Costantino HR, Langer R, Klibanov AM. Solid-phase aggregation of proteins under pharmaceutically relevant conditions. Pharm Res 1994; 11:21–29.
51. Schrier JA, Kenley RA, Williams R, Corcoran RJ, Kim Y, Northey RP Jr, D'Augusta D, Huberty M. Degradation pathways for recombinant human macrophage colony-stimulating factor in aqueous solution. Pharm Res 1993; 10(7):933–944.
52. Brange J, Langkjaer L, Havelund S, Volund A. Chemical stability of insulin. 1. Hydrolytic degradation during storage of pharmaceutical preparations. Pharm Res 1992; 9(6):715–726.

53. Vooter CEM, de Haard-Hoekman WA, van den Oetelaar PJM, Bloemendal H, de Jong WW. Spontaneous peptide bond cleavage in ageing a-crystallin through a succinimide intermediate. J Biol Chem 1988; 263(35):19020–19023.
54. Kenley RA, Warne NW. Acid-catalyzed peptide bond hydrolysis of recombinant human interleukin 11. Pharm Res 1994; 11(1):72–76.
55. Nursten HE. In: Macarthy D, ed. Maillard Browning Reactions in Dried Foods in Concentration and Drying of Foodstuffs. London: Elsevier Applied Science, 1986:53–87.
56. Ledl F, Schleicher E. New Aspects of the Maillard Reaction in Foods and in the Human Body. Angew Chem Int Ed Engl 1990; 29:565–594.
57. Quan CP, Wu S, Dasovich N, Hsu C, Patapoff T, Canova-Davis E. Susceptibility of rhDNase I to glycation in the dry-powder state. Anal Chem 1999; 71:4445–4454.
58. Pikal MJ, Dellerman KM, Roy ML, Riggin RM. The effects of formulation variables on the stability of freeze-dried human growth hormone. Pharm Res 1991; 8(4):427–436.
59. Shenoy B, Wang Y, Shan W, Margolin AL. Stability of crystalline proteins. Biotechnol Bioeng 2001; 73(5):358–369.
60. Pikal MJ, Rigsbee DR. The stability of insulin in crystalline and amorphous solids: observation of greater stability for the amorphous form. Pharm Res 1997; 14(10):1379–1387 (errata in Pharm Res 1998; 75(2):362–363).
61. Pikal MJ. Freeze-drying of proteins: process, formulation, and stability. In: Cleland JL, Langer R, eds. Formulation and Delivery of Proteins and Peptides. ACS Symposium Series 567. Washington, DC: American Chemical Society, 1994:20–33.
62. Carpenter JF, Prestrelski SJ, Anchordoguy TJ, Arakawa T. Interactions of stabilizers with proteins during freezing and drying. In: Cleland JL, Langer R, eds. Formulation and Delivery of Proteins and Peptides. ACS Symposium Series 567. Washington, DC: American Chemical Society, 1994:148–169.
63. Mumenthaler M, Hsu CC, Pearlman R. Feasibility study on spray-drying protein pharmaceuticals: recombinant human growth hormone and tissue-type plasminogen activator. Pharm Res 1994; 11(1):12–20.
64. Maa Y-F, Nguyen P-A, Sweeney T, Shire SJ, Hsu CC. Protein inhalation powders: spray drying vs. spray freeze drying. Pharm Res 1999; 16(2):249–254.
65. Inactive Ingredient Guide, Division of Drug Information Resources, Food and Drug Administration Center for Drug Evaluation and Research Office Management, Washington DC, 1996.
66. Physician's Desk Reference 2001, Medical Economics, Montvale, New Jersey, 2001.
67. Powell MF, Nguyen T, Baloian L. Compendium of excipients for parenteral formulations. PDA J Pharm Sci Technol 1998; 52(5):238–311.
68. Brazeau GA, Cooper B, Svetic KA, Smith CL, Gupta P. Current perspectives on pain upon injection of drugs. J Pharm Sci 1998; 87(6):667–677.
69. Napaporn J, Thomas M, Svetic KA, Shahrokh Z, Brazeau GA. Assessment of mytotoxicity of pharmaceutical buffers using an in vitro muscle model: effect of pH, capacity, tonicity, and buffer type. Pharm Dev Technol 2000; 5(1):123–130.
70. Helm VJ, Muller BW. Stability of gonadorelin and triptorelin in aqueous solution. Pharm Res 1990; 7(12):1253–1256.
71. Allen JD, Bentley D, Stringer RA, Lowther N. Hybrid (BDBB) interferon-a: preformulation studies. Int J Pharm 1999; 187:259–272.
72. Glasser L, Sternglanz PD, Combie J, Robinson A. Serum osmolality and its applicability to drug overdose. Am J Clin Pathol 1973; 60(5):695-699.
73. Hendry EB. Osmolarity of human serum and of chemical solutions of biologic importance. Clin Chem 1961; 7:154-164.
74. Hendry EB. The osmotic pressure and chemical composition of human body fluids. Clin Chem 1962; 8:246-265.
75. Keele CA, Armstrong D. Substances Producing Pain and Itch. Monographs of the Physiological Society. Vol.12. London: Edward Arnold, 1960:12:667–572.

76. Lam XM, Yang JY, Cleland JL. Antioxidants for prevention of methionine oxidation in recombinant monoclonal antibody HER2. J Pharm Sci 1997; 86(11):1250–1255.
77. Andya JD, Maa Y-F, Costantino HR, Nguyen P-A, Dasovich N, Sweeney TD, Hsu CC, Shire SJ. The effect of formulation excipients on protein stability and aerosol performance of spray-dried powders of a recombinant humanized anti-IgE monoclonal antibody. Pharm Res 1999; 16(3):350–358.
78. Cleland JL, Lam X, Kendrick B, Yang J, Yang T-H, Overcashier D, Brooks D, Hsu C, Carpenter J. A specific molar ratio of stabilizer to protein is required for storage stability of a lyophilized monoclonal antibody. J Pharm Sci 2001; 90(3):310–321.
79. Chang BS, Reeder G, Carpenter JF. Development of a stable freeze-dried formulation of recombinant human interleuking-1 receptor antagonist. Pharm Res 1996; 13(2): 243–249.
80. Bam NB, Randolph TW, Cleland JL. Stability of protein formulations: investigation of surfactant effects by a novel EPR spectroscopic technique. Pharm Res 1995; 12(1):2–11.
81. Kerwin BA, Heller MC, Levin SH, Randolph TW. Effects of tween 80 and sucrose on acute short-term stability and long-term storage at –20°C of a recombinant hemoglobin. J Pharm Sci 1998; 87(9):1062–1068.
82. Fatouros A, Sjostrom B. Recombinant factor VIII SQ—the influence of formulation parameters on structure and surface adsorption. Int J Pharm 2000; 194:69–79.
83. Osterberg T, Fatouros A, Mikaelsson M. Development of a freeze-dried albumin-free formulation of recombinant factor VIII SQ. Pharm Res 1997; 14(7):892–898.
84. Kibbe AH (Ed.) Handbook of Pharmaceutical Excipients. 3rd ed. Washington, DC: American Pharmaceutical Association and Pharmaceutical Press, 2000.
85. Honigman JL. Disinfectant ototoxicity [letter]. Pharm J 1975; 215:523.
86. United States Pharmacopeia. 24th ed. Rockville, MD: the United States Pharmacopeial Convention, Inc., 1999.
87. 1997 European Pharmacopeia. 3rd ed and 1999 Supplement. Strasbourg, France: European Department for the Quality of Medicine within the Council of Europe, 1996.
88. Lam XM, Patapoff TW, Nguyen TH. The effect of benzyl alcohol on recombinant human interferon-g. Pharm Res 1997; 14(6):725–729.
89. Remmele RL Jr, Nightlinger NS, Srinivasan S, Gombotz WR. Interleukin-1 receptor (IL-1R) liquid formulation development using differential scanning calorimetry. Pharm Res 1998; 15(2):200–208.
90. Eckhart BM, Oeswein JQ, Bewley TA. Effect of freezing on aggregation of human growth hormone. Pharm Res 1991; 8(11):1360–1364.
91. Cleland JL, Daugherty A, Mrsny RJ. Emerging protein delivery methods. Curr Opin Biotechnol 2001; 12:212–219.
92. Fix JA. Oral controlled release technology for peptides: status and future prospects. Pharm Res 1996; 13(12):1760–1764.
93. Humphrey MJ, Ringrose PS. Peptide and related drugs: a review of their absorption, metabolism, and excretion. Drug Metab Rev 1986; 17:283–310.
94. Lee VHL, Yamamoto A, Kompella UB. Mucasal penetration enhancers for facilitation of peptide and protein drug absorption. Crit Rev Ther Drug Carrier Syst 1991; 8:91–192.
95. Merkle HP. New aspects of pharmaceutical dosage forms for controlled drug delivery of peptides and proteins. Eur J Pharm Sci 1994; 2:19–21.
96. Lehr C. Bioadhesion technologies for the delivery of peptide and protein drugs to the gastrointestinal tract. Crit Rev Ther Drug Carrier Syst 1994; 11:119–160.
97. Sarciaux JM, Acar L, Sado PA. Using microemulsion formulations for oral drug delivery of therapeutic peptides. Int J Pharm 1995; 120:127–136.
98. Patton JS, Bukar J, Nagaranjan S. Inhaled insulin. Adv Drug Deliv Rev 1999; 35 (2–3):235–247.
99. Patton JS. Insulin inhalation (Pfizer/Inhale Therapeutic Systems). HMR 4006, Drugs RD 1999; 2(2):112–113.

100. Guy RH. Current status and future prospects of transdermal drug delivery. Pharm Res 1996; 13(12):1765–1769.
101. Miller MS. Use of topical recombinant human platelet-derived growth factor-BB (becaplermin) in healing of chronic mixed arteriovenous lower extremity diabetic ulcers. J Foot Ankle Surg 1999; 38(3):227–231.
102. Robertson KE, Glazer NB, Campbell RK. The latest developments in insulin injection devices. Diabetes Educ. 2000; 25:135–152.
103. Selam JL, Charles MA. Devices for insulin administration. Diabetes Care 1990; 13: 955–979.
104. Kennedy FP. Recent developments in insulin delivery techniques: current status and future potential. Drugs 1991; 42:213–227.
105. Thompson JS, Duckworth WC. Insulin pumps and glucose regulation. World J Surg 2001; 25:523–526.
106. Saudek CD, Duckworth WC, Giobbie-Hurder A, Henderson WG, Henry RR, Kelley DE, Edelman SV, Zieve FJ, Adler RA, Anderson JW, Anderson RJ, Hamilton BP, Dormer TW, Kirkman MS, Morgan NA. Implantable insulin pump vs. multiple-dose insulin for noninsulin dependent diabetes mellitus: a randomized clinical trial. J Am Med Assoc 1996; 276:1322–1327.
107. Haardt MJ, Selam JL, Slama G, Bethoux JP, Dorange C, Mace B, Ramaniche ML, Bruzzo F. A cost-benefit comparison of intensive diabetes management with implantable pumps versus multiple subcutaneous injections in patients with type I diabetes. Diabetes Care 1994; 17(8):847–851.
108. Tracy MA. Development and scale-up of a microsphere protein delivery system. Biotechnol Prog 1998; 74:108–115.
109. Clark AR. Medical aerosol inhalers: past, present, and future. Aero Sci Technol 1995; 22:374–391.
110. Clark AR, Shire SJ. Formulation of proteins for pulmonary delivery. In: McNally EL, ed. Protein and Formulation Delivery. New York: Marcel Dekker, Inc., 2000:201–234.
111. International Conference on Harmonisation of Technical Requirements for Registration of Pharmaceuticals for Human Use, ICH Harmonised Tripartite Guideline—Quality of Biotechnological Products: Stability Testing of Biotechnological/Biological Products, 1995.
112. Jones AJS. Analytical methods for the assessment of protein formulations and delivery systems. In: Cleland JL, Langer R, eds. Formulation and Delivery of Proteins and Peptides. ACS Symposium Series 567. Washington, DC: American Chemical Society, 1994:22–45.
113. Willig SH, Stoker JR. Good Manufacturing Practices for Pharmaceuticals: A Plan for Total Quality Control. 4th ed. New York: Marcel Dekker, Inc., 1997.
114. Devine RA. Licensing biotechnology facilities. In: Lubiniecki AS, Vargo SA, eds. Regulatory Practice for Biopharmaceutical Production. New York: Wiley-Liss, Inc., 1994:357–382.
115. Wu VL, Leo FN. Advances in blow/fill/seal technology: a case study in the qualification of a biopharmaceutical product. In: Avis KE, Wu VL, eds. Biotechnology and Biopharmaceutical Manufacturing, Processing, and Purification. Buffalo Grove, IL: Interpharm Press, Inc., 1996:265–292.
116. Sharp JR. Manufacturing of sterile pharmaceutical products using "blow-fill-seal" technology. Pharm J 1987; 239:106–108.
117. Sharp JR. Validation of a new form-fill-seal installation. Manufact Chem 1988; 55: 22–27.
118. Leo F. B/f/s aseptic packaging technology. In: Groves MJ, Olson WP, eds. Aseptic Pharmaceutical Manufacturing for the 1990s. Buffalo Grove, IL: Interpharm Press, Inc., 1989.

15

Validation of Cell Culture-Based Processes and Qualification of Associated Equipment and Facility

Chandra M. Dwivedi*
Bayer Corporation, Berkeley, California, U.S.A.

INTRODUCTION

Why validate? Though validation is a well-accepted and recognized cGMP requirement in today's Pharma business, this question is often posed during the product or process development (PD) activities in a start-up or even in an established company. In a nutshell, validation is not only a regulatory requirement, but it makes "good business sense." Validated processes assure production of quality product, batch after batch, and ultimately result in fewer headaches down the road in terms of fewer deviations during production, quality assurance (QA) discrepancy investigations, adverse events from the field, and regulatory observations (483s and its global equivalent) during regulatory inspections. In addition, they improve cost effectiveness in terms of preventing process failures, lot rejections, re-processing of salvageable lots, and attaining maximum plant capacity. Moreover, a sound and thorough validation strategy not only assures the production of top quality products, but also builds confidence and provides peace of mind to its customers. It also boosts the morale of the company employees and help build a sound and trustworthy relationship and track record with the regulatory agencies. The latter may come as a blessing for a company's future dealings with the regulatory agencies.

The term process validation originated in 1983 when the Food and Drug Administration (FDA) expanded the cGMP guidelines to cover demonstration of process consistency/reproducibility, but the guidelines were not finalized until 1987 (1). These guidelines were originally intended to be adopted by all drug product and biological manufacturers, but were later extended to the medical device and diagnostic manufacturers and to the blood collection/distribution/users and blood product manufacturers (2,3). Though originally intended only for the finished drug

---

*Currently at BIOGEN IDEC, Inc., Cambridge, MA, U.S.A.

product, these regulations have been recently extended to bulk drugs and bulk biologicals (4).

The original definition of the term process validation was described by the FDA as "Establishing a documented evidence which provides a high degree of assurance that a specific process will consistently produce a product meeting its predetermined specifications and quality attributes."

In practice, process validation (process performance qualification, PPQ) is more complicated than the simple definition stated above and is only one element of the overall validation process. It is a culmination of all other validation studies, such as equipment qualification (installation qualification, IQ; operational qualification, OQ; and performance qualification, PQ), computer qualification (IQ and OQ), utilities and facilities qualification (IQ, OQ, and PQ) cleaning validation (PQ), environmental qualification (PQ), and analytical qualification (PQ), all covered under a validation master plan (VMP) written for each new technology, process, or a product.

The invention of recombinant DNA technology in the late 1970s and its widespread application to eukaryotic and prokaryotic cells for developing unique medical applications/treatments resulted in the establishment of a new field known as "Genetic Engineering" today (5,6). These developments opened the floodgates for innovation that resulted in the establishment of many biotechnology companies worldwide. Of these, more than 50% of the biotechnology companies are working on cell culture technology for producing pharmaceutical and cellular therapies.

Due to continued innovation in this field the application of process validation concepts and guidelines are becoming increasingly complex, challenging, and difficult to understand by technical professionals, regulatory auditors, and cGMP compliance enforcers working in the pharmaceutical and biotechnology-related organizations. Since it is impossible to cover all aspects of process validation for the numerous biotechnology-derived products in this chapter, an attempt will be made only to provide a simplified version of the regulatory requirements that are needed for licensing cell culture-derived pharmaceuticals and cellular therapies. This chapter is intended to provide a bird's-eye view of the regulatory requirements for process validation to entrepreneurs before they plan for building a new manufacturing plant and expect to obtain licensure for a product (Product License Application, PLA) or a biologic (Biologic License Application, BLA) or a drug (New Drug Application, NDA) from regulatory agencies. This chapter is expected to prepare them well before they begin that challenging, eventful, exhausting, memorable, and ultimately rewarding journey.

APPROACH AND RATIONALE

The innumerable amount of research and development studies conducted on a large number of medical products has enabled us to understand that the quality attributes for any given product are not an unexpected output. But, are largely dependent on the process parameters used during their production. Therefore, the control of quality attributes for any biological or pharmaceutical product is in our hands; and with the development of new technologies, quality attributes for the new products can now be built into the manufacturing process. In this respect, the process design in relation to the respective product quality attributes has become crucial for the development and licensing of the medical and pharmaceutical products (7–10).

Since the breakthrough in genetic engineering a few decades ago, numerous medical, biological, pharmaceutical, and diagnostic products and applications based on cell culture technology have been invented. They are based on microbial fermentation (eukaryotic and prokaryotic), hybridoma technology, and tissue regeneration. Even plant cell technology is being evaluated to produce medical and therapeutic products for human use. The examples of the cells used for this purpose are: bacteria (*Escherichia coli*), fungi (*Aspergillus*, *Saccharomyces*), mammalian cells (CHO, BHK, myeloma, melanoma, hybridoma, etc.), insect cells (*Drosophila*), and plant cells (tobacco, spinach, etc.). The majority of the products are secreted by the cells in the spent medium (harvest) by applying the rDNA technologies and manipulation of respective genes in the cells. The examples of recombinant products derived from these technologies are: erythropoietin (rEPO), anti-hemophilic factor (rFVIII), tissue plasminogen activator (rTPA), growth factors (EGF, TGF, PDGF, TNF, etc.), hormones (Insulin, LH, FSH, etc.), interferons (IF-1, IF-2, etc.), interleukins (IL-2, IL-4, IL-6, etc.), monoclonal antibodies (mAbs), and other enzymes and proteins (cerezyme, galactosidase, etc.). Some of the products are expressed in the inclusion bodies within cells and the cells therefore must be lysed to extract the products out (insulin, EGF, etc.). Epithelial cells, neuroblastoma, osteoblastoma and cartilage cells are being grown in laboratories and used as medical devices for a number of treatments (burns, tissue implant, tissue regeneration, etc.). A number of monoclonal antibodies are being generated from bacterial, mammalian, and plant cell technologies for the treatment of cancer, autoimmune diseases, and other immunological disorders.

A general approach to streamline validation concepts and policies has been evolving over the last number of years. These efforts have resulted in better understanding of the requirements for the validation by the industry professionals. For the purposes of clarity and better understanding this article will employ the newly emerging approach on validation concepts (11,12). Accordingly, qualification of all equipment and systems (design qualification, DQ; installation qualification, IQ; operational qualification, OQ; and performance qualification, PQ) will be referred as "Equipment Qualification" and not as "Validation." The term "Validation" will be used only for "Process Validation" studies that are related with the studies (with or without active ingredient) at the small-scale (lab-scale) or full production scale (process validation, PV or PPQ).

The variety of cell culture technologies and many different approaches to use them as pharmaceutical products or medical devices makes the task of building the quality attributes in the manufacturing process very challenging. This also makes the task of process validation more difficult as generic models of process validation cannot be used, and every process validation study needs to be devised from scratch based on the technology being used. For example, the level of impurities (DNA, host cell contaminating proteins, etc.) may be substantially less in the starting material where the product is secreted out in the spent medium (harvest) as compared to the product that is expressed intracellularly such as in the inclusion bodies. Therefore, the design of the manufacturing process and the resultant process validation studies would be very different for the two approaches to isolate and purify the product(s).

The possible impurities and contaminants in a cell culture-based product are: intact cells, adventitious agents [bacteria, fungi, mycoplasma, viruses, transmissible spongiform encephalitis (TSE)/bovine spongiform encephalitis (BSE)], endogenous retroviruses, host cell nucleic acids and proteins, foreign proteins (from raw materials and microbial contaminations), endotoxins, and contaminating process chemicals (13). A validated process, therefore, must demonstrate effective removal, inactivation, or reduction of these impurities and contaminants to acceptable levels.

Though it is preferable to perform process validation studies at full-scale operational level, it is not always possible to perform them at manufacturing scale due to practical limitations (e.g., virus and nucleic acid reduction studies may require huge amounts of model viruses and nucleic acids). In such cases, scaled-down bench-level studies are acceptable as long as all process input parameters are kept the same as in the full-scale and the output parameters are comparable to the full-scale (14). Whenever this approach is used, demonstration and justification of the acceptability of the scaled-down model should be performed prior to formal process validation.

PROCESS DEVELOPMENT

Development of a Defined Process

The critical steps for the development of a defined process are outlined in Fig. 1. We will examine below the requirements for developing a reliable and reproducible process for a cell culture derived product. The definition of a defined process may be summarized as "a process that provides a high degree of assurance that it will consistently produce a product meeting its predetermined specifications and quality attributes." This definition seems simple and doable (in the beginning phase of a project) but becomes difficult to achieve when all the details for a cell culture-based product are brought into consideration. Adequate confidence must be built by doing sufficient experimentation and development work to demonstrate that the process can consistently produce a product of pre-specified quality. Range finding (feed stream) studies should be performed for every critical and noncritical process parameter (15), and operational set-points must be established after completion of the range finding studies. Worst-case studies (upper and lower ranges) should be performed during the development phase, (as it is much easier to do them during development than during actual production). Alert and action levels (limits) for out-put parameters (test results and specifications) must be established with adequate justification. In-process and final product specifications (acceptance criteria) must be defined clearly with sound scientific justifications.

The success of a well executed project depends on a well written process development (PD) report with sufficient details for every aspect of the process and a well executed transfer of technology from the R&D department to the operations department. The R&D personnel not only adequately transfer the technology, but must provide training to production personnel in every aspect of the process. The role of the R&D personnel does not end here, they should actively monitor the process after successful process validation by applying the statistical tools such as statistical process control. Post-validation process data must be analyzed to ensure that the process performs within the established boundaries. Process capability (Cpk) calculations must be performed on the post-validation process data to evaluate process performance. The process data should also be analyzed by applying other statistical tests, such as Student's *t*-test, to determine confidence intervals on process performance. A 95% confidence interval is generally acceptable for process validation studies. Many companies, however, run their production processes at 98% confidence interval or up to ± 6 SD of the validated process parameters. These analyses demonstrate whether the process is in control and build confidence for running the process on a consistent basis.

The importance and relevance of good PD work that eventually pays off many fold must be emphasized here. It is generally acknowledged that many pharmaceutical

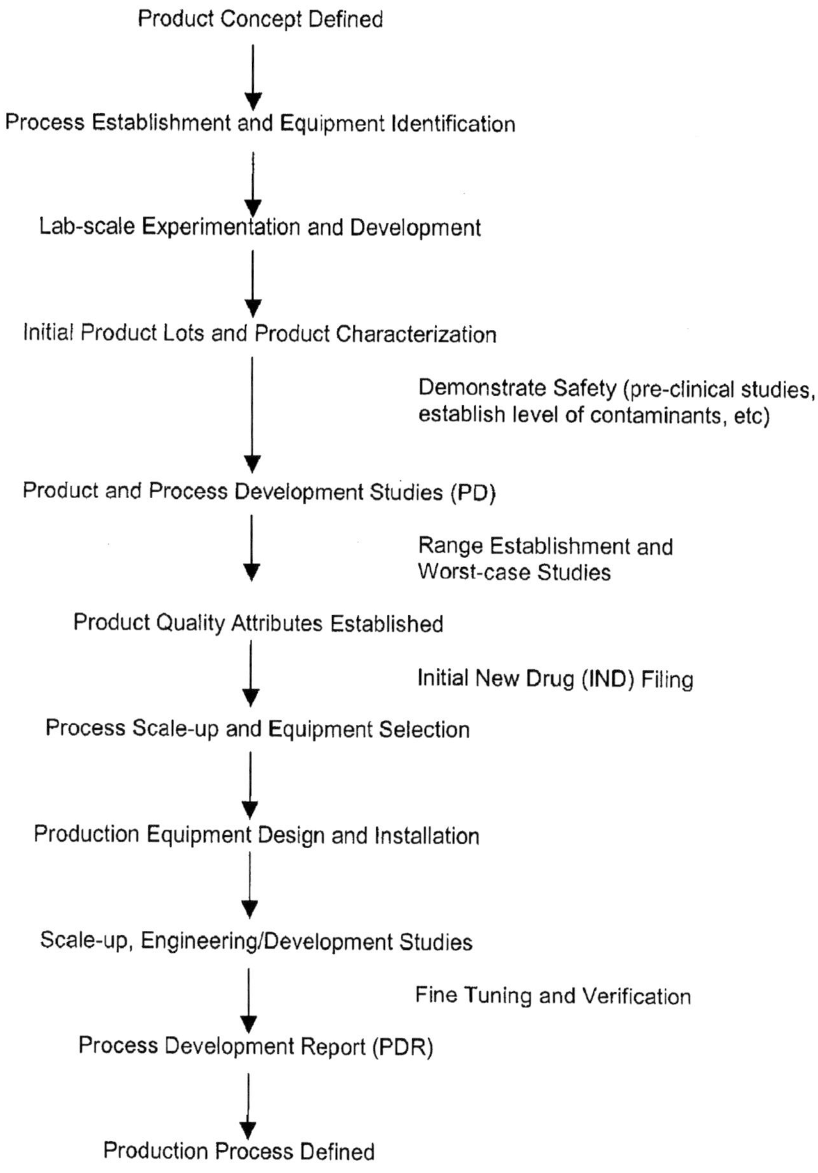

Figure 1 Critical steps for developing a defined process.

and biotechnology organizations shy away from doing comprehensive PD work as they are in a rush to reach the marketplace. In our competitive world of today, timing is key for making or breaking of an organization. Often what we do not realize is that there are no short cuts and eventually (sooner or later) we have to do the required PD work. The smart approach, therefore, would be to perform all required PD work before process validation, rather than during process validation or after completion of a process validation project. In the latter case, the validation projects generally become confusing, cost a great deal of money, and delay project completion (16).

A poorly developed process will typically allow only narrow ranges for operational parameters and may result in the rejection of large amounts of otherwise good in-process material produced slightly outside the narrow process ranges developed. Extension of the process ranges or scale-up of manufacturing processes after initial validation requires time consuming regulatory review and approvals, repetition of all the work performed previously, and almost always turns into a costly validation project. It prevents pioneering organizations from taking leadership positions in the marketplace due to limited product supplies. It may lead and encourage competitors to enter the field and snatch the leadership position from the organizations that developed the product at the first place. It is a lesson many organizations learn, albeit late.

Process Development Report

The importance of well-executed PD work and a well-written process development report (PDR) cannot be emphasized enough. PD and PDRs are the key components of a successful technology transfer from R&D to manufacturing (8). The success or failure of a process validation project greatly depends on the quality and details of the PD work performed and the quality of PDRs in terms of their content, clarity, and completeness. Poorly written reports often cause a great deal of frustration for all involved, result in unnecessary delays, impact project schedule, and even lead to ultimate failure of a project. Many organizations perform excellent PD work, but lack in writing clear and complete reports. Ideal PDRs should contain the following information in as much detail as possible:

- Objective and definition of a process/product
- Scope and rationale
- Process description
- Process flow chart
- Materials and methods
- Equipment and facilities
- Utilities and accessories
- HVAC and environmental requirements
- Process input and output parameters (critical and noncritical)
- In-process testing and acceptance criteria
- Product specifications
- Calibration and preventive maintenance
- Other process requirements
- Result and discussion
- Conclusion
- References

Process Parameters

It is paramount that all process operating parameters (input parameters) that affect product quality attributes (output parameters) are established clearly during the PD phase of a new process, product, or a technology. This is accomplished typically by performing studies at lower and upper limits of the operating ranges generally referred to as the worst-case studies, crash studies or feed-stream studies. Some studies are performed up to the edge of failure and then stepped back to the ranges where process performance is acceptable. These studies can be simulated or performed with active ingredient or product derived from starting material generated during PD phase of the project. These studies can also be performed by generating starting material by artificially setting the parameters to the upper and lower limits of the range. The process parameters are generally classified as critical process parameters and noncritical process parameters.

Critical Process Parameters

By definition the critical process parameters are "those operating parameters that directly influence the quality attributes of the product being produced." For example, temperature and pH in a fermenter are considered critical operating parameters as they have a direct influence on the viability of the organism and the chemical or biological activity of the product being produced. Other parameters that may be considered critical for fermentation processes are: cell viability, media conductivity, glucose concentration, oxygen and air uptake rates, and cell density in the production vessel or device.

Noncritical Process Parameters

The noncritical process parameters are "those operating parameters that have no direct influence on the quality attributes of the product being produced." For example, cell age and media flow rates in a fermenter are considered as noncritical operating parameters as they have no direct influence on the viability of the organism or the activity of the product being produced. Other parameters that may be considered noncritical for fermentation processes are: cell density in the inoculum, cell productivity, agitation rate, perfusion rate, and cell osmolality in the production vessel or device.

Cell Culture and Fermentation Process

A number of different approaches have been used to exploit cell culture technology and develop pharmaceutical products and medical devices, for example expression of the molecule of interest by cells through genetic manipulation or the use of cells as such for treating certain medical conditions. Of these, the technology based on product expression through genetic manipulation is most common. Commercial fermentation processes and bioreactor technologies have been developed in the last several decades to state of the art production of pharmaceutical agents of interest. The introduction of rEPO, rTPA, rFVIII, rInsulin, rHGH, rPDGF, etc. to treat many medical problems would have not been developed without these advances in the technologies. Figure 2 depicts a flow diagram for a typical fermentation process. We will discuss below the steps involved in the development of a commercial cell culture process in the light of process validation. Of special interest here is the establishment of critical and noncritical process parameters that will be verified during the process validation phase.

Cell Line Development

Once a clone has been selected for commercial development it is crucial that the nutritional requirements for the cell line must be defined. The cell line may need to be adapted for growth in certain cases, such as the expression and production of a product in a serum-enriched or serum-free media. The following nutritional requirements in terms of their concentration (% or molarity) or amounts (g/L or PPM or PPB as appropriate), and growth conditions must be established:

- Chemically defined growth medium
- Need for protein/serum/plasma or a protein-free media

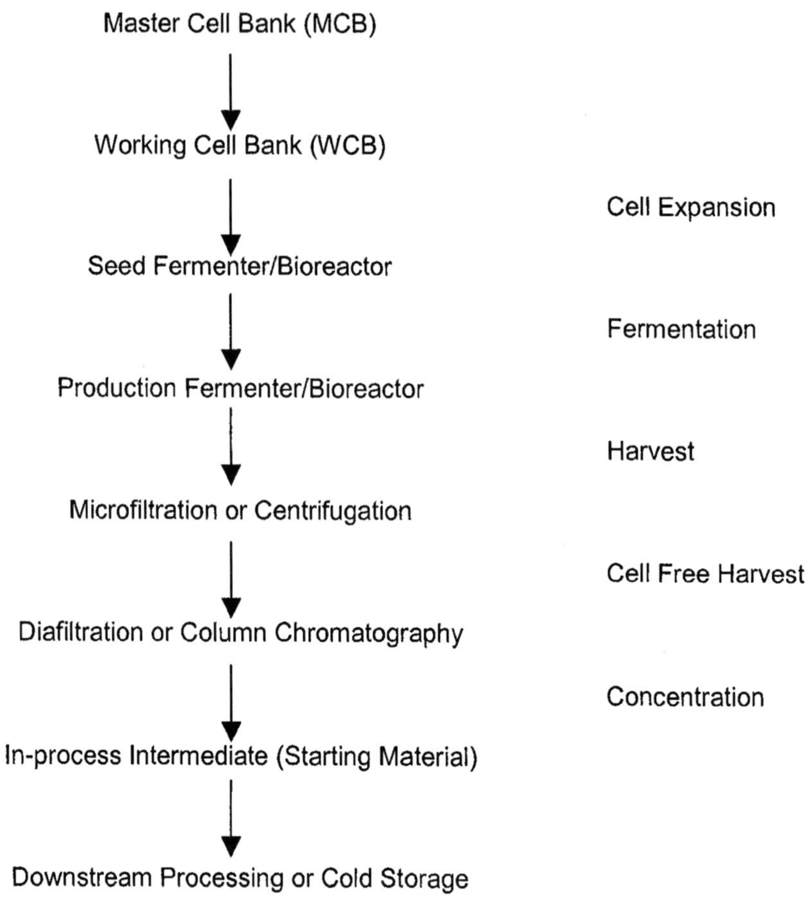

Figure 2 Typical fermentation process flow diagram.

- Requirements for vitamins or fatty acids
- Requirements of any special chemicals
- Requirements of growth factors or hormones, etc.
- Optimizatipon and maintenance of appropriate pH and ionic strength
- Requirements of oxygen, carbon dioxide or other gases
- Optimizatipon and maintenance of appropriate temperature
- Frequency of media changeover
- Frequency of harvesting of the cell line or product

Cell Line Characterization

The cell line must be fully characterized (17,18) for the absence of objectionable organisms or contaminants as follows:

- Absence of bacteria or spores
- Absence of fungi and mycoplasma
- Screening for adventitious viruses (nonretroviral)
- Screening for species-specific viruses (nonretroviral)
- Retrovirus contamination from other species

- Absence of other objectionable agents such as prions and TSE/BSE
- Phenotype/genotype characterization
- Identity and genetic stability

Master Cell Bank (MCB) Preparation

After selection of the appropriate clone for production purposes, the clonal cells should be expanded to appropriate fermentation scale, preferably to production scale fermentation, and used to prepare an MCB in appropriate sized vials for storage in liquid nitrogen. After preparation, the MCB should be scaled-up to evaluate its life cycle and productivity. The process parameters that need to be established at this stage are: cell concentration and viability for preparing MCB, cell volume in the vial, purity and amount of the preserving agent such as DMSO, storage temperature, and acceptable cell recovery and viability after thawing of the MCB vial.

Working Cell Bank (WCB) Preparation

The MCB is typically expanded up to seed or production scale fermentation to prepare the WCB in appropriate sized vials or bags suitable for storage at −70°C or colder. The WCB should also be scaled-up to production scale fermentation to evaluate its productivity and other growth conditions, and it should be fully characterized as outlined in "Approach and Rationale" to ascertain that it is free of objectionable organisms and contaminants. The process parameters that should be established at this stage are: cell concentration and viability for preparing WCB, cell volume in the vial or bag, purity and amount of the preserving agent such as DMSO, storage time and temperature, and acceptable cell recovery and viability after thawing of the WCB vial or bag.

Cell Expansion and Seed Preparation

Procedures (SOPs, BPRs, etc.) should be prepared that describe in detail all the steps for the expansion of cells starting from WCB through preparation of the seed for inoculation of the final-scale production device (fermenter, bioreactor, bag or bottle or vessel). The cell expansion procedure may require only a few steps or may have a number of steps before a seed is ready for inoculation of the production device. In addition, the seed may be used immediately to inoculate a production vessel or it may be stored further until use. Therefore, it is important to evaluate the process and identify the critical and noncritical process parameters for each process independently.

The examples of the process parameters that may be established at this stage are: thawing time and temperature, volume of media and size of flask or bottle for initial cell growth, time and temperature for initial cell growth, media pH, conductivity, temperature, glucose concentration, oxygen and air uptake rates, cell viability and cell recovery at different stages, cell density for scale-up to the final seed vessel (bottle or bag or fermenter or bioreactor), cell density and cell viability in the seed to be used for inoculation of the production device (bottle or bag or vessel or fermenter or bioreactor), and storage time and temperature for the seed (inoculum).

Production Scale Fermentation

Fermentation at the production scale may be carried out in a vessel (fermenter or bioreactor), bottle, or a bag depending on the product type. The product may be the cells themselves, for which efficient cell growth may be critical, or the product may be a biochemical entity (enzyme, protein, hormone, etc.) expressed either in the cells intracellularly retained in the inclusion body or secreted out of the cell in the spent medium. In the latter case, the stability of the molecule in the spent medium should be explored as storage time and temperature for the harvest will be critical for the stability of the product. The process parameters required to be established for the fermentation are: cell viability and cell density in the inoculum, media pH, conductivity, temperature, glucose concentration, oxygen and air uptake rates, cell productivity, cell life span, agitation rate, perfusion rate, cell density, cell viability, and cell osmolality in the production vessel or device.

Continuous fermentation (perfusion). Continuous fermentation, where the product is generally secreted in the spent medium, is the most efficient and commonly employed technology for the production of biopharmaceuticals today. The main concept of this technology is to keep the cells alive as long as they produce a quality product. The number of days the cells are kept in a fermenter (fermenter days) varies depending on the cell type and established time period (weeks or months) for producing a quality product. In this approach the cells are expanded to desired optimum concentration and induced to adhere to coated (with proteins such as collagen) or noncoated acrylic beads where they can survive for many months as long as their nutritional needs are met. Fresh medium is introduced (perfused) and spent medium (harvest) is removed from the fermenter on a continuous basis. A number of cell sedimentation devices (conical, incline or plate settlers) are used to separate the cells from harvest. The cells are returned to the fermenter and the harvest is collected in a harvest tank or bag. The fermenters used in this technology are typically smaller in size (50–2500 L), as continuous perfusion of media allows sufficient volume of harvest collected on a daily basis.

Since the equipment used is more complex and the fermenter cycle is typically long (months), the validation effort is more rigorous for this technology. Establishment of acceptable fermenter days requires full cell characterization (see Cell Line Characterization) at the beginning (early), middle, and end (late) of fermentation to demonstrate that the cell characteristics do not change over time. In addition, product quality attributes are evaluated for the product derived from early, middle, and late stages of fermentation. These activities are performed and established during PD phase and confirmed during formal process validation (PPQ).

Batch fermentation. This nonperfusion technology is employed for products that are either secreted in the spent medium or expressed intracellularly. The cells or the harvest is collected for the isolation and purification of the product depending on the expression of the product in the cells or in the spent medium. The fermenter cycle is generally short (days) for batch fermentation process than for continuous fermentation process (weeks or months). This technology is most efficient for products that are expressed intracellularly where cell mass expansion is critical for productivity. It is less efficient for cell secreted products as the cost of operation is high. The fermenters used in this technology are typically larger in size (500–25,000 L), as it is a batch operation that allows collection of cells or harvest only once per fermenter cycle. Since this technology does not use cell sedimentation devices and fermenter cycle is short (days), the validation effort for batch fermentation process is less rigorous.

Cell mass expansion. This technology is similar to batch fermentation process except the main objective of the fermentation is to expand cell mass. It is used mainly for the products that are expressed intracellularly or where the cells themselves are used for medical treatment (as a medical device), and the cell mass is critical for productivity. The fermenter cycle is short (days) for this technology than for continuous fermentation process (weeks or months). The fermenter sizes used in this technology vary depending on the requirements of the cell mass. Since the equipment used is simpler and the fermenter cycle is short (days), the validation effort for this fermentation process is less rigorous.

Though validation of fermentation process may be simpler for medical devices using biologically active cells, process validation for their formulation, storage, and delivery are more complex. Since mammalian cells are more fragile than protein molecules, their storage without impacting their quality attributes are more challenging. Demonstration of biological activity retention for a heterologus cell-based product during production, distribution, and storage is a daunting task. Cell characterization studies (see Cell Line Characterization) may have to be performed more rigorously after formulation, storage, and end of shelf life of these devices. In addition, the level of impurities and contaminants would also require rigorous investigation during these stages. Moreover, an assurance that the biological activity and safety do not impair and adverse reactions do not increase during these stages also needs to be demonstrated.

New approaches and future of cell-based therapies. The manipulation of cell culture technologies to generate unique therapies and medical treatments has just begun. Further development of these technologies would be essential for their impromptu use in new ways to treat diseases. Mammalian cells are being evaluated for grafting, transplantation, tissue regeneration, and organ culture. Stem cells are being developed into numerous cell-based treatments/cure for many diseases such as cancer, HIV, Alzheimer's, Parkinson's, etc. Gene therapy is expected to be the ultimate cure for many diseases in the 21st. century. This field is expected to grow exponentially in the next 25 years and bring numerous challenges for cell culture scientists. A number of microbial hosts (bacteria, plasmids, viruses, etc.) are being evaluated as carriers or vehicle for gene therapy products. For a successful gene therapy product it is crucial that it is free of any side effects, is long lasting, and is fully effective. To accomplish these goals the gene therapy products would have to be pure, free from undesirable components, easy to use, effectively targeted to desirable site, effective transformation and expression of desired genes, complete correction or deletion of defective genes, and they must prove to increase longevity. Cell culture scientists would definitely address all these issues and develop appropriate technologies to attain desired results. However, imagination for validation of all these diverse technologies and processes is mind-boggling today. New approaches to validate these technologies, production equipment, and production processes would have to be developed to meet yet to be established regulatory requirements.

Product Isolation, Separation, and Concentration

The procedures for the isolation of the product, its separation from impurities and contaminants, and its concentration by different technologies, depending on the product type, are performed after the fermentation process is complete. The cell separation techniques such as centrifugation or microfiltration may be applied to concentrate cells for product recovery or remove cells from the harvest that contains the product. The parameters that may be critical for process validation for this process are: cell concentration and viability in the fermenter effluent (spent media with or without cells), storage time and temperature for the effluent, centrifugation speed or microfiltration rate for cell separation, cell separation time and temperature, and storage time and temperature for the concentrated in-process intermediate (IPI, starting material).

In-Process Intermediate (IPI) or Product Preparation and Stabilization

The IPI may be the cell suspension or cell extract or concentrated harvest fluid depending on the product type. Some of the processes require stabilization of the IPI for storage prior to further processing of the product. In these cases a formulating agent may be added to the concentrated IPI prior to its storage in the cold. The parameters that should be considered for process validation here are: the purity and concentration of formulating agent, mixing of the formulating agent with the IPI, freezing time and temperature for the formulated IPI, and storage time and temperature for the formulated IPI.

For medical device applications, the IPI (cell suspension/tissue) may be the final product that may require cleaning and removal of impurities, formulation for stabilization, and preparation of the product for clinical use. In such cases, the parameters that need to be established for process validation are: amount of contaminants (DNA, proteins, etc.) in the final product, cell/tissue morphology, genetic characterization of the cells/tissue, amount of formulating agent, storage time and temperature, and shelf life of the product.

Downstream Process Development

The pharmaceutical or therapeutic proteins expressed in cells are further purified from the IPI generated during the fermentation. The purification steps may involve microfiltration/diafiltration, salt or solvent fractionation, and column chromatography (ion exchange, hydrophobic interaction, affinity, gel filtration, etc.). The examples of critical process parameters that should be established here are: pH, conductivity, salt or solvent residues, membrane life cycle, column operation parameters (equilibration, load, wash, elution, regeneration, storage, cleaning, life-cycle, etc.), impurities and contaminant clearance (DNA, microbes, viruses, proteins, etc.), and hold times and temperature for in-process material and equipment. There is a whole battery of process parameters that should be established for process validation for these operations. It is beyond the scope of this book chapter to go into details of these parameters, but they are discussed in more detail elsewhere (19).

Bulk Formulation, Stabilization, and Final Packaging

The purified in-process material (bulk) is typically formulated for stabilization and either stored until further processing or filled and/or freeze-dried depending on the mode of application for the product. These operations are generally performed in the clean room environments (class 100) that require detailed and cumbersome PQ studies. At this stage, the product is characterized in detail and every lot of the product is evaluated against pre-established specifications and quality attributes. These include: product amount (units or weight), purity, strength, pH, ionic strength, amount of trace metals, amount of impurities and contaminants (DNA, proteins, microbial load, endotoxins, etc.), amount of residual reagents (solvents, polymers, chemicals, etc.), amount of formulating agents (excipients, sugars, etc.), sterility, storage time and temperature, and product shelf life. The process parameters governing these quality attributes should be considered as critical parameters during process validation studies. The filled or lyophilized product vials are inspected for container closure integrity and labeled appropriately. The product coming from validation runs is placed on stability to demonstrate that the product is stable during its entire shelf life. In addition, shipping studies are performed to demonstrate that the packaging is shatter-proof and the product is stable during shipping. All appropriate process parameters covering these quality aspects of the process should be established and verified during the process validation studies.

Process Validation

Process validation projects are complex and cover a great deal of details. The critical steps for performing process validation studies are depicted in Fig. 3. The process validation begins with the receipt of a PD report from the R&D arm of an organization, and ends with the approval of a process validation report and applicable SOPs/BPRs by the QA arm of the organization. A well-written PD report with adequate details of the developed process greatly facilitates process validation. In addition, efficiency and success of a process validation depends heavily on the extent and quality of PD work and/or engineering runs performed on associated equipment (see Process Validation Protocols and Final Reports for details). Prior to formal process validation, processes should be run rigorously on associated equipment to gain valuable experience. The experience gained is well worth as it saves time and resources during the formal validation stage by efficiently resolving any unforeseen problems, discrepancies or deviations. In no event, the engineering runs or PD studies be left for performance during the formal process validation stage (PPQ phase). Successful completion of a process validation study requires good coordination among the responsible departments such as production/manufacturing, quality control, QA, engineering, R&D, regulatory affairs, etc. We will discuss below the types of processes that are generally covered under process validation projects (PV or PPQ).

- Manufacturing procedures for fermentation, product isolation, purification, formulation, sterilization, filling, and freeze drying of products
- Microfiltration, ultrafiltration, and sterile filtration
- Cleaning procedures for equipment and processes (clean-in-place, CIP)
- Lifecycle determinations for chromatographic resins, membranes and filters
- Impurity/contaminant clearance (DNA, viruses, host cell proteins, etc.) studies
- Impurity/contaminant inactivation (virus, endotoxins, TSE/BSE, etc.) studies
- Sterilization and steam-in-place (SIP) systems
- Critical utilities such as water for injection (WFI). (PPQ is performed to demonstrate process reproducibility/consistency and product quality)
- Environmental qualification for facilities (EQ). (EQ is a combination of air handling, equipment/facility cleaning, sanitation, gowning, and environmental monitoring)
- Re-processing of process intermediates, bulk, and final product
- Revalidation of processes due to major change control requests (CCRs)

Unlike equipment re-qualification, revalidation of a production process is currently not a mandatory cGMP requirement, but must be considered after implementation of a large number of CCRs or as soon as a process shift is noticed. The

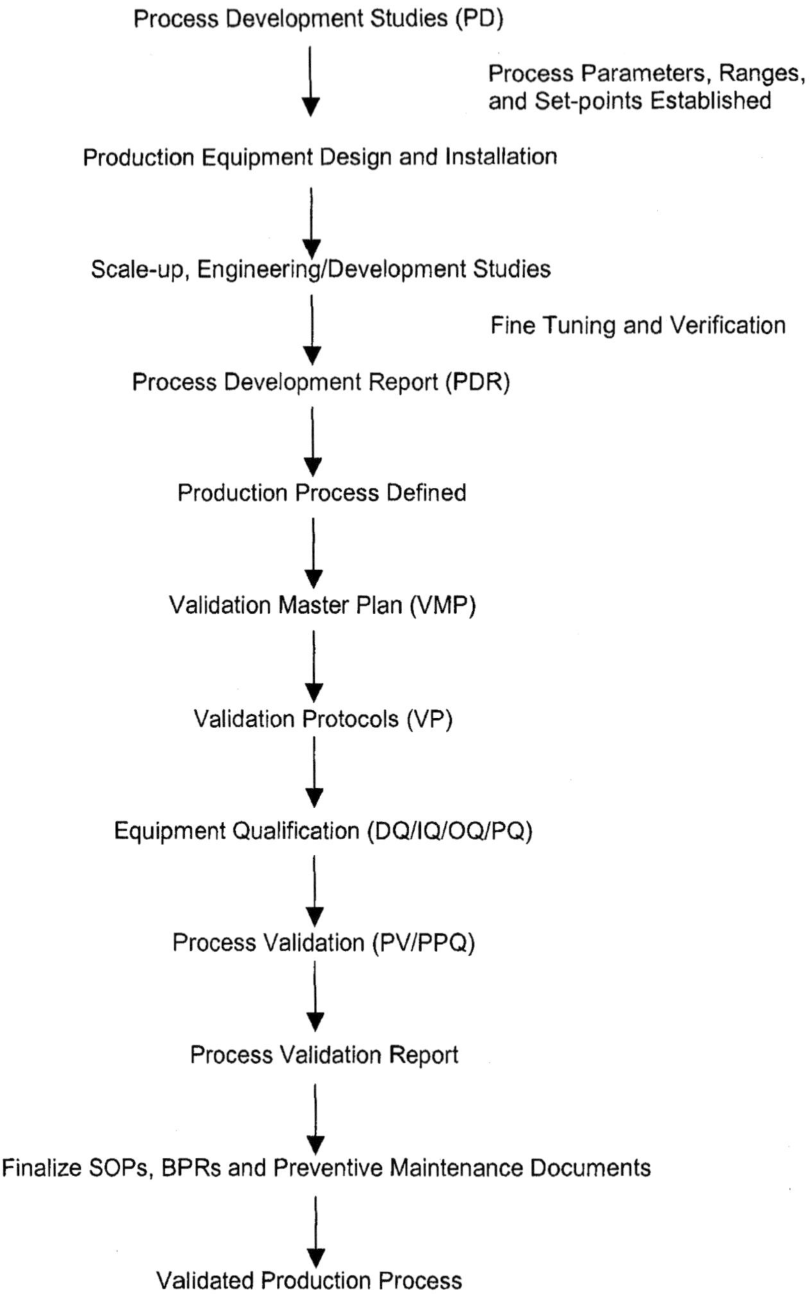

Figure 3 Critical steps leading to process validation.

purpose of the process revalidation should still be to demonstrate that the implementation of CCRs or an observed shift in the process has not affected quality attributes of the final product. Even if no major changes were implemented or process has not shifted, process revalidation should still be considered to demonstrate that the process is running within controlled limits, at some appropriate time intervals after original process validation (after 100/ 500/1000 lots or 2/5/10 years, as appropriate).

Life Cycle of Process Validation

Process validation is a continuous process, it does not end after the sign-off of the original process validation report. Figure 4 illustrates the life cycle of process validation by keeping the process and product quality attributes in mind. It is not uncommon for companies to initiate change control requests within days or weeks after the completion of the original process validation. Many companies have made major process changes within one year of the original process validation.

Since cell culture and fermentation processes are continually evolving, changes are being made to manufacturing processes on a regular basis by many companies. In fact, a number of second-generation products are derived from the same cell lines, except the cell lines have been adapted to produce the product in a protein-free (serum/plasma) media.

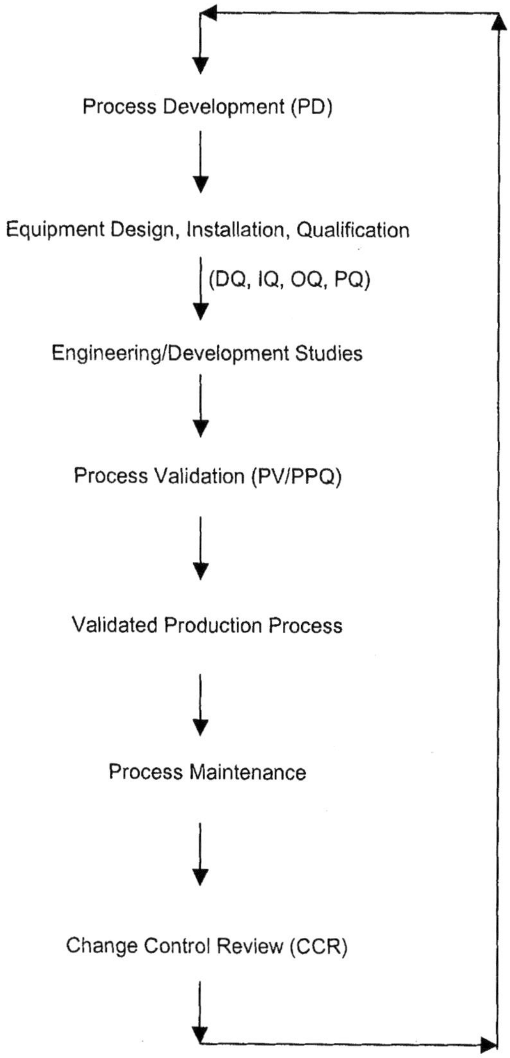

Figure 4 Life cycle of process validation.

Though such changes improve the product quality and safety tremendously, the process validation task is nevertheless the same. Such changes often lead to the construction of a new manufacturing plant and complete re-validation of the production process. Therefore, the life cycle of a manufacturing process after original validation is mainly dependent on the volume of change control activity.

Though regulatory agencies have not made a guideline yet for process re-validations, they have been advising drug/biological manufacturers for some time to evaluate the need for process re-validation at some regular intervals. Many companies do not feel the need for process re-validations based on the activities that are covered under their change control management programs. However, it may be prudent to perform process re-validation after a major process change or after a number of small changes to assess the cumulative effect of many change control requests. One possible mechanism may be to perform process re-validations after every 100/200/500/1000 production runs or, 2/5/10 years or sooner, as justified by the evaluation of the historical process input and output parameters, specifications, set-points, action limits, or other observations that suggest that the validated process may have shifted.

VALIDATION OF THE MANUFACTURING PROCESSES AND ASSOCIATED EQUIPMENT QUALIFICATION

Validation SOPs

Establishment of sound validation SOPs and strict adherence to them is crucial for the success of any validation project in a cell culture-derived drug/biological manufacturing organization. These SOPs are typically written by validation professionals and are approved by the responsible groups (e.g., operations, engineering, validation, research, PD, QC, QA, and regulatory affairs) that have a stake in the validation projects. The SOPs should be established for the required validation functions, as listed below:

- Site or facility validation policy and management
- Validation master plan and final report
- Validation requirements for DQ, IQ, OQ, PQ, PD, PPQ, EQ, and RQ validation protocols
- Design of worst-case studies for OQ, PD, and PPQ protocols
- Determination of acceptance criteria for validation protocols
- Design of prospective or concurrent validation studies, and retrospective data analysis
- Write a validation protocol and a validation final report
- Revision of a validation plan, protocol or a final report
- Execution of IQ and OQ protocols
- Execution of engineering and PD protocols (studies)
- Execution of PV or PPQ protocols (studies)
- Validation of analytical methods, assays, and procedures
- Performance of validation studies on clearance and/or inactivation of impurities contaminants
- Performance of filter qualifications and sterile filtration validations
- Performance of cleaning validation studies on equipment, accessories, and processes

- Performance of sanitation and sterilization validation studies
- Performance of validation studies on automated systems (computers, DCS, and PLCs)
- Performance of environmental qualification (EQ) studies
- Review, verification and analysis of validation data and documents
- Documentation of validation discrepancies and deviations
- Performance of re-qualification (RQ) studies on validated equipment, systems and processes
- Establishment of a validated lifecycle (for cells, resins, filters, membranes, etc.)
- Requirements for maintaining a equipment or a process in a validated state
- Display of validation status (labels) for validated equipment and systems
- Training and certification program for personnel involved in validation projects

In addition to these, many other SOPs may be established for performing specific validation functions depending on the need of the equipment, system or process (e.g., determination of equipment surface finish, calibration of instruments and thermocouples, determination of agitation rates, determination of HETP and Af on chromatography columns). The required SOPs are typically identified and established by qualified validation professionals depending on the need of an organization.

Validation Master Plan

It is paramount that a detailed VMP should be written before implementation of any new or unlicensed process for the production of a cell culture derived drug/biological product (11,12,20–22). The VMP provides details of an organization's plans for carrying out all validation activities on production equipment, systems, and processes. The plan should provide details of equipment, facilities, utilities, raw materials, storage times and temperatures, environmental requirements, production processes, critical and noncritical process parameters, process set points, SOPs and BPRs, in-process and final product sampling, analytical assays and methods, QC testing, in-process testing and acceptance criteria, and product release specifications. The VMPs should be approved by all responsible stakeholders or senior management of a company (operations/manufacturing, QA, QC, engineering, PD, and regulatory affairs), and should contain the following information at a minimum:

- Objective
- Scope and rationale
- Process and product design
- Description of manufacturing facility
- Equipment description and qualification (DQ/IQ/OQ/PQ)
- Process description and qualification (PQ/PPQ)
- Description of utilities and supplies
- Description of automated systems
- Equipment cleaning (CIP/COP) and sanitation (SIP)
- HVAC and EQ
- Process parameters and set points
- In-process testing and product specifications
- Analytical methods and procedures
- Manufacturing procedures and batch records

- Responsibilities
- Execution plan and schedule
- Documentation and training
- Modification and change control
- Preventive maintenance
- References
- Attachments
- Facility diagram
- Process flow diagram
- List of validation protocols
- List of PDRs
- List of engineering reports
- Project schedule and Gantt charts
- Other pertinent documents

Equipment Qualification

A great deal of information is available in the literature, web sites of the regulatory agencies (FDA, EMEA, ICH), professional societies (PDA, ISPE, AAPS, etc.), various seminars and symposia, and from professional consultants for the qualification of process equipment. The key components of equipment qualifications are covered in the following validation studies:

- Design qualification (DQ)
- Installation qualification (IQ)
- Operational qualification (OQ)
- Performance qualification (PQ)

Equipment qualification studies should be performed for all process equipment (fermenters/bioreactors, chromatography, microfiltration/ ultrafiltration systems, tanks and vessels, autoclaves, CIP and SIP systems, freeze dryers, etc.), utility equipment (WFI, clean steam, solvent delivery, etc.), supply equipment and accesories (gases, filters, raw materials, etc.), automated systems (computers, DCS, PLCs, etc.), and critical facility equipment (HVAC, warehouse, storage chambers, shipping containers, etc.). It is beyond the scope of this chapter to cover details of equipment qualifications. The readers can review selected references (11,12,22,23) to learn more about equipment qualifications.

Equipment Engineering Runs and Process Development Studies

After completion of the DQ, IQ, OQ and PQ studies on the equipment, it is prudent to perform equipment engineering runs and PD studies (pre process validation studies, also known as trial runs) to ascertain that the developed process is scalable to the production scale. These studies confirm that the process can perform effectively within the ranges established at the small-scale or previously developed scale. These studies also help in ensuring that the formal process validation studies would not run into major discrepancies, deviations, or failures. They also provide an opportunity to fix any problems that may have been ignored previously. Furthermore, they provide a chance to develop or fine tune acceptance criteria and specifications for the formal process validation studies down the road.

These studies are performed under approved PD protocols (see Cell Culture and Fermentation Process). Typically three runs are performed at the production scale by following the SOPs and BPRs written for the process. All sampling and testing is performed as in the process validation studies except that some tests may not be required (have no chance of failure due to scale of operation). PD final reports are prepared after review and analysis of all results and associated data. Appropriate conclusions are drawn and recommendations for formal process validation studies are made in the final reports. The examples of these studies are equipment load studies (autoclaves, depyrogenation ovens, viral inactivation tanks, pasteurization, column chromatography, cleaning validations, etc.), mock runs (process runs without active ingredient), partial load studies (part load with active ingredient and remaining load with an excipient), and full process runs with active ingredient. In conclusion, these studies provide peace of mind that the formal process validation would be un-eventful or would be completed with minimum difficulties.

Process Validation Protocols (PV or PPQ) and Final Reports

Process validation protocols are written to demonstrate that the production processes are reproducible, are in control, and consistently produce a product of predefined specifications and quality attributes. The protocols are also used to demonstrate impurity/contaminant clearance, validation of operating ranges, equipment cleaning, and establishment of lifecycles for chromatographic resins, filters, and membranes. It is indisputable that well written process validation protocols are instrumental for efficiency and success of process validation studies. The protocols should contain details of the production process and equipment to be used, SOPs and BPRs to be used, critical and noncritical process parameters, process set-points and action limits, sampling and QC testing, analytical methods and assays to be used, and in-process and final product specifications. The process validation protocols should be designed to incorporate worst-case studies based on the acceptable level of risk for the process and the product. It is important that all pre-requisites (e.g., DQ/IQ/OQ/PQ on equipment or a system, approved SOPs/BPRs, instrument calibration, and personnel training) are completed before execution of a process validation protocol.

After execution of the validation protocols and completion of all testing, validation final reports should be written with complete details of protocol execution (BPRs and sample tables), test results, discrepancies and deviations, modifications or change control requests, passage/failures against acceptance criteria, statistical data analysis, preventive measures and maintenance, validated critical and noncritical process parameters, process set-points and action/alert limits, supporting data, and conclusions. The validation protocols (and final reports after completion of the study, see below) should be reviewed and/or approved by all responsible parties such as manufacturing, PD, engineering, quality control, QA, validation, and regulatory affairs (optional), as appropriate, prior to its execution. A typical outline of a process validation protocol and final report (validation package) are given below:

- Approval signatures page
- Objective
- Scope and rationale
- Process description
- Acceptance criteria
- Responsibilities
- References

- Prerequisites
- Validation procedures or test functions
- Result, data analysis and discussion
- Validated critical and noncritical process parameters
- List of SOPs and BPRs
- List of supporting data and documentation
- List of discrepancies and deviations
- Conclusion

The validation final reports must contain all results obtained during the execution of a process validation protocol. All deviations should be described along with justifications for their acceptance. The results must be evaluated against the pre-established acceptance criteria, product specifications, and quality attributes. The data should also be evaluated statistically and confidence intervals for the data should be calculated to demonstrate process robustness. Process capability (Cpk) calculations should also be performed to demonstrate process reproducibility. The final report should draw a scientifically sound conclusion based on the results obtained during the study. Ranges and set-points for all validated critical and non-critical process parameters should be established. In addition, alert and action limits for validated process parameters should also be established wherever applicable. Moreover, a plan to monitor process parameters during production should be devised, and the historical data should be evaluated statistically at pre-established time periods (yearly or after every 100 lots or other suitable interval). Process capability (CpK) calculations should also be repeated on historical data to demonstrate process reproducibility. The in-process action and alert limits and product specifications should be re-evaluated and tightened, wherever possible, after complete data analysis at pre-established time periods.

Raw Materials

The quality of raw materials plays a major role in attaining quality attributes for the final product and indirectly affects success or failure of process validation. The critical raw materials for a cell-derived product during fermentation are: basal media, purified water or WFI, salts and buffering agents, oxygen and carbon dioxide, amino acids, vitamins, glucose, serum or plasma proteins (animal or human), and other nutrients such as hormones or growth factors. It is essential to establish specifications for all raw materials used in the production process, and ensure that quality attributes of all incoming raw material lots are met against the established specifications. Any changes made to the specifications for raw materials should be evaluated through change control management and process validation(s) performed wherever necessary. In addition, FIFO (first in first out) procedures should be established for approved/released incoming raw material lots.

Facilities

All equipment installed in a facility must be qualified (DQ, IQ, OQ and PQ, wherever applicable). The typical facility related equipment includes HVAC, cold/freezer freezer rooms, cooling towers and heat exchangers, chemical/solvent tanks and distribution system, and waste treatment and disposal systems. The facility cleaning procedures are established based on the requirements for each classification (class

100, 1000, 10,000, and 100,000; or grade A, B, C, or D), and must be validated under the facility EQ. Emergency power systems should also be qualified (DQ, IQ, and OQ) to demonstrate that uninterrupted power supply is available to critical production equipment (cold rooms, freezers, freeze dryers, etc.).

Utilities

The critical utility systems must be qualified (DQ, IQ, OQ, and PQ). The examples of utilities that are used in a typical cell culture based production facility are potable water, purified water, WFI, plant steam and clean steam systems, gas distribution systems, CIP and SIP systems, and electricity supply systems.

Production Equipment

Besides equipment qualification (DQ, IQ, OQ, and PQ), typical production equipment used in a cell culture-based facility must be validated (PPQ). The production equipment includes media and buffer preparation equipment, media filtration equipment, media storage tanks, media and buffer distribution system, seed fermenters or bioreactors, production-scale fermenters or bioreactors, aseptic transfer equipment, cell settlers, heat exchangers, pumps and agitation systems, gas sparging equipment, harvest tanks, microfiltration and ultrafiltration systems, centrifuges, and initial capture and concentration equipment. These validation studies may be performed separately or may be combined in the process validation study for a new process, product or technology.

Instruments

Though all instruments associated with equipment systems are generally covered in the equipment qualification, any stand-alone instrument must be qualified separately. In addition, all instruments must be calibrated at the time of their qualification and must remain on a regular calibration schedule after their qualification. The established calibration schedule must be justified and scientifically sound. All test instruments used to measure a specific parameter during validation must also be calibrated.

Analytical Equipment, Instruments, and Methods

All analytical equipment and instruments must be qualified (DQ, IQ, OQ, and PQ), and associated methods, procedures or assays must be validated (PPQ). During this phase of validation, the procedures must be evaluated for precision, accuracy, repeatability, and variability. The results must be evaluated statistically to demonstrate confidence intervals for each sampling condition. The instrument to instrument, operator to operator, and intra- or inter-assay variabilities must be established.

Distributed Control Systems (DCS) and Programmable Logic Controllers (PLCs)

All automated systems (computers, DCS and PLCs) must be qualified (DQ, IQ, OQ, and PQ), and associated software must be tested for its intended function, use, and its lifecycle (24–26). The validation aspects of these systems are verified and confirmed during the process validation phase of the validation activity.

Cleaning and Disinfecting

The cleaning of the equipment including the CIP systems must be qualified (DQ, IQ, OQ, and PQ), and the procedures must be validated (PD, PV, and PPQ). The cleaning validation studies must demonstrate that the process residues and cleaning agent residues are removed to the acceptable levels after the cleaning (27). The acceptable levels for the residues must be established by actual scientific data or sound scientific knowledge (28–31). Appropriately validated assays to test the residues must be used. In addition, all cleaning and disinfecting agents must be qualified or approved for use. The cleaning requirements for each step of the process must be established per appropriate guidelines established by regulatory agencies (27–29). Since the final rinse in cleaning processes for biological products is performed by WFI, many companies use the quality attributes of WFI as acceptance criteria for rinse samples. In addition to rinse sampling, cleaning agent residues and process residues (protein, fatty acids, nucleic acids, raw material components, etc.) are tested by surface swabbing and evaluated against pre-established acceptance criteria. The typical assays used for cleaning validations are pH, conductivity, TOC, microbial load, endotoxin, protein assays/analysis, and other assays for specific residues. A cleaning monitoring program should be established to maintain the equipment in a validated state (31).

Standard Operating Procedures (SOPs) and Batch Production Records (BPRs)

The appropriate SOPs and BPRs must be drafted, reviewed, and approved prior to beginning of process validation studies. These procedures may be modified, if needed, with appropriate justifications during validation as long as the last validation runs (three or more) are performed after making the modifications and all validation acceptance criteria are met. The modifications to the procedures during validation should be made through the change control system of the organization. It is important that all changes made during validation are fully incorporated in the SOPs and BPRs prior to approval of the final validation package.

Personnel and Training

The success of a process validation project depends solely on two things—first an effective and detailed technology transfer program, and second an effective and detailed training program for the new processes or technologies. Therefore, it is crucial that the organizations establish a very effective and practical technology transfer function, and an effective and practical training program for the operations personnel. The initial training is typically provided by the R&D arm of the organization (or whoever developed the process and has the most knowledge or experience). The training program and procedures must be well documented and must follow cGMP guidelines.

Documentation

The regulatory agencies have stated in no uncertain terms that the lack of documentation (even if the work was performed) would be interpreted as if no work was performed. Therefore, the importance of documentation cannot be emphasized enough. The following documents need to be in place for the purposes of process validation:

- An approved validation master plan/report, if applicable.
- An approved validation protocol/report/final package.
- Original or copies of all approved specifications or acceptance criteria.
- Original or copies of all PD reports.
- Original or copies of all approved process validation parameter documents.
- Original copies of all prerequisite sheets completed during validation.
- Original copies of all validation attachments/execution documents.
- Original or copies of all SOPs/BPRs employed during validation execution.
- Original or copies of all QC test reports.
- Original or copies of all other test reports.
- Original or copies of all raw data, or location of all archived raw data.
- Original or copies of all supporting data related to validation.
- Original or copies of all change control requests implemented during validation.
- Original or copies of all deviation reports encountered during validation.
- Original or copies of all corrective action reports, if applicable.
- Original or copies of all SOPs/BPRs/documents revised during or as a result of validation.
- Original or copies of all other documents related to validation.

Preventive Maintenance

Maintenance of validation post-licensure is as important as original validation, as it transforms into cGMP compliance after licensure of the production process. To maintain a process in the validated state, it is crucial that procedures for preventive maintenance (PM) be established prior to original validation, modified as appropriate during validation, and are followed thoroughly and timely after validation. In addition, the processes should be monitored regularly with respect to process parameters observed during production. The historical process parameters data should be evaluated at some fixed intervals (annually or sooner if needed) with respect to validated parameters and be tightened or loosened (in principle) as appropriate and justified by the data based on sound scientific principles and policies. The historical data must be evaluated by applying appropriate statistical methods and calculation for process capability (Cpk). The procedures for the following activities should be established as needed:

- Evaluation of production equipment, parts, and accessories.
- Calibration program for all instruments used in production and testing.
- Equipment re-qualification and process re-validation program.
- Preventive maintenance program for equipment, facilities, and processes.
- Change control program.
- Evaluation of validated process parameters.

Change Control

A change control system must be instituted to document all changes made to validated production processes (32). The change control activity begins after the installation of the equipment systems and continues throughout the lifecycle of the process or product. The change control requests (CCRs) should be initiated prior to making the change, except in emergency situations, and must be reviewed and approved by all involved groups. The impact on validated systems and processes must be assessed and any identified validation work must be completed before closing the CCR. If the change is minor, little or no validation work may be required, however if the change is major, full revalidation may be required. The US FDA published a guideline in July 1997 (33) that requires notification of all changes to the agency depending on the extent of the change in the following manner:

Major changes require submission and approval of a supplement prior to the distribution of the affected product. *Moderate* changes require submission of a supplement at least 30 days (CBE30) prior to the distribution of the affected product. *Minor* changes do not require any submission prior to the distribution of the affected product, but must be documented in the annual report.

MANUFACTURING PLANT QUALIFICATION

The manufacturing plant qualification and licensing requires additional validation, testing, and documentation besides equipment qualification, utilities/facilities qualification, process validation, and establishment of PM and change control programs. Successful completion of the following items is key to the licensure of a manufacturing plant for a cell culture-derived pharmaceutical or therapeutic or diagnostic product:

Plant Design and Construction

The manufacturing plant must be designed and constructed per appropriate local, state and federal regulations and bylaws. The plant must be built by keeping the product and personnel flow in mind. In general, the product flow should be unidirectional. All required essential utilities (power, water, waste disposal, sewer, etc.) and facilities (warehouse, receiving, shipping, etc.) must be planned for and built into the manufacturing plant. Location and installation of all production equipment, process utilities, facilities, and support systems must be designed and procured per requirement of the production process. Additional details for manufacturing plant design and construction are available elsewhere (8,20,21).

Validation Master Plan

A VMP must be written, approved by the management, executed, and revised/updated as needed starting with the site selection and plant construction and ending with the licensure of the manufacturing plant (see Process Development Report Section for the details that must be included in the VMP).

Equipment Installation

The production equipment, utilities/facilities equipment, and all support system equipment as designed for the process must be procured from quality manufacturers and installed per manufacturer's recommendations in the manufacturing plant. All essential utilities (power, water, disposal, etc.) and supplies (gases, solvents, etc.) must be available prior to the installation of equipment systems. Equipment checkouts (ECOs) must be performed to evaluate safe and normal operation of the equipment prior to equipment qualification. It is also essential that the equipment life cycle be established at this phase to ensure that the equipment performs as designed during the entire life of the equipment.

Equipment Qualification

Please refer to Equipment Engineering runs and Process Development Studies, and Process Validation Protocols and Final Reports Sections for details on equipment qualification. It is crucial that ECOs and dry runs (trial/engineering runs) are performed during the equipment qualification phase to ensure that the equipment operates within the ranges as designed for the process and certified by the manufacturer.

Equipment Performance Qualification

These studies are performed during the qualification phase of the equipment to ensure that the equipment delivers the desired output for the manufacturing process such as flow rates, temperature, pH, conductivity, agitation rates, sparging rates, cell retention, bioburden reduction, endotoxin removal, depyrogenation, sterile filtration, sterilization, etc. Equipment loading studies (autoclave loads, vial washer loads, depyrogenation loads, stopper processing loads, chemical inactivation tank loads, etc.) are performed during this phase to ensure that the equipment loads are processed appropriately to meet all quality attributes desired in the qualified loads. It is essential to evaluate which of the equipment needs to be placed on equipment re-qualification program to ensure that equipment delivers desired quality attributes in the processed loads during the entire life of the equipment.

Heat, Ventilation, and Air Conditioning (HVAC)

The qualification of HVAC system (DQ, IQ, OQ, and PQ) for the manufacturing plant must be completed during the validation phase. This includes qualification of the air handlers, HEPA filters, distribution piping, and associated equipment. The PQ on the HVAC system must demonstrate that the adequate air flow and particle levels (viable and nonviable) are achieved per specification (Federal Standard 209E or European ISO Standard or equivalent) for the different manufacturing environments (class 100, 1000, 10,000, and 100,000; or grade A, B, C, D by European standards).

Cleaning and Sanitation

Effective equipment and facility cleaning and sanitation procedures play a major role in maintaining equipment in a validated state and maintaining its life cycle, and by doing so, assures maintenance of product quality attributes (27–31). Cleaning and sanitation procedures should be developed that can effectively clean equipment product contact surfaces, working and processing areas, work surfaces, floor and wall surfaces, and drainage and disposal systems. The cleaning/sanitation procedures must be validated by using appropriate acceptance criteria. The specifications or acceptance criteria for cleaning and sanitation should be established on a case by case basis by consideration of the manufacturing process, processing time, and in-process materials used. Hold times for process equipment after cleaning must be established and demonstrated (qualified). PD studies should be performed to evaluate effectiveness of the cleaning agents and procedures and establishment of acceptance criteria before PPQ studies are performed. The typical tests used to demonstrate cleaning of equipment product contact surfaces are visual examination, microbial load, endotoxin level, pH, conductivity, level of residual process impurities, level of residual cleaning agents, TOC, and any other appropriate test for a residue. The typical tests used to demonstrate effective sanitation procedures are visual examination, levels of viable particles, level of nonviable particles, and absence of objectionable organisms. The following cleaning and sanitation procedures should be established at a minimum:

- Procedures for cleaning inner and outer surfaces of equipment.
- Procedures for cleaning of chromatography columns, filters and membranes
- Recipes and methods for CIP/COP systems.
- Manual cleaning procedures.
- Cleaning/sanitation procedures for work surfaces, floors, and walls.
- Cleaning/sanitation procedures for drains and disposal systems.
- Cleaning/sanitation procedures for utilities (WFI, gases, steam, etc.)

Gowning and Personal Safety

Appropriate and protective gowning is essential for preventing the product from getting contaminated by human interaction and also to safeguard humans from any undesirable effects due to exposure of finished products, in-process materials, process chemicals, and supplies or accessories used during production. The personal safety devices used are lab coats or gowns, eye/hearing protection devices, gloves, face shields, and chemical or solvent handling devices. Clear procedures should be written for effective use of gowning and personal safety devices.

Environmental Qualification

In addition to qualification of HVAC, equipment/facility cleaning, and gowning; an EQ should be performed to demonstrate clean manufacturing environment as a combination of an effective HVAC system, effective equipment cleaning procedures, effective gowning procedures, and effective facility cleaning procedures at dynamic and at rest conditions. The acceptance criteria for the EQ for class 100,000, 10,000, 1000, and 100 (grade A, B, C, and D by European Standard) are different and are derived from Federal Standard 209E and other US and European regulations. The typical documentation and testing performed during EQ are as follows:

- Number and gowning status of the operators present during dynamic conditions.
- Number and gowning status of the operators present during rest conditions.
- Swab testing of operator's gowns for presence of viable and nonviable particles.
- Verification of air flow changes/hour and air flow rates during testing.
- Testing of viable and nonviable particles in the air. Testing of viable and nonviable particles in swab samples from work surfaces, floor surfaces, and walls.
- Verification of all cleaning performed during the testing period.

- Presence/absence of objectionable organisms in the environment.
- Discrepancy investigations and implementation of corrective actions.
- Establishment of an environmental monitoring program.

Preventive procedures to check the presence of insects, pests, reptiles, and rodents should also be employed.

Capacity Evaluation

Though not a regulatory requirement yet, regulatory agencies sometimes request that a production capacity evaluation be performed for each manufacturing facility that requires licensure. Capacity evaluation should be performed by verifying the capacity of the equipment, utilities, supplies, facilities, raw materials, in-process materials, and final product production. The following items should be evaluated for the capacity assessment of the manufacturing facility:

- Capacity for manufacturing areas in terms of space and production for equipment such as seed and production fermenters, media/buffer tanks, harvest tanks, filtration/microfiltration, ultrafiltration/diafiltration, chromatography columns and skids, in-process material and bulk storage tanks, filling machines, lyophilizers, vial washers and processors, stopper washer and processors, cold rooms and freezer rooms, refrigerators and freezers, warehouse capacity, storage of raw materials and process intermediates, and storage of quarantine and released product.
- Capacity for processing/storage of potable water, purified water, WFI, plant steam, clean steam, CIP, SIP, cooling towers, heat exchangers, gases, etc.
- Capacity or production/storage of critical raw materials such as plasma, serum, hormones, peptides or proteins, affinity-matrix for columns, etc. used in the production.
- Capacity in terms of processing supplies such as autoclave loads, depyrogenation oven loads, vial/stopper processor loads, filters, etc.
- Capacity in terms of cleaning of equipment and turnaround time.
- Capacity for maintaining appropriate manufacturing environment
- Capacity in terms of personnel for working space, production schedule training, document archival, etc. performed in an orderly and a normal way.

A capacity report should be prepared and approved by responsible departments after evaluation of the above listed and any other requirements for capacity. The capacity report should clearly demonstrate the plant capacity in terms of product produced per day/week/month/year based on the capacity of each equipment, process, schedules, and trained personnel.

Access and Security

The manufacturing areas should be accessible only to qualified and authorized personnel. Adequate security procedures should be established to demonstrate that the manufacturing facility is secure and not accessible to unauthorized people. Regulatory agencies have been paying special attention to this issue lately due to recent product counterfeiting incidents in generic drug companies (34).

Building Licensure

A BLA is submitted to the FDA or its equivalent EMEA or other global regulatory agencies, after completion of the above listed and other required activities and documentation, for licensure of the manufacturing facility. A pre-approval inspection (PAI) is often performed by the regulatory agencies to verify that the information submitted to them is accurate and complete. The FDA and other global regulatory agencies have indicated recently that they may make PAI optional, if the previous compliance or system based inspections of the applying company have been satisfactory. Upon successful inspection or verification of the information submitted, a BLA may be approved by the regulatory agencies permitting shipping of the approved product produced from the manufacturing facility to its customers.

SUMMARY

In summary, process validation for a new cell culture derived-product, process, technology, or a new manufacturing facility should be carried out with a lot of careful planning and brainstorming along with a pinch of passion and dedication to details. Often process validation projects end up in tremendous delays and exorbitant costs due to poor planning, unrealistic goals and schedules, and inexperience of the assigned staff. It is a very expensive activity and should be carried out with good planning and caution. Validation activities take time on their own and often cannot be sped up no matter how many resources are poured into it. Albeit challenging and tough, it is an unforgettable journey that more often than not results in a joyful regulatory approval of a new product, process, technology, or a new manufacturing facility. Once taken successfully, one craves for this journey again and again and again.

REFERENCES

1. Guidelines on General Principles of Process Validation. FDA, May 1987.
2. Production and Testing of New Drugs and Biologicals Produced by Recombinant DNA Technology. FDA, April 1985.
3. Production and Testing of New Drugs and Biologicals Produced by Recombinant DNA Technology: Nucleic Acid Characterization and Genetic Stability. FDA, April 1992.
4. Sampling and Testing of In-process Materials and Drug Products, Code of Federal Regulations, Title 21, Part 211.110. FDA, April 1998.
5. Smith A. (ed.). Gene Expression in Recombinant Microorganisms. New York: Marcel Dekker, Inc., 1995.
6. Blanch, HW, Clark DS (eds). Biochemical Engineering. New York: Marcel Dekker, Inc., 1997.
7. Petrides DP, Cooney CL, Evans LB. An introduction to biochemical process design. In: Shuler ML, ed. Chemical Engineering Problems in Biotechnology. Vol. 1. New York: AICHE, 1989:351.
8. Dale J. Successful technology transfer: lessons from experience. BioPharm 2000; 13(10):48–56.
9. Ernst S, Garro OA, Winkler S, Venkataraman G, Langer R, Cooney CL, Sasisekharan R. Process simulation for recombinant protein production: cost estimation and sensitivity analysis for heparinase I expressed in *E. coli.* Biotechnol Bioeng 1997; 53:575–582.
10. Petrides, D. Bioprocess design: personal communication, Intelligen, Inc., 2000: www.intelligen.com.

11. European Commission. Validation Master Plan Design Qualification, Installation and Operational Qualification, Non-Sterile Process Validation, Cleaning Validation. [E-3D (99)]. 30 October 1999.
12. Proposed Validation Standard VS-1. Non-Aseptic Pharmaceutical Processes by Institute of Validation Technology Standards Committee. J Validation Technol 2000; 6(2): 502–520.
13. Lubiniecki AS, Wiebe ME, Builder SE. Process validation for cell culture-derived pharmaceutical proteins. In: Lubiniecki AS, ed. Large-Scale Mammalian Cell Culture Technology. New York: Marcel Dekker, 1990:515–541.
14. Gardner AR, Smith TM. Identification and establishment of operating ranges of critical process variables. In: Sofer G, Zabriskie DW, eds. Biopharmaceutical Process Validation. New York: Marcel Dekker, 2000:61–76.
15. Seely RJ, Hutchins HV, Luscher MP, Sniff KS, Hassler R. Defining critical variable in well-characterized biotechnology processes. BioPharm 1999; 12(4):33–36.
16. Bobrowicz G. The compliance costs of hasty process development. BioPharm 1999; 12(8):35–38.
17. Points to Consider in the Characterization of Cell Lines Used to Produce Biologicals, FDA, 1993.
18. Lao MS. Cell bank validation: comparing points to consider and ICH guidelines. BioPharm 1999; 12(1):48–51.
19. Sofer G, Zabriskie DW (eds). Biopharmaceutical Process Validation. New York: Marcel Dekker, Inc., 2000.
20. Agalloco J. Master planning and validation: Part 1. Anatomy of a validation master plan. J BioPharm 2000; 13(3):40–44.
21. Agalloco J. Master planning and validation: Part 2. System definition. BioPharm 2000; 13(4):34–38.
22. Vincent DW, Revie DR. Technical guide: the validation life cycle of a biotechnology derived drug. J Validation Technol 1999.
23. Global Harmonization Task Force. Final Document-Process Validation Guidance for Medical Device Manufacturers, GHTF/FD-99. Finalized 29 June 1999.
24. Guidance on General Principles of Software Validation. FDA, June 1997.
25. Budihandojo R, Bergeson DJ, D'Eramo P, Huber L, Lopez O, Perez A, Selby D, Wingate G, Wyrick M. The future state of computer validation, part I: increasing the efficiency of computer validation practices. Pharm Technol 2001; 25(7):86–93.
26. Budihandojo R, Bergeson DJ, D'Eramo P, Huber L, Lopez O, Perez A, Selby D, Wingate G, Wyrick M. The future state of computer validation, part ii: increasing the efficiency of computer validation systems. Pharm Technol 2001; 25(9):50–60.
27. Fourman GL. Mullen MV. Determining cleaning validation acceptance limits for pharmaceutical manufacturing operations. Pharm Technol 1993; 17(4):54–60.
28. Zeller AO. Cleaning validation and residue limits: a contribution to current discussions. Pharm Technol 1993; 17(10):70–80.
29. Leblanc CA. Establishing scientifically justified acceptance criteria for cleaning validation of finished drug products. Pharm Technol 1998; 23(10):136–148.
30. Hill L. Performance qualification of CIP system: lessons from experience in multiproduct biopharmaceutical manufacturing. BioPharm 2000; 13(4):24–29.
31. Hwang R. How to establish an effective maintenance program for cleaning validation. Pharm Technol 2000; 24:62–72.
32. Moore RE. Expanding FDA's concepts of change control and reworking/reprocessing as applied to APIs. Pharm Technol 1999; 24(2):54–66.
33. Guidance for Industry: Changes to an Approved Application for Specified Biotechnology and Specified Synthetic Biological Products, FDA, July 1997.
34. Nordenberg T. Inside FDA: barring people from the drug industry. FDA Consumer 2000; 31(2).

16
Facility Design

Kim Nelson
CRB Consulting Engineers, Plymouth Meeting, Pennsylvania, U.S.A.

INTRODUCTION

The last two decades have seen numerous cell culture -based production facilities come on-line as cell culture-based therapeutics have been approved and launched onto the biopharmaceutical market. One might expect that facility design for cell culture-based manufacturing has become a mature discipline during that period, but this is not entirely the case. New market pressures have forced companies and engineers to adapt and bring projects swiftly while attempting to keep the budgets manageable.

There is, at this point in 2003, a production capacity shortfall looming as a result of a combination of a lag in construction of new production facilities and product pipelines being full and a great many products entering or in late-phase clinical trials. Companies are planning, designing, and constructing in order to prepare for their production requirements. This chapter will discuss issues and paradigms in the design and construction of cell culture production facilities.

There is no single correct way to design, nor operate, a biopharmaceutical manufacturing facility, which is Good Manufacturing Practice (GMP) compliant. Product can be acceptably prepared using a variety of technologies, and at a range of scales. Sometimes the development time and investment required changing technologies and scale-up a process must be balanced against the expediency of suboptimal small-scale operations. An example of this trade-off was that of the launch of Epogen®, a recombinant erythropoietin cell culture-based product which has been (very) successfully produced in roller bottles (1).

GMP AND REGULATORY CONSIDERATIONS

Acceptable GMPs are based upon properly and reproducibly maintaining control over a process. Another aspect of GMPs requires that supporting data be examined before changes to a facility or process are implemented. While product quality is not negotiable, the methods and systems used to achieve a predetermined product specification are open to interpretations. Beyond minimal GMP compliance, manufacturers may choose to make discretionary investments in capital and operating practices for business reasons. These decisions are typically ones of economic tradeoff and do not compromise product quality.

Current Good Manufacturing Practices Compliance

Regulatory agencies require GMPs be followed for any drug product going into humans, including clinical trial materials. For example, the Food and Drug Administration's (FDA) "Guideline on the Preparation of Investigational New Drug Products" (2) states:

> When drug development reaches the stage where the drug products are produced for clinical trials in humans or animals, then compliance with GMP regulations is required.
>
> FDA, while recognizing the differences between the manufacture of investigational products and commercial products, believes that is nonetheless vital that investigational products be made in conformance with current good manufacturing practices.

While the EU's Annex 13 "Manufacture of investigational medicinal Products" states:

> Manufacture of investigational medicinal products should comply with the basic GMP for medicinal products, with applicable annexes, other relevant guidelines published by the European Commission (e.g., validation of virus activation/removal) and the guidelines in this annex. (3)

An early issue to be resolved before a facility design can proceed is the regulatory basis for the project. It is important to identify which countries a product will be licensed in, and the corresponding regulatory agencies, which must be satisfied. Mutual Recognition Treaties are improving international drug manufacture and distribution between some countries, but care must be taken to examine these issues while in the planning and design stages.

An emphasis in manufacturing steps should be to control bioburden, minimizing the introduction or growth of contaminants to acceptably low level such that they do not compromise the safety and efficacy of the product. Contamination control is achieved through a combination of operating practices and design approaches for the process and facilities. Cell culture processes inherently are produced in an environment where a contamination, such as a bacteria or adventitious agents, can get established and grow much more rapidly than the cell culture itself. Viable organisms introduced into the process can compromise the process, or worse, adversely affect the molecule that is ultimately being produced, or produce contaminants which may not be removed during the normal purification process. Depending on the stage of manufacture, open processes may likely be utilized, exposing the product to potential contamination. In such cases these open processes should be carried out in controlled room environments, protected by a combination of air quality control, personnel flow controls, material flow controls, room pressurization, etc.

Closed processes reduce the risk of contamination. An argument can be made that a truly closed process could be carried out in an uncontrolled environment—the so-called "parking-lot" case. Unfortunately, virtually all closed processes are opened at some point, such as for maintenance, cleaning, or the introduction of a raw material. They do as a result become susceptible to contamination. To maintain integrity, a closed process must either be accessed in a controlled manner (i.e., from a classified environment), or have a validated process for cleaning the system after opening and re-closure. The recognition of closed process systems is an important subject and is a central theme in current manufacturing facility designs for cell culture processes.

ICH Harmonization

The regulatory agencies of the United States, Japan and the EU and representatives of the pharmaceutical industries in these countries, have been working on harmonization of GMPs and specifications and have reached a consensus document for a GMP harmonization framework. The six ICH sponsors are the European Commission, the European Federation of Pharmaceutical Industries Associations, the Japanese Ministry of Health and Welfare, the Japanese Pharmaceutical Manufacturers Association, US FDA, and the Pharmaceutical Research and Manufacturers of America. Established in 1990, The International Conference on Harmonization of Technical Requirements for Registration of Pharmaceuticals for Human Use (ICH) has the goal of harmonizing across all three regions, the requirements for data submitted to support safety, efficacy and quality determinations in new drug applications, and to develop guidelines for the industry based on the harmonized requirements. The objective is to harmonize the requirements and registration of new medicinal products thus improving the efficiency of the development and licensure while not compromising each country's regulatory obligations for safety and effectiveness.

To achieve this, one of the goals of harmonization is to identify and then reduce differences in technical requirements for medical product development among regulatory agencies. The purpose is to achieve greater harmonization in the interpretation and application of technical guidelines and requirements for product registration in order to reduce or obviate the need to duplicate the testing carried out during the research and development of new medicines. The objective of such harmonization is a more economical use of human, animal and material resources, and the elimination of unnecessary delay in the global development and availability of new medicines while maintaining safeguards on quality, safety and efficacy, and each country's regulatory obligations to protect public health.

Since its inception in 1990, a series of harmonization documents have been issued and adopted by the participating countries. The first documents concerned such things as excipients in oral solid dosage drug, and more recently Q7A [REF] was issued for Good Manufacturing Practice Guidance for Active Pharmaceutical Ingredients. In the United States FDA published this as a Guidance document—"Guidance for Industry—Q7A Good Manufacturing Practice Guidance for Active Pharmaceutical Ingredients (August 2001)" (4). This document applies to bulk pharmaceutical chemicals as well as biological and fermentation products.

Facility and Process Changes Requiring Notification to the FDA

There was a time when the process, facility, and site for biologics manufacturing, once licensed, were almost never changed because of concerns that the FDA may have required repeat clinical trials, or that the changes may not have been approved. This is sometimes called the "product = process = facility" paradigm. This paradigm of product, process and facilities be inextricably linked arose from the early biologies processes, where the processes were sometimes more of an art, and fundamental knowledge was sometimes lacking. This situation has changed, in part due to the reproducibility of recombinant processes and advances in analytical technology. The FDA now approves changes in a process or facility, and in some cases quite significant changes.

Postapproval changes to a product's production process, equipment, facility, QC testing, or responsible personnel require notification to the FDA, and amendments to the product license under 21 CFR §601.12. FDA issued a Guidance document in 1995 regarding changes to be reported for product and establishment licenses (5). These notifications and supplements are termed a Prior Approval Supplement (PAS). The supplement is a request to the Director of CBER to approve a change to an approved license application. These postapproval changes fall into three broad categories:

> Category III Change under §601.12 (b) Category III changes are considered significant changes which could impact product quality, and require submission of a supplement and approval by the FDA prior to distribution of any product made using the changes. These are changes requiring submission of a supplement and approval by FDA before a product made using the change is distributed. These changes may be to the product, production process, quality controls, equipment, facilities, or responsible personnel that have a substantial potential to have an adverse effect on the product's identity, strength, quality, purity, or potency (as they may relate to the safety or effectiveness of the product). A few examples of some of these process and facility changes requiring supplement include but are not limited to such things as: scale-up requiring a larger fermentor, bioreactor, and/or purification equipment; extension of culture growth time leading to significant increase in number of cell doublings beyond validated parameters; new or revised recovery procedures; new or revised purification process, including a change in a column; changes in the sequence of processing steps or addition, deletion, or substitution of a process step; changes in the chemistry or formulation of solutions used in processing; conversion of production and related areas from single to multiproduct manufacturing.
>
> Category II Changes under §601.12 (c) These are changes requiring supplement submission at least 30 days prior to distribution of the product made using the change. A few examples of some of these process and facility changes requiring a supplement include, but are not limited to, such things as: automation of one or more process steps without a change in process methodology; addition of a duplicated process train or individual unit process, such as a fermentation process or duplicated purification columns, with no change in process parameters; addition or reduction in the number of pieces of equipment (e.g., centrifuges, filtration devices, blending vessels, columns, etc.) to achieve a change in purification scale not associated with a process change; modification of an approved manufacturing facility that is not likely to have an adverse effect on safety, sterility assurance, purity, or potency of product, such as adding new interior partitions or walls to increase control over the environment; automation of process control for steps to replace manual process control; downgrading of room or area environmental quality classification except for aseptic processing areas; manufacture of an additional product in a previously approved multiproduct manufacturing facility using the same equipment and/or personnel, provided there have been no changes to the approved and validated cleaning and changeover procedures and there are no additional containment requirements.
>
> Category I Changes under §601.12 (d) These are changes that can be described in an annual report (minor changes). A few examples of minor changes which need to be described in an annual report under Category I Changes §601.12(d) include,

> but are not limited to such things as: change in harvesting or pooling procedures provided they do not affect the method of manufacture, recovery, storage conditions, production scale, or sensitivity of detection of adventitious agents; changes in cell inoculum preparation; such as the mode of expansion (attached vs. suspension; bioreactor vs. spinner), the cell density at inoculum, or the staging of cultures (excluding viral products); replacement of equipment with like equipment of an identical design and operating principle and involving no change in the process parameters; addition of like equipment, which is identical to the primary system and serves as an alternate resource for the manufacturing processes within an approved production room or area; room upgrades, such as installation of improved finishes on floors/walls; upgrade to area HVAC cleanliness; upgrades to material, or personnel flows where the product specifications remain unchanged and the changes do not involve changes in equipment or physical structure of production rooms; relocation or rearrangement of equipment within an approved operating process room with no change in room air classification; change in the simple floor plan that does not affect production process or contamination precautions.

Comparability Protocols Under §601.12(e)

Prior Approval Supplement for a revised process or for a new site require Comparability Testing showing the equivalence of products produced by the revised process and that process approved by Agency in the original license. Some in the pharmaceutical industry have contrasted comparability with equivalence. It is contended that comparability is a tool, based on the innovator's long experience and analytical methods, which the innovator may use to modify their manufacturing process or facility, while equivalence is merely a marketing authorization recognition of therapeutic equivalence. While this is at its core, a regulatory agency decision, it may, in view of recent immunogenicity issues arising in biologies be relevant to the issues of biogenerics.

A comparability protocol, described in §601.12(e), establishes the tests to be done and acceptable limits to be achieved in order to demonstrate the lack of adverse effect on the safety and effectiveness of a product for specified types of manufacturing changes.

ISPE Baseline® Guides

Over the past decade, the cost of biopharmaceutical facilities has racketed up significantly, partially as a result of foverdesign. The International Society of Pharmaceutical Engineers (ISPE), in partnership with US FDA and industry representatives, is preparing a series of 10 guides for industry. This series of guides, called the Baseline® Pharmaceutical Engineering Guides, approach facility design by identifying features of the facility which could affect the ability to reliably and consistently produce quality pharmaceutical products or medical devices. The ISPE Guides seek to provide understanding and guides which may help avoid unnecessary spending on facility features that do not contribute to a consistent production of quality products.

Facility features are prioritized to focus attention upon those features, which are most critical to the production of safe and reliable products. Investment capital and manpower can then be allocated most productively through intelligent selection of simple and effective systems. The objective of the Baseline® Guides is to provide engineers and owners in the pharmaceutical industry with consistent guidance on the design, construction, and commissioning of new and renovated facilities, as well as information on equipment and systems to achieve regulatory acceptance.

It is important to understand that the Baseline® Guides are not regulatory documents. The intent of the Guides is to prioritize facility design features based upon the impact on product and process. The Guides are not intended to be used as standards or detailed design guides, and as they are not an FDA document they obviously are not intended to replace governing laws or regulations applying to facilities.

The Baseline® Guides series are organized into "Vertical" guides which address specific types of facilities, and "Horizontal" guides which address issues common to all facility types. The Guides drafted, issued, or planned thus far are:

Vertical Guides (addressing specific types of facilities):
Bulk Pharmaceutical Chemicals
Oral Solid Dosage Forms
Sterile Manufacturing Facilities
Biopharmaceuticals
R&D Facilities
Oral Liquids and Aerosols
Horizontal Guides (addressing issues applying to all facility types):
Water and Steam Systems
Commissioning and Qualification
Packaging and Warehousing

Because manufacturers and engineering firms have numerous approaches which may be equally satisfactory in facility and system designs, the Baseline® Guides address alternative designs and, where appropriate, advantages and disadvantages of design alternative. Alternatives can then be reviewed and manufacturers may elect to construct simpler, less capital intensive facilities and utilize procedural controls to maintain quality, or may invest more capital to construct highly automated operations and rely less upon operational procedures to assure product quality. This flexibility allows the Guides to be adapted to the needs of each pharmaceutical manufacturer. The most recent Guide, currently in final revisions, and due for release in 2004, is Volume 6: Biopharmaceuticals (6).

Multiproduct Facilities

Prior to a product being approved and proved successful on the market, it is difficult to justify the extremely large investment a dedicated manufacturing plant for a single product requires. For companies without existing manufacturing capacity, it is often strategically desirable to contract manufacture late clinical and production for early market launch. This is due to the long lead times of plant construction and commissioning, and the significant capital costs, as well as the very real risk of any given product failing during clinical trials. Because of these risks, it is desirable, where feasible, to design facilities with the flexibility to operate as multiproduct facilities able to accommodate a range of potential products or processes.

Multiproduct facilities may be operated in a campaigned mode (temporal separation) or in a concurrent manufacturing mode (physical segregation). For concurrent operations, the emphasis is on avoiding cross-contamination through segregation, procedural controls, and cleaning validation. In campaigned operations, the emphasis is primarily on cleaning validation and changeover procedures since there are not other products being manufactured in the area. In both cases, the emphasis is on product by focusing on systems and procedures, which protect against cross-contamination and procedural mix-ups. Critical to any multiproduct operation is cleaning validation and good analytical support. In some cases the cost of dedicated equipment is preferable to the cost and time of cleaning validation and QC release of the shared equipment. An important point is that QC must establish analytical endpoints for cleaning criteria. From an operational standpoint and training standpoint, multiproduct facilities require a higher discipline in operations—such as better training, procedures and controls, and particularly documentation.

Perhaps the biggest issue with multiproduct manufacturing is the perception by company regulatory personnel, and perhaps FDA inspectors, of a perceived risk of contamination. Such perceptions may, in some situations, result in an unnecessary overdesign to prevent hypothetical scenarios. The basis of design for multiproduct facilities must be sound engineering, procedural controls, and personnel training. Belts and suspenders may not provide an actual increase in product protection. For example, a closed sterile bioreactor located in a class 10,000 (ISO 7) does not provide any sterile advantage over one located in a lower classification, such as class 100,000 (ISO 8) or a controlled unclassified environment. Integrity of the system is dependent upon good cleaning, and proper SIP sterilization.

Viral Clearance

FDA has a high concern for viral clearance and inactivation, and the risk of undetected viral contamination of biologic products (7). The European Committee on Proprietary Medicinal Products (CPMP) has noted in their Notes for Guidance on Virus Validation Studies (8) that "...many instances of contamination in the past have occurred with agents whose presence was not known or even suspected at the time of manufacture." This latter fact is the driving force behind regulatory agencies requiring validated viral clearance (removal/inactivation) steps. The table below lists some of incidents involving viral contamination of biologics, which help form a historic perspective on why the FDA is concerned.

Product	When	Contaminant
Yellow fever vaccine	1940s	Avian leucosis virus, hepatitis B virus
Polio vaccine	1950s	SV-40
Human growth hormone	1980s	Creutzfeldt–Jakob agent
Blood-derived products (e.g., factor VIII)	1980s	HIV, HBV, HCV
Human monoclonals	1990s	LCM

Because they are derived from mammals, cell lines could potentially carry viral contaminants, although with proper screening this risk is manageable. An additional risk, which is perhaps more problematic and difficult to manage, is the risk of viral contamination introduced from raw materials or worker exposure. Animal-derived raw materials such as serum, growth factors, etc., are introduced either in cell line development, process development or manufacturing, or perhaps as additives used as stabilizers in formulation. Because of the concerns special testing and controls precautions must be implemented to minimize the risk of the products becoming contaminated and transmitting these viral infections to the recipient patient. The result of these raw material concerns is pressure from the FDA to have a serum-free process, and not to use animal-derived raw materials. In addition, the FDA requires viral clearance validation. The EU requires that when viral inactivation/removal is required in a clinical manufacturing process, that it be no less rigorous than for products authorized for marketing.

The starting point of viral clearance processes is a risk assessment of the process based on the potential for viral presence, load, and drug dose and regimen. These allow determination of the total process clearance factor which must be achieved for an acceptable level of risk of patient exposure. Viral clearance procedures should incorporate multiple viral clearance steps whose methods are orthogonal (that is acting by different mechanisms). The critical assumption is that "orthogonal" methods can be log additive.

Containment and Biosafety

Work with pathogens or work with recombinant DNA molecules is governed in the United States by the Center for Disease Control's monograph *Biosafety in Microbiological and Biomedical Laboratories* (BMBL) (9), and the National Institutes of Health's "Guidelines for Research Involving Recombinant DNA Molecules (NIH Guidelines)" (10). Both the NIH Guidelines and the CDC's BMBL describes combinations of standard and special microbiological practices, safety equipment, and facilities design features, which are recommended for work with a variety of infectious agents or GM organisms in various laboratory or large-scale settings. The BMBL predates the National Institutes of Health's Guidelines for Research Involving Recombinant DNA Molecules (NIH Guidelines) the NIH Guidelines purpose is to specify practices for constructing and handling: (i) recombinant deoxyribonucleic acid (DNA) molecules, and (ii) organisms and viruses containing recombinant DNA molecules.

The facility should be designed appropriate for the organism and product being produced. For multiproduct facilities an analysis of the probable organisms should be made and risk factors evaluated. The most stringent resulting containment level should serve as the basis of design for the facility. Biosafety central dogma is the overlapping use of (1) primary containment equipment (BSC, etc.), (2) secondary containment within a facility, and (3) use of good procedures and techniques (Fig. 1).

Risk assessment of bioprocesses must evaluate the nature of the organism, its expressed product if recombinant, routes of exposure, disease or toxic potential, and an equipment and facility evaluation. The nature of the agent includes whether it is infectious, a toxin, or whether it has pharmacological or immunological activity. Biological agents must be evaluated for disease potential in animals and humans, the potential host range, whether it is a pathogen, and whether it is excreted or cage-mate transmitted. The route of exposure must be considered—whether it is an inhalation, ingestion, ocular, or skin puncture exposure. Equipment and facilities are evaluated for the primary containment design of equipment, the secondary containment design of the facility, the program of activity, personnel qualifications and training, and any medical requirements such as vaccinations or medical surveillance. Facility design issues include the facility layout, HVAC design such as air handling unit zoning, room pressurization, exhaust filtration, and biowaste treatment for both liquid and solid waste.

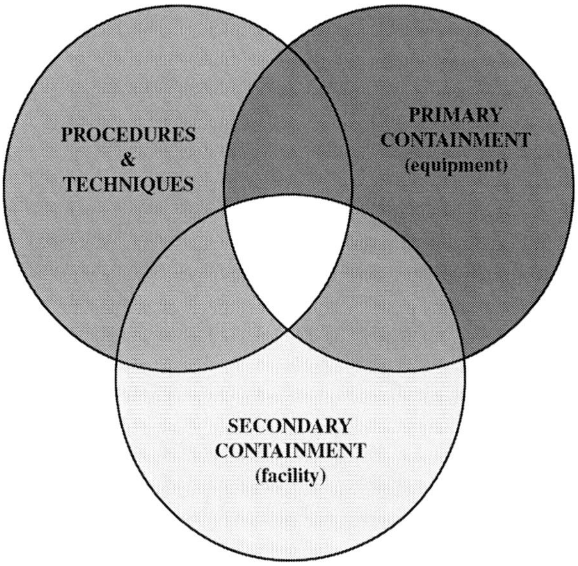

Figure 1 Successful Biocontainment is achieved through the overlap of the central principles of biosafety - the use of primary containment equipment, secondary containment by the facility, and the proper training and procedures used by the personnel.

Risk categories for biological agents, which would include cell culture and cell culture-derived products, are divided into four categories based on the disease or toxic potential, and the availability of vaccines or effective prophylaxis. The four risk categories range from Risk Group 1, where the biological agents are unlikely to cause human disease, to Risk Group 4 Biological Agents, which cause severe human disease and present a serious hazard to workers, and may present a risk of spreading to the community. There are usually no effective prophylaxis or treatments available for Group 4 Biological Agents.

For the most part cell culture-derived products will fall into the lowest levels of risk, and containment levels are generally GLSP, or BL1. The exception to this is vaccine production processes where the virus may present a risk potential.

Where required, a decontamination autoclave should be provided. This is best done in a separate room (Decon Autoclave Room) where items can be staged and unloaded for subsequent disposal. In high containment facilities, these decon autoclaves should be double door autoclaves for deconing items as they are passed out of the containment suite.

PROCESS TECHNOLOGIES AND FUNCTIONAL AREA REQUIREMENTS

Process Overview

Cell culture processes are used in production of a variety of products, including therapeutic enzymes, monoclonal antibodies, cytokines, vaccines, and gene therapy products. For protein products, the processes vary in detail, but are similar in concept. The processes include the steps of: (a) preparing an inoculum, (b) expanding the culture to production volumes, (c) harvesting the conditioned media (batchwise or

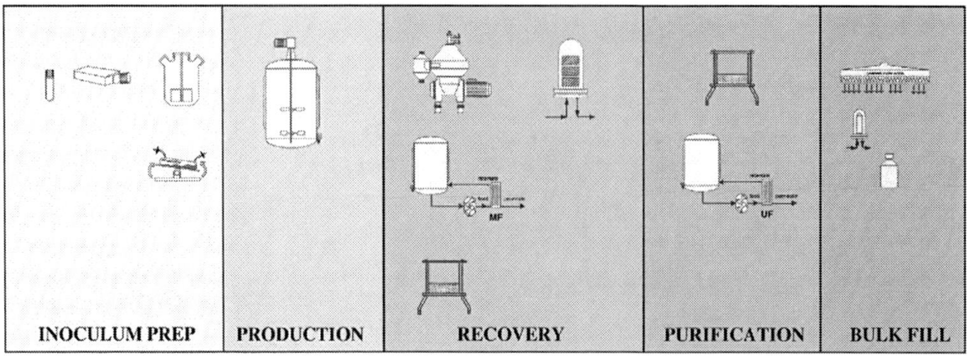

Figure 2 Basic steps in bulk manufacturing of a cell culture based therapeutic. (*See color insert p. 7.*)

perfusion), (d) clarifying the broth by removal of whole cells and cellular debris, (e) product capture and initial purification steps, (f) final purification steps, (g) a viral clearance step, and (h) formulation and bulk filtration (Fig. 2).

Bioreactor Technology

The scale of manufacturing operation depends upon the nature of the product and its potency, the market size, and the process which development has delivered to the manufacturing group. Once a product has been approved based on a given process, changing the process becomes much more difficult and expensive due to the testing and possibility that additional clinical trials may be required to show equivalency. Such expenses and development efforts may or may not be justified.

Available cell culture production technologies used in approved GMP production operations include several bioreactor systems presented in Fig. 3.

Technology	Usage in an approved GMP process
Roller bottles (manual manipulations or automated systems)	X
Hollow fiber systems	X
Cell cubes	X
Wave bioreactors	X
Stirred tank bioreactors	
Batch	X
Perfusion	X
Airlift bioreactors	X
Cell factory (NageneNunc)	X

The production bioreactor room generally includes both the seed bioreactors and the production bioreactors. The majority of products in the pipeline today are monoclonal antibody or related molecules. The bioreactors scale required for

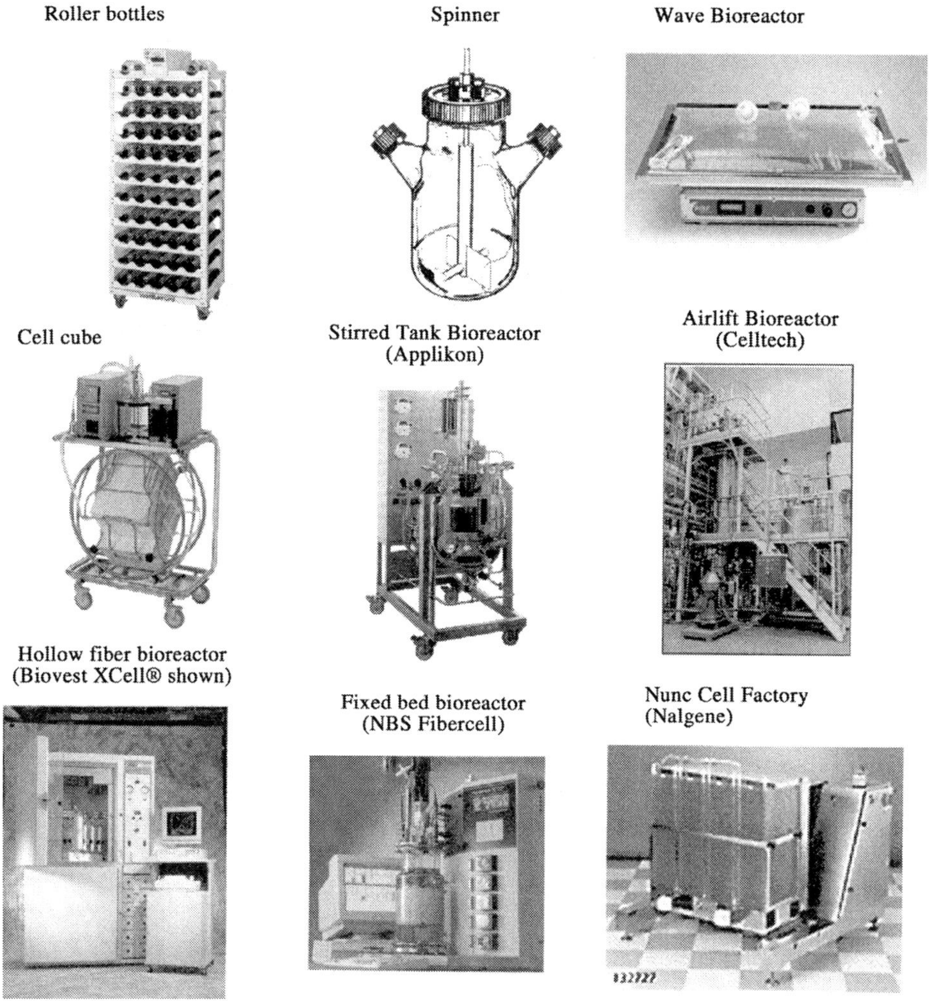

Figure 3 Representative cell culture systems successfully used in GMP licensed production systems. (*See color insert p. 8.*)

production of these types of molecules is typically in the 10,000–20,000-L bioreactor volume. Depending upon the inoculum prep approach the first stage in the seed bioreactor expansion may range from 15 L to as much as 1000 L. Bioreactors at the lower size range may require partial or full disassembly for proper cleaning. Bioreactors of 100 L working volume and above are almost always cleaned in place using a clean-in-place (CIP) system, although there are examples of the bioreactors at the smaller end of the scale (<1000 L) being cleaned out of place and autoclaved.

The number of seed bioreactors required to reach the production reactor depends upon the ultimate scale, and the split ratios. The initial seed bioreactor is inoculated using flexible sanitary tubing. Bioreactor to bioreactor transfers are done using air pressure to avoid pumps and pump seals. Bioreactor operations may require nutrient feeds, which at the smaller scales may require bench or floor scales, and at the large scales fixed nutrient feed tanks. Harvesting operations are based on specific criteria, usually culture age, at which point the culture is sent to harvesting and recovery operations.

Inoculum Preparation

The early steps in the inoculum build-up takes place in an Inoculum Prep Room. The inoculum operations include open manipulations and present the greatest risk of contamination. As a result the Inoculum prep room is segregated and generally classified as a ISO 7 class (Class 10,000) area. Vials of frozen working seed stock are taken in as required, thawed and used to initiate T-Flask or shaker cultures. These cultures are expanded into any number of culture systems including spinners, roller bottles, wave reactors, or stirred tank bioreactors. There are no limitations on the technology which may be used, other than it must be validated.

Small-scale transfers are usually done in Class II biosafety cabinets which provide product protection and worker protection, although for low biosafety level work, laminar flow hoods are also used. For inoculum operations, flask, roller bottle, and spinner cell cultures are incubated in CO_2 incubators. The culture is progressively expanded into larger vessels, ultimately going into a bioreactor (seed or production) using aseptic transfer. After the culture has reached the target transfer density (generally mid log phase growth), the next larger fermentor is inoculated.

A separate room, outside of the production suite, is often provided in cell culture manufacturing buildings to prepare master and working seeds. Because of the critical nature of preparing the cell bank, the HVAC classification is a Class 10,000 (ISO 7) provided with a local Class 100 biosafety hoods. The room is generally equipped similar to an inoculum prep suite with Biosafety Cabinent, incubator(s), centrifuge, refrigerator/freezer, and in addition, with a controlled rate freezer and instrumentation. A small $-70°C$ freezer may also be provided if a liquid nitrogen dewer is not available when the vials are withdrawn from the controlled rate freezer. Some companies do not use controlled rate freezers, although this generally increase the viability at recovery.

Following initial clearance by QC, a vial of the master cell bank (MCB) will be subcultured and a working cell bank frozen down. The working cell bank (WCB) is defined as a quantity of cells derived from one or more ampules of the MCB, stored cryogenically and used to initiate the production batch. The master cell banks are prepared prior to the initiation of clinical manufacturing process, and a portion of the master cell bank will be transferred to the manufacturing site. The working cell bank are to be prepared for the actual production runs. All master and working cell banks are divided and stored in liquid nitrogen dewers located in separate facilities in order to preclude destruction of the entire cell bank in the event of a disaster. Seed stocks from different cell lines are sometimes stored in separate dewers to prevent potential cross-contamination.

Dewers should be equipped with alarms and auto fill valve systems with the liquid nitrogen usually being supplied from a portable storage tank either within the Cell Bank Storage Room or from an adjoining closet.

Scale of Operations

The major impact of process scale on facility design is the resulting facility size (height as well as area) and the utilities required to support the operations. As tank size goes up, the room height increases dramatically since tanks should have a clear-

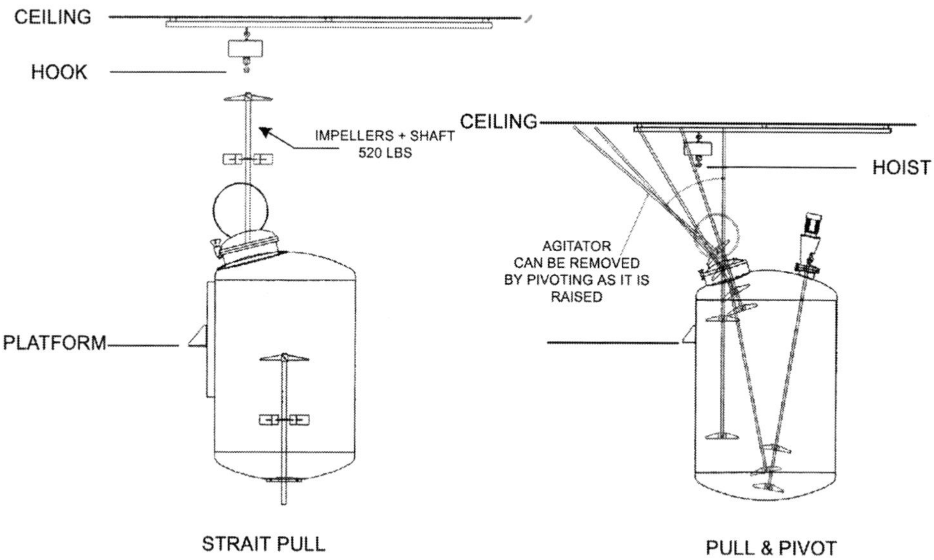

Figure 4 The need to consider pull paths and pull lengths in establishing room heights.

ance above the tank approximately equal to the straight side of the tank. This allowance provides space for a hoist to remove internals such as impeller shafts, as well as provide safety lift points for extracting workers when they are in the vessel. In rooms with limited clearances, long items such as impeller shafts, may be pivoted as they are brought out of the vessel, dramatically decreasing the height which might otherwise be required for a straight pull (Fig. 4).

Recovery

Cell culture-derived products may be soluble, such as most recombinant therapeutics, or viral products including gene therapy vectors. In the case of the soluble products, recovery operations clarify and remove cell debris. Viral production recovery processes remove the heavier cellular debris and proceed with the viral purification.

Cell culture processes may, in some cases, directly load the initial capture chromatography column without removal of cellular debris, but clarification is usually done to remove such debris. Clarification operations may be done using centrifugation, depth filtration, dead-end sterile filtration, or tangential flow microfiltration. For large-volume systems, disk stack continuous flow centrifugation is generally used to remove the bulk of cell debris prior to depth filter followed by an absolute rated membrane filter.

Prior to purification, the harvest may have a volume reduction UF concentration step to improve the processing times of the initial capture column. This eliminates the need to handle large volumes of liquid which might otherwise require large columns or extended numbers of load/elute cycles . The properties of each culture broth and its debris load is unique to the cell line and media components used. For this reason, any clarification filtration steps in a flexible multiproduct facility must be sized to handle a range of filter medias and areas. Depending upon the resin and buffer systems, it may also be necessary to adjust the pH and/or the conductivity prior to the first chromatographic column and sometimes between chromatography steps in order to optimize the column loading. Volume reduction and product concentration are usually combined with diafiltration to adjust the conductivity, and sometimes to remove contaminating small molecules or salts which could adversely affect the subsequent chromatography step. In recent years, the introduction of more robust resins has allowed direct product capture without the need for a prior ultrafiltration or diafiltration step. Direct product capture offers the opportunity for fewer steps and a resulting simpler process, and a shortened total purification cycle time. It eliminates the need for expensive concentration equipment, and in some cases, such as expanded bed chromatography, eliminates the need for prior clarification such as centrifugation, filtration or tangential flow microfiltration.

Purification

The purification suite should be designed for flexibility to accommodate the operation of a variety of purification steps, typically a mixture of chromatography and ultrafiltration.

At manufacturing scale, load flow-through from the chromatography operations are seldom collected into vessels and retained until the chromatography step has been determined to be successful and product not lost. This contrasts to development facilities where this temporary retention may be done until the process is well characterized and robust. The practicality of such a retention of flow-through depends, of course, upon the size of the operation. At manufacturing scales it is not practical, nor should it necessary for a validated process where the performance is reproducible.

Whenever possible, manufacturing processes are best designed for ambient temperature processing unless the stability of the product is demonstrated to require chilled conditions. Chilled processing introduces operational inconveniences and equipment costs. Where necessary, the location of the equipment and approach to temperature control is dependent upon scale. Large volume buffers are stored in jacketed tanks, while small volume systems may have the chromatography system, column and buffer containers all located in a walk-in cold room. In some cases large volume chromatography systems may be located in a cold room in order to avoid insulation of the columns and skid piping and the associated condensate sweating of lines.

Chromatography is the workhorse of the biopharm industry, providing an effective separation step. Chromatography methods include: ion exchange (cation and anion), affinity, hydrophobic interaction, reverse phase, size exclusion (also called gel chromatography), and hydroxyappitite. A prime example of affinity chromatography is Protein A, which preferentially binds monoclonal antibodies, and is often used as the capture step in large-scale antibody production process.

The scale-up of chromatography operations is based upon using the same resin and maintaining similar residence times in the column. It is therefore critical that the pilot plant scale that purification processes be developed with resins which are suitable for commercialization (cost, availability, scalability, and process characteristics). Some resins are constrained by the allowable total pressure drop of the column beyond which they may compress. This translates into upper limits on the linear velocity through the column. Many newer resins have a rigid structures and are rated for substantially higher flow rates and pressures, thus offering improved efficiencies in the separation process by cutting down the cycle time.

Chromatography columns require routine re-packing. This may be done manually, or using a packing skid. Columns larger than approximately 60 or 80 cm are generally packed-in-place using a packing skid. Space must be provided for a resin slurry tank for preparation of column resins, and the storage of the packing equipment. Generally a dedicated packing room is provided in facilities for purification support.

Interspersed in the purification process are ultrafiltration/diafiltration steps to adjust the conductivity or switch buffers for subsequent chromatography steps, or reduce the volumes. Although an alternative to ultrafiltration/diafiltration for buffer exchange is gel filtration, this is not generally used due to the inherent difficulties associated with gel filtration. Diafiltration is also used for the final bulk product formulation in order to exchange the product into the formulation buffer. Diafiltration operations start with an over concentration of the product stream, followed by a buffer wash/exchange, either as a continuous diafiltration wash, or as batchwise washes. Diafiltrations generally use approximately 10 volumes of buffers relative to the over-concentrated product stream volume, with continuous diafiltration wash being more efficient.

Product vessels are usually jacketed to allow chilling in the event that a problem arises and processing must be delayed. These tanks will also require the ability to be warmed back up to the proper processing temperature using a tempered water system.

Viral Clearance

Viral clearance is generally expressed as a log reduction factor. Clearance in multiple orthogonal steps (where viral inactivation or clearance operate on independent principles) are calculated by adding log reduction factors together. Viral clearance is composed of a combination of orthogonal viral burden reduction steps which may be divided broadly into viral partitioning or viral inactivation. Removal, or partitioning, is the physical separation of the virus or viral particles from the therapeutic component. Examples of partition methods include filtration, some chromatography operations, diafiltration, and polyethylene glycol fractionation.

In order to be effective, viral inactivation techniques must destroy at least one viral element essential to replication. Inactivation processes destroys the virus so that the remaining viral fragments lack the structure and components needed to infect an individual receiving the product. Viral inactivation may be achieved by:

- by process buffers
- by specific inactivation steps
  - detergents
  - solvents
  - pH extremes
  - chemicals (β-propiolactone, alcohol)
  - heat
  - radiation
- by partition methods
  - chromatographic separations
  - virus filtration
  - diafiltration
  - precipitation

Heating and solvent detergent treatments are examples of processes for viral inactivation. Heat treatment denatures viral proteins and nucleic acids, rendering viruses incapable of replication. Irradiation processes destroy viral nucleic acids by inducing breaks and linkages. Solvent detergent methods destroy the viral envelope in lipid-enveloped viruses.

Some processes, such as ethanol fractionation (used in blood fractionation, but not recombinant cell culture-derived product purification), both remove and inactivate viruses. There are also a number of new methods under development, such as irradiation, photoinactivation and treatment with a variety of chemicals. Photoinactivation techniques use light-activated dyes which when irradiated, cause the dyes to convert to molecules that can react with DNA or membrane lipoproteins.

All processes which claim to remove virus must be fully validated. Viral validation studies are performed on scaled down laboratory models of the production process, as it is an unacceptable risk to introduce model viruses into a production area in order to validate a viral clearance step. The validation studies should be designed to represent the worst case conditions that could be encountered during full-scale production, i.e., those conditions under which the removal or inactivation of viruses would be expected to be the least effective. Viral clearance validation studies usually take the form of "spiking" experiments in which large amounts of a model virus is added to the feed stream of the manufacturing step being validated. The reduction in the amount of added virus by the manufacturing step in question is then measured. It should be noted that the viral load is far in excess of what would be expected in a "contaminated" product stream.

These validation studies are performed when a new product is under development or when a new step is introduced into a preexisting manufacturing process. In the latter cases of introducing a new step in the purification process, the validation should be designed along with the process so as to optimize for viral clearance. It is very important to consider scale-up issues and the possibility that pilot scale validation studies may not be relevant to the final manufacturing process. Where scale-up involves changes in the manufacturing process which could adversely affect the viral clearance validated at the pilot stage, then the viral clearance would need to be revalidated for the altered process.

Bulk Formulation

The final step of the purification process is the bulk filtration of the final product in preparation for filling or storage. This step should be done under Class 100 laminar flow, such as is provided by a laminar flow booth with soft wall curtains. The laminar flow zone should be surrounded by a Grade B area (equivalent to a class 10,000 active classification). Typically, the bulk formulation will involve adjusting the concentration and exchanging the buffers for the formulation buffer which would also contain stabilizers. This is usually done via UF diafiltration. The product will be overconcentrated and diafiltered against the formulating buffer. The UF system is then emptied, flushed to recover any residual product held up in the UF system and membranes. A sample will be withdrawn and assayed for product concentration, and the batch adjusted to the target concentration by dilution. The batch is then sterile filtered into the bulk container and stored. Smaller volumes are often dispensed into bottles or bags. Freezing of bags is problematic if the required temperatures are too low. Below a certain temperature, the plastic materials used in bags will go through a glass transition, and become fragile and liable to fracture. As freezing is required for large containers, a system has been developed for rapid cryogenic freezing of portable tanks (11). The tanks, once frozen, are typically stored in a walk-in cold room at −20°C or −40°C. Cryogenic freezing systems are available from Integrated Biosystems. Facility design for this type of freezing operation requires additional space for the equipment and back-up units, as well as substantial storage and maintenance space for these very specialized tanks. The rooms where the cryogenic freezing units are located require electrical classification due to the heat transfer fluids used.

Buffer and Media Preparation

Buffer and Media Prep areas are generally segregated operations because of the growth promoting nature of media powders, and the fear that this nutrient rich materials may either allow higher environmental bioburden, or if introduced unknowingly into buffers, allow growth of contaminants which may produce endotoxins. Because of the nature of the solution prep process (adding salts and components into open mixing vessels), Buffer and Media Prep operations are invariably located in classified spaces. In Class 100,000 (ISO 8) HVAC areas classified spaces, Buffers and Media can be prepared OPEN, with proper segregation and dust control.

Weighing and charging of powders should be done in a manner which minimizes dust dispersion into the room. Dust control for additions to the mix vessel can be done using either a dust extraction elephant trunk or hood, or using a closed powder addition system. Preweighed powders for media or buffer make-up are charged into small containers or bags, often in a laminar flow dust extraction hood. There are methods of adding solids to mix vessels in a closed system manner, however there usually still a need to open up the system for minor component additions or pH adjustments. In the industry as a whole, prepared medias or buffers are generally sterilized by membrane filtration. However, there are manufacturers who sterilize their cell culture media by high temperature short time continuous sterilization in order to ensure that there is no viral carryover from raw materials.

The cleaning of media mix tanks and holding tank cleaning must be given special consideration because of the potential of bioburden growth. By contrast, buffer tanks, which hold only simple salt buffers, can often be acceptably cleaned by rinsing with hot purified water. Cleaning validation is necessary, but simple mineral salts are highly soluble in water, and generally rinse clean without the need for caustic or acid washes. These reduced CIP cycles can be important in minimizing water usage and speeding turnaround of the equipment.

A recent innovation in buffer and media preparation is a totally disposable mixing system introduced by HyNetics. The system utilizes presterilized bags provided complete with mixing impellers. The systems provide for powder component additions introduced in a totally closed manner using bags.

Raw Material Weigh/Dispense

Raw material weigh and dispensing operations are either located adjacent to the warehouse operations as a support function to the entire facility, or are provided as ancillary rooms to the area requiring the weigh/dispense, such as Media Prep, or Buffer Prep. Where centralized weigh operations are done, it is generally better to isolate weigh operations for handling growth promoting materials (media) from those for nongrowth promoting materials (buffers). This would usually be done using dedicated weigh booths, although in small facilities this could be handled procedurally with wipe downs when switching from growth promoting to nongrowth promoting weighing operations.

Chronic exposure to many media components may be allergenic, for example, yeast extracts. A review of all raw materials should be done by the firm's Environmental Health and Safety group. Weigh operations with potentially allergenic materials should be done in a dust containment booth. In cases where the materials are hazardous, it may be necessary for operators to wear breathing protection equipment, and in the case of highly hazardous, such operations should be done in isolators.

Dispensing of powdered raw material are best conducted in an dust control weigh/dispense booth. These booths are designed to contain material powders and dusts, thus minimizing operator exposure and room contamination, as well as providing product protection via an HEPA laminar flow air curtain which is circulated down over the weigh operation (and operator) before being collected, filtered, and exhausted. The front of the weigh booth is open, and multiple weigh booths can be placed in the same room without cross-contamination risk (Fig. 5).

The size of the bioreactor operations will dictate the scales provided, but generally a combination of benchtop balances and floor scales are necessary. If operations require large quantities of media, considerations for drum hoist equipment may be necessary. For example, for a 15,000 L batch of cell culture media will require approximately 200–300 kg of powdered media.

Partials from raw material weigh and dispensing operations may be returned to the warehouse or stored in a small in-use chemical storage area adjacent to the weigh operations to streamline the work-flow.

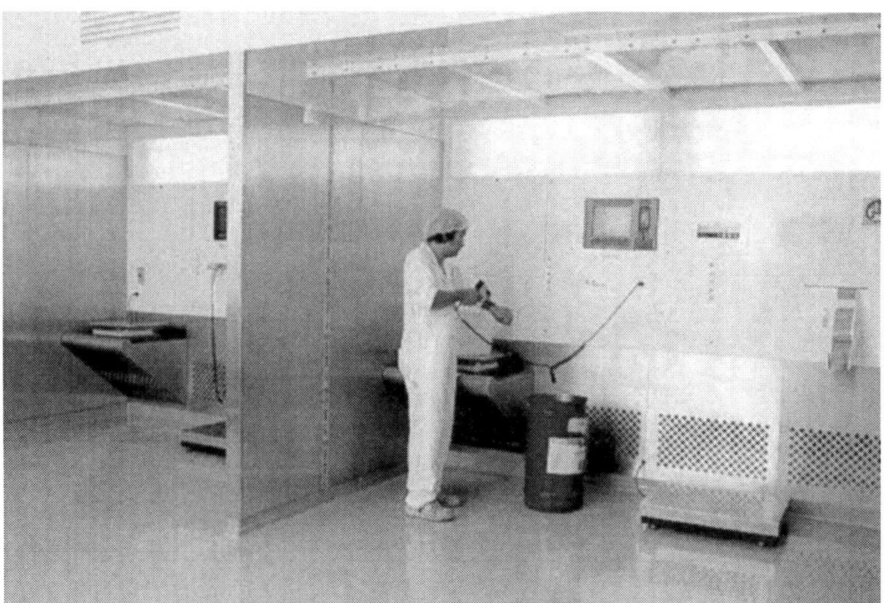

Figure 5 Modular weigh booth providing product protection with HEPA air filtration coverage as well as dust extraction during raw material weighing and dispensing operations. Photo courtesy of Extract Technology.

Warehouse

Receiving and dispatch bays of a warehouse should protect materials and products from the weather during the unloading. This is done by having the loading dock either recessed into the building, or having a exterior awning covering the receiving bay doors. In either case, dock seals should be provided to minimize wind and debris infiltration during loading or unloading. Loading dock areas should be designed and equipped to allow containers of incoming materials to be cleaned, where necessary, before storage. Ideally, pallets used in the warehouse are of plastic or aluminum, and captive to the facility. Incoming supplies are transferred to these pallets upon receipt and the pallets are cleaned before reuse in a pallet CIP booth.

All starting materials must be sampled and tested for identity and conformance to purity specifications. Such sampling is normally done in a separate sampling room. If sampling is performed in the storage area, it should be conducted in such a way as to prevent contamination or cross-contamination. Storage areas should be of sufficient capacity to allow orderly storage of raw materials, intermediate, bulk and finished products, products in quarantine, released.

The storage areas should be designed to ensure good storage conditions. In particular, they must be maintained within acceptable temperature limits and kept clean and dry. Where materials require special storage conditions such as temperature or humidity, these should be provided, checked, and monitored. Sometimes if the volumes justify it, receiving areas may have small walk-in cold rooms or freezer rooms for temporary staging of refrigerated or frozen materials.

Segregated areas should be provided for the storage of rejected, recalled or returned materials or products.

Additional spaces the warehouse will generally require include: Shipping/ Receiving Office; battery charging room for fork lifts; a small waiting room for delivery drivers; a toilet (for delivery drivers so they do not need to go through the building to reach a restroom); a packaging area for outgoing shipments; a chemical storage area (sometimes including an outdoor tank farm for caustic or other chemicals). Some areas, which may be required depending upon the manufacturing process, may include solvent storage areas.

Where quarantine status is ensured by storage in physically segregated areas, these areas must be clearly marked and access restricted to authorized personnel only. Any system replacing the physical quarantine, such as electronic bar coding or automated storage and retrieval systems, should give equivalent security.

Final bulk product must be stored in a secure storage area prior to being sent to filling and finishing. Depending upon the quantity and storage conditions of the product, this may be a caged area with upright -70°C freezers, or walk-in freezers, or refrigerators. For larger quantities of frozen bulk product, cryogenic storage tanks are often used. These tanks offer capacities of 100–300 L, with rapid freezing and thawing will be located in the secured raw materials warehouse area to store finished product before delivery to the client.

Support Functions

Support functions in a cell culture manufacturing facility include a myriad of laboratories and storage spaces. A few of these support functions are listed below:

- In-process Testing
- Qc Testing (Raw Material and Final Product) Laboratory

- Environmental Monitoring Laboratory
- Offices
- Qc Retention Samples
- Metrology (Calibration) Laboratory
- Maintenance Shop for GMP equipment
- Batch Record Storage Archive
- Janitorial Closets and Supplies Storage
- Trash and Waste Disposal Dock
- Equipment Storage Rooms
- Cell Line Quarantine Laboratory
- Data Center
- DCS Control Room (not required unless a DCS system used)

The particular facility may or may not require all of these spaces, depending upon whether some of these functions are provided elsewhere, e.g., the Cell Line Quarantine Laboratory.

A storage location for QC retention samples is necessary. QC retention samples will include raw material samples at 4°C and −70°C. QC sample retention areas generally have a bank of refrigerators and freezers each assigned to separate processes rather than shared cold boxes for all the processes.

Office areas are always necessary for support of the operations. Unless these are available in adjoining buildings, a manufacturing facility must include these suitable quantities to allow the proper supervision and batch record preparation for the manufacturing process.

Maintenance operations for a biopharmaceutical manufacturing plant require some level of mechanical shop support. It is recommended to have a separate Maintenance Shop for GMP process equipment from the building mechanical systems. This is because it is best not to utilize tools used for nonstainless steel when working with stainless steel. Any carry-over of iron filings will result in localized rust. Spare parts may also be stored in this area, as well as storage of calibration equipment.

A GMP documentation store room is required to house documentation on all equipment validation, air monitoring, facility operation, and all batch records for processing. This documentation area is constructed as a fire-rated book vault.

Separate storage room for janitorial supplies should be provided. Janitorial Closets with mop basins are distributed throughout the facility. If a facility uses clean/return corridor systems, there should be Janitorial Closets on both the Clean and Return corridors. In high containment facilities, there is often a JC within the containment suite in order to minimize the chances of a breach of containment.

There should be a separation of raw materials and waste or trash. At the loading docks, this separation is achieved by a separate trash room which is the loading point for the facility trash compactor, and if necessary a dock with recycle dumpsters. Trash should not be taken out through the GMP warehouse.

Storage rooms in manufacturing facilities are the most visible and vulnerable rooms subject to cuts during the design stage. This is unfortunate, since these storage rooms are some of the least expensive in the facility and aid substantially in organizing a GMP facility and keeping un-needed equipment out of operating areas. For example, cutting out an equipment storage room of 600 sq. ft (20′ × 30′) at an approximate cost of \$250–\$350/sq. ft may reduce the project budget by \$150,000–\$210,000 but as a result make the operations areas and corridors into pseudo store rooms thus compromising GMPs. Providing adequate storage space in manufacturing buildings, and within, the clean manufacturing envelope is extremely important from an overall GMP standpoint.

Smaller skids and other portable equipment not in use for a particular campaign should be removed from the manufacturing suites and housed in store rooms. When equipment is returned to service, it should be CIP'd and cleaned externally by a wiped-down as necessary.

Cell lines should be isolated for testing and clearance for the presence of adventitious agents, such as bacteria, fungi, or viruses, prior to being allowed into the manufacturing area. A Cell Line Quarantine Laboratory is often provided to isolate incoming cell lines destined for manufacturing in the GMP facility. Such a Cell Line Quarantine Laboratory is usually quite small and outside of the manufacturing zone. It would require airlocks and appropriate HVAC classification.

Some organizations have intermediate hold steps with a required QC release prior to further processing. Work in Progress (WIP) storage is required for these processes while in-process QC testing is completed. If WIP storage quantities or storage durations are substantial, dedicated space is recommended for intermediate storage of these process product intermediates. Storage conditions can range from refrigerated walk-in cold rooms (2–8°C) to frozen material in upright freezers, or walk-in −20°C freezers to −70°C freezers. The facility design must make assumptions and estimate the quantity and type of storage units. While the locations may be within the clean manufacturing areas, or out in unclassified spaces depending upon the frequency of access required, it is more convenient from a work flow standpoint to have it within the clean envelope and near to the process area.

Process Simulation

Process simulators are software tools that enable the representation and analysis of integrated processes. The use of simulator dates back to the since the early 1960s in the petrochemical industries and the mid 1980s in the biotech industry. Established simulators for the petrochemical industries have generally been found to be unacceptable for application to the biotech industry, primarily due to the batch-wise manufacturing approach. Biopharm process simulators are generally designed to handle batch processes and the ability to model the unit operations that are specific to bioprocessing. Additionally, it should provide process equipment timelines (Gantt charts) for equipment utilization scheduling. Depending upon the simulation approach, detailed mass balances may not be as important as the equipment utilization timelines and water utilization timelines. Simulating the mass balances provides valuable information, but is much more involved and can be generated simply in spreadsheets for warehouse sizing. Depending upon the process it may be important to estimate the waste discharge, including flowrates and compositions, although in simpler situations this can also be estimated using spreadsheets.

Several simulators used in the biopharmaceutical industry include SuperPro Designer (from Intelligen), and BioProcess Simulators (BPS) (from Aspen Technology, Inc.) and Extend (from Imagine That). SuperPro handles material and energy balances, equipment sizing and costing, economic evaluation, environmental impact assessment, process timeline scheduling, and debottlenecking of batch and continuous processes.

The use of process simulators has gained acceptance in the biopharmaceutical industry as an important tool in process optimization, debottlenecking, and utility demand estimates. The specific tasks handled by these process simulators depend upon the questions being asked, but might include process timeline scheduling, utility system sizing, throughput analysis and debottlenecking and equipment sizing, material and energy balances, as well as cost analysis, and in some cases even environmental impact assessment, etc. Process simulation tools can be used throughout the life cycle of a product's development and commercialization.

Frequently, the total throughput of a batch plant is limited (bottlenecked) by an undersized unit operation, equipment availability, or utility limitations. Simulators provide tools to aid in debottlenecking processes in order to optimize a throughput or equipment utilizations. If the goal is to increase plant throughput, changes that increase the batch size and/or reduce the effective batch time (plant cycle time) should be made. First the simulation model batch size is increased until at least one cyclical step operates at 100% capacity utilization. If its uptime is low, then one increases its number of cycles per batch. This may create opportunities for additional increases in batch size. A side benefit of increased batch size is the reduced cost for quality control (QC) and quality assurance (QA), which depends on the number and not the size of batches. Once the process operates at its maximum batch size, attention is then focused on elimination of time bottlenecks. Long process steps and equipment sharing are the causes of time bottlenecks. Equipment sharing causes time bottlenecks when an operation must wait until the shared equipment item is available. Equipment sharing time bottlenecks are eliminated by installing extra equipment that reduces sharing. If the time bottleneck is caused by a step that has a very long cycle time, additional new equipment should installed and operated in a staggered mode based on the cycle time of the next time bottleneck. Optimization of a processes using a simulator must be careful to select robust operating range solutions. Figure 6 shows an example of this, the relative stagger between unit operations can result in high shifts in water for injection (WFI) demand. It is necessary to avoid an unstable operating points where, for example, the delay of an operation could inadvertently result in a cascading of problems excessive demands, or unavailable equipment.

DESIGN AND OPERATIONAL PARADIGMS

Facility design features need to be prioritized in order to focus attention upon those which are most critical to the reliable production of safe products. Prioritizing such facility features allows capital budget and manpower to be intelligently allocated through the appropriate selection of simple and effective systems. Discretionary spending can be made to incrementally provide better features, but such spending is above that which is necessary to satisfy basic GMPs. Some of the design paradigms which have an influence (sometimes great) on the design approach and ultimately the facility operations include but are certainly not limited to:

- Bioburden control vs. aseptic processing
- Segregation vs. closed systems
- Transfer lines vs. portable vessels
- Classified vs. controlled unclassified spaces
- Transfer panels vs. valve arrays
- Stainless steel tanks vs. bags

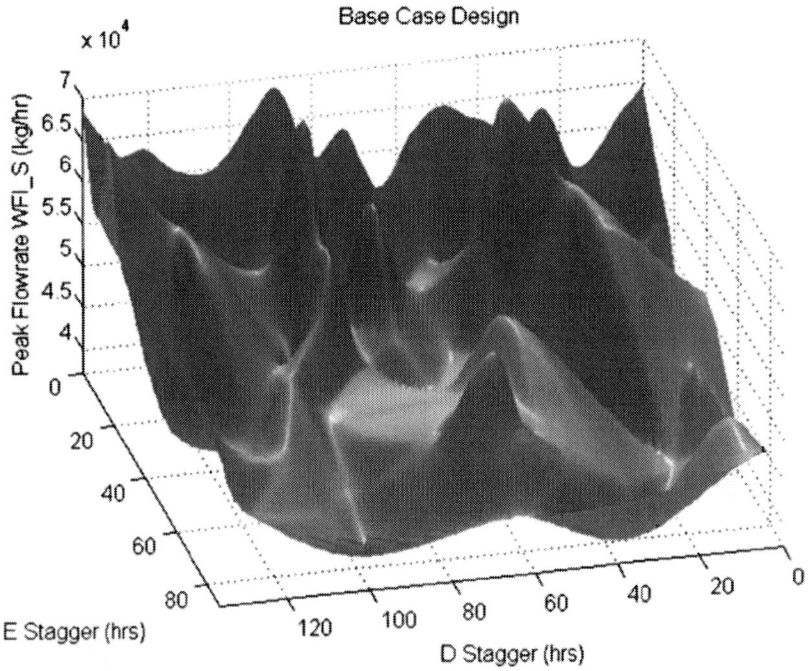

Figure 6 Optimization of processes must be careful to select robust operating range solutions. This is necessary to avoid an unstable points where, for example, the delay of an operation could inadvertently result in a cascading of problems excessive demands, or unavailable equipment. (Response surface plot courtesy of CRB Consulting Engineers, Inc.) (*See color insert p. 8.*)

- Clean/dirty corridors system vs. mono-corridor
- Single product vs. multiproduct
- Stick-built vs. modular construction
- Mechanical chases and interstitial spaces

Bioburden Control vs. Aseptic Processes

Bioburden control and endotoxin control are key concepts which impact process and facility design, particularly when they are not clearly distinguished from sterile or aseptic processing. Most bulk biopharm products are not produced via aseptic or sterile processes, but do control bioburden in the processing by a combination of equipment specification, piping design, and procedures. Some aspects of aseptic design or operational procedures may be employed by but an aseptic process is not a requirement. The focus in design is on protecting the product and controlling bioburden through appropriate facility design and operation, while better optimizing life cycle cost of the facility.

The issue is establishing the appropriate level of control. Product manufacturers must establish in-process bioburden limits. Sterilizing every vessel does not necessarily provide a "better" product. For example, in some ultrafiltration operations the membranes and piping systems are sanitized only (e.g., using dilute caustic). A recirculation reservoir used by such a system does not necessarily need to be SIP'd, provided parameters are set for the cleaning and allowable times prior to use. Aseptic processing applies to parenteral products being filled into final dosage containers.

Segregation vs. Closed Processes

Segregation is a traditional concept in biopharmaceutical production operations and includes segregation of, and between, the following:

- Organisms or products
- Processing steps
- Materials at different stages of quality control release
- Components or equipment at different stages of cleanliness, etc.

Segregation is a central paradigm in process and facility design which ensures product protection in the biopharmaceutical operations. Segregation can be accomplished by space (physical), by time (temporal), by environmental controls (e.g., biosafety cabinet), or by process design (system closure). Procedural approaches use one or more of these methods to achieve segregation.

Closed processes greatly reduce risks to the product from both the facility and from operator interventions. For a validated closed process which protects the product, the facility becomes of secondary importance to GMP, and it is possible to focus on the process rather than the facility. Manufacturing areas need not be classified where there is no product exposure to the room environment, and the facility may begin to look similar to facilities for bulk pharmaceutical chemicals or oral solid dosage products. Indeed, the International Conference on Harmonization went so far as to state in the ICH Q7A (4):

> Where the equipment itself (e.g., closed or contained systems) provides adequate protection of the material, such equipment can be located outdoors. (ICHQ7A, August 2001)

This document applies to biological and fermentation products as well. Logically, multiple products may be handled in the same area if processes are closed. While no manufacturer has implemented this approach to its logical extension, it does clearly imply the agencies attitudes toward closed processes and unclassified spaces have changed.

It is always good engineering practice to keep the process stream as contained as is practical. This approach reduces risk of product contamination from the environment, as well as protects workers from product exposure where potent compounds are involved. It should be noted that each process must be individually considered, not all process or products are appropriate for a fully closed system approach. For example, clinical trial manufacturing processes are more difficult to operate in a fully closed manner, in part due to the size of the equipments, and the fact that the process may not be fully defined (or locked) and a degree of flexibility in the operations is required. The selection of the appropriate HVAC classification for a particular process step must take into consideration not only the degree of system "closure," but also operational considerations. Risk management is an important business decision. The use of a classified environment around a particular step may be justified if it allows recovery from a system breach or process disturbance.

Closed systems allow a reduction in the facility cost, which allows spending moneys on the process and equipment rather than on architectural finishes and higher HVAC classifications. The impact trickles down to gowning, environmental monitoring, and cleaning/housekeeping requirements which correspond with the area classifications. The area obviously should be kept clean and be easily maintained, and process piping can be exposed providing convenient access for maintenance.

Stainless Steel Tanks vs. Disposable Bags

One of the major cost factors in a biopharmaceutical plant is the costs associated with tanks, stainless process piping-associated instrumentation and CIP systems. Sterile bags have emerged as a proven alternative to fixed stainless steel tanks in the past decade.

Stainless steel tanks, whether jacketed or unjacketed, represent the traditional (and proven) technology used by the biopharmaceutical industry at medium and large scales. The cleaning, sterilizing, and filling of fixed tanks is a proven technology. System cleaning can be automated to minimize the amount of operator time needed. Tank rooms are at ambient temperature with jacketing for vessel cool-down following SIP, and for any chilled holding of solutions. Tank dimensions can be customized to optimize floor space in the hold room and operational procedures. There are also no limitations to the volume of the tanks other than the room within which they are to be placed.

On the downside, fixed stainless steel tanks represent the most expensive first cost. They are mechanically complex with many valves, fittings and instrumentation, all of which present potential sites of failure or contamination. As a result tank and transfer line systems have the highest odds of sterility failure, and require careful design, installation, and validation. If sterility failures do occur, their causes may be difficult to trace, since both mechanical failure and operator error will need to be considered as the potential causes. Performing a pressure leak test of the vessel prior to steaming is a critical part of the SIP protocol. These tests will find most leaks attributable to gaskets or mechanical fittings.

Automated procedures are more difficult to troubleshoot if a problem occurs. In addition, in the normal startup situation, schedule is everything, and each automated system will require extensive validation. Where the complexity and scale are appropriate lower levels of automation may be favored for achieving the overall project schedule till "product out the door" is achieved. Because of the extensive piping, and automation around most process tanks, more maintenance is needed relative to options using disposable bags. Additional operating costs will also be incurred for instrument calibration and system maintenance.

The use of bags in fixed or portable drums or totes represent a proven and preferred method of small and intermediate volume media storage for many biopharmaceutical firms (Fig. 7). The initial equipment costs are the lowest of the various options and the use of bags is flexible. The bag technology is proven, and sterility problems may be reduced.

Validation is simplified and can be performed quickly for a fast start-up. Changing processes may have more or larger volume solution containers without facility modifications. Sterile connections are made using simple tubing welder techniques—an operation which is quick compared to steam sterilizing connections.

The main drawback this method of media hold is that where the bins need to be moved from an ambient hold area to a cold room for storage, the volume is limited by reasonable or practical volume to be moved by pallet jacks or pushed by an operator. This is dependent upon the frequency of such movement. Storage requires a large cold room which increases the initial costs. Another drawback is that extra floor space is needed in order to increase the hold capacity although bins can be stacked. Handling and disposal of large bags must also be considered, particularly if these have any biological contact which would require disposal as a medical waste.

Cost comparison studies indicate potential cost savings in favor of single-use system when compared with reusable (i.e., stainless steel tank) systems (12). This

Figure 7 Stainless steel pallet tank used for holding media and buffer bags (photos courtesy of Stedim). (*See color insert p. 9.*)

is due to the savings in first cost relative to fixed tankage, and savings in CIP and SIP relative to stainless steel systems. There are also savings in production schedule due to the quick changes and prep time to be ready to fill a container—thus saving labor and improving productivity.

Filling sterile bags in portable drums or bins is currently a proven technology and failures in sterility are rare. Where failures do occur, the problem is traced quickly through visual observation to a bag with a hole, a leaky tubing or fitting, or bad tubing weld. The bags themselves, along with any sterile filters for filling, are gamma radiation sterilized, and have very long shelf lives. Filter failures are detected through postuse integrity testing. Sterile connections are made using sterile tubing connector devices available from several companies. Finally, validation of the hold bags is relatively simple and quick compared to a large sterile hold tank with the associated piping.

In considering which approach is most appropriate for a particular application, the single-most important factor one needs to take into account is process scale and required solution volumes. Bags are currently limited to approximately 1500–2500 L, with larger bags up to 5000 and 10,000 L in development. While multiple bags could be used for larger volumes this becomes undesirable if overdone. The second factor which drives the economic decision is the frequency of bag changes, and whether a bag can be used multiple times.

Bags can be customized. Many bag manufacturers will design and customize the bags to meet the customer's specific process requirements. In addition to holding bags, tank liners are available for small volume mixing applications. Again, once used the liner is disposed of and a fresh one inserted. The containers for such tank liners or bags are generally inexpensive and readily available polypropylene. For larger bags, bins or totes are used. Such totes are approximately the size of a pallet and vary in height. As an example, a 1500-L pallet tank is approximately 7 ft tall. Materials of construction are usually stainless steel for larger pallet tanks, and polypropylene drums for smaller bags.

Fixed bins for bags offer a combination of advantages of the portable bins and fixed tanks. They use the bag technology and therefore minimize sterility problems and eliminate CIP requirements. Validation of the system is simplified. The fixed bins can offer additional hold capacity over portable bag containers, as well as the opportunity for temperature control using a jacketed tote. Floor space requirements can be optimized with the design of custom bins, including stackable totes.

Steam Sterilized Transfer Lines vs. Flex Lines and Tubing Welders or Aseptic Connectors

Sterile transfers may be made using a steam-sterilized lines with double block and bleed valve assemblies at each end, or where bags and tubing systems are used one may alternately use sterile tubing welder techniques. For smaller tubing, sterile tubing welding/fusing techniques are quick compared to steam sterilizing connections. They are also very convenient for use at small scales in inoculum laboratories and small-scale bioreactors. Tubing welders are available from Terumo and Wave Biotech. The Terumo unit was developed for use in home dialysis and is limited to small bore tubing. The more recent Wave Biotech unit is capable of fusing tubing up to 7/8 in. tubing, and was purpose-designed for industrial use in pharmaceutical manufacturing. Recently, aseptic connectors have been introduced by Pall to provide aseptically join sterile sections of tubing without tubing welds or steam sterilization. For larger lines, there is no alternative to steam-sterilized transfer lines. This approach is reliable, and robust, but requires careful design and validation.

Transfer Panels vs. Valve Arrays

The routing and directing of fluid transfers within a manufacturing plant represents a significant cost in terms of stainless steel piping and transfer panels and valve arrays (Figs. 8 and 9). Transfer systems for use with media, buffers, culture, harvest, or in-process materials may be designed around portable vessels, transfer panels, or dedicated transfer lines. CIP is usually routed using fixed CIP supply and CIP return lines which are piped to the various pieces of equipment and transfer panels. The CIP supply and return systems can be very complicated in large-scale manufacturing. Two fundamental approaches may be taken—transfer panels with U-bends, or fixed valve arrays. There are advantages to both, however, the costs may not be that substantially different when installed. Transfer panels are operator-intensive, and automation-nonintensive, while valve arrays are the reverse. When transfer panels handle dangerous materials or pressurized contents, they should be equipped with proximity switches and be tied into a control system which only allows transfers of CIP

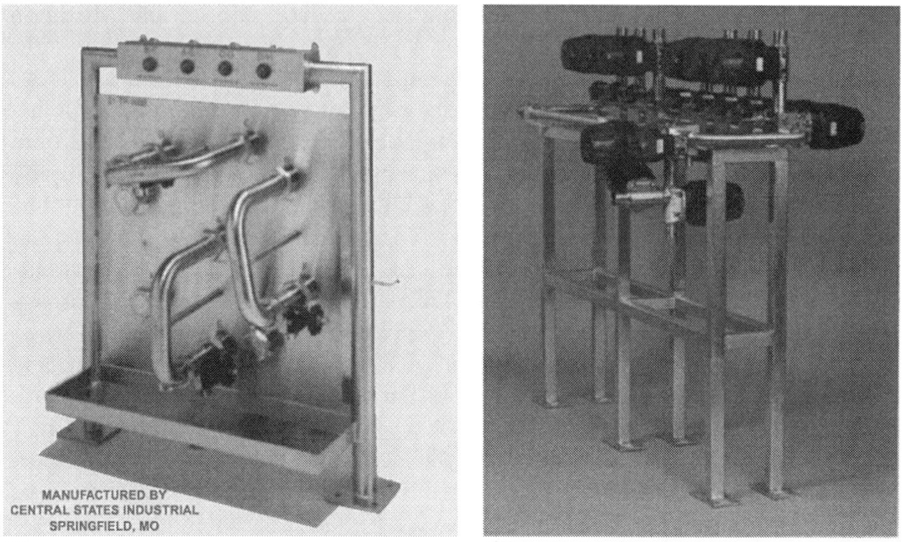

(A) (B)

Figure 8 (A) Transfer panel. (Courtesy of Central States Industrial.) (B) Valve array. (Courtesy of ITT Sherotec.)

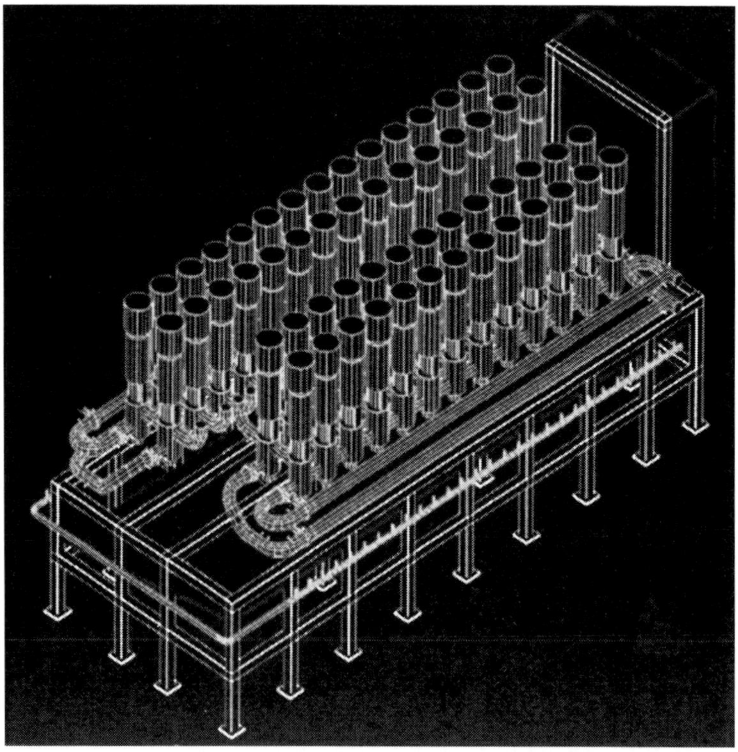

Figure 9 3D design renderings of equivalent transfer panel and mix-proof valve arrays. In this case, when bids were received, the capital cost of the transfer panel and the valve array were nearly identical. (Courtesy of CRB Consulting Engineers.) (*See color insert p. 9.*)

solutions on a permissive basis after an allowable flow path has been configured. In addition, lines should be equipped with pressure gauges and bleed valves to allow depressurization of lines prior to loosening of clamped nozzle caps.

For small-scale operations, in-process materials or cultures are generally hand-transferred in portable stainless steel vessels or bags between operation steps. When transferring materials in portable vessels or bags, pass-throughs between adjacent process rooms are preferred as a means to improve the material flows through the facility. Without pass-throughs between adjacent process rooms, the intermediate product in portable vessel or bag would need to exit to a corridor and enter the next processing suite.

Solutions are usually transferred by overpressure wherever possible, in order to eliminate a potential contamination source from pump seals, as well as simplifying the design which reduces initial cost and ongoing maintenance (by eliminating the pump), and reduce potential shear damage to the product or organism.

Clean-in-Place Strategies

Reproducible cleaning of equipment is critical to prevent contamination of the product which could alter its safety, identity, strength, quality, or purity. The basic requirement for suitable cleaning of equipment in GMP manufacturing operations is put forth in 21 CFR 211.067(a) which states "Equipment and utensils shall be cleaned, maintained, and sanitized at appropriate intervals to prevent malfunctions or contamination that would alter the safety, identity, strength, quality, or purity of the drug product beyond the official or other established requirements." Additionally, GMPs (21 CFR 211.063) require that there should be a written procedure detailing the equipment cleaning and the cleaning materials to be utilized. 21 CFR 211.182 requires written records of major equipment cleaning.

Stationary tanks are usually cleaned using a CIP system. CIP cleaning of a process vessel is more complicated than merely cleaning the tank itself. During the CIP cycle, all inlet and outlet lines must also be cleaned. For a full understanding of the CIP of complicated systems, piping and instrument diagrams (P&IDs) highlighting each of the cleaning paths will be necessary.

An important aspect of the design of a facility is the development of the integrated concepts and approaches for cleaning all of the equipment and transfer line systems. This is particularly important in multiproduct GMP pilot plants because of the concern for cross-contamination between processes. In order to avoid cross contamination in multiproduct facilities concurrently manufacturing multiple products, the CIP systems should be dedicated to a single product, or be once-through. For the same reason, the media and buffer prep, as well as upstream and downstream operations should be segregated or be once-through in order to avoid potential cross contamination.

The approach recommended by this author is to develop a single (or several) drawing(s) showing a schematic overview of the entire process which identifies all of the major equipment requiring CIP, all transfer lines and transfer panels involved in a process, and the flow paths which will be used to CIP each line and piece of equipment. This type of drawing is schematic in that it also shows relative elevations within the facility. This design approach incorporates an agreed upon basis of design for whether CIP skids will be shared between suites, whether CIP systems are to be once-thru or recirculated, and a resulting analysis of the required number of CIP skid systems. An important part of the CIP system analysis is a simulation of the process which will identify potential conflicts with CIP skid availability.

Validation of the cleaning procedures for the process equipment and piping, including chromatography columns, must be carried out. This is especially critical for a multiproduct facility where carry over cross-contamination is of major concern. The cleaning should remove endotoxins, bacteria, toxic elements, and contaminating proteins, while not adversely affecting the performance of the equipment. The manufacturer should have validated the effectiveness of the cleaning procedure for each bulk drug substance or intermediate used in a particular piece of equipment. It is recognized that the sensitivity of assays may make it impossible to reduce contaminants below detectable limits, even with a reasonable number of cleaning cycles. The validation data should verify that the cleaning process will reduce the specific residues to an acceptable level, generally expressed in parts per million (ppm) or parts per billion (ppb), and these levels should be scientifically justified by the manufacturer.

FDA does not specify analytical methods nor set acceptance criteria for determining whether a cleaning process is suitable or validated. The rationale for the residual contaminant limits for each piece of equipment should be scientifically sound based on the manufacturer's knowledge of the materials involved, and be practical, achievable, and verifiable. It is important to define the sensitivity of the analytical methods in order to set reasonable limits. Some limits that have been mentioned in the literature or in presentations by industry representatives include analytical detection levels such as 10 ppm, biological activity levels such as 1/1000 of the normal therapeutic dose, and organoleptic levels such as no visible residue. If an analytical method shows results below detectable limits, the assay should be revised or substituted with a more sensitive level. Essential to CIP validation, as well as routine monitoring, is a sampling plan which assures that the surface has been cleaned to the validated level. One common method is the analysis of the final rinse water for the presence of the cleaning agents from the last cleaning of that piece of equipment. There should also always be a direct determination of the residual substance. Another method is swab sampling of surfaces.

Classified vs. Controlled Unclassified Spaces

Traditionally cell culture manufacturing facilities have been designed around classified HVAC clean spaces for the manufacturing operations, such as Class 100,000 or Class 10,000. This was due to the need to minimize or prevent contamination of the culture process, or contamination of the product as it was being purified. Classification of clean room in the United States were, in the past, established by Federal Standard 209E (13), but this has been superceded by ISO-14644–1 (14). The International Standards Organization has established a new international standard in ISO 14644–1 "Clean Rooms and Associated Controlled Environments—Part 1: Classification of air cleanliness." This ISO standard is being reviewed by British and European organization in an effort to harmonize the classification of air cleanliness levels in clean rooms. The table below gives U.S. clean room classification, equivalent to the ISO classes and European Union standards, as well as recommended viable counts, and typical air changes. It is important to note that the EC classifications provide distinctions of "in operation" and "at rest" while the United States recognizes "in operation" only.

United States vs. ISO Clean Room Classifications

ISO class 14644	US 209E classification (superceded by ISO 14644)		EC grade[a]	Typical air charges[b]		Recommended viables	
	Traditional	Metric	In operation	Operational	Design	US CFU/ 10ft³	EU
ISO class 5	Class 100	M3.5	Grade A	90–100 ft per minute over entire area		1	<1
ISO class 6	Class 1000	M4.5		60 ACH, min	70 ACH	5	10
ISO class 7	Class 10,000	M5.5	Grade B	35 ACH, min	40 ACH	25	100
ISO class 8	Class 100,000	M6.5	Grade C	25 ACH, min	30 ACH	25	200
	"Pharmaceutical"		Grade D				

[a] EC classifications provide distinctions of "in operation" and "at rest".
[b] Specific rooms may require higher air change rates due to internal cooling loads, exhaust air requirements, particulate generating operations, etc.

Particle Count vs. Particle Size Chart in Different Classes

ISO Class 14644-1	Traditional (Fed Std 209E)	Permissible particle counts of given size (μm)					
		0.1	0.2	0.3	0.5	1	5
ISO class 1		10	2				
ISO class 2		100	24	10	4		
ISO class 3	Class 1	1000	237	102	35	8	
ISO class 4	Class 10	10,000	2370	1020	352	83	
ISO class 5	Class 100	100,000	23,700	10,200	3520	832	29
ISO class 6	Class 1000	1,000,000	237,000	102,000	35,200	8320	293
ISO class 7	Class 10,000				352,000	83,200	2930
ISO class 8	Class 100,000				3,520,000	832,000	29,300
ISO class 9					35,200,000	8,320,000	293,000

In addition to these recognized clean room classifications, the biopharmaceutical industry has also HEPA filtered but unclassified spaces, sometimes call them "clean manufacturing" or "controlled unclassified" spaces.

Air handler sizes, and corresponding energy costs, are proportional to several factors: air changes per volume; percent outside air; ambient temperature and (more importantly) humidity; and finally, the operating space conditions. Air changes per hour are related to the area classification, the room dimensions and the room height:

Air Changes per Hour: Relationship to Physical Specs

Class	Air changes per hour 8′ ceiling	9′ ceiling	10′ ceiling	12′ ceiling	Design range
100	675	600	540	450	Laminar flow 90 fpm at 12 in. above the point of product exposure
1000	150	133	120	100	50–80 AC/H
10,000	75	67	60	50	40–60 AC/H
100,000	38	33	30	25	20–30 AC/H

Air changes affect the rate of dilution and wash-out of particulates in the air. The higher the air changes per hour, the quicker a room will return to a state of cleanliness after an introduction of particulates. The table below illustrates this:

675 Air changes per hour Particles	Time (sec)	30 Air changes per hour Particles	Time (sec)
1,000,000	0	1,000,000	0
10,000	5.3	10,000	120
10	10.7	10	240
1	16	1	360

In the United States, HEPA filters are almost exclusively of the laminar flow HEPA module type, which provides a laminar flow washing of the room. Such a laminar flow washing works best in conjunction with low-level returns, but is subject to stagnant areas due to equipment placement, HEPA module placement, or return grill locations. In Europe, by contrast, a turbulent mixing diffuser is commonly used in non-Grade A/B (Class 100) areas. These turbulent mixing diffusers, sometimes called radial vane diffusers, provide turbulent mixing of the room resulting in a so-called "well-mixed" condition. Any particulate contaminates are assumed to be more homogenously distributed, and washout uniformly. Areas where turbulent radial vane diffusers have obvious application include areas of ISO 8 (Class 100,000) or controlled unclassified areas where the HEPA coverage is lower, such as high bay areas, manufacturing corridors with ceiling returns, and support areas such as media and buffer prep. Suppliers of these radial vane diffusers include TROX in Europe, and E.H. Price in the United States.

Glassware and Equipment Washing and Sterilization

Soiled equipment such as flasks, bottles, tanks, and glassware must be cleaned in a validated fashion. The first choice is often to use disposable items where ever possible. Beyond this, a choice must be made where to clean the glassware and equipment—either at a centralized Equipment Wash Area, or in a distributed fashion within the suites where the glassware or equipment are used.

Glassware and equipment wash areas are preferably organized around a flow from dirty to clean to sterile (Fig. 10). Generally this is achieved by organizing the glassware and equipment wash suite, with a staging and dirty wash area, an area for the wrapping and assembly of clean glassware and equipment and loading into the sterilizing autoclave, and an area where the autoclaves are unloaded into a separate room. Autoclave unload area generally has an HEPA filter area above the unload zone of the autoclave. It is important to provide space for adequate staging space for the dirty glassware, carts, and equipment. Ideally, double door pass-through washers are used so that only clean glassware and equipment is introduced to the clean assembly area. Where quantities justify, a separate sterile glassware storage room, or sterile tank storage room may be justified. Sterile-glassware and tanks are transported as required to the various operating areas.

When necessary, equipment from the a biocontainment area may require decontamination via SIP or an autoclave (or for lower containment levels, perhaps chemical decontamination) prior to disassembly and cleaning. Decon autoclaves for situations other than high containment may be located centrally adjacent to the dirty wash area. The decon autoclave should be designed using a double-door autoclave which unloads outside of the manufacturing area. In higher containment situations, the decon autoclave should load directly from within the bioreactor suite, so all materials are decon'd prior to removal from the containment envelope.

In situations where bags are used for buffer and media, the drums are preferably washed using a large equipment washer with a drying cycle, which unloads in a clean drum storage area. While there is not product contact with the bag holder drums or totes, and they could be manually wiped down, this is very labor-intensive and not very reproducible. A pass-through from the dirty to the clean side will be required to move any bulk items which cannot be washed nor autoclaved, such as carts.

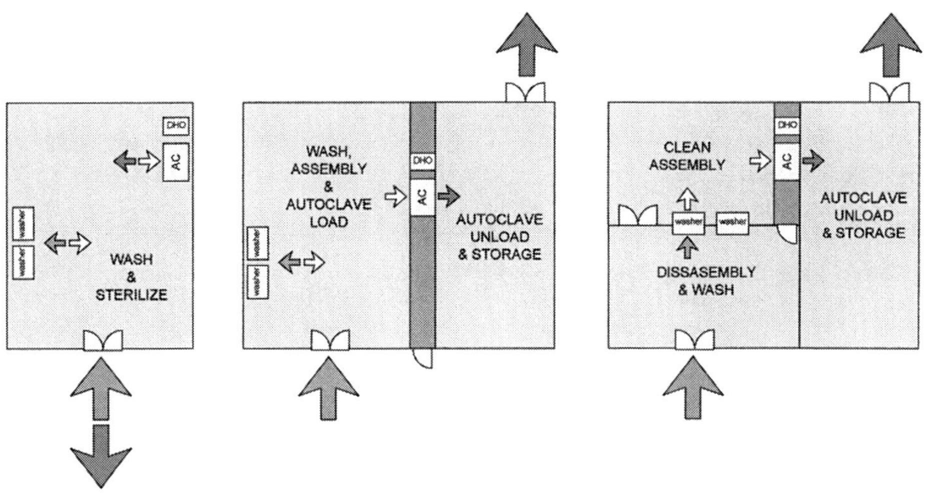

Figure 10 Organization options for wash areas. The area shown on the left depicts a small, single room, dedicated wash area may be appropriate for support of some operations where cleaning out of place is necessary (e.g., centrifuge operations), while the multiroom wash suite depicted at the far right is appropriate for supporting a full train of process suites. It provides distinct rooms for disassembly and washing, assembly and autoclave loading, and the autoclave unload and sterile storage.

Gowning Requirements

Gowning and Plant Uniform requirements in a facility should be discussed and agreed upon very early in the design process. While seemingly minor, this is a major driver in corridor and airlock requirements, and has a major impact on layouts. One method of examining this issue is to develop a table of gowning level requirements, and then assign each operational area to these. This is an invaluable tool in gaining consensus agreements between manufacturing, regulatory, and any other interested stakeholders. The table below shows an example of gowning levels (arbitrary numbered 1–4), and typical gowning:

Bulk production gowning requirements

Level 1 gowning—Plant uniform (e.g., done in locker room)	Sanitize hands Plant uniform/scrubs Plant shoes
Level 2 (e.g., just prior to entry into the Production supply hallways)	Booties Hair covers Beard covers Safety glasses Gloves Sanitize gloved hands
Level 3 (e.g., just prior to entry into Class 100,000 areas)	Labcoats or gowns Another set of booties Re-Sanitize gloved hands
Level 4 (e.g., just prior to entry into Class 10,000 areas)	Overgown w/coveralls Overgown w/booties Optional full headcover Second set of gloves Re-sanitize gloved hands
Level 5 (e.g., just prior to entry into Class 100 areas)	Overgown w/sterile coveralls Overgown w/sterile booties Overgown with full sterile headcover Second set of sterile gloves Re-sanitize gloved hands

Flexibility and Expansion Options

Any project requirement for future facility expansion must be defined early so that feasible expansion approaches may be considered during site selection and schematic building layout development. This is a critical layout issue, since areas set for expansion typically are best located on outside walls, or otherwise built out as unfinished shell space or user space identified for future renovation (such as warehouse space). One approach is to arrange the building with one of the corridors along the outside wall where expansion may take place in order to allow expansion without disrupting ongoing operations within the suites.

Beyond the basic question of expansion capability, the type of expansion required must be defined. Is it to be expanded capacity for the same line (i.e., larger equipment)? An existing line producing a new product (i.e., possibly requiring different equipment)? Or a new production suite with the same product? Obviously, these scenarios could result in different layouts. The process and architectural programming will identify expansion requirements and if required, approaches addressing these requirements will be developed during the early layout development.

Gravity Flow Vertical Integration of a Process

In larger manufacturing operations, it is often desirable to utilize gravity flow to enhance the organization of product and raw material flows through a building, as well offer improved CIP and SIP design. The basic premise is to place the buffer and media prep operations at the top of the building, with buffer hold tanks and the bioreactors on the floor below, and the recovery operations and purification operations on the floor below, as shown in Fig. 11. This concept can be carried a step further with a mechanical basement or floor below the lowest process area for CIP and biowaste systems. It should be remembered that the building will still require adequate interstitial space for the HVAC and mechanical systems. Organizationally, this stacking is the recommended approach to organizing the building. The challenge to this vertical organization of the building is the resultant overall height of the building, and the increased structural steel (and resulting cost). An early sitting issue to be resolved is the applicable zoning restrictions on total building height.

Corridor System Design and GMP Flows

One of the biggest contentions in facility design is maximizing production space and minimizing "nonproductive" spaces. This often results in storage rooms being eliminated, and often challenges the use of dual corridor systems. One of the fundamental paradigms which must be addressed early in design is the approach to satisfy GMP flows for people, materials, product, and waste. Raw materials and product should circulate through the operation in a logical fashion. Indeed, the Europeans actually call this out explicitly:

Chapter 3, paragraph 7 reads:

> "Premises should preferably be laid out in such a way as to allow the production to take place in areas connected in a logical order corresponding to the sequence of the operations and to the requisite cleanliness levels." (15)

This maybe achieved by a combination of engineered solutions, such as the dual corridor system, or by temporal and procedural controls (Fig. 12). What should be discussed early is the approach to be taken with an agreement by operations and regulatory groups regarding GMP flows. As an example, a table can be prepared to better focus the discussion on potential crossovers and when these are acceptable or unacceptable, or when procedural controls are adequate.

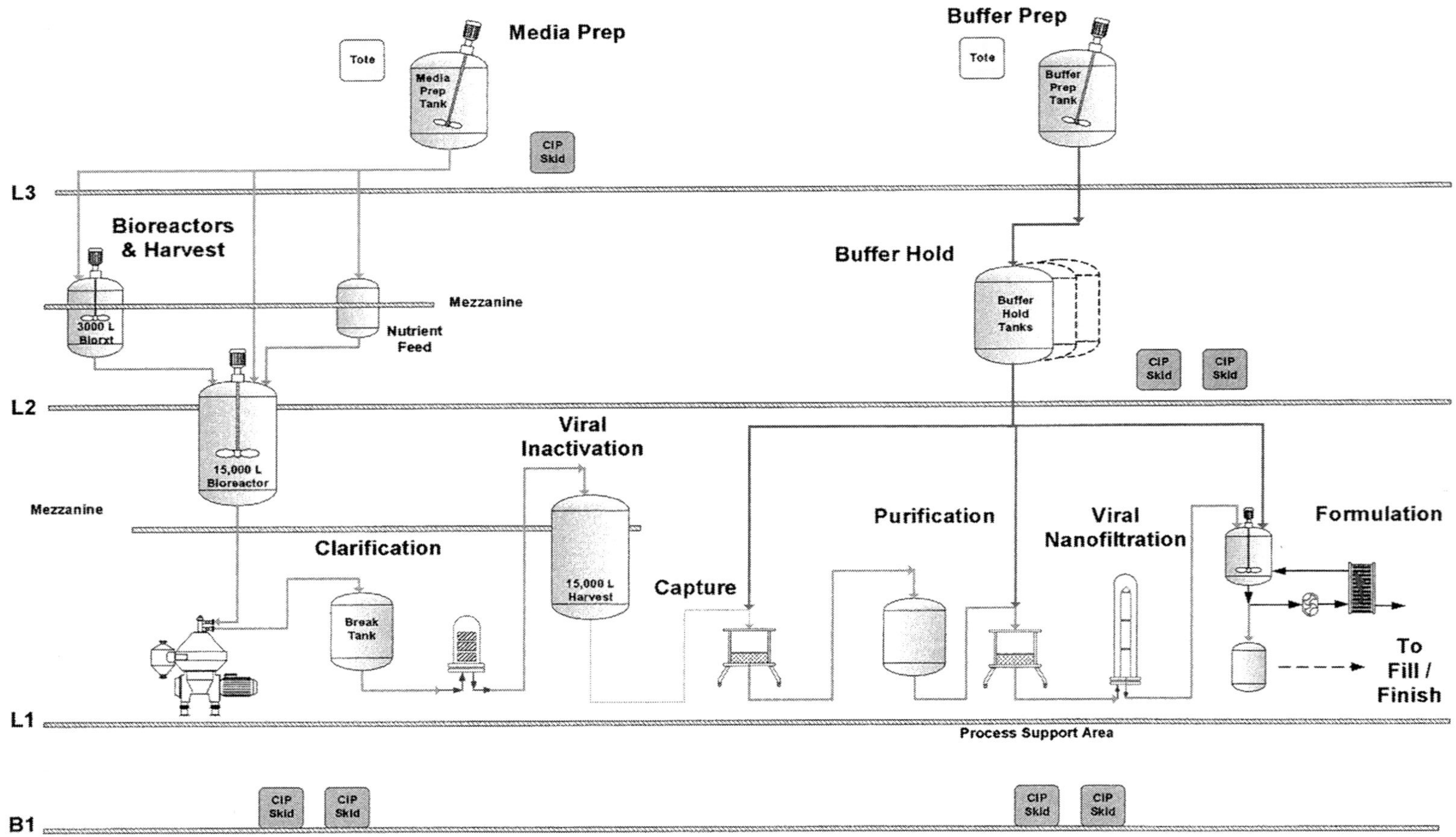

Figure 11 Schematic organization of a process building to provide gravity flow of solutions from the media and buffer prep areas down through the purification areas.

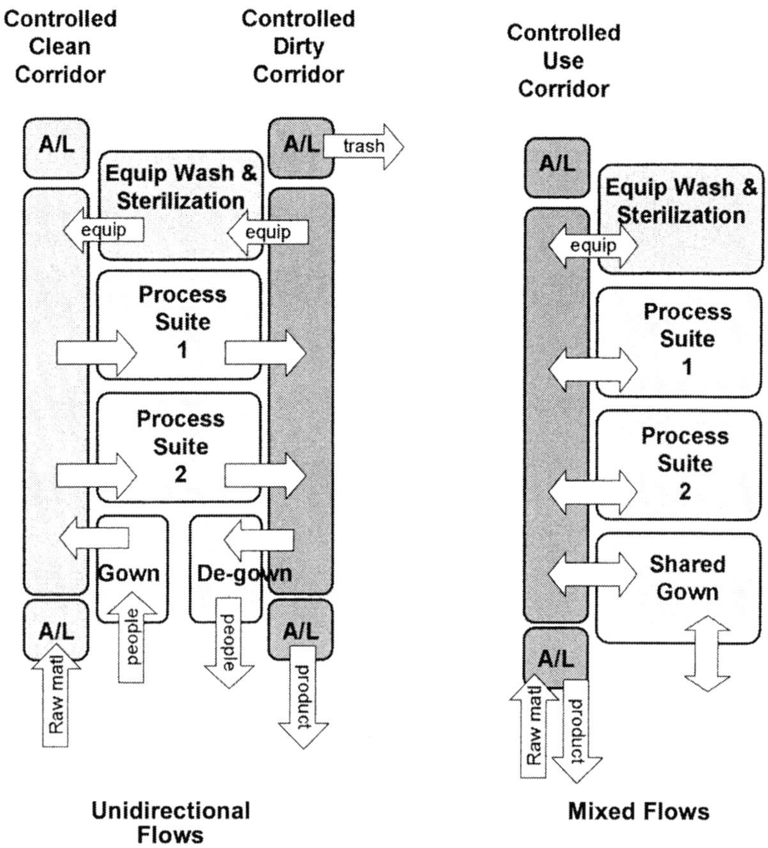

Figure 12 Dual and mono-corridor facility arrangements.

	People	Supplies	Raw materials	Product
People	—			
Supplies	Minimize	—		
Raw materials	Minimize	OK	—	
Product	Minimize	Minimize	Minimize	—
Waste and trash	Never	Never	Never	Never

If dirty equipment is to be washed within production suites in a distributed fashion, the mono-corridor becomes more feasible, although still not recommended by this author due to the remaining issues around flows. If a common equipment wash area is used, a dual corridor system is usually better because of the higher amount of traffic to the equipment wash area. For closed system processing this table must be revised to reflect the fact that beyond the bulk mixing stages for medias and buffers, the entire process may be closed, and thus present no danger to the product. Under those circumstances the table will look quite different, and the conclusion must be that dual corridors are entirely optional.

Personnel flows and access to Production areas must be controlled. Access control is generally done by electronic access control applied at the outside entrance into a manufacturing building, and also at the entrance to the lockers or gowning rooms. In multiproduct facilities an additional level of access control may be applied at the airlocks entering into different products manufacturing suites.

Personnel should enter the GMP production areas via plant locker rooms, where personnel change from street clothing to plant uniforms. This reduces the environmental burden carried by personnel. Prior to entering a manufacturing corridor, personnel usually go through a gowning stage where they don booties and hair nets, and proceed into the controlled corridor. Even gowned, there is a certain amount of billowing in and out of the loose fabric, which the gowns undergo as a person walks. Gowning over street clothing allows this shedding to introduce more particulates and bioburden to the clean environment, relative to gowning over clean plant uniforms.

From the controlled corridor, personnel may access the manufacturing areas, usually through an airlock. Flow in these areas is an uncontrolled mixed flow with-personnel entering and exiting each suit through the same airlock and controlled corridor. As personnel enter each suite, they can don lab coats. When personnel exit suites, they remove the lab coats and exchange shoe cover booties, thus minimizing the chances of carrying contaminants out into the controlled corridor.

Equipment flows through a facility is a cyclic nature. Clean equipment is distributed, used and dirty equipment is returned to Equipment Wash for cleaning. Where the equipment wash areas are distributed and located within the suites, this merely is internal circulation within the suite. Where a shared Equipment Wash is used, this requires circulation through the corridors. In a dual corridor system (clean/dirty) this presents no problems.

When a mono-corridor exists, it may require that the corridor traffic be managed with a time period for clean items and a time period for dirty items.

Trash and disposables should be taken out of the manufacturing area in such a manner as to not put the product at risk. In a clean/dirty corridor system this is straightforward. In a mono-corridor this requires adherence to SOPs with temporal separation of traffic with intervening cleaning. In the ideal case, trash is taken to an dedicated trash airlock exiting the dirty corridor. Trash should not be taken out through the raw materials warehouse, and a separate loading dock with appropriate staging is recommended.

Interstitial Spaces

Interstitial space is that volume of space above the operating area's ceilings where the mechanical systems such as HVAC duct, electrical, and piping are run. Because these mechanical systems have valves, instruments and items which require periodic service, it is necessary to have access to those areas. Approaches to this include fully walkable ceilings, limited catwalk systems, and in some circumstances, the use of access panels from the clean area. The use of access panels from the clean area is strongly discouraged since when the panels are opened, the operating space must be cleaned and recertified. If routine access is required in these situations, the relevant valves or items should be boxed in so that opening the panel does not expose the room to the dirty interstitial space, but rather to a sealed, enclosed volume.

A major challenge in construction of biopharmaceutical plants is the sheer quantity of HVAC duct, piping and electrical which must be run through the interstitial space. This can result in difficulties (and expense) in construction, and ongoing difficulties with maintenance. For these reasons, where ever possible, a generous interstitial space is recommended. While rules of thumb are difficult to arrive at, and often open to criticism, for interstitial space approximately 1 ft above ceiling is recommended for each foot of process area room height (excluding any local high-hat areas, such as above a tall tank or for agitator pulls). This allows room for the HEPA filters, duct work, catwalks, pipe racks, cable raceways, and building structural steel for the floor above. The downside of the generous interstitial space is greater cost due to added building steel, and possible issues with the total building height. Obviously for less complex or less extensive systems this guidance may be reduced, and non-GMP areas of the building, such as laboratories or offices, need only have interstitial space appropriate for these types of spaces, with access provided by removable drop ceiling tiles.

Walkable ceilings are constructed of self-supporting panels suspended from the building structural steel above or a self-supporting steel structure erected above the clean room (Fig. 13). Walkable ceilings allow service of the valves and instruments, as well as the servicing of the fluorescent lights. While more expensive, walkable ceilings provide one of the better maintenance access approaches. A downside to the walkable ceiling are the many joints between the structural support grid and the panels. While these are gasketed, there is still a resulting crevice which will be more difficult to clean as compared to a monolithic ceiling such as gypsum board.

Grey Space and Mechanical Corridors

A goal in many biopharmaceutical facility designs is to remove as much equipment from the clean room spaces as possible, locating this equipment in so-called "gray space" or controlled unclassified space. This allows for easier cleaning and smaller clean rooms. By reducing the clean room area, the construction cost is reduced, and perhaps as importantly, the ongoing operating cost is reduced due to lower energy usage, as well as possibly reducing the clean room operating staff. The

Figure 13 Walkable ceiling interstitial space. The air handling units are on the grating floor above. The walkable ceiling allows easy access to HVAC control instruments and dampers, as well as allowing for the servicing of lights from outside the clean envelope. (Courtesy Clestra Cleanrooms Inc.) (*See color insert p. 10.*)

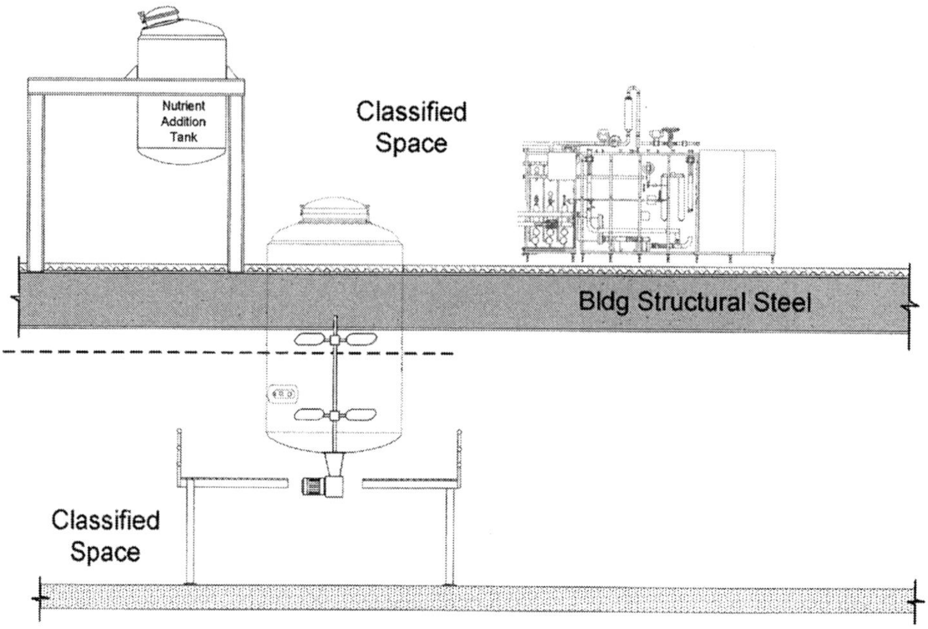

Figure 14 Critical process equipment within the clean envelope.

various levels of usage of the grey-space are illustrated in the series of diagrams which follow. Figure 14 shows all critical process equipment within the clean envelope. This would include process tanks, skids, bioreactors, etc. The use of gray space would be limited to utility systems or CIP systems with no process tanks or skids located in utility or gray spaces. This is representative of smaller biotech operations, and a few of the larger operations. At smaller scales, it is more cost-effective to have a single bioreactor skid rather than a split skid system, and the cost in having a larger clean classified area is offset by the lower staffing requirement. Some biopharm companies feel that having the bioreactor skid within the bioreactor room allows operating staff to troubleshoot the equipment much more efficiently. There is an efficiency loss when skids are located in an adjoining room due to the redundancy of staff, and the communication which must take place.

The use of controlled unclassified (gray space) is illustrated in Fig. 15. This figure shows such an adjacent mechanical area supporting the bioreactors, with a split skid design.

This approach has become widely accepted, and removes a great deal of piping, valves, and instrumentation from the clean envelope, allowing for easier wipe downs and sanitization.

Finally, Fig. 16 shows such an approach where the use of grey space or controlled unclassified space is maximized. This approach has the bioreactor support skids, and the majority of the vessel outside of the clean envelope, with only the probes and sample points within a clean room. This approach has been implemented on several very large-scale cell culture facilities, and the design does minimize the clean envelope space.

Construction Approach and Modularization

One of the biggest opportunities in current design of large-scale cell culture facilities is the use of modularization during the design and construction phase. Modulariza-

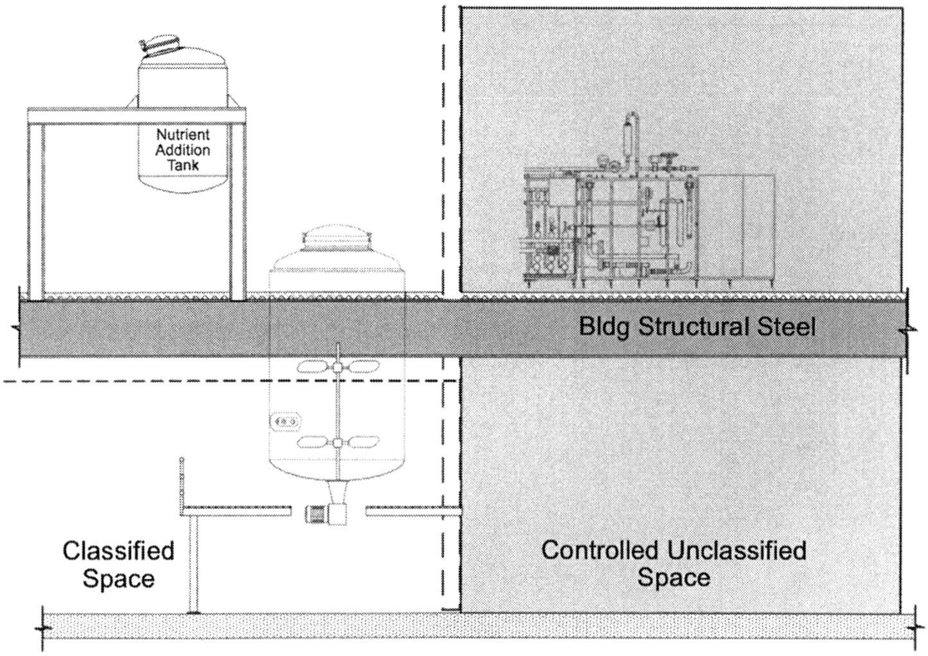

Figure 15 An adjacent mechanical area supports the bioreactors, using a split skid design to remove piping and utility skids outside the area.

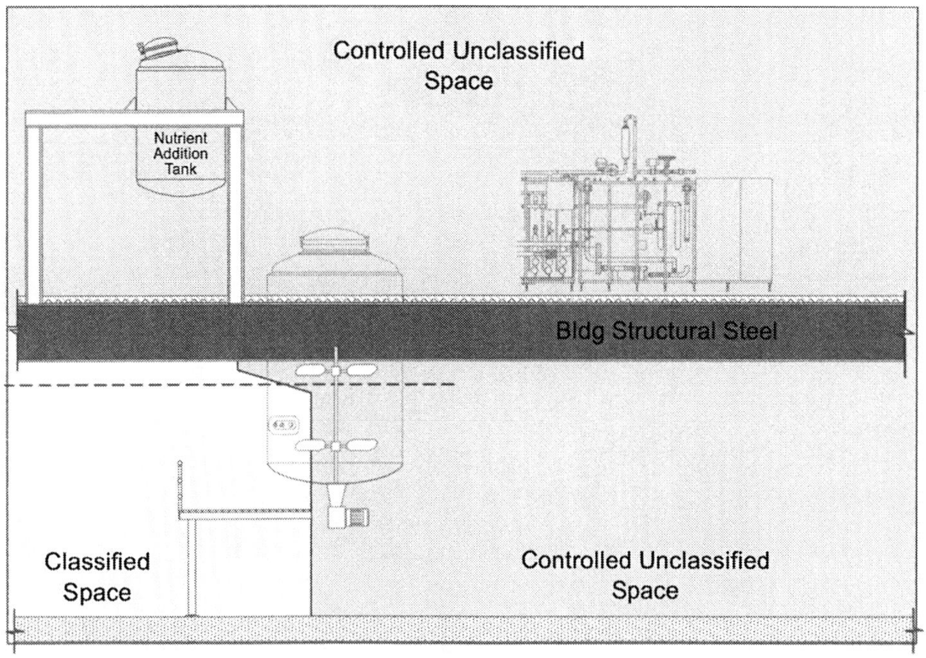

Figure 16 Use of gray space or controlled unclassified space, with the bioreactor support skids, and the majority of the vessel outside of the clean envelope.

Figure 17 Rendering of buffer hold module for Biogen-IDEC's NIMO facility in Oceanside, CA. (Courtesy of Biogen Idec.) (*See color insert p. 10.*)

tion offers significantly accelerated schedules, and significantly greater quality control due to the controlled fabrication in a shop environment as opposed to the field conditions at a construction site. To maximize its benefits, modularization must be integrated into the design phase, such that the design and fabrication of the modules are accelerated and they are delivered shortly after the building is closed in. In a traditional construction schedule, the platforms and tanks would normally be delivered at this same point, and all of the hygienic piping would then be field-fabricated. Modularization looks to fabricate the biggest pieces of the project feasible, with partial disassembly to accommodate over the road shipping. While modular construction has been very successfully applied in the petrochemical industry for years, the biopharmaceutical industry's use was essentially nonexistent till just a few years ago. The first significant projects to use modularization widely was the Biogen's Large Scale Manufacturing facility in RTP, North Carolina. Figure 17 shows a photograph of the buffer hold modules from another project, Biogen-IDEC's new manufacturing facility located in Oceanside, California (photo courtesy of Biogen-IDEC and Paul Mueller Co.). This module featured 22 tanks on four platforms with all interconnecting piping.

EQUIPMENT AND STERILE PIPING DESIGN

Cleanability and Deadlegs

The proper installation of sterile or hygienic piping is critical to the ability to clean and sterilize that piping. As is often the case, the devil is in the details, and details in piping systems can make or break the projects operational success and/or construction budget. Care must be taken in the selection of the hygienic valves, and extreme care must be taken in the design and installation of the piping. Deadlegs are sections of pipe without throughflow during CIP or SIP, such as branches off of main lines. The 3A Sanitary Standards (16), developed for dairy industry, provide guidelines and standards for hygienic equipment for the food and dairy industries. These had been followed by the biopharmaceutical industry in its early years from a lack of a more appropriate standard. The American Society of Mechanical Engineers (ASME) has developed and published such standards in their ASME Bioprocessing Equipment (BPE), with the most recent revision being published in 2002 (17). The development of the ASME BPE represents a very significant advancement in presenting good design practices, and standards, and is now an American National Standard. These standards, based on best technology and practices of the day, established a guideline of deadlegs being less than six-pipe diameters from the center of the main line to the end of the deadleg such as the clamped cap as shown below, or weir of a diaphragm valve. In the biopharm industry experience showed this to be too great of a deadleg length and companies would set their own internal standards to a closer dimension. Subsequent research laboratory studies and computational fluid dynamics studies, have found that deadlegs must be considerably shorter to ensure cleanability. The ASME BPE-2002 suggests, in section SD-3.11.1, a target L:D ratio of 2:1 from the inside wall of the mainline as the recommended practice (Fig. 18). This is felt to be achievable in most, although perhaps not all, situations.

The design of valves have progressed tremendously in the last decade, with the wide use of computer controlled machining. This has allowed multiple valves to be machined in a very tight configurations which minimize deadlegs. As an example, a traditional ported valve assembly would result in significant deadlegs. The more recent machined versions of this assembly would have the weir of the branch valve right at the edge of the main line, eliminating the deadleg. These valves can be very important in critical sterility systems, chromatography systems and in process systems where product loss may be a significant issue. It should be remembered that

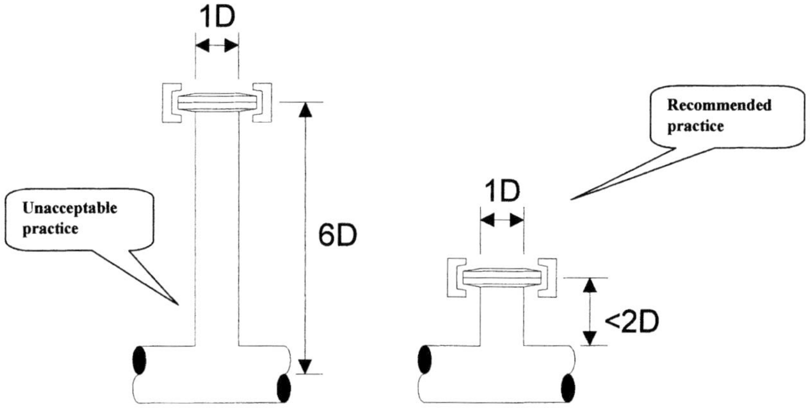

Figure 18 Traditional recommendation of maximum Deadlegs (< 6D)—now regarded as unacceptable practice, and ASME BPE's recommendation of a target of less than two-pipe diameters—which is the current recommended practice.

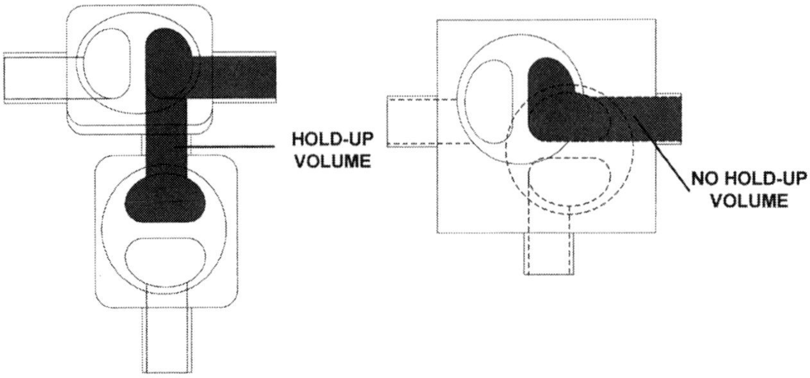

Figure 19 Hold-up of branch valves.

some of these therapeutics have values in the thousands of dollars per dose (typically only a few milliliters or less).

Figure 19 shows hold-up of branch valves. The traditional approach using ported valves, or branch Tees, results in substantial deadlegs, while newer approaches use a so-called "zero-static" machined valves have the valve weir located flush with the interior of the pipe. This results in no deadleg.

The flowrate required to clean transfer lines is often a point of discussion. Initial thoughts would suggest that the flowrate only needs to be sufficient to be in the turbulent regime. This may only require a flowrate of <1 ft/sec in larger pipelines. The problem which arises is removing entrained air, particularly in any vertical lines. During CIP, it is important to have a high enough flowrate to overcome the buoyancy of entrained air in vertical lines. It is also helpful to have the flow from lowest to highest so that the lines are fully flooded, and entrained air goes with flow rather than trying to buoyantly rise against the flow direction in vertical segments.

In an interesting study by Haga et al. (18), the cleanability of deadlegs was examined, and parameters were established to provide an approximate guide for the flowrates required to clean branch deadlegs. Figure 20 shows this. It might be noted that at the six-pipe diameter for a deadleg, the required flowrate is 1.5 m/s which is approximately 5 ft/sec.

Drainability and Sterilization

Dead legs are segments of piping, typically located at a branch point, where the product or cleaning solution do not flow through, and as a result are stagnant pockets. If these sections are to be cleaned, the cleaning solution must develop sufficient turbulence at the branch point that even the farthest point in the dead leg section will see adequate mixing.

The proper installation of sterile or hygienic piping is critical to the cleanability and sterilizability of the piping system. All piping must be pitched to a low-point drain with sufficient pitch to allow condensate to roll down the pipe during sterilization processes. The routing of the piping must ensure that any low points are provided with low-point valves and traps as necessary. The ASME BPE-2002 (SD-3.12.1) lists slopes for process lines which vary from 1/16 in. per foot for very long runs (such as in a pipe-rack), to $^1/_8$ in. to $^1/_4$ in. per foot for short runs. ASME

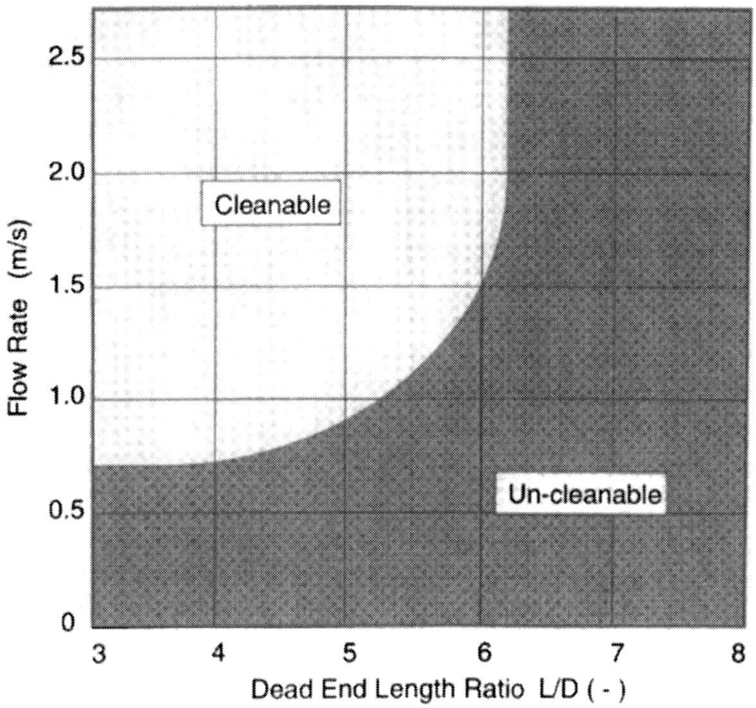

Figure 20 Flow rates required to clean deadlegs of different L/D ratios [From Ref. (18).].

further recommends $^1/_4$ in. per foot for slopes on process pipe runs on skidded equipment.

One thing which is sometimes overlooked is the care with which valves must be installed. The design of diaphragm valves requires their installation at what is called their "Self-Draining Angle." This is sometimes indicated by a "hash" mark or engraving at the top dead center of the exit or entrance tubing. These markings should not be trusted, because it requires the welder to orient the valve by eye, and no human has an eye which can accurately spot that orientation exactly at top dead center. Rolling the valves too far results in the widening of the diaphragm to form a puddle. Figure 21 demonstrates valve angles and resulting hold-up at weir or behind weir. The proper angle is critical to prevent either puddling at the valve if it is overrotated, or pooling behind the weir if the valve is under rotated and acts as a dam in holding liquid from draining. Of the two, under rotating and backing up condensate or material in the line is the more serious situation. It is important to note that the angle is valve model-specific, size-specific, and manufacture-specific. These angles may also conceivably change from year to year with minor manufacturing or valve design changes, and it is critical that installers be provided with a table of the relevant valve models, size, and required installation angles. A digital protractor is the most accurate means of measuring the angle, and the accuracy should be one digit beyond the required installation accuracy. The author has personally seen hundreds of valves cut out because the welders were unaware that hash marks were not sufficiently accurate for installation angles, or their means of measurement were inadequate. Hash marks should not to be trusted.

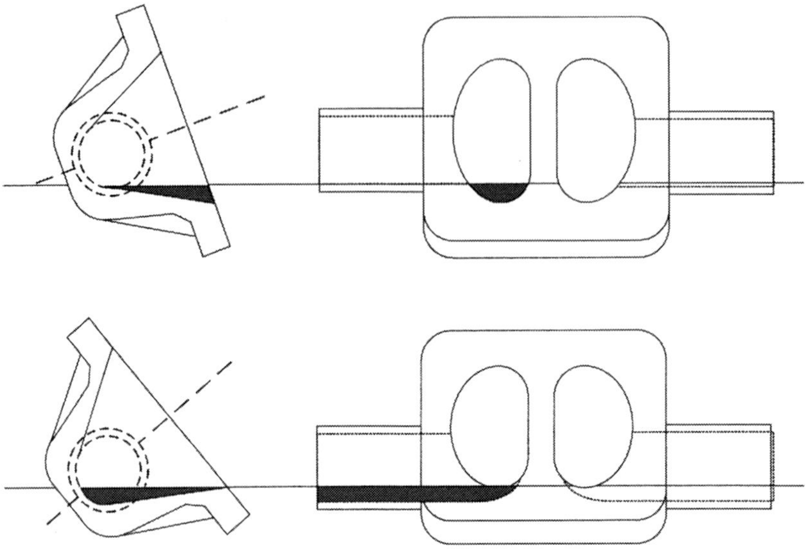

Figure 21 Valve angles and resulting hold-up at weir or behind weir.

As in cleanability, the length of deadlegs has a significant effect on the ability to sterilize a line. The issue in this is the heat transfer from the main line to the most distant point of the deadleg branch, and the heating up of the stainless steel at the far end of the deadleg (19).

Passivation

Passivation is a chemical treatment/cleaning process that removes exogenous iron or iron compounds from the surface of stainless steel piping and equipment by the use of a mild oxidant, such as a nitric acid solution, or a chelating solution. The purpose of passivation is to restore or enhance the formation of the chemically inert surface or protective passive film. A passive oxide film forms naturally on a stainless steel surface when exposed to air or similar oxidizing environment. This passive oxide film protects the underlying metal from corrosion. The ratio of chromium to iron is a measure of corrosion resistance.

Welding disturbs the passive layer by reducing the chromium and increasing the iron, thus altering the chromium/iron ratio and the weld and adjacent weld affected area must be repassivated (20).

A surface which has not been passivated may have active sites where corrosion can occur. Such active sites are caused by metal contaminants. Surface contaminants can be readily removed or treated, however, subsurface contaminants or severely embedded surface contaminants will require physical removal by mechanical polishing prior to passivation. In order to be effective the piping or equipment must be thoroughly cleaned and degreased prior to passivation.

Normal operating conditions in typical WFI, clean steam, reverse osmosis, deionized water, CIP, and process piping systems often lead to formation of a mild corrosion called rouge. Rouge is the most common form of self catalyzing corrosion and is a colloidal form of rust containing iron oxide, chromium and nickel in various forms. Rouge formation is accentuated by high temperature. Rouge (French for "red") varies in color from pink to red, brown, golden, or colors in between. Rouge formed in one spot in a system, such as a pump impeller, can be spread throughout the system.

PROCESS UTILITY SYSTEMS

Purified Water (Water for Injection and USP Purified Water)

The water used in biopharmaceutical manufacturing must be appropriate for the process step. There is a central theme of the ISPE's Baseline Guide for Purified Water Systems (21), which emphasizes the use of an appropriate grade of water for the particular task or process requirements. There is not an absolute list of what should be used where, and this must established by the manufacturer. Overly conservative use of WFI for all users may result in a more expensive system, and higher operating and QC monitoring costs. In smaller facilities there is often a trade-off where the greater cost of WFI is traded off with having a single distribution loop rather than, for example, have an additional USP purified water loop. The reader is strongly encouraged to read this Baseline Guide for Purified Water Systems, as well as the ISPE's other Baseline Guides.

Water for injection may be generated by a variety of means including multiple effect still, vapor compression stills, or reverse osmosis. The European Union has recently provided guidance that WFI may only be produced by distillation. The use of reverse osmosis, while not disallowed by the FDA, is subject to much higher scrutiny due to the risk of contamination in these systems unless they can be heat sanitized routinely. There is no requirement that USP Purified Water be used as a feed for the WFI generation equipment. The choice of feedwater pretreatment should be based on the WFI generation equipment's specific requirements. The high temperatures of multieffect stills requires a suitable pretreatment to remove water contaminants which might otherwise scale or foul the heat transfer surfaces. Vapor compression stills require only softened water as a feed due to the lower temperatures in those stills. The outlet from the generator unit should include instrumentation for monitoring of pH and conductivity. During the start-up period, water not meeting specifications is diverted to the drain.

In addition to the WFI generation equipment, the WFI system includes an insulated storage tank(s), pump(s), distribution piping, loop reheat heat exchanger, and controls. All metal WFI contact surfaces is generally 316L SS with a surface finish of 15–20/μin. Ra. All elastomeric components need to comply with USP Class VI requirements for extractables.

Figure 22 shows typical hot WFI loop using a jacket for temperature control. The temperature of the WFI system should be based on the process requirements. Cleaning operations require hot final rinses, while buffer or media prep will require ambient WFI. Ambient systems must be sanitized regularly to control the microbial counts. Typically ambient WFI systems are sanitized once a day at temperatures above 80°C. Hot WFI systems are typically maintained above 80°C.

Design of the storage and distribution systems must include sanitizing capability, sloped piping, an absence of deadlegs through the extensive use of zero deadleg valves, a recirculating piping loop, low-point drains and sample valves. Design flow rates are set in order to maintain turbulence, and generally are approximately 5 ft/sec. Distribution pumps are usually controlled via a variable frequency drive (VFD) by a loop mounted pressure transmitter. Flow control is usually achieved through the use of a dual range-controlled diaphragm valve. Alarms are provided

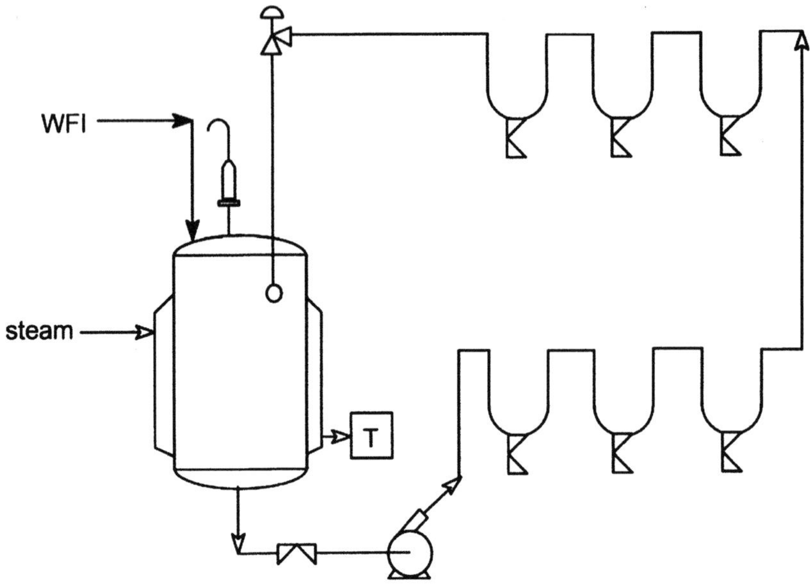

Figure 22 Typical hot WFI loops using a jacket for temperature control.

for temperature, pressure, low flow, and conductivity. All critical parameters should be monitored and recorded. The United States Pharmacopia (22) requires the measurement of total organic carbon (TOC) as one of the critical quality parameters for purified water and WFI. In one approach, a manifolded TOC monitor is used to allow multiple sanitary systems (USP, WFI and Clean Steam) to be monitored on-line. Alternately, samples are often be taken and measured off-line.

WFI must be sampled during validation and on an ongoing basis. The system should be designed to allow the collection of samples for this purpose. Equipment that has been directly connected to the WFI system, i.e., Glassware Washers, Stopper Processors, etc. must be designed with an integral sample valve as part of the point of use valve. This may present difficulties due to the height at which these WFI valves are usually located above these and other pieces of equipment. Provisions should be taken to route the sample line to a reasonable height for sampling, although more extensive flushing will be required. Movable equipment is connected to the WFI system by flexible hose, in which case a separate sample valve is not required. Sampling at these points shall be done by removing the connection at the equipment end, flushing the hose for a period of time and then collecting a sample. It is usually advisable to minimize WFI points of use in order to control the amount of environmental water sampling routinely required. Unless multiple drops are absolutely required in a particular room, it is not like adding a simple sink—there are long-term implications, and thought needs to be taken.

Purified Water standards (along with WFI) are dictated in the United States by the U.S. Pharmacopia (USP). USP Purified Water can be generated by a several methods, but is usually generated using standard water pretreatment systems, reverse osmosis, and deionization. The pretreatment system design depends upon the inlet source water, and usually consists of multimedia filters, softeners, a heat exchanger, and a chlorine removal system. Chlorine removal may be done using activated carbon beds or using sodium metabisulfite addition. The use of activated carbon can be problematic due to issues with microbial growth on the activated carbon bed and the resulting need to regularly sanitize the bed. The use of sodium metabisulfite avoids this problem and any residual chemicals are rejected by the reverse osmosis membranes.

Continuous electrodeionization (CDI) has gained wider acceptance and is now often used in place of traditional ion exchange deionizers. The choice of CDI is often driven as a result of the regeneration not requiring chemicals.

The outlet from the deionization system includes monitoring of pH and conductivity with water not meeting requirements being diverted to the sanitary waste system. The tank and loop for a USP water system could be stainless steel or plastic, but the materials of construction must not compromise the quality of USP water. Although not required, all contact surfaces usually 316L SS with a suitable surface finish such as 20–35 μ in. Ra. Stainless steel distribution systems are preferred due to the need to periodically sanitize the loops to control bioburden.

Design of the distribution systems for USP Purified water is similar to that of WFI, and should include sanitization capability, sloped piping, absence of deadlegs, recirculating piping, low-point drains, and sample valves.

The system should be designed for sanitization. Usually this is done heating the loop to elevated temperature (80°C) for sanitization at regular intervals. Either the storage tank should be jacketed to allow heating, or a heat exchanger should be provided in the USP circulating loop. The USP Purified Water storage tank should be sized to allow selected utilities such as the clean steam system to operate for perhaps 12 hr without any USP Purified water production. This allows down time and maintenance and servicing of the USP generation equipment without shutting down the facility.

Biowaste Inactivation System

Biopharm facilities may or may not require a biowaste inactivation system depending upon the biohazard risk and corresponding biocontainment level. Most simple cell culture-based biotherapeutic processes which utilize cell lines such as CHO, will have a biosafety level of Good Large Scale Practices (GLSP). This does not require inactivation prior to disposal. However, some local municipalities or state plumbing codes may require inactivation prior to disposal. While it is resisted from a cost standpoint, it is usually a very good design practice to design the waste drain systems with this possible need in mind, even if the lines may be tied together just prior to exiting the building. Usually a deep pit should be provided for a collection tank, as well as a location for the actual treatment inactivation system. Inactivation is based upon the organism sensitivity. For a simple CHO cell line this is quite low, and for a short time period. If it were for a viral vaccine process, it may be for a quite elevated temperature for a longer time period.

As an example of the reasons for anticipating, a large-scale manufacturing facility had a last-minute change which would introduce a new product to the facility. The facility was originally designed as GLSP, and the new product was nominally a BL1-LS process. This required some modifications of the drain collection systems to accommodate the segregation of the contained waste lines form the bioreactors and recovery areas from the remainder of the facility waste streams. A lesson from this should be to examine the range of process and organisms, and design a little beyond where you feel you will be operating; designing in some flexibility to be able to modify the facility without major problems (which underground drain systems would be!).

Backup or Standby Power

Standby power should be considered distinct from emergency (life-safety related) power. Standby power is a backup system which is provided for economic reasons in order to minimize the losses which might otherwise occur with a power outage. Standby power is used to provide back-up for critical equipment such as freezers, refrigerators, bioreactors, control systems, incubators, as well as HVAC pressurization. In a manufacturing facility the difficulty comes in when users want the entire facility on back-up power. It is often impractical to have building chillers on standby power due to their huge electrical draws. Standby generators are generally provided as diesel or natural gas driven electrical generator, depending upon the size. In some situations a facility is located where dual electrical feeds may be available, and a standby generator may be either unnecessary or significantly smaller. In other situations companies having large sites may utilize cogeneration of steam and electricity. The availability of cogeneration provides the electrical back-up, since the building would have a normal electrical hookup as well.

Uninterruptible Power Supply Power

Uninterruptible power supply (UPS) power supplies utilize battery systems to provide fully uninterruptible power. UPS is provided to critical control systems and process equipment which cannot tolerate any power outages, as well as any data recording systems and chart recorders which document temperatures in freezers and refrigerators. UPS power covers the few seconds of power outage while the back-up generators come up to speed. It does not replace backup or standby power. If a skid has a large motor and control panel, the UPS power may be connected to the control panel, and stand-by power may (or may not) be connected to the skid motor. The important issue is having uninterrupted recording of the process parameters and environmental conditions.

REFERENCES

1. Kunitake R, Suzuki A, Ichihashi H, Matsuda S, Hirai O, Morimoto K. Fully automated roller bottle handling system for large scale culture of mammalian cells. J Biotechnol 1997; 52:289.
2. FDA's "Guideline on the Preparation of Investigational New Drug Products", March 1991.
3. The rules governing medicinal products in the European Union, Vol. 4. Good Manufacturing Practices, ANNEX 13 Manufacture of Investigational Medicinal Products, November 2001.
4. FDA/ICH Harmonized Tripartite Guideline "Guidance for Industry—Q7A Good Manufacturing Practice Guidance for Active Pharmaceutical Ingredients", August 2001.
5. FDA. "Changes to be Reported for Product and Establishment License Applications; Guidance", 1995.
6. ISPE Baseline Pharmaceutical Engineering Guides—Volume 6—Biopharmaceuticals Baseline® Guide. Tampa, FL: International Society of Pharmaceutical Engineering (anticipated release 2004).
7. FDA/ICH Harmonized Tripartite Guideline. Guidance on viral safety evaluation of biotechnology products derived from cell lines of human or animal origin. Fed Reg 1998; 63(185):51074–51084.

8. EUDRA Committee on Proprietary Medicinal Products (CPMP). Notes for Guidance on Virus Validation Studies: The Design, Contribution and Interpretation of Studies Validating the Inactivation and Removal of Viruses. CPMP/BWP/268/9, 1996.
9. Center for Disease Control's monograph Biosafety in Microbiological and Biomedical Laboratories (BMBL) Washington: U.S. Government Printing Office, 1999.
10. National Institutes of Health. Guidelines for Research Involving Recombinant DNA Molecules (NIH Guidelines), 2002.
11. Wisniewski R. Developing large-scale cryopreservation systems for biopharmaceutical products. BioPharm 1998; 11(6):50.
12. Sinclair A, Monge M. Quantitative economic evaluation of single use disposables in bio-processing. Pharm Eng 2002; 22(3):20.
13. FS209E. Airborne Particulate Cleanliness Classes in Clean Rooms and Clean Zones. Mount Prospect, IL: Institute of Environmental Sciences, 1992.
14. ISO 14644–1 Clean Rooms and Associated Controlled Environments—Part 1: Classification of air cleanliness.
15. EU Ch 7 paragraph 7 ... Chapter 3, paragraph 7 reads: "Premises should preferably be laid out in such a way as to allow the production to take place in areas connected in a logical order corresponding to the sequence of the operations and to the requisite cleanliness levels." [REF].
16. "3-A Accepted Practices for permanently installed product and solution pipelines and cleaning systems used in milk and milk product processing plants, Number 605–04." Diary, Food and Environmental Sanitation, 12(2).
17. ASME BPE-2002. Bioprocessing Equipment. New York: The American Society of Mechanical Engineers, 2002.
18. Haga R, Sei M, Ostrove S, Weiss S. Cleaning mechanism study for biopharmaceutical plant design. Pharm Eng 1997; 17(5):8.
19. Young JH, Lasher WC, Gaber RP. Transport processes during sterilization of vertical and 5 degree horizontal deadlegs. Bioprocess Eng 1995; 12:293.
20. Banes PH. Passivation: understanding and performing procedures on austenitic stainless steel systems. Pharm Eng 1990; 10(6):41.
21. ISPE Baseline Pharmaceutical Engineering Guides—Vol. 4 Water and Steam Systems Baseline® Guide. Tampa, FL: International Society of Pharmaceutical Engineering, 2001.
22. United States Pharmacopeia XXVII, Water for Pharmaceutical Purposes, section <1231>, 2003.

17
Production of Proteins by Transient Expression

Alain R. Bernard
Serono Pharmaceutical Research Institute, Geneva, Switzerland

INTRODUCTION

Historically, the first observation that transfected heterologous DNA may translate into protein in a controlled way in a mammalian cell environment was by transient expression (1,2). Following this achievement, the search for novel genes and their associated gene product has been the major driving force for the development of several powerful transient expression systems. The complete sequencing of the human genome has then started the expected race to the next, more interesting level: the complete human proteome, comprising tens of thousands of gene products. Clearly, today, only a minority of these gene products is fully understood in terms of their biochemical and biological function. Transient gene expression bridges these two worlds in the most efficient manner. Nowadays, several transient expression systems have matured and are as diverse as direct DNA transfection into mammalian cells, recombinant virus infection, injection of *Xenopus* oocytes with mRNA or complete in-vitro coupled transcription–translation. Whereas functional expression of large cDNA libraries has allowed the cloning of many new genes by their function, the analysis of the gene products themselves, i.e., the corresponding proteins, however has lagged behind because of the lack of efficient, quick and easy ways of producing sufficient amounts.

All transient expression methods have historically been introduced and designed for small-scale operations (petri dish or tissue culture flasks). However, the emphasis of recent developments has been on the throughput, scale-up and the application of these laboratory methods to large-scale production systems. The most developed methods can now be operated at pilot bioreactor scale.

The ability to express rapidly and reproducibly one or several recombinant genes in an eukaryotic cell environment and to produce large amounts of the gene product is now established. This has prompted several laboratories to implement ambitious programs to express thousands of unknown DNA sequences in a mammalian cell host in order to obtain sufficient protein amount to carry out a functional biological analysis. This article reviews the most advanced transient expression methods that can be used on a large scale for the production of recombinant proteins of pharmaceutical interest for in vitro biological, biochemical, animal and human studies. A recent review also describes experimental protocols in more detail (3). The various methods presently available are classified according to the vector used: plasmid DNA, or viruses. Several of the vectors used in transient expression are also used in gene therapy applications, but we will restrict the scope of this review to the production of recombinant proteins in cell cultures.

STABLE VS. TRANSIENT EXPRESSION

Stable Expression

Methods for stable expression of heterologous genes in mammalian cells have been available and optimized for many years and presently form the basis for the development and manufacturing of most novel biopharmaceuticals production in cell culture.

Expression levels achieved by stable expression can reach values in the gram/liter range of the target protein in particular when expressing antibodies (4–7). These levels are much higher than those reported by transient expression (8–13). This is due to several combined actions such as amplification of the heterologous gene to high copy numbers, integration into targeted, transcriptionally active regions, several rounds of selection of high producer clones.

Nevertheless, the time required for the construction and isolation of an optimal, stable recombinant cell line remains as a major drawback and bottleneck of the methodology. By definition, a stable cell line, in which the heterologous DNA has integrated into the host genome and is maintained throughout many generations, requires an extensive time (6 months to a year) to generate a stable clone and to demonstrate that it has the required properties of clonality, integration and stability of high expression.

Transient Expression

On the contrary, the short time lines of most transient expression systems have made them attractive for producing human or mammalian recombinant proteins, mainly for research purposes (Fig. 1) In this type of methodology, the expression is exploited immediately after introduction of the foreign DNA into the cell. The cell line usually loses its ability to express the target protein with time. The transfected or infected cells are used for a limited life span. The heterologous DNA is usually not integrated into the host genome. It may be maintained inside the cell as an autonomous replicon or as part of a virus. It may also be delivered as a nonreplicating entity (see Alphavirus Vectors), in which case it is rapidly lost by the cell. Therefore, the two systems (stable vs. transient) are really by nature very different. Table 1 summarizes the main similarities and differences. Nevertheless the applicability of transient gene expression to the production of therapeutic proteins for clinical purposes should, technically, be feasible but remains still to be demonstrated and the associated regulatory issues raised by such a different concept need to be discussed.

CELL HOSTS

Three cell types have been most frequently used for transient expression: HEK-293, COS and BHK cells. More work on CHO is currently in progress. COS cells are the most widely used host for transient expression at small scale (petri dish). Derived

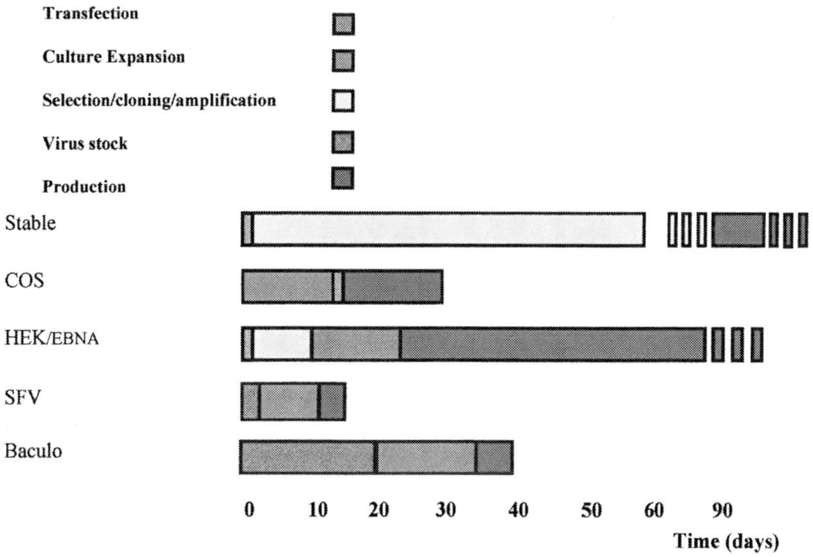

Figure 1 Transient expression timings. The average times for the various step operations have been estimated for each system. Time 0 corresponds to the availability of the expression vector (whether virus or nucleic acid). Repetitive operations are indicated by leaving some empty time between each of them. Baculovirus technology is used for comparative purposes. (*See color insert p. 11.*)

from green monkey kidney cells (CV-1) they were transformed with an origin defective SV40 virus (14). Historically, three similar cell lines: COS-1, COS-3, COS-7 were generated. The cells express high levels of the wild type SV40 large tumor (T) antigen, an early gene product of 100 bp, which binds to the vector DNA at the origin of replication and allows the host-cell polymerase to run repetitive cycles of DNA replication. As a consequence, plasmids carrying the SV40 origin are maintained in the cell at high copy number (10^4–10^5 copies per cell), giving rise to abundant mRNA and yielding high level recombinant gene expression. COS cells are known to express for up to 1 week post-transfection, with peak protein production between 48 and 72 hr post-transfection.

Table 1 Stable and Transient Expression Features

	Stable cell lines	Transient
Time	Months	Days
Stability of expression	Good	Variable
Host cells	CHO, NS0	CHO, HEK293, BHK, COS, HKB11
Expression level	Gram/L	1–10 mg/L
Post-translational modifications	Rodent-like	Rodent, simian or human
Main applications	Therapeutic proteins	Biochemical research and functional genomics

Lusky and Botchan (15) were the first to use the large T antigen expressing COS in combination with an SV40 origin containing plasmid for the identification of DNA transcription sequences. In 1982, the COS cell expression system was used for the first time by Rose and Bergmann to express cell surface and secreted proteins such as vesicular stomatitis virus glycoprotein (16). Subsequent applications by numerous other investigators demonstrated that the COS cell expression system generally yields biologically active proteins. The cells are readily available from public collections (ATCC, ECACC) and are efficiently transfected. Cultivation of these adherent cells does not pose any major problem, even for large-scale cultures. HEK-293 cells (human embryonic kidney fibroblasts, ATCC CRL-1573) provide an alternative to COS cells (17). It is a permanent line of primary human embryonic kidney transformed by sheared human adenovirus type 5 (Ad 5) DNA. The 293/SF isolate (ATCC CRL-1573-1) was derived from its parent after adaptation to growth in serum-free medium. Their human origin and robustness has motivated the development of several specific tools such as expression vectors and culture media and conditions. A number of HEK-293 cell isolates, some of them modified genetically, are useful for transient expression: HEK-293 T are cells that express the SV-40 large T antigen (18), HEK 293 EBNA express the nuclear antigen from Epstein–Barr virus EBNA (10). To facilitate large-scale operations, HEK-293 have been adapted to suspension (19,20). However, frequently, these cells grow as small aggregates in suspension. Adapting these "difficult" cells to suspension growth is due to the development of specialized media (21–23) in which cell aggregation can be controlled to less than 20 cells per aggregate.

BHK cells were originally isolated by polyoma transformation of hamster cells (24,25) and were extensively used as substrates for virus propagation for vaccine and more generally for viral mediated expression. It is also considered as a useful host cell for stable expression (26,27).

Surprisingly, CHO cells, while probably the most widely used host for stable expression of proteins, have not been used frequently so far mostly because of low transfection efficiencies and absence of robustness during transfection. However, this was probably due more to a lack of specific tools and efforts to understand the requirements of these cells when used for transient gene expression rather than to an intrinsic cellular property. In fact by engineering an anti-apoptosis dominant negative mutation, caspase 9-DN into a standard CHO host cell, it was recently shown that cell viability is no longer adversely affected by high levels of transfection reagents (11).

Other recent studies have been done with a new host cell line HKBl1 is a somatic human hybrid cell line of HEK-293 and 2B8 (a derivative of Burkitt's lymphoma cell line). HKBl1 cells grow well as a suspension culture with small loose aggregates under serum-free conditions. HKBl1 cells can be transfected with high efficiency and secrete high levels of proteins with human-specific glycosylation profiles (28). Recent studies suggest that host cells such as COS, CV-1, and HEK-293 cells demonstrate higher expression of the target gene when overexpressing wild-type PKB concomitantly. This results in an improved recombinant protein titer, up to threefold when using three different secreted proteins as reporters (29).

VECTORS FOR TRANSIENT EXPRESSION

Plasmid Expression Vectors

Introducing the heterologous target gene into the cell via plasmid DNA is the most straightforward approach (30). The best transfection methods commonly achieve transfection efficiencies of more than 70% and thus guarantee satisfactory expression levels.

Attempts to transiently express recombinant proteins from plasmid vectors were originally performed with "conventional" expression vectors that contain strong viral promoters, such as SV40 or CMV (8,10,31,32). More recently, at least one nonviral promoter, the EF-1 promoter (elongation factor 1) (33,34) has demonstrated its utility and efficiency (35–37).

Appropriate vectors can provide extended maintenance of plasmids in cells. These vectors are used in combination with hosts that are themselves previously engineered to express elements that promote episomal plasmid replication, such as the large T-antigen from SV40 (see Cell Hosts). In this case, the plasmid carries the origin of replication from SV40. In another approach, the transfected plasmid may carry a second expression cassette (in addition to the gene of interest) for the large T-antigen. Experience in some laboratories (8) indicates that this approach improves recombinant protein production three- to fivefold.

Today, many SV40 T antigen dependent vectors (e.g., ATCC 37192, 37193, 53100) are available for gene expression with COS cells. Typical plasmids for COS transient expression carry the following major components:

- the SV40 origin for plasmid replication
- appropriate promoter, enhancer, and polyadenylation sequences
- a prokaryotic origin of replication
- a prokaryotic selectable marker
- unique restriction sites for target gene subcloning.

Epstein–Barr virus transformed cell lines expressing the EBV nuclear antigen-1 (EBNA-1) replicate transfected plasmids carrying the replication origin region (OriP) from EBV. As a consequence, they express the subcloned gene, in a similar way to the COS cell-based system. Both genomic and cDNA sequences can be expressed in this system. EBV vectors replicate episomally in the nucleus, and, under certain conditions, can be maintained by the cells upon prolonged passage, thus providing a way of obtaining either transient or stable expression of the recombinant protein. In addition, and just like with the SV40 T antigen expression described above, incorporation of the coding sequence for EBNA-1 on the expression vector yields a self-replicating expression plasmid, which can be autonomously maintained under selection pressure. Newly developed improvements include a vector where the CMV enhancer/promoter (CMVe/p) is controlled by transactivating elements. This vector was designated as TT-oriP expression vector. The expression vector comprises two key functional elements, oriP for plasmid maintenance and HIV-1 Tat/TAR transactivator element for protein expression. This expression vector, when used in HKBl 1 cells (see Cell Hosts), has been shown to sustain high levels of protein production over up to 10 days (28,38).

Alphavirus Vectors

Sindbis Virus (SIN) or the Semliki Forest Virus (SFV) are the alphaviruses most commonly used as vectors for heterologous expression (39). They are very small (40–50 nm) enveloped viruses containing single stranded RNA of positive polarity. After entering the cell, the 5′ two-thirds of the RNA is translated into precursors of four nonstructural proteins (nsPl–nsP4). The transcriptase, a key nonstructural protein produces a negative strand copy of the RNA genome. This negative strand then serves as a template for the generation of positive strand RNA copies. The 3′ third of the genome codes for the viral structural proteins, i.e., the capsid (C) and two membrane proteins p62 and El. Once expressed, the capsid protein packages the genomic RNA within the cytoplasm. The membrane proteins are translocated into the ER, where they form heterodimers and allow virus budding from the cell surface (39,40). The released infectious virus particles can then be taken up by other cells via receptor mediated endocytosis. This type of virus infects a wide range of mammalian cells and eventually leads to cell death within 24–48 hr post-infection.

The high level expression that these viruses allow and their broad host range have made them particularly attractive for functional studies and investigation of the biological role of proteins (41). The exploitation of such viral vectors was originally developed by Liljeström and Garoff (Fig. 2) who split the SFV genome into two plasmids: pSFV1 and pSFV-Helper (42). pSFV1, the cloning vector, harbors the replicase genes and a multiple cloning site where the heterologous gene is inserted, under control of the 26S promoter. pSFV-Helper, the helper vector, encodes the structural proteins of the virus. In a variant version of the helper vector pSFV-HelperII, three point mutations were introduced into the spike protein encoding gene p62. This modification was made to increase the level of biological safety (43). The mutation prevents the natural proteolytic processing of the spike protein into an active protein that mediates virus entry into the cell thereby rendering the virus non-infectious. After packaging, the spike protein can then be processed in a

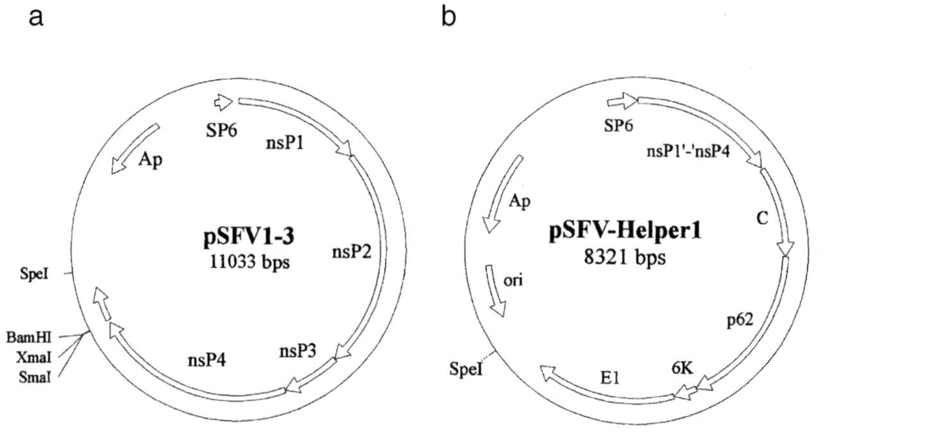

Figure 2 Semliki Forest virus expression vectors pSFVl-3 (panel a) and pSFV-Helperl plasmids (panel b). Construction of the SFV expression plasmids were done by deleting most of the coding regions for the structural proteins (C, p62, 6K, El) from the cDNA clone which is spanning the complete SFV genome. The vectors retain the promoter region of the 26S subgenomic RNA (at the end of the nsP4 region) and the last 49 amino acids of the El protein as well as the complete noncoding 3′ end of the SFV genome. Downstream of the 26S promoter a polylinker (*Bam*HI, *Xma*I, *Sma*I) was introduced. The SP6 promoter, which is followed by the replicase (nsPl–nsP4) allows in vitro generation of full-length SFV RNA transcripts. The eukaryotic origin of replication (ori) and the antibiotic resistance (Ap) allow for plasmid production in *Escherichia coli.* A unique *Spe*I restriction site was inserted for plasmid linearization. The pSFV-Helperl represents the SFV genome cDNA being deleted of most of the nonstructural regions but carrying all the structural genes (C, p62, 6K, El).The pSFV-Helper2 code for a cleavage-deficient variant of the spike protein. Reprinted from Ref 39 with permission of *Nature-Biotechnology*, NY.

controlled manner by chymotrypsin cleavage thus restoring full infectivity of the virus stock (43).

The extremely efficient SFV 26S promoter and the high number of copies of RNA (approximately 200,000) per infected cell lead to high expression levels. Recombinant proteins are produced with post-translational modifications conferred by the specific processing capacities of the host cell that the virus infects (44–47) and the system has been scaled up to stirred tank reactor scale (48).

RNA generated in vitro from the cloning vector has also been used as a stand-alone expression vector by direct transfection of cells. This method was shown to lead to subsequent expression of the heterologous gene of interest (49). This approach, although less efficient than recombinant packaged SFV particles, is well suited to the expression of oncogenes and other hazardous genes and also potentially to cell or gene therapy.

Adenovirus Vectors

Isolated in the early 1950s, adenoviruses have more recently received a renewed interest for their use as vectors for DNA delivery to cells in vivo in gene therapy applications, as potential vaccine vectors or as expression vectors for heterologous genes expression in vitro.

The extensive knowledge accumulated over the years on adenovirus transcription regulation allowed the engineering of adenovirus vectors modified for heterologous expression (50,51). The adenovirus life cycle is composed of several phases. In the early phase, regions E1–E4 and L1 of the linear double-stranded DNA (36 kb total length) are actively transcribed from the early (E) and major late promoters (MLP), respectively (50). In the following phase, infection is followed by viral replication 8 hr later, and host transcription and translation are gradually shut-off as the cycle progresses to the late phase of infection. In this phase, late transcripts from the MLP, L2–L5 are synthesized, together with structural proteins. The MLP promoter drives high levels of expression of these late proteins and they can account for up to 50% of cellular protein mass. The cycle then continues with cell death and release of high virus titers (10^8–10^9 pfu/mL) in the extracellular medium. These infectious particles can in turn infect more cells and the cycle repeats.

For recombinant expression purposes, the early regions El and E3 were deleted. This renders the virus replication incompetent and this function must then be provided in trans by the host cell. Replication competent cell lines have been engineered from HEK-293 specifically to provide that function (17,52). An expression cassette was set in place of the deleted El region. In the cassette, the recombinant gene is placed under control of an additional (or ectopic) MLP or under control of an exogenous promoter such as CMV (53). More recently, a new series of vectors, pAdBM5 was constructed (54) with enhancers added to the ectopic MLP region. Late promoters are preferred for yielding high level of protein expression whereas an early or compound early/late promoter may be chosen when the infection cytopathic effect has to be minimized (for functional studies for example) (51).

Vaccinia Vectors

The vaccinia virus genome is composed of a double-stranded DNA of nearly 200,000 bp and is replicating in the cytoplasm of the host cell (55). Infection of cells with the vaccinia virus leads to the production of up to 5000 virus particles per cell.

Table 2 Vaccinia Large-scale Productions

Scale (L)	Host cells	Company	Product	Reference
150	BHK	Pasteur Merieux	HIV-1 rgpl60	96
1200	VERO	Immuno AG	HIV-1 gpl60	97
1200	VERO	Immuno AG	Human prothrombin	98

[From Ref. (3).]

For heterologous gene expression applications, the recombinant gene of interest is placed under the control of a vaccinia promoter.

Integration of the foreign gene can be accomplished by one of three methods: homologous recombination, direct cloning or by constructing a hybrid of vaccinia with the bacteriophage T7 RNA polymerase.

The generation of a recombinant vaccinia virus by homologous recombination is usually done in three steps: (1) Subcloning cDNA into a plasmid next to a vaccinia promoter, flanked by nonessential vaccinia gene sequences, (2) Recombining into the vaccinia genome: cell infection with wild type vaccinia virus and transfection of recombinant plasmid, and (3) Screening for recombinant virus.

Direct cloning is performed by inserting the target cDNA into a unique restriction site of the vaccinia virus genome. The genome is then packaged by transfecting helper cells (fowl pox virus-infected cells).

High level expression can also be achieved with the vaccinia virus-T7 RNA polymerase hybrid system. This system relies on the cytoplasmic expression of the bacteriophage T7 RNA polymerase, e.g., by a stable transfected cell line. The recombinant gene is inserted into a vaccinia virus under the control of the T7 RNA polymerase promoter. To provide the T7 polymerase, either constitutively expressing cells are infected or cells are co-transfected with a polymerase expressing vaccinia virus.

The vaccinia expression system has been used extensively for HIV vaccine preparations and human prothrombin (Table 2).

PLASMID DNA-MEDIATED EXPRESSION

Vehicle Systems

Calcium phosphate (56), polyethyleneimine/DNA complexes (PEI) (57), electroporation (58,59), DEAE Dextran (60) and lipofection (61,62) are some of the most conventional vehicles/approaches for cell transfection and subsequent transient gene expression. In addition, other methods have been described such as receptor-mediated endocytosis, protoplast fusion, poly-lysine and polybrene-mediated transfection. Direct DNA injection into cells with the gene gun (bombardment with DNA-coated gold particles) is used in special applications where cells are difficult to transfect or for certain electrophysiology studies. DNA complexed in some form to cationic molecules such as PEI is convenient and cost effective for large-scale operations. A combination of dioleoylmellitin (DOM) and PEI increases stability of the DNA–PEI complexes and subsequently increases expression levels (23,63,64). New transfection reagents such as RO1539 are also regularly designed and tested (65). The RO1539 reagent, reportedly not toxic to cells, generates high transfection efficiency in different cell lines at low DNA concentration (0.2 μg/mL) (66). Lipofectamine2000™ and RO1539 are reported to result in highly efficient transfections. Both reagents show very similar high expression levels of secreted alkaline phosphatase used as a model reporter gene, however the cell toxicity is minimized with RO1539.

Generally, medium composition, environment and cell physiological states need to be optimized for each transfection reagent in order to maximize the transfection and expression efficiency.

The transfection methods described above can be applied to cells grown either in adherent or suspension mode. For cells grown in suspension, they have been used at scales up to about 100 L bioreactors (67). Expression levels of 1–10 mg/L of product have been reported for secreted proteins, corresponding to specific productivities of about 0.1–1 pg/cell/day.

Elegant studies elucidate the complex and numerous mechanisms involved in the rather obscure process of DNA transfer from a test tube into the nucleus of a target cell when using cationic polymers/DNA complexes (68–70).

After DNA complex formation, complexes are taken up by cells through endocytosis (71,72). After cell penetration, DNA complexes migrate inside the cytoplasm via endosomes. Polyethyleneimine complexes induce endosome osmotic swelling and rupture via the proton sponge mechanism (73). Finally, the last barrier to overcome is the nuclear membrane (74). It is estimated that large quantities of DNA are actually degraded during the complete process (68). However, this very low yield is balanced by putting into contact with the cells a large molar excess of plasmid molecules ($1–10 \times 10^6$ per cell), which ensures that a sufficient number does reach the nucleus.

Calcium Phosphate

DNA uptake by mammalian cells can be achieved by exposing cells to condensed forms of DNA. Calcium phosphate-DNA co-precipitation provides one amongst many other methods to achieve DNA condensation. A solution of DNA in $CaCl_2$ is diluted drop wise into a concentrated HEPES-balanced salt buffer. The co-precipitate is formed within a few minutes. Several parameters for transfection with calcium phosphate are critical, including the time allowed for formation of the calcium phosphate–DNA co-precipitate and the pH at which this operation is performed. Even though routinely used this method is prone to large variability in the final results and detailed studies report on the optimization of the parameters involved (75–77). Calcium can be substituted by strontium and this substitution has been reported to be beneficial at least in some cases (78). After formation, the DNA–$CaPO_4$ co-precipitate is then added to the cells in adherent cultures and some investigators recommend replacing the medium after about 10 min with fresh medium. In all cases, the transfection mixture is removed from the culture within one day after precipitate addition and replaced by fresh culture medium, allowing 2–3 days for gene expression.

Calcium phosphate transfection is also applicable to cells grown in suspension. After a wash with PBS, the cells are re-suspended and incubated in a solution containing the calcium phosphate–DNA co-precipitate. After adding fresh medium in order to dissolve remaining calcium phosphate precipitates, the cells are incubated for 24 hr, then centrifuged and transferred into complete fresh medium.

Scaling-up this method requires that the standard protocols be adapted to reduce or avoid the incubation, washing- and separation steps, since these steps are not compatible with large volumes. Significant progresses have been made quite recently and it looks like this methodology is truly adaptable to large-scale operations (67).

Lipofection

DNA can be delivered to cells by liposome vehicles (61,62,79,80). Positively charged liposomes are formed with various cationic lipids such as *N*-[1(2,3-diolexyloxy) propyl]-*N*, *N*,*N*-trimethylammonium chloride (DOTMA). These liposomes, when mixed in appropriate conditions with DNA produce DNA-liposome complexes. These complexes are adsorbed to and subsequently fuse with the lipid bilayer of cell membranes, delivering DNA into the cytoplasm.

The experimental protocols are relatively simple (62): plasmid DNA is diluted into serum free medium and mixed with pre-formed cationic liposomes. For the actual transfection, cells are overlaid with this mixture for several hours. Finally, fresh medium is added and the culture is continued for a few days during which expression of the target gene occurs.

No developments in large-scale transfections have been reported to date, probably because of the high costs of the liposome starting material required for lipofection. However, lipofection has been successfully carried out in spinner systems and this indicates the potential for scale-up.

Electroporation

Transfecting cells by electroporation is achieved by imposing a short, powerful electrical impulse onto the cells. This method has been applied to DNA transfer into mammalian (58,59,81,82) as well as plant cells (83,84). The pulse is calculated to prevent excessive damage of the membrane structure. It is generated by discharging a capacitor and, as a consequence, small pores are created into the cell membrane by interaction between the electrical field and the membrane lipid dipoles. These pores allow DNA to enter the cell by diffusion within 10 min after the electrical pulse (58,59). A few pulses (two to five) are usually applied to the same cell suspension in order to maximize transfection efficiency.

Practically, cells are grown to sub-maximal density, pelleted by centrifugation and suspended into a conductive medium (e.g., normal culture medium or PBS) together with the plasmid DNA. The mixture is transferred to a sterile electroporation cuvette. The electrical pulses (typically 230 V, 960 μF, for 250 μL) are applied and the cells are then diluted back into fresh medium.

Parameters which determine the transfection efficiency are:

- the field strength (V/cm) defined by the voltage applied and the distance between the cuvette electrodes
- the pulse shape which can be square or with a logarithmic decay
- the time constant which is directly related to the type of the electroporation cuvette
- the conductivity of the medium and the cell density during electroporation
- the concentration of DNA
- the number of electrical pulses applied

The optimal settings of these parameters vary between different cell types. Other variables like the pre and post transfection incubation times with the DNA, the temperature at which the electroporation is performed, and the physiological state of the cells also influence the final result. Blasey (personal communication) adapted the electroporation method to moderate scale operations. A drawback of the electroporation method is that a very large proportion of cells is killed by the electric discharge. Hence only a small percentage (<10%) of surviving clones out of several million transfected cells can be expanded further (59). Electroporation has been used to transfect COS cells at a liter scale, however, the process was carried out in batch mode, i.e., by electroporating cells in a large number of small volume cuvettes and this operation mode clearly limits further scale-up (13). Commercially available continuous electroporation systems are available but only few results have not been reported using these systems (85).

Large-Scale Transfection

Culture Scale-Up

COS cells can be grown up to large quantities in any appropriate culture vessel such as tissue culture flasks, roller bottles, microcarrier cultures or larger multilayer systems (Cellfactory(TM) or Cellcube(TM) for example), but clearly, only the last two provide an efficient means of large-scale cell productions.

For scale-up purposes, some HEK-293 lines have been adapted to suspension (19,20,86). However, frequently, these cells form aggregates when grown in suspension. The ability to adapt these cells to suspension growth is due, by large, to the development of specially designed media (22,23,87). In these media, cell aggregation can be limited to a small size (<20 cells).

For the well known CHO and BHK cell lines, large-scale culture methods are now well established and a large variety of technological options are available (see Chapter 2 in this book). For mammalian cells that grow in anchorage dependence mode at intermediate scales (1–101), the use of stacked, multilayer systems has become very popular. Cells can be maintained in these production vessels for a limited period in static, batch-type culture or, for extended periods, in a perfused mode. The major disadvantage of stacked multilayer systems is cost. However, they are usually supplied as sterile, ready-to-use and disposable units. In addition, they have become a first choice approach in the field of somatic gene therapy for mass production of virus-based DNA vectors.

Transfection and Protein Production

The first step of large-scale transfection processes is the preliminary preparation of plasmid DNA in sufficient amount and quality. The quantities of required plasmid DNA range from 0.2 to 2 mg per 10^9 cells. DNA is generally thought to be required in the purest form, after extraction from *Escherichia coli* although no definite study demonstrates this requirement. This purity requirement stems from developments of large-scale DNA preparations to be used in gene therapy applications (88–90) but in vitro cell transfection does not necessarily require the same DNA purity. After transfection, cells are usually maintained under production conditions for a limited time span, depending on the vector used. Viral vectors eventually lead to cell death in a short time while DNA plasmid vectors may allow longer production times, e.g., 5–10 days post-transfection. Even longer product accumulation periods can be reached with combinations of vectors and hosts that are competent for episomal replication. Plasmid DNA transfections of cells grown in suspension are most suited to process scales beyond 10 L. However this scale limit is constantly pushed up and a number of articles show that the basic principles of the technology are established (9,67). Nevertheless, a lot more work is required to make this process reliable and robust.

The transfected culture is allowed to recover for 1 day in FCS containing medium, which is then exchanged for low protein, serum-free medium for the protein production phase. Daily medium exchanges allow the extension of the culture to up to 10 days post transfection. A typical run yields several milligrams of pure protein (13).

More specific examples are illustrated in the following section.

COS Transfections

After growing COS cells in roller bottles, the culture is transfected by the DEAE dextran method. For this purpose, it is overlaid with a DEAE transfection mix and after a DMSO shock, medium is added and the transfected culture is further incubated for several days This method was reported for producing, at a liter scale, 1–2 mg/L of the scFv::IgG-kappa fusion protein (60).

HEK293 Transfections

Two distinct strategies can be followed for large-scale transient expression with HEK293 cells. One is based on cell expansion up to reactor stage and in-situ transfection, the other consists in transfecting the cells at very small scale and successive expansion of the transfected cell population to reactor scale (Fig. 3).

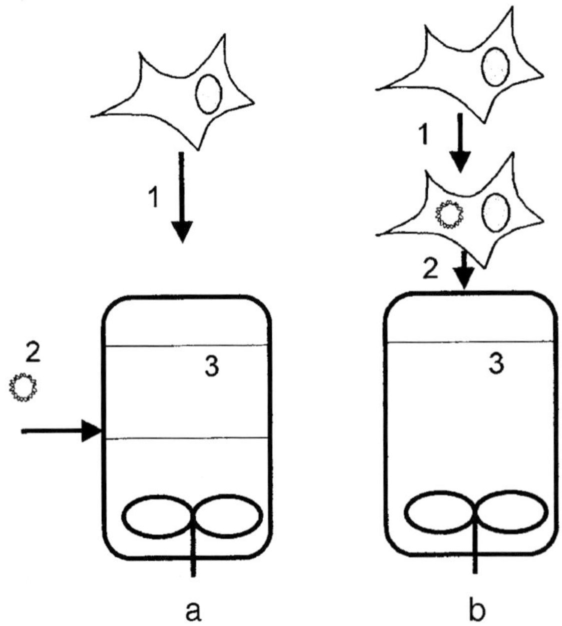

Figure 3 Large-scale expression by plasmid DNA transfection. The two possible approaches are illustrated in the two panels (a, b). In panel a, cells are first grown to a reasonable density in a large bioreactor (step 1). They are then transfected in situ (step 2) by addition of DNA complexed in some form (CaPO4, liposomes, PEI, . . .). Shortly after transfection, the culture volume is raised to its final value by addition of fresh medium (step 3). Harvest of the protein can be made a few days later. In panel (b), cells are first transfected with the expression vector at very small scale (step 1) by the same transfection methods used in panel (a), or by electroporation. The surviving transfection pools are then put into gradually expanding culture vessels (step 2) until the final bioreactor stage (step 3). Like in panel (a), harvest can be operated a few days later.

In situ transfection. HEK cells previously adapted to suspension growth are propagated in a stirred bioreactor and transfected by the calcium phosphate method. Once the cell density reaches 3×10^5 cells/mL, the calcium concentration in the medium is elevated to 10 mM by addition of $CaCl_2$. After 30 min transfection is initiated by addition of 20 mL/L of the calcium phosphate/DNA precipitate (at 500 μg DNA/L). Four hours later, calcium phosphate is diluted by simple addition of fresh culture medium, thus avoiding a separation step. In a typical example, 0.5 mg/L recombinant tPA was produced during a 5-day production period (86). HEK293 and CHO cells have also been transfected at large scale by using the transfection reagent, RO1539 (9). SEAP was used as a model protein and the average expression level after 3–4 days was 30 μg /mL with HEK293EBNA and 12 μg/mL with CHO cells (9,66). A transfection protocol in a final culture volume of more than 100 L was recently reported (67). Suspension-adapted HEK 293 EBNA cells were transfected in situ in a large bioreactor by the calcium phosphate method. More than half a gram of a monoclonal antibody (IgG) was produced in <10 days.

Expansion of transfected cell pools. An alternative approach consists in transfecting cells at very small scale and thereafter expanding the culture volume gradually. In addition, transfected cell pools can be maintained under selective pressure with hygromycin. Under these conditions, HEK-293 EBNA-1 cells can be subcultured for up to 6 months while still expressing recombinant proteins at a high level. For example, human placental alkaline phosphatase, tissue plasminogen-activator, vascular endothelial cell growth factor, hepatocyte growth factor, prolactin, neurotrophins and soluble tumor necrosis factor receptors were expressed in these conditions at levels of 0.2–10 μg/mL (10).

VIRUS-MEDIATED EXPRESSION

This section will focus on the use of Semliki Forest Virus (SFV) in large-scale productions (91) (Fig. 4). With SFV, levels up to 25% of total cell protein have been reported for β-galactosidase (39) and several million neurokinin-1 receptors per cell (92). SFV can infect a broad range of host cells, however the level of expression differs between cell lines: BHK, CHO and COS cells yield higher levels of expression than other cells such as NS0, RPMI 8226 and HeLa. This was clearly demonstrated for different types of proteins: multi-transmembrane receptors (such as neurokinin, serotonin, dopamine, adenosine or purinergic receptors), secreted proteins (such as soluble placental alkaline phosphatase, lysozyme, viral glycoproteins) and some cytosolic proteins (such as β-gal, cyclooxygenase, DHFR).

At least in BHK cells it is known that the optimal pH is different for SFV infection (pH 6.9) and protein production (pH 7.3). The multiplicity of infection (MOI) also strongly modulates the expression level (91), i.e., a higher MOI (above 10) results in elevated expression levels.

Virus Generation

After subcloning the gene of interest into the multiple cloning site of pSFV (see Alphavirus Vectors), the recombinant and helper plasmids are linearized by an *Spe*I digestion to prepare for in vitro transcription and RNA production by the SP6 RNA polymerase. BHK-21 cells are then co-transfected with the recombinant and helper RNAs which initiates the in vivo packaging process. It is important to note that the packaging signal is present only on pSFV-1 and absent from the pSFV-Helper

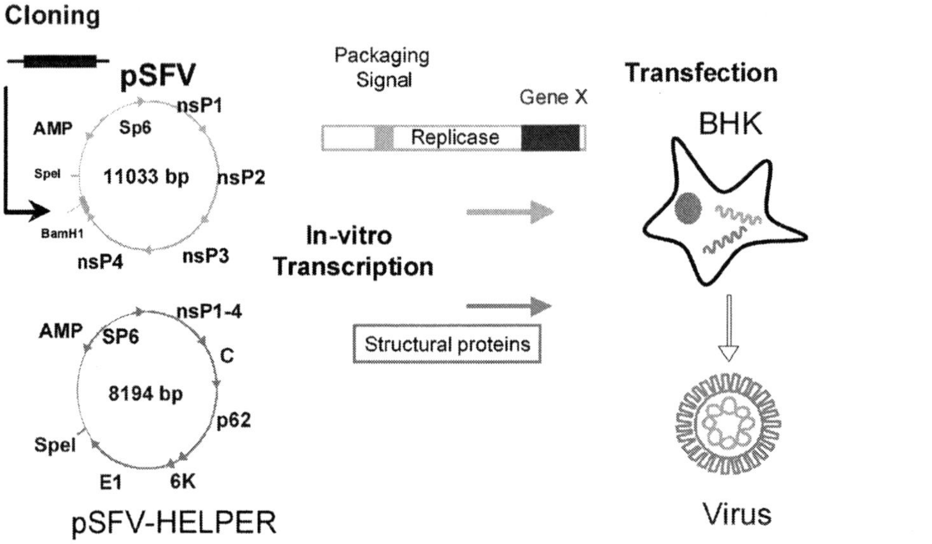

Figure 4 Semliki Forest virus expression system. In vivo packaging of recombinant SFV RNA transcripts into infectious particles. After co-transfection of RNA transcipts from the expression and the helper vector, the replicase, coded on the recombinant vector, drives RNA replication and transcription for both plasmids. Structural genes from the helper are translated into proteins. While the capsid protein stays cytosolic and binds RNA to form the nucleocapsid, the membrane proteins undergo post-translational modifications. El and p62 are dimerized in the ER and are transported to the plasma membrane where virus budding occurs. Since the viral packaging signal is only found on the recombinant RNA, selective packaging of the pSFVdhfr occurs and as a result yields nonreplicative infectious SFV particles. [Adapted from Ref. (39).] (*See color insert p.11.*)

vector. As a consequence, infectious but non-replicative, recombinant virus particles are produced and a high titer (approximately 10^9 virus particles/mL) is harvested 24hr after transfection.

Protein Expression

As a first step, BHK cells are grown to confluence in T-flasks. In parallel, the non-infectious virus stock is activated by incubation with chymotrypsin, which is then inactivated after 20 min by addition of aprotinin. For large-scale applications, BHK cells are grown in single cell suspension to 10^6 cells/mLin a stirred tank reactor. Before infection, 90% of the medium is renewed either by centrifugation or by external cross-flow filtration. The virus culture is infected by addition of the activated virus stock at a high MOI (>10). Within few hours, the host cell protein synthesis is dramatically reduced or shut down and cell growth is arrested. Maximum product concentration is usually reached within 1 day post-infection, but may vary depending on the proteins expressed. Typically, 16 hr post-infection the culture is harvested by centrifugation and the recombinant protein purified. The above procedure can also be applied to other host cell lines after fine tuning the conditions for each specific one.

Results of large-scale protein production with SFV have recently been described with cyclooxygenase-2 [16 mg microsomal protein per 10^9 cells, (91)], or

Table 3 Transient Expression Systems

	Time scale	Advantages	Disadvantages	Applications	Yields
Large T-expressing cell lines (e.g., COS)	3–4 weeks	Established technology Simple to execute Liter scale demonstrated	Large DNA and cell quantities required for transfection Large-scale transfection Short expression time after transfection (1–2 weeks), overgrowth by nonproducers	Functional and structural studies	0.1–3 mg/L Up to 10 mg protein
HEK293 EBNA	4–8 weeks			Functional and structural studies, target validation, pre-clinical studies	0.1–10 mg/L 100 mg protein
In situ transfection		Speed Scale-up of host cell culture only	Large amounts of DNA needed Reproducibility?		
Expansion of transfected cells		Selective pressure available Transfection requires only small quantities of DNA and cells Potential selection of clones Seed bank possible			
Baculovirus	6–8 weeks	Established technology Easy scale-up Safe viral system	Lytic system Non-mammalian glycosylation	Functional and structural studies, target, validation, preclinical studies	0.5–30 mg/L Gram quantities

(*Continued*)

Table 3 Transient Expression Systems (*Continued*)

	Time scale	Advantages	Disadvantages	Applications	Yields
Semliki Forest Virus	2–6 weeks	Exceptionally short time from gene to product High yields Very broad host range	Safety category 2 for activated virus Limited virus scale-up Lytic system	Functional and structural studies, target validation	Up to 8×10^6 receptors per cell Up to 100 mg protein
Vaccinia	2–8 weeks	Easy scale up to very large scale Virus simple to scale-up Broad viral host range	Pathogenic: safety category 2 Lytic system	Functional and structural studies, target validation, vaccines	Around 0.1–5 mg/L Gram quantities

5HT3 receptor (1–2 mg 5HT3 receptor protein/L culture, equivalent to 25–50 pmol binding sites per cell (93,94)).

Virus generation for large-scale infection is still labor intensive, in particular the production of RNAs in vitro and co-transfection of cells with the two RNAs. This constitutes the main difference with other viral-mediated expression systems. Biological safety of the SFV expression system has also been given significant attention. The virus has now been made nonreplicative and only conditionally infective. However, since it is not possible to completely exclude that the culture contains active virus at the time of harvest, it is recommended to consider that this expression system and the handling of products obtained by this technology is operated under safety category 2 or higher.

CONCLUSIONS

Transient expression technologies have evolved in the recent years and have found an increasing number of applications. Table 3 summarizes the essential features of each technology. Large-scale use of these technologies for producing recombinant proteins in amounts sufficient for biochemical and biological studies (100–1000 mg) is becoming routine. Speed is the principal motivation. Starting from a cDNA, current protocols allow the protein to be produced within 10–15 days. By contrast, more classical technologies such as stable cell lines require selection and amplification of the recombinant gene and therefore considerably extend the time required for product generation. The eukaryotic environment in which most of these transient expression systems operate ensures that elaborate post-translational modifications are properly carried out and that the protein product obtained provides an authentic copy of its native human counterpart. In addition, novel routes to pharmaceutical discovery such as human genome sequencing create a serious need for high-throughput, medium-scale (10–100 mg of protein) expression systems in order to quickly characterize and evaluate the function of novel gene products. This is the next major challenge, which should be overcome in the coming few years.

REFERENCES

1. Wu G, Sharp PA. SV40 DNA transfection of cells in suspension: analysis of efficiency of transcription and translation of T-antigen. Gene 1981; 13:197.
2. Capecchi MR. High efficiency transformation by direct microinjection of DNA into cultured mammalian cells. Cell 1980; 2:479.
3. Bernard AR, Blasey HD. Transient expression systems. In: Flickinger MC, Drew SW, eds. Encyclopedia of Bioprocess Technology: Fermentation, Bio catalysis and Bioseparation. New York: John Wiley, 1999:2580.
4. Barnes LM, Bentley CM, Dickson AJ. Advances in animal cell recombinant protein production: GS-NS0 expression system. Cytotechnology 2000; 32:109.
5. Sauer PW, Burky JE, Wesson MC, Sternard HD, Qu L. A high-yielding, generic fed-batch cell culture process for production of recombinant antibodies. Biotechnol Bioeng 2000; 67:585.
6. Zhou W, Chen CC, Buckland B, Aunins J. Fed-batch culture of recombinant NS0 myeloma cells with high monoclonal antibody production. Biotechnol Bioeng 1997; 55:783.
7. Bebbington CR, Renner G, Thomson S, King D, Abrams D, Yarranton GT. High-level expression of a recombinant antibody from myeloma cells using a glutamine synthetase gene as an amplifiable selectable marker. BioTechnology 1992; 10:169.

8. Geisse S, Patoux A, Rohr F, Glueck D, Hagen S, Rinaldi D, Gram H. EBNA-1/ori P Driven Expression in HEK293 Cells, Transient Gene Expression in Animal Cells, Island of Jersey, United Kingdom, Hotel de France (Chairs: Alain Bernard, Ares-Serono Pharmaceutical Research Institute, Geneva, Switzerland and Robert Posse, NERC Oxford, UK), May 11–14, 1997.
9. Schlaeger EJ, Legendre JY, Trzeciak A, Kitas EA, Christensen K, Deuschle U, Supersaxo A. Transient transfection in mammalian cells. In: Merten O-W, Perrin P, Griffiths B, eds. New Developments and New Applications in Animal Cell Technology. Dordrecht: Kluwer Academic Publishers, 1998:105–112.
10. Cachianes G, Ho C, Weber RF, Williams SR, Goeddel DV, Leung DW. Epstein–Barr virus-derived vectors for transient and stable expression of recombinant proteins. Biotechniques 1993; 15:255.
11. Etcheverry T, Closkey L, Lee E, Tezare R. Protein production in the era of gene discovery: transient transfections in 10 liter Bioreactors. Conference Abstract, Current Status and Trends in Mammalian Transient Protein Expression, Ittingen, Switzerland, Oct 1–3, 2000.
12. Wurm F, Bernard A. Large-scale transient expression in mammalian cells for recombinant protein production. Curr Opin Biotechnol 1999; 10:156.
13. Blasey HD, Aubry JP, Mazzei GJ, Bernard AR. Large scale transient expression with COS cells. Cytotechnology 1996; 18:183.
14. Glutzman Y. SV40-transformed simian cells support the replication of early SV40 mutants. Cell 1981; 23:175.
15. Lusky M, Botchan M. Inhibition of SV40 replication in simian cells by specific pBR322 DNA sequences. Nature 1981; 27:279.
16. Rose JK, Bergmann JE. Expression from cloned cDNA of cell-surface secreted forms of the glycoprotein of vesicular stomatitis virus in eukaryotic cells. Cell 1982; 30:753.
17. Graham FL, Smiley J, Russell WC, Nairn R. Characteristics of a human cell line transformed by DNA from human adenovirus type 5. J Gen Virol 1977; 36:59.
18. Kim CH, Oh Y, Lee TH. Codon optimization for high-level expression of human erythropoietin (EPO) in mammalian cells—human or yeast codon usage effect on overexpression in 293T cell culture. Gene 1997; 199:293.
19. Graham FL. Growth of 293 cells in suspension culture. J Gen Virol 1987; 68:937.
20. Garnier A, Cote J, Nadeau I, Kamen A, Massie B. Scale-up of the adenovirus expression system for the production of recombinant protein in human 293s cells. Cytotechnology 1994; 15:145.
21. Tolbert WR, Hitt MM, Feder J. Cell aggregate suspension culture for large-scale production of biomolecules. In Vitro 1980; 16:486.
22. Peshwa MV, Kyung YS, McClure DB, Hu WS. Cultivation of mammalian cells as aggregates in bioreactors: effect of calcium concentration on spatial distribution of viability. Biotechnol Bioeng 1993; 41:179.
23. Schlaeger EJ, Christensen K. Transient gene expression in mammalian cells grown in serum-free suspension culture. Cytotechnology 1999; 30:71.
24. Macpherson I, Stoker M. Polyoma transformation of hamster cell clones—an investigation of genetic factors affecting cell competence. Virology 1962; 16:147.
25. Macpherson I. Characteristics of a hamster cell clone transformed by polyoma virus. J Natl Cancer Inst 1963; 30:795.
26. Capon DJ, Lawn RM, Levinson AD, Vehar GA, Wood WI. Preparation of functional human factor VIII in mammalian cells using methotrexate based selection. U.S. Pat. 4,965,199, 1990.
27. Schmid G, Johannsen G. Metabolic quotients for recombinant CHO and BHK cell lines producing human antithrombin III. Biotechnol Lett 1990; 12:317.
28. Cho M-S, Yee H, Brown C, Mirenda C, Mei B, Chan S. Mammalian cell expression system for high throughput transient transfection and large scale transfection for the

production of research proteins. Conference Abstract, 17th ESACT Meeting, Tylosand, Sweden, June 10–14, 2001.
29. Ettehadieh E, Wong-Madden S, Aldrich T, Morris AE. Overexpression of protein kinase B (PKB) enhances recombinant protein expression in transient systems. Conference Abstract, 17th ESACT meeting, Tylosand, Sweden, June 10–14, 2001.
30. Chen C, Okayama H. High efficiency transformation of mammalian cells by plasmid DNA. Mol Cell Biol 1987; 7:2745.
31. Paborsky LR, Fendly BM, Fisher KL, Lawn RM, Marks BJ, McCray G, Tate KM, Vehar GA, Gorman GA. Mammalian cell transient expression of tissue factor for the production of antigen. Protein Eng 1990; 3:5473.
32. Gorman CM, Gies DR, McCray G. Transient production of proteins using an adenovirus transformed cell line. DNA Protein Eng Technol 1990; 2:3.
33. Kim DW, Uetsuki T, Kaziro Y, Yamaguchi N, Sugano S. Use of the human elongation factor 1 alpha as a versatile and efficient expression system. Gene 1990; 91:217.
34. Kurth R, Philippsen P, Steiner S, Wright MC. Isolated promoter and terminator of elongation factor 1-alpha. U.S. Pat. 5,650,294, 1997.
35. Gill DR, Smyth SE, Goddard CA, Pringle IA, Higgins CF, Colledge WH, Hyde SC. Increased persistence of lung gene expression using plasmid containing the ubiquitin C or elongation factor-1 alpha promoter—plasmid pCI expression in mouse HEK293T cell and cationic liposome useful in gene therapy. Gene Ther 2001; 8:1539.
36. Kobayashi M, Tanaka A, Hayashi Y, Shimamura S. The CMV enhancer stimulates expression of foreign genes from the human EF-1 alpha promoter. Anal Biochem 1997; 247:179.
37. Mizushima S, Nagata S. pEF-BOS, a powerful mammalian expression vector. Nucleic Acids Res 1990; 18:532.
38. Cho M-S, Yee H, Brown C, Chan S. Transient transfection for protein production (up to 100 mg) using HKB11. Conference Abstract, 17th ESACT Meeting, Tylosand, Sweden, June 10–14, 2001.
39. Liljeström P, Garoff H. A new generation of animal cell expression vectors based on the Semliki Forest virus system. Biotechnology 1991; 9:1356.
40. Rice CM. Examples of expression systems based on animal RNA viruses: alphaviruses and influenza virus. Curr Opin Biotechnol 1992; 3:523.
41. Lundstrom K. Alphaviruses as expression vectors. Curr Opin Biotechnol 1997; 8:578.
42. Liljeström P. Alphavirus expression systems. Curr Opin Biotechnol 1994; 5:495.
43. Berglund P, Sjöberg M, Garoff H, Atkins GJ, Sheahan BJ, Liljeström P. Semliki Forest virus expression system: production of conditionally infectious recombinant particles. BioTechnology 1993; 11:916.
44. Blasey HD, Lundström K, Tate S, Bernard AR. Recombinant proteins production using the Semliki Forest virus expression system. Cytotechnology 1997; 24:65.
45. Bikker H, Baas F, De-Vijlder JJM. Molecular analysis of mutated thyroid peroxidase detected in patients with total iodide organification defects. J Clin Endocrinol Metab 1997; 82:649.
46. Werner P, Kawashima E, Reid J, Hussy N, Lundström K, Buell G, Humbert Y, Jones KA. Organization of the mouse 5-HT3 receptor gene and functional expression of two splice variants. Mol Brain Res 1994; 26:233.
47. Paul NL, Marsh M, McKeating JA, Schulz TF, Liljeström P, Garoff HA, Weiss R. Expression of HIV-1 envelope glycoproteins by Semliki Forest virus vectors. AIDS Res Hum Retroviruses 1993; 9:9630.
48. Blasey HD, Brethon B, Hovius R, Lundström K, Rey L, Vogel H, Tairi AP, Bernard AR. Large scale application of the Semliki Forest virus system: 5-HT3 receptor production. In: Merten O-W, Perrin P, Griffiths B, eds. New Developments and New Applications in Animal Cell Technology. Dordrecht: Kluwer Academic Publishers, 1998:449–455.

49. Tubulekas I, Berglund P, Fleeton M, Liljestrom P. Alphavirus expression vectors and their use as recombinant vaccines: a minireview. Gene 1997; 190:191.
50. Graham FL, Prevec L. Methods for construction of adenovirus vectors. Mol Biotechnol 1995; 3:207.
51. Berkner KL. Expression of heterologous sequences in adenoviral vectors. Curr Top Microbiol Immunol 1992; 158:39.
52. Berkner KL. Development of adenovirus vectors for the expression of heterologous genes. Biotechniques 1988; 6:616.
53. Fooks AR, Schadeck E, Liebert UG, Dowsett AB, Rima BK, Steward M, Stephenson JR, Wilkinson GWG. High-level expression of the measles virus nucleocapsid protein by using a replication-deficient adenovirus vector: induction of an MHC-1-restricted CTL response and protection in a murine model. Virology 1995; 210:456.
54. Massie B, Dionne J, Lamarche N, Fleurent J, Langelier Y. Improved adenovirus vector provides herpes simplex virus ribonucleotide reductase Rl and R2 subunits very efficiently. BioTechnology 1995; 13:602.
55. Zhang Y, Moss B. Inducer dependant conditional lethal mutant animal viruses. Proc Natl Acad Sci USA 1991; 88:1511.
56. Graham FL, Van der Eb AJ. A new technique for the assay of infectivity of human adenovirus 5 DNA. Virology, 1973; 52:456.
57. Boussif O, Lezoualc'h F, Zanta M, Mergny M, Scheman D, Demeneix B, Behr JP. A versatile vector for gene and oligonucleotide transfer into cells in culture and in vivo: polyethyleneimmine. Proc Natl Acad Sci USA 1995; 9:297.
58. Chu G, Hayakawa H, Berg P. Electroporation for the efficient transfection of mammalian cells with DNA. Nucleic Acids Res 1987; 15:1311.
59. Neumann E, Schäfer-Ridder M, Wang Y, Hofschneider PH. Gene transfer into mouse lyoma cells by electroporation in high electric fields. EMBO J 1982; 1:841.
60. Ridder R, Geisse S, Kleuser B, Kawalleck P, Gram H. A COS-cell-based system for rapid production and quantification of scFv: IgC kappa antibody fragments. Gene 1995; 166:273.
61. Fraley R, Subrami S, Berg P, Papahadjopoulos D. Introduction of liposome-encapsulated SV40 DNA into cells. J Biol Chem 1980; 255:10431.
62. Felgner PL, Gadek TR, Holm M, Roman R, Chan HW, Wenz M, Northrop JP, Ringold GM, Danielsen M. Lipofection: a highly efficient lipid mediated DNA transfection procedure. Proc Natl Acad Sci USA 1987; 84:7413.
63. Legendre JY, Trzeciak A, Bohrmann B, Deuschle U, Kitas EA, Supersaxo A. Dioleolylmelittin as a novel serum insensitive reagent for efficient transfection of mammalian cells. Bioconjug Chem 1996; 8:57.
64. Schlaeger EJ, Legendre JY, Trzeciak A, Kitas EA, Christensen K, Deuschle U, Supersaxo A. Synergistic enhancement of transient expression by dioleolylmelittin (DOM) and polyethylenimine (PEI) in mammalian cells in suspension culture. In: Merten O-W, Perrin P, Griffiths B, eds. New Developments and New Applications in Animal Cell Technology. Dordrecht: Kluwer Academic Publishers, 1998:117–120.
65. Regelin AE, Fernholz E, Krug HF, Massing U. High throughput screening method for identification of new lipofection reagents. J Biomol Screen 2001; 4:245.
66. Schlaeger E-J, Guthy D, Kitas E. Optimal transfection complex formation depends on the medium composition. Conference Abstract, Current Status and Trends in Mammalian Transient Protein Expression, Ittingen, Switzerland, Oct 1–3, 2000.
67. Girard P, Derouazi M, Baumgartner G, Bourgeois M, Jordan M, Wurm FM. 100 liter transfection. Conference Abstract, 17th ESACT Meeting, Tylosand, Sweden, June 10–14, 2001.
68. Behr JP. Towards artificial viruses. Conference Abstract, Current Status and Trends in Mammalian Transient Protein Expression, Ittingen, Switzerland, Oct 1–3, 2000.
69. Blessing T, Remy JS, Behr JP. Template oligomerization of DNA-bound cations produces calibrated nanometric particles. J Am Chem Soc 1998; 120:8519.

70. Blessing T, Remy JS, Behr JP. Monomolecular collapse of plasmid DNA into stable virus-like particles. Proc Natl Acad Sci USA 1998; 95:1427.
71. Zanta MA, Boussif O, Adib A, Behr JP. In vitro gene delivery to hepatocytes with galactosylated polyethylenimine. Bioconj Chem 1997; 8:841.
72. Erbacher P, Remy JS, Behr JP. Gene transfer with synthetic virus-like particles via the integrin-mediated endocytosis pathway. Gene Ther 1999; 6:138.
73. Boussif O, Lezoualch F, Zanta MA, Mergny MD, Scherman D, Demeneix B, Behr JP. A versatile vector for gene and oligonucleotide transfer into cells in culture and in vivo: polyethylenimine. Proc Natl Acad Sci USA 1995; 92:7297.
74. Zanta MA, Belguise P, Behr JP. Gene delivery: A single nuclear localization signal (NLS) peptide is sufficient to carry DNA to the cell nucleus. Proc Natl Acad Sci USA 1999; 96:91.
75. Jordan M, Wurm FM. Measurable parameters of cells and precipitate predict transfectability with calcium phosphate. In: Merten O-W, Perrin P, Griffiths B, eds. New Developments and New Applications in Animal Cell Technology. Dordrecht: Kluwer Academic Publishers, 1998:125–128.
76. Jordan M, Köhne C, Wurm FM. Principles of a scaleable transient transfection and expression technology for mammalian cells. In: Carrondo MJ, et al., eds. Animal Cell Technology—From Vaccines to Genetic Medicine. Dordrecht: Kluwer Academic Publishers, 1997:47–50.
77. Jordan M, Schallhorn A, Wurm FM. Transfecting mammalian cells: optimization of critical parameters affecting calcium-phosphate precipitate formation. Nucleic Acids Res 1996; 24:596.
78. Brash DE, Reddel RR, Qanrud M, Yank K, Farrell MP, Harris CC. Strontium phosphate transfection of human cells in primary culture: stable expression of the simian virus 40 target-T-antigen gene in primary human bronchial epithelial cells. Mol Cell Biol 1987; 7:2031.
79. Yang JP, Huang L. Overcoming the inhibitory effect of serum on lipofection by increasing the charge ratio of cationic liposome to DNA. Gene Ther 1997; 4:950.
80. Schäfer-Ridder M, Wang Y, Hofschneider PH. Liposomes as gene carriers: efficient transformation of mouse L cells by thymidine kinase gene. Science 1982; 215:166.
81. Wong TK, Neumann E. Electric field mediated gene transfer. Biochem Biophys Res Commun 1982; 107:584.
82. Potter H, Weir L, Leder P. Enhancer-dependent expression of human kappa immunoglobulin genes introduced into mouse pre-B lymphocytes by electroporation. Proc Natl Acad Sci USA 1984; 81:7161.
83. Fromm M, Taylor LP, Walbot V. Expression of genes transferred into monocot and dicot plant cells by electroporation. Proc Natl Acad Sci USA 1985; 82:5824.
84. Fromm M, Taylor LP, Walbot V. Stable transformation of maize after gene transfer by electroporation. Nature 1986; 319:791.
85. Rols MP, Coulet D, Teissie J. Highly efficient transfection of mammalian cells by electric field pulses: application to large volumes of cell culture by using a flow system. Eur J Biochem 1992; 206:115.
86. Jordan M, Köhne C, Wurm FM. Calcium-phosphate mediated DNA transfer into HEK-293 cells in suspension: control of physicochemical parameters allows transfection in stirred media. Cytotechnology 1998; 26:39.
87. Tolbert WR, Hitt MM, Feder J. Cell aggregate suspension culture for large-scale production of biomolecules. In Vitro 1980; 16:486.
88. Horn NA, Meek JA, Budahazi G, Marquet M. Cancer gene therapy using plasmid DNA: purification of DNA for human clinical trials. Hum Gene Ther 1995; 6:565.
89. Schluep T, Cooney CL. Purification of plasmids by triplex affinity interaction. Nucleic Acids Res 1998; 26:4524.

90. Levy MS, Collins IJ, Tsai JT, Shamlou PA, Ward JM, Dunnill P. Removal of contaminant nucleic acids by nitrocellulose filtration during pharmaceutical-grade plasmid DNA processing. J Biotechnol 2000; 76:197.
91. Blasey HD, Lundström K, Tate S, Bernard AR. Recombinant protein production using the Semliki Forest virus expression system. Cytotechnology 1997; 24:65.
92. Lundström K, Mills A, Buell G, Allet E, Adami N, Liljeström P. High-level expression of the human neurokinin-1 receptor in mammalian cell lines using the Semliki Forest virus expression system. Eur J Biochem 1994; 224:917.
93. Hovius R, Schmid EL, Tairi AP, Blasey H, Bernard AR, Lundstrom K, Vogel, H. Fluorescence techniques for fundamental and applied studies of membrane protein receptors: the 5-HT3 serotonin receptor. J Recept Signal Transduct Res 1999; 19:533.
94. Blasey HD, Hovius R, Vogel H, Bernard AR. Transient-expression technologies, their application and scale-up: 5-HT3 serotonin receptor case study. Biochem Soc Trans 1999; 27:955.
95. Pialoux G, Excler JL, Riviere Y, Gonzalez-Canali G, Feuillie V, Coulaud P, Gluckman JC, Matthews TJ, Meignier B, Kieny MP, Gonnet P, Diaz I, Meric C, Paoletti E, Tartaglia J, Salomon H, Plotkin S. The Agis Group and L'Agence Nationale de Recherche sur le Sida, a prime-boost approach to HIV preventive vaccine using a recombinant canarypox virus expressing glycoprotein 160 (MN) followed by a recombinant glycoprotein 160 (MN/LAI). AIDS Res Hum Retroviruses 1995; 11:373.
96. Barrett N, Mitterer A, Mundt W, Eibl J, Eibl M, Gallo RC, Moss B, Dorner F. Large-scale production and purification of a vaccinia recombinant-derived HIV-1 gpl60 and analysis of its immunogenicity. AIDS Res Hum Retroviruses 1989; 5:159.
97. Falkner FG, Turecek PL, MacGillivray RT, Bodemer W, Scheiflinger F, Kandels S, Mitterer A, Kistner O, Barrett N, Eibl J, Dorner F. High level expression of active human prothrombin in a vaccinia virus expression system. Thromb Haemost 1992; 68:119.

18

Principles and Applications of the Insect Cell-Baculovirus Expression Vector System

Laura A. Palomares, Sandino Estrada-Mondaca, and Octavio T. Ramírez
Departamento de Medicina Molecular y Bioprocesos, Instituto de Biotecnología, Universidad Nacional Autónoma de México, Morelos, Mexico

INTRODUCTION

Animal cell culture has gained a paramount position in modern biotechnology, as evidenced by the large market share and important applications that proteins derived from this technology have in therapy, prophylaxis, and diagnosis. In fact, at least 16 recombinant proteins (r-proteins) derived from cell cultures, which account for about half of the r-DNA products approved to date, have reached the market (1). Furthermore, over 100 proteins produced from animal cell culture are presently undergoing clinical trials, and it is expected that more than half of the new protein drugs will require animal cells for their production. Gene expression in animal cells is required since the in vivo activity of many r-proteins strongly depends on complex posttranslational modifications that can only be performed by higher eukaryotic cells (2). In particular, mammalian cells are preferred for human applications since posttranslational modifications are closer to those found in human proteins. Mammalian cell culture technology has evolved significantly during the last 20 years; however, it still faces important challenges. For instance, transfection, selection, and amplification of stable and high-producing cell lines can be a laborious and time-consuming task. Furthermore, product concentrations and productivities are in general lower, whereas overall bioprocessing costs are usually higher than for other expression systems, such as prokaryotes, lower eukaryotes, and transgenic plants and animals. Accordingly, there exists a permanent interest in better expression systems with the advantages of mammalian cell cultures but without their drawbacks. Among the various options, the insect cell-baculovirus expression vector system (IC-BEVS) represents a promising alternative.

Production of heterologous proteins by the IC-BEVS consists of a two-stage process where insect cells are first grown to a desired concentration and then infected with a recombinant (r) -baculovirus containing the gene that codes for the protein of interest. The origins of insect cell culture date from the early 1960s when the first insect cell lines were established (3). In turn, wild-type baculoviruses were originally propagated in insect larvae (in vivo) with the purpose of producing bioinsecticides.

During the 1950s, extensive safety testing of baculoviruses had already started and in 1975, the U.S. Environmental Protection Agency registered the first baculovirus for use as a pesticide (4–8). By 1983, the first r-baculovirus was constructed (9,10). This development opened whole new fields of activity. The initial interest placed on wild-type bioinsecticides was expanded to include improved r-bioinsecticides as well as other new applications, particularly in the pharmaceutical and medical areas. At present, genetically modified baculoviruses have become powerful vectors for expressing heterologous proteins in insect cells, both in vivo and in vitro. Furthermore, it has been recently demonstrated that r-baculoviruses can be efficiently taken up by human hepatocytes and subsequently express the heterologous genes (11). This has uncovered new applications of baculovirus vectors for gene therapy; however, it has also cautioned on the potential risks that should be considered when using r-baculoviruses in agriculture (11,12). Interesting perspectives of insect cell culture still lie ahead as the unique and diverse molecular pathways and metabolic capabilities of insects should represent a rich source of new applications in very diverse areas (13).

Compared to other expression systems, the IC-BEVS has important advantages, including speed, simplicity, reliability, versatility, high-level expression, capacity for posttranslational processing of proteins, and various convenient features for bioprocessing (see Table 1 for details). As a result, increasing attention has been placed on the IC-BEVS, as observed from the increasing number of expressed products, published papers, and patents that have appeared during the last decade (15,17). Nonetheless, there is still much to learn and improve if the IC-BEVS is to become a prevalent commercial technology in agricultural, pharmaceutical, and medical applications. As with other animal cell cultures, overall bioprocess productivities must increase and costs decrease in order to compete with other simpler expression systems. Generation of protein posttranslational modifications, particularly glycosylation, similar to those found in human proteins is needed before the IC-BEVS is considered for the production of many therapeutic products. Finally, safety and potency issues of promising new applications, such as r-pesticides and gene therapy, must be carefully addressed. All such issues are covered in the present chapter, along with a revision of fundamental aspects of insect cell lines and baculoviruses. A detailed description is presented covering bioengineering issues of in vitro insect cell culture, product characteristics, and applications of the IC-BEVS. A brief description is also included of stable expression systems based on insect cells, such as the *Drosophila* metallothionein system, as well as in vivo (larval) production methods.

INSECT CELL LINES

Sources and Isolation

The first continuous invertebrate cell line was established in 1962 by Grace (3) from ovaries of the Australian Emperor gum moth, *Antheraea eucalypti*. Since then, several insect species from the orders Lepidoptera, Diptera, Homoptera, Hymenoptera, Orthoptera, Hemiptera, and Coleoptera have been the source of cell lines isolated from eggs, larvae, embryos, or adult tissues (20–22). To date, at least 450 continuous cell lines have been established from more than 100 species of insects (21,22). Interestingly, only a relatively small fraction of those cell lines are listed in the two main cell culture depositories in the world (The American Type Culture Collection and the European Collection of Animal Cell Cultures). A list of representative insects and tissues used for generating cell lines is given in Table 2 along with baculoviruses able

Table 1 Relevant Characteristics of the IC-BEVS

Safety
Baculoviruses exhibit a narrow range of infectivity, generally restricted to arthropods
Insect cell lines are not considered as transformed, thus risk of transmission of cancer-promoting compounds is reduced
There exists a low probability that insect cells contain viruses hazardous to humans or other mammals
Speed, simplicity, and reliability
Construction of reliable recombinant baculovirus is simple and rapid (in about 5 days)
No labor- and time-intensive cell line transfection, selection, screening or amplification procedures are required
As a transient expression system, only short time periods are required for generating substantial quantities of heterologous protein
Versatility
Approximately 500 different functional recombinant proteins have been produced
Baculoviruses allow packaging and expression of very large genes
Multiple genes can be cloned into a single baculovirus and expressed simultaneously to produce properly assembled multimeric proteins
Coinfection with various baculoviruses containing different genes can be used to produce properly assembled multimeric proteins
Baculoviruses can perform intron/exon splicing and, thus, can express unspliced genes
Heterologous protein can be transported to natural cellular locations (organelles, membranes, secreted)
High-level expression
Heterologous protein can be as high as 50% of the total cellular protein, which corresponds to $\approx$1 ng recombinant protein per cell
Compared to mammalian cells, heterologous protein yields can be 20–250 times higher
High maximum concentrations of heterologous protein can be obtained (>500 mg/L)
Posttranslational modifications
Heterologous proteins can undergo N- and O-linked glycosylation, phosphorylation, fatty acid acylation, α-amidation, N-terminal amidation, carboxymethylation, isoprenylation, folding, signal peptide cleavage, and proteolytic cleavage, although differences can exist with respect to the authentic proteins
Convenient features for bioprocessing
Insect cell lines can be subcultured indefinitely
Insect cell lines can be readily adapted to grow in suspension
High cell concentrations can be attained in batch ($\sim 10 \times 10^6$ cells/mL) and perfusion or fed-batch cultures ($>50 \times 10^6$ cell/mL)
Straightforward scale-up (cell culture volumes as high as 1000 L have been reported)
Expression heterologous protein can be simply controlled by manipulating time and multiplicity of infection
Fusion proteins can be easily constructed for facilitating purification and/or monitoring of heterologous protein
Very late promoters allow expression of cytotoxic recombinant gene products

[Adapted from Refs. (14–19).]

to infect lepidopteran cell lines. Lepidopteran and Dipteran cell lines are the most abundant, and have been extensively used not only for production purposes but also for better understanding of their relationship with viruses that specifically infect them. The most popular cell lines have been derived from *Spodoptera frugiperda* (fall armyworm) (see Fig. 1), *Trichoplusia ni* (cabbage looper), and *Mamestra brassicae* (cabbage moth). In particular, Sf9 and Sf21, clones obtained from the pupal ovarian

Table 2 List of Some Insect Species Used for Cell Line Isolation and the Tissues from which They Were Isolated

Species of origin	Tissue of origin	Infectivity by NPVs
Lepidoptera		
Adoxophyles orana	NAD	*Ao*NPV, *Ac*MNPV, *Bb*NPV
Antheraea eucalypti	Trypsinized pupal ovaries	*Bm*NPV
Antheraea pernyi	Pupal ovary	*Ap*NPV
Anticarsia gemmatalis	Embryo	*Ag*MNPV, *Ac*MNPV
Amyelois transitella	NAD	*Ac*MNPY
Bombyx mori	Larval ovaries	*Bm*NPV, *Ac*MNPV, *Ha*NPV
Chilo suppressalis	Larval hemocytes	NAD
Choristoneura disstria	Neonate larvae	NAD
C. fumiferana	Minced tripsinized larvae	*Ac*MNPV
Cydia pomonella	NAD	*Cm*NPV
Diatrea grandiosellla	Hemocytes and fat body	NAD
Estigmene acrea	Larval hemocytes	*Ac*MNPV
Euxoa scandens	Ovary	*Ac*MNPV, *Bm*MNPV, *Dw*NPV, *Gm*NPV
Heliothis virescens	Larval fat body	*Ac*MNPV, *Hz*SNPV
Helicoverpa zea	Pupal ovary	*Hz*-1, *Tn*NPV, *Ac*MNPV, *Hz*SNPV
Laspeyresia pomonella	Minced embryo	NAD
Leucania separata	Pupal ovary, embryo	*Ac*MNPV
Lymantria dispar	Hemocytes	*Bm*NPV, *Ac*MNPV, *Ap*NPV, *Gm*NPV, *Ld*MNPV, *Op*NPV
Malocosoma disstria	Blood cells, larval fat body	*Lfs*NPV, *Cf*NPV, *Ac*MNPV
Mamestra brassicae	Neonate larvae	*Ac*MNPV, *Mb*NPV
Manduca Sexta	Neonate larvae, pupal ovaries	*Ac*MNPV
Ostrinia nubilalis	Pupal ovary	NAD
Papilio xuthus	Pupal ovaries	NAD
Plutella xylostella	NAD	*Ac*MNPV, *Pcr*NPV
Samia Cynthia	Pupal hemocytes	NAD
Spodoptera exigua	Pupal ovary	*Ac*MNPV, SexiNPV
S. frugiperda	Adult ovary, embryo	*Sf*NPV, *Ac*MNPV, *Tn*NP, *Gm*NPV, *Sf*MNPV, *Sl*NPV, SexiNPV, SexeNPV, *Xc*NPV
S. littoralis	NAD	*Ac*MNPV, *Sl*NPV, *Sf*MNPV, SexiNPV
Trichoplusia ni	Adult ovary, embryo	*Ac*MNPV, *Tn*NPV, *Gm*NPV, *Ag*MNPV, *Hc*MNPV

(*Continued*)

Table 2 List of Some Insect Species Used for Cell Line Isolation and the Tissues from which They Were Isolated (*Continued*)

Species of origin	Tissue of origin	Infectivity by NPVs
Diptera		
Aedes aegypti	Minced larvae; homogenized embryos	
A. albopictus	Minced trypsinized larvae, ovarian tissue	
A. malayensis	Minced larvae	
A. novo-albopictus	NAD	
A. pseudoscultellaris	Minced larvae	
A. taeniorhynchus	First-stage larvae	
A. vexans	Minced pupae	
A. w-albus	Newly hatched trypsinized larvae	
Anopheles gambiae	Minced trypsinized larvae	
A. stephensi	Minced larvae	
Armigeries subalbatus	First-stage larvae	
Culex molestus	Adult ovaries	
C. quinquefasciatus	Adult ovaries	
C. salinarius	First-stage larvae	
C. tarsalis	Embryos	
C. tritaeniorhynchus	Ovarian tissue, first-stage larvae	
Culiseta inornata	Minced adult	
Drosophila immigrans	Late embryos and first -stage larvae	
D. melanogaster	Embryos	
Musca domestica	Embryos	
Orthoptera		
Blabera fusca	Minced trypsinized	
Blattella germanica	Embryos, dorsal closure	
Leucophaea maderae	Nymphal dorsal vessel; nymphal ovaries	
Periplaneta americana	Minced trypsinized embryos; hemocytes from nymphal male	
Homoptera		
Aceratagallia sanguinolenta	Embryos	
Agallia constricta	Minced embryos	
A. quadripunctata	Minced embryos	
Agalliopsis novella	NAD	
Colladonus montanus	Fragmented trypsinized embryos	
Dalbulus maidis	Embryos	
Macrosteles sexnotatus	Embryos	
Nephotettix apicalis	Embryonic tissue	

(*Continued*)

Table 2 List of Some Insect Species Used for Cell Line Isolation and the Tissues from which They Were Isolated (*Continued*)

Species of origin	Tissue of origin	Infectivity by NPVs
N. cincticeps	Embryonic tissue	
Hemiptera		
Triatoma infestans	Embryonic tissues	
Coleoptera		
Xylotrechus pyrrhoderus	Larval fat body tissue	

Note: Rightmost column indicates NPVs able to infect lepidopteran cell lines. NAD, for nonavailable data.
[Adapted from Refs. (14,17,21,23,24).]

tissue of *S. frugiperda*, are the most widely used cell lines for r-protein production (26), although other cell lines show better potential. For instance, the yield of an r-protein produced in *T. ni* or *M. brassicae* clones is in general up to 2.4 or 1.5 times higher, respectively, than in Sf9 cells. The yields of various r-proteins produced in different cell lines, compared to Sf9, have been reviewed by Palomares and Ramírez (17). The performance of the commercial cell line High Five® (*T. ni* BTI-TN-5bl-4) is among the best since high r-product and cell (8.5×10^6 cells/mL in batch) concentrations can be obtained. It must be noted that the relative yield of r-protein not only depends on the particular cell line, but also on the r-protein itself. For instance, 23.5 higher yields of r-SeAP (secreted alkaline phosphatase) were obtained in High Five® compared to Sf9 cells, whereas only 2.0 higher yields were obtained for r-tumor necrosis factor (TNF) (27,28). Caution should be taken with data showing very large differences in r-protein yields between different cell lines.

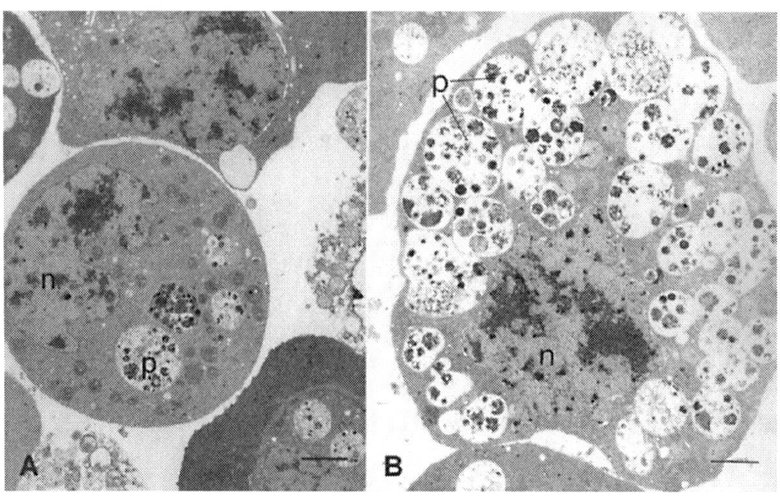

Figure 1 Transmission electron microscopy of viable uninfected Sf9 cells. Note the large amount of phagosomes (p), particularly in the cell shown in panel B, which result from a characteristic active phagocytic process detailed in Ref. 25. (n, nucleous.)

A cell line is established by mechanically or enzymatically dissociating the tissue of interest. The dissociated cells are maintained in rich medium until growth is observed; thereafter, cells are subcultured until a morphologically homogenous culture is obtained. For instance, the cell line obtained by Grace (3) originated from ovaries dissociated with trypsin, which were kept in culture medium at 27–29°C for 10 months until growth was detected. A detailed guideline for the isolation and maintenance of insect cells can be found elsewhere (29,30). The choice of the tissue is important for establishing a cell line. Large insects facilitate the isolation of cells from organs, like ovaries. However, larvae and embryos of small insects have also been the source of several cell lines. Various features must be considered when selecting a cell line for a particular application. These include optimum stoichiometric and kinetic parameters, such as maximum growth rate, cell concentration, yield and production rates of r-protein, as well as other characteristics such as cell line stability, virus susceptibility, and protein posttranslational processing (20).

Cell Line Characterization

Cell lines can be initially identified by growth characteristics and morphology; however, these can vary with culture time and conditions. Due to the increasing number of established cell lines, there exists an even more urgent need to identify them by additional criteria. Accordingly, some or all of the following characterization procedures must be implemented in laboratories where cultivation of several cell lines is a regular practice: karyology, immunologic and enzymatic analyses, flow cytometry, and DNA amplification fingerprinting (DAF).

Karyotyping. This common method aims to accurately count the chromosomal complement of the chosen organism. Different insect species have different haploid chromosome numbers ranging from 1 to 221 (31). Karyotyping can be accurately carried out on dipteran cell lines, whose diploid chromosome numbers are low (6 for mosquitoes and 8 for *Drosophila* species) and have low poliploidy (i.e., 3n for *Aedes aegypti* and for *Anopheles stephensi*). In hemipteran and orthopteran cells, it is also easy to perform karyology characterization studies. However, karyotyping can be very difficult for lepidopteran cells due to the presence of tiny microchromosomes, to the fact that macrochromosome centromeres are rather diffuse, and to high poliploidy. For instance, the haploid number of a cell culture derived from *A. eucalypti* is 25, but the ploidy may range from 2n to 128n (32). Procedures for enumerating chromosomes in insect cells in general and in *Drosophila* cells in particular are described in Schneider (32) and Echalier (30), respectively.

Isoenzymes. The presence of characteristic enzymes in cell extracts, as assayed by electrophoretic methods, can be used as the basis for discriminating between different cell lines. In this way, specific isoenzyme profiles from lepidopteran cell lines belonging to four different genera were identified by isoelectrofocusing and phosphoglucoisomerase staining in starch gels (33). This method showed that differences in cell passage do not affect the isoenzyme band pattern. Furthermore, it is equally reliable for identification of intergenus as well as intragenus differences. Other enzymes can also be assayed, such as aspartate aminotransferase, esterase, isocitrate dehydrogenase, lactate dehydrogenase, leucine aminopeptidase, malate dehydrogenase, malic enzyme, phosphoglucoisomerase, and phosphoglucomutase. For instance, by comparing such enzymes to last-instar larvae, 28 continuous insect cell lines were readily identified (34). Using such an analysis, the authors were able to detect contaminations between insect cell lines. Other authors report similar results of isoenzyme analysis using a cellulose acetate technique (35). In this case, differences between cell lines belonging to different species were detected.

Serologic studies. These types of analyses are based on the identification of cell line-specific antigens, by techniques such as complement fixation, hemagglutination, immunodiffusion, and immunoelectrophoresis. Antigens are usually obtained from the sera of mice immunized by intraperitoneal injections of cell extracts. For instance, Aldridge and Knudson (36) were able to distinguish up to the family level differences in four out of five lepidopteran cell lines, with immunoelectrophoresis being the technique that yielded the best results. Improved specificity could be obtained with the use of monoclonal antibodies and ELISA or RIA techniques. Nonetheless, serologic studies are laborious, time-consuming, and can suffer from lack of specificity due to sera cross-reactivity between cell lines. Unless such drawbacks can be overcome, other methods should be preferred.

Flow cytometry. Common flow cytometry protocols for analyzing proliferation and ploidy of mammalian cells can also be used to characterize nuclear DNA of insect cells (25,37–40). Flow cytometry allows the identification of specific DNA peak profiles, which indicate ploidy levels for individual cell lines. Peaks can be objectively quantified and in general are not affected by passage number or some other culture conditions such as temperature (39). Léry et al. (39) carried out a study on eight insect cell lines, each one displaying different DNA profiles that were characterized by a specific number of peaks and intensity. Moreover, by using the cell size discrimination option of this technique, they were able to distinguish a mixture of cell lines. Interestingly, the most homogenous cell line turned out to be Sf9, the only cloned cell line used in their study (39). Flow cytometry can also be a powerful technique for characterizing cell cycle phases and apoptotic death (25).

DNA amplification fingerprinting (DAF). Due to the increasing number of insect cell lines, mislabeling and cross-contamination becomes a common problem that can be neglected unless adequate methods are used. With the advent of PCR-based techniques, fingerprinting of cell lines is now a reality. DAF is a PCR-based method, which uses arbitrarily selected primers that allow amplification of a DNA sequence in a species-specific fashion, resulting in characteristic patterns useful for identification. DAF is a trustworthy method. McIntosh et al. (41) used primer couples, originally designed for PCR amplification on mammalian aldolase, prolactin receptor, and interleukin-1β, in a PCR reaction on DNA isolated from 20 insect cell lines. The amplification patterns obtained differed according to the species whose DNA was used as a template. DAF has allowed the identification of cell lines at the species level, whereas other methods, such as isoenzyme analysis, are only useful to distinguish up to the genus level (35). A similar technology, also PCR-based, is random amplification of polymorphic DNA (RAPD) in which small six to nine oligonucleotide primers randomly selected are used for a PCR amplification of an unknown DNA (42). This method amplifies DNA sequences whose length can differ with the species used for isolating the genetic material. RAPD has been used for identifying species and clones of aphids (43) and for distinguishing among biotypes of the white fly *Bemisia tabaci* (44). Thus, RAPD should also be a useful technique for identification of insect cell lines in the future.

Ribosomal RNA genes. Throughout evolution, ribosomes along with the primary, secondary and tertiary structures of ribosomal RNA molecules (rRNA), have been conserved. Both, ribosomes and rRNAs are part of the translational machinery, thus vital for the survival of the cell. These latter molecules exist as different subunits (18S , 5.8S, and 28S rRNAs) and are encoded by genes contained in tandemly repeated clusters. The genes are separated by spacers, namely, intergenic spacer (IGS) or nontranscribed spacer (NTS) and the internal transcribed spacer (ITS). Such ITS sequences are a tool for phylogenetic analysis since they are conserved in all eukaryotes and show a high rate of divergence, which makes them a useful tool for the identification of closely related species. Evolutionary studies based on the analysis of such ITS sequences have been carried out by Schlötterer et al. (45) on *Drosophila* species and by Mukha et al. (46) on cockroaches. The reported characterization of dipteran (47) and lepidopteran (48) cell lines by using rRNA gene sequences consisted of the use of specific primers which were used with DNA extracted from the cell lines as template in a PCR reaction. This reaction amplified a fragment of a gene that codes for the mitochondrial 16s rRNA. By denaturing and suddenly renaturing the PCR product, a single band corresponding to a homoduplex DNA molecule indicated that the DNA donor cells belong to a single lineage. In contrast, a mixed population was identified if multiple bands (heteroduplex) were obtained.

BACULOVIRUS

Baculoviruses are large (80–230 kbp) double-stranded, circular, and supercoiled DNA viruses (4). The name baculovirus is derived from their rod-shaped capsid, which measures 40–50 nm in diameter and 200–400 nm in length. The baculoviridae family includes two genera of lytic viruses: Granulovirus (GNV) and Nucleopolyhedrovirus (NPV) (49). The GNVs are characterized for having only one virion (enveloped nucleocapsid) occluded within an ovicylindrical crystal matrix made of granulin. In contrast, NPVs are characterized by the presence of multiple virions embedded in occlusion bodies formed in the nucleus of infected cells. In this case, occlusion bodies are made of a polyhedral crystalline matrix composed of a 29 kDa protein known as polyhedrin. NPVs are classified as multiple (MNPV) or single (SNPV) nucleocapsid polyhedroviruses, depending on whether a membrane unit envelopes multiple (i.e., two or more) or individual nucleocapsids, respectively. At least 498 NPV and 135 GNV species have been identified (49). Some common NPVs and their insect hosts are listed in Table 2.

Baculoviruses are usually only pathogenic to arthropods, particularly holometabolous insects from the orders Lepidoptera, Diptera, Hemiptera, Hymenoptera, Coleoptera and Crustaceans from the order Decapoda. Furthermore, baculoviruses can only replicate in a few closely related species within a single genus or within a family, with larval stages being the most susceptible (4,14). For instance, the NPV from *Bombyx mori* (*Bm*NPV) has a host range limited to *B. mori* cells. Due to such a narrow host range, wild-type baculoviruses have received considerable attention since the 1940s as remarkably specific pesticides. As described in the section on heterologous gene products, it has recently been shown that genetic material can be delivered to mammalian cells through baculoviruses. During the past two decades, baculoviruses have also proven to be powerful expression vectors for the production of heterologous proteins in insect cells. In this case, the NPV from *Autographa californica* (*Ac*MNPV) has been used as the vector of choice since its host range is not as limited as that of other baculoviruses. *Ac*MNPV was first isolated from the alfalfa looper but it can readily replicate in over 30 cell lines isolated from several lepidopteran species (15,50). *Ac*MNPV is by far the NPV most utilized for production of r-proteins, although others such as *Bm*NPV or *Tn*SNPV are also used.

Furthermore, the entire DNA sequence of *Ac*MNPV is already available (51). Accordingly, in the following sections *Ac*MNPV is used as the model for describing the basic structure and life cycles of wild-type and r-baculoviruses.

Life Cycle and Structure of Wild-Type Baculovirus

The life cycle of baculoviruses in nature (schematically represented in Fig. 2) is temporally divided into four phases of gene expression: immediate early (α), delayed early (β), late (γ), and very late (δ) (15,26,52,53). Occluded virions of wild-type NPV are found in the soil and on surfaces of plants, where they are protected from the harsh environmental conditions by means of the polyhedrin matrix. Upon ingestion by a host larva, the polyhedrin matrix dissolves in the alkaline midgut ($pH > 10$) of the insect. Subsequently, virions are released from the polyhedrin crystals and adsorbed specifically onto microvilli of columnar epithelial cells of the gut (52,53). At this moment, the α phase of the infection cycle begins and lasts until $\approx$4 hr post-infection (hpi). During the first hour of being ingested, virions migrate to the cell nucleus where they get uncoated. Throughout the α phase, viral genes are transcribed and translated using host cell factors. Viral RNA can be detected as early as 0.5 hpi. The β phase of infection spans from 4 to 7 hpi and is characterized by cytoskeletal and nuclear rearrangements, and ultimately by degradation of the host genome. The first round of viral DNA replication starts about 7 hpi, marking the onset of the γ phase, which extends to $\approx$20–24 hpi. At 12 hpi, nucleocapsid assembly begins. Nucleocapsids then bud through the nuclear membrane and move through the

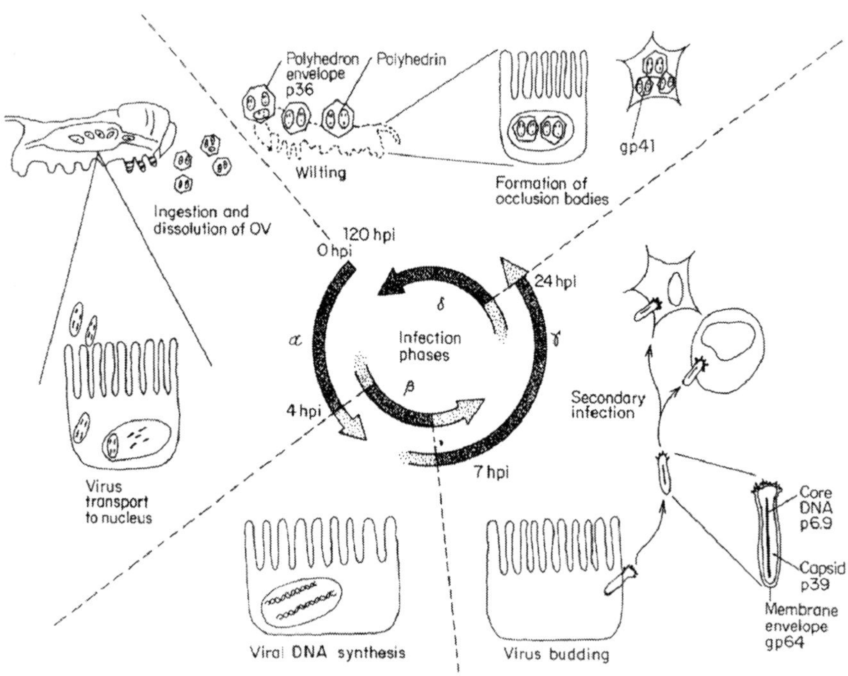

Figure 2 In vivo life cycle of baculovirus. The main structural baculovirus proteins are shown.

cytoplasm where they bud from the cell acquiring their envelope. Such virions are known as budded viruses or nonoccluded viruses and, together with the occluded viruses, make up the two phenotypes of NPV. Budded viruses are covered with filaments known as peplomers that are important for virus attachment to cells and tissues within the insect hemocel. After being released, budded viruses are transported through the hemolymph to infect other tissues, including fat body, muscle, tracheal matrix, hemocytes, and epithelium (49). This process is known as secondary infection. Finally, the very late phase of the infection cycle begins with transcription of the so-called "hyperexpressed" genes, such as those for polyhedrin and p10. Budded virus release ceases and, instead, virions accumulate on the nucleus where they become occluded in a polyhedrin matrix. During this phase, p10 accumulates in the cytoplasm and nucleus, forming large fibrous networks believed to participate in the correct synthesis of polyhedrin (54–56). Accumulation of polyhedral crystals continues for the next 3–5 days until the larva eventually disintegrates (wilts), liberating the occluded viruses to the environment where other larvae can ingest them to initiate again the infection cycle.

A relevant genetic feature of baculoviruses is the presence of regions containing two to eight imperfect palindromic sequences, which are homologous to each other and harbor an EcoRI site. These homologous regions act as enhancers for the expression of early (immediate and delayed) and late genes (26,50). Although many genes can be assigned to a single phase of the infection cycle, some are transcribed in more than one. During the early phase, preparation for replication and expression of viral DNA takes place. The difference between the α and β phases is given by the dependence on the regulatory activity of IE-1 or IE-0 (4,50). These are multifunctional viral regulatory proteins, products of splicing of the IE-1 gene (57), which are expressed throughout the virus life cycle. Genes of the α phase, such as *ie*-0, *ie*-1, and *ie*-n, are less dependent on the activity of IE-1, whereas genes of the β phase are strongly transactivated by IE-1 (4). Interestingly, the genes of the α phase are the only ones clustered together in the baculovirus genome. Genes expressed in the early phase include those coding for DNA polymerases, helicases, an apoptosis inhibitor, and a molting inhibitor.

r-Baculovirus and Infection of Cells in Culture

Polyhedrin is necessary for the survival of NPV in nature, but not in cell culture (9). Furthermore, budded viruses can readily infect and replicate in cultured insect cells. Accordingly, a gene of interest can be placed instead of the polyhedrin gene (*polh*) and its expression controlled by the *polh* promoter without detriment to the budded virus life cycle. By infecting insect cells with such a type of r-baculoviruses, very high yields of the r-protein can be obtained since the *polh* promoter is very strong. Such a strategy was originally proposed by Smith et al. (10) and applied to the production of recombinant human interferon-β in insect cells. Today, most r-genes expressed by IC-BEVS are located under the *polh* promoter. Since *polh* is hyperexpressed during the very late phase of infection, the r-protein is not detected in culture until about 18 hpi, whereas its concentration usually peaks between 48 and 120 hpi. An r-baculovirus in cell culture follows a similar life cycle to that shown in Fig. 2 for a wild type, except for some differences. Namely, cultures are directly infected with budded viruses and virus progeny will not be occluded in a polyhedrin matrix if the polyhedrin gene has been deleted. In this case, the r-protein is produced instead of polyhedrin. A secondary infection can occur as well. In such a process, virus progeny directly infect cells that were not infected during the primary infection. Wild-type baculoviruses with an intact polyhedrin gene can also be propagated in cell culture to produce occluded virus (58,59). Their life cycle is also similar to that shown in Fig. 2 for an in vivo infection, except that cultures must be directly infected with nonoccluded viruses. The in vitro infection process can be followed in various ways: by using antibodies against baculovirus-infected Sf9 cells (60), by agar plaque assays to determine the number of infectious particles present in a sample, or by amplifying polyhedrin conserved sequences using PCR-based techniques (26,61). Figure 3(C) and (D) shows r-baculoviruses accumulated in the nucleus of Sf9 cells.

Apart from polyhedrin, heterologous genes have also been placed under the control of other strong promoters, such as that for *p*10, which can allow easier vector construction and screening (26,62–64). Other benefits of using the *p*10 promoter have been reported. For instance, by simultaneously placing copies of the heterologous gene behind the *p*10 and *polh* promoters, DiFalco et al. (65) obtained a higher concentration of r-protein than when using each promoter alone. If larvae production is preferred, polyhedra should usually remain intact for an efficient primary infection. This can be achieved by placing the r-gene under the control of the *p*10 promoter instead of *polh* (66,67). Improved secretion and complete complex glycosylation, including addition of sialic acid residues, have also been reported when *p*10

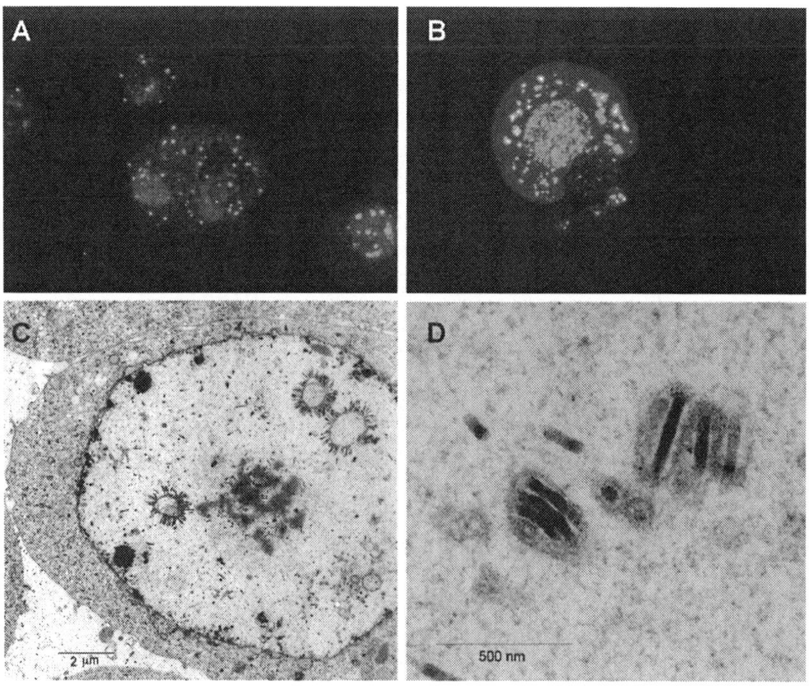

Figure 3 Micrographs of infected insect cells. Epifluorescence micrographs: (A) Sf9 cells at 72 hpi (× 560); (B) Tn5Bl-4 cells at 72 hpi (× 560). Cells in A and B are expressing a fusion protein containing GFP and VP2 of rotavirus. (Baculovirus provided by Prof. J. Cohen, Institut National de la Recherche Agronomique (INRA), France.) (C) Transmission electron microscopy of Sf9 cells infected with a recombinant AcMNPV. (D) Enlarged section of the nucleus. Note the large nucleus in C and enveloped multiple virions in D. (*See color insert p. 12.*)

promoter is used instead of *polh* promoter (68). Such a behavior was attributed to the relatively weaker strength of *p*10 promoter compared to *polh*, which did not overwhelm the processing capacity of the endoplasmic reticulum (ER). The fibrous networks composed by *p*10 are also thought to participate in cell lysis and larvae disintegration since these are delayed in *p*10 mutants (54–56,64). Thus, expression vectors based on *p*10 promoters could be useful for extending the r-protein productive phase, although this must still be thoroughly tested.

A disadvantage of gene expression based on the very late *polh* or *p*10 promoters is that r-protein production occurs toward the end of the culture when cells are in their death phase. Accordingly, secretion and correct posttranslational processing can be jeopardized (69) (see also Posttranslational Modifications). Very late expression can also be a drawback when r-baculoviruses are used as pesticides since the toxic effect is delayed, allowing prolonged feeding of the pests (see also Posttranslational Modifications). To prevent such problems, heterologous gene expression under immediate early (α phase), delayed early (β phase), or late (γ phase) promoters has been explored (70–73). For instance, r-protein expression can be obtained as early as 12 hpi with late promoters, such as those for vp39 (major capsid protein), p6.9 (core-associated protein), or lef-7 (26,74–78). With delayed early-phase promoters, such as *etl*, *pe38*, *me53*, *p35*, *egt*, and *da26*, r-proteins are expressed up to 6 hr earlier than with late promoters (26,78–83). Expression is even earlier with α-phase promoters such as *iel* (72,84). Furthermore, α-phase promoters allow a better and more extensive posttranslational processing of r-glycoproteins (72,73,84,85) (see also Posttranslational Modifications). Nonetheless, early and late promoters can present some drawbacks. First, they are not as strong as very late promoters, and thus lower r-protein concentrations are attained. For instance, expression of r-chloramphenicol acetyl transferase from a very late promoter was two- to threefold higher than from a late promoter and 10- to 20-fold higher than expression from an earlier promoter (86). Second, early and late promoters can interfere with the formation of budded virus, which in turn can affect secondary infection, and make difficult r-virus identification, isolation, and propagation necessary for generating viral stocks.

Success in enhancing the expression of r-proteins in the baculovirus-insect cell system has been accomplished by modifying the length of the 5′ untranslated region –the sequence that precedes the ATG codon. Sano et al. (87) were capable of enhancing the expression of luciferase and rabbit skeletal muscle α-tropomyosin 7- and 20-fold, respectively, by using a 21-base leader sequence rich in adenosine in positions -15 to -9 and the Kozac sequence in the -8 to -1 positions.

Generation of r-Baculoviruses

Construction of r-baculovirus involves three main steps: construction of a transfer vector; recombination between transfer vector and baculovirus DNA, and screening for the resulting r-baculovirus; and virus titration and elaboration of a viral stock (denoted as A, B, and C in Fig. 4). In the first step, the gene of interest is inserted into a transfer vector (plasmid). The gene is placed under the control of a promoter and flanked by sequences homologous to parental DNA in a locus not essential for the virus. A wide variety of baculovirus transfer vectors are available from commercial sources; some of them are sold in kits for readily expressing r-proteins in insect cells. The choice of a particular vector is based on the promoter to be used and on the number and form (e.g., fusion constructs) of proteins to be expressed. In general, transfer vectors contain an *Escherichea coli* origin of replication, an antibiotic

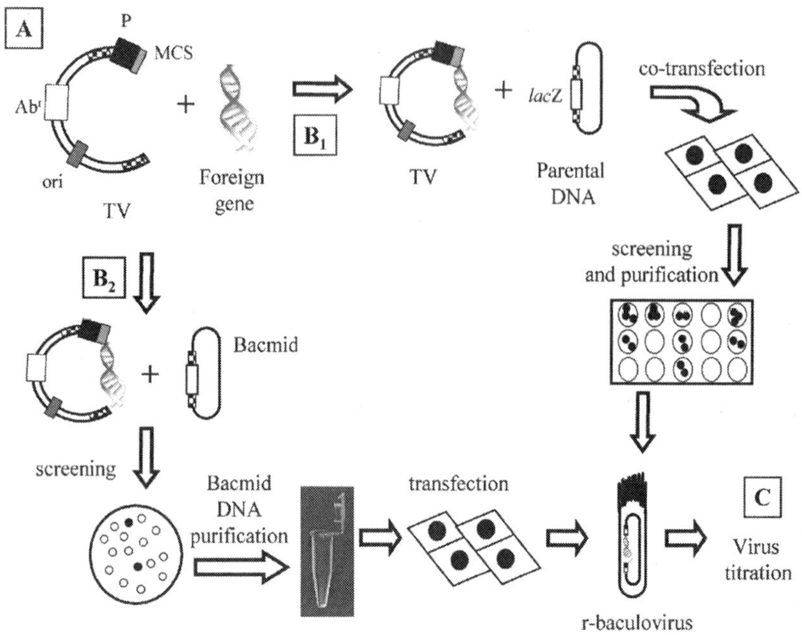

Figure 4 Main steps for obtaining an r-baculovirus. (A) The transfer vector (TV) is depicted with its main elements (ori, bacterial replication origin; Ab^r, antibiotic resistance marker; P, promoter; MCS, multiple-cloning site; patterned areas, homologous recombination sites). Depending on the transfer vector used, two routes can be chosen: (B1) Traditional method in which recombination occurs in the insect cells; (B2) Bacmid-transposition method. (*See color insert p. 12.*)

resistance marker, a promoter, a multiple-cloning site, and a large stretch of *Ac*MNPV genome homologous sequences flanking the cloning site. The cloning site is located downstream from the promoter for inserting the gene to be expressed.

The transfer vector and the parental DNA are then cotransfected into insect cells by any of the conventional methods available (B1 in Fig. 4). Recombination occurs between homologous sequences of the vector and parental DNA, resulting in a recombinant viral progeny that will usually not exceed 1%. To optimize this step, Kitts et al. (88) linearized the parental virus near the recombination site, thus improving r-yield by 30%. The parental DNA carries a marker gene, commonly *lacZ* from *E. coli*, which is replaced upon recombination. Therefore, selection of recombinants is achieved by using a color-screening test for β-galactosidase. This method requires further purification of β-gal$^-$ viruses by plaque assay. A relevant improvement for generating r-baculovirus consists in performing within *E. coli* a site-specific transposition between a baculovirus shuttle vector (known as bacmid) and an expression cassette (89) (B2 in Fig. 4). Upon transposition, white *E. coli* colonies (β-gal$^-$), which correspond to those where insertion of the foreign gene in the bacmid occurred, can be directly selected without the need for virus plaque assays. Such colonies are amplified and the bacmid DNA is purified and transfected into the insect cells to produce the r-baculovirus.

r-Protein productivity and emergence of defective interfering particles (see later) will closely depend on the number of infective viral particles per cell [or multiplicity of infection (MOI)] added to the culture (see Bioengineering Issues in Insect

Cell Culture). Accordingly, once the r-baculovirus has been identified it must be quantified (C in Fig. 4). There exist two main titration methodologies, plaque assay and end-point dilution (26). The first relies on the formation of characteristic plaques (lytic halo) upon infection of insect cells fixed in multiwell plates with a serially diluted viral stock. By directly counting the plaques, viral titers can be calculated. This method assumes that each plaque is derived from the infection of a single viral particle. In the end-point dilution assays, viral stock is also serially diluted, but cells are not necessarily fixed. In this case, a characteristic change, such as in morphology, viability, or even plaque formation, induced by viral infection is determined. The tissue culture infectious dose 50 ($TCID_{50}$) can then be calculated, which is the viral concentration that infects 50% of the insect cell population. $TCID_{50}$ can easily be converted to plaque-forming units (pfu) by simply multiplying the $TCID_{50}$ by 0.693 (26). Dee and Shuler (90) proposed a modification of the end-point dilution assay to improve its accuracy. They were able to reduce the experimental variation from one order of magnitude to only 10–40% by including a centrifugation step to enhance virus attachment to cells. Other modifications have been aimed at shortening the assay time and reducing errors due to the subjective interpretation of changes caused by infection. For instance, viruses marked with reporter genes, such as β-galactosidase (91,92) or green fluorescent protein (GFP) (93), have been used. In these cases, either absorbance or fluorescence is proportional to the amount of infective virus particles and therefore $TCID_{50}$ can be readily and objectively determined (93). Mena et al. (94) recently proposed an assay based on viability that allows the objective determination of titers without a reporter protein. Alternatively, antigens produced upon baculoviral infection can be detected by polyclonal antibodies. Detection occurs even before any plaque can be detected, thus representing an interesting advantage over other methods (95,96). A modification to such a method for further shortening the time required to obtain a result was reported by Kitts and Green (97).

Both virus titering methods have important limitations, although the plaque assay is more laborious and time-consuming than the end-point method. In the plaque assay, culture medium is commonly replaced after 1–2 hr of infection, which may not be a sufficient time for virus attachment to Sf cells (90,98). Furthermore, difficulty in counting plaques can lead to errors. In contrast, operator-independent methods for assessing cell changes due to infection can be implemented in the end-point dilution (94). Likewise, the number of infecting viruses is estimated from the presence or absence of at least one infective particle, and the culture medium is not removed from the plate. Therefore, the end-point dilution method should be preferred to the plaque assay.

Finally, viral stocks are prepared by infecting either larvae or cultured cells and recovering the virus. Viruses are commonly resuspended in culture medium, titrated, and maintained for long-term storage at 4°C, −80°C, or in liquid nitrogen; however, virus instability problems can occur even under such conditions. Thus, it is advisable to retiter the virus after long-term storage since inactivation can occur. Fetal bovine serum (FBS) is usually added to viral stocks to increase their stability during storage.

Defective Interfering Particles

A well-known phenomenon, that occurs when viral stocks are serially passaged for many times, is the appearance of mutant viruses, which alter the infection process. A type of such mutants is known as defective interfering particles (DIP). DIPs are viral particles bearing genetic deletions, usually unable to form plaques in appropriate assays, and physically characterized by being smaller than normal viral particles. For instance, in a normal stock more than 95% of viral particles have an average size of 330 nm, whereas a high-passage virus sample can be composed of up to 60% DIP with a size of 200 nm (99). DIPs are incapable of replicating by themselves, and require a complete virus as a helper. A high MOI increases the probability that both DIP and a complete virus enter the same cell. Therefore, DIPs are generated when stocks are passaged at a high multiplicity, and will compete with normal viruses for components of the replicative machinery (100). Since DIPs can replicate faster, the yield of infective viruses will decline. Accordingly, viral titers of high-passage stocks containing DIPs can be several-fold lower than titers of stocks of low-passage number (99). Some other mutant viruses can have deletions of the polyhedrin gene or other genes relevant for virus replication (101). A class of mutant viruses affects the formation of occluded bodies since it produces few polyhedra (FP), whereas another class producing many polyhedra has also been described (102). FP viruses can form plaques in in vitro assays, leading to false positives when screening for r-baculoviruses. To avoid the negative effects of DIP or other mutant viruses, multiple passaging of virus stocks must be avoided and low MOI must be employed (see Bioengineering Issues in Insect Cell Culture).

INSECT CELL CULTURE

In vitro culture provides an environment very different from that in a physiologically regulated complete organism. Nutrient levels are not constant or regulated, and by-products are not eliminated. Cells can be subjected to hydrodynamic, osmotic, or pH stresses. All these factors can affect the ability of cells to produce r-proteins or baculoviruses. Culture techniques and flasks commonly used for culturing mammalian cells are also adequate for insect cell culture. However, culture conditions are different (see Table 3). Insect cells do not require an external CO_2 provision for growth and in general are not anchorage dependent. Most insect cell lines form a monolayer in static culture, but some do not attach to surfaces, such as the BTI-Pu-A7S cell line (104). Insect cells can be cultured in suspension with minimum adaptation. Suspension cultures are preferred for r-protein or baculovirus production, while attached cultures are convenient for virus tittering and isolation. Some cell lines, such as Tn5Bl-4, may form clumps when cultured in suspension. Such a problem can be reduced by supplementing serum-free medium (SFM) with 40 μg/mL of dextran

Table 3 Optimal Conditions for Insect Cell Cultures

pH	6.2–6.5
Osmolarity (mOsm)	300–380
Temperature (°C)	27–29
Temperature (°C)[a]	22–29
Laminar shear stress (N/m^2)	< 0.59

[a]For recombinant protein production.
[Adapted from Refs. (14,17,21,23,24).]

sulfate (105). However, no difference between the amount of r-protein produced by individual cells and aggregates has been found (106).

The age of a cell line in culture has a drastic effect on its performance. Donaldson and Shuler (107) compared Tn5Bl-4 cells passaged 130 or 360 times. They found a decrease of 2.5 times in the concentration of r-product produced by the high-passage cells. Furthermore, high-passage cells were 4 μm larger and reached maximum cell concentrations of about half of those of low-passage cultures. It is therefore necessary to create a master cell bank and renew a working stock approximately every 3 months or 30 passages. The master cell bank should be stored in liquid nitrogen as described elsewhere (26). Inoculum size and its metabolic activity are also important. The minimum inoculum size for insect cell growth depends on the age of the inoculum, cell line, and culture medium employed. The inoculum size determines the concentration of growth-promoting factors produced by cells, as well as those carried over from the seed culture (108). Furthermore, Sf9 cultures can be seeded in serum-containing medium with as little as 0.7×10^5 cells/mL if the inoculum comes from an exponentially growing culture (109). In SFM, a minimum inoculum of 0.5×10^6 cells/mL is recommended (110).

Nutritional Requirements of Insect Cells in Culture

Insect cells require many compounds for growth, including carbon and energy sources, amino acids, growth factors, hormones, lipids, vitamins, and various trace elements. In general, nutritional requirements of infected insect cells have been reported to be similar to those of uninfected cultures (111,112). The nutritional requirements of cell lines from *S. frugiperda* (Sf) and *T. ni* (Tn), two of the most utilized insect cell lines for r-protein production, are discussed below. Typical nutrient consumption rates of both cell lines are listed in Table 4. The most obvious difference between Sf9 and Tn5Bl-4 is their metabolic activity. Nutrient uptake rates of Tn5Bl-4 cells are 1.4–2.5 times faster than in Sf9 cell cultures. Additionally, the metabolisms of these cell lines are regulated in different ways.

Carbohydrates are the main energy source for insect cells (111,114,118). Insect cells can survive for at least 96 hr without a carbohydrate provision, but no growth will occur (113–115,119). Fructose, maltose, and glucose (Glc) can support insect cell growth, but Glc is preferred (113,115). Meneses-Acosta et al. (25) have shown that Glc depletion marks the appearance of apoptotic Sf9 cells in culture. Carbohydrate metabolism is regulated in a similar way as in other eukaryotic cells. An excessive provision of Glc (>1 g/L) results in the accumulation of by-products, mostly lactate and alanine (114,115,120). In contrast, cells grown under Glc limitation or in a different carbohydrate, such as fructose, reduce their monosaccharide uptake rate and production of by-products (114). Sucrose, which is added to adjust medium osmolarity, is usually not consumed by Sf9 or Tn5Bl-4 cells (111,113,115). However, sucrose can be utilized by other cell lines if Glc is not available (121).

Amino acids fulfill the nitrogen requirements of insect cells. Amino acid requirements of various insect cell lines have been summarized by Mitsuhashi (122). Aspartic acid, cystine, glutamine, and methionine are the fastest-consumed amino acids. Asparagine is also highly consumed in cultures of Tn5Bl-4 cells and in glutamine-limited Sf9 cell cultures (111,120,123). Glutamine is mainly utilized for biosynthesis by Sf cells (17,118). Sf9 cells can grow in the absence of glutamine, although their growth rate may be reduced (114,124). Interestingly, Sf9 cells, but not Tn5Bl-4 cells, are capable of synthesizing glutamine if ammonium is available (124).

Table 4 Metabolic Coefficients for Selected Nutrients in Sf9 and BTI-Tn-5Bl-4 Cell Cultures in TnM-FH Medium Supplemented with 10% FBS

	Sf9	BTI-Tn-5Bl-4	References
Uninfected cultures			
Maintenance coefficient, $\times 10^{-10}$ mmol ATP/cell/hr	2.53	—	(113)
Glucose consumption rate, $\times 10^{-10}$ mmol/cell/hr	0.52–0.78	1–2	(111,114,115)
Glutamine consumption rate, $\times 10^{-10}$ mmol/cell/hr	0.37–0.4	0.54	(111,114)
Oxygen consumption rate, $\times 10^{-10}$ mmol/cell/hr	2–4	2.8–5.4	(111,113)
Infected cultures			
Glucose consumption rate, $\times 10^{-10}$ mmol/cell/hr	0.61	2.6–1.8	(111,116)
Glutamine consumption rate, $\times 10^{-10}$ mmol/cell/hr	0.18	0.58–0.78	(111,116)
Oxygen consumption rate, $\times 10^{-10}$ mmol/cell/hr	2–7.6	3.2–5.8	(111,117)

Note: All values were obtained under no-nutrient limiting conditions.

Cystine and glycine are essential for Sf cells, which do not express the methylene-tetrahydrofolate synthetase required for the synthesis of both amino acids (125). This is a unique case in eukaryotic cells. The rest of the amino acids are slightly, or not at all, consumed by cells in culture (113,116).

Insect cells have a safe way to dispose of by-products, especially ammonia, in the form of alanine. Alanine is a by-product of Glc and glutamine metabolism that is not toxic to insect cells at concentrations of up to 100 mM (120). It is produced through the transamination of pyruvic and glutamic acids and the degradation of other amino acids and serum proteins. Alanine production is finely regulated, and both Glc and glutamine are required for its formation (114,119,120). An excesxsive supply of any of these two nutrients, or the absence of any one, will block alanine formation, cause ammonium accumulation, and reduce growth rate (114,120).

Insect cells have a limited capacity of synthesizing, desaturating, and elongating fatty acids (120). Lipid privation causes cell degeneration and the production of noninfective baculoviruses (126). Moreover, insect cells cannot synthesize cholesterol, which is required for synthesis of hormones and membranes (122). Therefore, lipids and cholesterol must be provided to insect cells cultured in vitro. Vitamins of the B complex are important for maintaining insect cell lines. Thiamine, riboflavin, pyridoxine, folic acid, niacinamide, inositol, choline, carnitine, and cobalamine are essential for insect cells (122). Trace elements, such as zinc, iron, and copper, are also required for cell growth.

Insect cells are strict aerobes. Optimum dissolved oxygen (DO) concentrations for their growth range between 30% and 100% with respect to air saturation, and the Monod saturation constant for oxygen (K_{O_2}) has been reported to be 1.9% (17,113). In contrast, r-protein production is more senstive to DO. r-Protein yield is reduced 50% at DO of 10% or 50%, compared with infections effected at 25% DO (127). Moreover, DO can affect r-protein glycosylation (128). Typical oxygen uptake rate (OUR) of insect cells is similar to those of mammalian cells (17,113). However, when insect cells are infected, their OUR can increase up to 100% (117,127,129,130). Interestingly, such an increase in OUR has not been correlated with a higher consumption of other nutrients (111,112).

Toxic by-product accumulation is one of the main problems that limit in vitro culture of high eukaryotic cells. However, insect cells are less sensitive to by-products than mammalian cells. Ammonium is produced from the metabolism of glutamine and other amino acids. As already mentioned, alanine is a by-product through which ammonium is safely eliminated. Sf and *B. mori* cells under optimal culture conditions do not produce ammonium and are capable of using it if present in the culture medium (115,124,131). However, ammonium does accumulate in cultures of Tn5Bl-4 cells (up to 25 mM), which cannot use it (111,112,124). r-Protein production by Tn5Bl-4 cells continues even at ammonium concentrations of 15 mM, but 20 mM can reduce cell and r-product concentrations 20% and 73%, respectively (112,132). In contrast, ammonium concentration as low as 4 mM can reduce the growth of mammalian cells by 50% (133). Such a difference in ammonium sensitivity may be a result of the lower pH of insect cell culture media (6.0–6.5), which increases the concentration of ammonium ion with respect to ammonia. Dissolved ammonia is soluble in cell membranes and can freely enter cells and increase the pH of lysosomes (134,135). Lactate is another extensively studied by-product that is produced by insect cells under oxygen limitation (113). It seldom accumulates in Sf9 cultures, but it can reach concentrations of up to 40 mM in cultures of Tn5Bl-4 cells (111). The effect of lactate at 40 mM has not been determined for Tn5Bl-4 cells, although concentrations of up to 10 mM do not affect the growth of Sf9 cells (113). Lactate is consumed by Sf9 cells after carbohydrate depletion (113,136), causing a characteristic increase of pH toward the end of the cultures (17). Metabolic engineering has been utilized for increasing the efficiency of nutrient utilization by Tn5Bl-4 cells. For instance, yeast's pyruvate carboxylase was overexpressed by Elias et al. (137), which resulted in a reduction of lactate and ammonia production.

Culture Media

The initial medium formulations for insect cell culture were based on the composition of insect hemolymph. Grace (3) formulated the first widely used culture medium for insect cells. Grace's medium was originally made for culturing *A. eucalypti* cells, but it is now utilized for culturing cells of many insect species. It contains all the proteic amino acids, except cysteine, various organic acids, inorganic salts, vitamins, Glc, fructose, and sucrose. Another medium for insect cell culture was designed by Schneider (138), specifically for the in vitro culture of cells from *Drosophila*. Both media require supplementation of insect hemolymph, and both are still extensively used.

Grace's medium has been modified many times, the most popular being the modification by Hink (139), in which hemolymph was substituted by bovine serum and medium was supplemented with lactoalbumin hydrolysate and yeastolate. This modification is called TnM-FH, and was originally formulated for the establishment of cell lines from *T. ni*. TnM-FH is widely utilized for laboratory-scale applications of the IC-BEVS with many different cell lines. The replacement of insect hemolymph with bovine serum was an important development for insect cell culture, as it reduced the cost of culture media and allowed an increase in the scale of cultures.

The empirical formulation of Grace's medium is very complex. The monosaccharide concentration is very low and only supports the growth of 2×10^6 cells/mL to 3×10^6 cells/mL (110,113,115). Glc is the first nutrient to be depleted in batch cultures and increasing its concentration to 2.1 g/L increases cell concentration 57% (115). The nutrients to be depleted next from Grace's medium are cysteine, glutamine, and glutamate (113–115). Arginine is also a limiting nutrient for Tn5Bl-4 cultures. α-alanine, which is a by-product of metabolism, is included in the formulation. In addition, various organic acids are included in Grace's medium, although they are not required for cell growth (120). Recently, Palomares et al. (140) showed that elimination of fructose (if an equivalent amount of Glc is added instead), sucrose (if osmolarity is adjusted with NaCl), FBS, and lactoalbumin hydrolysate did not affect the maximum concentration of r-protein produced by Sf9 cells in TnM-FH. However, yeastolate is required for r-protein production, while FBS promotes cell growth after infection (140).

Gardiner and Stockdale (141) proposed a simpler version of Grace's medium, TC-100, for the propagation of baculovirus in Sf cells. In this formulation fructose was substituted by Glc, sucrose was eliminated, sodium chloride was added to adjust osmolarity, organic acids were omitted, and 2.6 g/L of tryptose extract was included. These changes resulted in a less complex formulation. TC-100, however, requires supplementation with FBS.

In the early 1980s, a more complex and complete FBS-supplemented medium, IPL-41, was formulated for large-scale production of baculoviruses. It includes more trace elements than Grace's medium, such as Cu, Co, Fe, Mo, and Zn. Glc concentration was increased to 2.5 g/L, fructose was eliminated, sucrose concentration was reduced, and maltose was added (1 g/L). The concentration of most amino acids was increased, α-alanine was eliminated, and hydroxyproline was included. The concentration of organic acids was reduced more than 10 times. Still, IPL-41 did not perform better than TnM-FH in Sf9 or *Estigmena acrea* cultures (115). It is known that Glc concentration in IPL-41 medium does not limit growth; however, the limiting nutrient has not been identified. The performance of IPL-41 increases when supplemented with yeastolate or tryptose phosphate broth.

All media described in this section require the supplementation of FBS to support insect cell growth. Interestingly, FBS is an effective substitute for insect hemolymph even when no response to vertebrate growth factors has been observed in insect cells. FBS provides growth factors, hormones, and lipids, inactivates toxic materials, provides carrier proteins, contains protease inhibitors, affects the solubility of nutrients, has buffering capacity, and modifies some physical properties of culture medium (such as surface tension and osmolarity) (142). Serum provides protection to cells from hydrodynamic stress, as it decreases plasma membrane fluidity and cell surface hydrophobicity (143,144). Lower plasma membrane fluidity results in stronger cells, whereas lower cell surface hydrophobicity prevents cell attachment to bursting bubbles. Serum addition can have various effects on cell physiology. Joshi et al. (145) reported an increase in complex glycan structures present in an r-glycoprotein expressed by Tn cells when FBS was added to culture medium. Moreover, it has been demonstrated that FBS reduces baculovirus attachment to Sf9 cells (98,146). Bovine serum albumin is, at least in part, responsible for this effect (147). Serum composition is not fully defined and it is frequently impossible to identify the compounds responsible for an observed effect. In spite of its advantages, not all of the properties of hemolymph are present in FBS. Addition of 10% hemolymph instead of FBS can delay cell death after infection by 164 hr, decrease the specific death rate, prevent apoptosis, and increase r-protein concentration 4.5 times (146,148). Similar results were obtained by Maranga et al. (149). However, the high cost and low availability of insect hemolymph makes its use impractical. In addition, serum fails to provide some lipids required for growth, but this deficiency has been overcome by the addition of yeastolate. Yeastolate not only provides lipids, but also amino acids, B vitamins, nucleotides, and growth-promoting factors (150).

FBS is the most expensive additive in cell culture media; its price ranges between US$400 and US$500 per liter. Moreover, serum contains a high concentration of proteins, it is subjected to unpredictable lot-to-lot variations, it can be contaminated with adventitious agents, and it complicates downstream processing. These drawbacks make serum an undesirable supplement to cell culture media and efforts to replace it with several additives are abundant. Most SFMs contain Pluronic F-68 (PF68), a nonionic copolymer. PF68 is a shear protective agent, utilized in both mammalian and insect cell cultures. In a similar way to FBS, PF68 can decrease plasma membrane fluidity, increasing cellular resistance to shear (151). PF68 is added at concentrations between 0.05% and 0.5% (w/v). Lipids are usually added to SFM as lipid–sterol emulsions at 0.1 to 2.5 % (v/v) (116). Stock lipid emulsions usually contain cholesterol (4.5 g/L), cod liver oil fatty acids (methyl esters, 10 g/L), α-tocopherol (2 g/L), and a detergent as an emulsifier, usually Tween 80 (25 g/L). PF68 can also be utilized as an emulsifier. Yeastolate has been found to be the most critical supplement for Tn5Bl-4 cells in SFM, with optimum concentrations between 4 and 8 g/L (116). Sf9 cells also require yeastolate for growth on SFM. Yeastolate is the only chemically undefined supplement in commercial insect cell SFM. Yeastolate presents some drawbacks similar to FBS, such as lot-to-lot variations, but despite many efforts, it has not been possible to replace it with chemically defined compounds.

Various SFMs are commercially available and many of them are also protein-free. Some formulations have been specifically designed to support the growth of a particular cell line. SFMs for culturing Tn5Bl-4, Sf9, and D.mel 2 (from *Drosophila melanogaster*) cells are available. Most SFMs have been optimized and their performance is better than serum-supplemented formulations. SFM contain from 5 to 10 g/L of Glc and support from 1.5 to 2 times higher cell densities than traditional media (110). However, media are often optimized for cell growth and not for protein or virus production. In some cases serum-containing media may perform better than serum-free formulations, especially when glycosylated and secreted proteins are produced (110). This result agrees with the existence of compounds in FBS that promote protein glycosylation (145). It is clear that media designed to promote cell growth are not necessarily the best for r-protein production, and vice versa. Moreover, media can perform differently depending on the r-protein being produced. It should be noted that media formulations could also have an important effect on protein quality (see Posttranslational Modifications). No effect of extended passaging of cells in SFM has been found on the number of occlusion bodies obtained (22).

Commercially available SFMs have two disadvantages: their composition is proprietary, which prevents further medium optimization, and they are expensive, with prices between US$30 and US$43 per liter (17). Note that the cost of such media would only be enough to pay for the fraction of FBS needed to supplement serum-containing media. Accordingly, prices are still too high for large-scale applications of low-value products such as bioinsecticides. To solve both problems, many groups have successfully developed SFM from inexpensive compounds, such as protein hydrolysates (116,150,152).

BIOENGINEERING ISSUES IN INSECT CELL CULTURE

The technological utilization of IC-BEVS can be a complicated task as it involves two different processes. Insect cells are first grown to a desired concentration and then infected by a baculovirus to obtain the product of interest (r-protein or baculovirus progeny). After infection cell division stops and the synthesis of viral proteins begins. The most favorable conditions for r-protein production are quite different from the conditions for optimal cell growth (110,153). Baculovirus and r-protein production is a transient process that ends when cells die as a consequence of infection. Typical kinetics of an infected insect cell culture are shown in Fig. 5. Several strategies have been utilized for increasing r-protein and baculovirus production at higher scales. An overview of such efforts is given in this section.

Overcoming the "Cell Density Effect"

The "cell density effect" consists in the reduction of specific productivity as cell concentration increases above 3×10^6 cells/mL or 4×10^6 cells/mL (155,156). After several years of research, it has been concluded that the cell density effect is mainly a consequence of nutrient limitation at high cell densities (112,155–157). The first approach utilized to avoid the cell density effect was replacing the medium at the time of infection (115,158–161). However, for large-scale applications this option is neither economical nor practical. Nutrient feeding has proven to be an efficient strategy for avoiding the cell density effect (155,157). Nonetheless, nutrient feeding or medium replacement must be performed before autoregulatory events leading to programmed cell death or cycle arrest occur (25,108). Various research groups have analyzed the nutrient requirements of insect cells and designed feeding strategies to overcome the cell density effect [reviewed in Ref. (17)]. Such efforts have resulted in cell concentrations as high as 5.2×10^7 cells/mL and constant specific productivities at a cell concentration of 1.5×10^7 cells/mL (157). To achieve such results, Elias et al. (157) utilized a semicontinuous nutrient addition that included Glc, amino acids, yeastolate, lipids, vitamins, and trace elements. Overall, nutrient feeding increased three times the concentration of β-galactosidase (Table 5). Oxygen is also a key element to avoid the cell density effect (155). It should be provided in sufficient quantities to satisfy the increase of OUR at the time of infection.

Another alternative for avoiding the cell density effect is by operating the culture in perfusion mode. Perfusion is usually not economically attractive for the production of intracellular or membrane proteins but can be convenient for secreted proteins. Secreted proteins are usually glycosylated and occasionally can be produced at low concentrations by insect cells. Their production can be more energy

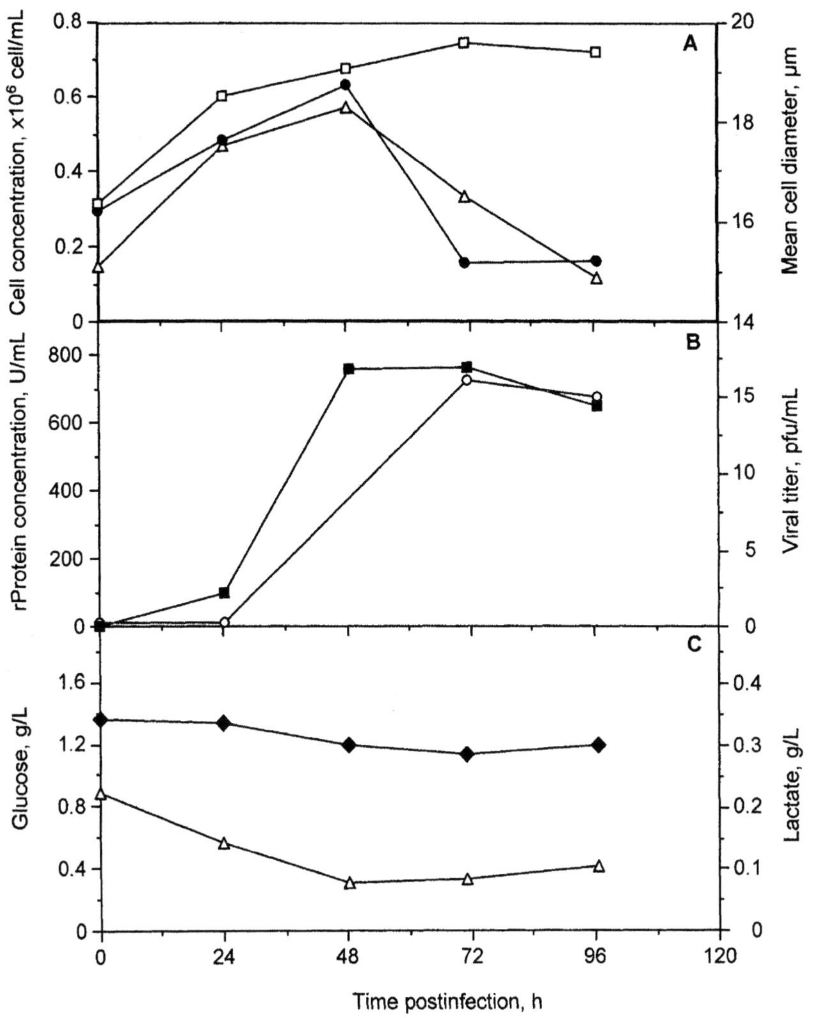

Figure 5 Typical kinetics of an infected insect cell culture with an MOI of 1 pfu/cell. Sf9 cells were infected with a recombinant baculovirus coding for rotavirus VP7 [for details, see Ref. (154)]. (A) Total cell concentration (□), viable cell concentration (•), cell diameter (Δ). (B) r-protein concentration (■), viral titer (○). (C) glucose (◆), lactate (Δ).

and nutrient demanding than that of intracellular proteins. The cell density effect for the production of a glycosylated protein (β-trace glycoprotein) has been observed at cell concentrations much lower than for intracellular proteins ($>1\times10^6$ cells/mL) (132). Utilizing perfusion, Chico and Jager (132) maintained constant the specific production rate of β-trace glycoprotein up to cell densities of 3×10^6 cells/mL. A productivity of 1.2 g/L/day and a concentration of 0.5 g/L of r-protein were obtained (Table 5). Such values are very high for a glycosylated protein produced through any expression system. Even when promising results have been obtained by increasing insect cell concentrations, more work is still needed to fully define the nutritional requirements of insect cells for growth and r-protein or virus production (see Insect Cell Culture).

Table 5 Bioreactors Utilized for r-Protein Production by Insect Cells

Bioreactor	Culture medium	Scale	Cell line	Maximum r-protein concentration	References
Batch STR	TC100 5% FBS SF900II IPL41 5% FBS	1.4 L 3.5 L 2 L	Sf9	140–150 U/mL, β-gal[a]	157,162,163
Fed-batch STR	SF900II	3.5 L	Sf9	430 U/mL, β-gal[a]	157
Air lift	SF900	14 L	Sf9	104 U/mL, β-gal[a]	164
Tubular segmented-flow	ExCell 401	0.3 L	Sf9	36 U/mL, β-gal[a]	165
Helical ribbon STR	IPL41 SFM	11 L	Sf9	70 μg/mL, VP6[b]	136
HARV[c]	Express 5	50 mL	Tn5Bl-4	2.7 μg/mL, SeAP	166
Split-flow air lift (packed-bed)	ExCell 400	0.5 L	Tn5Bl-4	10.7 μg/mL, SeAP	167
Two-stage STR	SF900II	1 L × 2	Sf9	5 mg/10^6 cells, EBNA1[d]	168
Two-stage STR	IPL41 10% FBS	1.5 L × 2	Bm5	250 mg/L, CAT[e]	161
Perfused reactor membrane aerated	Excell 401	1.7 L	Tn5Bl-4	0.5 g/L, 1.2 g/L/day, β-trace protein	132
Packed-bed	TnM-FH 10% FBS, 0.2% PF68	0.25 L	Sf21	4×10^5 U/mL, IL-5[f]	169

[a] β-Galactosidase.
[b] Structural protein of rotavirus.
[c] High-aspect rotating-wall vessel.
[d] Epstein–Barr virus nuclear antigen 1.
[e] Bacterial chloramphenicol acetyltransferase.
[f] Human interleukin 5.

Infection Strategies

An efficient infection requires that all cells in culture are infected and that culture conditions are optimal for baculovirus and r-protein production. Overall nutrient demand, cell growth and death after infection, and product concentrations are all determined by infection conditions. A simultaneous infection of all cells in culture results in synchronized protein and virus production. Alternatively, if only a fraction of the population is initially infected, the remaining cells will continue to grow until infected by the viral progeny of the infected cells (secondary infection, see Baculovirus). Infection can be affected by media composition, cell growth phase, cell concentration, mode of culture, and MOI (90,98,146,170,171; reviewed in 17).

Insect cell–baculovirus interactions have been studied and mathematical models for virus attachment and trafficking have been proposed (90,170). Baculovirus binding to cells is the limiting step for infection, as the mean time for binding is at least one order of magnitude higher than those for each of the subsequent intracellular steps (90). Moreover, a direct correlation between the amount of attached virus and r-protein concentration has been found (98). Accordingly, infection strategies should be aimed at increasing the rate of virus binding. Brownian diffusion governs baculovirus encounter with cells (90,162,170). Thus, virus attachment for fully dispersed cells is affected by temperature, viscosity, and the cell surface available for infection, but not by hydrodynamic conditions (98,162,170). A first-order dependence of baculovirus attachment on cell concentration (between 0.5×10^6 cells/mL and 20×10^6 cells/mL) has been found (98,170). Nonetheless, virus concentration only affects virus binding at low cell to virus ratios (below 1 pfu/cell) (172). Medium composition can also affect virus binding. Baculovirus binding to Sf9 cells is reduced three times if medium is supplemented with 10% FBS (98). Rhee et al. (146) also observed a decrease of infection when FBS was added. Palsson et al. (173) demonstrated in mammalian cells that increasing virus attachment to cells increases gene delivery, the ultimate objective of baculovirus infection for r-protein production. They utilized a flow-through procedure for the infection of various mammalian cells with retrovirus. Gene transfer was increased by one to two orders of magnitude. However, only Wood (174) and Dee and Shuler (90) have proposed a strategy (centrifugation) to increase baculovirus binding to insect cells. Nonetheless, both centrifugation and flow-through devices are difficult to operate at large scales.

The rate of infection and the fraction of the population initially infected are a direct function of MOI (90,159,175,176). MOI is defined as the number of infectious particles (pfu) added per cell (see Baculovirus). The effect of MOI on the infection of a culture can be modeled by a Poisson distribution (159,175). The probability (p) of a cell being infected by (w) virus would be given by:

$$p(w) = \left[\left(\frac{\mathrm{MOI}^{w}}{w!}\right)e^{-\mathrm{MOI}}\right] \tag{1}$$

The solution to Eq. (1) is shown in Fig. 6. It can be seen that at low MOI (below 2) the fraction of infected cells is very sensitive to differences in MOI. As MOI increases above 4 pfu/cell, practically all cells are initially infected by at least one virus. It should be noted that the MOI utilized in Eq. (1) should reflect the actual infection conditions in culture. That is, it should include factors that may delay or reduce virus attachment. This is seldom the case, as virus titering is performed in static cultures under conditions substantially different from those in production cultures (see Baculovirus).

Although MOI appears to be a parameter easily manipulated, two important factors should be considered (*see Baculovirus*). First, traditional methods for baculovirus titering are in general inaccurate, and virus stocks are subject to inactivation with time. Thus, the actual MOI utilized can be different from that planned, especially if calculated from outdated titers (older than 6 months). Second, infection requires a viral stock, which needs to be previously produced in insect cell cultures. Such viral stock must be free of defective particles and should have a high titer to prevent spent medium carryover. Defective particles reduce r-protein or baculovirus

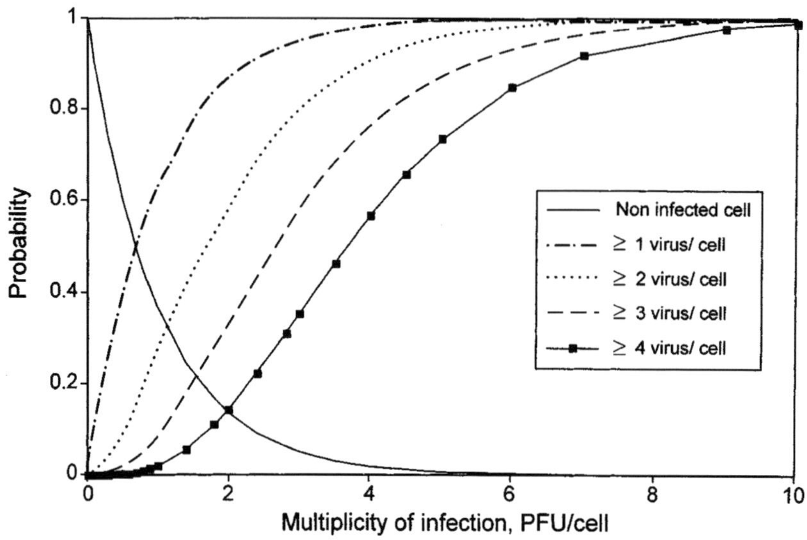

Figure 6 Probability of infection of a population according to the Poisson distribution. [From Refs. (159,175).]

yields at high MOI (99). To avoid defective particles viral stocks should be produced utilizing low MOI (< 1 pfu/cell), and the r-virus should be plaque-purified every 7–10 passages (see Baculovirus) (99). Concentrating viral stocks through centrifugation or ultrafiltration may reduce spent medium carryover.

Many research groups have studied the effect of MOI on r-protein and baculovirus production [reviewed in Ref. (17)]. Such an effect depends directly on the time of infection (TOI). A simulation by Licari and Bailey (159) of the effect of MOI and TOI is shown in Fig. 7. It can be seen that as TOI (the age of the culture) increases, a higher MOI should be utilized for maximizing the r-protein yield. This is explained by the fact that due to infecting at low cell concentrations (culture in the early exponential growth phase) a low MOI (< 4 pfu/cell) results in only a fraction of initially infected cells (Fig. 6). The remaining cells are then capable of growing until progeny viruses infect them. Such a strategy has been used for obtaining high product yields with minimum virus requirements (156,175). As the cell concentration increases (close to the stationary growth phase), higher MOI should be utilized since the medium nutritional capacity is almost exhausted and will not support additional cell growth. Yang et al. (112) determined the medium nutritional capacity (or nutritional depth) and used it to estimate optimal MOI and TOI combinations. Infection strategies should also be designed considering the stability of the r-product. The use of low MOI and TOI results in a long production phase, which increases the time of exposure of the product to proteases (171).

Among its advantages, the IC-BEVS can be used to simultaneously produce several r-proteins. The most complex case is the production of virus-like particles (VLPs), which are conformed by hundreds of various molecule kinds (up to five different proteins). MOI can be utilized to manipulate the stoichiometric relation between protein molecules that are encoded by genes in different baculoviruses (175–177). In turn, different stoichiometric relations between molecules can be used to manipulate the composition of VLP (175,177). Likewise, Palomares et al. (176) utilized sequential infections for manipulating the stoichiometry between rotavirus

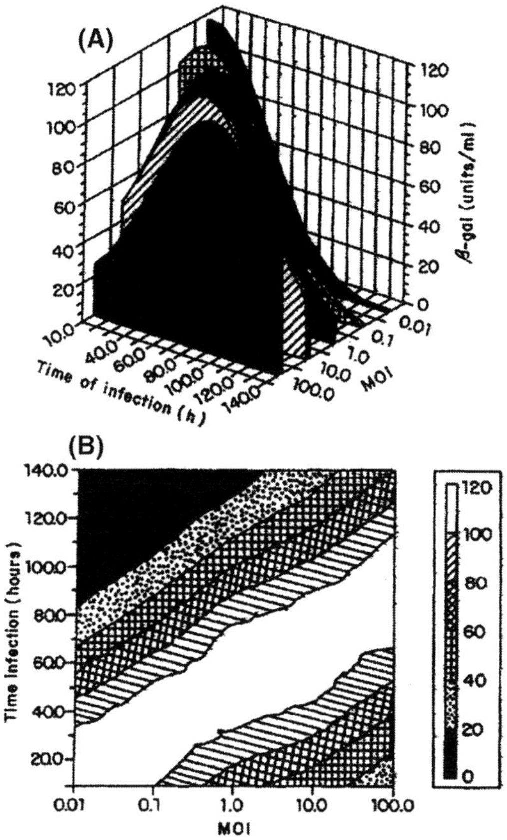

Figure 7 (A) Final-state β-galactosidase titers (activity units/mL) as a function of both time of infection (hours after cell inoculation) and MOI. (B) Contour plot for optimizing infection strategies. Domains of time of initial infection (hr) and MOI corresponding to different ranges of final β-galactosidase titers are shown. The legend to the right indicates the final product titers (activity units/mL). [From Ref. (159).]

structural proteins. MOI and TOI are valuable tools to design rational infection strategies for the production of VLPs. Such rational strategies can increase the yield of functional multimeric molecules (175).

Shear Stress in Bioreactors

One of the main issues when scaling up animal cell cultures is cellular fragility. Cells cultured in bioreactors are subjected to shear stress caused by agitation and sparging. The damage that a cell suffers is determined by the type, duration, and magnitude of the hydrodynamic force (178,179). Insect cells have been found susceptible to laminar shear above $0.59\ N/m^2$ (103) or to energy dissipation rates above $2.25 \times 10^4\ W/m^3$ (180). Energy dissipation rates in stirred tanks are in the range of $2 \times 10^3\ W/m^3$ to $3.5 \times 10^3\ W/m^3$, and in bubble columns are between $0.2 \times 10^3\ W/m^3$ and $2 \times 10^3\ W/m^3$ (181). Moreover, cells will not be subjected to turbulence at eddy scale since the Kolmogoroff eddy size in reactors utilized for animal cell culture is usually larger than the insect cell diameter (127,161). Therefore, insect cell damage from agitation in typical reactors is unlikely. However, it has been found that bubble rupturing can liberate much more energy than agitation. Namely, shear stress originated from bubble bursting is equivalent to $\approx 10.4\,N/m^2$ (181) and energy dissipation rates have been calculated to be in the order of 10^4 to $10^8\,W/m^3$ (180). As a result, almost every cell surrounding the area where a bubble bursts will be killed (180). Accordingly, Kioukia et al. (162) found that Sf9 cells suffer little damage in a reactor agitated at 400 rpm with a Rushton turbine. The observed damage was attributed to bubble entrainment. However, sparging at a rate of only 0.007 volumes of gas per volume of liquid per minute (vvm) reduced significantly the cell concentration and reduced 50% viral titer and polyhedra yield. The energy liberated by bubble bursting is inversely proportional to bubble size, so if sparging is required large bubbles should be utilized (181,182). However, the oxygen transfer coefficient is reduced as the area per volume of sparged gas decreases. Accordingly, an alternative would be the use of microbubbles, which will completely dissolve before reaching the liquid surface.

Damage to cultures from hydrodynamic forces can occur even when there is no appreciable decrease in cell concentration or growth rate. Cells may die during exposure to shear (lethal effect) or survive and experience physiological effects (sublethal damage). The sublethal effects of hydrodynamic stress comprise a variety of metabolic or structural changes that do not appear to damage cells, but that can affect many of their functions (183). Such physiological effects can include apoptosis triggering, changes in the protein synthesis and nutrient consumption rates, cell cycle arrest, and changes in intracellular pH (40,184; reviewed in 178,179). Sublethal effects reduce the concentration of r-protein as well as the assembly of VLPs (127).

The high specific OUR of infected insect cells (almost twice the typical OUR of animal cells) requires higher oxygen transfer rates (OTRs) than those usually found in animal cell reactors (185). Furthermore, as scale and concentration increase, higher OTRs are required. Thus, in many cases OUR can only be satisfied through sparging. Hydrodynamic damage of sparged insect cell cultures has been prevented by the addition of PF68 (186). Insect cell cultures in TnM-FH medium supplemented with 0.05% PF68 can tolerate a shear stress of at least $4.5\,N/m^2$ (154). In Fig. 8 the effect of aeration and agitation rates on the growth rate of Sf9 cells in SFM supplemented with 0.2% of PF68 is shown. Agitation rates of 275 rpm and aeration rates of 0.04 vvm can be utilized without reducing the cell growth rate. It has been shown that PF68 addition reduces plasma membrane fluidity (151) and yields stronger hybridoma cells (187). PF68 has also been reported to reduce the attachment of insect cells to bubbles by two mechanisms, reducing their hydrophobicity (144) and reducing the medium surface tension (103,188). Palomares et al. (154) found that the protective effect of PF68 remains up to 24 hr, even when the additive is eliminated from the culture medium. This indicates that PF68 interacts with insect cells in a strong and stable manner. Moreover, cells are protected from shear immediately upon PF68 addition, ruling out a slower biological effect (154,189). However, it has been found that PF68 has physiological effects on animal cells. Even under conditions of low shear PF68 promotes the growth of insect cells (162), increases r-protein yield, and decreases baculovirus production (154).

Scale-up

Several considerations should be made when scaling-up animal cell cultures [reviewed in Refs. (185,190)]. A bioprocess can be scaled up either by increasing the cell concentration or by increasing the culture volume. An efficient scale-up

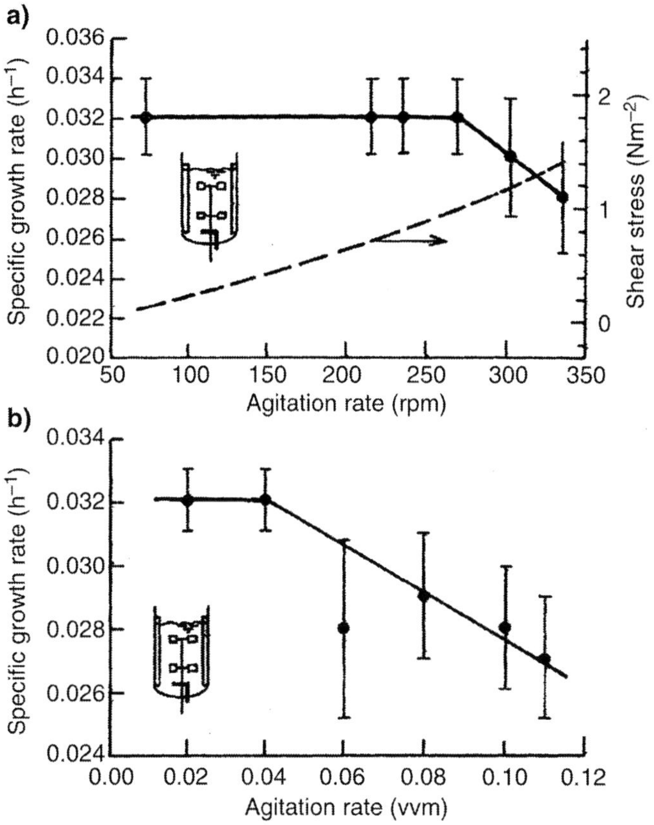

Figure 8 Critical agitation and aeration rates for Sf9 cells in Sf900II medium supplemented with 0.2% PF68. Cells were cultured in a 2 L bioreactor agitated with two Rushton turbines with standard geometry (127). (a) Aeration was kept constant at 0.03 vvm. Shear stress originated from agitation was calculated as described in Ref.(127). (b) Agitation was kept constant at 200 rpm. [From Ref. (181).]

should consider both possibilities simultaneously and aim at large volumes with high cell densities. Such an approach should result in the most efficient use of resources. Of course, the large-scale process should have similar yields and quality as the small-scale process. However, both the fragility and the nutrient demand of insect cells make scaling up a nontrivial task.

Bioreactors for Insect Cell Culture

Commonly used reactors for producing r-proteins by IC-BEVS are summarized in Table 5. Cells can be either freely suspended or entrapped in packed beds, microspheres, or microcarriers [reviewed in Ref. (17)]. Immobilized cultures have been utilized in small scales (0.25–0.5 L) (167,169). Packing materials include glass beads (167), and polyurethane and cellulose foams (169). Cell immobilization is particularly useful for attachment-dependent cells. In general, suspension cultures are preferred for scaling up, as higher OTR and homogenous conditions can be obtained. Stirred-tank reactors (STR) have been operated in batch, perfusion, or continuous modes. It is difficult to compare between various reactor configurations and operation modes because of the variety of r-proteins expressed, culture media, and cell lines employed. By comparing only Sf9 cultures producing β-galactosidase, it can be seen that STRs yield the highest concentration of r-product. Interestingly, β-galactosidase yields obtained by three different research groups in three different STRs are very similar (Table 5). Reactor design for insect cell culture has focused on preventing shear damage. Impellers that provide high mass transfers at low shears, such as helical ribbons, have been utilized. If possible, surface aeration is preferred, although it does not provide enough oxygen for pilot plant or larger-scale cultures. Strategies for increasing oxygen transfer without exposing cells to bubbles include permeable tubing (132) or perfluorocarbon-mediated aeration (191). However, large-scale application of these alternatives is either impractical or too expensive. Sparging must then be utilized.

Continuous and perfusion cultures have also been employed for insect cells. High cell concentrations (3×10^7 cells/mL) can be attained in perfusion cultures (192). Cells can then be diluted with fresh medium and transferred to another reactor for infection. Product concentrations as high as 0.5 g/L have been reported using the perfusion mode (132). Cell separation from the culture medium is often a challenge, especially at very high cell densities ($>10^6$ cells/mL) or at large scales. Several alternatives have been proposed to separate cells from the culture medium, including gravitational, electronic, or sonic settlers (192).

Continuous or semicontinuous cultures require two tanks, one for cell growth and another for infection (Table 5). However, continuous cultures are unstable due to the generation of defective particles in the infection reactor (see Baculovirus). Interference of defective particles has been detected after 1 month of continuous operation (193). The passage effect can be reduced, but not eliminated, by semicontinuous operation (194). Hu et al. (165) proposed the utilization of a tubular segmented-flow bioreactor for infection, where virus is added at the entrance. A segmented plug flow ensures that only virus from a low-passage viral stock infects the cells, which overcomes the passage effect. After 10 hr of residence in the tubular reactor, the infected cells are placed in an STR for baculovirus or r-protein production.

Scale-up of insect cell cultures can be performed in bioreactors designed for mammalian cell culture, although mass transfer coefficients may need to be increased (185). Insect cell cultures have been scaled up to 150 L in traditional sparged STRs with similar baculovirus yields than at smaller scales (182). Fermenters originally designed for microbial cultures can be adapted for the IC-BEVS if Rushton turbines and side baffles are replaced by low-shear pitch blade impellers and surface baffles (153). It is expected that the scale of insect cell cultures will increase since agricultural application of baculoviruses requires large scale to be economically attractive. Competitive bioinsecticides can only be produced in 20,000–50,000 L tanks (182). Scaling up to those volumes is actually in progress (182). If successful, such reactors will become the largest tanks for any animal cell culture, which is currently performed in scales of up to 10,000 L (185).

Scaling up may result in concentration gradients due to deficient mixing, as the fragility of insect cells impedes the utilization of high stirring and aeration rates (190). Concentration gradients of carbon dioxide, microcarrier concentration, pH, oxygen, and culture segregation have been found in animal cell cultures (153,185). Environmental heterogeneities often lead to poor culture performance (185). However, very little work has been done for assessing the effect of such heterogeneities in the production of r-protein or baculovirus by insect cells. One of the consequences of scaling up insect cell cultures is the accumulation of CO_2. This is caused by insufficient CO_2 desorption in cultures that are only surface aerated, membrane aerated, or when very low sparging rates or microbubbles are used. Garnier et al. (153) found that CO_2 accumulates to 0.18 atm in 150 L insect cell culture tanks. CO_2 accumulation did not affect Sf9 growth, but it reduced three times the concentration of r-TGFβ receptor. In contrast, Mitchell-Logean and Murhammer (195) found that CO_2 (24 mM) reduces the growth rates of Sf9 and Tn5Bl-4 cells. They propose head-space purging as a solution, which proved effective for reducing dissolved CO_2 concentration to 6 mM. CO_2 accumulation may cause other unknown effects on cell physiology, protein quality, and baculovirus production.

As a consequence of concentration gradients cells are exposed to various oscillating environmental conditions as they move through different zones in a reactor. Such changing conditions can be reproduced at a small scale through the so-called "scale-down" methodologies (190). Scale-down is utilized for diagnosing problems in large-scale cultures, or for predicting the behavior of a culture in a large-scale vessel. To our knowledge, only Rhiel and Murhammer (196) have assessed the effect of oscillating DO conditions in the IC-BEVS, but under conditions that do not represent those of large-scale cultures. They found that oscillations between 0% and 15% DO concentration reduced growth and r-protein production. It can be expected that other gradients with various frequencies, amplitudes, and oscillation axes may affect in several ways the IC-BEVS, but this remains to be studied.

Process Monitoring and Control

Commercial applications of insect cells require highly productive cultures that can only be obtained through optimal process monitoring and control strategies. Specific issues regarding the IC-BEVS are discussed here. Parameters that can be measured on-line are of special interest as they allow real-time control actions. Such parameters include OUR and carbon dioxide production rate (CPR) (113,117,136). On-line estimation of uninfected viable cell concentration and of the time of monosaccharide depletion is possible through OUR measurements (113,117). Moreover, an effective infection is accompanied by an increase of OUR, which depends on the MOI utilized (117,127,129,130). The decrease in CPR after infection has been observed to coincide with virus release (197). Another simple and reliable parameter for monitoring the IC-BEVS is cell size. Cell size has been found to correlate with the phases of the insect cell cycle and with cell death (25). The size of uninfected insect cells has been found to correlate with their ability to produce r-proteins (28). Moreover, the optimal time of infection can be determined from cell size (198). Cell size increases after infection. Such an increase has been found to correlate with the efficiency of virus replication (155) and with the amount of r-protein produced (140). An important issue for the IC-BEVS is to accurately determine the optimal harvest time since labile proteins can be rapidly degraded after cell lysis has started. The times of maximum OUR and cell size have been correlated with the time of maximum r-protein concentration (117,140,199). Another parameter that has been utilized for monitoring insect cell cultures is relative permittivity (157,197).

A novel and interesting approach for monitoring r-protein production is the use of GFP (see Fig. 3C). Chimeras of the gene of interest and the gene for GFP have allowed the visualization and relative quantification of an r-protein in larvae or in bioreactors (191,200). This can also be utilized for the rapid screening of the most productive r-virus or cell lines (200). GFP can also be utilized for studying protein production and posttranslational modification pathways (201).

POSTTRANSLATIONAL MODIFICATIONS

The ideal system for r-gene expression should produce proteins at high productivities and identical to their native counterparts. This is particularly important for therapeutic applications. Differences between native and r-products are the result of different posttranslational processes. Such differences allow organisms to identify and eliminate foreign proteins. Therefore, posttranslational modifications can have an important effect on protein pharmokinetics. Moreover, unusual modifications in a therapeutic protein can result in allergic reactions. Posttranslational modifications can also affect protein solubility, stability, or specific activity. It is clear that for some applications the characteristics of the product (quality) are much more important than the quantity (202).

The extent of posttranslational processing is highly dependent on the protein being expressed and on specific factors of the host cell, and in general cannot be predicted. Some posttranslational processes are specific to each organism, in some cases even to each particular tissue. In these cases, the use of more complex and costly expression systems, such as those utilizing animal cells, is justified. Insect cells are far more capable of performing the complex posttranslational processing present in human proteins than other eukaryotic hosts, such as yeast, fungi, or plant cells (1). The capability of insect cells for posttranslational processing is discussed below.

Protein Folding and Assembly

The first posttranslational event is folding. Improperly folded proteins may form aggregates in the ER or in the cytosol. The recovery of aggregated proteins is only possible by renaturing them through a costly and labor-intensive task (1). Proteins directed to the ER may be retained there if improperly folded, and will neither be subjected to further processing nor reach their final destination. Cells have proteins that prevent protein aggregation (chaperones), catalyze folding, and retain improperly folded proteins in the ER.

Most of the proteins expressed in insect cells are correctly folded. However, very high levels of r-protein may saturate the cellular folding machinery, and consequently cause protein aggregation. Moreover, the utilization of tags for protein purification, such as polyhistidine and glutathione-*S*-transferase (GST) tails, may result in incorrectly folded proteins (203). Some research groups have reduced aggregation by coexpressing chaperones with the protein of interest. Whiteley et al. (204) and Ailor and Betenbaugh (205) coexpressed r-IgG with the chaperones BiP or cytosolic hsp70. Using this approach they reduced protein aggregation and doubled the concentration of secreted r-IgG. Coexpression of hsp70 with its cofactors hsdj or hsp40 further increased the solubility of the product of interest (206). Other chaperones, such as hsp47, may facilitate secretion, although they do not increase protein solubility (207). Folding catalysts have also been overexpressed simultaneously with the gene of interest. Hsu et al. (208) overexpressed the ER enzyme protein disulfide isomerase (PDI), which catalyzes oxidation, reduction, and isomerization of disulfide bonds. Folding and secretion of heterologous IgG was increased. Interestingly, PDI was even capable of resolubilizing aggregated IgGs both in vivo and in vitro. Likewise, Lenhard and Reiländer (209) stably transformed insect cells with the *ninaA* gene from *D. melanogaster*, which codes for a peptidyl–prolyl *cis*/*trans* isomerase. This improved the folding and activity of heterologous human dopamine transporter.

Protein assembly is also efficient in insect cells. Very complex structures can assemble by coexpressing their components. In the case of VLPs, the structural proteins of nonenveloped virus are simultaneously expressed in insect cells. Virus structural proteins self-assemble to generate structures almost identical to the original virion, but without the genetic material. Up to five genes have been simultaneously expressed in insect cells to obtain VLPs of rotavirus (210), parvovirus (175), and bluetongue virus (211), among others. Such structures have proven to be useful as vaccines and as tools for studying protein–protein and protein–cell interactions (212,213).

Proteolytic Processing

The signal peptide utilized by many eukaryotic cells to direct proteins to the ER is usually cleaved properly by insect cells. However, in some cases signal cleavage may be the limiting step for secretion (26,214). It is also possible that a foreign signal peptide is not fully recognized by insect cells. Both problems may be reduced if the original signal peptide is replaced with one that is commonly found in insect proteins, such as signal peptides from honeybee melittin, cecrobin B, or gp64 from baculovirus (203,215,216). However, in some cases the utilization of such signal peptides has no effect on r-protein yield (217). This strategy should only be utilized when evidence of inefficient cleavage or inefficient signal recognition exists (26).

Insect cells are usually capable of performing other types of proteolytic processing than signal peptide cleavage. Some processing requires tissue-specific proteases, which seldom exist in host cells, even in those of mammalian origin. It is possible to simultaneously produce such proteases with the protein of interest. Another alternative is to express the mature form of the protein instead of its precursor. Again, excessive amounts of the r-product may reduce the efficiency of proteolytic cleavage. Overexpression of proteases simultaneously to the protein of interest can increase precursor processing and extracellular concentrations of the processed protein (218,219).

Glycosylation

Glycosylation is the most complex posttranslational modification, and consists of the addition of carbohydrate groups. Protein glycosylation occurs in several steps through the ER and Golgi complex and involves dozens of enzymes. Most of the secreted and membrane proteins produced by mammalian cells have complex glycosylation (Fig. 9). In contrast, most of the proteins produced by insect cells have high-mannose (Man) or hybrid glycosylation, which are not typical in native human proteins (202). Due to such different glycosylation profiles, the commercial utility of IC-BEVS for producing human therapeutic proteins is at present limited. However, this situation can change if the ability of the IC-BEVS for producing complex glycosylated proteins is improved. In the following section, factors determining posttranslational processing by insect cells are described, and approaches taken for obtaining proteins with complex glycosylation from insect cells are discussed.

N-Glycosylation

N-glycosylation starts in the ER when the oligosaccharide $Glc_3Man_9GlcNAc_2$ is transferred from a lipid complex to an asparagine in a protein (GlcNAc is *N*-acetylgluco-

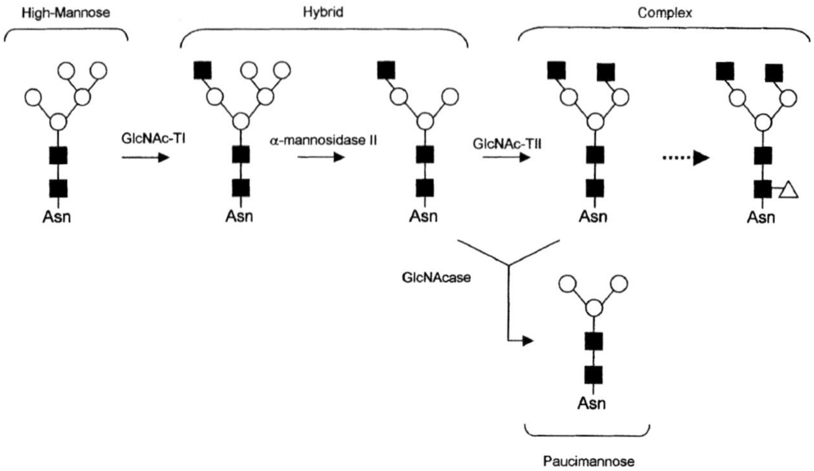

Figure 9 *N*-glycan diversification in the Golgi apparatus. The structure shown may be further diversified by additional branch formation. (■), *N*-Acetylglucosamine; (○) mannose; (Δ) fucose. GlcNAc-T, GlcNAc-transferase. Fucose can be added earlier than indicated.

samine) (216). This initial oligosaccharide is trimmed and processed in the ER and Golgi complex. Several research groups have analyzed the capability of insect cells for performing N-glycosylation. Various insect enzymes involved in glycan processing have been detected and some have been cloned. It is clear that insects contain the enzymes required for the production of complex glycosylated proteins (220,221; see discussion in 222). However, complex glycoforms are seldom found in proteins expressed by insect cells (223). One of the reasons for the absence of complex glycoforms may be the low activity of some glycosyltranferases in insect cells. β1-4 galactosyltransferase (GalT) activity, required for the production of galactosylated forms, is almost undetectable in Sf9 cells (224). However, detectable GalT activities have been reported in insect cells capable of producing complex glycosylated proteins (221,225). Sialyltransferase (SialT) activity, required for glycan sialylation, is usually very low or not found at all in insect cells (226–228). In spite of this, some groups have reported the existence of sialylated glycoforms (with tri and tetra-antennary structures) in proteins expressed by insect cells (68,225)(228–230). It appears that sialylation is only performed by insects in particular developmental stages. For instance, larvae of *Galleria mellonella* contained 16 times more sialic acid (NeuAc) than adults (231). Moreover, baculovirus infection reduces up to 80 times the activity of glycosyltransferases (221).

β-*N*-Acetylglucosaminidase (GlcNAcase) is active in cells and culture supernatants of Tn5Bl-4, Mb0503, DpNl, A7S, and Sf9 (221,225,232,233). GlcNAcase specifically cleaves the GlcNAcMan$_3$GlcNAc2 oligosaccharide, which is required for further glycan diversification (232). Inhibition of GlcNAcase results in an increase of terminal GlcNAc in hemagglutinin produced by Sf9 cells (232), and in a sialylated glycoprotein produced by Tn5Bl-4 cells (234). Sialidase activity has also been found in Tn4h and Tn4s cells (derived from the Tn5Bl-4 cell line) (228,233). Both the low activity of glycosyltransferases and the presence of glycosydases explain the fact that glycans attached to proteins expressed by insect cells usually have a truncated trimannosyl form (221,232). An increase in GalT activity reduces the percentage of pauci-Man forms, indicating a reduction in GlcNAcase substrate concentration (235). Active β1-4-*N*-acetylgalactosaminyltransferase (GalNAcT) has been found in

Tni, Sf, and *M. brassicae* cells (221). GalNAcT is responsible for attaching *N*-acetylgalactosamine (GalNAc) residues to terminal *β*-linked GlcNAc in N-linked glycans.

The activity of α-1–3-fucosyltransferase in insect cells is another source of difference between glycoproteins produced by insect and mammalian cells (223). This enzyme attaches a fucose (Fuc) in position 3 of the GlcNAc residue proximal to the Asn. The resulting α-1,3-fucosylated core is immunogenic to mammals, and thus is undesirable in proteins with therapeutic applications, α1–3-fucosylation has been observed in proteins produced by Tn5Bl-4, Sf, *M. brassicae*, and *B. mori* cells (222; reviewed in 223).

Another factor influencing protein glycosylation is the availability of nucleotide sugars for glycan buildup. Tomiya et al. (236) found that UDP–sugar concentrations [UDP–GlcNac, UDP–galactose (Gal), UDP–Glc, UDP–Fuc, UDP–Man] are similar or higher in Sf9 and H5 than in CHO cells. Moreover, Ea cells have a higher content of UDP–GlcNAc than Sf cell lines (2.8 mmol per 10^6 Ea cells) (237). NeuAc has been found in *D. melanogaster* (238), larvae of *Philaenus spumarius* (239), and *G. mellonella* (231), and in Sf9 cells (240). However, no CMP–NeuAc, required for sialylation, has been found in Sf9, Sf21, or Ea4 cells (236,237,241). Substrate concentration in culture medium can also affect protein glycosylation. Mannosamine feeding increased the amount of terminal GlcNAc in SeAP produced by Sf21 and Tn5Bl-4 (242). The size of the oligosaccharides obtained indicated that tri- and tetra-antennary structures were present. The mechanism by which this happens is not known. In turn, Lawrence et al. (241) found that *N*-acetylmannosamine (ManAc) was required for NeuAc synthesis by insect cells that were expressing human genes for NeuAc phosphate synthase and CMP–NeuAc synthase. Both results underscore the importance of medium composition in protein glycosylation.

Different enzymatic activities and sugar nucleotide pools result in cell lines with different capacities for performing glycosylation. In Table 6, some of the quantitative information available on the structure of glycans attached to proteins produced by various insect cell lines is shown. Three insect cell lines, Ea4, Tn5Bl-4, and DpNl, can be distinguished for their higher ability to perform hybrid or complex glycosylation. Up to 48% of the glycans produced by these cell lines were complex, some with terminal Gal (Table 6) (123,145,222,225,243,244). This percentage of complex glycans is similar to that in proteins expressed by mammalian cell lines (222). GalT activity detected in Ea4 and Tn5Bl-4 cell lines is similar, and more than double of the activity is found in Sf9 cells (221). Interestingly, no activity of GlcNAcase has been observed in Ea cells, which may explain their increased capacity to produce complex glycoforms (232,243). In contrast, most glycoproteins produced by Sf9 cells have high-Man and pauci-Man forms (223), although a few exceptions have been reported (229,243,244). The profiles obtained in glycoproteins of three additional cell lines and larvae from five different species have been compared by Kulakosky et al. (245). Only pauci-Man and high-Man glycoforms were detected in all cases. The utilization of a different baculovirus from the traditionally used *Ac*MNPV can also affect protein glycosylation. When Joshi et al. (145) infected Tn5Bl-4 cells with recombinant *T. ni* NPV instead of *Ac*MNPV the percentage of complex glycans attached to SeAP decreased five times. As described previously (see Baculovirus), replacing the very strong, very late *polh* promoter by earlier promoters can also result in improved capability for complex glycosylation (69,72,73,84,85).

Different r-proteins may be glycosylated in different ways. Protein sequence and three-dimensional structure can influence its glycosylation. For example, Sf9 cells are capable of producing hybrid glycoforms in interferon-γ, but not in SeAP

Table 6 Comparison of Different Cell Lines for Performing Protein Glycosylation

Cell line	Medium	High-mannose (%)	Paucimannose[a] (%)	Hybrid (%)	Complex (%)
DpNl[b]	TnM-FH[c]	24	44	6	26
Ea4[d]	TC-100[c]	ND	22	36	42
Sf9[d]	TC-100[c]	ND	81	19	ND
Sf9[b]	TnM-FH[c]	16	84	ND	ND
Tn4h[be]	TnM-FH[c]	6	43	3	48
Tn4h[be]	ExCell 405	13	86	1	ND
Tn5Bl-4[f]	ExCell 405	ND	35	30	35
Tn5Bl-4[g]	ExCell 405	31	55	14	ND

[a]Forms with three or less mannose residues.
[b]Profile of glycans attached to SeAP (145,225).
[c]Supplemented with 10% FBS.
[d]Profile of glycans attached to Asn25 of interferon g (243).
[e]Tn4h cells are derived from the Tn5Bl-4 cell line.
[f]Profile of glycans attached to IgG (222).
[g]Profile of glycans attached to human transferrin (235)
Note: ND, not detected.

(Table 6) (145,243). Moreover, Tn5Bl-4 cells produce complex glycoforms attached to SeAP and IgG, but not attached to human transferrin (145,222,235). Enzymes may be more active in one glycosylation site than in another of the same protein. Ogonah et al. (243) observed complex glycoforms in 42% of the glycans attached to the Asn25 of interferon-γ, but only 19% of the glycans attached to Asn97 of the same protein were complex. It is known that different consensus sequences can result in different glycan profiles, which may explain this result (243). In addition, very high protein concentrations may overwhelm the glycosylation machinery and result in incomplete glycosylation (68).

Engineering N-Glycosylation Pathways

Genetic engineering has been utilized for increasing the activity of glycosyltransferases in insect cells (Table 7). Genes coding for glycosyltransferases have been cloned into baculoviruses under early promoters (85,235). Stably transfected Sf9 and Tn5Bl-4 cells have also been generated which have higher GalT and SialT activities than mammalian COS cells (224). Marchal et al. (247) took a different approach. They cloned into *Ac*MNPV a chimeric gene containing DNA coding for the catalytic domain of *Tripanosoma cruzi trans*-sialidase, the transmembrane domain of baculovirus gp64, and a signal peptide. When insect cells were infected with this virus, they produced a functional *trans*-sialidase which sialylated a galactosylated acceptor added to culture medium in the presence of a sialylated donor. This approach does not require a CMP–NeuAc pool in insect cells. Lawrence et al. (241) have gone one step further as they engineered the metabolic pathway to produce NeuAc in insect cells by introducing the human genes for NeuAc phosphate synthase and CMP–NeuAc synthase. Utilizing this approach the NeuAc pool in insect cells exceeded that typical of mammalian cell lines.

Interestingly, the overexpression of glycosyltransferases alone has been sufficient for obtaining sialylated proteins. This means that the r-glycosyltransferases were located in the correct position with respect to other enzymes in the Golgi com-

Table 7 Metabolic Engineering of Insect Cells to Improve Their Glycosylation Potential

Glycosyltransferases	Cell line/protein	Result	References
Human β1,2-*N*-GlcNAc-transferase I	Sf9/r-fowl plague virus hemmaglutinin	Fourfold increase in terminal GlcNAc residues	232
Bovine GalT	Sf9/gp64	Galactosylated gp64	72
Bovine GalT	Stably transformed Sf9/r-TPA	Galactosylated TPA	246
Bovine GalT	Tn5B1-4/r-transferrin	Hybrid galactosylated glycans Increase of hybrid structures from 14% to 16.4% Decrease of paucimannose structures from 54% to 18%	235
Bovine GalT and rat α2,6-sialT	Stably transformed Sf9/gp64 and r-GST-SfManI[a]	Galactosylated and sialylated gp64 and r-GST-SfManI	224
Bovine GalT and rat α2,6-sialT	Sf9/gp64	Galactosylated and sialylated gp64	85
Bovine GalT and rat α2,6-sialT	Stably transformed Tn5Bl-4/gp64	Galactosylated and sialylated gp64	70
Tripanosoma cruzi trans-sialidase	Sf9	Culture with sialyltransferase activity	247

[a]GST-SfManI, GST-tagged Sf9 cell class I α-mannosidase.

plex, and that insect cells were able to synthesize CMP–NeuAc and to transport it into the Golgi (85). Most of the studies that employ genetic engineering for improving glycosylation by insect cells only give qualitative results of the presence or absence of Gal or NeuAc in glycoproteins. One exception is the work of Ailor et al. (235), who identified the structures of glycans attached to r-transferrin, as well as their relative amounts. Upon coexpression of GalT, they observed an increase in the percentage of hybrid glycans and detected galactosylated forms, which were 12.6% of all glycans attached to r-transferrin. They also observed a reduction in the percentage of pauci-Man forms from 58 to 18, which indicates that GalT overexpression reduces the amount of glycans that can be a substrate for GlcNAcase. Unfortunately, publications describing the use of genetic engineering for improving the glycosylation potential of insect cells do not report if the coexpression of the r-glycosyltransferase has any effect on the yield of r-protein. Future efforts on metabolic engineering of the glycosylation pathways of insect cells may be to knock out the α-1–3-fucosyltransferase gene to prevent α-1–3-fucosylation.

O-Glycosylation

O-linked glycosylation consists in the initial addition of a GalNAc residue to a serine or threonine residue. O-linked glycans are less branched than N-linked glycans and are usually biantennary structures (248). r-Proteins produced in insect cells are usually O-glycosylated in the same position as native proteins. Mammalian O-linked oligosaccharides commonly have multiple lactosamine units (Galβ1-4GlcNAc), and are sialylated and fucosylated. In contrast, proteins expressed by insect cells have truncated glycans with only the core GalNAc residue or Galβ1-3GalNAc (227,249,250). The potentially immunogenic *O*-glycan Galα1–4Galβ1–3GalNAcα1-*O*-Ser/Thr has been detected in proteins produced by *M. brassicae* cells (227). Little effort has been put on increasing the O-glycosylation potential of insect cells, although overexpression of GalT and SialT should also promote the formation of galactosylated and sialylated O-glycans.

Other Posttranslational Modifications

Other posttranslational modifications have not been studied to the same extent as glycosylation. It is known that myristoylation, palmitoylation, isoprenylation, phosphorylation, sulfation, acetylation, C-terminal amidation, β-hydroxylation, and methylation occur in insect cells (203,251; reviewed in 26,223). It has been found that fatty acid acylation (palmitoylation and myristoylation) is only performed in non-aggregated proteins, but occurs even when glycosylation is inhibited (203). In some cases, isoprenylation may be heterogenous and incomplete (251). α-amidation at the carboxyl terminus does not always occur and is difficult to predict (26).

Other Parameters that affect Posttranslational Processing by Insect Cells

In addition to the factors discussed previously, several other parameters can affect posttranslational processing of proteins produced through the IC-BEVS. Baculovirus infection causes multiple changes in the intracellular environment. Moreover, due to the lytic nature of baculovirus infection, the functionality of infected cells can be expected to decline as infection proceeds. Infection may reduce the synthesis of many cellular proteins required for proper posttranslational processing, such as chaperones, folding catalysts, glycosyltransferases, and other enzymes. In addition, infected cells may die before fully processing the r-product. In fact, it has been found that phosphorylation and acylation are less efficient as infection progresses [reviewed in Ref. (26)]. Moreover, van Die et al. (221) observed a reduction of 80 times in the activity of a glycosyltransferase at 48 hpi. Others have not observed any effect of infection on GlcNAcT I activity (252) or any difference between glycans attached to interferon-γ analyzed 48 and 144 hpi (243). In contrast, Davidson and Castellino (253) and Donaldson et al. (123) found that SeAP and human tissue plasminogen activator (TPA), respectively, produced very late after infection are more extensively glycosylated than proteins produced earlier. It is possible that different culture conditions were responsible for such contrasting results. In any case, the utilization of promoters earlier than *polh* (see Baculovirus) has been useful for obtaining more efficiently processed proteins, which were not functional or soluble otherwise (68,73). An alternative approach, which was exploited by Jarvis et al. (254) for increasing the concentration of the extensively glycosylated TPA molecule, is the utilization of stably transfected insect cells (see the next section).

The concentration of immature r-product can also affect its posttranslational modification. Posttranslational processing is often the limiting step for obtaining r-proteins, especially secreted and membrane proteins (69). Very high concentrations of protein may overwhelm the cellular posttranslational machinery. Consequently, a protein that should be secreted may accumulate intracellularly and heterogenous protein populations may be obtained. Proteins accumulated intracellularly can contain immature high-Man forms, or not be glycosylated at all (203,222). The presence of high-Man forms indicates that glycoproteins are not being processed in the Golgi complex. Posttranslational processing may be improved by reducing the r-protein synthesis rate. This can be achieved by decreasing the culture temperature (123,206), reducing MOI (162,176), or utilizing a weaker promoter than *polh* (68). A reduction in the concentration of immature protein may result in an overall increase of mature, secreted, and functional molecules. Proper intracellular localization of nonsecreted proteins may also be obtained by reducing the r-protein synthesis rate. Improved posttranslational modifications can also be obtained by increasing the concentration of chaperones, folding catalysts, or glycosyltransferases (205,224). The metabolic activity of cells at the moment of infection can also affect the synthesis of r-product and the extent of posttranslational processing (28). A higher metabolic activity can be expected to result in better posttranslational processing.

Environmental and culture conditions also affect posttranslational processing. However, their effect has been scarcely studied for r-proteins produced by insect cells. Davidson and Castellino (230) found that only insect cells cultured in static mode are able to perform complex glycosylation. Joshi et al. (228) found glycoproteins with tri-and tetra-antenary N-linked oligosaccharides in SeAP produced under simulated microgravity. Sialylated glycoproteins were also found in T-flask cultures when supplemented with 5 mM ManAc, a precursor of CMP–NeuAc. Cruz et al. (127) found that low DO (10%) decreases protein yield, but increases the amount of protein assembled in VLPs. Moreover, Zhang et al. (128) found a decrease of high-Man forms in cultures at 50% DO, compared to cultures at 10% and 190%. Ammonia is a by-product that has an important effect on glycosylation of proteins produced by mammalian cell lines. However, Donaldson et al. (123) did not find a significant effect of ammonium sulfate added at concentrations of up to 40 mM on glycosylation of SeAP. Other factors that are known to affect posttranslational processing of proteins produced by mammalian cells can also have an effect on glycosylation by insect cells, although this has not been studied. Some of these factors are the concentration of saccharides, nucleotides, DO, dissolved carbon dioxide, lipids, amino acids, osmolarity, and pH [reviewed in Ref. (255)].

Evidence indicating that insect cells possess all the enzymes required for performing posttranslational modifications similar to those performed by mammalian cells, including complex glycosylation, is continuously increasing. However, as reviewed here it appears that the posttranslational machinery of insect cells is only functional under very specific circumstances, which are still scarcely known.

LARVAL PRODUCTION AND STABLE EXPRESSION SYSTEMS

Larval Production

The traditional alternative approach to in vitro cell culture is larval production, which consists in directly infecting caterpillars with wild-type or r-baculoviruses for expressing either r-proteins or baculoviruses (15,26). *Ac*MNPV and *Bm*NPV are the two most commonly used baculoviruses for larvae expression. *Ac*MNPV has broad host specificity, including *S. frugiperda, T. ni, Heliothis virescens, Man-*

duca sexta, and *Hyalaphora cecropia*, among others. In contrast, *Bm*NPV exclusively infects *B. mori*. For instance, *Ac*MNPV has been used for the functional expression of interleukin-2 in *T. ni* (67) and α-anti-insect toxin in *S. litoralis* and *Heliothis armigera* larvae (256). Likewise, *Bm*NPV and larvae have been used for production of bovine interferon-γ (257). A hybrid baculovirus, which was generated by heterologous recombination between *Ac*MNPV and *Bm*NPV, exhibits a wider host range and allows the production of r-baculoviruses in Sf cells for infection on *B. mori* larvae (258). Such a virus has been used for expressing larva firefly luciferase and for obtaining fibroblast growth factors under control of the polyhedrin promoter (259).

Efficient in vivo production requires a homogenous larvae population. This can only be accomplished by rearing caterpillars born from eggs that hatch in a controlled environment (25–27°C in humidified chambers). The resulting larvae are maintained in small containers (typically 180 mL cups, each holding ca. 10–15 caterpillars) and fed on controlled diets. Larvae must be infected within 1 or 2 days after the sixth molting event in order to insure it is early enough before preparation for pupation (26). Under such conditions, fat body is the major tissue undergoing active metabolism. Infection must be as synchronous as possible and it is performed either by feeding larvae with preadsorbed virus or by directly injecting viruses through their cuticle. Occluded virions are preferred when infecting via diet since nonoccluded virions rapidly decay unless formulated with protective agents (260–262). Accordingly, transfer plasmids for r-protein or r-baculovirus production are commonly constructed with the heterologous gene under control of the *p*10 promoter, whereas *polh* is kept intact for proper occlusion of viral particles (26,67). Some recent studies report up to a 2.7-fold increase in infectivity when using occluded virions produced in vivo compared to in vitro (263). In contrast, others have found no differences in production between both systems (5,58,59). Injecting budded viruses into the hemolymph is a good choice when few insects are to be used; however, it is a time-consuming procedure and becomes impractical at large scales.

As with in vitro culture, TOI, MOI, and harvesting time are three fundamental parameters for attaining an efficient process. A good timing of inoculation permits infection of a large number of fat body cells when the tissue is still actively synthesizing proteins. Either too early or too late infection will result in low yields of r-protein. An early infection will prematurely arrest cell division and will consequently limit larval growth and reduce the number of infected cells. In contrast, if a late infection is employed then production will occur near pupation and part of the rapid protein expression phase will be missed (67). Control of MOI can also be used for improving r-protein production. A strong dependence between MOI and larval growth and time of maximum protein yield has been observed. For instance, in per os infected larvae Pham et al. (67) showed that r-protein yield was maximum at intermediate virus loading (3.8×10^7 pfu/cup), whereas it decreased 77% and 10% at higher (1.5×10^8 pfu/cup) or lower (1.9×10^7 pfu/cup) MOI, respectively. Such a behavior was attributed to a proper balance between infection rate, larval growth, and protease activity. Harvest time must also be carefully selected. Maximum product production can be missed by an early harvest, but proteases released during wide-spread cell lysis can degrade the protein if harvest is delayed for too long (67,264). To help determine the maximum protein expression time, a baculovirus expressing a GFP variant optimized for UV light stimulation was used by Cha et al. (264) as a reporter gene. Such a strategy allowed a noninvasive estimation of the optimal harvest time at ≈3 days postinfection. Anticipating the harvest time by fluorescence measurements can be further simplified by expressing the protein of interest fused to GFP (200). An additional advantage of using such fusion proteins is the possibility of simple product quantification since fluorescence intensity is proportional to the amount of r-protein.

The method of recollection of r-protein will depend on whether it is secreted or not. Secreted proteins are easily collected since they are found in the hemolymph. Therefore, bleeding the caterpillar by clipping its prolegs is the best way to recover the product. However, as warned by O'Reilly et al. (26), when working with hemolymph, melanization should be avoided by addition of DTT, an inhibitor of polyphenol oxidases that catalyze the formation of quinones which irreversibly bind to proteins and make their recovery difficult. In general, r-protein yields of between 0.1 and 25 μg/larva have been reported (67,200). A direct comparison of expression levels between in vivo and in vitro is difficult. Nonetheless, in some cases higher levels have been reported for proteins secreted into the hemolymph compared to an equivalent number of cultured cells, whereas the opposite has been found for some membrane-bound proteins (15). In the case of nonsecreted proteins, larvae should be milled for recovering the r-product. The abundant presence of proteases and hydrolases can degrade the r-protein. Thus, it is recommended to remove the gut before homogenization. For large insects, a good option is to remove the fat body where a large proportion of the r-protein can be found (26,259). Nonetheless, these manipulations, as with infection by injection, are not practical for a large-scale operation.

In addition to the limitations already mentioned, larval production can present other drawbacks. It is very difficult to guarantee a parasite- and virus-free process. Some larvae are aggressive and display cannibalistic behavior, which complicate insect rearing. Furthermore, adequate control of process parameters, including MOI and TOI, can be difficult, particularly when infecting via diet supply. Nonetheless, under certain situations larval production can present advantages over in vitro: capital investment is lower, cell lines and aseptic infrastructure are only required for viral stock preparation, and highly skilled personnel trained in bioreactor operation are not required. Accordingly, larval production is still preferred for several applications, in particular for the production of bioinsecticidal wild-type baculovirus (see the next section). For such an application, in vivo production can be justified as long as the market size does not exceed 1 million ha. Each larva produces an average of ca. 10^9 polyhedra; accordingly, treating 1 ha would require 1000 to 10,000 infected caterpillars (at 10^{12}–10^{13} polyhedra/ha) (5,7,265). In vivo methods are also used in research and industry for rapid and simple production of r-proteins. Larval production is a mature technology and companies already exist that provide larvae, baculovirus, feed stock, and all the materials required for readily establishing an in vivo operation.

Stable (Nonlytic) Expression Systems

Much has been learned and improved during the past 20 years of use of the IC-BEVS for production of heterologous proteins. A significant improvement has been the development of nonlytic and stable expression of r-proteins in insect cell lines by integration of foreign DNA sequences in their genome. By establishing insect cell lines capable of producing r-proteins in a continuous manner, some inherent limitations of a transient and lytic expression system can be overcome (19,266). Traditional r-baculoviruses rely on the use of promoters that become active very late during the infection cycle. Accordingly, most of the baculovirus genetic information must be preserved for late promoters to work. Moreover, due to the lytic nature of baculovirus, cellular integrity is compromised soon after r-protein synthesis begins when very late promoters are used. With stable expression, the yields of membrane bound and secreted r-proteins can improve (267). For instance, Farrell et al. (268) obtained yields of TPA from stably transformed Bm5 cells up to 10 times higher than those of CHO cells, and comparable to those of *S. cerevisiae*.

Stable transformation is possible both in dipteran, such as mosquito, S2, S3, and Kc (30), and in lepidopteran, such as Sf9, Bm5, and TN368 (254,269,270), cell lines. Genetic transformation has been achieved by transfecting into an insect cell, by any of the standard methods, a vector plasmid containing a selection marker and the foreign gene under the control of a strong promoter (30,266,267). Since successful genomic integration is a rare event, two selection systems have been commonly used to preserve the few transformants. One relies on transferring dominant bacteria-derived selective markers, which give cells the advantage of employing unusual metabolites not present in the culture medium. Another is based on cytotoxicity of drugs added to the medium (267). In the former case xanthine instead of hypoxanthine has been used for purine nucleotide synthesis, whereas in the latter case the most commonly used drugs are G418, hygromycin B, methotrexate, and α-amanitin (30).

Both, constitutive and inducible promoters have been used for stable expression in insect cells. The immediate early (IE-1) promoter from *Ac*MNPV (84,254) and the IE-2 promoter from *Orgyia pseudotsugata* MNPV (271) are among the preferred constitutive promoting sequences for stable expression. Both are active in uninfected lepidopteran and dipteran cells since they use RNA polymerase of the host. Several homologous constitutive promoters for use on *Drosophila* cells have been tested (30). These include the long-terminal repeat of the retrotransposon *copia* (a mobile genetic element present in the *D. melanogaster* genome that encodes most of the transcriptional regulatory sequences), the alcohol dehydrogenase gene promoter, and the *actin 5c* gene promoter. Alcohol dehydrogenase is believed to participate in tolerance to alcohol since flies feed on fermenting fruits, whereas the *actin 5c* gene promoter is involved in the expression of a widely expressed cytoskeletal protein. The use of inducible promoters allows the attainment of high growth rates of insect cells without the metabolic burden of expressing an r-protein. Once a desired cell concentration has been reached, expression can then be triggered by various means, including rise in temperature (*hsp* genes) or addition of divalent cations (such as Cd, Zn, or Cu for metallothionein genes). Synthesis of heat shock proteins can be triggered by an increase in temperature as small as 2°C (from 25°C to 27°C). The best-known promoter of this group is that of the protein HSP70. Metallothioneins are detoxifying proteins that bind heavy metals and are a valuable tool for stable transformation of insect cells since induction can be simply accomplished by the presence of heavy metals in the culture medium (30).

Stable transformation of easy-to-handle insect cell lines is a promising avenue for the continuous expression of adequately processed r-proteins at the milligram per liter level (246,266). A further improvement that should push this technology ahead is the development of transposon-based transformation. The use of P-element mediated transposition for the stable genetic transformation of *Drosophila* cells has been reported (272). Such a strategy should allow integration of multiple copies of the gene of interest into the genome, thus leading to higher yields in stably transformed insect cell lines. As reviewed in the next section, stable expression based on r-baculoviruses has also been achieved on vertebrate cells (273,274).

PRODUCTS AND APPLICATIONS OF IC-BEVS

Bioinsecticides

Wild-Type Baculoviruses

The first practical application of baculoviruses was in agriculture as bioinsecticides. In fact, there exist reports as early as 1892 documenting attempts to use baculoviruses as bioinsecticides (8,261), although it was not until the 1940s that formal demonstrations of their utility for pest control appeared (12). Extensive field trials were performed during the following 30 years, culminating in 1975 with registration by the U.S. Environmental Protection Agency (EPA) of *Helicoverpa zea* NPV (under the commercial name Elcar) as the first baculovirus for use as a pesticide (4–8). To date at least 29 different baculoviruses have been tested in detail for pathogenicity, and more than 40 (both NPV and GNV) have been used extensively in many countries around the world without adverse effects on nontargeted organisms (Table 8). For instance, in Brazil *Anticarsia gemmatalis* NPV is successfully used on approximately 1 million ha annually to protect soybean crops; in many European countries *Cydia pomonella* GNV is used against the codling moth on apples, pears, and walnuts; and *Helicoverpa armigera* NPV is used in China on ≈100,000 ha annually to protect cotton, tobacco, and tomato plants, just to mention a few examples (261). In the United States, the EPA has registered seven NPVs and one GNV (Table 8). A detailed review of commercial products and the countries and crops where they are applied can be found elsewhere (261).

Ironically, it is the narrow host specificity that characterizes wild-type baculoviruses as environmentally robust pesticides that has limited their market and hampered their economic competitiveness with respect to broader-spectrum synthetic chemical insecticides (7,8). For instance, due to economic considerations, the production of Elcar is at the moment suspended (7,261). Yet, treatment costs for various applications are competitive with respect to conventional ones. For instance, the treatment cost for cotton and vegetables with either *H. zea* or *Spodoptera exigua* NPVs produced in larvae or with chemical pesticides is about the same (between US$12 and US$25/ha). Likewise, *A. gemmatalis* NPV has been successful since it is highly virulent to host larvae and therefore fewer applications are needed to control pests compared to chemical insecticides (261). Baculoviruses are typically applied in the field at 10^{12}–10^{13} polyhedral inclusion bodies (PIBs) per hectare. Commercial pesticide products contain occluded virions as the active ingredient (at $\approx 1 \times 10^{10}$ PIBs/g, 7% humidity), are usually sold as liquid or wettable powders, are applied as foliar sprays, and all target moth larvae. In vitro processes are well suited for the production of occluded virions, even though their biological activity is not always similar to the activity of those produced in larvae (5,58,59,263). Nonetheless, all baculoviruses used for pesticides are presently produced in larvae since manufacturing costs are lower than in in vitro production (8). In particular, development of inexpensive culture media (see Insect Cell Culture), reduction of capital investments, and productivity improvements are needed for attaining an in vitro competitive bioprocess. For instance, it has been estimated that culture medium must be less than US$1/L for successfully producing a biocontrol agent at a commercial level (276). Chakraborty et al. (5) reported a maximum yield of 100 *H.*

Table 8 Baculoviruses used as Active Ingredients in Commonly Employed Commercial Biopesticides

Baculovirus	Insect's common name	Harvest	Commercial name	Country of use
Adoxophyes orana GNV	Summer fruit tortrix	Apple	Capex 2	Switzerland
Adoxophyes sp. GNV	*Adoxophyes* sp.	Tea		Japan
Agrotis segetum GNV	Turnip moth, winter cutworm	Cucurbits		China, USA
Anagrapha falcifera NPV[a]	Celery looper	Cotton, vegetables		USA
Anticarsia gemmatalis NPV	Soybean caterpillar	Soybean	Baculoviron, Baculovirus Nirtal, Coopervirus, Protégé	Brazil, Argentina
Autographa californica NPV[a]	Alfalfa semilooper	Cabbage, cotton, ornamentals	VPN80	Guatemala
Buzura suppressaria NPV	*B. suppressaria*	Tea, tung oil tree		China
Cydia pomonella GNV[a]	Codling moth	Apples, pears, walnuts	Carpovirusine	
			CYD-X	France
			Granusal	USA
			Virin-GyAp	Germany
				Russia
Erinnyis ello GNV	Ello sphinx	Cassava		Brazil
Helicoverpa (=*Heliothis*) *armigera* NPV	American bollworm	Cotton	Virin-HS	Russia

				China
H. virescens NPV	Tobacco budworm	Cotton	Elcar, Gemstar	USA
H. (=*Heliothis*) *zea* NPV[a]	Cotton bollworm, tomato fruitworm, soybean podworm, sorghum headworm, corn earworm, tobacco budworm	Vegatables, corn, cotton, peanuts, sorghum, tobacco, ornamentals, tropical fruits	Elcar, GemStar	USA
Homona magnanima GNV	Oriental tea tortrix	Tea		Japan
Hypantrea cunea NPV	*H. cunea*	Forest, mulberry	Virin-ABB	Russia
Lymantria dispar NPV[a]	Gypsy moth	Forest trees, ornamental or noncommercial trees	Gypcheck, Disparvirus, Virin-ENSH	USA
Mamestra brassicae NPV	Cabbage looper	Cabbage	Mamestrin	France
			Virin-EKS	Russia
Mamestra configurata NPV[b]	Bertha armyworm	Canola	VirosoftBA3	Canada
Neodiprion lecontei NPV	Redheaded pine sawfly	Pine trees		Canada
Neodiprion sertifer NPV	European pine sawfly	Pine trees		

(*Continued*)

Table 8 Baculoviruses used as Active Ingredients in Commonly Employed Commercial Biopesticides (*Continued*)

Baculovirus	Insect's common name	Harvest	Commercial name	Country of use
Orgyia pseudotsugata NPV[a]	Douglas fir tussock moth	Forest trees, ornametal or noncommercial trees	TM Biocontrol-1, Virtuss	USA
Orytes rhinoceros	Coconut beetles	Coconut palm		Southeast Asia
Panolis flammea NPV	Pine beauty moth	Pine trees		
Phtorimae opoerculella GNV	*P. operculella*	Field and stored potatoes	PTM baculovirus	Peru, Egypt
Pieris rapae GNV	Cabbage white	Cabbage		China
Plodia interpunctell GNV	Indian-meal moth	Stored almonds and raisins		USA
Plutella xylostella GNV	Diamond back or cabbage moth	Cabbage		China
S. exigua NPV[a]	Beet or Lesser armyworm	Vegetables, ornamentals, cotton, corn, and peanuts	Spod-X	USA
S. frugiperda NPV	Fall armyworm	Maize		Brazil
S. littorallis NPV	Egyptian cotton leafworm	Cotton	Spodopterin	Africa
S. litura NPV	Cluster caterpillar, tobacco cutworm, rice cutworm	Vegetables, cotton, rice, peanuts, tropical fruits, cucurbits		China
S. sunia NPV	Legume caterpillar		VPN 82	Guatemala

[a] Registered by the EPA (USA).
[b] EPA received pesticide petition.
[Adapted from Refs. (6,8,261,275) and http://www.epa.gov.]

zea PIBs per cell at 2×10^6 cells/mL in Sf900II medium. At such productivities, 5 L of fermentation product per hectare would be required. Accordingly, their process would only need a twofold improvement in PIB productivity and a low-cost medium to be economical (5).

In addition to cost considerations, other factors have limited commercialization of baculovirus pesticides. Environmental conditions, such as solar radiation and rainfall can rapidly inactivate baculoviruses even in their occluded form. To prevent such a problem, various materials that act as UV-shielding agents, including lignin derivatives, starch, and fluorescent brighteners, have been employed in product formulations to further protect the virions (260,261). Temperature, application timing, leaf distribution, plant architecture, and virulence differences due to genetic variability must also be considered for proper performance of baculoviruses in the field (261).

r-Baculoviruses

One of the main limitations of baculovirus pesticides is their slow lethal action, which can be delayed between 5 and 15 days postinfection before killing the insect pest. During such time, larvae can continue feeding and consequently cause substantial damage to crops (6,8,275). A solution to such a problem, initially attempted by Carbonell et al. (277), has been the development of genetically modified baculoviruses. In r-baculoviruses, genes that exert a deleterious effect on the insect are placed under the control of the *polh* promoter or others. To date, various genes have been tested (81,256,275,278,279). These include those for toxins of several scorpions (*Buthus eupeus* and *Androctonus australis*), mite (*Pyemotes tritici*), wasp, and bacteria (*Bacillus thuringiensis* delta-endotoxin), as well as hormones (*Manduca sexta* diuretic hormone), enzymes (*H. virescens* juvenile hormone esterase and ecdysteroid UDP–glucosyl transferase), and other proteins (viral enhancing factor). Depending on the gene selected, diverse effects on the larvae are obtained, including reduction of hemolymph volume, feeding cessation, blackening, paralysis, tremors, dorsal arching, premature melanization, and low weight gain (6). In the most effective cases, such as with the *A. australis* toxin, reductions of more than 40% in insect survival time after infection, with respect to wild-type baculoviruses, have been obtained. However, a more modest effect or no effect at all has been found for other genes such as that for *B. thuringiensis* delta-endotoxin. Detailed reviews of the approaches taken for constructing more effective r-baculoviruses and the results obtained can be found elsewhere (6,8,261,280).

Baculoviruses have also been engineered to broaden their host specificity (258,281). Accordingly, broad-host-range virus strains could be modified with one of the previously mentioned genes and target larger markets, which would increase the economic viability of the pesticide. However, such an approach has to be carefully analyzed since nontargeted beneficial insects could be affected (7).

Nonoccluded virions have been found more potent than occluded virions, and should be more efficiently produced in vitro since metabolic energy is not wasted in the synthesis of high amounts of polyhedrin (262,265,282). Furthermore, an effective means to contend with the possible environmental risks of r-pesticides is to use polyhedrin gene-deleted strains, which will act as suicide delivery systems. By proper formulation, the nonoccluded virions can be stabilized for adequate performance in the field but their unprotected nonoccluded progeny will rapidly decay in the environment (262). Finally, by placing a strong toxic gene under a very early promoter the larvae will die before being able to produce viral progeny, which will further guarantee that r-baculoviruses will not persist in the environment (12). From an environmental perspective, such a trait is desirable; however, it prevents the use of in vivo methods for producing rapidly acting r-baculoviruses. Thus, in this case baculovirus production by cell culture is necessary.

In order to address the environmental impact of releasing r-baculoviruses, contained greenhouse and open field trials have been conducted in the United Kingdom and the United States since 1986 (12,280). In general, the trials have demonstrated improved plant protection and reduced secondary transmission and persistence of the r-baculoviruses compared with the wild type (278,279). From the positive safety results obtained so far, it can be anticipated that commercial pesticides based on r-baculoviruses will soon be available. Carbonell and Miller (283) corroborated the safety of r-baculovirus, with respect to vertebrates, by showing their inability to express the heterologous gene in two model mammalian cell lines. However, it was recently demonstrated that human cells could efficiently uptake r-baculovirus and express the heterologous gene (11). Such demonstration underlies the necessity of performing detailed risk-assessment studies (11,12,280) and integral pest management policies for new r-baculovirus pesticides.

Heterologous Gene Products

Therapeutics and Vaccines

The advantages of IC-BEVS have made it the eukaryotic expression system of choice for research. More than 500 genes of different origins have been expressed through the IC-BEVS for a large variety of applications (15,16,26). Furthermore, IC-BEVS is utilized by almost 50% of the scientific papers that use a high eukaryotic expression system (17). The pharmaceutical industry extensively employs such an expression system to generate material for preliminary testing of candidate drugs. However, utilization of IC-BEVS for industrial production of medicines has been limited by differences in glycosylation between proteins produced by human and insect cells (see Posttranslational Modifications). Still, several pharmaceuticals for veterinary and human use, such as Lysodase (PEGylated-glucocerebrosidase), are being produced by insect cells and are now in clinical trials. The IC-BEVS has been particularly suitable for producing vaccines, which may be composed of individual antigens or by VLPs. In this case, the lack of mammalian-like glycosylation may represent a benefit since it can enhance the immune response to the vaccine. VLPs have the advantage of being structurally identical to the native virus, but without its genetic material. This makes them innocuous. In addition, the immunogenicity of VLPs can be manipulated by varying the ratio between structural proteins (175), or by simultaneously utilizing proteins from different virus strains (polyvalent VLPs). VLPs are potential vaccines against rotavirus (212), human immunodeficiency virus (HIV) (284), blue tongue virus (211), human parvovirus (175), and human papillomavirus (285). Clinical trials have been conducted for most of them. VLPs are also valuable research tools.

r-Proteins produced by insect cells have found various commercial applications. Several diagnostic kits (e.g., detection of herpes simplex virus type 2 by ELISA, Centocor, Malvern, CA) employ as standards r-proteins produced by insect cells. Protein Sciences Corp. (www.proteinsciences.com, Meriden, CT) sells various structural proteins from the HIV, human CD4, and human carcinoembryonic antigen (CEA) for laboratory use or in vitro testing. The company offers for licensing the production of erythropoietin using BEVS technology. It is also producing r-hemagglutinin, r-neuraminidase, and r-CEA using IC-BEVS, and insect cells or cell fractions as adjuvants for antigens. Such products are employed in vaccines against human influenza, malaria, and cancer; a bird-flu vaccine should be in the market soon. Various r-glycosyltransferases in the market are produced by the IC-BEVS (Calbiochem catalog, 2000). Other selected pharmaceutical products are described in other sections of this chapter and have been reviewed elsewhere (17).

Several novel applications of the IC-BEVS have been developed, such as the technology for protein display that uses baculoviruses [reviewed in Ref. (286)]. It has the advantage of employing a eukaryotic virus that can display proteins or peptides too complex for phages. Such proteins often require posttranslational processing. Protein display is accomplished by fusing the baculovirus coat protein gp64 with a protein of interest. As a result, the r-protein is displayed on the surface of infected cells and sometimes in the viral progeny. Membrane-anchor domains of surface proteins of other viruses can also be utilized with good results (286). It is estimated that a library of 2×10^6 individual clones can be obtained using baculoviruses, which is enough to consider the system useful (286). Baculovirus display has been successfully utilized for screening a library of variants of an epitope of a monoclonal antibody (Mab) (287) and a library of HIV-1 gpl20 fragments (286). The baculovirus display system is also useful for generating antibodies when other methods have failed (288). Generation of Mabs through baculovirus display does not require a purified protein. In fact, Mabs can be generated from the partial DNA sequence of a gene.

Gene Transfer into Vertebrate Cells and Gene Therapy

One of the most exciting novel applications of baculoviruses is gene delivery. Although it was thought that genes delivered by baculoviruses could not be expressed in mammalian cells (283,289), Hofmann et al. (11) demonstrated otherwise. They observed high transient expression of the r-luciferase gene after the infection of human and rabbit hepatocytes with an r-baculovirus. The r-virus contained a reporter gene under the immediate early promoter of cytomegalovirus. When the *polh* promoter was utilized instead, no expression was observed, indicating that such a promoter is inactive in the mammalian cell lines tested. These results pointed out the utility of r-baculoviruses as a convenient vector for gene transfer. Since then, it has been found that baculoviruses can also deliver genes to a variety of vertebrate cells. These include porcine and monkey kidney cells, human cervix carcinoma cells, human gastric cancer cells, human keratinocytes, primary human and mouse pancreatic islet cells, bone marrow fibroblasts, human neural cells, and human osteogenic sarcoma cells (273,290–293; reviewed in 294). Stably transformed cell lines can also be obtained utilizing r-baculoviruses (273,274). In particular, the constructs of Condreay et al. (273), which coexpress GFP, represent a significant improvement since a selectable marker for stable transformed cells is included. Baculovirus has also been utilized for launching viral infection in mammalian cells. The DNA of the virus of interest is introduced in the baculovirus, and is expressed by the host. This can be useful to produce viruses that cannot be expanded in cultured cells, such as hepatitis B and C viruses, or for the production of viral vectors without a helper virus [reviewed in Ref. (294)].

Baculoviruses have various characteristics that make them useful vectors for gene therapy. They are easy to construct and propagate, they can deliver genes into nondividing mammalian cells (295), preexisting immunity in humans is not likely

(296), and they are not pathogenic for mammals. When internalized by a mammalian cell, their promoters remain silent and they cannot replicate (11). A problem with the utilization of baculoviruses for gene therapy is that they are inactivated by the complement system in blood or serum. This has been overcome by direct injection in tissues, such as skeletal muscle and brain (293,297). Airenne et al. (298) delivered a gene to rabbit carotid arteries using a silastic collar to avoid direct contact of the viruses with blood. Such a proprietary device could be utilized in human clinical trials in the near future (www.arktherapeutics.com). Hüser et al. (296) constructed complement-resistant baculoviruses by incorporating into them the complement-regulatory protein human decay accelerating factor. Such approaches have made possible the in vivo use of baculoviruses as gene transfer vehicles.

Baculovirus gene vectors can be employed for targeted gene therapy using the baculovirus display system. Ojala et al. (299) displayed specific antibody fragments against CEA, which resulted in specific gene delivery to cells with CEA on their surface. Moreover, the Z domain of protein A has been displayed in baculoviruses. Such a domain binds nonspecifically to any IgG. This allows a versatile and simple construction of very specific delivery vectors. VLPs can also be utilized for gene delivery or DNA immunization. VLPs can package unrelated DNA in vitro, and deliver it to target cells (300). This permits the possibility of designing a very specific gene delivery system. Another application of baculoviruses in gene therapy is the production of fully deleted vectors that require a helper virus for their replication. Cheshenko et al. (301) introduced a DNA fragment of 38 kb that encoded the genome of a packaging defective adenovirus genome. Infection of insect cells with this r-baculovirus resulted in good titers of the vector that could transduce 293 cultured cells, although replication-competent adenovirus was obtained after subsequent passaging.

ACKNOWLEDGMENTS

We are grateful to A. Ortega, V. Hernández, P. Saucedo, S. Ainsworth, R. Ciria, A. Martínez, and J. A. Mena for help during the elaboration of the manuscript. We thank A. Meneses and H. Merchant for help in electron microscopy. Support by DGAPA-UNAM IN-216100, ES-218202, CONACYT 33348-B, I33104-B, and NC-230 is acknowledged.

REFERENCES

1. Palomares LA, Kuri-Breña F, Ramírez OT. Industrial production. In: Al Gobaisi D, ed. The Encyclopedia of Life Support Systems. Oxford: EOLSS Publishers, 2002.
2. Gooche CF, Gramer MJ, Andersen DC; Bahr JB, Rasmussen J. The oligosaccharides of glycoproteins. Bio/Technology 1991; 9:1347–1355.
3. Grace TDC. Establishment of four strains of cells from insect tissue grown in vitro. Nature 1962; 195:788–789.
4. Miller LK. Insect viruses. In: Fields BN, Knipe DM, Howley PM, et al., eds. Fields Virology. Philadelphia, PA: Lippincott-Raven, 1996:533–556.
5. Chakraborty S, Kanhaisingh A, Greenfield PF, Reid S, Monsour C, Teakle R. In vitro production of wild type Heliothis baculo viruses for use as biopesticides. Aust Biotechnol 1995; 5:82–86.
6. Maeda S. Further development of recombinant baculovirus insecticides. Curr Opin Biotechnol 1995; 6:313–319.

7. Teakle RE. Prospects for the use of baculoviruses as bioinsecticides. Aust Biotechnol 1995; 5:345–348.
8. Wood HA. Development and testing of genetically improved baculovirus insecticides. In: Shuler ML, Wood HA, Granados RR, Hammer DA, eds. Baculovirus Expression Systems and Biopesticides. New York: Wiley-Lyss, 1995:91–102.
9. Smith GE, Fraser MJ, Summers MD. Molecular engineering of the *Autographa californica* nuclear polyhedrosis virus genome: deletion mutations within the polyhedrin gene. J Virol 1983; 46:584–593.
10. Smith GE, Summers MD, Fraser MJ. Production of human β-interferon in insect cells infected with a baculovirus expression vector. Mol Cell Biol 1983; 3:2156–2165.
11. Hofmann C, Sandig V, Jennings G, Rudolph M, Schlag P, Strauss M. Efficient gene transfer into human hepatocytes by baculovirus vectors. Proc Natl Acad Sci USA 1995; 92:10099–100103.
12. Richards A, Matthews M, Christain P. Ecological considerations for the environmental impact evaluation of recombinant baculovirus insecticides. Annu Rev Entomol 1998; 43:493–517.
13. Dall DJ, East PD, Oakeshott JG. Perspective on bioprocessing. Aust Biotechnol 2001; 11:27–28.
14. Vaughn J. Insect cell culture, protein expression. In: Flickinger MC, Drew SW, eds. Encyclopedia of Bioprocess Technology, Fermentation, Biocatalysis and Bioseparation. New York: John Wiley and Sons, 1999; 3:1444–1457.
15. Luckow VA. Cloning and expression of heterologous genes in insect cells with baculovirus vectors. In: Prokop A, Bajpai R, Ho C, eds. Recombinant DNA Technology and Applications. New York: McGraw-Hill, Inc., 1991:97–152.
16. Luckow VA. Protein production and processing from baculovirus expression vectors. In: Shuler ML, Wood HA, Granados RR, Hammer DA, eds. Baculovirus Expression Systems and Biopesticides. New York: Wiley-Lyss, 1995:51–90.
17. Palomares LA, Ramírez OT. Insect cell culture: recent advances, bioengineering challenges and implications in protein production. In: Galindo E, Ramirez OT, eds. Advances in Bioprocess Engineering II. Dordrecht: Kluwer Academic Publishers, 1998:25–52.
18. Crossen R, Grüenwald S. Baculovirus Expression Vector System Manual. 4th ed. San Diego, CA: Pharmingen, 1997.
19. Bernard AR, Blasey HD. Transient expression systems. In: Flickinger MC, Drew SW, eds. Encyclopedia of Bioprocess Technology, Fermentation, Biocatalysis and Bioseparation. Vol. 3. New York: John Wiley & Sons, 1999:2580–2597.
20. Betenbaugh MJ, Ailor E, Whiteley EM, Hsu T-A. Insect cells and larvae, gene expression systems. In: Flickinger MC, Drew SW, eds. Encyclopedia of Bioprocess Technology. Fermentation, Biocatalysis and Bioseparation. Vol. 3. New York: John Wiley & Sons, 1999:1457–1476.
21. Lynn DE. Development of insect cell lines: virus susceptibility and applicability to prawn cell culture. Methods Cell Sci 1999; 21:173–181.
22. Lynn DE. Comparison of cell line maintenance procedures on insect cells used for producing baculoviruses. In Vitro Cell Dev Biol Anim 1999; 55:248–251.
23. Iwabuchi K. An established cell line from the beetle, *Xilotrechus pyrrhoderus* (Coleoptera: cerambycidae). In Vitro Cell Dev Biol Anim 1999; 35:612–615.
24. Trisyono A, Goodman CL, Grasela JJ, Mcintosh AH, Chippendale GM. Establishment and characterization of an *Ostrinia nubilalis* cell line, and its response to ecdysone agonists. In Vitro Cell Dev Biol Anim 2000; 36:400–404.
25. Meneses-Acosta A, Mendonça RZ, Merchant H, Covarrubias L, Ramírez OT. Comparative characterization of cell death between Sf9 insect cells and hybridoma cultures. Biotechnol Bioeng 2001; 72:441–457.
26. O'Reilly DR, Miller LK, Luckow VA. Baculovirus Expression Vectors. A Laboratory Manual. New York: Oxford University Press, 1994.

27. Davis TR, Wickham K, McKenna K, Granados RR, Shuler ML, Wood HA. Comparative recombinant protein production of eight insect cell lines. In Vitro Cell Dev Biol 1993; 29A:388–390.
28. Chai H, Al-Rubeai M, Chu K, Oh S, Yap M. Insect cell line dependent gene expression of recombinant human necrosis factor. Enzyme Microb Technol 1996; 18:126–132.
29. Lynn DE. Methods for the development of cell lines from insects. J Tissue Culture Methods 1989; 12:23–29.
30. Echalier G. *Drosophila* Cells in Culture. San Diego, CA: Academic Press, 1997:131–187, 439–511.
31. Hoy MA. Chromosomal and Extrachromosomal Organization of DNA in Insects. Insect Molecular Genetics. An Introduction to Principles and Applications. San Diego, CA: Academic Press Inc., 1994:45–122.
32. Schneider I. Karyology of cells in culture. F. Characteristics of insect cells. In: Kruse PF, Patterson MK, eds. Tissue Culture Methods and Applications. New York: Academic Press, 1973:788–796.
33. McIntosh AH, Ignoffo C. Characterization of lepidopteran cell lines by isoelectric focusing and phosphoglucoisomerase. In: Mitsuhashi J, ed. Invertebrate Cell System Applications. Vol. 1. Boca Raton, FL: CRC Press, 1989:65–70.
34. Harvey GT, Sohi SS. Isozyme characterization of 8 hymenopteran and 20 lepidopteran cell lines. In: Mitsuhashi J, ed. Invertebrate Cell System Applications. Vol. 1. Boca Raton, FL: CRC Press, 1989:71–75.
35. Brown SE, Knudson DL. Characterization of invertebrate cell lines. IV. Isozyme analyses of dipteran and acarine cell lines. In Vitro 1982; 18:347–350.
36. Aldridge CA, Knudson DL. Characterization of invertebrate cell lines. I. Serologic studies of selected lepidopteran lines. In Vitro 1980; 16:384–391.
37. Fertig G, Klöppinger M, Miltenburger HG. Cell cycle kinetics of insect cell cultures compared to mammalian cell cultures. Exp Cell Res 1990; 189:208–212.
38. Schopf B, Howaldt MW, Bailey JE. DNA distribution and respiratory activity of *Spodoptera frugiperda* populations infected with wild-type and recombinant *Autographa californica* nuclear polyhedrosis virus. J Biotechnol 1990; 15:169–186.
39. Léry X, Charpentier G, Belloncik S. DNA content analysis of insect cell lines by flow cytometry. Cytotechnology 1999; 29:103–113.
40. Cowger NL, O'Connor KC, Hammond TG, Lacks DJ, Navar GL. Characterization of bimodal cell death of insect cells in a rotating-wall vessel and shaker flask. Biotechnol Bioeng 1999; 64:14–26.
41. McIntosh AH, Grasela JJ, Matteri RL. Identification of insect cell lines by DNA amplification fingerprinting (DAF). Insect Mol Biol 1996; 5:187–195.
42. Tagu D. RAPD: polymorphisme d'ADN par amplification aléatoire (Random amplified polymorphic DNA). In: Tagu D, ed. Principes des Techniques de Biologie Moléculaire. Paris, INRA 1999:113–114.
43. Cenis JL, Perez P, Fereres A. Identification of aphid (Homoptera: Aphididae) species and clones by random amplified polymorphic DNA. Ann Entomol Soc Am 1993; 86:545–550.
44. Moya A, Guirao P, Cifuentes D, Beitias F, Cenis JL. Genetic diversity of Iberian populations of *Bemisia tabaci* (Hemiptera: Aleyrodidae) based on random amplified polymorphic DNA-polymerase chain reaction. Mol Ecol 2001; 10:891–897.
45. Schlötterer C, Hauser M-T, von Haesler A, Tautz D. Comparative evolutionary analysis of rDNA ITS regions in *Drosophila*. Mol Biol Evol 1994; 11:513–522.
46. Mukha D, Wiegmann BM, Schal C. Evolution and phylogenetic information content of the ribosomal DNA repeat unit in the Blattodea (Insecta). Insect Biochem Mol Biol 2002; 32:951–960.
47. Kshirsagar SG, Patole MS, Shouche YS. Characterization of insect cell lines: heteroduplex analysis employing a mitochondrial 16S ribosomal RNA gene fragment. Anal Biochem 1997; 253:65–69.

48. Sudeep AB, Mourya DT, Shouche YS, Pidiyar V, Pant U. A new cell line from the embryonic tissue of *Helicoverpa armigera* Hbn. (Lepidoptera: Noctuidae). In Vitro Cell Dev Biol Anim 2002; 38:262–264.
49. Volkman LE, Blissard GW, Friesen P, Keddie BA, Possee R, Theilmann DA. Family Baculoviridae. Virus taxonomy. Classification and nomenclature of viruses. In: Murphy FA, Fauquet CM, Bishop DHL, Ghabrial SA, Jarvis AW, Martelli GP, Mayo MA, Summers MD, eds. Sixth Report of the International Committee on Taxonomy of Viruses. Arch Virol 1995; suppl 10:104–113.
50. Vialard JE, Arif BM, Richardson C. Introduction to the molecular biology of baculoviruses. In: Richardson CD, ed. Methods in Molecular Biology: Baculovirus Expression Protocols. Vol. 39. Totowa, NJ: Humana Press Inc., 1995:1–24.
51. Ayres MD, Howard SC, Cuzio J, Ferber ML, Possee RD. The complete DNA sequence of *Autographa californica* nuclear polyhedrosis virus. Virology 1994; 202:586–605.
52. Blissard GW, Rohrmann GF. Baculovirus diversity and molecular biology. Annu Rev Entomol 1990; 35:127–155.
53. Granados RR. Infectivity and mode of action of baculoviruses. Biotechnol Bioeng 1980; 25:1377–1405.
54. van Oers MM, Vlak JM. The baculovirus 10-kDa protein. J Invert Pathol 1997; 70: 1–17.
55. Alaoui-Ismaili MH, Richardson CD. Insect virus proteins (FALPE and p10) self-associate to form filaments in infected cells. J Virol 1998; 72:2213–2223.
56. Bianchi FJJA, Snoeijing I, van der Werf W, Mans RMW, Smits PH, Vlak JM. Biological activity of SeMNPV, AcMNPV, and three AcMNPV deletion mutants against *Spodoptera exigua* larvae (Lepidoptera: Noctuidae). J Invert Pathol 2000; 75:28–35.
57. Chisholm GE, Henner DJ. Multiple early transcripts and splicing of the *Autographa californica* nuclear polyhedrosis virus IE-1 gene. J Virol 1988; 62:3193–3200.
58. Visnovsky G, Claus J. Influence of time and multiplicity of infection on the batch production of *Anticarsia gemmatalis* nuclear polyhedrosis virus in lepidopteran insect cell cultures. In: Galindo E, Ramírez OT, eds. Advances in Bioprocess Engineering. Dordrecht: Kluwer Academic Publishers, 1994:123–128.
59. Claus JD, Remondetto GE, Guerrero SA, Demonte AM, Murguía M, Marcipar AJ. *Anticarsia gemmatalis* nuclear polyhedrosis virus replication in serum-free and serum-reduced insect cell cultures. J Biotechnol 1993; 31:1–15.
60. Mariani SM, Krammer PH. Detection of active infection of Sf9 insect cells by recombinant baculoviruses. J Virol Methods 1997; 69:19–28.
61. de Moraes RR, Maruniak JE. Detection and identification of multiple baculoviruses using the polymerase chain reaction (PCR) and restriction endonuclease analysis. J Virol Methods 1997; 63:209–217.
62. Vlak JM, Klinkenberg FA, Zaal KJM, Usmany M, Klinge-Roode EC, Geervliet JBF, Roosien J, van Lent JWM. Functional studies on the p10 gene of *Autographa californica* nuclear polyhedrosis virus using a recombinant expressing a p10-β-galactosidase fusion gene. J Gen Virol 1988; 69:765–776.
63. Vlak JM, Schouten A, Usmany M, Belsham GJ, Klinge-Roode EC, Maule AJ, van Lent JWM, Zuidema D. Expression of cauliflower mosaic virus gene I using a baculovirus vector based upon the p10 gene and a novel selection method. Virology 1990; 779:312–320.
64. Williams GV, Rohel DZ, Kuzio J, Faulkner P. A cytopathological investigation of *Autographa californica* nuclear polyhedrosis virus p10 gene function using insertion-deletion mutants. J Gen Virol 1989; 70:182–202.
65. DiFalco MR, Bakopanos E, Patricelli M, Chan G, Congote LF. The influence of various insect cell lines, p10 and polyhedrin promoters in the production of secreted insulin-like growth factor-interleukin-3 chimeras in the baculovirus expression system. J Biotechnol 1997; 56:49–56.

66. Heldens JGM, Kester HA, Zuidema D, Vlak JM. Generation of a p10-based baculovirus expression vector in yeast with infectivity for insect larvae and insect cells. J Virol Methods 1997; 68:57–63.
67. Pham MQ, Naggie S, Wier M, Cha HJ, Bentley WE. Human interleukin-2 production in insect (*Trichoplusia ni*) larvae: effects and partial control of proteolysis. Biotechnol Bioeng 1999; 62:175–182.
68. Pajot-Augy E, Bozon V, Remy JJ, Couture L, Salesse R. Critical relationship between glycosylation of recombinant lutropin receptor ectodomain and its secretion from baculovirus-infected insect cells. Eur J Biochem 1999; 260:635–648.
69. van Oers MM, Thomas AAM, Moormann RJM, Vlak JM. Secretory pathway limits the expression of classical swine virus E2 glycoprotein in insect cells. J Biotechnol 2001; 86:31–38.
70. Brietbach K, Jarvis DL. Improved glycosylation of a foreign protein by Tn-5BI-4 cells engineered to express mammalian glycosyltransferases. Biotechnol Bioeng 2001; 74:230–239.
71. Zhao YG, Eggleston P. Comparative analysis of promoters for transient gene expression in cultured mosquito cells. Insect Mol Biol 1999; 1:31–38.
72. Jarvis DL, Finn EE. Modifying the insect cell N-glycosylation pathway with immediate early baculovirus expression vectors. Nat Biotechnol 1996; 14:1288–1292.
73. Chazenbalk GD, Rapoport B. Expression of the extracellular domain of the thyrotropin receptor in the baculovirus system using a promoter active earlier than the polyhedrin promoter. J Biol Chem 1995; 270:1543–1549.
74. Wilson ME, Mainprize TH, Friesen PD, Miller LK. Location, transcription, and sequence of a baculovirus gene encoding a small arginine-rich polypeptide. J Virol 1987; 61:661–666.
75. Thiem SM, Miller LK. Identification, sequence, and transcriptional mapping of the major capsid protein gene of the baculovirus *Autographa californica* nuclear polyhedrosis virus. J Virol 1989; 63:2008–2018.
76. Hill-Perkins MS, Possee RD. A baculovirus expression vector derived from the basic promoter of *Autographa californica* nuclear polyhedrosis virus. J Gen Virol 1990; 71:971–976.
77. Thiem SM, Miller LK. Differential gene expression mediated by late, very late, and hybrid baculovirus promoters. Gene 1990; 91:87–94.
78. Morris TD, Todd JW, Fisher B, Miller LK. Identification of lef-7: a baculovirus gene affecting late gene expression. Virology 1994; 200:360–369.
79. Crawford M, Miller LK. Characterization of an early gene accelerating expression of late genes of the baculovirus *Autographa californica* nuclear polyhedrosis virus. J Virol 1988; 62:2773–2781.
80. Nissen MS, Friesen PD. Molecular analysis of the transcriptional regulatory region of an early baculovirus gene. J Virol 1989; 63:493–503.
81. O'Reilly DR, Miller LK. A baculovirus blocks insect molting by producing ecdysteroid UDP-glucosyl transferase. Science 1989; 245:1110–1112.
82. O'Reilly DR, Passarelli AL, Goldman IF, Miller LK. Characterization of the DA26 gene in a hypervariable region of the Autographa californica nuclear polyhedrosis virus genome. J Gen Virol 1990; 71:1029–1037.
83. Knebel-Morsdorf D, Flipsen JT, Roncarati R, Jahnel F, Kleefsman AW, Vlak JM. Baculovirus infection of *Spodoptera exigua* larvae: lacZ expression driven by promoters of early genes pe38 and me53 in larval tissue. J Gen Virol 1996; 77:815–824.
84. Jarvis DL, Wemkauf C, Guarino LA. Immediate-early baculovirus vectors for foreign gene expression in transformed or infected insect cells. Protein Expr Purif 1996; 8: 191–203.
85. Jarvis DL, Howe D, Aumiller JJ. Novel baculovirus expression vectors that provide sialylation of recombinant glycoproteins in lepidopteran insect cells. J Virol 2001; 75: 6223– 6227.

86. Morris TD, Miller LK. Promoter influence on baculovirus-mediated gene expression in permissive and nonpermissive insect cell lines. J Virol 1992; 66:7397–7405.
87. Sano K-I, Maeda K, Oki M, Maeda Y. Enhancement of protein expression in insect cells by a lobster tropomyosin cDNA leader sequence. FEBS Lett 2002; 532:143–146.
88. Kitts PA, Ayres MD, Possee RD. Linearization of baculovirus DNA enhances the recovery of recombinant virus expression vectors. Nucleic Acids Res 1990; 18: 5667–5672.
89. Luckow VA, Lee SC, Barry GF, Olins PO. Efficient generation of infectious recombinant baculoviruses by site-specific transposon-mediated insertion of foreign genes into a baculovirus genome propagated in *Escherichia coli*. J Virol 1993; 67:4566–4579.
90. Dee KU, Shuler ML. Optimization of an assay for baculovirus titer and design of regimens for the synchronous infection of insect cells. Biotechnol Prog 1997; 13:14–24.
91. Sussman DJ. 24-Hour assay for estimating the titer of beta-galactosidase-expressing baculovirus. BioTechniques 1995; 18:50–51.
92. Yahata T, Andriole S, Isselbacher KJ, Shioda T. Estimation of baculovirus titer by β-galactosidase activity assay of virus preparations. BioTechniques 2000; 29:214–215.
93. Cha HJ, Gotoh T, Bentley WE. Simplification of titer determination for recombinant baculovirus by green fluorescent protein marker. BioTechniques 1997; 23:782–786.
94. Mena JA, Ramírez OT, Palomares LA. Titration of non-occluded baculovirus using a cell viability assay. BioTechniques 2003; 34:260–264.
95. Volkman LE, Goldsmith PA. Baculovirus bioassay not dependent upon polyhedra production. J Gen Virol 1981; 56:203–206.
96. Volkman LE, Goldsmith PA. Generalized immunoassay for *Autographa californica* nuclear polyhedrosis virus: infectivity in vitro. Appl Environ Microbiol 1982; 44: 227–233.
97. Kitts PA, Green G. An immunological assay for determination of baculovirus titers in 48 hours. Anal Biochem 1999; 268:173–178.
98. Petricevich VL, Palomares LA, González M, Ramírez OT. Parameters that determine virus adsorption kinetics: toward the design of better infection strategies for the insect cell-baculovirus expression system. Enzyme Microb Technol 2001; 28:52–61.
99. Wickham TJ, Davis T, Granados RR, Hammer DA, Shuler ML, Wood HA. Baculovirus defective interfering particles are responsible for variations in recombinant protein production as a function of multiplicity of infection. Biotechnol Lett 1991; 13:483–488.
100. Coen DM, Ramig RF. Viral Genetics. In: Fields BN, Knipe DM, Howley PM, et al., eds. Fields Virology. Philadelphia, PA: Lippincott-Raven, 1996:113–151.
101. Kool M, Voncken JW, van Lier FLJ, Tramper J, Vlak JM. Detection and analysis of *Autographa californica* nuclear polyhedrosis virus mutants with defective interfering properties. Virology 1991; 183:739–746.
102. Slavicek JM, Mercer MJ, Kelly ME, Hayes-Plazolles N. Isolation of a baculovirus variant that exhibits enhanced polyhedra production stability during serial passage in cell culture. J Invert Pathol 1996; 67:153–160.
103. Goldblum S, Bae Y, Hink F, Chalmers J. Protective effect of methylcellulose and other polymers on insect cells subjected to laminar shear stress. Biotechnol Prog 1990; 6: 383–390.
104. Wang P, Toung R, Granados RR. The establishment of new cell lines from *Pseudaletia unipuncta* with differential responses to baculovirus infection. In Vitro Cell Dev Biol-Anim 1999; 35:333–338.
105. Dee KU, Shuler ML, Wood HA. Inducing single-cell suspension of BTI-TN5B1-4 insect cells: I. The use of sulfated polyanions to prevent cell aggregation and enhance recombinant protein production. Biotechnol Bioeng 1997; 54:191–205.
106. Saarinen MA, Troutner KA, Gladden SG, Mitchell-Logean CM, Murhammer DW. Recombinant protein synthesis in *Trichoplusia ni* BTI-Tn-5Bl-4 insect cell aggregates. Biotechnol Bioeng 1999; 63:612–617.

107. Donaldson MS, Shuler ML. Effects of long-term passaging of BTI-Tn5Bl-4 insect cells on growth and recombinant protein production. Biotechnol Prog 1998; 14:543–547.
108. Doverskog M, Bertram E, Ljunggren J, Öhman L, Sennerstam R, Häggström L. Cell cycle progression in serum-free cultures of Sf9 insect cells: modulation by conditioned medium factors and implications for proliferation and productivity. Biotechnol Prog 2000; 16:837–846.
109. Wu J, King G, Daugulis AJ, Faulkner P, Bone DH, Goosen MFA. Adaptation of insect cells to suspension culture. J Ferment Bioeng 1990; 70:1–5.
110. Taticek RA, Choi C, Phan SE, Palomares LA, Shuler ML. Comparison of growth and recombinant protein expression in two different insect cell lines in attached and suspension culture. Biotechnol Prog 2001; 117:676–684.
111. Rhiel M, Mitchell-Logean CM, Murhammer DW. Comparison of *Trichoplusia ni* BTI-Tn-5Bl-4 (High Five®) and *Spodoptera frugiperda* Sf-9 insect cell line metabolism in suspension cultures. Biotechnol Bioeng 1997; 55:909–920.
112. Yang JD, Gecik P, Collins A, Czarnecki S, Hsu HH, Ladsun A, Sundaram R, Muthukumar G, Silberklang M. Rational scale-up of a baculovirus-insect cell batch process based on medium nutritional depth. Biotechnol Bioeng 1996; 52:696–706.
113. Palomares LA, Ramírez OT. The effect of dissolved oxygen tension and the utility of oxygen uptake rate in insect cell culture. Cytotechnology 1996; 22:225–237.
114. Mendonca RZ, Palomares LA, Ramírez OT. An insight into insect cell metabolism through selective nutrient manipulation. J Biotechnol 1999; 72:61–75.
115. Bédard C, Tom R, Kamen A. Growth, nutrient consumption and end-product accumulation in Sf-9 and BTI-EAA insect cell cultures: insights into growth limitation and metabolism. Biotechnol Prog 1993; 9:615–624.
116. Donaldson MS, Shuler ML. Low-cost serum-free medium for the BTI Tn5Bl-4 insect cell line. Biotechnol Prog 1998; 14:573–579.
117. Kamen AA, Bédard C, Tom R, Perret S, Jardin B. On-line monitoring of respiration in recombinant-baculovirus infected and uninfected insect cell bioreactor cultures. Biotechnol Bioeng 1996; 50:36–48.
118. Neerman J, Wagner R. Comparative analysis of glucose and glutamine metabolism in transformed mammalian cell lines, insect and primary liver cells. J Cell Physiol 1996; 166:152–169.
119. Drews M, Doverskog M, Öhman L, Chapman BE, Jacobson U, Kuchel PW, Häggstrom L. Pathways of glutamine metabolism in *Spodoptera frugiperda* (Sf9) insect cells: evidence for the presence of the nitrogen assimilation system, and a metabolic switch by 1H/15N NMR. J Biotechnol 2000; 78:23–37.
120. Öhman L, Ljunggren J, Haggstrom L. Introduction of a metabolic switch in insect cells by substrate-limited fed batch cultures. Appl Microbiol Biotechnol 1995; 43:1–8.
121. Clements A, Grace TDC. The utilization of sugars by insect cells in culture. J Insect Physiol 1967; 13:1327–1332.
122. Mitsuhashi J. Nutritional requirements of insect cells in vitro. Mitsuhashi J, ed. Invertebrate Cell System Applications. Boca Raton, FL: CRC Press, 1989:3–20.
123. Donaldson M, Wood HA, Kulakosky PC, Shuler ML. Glycosylation of a recombinant protein in the Tn5Bl-4 insect cell line: Influence of ammonia, time of harvest, temperature, and dissolved oxygen. Biotechnol Bioeng 1999; 63:255–262.
124. Öhman L, Alarcon M, Ljunggren J, Ramqvist AK, Häggstrom L. Glutamine is not an essential amino acid for Sf-9 insect cells. Biotechnol Lett 1996; 18:765–770.
125. Tremblay G, Mejia N, MacKenzie R. The NADP-dependent methylenetetrahydrofolate synthetase is not expressed in *Spodoptera frugiperda* cells. J Biol Chem 1992; 267:8281–8285.
126. Goodwin H. Replacement of vertebrate serum with lipids and other factors in the culture of invertebrate cells, tissue, parasites and pathogens. In Vitro Cell Dev Biol 1991; 24A:470–478.

127. Cruz PE, Cuhna A, Peixoto CC, Clemente J, Moreira JL, Carrondo MJT. Optimization of the production of virus-like particles in insect cells. Biotechnol Bioeng 1998; 60: 408–418.
128. Zhang F, Saarinen MA, Itle LJ, Lang SC, Murhammer DW, Linhardt RJ. The effect of dissolved oxygen (DO) concentration on the glycosylation of recombinant protein produced by the insect cell-baculovirus expression system. Biotechnol Bioeng 2002; 77: 219–224.
129. Hensler W, Agathos S. Evaluation of monitoring approaches and effects of culture conditions on recombinant protein production in baculovirus-infected insect cells. Cytotechnology 1994; 15:177–186.
130. Wong TKK, Nielsen LK, Greenfield PF, Reid S. Relationship between OUR and time of infection of Sf9 insect cells infected with a recombinant baculovirus. Cytotechnology 1994; 15:157–167.
131. Stavroulakis D, Kalogerakis N, Behie L, Iatrou K. Kinetic data for the BM-5 insect cell line in repeated-batch suspension cultures. Biotechnol Bioeng 1991; 38:116–126.
132. Chico E, Jager V. Perfusion culture of baculovirus-infected BTI-Tn-5Bl-4 insect cells: a method to restore cell specific β-trace glycoprotein productivity at high cell density. Biotechnol Bioeng 2000; 70:574–586.
133. Ozturk S, Riley M, Palsson BO. Effects of ammonia and lactate on hybridoma growth, metabolism, and antibody production. Biotechnol Bioeng 1992; 39:418–431.
134. Doyle C, Butler M. The effect of pH on the toxicity of ammonia to a murine hybridoma. J Biotechnol 1990; 15:91–100.
135. Martinelle K, Haggstrom L. Mechanisms of ammonia and ammonium ion toxicity in animal cells: transport across cell membranes. J Biotechnol 1993; 30:339–350.
136. Kamen AA, Tom R, Caron A, Chavarie C, Massie B, Archambault J. Culture of insect cells in a helical ribbon impeller bioreactor. Biotechnol Bioeng 1991; 38:619–628.
137. Elias CB, Carpentier E, Durocher Y, Bisson L, Wagner R, Kamen AA. Improving glucose and glutamine metabolism of human HEK293 and *Trichoplusia ni* insect cells engineered to express a cytosolic pyruvate carboxylase enzyme. Biotechnol Prog 2003; 19:90–97.
138. Schneider I. Differentiation of larval *Drosophila* eye-antennal discs in vitro. J Exp Zool 1964; 156:91–100.
139. Hink WF. Established cell line from the cabbage looper, *Trichoplusia ni*. Nature 1970; 226:466–467.
140. Palomares LA, Pedroza JC, Ramírez OT. Cell size as a tool to predict the production of recombinant protein by the insect-cell baculovirus expression system. Biotechnol Lett 2001; 23:359–364.
141. Gardiner GR, Stockdale H. Two tissue culture media for production of Lepidopteran cells and nuclear polyhedrosis viruses. J Invert Pathol 1975; 25:363–370.
142. Jayme W. Nutrient optimization for high density biological production applications. Cytotechnology 1991; 5:15–30.
143. Ramírez OT, Mutharasan R. Effect of serum on the plasma membrane fluidity of hybridomas: an insight into its shear protective mechanism. Biotechnol Prog 1992; 8:40–50.
144. Wu J, Ruan Q, Lam HYP. Effects of surface-active medium additives on insect cell surface hydrophobicity relating to cell protection against bubble damage. Enzyme Microb Technol 1997; 21:341–348.
145. Joshi L, Davis TR, Mattu TS, Rudd PM, Dwek RA, Shuler ML, Wood HA. Influence of baculovirus-host cell interactions on complex N-linked glycosylation of a recombinant human protein. Biotechnol Prog 2000; 7:9–14.
146. Rhee WJ, Kim EJ, Park TH. Kinetic effect of silkworm hemolymph on the delayed host cell death in an insect cell-baculovirus system. Biotechnol Prog 1999; 15:1028–1032.
147. Maranga L, Coroadinha AS, Carrondo MJ. Insect cell culture medium supplementation with fetal bovine serum and bovine serum albumin: effects on baculovirus adsorption and infection kinetics. Biotechnol Prog 2002; 18:855–861.

148. Kim EJ, Rhee WJ, Park TH. Isolation and characterization of an apoptosis-inhibiting component from the hemolymph of *Bombyx mori*. Biochem Biophys Res Commun 2001; 285:224–228.
149. Maranga L, Mendonca RZ, Bengala A, Peixoto CC, Morales RHP, Pereira CA, Carrondo MJ. Enhancement of Sf9 cell growth and longevity through supplementation of culture medium with hemolymph. Biotechnol Prog 2003; 19:58–63.
150. Ferrance J, Goel A, Ataai M. Utilization of glucose and amino acids in insect cell cultures: quantifying the metabolic flows within the primary pathways and medium development. Biotechnol Bioeng 1993; 42:697–707.
151. Ramírez OT, Mutharasan R. The role of plasma membrane fluidity on the shear sensitivity of hybridomas grown under hydrodynamic stress. Biotechnol Bioeng 1990; 36:911–920.
152. Schlaeger E. Medium design for insect cell culture. Cytotechnology 1996; 20:57–70.
153. Garnier A, Voyer R, Tom R, Perret S, Jardin B, Kamen A. Dissolved carbon dioxide accumulation in a large scale and high density production of TGFD receptor with baculovirus infected Sf-9 cells. Cytotechnology 1996; 22:53–63.
154. Palomares LA, González M, Ramírez OT. Evidence of Pluronic F-68 direct interaction with insect cells: impact on shear protection, recombinant protein and baculovirus production. Enzyme Microb Technol 2000; 26:324–331.
155. Taticek RA, Shuler ML. Effect of elevated oxygen and glutamine levels on foreign protein production at high cell densities using the insect cell-baculovirus expression system. Biotechnol Bioeng 1997; 54:142–152.
156. Wong KTK, Peter CH, Greenfield PF, Reid S, Nielsen LK. Low multiplicity infection of insect cells with a recombinant baculovirus: the cell yield concept. Biotechnol Bioeng 1996; 49:659–666.
157. Elias CB, Zeiser A, Bédard C, Kamen AA. Enhanced growth of Sf-9 cells to a maximum density of 5.2×10^7 cells per mL and production of β-galactosidase at high cell density by fed batch culture. Biotechnol Bioeng 2000; 68:381–388.
158. Kioukia N, Nienow A, Emery A, Al-Rubeai M. Physiological and environmental factors affecting the growth of insect cells and infection with baculovirus. J Biotechnol 1995; 38:243–251.
159. Licari P, Bailey JE. Modeling the population-dynamics of baculovirus-infected insect cells—optimizing infection strategies for enhanced recombinant protein yields. Biotechnol Bioeng 1992; 39:432–441.
160. Wang MY, Kwong S, Bentley WE. Effects of oxygen/glucose/glutamine feeding on insect cell baculovirus protein expression: a study on epoxide hydrolase production. Biotechnol Prog 1993; 9:355–361.
161. Zhang J, Kalogerakis N, Iatrou K. A two-stage bioreactor system for the production of recombinant proteins using a genetically engineered baculovirus/insect cell system. Biotechnol Bioeng 1993; 42:357–366.
162. Kioukia N, Nienow A, Al-Rubeai M, Emery AN. Influence of agitation and sparging on the growth rate and infection of insect cells in bioreactors and a comparison with hybridoma culture. Biotechnol Prog 1996; 12:779–785.
163. Power J, Reid S, Radford K, Greenfield P, Nielsen L. Modeling and optimization of the baculovirus expression vector system in batch suspension culture. Biotechnol Bioeng 1994; 44:710–719.
164. King GA, Dougulis A, Faulkner P, Goosen M. Recombinant β-galactosidase production in serum free medium by insect cells in a 14-L airlift bioreactor. Biotechnol Prog 1992; 8:567–571.
165. Hu YC, Wang MY, Bentley WE. A tubular segmented-flow bioreactor for the infection of insect cells with recombinant baculovirus. Cytotechnology 1997; 24:143–152.
166. Saarinen MA, Murhammer DW. Culture in the rotating-wall vessel affects recombinant protein production capability of two insect cell lines in different manners. In Vitro Cell Dev Biol Anim 2000; 36:362–366.

167. Chung IS, Taticek RA, Shuler ML. Production of human alkaline phosphatase, a secreted, glycosylated protein, from a baculovirus expression system and the attachment-dependent cell line *Trichoplusia ni* BTI-Tn 5B1-4 using a split-flow, air-lift bioreactor. Biotechnol Prog 1993; 9:675–678.
168. Meij P, Vervoort MBHJ, de Gooijer K, Bloemena E, Meijer CJLM, Middeldorp JM. Bioreactor-scale production and one-step purification of Epstein-Barr nuclear antigen 1 expressed in baculovirus-infected insect cells. Protein Expr Purif 2000; 20: 324–333.
169. Chiou TW, Wang YC, Liu HS. Utilizing the macroporous packed bed for insect cell/baculovirus expression. Bioprocess Eng 1998; 18:91–100.
170. Dee KU, Shuler ML. A mathematical model of the trafficking of acid-dependent enveloped viruses: application to binding, uptake, and nuclear accumulation of baculovirus. Biotechnol Bioeng 1997; 54:468–490.
171. Radford KM, Cavegn C, Bertrand M, Bernard AR, Reid S, Greenfield P. The indirect effects of multiplicity of infection on baculovirus expressed proteins in insect cells: secreted and non-secreted products. Cytotechnology 1997; 24:73–81.
172. Wickham TJ, Shuler ML, Hammer DA, Granados RR, Wood HA. Equilibrium and kinetic analysis of *Autographa californica* nuclear polyhedrosis virus attachment to different insect cell lines. J Gen Virol 1992; 73:3185–3194.
173. Palsson BO, Chuck AS, Huang G. Gene therapy: the importance of random Brownian motion (abstract). Cytotechnology 1996; 17:22.
174. Wood HA. An agar overlay plaque assay method for *Autographa californica* nuclear polyhedrosis virus. J Invert Pathol 1977; 29:304–307.
175. Tsao E, Mason M, Cacciuttolo M, Bowen S. Production of parvovirus B19 vaccine in insect cells coinfected with double baculovirus. Biotechnol Bioeng 1996; 49: 130–138.
176. Palomares LA, López S, Ramírez OT. Strategies for manipulating the relative concentration of recombinant rotavirus structural proteins produced by insect cells. Biotechnol Bioeng 2002; 78:635–644.
177. Hu Y-C, Bentley WE. Effect of MOI ratio on the composition and yield of chimeric infectious bursal disease virus-like particles by baculovirus co-infection: deterministic predictions and experimental results. Biotechnol Bioeng 2001; 75:104–119.
178. Chisti Y. Hydrodynamic damage to animal cells. Crit Rev Biotechnol 2001; 21:67–110.
179. Thomas CR, Zhang Z. The effect of hydrodynamics on biological materials. In: Galindo E, Ramírez OT, eds. Advances in Bioprocess Engineering II. Dordrecht: Kluwer Academic Publishers, 1998:137–170.
180. Garcia-Briones MA, Brodkey RS, Chalmers JJ. Computer simulation of the rupture of a gas bubble at a gas-liquid interface and its implications in animal cell damage. Chem Eng Sci 1994; 49:2301–2320.
181. Chisti Y. Animal-cell damage in sparged reactors. Trends Biotechnol 2000; 18:420–432.
182. Jem KJ, Gong T, Mullen J, Georgis R. Development of an industrial insect cell culture process for large scale production of baculovirus pesticides. In: Maramorosch K, Mitsuhashi J, eds. Invertebrate Cell Culture: Novel Direction and Biotechnology Applications. Enfield, NH: Science Publishers, 1997:173–180.
183. Traub O, Berk BC. Laminar shear stress—mechanisms by which endothelial cells transduce an atheroprotective force. Arterioscler Thromb Vasc 1998; 18:677–685.
184. Al-Rubeai M. Apoptosis and cell culture technology. Adv Biochem Eng Biotechnol 1998; 59:225–249.
185. Palomares LA, Ramírez OT. Bioreactor scale-up. In: Spier RE, ed. Encyclopedia of Cell Technology. Vol.1.New York: John Wiley & Sons, 2000:183–201.
186. Murhammer D, Goochee C. Scale up of insect cell cultures: protective effects of Pluronic F-68. Biotechnology 1988; 6:1411–1418.
187. Zhang Z, Al-Rubeai M, Thomas CR. The effect of Pluronic F68 on mechanical properties of mammalian cells. Enzyme Microb Technol 1992; 14:980–983.

188. Bavarian F, Fan L, Chalmers J. Microscopic visualization of insect cell-bubble interactions. I: rising bubbles, air medium interface, and the foam layer. Biotechnol Prog 1991; 7:140–150.
189. Michaels JD, Petersen JF, McIntire LV, Papoutsakis ET. Protection mechanisms of freely suspended animal cells CLR 8018 from fluid-mechanical injury. Viscometric and bioreactor studies using serum, Pluronic F68 and polyethylene glycol. Biotechnol Bioeng 1991; 38:169–180.
190. Palomares LA, Ramírez OT. Bioreactor scale-down. In: Spier RE, ed. Encyclopedia of Cell Technology. Vol.1.New York: John Wiley & Sons, 2000:174–183.
191. Gotoh T, Mochizuki G, Kikuchi K. Perfluorocarbon-mediated aeration applied to recombinant protein production by baculovirus-infected insect cells. Biochem Bioeng J 2001; 7:69–78.
192. Zhang J, Collins A, Chen M, Knyazev I, Gentz R. High-density perfusion culture of insect cells with a BioSep ultrasonic filter. Biotechnol Bioeng 1998; 55:351–359.
193. Kompier R, Tramper J, Vlak JM. A continuous process for the production of baculovirus using insect cell cultures. Biotechnol Lett 1988; 10:849–854.
194. van Lier F, van den Hombergh JP, de Gooijer CD, den Boer MM, Vlak JM, Tramper J. Long term semi-continuous production of recombinant baculovirus protein in a repeated (fed-) batch two-stage reactor system. Enzyme Microb Technol 1996; 18:460–466.
195. Mitchell-Logean C, Murhammer DW. Bioreactor headspace purging reduces carbon dioxide accumulation in insect cell cultures and enhances cell growth. Biotechnol Prog 1997; 13:875–877.
196. Rhiel M, Murhammer D. The effect of oscillating dissolved oxygen concentrations on the metabolism of a *Spodoptera frugiperda* IPLB-Sf21-AE clonal isolate. Biotechnol Bioeng 1995; 47:640–650.
197. Zeiser A, Elias CB, Voyer R, Jardin B, Kamen AA. On-line monitoring of physiological parameters of insect cell cultures during the growth and infection process. Biotechnol Prog 2000; 16:803–808.
198. Ljunggren J, Alarcon M, Ramqvist A-K, Westlund A, Öhman L. A method to determine the optimal time to infect cells with the baculovims expression system. In: Bernard A, Griffiths B, Noé W, Wurm F, eds. Animal Cell Technology: Products from Cells, Cells as Products. Dordrecht: Kluwer Academic Publishers, 1999:359–361.
199. Jain D, Ramasumbramanyan K, Gould S, Seamans C, Wang S, Lenny A, Silberklang M. Production of Antistasin using the baculovirus expression system. In: Hatch R, Gooche C, Moreira A, Alroy Y, eds. Expression Systems and Processes for r-DNA Products. ACS Symposium Series. Vol.477. Washington, DC: American Chemical Society, 1991:97–110.
200. Cha HJ, Dalal NG, Pham MQ, Vakharia VN, Rao G, Bentley W. Insect larval expression process is optimized by generating fusions with green fluorescent protein. Biotech Bioeng 1999; 65:316–324.
201. LeDuc PR, Whiteley EM, Bao G, Betenbaugh MJ. Investigating the secretory pathway of the baculovirus-insect cell system using a secretory green fluorescent protein. Biotechnol Prog 2000; 16:716–723.
202. Jenkins N, Parekh RB, James DC. Getting the glycosylation right: implications for the biotechnology industry. Nat Biotech 1996; 14:975–981.
203. Grünewald S, Haase W, Reiländer H, Michel H. Glycosylation, palmitoylation, and localization of the human D_{2S} receptor in baculovirus-infected insect cells. Biochemistry 1996; 35:15149–15161.
204. Whiteley EM, Hsu TA, Betenbaugh MJ. Modeling, assembly, aggregation, and chaperoning of immunoglobulin G production in insect cells. Biotechnol Bioeng 1997; 56:106–116.
205. Ailor E, Betenbaugh MJ. Overexpression of a cytosolic chaperone to improve solubility and secretion of a recombinant IgG protein in insect cells. Biotechnol Bioeng 1998; 58:196–203.

206. Yokoyama N, Hirata M, Ohtsuka K, Nishiyama Y, Fujii K, Fujita M, Kuzushima K, Kiyono T, Tsurumi T. Co-expression of human chaperone Hsp70 and Hsdj or Hsp40 co-factor increases solubility of overexpressed target proteins in insect cells. Biochim Biophys Acta 2000; 1493:119–124.
207. Tomita M, Yoshizato K, Jagata K, Kitajima T. Enhancement of secretion of human procollagen I in mouse HSP47-expressing insect cells. J Biochem 1999; 126: 1118–1126.
208. Hsu TA, Watson S, Eiden JJ, Betenbaugh MJ. Rescue of immunoglobulins from insolubility is facilitated by PDI in the baculovirus expression system. Protein Expr Purif 1996; 7:281–288.
209. Lenhard T, Reiländer H. Engineering the folding pathway of insect cells: generation of a stably transformed insect cell line showing improved folding of a recombinant membrane protein. Biochem Biophys Res Commun 1997; 238:823–830.
210. Crawford SE, Labbé M, Cohen J, Burroughs M, Zhou Y, Estes MK. Characterization of virus-like particles produced by the expression of rotavirus capsid proteins by insect cells. J Virol 1994; 68:5945–5952.
211. French T, Marshall J, Roy P. Assembly of double-shelled, virus-like particles of bluetongue virus by simultaneous expression of four structural proteins. J Virol 1990; 64:5695–5700.
212. Crawford SE, Estes MK, Ciarlet M, Barone C, O'Neal CM, Cohen J, Conner ME. Heterotypic protection and induction of a broad heterotypic neutralization response by rotavirus-like particles. J Virol 1999; 73:4813–4822.
213. Zeng CQY, Wentz MJ, Cohen J, Estes MK, Ramig RF. Characterization and replicase activity of double-layered and single-layered rotavirus-like particles expressed from baculovirus recombinants. J Virol 1996; 70:2736–2742.
214. Vernet T, Tessier DC, Richardson C, Laliberte F, Khouri HE, Bell AW, Storer AC, Thomas DY. Secretion of functional papain precursor from insect cells. Requirement for N-glycosylation of the pro-region. J Biol Chem 1990; 256:16661–16666.
215. Tessier DC, Thomas DY, Khouri HR, Laliberte F, Vernet T. Enhanced secretion from insect cells of a foreign protein fused to the honeybee melittin signal peptide. Gene 1991; 98:177–183.
216. Ailor E, Betenbaugh MJ. Modifying secretion and post-translational processing in insect cells. Curr Opin Biotechnol 1999; 10:142–145.
217. Jarvis DL, Summers MD, Garcia A, Bohlmeyer DA. Influence of different signal peptides and prosequences on expression and secretion of human plasminogen activator in the baculovirus system. J Biol Chem 1993; 268:16754–16762.
218. Ailor E, Pathmanathan J, Jongbloed JDH, Betenbaugh MJ. A bacterial signal peptidase enhances processing of a recombinant single chain antibody fragment in insect cells. Biochem Biophys Res Commun 1999; 255:444–450.
219. Laprise MH, Grodin F, Dubois CM. Enhanced TGFβ1 maturation in High Five cells coinfected with recombinant baculovirus encoding the convertase furin/pace: improved technology for the production of recombinant proteins in insect cells. Biotechnol Bioeng 1998; 58:85–91.
220. Palomares LA, Ramírez OT. Complex N-glycosylation of recombinant proteins by insect cells. Bioprocessing 2002; 1:70–73.
221. van Die I, van Tetering A, Bakker H, van den Eijden DH, Joziasse DH. Glycosylation in lepidopteran insect cells: identification of a β-1–4-N-acetylgalactosaminyltransferase involved in the synthesis of complex-type oligosaccharide chains. Glycobiology 1996; 6:157–164.
222. Hsu TA, Takahashi N, Tsukamoto Y, Kato K, Shimada I, Masuda K, Whiteley M, Fan JQ, Lee YC, Betenbaugh MJ. Differential N-glycan patterns of secreted and intracellular IgG produced in *Trichoplusia ni* cells. J Biol Chem 1997; 272:9062–9070.
223. Altmann F, Staudacher E, Wilson IBH, März L. Insect cells as hosts for the expression of recombinant glycoproteins. Glycoconjug J 1999; 16:109–123.

224. Hollister JR, Jarvis DL. Engineering lepidopteran insect cells for sialoglycoprotein production by genetic transformation with mammalian β1,4-galactosyltransferase and α2,6-sialyltransferase genes. Glycobiology 2001; 11:1–9.
225. Palomares LA, Joosten CE, Hughes PR, Granados RR, Shuler ML. Novel insect cell line capable of complex N-glycosylation and sialylation of recombinant proteins. Biotechnol Prog 2003; 19:185–192.
226. Seo N-S, Hollister JR, Jarvis DL. Mammalian glycosyltransferase expression allows sialoglycoprotein production by baculovirus-infected insect cells. Protein Expr Purif 2001; 22:234–241.
227. Lopez M, Tetaert D, Juliant S, Gazon M, Cerruti M, Verbert A, Delannoy P. O-glycosylation potential of lepidopteran insect cell lines. Biochim Biophys Acta 1999; 1427: 49–61.
228. Joshi L, Shuler ML, Wood HA. Production of a sialylated N-linked glycoprotein in insect cells. Biotechnol Prog 2001; 17:822–827.
229. Davidson DJ, Fraser MJ, Castellino FJ. Oligosaccharide processing in the expression of human plasminogen cDNA by lepidopteran insect (*Spodoptera frugiperda*) cells. Biochemistry 1990; 29:5584–5590.
230. Davidson DJ, Castellino FJ. Structures of the asparagine-289-linked oligosaccharides assembled on recombinant human plasminogen expressed in a *Mamestra brassicae* cell line (IZD-MBO503). Biochemistry 1991; 30:6689–6696.
231. Karacali S, Kirmizigul S, Daveci R. Sialic acid in the developing testis of *Galleria mellonella* (Lepidoptera). Invert Reprod Dev 1999; 35:225–229.
232. Wagner R, Geyer H, Geyer R, Klenk H-D. N-acetyl-D-glucosaminidase accounts for differences in glycosylation of influenza virus hemagglutinin expressed in insect cells from baculovirus vector. J Virol 1996; 70:4103–4109.
233. Joosten CE, Shuler ML. Production of sialylated N-linked glycoproteins in insect cells: role of glycosidases and effect of harvest time on glycosylation. Biotechnol Prog 2003; 19:193–201.
234. Watanabe S, Kokuho T, Takahashi H, Takahashi M, Kubota T, Inamuru S. Sialylation of N-glycans on the recombinant proteins expressed by a baculovirus-insect cell system under β-N-acetylglucosaminidase inhibition. J Biol Chem 2002; 277:5090–5093.
235. Ailor E, Takahashi N, Tsukamoto Y, Masuda K, Rahman BA, Jarvis DL, Lee YC, Betenbaugh MJ. N-glycan patterns of human tranferrin produced in *Trichoplusia ni* insect cells: effects of mammalian galactosyltransferase. Glycobiology 2000; 10:837–847.
236. Tomiya N, Ailor E, Lawrence SM, Betenbaugh MJ, Lee YC. Determination of nucleotides and sugar nucleotides involved in protein glycosylation by high-performance anion-exchange chromatography: sugar nucleotide contents in cultured insect cells and mammalian cells. Anal Biochem 2001; 293:129–137.
237. Hooker AD, Green NH, Baines AJ, Bull AT, Jenkins N, Strange PG, James DC. Constraints on the transport and glycosylation of recombinant IFN-γ in Chinese hamster ovary and insect cells. Biotechnol Bioeng 1999; 63:559–572.
238. Roth J, Kemp A, Reuter G, Schauer R, Gehring WJ. Occurrence of sialic acids in *Drosophila melanogaster*. Science 1992; 256:673–675.
239. Malykh YN, Krisch B, Gerardy-Schahn R, Lapina EB, Shaw L, Schauer R. The presence of N-acetylneuraminic acid in Malpighian tubules of larvae of the cicada *Philaenus spumarius*. Glyconjug J 1999; 16:731–739.
240. Lawrence SM, Huddleston KA, Pitts LR, Nguyen N, Lee YC, Vann WF, Coleman TA, Betenbaugh MJ. Cloning and expression of the human N-acetylneuraminic acid phosphate synthase gene with 2-keto-3-deoxy-D-glycero-D-galacto-nononic acid biosynthetic ability. J Biol Chem 2000; 275:17869–17877.
241. Lawrence SM, Huddleston KA, Tomiya N, Pitts L, Nguyen N, Lee YC, Vann WF, Coleman TA, Betenbaugh MJ. Expression of human sialic acid pathway genes to generate CMP-sialic acids in insect cells. Glycoconjug J 2001; 18:205–213.

242. Donaldson M, Wood HA, Kulakosky PC, Shuler ML. Use of mannosamine for inducing the addition of outer arm N-acetylglucosamine onto N-linked oligosaccharides of recombinant proteins in insect cells. Biotechnol Prog 1999; 15:168–173.
243. Ogonah OW, Freedman RB, Jenkins N, Patel K, Rooney BC. Isolation and characterization of an insect cell line able to perform complex N-linked glycosylation on recombinant proteins. Bio/Technology 1996; 14:197–202.
244. Rudd PM, Downing AK, Cadene M, Harvey DJ, Wormland MR, Weir I, Dwek RA, Rifkin DB, Gleizes P-E. Hybrid and complex glycans are linked to the conserved N-glycosylation site of the third eight-cysteine domain of LTPB-1 in insect cells. Biochemistry 2000; 39:1596–1603.
245. Kulakosky PC, Hughes PR, Wood HA. N–linked glycosylation of a baculovirus-expressed recombinant glycoprotein in insect larvae and tissue culture cells. Glycobiology 1998; 8:741–745.
246. Hollister JR, Shaper JH, Jarvis DL. Stable expression of mammalian β1,4-galactosyltransferase extends N-glycosylation pathway in insect cells. Glycobiology 1998; 8:473–480.
247. Marchal I, Cerutti M, Mir AM, Juliant S, Devauchelle G, Cacan R, Verbert A. Expression of a membrane-bound form of *Trypanosoma cruzi* trans-sialidase in baculovirus-infected insect cells: a potential tool for sialylation of glycoproteins produced in the baculovirus-insect cells system. Glycobiology 2001; 11:593–603.
248. Varki A, Cummings R, Esko J, Freeze H, Hart G, Marth J. Essentials of Glycobiology. Vol.92. New York, Cold Spring Harbor Press 1999: pp. 101–113.
249. Wathen MW, Aeed PA, Elhammer AP. Characterization of oligosaccharide structures on a chimeric respiratory syncytial virus protein expressed in insect cell line Sf9. Biochemistry 1991; 30:2863–2868.
250. Thomasen DR, Post LE, Elhammer AP. Structure of O-glycosidically linked oligosaccharides synthesized by the insect cell line Sf9. J Cell Biochem 1990; 43:67–79.
251. Cha K, Bruel C, Inglese J, Khorana HG. Rhodopsin kinase: expression in baculovirus-infected insect cells, and characterization of posttranslational modifications. Proc Natl Acad Sci USA 1997; 94:10577–10582.
252. Altmann F, Kornfeld G, Dalik T, Staudacher E, Glössl J. Processing of asparagine-linked oligosaccharides in insect cells. N-Acetylglucosaminyltransferase I and II activities in cultured lepidopteran cells. Glycobiology 1993; 5:619–625.
253. Davidson DJ, Castellino FJ. Asparagine-linked oligosaccharide processing in lepidopteran insect cells-temporal dependence of the nature of oligosaccharides assembled on Asparagine-289 of recombinant human plasminogen produced in baculovirus vector infected *Spodoptera frugiperda* (IPLB-Sf21ae) cells. Biochemistry 1991; 30:6167–6174.
254. Jarvis DL, Fleming JGW, Kovacs GR, Summers MD, Guarino LA. Use of early baculovirus promoters for continuous expression and efficient processing of foreign gene products in stably-transformed lepidopteran cells. BioTechnology 1990; 5:950–955.
255. Andersen DC, Gooche CF. The effect of cell culture conditions on the oligosaccharide structures of secreted glycoproteins. Curr Opin Biotechnol 1994; 5:546–549.
256. Chejanovsky N, Zilberberg N, Rivkin H, Zlotkin E, Gurevitz M. Functional expression of an alpha anti-insect scorpion neurotoxin in insect cells and lepidopterous larvae. FEBS Lett 1995; 376:181–184.
257. Murakami K, Uchiyama A, Kokuho T, Mori Y, Sentsui H, Yada T, Tanigawa M, Kuwano A, Nagaya H, Ishiyama S, Kaki H, Yokomizo Y, Inumaru S. Production of biologically active recombinant bovine interferon-γ by two different baculovirus gene expression systems using insect cells and silkworm larvae. Cytokine 2001; 13:18–24.
258. Mori H, Nakazawa H, Shirai N, Shibata N, Sumida M, Matsubara F. Foreign gene expression by a baculovirus vector with an expanded host range. J Gen Virol 1992; 73:1877–1880.
259. Wu X, Kamei K, Sato H, Sato S, Takano R, Ichida M, Mori H, Hara S. High-level expression of human acidic fibroblast growth factor and basic fibroblast growth factor

in silkworm (*Bombyx mori* L.) using recombinant baculovirus. Protein Expr Purif 2001; 21:192–200.

260. Tamez-Guerra P, McGuire MR, Behle RW, Hamm JJ, Sumner HR, Shasha BS. Sunlight persistence and rainfastness of spray-dried formulations of baculovirus isolated from *Anagrapha falcifera* (Lepidoptera: Noctuidae). J Econ Entomol 2000; 93:210–218.
261. Moscardi, F. Assesment of the application of baculoviruses for control of lepidoptera. Annu Rev Entomol 1999; 44:257–289.
262. Behle RW, McGuire MR, Tamez-Guerra P. Effect of light energy on alkali-released virions from *Anagrapha falcifera* nucleopolyhedrovirus. J Invertebr Pathol 2000; 76: 120–126.
263. Chakraborty S, Monsour C, Teakle R, Reid S. Yield, biological activity, and field performance of a wild-type Helicoverpa nucleopolyhedrovirus produced in *H. zea* cell cultures. J Invertebr Pathol 1999; 73:199–205.
264. Cha HJ, Pham MQ, Rao G, Bentley W. Expression of green fluorescent protein in insect larvae and its application for heterologous protein production. Biotechnol Bioeng 1997; 56:239–247.
265. Hughes PR, Wood HA. In vivo production, stabilization, and infectivity of baculovirus preoccluded virions. Appl Environ Microbiol 1996; 62:105–108.
266. Pfeifer TA. Expression of heterologous proteins in stable insect cell culture. Curr Opin Biotechnol 1998; 9:518–521.
267. McCarroll L, King LA. Stable insect cell cultures for recombinant protein production. Curr Opin Biotechnol 1997; 8:590–594.
268. Farrell PJ, Behie LA, Iatrou K. Transformed lepidopteran insect cells: new sources of recombinant human tissue plasminogen activator. Biotechnol Bioeng 1999; 64: 426–433.
269. Cartier JL, Hershberger PA, Friesen PD. Suppression of apoptosis in insect cells stably transfected with baculovirus p35: dominant interference by N-terminal sequences p351–76. J Virol 1994; 68:7728–7737.
270. Farrel PJ, Lu M, Prevost J, Brown C, Behie L, Iatrou K. High level expression of secreted glycoproteins in transformed lepidopteran insect cells using a novel expression vector. Biotechnol Bioeng 1998; 60:656–663.
271. Li E, Brown SL, Dolman CS, Brown GB, Nemerow GR. Production of functional antibodies generated in a non lytic insect cell expression system. Protein Expr Purif 2001; 21:121–128.
272. Segal D, Cherbas L, Cherbas P. Genetic transformation of *Drosophila* cells in culture by P element-mediated transposition. Somat Cell Mol Genet 1996; 22:159–165.
273. Condreay JP, Whiterspoon SM, Clay WC, Kost TA. Transient and stable gene expression in mammalian cells transduced with a recombinant baculovirus vector. Proc Natl Acad Sci USA 1999; 96:127–132.
274. Palombo F, Monciotti A, Recchia A, Cortese R, Ciliberto G, La Monica N. Site-specific integration in mammalian cells mediated by a new hybrid baculovirus-adeno-associated virus vector. J Virol 1998; 72:5025–5034.
275. Chen X, Sun X, Hu Z, Li M, O'Reilly DR, Zuidema D, Vlak JM. Genetic engineering of Helicoverpa single-nucleocapsid nucleopolyhedrovirus as an improved pesticide. J Invertebr Pathol 2000; 76:140–146.
276. Kennedy MJ, Davies RJ, Surrey MR, Reader SL, Hoefakker PC. Process development for biological control agent production in New Zealand. Aust Biotechnol 1995; 5: 349–354.
277. Carbonell LF, Hodge MR, Tomalski MD, Miller LK. Synthesis of a gene coding for an insect-specific scorpion neurotoxin and attempts to express it using baculovirus vectors. Gene 1988; 73:409–418.
278. Treacy MF, All JN, Ghidiu GM. Effect of ecdysteroid UDP-glucosyltransferase gene deletion of efficacy of a baculovirus against *Heliothis virescens* and *Trichoplusia ni* (Lepidoptera: Noctuidae). J Econ Entomol 1997; 90:1207–1214.

279. Mulock B, Faulkner P. Field studies of a genetically modified baculovirus, *Autographa californica* nuclear polyhedrosis virus. J Invertebr Pathol 1997; 70:33–40.
280. Bonning BC, Hammock BD. Development of recombinant baculoviruses for insect control. Annu Rev Entomol 1996; 41:191–210.
281. Thiem SM. Prospects for altering host range for baculovirus bioinsecticides. Curr Opin Biotechnol 1997; 8:317–322.
282. Ignoffo CM, Mclntosh AH, Garcia C. In vitro and in vivo activity of a polyhedrin gene-deleted and wild strain of the nucleopolyhedrosis virus of *Autographa californica*. J Invertebr Pathol 1998; 72:90–92.
283. Carbonell LF, Miller LK. Baculovirus interaction with nontarget organisms: a virus-borne reporter gene is not expressed in two mammalian cell lines. Appl Environ Microbiol 1987; 55:1412–1417.
284. Yamschikov GV, Ritter GD, Vey M, Compans RW. Assembly of HIV virus-like particles containing envelope proteins using a baculovirus expression system. Virology 1995; 214:50–58.
285. Kirnbauer RJ, Taub H, Greenstone R, Roden M, Durst L, Gissmann DR, Lowry R, Schiller JT. Efficient self-assembly of human papilloma type 16 LI and L1-L2 into virus like particles. J Virol 1993; 67:6929–6936.
286. Grabherr R, Ernst W, Oker-Blom C, Jones I. Developments in the use of baculoviruses for the surface display of complex eukaryotic proteins. Trends Biotechnol 2001; 19:231–236.
287. Ernst W, Grabherr R, Wegner D, Borth N, Grassauer A, Katinger H. Baculovirus surface display: construction and screening of a eukaryotic epitope library. Nucleic Acids Res 1998; 26:1718–1823.
288. Lindley KM, Su JL, Hodges PK, Wisely GB, Bledsoe RK, Condreay JP, Winegar DA, Hutchins JT, Kost TA. Production of monoclonal antibodies using recombinant baculovirus displaying gp64-fusion proteins. J Immunol Methods 2000; 234:123–135.
289. Volkman LE, Goldsmith PA. In vitro survey of *Autographa californica* nuclear polyhedrosis virus interaction with nontarget vertebrate host cells. Appl Environ Microbiol 1983; 45:1085–1093.
290. Shoji I, Aizaki H, Tani H, Ishii K, Chiba T, Saito I, Miyamura T, Matsuura Y. Efficient gene transfer into various mammalian cells, including non-hepatic cells, by baculovirus vectors. J Gen Virol 1997; 78:2657–2664.
291. Ma L, Tamarina N, Wang Y, Kuznetsov A, Patel N, Kending C, Hering BJ, Philipson LH. Baculovirus-mediated gene transfer into pancreatic islet cells. Diabetes 2000; 49:1986–1991.
292. Song SU, Boyce FM. Combination treatment for osteosarcoma with baculoviral vector mediated gene therapy (p53) and chemotherapy (adriamycin). Exp Mol Med 2001; 35:46–53.
293. Sarkis C, Serguera C, Petres S, Buchet D, Ridet J-L, Edelman L, Mallet J. Efficient transduction of neural cells in vitro and in vivo by a baculovirus-derived vector. Proc Natl Acad Sci USA 2000; 97:14638–14643.
294. Kost TA, Condreay JP. Recombinant baculoviruses as mammalian cell gene delivery vectors. Trends Biotechnol 2002; 20:173–180.
295. van Loo ND, Fortunati E, Ehlert E, Rabelink M, Grosveld F, Scholte BJ. Baculovirus infection of nondividing mammalian cells: Mechanisms of entry and nuclear transport of capsids. J Virol 2001; 75:951–970.
296. Hüser A, Rudolf M, Hofmann C. Incorporation of decay-accelerating factor into the baculovirus envelope generates complement-resistant gene transfer vectors. Nat Biotechnol 2001; 19:451–455.
297. Pieroni L, Maione D, La Monica N. In vivo gene transfer in mouse skeletal muscle mediated by baculovirus vectors. Human Gene Ther 2001; 12:871–881.
298. Airenne KJ, Hiltunen MO, Turunen MP, Turunen AM, Laitinen OH, Kulomaa MS, Yla-Herttuala S. Baculovirus-mediated periadventitial gene transfer to rabbit carotid artery. Gene Ther 2000; 7:1499–1504.

299. Ojala K, Mottershead DG, Suokko A, Oker-Blom C. Specific binding of baculoviruses displaying gp64 fusion proteins to mammalian cells. Biochem Biophys Res Commun 2001; 284:777–784.
300. Touze A, Coursaget P. In vitro gene transfer using human papillomavirus-like particles. Nucleic Acids Res 1998; 26:1317–1323.
301. Cheshenko N, Krougliak N, Eisensmith RC, Krougliak VA. A novel system for the production of fully deleted adenovirus vectors that does not require helper adenovirus. Gene Ther 2001; 8:846–854.

19
Advances in Adult Stem Cell Culture

E. S. Tzanakakis
School of Medicine, Diabetes Center, University of California-San Francisco, San Francisco, U.S.A.

C. M. Verfaillie
Stem Cell Institute, University of Minnesota, Minneapolis, Minnesota, U.S.A.

INTRODUCTION

Stem cells are defined as cells that have extensive, some would say indefinite, proliferation potential, differentiate into several cell lineages, and repopulate tissues upon transplantation. Embryonic stem (ES) cells, which are derived from the inner cell mass of a blastocyst, are by all accepted criteria true stem cells capable of generating mature progeny of all cell types. ES cell isolation was initially reported in mouse embryos (1) and more recently in nonhuman primates and humans (2,3). When introduced into mouse blastocysts, ES cells can contribute to all tissues of the mouse (4). Following transplantation in postnatal animals, ES cells generate teratomas (3,5), which again demonstrates their pluripotency, but at the same time raises serious concerns when clinical applications of ES cells are considered. Additional legal, ethical, and moral problems accompany the derivation of ES cells from human blastocysts.

Stem cells have also been identified in most adult tissues including bone marrow, brain, skeletal muscle, liver, and pancreas. These findings raise the question of whether adult stem cells preferentially reside within certain tissues or whether they are widely distributed. Stem cells residing in a specific tissue have been traditionally considered as precursors of mature cells of only this tissue, and therefore have been named multipotent, but not pluripotent. Thus, their differentiation potential has been described as limited or tissue-restricted. However, this dogma has recently been challenged and there is growing evidence that adult stem cells may possess a greater plasticity than previously thought.

This chapter is focused on somatic tissue-derived stem cells, their in vitro expansion and differentiation, potential applications as well as concerns associated with recent discoveries and reports related to adult stem cells.

SOMATIC STEM CELLS

A plethora of recent reports has dealt with the identification and isolation of stem cells from various postnatal tissues. Although the majority of papers have described their tissue-specific differentiation potential, in more recent reports, investigators have claimed direct demonstration of tissue-committed stem cells differentiating toward a distant cell type or even de-differentiating to become stem cells for other tissues (transdifferentiation). However, one serious pitfall of many of these studies is that the experiments are rarely performed with single cells. Picking single-stem cells and showing that they can generate progeny of various cell types in vivo or in vitro is the most widely accepted criterion for proving that they are truly multipotent stem cells. True stem cells are clonogenic by definition and this property is the basis for clonality assays proposed to be standard tests when claims regarding transdifferentiation of adult stem cells come into view. Demonstration of clonality should be accompanied by evidence of functional progeny. It is not sufficient to show that the differentiated progeny derived from putative stem cells exhibits certain morphological features or is identified within a specific tissue, e.g., after transplantation. More importantly, the differentiated cells must assume the functional role of the expected cell type. Unfortunately, currently most but a few studies do not provide convincing evidence satisfying these criteria. Therefore, it is prudent—if not essential—to exercise caution before reaching conclusions, as many issues pertinent to adult stem cells are still unsettled. With this in mind, we will mention only a few categories of tissue-specific adult stem cells, as it would be beyond the scope and space limits of this chapter to give an extensive account of studies on all different types of somatic stem cells.

Hematopoietic Stem Cells

Hematopoietic stem cells (HSCs) are probably the best characterized somatic stem cells. Their isolation is based on cell surface markers and functional characteristics. The sialomucin CD34, whose normal function in hematopoiesis remains enigmatic, has become the distinguishing feature for isolation and manipulation of HSCs, although for murine cells that is not necessarily the case. In addition to its expression in stem cells and early progenitors during hematopoiesis, CD34 is also found in vascular endothelial cells and some fibroblasts (6–8), necessitating more antigenic determinants for HSC selection. The vascular endothelial receptor 2 (VEGFR2 or KDR) (9), CD133 (AC133) (10,11), CD90, CD117 and CD38 (12,13) are some of the markers that have been proved useful in generating functionally homogeneous populations of HSCs. Enriched populations of HSCs have also been obtained via a cell sorting method exploiting the ability of HSCs, that express the BCRP1 membrane pump, to efflux the fluorescent dye Hoechst 33342 (14). This procedure results in a so-called side population (SP), which has recently been identified in bone marrow-derived cells as well as in samples from other tissues.

After transplantation of HSCs into lethally irradiated animals or humans, the neutrophile-macrophage, megakaryote, erythroid and lymphoid cell pool are repopulated (12,14–16). In vitro, HSCs can be induced to undergo at least some self-renewing cell divisions and can be induced to differentiate to the same lineages as seen in vivo.

Recent studies have suggested that HSC may also differentiate to nonhematopoietic cell types. For example, Lagasse et al. (17) demonstrated that bone marrow-derived HSCs were capable to give rise to hepatocytes that could rescue animals with hereditary tyrosinemia. Grant et al. (18) showed that HSC may give rise to endothelial cells that repopulate retinal blood vessels following retinal injury. Other reports have indicated that HSCs can acquire gastrointestinal epithelium (19), skin (20), lung epithelium and muscle (21) as well as neural characteristics (22). Except for the study by Lagasse et al. (17) and the study by Grant et al. (18), the majority of studies did not demonstrate acquisition of functional characteristics of the distant tissue cells. In addition, aside from the studies by Grant et al. (18) and Krause et al. (21), most studies did not use single cells for transplantation, therefore making it impossible to prove that an HSC acquired the phenotype of a nonhematopoietic cell type. Furthermore, Wagers et al. (23) found levels of transdifferentiation that were significantly lower than what was demonstrated by Krause et al. (21). Therefore, most of these studies only suggest the possibility of "transdifferentiation," without full proof.

Mesenchymal Stem Cells

HSCs are not the only stem cells isolated from bone marrow. Mesenchymal stem cells (MSCs), initially described by Friedenstein et al. (24), are also known to have remarkable differentiation potential as several investigators have reported. MSCs can form osteoblasts, chondrocytes, adipocytes, and skeletal muscle cells (25–27). Purification of MSCs is predominantly based on their ability to adhere to culture dishes when incubated with fetal calf serum, but selection with respect to cell surface markers has also been employed for MSC isolation. MSCs are negative for the hematopoietic markers CD34, the leukocyte common antigen CD45, and the lipopolysaccharide receptor CD 14 (25,28). Conversely, cultured MSCs exhibit CD44, CD29, and the transferrin receptor CD71. Furthermore, MSCs express various extracellular matrix components and their receptors such as proteoglycan, collagen (type I, III, IV, V, and VI), laminin, the hyaluronide receptor, and ICAM-1 and -2. Typical marker genes expressed by bone marrow-derived MSCs also include cytokines and cytokine receptors such as interleukin (IL)-6 and IL-6 receptor (IL-6R), IL-7 and IL-7R, tumor growth factor (TGF)-β1R and TGF-β2R, basic fibroblast growth factor receptor (bFGFR), and platelet derived growth factor receptor (PDGFR) (29–31). The expression of cytokines, growth factors, matrix molecules, and receptors is associated with the notion that mesenchymal stem and progenitor cells participate in the organization and function of the stromal microenvironment that promotes differentiation of mesenchymal and hematopoietic cells. Cultured MSCs maintain normal karyotype even after several passages, without differentiating spontaneously.

Depending on the culture conditions, MSCs have been shown to commit toward adipocytes, osteocytes (25), chondrocytes (26), skeletal muscle cells (32), cell types derived from limb-bud mesoderm. In addition, some studies have suggested that MSC may also differentiate into cells with morphological and antigenic characteristics of cardiac muscle cells (33) and neuroectodermal cells (34). Again, no proof of function has been provided either in vitro or in vivo for differentiation to cells other than limb-bud mesoderm. Donor human MSCs have been detected in liver lung, marrow, spleen, and thymus after transplantation in nonirradiated immunodeficient (severe combined immunodeficiency, SCID) animals, even though it was not shown that the MSCs acquired morphologic, phenotypic, or functional characteristics of the tissue they engrafted in (35,36).

Muscle Cells

Several classes of stem cells exist in skeletal muscle, including muscle satellite cells, and a population of cells that may precede the satellite population. Satellite cells are unipotent myogenic progenitors found in skeletal muscle, characterized by expression of *pax7* (37), *myf5*, *C34* and *M-cadherin* (38), and integrin α-7, that are required for postnatal muscle growth and repair. Satellite cells give rise to large numbers of daughter myoblasts and repopulate the satellite compartment (39). Recent studies have demonstrated that satellite cells can also differentiate into adipocytes and osteocytes in vitro (40,41), indicating mesenchymal differentiation potential of satellite cells. One of the shortcomings of satellite cells for clinical therapy is the limited number of cells that can be harvested, and the low viability following transplantation. In addition, there is no evidence that satellite cells home to and engraft in muscles when administered systemically. Therefore, localized therapies in muscle would be needed.

A second population of cells that, in contrast to satellite cells does not express M-cadherin and can be expanded extensively ex vivo, was described by Qu-Petersen et al. (42). These muscle-derived stem cells (MDSCs) are selected from cultures as cells that can undergo multiple self-renewing cell divisions. MDSC are $CD34^+$, $Sca\text{-}1^+$, and $MyoD^+$, but $CD45^-$, $c\text{-}kit^-$, and $M\text{-}Cadherin^-$. Following transplantation in skeletal muscle, MDSC can give rise to long-term persisting grafts of cells that contribute to the satellite cell-like compartment as well as the mature myocyte compartment. In addition, MDSC may also give rise to endothelial cells and neuron-like structures in vivo. Not known is whether these cells will home to muscle when injected systemically.

According to Asakura et al. (43), the SP fraction of skeletal muscle, i.e., the cells that do not label with Hoechst 33342, are progenitors for satellite cells. SP cells in muscle contain both HSCs as well as a population of cells that upon transplantation in skeletal muscle or co-culture with skeletal myoblasts in vitro acquires a muscle fate, including both mononuclear myoblasts and multinucleated myotubes. Mouse SP (mSP) cells capable of differentiation into muscle cells are $Sca\text{-}1^+$, but $Myf5\text{-}nlacZ^-$, $desmin^-$ and $Pax7^-$. In contrast to satellite cells, mSP cells are located outside of muscle fibers. When mSP cells are transplanted in skeletal muscle, they can however differentiate into myocytes, and contribute, albeit at low levels, to muscle in *mdx* animals when injected systemically (44). Jackson et al.(45) published that skeletal muscle SP cells can contribute to the hematopoietic system when transplanted in lethally irradiated mice. Subsequent studies from the same group demonstrated that the hematopoietic potential may be derived from the $CD45^+$ subpopulation of SP cells in muscle (46). Recently, the $Sca\text{-}l^+CD45^-c\text{-}kit^-$ murine muscle fraction has also been shown to exhibit hematopoietic reconstitution activity following intravenous transplantation into lethally irradiated mice (47). These putative stem cells derived from muscle could be expanded in vitro without loss of differentiation potential or longevity (47).

Hepatic Stem Cells

Though normally senescent, hepatocytes exhibit rapid proliferation, driving liver regeneration after partial hepatectomy. This restoration is largely achieved by hepatocyte self-replication leading many groups to suggest that hepatocytes may function as liver stem cells. Others have identified so-called "oval cells" residing within the smallest branches of the intrahepatic biliary network (48,49). Oval cells are $Thyl^+$, $c\text{-}kit^+$, and $flt3^+$ and their number increases following hepatocytes loss due to liver disease (50,51). The origin of oval cells has long been debated as some groups theorize that oval cells are derived directly from intrahepatic proliferation of cells residing in the biliary tree whereas others support the notion that oval cells have precursors of bone marrow origin. To that end, a number of reports demonstrated that bone marrow hepatopoietic cells could give rise to hepatocytes although this claim has recently been challenged (23). From an embryo development viewpoint, hepatocytes share their endodermal origin with pancreatic cells and this relation may prove to be important for the generation of hepatocytes from pancreatic tissue cells and vise versa. It has long been known that cells within the pancreas acquire hepatocyte phenotype and morphology in rats placed on a copper-deficient diet. Krakowski et al. (52) found hepatocytes in the islets of Langerhans in insulin promoter-keratinocyte growth factor (KGF) transgenic mice, and similar results have been obtained in vitro following treatment of pancreatic exocrine cells with a combination of dexamethasone and oncostatin M (53,54).

Neural Stem Cells

The existence of stem cells in the central nervous system (CNS) was reported in the early 1990s (55–57). Employing clonal analysis, labeling and transplantation experiments, rodent neural cell precursors have been shown to differentiate into all three neural cell types: neurons, oligodentrocytes, and astrocytes. Interestingly, neural cell precursors isolated from different regions of the adult CNS and brain, vary in their growth characteristics and differentiation pathways. For example, A2B5 immunoreactive cells from the rat optic nerve give rise to astrocytes but not oligodentrocytes (58). Analogously, glial- and oligodentrocyte-restricted precursors have been isolated from other areas (59,60). In many instances, neural stem cells are maintained in culture as floating cell aggregates, or neurospheres, in the presence of epidermal growth factor (EGF) (61), but adherent cultures of FGF-dependent neural stem cells have also been reported (62). In a study by Uchida et al. (63), brain cells that were $CD133^+$, $CD24^{lo}$, $CD34^-$, $CD45^-$, and $5E12^+$, were shown to self-renew in neurospheres and differentiate into neurons and glia. Despite the exciting results reported in recent studies on neural stem cells, many issues require scrutiny especially if one considers the use of heterogeneous cell populations in many of these reports.

Multipotent Adult Progenitor Cells

We identified a rare cell within human, mouse, and rat bone marrow MSC cultures that can be expanded for >100 population doublings (64,65). This cell differentiates not only into mesenchymal lineage cells but also cells with phenotypic, morphologic, and functional characteristics of endothelium, neuroectoderm, and endoderm (64–67). We termed this cell "multipotent adult progenitor cell" (MAPC). Quantitative reverse transcription-polymerase chain reaction (RT-PCR) showed that a transcription factor important in maintaining ES cells undifferentiated, Oct-4 (68), was expressed in MAPC (64). Furthermore, like ES cells, mouse MAPC, but not human MAPC require leukemia inhibitory factor (LIF) to establish (3,64). Mouse MAPC expresses SSEA1 and human MAPC SSEA4 (69). MAPCs are also $Flk\text{-}1^+$, $Sca\text{-}1^+$, $Thy\text{-}1^+$, and $CD13^+$ (64), but CD105 and CD44 negative. Single-mouse MAPC injected in a blastocyst contributes to most, if not all, somatic cell types (64). Upon transplantation in a nonirradiated host, MAPC engraft and differentiate to the hematopoietic lineage, and epithelium of liver, lung, and gut (64).

ADULT STEM CELL CULTURE

Somatic stem cell proliferation and differentiation in vivo is controlled by the milieu within which these cells reside. Schofield (70) coined the term "niche" to describe tissue regions, which provide a suitable microenvironment to control the fate of HSCs. Such microenvironment can effect both symmetric and asymmetric stem cell division, allowing for normal tissue development and homeostasis while maintaining a pool of cells with extended neogenic potential. Stem cell proliferation and differentiation may be regulated by means of diffuse signaling, cell–cell, and cell–matrix interactions within these regions. The different developmental programs acting in concert during multicellular organization in vivo are mediated by cytokines, transcriptional regulators, and adhesion molecules. Many recent reports indicated the instrumental role in multiple differentiation pathways of growth factors such as VEGF, FGF, HGF and their receptors, transcription factors like members of the GATA, Nkx and LIM families and adhesion molecules that include members of the integrin family. Currently, models depicting the action of these factors in determining the stem cell proliferation and commitment are not complete but can serve as a basis for establishing regimens for directional and effective stem cell differentiation in vitro.

Strategies to induce in vitro differentiation of stem cells consist of:

1. *Epigenetic mechanisms*: these take action with the addition of extracellular signals such as growth factors and/or nonphysiological factors (e.g., dimethylsulfoxide, DMSO) and extracellular matrix components. The presence of these factors aims at modulating the gene expression creating a profile akin to what is observed during developmental processes in vivo.

2. *Genetic mechanisms*, which are mediated by the engineered expression of particular fate-deciding genes.

Epigenetic methods are easier and faster to implement but manipulation of the stem cell phenotype is often incomplete and yields heterogeneous cell populations. Conversely, genetic methods can be more demanding in terms of time and effort but allow for the generation of more pure cell populations. This however may be associated with increased risk for genomic aberrations and malignant transformations. Genetic methods have been applied extensively for directing the differentiation of ES cells. Using for example gene targeting, a drug resistance/toxic gene can be introduced under a cell-type-specific promoter resulting in the selective survival/ablation of a particular group of committed cells. In a similar manner, somatic stem cell differentiation can be directed with the insertion and/or overexpression of a gene that is critical for specification to a certain phenotype. The efficacy of genetic methods for stem cell differentiation depends on a number of factors including how amenable the particular type of stem cells is to genetic manipulation and the targeted gene(s). More information on genetic methods for stem cell differentiation can be found elsewhere (71) and references therein.

Extracellular Signals

Epigenetic methods, in the context of directing stem cell commitment, consist of incubating stem cells with a combination of growth factors and/or nonphysiological agents in order to modulate gene expression in favor of a desired cell type. In most instances, the premise for adding particular growth factors to stem cell cultures for differentiation is that these factors are present in the developing tissue in vivo.

The mechanism(s) of action of nonphysiological agents such as DMSO, retinoic acid, and 5-azacytidine in the differentiation of stem cells in the laboratory is even more puzzling. A list of factors used for in vitro differentiation of somatic stem cells is given in Table 1. It should be mentioned that studies dealing with in vivo differentiation of adult stem cells (e.g., via transplantation) have not been included in this list.

Besides the factor(s) used for differentiation, other related parameters are also important. These include the factor concentration, time point of stimulus introduction to the culture and duration of cell exposure to the agent(s). For example, different serum concentrations effect skin-derived precursor (SKP) differentiation toward

Table 1 Factors Mediating In Vitro Differentiation of Adult Stem Cells

Adult stem cell differentiation to	Factor(s)	Reference(s)
Neural cells (neurons, astrocytes, oligodendrocytes)	Basic FGF (FGF-2), β-mercaptoethanol, butylated hydroxyanisole, EGF, Brain-derived neurotrophic factor, glial-derived neurotrophic factor, PDGF, neural survival factor-1	34, 63, 64, 134, 135
Endothelial cells	VEGF	65
Hepatocytes	HGF, FGF-4, oncostatin M, dexamethasone, KGF	52, 53, 67
Cardiac muscle cells	Co-culture with neonatal cardiomyocytes, 5-azacytidine, TGF-β1	33, 136, 137
Skeletal muscle cells	5-Azacytidine, amphotericin B, dexamethasone, hydrocortisone, co-culture with myoblasts	32, 65, 138, 139
Adipocytes	Hydrocortizone, isobutyl-methylxanthine, indomethacin, dexamethasone, insulin, co-culture with adipocytes, serum	25, 65, 139, 140
Bone and cartilage cells (chondrocytes, osteocytes, etc.)	Dexamethasone, β-glycerolphosphate, TGF-β3, ascorbate-2-phosphate, co-culture with osteoblasts, indomethacin, TGF-β1	25, 65, 139–141
Blood cells	Stem cell factor, IL-3, IL-6, erythropoietin, methylcellulose culture	142, 143

distinct types (72). In the absence of serum, SKPs acquire a neuron and glial cell phenotype, whereas adipocytes are generated after incubation with 10% fetal bovine serum (FBS). Moreover, when SKPs differentiate in 3% rat serum they generate a small subpopulation of smooth muscle cells. Like the factor concentration, similarly important for the outcome of the differentiation, are the time point at which a factor is added to the cultured cells and the duration of the treatment. This dependence is expected as many developmental programs for cell commitment and maturation rely on presentation of cytokines to the cells by the extracellular milieu in a temporal manner. Many protocols try to mimic this temporal profile of cytokines. For example, retinoic acid applied to mouse ES cells induces time- and concentration-dependent differentiation toward neuronal, cardiac, myogenic, adipogenic, and vascular endothelial smooth muscle cells (73).

Except from differentiation, various combinations of factors and adhesion molecules have been used to maintain stem cells in a proliferative and undifferentiated state in culture. Thus far, the most commonly used factor for maintaining and expanding murine ES cells is the LIF (74). Similarly, mouse-derived somatic stem cells, for example mMAPCs (64), require the addition of LIF to the culture medium for expansion. If a murine stem cell-based system is considered, such as for production of a certain metabolite, then the cost arising from the use of LIF can be an important economic parameter. So far, the mechanism(s) by which LIF prevents adult stem cells from differentiating is unclear, but it is postulated that a modus operandi similar to that for ES cells exists. In ES cells, LIF binds to a receptor complex consisting of the LIF receptor and the gpl30 receptor. Upon binding of LIF to the receptor complex, activation of the STAT3 transcription factor ensues. STAT3 may interact with Oct4, another transcription factor whose expression is one of the hallmarks of undifferentiated pluripotent cells (75–77). Oct4 expression is evident in murine and human MAPCs albeit in lower levels compared to ES cells (64).

Oct4 is not the only ES cell marker found in MAPCs or other somatic stem cells. A partial list of stem cell markers is given in Table 2. The marker expression pattern depends on a number of factors including the tissue and the species from

Table 2 Markers Expressed in Undifferentiated Stem Cells

Marker	Comments
Oct-3/4	Octamer-specific protein present during mouse embryogenesis (144)
Rex-1	Transcription factor expressed in early embryo (145)
Telomere length	Human: ~11–15 kb (65), mouse: ~27 kb (64)
Telomerase activity	Remains high in undifferentiated cells
SSEA-1	Detected in mouse stem cells
SSEA-3 and -4	Detected in human but not mouse stem cells
Sca-1	Member of the Ly-6 family of GPI-linked surface proteins (146)
Thy-1, Thy-1.1, Thy-1.2	Also known as CD90 (147), CD90.1 and CD90.2, respectively

which the cells were isolated. For example, mMAPCs express stage-specific embryonic antigen 1 (SSEA-1) but no SSEA-3, and the same is true for murine ES cells (unpublished results and 64). A set of stem cell markers should be established for a certain type of tissue-specific stem cells [e.g., nestin has been postulated to be a marker of brain- and pancreas-residing stem cells (78,79)]. This can serve as quality control, or more specifically to assess whether the cell population retains its multipotency.

Most reports do not indicate the presence of feeder layers as a requirement for adult stem cell cultures. The presence of feeder layers if needed, would pose a considerable problem for the development of large-scale cultures. Murine somatic stem cells though require the addition of LIF to the culture medium for undifferentiated expansion.

Physical Conditions and Basal Medium

As with any other cell culture system, ex vivo expansion of stem cells in bioreactors requires suitable physiological conditions such as temperature, oxygen and carbon dioxide tension, pH, serum, and medium composition. In addition, some of the above parameters should be adjusted to levels that are essential for in vitro cell growth and depend on the specific stem cell type and its source. Nevertheless, adult stem cells are cultured under the same conditions as most common tissue cell cultures, i.e., at 37°C, 5% CO_2 in medium with pH at 7–7.2. Basal medium formulations may comprise, but not limited to, minimum essential medium (MEM), Dulbecco's modified essential medium (DMEM) and various MCDBs, providing necessary nutrients such as amino acids and salts. Supplementation with insulin, transferrin, and antibiotics is also common.

Serum

Like ES cells, adult stem cells are normally maintained in medium supplemented with certain percentage of serum. The use of serum is associated with a number of potential problems including physiological variability, lot consistency, availability and contamination with growth inhibitors and viruses. These problems can be minimized with the use of high-quality ES cell-screened serum but its high cost may prove prohibitive for large-scale cultures. To that end, low-serum or serum-free media formulations are greatly desirable. We are currently growing MAPCs in 2% fetal calf serum (64), which is advantageous compared to the 10–15% serum content required for ES cell culture. Furthermore, serum is removed during in vitro differentiation of MAPCs into endothelial, hepatic, and neuronal cells. Therefore, the development of serum-free medium for stem cell culture is feasible.

Cell Density

Another important determinant in a culture system is the cell density and this appears to be especially crucial in the case of stem cells. From our experience with ES cell cultures, we know that once ES cells reach a certain density, they lose their self-renewal capacity and start to differentiate uncontrollably necessitating frequent culture splitting. Differentiation is also prominent when ES cells are cultured on a substratum promoting the formation of aggregates, or embryoid bodies. A similar link between cell density and differentiation appears to exist for tissue-specific stem cells. Densities above 1500–2000 cells/cm^2 favor commitment of cultured mMAPCs to various cell types instead of undifferentiated proliferation. Regardless of whether cell–cell contact or some unknown diffusible signal (or both) may promote differentiation, such low density is an obvious impediment for the use of MAPCs in clinical applications requiring large cell numbers. Higher cell densities have been reported, for example, for cultures of human CNS stem cells (63). Single cells can initiate the formation of aggregates, or neurospheres, which increase in size as the cells divide. The cultured neurospheres are gently dissociated to a single-cell suspension and passaged every 2–3 weeks on plastic tissue-culture substrata. However, as the neurospheres grow in size, oxygen and nutrient transfer limitations become important, prohibiting in turn higher cell densities. A solution to this problem may be the use of stirred suspension bioreactors. When neurospheres are cultured in suspension (80), the maximum cell density is improved compared to T-flask cultures, possibly due to the superior mass transfer commonly achieved in suspension bioreactors. The final cell density is boosted even further by increasing the oxygen tension (from 5% to 20%). Upon subculturing, cells retain their capacity to generate new aggregates and to differentiate into all of the primary cell phenotypes in the CNS (81).

QUALITY CONTROL IN ADULT STEM CELL CULTURE

Quality control assays are essential in adult stem cell cultures as in any other type of cell culture and must be performed frequently and on a regular basis. Of course, quality control assays related to sterility, culture medium, and other parameters are the norm. However, here we would like to focus on assays related to the quality of the cultured stem cells per se. We will limit our discussion on three assays but it is obvious that performing additional assays will help to ensure high culture quality while preventing potential costly and time-consuming problems down the road.

Mycoplasma Testing

Mycoplasma are common contaminants of tissue culture cells and can be transmitted from culture to culture by contaminated tissue culture facilities and/or poor sterility techniques. In fact, mycoplasma contamination appears to be endemic in many laboratories. Part of the reason is that mycoplasma can be latent for long periods making its detection difficult in asymptomatic cultures. Sudden surges of mycoplasma contamination result in changes in cell morphology/function and eventually cell death. Furthermore, it can be extremely challenging to rescue contaminated cells. Treatment with antibiotics such as gentamycin may be effective against mycoplasma. In general, however, the uncertainty about the effect of the mycoplasma or treatment on the pluripotency of the cells makes as the best solution the immediate disposal of all contaminated cultures and frozen cells. Therefore, it is important to monitor cultures for mycoplasma, for example, once every month. For routine monitoring, nucleic acid-staining dyes may be used allowing for visualization of mycoplasma (disperse punctate cytoplasmic pattern). Other mycoplasma tests with higher sensitivity are based on amplification and detection of mycoplasma-specific ribosomal RNA in tissue culture medium. Along with mycoplasma, provisions should be made for testing possible contamination from other infection agents due to animal products (e.g., serum, animal cells).

Cytogenetic Analysis

By cytogenetic analysis, we refer to chromosome count and karyotype determination. Analysis of the chromosome complement is commonly performed by inducing the cells in metaphase arrest (by incubating with colcemid). Subsequently, the cells are suspended in hypotonic solution causing them to "burst" and spread their nuclear contents. Chromosome spreads are visualized by staining with Giemsa or quinacrine. Ideally, the chromosome number is counted in 30–70 spreads. For karyotyping, images are collected from 10 to 15 chromosome spreads and individual chromosomes are sorted into sequence and compared to the expected karyotype of a particular species.

Recently, more sensitive methods have been developed for the detection and analysis of chromosomal imbalances. Comparative genomic hybridization, which is performed on microarray slides, has greatly contributed to the current knowledge of genomic alterations associated with constitutional and sporadic genetic diseases (82–84). Furthermore, a novel CGH approach, termed matrix-CGH (85), where chromosome preparations are substituted by sets of well-defined genomic DNA fragments, exhibits superior resolution and potential for automation.

These methods can be implemented to address putative chromosomal alterations in stem cells and to obtain information about the species and sex from which the cells were derived. This is especially important if adult stem cells from more than one donor or species are simultaneously cultured in the same laboratory. In addition, there is always a possibility that the karyotype may be altered during cell culture. Cytogenetic analysis provides the means to monitor whether the cells retain their original karyotype even after multiple rounds of expansion. The presence of malignant cells can be easily diagnosed, for example by CGH, because their karyotype and/or chromosome number is altered compared to normal cells, and this assessment is imperative if the use of cells is intended for therapeutic applications.

Differentiation Capacity

One of the major characteristics of stem cells is their ability to differentiate toward specific phenotype(s). Cell differentiation protocols, which are established in the laboratory and/or published in the literature, should be employed frequently for testing of the neogenic potential of the cells. In our laboratory, we often test the potential of MAPCs to acquire phenotypes, each of which is associated with a different germ layer. Following are the most common differentiations we perform with MAPCs:

1. Endothelial cells (mesoderm): cells are plated at a density of 20×10^3 cells/cm^2 and treated with 10 ng/mL VEGF (64). After 10–14 days, cell morphology is akin to endothelial cells and distinct from undifferentiated MAPCs. Furthermore, endothelial markers such as von Willebrand factor, CD62E, CD31, and vascular endothelial (VE)-cadherin can be detected by immunocytochemistry.

2. Hepatocyte-like cells (endoderm): in this case, the plating density is approximately 18×10^3 cells/cm^2 while 20 ng/mL HGF and/or 10 ng/mL FGF4 are added for incubation (67). As with endothelial cells, cell morphology is changing drastically over the next 2 weeks accompanied by the presence of liver-specific markers such as albumin, cytokeratin 18 (CK18), and α-fetoprotein.

3. Neural cells (ectoderm): incubation of $3–5 \times 10^3$ MAPCs/cm^2 with 10 ng/mL bFGF results in phenotypes resembling oligodentrocytes, astrocytes, and neurons

(64). The resemblance is morphological and functional as many neural markers are detected including neuron-specific enolase, GABA, Tau, microtubule-associated protein (MAP) -2 and serotonin.

The protocols presented here have been applied with modifications for differentiations of other non-MAPC stem cells.

Except from these assays, other methods can be implemented for quality control of the undifferentiated stem cells/committed progeny (e.g., restriction fragment length polymorphism (RFLP) technique can be employed when considering the donor in stem cells of human origin, etc.).

PHENOTYPIC CHARACTERIZATION DURING IN VITRO DIFFERENTIATION OF STEM CELLS

The phenotype in stem cells undergoing in vitro differentiation is commonly assessed by detecting markers specific to a particular cell type. Identification of a single marker does not provide sufficient evidence for determining cell commitment. Normally, cultured cells should exhibit a group of markers related to a certain phenotype (Table 3).

Obviously, cell morphology by itself is not sufficient to draw conclusions regarding cell differentiation but it should have a complementary role. It is also a good practice to use multiple techniques such as immunocytochemistry, flow cytometry, RT-PCR and microarrays in order to minimize possible artifacts associated with the implementation of a single method. Determination of the final phenotype should not be limited to the identification of particular markers but should include functional assays such as substrate metabolism by tissue-specific enzymes, electrophysiological measurements etc., depending on the progeny. Lastly, it should be kept in mind that most in vitro differentiations involve starting cell populations which are not totally homogeneous and probably not all the cells will differentiate at the same pace or direction. Therefore, selection and quantitation (e.g., percentage) may be necessary when in vitro stem cell differentiation is assessed.

LARGE-SCALE STEM CELL CULTURE SYSTEMS

The rapid progress in stem cell research fueled great hope for effective treatment of a number of pathological conditions. In most studies, the therapeutic potential of stem cells has been demonstrated in animal models, usually rodents, via cell transplantation and subsequent amelioration of conditions such as diabetes, heart infarction, and dopamine deficiency. However, one of the major obstacles for translating such therapies to human patients is the requirement for large numbers of cells. This prerequisite emphasizes the importance of engineering culture systems of sufficiently large scale for ex vivo generation, maintenance, and differentiation of stem cells. A solution to this problem may be the use of bioreactors, which have been used for long time for the generation of large quantities of cells and cellular products. Bioreactors with capacities between 1 and 1000 L may be adequate for the expansion and differentiation of stem cells in clinically relevant numbers (86), although the size of the cell culture system will eventually depend on the specific end-application as well as on cost-based criteria. The size of the bioreactor is also intimately related

Table 3 Markers Associated with Specific Cell Phenotypes

Cell type	Markers
Neural cells (astrocytes, neurons, oligodendrocytes)	Nestin, microtubule-associated protein-2 (MAP2), Tau, neurofilament-200[a], neuron-specific enolase, GABA, serotonin, glial acidic fibrillary protein (GFAP)[b], galactocerebrocide (GalC)[c]
Endothelial cells	Vascular endothelial (VE)-cadherin, platelet endothelial cell adhesion molecule-1 (PECAM1, a.k.a. CD31), Flt-1, CD62E, Tek, von Willebrand factor (vWF), vinculin, vascular endothelial growth receptor factor-2 (VEGFR-2)[d]
Hepatocytes	Hepatocyte nuclear factors (HNFs), α-fetoprotein, cytokeratin 18 (CK18), transyretin, transferrin, albumin, urea, dipeptidyl peptidase IV (DPPIV)
Pancreatic cells	Nestin, neogenin3 (ngn3), PDX1, Nkx6.1, Isl-1, insulin, Glut-1, Glut-2
Cardiac muscle cells	Nkx2.5/Csx, α-myosin heavy chain, atrial natriuretic peptide (ANP)[e], a-actinin, cardiac troponin I, desmin
Skeletal muscle cells	Dystrophin, MyoD, myosin, myogenin, MRF4, Myf-5
Adipocytes	Neutral lipid vacuoles[f], peroxisome proliferation-activated receptor γ2 (PPARγ2), lipoprotein lipase (LPL), fatty acid-binding protein αP2
Bone and cartilage cells (chondrocytes, osteocytes, etc.)	Glycosaminoglycans[g], collagen type I and II[h], alkaline phosphatase, calcium accumulation, osteopontin, osteocalcin, osteonectin

[a] Expressed in neurons.
[b] Expressed in astrocytes.
[c] Expressed in oligodendrocytes.
[d] VEGFR-2 is the mouse ortholog of FLK-1 and human KDR.
[e] Also known as atrial natriuretic factor.
[f] Detected by oil red staining.
[g] Detected by toluidine blue staining.
[h] Alizarin Red S stains both collagen type I and II. Antibodies are also available for specific collagen types.

to the reactor type. Of course, the development of large-scale bioprocesses assumes solid understanding of aspects of stem cell biology, which are currently the focus of intense investigation. Although a wide variety of bioreactor configurations is available for cell culture, we will limit our discussion to reactor types that have already been employed for the expansion and/or differentiation of stem cells, particularly bone marrow-derived stem cells.

Stirred Suspension Bioreactors

In the laboratory, stem cells are typically cultured in static configurations including multiwell plates, T-flasks, and blood bags equipped with gas-permeable membranes. These static culture systems are not easily amenable to scale-up, on-line monitoring, and control of culture parameters. Conversely, stirred suspension bioreactors are ideal for high-volume cultures of stem cells (Fig. 1). These types of culture vessels are readily scalable, simple in their operation, and their technology is mature. In addition, they provide a homogeneous environment allowing for better control of culture conditions. Several groups have tried to expand stem cells, primarily hematopoietic, in stirred suspension bioreactors (80,87,88). Compared to static cultures, stirred suspension cultures fared better in terms of hematopoietic cell proliferation in the presence of Steel factor and IL-3 (87).

Suspension bioreactors may be used not only for stem cells expansion but also for evaluating possible mechanisms related to stem cell physiology. In studies regarding the effects of different culture variables on the propagation and differentiation of bone marrow- and umbilical cord blood-derived cells, the rates of progenitor expansion and cytokine depletion were measured and compared among cultures of various bone marrow fractions in the presence of IL-3, IL-6, IL-11, and stem cell factor (SCF) (89,90). The results suggested the possibility that a feedback mechanism based on cytokine depletion controls the proliferation and commitment of hematopoietic progenitors. In addition to cytokine depletion, agitation, and inoculum density (91), oxygen uptake and glucose metabolism (92,93) were shown to affect the rate of blood progenitor cell expansion in stirred suspension bioreactors.

These reactors have also been used for the propagation of neural stem cells in cell aggregates or neurospheres (80,94). After growing the cells for over a month, 10^7-fold higher cell number was obtained without reduction in growth rate and viability. The cells retained a normal karyotype throughout 25 population doublings and were able to differentiate toward neurons, oligodendrocytes, and astrocytes. The effect of agitation and medium viscosity were also investigated in these studies.

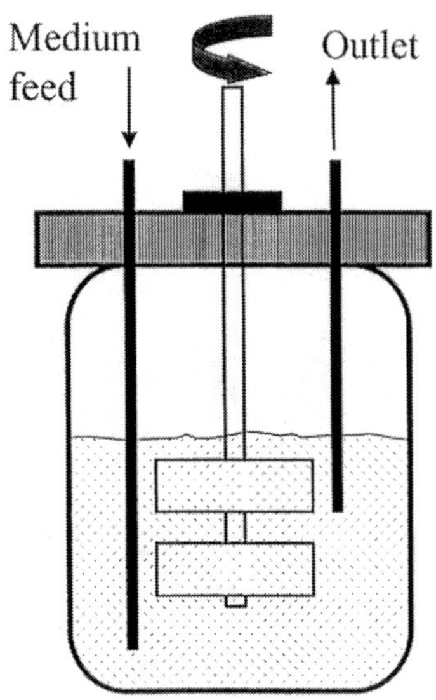

Figure 1 Stirred tank bioreactor.

Perfusion Chamber Bioreactors

Another bioreactor system that has been studied for large-scale propagation and differentiation of stem cells, primarily hematopoietic, is the perfusion chamber. In one report (95), 10–30-fold expansion of human bone marrow mononuclear cells (BMMCs) was achieved over 2 weeks of culture starting with inocula of $\sim 10^7$ cells. The extent of increase in cell number was linked to the seeding density, the time of cell harvest and the oxygen tension maintained within the system. In a later study (96), oxygen and growth factor transfer were found to be limiting for the generation of large amounts of cells in perfusion chambers. Compared to unprocessed whole bone marrow cells (WBMC), CD34-enriched cells on stroma exhibited a markedly increased expansion (1300-fold vs. 13-fold for BMMCs) over the same period (97). Interestingly however, the WBMCs generated five times higher concentrations of colony-forming units of granulocyte-macrophage (CFU-GM/mL of bone marrow aspirate). A possible explanation may be the increased amounts of EGF and platelet-derived growth factor (PDGF) present in the WBMC cultures. As in the case of stirred bioreactors, the effects of cytokine consumption kinetics have been studied in perfusion chamber cultures of hematopoietic cells with the intention of elucidating interactions among pathways regulating stem cell proliferation and differentiation (98).

Furthermore, the geometry and transport phenomena in various perfusion bioreactor configurations have been studied in order to increase the efficiency of these reactors with respect to HSC proliferation. A flat-bed perfusion reactor (Fig. 2) was modified with the addition of grooves perpendicular to the direction of flow at the chamber bottom (99). Through a combination of experimental results and numerical simulations, the authors showed that a decrease in the spacing between grooves resulted in increase in reactor efficiency by as much as 25% without growth restriction. In an earlier report (100), after analysis of four geometries (slab, gondola, diamond, and radial shapes) for a parallel-plate bioreactor, the radial-flow-type bioreactor appeared to provide the most uniform environment for parenchymal

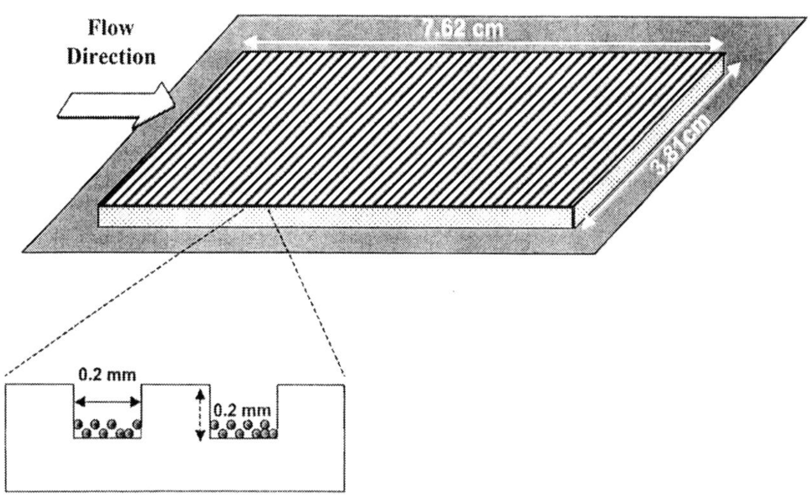

Figure 2 Flat-bed bioreactor with grooved bottom. [Adapted from Ref. (99).]

cell growth and differentiation ex vivo due to the absence of walls that are parallel to the flow paths creating slow flowing regions. Such studies demonstrate the flexibility as well as the large margin for optimization on variations of perfusion chambers employed for HSC culture.

More recently (101), hematopoietic progenitors expanded in a perfusion bioreactor were employed in a clinical trial for the treatment of breast cancer patients. Cells were collected from patients, expanded and infused back to the patients, who in the mean time had undergone chemotherapy. The investigators observed platelet and neutrophile recovery after re-infusion suggesting the potential applicability of perfusion systems for stem cell culture in a clinically pertinent scale.

Although the results are encouraging, certain limitations of perfusion-based systems are also pointed out by these studies. Growth factor or oxygen transport constraints may curtail stem cell expansion. Surface area for cell growth poses a considerable limitation, a problem that can be exacerbated by the fact that high cell density favors cell differentiation instead of uncommitted propagation.

Other Bioreactor Systems

Except from stirred suspension vessels and perfusion chambers, a variety of other cell culture configurations has also been studied with the aim of large-scale stem cell expansion and differentiation. Sardonini and Wu (102) examined static culture, suspension culture, microcarrier culture, airlift reactor, and hollow fiber culture. Using identical media, cytokines, and feed schedules, low density mononuclear cells in the suspension bioreactor expanded to a value of 1.6 compared to a normalized value of 1.0 for static cultures for the two runs investigated. Conversely, microcarrier and airlift bioreactor cultures averaged 0.75- and 0.70-expansion, respectively, compared to static cultures. However, in a later study (103), airlift packed bed bioreactor (Fig. 3) culture yielded promising results in terms of ex vivo hematopoiesis. In this case, stromal cells were grown on a fiberglass matrix packed in the bottom of an airlift reactor, imitating the environment in the native bone marrow. Once the stromal cell layer was established, fresh bone marrow was inoculated and over an 11-week culture period approximately 3.6×10^8 cells were obtained. Modifying a continuously perfused bioreactor, Meissner et al. (104) developed a fixed bed reactor (Fig. 4) with stromal cells immobilized on porous glass carriers and cocultivated human hematopoietic progenitor cells for several weeks. Early progenitor cells expanded up to 4.2-fold whereas colony-forming unit-granulocyte monocyte and burst-forming unit-erythrocyte up to 7- and 1.8-fold, respectively. Extensive production of lymphoid cells and all stages of committed cells (105) as well as erythropoiesis (106) were also achieved in bioreactors packed with porous beads in the presence of various cytokines.

These observations suggest that these systems have the potential to play a major role in realizing the promise of stem cell-based therapies. At the same time, they indicate that the design of bioreactor systems is crucial for efficient stem cell expansion.

RECENT ISSUES IN ADULT STEM CELL RESEARCH

There is now a large body of evidence suggesting that tissue-specific stem cells are capable of adopting multiple fates. Claims of stem cells "transdifferentiation" or

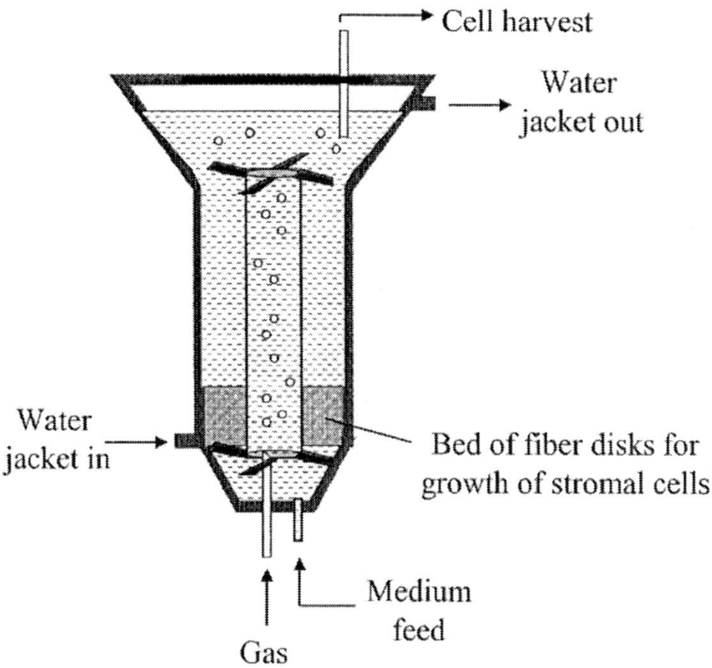

Figure 3 Airlift packed bed bioreactor. [Adapted from Ref. (103).]

"plasticity" have been met with skepticism—if not criticism—as they challenge decades of experiments that have proved that animal cells once differentiated show little sign of haphazard fate-switching. Both proponents and opponents have come forward with various interpretations of the observed "transdetermination" phenomena. The former believe that the great plasticity of adult somatic stem cells may be

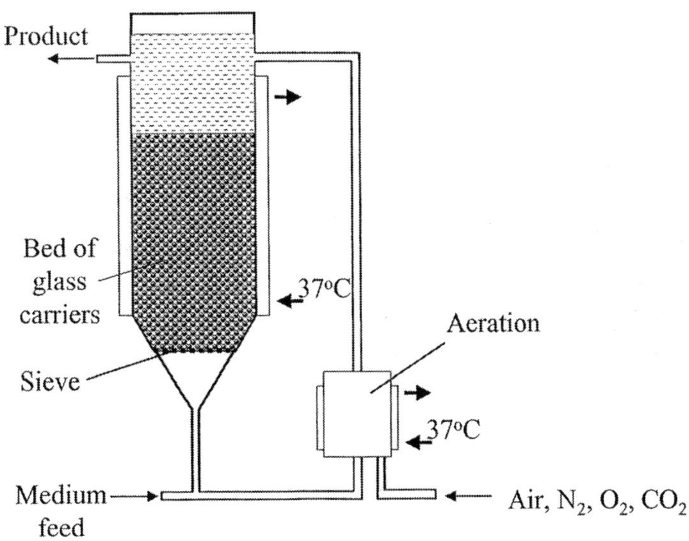

Figure 4 Fixed bed bioreactor. [Adapted from Ref. (104).]

expected because different niches share cytokine requirements, extracellular matrix components and transcriptional regulators. Considering this concept, many studies have illustrated the pleiotropic functions of growth factors and their receptors, adhesion molecules, and transcription factors.

However, two recent publications suggested that the apparent reprogramming of adult stem cells might instead be due to cell fusion. When bone marrow from GFP transgenic mice was cultured with ES cells, about 2–11 cells/10^6 cells fused with ES cells and these cells could subsequently adopt some of the phenotypes typical of ES cell differentiation (107). The cell fusion frequency was not greater in the hematopoietic fraction of bone marrow and this finding raises concerns since these cells were thought to be responsible for liver engraftment (17). In a parallel study (108), pyromycin-resistant GFP-labeled mouse neural stem cells were cocultured with hygromycin-resistant ES cells. Following selection in ES cell growth medium, the remaining cells expressed GFP and were resistant to both pyromycin and hygromycin. Moreover, these cells had double the normal DNA content and showed many characteristics of ES cells. The data supported the occurrence of fusion at low frequency (10^{-5}) between CNS stem cells with ES cells and these fused cells exhibited multilineage potential when injected into blastocysts, most prominently into liver.

These studies were a wake-up call for several groups and fueled efforts to establish rigorous standards for proving plasticity. By definition, stem cells are clonal precursors of more stem cells (symmetric division) and differentiated progeny (asymmetric division). Therefore, putative stem cells, which may be marked, should give rise to more stem cells starting from a single cell (clonality assay). Moreover, such cells should have the ability to repopulate and regenerate stem cells, progenitors and differentiated and functional progeny (109). Demonstrating the presence of tissue-specific proteins or morphology is not sufficient; the cells must contribute to the functions of the host tissue, e.g., detoxification functions for liver cells or electrophysiological responses for heart or brain cells.

With these criteria in mind, many investigators are re-examining previously published results. For example, what was thought were muscle cells turning into blood-forming cells were actually hematopoietic cells already residing in muscle (46). To that end, little evidence was found of HSCs contributing appreciably, if any, to nonhematopoietic tissues including brain, kidney, gut, liver, and muscle (23). It is becoming clear that isolation and characterization of somatic stem cells are challenging tasks. Yet, dealing with these tasks may turn out to be easier through continuous advances in stem cell biology as well as remarkable progress in various cell-sorting techniques.

POTENTIAL THERAPEUTIC APPLICATIONS

The demand for stem cell-based therapies is great and grows continuously among different groups of patients. Degenerative diseases are developing into a frequent theme among the elderly as the average life expectancy has increased rapidly over the last decades. Cardiovascular and musculoskeletal diseases, diabetes, Alzeimer's, and Parkinson's pathologies are some of the most prominent conditions affecting millions of people even in western countries and future projections are bleak. The solution to these and other diseases may lay on stem cells in conjunction with novel bioengineering approaches such as tissue engineering. The moment when stem cells will be routinely used for treating human patients is probably not in the immediate future since numerous issues pertinent to stem cell biology are still under intense investigation.

Even so, results from a small number of trials have been reported while still others are underway and a large body of evidence exists from studies on various animal models. MSCs isolated from a bone marrow aspirate were expanded ex vivo and re-infused in breast cancer patients undergoing bone marrow transplantation (110,111). The results from a group of 28 patients indicated that the autologous MSCs effected early hematopoietic recovery based on platelet and neutrophil counts. In another study, MSC transplantation was employed to rectify skeletal abnormalities. Initially, transgenic mice with fragile bones due to osteogenesis imperfecta, a disease resulting from mutations in the type I collagen gene, underwent transplantation of wild-type MSCs. Bone collagen and minerals were restored partially in the recipient mice (112). Later, Horwitz et al. (113) performed bone marrow transplantations from HLA-matched donors in three children with osteogenesis imperfecta. Engraftment was confirmed and was concomitant with new dense bone formation, increase in total body mineral content, and reduction of bone fracture frequency.

Except from MSCs, fetal neural cells which contain multipotent NSCs as well as more restricted neural precursors and terminally differentiated cells, have been employed for the treatment of Parkinson's and Huntington's syndromes. Fetal mesencephalic tissues were transplanted into the striatum of human Parkinson's patients. The grafted tissue survived for a long period in the human brain. Dopaminergic innervation to the striatum was restored and the patient had sustained improvement in motor function (114). In a similar fashion, the outcome of a clinical study with patients with Huntington's disease indicated increased scores on some measures of cognitive functioning following bilateral intrastriatal implantation of fetal striatal tissue (115). Although further investigation in this direction is necessary, the results suggest the possibility that purified populations of adult NSCs may be the key in reversing neurodegenerative conditions.

Stem cells can also be used as therapeutic agents in conjunction with gene delivery methods. The cells can be genetically modified to confer specific properties on their progeny. The ex vivo gene modification of stem cells intended for cell therapy reduces the possibility of unintentional carry-over of viral or other pathogenic vectors. Gene targeting methods, which are widely used to manipulate in a specific manner different chromosomal loci in ES cells, may be applied to adult stem cells as well. Furthermore, retroviruses, adenoviruses and adeno-associated viruses, which are typically used as gene transfer vehicles, exhibit tropism for various tissue-specific stem cells and provide additional means for ex vivo genetic manipulation of stem cells (116,117). MDSCs, or myoblasts, have been genetically engineered to deliver dystrophin to mdx mouse muscles (118), bone morphogenetic protein 2 (BMP2) for healing bone defects (119 and references therein), growth hormone (120), erythropoietin (121), and factor IX (122). Human MSCs can be transduced with viruses and maintain transgene expression during expansion in culture and differentiation into cartilage, osteocytes, and adipocytes (123). Human skin stem cells have also been modified genetically to correct the expression of laminin in patients with severe junctional epidermolysis bullosa (124). Transplantation of neural stem cells retrovirally transduced to secrete nerve growth factor (NGF) was able to ameliorate the death of striatal projection neurons caused by transient focal ischemia in the adult rat (125).

A handful of gene therapy trial involved the use of hematopoietic cells for treating patients with conditions such as adenosine deaminase deficiency (126) and

Gaucher disease (127). Recently, Hacein-Bey-Abina et al. (128) reported the use of genetically modified $CD34^+$ bone marrow cells for the correction of X-linked severe combined immunodeficiency due to mutation in the common γ chain (γc) gene. The study was conducted over a period of 2.5 years with the participation of five patients. The γc gene was transferred ex vivo via a retroviral vector to $CD34^+$ cells from these patients. Transduced T cells and natural killer cells appeared in the blood of four of the five patients within 4 months. Although the frequency of transduced B cells was low, serum immunoglobulin levels and antibody production after immunization were sufficient to avoid the need for intravenous immunoglobulin. Correction of the immunodeficiency eradicated established infections and allowed patients to have a normal life.

In a later correspondence (129), however, the authors commented that although three of the four patients continued to do well, 3.6 years after gene therapy, a serious adverse event occurred in the fourth patient. At a routine checkup 30 months after gene therapy, lymphocytosis consisting of a monoclonal population of $V\gamma9/V\delta1$, γ/δ T cells of mature phenotype was detected. Upon further examination, one proviral integration site within the *LMO-2* locus of chromosome 11 was found resulting in aberrant expression of *LMO-2*, which is linked to acute lymphoplastic leukemia (130). The increase in patient's lymphocyte count was accompanied by the development of hepatosplenomegaly. Although other possible explanations have not been excluded, these findings were interpreted as the consequence of the insertional mutagenesis event associated with the gene transfer using recombinant retrovirus. So far, the risk from such an event has been considered very low in humans (131).

As this and other studies on the development of stem cell therapies show, despite the promising results, important issues need to be addressed for successful application of these therapies in the clinical setting. The availability and expansion of the cells is obviously a major issue. Not only more refined methodologies for the isolation and purification of somatic stem cells are needed but the development of systems for in vitro maintenance, increase in cell number and directed differentiation of stem cells is essential. In the case of ex vivo gene delivery to stem cells, the gene transfer efficiency, the persistence of transgene expression over time and potential immunogenicity and/or oncogenicity of cells undergoing gene transfer are parameters which should be scrutinized extensively. Thus far, typical viral and nonviral vehicles used for gene therapy appear to be suitable for transduction/ transfection of adult stem cells although further investigation is forthcoming. Transgene expression on the other hand is frequently dampened, for example, after retroviral transduction of stem cells. Silencing of a retroviral long-term repeat (LTR) promoter, which can result from methylation of the LTR sequence or positional inactivation can be addressed by substituting LTR elements with tissue-specific promoters/enhancers (132,133). In vivo, silencing may be linked to factors/signals originating in the extracellular milieu where the cells are homing. Not only cell homing but also the requirement for local or systemic expression of the gene product is an important determinant for choosing the route of cell delivery. For systemic expression, cells can be infused in the general circulation whereas examples of delivery for local protein expression include subcutaneous implantation of encapsulated or scaffold-attached cells. Regardless of the way the cells are delivered, thorough assessment of potential immunogenic or tumorgenic properties of stem cells intended for transplantation is required. As the aforementioned study by Hacein-Bey-Abina et al. exemplifies, the risk for insertional mutagenesis associated with retroviral gene delivery should not be overlooked as improbable. Such risk may be eliminated with the use of homologous recombination methodologies already applied for the genetic modification of ES cells. Homologous recombination allows for better control over the site of integration but is hampered by low transfection efficiency, a problem amenable to optimization.

It is obvious that most studies involve animal models whereas data from stem cell therapy trials on human patients are sparser. Animal models certainly have limitations but they provide the best proving arena for potential cell therapies. Regardless of intrinsic differences between animal and human pathophysiology, data gathered from animal studies have tremendous predictive value for clinical translation of stem cell therapeutics in patients. In the clinical context, one envisions that the donor and the recipient of the adult stem cells are the same person and the importance of this is realized considering the minimization of serious immunorejection problems. Furthermore, no study to this day has proved that adult stem cells are tumorogenic as ES cells which form teratomas upon transplantation (3,5). So far, however, the isolation and purification of adult stem cells is difficult and unlike ES cells, most tissue-specific stem cells do not appear to be pluripotent. Finally, the quality and characteristics of somatic stem cells may depend on a number of factors including the age of the donor.

REFERENCES

1. Martin GR. Isolation of a pluripotent cell line from early mouse embryos cultured in medium conditioned by teratocarcinoma stem cells. Proc Natl Acad Sci USA 1981; 78:7634–7638.
2. Thomson JA, Kalishman J, Golos TG, Durning M, Harris CP, Becker RA, Hearn JP. Isolation of a primate embryonic stem cell line. Proc Natl Acad Sci USA 1995; 92:7844–7848.
3. Thomson JA, Itskovitz-Eldor J, Shapiro SS, Waknitz MA, Swiergiel JJ, Marshall VS, Jones JM. Embryonic stem cell lines derived from human blastocysts. Science 1998; 282:1145–1147.
4. Orkin SH. Embryonic stem cells and transgenic mice in the study of hematopoiesis. Int J Dev Biol 1998; 42:927–934.
5. Wobus AM, Holzhausen H, Jakel P, Schoneich J. Characterization of a pluripotent stem cell line derived from a mouse embryo. Exp Cell Res 1984; 152:212–219.
6. Fina L, Molgaard HV, Robertson D, Bradley NJ, Monaghan P, Delia D, Sutherland DR, Baker MA, Greaves MF. Expression of the CD34 gene in vascular endothelial cells. Blood 1990; 75:2417–2426.
7. Baumhueter S, Dybdal N, Kyle C, Lasky LA. Global vascular expression of murine CD34, a sialomucin-like endothelial ligand for L-selectin. Blood 1994; 84:2554–2565.
8. Brown J, Greaves MF, Molgaard HV. The gene encoding the stem cell antigen, CD34, is conserved in mouse and expressed in hematopoietic cell lines, brain and embryonic fibroblasts. Int Immunol 1991; 3:175–184.
9. Ziegler BL, Valtieri M, Porada GA, De Maria R, Muller R, Masella B, Gabbianelli M, Casella I, Pelosi E, Bock T, Zanjani ED, Peschle C. KDR receptor: a key marker defining hematopoietic stem cells. Science 1999; 285:1553–1558.
10. Wuchter C, Ratei R, Spahn G, Schoch C, Harbott J, Schnittger S, Haferlach T, Creutzig U, Sperling C, Karawajew L, Ludwig WD. Impact of CD133 (AC133) and CD90 expression analysis for acute leukemia immunophenotyping. Haematologica 2001; 86:154–161.
11. Bhatia M. AC133 expression in human stem cells. Leukemia 2001; 15:1685–1688.

12. Bhatia M, Wang JC, Kapp U, Bonnet D, Dick JE. Purification of primitive human hematopoietic cells capable of repopulating immune-deficient mice. Proc Natl Acad Sci USA 1997; 94:5320–5325.
13. Thomas TE, Miller CL, Eaves CJ. Purification of hematopoietic stem cells for further biological study. Methods 1999; 17:202–218.
14. Goodell MA, Brose K, Paradis G, Conner AS, Mulligan RC. Isolation and functional properties of murine hematopoietic stem cells that are replicating in vivo. J Exp Med 1996; 183:1797–1806.
15. Tricot G, Gazitt Y, Leemhuis T, Jagannath S, Desikan KR, Siegel D, Fassas A, Tindle S, Nelson J, Juttner C, Tsukamoto A, Hallagan J, Atkinson K, Reading C, Hoffman R, Barlogie B. Collection, tumor contamination, and engraftment kinetics of highly purified hematopoietic progenitor cells to support high dose therapy in multiple myeloma. Blood 1998; 91:4489–4495.
16. McCune JM, Namikawa R, Kaneshima H, Shultz LD, Lieberman M, Weissman IL. The SCID-hu mouse: murine model for the analysis of human hematolymphoid differentiation and function. Science 1988; 241:1632–1639.
17. Lagasse E, Connors H, Al-Dhalimy M, Reitsma M, Dohse M, Osborne L, Wang X, Finegold M, Weissman IL, Grompe M. Purified stem cells can differentiate into hepatocytes in vivo. Nat Med 2000; 6:1229–1234.
18. Grant MB, May WS, Caballero S, Brown GA, Guthrie SM, Mames RN, Byrne BJ, Vaught T, Spoerri PE, Peck AB, Scott EW. Adult hematopoietic stem cells provide functional hemangioblast activity during retinal neovascularization. Nat Med 2002; 8:607–612.
19. Potten CS. Stem cells in gastrointestinal epithelium: numbers, characteristics and death. Philos Trans R Soc Lond B Biol Sci 1998; 353:821–830.
20. Watt F. Epidermal stem cells: markers, patterning and the control of stem cell fate. Philos Trans R Soc Lond B Biol Sci 1997; 353:831–837.
21. Krause DS, Theise ND, Collector MI, Henegariu O, Hwang S, Gardner R, Neutzel S, Sharkis SJ. Multi-organ, multi-lineage engraftment by a single bone marrow-derived stem cell. Cell 2001; 105:369–377.
22. Eglitis MA, Mezey E. Hematopoietic cells differentiate into both microglia and macroglia in the brains of adult mice. Proc Natl Acad Sci USA 1997; 94:4080–4085.
23. Wagers AJ, Sherwood RI, Christensen JL, Weissman IL. Little evidence for developmental plasticity of adult hematopoietic stem cells. Science 2002; 297:2256–2259.
24. Friedenstein AJ, Gorskaja JF, Kalugina NN. Fibroblast precursors in normal and irradiated mouse hematopoietic organs. Exp Hematol 1976; 4:267–274.
25. Pittenger MF, Mackay AM, Beck SC, Jaiswal RK, Douglas R, Mosca JD, Moorman MA, Simonetti DW, Craig S, Marshak DR. Multilineage potential of adult human mesenchymal stem cells. Science 1999; 284:143–147.
26. Johnstone B, Hering TM, Caplan AI, Goldberg VM, Yoo JU. In vitro chondrogenesis of bone marrow-derived mesenchymal progenitor cells. Exp Cell Res 1998; 238:265–272.
27. Pereira RF, Halford KW, O'Hara MD, Leeper DB, Sokolov BP, Pollard MD, Bagasra O, Prockop DJ. Cultured adherent cells from marrow can serve as long-lasting precursor cells for bone, cartilage, and lung in irradiated mice. Proc Natl Acad Sci USA 1995; 92:4857–4861.
28. Conget PA, Minguell JJ. Phenotypical and functional properties of human bone marrow mesenchymal progenitor cells. J Cell Physiol 1999; 181:67–73.
29. Majumdar MK, Thiede MA, Mosca JD, Moorman M, Gerson SL. Phenotypic and functional comparison of cultures of marrow-derived mesenchymal stem cells (MSCs) and stromal cells. J Cell Physiol 1998; 176:57–66.
30. Haynesworth SE, Baber MA, Caplan AI. Cytokine expression by human marrow-derived mesenchymal progenitor cells in vitro: effects of dexamethasone and IL-1 alpha. J Cell Physiol 1996; 166:585–592.

31. Minguell JJ, Erices A, Conget P. Mesenchymal stem cells. Exp Biol Med 2001; 226:507–520.
32. Wakitani S, Saito T, Caplan A. Myogenic cells derived from rat bone marrow mesenchymal stem cells exposed to 5-azacytidine. Muscle Nerve 1995; 18:1417–1426.
33. Makino S, Fukuda K, Miyoshi S, Konishi F, Kodama H, Pan J, Sano M, Takahashi T, Hori S, Abe H, Hata J, Umezawa A, Ogawa S. Cardiomyocytes can be generated from marrow stromal cells in vitro. J Clin Invest 1999; 103:697–705.
34. Woodbury D, Schwarz EJ, Prockop DJ, Black IB. Adult rat and human bone marrow stromal cells differentiate into neurons. J Neurosci Res 2000; 61:364–370.
35. Sandhu JS, Clark BR, Boynton EL, Atkins H, Messner H, Keating A, Hozumi N. Human hematopoiesis in SCID mice implanted with human adult cancellous bone. Blood 1996; 88:1973–1982.
36. Kittler EL, McGrath H, Temeles D, Crittenden RB, Kister VK, Quesenberry PJ. Biologic significance of constitutive and subliminal growth factor production by bone marrow stroma. Blood 1992; 79:3168–3178.
37. Seale P, Sabourin LA, Girgis-Gabardo A, Mansouri A, Grass P, Rudnicki MA. Pax7 is required for the specification of myogenic satellite cells. Cell 2000; 102:777–786.
38. Beauchamp JR, Heslop L, Yu DS, Tajbakhsh S, Kelly RG, Wernig A, Buckingham ME, Partridge TA, Zammit PS. Expression of CD34 and Myf5 defines the majority of quiescent adult skeletal muscle satellite cells. J Cell Biol 2000; 151:1221–1234.
39. Zammit PS, Heslop L, Hudon V, Rosenblatt JD, Tajbakhsh S, Buckingham ME, Beauchamp JR, Partridge TA. Kinetics of myoblast proliferation show that resident satellite cells are competent to fully regenerate skeletal muscle fibers. Exp Cell Res 2002; 281:39–49.
40. Asakura A, Komaki M, Rudnicki M. Muscle satellite cells are multipotential stem cells that exhibit myogenic, osteogenic, and adipogenic differentiation. Differentiation 2001; 68:245–253.
41. Wada MR, Inagawa-Ogashiwa M, Shimizu S, Yasumoto S, Hashimoto N. Generation of different fates from multipotent muscle stem cells. Development 2002; 129:2987–2995.
42. Qu-Petersen Z, Deasy B, Jankowski R, Ikezawa M, Cummins J, Pruchnic R, Mytinger J, Cao B, Gates C, Wernig A, Huard J. Identification of a novel population of muscle stem cells in mice: potential for muscle regeneration. J Cell Biol 2002; 157:851–864.
43. Asakura A, Seale P, Girgis-Gabardo A, Rudnicki MA. Myogenic specification of side population cells in skeletal muscle. J Cell Biol 2002; 159:123–134.
44. Gussoni E, Soneoka Y, Strickland CD, Buzney EA, Khan MK, Flint AF, Kunkel LM, Mulligan RC. Dystrophin expression in the mdx mouse restored by stem cell transplantation. Nature 1999; 401:390–394.
45. Jackson K, Mi T, Goodell MA. Hematopoietic potential of stem cells isolated from murine skeletal muscle. Proc Natl Acad Sci USA 1999; 96:14482–14486.
46. McKinney-Freeman SL, Jackson KA, Camargo FD, Ferrari G, Mavilio F, Goodell MA. Muscle-derived hematopoietic stem cells are hematopoietic in origin. Proc Natl Acad Sci USA 2002; 99:1341–1346.
47. Howell JC, Yoder MC, Srour EF. Hematopoietic potential of murine skeletal muscle-derived CD45(-)Sca-l(+)c-kit(-) cells. Exp Hematol 2002; 30:915–924.
48. Yaswen P, Hayner NT, Fausto N. Isolation of oval cells by centrifugal elutriation and comparison with other cell types purified from normal and preneoplastic livers. Cancer Res 1984; 44:324–331.
49. Alison MR, Golding M, Sarraf CE, Edwards RJ, Lalani EN. Liver damage in the rat induces hepatocyte stem cells from biliary epithelial cells. Gastroenterology 1996; 110:1182–1190.
50. Petersen BE, Goff JP, Greenberger JS, Michalopoulos GK. Hepatic oval cells express the hematopoietic stem cell marker Thy-1 in the rat. Hepatology 1998; 27: 433–445.

51. Baumann U, Crosby HA, Ramani P, Kelly DA, Strain AJ. Expression of the stem cell factor receptor c-kit in normal and diseased pediatric liver: identification of a human hepatic progenitor cell?. Hepatology 1999; 30:112–117.
52. Krakowski ML, Kritzik MR, Jones EM, Krahl T, Lee J, Arnush M, Gu D, Sarvetnick N. Pancreatic expression of keratinocyte growth factor leads to differentiation of islet hepatocytes and proliferation of duct cells. Am J Pathol 1999; 154:683–691.
53. Shen CN, Slack JM, Tosh D. Molecular basis of transdifferentiation of pancreas to liver. Nat Cell Biol 2000; 2:879–887.
54. Kinoshita T, Sekiguchi T, Xu MJ, Ito Y, Kamiya A, Tsuji K, Nakahata T, Miyajima A. Hepatic differentiation induced by oncostatin M attenuates fetal liver hematopoiesis. Proc Natl Acad Sci USA 1999; 96:7265–7270.
55. Reynolds BA, Weiss S. Generation of neurons and astrocytes from isolated cells of the adult mammalian central nervous system. Science 1992; 255:1707–1710.
56. Richards LJ, Kilpatrick TJ, Bartlett PF. De novo generation of neuronal cells from the adult mouse brain. Proc Natl Acad Sci USA 1992; 89:8591–8595.
57. Kirschenbaum B, Nedergaard M, Preuss A, Barami K, Fraser RA, Goldman SA. In vitro neuronal production and differentiation by precursor cells derived from the adult human forebrain. Cereb Cortex 1994; 4:576–589.
58. Mi H, Barres BA. Purification and characterization of astrocyte precursor cells in the developing rat optic nerve. J Neurosci 1999; 19:1049–1061.
59. Rao MS, Noble M, Mayer-Proschel M. Glial restricted precursors are derived from multipotent neuroepithelial stem cells. Proc Natl Acad Sci USA 1998; 95:3996–4001.
60. Raff MC, Miller RH, Noble M. A glial progenitor cell that develops in vitro into an astrocyte of an oligodendrocyte depending on culture medium. Nature 1983; 303:390–396.
61. Kalyani A, Hobson K, Rao MS. Neuroepithelial stem cells from the embryonic spinal cord: isolation, characterization, and clonal analysis. Dev Biol 1997; 186:202–223.
62. Palmer TD, Takahashi J, Gage FH. The adult rat hippocampus contains primordial neural stem cells. Mol Cell Neurosci 1997; 8:389–404.
63. Uchida N, Buck DW, He D, Reitsma MJ, Masek M, Phan TV, Tsukamoto AS, Gage FH, Weissman IL. Direct isolation of human central nervous system stem cells. Proc Natl Acad Sci USA 2000; 97:14720–14725.
64. Jiang YH, Jahagirdar BN, Reinhardt RL, Schwartz RE, Keene CD, Ortiz-Gonzalez XR, Reyes M, Lenvik T, Lund T, Blackstad M, Du JB, Aldrich S, Lisberg A, Low WC, Largaespada DA, Verfaillie CM. Pluripotency of mesenchymal stem cells derived from adult marrow. Nature 2002; 418:41–49.
65. Reyes M, Lund T, Lenvik T, Aguiar D, Koodie L, Verfaillie CM. Purification and ex vivo expansion of postnatal human marrow mesodermal progenitor cells. Blood 2001; 98:2615–2625.
66. Reyes M, Dudek A, Jahagirdar B, Koodie L, Marker PH, Verfaillie CM. Origin of endothelial progenitors in human postnatal bone marrow. J Clin Invest 2002; 109:337–346.
67. Schwartz RE, Reyes M, Koodie L, Jiang Y, Blackstad M, Lund T, Lenvik T, Johnson S, Hu WS, Verfaillie CM. Multipotent adult progenitor cells from bone marrow differentiate into functional hepatocyte-like cells. J Clin Invest 2002; 109:1291–1302.
68. Nichols J, Zevnik B, Anastassiadis K, Niwa H, Klewe-Nebenius D, Chambers I, Scholer H, Smith A. Formation of pluripotent stem cells in the mammalian embryo depends on the POU transcription factor Oct4. Cell 1998; 95:379–391.
69. Kannagi R, Cochran NA, Ishigami F, Hakomori S, Andrews PW, Knowles BB, Solter D. Stage-specific embryonic antigens (SSEA-3 and -4) are epitopes of a unique globo-series ganglioside isolated from human teratocarcinoma cells. EMBO J 1983; 2:2355–2361.
70. Schofield R. The relationship between the spleen colony-forming cell and the haemopoietic stem cell. Blood Cells 1978; 4:7–25.

71. O'Shea KS. Directed differentiation of embryonic stem cells: genetic and epigenetic methods. Wound Repair Regen 2001; 9:443–459.
72. Toma JG, Akhavan M, Fernandes KJL, Barnabe-Heider F, Sadikot A, Kaplan DR, Miller FD. Isolation of multipotent adult stem cells from the dermis of mammalian skin. Nat Cell Biol 2001; 3:778–784.
73. Rohwedel J, Guan K, Wobus AM. Induction of cellular differentiation by retinoic acid in vitro. Cells Tissues Organs 1999; 165:190–202.
74. Rathjen PD, Toth S, Willis A, Heath JK, Smith AG. Differentiation inhibiting activity is produced in matrix-associated and diffusible forms that are generated by alternate promoter usage. Cell 1990; 62:1105–1114.
75. Burdon T, Chambers I, Stracey C, Niwa H, Smith A. Signaling mechanisms regulating self-renewal and differentiation of pluripotent embryonic stem cells. Cells Tissues Organs 1999; 165:131–143.
76. Matsuda T, Nakamura T, Nakao K, Arai T, Katsuki M, Heike T, Yokota T. STAT3 activation is sufficient to maintain an undifferentiated state of mouse embryonic stem cells. EMBO J 1999; 18:4261–4269.
77. Niwa H, Burdon T, Chambers I, Smith A. Self-renewal of pluripotent embryonic stem cells is mediated via activation of STAT3. Genes Dev 1998; 12:2048–2060.
78. Lendahl U, Zimmerman LB, McKay RD. CNS stem cells express a new class of intermediate filament protein. Cell 1990; 60:585–595.
79. Hunziker E, Stein M. Nestin-expressing cells in the pancreatic islets of Langerhans. Biochem Biophys Res Commun 2000; 271:116–119.
80. Kallos MS, Behie LA. Inoculation and growth conditions for high-cell-density expansion of mammalian neural stem cells in suspension bioreactors. Biotechnol Bioeng 1999; 63:473–483.
81. Sen A, Kallos MS, Behie LA. Passaging protocols for mammalian neural stem cells in suspension bioreactors. Biotechnol Prog 2002; 18:337–345.
82. DeRisi J, Penland L, Brown PO, Bittner ML, Meltzer PS, Ray M, Chen Y, Su YA, Trent JM. Use of a cDNA microarray to analyse gene expression patterns in human cancer. Nat Genet 1996; 14:457–460.
83. Forozan F, Karhu R, Kononen J, Kallioniemi A, Kallioniemi OP. Genome screening by comparative genomic hybridization. Trends Genet 1997; 13:405–409.
84. Knuutila S, Bjorkqvist AM, Autio K, Tarkkanen M, Wolf M, Monni O, Szymanska J, Larramendy ML, Tapper J, Pere H, El-Rifai W, Hemmer S, Wasenius VM, Vidgren V, Zhu Y. DNA copy number amplifications in human neoplasms: review of comparative genomic hybridization studies. Am J Pathol 1998; 152:1107–1123.
85. Wessendorf S, Fritz B, Wrobel G, Nessling M, Lampel S, Goettel D, Kuepper M, Joos S, Hopman T, Kokocinski F, Dohner H, Bentz M, Schwaenen C, Lichter P. Automated screening for genomic imbalances using matrix-based comparative genomic hybridization. Lab Invest 2002; 82:47–60.
86. Zandstra PW, Nagy A. Stem cell bioengineering. Annu Rev Biomed Eng 2001; 3:275–305.
87. Zandstra PW, Eaves CJ, Piret JM. Expansion of hematopoietic progenitor cell populations in stirred suspension bioreactors of normal human bone marrow cells. Biotechnology 1994; 12:904–914.
88. Sandstrom CE, Collins PC, McAdams TA, Bender JG, Papoutsakis ET, Miller WM. Comparison of whole serum-deprived media for ex vivo expansion of hematopoietic progenitor cells from cord blood and mobilized peripheral blood mononuclear cells. J Hematother 1996; 5:461–473.
89. Petzer AL, Zandstra PW, Piret JM, Eaves CJ. Differential cytokine effects on primitive ($CD34^+CD38^-$) human hematopoietic cells: novel responses to Flt3-ligand and thrombopoietin. J Exp Med 1996; 183:2551–2558.

90. Zandstra PW, Petzer AL, Eaves CJ, Piret JM. Cellular determinants affecting the rate of cytokine depletion in cultures of human hematopoietic cells. Biotechnol Bioeng 1997; 54:58–66.
91. Collins PC, Miller WM, Papoutsakis ET. Stirred culture of peripheral and cord blood hematopoietic cells offers advantages over traditional static systems for clinically relevant applications. Biotechnol Bioeng 1998; 59:534–543.
92. Collins PC, Nielsen LK, Patel SD, Papoutsakis ET, Miller WM. Characterization of hematopoietic cell expansion, oxygen uptake, and glycolysis in a controlled, stirred-tank bioreactor system. Biotechnol Prog 1998; 14:466–472.
93. Collins PC, Nielsen LK, Wong CK, Papoutsakis ET, Miller WM. Real-time method for determining the colony-forming cell content of human hematopoietic cell cultures. Biotechnol Bioeng 1997; 55:693–700.
94. Sen A, Kallos MS, Behie LA. Expansion of mammalian neural stem cells in bioreactors: effect of power input and medium viscosity. Brain Res Dev Brain Res 2002; 134: 103–113.
95. Palsson BO, Paek SH, Schwartz RM, Palsson M, Lee GM, Silver S, Emerson SG. Expansion of human bone marrow progenitor cells in a high cell density continuous perfusion system. Biotechnology 1993; 11:368–372.
96. Oh DJ, Koller MR, Palsson BO. Frequent harvesting from perfused bone marrow cultures results in increased overall cell and progenitor expansion. Biotechnol Bioeng 1994; 44:609–616.
97. Koller MR, Manchel I, Newsom BS, Palsson MA, Palsson BO. Bioreactor expansion of human bone marrow: comparison of unprocessed, density-separated, and CD34-enriched cells. J Hematother 1995; 4:159–169.
98. Koller MR, Bradley MS, Palsson BO. Growth factor consumption and production in perfusion cultures of human bone marrow correlate with specific cell production. Exp Hematol 1995; 23:1275–1283.
99. Horner M, Miller WM, Ottino JM, Papoutsakis ET. Transport in a grooved perfusion flat-bed bioreactor for cell therapy applications. Biotechnol Prog 1998; 14:689–698.
100. Peng CA, Palsson BO. Cell growth and differentiation on feeder layers is predicted to be influenced by bioreactor geometry. Biotechnol Bioeng 1996; 50:479–492.
101. Chabannon C, Blache JL, Sielleur I, Douville J, Faucher C, Gravis G, Arnoulet C, Oziel-Taieb S, Blaise D, Novakovitch G, Camerlo J, Chabbert I, Genre D, Appel M, Armstrong D, Maraninchi D, Viens P. Production of ex vivo expanded hematopoietic cells and progenitors in a closed bioreactor, starting with a small volume marrow collection: a feasibility study in patients with poor-risk breast cancer and receiving high-doses of cyclophosphamide. Int J Oncol 1999; 15:511–518.
102. Sardonini CA, Wu YJ. Expansion and differentiation of human hematopoietic cells from static cultures through small-scale bioreactors. Biotechnol Prog 1993; 9:131–137.
103. Highfill JG, Haley SD, Kompala DS. Large-scale production of murine bone marrow cells in an airlift packed bed bioreactor. Biotechnol Bioeng 1996; 50:514–520.
104. Meissner P, Schröder B, Herfurth C, Biselli M. Development of a fixed bed reactor for the expansion of human hematopoietic progenitor cells. Cytotechnology 1999; 30:227–234.
105. Wang TY, Brennan JK, Wu JH. Multilineal hematopoiesis in a three-dimensional murine long-term bone marrow culture. Exp Hematol 1995; 23:26–32.
106. Mantalaris A, Keng P, Bourne P, Chang AY, Wu JH. Engineering a human bone marrow model: a case study on ex vivo erythropoiesis. Biotechnol Prog 1998; 14:126–133.
107. Terada N, Hamazaki T, Oka M, Hoki M, Mastalerz DM, Nakano Y, Meyer EM, Morel L, Petersen BE, Scott EW. Bone marrow cells adopt the phenotype of other cells by spontaneous cell fusion. Nature 2002; 416:542–545.
108. Ying QL, Nichols J, Evans EP, Smith AG. Changing potency by spontaneous fusion. Nature 2002; 416:545–548.

109. Weissman IL, Anderson DJ, Gage F. Stem and progenitor cells: origins, phenotypes, lineage commitments and transdifferentiations. Annu Rev Cell Dev Biol 2001; 17:387–403.
110. Lazarus HM, Haynesworth SE, Gerson SL, Rosenthal NS, Caplan AI. Ex vivo expansion and subsequent infusion of human bone marrow-derived stromal progenitor cells (mesenchymal progenitor cells)—implications for therapeutic use. Bone Marrow Transplant 1995; 16:557–564.
111. Koc ON, Gerson SL, Cooper BW, Dyhouse SM, Haynesworth SE, Caplan AI, Lazarus HM. Rapid hematopoietic recovery after coinfusion of autologous-blood stem cells and culture-expanded marrow mesenchymal stem cells in advanced breast cancer patients receiving high-dose chemotherapy. J Clin Oncol 2000; 18:307–316.
112. Pereira RF, Ohara MD, Laptev AV, Halford KW, Pollard MD, Class R, Simon D, Livezey K, Prockop DJ. Marrow stromal cells as a source of progenitor cells for nonhematopoietic tissues in transgenic mice with a phenotype of osteogenesis imperfecta. Proc Natl Acad Sci USA 1998; 95:1142–1147.
113. Horwitz EM, Prockop DJ, Fitzpatrick LA, Koo WWK, Gordon PL, Neel M, Sussman M, Orchard P, Marx JC, Pyeritz RE, Brenner MK. Transplantability and therapeutic effects of bone marrow-derived mesenchymal cells in children with osteogenesis imperfecta. Nat Med 1999; 5:309–313.
114. Kordower JH, Freeman TB, Snow BJ, Vingerhoets FJG, Mufson EJ, Sanberg PR, Hauser RA, Smith DA, Nauert GM, Perl DP, Olanow CW. Neuropathological evidence of graft survival and striatal reinnervation after the transplantation of fetal mesencephalic tissue in a patient with Parkinson's disease. N Engl J Med 1998; 332:1118–1124.
115. Philpott LM, Kopyov OV, Lee AJ, Jacques S, Duma CM, Caine S, Yang M, Eagle KS. Neuropsychological functioning following fetal striatal transplantation in Huntington's chorea: three case presentations. Cell Transplant 1997; 6:203–212.
116. Pittenger M, Vanguri P, Simonetti D, Young R. Adult mesenchymal stem cells: potential for muscle and tendon regeneration and use in gene therapy. J Musculoskelet Neuronal Interact 2002; 2:309–320.
117. Toma C, Pittenger MF, Cahill KS, Byrne BJ, Kessler PD. Human mesenchymal stem cells differentiate to a cardiomyocyte phenotype in the adult murine heart. Circulation 2002; 105:93–98.
118. Floyd SS, Clemens PR, Ontell MR, Kochanek S, Day CS, Yang J, Hauschka SD, Balkir L, Morgan J, Moreland MS, Feero GW, Epperly M, Huard J. Ex vivo gene transfer using adenovirus-mediated full-length dystrophin delivery to dystrophic muscles. Gene Therapy 1998; 5:19–30.
119. Young BH, Peng H, Huard J. Muscle-based gene therapy and tissue engineering to improve bone healing. Clin Orthopaed Relat Res 2002; 403:S243–S251.
120. Dhawan J, Pan LC, Pavlath GK, Travis MA, Lanctot AM, Blau HM. Systemic delivery of human growth hormone by injection of genetically engineered myoblasts. Science 1991; 254:1509–1512.
121. Hamamori Y, Small B, Tian J, Kedes L. Myoblast transfer of human erythropoietin gene in a mouse model of renal failure. J Clin Invest 1995; 95:1808–1813.
122. Wang JM, Zheng H, Blaivas M, Kurachi K. Persistent systemic production of human factor IX in mice by skeletal myoblast-mediated gene transfer: feasibility of repeat application to obtain therapeutic levels. Blood 1997; 90:1075–1082.
123. Lee K, Majumdar MK, Buyaner D, Hendricks JK, Pittenger MF, Mosca JD. Human mesenchymal stem cells maintain transgene expression during expansion and differentiation. Mol Ther 2001; 3:857–866.
124. Dellambra E, Pellegrini G, Guerra L, Ferrari G, Zambruno G, Mavilio F, De Luca M. Toward epidermal stem cell-mediated ex vivo gene therapy of junctional epidermolysis bullosa. Hum Gene Ther 2000; 11:2283–2287.

125. Andsberg G, Kokaia Z, Bjorklund A, Lindvall O, Martinez-Serrano A. Amelioration of ischaemia-induced neuronal death in the rat striatum by NGF-secreting neural stem cells. Eur J Neurosci 1998; 10:2026–2036.
126. Blaese RM, Culver KW, Miller AD, Carter CS, Fleisher T, Clerici M, Shearer G, Chang L, Chiang Y, Tolstoshev P, Greenblatt JJ, Rosenberg SA, Klein H, Berber M, Mullen CA, Ramsey WJ, Muul L, Morgan RA, Anderson WF. T lymphocyte-directed therapy for ADA-SCID: initial trial results after 4 years. Science 1995; 270:475–480.
127. Dunbar CE, Kohn DB, Schiffmann R, Barton NW, Nolta JA, Esplin JA, Pensiero M, Long Z, Lockey C, Emmons RV, Csik S, Leitman S, Krebs CB, Carter C, Brady RO, Karlsson S. Retroviral transfer of the glucocerebrosidase gene into CD34+ cells from patients with Gaucher disease: in vivo detection of transduced cells without myeloablation. Hum Gene Ther 1998; 9:2629–2640.
128. Hacein-Bey-Abina S, Le Deist F, Carlier F, Bouneaud C, Hue C, De Villartay JP, Thrasher AJ, Wulffraat N, Sorensen R, Dupuis-Girod S, Fischer A, Davies EG, Kuis W, Leiva L, Cavazzana-Calvo M. Sustained correction of X-linked severe combined immunodeficiency by ex vivo gene therapy. N Engl J Med 2002; 346:1185–1193.
129. Hacein-Bey-Abina S, von Kalle C, Schmidt M, Le Deist F, Wulffraat N, Mcntyre E, Radford I, Villeval JL, Fraser CC, Cavazzana-Calvo M, Fischer A. A serious adverse event after successful gene therapy for X-linked severe combined immunodeficiency. N Engl J Med 2003; 348:255–256.
130. Rabbitts TH, Bucher K, Chung G, Grutz G, Warren A, Yamada Y. The effect of chromosomal translocations in acute leukemias: the LMO2 paradigm in transcription and development. Cancer Res 1999; 59:1794s–1798s.
131. Stocking C, Bergholz U, Friel J, Klingler K, Wagener T, Starke C, Kitamura T, Miyajima A, Ostertag W. Distinct classes of factor-independent mutants can be isolated after retroviral mutagenesis of a human myeloid stem cell line. Growth Factors 1993; 8:197–209.
132. Lund AH, Duch M, Pederson FS. Transcriptional silencing of retroviral vectors. J Biomed Sci 1996; 3:365–378.
133. Miller N, Whelan J. Progress in transcriptionally targeted and regulatable vectors for genetic therapy. Hum Gene Ther 1997; 8:803–815.
134. Kondo T, Raff M. Oligodendrocyte precursor cells reprogrammed to become multipotential CNS stem cells. Science 2000; 289:1754–1757.
135. Sanchez-Ramos J, Song S, Cardozo-Pelaez F, Hazzi C, Stedeford T, Willing A, Freeman TB, Saporta S, Janssen W, Patel N, Cooper DR, Sanberg PR. Adult bone marrow stromal cells differentiate into neural cells in vitro. Exp Neurol 2000; 164:247–256.
136. Condorelli G, Borello U, De Angelis L, Latronico M, Sirabella D, Coletta M, Galli R, Balconi G, Follenzi A, Frati G, De Angelis MGC, Gioglio L, Amuchastegui S, Adorini L, Naldini L, Vescovi A, Dejana E, Cossu G. Cardiomyocytes induce endothelial cells to trans-differentiate into cardiac muscle: implications for myocardium regeneration. Proc Natl Acad Sci USA 2001; 98:10733–10738.
137. Tomita S, Li RK, Weisel RD, Mickle DAG, Kim EJ, Sakai T, Jia ZQ. Autologous transplantation of bone marrow cells improves damaged heart function. Circulation 1999; 100:247–256.
138. Galli R, Borello U, Gritti A, Minasi AG, Bjornson C, Coletta M, Mora M, De Angelis MGC, Fiocco R, Cossu G, Vescovi AL. Skeletal myogenic potential of human and mouse neural stem cells. Nat Neurosci 2000; 3:986–991.
139. Zuk PA, Zhu M, Mizuno H, Huang J, Futrell JW, Katz AJ, Benhaim P, Lorenz HP, Hedrick MH. Multilineage cells from human adipose tissue: implications for cell-based therapies. Tissue Eng 2001; 7:211–228.
140. Erices A, Conget P, Minguell JJ. Mesenchymal progenitor cells in human umbilical cord blood. Br J Haematol 2000; 109:235–242.
141. Gronthos S, Graves S, Ohta S, Simmons P. The STRO-1+ fraction of adult human bone marrow contains the osteogenic precursors. Blood 1994; 84:4164–4173.

142. Asakura A, Rudnicki MA. Side population cells from diverse adult tissues are capable of in vitro hematopoietic differentiation. Exp Hematol 2002; 30:1339–1345.
143. Berardi AC, Wang A, Levine JD, Lopez P, Scadden DT. Functional isolation and characterization of human hematopoietic stem cells. Science 1995; 267:104–108.
144. Scholer HR, Hatzopoulos AK, Balling R, Suzuki N, Gruss P. A family of octamer-specific proteins present during mouse embryogenesis: evidence for germline-specific expression of an Oct factor. EMBO J 1989; 8:2543–2550.
145. Ben-Shushan E, Thompson JR, Gudas LJ, Bergman Y. Rex-1, a gene encoding a transcription factor expressed in the early embryo, is regulated via Oct-3/4 and Oct-6 binding to an octamer site and a novel protein, Rox-1, binding to an adjacent site. Mol Cell Biol 1998; 18:1866–1878.
146. Ortega G, Korty PE, Shevach EM, Malek TR. Role of Ly-6 in lymphocyte activation. I. Characterization of a monoclonal antibody to a nonpolymorphic Ly-6 specificity. J Immunol 1986; 137:3240–3246.
147. Gunter KC, Malek TR, Shevach EM. T cell-activating properties of an anti-Thy-1 monoclonal antibody. Possible analogy to OKT3/Leu-4. J Exp Med 1984; 159:716–730.

20

Ex Vivo Culture of Hematopoietic and Mesenchymal Stem Cells for Tissue Engineering and Cell-Based Therapies

A. Mantalaris
Department of Chemical Engineering and Chemical Technology, Imperial College of Science, Technology, and Medicine, London, U.K.

J. H. D. Wu
Department of Chemical Engineering, Biomedical Engineering, and Microbiology and Immunology, University of Rochester, Rochester, New York, U.S.A.

INTRODUCTION

Hematopoietic stem cell (HSC) cultures have generated intense clinical, commercial, and academic interest due to their potential applications in cell-based therapies, including genetic, cellular and tissue engineering, transfusion support, and immunotherapy. In addition, HSC cultures have proved useful for leukemia and cancer cell purging and serving as models for drug efficacy, toxicity, and pathogenicity testing, as well as a tool for discovering and elucidating the roles of the many factors modulating hematopoietis. However, there still exist numerous problems to overcome and new methodologies to develop in order to achieve the consistent, controlled, and reproducible growth of clinically relevant numbers of HSCs and their derivatives so that this technology can expand its, for now, limited applicability. This chapter will focus on the essential bioprocess parameters involved in controlling the complex kinetics encountered in the ex vivo hematopoietic cell cultures, including cell source, the desired characteristics of the cultured cells, and culture conditions, including media, scaffolds, and bioreactors, for the development of clinically desirable "designer transplants" (1).

HEMATOPOIESIS AND BONE MARROW

Bone marrow (BM), located in the medullary cavity of bone, is the hematopoietic tissue and a primary lymphoid organ. In normal adults, it produces about 2.5 billion red blood cells (RBC), 2.5 billion platelets, and 1 billion granulocytes per kilogram of body weight per day. Furthermore, the rate of production is adjusted to actual

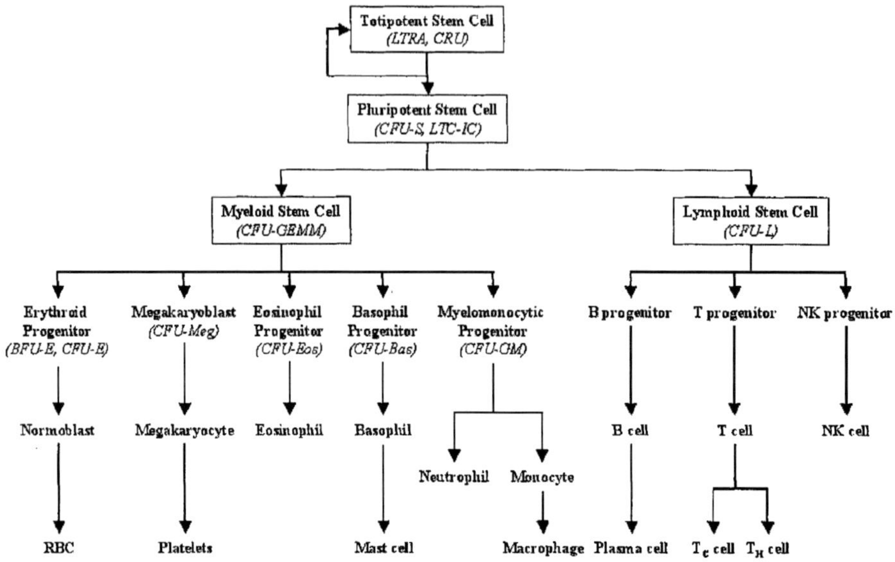

Figure 1 The hematopoietic system hierarchy, including stem and progenitor cell assays. Hematopoiesis is a three-stage process, which includes proliferation, commitment, and differentiation.

needs and can vary from nearly zero to many times normal (2). Hematopoiesis is a three-stage continual process that is the result of proliferation, commitment, and differentiation of the hematopoietic cells. All blood cells are derived from a common precursor, termed the HSC (Fig. 1). Self-renewal of the HSC ensures maintenance of blood formation throughout life. Commitment of the HSC to differentiation is the first step in a series of cell divisions through which specialization and differentiation along a single lineage occur resulting in the production of mature blood cells. Within these three stages, extensive expansion and maturation occurs, and a single HSC has been proposed to be capable of more than 50 cell divisions to generate up to 10^{15} cells (3).

Bone marrow is a complex tissue with an elaborate architecture that contains, besides the hematopoietic cells (<30% of the cellular elements within normal BM), the supporting stroma consisting of reticular cells, osteocytes, adipocytes, vascular endothelium, and the extracellular matrix (ECM) (4). The concept of the hematopoietic inductive microenvironment (HIM) was introduced in the 1960s to explain the restriction of adult hematopoiesis to the BM (5). The HIM (Fig. 2) consists of the stromal cells, which through their intimate physical contact with the hematopoietic cells, the ECM, and the growth factors they secrete, create this unique three-dimensional structural and chemical microenvironment that regulates, through positive and negative factors, the survival, proliferation, and differentiation of the HSCs.

A model for hematopoietic regulation has been proposed in which specific microenvironmental niches determine the fate of HSCs (6). According to the model, rare niches within the HIM specifically promote the self-renewal of HSCs. Space within each niche acts as the limiting factor for stem cell renewal. When sufficient numbers of self-renewing divisions have occurred and the niche is filled with stem cells, the excess cells are pushed into a neighboring niche, which fosters the production of differentiated progeny by these cells. Eventually, mature blood cells exit the marrow space and enter the blood circulatory system. The existence of niches

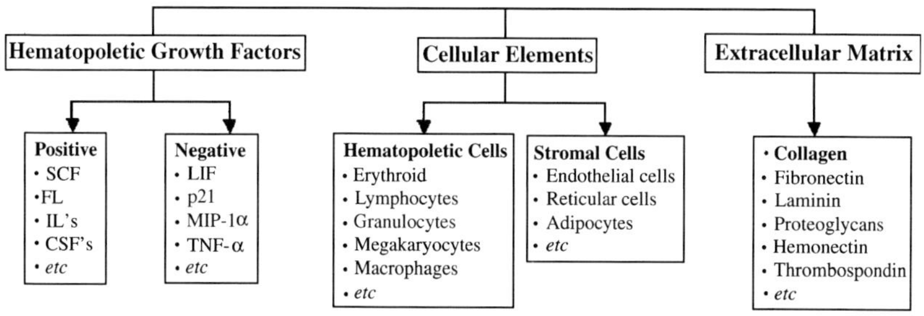

Figure 2 The components of the hematopoietic inductive microenvironment.

as distinct microenvironments that possess specialized functions to control the selective regulation of cell populations would probably be due to the presence and/or absence of particular positive or negative factors, as well as concentration gradients of membrane-bound cytokines that are thought to be important for instructing HSCs to differentiate or remain quiescent (7,8). Unfortunately, much of the lack of understanding of these molecules is due to the considerable difficulty in recapitulating these specialized microenvironments in vitro, although several groups are now working toward this direction recently (9).

HEMATOPOIETIC STEM/PROGENITOR CELL ASSAYS

The biggest impediment to the successful development of methods for the ex vivo expansion of HSCs has been the lack of accurate methods to enumerate them (Table 1). An HSC is defined as a cell with self-renewal and proliferative potential, coupled with the capacity to differentiate into the progenitors of all blood cell lineages. The "gold" standard for assaying the HSC is, therefore, the ability of a single cell to reconstitute long-term hematopoiesis in the recipient host (10).

Since the long-term repopulation assay (LTRA) is both time-consuming and costly, various alternative in vitro and in vivo assays have been developed to identify stem cells and to learn more about their properties. Phenotypic characterization using flow cytometry is useful not only for the initial isolation of cells that are enriched for HSCs, but also because they have the potential to allow the rapid, relatively inexpensive determination of culture status (1). It is widely accepted that the cell-surface glycoprotein CD34 is expressed on almost all human cells with discernable hematopoietic potential (11). In more recent years, additional markers have been used to fractionate the $CD34^+$ cells further and to identify more "primitive" stem cells (12). In particular, cells expressing $CD34^+/CD38^-$ or $CD34^+/Thy\text{-}1^+$ phenotype are highly enriched for very primitive hematopoietic cells (13,14). However, the identification of $CD34^-$ cells with significant hematopoietic potential has also been reported (15). On the contrary, the murine HSCs are identified by cells lacking the markers of the mature blood cells (lineage minus or Lin^-) and the expression of Sca-1 and c-kit surface markers (16). Recently, the combination of these cell-surface markers with fluorescent dyes that indicate cell proliferative status has become a method of choice for isolating HSCs. Hoechst-33342 staining, followed by flow cytometric analysis, has been used to define a small subset of murine and human BM cells (side population, SP) , which constitute approximately 0.1% of the whole BM and are highly enriched for long-term reconstituting HSCs (17),

Table 1 Hematopoietic Stem (HSC)/Progenitor Cell (HPC) Assays

Assay	Species	Target cell population/comments
Phenotypic markers (flow cytometry)		
$CD34^+$, Thy-1^+, Lin^-, c-kit^{lo}, $CD38^-$	Human	Subset of HSCs; questionable dependability of phenotypic markers as a predictor of biological activity
Lin^-, Sca-l^{hi}, Thy-l.l^{lo}, c-kit^+	Murine	Same as above
Side population defined by Hoechst-33342 staining	Human/ murine	Enriched HSCs
In vitro (long-termculture)		
Colony-forming assay (CFU, BFU, HPP-CFC)	Human/ murine	HPC
LTC-IC	Human/ murine	Primitive HPCs; do not enumerate HSCs
In vivo (animal models)		
CFU-S	Murine	Primitive HPC and subset of HSCs (depending on the assay endpoint) forming colonies on the host spleen upon transplantation; exhibition of high variable self-renewal and differentiation of even the most phenotypically homogeneous CFU-S populations
Long-term repopulation assay (syngeneic/ xenogeneic transplantation of HSC into SCID, NOD-SCID mice, or sheep, etc.)	Human/ murine	HSC; gold standard of HSC assay; xenogeneic microenvironment may produce a less conducive environment for homing, growth, and differentiation of human HSCs

whereas the mitochondrial dye rhodamine-123 has been used to further subdivide primitive stem cells (18).

The advantages of flow cytometric analysis of the stem cell phenotypic markers are speed and convenience. Unfortunately, the relationship between the HSC phenotype and their function or genotype is based on uncultured cells and this relationship rapidly dissociates in in vitro culture environments (19); making the dependability of these phenotypic markers as a predictor of biological potential doubtful (12).

In vitro functional assays have been developed to allow specific lineage-restricted and multi-, bi-, and mono-potent subpopulations of hematopoietic progenitor cells to be detected and quantified. These methods require the provision of appropriate soluble, or cell, or matrix-bound factors essential to the support and stimulation of the progenitors being assessed. The colony-forming unit (CFU), burst-forming unit (BFU), and high-proliferative potential colony-forming cell (HPP-CFC) assays enumerate committed hematopoietic progenitor cells, which form colonies of differentiated cells on semisolid culture media (20,21). These progenitors are therefore operationally defined as CFUs with the prefix indicating their lineage commitment. For example, CFU granulocyte/macrophage (CFU-GM) is a progenitor committed to the granulocytic and macrophage lineages. Although some progenitors with multilineage differentiation potential can be detected, these progenitors display only limited self-renewal capacity in vitro.

An in vitro assay that has shown greater specificity for quantifying hematopoietic cells that appear to be more primitive than most CFCs is the long-term culture-initiating cell (LTC-IC) assay in which cells are co-cultured with hematopoietic supportive stromal feeders, with or without exogenous cytokines (13). The LTC-IC assay fulfills the HSC criterion of being able to initiate and sustain myelopoiesis, but not the two other criteria of self-renewal and long-term lymphopoiesis (22).

The colony-forming unit-spleen (CFU-S) assay is an in vivo clonal assay in which BM or spleen cells are transplanted into lethally irradiated recipients whose spleens are harvested and analyzed after 8 (CFU-S_8) and 12 (CFU-S_{12}) days (23). The colonies of hematopoietic cells appearing on the spleen surface represent different populations of cells with CFU-S_8 cells being predominantly unipotential, whereas CFU-S_{12} cells being multipotential. While transplanted HSCs do not generate CFU-S_8, at least a subset of HSCs produce CFU-S_{12}.

The most stringent assay for measuring HSC activity is the long-term repopulating assay because it requires donor cells to fulfill the criteria that define HSCs—the abilities to self-renew and differentiate and to reconstitute hematopoiesis in the transplant host. In the murine system, typically syngeneic mice are used as donors and hosts to avoid rejection. For human, animals are used as hosts for obvious reasons. In these xenogeneic transplant systems, human HSCs are transplanted in immune-incompetent animals, such as severe combined immunodeficiency (SCID) mice (24), nonobese diabetic (NOD)-SCID mice (25), or preimmune fetal lambs (26), allowing a demonstration of not only multilineal differentiation, but also self-renewal and repopulating ability. However, as human HSCs have to repopulate a xenogeneic microenvironment that may be less conducive to HSC homing, growth and differentiation than the human microenvironment, it remains to be proved that these assays enumerate all human HSCs (22). Studies utilizing these models demonstrated that the frequency of in vivo repopulating HSCs within a phenotypically defined population was significantly lower (10–100-fold) than previously estimated by in vitro (CFU and LTC-IC) assays (27). Furthermore, the idea of "phenotypic fidelity" has been challenged by groups that demonstrated that $CD34^+/CD38^-$ cells are not identical before and after ex vivo culture (12).

Hematopoietic Cell Source

The two major issues from a clinical and bioprocess aspect for the ex vivo expansion are the source of the hematopoietic cells and the extent of purification. Several sources of hematopoietic stem and progenitor cells are used for clinical therapies, including BM, peripheral blood, and umbilical cord from the patient or from human leukocyte antigen (HLA)-compatible donors. There are significant differences in the quantity and quality (in terms of their proliferative and engraftment abilities and responses to hematopoietic growth factors) of the hematopoietic sample collected from these tissues, which also serve as the starting material for ex vivo expansion, although there is currently no clear consensus as to the best starting population (28).

Bone marrow remains the best-established source of stem and progenitor cells for HSC transplantation. A total of 0.4–1 L of fluid containing BM and blood cells can be collected under spinal or general anesthesia from the iliac crest (occasionally from the sternum). The typical $CD34^+$ cell content in BM is 1.7% (1–5% of the $CD34^+$ population is $CD38^-$ and 25% is Thy-1^+) (28).

Stem and progenitor cells have the capacity to migrate from the BM to the peripheral circulation and vice versa. Therefore, stem and progenitor cells from PB, collected through aphaeresis—a process that separates nucleated cells from RBC and platelets—are used for transplantation. However, the primitive cells that circulate within the PB are 2% of those in the BM. Specifically, the $CD34^+$ content in the PB is 0.15% (<3.3% of the $CD34^+$ population is $CD38^-$) (28). Initially, PB transplants were used as an alternative to BM, when patients had undergone irradiation to the pelvic area or when the BM had extensive cancer-cell contamination. Through the discovery that stem and progenitor cell content increases following cytokine-aided postchemotherapy recovery, the practice of "mobilizing" PB (MPB) stem and progenitor cells by administering growth factors such as GM-CSF has become popular. GM-CSF increases the $CD34^+$ content in MPB to 0.6–2% (5% of the $CD34^+$ population is $CD38^-$ and 8% is Thy-1^+) (28). As a result of this practice, fewer aphaeresis sessions are required (1–3 as opposed to 7–10) to collect sufficient PB for transplantation.

Over the last 10 years, umbilical cord blood (CB) has clinically been investigated as an alternative source of HSCs for allogeneic transplantation for patients lacking HLA-matched marrow donor (29). The ease of collection and potential availability to groups underrepresented in transplant registries are advantages of CB compared to BM. In addition, it has been suggested that CB contains fewer T cells and/or more naïve T cells than BM and may permit a greater degree of mismatch with less graft vs. host disease (GVHD) (30). Furthermore, CB has lower levels (compared to BM and MPB) of contamination with common viral pathogens (28). Although, CB is rich in stem and progenitor cells (the $CD34^+$ content in CB is 0.8%; 2–25% of the $CD34^+$ population is $CD38^-$ and 25% is Thy-1^+), the total quantity of stem cells due to the small volume collected ($\sim$100 mL) is low compared to the numbers of BM or MPB HSCs required for adult engraftment. Hence the vast majority of CB recipients have been children with an average weight of 20 kg. The progression-free survival rates reported thus far are comparable to the results achieved with allogeneic BM transplantation with a suggestion of decreased GVHD, although the time to neutrophil and platelet engraftment in CB recipients has been delayed compared to BM (31). However, the future clinical potential for CB HSCs is enormous if methods for successfully expanding the HSCs ex vivo without compromising their engraftment ability could be developed, especially since CB is more responsive to growth factors and is capable of greater expansion than the BM or MPB HSCs.

It is unclear what is the best starting population, in terms of the extent of cell purification, for inoculating ex vivo expansion cultures. Most clinical trials have used only CD34-enriched populations and there are advantages and disadvantages to this approach. In terms of clinical-scale ex vivo expansion and volumetric culture requirements, a bioreactor will support the same number of total cells and CFU-GMs and will have the same volumetric culture requirements, regardless of the

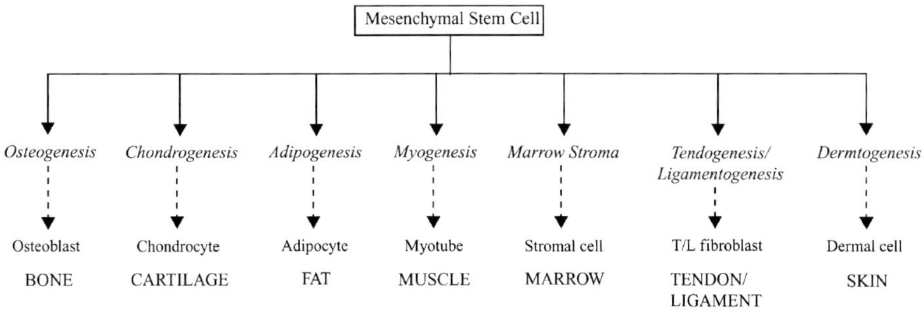

Figure 3 Mesengenic differentiation pathway. The differentiation process involves stepwise cellular transitions from the MSC to highly differentiated end-stage phenotypes. [From Refs. (37,39).]

extent of purification of the input cells (32,33). It has been shown that transplantation of BM mononuclear cells (BMMNCs) support rapid neutrophil and platelet engraftment and long-term stem cell engraftment, presumably because of the presence of accessory cells such as stromal cells in the MNCs (34,35). However, despite the 10-fold expansion of CFU-GM observed in this study, the LTC-IC and $CD34^+/Lin^-$ cells were reduced to half or less of the input amount (34,35).

Bone marrow stroma, as has been discussed earlier, provides the structural and instructional support for hematopoiesis. Considerable progress has been made in the identification and characterization of precursors for the stromal tissue of BM, although this very small (0.01–0.0001% of the total nucleated cells) and heterogeneous population still lacks conclusive morphologic and molecular characterization (36,37). Nevertheless, evidence for a developmental hierarchy of stromal cells within the BM is strong, pointing to the existence of multipotent mesenchymal stem cells, capable of forming a spectrum of specialized mesenchymal and a variety of connective tissues (Fig. 3) (37–39). In vivo analysis of MSCs has been augmented by the development of in vitro functional assays (CFU-F), which demonstrate the ability of clonogenic progenitor stromal cells to form a variety of differentiated cell types (40,41). Despite the fact that antibodies to several cell surface antigens that recognize MSCs exist, there is a lack of monospecific and unique molecular probes to identify these cells unequivocally in situ. The monoclonal antibody STRO-1, which identifies an as yet uncharacterized cell-surface antigen expressed by a small, heterogeneous population of adult BMMNC, appears to be a suitable candidate (42). Further research is clearly required before we can fully exploit the biological potential of MSCs for therapeutic purposes. However a number of investigators are already using these cells in a range of therapeutic strategies, including the replacement of marrow stromal tissue damaged by chemotherapy as part of cancer treatment (43), as vesicles for gene therapy in the treatment of mesenchymal tissues (44), and for the ex vivo expansion for autologous repair of bone and cartilage defects as well as osteoporosis (37,45).

Hematopoietic Culture Parameters

The normal BM in vivo state is highly dynamic, including the state of the cell populations present and the physical rate processes of supply and removal of nutrients, growth factors, and metabolites. In addition, the local geometry of the microenvironment may not be static (46). The establishment of optimal culture parameters for the ex vivo expansion of stem and progenitor cells is therefore a major challenge, especially since most studies ignore the complex kinetics, transient nature, intricate interactions between parameters, and the lack of invariant measures (47).

Inoculum density is a well-known parameter in cell culture research. In particular, in BM cultures, it is not only the total number of cells inoculated that is important, but also the composition of stem, progenitor, and accessory cells. This is because the starting population varies widely in terms of degree of differentiation. Even cell populations that have been purified, as in the case of primitive erythroid cells by three-color cell sorting, vary in expansion potential (1000-fold in this case) (48). Furthermore, most primitive cells will be in quiescence at the onset of the culture (49). It is important that individual laboratories determine the initial inocula concentrations that generate optimal cell-outputs of the desired cells from a given starting population under the specific experimental conditions. Large inocula of BM-derived cells in traditional static cultures fed periodically by partially replacing the culture medium can be limited by nutrient and growth factor depletion and exposure to certain metabolites (50). Increased medium exchange results in increased total and progenitor cell numbers (32,51), increased specific glucose consumption and lactate production (52), as well as dilution of metabolites and inhibitory factors. In the absence of feeding, studies have shown that $CD34^+$ selection is required for optimal expansion (53), however when cultures are frequently fed, unpurified cell samples perform as well or better than $CD34^+$ samples (47). In addition, the initial concentration of $CD34^+$ cells can affect the expansion of all hematopoietic compartments. Hence, reduction in the inoculum size from 3×10^4 to 1.5×10^3 cells/mL resulted in a 20-fold increase in total cell expansion and a threefold increase in progenitor expansion (54).

Hematopoietic cultures can experience pH variations of up to 0.5 pH units, unless controlled. It is well established that medium acidification causes growth inhibition. Furthermore, differences in the survival and differentiation of myeloid progenitors (optimally in the range of pH 7.2–7.4) vs. erythroid progenitors (optimally around pH 7.6) have been observed (55,56). In general, the effect of pH can be very complex and recent attempts to elaborate the effects of lactate production, pH, and medium utilization have shown that each of these parameters likely influences stem and progenitor cell responses (57).

Dissolved oxygen concentration varies substantially over the span of a culture if left uncontrolled. In BM, oxygen concentrations are on the order of 2–7% saturation. Several reports have demonstrated that oxygen partial pressure ranging from 1% to 10% in hematopoietic cultures can enhance the size and number of colonies obtained in semisolid media when compared to cultures maintained in atmospheric oxygen pressure (58,59). The benefit of low oxygen concentration could, at least in part, be attributed to lower levels of reactive oxides and to increased responsiveness to cytokines (60). Furthermore, there may exist an "optimal" oxygen level based on the cytokines used and on the developmental state/potential of the cells so that under higher oxygen conditions there is an expansion of HSCs (51) whereas at low oxygen conditions there is no net expansion of HSCs (61). It is important to note that the local concentration of oxygen that the cells experience is more important than the gas phase oxygen level (46), and thus additional research on the effect of oxygen level along with measurements of local oxygen concentrations should be very informative.

The duration of hematopoietic cultures is another parameter of significance, especially for clinically relevant applications. In $CD34^+$ ex vivo expansion from

MPB, a two-phase growth was observed with a pre-CFU phase during the first 7–10 days and a phase past day 10 where CFU-GM decreased and mature myeloid cell numbers increased (62). Many clinically relevant studies have reported a decrease in progenitor production during days 12–15 of culture, indicating that expansion in static liquid cultures should be limited to 7–14 days (63), probably due to inadequate feeding protocols and high cell concentrations. However, several preclinical studies have reported continuously increasing levels of expansion of all hematopoietic lineages for several months (64).

Hematopoietic Culture Media

Hematopoiesis is the result of complex, dynamic interactions of HSCs with cytokines, the ECMs, and the stromal cells (8). The successful ex vivo expansion and maintenance of transplantable stem and progenitor cells require the regulation between self-renewal and differentiation of HSCs as well as avoidance of apoptosis and sustaining the ability to home and engraft. The regulation of this intricate environment is due, in part, to the effects ofgrowth-promoting and inhibitory cytokines and their receptors interacting with each other in complex networks (Fig. 4) (1).

Traditional media for hematopoietic cultures contain xenogenic supplements such as fetal calf and horse sera (about 25% v/v) or bovine serum albumin (65). However, use of these additives is problematic for clinical applications due to the risk of contamination by infectious agents and the lack of well-defined and standardized media components. A number of serum-free alternatives have been developed (66,67). Serum-free media promote greater expansion of the erythroid and megakaryocytic lineages, mainly due to the fact that serum contains transforming growth factor-β (TGF-β) (48,68), whereas serum-containing media are superior for the expansion of granulocytic and monocytic lineages (69). Autologous serum and plasma have been used as substitutes to animal sera and some studies have reported that even low concentrations (1–2%) of autologous plasma are sufficient to support HSC expansion (70,71).

The first demonstration of the use of growth factors to generate increased numbers of specific cell populations was performed in the early 1980s using crude conditioned media as a source of hematopoietic growth factors. From in vitro studies, multiple factors have been identified to have the capacity to stimulate or potentiate the proliferation of hematopoietic cells in the various lineages (Table 2). Nowadays a key feature of any hematopoietic culture system is the combination of cytokines it delivers to the cells and how the concentration of these cytokines are maintained over time, since both these aspects will influence the cumulative response obtained (72). HSCs and HPCs are thought to express most cytokine receptors, however what dictates whether cells respond to some but not all cytokines is not known (22).

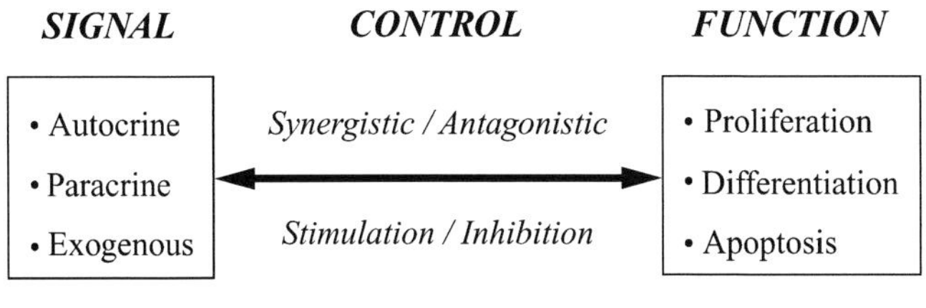

Figure 4 Regulation of hematopoiesis by cytokine networks.

Table 2 Cytokines with Stimulating Action on Hematopoietic Cell Proliferation

Cell type	Cytokines
B-cell	IL-7, SCF, IL-2, IL-5, FL
T-cell	IL-2, SCF, FL
NK-cell	IL-2, IL-12
Erythroid	Epo, SCF, IL-3
Megakaryocytes	TPO, CSF, LIF, Epo, IL-3, IL-6, IL-11
Neutrophil	G-CSF, SCF, GM-CSF, M-CSF, IL-3, IL-6
Eosinophil	IL-5, GM-CSF
Basophil	IL-3, SF
Macrophage	M-CSF, GM-CSF, IL-3

Cytokine receptors belong to three general families: the tyrosine kinase receptors, the cytokine receptors, and the gpl30 receptor family, c-kit, the receptor for stem cell factor (SCF) (73) and flk2/flt3, the receptor for flt3 ligand (FL) (74), are examples of the tyrosine kinase receptors. Both are expressed on very primitive hematopoietic cells and are implicated in the regulation of early stages of hematopoiesis. SCF promotes the survival and growth of primitive hematopoietic progenitors when used in combination with other stimulatory cytokines (75), whereas little expansion is observed in cultures supplemented with SCF alone (76). FL, a factor secreted by stromal cells, has also been investigated as a potential mediator of HSC expansion and maintenance (77) by recruiting HSCs into the cell cycle (78) and by promoting survival, presumably through inhibition of apoptosis (79).

The cytokine receptor family lacks endogenous tyrosine kinase activity. Ligands for these receptors include interleukin-3 (IL-3) and thrombopoietin (TPO), both of which are active on primitive progenitors. IL-3 is required, along with SCF and FL, to maximize the expansion of human LTC-IC from $CD34^+/CD38^-$ cells isolated from BM (80). However, IL-3 may be detrimental to the pluripotentiality of murine progenitors (81), and when added in inappropriately high concentration to human progenitors it induces terminal differentiation and loss of primitive HPCs (19). TPO supports the proliferation and long-term maintenance of primitive hematopoietic progenitors and repopulating HSCs when used in combination with other stimulatory cytokines (82,83). However, when used alone, TPO has little effect on the expansion of HSCs in suspension cultures (84). In the presence of stromal cells, TPO may promote sustained HSC expansion (85).

The gpl30 receptor family includes receptors for IL-6, IL-11, Oncostatin-M, and leukemia inhibitory factor (LIF). Similar to the cytokine receptor family, these receptors do not have intrinsic tyrosine kinase activity; instead they interact with a receptor complex that contains the gpl30 signal molecule. IL-6 has direct proliferative effects on hematopoietic cells and interacts synergistically with other growth factors to stimulate myeloid proliferation (86). IL-11 is produced by stromal cells and has been found, in synergy with SCF and FL, to stimulate the ex vivo expansion of adult murine stem cells (87). The addition of LIF to purified $CD34^+/Thy\text{-}1^+$ cells on a murine BM-derived stromal cell line caused the expansion of the $CD34^+/Thy\text{-}1^+$ cells, which in synergy with other growth factors such as IL-3, IL-6, SCF, and GM-CSF resulted in a 150-fold expansion of cells retaining the $CD34^+/Thy\text{-}1^+$ phenotype (88). These studies indicated that the action of LIF is indirect and mediated via stromal cells.

Although most studies have focused on hematopoietic-related cytokines, novel factors, not originally identified as a result of their effects on hematopoietic cells, have been investigated. Bone morphogenetic proteins can modulate the proliferation and differentiation of primitive human hematopoietic cells ($CD34^+/CD38^-/Lin^-$) (89). Recently, the addition of the Notch ligand, Jagged-1, to cultures of primitive progenitor cells induced the survival and expansion of repopulating HSCs (90).

Aside from growth-promoting cytokines, a number of growth-inhibitor cytokines, such as TGF-β, macrophage inflammatory protein-1α (MIP-1α) , and tumor necrosis factor-α (TNF-α) , have been shown to affect hematopoiesis (22). Although the role of the growth-inhibitory cytokines in vivo is not known, in vitro production of inhibitory proteins may be one reason for the transient expansion observed in many cultures (1). In long-term cultures, TGF-β is responsible for the quiescent state of the progenitor 5–7 days after the last medium exchange (91). Similarly, addition of MIP-1α to cultures prevents cells from cycling (92). Hence, addition of MIP-lα and platelet factor-4 to IL-3 containing cultures of human hematopoietic cells resulted in the preservation of LTC-IC for at least 8 weeks (93). The presence of TNF-α in cultures supplemented with stimulatory cytokines, such as SCF and FL, can potentially inhibit the proliferation of progenitor cells, likely by promoting apoptosis (94,95).

Significant attention has been given toward the elucidation of suitable cytokine combinations for the ex vivo expansion of hematopoietic cells. However, the cytokine microenvironment in hematopoietic culture systems is dynamic (50), with multiple cell types (at different differentiation and cell cycle stages) competing for cytokines that each influence cell fate directly or indirectly, making the cytokine composition of HSC expansion medium particularly challenging to optimize (96). Factorial and composite design have been used to increase the efficiency of investigations of the effects of stimulatory factors on HSCs and to reveal unexpected interactions that would otherwise be missed by conventional dose–response analyses (19,97).

Hematopoietic Culture Systems

Bone marrow cells have traditionally been cultured on the flat, two-dimensional surface of culture dishes following the pioneering work of Dexter et al. who developed a stroma-dependent long-term murine BM culture system, which primarily produced granulocytes (98). The major challenge for the ex vivo expansion of hematopoietic stem and progenitor cells is to provide a suitable culture environment that would support the expansion, from relatively small donor samples, of clinically relevant cell numbers of the desired cell type(s) (99). Furthermore, regulatory requirements require that the culture devices used in the ex vivo expansion of hematopoietic cells comply with good manufacturing practice rules.

In general, there exist two main culture approaches for hematopoietic cultures, namely stroma-free and stroma-dependent cultures. From a clinical standpoint, stroma-free cultures are obviously more easily standardized and maintained (22). However, stroma-based cultures provide a more physiologically relevant microenvironment to the hematopoietic stem and progenitor cells. In addition, in the clinical setting, stromal feeders have been used to support the retroviral transduction of autologous HSCs (100) and to purge chronic myelogenous leukemia (CML) (101) and acute myelogenous leukemia (AML) (102) autografts ex vivo. The advantage of such autologous feeders in the clinical setting is that no allogeneic or xenogeneic feeder cells contaminate the transplant stem cell population (22). However, hematopoietic support may be defective due to exposure of the stroma-donor in vivo to chemotherapy and radiation (103) or due to defects in the stromal cells as a result of the disease (104). To circumvent such problems, a number of murine or human stromal cell feeders have been cloned from a variety of sources (22).

Several different culture devices have been developed, including culture vessels for static cultures and the use of bioreactors. Static culture systems, such as well-plates, T-flasks, and gas permeable blood bags (most published clinical results have been obtained using the gas-permeable bags) , are the most widely used culture systems for the ex vivo expansion of hematopoietic cells (105,106). Despite their widespread usage, however, static systems have several limitations, including concentration gradients (pH, dissolved oxygen, metabolites, etc.) in the culture medium, the inability to be readily monitored and controlled on-line, the requirement for repeated handling for feeding, and the limited productivity by the number of cells that can be supported by a given surface area (106). The use of bioreactors is an appealing alternative approach because they offer closed culture conditions, support high cell densities and accessory cells, and can be automated to perform feeding strategies, thus reducing handling and contamination (47). However, to date, no bioreactor has obtained market approval from the FDA in the United States, and only one has obtained a CE mark in Europe (51).

Perfusion bioreactors have been developed to continually replace culture medium (thus removing any harmful metabolites and maintaining a constant supply of growth nutrients) yet not over-dilute endogenously produced factors or mechanically damage cells. Flat-bed perfusion systems maintain relatively homogeneous bulk culture medium conditions, although they do not completely eliminate gradients in culture conditions (106). These systems are typically medium intensive and scale-up to larger flat-bed reactors is not straightforward due to the geometry of the system. Perfusion systems have demonstrated benefits in the expansion of several hematopoietic cell types when compared to static controls (32,51,107,108) and have been used for clinical trials (105,109). These types of bioreactors have been used to expand unselected BMMNC, in which a stromal layer developed from the inoculum and thus no preformed stroma was required (51,110). CFU-GM progenitor cells expanded 10–30-fold in this system. In addition, a 7.5-fold expansion of LTC-IC was obtained and more than 3 billion cells containing 12 million CFU-GM were generated from the equivalent of a 10–15-mL BM aspirate (51). BM $CD34^{-}$ enriched cell expansion (1300-fold) on stroma with a 41-fold expansion of CFU-GM has also been reported in such a bioreactor (33). Continuous perfusion of the stroma-dependent cultures may improve HSC and HPC support from umbilical CB and MPB (111). This is in part due to the removal of toxic metabolic products and in part due to the induction of certain growth factors in the stromal feeder. When supplemented with cytokines and under the best conditions tested, this bioreactor supports expansion of the LTC-IC, CFU-GM, and indicate that adult-size grafts from CB and MPB could be obtained. These culture systems are currently being clinically evaluated and initial results suggest that human repopulating stem cells can be preserved (22). Perfusion bioreactors have also been used to analyze the consumption and production of growth factors and their relation with specific cell production (112) and to demonstrate that changes in the medium perfusion schedule significantly influence growth factor secretion by stromal cells (113). This suggests that it may not be necessary to add large amounts of growth factors to the cultures, but instead to provide culture conditions in which the stroma is maintained in a state that supports hematopoiesis. In addition, perfusion cultures have been used to study the influence of preformed irradiated stroma (a modified culture chamber with groves perpendicular to the direction of flow was developed for stroma-free cultures) on peripheral blood and marrow mononuclear cells (7), although it is not yet clear whether HSCs need to remain in direct contact with stromal cells to survive and/or expand (22,93,110,111).

Stirred culture systems are well characterized and widely used for the culture of both microbial and mammalian cells. They provide a homogeneous, well-controlled, closed, and reproducible culture environment and are superior to most static and perfusion systems in terms of sampling, data collection, and control of medium conditions. Furthermore, stirred bioreactors could prove advantageous for clinical applications in terms of the scale anticipated for clinical cultures (96). BMMNC have been successfully propagated in serum-containing stirred culture systems (114,115), whereas MPB and CB mononuclear cells and $CD34^+$ cells have been propagated in both serum and serum-free stirred cultures (106). Natural killer (NK) cells isolated from peripheral blood mononuclear cells have also been cultivated in stirred culture bioreactors (116). These studies indicated that the performance was improved, presumably because mixing overcame the diffusion limitations observed in static cultures. However, suspension culture systems do not mimic the complex three-dimensional microenvironment found in BM. Although macrophages tolerate suspension, fibroblasts and endothelial cells are unlikely to perform properly in stirred cultures (47). Whether extensive stem cell self-renewal can be sustained in the stirred culture remains to be investigated. In addition, since hematopoietic cells are relatively sensitive to shear, low shear rates are required to avoid physical damage (106), especially since it has been demonstrated that shear stress can affect surface marker expression, including cytokine receptors, and metabolism in other experimental systems (117). Thus agitation could have a profound effect on the type of cells that will expand and the extent of the expansion.

In vivo, the marrow three-dimensional microenvironment, as has been discussed earlier, plays a critical role in the regulation of proliferation and differentiation of hematopoietic stem and progenitor cells. Whereas in vivo hematopoiesis is regulated by cell–cell, cell–matrix, and growth factors, most ex vivo culture systems have focused on the addition of cytokines alone, which is insufficient to promote optimal development of blood cell development and retain long-term generative capacity (118,119). Most traditional stroma-dependent culture systems are not designed to enhance cell–cell and cell–matrix interactions, create a spatial distribution of stromal and hematopoietic cells, and mimic the in vivo three-dimensional configuration. Furthermore, comparisons between signal transduction in three-dimensional matrices (3-D signaling) with signaling events on two-dimensional coated surfaces reveal some striking differences (120). On the basis of BM physiology, novel three-dimensional culture systems were developed based on the hypothesis that growth in 3-D culture may more faithfully replicate the in vivo BM HIM required for self-renewal and differentiation of stem and progenitor cells. A collagen-coated nylon mesh has been used as a template to grow BM cells, maintaining multilineage differentiation with, however, declining CFU numbers (121–123). Porous microspheres have been used in a packed bed bioreactor to grow both murine and human BM cells, which supported multilineal differentiation and a 3-D growth configuration that mimicked that of BM in vivo in the absence of exogenous growth factors (124,125). In the presence of cytokines, including erythropoietin (Epo), this system supported extensive erythropoiesis when compared to a traditional flask culture under lower, more physiological, concentrations of Epo (125). A tantalum-coated porous biomaterial was found to enhance the maintenance and retroviral transduction of HPCs, improve the cell pluripotency (126), and support marrow $CD34^+$ cell expansion up to 1.5-fold in the absence of exogenous cytokines, due to the 3-D culture environment (127). More recently, 3-D nonwoven fibrous matrices have been used for the culture of human CB cells, which, when compared to a 2-D film, prolonged the proliferation process and supported higher total $CD34^+$ cell numbers (128). The 3-D culture may provide a better system for stem and progenitor cell culture in an environment that resembles the BM in vivo.

CONCLUSIONS

Although significant progress has been made in the characterization of the factors that govern hematopoiesis, ex vivo expansion appears to be a technique still in its infancy. Nevertheless, the ex vivo expansion of hematopoietic cells for clinical use is now recognized to be a very promising method. The engineering of optimal hematopoietic cell culture systems will require the design of new expansion systems mimicking the in vivo self-regulating BM environment that operate under reliable and reproducible conditions and offer a broad spectrum of possibilities for different culture strategies and the cultivation of various cell types—from stem cells to differentiated cells for gene, cellular, and tissue therapies.

REFERENCES

1. Madlambayan GJ, Rogers I, Casper RF, Zandstra PW. Controlling culture dynamics for the expansion of hematopoietic stem cells. J Hematother Stem Cell Res 2001; 10:481–492.
2. Abboud CN, Lichtman MA. Structure of the marrow and the hemopoietic microenvironment. In: Seligsohn U, ed. Hematology. New York: McGraw-Hill, 2001:29–58.
3. McNiece I, Briddell R. *Ex vivo* expansion of hematopoietic progenitor cells. Exp Hematol 2001; 29:3–11.
4. Charbord P. Microenvironmental cell populations essential for the support of hematopoietic stem cells. In: Zon LI, ed. Hematopoiesis: A Developmental Approach. New York: Oxford University Press, 2001:691–701.
5. Trentin JJ. Influence of hematopoietic organ stroma (hematopoietic inductive microenvironments) on stem cell differentiation. In: Gordon AS, ed. Regulation of Hemopoiesis. Vol.I. New York: Appleton-Century-crofts, 1970:161–186.
6. Weissman I. Developmental switches in the immune system. Cell 1994; 76:207–218.
7. Aiuti A, Friedrich C, Sieff C, Gutierrez-Ramos J. Identification of distinct elements of the stromal microenvironment that control human hematopoietic stem/progenitor cell growth and differentiation. Exp Hematol 1998; 26:143–157.
8. Quesenberry P, Crittenden R, Lowry P, Kittler E, Rao S, Peters S, Ramshaw H, Stewart F. *In vitro* and *in vivo* studies of stromal niches. Blood Cells 1994; 20:97–106.
9. Pohlmann SJ, Slayton WB, Spangrude GJ. Stem cell populations: purification and behavior. In: Zon LI, ed. Hematopoiesis: A Developmental Approach. New York: Oxford University Press, 2001:35–47.
10. Quesenberry P, Colvin GA. Hematopoietic stem cells, progenitor cells, and cytokines. In: Seligsohn U, ed. Hematology. New York: McGraw-Hill, 2001:153–174.
11. Krause DS, Fackler MJ, Civin CI, May WS. CD34: structure, biology, and clinical utility. Blood 1996; 87:1–13.
12. Shih C-C, DiGiusto D, Forman SJ. *Ex vivo* expansion of transplantable human hematopoietic stem cells: where do we stand in the year 2000? J Hematother Stem Cell Res 2002; 9:621–628.

13. Sutherland HJ, Lansdorp PM, Henkelman DH, Eaves AC, Eaves CJ. Functional characterization of individual human hematopoietic stem cells cultured at limiting dilution on supportive marrow stromal layers. Proc Natl Acad Sci USA 1990; 87:3584–3588.
14. Craig W, Kay R, Cutler RL, Lansdorp PM. Expression of Thy-1 on human hematopoietic progenitor cells. J Exp Med 1993; 177:1331–1342.
15. Osawa M, Hanada K, Hamada H, Nakauchi H. Long-term lymphohematopoietic reconstitution by $CD34^-$ low/negative hematopoietic stem cells. Science 1996; 273:242–245.
16. Randall T, Weissman I. Characterization of a population of cells in the bone marrow that phenotypically mimics hematopoietic stem cells: resting stem cells or mystery population? Stem Cells 1998; 16:34–48
17. Goodell M, Brose K, Paradis G, Conner A, Mulligan A. Isolation and functional properties of murine hematopoietic stem cells that are replicating *in vivo*. J Exp Med 1996; 183:1797–1806.
18. Spangrude GJ, Johnson GR. Resting and activated subsets of mouse multipotent hematopoietic stem cells. Proc Natl Acad Sci USA 1990; 87:7433–7477.
19. Zandstra PW, Conneally E, Petzer AL, Piret JM, Eaves CJ. Cytokine manipulation of primitive human hematopoietic cell self-renewal. Proc Natl Acad Sci USA 1997; 94:4698–4703.
20. Williams DE, Hangoc G, Cooper S, Boswell HS, Shadduck RK, Gillis S, Waheed A, Urdal D, Broxmeyer HE. The effects of purified recombinant murine interleukin-3 and/or purified natural murine CSF-1 *in vivo* on the proliferation of murine hifg- and low-proliferative potential colony-forming cells: demonstration of *in vivo* synergism. Blood 1987; 70:401–403.
21. Eaves CJ, Eaves AC. Stem cell kinetics. Baillieres Clin Haematol 1997; 10:233–257.
22. Verfaillie CM. *Ex vivo* expansion of hematopoietic stem cells. In: Zon LI, ed. Hematopoiesis: A Developmental Approach. New York: Oxford University Press, 2001: 119–129.
23. Till JE, McCulloch EA. A direct measurement of the radiation sensitivity of normal mouse bone marrow cells. Radiat Res 1961; 14:213–222.
24. McCune JM, Namikawa R, Kaneshima H, Shultz LD, Lieberman M, Weissman IL. The SCID-hu mouse: murine model for the analysis of human hematolymphoid differentiation and function. Science 1998; 24:1632–1639.
25. Larochelle A, Vormoor J, Hanenberg H, Wang JC, Bhatia M, Lapidot T, Moritz T, Murdoch B, Xiao XL, Kato I, Williams DA, Dick JE. Identification of primitive human hematopoietic cells capable of repopulating NOD/SCID mouse bone marrow: implications for gene therapy. Nat Med 1996; 2:1329–1337.
26. Civin CI, Almeida-Porada G, Lee MJ, Olweus J, Terstappen LW, Zanjani ED. Sustained, retransplantable, multilineage engraftment of highly purified adult human bone marrow stem cells *in vivo*. Blood 1996; 88:4102–4109.
27. Gan OI, Murdoch B, Larochelle A, Dick JE. Differential maintenance of primitive human SCID-repopulating cells, clonogenic progenitors, and long-term culture-initiating cells after incubation on human bone marrow stromal cells. Blood 1997; 90:641–650.
28. McAdams TA, Winter JN, Miller WM, Papoutsakis ET. Hematopoietic cell culture therapies (Part II): clinical aspects and applications. Trends Biotechnol 1996; 14:388–396.
29. Cairo MS, Wagner JE. Placental and/or umbilical cord blood: an alternative source of hematopoietic stem cells for transplantation. Blood 1997; 90:4664–4671.
30. De La Selle V, Gluckman E, Bruley-Rosset M. Newborn blood can engraft adult mice without inducing graft versus host disease across non-H-2 antigens. Blood 1996; 87:3977–3984.
31. Rubinstein P, Carrier C, Scaradavou A, Kutzberg J, Adamson J, Migliaccio AR, Berkowitz RL, Cabbad M, Dobrila NL, Taylor PE, Rosenfield RE, Stevens CE. Outcomes among 562 recipients of placental-blood transplants from unrelated donors. N Engl J Med 1998; 339:1565–1577.

32. Sandstrom CE, Bender JG, Papoutsakis ET, Miller WM. Effects of $CD34^+$ cell selection and perfusion on *ex vivo* expansion of peripheral blood mononuclear cells. Blood 1995; 86:958–970.
33. Roller MR, Manchel I, Newsom BS, Palsson MA, Palsson BO. Bioreactor expansion of human bone marrow: comparison of unprocessed, density-separated and $CD34^-$ enriched cells. J Hematother 1995; 4:159–169.
34. Roller MR, Oxender M, Jensen TC, Goltry KL, Smith AK. Direct contact between CD34+/Lin-cells and stroma induces a soluble activity that specifically increases primitive hematopoietic cell production. Exp Hematol 1999; 27:734–741.
35. Bachier CR, Gokmen E, Teale J, Lanzkron S, Childs C, Franklin W, Shpall E, Douville J, Weber S, Muller T, Armostrong D, LeMaistre CF. *Ex-vivo* expansion of bone marrow progenitor cells for hematopoietic reconstitution following high-dose chemotherapy for breast cancer. Exp Hematol 1999; 27:615–623.
36. Simmons PJ, Gronthos S, Zanneting ACW. The development of stromal cells. In: Zon LI, ed. Hematopoiesis: A Developmental Approach. New York: Oxford University Press, 2001:718–726.
37. Haynesworth SE, Reuben D, Caplan AI. Cell-based tissue engineering therapies: the influence of whole body physiology. Adv Drug Deliv Rev 1997; 33:3–14.
38. Owen ME. The marrow stromal cell system. In: Owen ME, ed. Marrow Stromal Cell in Culture. Cambridge, UK: Cambridge University Press, 1988:88–110.
39. Caplan AI, Bruder SP. Mesenchymal stem cells: building blocks for molecular medicine in the 21st century. Trends Mol Med 2001; 7:259–264.
40. Owen ME. Lineage of osteogenic cells and their relationship to the stromal system. Bone Miner Res 1985; 3:1–25.
41. Friedenstein AJ, Gorskaja JF, Kulagina NN. Fibroblast precursors in normal and irradiated mouse hematopoietic organs. Exp Hematol 1976; 4:267–274.
42. Simmons PJ, Torok-Storb B. Identification of stromal cell precursors in human bone marrow by a novel monoclonal antibody, STRO-1. Blood 1991; 78:55–62.
43. Gronthos S, Simmons PJ. The biology and application of human bone marrow stromal cell precursors. J Hematother 1996; 5:15–23.
44. Onyia JE, Clapp DW, Long H, Hock JM. Osteoprogenitor cells as targets for *ex vivo* gene transfer. J Bone Miner Res 1998; 13:2–30.
45. Young RG, Butler DL, Weber W, Caplan AI, Gordon SL, Fink DJ. Use of mesenchymal stem cells in collagen matrix for Achilles tendon repair. J Orthop Res 1998; 16:406–413.
46. Koller MR, Palsson BO. Tissue engineering: reconstitution of human hematopoiesis *ex vivo*. Biotechnol Bioeng 1993; 42:909–930.
47. Nielsen LK. Bioreactors for hematopoietic cell culture. Annu Rev Biomed Eng 1999; 1:129–152.
48. Krystal G, Lanm V, Dragowska W, Takahashi C, Appel J. Transforming growth factor beta 1 is an inducer of erythroid differentiation. J Exp Med 1994; 180:851–860.
49. Metcalf D. Some general aspects of hematopoietic cell development. In: Zon LI, ed. Hematopoiesis: A Developmental Approach. New York: Oxford University Press, 2001:3–14.
50. Zandstra PW, Petzer AL, Eaves CJ, Piret JM. Cellular determinants affecting the rate of cytokine depletion in cultures of human hematopoietic cells. Biotechnol Bioeng 1997; 54:58–68.
51. Koller MR, Emerson SG, Palsson BO. Large-scale expansion of human stem and progenitor cells from bone marrow mononuclear cells in continuous perfusion cultures. Blood 1993; 82:378–384.
52. Caldwell J, Locey B, Clarke MF, Emerson SG, Palsson BO. Influence of medium exchange schedules on metabolic, growth, and GM-CSF secretion rates of genetically engineered NIH-3T3 cells. Biotechnol Prog 1991; 7:1.

53. Briddell RA, Kern BP, Zilm KL, Stoney GB, McNiece IK. Purification of $CD34^+$ cells is essential for optimal *ex vivo* expansion of umbilical cord blood cells. J Hematother 1997; 6:145–150.
54. Poloni A, Giarratana MC, Kobari L, Firat H, Bouchet S, Gorin NC, Douay L. The *ex vivo* expansion capacity of normal human bone marrow cells is dependent on experimental conditions: role of cell concentration, serum and $CD34^+$ cell selection in stroma-free cultures. Hematol Cell Ther 1997; 39:49–58.
55. McAdams TA, Miller WM, Papoutsakis ET. pH is a potent modulator of erythroid differentiation. Br J Haematol 1998; 103:317–325.
56. McAdams TA, Miller WM, Papoutsakis ET. Variations in culture pH affect the cloning efficiency and differentiation of progenitor cells in *ex vivo* haematopoiesis. Br J Haematol 1997; 97:889–895.
57. Patel SD, Papoutsakis ET, Winter JN, Miller WM. The lactate issue revisited: novel feeding protocols to examine inhibition of cell proliferation and glucose metabolism in hematopoietic cell cultures. Biotechnol Prog 2000; 16:885–892.
58. Bradley TR, Hodgson GS, Rosendaal M. The effect of oxygen tension on haematopoietic and fibroblast cell proliferation. J Cell Physiol 1978; 97:517–522.
59. Broxmeyer HE, Cooper S, Lu L, Miller ME, Langefeld CD, Ralph P. Enhanced stimulation of human bone marrow macrophage colony formation *in vitro* by recombinant human macrophage colony-stimulating factor in agarose medium and at low oxygen tension. Blood 1990; 76:323–329.
60. Pick E, Mizel D. Rapid microassays for the measurement of superoxide and hydrogen peroxide production by macrophages in culture using an automatic enzyme immunoassay reader. J Immunol Methods 1981; 46:211–226.
61. Cipolleschi MG, Dello Sbarba P, Olivotto M. The role of hypoxia in the maintenance of hematopoietic stem cells. Blood 1993; 82:2031–2037.
62. Haylock DN, To LB, Dowse TL, Juttner CA, Simmons PJ. *Ex vivo* expansion and maturation of peripheral blood $CD34^+$ cells into the myeloid lineage. Blood 1992; 80:1405–1412.
63. Bregni M, Magni M, Siena S, Di Nicola M, Bonadonna G, Gianni AM. Human peripheral blood hematopoietic progenitors are optimal targets of retroviral-mediated gene transfer. Blood 1992; 80:1418–1422.
64. Piacibello W, Sanavio F, Garetto L, Severino A, Dane A, Gammaitoni I, Agglietta M. Differential growth factor requirement of primitive cord blood hematopoietic stem cells for self-renewal and amplification versus proliferation and differentiation. Leukemia 1998; 12:718–727.
65. Douay L. Experimental culture conditions are critical for *ex vivo* expansion of hematopoietic cells. J Hematother Stem Cell Res 2001; 10:341–346.
66. Koller MR, Maher RJ, Manchel I, Oxender M, Smith AK. Alternatives to animal sera for human bone marrow cell expansion: human serum and serum free media. J Hematother 1998; 7:413–421.
67. Lebowski JS, Schain LR, Okarma TB. Serum-free culture of hematopoietic stem cells: a review. Stem Cells 1995; 13:607–612.
68. Dybedal I, Jacobsen SE. Transforming growth factor beta (TGF-beta), a potent inhibitor of erythropoiesis: neutralizing TGF-beta antibodies show erythropoietin as a potent stimulator of murine burst-forming unit erythroid colony formation in the absence of a burst-promoting activity. Blood 1995; 86:949–957.
69. Lill MC, Lynch M, Fraser JK, Chung GY, Schiller G, Glaspy JA, Souza L, Baldwin GC, Gasson JC. Production of functional myeloid cells from CD34-selected hematopoietic progenitor cells using a clinically relevant *ex vivo* expansion system. Stem Cells 1994; 12:626–637.
70. Rosenzwajg M, Canque B, Gluckman JC. Human dendritic cell differentiation pathway from $CD34^+$ hematopoietic precursor cells. Blood 1996; 87:535–544.

71. Brugger W, Mocklin W, Heimfeld S, Berenson RJ, Mertelsmann R, Kanz L. *Ex vivo* expansion of enriched peripheral blood $CD34^+$ progenitor cells by stem cell factor, interleukin-1 (IL-1), IL-6, IL-3, interferon y and erythropoietin. Blood 1993; 81:2579–2584.
72. Audet J, Zandstra PW, Eaves CJ, Piret JM. Advances in hematopoietic stem cell culture. Curr Opin Biotechnol 1998; 9:146–151.
73. Broudy VC. Stem cell factor and hematopoiesis. Blood 1997; 90:1354–1364.
74. Lyman SD, Williams DE. Biology and potential clinical applications for flt3 ligand. Curr Opin Hematol 1995; 2:177–181.
75. Brandt JE, Bhalla K, Hoffman R. Effects of interleukin-3 and c-kit ligand on the survival of various classes of human hematopoietic progenitor cells. Blood 1994; 55:1507–1514.
76. Ohmizono Y, Sakaba H, Kimura T, Tanimukai S, Matsumura T, Miyazaki H, Lyman SD, Sonoda Y. Thrombopoietin augments *ex vivo* expansion of human cord blood-derived hematopoietic progenitors in combination with stem cell factor and flt3 ligand. Leukemia 1997; 11:524–530.
77. Lyman SD, Jacobsen SE. c-kit ligand and Flt3 ligand: stem/progenitor cell factors with overlapping yet distinct activities. Blood 1998; 91:1101–1134.
78. Shah AJ, Smogorzewska EM, Hannum C, Crooks GM. Flt3 ligand induces proliferation of quiescent human bone marrow $CD34^+CD38^-$ cells and maintains progenitor cells *in vitro*. Blood 1996; 87:3563–3570.
79. Meyer C, Drexler HG. FLT3 ligand inhibits apoptosis and promotes survival of myeloid leukemia cell lines. Leuk Lymphoma 1999; 32:577–581.
80. Petzer AL, Zandstra PW, Piret JM, Eaves CJ. Differential cytokine effects on primitive ($CD34^+$ $CD38^-$) human hematopoietic cells: novel responses to flt3-ligand and thrombopoietin. J Exp Med 1996; 183:2551–2558.
81. Matsunaga T, Hirayama F, Yonemura Y, Murray R, Ogawa M. Negative regulation by interleukin-3 (IL-3) of mouse early B-cell progenitors and stem cells in culture: transduction of the negative signals by betac and betaIL-3 proteins of IL-3 receptor and absence of negative regulation by granulocyte–macrophage colony-stimulating factor. Blood 1998; 92:901–907.
82. Won JH, Cho SD, Park SK, Lee GT, Baick SH, Suh WS, Hong DS, Park HS. Thrombopoietin in synergy with other cytokines for expansion of cord blood progenitor cells. J Hematother Stem Cell Res 2000; 9:465–473.
83. Kobayashi M, Laver JH, Kato T, Miyazaki H, Ogawa M. Thrombopoietin supports proliferation of human primitive hematopoietic cells in synergy with steel factor and/or $orinterleukin - 3. Blood 1996; 88 : 429–436$.
84. Kobayashi M, Laver JH, Lyman SD, Kato T, Miyazaki H, Ogawa M. Thrombopoietin, steel factor and the ligand for flt3/flk2 interact to stimulate the proliferation of human hematopoietic progenitors in culture. Int J Hematol 1997; 66:423–434.
85. Yagi M, Ritchie KA, Sitnicka E, Storey C, Roth GJ, Bartelmez S. Sustained *ex vivo* expansion of hematopoietic cells mediated by thrombopoietin. Proc Natl Acad Sci USA 1999; 96:8126–8131.
86. Ogawa M, Clark SC. Synergistic interaction between interleukin-6 and interleukin-3 in support of stem cell proliferation in culture. Blood Cells 1988; 14:329–335.
87. Holyoake TL, Freshney MG, McNair L, Parker AN, McKay PJ, Steward WP, Fitzsimons E, Graham GJ, Pragnell IB. *Ex vivo* expansion with stem cell factor and interleukin-11 augments both short-term recovery posttransplant and the ability to serially transplant marrow. Blood 1996; 95:4589–4595.
88. Shih C-C, Hu MCT, Hu J, Wenh Y, Yazaki P, Medeiros J, Forman SJ. A secreted and LIF-mediated stromal cell-derived activity that promotes *ex vivo* expansion of human hematopoietic stem cells. Blood 2000; 95:1957–1966.

89. Bhatia M, Bonnet D, Wu D, Murdoch B, Wrana J, Gallacher L, Dick JE. Bone morphogenetic proteins regulate the developmental program of human hematopoietic stem cells. J Exp Med 1999; 189:1139–1148.
90. Karanu FN, Murdoch B, Gallacher L, Wu DM, Koremoto M, Sakano S, Bhatia M. The notch ligand jagged-1 represents a novel factor of human hematopoietic stem cells. J Exp Med 2000; 192:1365–1372.
91. Eaves CJ, Cashman JD, Kay RJ, Dougherty GJ, Otsuka T, Gaboury LA, Hogge DE, Lansdorp PM, Eaves AC, Humphries RK. Mechanisms that regulate the cell cycle status of very primitive hematopoietic cells in long-term human marrow cultures. II. Analysis of positive and negative regulators produced by stromal cells within the adherent layer. Blood 1991; 78:110–117.
92. Broxmeyer HE, Sherry B, Cooper S, Lu L, Maze R, Beckmann MP, Cerami A, Ralph P. Comparative analysis of the human macrophage inflammatory protein family of cytokines (chemokines) on proliferation of human myeloid progenitor cells. Interacting effects involving suppression, synergistic suppression, and blocking of suppression. J Immunol 1993; 150:3448–3458.
93. Verfaillie CM, Hurley R, Bhatia R, McCarthy JB. MIP-loc combined with IL-3 conserves primitive human LTBMC-IC for at least 8 weeks in *ex vivo* "stroma-non-contact" cultures. J Exp Med 1994; 179:643–649.
94. Veiby OP, Jacobsen FW, Cui L, Lyman SD, Jacobsen SE. The flt3 ligand promotes the survival of primitive hemopoietic progenitor cells with myeloid as well as B lymphoid potential. Suppression of apoptosis and counteraction by TNF-alpha and TGF-beta. J Immunol 1996; 157:2953–2960.
95. Rusten LS, Smeland EB, Jacobsen FW, Lien E, Lesslauer W, Loetscher H, Dubois CM, Jacobsen SE. Tumor necrosis factor-alpha inhibits stem cell factor-induced proliferation of human bone marrow progenitor cells *in vitro*. Role of p55 and p75 tumor necrosis factor receptors. J Clin Invest 1994; 94:165–172.
96. Zandstra PW, Nagy A. Stem cell bioengineering. Annu Rev Biomed Eng 2001; 3: 275–305.
97. Zandstra PW, Conneally E, Piret JM, Eaves CJ. Ontogeny-associated changes in the cytokine responses of primitive human haematopoietic cells. Br J Haematol 1998; 101:770–778.
98. Dexter TM, Allen TD, Lajtha LG. Conditions controlling the proliferation of haematopoietic stem cells *in vivo*. J Cell Physiol 1977; 91:335–344.
99. Cabral JMS. *Ex vivo* expansion of hematopoietic stem cells in bioreactors. Biotechnol Lett 2001; 23:741–751.
100. Emmons RV, Doren S, Zujewski J, Cottier-Fox M, Carter CS, Hines K, O'Shaughnessy JA, Leitman SF, Greenblatt JJ, Cowan K, Dunbar CE. Retroviral gene transduction of adult peripheral blood or marrow-derived $CD34^+$ cells for six hours without growth factors or on autologous stroma does not improve marking efficiency assessed *in vitro*. Blood 1997; 89:4040–4046.
101. Barnett MJ, Eaves CJ, Phillips GL, Gascoyne RD, Hogge DE, Horsman DE, Humphries RK, Klingemann HG, Lansdorp PM, Nantel SH, Eaves AC. Autografting with cultured marrow in chronic myeloid leukemia: results of a pilot study. Blood 1994; 84:724–732.
102. Coutinho LH, Testa NG, Chang J, Morgenstern G, Harrison C, Dexter TM. The use of cultured bone marrow cells in autologous transplantation. Prog Clin Biol Res 1990; 333:415–432.
103. Mayani H. Composition and function of the hematopoietic microenvironment in human myeloid leukemia. Leukemia 1996; 10:1041–1047.
104. Bhatia ML, McGlave PB, Dewald GW, Blazar BR, Verfaillie CM. Abnormal function of the bone marrow microenvironment in chronic myelogenous leukemia: role of malignant stromal macrophages. Blood 1995; 85:3636–3645.

105. Chabannon C, Olivero S, Blaise D, Maraninchi D, Viens P. *Ex vivo* expansion of human hematopoietic progenitors and cells to support high-dose chemoradiation therapy: five years of clinical experience. Cytokines Cell Mol Ther 2000; 6:97–108.
106. Collins PC, Miller ME, Papoutsakis ET. Stirred culture of peripheral and cord blood hematopoietic cells offers advantages over traditional static systems for clinically relevant applications. Biotechnol Bioeng 1998; 59:534–543.
107. Sandstrom CE, Bender JG, Miller ME, Papoutsakis ET. Development of a novel perfusion chamber to retain nonadherent cells and its use for comparison of human "mobilized" peripheral blood mononuclear cell cultures with and without irradiated bone marrow stroma. Biotechnol Bioeng 1996; 50:493–504.
108. Koller MR, Bender JG, Miller ME, Papoutsakis ET. Expansion of primitive human hematopoietic progenitors in a perfusion bioreactor system with IL-3, IL-6, and stem cell factor. Biotechnology 1993; 11:358–363.
109. Champlin R, Mehra R, Gajewski J, Khouri I, Geisler D, Davis M, Oba K, Thomas MRD, Douville J, Weber S, Silver S, Muller T, Deisseroth A. *Ex vivo* expanded progenitor cell transplantation in patients with breast cancer. Blood 1995; 86:1166–1166.
110. Palsson BO, Paek S-H, Schwartz RM, Palsson MA, Lee G-M, Silver S, Emerson SG. Expansion of human bone marrow progenitor cells in a high cell density continuous perfusion system. Biotechnology 1993; 11:368–372.
111. VanZant G, Rummel SA, Koller MR, Larson DB, Drubachevsky I, Palsson MA, Emerson SG. Expansion in bioreactors of human progenitor populations from cord blood and mobilized peripheral blood. Blood Cells 1994; 20:482–491.
112. Koller MR, Bradley TR, Palsson BO. Growth factor consumption and production in perfusion cultures of human bone marrow correlate with specific cell production. Exp Hematol 1995; 23:1275–1283.
113. Caldwell J, Palsson BO, Locey B, Emerson SG. Culture perfusion schedules influence the metabolic activity and granulocyte–macrophage colony-stimulating factor production rates of human bone marrow stromal cells. J Cell Physiol 1991; 147:344–353.
114. Sardonini CA, Wu YJ. Expansion and differentiation of human hematopoietic cells from static cultures through small-scale bioreactors. Biotechnol Prog 1993; 9:131–137.
115. Zandstra PW, Eaves CJ, Piret JM. Expansion of hematopoietic progenitor cell populations in stirred suspension bioreactors of normal human one marrow cells. Biotechnology 1994; 12:909–914.
116. Pierson BA, Europa AF, Hu WS, Miller JS. Production of human natural killer cells for adoptive immunotherapy using a computer-controlled stirred-tank bioreactor. J Hematother 1996; 5:474–483.
117. McDowell CL, Papoutsakis ET. Increased agitation intensity increases CD13 receptor surface content and mRNA levels, and alters the metabolism of HL60 cells cultures in stirred tank bioreactors. Biotechnol Bioeng 1998; 60:239–250.
118. Dao MA, Hashimo K, Kato I, Nolta JA. Adhesion to fibronectin maintains regenerative capacity during *ex vivo* culture and transduction of human hematopoietic stem and progenitor cells. Blood 1998; 92:4612–4621.
119. Chute JP, Saini AA, Kampen RL, Wells MR, Davis TA. A comparative study of the cell cycle status and primitive cell adhesion molecule profile of human $CD34^+$ cells cultured in stroma-free versus porcine microvascular endothelial cell cultures. Exp Hematol 1999; 27:370–379.
120. Cukierman E, Pankov R, Yamada KM. Cell interactions with three-dimensional matrices. Curr Opin Cell Biol 2002; 14:1–7.
121. Naughton BA, Naughton GK. Hematopoiesis on nylon mesh templates—comparative long-term bone-marrow culture and the influence of stromal support cells. Ann NY Acad Sci 1989; 554:125–140.
122. Naughton BA, Jacob L, Naughton GK. A three-dimensional culture system for the growth of hematopoietic cells. Prog Clin Biol Res 1990; 333:435–445.

123. Naughton BA, Tjota A, Sibanda B, Naughton GK. Hematopoiesis on suspended nylon screen-stromal cell microenvironments. J Biomech Eng Trans ASME 1991; 113:171–177.
124. Wang T-Y, Brennan JK, Wu JHD. Multilineal hematopoiesis in a three-dimensional murine long-term bone marrow culture. Exp Hematol 1995; 23:26–32.
125. Mantalaris A, Keng P, Bourne P, Chang AYC, Wu JHD. Engineering a human bone marrow model: a case study on *ex vivo* erythropoiesis. Biotechnol Prog 1998; 14:126–133.
126. Rosenzweig M, Pykett MJ, Marks DF, Johnson RP. Enhanced maintenance and retro viral transduction of primitive hematopoietic progenitor cells using a novel three-dimensional culture system. Gene Ther 1997; 4:928–936.
127. Bagley J, Rosenzwajg M, Marks DF, Pykett MJ. Extended culture of multipotent hematopoietic progenitors without cytokine augmentation in a novel three-dimensional device. Exp Hematol 1999; 27:496–504.
128. Li Y, Kniss DA, Yang S-T, Lasky LC. Human cord cell hematopoiesis in three-dimensional nonwoven fibrous matrices: in vitro simulation of the marrow micro-environment. J Hematother Stem Cell Res 2001; 10:355–368.

Index